APPLET CORRELATION

AF861507

Applet	Concept Illustrated	Description	Applet Activity
Sample from a population	Assesses how well a sample represents the population and the role that sample size plays in the process.	Produces random sample from population from specified sample size and population distribution shape. Reports mean, median, and standard deviation; applet creates plot of sample.	**4.4**, 196; **4.6**, 216; **4.7**, 235
Sampling distributions	Compares means and standard deviations of distributions; assesses effect of sample size; illustrates undbiasedness.	Simulates repeatedly choosing samples of a fixed size n from a population with specified sample size, number of samples, and shape of population distribution. Applet reports means, medians, and standard deviations; creates plots for both.	**4.9**, 251; **4.10**, 251
Random numbers	Uses a random number generator to determine the experimental units to be included in a sample.	Generates random numbers from a range of integers specified by the user.	**1.1**, 22; **1.2**, 22; **3.6**, 164; **4.1**, 183; **4.8**, 245
Long-run probability demonstrations illustrate the concept that theoretical probabilities are long-run experimental probabilities.			
Simulating probability of rolling a 6	Investigates relationship between theoretical and experimental probabilities of rolling 6 as number of die rolls increases.	Reports and creates frequency histogram for each outcome of each simulated roll of a fair die. Students specify number of rolls; applet calculates and plots proportion of 6s.	**3.1**, 128; **3.3**, 138; **3.4**, 138; **3.5**, 152
Simulating probability of rolling a 3 or 4	Investigates relationship between theoretical and experimental probabilities of rolling 3 or 4 as number of die rolls increases.	Reports outcome of each simulated roll of a fair die; creates frequency histogram for outcomes. Students specify number of rolls; applet calculates and plots proportion of 3s and 4s.	**3.3**, 138; **3.4**, 138
Simulating the probability of heads: fair coin	Investigates relationship between theoretical and experimental probabilities of getting heads as number of fair coin flips increases.	Reports outcome of each fair coin flip and creates a bar graph for outcomes. Students specify number of flips; applet calculates and plots proportion of heads.	**3.2**, 128; **4.2**, 183
Simulating probability of heads: unfair coin ($P(H) = .2$)	Investigates relationship between theoretical and experimental probabilities of getting heads as number of unfair coin flips increases.	Reports outcome of each flip for a coin where heads is less likely to occur than tails and creates a bar graph for outcomes. Students specify number of flips; applet calculates and plots the proportion of heads.	**4.3**, 196
Simulating probability of heads: unfair coin ($P(H) = .8$)	Investigates relationship between theoretical and experimental probabilities of getting heads as number of unfair coin flips increases.	Reports outcome of each flip for a coin where heads is more likely to occur than tails and creates a bar graph for outcomes. Students specify number of flips; applet calculates and plots the proportion of heads.	**4.3**, 196
Simulating the stock market	Theoretical probabilities are long run experimental probabilities.	Simulates stock market fluctuation. Students specify number of days; applet reports whether stock market goes up or down daily and creates a bar graph for outcomes. Calculates and plots proportion of simulated days stock market goes up.	**4.5**, 196
Mean versus median	Investigates how skewedness and outliers affect measures of central tendency.	Students visualize relationship between mean and median by adding and deleting data points; applet automatically updates mean and median.	**2.1**, 61; **2.2**, 61; **2.3**, 61

(Continued)

Applet	Concept Illustrated	Description	Applet Activity
Standard deviation	Investigates how distribution shape and spread affect standard deviation.	Students visualize relationship between mean and standard deviation by adding and deleting data points; applet updates mean and standard deviation.	**2.4**, 68; **2.5**, 68; **2.6**, 69; **2.7**, 81
Confidence intervals for a proportion	Not all confidence intervals contain the population proportion. Investigates the meaning of 95% and 99% confidence.	Simulates selecting 100 random samples from the population and finds the 95% and 99% confidence intervals for each. Students specify population proportion and sample size; applet plots confidence intervals and reports number and proportion containing true proportion.	**5.5**, 294; **5.6**, 295
Confidence intervals for a mean (the impact of confidence level)	Not all confidence intervals contain the population mean. Investigates the meaning of 95% and 99% confidence.	Simulates selecting 100 random samples from population; finds 95% and 99% confidence intervals for each. Students specify sample size, distribution shape, and population mean and standard deviation; applet plots confidence intervals and reports number and proportion containing true mean.	**5.1**, 277; **5.2**, 277
Confidence intervals for a mean (not knowing standard deviation)	Confidence intervals obtained using the sample standard deviation are different from those obtained using the population standard deviation. Investigates effect of not knowing the population standard deviation.	Simulates selecting 100 random samples from the population and finds the 95% z-interval and 95% t-interval for each. Students specify sample size, distribution shape, and population mean and standard deviation; applet plots confidence intervals and reports number and proportion containing true mean.	**5.3**, 287; **5.4**, 287
Hypothesis tests for a proportion	Not all tests of hypotheses lead correctly to either rejecting or failing to reject the null hypothesis. Investigates the relationship between the level of confidence and the probabilities of making Type I and Type II errors.	Simulates selecting 100 random samples from population; calculates and plots z-statistic and P-value for each. Students specify population proportion, sample size, and null and alternative hypotheses; applet reports number and proportion of times null hypothesis is rejected at 0.05 and 0.01 levels.	**6.5**, 354; **6.6**, 354
Hypothesis tests for a mean	Not all tests of hypotheses lead correctly to either rejecting or failing to reject the null hypothesis. Investigates the relationship between the level of confidence and the probabilities of making Type I and Type II errors.	Simulates selecting 100 random samples from population; calculates and plots t statistic and P-value for each. Students specify population distribution shape, mean, and standard deviation; sample size, and null and alternative hypotheses; applet reports number and proportion of times null hypothesis is rejected at both 0.05 and 0.01 levels.	**6.1**, 329; **6.2**, 334; **6.3**, 334; **6.4**, 334
Correlation by eye	Correlation coefficient measures strength of linear relationship between two variables. Teaches user how to assess strength of a linear relationship from a scattergram.	Computes correlation coefficient r for a set of bivariate data plotted on a scattergram. Students add or delete points and guess value of r; applet compares guess to calculated value.	**10.2**, 597
Regression by eye	The least squares regression line has a smaller SSE than any other line that might approximate a set of bivariate data. Teaches students how to approximate the location of a regression line on a scattergram.	Computes least squares regression line for a set of bivariate data plotted on a scattergram. Students add or delete points and guess location of regression line by manipulating a line provided on the scattergram; applet plots least squares line and displays the equations and the SSEs for both lines.	**10.1**, 572

MyStatLab

Dear Temple Student:

In this course you will be using MyStatLab – an online resource to help you assess your math skills, do homework, take tests, view tutorial videos, and more!

What you need to begin:

- **A Student Access Kit** (packaged inside a new textbook, purchased online in step 3, or with a standalone access kit).
- **A valid email address**
- **Your school's zip code:**
  Main Campus: 19122
  Ambler: 19002
  TUCC: 19102

In addition, your instructor will provide you with:

- **A "Course ID" Number:** ______________________________

Student Registration:

1. Go to www.coursecompass.com.
2. Click on the **Register** button below Students
3. Review the **Before You Start** information. If you have everything you need to register, click **Next.**
4. Enter your **Course ID** exactly as provided by your instructor and click **"Find Course."** *Your course information should appear. If not, contact your instructor to verify the correct Course ID.*
5. Select **Access Code** and type your **Access Code** in the fields provided (one word per field) and click **Next.**
6. Read the License Agreement and Privacy Policy and click **"I Accept."**
7. On the Access Information Screen, you'll be asked whether you already have a Pearson Education Account. Click:
   - **"YES"** if you have registered for other Pearson online products and already have a login name and password. Boxes will appear for you to enter your existing login information.
   - **"NO"** if this is the first time you have registered for a Pearson online product. Boxes will appear for you to create your login name and password.
   - **"NOT SURE"** if you want to check for a pre-existing account and receive an email with your login name and password.
8. Simply follow the registration screens and enter your information as prompted. You will enter your name, email address, school information, and provide a security question/answer to ensure the privacy of your account.

Once your registration is complete, you will see a **Confirmation** screen (this information will also be emailed to you). Simply print your confirmation and you are now ready to Log in and access your resources!

Important: You **MUST** install all required software **plugins** to use this website from a home or work computer. A link to these plugins is located from within the MyStatLab – Look for the **Installation Wizard** or **Browser Check**, often found under Announcements or as a button on the left. AOL users must minimize that browser and use Internet Explorer. You must also disable any pop-up blockers for this site only.

Technical Support

Contact Product Support at **http://www.mymathlab.com/contactus.html** for student "getting started" information, to fill out an online request form, or use Pearson's live Chat feature to talk with a member of the support team.

EXCEL 2007 Statistical Functions

CONTENTS

I. Getting Started with MS Excel

Introduction:
Microsoft Excel is a spreadsheet application and is a powerful computation and presentation tool used extensively in just about any desk job. The format is so easy to use and flexible, that everyone uses it daily from administrators to fluid dynamic engineers and yes, even by those who do data analysis and statistics. Below are instructions for one possible way to do things. In Excel there are many ways to do any one thing. Over time, you will become familiar with more and more ways. Use the Help Menu (click the white question mark in the blue circle on the top right of the screen or hit F1) if you need any additional information as this document only covers the very basic functions of Excel. Follow directions labeled "*Tutorial*" for a simple example.

Open a File/Workbook:
To open an Excel file in Excel, either double click on the file icon itself. Alternatively, you can start Excel then open the file. There are several ways to do this.

There are two ways to open Excel:
1. Double click on the Excel 2007 icon
2. Go to the *Start* menu -> *Programs* -> Microsoft Office 2007 -> Excel

Once Excel is open, you need to open the file. This can be done by one of several ways:

1. Click the *Office Button* (top left corner of the screen) -> *Open* to browse folders
2. Use the keystrokes "Control+O"
3. Select the open folder icon from the tool bar to the right of the office button (Warning: this may not be displayed on your copy).

Tutorial: Start Excel.

Create and Save a New File/Workbook:
There are several ways to create a new file in Excel:
1. Click the *Office Button -> New -> Blank Workbook.*
2. Use the keystrokes "Control+N"
3. Select the new file icon from the tool bar to the right of the office button (Warning: this may not be displayed on your copy)

You can save the new spreadsheet by:
1. Click the *Office Button -> New*

2. Use keystroke "Control+S"
3. Select the save file icon from the tool bar to the right of the office button.

Tutorial: Create a new file (known as a spreadsheet) in Excel. Save the new file on your desktop as "x*tutorial*".

Selecting Cells:
To select a cell in Excel, use the mouse to click on one. By holding the left mouse button down as you move your mouse, one can select many cells at once. For complicated non-connected cell selection, experiment with holding down the Shift key and the Control key as you select your cells. Cell selection can also be done without the mouse using the Shift key in concert with the arrow keys.

Notice if you select the very first cell (in upper left hand corner of spreadsheet) the "address" of the cell is displayed in the name box (box on left on line above cells). The address of the very first cell is "A1". Notice the labels for rows are numbered and the labels for the columns are lettered. These addresses are very important.

You can also select whole rows or whole columns with a single mouse click by mouse clicking on the name of the row or column. You can also select groups of rows or groups of columns. You can also change the width of any row or column by moving the mouse curser in between two labels for either the rows or the columns. When you find the right spot, the curser will change to a double arrow. At that point, hold the mouse button down and more the mouse to specify what size you want your row or column.

Tutorial: Select cells A1, A2, A3. Deselect them by clicking on a different cell. Select row 3. Now select column D. Deselect the column.

Enter Values or Text Into a Cell:
Simply select a cell and type numbers or words into a cell. The contents of the cell will display in the formula bar (to the right of the name box). To empty a cell, select the cell and type "Backspace" or "Delete". To empty a group of cells, select the group, and right click to select "Clear Contents".

If you don't like where something you entered is located, you can select the cells you wish to move and carefully move the mouse curser over the border of the selected region. When the mouse curser turns to an arrow, press and hold the mouse button and simply move the mouse around until you fine where you want your section of cells to go!

Sometimes the Format of a cell will change the visual appearance of cell contents. For example, if you enter "3/25/5", Excel may change it to "March 25, 2005" automatically. To specify what Format you wish a cell to have, select the cell of group of cells and either right click on your selection to get to "Format Cells..." or use the Menu Item: *Home* Tab -> *Format* (in second box from right on Home ribbon) -> *Format Cells...* or use keystroke "Control+1". Then specify your desired format – use Text format to have Excel leave your character entry exactly as entered.

Tutorial: Enter into cell "A1" the value 5, into cell "A2" the value 7, and in cell "A3" the value 9. In cell "D1" type," This is my first Excel Spreadsheet". Notice the contents overrun the cell you typed in. Resize the column width for column D (hint: one you get the double arrow, you can simply double click and Excel automatically sizes the column to the largest entry in that column). Move the contents of cell "D1" to cell "D2".

Perform Calculations:
Excel is capable of doing just about any calculation you wish. Also, it is programmable and will dynamically update values as you change things in your spreadsheet. If you wish to add two numbers together such as 5 + 7, select a cell that is not used and type, "=5+7" and then press "Enter" (a 12 will appear in the cell). Or instead of typing the 5 and 7 in your expression, you can replace them with a cell address where you may already have numbers stored.

Tutorial: Select cell "C2". Then enter "=A1 + A2" and press "Enter" (a 12 will appear in cell "C2").

If you wish to calculate a value based on a column or row of numbers already entered in your spreadsheet (or any subset of numbers anywhere), simply select a cell where you wish your calculated value to go. Then either click the f_x (left of the formula bar) or use the Menu Item: *Formulas* Tab -> *Insert Function* (large f_x on left). Either type a search term in the first box or select a category (e.g. financial, statistical, etc.) and a function (e.g. sum, average, minimum, etc.) A dialogue box will most likely appear. Click on the blue and red button to define what cells to use for your function. Use the mouse to physically select the cells with the values you wish your calculation to use. Press "Enter" when selected. Notice the sequence of addresses of your selected cells (not the values of the cells) is recorded. Then click on "OK" and the calculation value will appear in the cell. If you know the exact typing needed for your function as you use it often, you don't need to use the menu at all. You can type it in directly. For example, to sum the first three values in the first column (A), select the fourth cell in column A (cell A4) or any unused cell and type,"=sum(A1:A3)". This will sum the values starting at cell "A1" through "A3".

Tutorial: Select cell "A4" and sum the first three values of column A into that cell.

Automatic Propagation of Calculations:
One of the most powerful aspects of Excel is its ability to automatically propagate your calculations.

For example, lets say you teach a history course and you have 40 students whose names all appear in column A of your spreadsheet. In column B you have each students Homework 1 score, in column C you have all the Homework 2 scores, etc. up to column K for 10 assignments. You wish to use Excel to calculate each students score for the course overall. Since all the homeworks are scored as a percentage, you simply want to average each students scores to determine their final percentage score for the course. Lets say the first student information is listed in the second row since the first row has column heading information such as "Name, HW1, HW2, HW3,..., HW10, Final Score". You would select cell "L2" which is at the end of the row for the first student and type in, "=average (B2:K2). Then press "Enter". This will calculate the average score for the first student whose scores appear in cells "B2" through "K2" and display it in cell "L2".

Now, we wish to automatically propagate this average calculation to the other 39 students without programming for each and every student. To do this, simply select cell "L2" as your cell to propagate. Notice the cell is indicated with a border around the cell as usual. However, notice that the lower right hand corner of the cell has a small square on it. Move the mouse curser over this point until it changes to a small line cross. Once the mouse curser is a line cross, hold the mouse button down and drag downward until the 40th student cell is reached (row 41). Release the mouse. Averages for each student are automatically calculated in one move!

Tutorial: Select cell "B1" and type, "=A1 * 5". Press "Enter". This will multiply the value in cell "A1" by 5 and place it in cell "B1" (should get 25). Propagate this new calculation by selecting the cell "B1" again, moving the mouse curser until it is a line cross, and dragging down to cell "B3" (should get 35, and 45). To calculate these new values times 5 again, select all three new values with the mouse, get the line cross, and drag one column to the right, and release. You should have the column 125, 175, 225. Now propagate the sum at the bottom by selecting cell "A4" and propagating it two cells to the right (should get 105, 525). This is a very important function of Excel.

Charts and Graphs:

Another one of the most commonly used aspects of Excel is its ability to create all types of charts and graphs. They could be pie charts, bar charts, line charts, 3-D charts, etc. The complexity and variations possible are almost limitless. This document only shows how to make the simplest versions to get you started.

To create a chart or graph of data, select the data using the mouse. The data should be arranged so that a single variable is in one column (or row). If you have multiple variables you wish to graph all at once, select all of them (but each variable should be in respective columns). Then choose the Menu Item: *Insert* Tab -> select chart type. For other chart options, you can either select *Other Charts -> All Chart Types* or click on the small arrow in a square that is located in the bottom right corner of the charts box. Once you select the type of chart you are interested in, you will be given opportunities to choose what type of chart you want, to add labeling to your axes, as well as a host of other options to use.

Tutorial: To make a bar chart of the three columns of three values we created so far (not including the sums), select cells "A1", "A2", "A3", "B1", "B2", "B3", "C1", "C2", "C3" all at once. We wish to create a chart, which treats each of the three columns as a different variable. Select the Menu Item: *Insert* Tab -> *Column -> 2-D Clustered Column* (first chart in the top row). You can move the chart location on the existing sheet by clicking on the chart, dragging it to a new location and dropping it there, or you can cut and paste it in a new location. To move it to a different sheet or a new sheet, click on *Move Chart Location*. This brings up a dialog box, and you can specify which sheet in which you would like it placed. Now you can add labels to your axes and select many other options (e.g. legend, data labels, grid lines, etc.). To do this, click on the *Layout* tab and the preferred option(s). Congratulations, you have created a chart.

To Install Analysis Toolpak:
For the Statistics courses at Temple University, each student will have to install the Analysis Toolpak on their own computer at home or at work. The computers in the basement of Speakman Hall (available to all students) already have this Toolpak installed. This Toolpak allows for the complex analysis necessary for these courses. It is fortunate that Excel can also perform these high level analyses, so that when you are at work, you will have these techniques available to you. Not all work peers will be familiar with all the high level analyses you will be trained on from these two courses, so this is an important aspect of your skill set which will set you apart from others in the workforce. Learn the material in these courses well.

To install the Analysis Toolpak: *Office Button -> Excel Options* (bottom of box) -> *Add-Ins* (3rd item from bottom on the menu on the left). If the analysis toolpak is listed under "Active Applications", it is installed and you can hit "Cancel". If not, it will be listed under "Inactive Applications." Highlight it, and click *Go*. This brings up a dialog box with available Add-Ins. Put a checkmark next to "Analysis Toolpak," and check *OK*. This should prompt a message indicating the Add-in is not currently installed and asking if you would like to install it. Click yes, and follow the directions on the screen. Depending on the original Excel installation, you may have to insert the original Microsoft Office installation disks if your computer requests them (so be prepared). If you are at work, you may need to get assistance from your IT representative (so work on this issue in advance of your needing the Analysis Toolpak). There you have it!

II Statistical Functions

This section presents some select statistical functions for Excel for Windows to make it easier for instructors and students to use Excel. Data files are provided in Excel format in the CD for most of the examples, exercises and cases in the textbook. Your instructor may cover additional functions in class.

Histogram

1. Type or import the data into column A.
2. In column B type "Bins" in the first cell and then enter the upper limit of each class (optional).
3. Click **Data** tab -> **Data Analysis** -> **Histogram**.
4. Type the **Input Range** of the data. Include the cell containing the name of the variable (if applicable).
5. Type the **Input Range** of the bins.
6. If the names of the variables have been included in the first row of the ranges click **Labels**.
7. Click **Chart Output** and **OK**. This will create a chart with gaps between the rectangles.
8. To remove the gaps between the bars place the cursor over one of the rectangles and click the right button of the mouse.
9. Click (with the left button) **Format Data Series...**
10. Click **Options**, move the pointer to **Gap Width**: and change the number from 150 to 0. Click **OK**.

Bar Chart

There are two ways to make a bar chart in Excel:

A. Proceed through the first seven steps in constructing a histogram as described above. Use the codes representing the categories as the upper limits of the bins. The result is a bar chart.

B. If you have determined the frequency with which each value occurs follow the instructions below.
   1. Type or import the frequencies into a column.
   2. Highlight the frequencies, click the **Insert** tab -> **Bar** -> **2-D Clustered Bar** (first chart in top row).
   3. If you wish you can include the name of the category in a cell at the top of the column.

Pie Chart

1. Type or import the frequencies into a column.
2. Highlight the frequencies, click the **Insert** tab -> **Pie**, -> **Pie** (first chart in top row).

Scatter Diagram

1. Type or import the data into two adjacent columns. Store variable X in the first column and variable Y in the next column.
2. Highlight the data
3. Click the **Insert** tab -> **Scatter** -> **Scatter Only with Markers** (first chart in top row).
4. If you wish to label the chart and axis, click **Layout** tab -> **Chart Title** or **Axis Titles** (*Primary Horizontal* or *Primary Vertical*) and fill in the boxes.
5. To eliminate the horizontal lines that automatically appear, click **Layout** tab -> **Gridlines** -> **Primary Horizontal Gridlines** -> **None**.

Descriptive Statistics

1. Type or import the data.
2. Click **Data** tab -> **Data Analysis** -> **Descriptive Statistics.**
3. Specify the **Input Range.**
4. Click **Labels in First Row** if applicable, check **Summary Statistics** and click **OK.**

Covariance/Correlation

1. Type or import the data into two adjacent columns.
2. Click **Data** tab -> **Data Analysis** -> **Covariance.**
3. Type the **Input Range.**
4. Click **Labels in First Row** if applicable, and click **OK.**
5. For correlation, click **Correlation** instead of **Covariance** in the instructions above.

Binomial Probabilities

1. Click **Formulas** tab - > f_x, and either enter "binomial" in the search box or pick **Category: Statistical**, and **Function Name: BINOMDIST**. Click **OK.**
2. Type the value of x (**Number_s**), the number of trials, n (**Trials**), the probability of success, p (**Probability_s**), and "**true**" (**Cumulative**) to yield a cumulative probability or

"false" for an individual probability. The probability appears on the right side of the dialog box. Clicking **OK** will print the probability in the active cell.

Normal Probabilities

To compute cumulative normal probabilities P(X<x), proceed as follows:

1. Click **Formulas** tab - > f_x, and either enter "normal" in the search box or pick **Category: Statistical**, and **Function Name: NORMDIST**. Click **OK**. (Note: do not confuse NORMDIST with NORMSDIST. They are two different functions).

2. Type the value of x (**X**), the mean (**Mean**), the standard deviation (**Standard_dev**), and **"true" (Cumulative**) to yield a cumulative probability. The probability appears on the right side of the dialog box. Clicking **OK** will print the probability in the active cell.

To determine a value of X given a cumulative probability, follow these instructions:

1. Click **Formulas** tab - > f_x, and either enter "inverse normal" in the search box or pick **Category: Statistical**, and **Function Name: NORMINV**. Click **OK**.

2. Type the cumulative probability (**Probability**), the mean (**Mean**), the standard deviation (**Standard_dev**). The value of x will appear in the dialog box. Clicking **OK** will print the probability in the active cell.

Testing Two Independent Means

1) Retrieve and open data file from Folder or CD-ROM
2) Click: **Data** tab -> **Data Analysis - > t-Test: Two-Sample Assuming Equal Variances**
3) Input: Variable 1 Range
4) Input: Variable 2 Range
5) Enter Hypothesized Mean Difference (e.g., zero)
6) Check "labels" box if variable name used
7) Enter Desired Alpha Level (to obtain 1 & 2 tailed t critical values in output)
8) Output Display Options – select location to display output: cell range (select staring point), new worksheet page (the default), or new workbook (i.e. a new Excel file)
9) Click "OK"

Conduct F test for Homogeneity of Variance

1) Retrieve and open data file from Folder or CD-ROM
2) Click: **Data** tab -> **Data Analysis - > F-Test Two-Sample For Variances**
3) Input: Variable 1 Range
4) Input: Variable 2 Range
5) Check "labels" box if variable name used
6) Enter Desired Alpha Level (to obtain F critical value in output)
7) Output Display Options – select location to display output: cell range (select staring point), new worksheet page (the default), or new workbook (i.e. a new Excel file)
8) Click "OK"

Testing Two Dependent Means (Matched-Pairs)

1) Retrieve and open data file from Folder or CD-ROM

2) Click: **Data** tab -> **Data Analysis - > t-Test: Paired Two Sample for Means**
3) Input: Variable 1 Range
4) Input: Variable 2 Range
5) Enter Hypothesized Mean Difference (e.g., zero)
6) Check "labels" box if variable name used
7) Enter Desired Alpha Level (to obtain 1 & 2 tailed t critical values in output)
8) Output Display Options – select location to display output: cell range (select staring point), new worksheet page (the default), or new workbook (i.e. a new Excel file)
9) Click "OK"

Testing ANOVA SINGLE FACTOR

1) Retrieve and open data file from Folder or CD-ROM
2) Click: **Data** tab -> **Data Analysis - > Anova: Single Factor**
3) Input: Range
6) Check "labels in first row" box if variable name used
7) Enter Desired Alpha Level (to obtain F critical value in output)
8) Output Display Options – select location to display output: cell range (select staring point), new worksheet page (the default), or new workbook (i.e. a new Excel file)
9) Click "OK"

Obtain Simple Regression Analysis and Plots

1) Retrieve and open data file from Folder or CD-ROM
2) Click: **Data** tab -> **Data Analysis - > Regression**
3) Input: Y Range
4) Input: X Range
5) Check "labels" box if variable names used
6) Enter degree of confidence (95% is default value)
7) Output Display Options – select location to display output: cell range (select staring point), new worksheet page (the default), or new workbook (i.e. a new Excel file)
8) Other Output Options – select from the following as desired: residuals, standardized residuals, residual plots, line fit plots, or normal probability plots
9) Click "OK" (provides Regression Statistics, ANOVA, etc.)

Obtain Intercorrelations Among all Regression Variables (Correlation Matrix)

1) Retrieve and open data file from Folder or CD-ROM
2) Click: **Data** tab -> **Data Analysis - > Correlation**
3) Data Range: Select Y and all X variables simultaneously using click & drag method (Note: columns must be adjacent to each other.)
4) Check either **column** or **row** to indicate data grouping (column is the default).
5) Check "labels in first row" box if variable names used
6) Output Display Options – select location to display output: cell range (select staring point), new worksheet page (the default), or new workbook (i.e. a new Excel file)
7) Click "OK" (note, does not provide p-value(s) or sample size)

Obtain Multiple Regression Analysis and Plots

1) Retrieve and open data file from Folder or CD-ROM
2) Click: **Data** tab -> **Data Analysis - > Regression**
3) Input: Y Range
4) Input: X Range – select all columns with predictors simultaneously (Note: columns must be adjacent to each other.)
5) Check "labels" box if variable names used
6) Enter degree of confidence (95% is default value)
7) Output Display Options – select location to display output: cell range (select staring point), new worksheet page (the default), or new workbook (i.e. a new Excel file)
8) Other Output Options – select from the following as desired: residuals, standardized residuals, residual plots, line fit plots, or normal probability plots
9) Click "OK"

Statistics
for Business and Economics

Statistics for Business and Economics

Custom Edition for Temple University

Fox School of Business

James T. McClave
P. George Benson
Terry Sincich

Taken from:
Statistics for Business and Economics
Eleventh Edition
by James T. McClave, P. George Benson, and Terry Sincich

Learning Solutions

New York Boston San Francisco
London Toronto Sydney Tokyo Singapore Madrid
Mexico City Munich Paris Cape Town Hong Kong Montreal

The credit for this photo is Temple University Photography.

Excerpts taken from:

Statistics for Business and Economics, Eleventh Edition
by James T. McClave, P. George Benson, and Terry Sincich

Published by Prentice Hall
Upper Saddle River, New Jersey 07458

Pearson Learning Solutions, 501 Boylston Street, Suite 900, Boston, MA 02116
A Pearson Education Company
www.pearsoned.com

Printed in the United States of America

1 2 3 4 5 6 7 8 9 10 V303 15 14 13 12 11 10

000200010270592891

IW

PEARSON

ISBN 10: 0-558-83312-8
ISBN 13: 978-0-558-83312-1

Contents

*Indicates an optional chapter.

*Indicates an optional chapter.

SAVE 50¢
ON TWO
$5.00 OFF

SINGLE
LANE
AHEAD

Appendices

Preface

Incorporating extensive feedback from reviewers, we refined the eleventh edition of *Statistics for Business and Economics* to build on its reputation as the textbook for professors seeking a balanced and comprehensive approach. As in earlier editions, this revision contains an abundance of real-world business applications and many new and updated exercises. We focus on the data collection and analysis techniques students will need to evaluate reports of statistical studies and make informed business decisions. We emphasize inference, and we assume a mathematical background of basic algebra. Because it is so important, we continue to stress the development of statistical thinking, the assessment of credibility, and the value of inferences made from data, both by those who consume and those who produce them.

The text incorporates the following strategies, developed from the American Statistical Association (ASA)-sponsored conferences on *Making Statistics More Effective in Schools of Business* (MSMESB) and ASA's *Guidelines for Assessment and Instruction in Statistics Education* (GAISE).

1. Emphasize statistical literacy and develop statistical thinking
2. Use real data in applications
3. Use technology for developing conceptual understanding and analyzing data
4. Foster active learning in the classroom
5. Stress conceptual understanding rather than mere knowledge of procedures
6. Emphasize intuitive concepts of probability

New to This Edition

In this eleventh edition, we've made changes to strengthen our real-world focus, make use of the latest technology, and improve readability.

- **Updated exercises.** Thirty percent of the book's exercises have been updated. The text contains more than 1,200 exercises, and many are based on contemporary business-related studies and use real data. Most exercises foster and promote critical thinking skills. We also provide a few simpler exercises with smaller data sets in the first few exercises in the "Applying the Concepts—Basic" sections.
- **New Statistics in Action and Making Business Decisions cases.** Five of the 14 Statistics in Action cases are new, each based on real data from a recent business study. We've also added a new Making Business Decisions case after Chapter 12 (these were titled Real-World Case in previous editions).
- **Emphasis on ethics.** Where appropriate, Ethics in Statistics discussions have been added in every chapter to emphasize the importance of ethical behavior when collecting, interpreting, and reporting on data with statistics. These boxes provide the student with reminders of how to apply sound statistical methodology and to recognize when unethical statistics are clearly employed in a study.
- **Updated examples.** Twenty percent of the book's examples have been updated with the most current business-related data to be more timely and relevant.
- **Updated technology.** The Using Technology sections at the end of each chapter have been streamlined into easy-to-use steps. All output from statistical software (Excel®/DDXL, SPSS/PASW®, and Minitab®, plus the TI-84 graphing calculator) and corresponding instructions have been revised to reflect the latest versions of the technology.

- **Additional probability coverage.** Hypergeometric and exponential probability distributions are now introduced in Chapter 4 to give instructors added flexibility when covering probability.
- **Redesigned end-of-chapter summaries.** End-of-chapter notes have been redesigned for improved readability and reference. Important points are reinforced through flowcharts (which aid in selecting the appropriate statistical method), and Key Terms, Key Symbols or Formulas, and Key Ideas reinforce the chapter's concepts in a linear, easy-to-read way.
- **Updated data sets.** All data sets have been updated to include data from the year 2000 and later (with the exception of a few cases where the study is considered a "classic").

Content-Specific Changes to This Edition

- **Chapter 4 (Random Variables and Probability Distributions).** At the request of reviewers, we've added back material on two probability distributions—the hypergeometric and the exponential distributions. For pedagogical purposes, we've organized the chapter into three parts: "Discrete Random Variables" (the binomial, Poisson, and hypergeometric distributions), "Continuous Random Variables" (the normal, uniform, and exponential distributions), and "Random Variables Based on Sampling Distributions."
- **Chapter 5 (Confidence Intervals).** The methodology for finding a confidence interval for a population mean is developed based on using either a normal (z) statistic (Section 5.2) or a Student's t-statistic (Section 5.3). This change allows instructors to emphasize the theory behind confidence intervals while providing the flexibility of focusing on whether the sample is large or small.
- **Chapter 6 (Tests of Hypotheses).** We've added a new section emphasizing the formulation of the null and alternative hypotheses (Section 6.2).
- **Chapter 10 (Simple Linear Regression).** To streamline the chapter, we've combined and refined several sections from the previous edition. We now include the section on estimating σ^2 in Section 10.3 ("Model Assumptions") and combine the sections on the coefficients of correlation and determination into a single section (Section 10.5).
- **Chapter 11 (Multiple Regression and Model Building).** For pedagogical purposes, the chapter is now divided into three parts: "First-Order Models with Quantitative Independent Variables," "Model Building in Multiple Regression," and "Multiple Regression Diagnostics."

Traditional Strengths and Features

We have maintained or strengthened the pedagogical features that make *Statistics for Business and Economics* unique among introductory business statistics texts. These features assist the student in achieving an overview of statistics and an understanding of its relevance in both the business world and everyday life.

- **Use of examples as a teaching device.** Almost all new ideas are introduced and illustrated by data-based applications and examples. We believe that students understand definitions, generalizations, and theoretical concepts better *after* seeing an application. All examples have three components: (1) Problem, (2) Solution, and (3) Look Back. This step-by-step process provides students with a defined structure to approach problems, and it enhances their problem-solving skills. The Look Back feature often gives helpful hints to solving the problem and/or provides a further reflection or insight into the concept or procedure that is covered. At least 20 percent of the Examples have been updated.
- **Now Work exercises.** Now Work exercise suggestions follow most examples. The Now Work exercise (marked with the NW icon NW in the exercise sets) is similar in style and concept to the text example. This provides students with an opportunity to immediately test and confirm their understanding.

- **Statistics in Action.** Each chapter begins with a business case based on an actual contemporary, controversial, or high-profile issue. We present relevant research questions and data from the study and demonstrate the proper analysis in short Statistics in Action Revisited boxes throughout the chapter. These motivate students to critically evaluate the findings and think through the statistical issues involved.
- **Hands-on activity features.** In each chapter, we provide students with several opportunities to participate in hands-on classroom activities, ranging from data collection to formal statistical analysis. Students can perform these activities individually or as a class.
- **Applet exercises.** The text is accompanied by a CD containing applets (short JAVA computer programs). These point-and-click applets allow students to easily run simulations that visually demonstrate some of the more difficult statistical concepts (such as sampling distributions and confidence intervals). Each chapter contains several optional applet exercises in the exercise sets. They are denoted with the following icon:
- **Making Business Decisions cases.** Seven extensive business problem-solving cases, with real data and assignments for the student, cover two or three chapters. Each case serves as a capstone and review of the material that precedes it and requires the student to apply the methods presented in these chapters.
- **Real data-based exercises.** The text includes more than 1,200 exercises based on a wide variety of business applications. Nearly all of the applied exercises use current, real data extracted from newspapers, magazines, current business journals, and the Internet. Some students have difficulty learning the mechanics of statistical techniques when all problems are couched in terms of realistic applications. For this reason, most exercise sections are divided into at least four parts:

  Learning the Mechanics. Designed as straightforward applications of new concepts, these exercises test students' ability to comprehend a mathematical concept or a definition.

  Applying the Concepts—Basic. Based on applications taken from a wide variety of journals, newspapers, and other sources, these short exercises with smaller data sets help students begin developing the skills necessary to diagnose and analyze real-world problems.

  Applying the Concepts—Intermediate. Based on more detailed real-world applications, these exercises require students to apply their knowledge of the technique presented in the section.

  Applying the Concepts—Advanced. These more difficult real-data exercises require students to use their growing knowledge and developing critical thinking skills.

  Critical Thinking Challenges. These appear at the end of the "Supplementary Exercises" section and present students with one or two challenging real-life problems.

- **Exploring data with statistical software and the graphing calculator.** We demonstrate each statistical analysis method presented using output from three leading statistical software packages: Excel (with DDXL—an Excel add-in included on the companion CD), SPSS (PASW), and Minitab. These outputs appear throughout the text in examples and exercises, exposing students to the output they will encounter in today's high-tech business world. In addition, we provide output and keystroke instructions for the TI-83/TI-84 Graphing Calculator in the Using Technology section at the end of each chapter.
- **Using Technology tutorials.** At the end of each chapter we've included statistical software tutorials with instructions and screen shots for Minitab, SPSS (PASW), Excel (using the DDXL add-in), and the TI-83/TI-84 Graphing Calculator. These step-by-step tutorials, which have been streamlined for this edition, are easily located and show students how to best use statistical software.

- **Profiles of statisticians in history (Biography).** Brief descriptions of famous statisticians and their achievements are presented in side boxes with our hope that students will develop an appreciation of the statistician's efforts and the discipline of statistics as a whole.
- **Companion CD-ROM.** New copies of the text are accompanied by a CD that contains data set files for all of the text examples, exercises, Statistics in Action, and Making Business Decisions cases marked with a . Data sets are formatted for Excel, Minitab 14, Minitab 15, JMP, and SPSS (PASW), and are available as .txt files. The CD also contains DDXL (an Excel add-in), Chapters 13 and 14, and a set of applets that illustrate statistical concepts.

Flexibility in Coverage

We have designed the text to allow the instructor flexibility in coverage of topics. Here are suggestions for two topics, probability and regression:

- **Probability and counting rules.** One of the most troublesome aspects of an introductory Business Statistics course is the study of probability. Probability poses a challenge for instructors because they must decide on the level of presentation, and students find it a difficult subject to comprehend. We believe one cause of these problems is the mixture of probability and counting rules that occurs in most introductory texts. Consequently, we have included the counting rules (with examples) in an appendix (Appendix A) rather than in the body of Chapter 3. Thus, the instructor can control the level of coverage.
- **Multiple regression and model building.** This topic represents one of the most useful statistical tools for the solution of applied problems. Although an entire text could be devoted to regression modeling, we present coverage that is understandable, usable, and much more comprehensive than the presentations in other introductory Business Statistics texts. We devote two full chapters to discussing the major types of inferences that can be derived from a regression analysis, showing how these results appear in the output from statistical software, and, most importantly, selecting multiple regression models to be used in an analysis. Thus, the instructor has the choice of a one-chapter coverage of simple linear regression (Chapter 10), a two-chapter treatment of simple and multiple regression (excluding the sections on model building in Chapter 11), or complete coverage of regression analysis, including model building and regression diagnostics. This extensive coverage of such useful statistical tools provides added evidence to the student of the relevance of statistics to real-world problems.
- **Footnotes.** Although the text is designed for students with a noncalculus background, footnotes explain the role of calculus in various derivations. Footnotes are also used to inform the student about some of the theory underlying certain methods of analysis. These footnotes allow additional flexibility in the mathematical and theoretical level at which the material is presented.

Supplements

For the Student

Student's Solutions Manual, by Nancy Boudreau (Bowling Green State University), provides detailed, worked-out solutions to all odd-numbered text exercises. (ISBN-10: 0-321-64175-2; ISBN-13: 978-0-321-64175-5)

Excel Technology Manual, by Mark Dummeldinger (University of South Florida), provides tutorial instruction and worked-out text examples for Excel. (ISBN-10: 0-321-64180-9; ISBN-13: 978-0-321-64180-9)

Business Insight Videos. This series of videos, each about a well-known business and the challenges it faces, focuses on statistical concepts as they pertain to the real world. The videos can be downloaded to Video iPods® from within MyStatLab. Contact your Pearson representative for details.

Study Cards for Business Statistics Software. This series of study cards, available for Excel (0-321-64191-4), Minitab (0-321-64421-2), JMP (0-321-64423-9), SPSS/PASW (0-321-64422-0), R (0-321-64469-7), StatCrunch (0-321-62892-6), and TI-83/84 graphing calculators (0-321-57077-4) provides students with easy step-by-step guides to the most common business statistics software.

For the Instructor

Annotated Instructor's Edition contains answers to text exercises. Annotated marginal notes include Teaching Tips, suggested exercises to reinforce the statistical concepts discussed in the text, and short answers to the exercises within the exercise sets. (ISBN-10: 0-321-64176-0; ISBN-13: 978-0-321-64176-2)

Instructor's Solutions Manual, by Nancy Boudreau (Bowling Green State University), provides detailed, worked-out solutions to all of the book's exercises. Careful attention has been paid to ensure that all methods of solution and notation are consistent with those used in the core text. Available for download at www.pearsonhighered.com/irc.

TestGen® (www.pearsonhighered.com/testgen) enables instructors to build, edit, print, and administer tests using a computerized bank of questions developed to cover all the objectives of the text. TestGen is algorithmically based, allowing instructors to create multiple but equivalent versions of the same question or test with the click of a button. Instructors can also modify test bank questions or add new questions. Tests can be printed or administered online. The software and testbank are available for download from Pearson Education's online catalog (www.pearsonhighered.com/irc).

Technology Resources

A companion **CD-ROM** is bound in new copies of *Statistics for Business and Economics.* The CD holds a number of support materials, including:

- **Data sets** formatted for Minitab 14 and 15, SPSS (PASW), JMP, text, and Excel.
- **DDXL,** an Excel add-in, adds sound statistics and statistical graphics capabilities to Excel. Among other capabilities, DDXL adds box plots, histograms, statistical scatterplots, normal probability plots, and statistical inference procedures not available in Excel's "Data Analysis" pack.
- **Applets** (short JAVA computer programs). These point-and-click applets allow students to run simulations that visually demonstrate statistical concepts. (See page xv for a complete description.)
- **Chapters 13 and 14**

ActivStats® for Business Statistics (Mac and PC) by Data Description and Paul Velleman is an award-winning multimedia program that supports learning chapter-by-chapter with the book. It complements the book with videos of real-world stories, worked examples, animated expositions of all major statistics topics, and tools for performing simulations, visualizing inference, and learning to use statistics software. ActivStats includes 15 short video clips; 183 animated activities and teaching applets; 260 data sets; interactive graphs, simulations, and visualization tools; and much more. ActivStats (Mac and PC) is available in an all-in-one version for Data Desk, Excel, JMP, Minitab, and SPSS/PASW. (ISBN-10: 0-321-57719-1; ISBN-13: 978-0-321-57719-1)

MathXL® for Statistics Online Course (access code required) MathXL® for Statistics is a powerful online homework, tutorial, and assessment system that accompanies Pearson textbooks in statistics. With MathXL for Statistics, instructors can

- Create, edit, and assign online homework and tests using algorithmically generated exercises correlated at the objective level to the textbook.
- Create and assign their own online exercises and import TestGen tests for added flexibility.
- Maintain records of all student work, tracked in MathXL's online gradebook.

With MathXL for Statistics, students can:

- Take chapter tests in MathXL and receive personalized study plans based on their test results.

- Use the student plan to link directly to tutorial exercises for the objectives they need to study and retest.
- Students can also access supplemental animations and video clips directly from selected exercises.

MathXL for Statistics is available to qualified adopters. For more information, visit our website at www.mathxl.com, or contact your Pearson sales representative.

MyStatLab™ Online Course (access code required)—part of the MyMathLab® product family—is a text-specific, easily customizable online course that integrates interactive multimedia instruction with textbook content. MyStatLab gives you the tools you need to deliver all or a portion of your course online, whether your students are in a lab setting or working from home.

- **Interactive homework exercises,** correlated to your textbook at the objective level, are algorithmically generated for unlimited practice and mastery. Most exercises are free response and provide guided solutions, sample problems, and learning aids for extra help.
- **Personalized study plan,** generated when students complete a test or quiz, indicates which topics have been mastered and links to tutorial exercises for topics students have not mastered.
- **Multimedia learning aids,** such as video lectures, animations, and a complete multimedia textbook, help students independently improve their understanding and performance.
- **Statistics tools** MyStatLab includes built-in tools for statistics, including StatCrunch. Students also have access to statistics animations and applets that illustrate key ideas for the course. For those who use technology in their course, technology manual PDFs are included.
- **Assessment manager** lets you create online homework, quizzes, and tests that are automatically graded. Select just the right mix of questions from the MyStatLab exercise bank, instructor-created custom exercises, and/or TestGen test items.
- **Gradebook,** designed specifically for mathematics and statistics, automatically tracks students' results and gives you control over how to calculate final grades. You can also add offline (paper-and-pencil) grades to the gradebook.
- **MathXL Exercise Builder** allows you to create static and algorithmic exercises for your online assignments. You can use the library of sample exercises as an easy starting point.
- **Pearson Tutor Center** (www.pearsontutorservices.com) access is automatically included with MyMathLab. The Tutor Center is staffed by qualified math instructors who provide textbook-specific tutoring for students via toll-free phone, fax, e-mail, and interactive Web sessions.

MyStatLab is powered by CourseCompass™, Pearson Education's online teaching and learning environment, and by MathXL®, our online homework, tutorial, and assessment system. MyStatLab is available to qualified adopters. For more information, visit our Web site at www.mystatlab.com or contact your Pearson sales representative.

PowerPoint lecture slides provide an outline to use in a lecture setting, presenting definitions, key concepts, and figures from the text. These slides are available within MyStatLab or at www.pearsonhighered.com/irc.

StatCrunch is an online statistical software website that allows users to perform complex analyses, share data sets, and generate compelling reports of their data. Developed by Webster West, Texas A&M University, StatCrunch already has more than ten thousand data sets available for students to analyze, covering almost any topic of interest. Interactive graphics are embedded to help users understand statistical concepts and are available for export to enrich reports with visual representations of data. Additional features include:

- A full range of numerical and graphical methods that allow users to analyze and gain insights from any data set.
- Flexible upload options that allow users to work with their .txt or Excel® files, both online and offline.
- Reporting options that help users create a wide variety of visually-appealing representations of their data.

StatCrunch is available to qualified adopters. For more information, visit our website at www.statcrunch.com, or contact your Pearson sales representative.

The Student Edition of Minitab is a condensed edition of the professional release of Minitab statistical software. It offers the full range of statistical methods and graphical capabilities, along with worksheets that can include up to 10,000 data points. Individual copies of the software can be bundled with the text. (ISBN-10: 0-321-11313-6; ISBN-13: 978-0-321-11313-9)

JMP Student Edition is an easy-to-use, streamlined version of JMP desktop statistical discovery software from SAS Institute, Inc., and is available for bundling with the text. (ISBN-10: 0-321-67212-7; ISBN-13: 978-0-321-67212-4)

SPSS (PASW), a statistical and data management software package, is also available for bundling with the text. (ISBN-10: 0-321-67537-1; ISBN-13: 978-0-321-67537-8)

Acknowledgments

This book reflects the efforts of a great many people over a number of years. First, we would like to thank the following professors, whose reviews and comments on this and prior editions have contributed to the eleventh edition:

Reviewers of the Eleventh Edition of *Statistics for Business and Economics*

Sukhwinder Bagi, *Bloomsburg University*
John Beyers, *University of Maryland–University College*
Leszek Gawarecki, *Kettering University*
Katarina Jegdic, *University of Houston–Downtown*
Kenneth Leong, *College of New Rochelle*
Mark R. Marino, *Niagara University/Erie Community College*
Richard N. McGrath, *Bowling Green State University*
Keith Ord, *Georgetown University*
Leonard Presby, *William Paterson University*
James Wright, *Chadron State College*

Reviewers of Previous Editions

CALIFORNIA Joyce Curley-Daly, Jim Daly, Robert K. Smidt, *California Polytechnic State University* • Jim Davis, *Golden Gate University* • Carol Eger, *Stanford University* • Paul W. Guy, *California State University, Chico* • Judd Hammack, P. Kasliwal, *California State University, Los Angeles* • Mabel T. King, *California State University, Fullerton* • James Lackritz, *California State University, San Diego* • Beth Rose, *University of Southern California* **COLORADO** Rick L. Edgeman, Charles F. Warnock, *Colorado State University* • Eric Huggins, *Fort Lewis College* • William J. Weida, *United States Air Force Academy* **CONNECTICUT** Alan E. Gelfand, Joseph Glaz, Timothy J. Killeen, *University of Connecticut* **DISTRICT OF COLUMBIA** Phil Cross, Jose Luis Guerrero-Cusumano, *Georgetown University* Gaminie Meepagala, *Howard University* **FLORIDA** John M. Charnes, *University of Miami* • C. Brad Davis, *Clearwater Christian College* • Fred Leysieffer, Pi-Erh Lin, Doug Zahn, *Florida State University* • P. V. Rao, *University of Florida* • Jeffrey W. Steagall, *University of North Florida* • Edna White, *Florida Atlantic University* **GEORGIA** Robert Elrod, *Georgia State University* **HAWAII** Steve Hora, *University of Hawaii, Hilo* **ILLINOIS** Edward Minieka, *University of Illinois at Chicago* • Don Robinson, *Illinois State University* • Chipei Tseng, *Northern Illinois University* • Pankaj Vaish, *Arthur Anderson & Company* **IOWA** Dileep Dhavale, *University of Northern Iowa* • William Duckworth II, William Q. Meeker, *Iowa State University* • Tim E. McDaniel, *Buena Vista University* **KANSAS** Paul I. Nelson, *Kansas State University* • Lawrence A. Sherr, *University of Kansas* **LOUISIANA** James Willis, *Louisiana State University* **MARYLAND** John F. Beyers, Michael Kulansky, *University of Maryland–University College* • Glenn J. Browne, Mary C. Christman, *University of Maryland* **MASSACHUSETTS** Warren M. Holt, *Southeastern Massachusetts University* **MICHIGAN** Atul Agarwal, Petros Ioannatos, *GMI Engineering and Management Institute* • Richard W. Andrews, Peter Lenk, Benjamin Lev, *University of Michigan* • Toni M. Somers, *Wayne State University* • William Welch, *Saginaw Valley State University* • T. J. Wharton, *Oakland University* **MINNESOTA** Gordon J. Alexander, Donald W. Bartlett, David M. Bergman, Atul Bhatia, Benny Lo, Karen Lundquist, Vijay Pisharody, Donald N. Steinnes, Robert W. Van Cleave, Steve Wickstrom, *University of Minnesota* • Daniel G. Brick, Leigh Lawton, *University of St. Thomas* • Susan Flach, *General Mills, Inc.* • David D. Krueger, Ruth K. Meyer, Jan Saraph, Gary Yoshimoto, *St. Cloud State University* • Paula M. Oas, *General Office Products* • Fike Zahroom, *Moorhead State University* **MISSISSIPPI** Eddie M. Lewis, *University of Southern Mississippi* • Alireza Tahai, *Mississippi State University* **MISSOURI** James Holstein, Lawrence D. Ries, *University of Missouri, Columbia* • Marius Janson, *University of Missouri, St. Louis* • Farroll Tim Wright, *University of Missouri* **NEW HAMPSHIRE** Ken Constantine, *University of New Hampshire* **NEW JERSEY** Lewis Coopersmith, *Rider University* • Cengiz Haksever, *Rider University* • Lei Lei, Xuan Li, Zina Taran,

Rutgers University • Philip Levine, *William Paterson University* **NEW MEXICO** S. Howard Kraye, *University of New Mexico* **NEW YORK** James Czachor, *Fordham-Lincoln Center, AT&T* • Bernard Dickman, *Hofstra University* • Joshua Fogel, *Brooklyn College of City University of New York* • Martin Labbe, *State University of New York, College at New Paltz* • G. E. Martin, *Clarkson University* • Thomas J. Pfaff, *Ithaca College* • Gary Simon, *New York University, Stern School of Business* • Rungrudee Suetorsak, *SUNY-Fredonia* **NORTH CAROLINA** Golam Azam, *North Carolina Agricultural & Technical University* • Edward Carlstein, Douglas A. Elvers, *University of North Carolina at Chapel Hill* • Barry P. Cuffe, *Wingate University* • Don Holbert, *East Carolina University* • J. Morgan Jones, *University of North Carolina* • Douglas S. Shafer, *University of North Carolina, Charlotte* **OHIO** William H. Beyer, *University of Akron* • Michael Broida, Tim Krehbiel, *Miami University of Ohio* • Chih-Hsu Cheng, Douglas A. Wolfe, *Ohio State University* • Ronald L. Coccari, *Cleveland State University* • Richard W. Culp, *Wright-Patterson AFB, Air Force Institute of Technology* **OKLAHOMA** Larry Claypool, Brenda Masters, Rebecca Moore, *Oklahoma State University* • Robert Curley, *University of Central Oklahoma* **PENNSYLVANIA** Mohammed Albohali, *Indiana University of Pennsylvania* • Carl Bedell, *Philadelphia College of Textiles and Science* • Douglas H. Frank, *Indiana University of Pennsylvania* • Ann Hussein, *Philadelphia University* • Behnam Nakhai, *Millersville University* • Rose Prave, *University of Scranton* • Farhad Saboori, *Albright College* • Kathryn Szabet, *LaSalle University* • Christopher J. Zappe, *Bucknell University* **SOUTH CAROLINA** Iris Fetta, Robert Ling, *Clemson University* • Kathleen M. Whitcomb, *University of South Carolina* **TENNESSEE** Francis J. Brewerton, *Middle Tennessee State University* **TEXAS** Larry M. Austin, *Texas Tech University* • Jim Branscome, Robert W. Brobst, Mark Eakin, Grace Esimai, Michael E. Hanna, Craig W. Slinkman, *University of Texas at Arlington* • Virgil F. Stone, *Texas A & M University* **VIRGINIA** Edward R. Clayton, *Virginia Polytechnic Institute and State University* **WASHINGTON** June Morita, Kim Tamura, *University of Washington* **WISCONSIN** Ross H. Johnson, *Madison College* **CANADA** Clarence Bayne, *Concordia University* • Edith Gombay, *University of Alberta* **TURKEY** Dilek Onkal, *Bilkent University, Ankara* **OTHER** Michael P. Wegmann, *Keller Graduate School of Management.*

Other Contributors

Special thanks are due to our supplements authors, Nancy Boudreau and Mark Dummeldinger, who have worked with us for many years. C. Brad Davis provided the excellent student activities and applet exercises, while accuracy checkers Jackie Miller and Ann Cannon helped ensure a highly accurate, clean text. Finally, the Pearson Education staff of Deirdre Lynch, Chere Bemelmans, Christina Lepre, Dana Jones, Alex Gay, Kathleen DeChavez, Jean Choe, Tracy Patruno, Raegan Heerema, Thomas Benfatti, and Elm Street Publishing Services' Amanda Zagnoli all helped greatly with various stages of the book and media.

Statistics
for Business and Economics

CONTENTS

1 Statistics, Data, and Statistical Thinking

Where We're Going

- Introduce the field of statistics.
- Demonstrate how statistics applies to business.
- Establish the link between statistics and data.
- Identify the different types of data and data-collection methods.
- Differentiate between population and sample data.
- Differentiate between descriptive and inferential statistics.

Statistics IN Action A *20/20* View of Surveys: Fact or Fiction?

"Did you ever notice that, no matter where you stand on popular issues of the day, you can always find statistics or surveys to back up your point of view—whether to take vitamins, whether daycare harms kids, or what foods can hurt you or save you? There is an endless flow of information to help you make decisions, but is this information accurate, unbiased? John Stossel decided to check that out, and you may be surprised to learn if the picture you're getting doesn't seem quite right, maybe it isn't."

Barbara Walters gave this introduction to a segment of the popular prime-time ABC television program *20/20*. The story was titled "Fact or Fiction?—Exposés of So-Called Surveys." One of the surveys investigated by ABC correspondent John Stossel compared the discipline problems experienced by teachers in the 1940s and those experienced today. The results: In the 1940s, teachers worried most about students talking in class, chewing gum, and running in the halls. Today, they worry most about being assaulted! This information was highly publicized in the print media—in daily newspapers, weekly magazines, Ann Landers's column, the *Congressional Quarterly,* and the *(continued)*

Statistics IN Action *(continued)*

Wall Street Journal, among others—and referenced in speeches by a variety of public figures, including former first lady Barbara Bush and former Education Secretary William Bennett.

"Hearing this made me yearn for the old days when life was so much simpler and gentler, but was life that simple then?" asks Stossel. "Wasn't there juvenile delinquency [in the 1940s]? Is the survey true?" With the help of a Yale School of Management professor, Stossel found the original source of the teacher survey—Texas oilman T. Colin Davis—and discovered it wasn't a survey at all! Davis had simply identified certain disciplinary problems encountered by teachers in a conservative newsletter—a list he admitted was not obtained from a statistical survey, but from Davis's personal knowledge of the problems in the 1940s. ("I was in school then") and his understanding of the problems today ("I read the papers").

Stossel's critical thinking about the teacher "survey" led to the discovery of research that is misleading at best and unethical at worst. Several more misleading (and possibly unethical) surveys, conducted by businesses or special interest groups with specific objectives in mind, were presented on the ABC program. These are listed below. A sixth study, published in the *New York Times*, is also discussed.

The *20/20* segment ended with an interview of Cynthia Crossen, author of *Tainted Truth*, an exposé of misleading and biased surveys. Crossen warns, "If everybody is misusing numbers and scaring us with numbers to get us to do something, however good [that something] is, we've lost the power of numbers. Now, we know certain things from research. For example, we know that smoking cigarettes is hard on your lungs and heart, and because we know that, many people's lives have been extended or saved. We don't want to lose the power of information to help us make decisions, and that's what I worry about."

Reported Information (Source)	Actual Study Information
1. Eating oat bran is a cheap and easy way to reduce your cholesterol. (Quaker Oats)	Diet must consist of nothing but oat bran to reduce your cholesterol count.
2. 150,000 women a year die from anorexia. (Feminist group)	Approximately 1,000 women a year die from problems that were likely caused by anorexia.
3. Domestic violence causes more birth defects than all medical issues combined. (March of Dimes)	No study—false report.
4. Only 29% of high school girls are happy with themselves, compared to 66% of elementary school girls. (American Association of University Women)	Of 3,000 high school girls, 29% responded "Always true" to the statement "I am happy the way I am." Most answered "Sort of true" and "Sometimes true."
5. One in four American children under age 12 is hungry or at risk of hunger. (Food Research and Action Center)	Based on responses to questions: "Do you ever cut the size of meals?" "Do you ever eat less than you feel you should?" "Did you ever rely on limited numbers of foods to feed your children because you were running out of money to buy food for a meal?"
6. There is a strong correlation between a CEO's golf handicap and the company's stock performance: The lower the CEO's handicap (i.e., the better the golfer), the better the stock performs. (*New York Times*, May 31, 1998)	Survey sent to CEOs of 300 largest U.S. companies; only 74 revealed their golf handicaps. Data for several top-ranking CEOs were excluded from the analysis.

In the following *Statistics in Action Revisited* sections, we discuss several key statistical concepts covered in this chapter that are relevant to misleading surveys like those exposed in the *20/20* program.

Statistics IN Action Revisited

- Identifying the population, sample, and inference (p. 10)
- Identifying the data-collection method and data type (p. 17)
- Critically assessing the ethics of a statistical study (p. 20)

1.1 The Science of Statistics

What does *statistics* mean to you? Does it bring to mind batting averages? Gallup polls, unemployment figures, or numerical distortions of facts (lying with statistics!)? Or is it simply a college requirement you have to complete? We hope to persuade you that statistics is a meaningful, useful science whose broad scope of applications to business, government, and the physical and social sciences is almost limitless. We also want to show that statistics can lie only when they are misapplied. Finally, we wish to demonstrate the key role statistics play in critical thinking—whether in the classroom, on the job, or in everyday life. Our objective is to leave you with the impression that the time you spend studying this subject will repay you in many ways.

The *Random House College Dictionary* defines *statistics* as "the science that deals with the collection, classification, analysis, and interpretation of information or data." Thus, a statistician isn't just someone who calculates batting averages at baseball games or tabulates the results of a Gallup poll. Professional statisticians are trained in *statistical science*—that is, they are trained in collecting numerical information in the form of **data,** evaluating it, and drawing conclusions from it. Furthermore, statisticians determine what information is relevant in a given problem and whether the conclusions drawn from a study are to be trusted.

Statistics is the science of data. It involves collecting, classifying, summarizing, organizing, analyzing, and interpreting numerical information.

In the next section, you'll see several real-life examples of statistical applications in business and government that involve making decisions and drawing conclusions.

1.2 Types of Statistical Applications in Business

Statistics means "numerical descriptions" to most people. Monthly unemployment figures, the failure rate of startup companies, and the proportion of female executives in a particular industry all represent statistical descriptions of large sets of data collected on some phenomenon. Often the data are selected from some larger set of data whose characteristics we wish to estimate. We call this selection process *sampling*. For example, you might collect the ages of a sample of customers at a video store to estimate the average age of *all* customers of the store. Then you could use your estimate to target the store's advertisements to the appropriate age group. Notice that statistics involves two different processes: (1) describing sets of data and (2) drawing conclusions (making estimates, decisions, predictions, etc.) about the sets of data based on sampling. So, the applications of statistics can be divided into two broad areas: *descriptive statistics* and *inferential statistics*.

Descriptive statistics utilizes numerical and graphical methods to look for patterns in a data set, to summarize the information revealed in a data set, and to present the information in a convenient form.

Inferential statistics utilizes sample data to make estimates, decisions, predictions, or other generalizations about a larger set of data.

Although we'll discuss both descriptive and inferential statistics in the following chapters, the primary theme of the text is **inference.**

BIOGRAPHY

FLORENCE NIGHTINGALE (1820–1910)

The Passionate Statistician

In Victorian England, the "Lady of the Lamp" had a mission to improve the squalid field hospital conditions of the British army during the Crimean War. Today, most historians consider Florence Nightingale to be the founder of the nursing profession. To convince members of the British Parliament of the need for supplying nursing and medical care to soldiers in the field, Nightingale compiled massive amounts of data from the army files. Through a remarkable series of graphs (which included the first "pie chart"), she demonstrated that most of the deaths in the war were due to illnesses contracted outside the battlefield or long after battle action from wounds that went untreated. Florence Nightingale's compassion and self-sacrificing nature, coupled with her ability to collect, arrange, and present large amounts of data, led some to call her the "Passionate Statistician." ■

Let's begin by examining some business studies that illustrate applications of statistics.

Study 1 "U.S. Market Share for Credit Cards" (*Nilson Report,* May 2008)

The *Nilson Report* is the world's leading source of news and research on the consumer payment industry. The company tracked all credit card purchases in the United States during 2007. The amount of each purchase was recorded and classified according to type of card used (American Express, Discover, MasterCard, or VISA). The results are shown in Figure 1.1. From the graph, you can clearly see that nearly half of the purchases were made with a VISA card and more than 35% with a MasterCard. Because Figure 1.1 *describes* the type of card used in all credit card purchases in 2007, the graphic is an example of *descriptive statistics.*

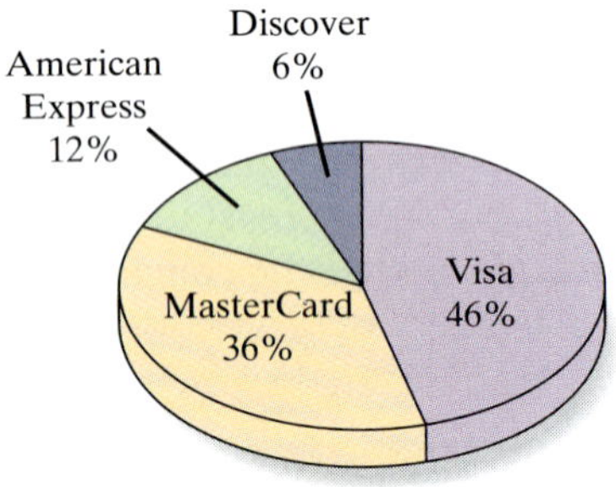

Figure 1.1
U.S. credit card market shares
Source: www.carddata.com

Study 2 "The Executive Compensation Scoreboard" (*Forbes,* April 30, 2008)

How much are the top corporate executives in the United States being paid, and are they worth it? To answer these questions, *Forbes* magazine compiles the "Executive Compensation Scoreboard" each year based on a survey of executives at the 500 largest U.S. companies. To determine which executives are worth their pay, *Forbes* also records the performance of shareholders' stock for the CEO's company over the previous 6 years. Based on a comparison of the stock's performance and the CEO's annual salary, each CEO is given an "efficiency" rating—the lower the rating, the more the CEO is worth his/her pay. An analysis of the scoreboard data in 2008 is summarized in Table 1.1. The table reveals that CEOs in the business services industry have the lowest average* efficiency rating (40.67), thus, they are the best performers. The CEOs in the food markets industry are the worst performers, with the highest efficiency rating (167). Armed with this sample information, *Forbes* might *infer* that, from the shareholders' perspective, typical chief executives in food markets are overpaid relative to CEOs in business services. Thus, this study is an example of *inferential statistics.*

Study 3 "The Consumer Price Index" (*U.S. Department of Labor*)

A data set of interest to virtually all Americans is the set of prices charged for goods and services in the U.S. economy. The general upward movement in this set of prices is referred to as *inflation;* the general downward movement is referred to as *deflation.* In order to *estimate* the change in prices over time, the Bureau of Labor Statistics (BLS) of the U.S. Department of Labor developed the Consumer Price Index (CPI). Each month, the BLS collects price data about a specific collection of goods and services

Table 1.1 Average Pay-for-Performance Efficiency Ratings of CEOs, by Industry

Industry	Average Efficiency Rating	Industry	Average Efficiency Rating
Aerospace and Defense	95.50	Hotels, Restaurants	100.00
Banking	97.08	Household Products	94.33
Business Services	40.67	Insurance	68.50
Capital Goods	88.44	Materials	69.62
Chemicals	127.00	Media	60.67
Conglomerates	97.00	Oil and Gas Operations	96.48
Construction	114.00	Retailing	83.00
Consumer Durables	108.33	Semiconductors	106.00
Diversified Financial	67.89	Software and Services	106.75
Drugs and Biotechnology	100.00	Technology Hardware	103.50
Food, Drink, and Tobacco	112.00	Telecommunications	95.67
Food Markets	167.00	Transportation	51.33
Health Care Equipment	102.71	Utilities	79.10

Source: Analysis of data in "Executive Compensation Scoreboard," *Forbes,* April 30, 2008.

*Although we will not formally define the term *average* until Chapter 2, *typical* or *middle* can be substituted here without confusion.

(called a *market basket*) from 85 urban areas around the country. Statistical procedures are used to compute the CPI from this sample price data and other information about consumers' spending habits. By comparing the level of the CPI at different points in time, it is possible to *estimate* (make an inference about) the rate of inflation over particular time intervals and to compare the purchasing power of a dollar at different points in time.

One major use of the CPI as an index of inflation is as an indicator of the success or failure of government economic policies. A second use of the CPI is to escalate income payments. Millions of workers have *escalator clauses* in their collective bargaining contracts; these clauses call for increases in wage rates based on increases in the CPI. In addition, the incomes of Social Security beneficiaries and retired military and federal civil service employees are tied to the CPI. It has been estimated that a 1% increase in the CPI can trigger an increase of over $1 billion in income payments. Thus, it can be said that the very livelihoods of millions of Americans depend on the behavior of a statistical estimator, the CPI.

Like Study 2, this study is an example of *inferential statistics*. Market basket price data from a sample of urban areas (used to compute the CPI) are used to make inferences about the rate of inflation and wage rate increases.

These studies provide three real-life examples of the uses of statistics in business, economics, and government. Notice that each involves an analysis of data, either for the purpose of describing the data set (Study 1) or for making inferences about a data set (Studies 2 and 3).

1.3 Fundamental Elements of Statistics

Statistical methods are particularly useful for studying, analyzing, and learning about *populations* of *experimental units*.

> An **experimental unit** is an object (e.g., person, thing, transaction, or event) upon which we collect data.

> A **population** is a set of units (usually people, objects, transactions, or events) that we are interested in studying.

For example, populations may include (1) *all* employed workers in the United States, (2) *all* registered voters in California, (3) *everyone* who has purchased a particular brand of cellular telephone, (4) *all* the cars produced last year by a particular assembly line, (5) the *entire* stock of spare parts at United Airlines' maintenance facility, (6) *all* sales made at the drive-through window of a McDonald's restaurant during a given year, and (7) the set of *all* accidents occurring on a particular stretch of interstate during a holiday period. Notice that the first three population examples (1–3) are sets (groups) of people, the next two (4–5) are sets of objects, the next (6) is a set of transactions, and the last (7) is a set of events. Also notice that *each set includes all the experimental units in the population* of interest.

In studying a population, we focus on one or more characteristics or properties of the experimental units in the population. We call such characteristics *variables*. For example, we may be interested in the variables age, gender, income, and/or the number of years of education of the people currently unemployed in the United States.

> A **variable** is a characteristic or property of an individual experimental unit.

The name *variable* is derived from the fact that any particular characteristic may vary among the experimental units in a population.

In studying a particular variable, it is helpful to be able to obtain a numerical representation for it. Often, however, numerical representations are not readily available, so the process of measurement plays an important supporting role in statistical studies. **Measurement** is the process we use to assign numbers to variables of individual population units. We might, for instance, measure the preference for a food product by asking a consumer to rate the product's taste on a scale from 1 to 10. Or we might measure workforce age by simply asking each worker, "How old are you?" In other cases, measurement involves the use of instruments such as stopwatches, scales, and calipers.

If the population we wish to study is small, it is possible to measure a variable for every unit in the population. For example, if you are measuring the starting salary for all University of Michigan MBA graduates last year, it is at least feasible to obtain every salary. When we measure a variable for every experimental unit of a population, the result is called a **census** of the population. Typically, however, the populations of interest in most applications are much larger, involving perhaps many thousands or even an infinite number of units. Examples of large populations include the 7 listed above, as well as all invoices produced in the last year by a *Fortune* 500 company, all potential buyers of a new fax machine, and all stockholders of a firm listed on the New York Stock Exchange. For such populations, conducting a census would be prohibitively time-consuming and/or costly. A reasonable alternative would be to select and study a *subset* (or portion) of the units in the population.

A **sample** is a subset of the units of a population.

For example, suppose a company is being audited for invoice errors. Instead of examining all 15,472 invoices produced by the company during a given year, an auditor may select and examine a sample of just 100 invoices (see Figure 1.2). If he is interested in the variable "invoice error status," he would record (measure) the status (error or no error) of each sampled invoice.

After the variable(s) of interest for every experimental unit in the sample (or population) is measured, the data are analyzed, either by descriptive or inferential statistical methods. The auditor, for example, may be interested only in *describing* the error rate in the sample of 100 invoices. More likely, however, he will want to use the information in the sample to make *inferences* about the population of all 15,472 invoices.

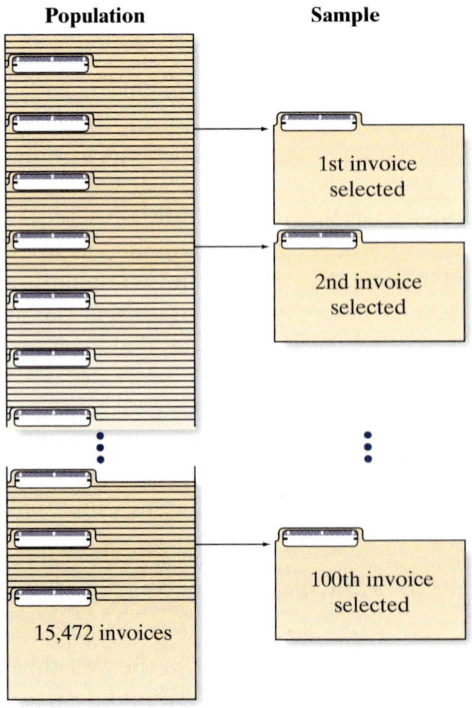

Figure 1.2
A sample of all company invoices

A **statistical inference** is an estimate or prediction or some other generalization about a population based on information contained in a sample.

*That is, we use the information contained in the sample to learn about the larger population.** Thus, from the sample of 100 invoices, the auditor may estimate the total number of invoices containing errors in the population of 15,472 invoices. The auditor's inference about the quality of the firm's invoices can be used in deciding whether to modify the firm's billing operations.

Example 1.1
Key Elements of a Statistical Problem

Problem According to *Variety* (June 29, 2008), the average age of viewers of television programs broadcast on ABC is 50 years. Suppose a rival network (e.g., FOX) executive hypothesizes that the average age of FOX viewers is less than 50. To test her hypothesis, she samples 200 FOX viewers and determines the age of each.

a. Describe the population.

b. Describe the variable of interest.

c. Describe the sample.

d. Describe the inference.

Solution

a. The population is the set of units of interest to the TV executive, which is the set of all FOX viewers.

b. The age (in years) of each viewer is the variable of interest.

c. The sample must be a subset of the population. In this case, it is the 200 FOX viewers selected by the executive.

d. The inference of interest involves the *generalization* of the information contained in the sample of 200 viewers to the population of all FOX viewers. In particular, the executive wants to *estimate* the average age of the viewers in order to determine whether it is less than 50 years. She might accomplish this by calculating the average age in the sample and using the sample average to estimate the population average.

Look Back A key to diagnosing a statistical problem is to identify the data set collected (in this example, the ages of the 200 FOX TV viewers) as a population or sample.

Example 1.2
Key Elements of a Statistical Problem

Problem *Cola wars* is the popular term for the intense competition between Coca-Cola and Pepsi displayed in their marketing campaigns. Their campaigns have featured movie and television stars, rock videos, athletic endorsements, and claims of consumer preference based on taste tests. Suppose, as part of a Pepsi marketing campaign, 1,000 cola consumers are given a blind taste test (i.e., a taste test in which the two brand names are disguised). Each consumer is asked to state a preference for brand A or brand B.

*The terms *population* and *sample* are often used to refer to the sets of measurements themselves, as well as to the units on which the measurements are made. When a single variable of interest is being measured, this usage causes little confusion. But when the terminology is ambiguous, we'll refer to the measurements as *population data sets* and *sample data sets,* respectively.

a. Describe the population.

b. Describe the variable of interest.

c. Describe the sample.

d. Describe the inference.

Solution

a. Because we are interested in the responses of cola consumers in a taste test, a cola consumer is the experimental unit. Thus, the population of interest is the collection or set of all cola consumers.

b. The characteristic that Pepsi wants to measure is the consumer's cola preference as revealed under the conditions of a blind taste test, so cola preference is the variable of interest.

c. The sample is the 1,000 cola consumers selected from the population of all cola consumers.

d. The inference of interest is the *generalization* of the cola preferences of the 1,000 sampled consumers to the population of all cola consumers. In particular, the preferences of the consumers in the sample can be used to *estimate* the percentage of all cola consumers who prefer each brand.

Look Back In determining whether the inference is inferential or descriptive, we assess whether Pepsi is interested in the responses of only the 1,000 sampled customers (descriptive statistics) or in the responses for the entire population of consumers (inferential statistics).

Now Work Exercise 1.14b

The preceding definitions and examples identify four of the five elements of an inferential statistical problem: a population, one or more variables of interest in a sample, and an inference. But making the inference is only part of the story. We also need to know its **reliability**—that is, how good the inference is. The only way we can be certain that an inference about a population is correct is to include the entire population in our sample. However, because of *resource constraints* (e.g., insufficient time and/or money), we usually can't work with whole populations, so we base our inferences on just a portion of the population (a sample). Consequently, whenever possible, it is important to determine and report the reliability of each inference made. Reliability, then, is the fifth element of inferential statistical problems.

The measure of reliability that accompanies an inference separates the science of statistics from the art of fortune-telling. A palm reader, like a statistician, may examine a sample (your hand) and make inferences about the population (your life). However, unlike statistical inferences, the palm reader's inferences include no measure of reliability.

Suppose, like the TV executive in Example 1.1, we are interested in the *error of estimation* (i.e., the difference between the average age for the population of TV viewers and the average age of a sample of TV viewers). Using statistical methods, we can determine a *bound on the estimation error*. This bound is simply a number that our estimation error (the difference between the average age of the sample and the average age of the population) is not likely to exceed. We'll see in later chapters that this bound is a measure of the uncertainty of our inference. The reliability of statistical inferences is discussed throughout this text. For now, we simply want you to realize that an inference is incomplete without a measure of its reliability.

A **measure of reliability** is a statement (usually quantified) about the degree of uncertainty associated with a statistical inference.

Let's conclude this section with a summary of the elements of both descriptive and inferential statistical problems and an example to illustrate a measure of reliability.

Four Elements of Descriptive Statistical Problems

1. The population or sample of interest
2. One or more variables (characteristics of the population or sample units) that are to be investigated
3. Tables, graphs, or numerical summary tools
4. Identification of patterns in the data

Five Elements of Inferential Statistical Problems

1. The population of interest
2. One or more variables (characteristics of the population units) that are to be investigated
3. The sample of population units
4. The inference about the population based on information contained in the sample
5. A measure of reliability for the inference

Example 1.3 Reliability of an Inference

Problem Refer to Example 1.2, in which the cola preferences of 1,000 consumers were indicated in a taste test. Describe how the reliability of an inference concerning the preferences of all cola consumers in the Pepsi bottler's marketing region could be measured.

Solution When the preferences of 1,000 consumers are used to estimate the preferences of all consumers in the region, the estimate will not exactly mirror the preferences of the population. For example, if the taste test shows that 56% of the 1,000 consumers chose Pepsi, it does not follow (nor is it likely) that exactly 56% of all cola drinkers in the region prefer Pepsi. Nevertheless, we can use sound statistical reasoning (which is presented later in the text) to ensure that our sampling procedure will generate estimates that are almost certainly within a specified limit of the true percentage of all consumers who prefer Pepsi. For example, such reasoning might assure us that the estimate of the preference for Pepsi from the sample is almost certainly within 5% of the actual population preference. The implication is that the actual preference for Pepsi is between 51% [i.e., $(56 - 5)\%$] and 61% [i.e., $(56 + 5)\%$]—that is, $(56 \pm 5)\%$. This interval represents a measure of reliability for the inference.

Look Back The interval 56 ± 5 is called a *confidence interval,* because we are "confident" that the true percentage of customers who prefer Pepsi in a taste test falls into the range (51, 61). In Chapter 5, we learn how to assess the degree of confidence (e.g., 90% or 95% confidence) in the interval.

Statistics IN Action Revisited — Identifying the Population, Sample, and Inference

Consider the study on the link between a CEO's golf handicap and the company's stock performance, reported in the *New York Times* (May 31, 1998). The newspaper gathered information on golf handicaps of corporate executives obtained from a *Golf Digest* survey sent to CEOs of the 300 largest U.S. companies. (A golf handicap is a numerical "index" that allows golfers to compare skills; the lower the handicap, the better the golfer.) For the 51 CEOs who reported their handicaps, the *New York Times* then determined each CEO's company stock market performance over a 3-year period (measured as a rate-of-return index, from a low value of 0 to a high value of 100). Thus, the experimental unit for the study is a corporate executive, and the two variables measured are golf handicap and stock performance index. Also, the data for the 51 CEOs represent a sample selected from the much larger population of all corporate executives in the United States. (These data are available in the **GOLFCEO** file.)

The *New York Times* discovered a "statistical correlation" (a method discussed in Chapter 10) between golf handicap and stock performance. Thus, the newspaper inferred that the better the CEO is at golf, the better the company's stock performance.

Data Set: GOLFCEO

1.4 Processes*

Sections 1.2 and 1.3 focused on the use of statistical methods to analyze and learn about populations, which are sets of *existing* units. Statistical methods are equally useful for analyzing and making inferences about *processes*.

> A **process** is a series of actions or operations that transforms inputs to outputs. A process produces or generates output over time.

The most obvious processes of interest to businesses are those of production or manufacturing. A manufacturing process uses a series of operations performed by people and machines to convert inputs, such as raw materials and parts, to finished products (the outputs). Examples include the process used to produce the paper on which these words are printed, automobile assembly lines, and oil refineries.

Figure 1.3 presents a general description of a process and its inputs and outputs. In the context of manufacturing, the process in the figure (i.e., the transformation process) could be a depiction of the overall production process or it could be a depiction of one of the many processes (sometimes called *subprocesses*) that exist within an overall production process. Thus, the output shown could be finished goods that will be shipped to an external customer or merely the output of one of the steps or subprocesses of the overall process. In the latter case, the output becomes input for the next subprocess. For example, Figure 1.3 could represent the overall automobile assembly process, with its output being fully assembled cars ready for shipment to dealers. Or, it could depict the windshield assembly subprocess, with its output of partially assembled cars with windshield ready for "shipment" to the next subprocess in the assembly line.

Besides physical products and services, businesses and other organizations generate streams of numerical data over time that are used to evaluate the performance of

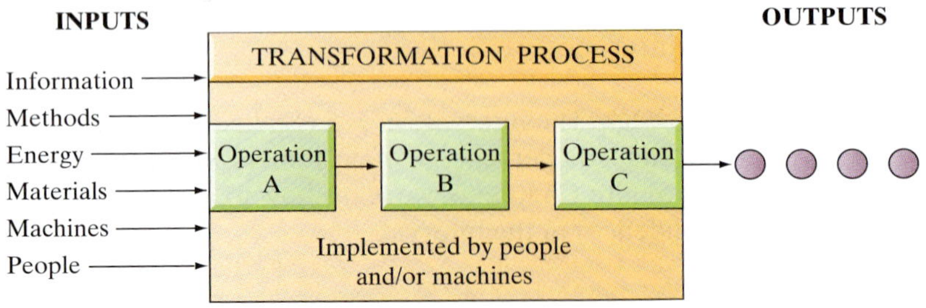

Figure 1.3
Graphical depiction of a manufacturing process

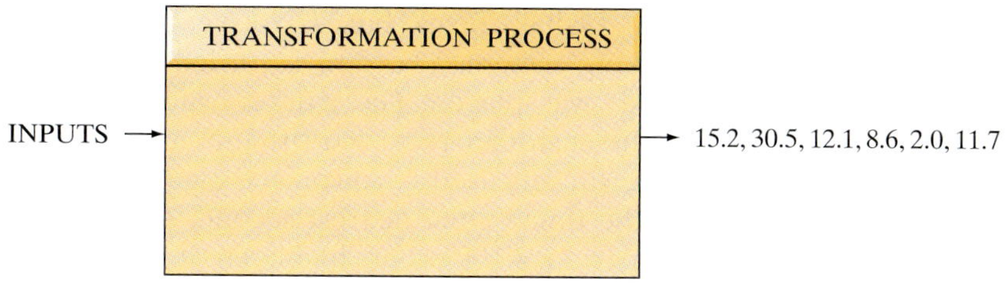

Figure 1.4
A black box process with numerical output

the organization. Examples include weekly sales figures, quarterly earnings, and yearly profits. The U.S. economy (a complex organization) can be thought of as generating streams of data that include the gross domestic product (GDP), stock prices, and the Consumer Price Index (see Section 1.2). Statisticians and other analysts conceptualize these data streams as being generated by processes. Typically, however, the series of operations or actions that cause particular data to be realized are either unknown or so complex (or both) that the processes are treated as *black boxes*.

> A process whose operations or actions are unknown or unspecified is called a **black box.**

Frequently, when a process is treated as a black box, its inputs are not specified either. The entire focus is on the output of the process. A black box process is illustrated in Figure 1.4.

In studying a process, we generally focus on one or more characteristics, or properties, of the output. For example, we may be interested in the weight or the length of the units produced or even the time it takes to produce each unit. As with characteristics of population units, we call these characteristics *variables*. In studying processes whose output is already in numerical form (i.e., a stream of numbers), the characteristic, or property, represented by the numbers (e.g., sales, GDP, or stock prices) is typically the variable of interest. If the output is not numeric, we use *measurement processes* to assign numerical values to variables.* For example, if in the automobile assembly process the weight of the fully assembled automobile is the variable of interest, a measurement process involving a large scale will be used to assign a numerical value to each automobile.

As with populations, we use sample data to analyze and make inferences (estimates, predictions, or other generalizations) about processes. But the concept of a sample is defined differently when dealing with processes. Recall that a population is a set of existing units and that a sample is a subset of those units. In the case of processes, however, the concept of a set of existing units is not relevant or appropriate. Processes generate or create their output *over time*—one unit after another. For example, a particular automobile assembly line produces a completed vehicle every four minutes. We define a sample from a process in the box.

> Any set of output (object or numbers) produced by a process is called a **sample.**

Thus, the next 10 cars turned out by the assembly line constitute a sample from the process, as do the next 100 cars or every fifth car produced today.

Example 1.4 Key Elements of a Process

Problem A particular fast-food restaurant chain has 6,289 outlets with drive-through windows. To attract more customers to its drive-through services, the company is considering offering a 50% discount to customers who wait more than a specified number of minutes to receive their order. To help determine what the time limit should be, the company decided to estimate the average waiting time at a particular drive-through

*A process whose output is already in numerical form necessarily includes a measurement process as one of its subprocesses.

window in Dallas, Texas. For 7 consecutive days, the worker taking customers' orders recorded the time that every order was placed. The worker who handed the order to the customer recorded the time of delivery. In both cases, workers used synchronized digital clocks that reported the time to the nearest second. At the end of the 7-day period, 2,109 orders had been timed.

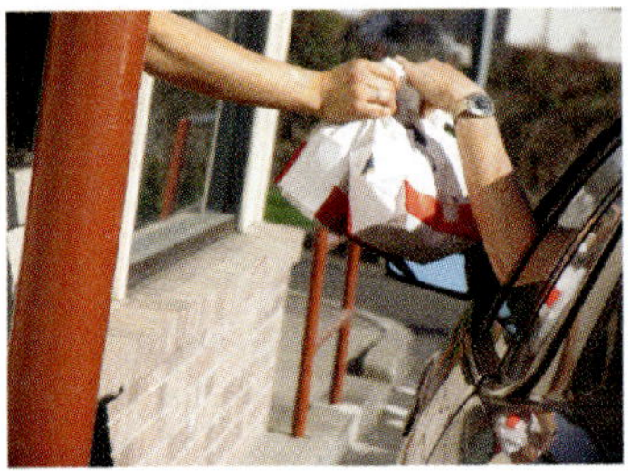

a. Describe the process of interest at the Dallas restaurant.

b. Describe the variable of interest.

c. Describe the sample.

d. Describe the inference of interest.

e. Describe how the reliability of the inference could be measured.

Solution

a. The process of interest is the drive-through window at a particular fast-food restaurant in Dallas, Texas. It is a process because it "produces," or "generates," meals over time—that is, it services customers over time.

b. The variable the company monitored is customer waiting time, the length of time a customer waits to receive a meal after placing an order. Because the study is focusing only on the output of the process (the time to produce the output) and not the internal operations of the process (the tasks required to produce a meal for a customer), the process is being treated as a black box.

c. The sampling plan was to monitor every order over a particular 7-day period. The sample is the 2,109 orders that were processed during the 7-day period.

d. The company's immediate interest is in learning about the drive-through window in Dallas. They plan to do this by using the waiting times from the sample to make a statistical inference about the drive-through process. In particular, they might use the average waiting time for the sample to estimate the average waiting time at the Dallas facility.

e. As for inferences about populations, measures of reliability can be developed for inferences about processes. The reliability of the estimate of the average waiting time for the Dallas restaurant could be measured by a bound on the error of estimation—that is, we might find that the average waiting time is 4.2 minutes, with a bound on the error of estimation of 0.5 minutes. The implication would be that we could be reasonably certain that the true average waiting time for the Dallas process is between 3.7 and 4.7 minutes.

Look Back Notice that there is also a population described in this example: the company's 6,289 existing outlets with drive-through facilities. In the final analysis, the company will use what it learns about the process in Dallas and, perhaps, similar studies at other locations to make an inference about the waiting times in its population of outlets.

Now Work Exercise 1.30

Note that output already generated by a process can be viewed as a population. Suppose a soft-drink canning process produced 2,000 twelve-packs yesterday, all of which were stored in a warehouse. If we were interested in learning something about those 2,000 twelve-packs—such as the percentage with defective cardboard packaging—we could treat the 2,000 twelve-packs as a population. We might draw a sample from the population in the warehouse, measure the variable of interest, and use the sample data to make a statistical inference about the 2,000 twelve-packs, as described in Sections 1.2 and 1.3.

In this optional section, we have presented a brief introduction to processes and the use of statistical methods to analyze and learn about processes. In Chapters 12 and 13 we present an in-depth treatment of these subjects.

1.5 Types of Data

You have learned that statistics is the science of data and that data are obtained by measuring the values of one or more variables on the units in the sample (or population). All data (and hence the variables we measure) can be classified as one of two general types: *quantitative* and *qualitative*.

Quantitative data are data that are measured on a naturally occurring numerical scale.* The following are examples of quantitative data:

1. The temperature (in degrees Celsius) at which each unit in a sample of 20 pieces of heat-resistant plastic begins to melt
2. The current unemployment rate (measured as a percentage) for each of the 50 states
3. The scores of a sample of 150 MBA applicants on the GMAT, a standardized business graduate school entrance exam administered nationwide
4. The number of female executives employed in each of a sample of 75 manufacturing companies

Quantitative data are measurements that are recorded on a naturally occurring numerical scale.

In contrast, qualitative data cannot be measured on a natural numerical scale; they can only be classified into categories.† Examples of qualitative data are as follows:

1. The political party affiliation (Democrat, Republican, or Independent) in a sample of 50 CEOs
2. The defective status (defective or not) of each of 100 computer chips manufactured by Intel
3. The size of a car (subcompact, compact, midsize, or full-size) rented by each of a sample of 30 business travelers
4. A taste tester's ranking (best, worst, etc.) of four brands of barbecue sauce for a panel of 10 testers

Often, we assign arbitrary numerical values to qualitative data for ease of computer entry and analysis. But these assigned numerical values are simply codes: They cannot be meaningfully added, subtracted, multiplied, or divided. For example, we might code Democrat = 1, Republican = 2, and Independent = 3. Similarly, a taste tester might rank the barbecue sauces from 1 (best) to 4 (worst). These are simply arbitrarily selected numerical codes for the categories and have no utility beyond that.

Qualitative data are measurements that cannot be measured on a natural numerical scale; they can only be classified into one of a group of categories.

Example 1.5
Types of Data

Problem Chemical and manufacturing plants sometimes discharge toxic-waste materials such as DDT into nearby rivers and streams. These toxins can adversely affect the plants and animals inhabiting the river and the riverbank. The U.S. Army Corps of Engineers conducted a study of fish in the Tennessee River (in Alabama) and its three tributary

*Quantitative data can be subclassified as either *interval* or *ratio*. For ratio data, the origin (i.e., the value 0) is a meaningful number. But the origin has no meaning with interval data. Consequently, we can add and subtract interval data, but we can't multiply and divide them. Of the four quantitative data sets listed, (1) and (3) are interval data, while (2) and (4) are ratio data.

†Qualitative data can be subclassified as either *nominal* or *ordinal*. The categories of an ordinal data set can be ranked or meaningfully ordered, but the categories of a nominal data set can't be ordered. Of the four qualitative data sets listed, (1) and (2) are nominal and (3) and (4) are ordinal.

creeks: Flint Creek, Limestone Creek, and Spring Creek. A total of 144 fish were captured, and the following variables were measured for each:

1. River/creek where each fish was captured
2. Species (channel catfish, largemouth bass, or smallmouth buffalo fish)
3. Length (centimeters)
4. Weight (grams)
5. DDT concentration (parts per million)

These data are saved in the **DDT** file. Classify each of the five variables measured as quantitative or qualitative.

Solution The variables length, weight, and DDT are quantitative because each is measured on a numerical scale: length in centimeters, weight in grams, and DDT in parts per million. In contrast, river/creek and species cannot be measured quantitatively: They can only be classified into categories (e.g., channel catfish, largemouth bass, and smallmouth buffalo fish for species). Consequently, data on river/creek and species are qualitative.

Look Back It is essential that you understand whether data are quantitative or qualitative in nature because the statistical method appropriate for describing, reporting, and analyzing the data depends on the data type (quantitative or qualitative).

Now Work Exercise 1.16

We demonstrate many useful methods for analyzing quantitative and qualitative data in the remaining chapters of the text. But first, we discuss some important ideas on data collection.

1.6 Collecting Data

Once you decide on the type of data—quantitative or qualitative—appropriate for the problem at hand, you'll need to collect the data. Generally, you can obtain the data in four different ways:

1. Data from a *published source*
2. Data from a *designed experiment*
3. Data from a *survey*
4. Data collected *observationally*

Sometimes, the data set of interest has already been collected for you and is available in a **published source,** such as a book, journal, newspaper, or Web site. For example, you may want to examine and summarize the unemployment rates (i.e., percentages of eligible workers who are unemployed) in the 50 states of the United States. You can find this data set (as well as numerous other data sets) at your library in the *Statistical Abstract of the United States,* published annually by the U.S. government. Similarly, someone who is interested in monthly mortgage applications for new home construction would find this data set in the *Survey of Current Business,* another government publication. Other examples of published data sources include the *Wall Street Journal* (financial data) and the *The Sporting News* (sports information).*

A second method of collecting data involves conducting a **designed experiment,** in which the researcher exerts strict control over the units (people, objects, or events) in the study. For example, a medical study investigated the potential of aspirin in preventing heart attacks. Volunteer physicians were divided into two groups—the *treatment* group and the *control* group. In the treatment group, each physician took one aspirin tablet a day

*With published data, we often make a distinction between the *primary source* and *secondary source.* If the publisher is the original collector of the data, the source is primary. Otherwise, the data are secondary source.

for 1 year, while each physician in the control group took an aspirin-free placebo (no drug) made to look like an aspirin tablet. The researchers, not the physicians under study, controlled who received the aspirin (the treatment) and who received the placebo. As you will learn in Chapter 10, a properly designed experiment allows you to extract more information from the data than is possible with an uncontrolled study.

Surveys are a third source of data. With a **survey,** the researcher samples a group of people, asks one or more questions, and records the responses. Probably the most familiar type of survey is the political polls conducted by any one of a number of organizations (e.g., Harris, Gallup, Roper, and CNN) and designed to predict the outcome of a political election. Another familiar survey is the Nielsen survey, which provides the major television networks with information on the most watched TV programs. Surveys can be conducted through the mail, with telephone interviews, or with in-person interviews. Although in-person interviews are more expensive than mail or telephone surveys, they may be necessary when complex information must be collected.

Finally, observational studies can be employed to collect data. In an **observational study,** the researcher observes the experimental units in their natural setting and records the variable(s) of interest. For example, a company psychologist might observe and record the level of "Type A" behavior of a sample of assembly line workers. Similarly, a finance researcher may observe and record the closing stock prices of companies that are acquired by other firms on the day prior to the buyout and compare them to the closing prices on the day the acquisition is announced. Unlike a designed experiment, an observational study is one in which the researcher makes no attempt to control any aspect of the experimental units.

Regardless of the data-collection method employed, it is likely that the data will be a sample from some population. And if we wish to apply inferential statistics, we must obtain a *representative sample*.

> A **representative sample** exhibits characteristics typical of those possessed by the population of interest.

For example, consider a political poll conducted during a presidential election year. Assume the pollster wants to estimate the percentage of all 140,000,000 registered voters in the United States who favor the incumbent president. The pollster would be unwise to base the estimate on survey data collected for a sample of voters from the incumbent's own state. Such an estimate would almost certainly be *biased* high.

The most common way to satisfy the representative sample requirement is to select a random sample. A **random sample** ensures that every subset of fixed size in the population has the same chance of being included in the sample. If the pollster samples 1,500 of the 140,000,000 voters in the population so that every subset of 1,500 voters has an equal chance of being selected, he has devised a random sample. The procedure for selecting a random sample is discussed in Chapter 3. Here, however, let's look at two examples involving actual sampling studies.

> A **random sample** of n experimental units is a sample selected from the population in such a way that every different sample of size n has an equal chance of selection.

Example 1.6 Method of Data Collection

Problem How do consumers feel about using the Internet for online shopping? To find out, a customer-experience software company commissioned a nationwide survey of 1,859 U.S. adults who had conducted at least one online transaction at a banking, shopping, travel, or insurance Web site in the past year. The findings, reported on *BusinessWeek.com* (January 5, 2006), revealed that 1,655 respondents, or 89%, experienced technical problems with an online transaction. Also, more than one-third of the consumers go to a competitor's Web site when a glitch in the online transaction occurs.

a. Identify the data-collection method.

b. Identify the target population.

c. Are the sample data representative of the population?

Solution

a. The data-collection method is a survey: 1,859 adults completed the questionnaire.

b. Presumably, the software company who commissioned the survey is interested in all consumers who have made at least one online transaction in the past year. Consequently, the target population is *all* consumers who use the Internet for online transactions.

c. Because the 1,859 respondents clearly make up a subset of the target population, they do form a sample. Whether or not the sample is representative is unclear because *BusinessWeek.com* provided no detailed information on how the 1,859 respondents were selected. If the respondents were obtained using, say, random-digit telephone dialing, then the sample is likely to be representative because it is a random sample. However, if the questionnaire was made available to anyone surfing the Internet, then the respondents are *self-selected* (i.e., each Internet user who saw the survey chose whether or not to respond to it). Such a survey often suffers from *nonresponse* bias. It is possible that many Internet users who chose not to respond (or who never saw the questionnaire) would have answered the questions differently, leading to a lower (or higher) sample percentage.

Look Back Any inferences based on survey samples that employ self-selection are suspect due to potential nonresponse bias.

Example 1.7 Data-Collection Method; Representative Data

Problem Marketers use wording such as "was \$100, now \$80" to indicate a price promotion. The promotion is typically compared to the retailer's previous price or to a competitor's price. A study in the *Journal of Consumer Research* investigated whether between-store comparisons result in greater perceptions of value by consumers than within-store comparisons. Suppose 50 consumers were randomly selected from all consumers in a designated market area to participate in the study. The researchers randomly assigned 25 consumers to read a within-store price promotion advertisement ("was \$100, now \$80") and 25 consumers to read a between-store price promotion ("\$100 there, \$80 here"). The consumers then gave their opinion on the value of the

Activity 1.1 *Keep the Change:* Collecting Data

Recently, Bank of America introduced a savings program called *Keep the Change.* Each time a customer enrolled in the program uses his or her debit card to make a purchase, the difference between the purchase total and the next higher dollar amount is transferred from the customer's checking account to a savings account. For example, if you were enrolled in the program and used your debit card to purchase a latte for \$3.75, then \$0.25 would be transferred from your checking to your savings account. For the first 90 days that a customer is enrolled in the program, Bank of America matches the amounts transferred up to \$250. In this and subsequent activities, we will investigate the potential benefit to the customer and cost to the bank.

1. Simulate the program by keeping track of all purchases that you make during one week that could be made with a debit card, even if you use a different form of payment. For each purchase, record both the purchase total and the amount that would be transferred from checking to savings with the *Keep the Change* program.
2. You now have two sets of data: *Purchase Totals* and *Amounts Transferred*. Both sets contain quantitative data. For each data set, identify the corresponding naturally occurring numerical scale. Explain why each set has an obvious lower bound but only one set has a definite upper bound.
3. Find the total of the amounts transferred for the one-week period. Because 90 days is approximately 13 weeks, multiply the total by 13 to estimate how much the bank would have to match during the first 90 days. Form a third data set, *Bank Matching,* by collecting the 90-day estimates of all the students in your class. Identify the naturally occurring scale, including bounds, for this set of data.

Keep the data sets from this activity for use in other activities. We suggest you save the data using statistical software (e.g., Minitab) or a graphing calculator.

discount offer on a 10-point scale (where 1 = lowest value and 10 = highest value). The value opinions of the two groups of consumers were compared.

a. Identify the data-collection method.

b. Are the sample data representative of the target population?

Solution

a. Here, the experimental units are the consumers. Because the researchers controlled which price promotion ad—"within-store" or "between-store"—the experimental units (consumers) were assigned to, a designed experiment was used to collect the data.

b. The sample of 50 consumers was randomly selected from all consumers in the designated market area. If the target population is all consumers in this market, it is likely that the sample is representative. However, the researchers warn that the sample data should not be used to make inferences about consumer behavior in other, dissimilar markets.

Look Back By using randomization in a designed experiment, the researcher is attempting to eliminate different types of bias, including self-selection bias.

Now Work Exercise 1.15

Statistics IN Action Revisited Identifying the Data-Collection Method and Data Type

Refer to the *New York Times* study on the link between a CEO's golf handicap and the company's stock performance. Recall that the newspaper gathered information on golf handicaps of corporate executives obtained from a *Golf Digest* survey that was sent to 300 corporate executives. Thus, the data-collection method is a survey. In addition to golf handicap (a numerical "index" that allows golfers to compare skills), the *Times* measured the CEO's company stock market performance over a 3-year period on a scale of 0 to 100. Because both variables, golf handicap and stock performance, are numerical in nature, they are quantitative data.

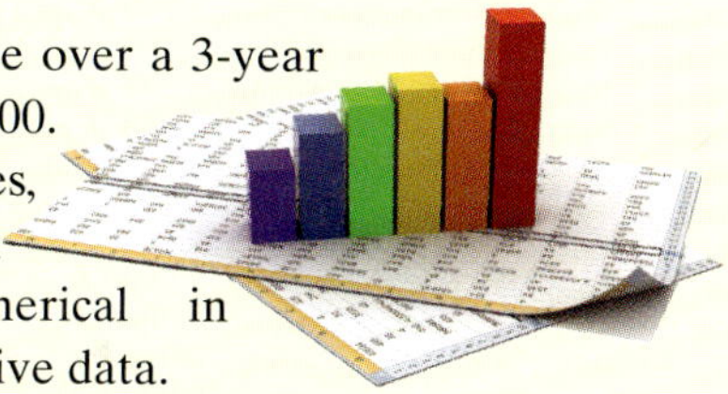

Data Set: GOLFCEO

1.7 The Role of Statistics in Managerial Decision Making

According to H. G. Wells, author of such science-fiction classics as *The War of the Worlds* and *The Time Machine*, "*Statistical thinking* will one day be as necessary for efficient citizenship as the ability to read and write." Written more than a hundred years ago, Wells's prediction is proving true today.

BIOGRAPHY H. G. WELLS (1866–1946)

Writer and Novelist

English-born Herbert George Wells published his first novel, *The Time Machine,* in 1895 as a parody of the English class division and as a satirical warning that human progress is inevitable. Although most famous as a science-fiction novelist, Wells was a prolific writer as a journalist, sociologist, historian, and philosopher. Wells's prediction about statistical thinking (see above) is just one of a plethora of observations he made about life on this world. Here are a few more of H. G. Wells's more famous quotes:

"Advertising is legalized lying."

"Crude classification and false generalizations are the curse of organized life."

"The crisis of today is the joke of tomorrow."

"Fools make researchers and wise men exploit them."

"The only true measure of success is the ratio between what we might have done and what we might have been on the one hand, and the thing we have made and the things we have made of ourselves on the other." ■

The growth in data collection associated with scientific phenomena, business operations, and government activities (quality control, statistical auditing, forecasting, etc.) has been remarkable in the past several decades. Every day the media present us with the published results of political, economic, and social surveys. In increasing government emphasis on drug and product testing, for example, we see vivid evidence of the need for *quantitative literacy,* i.e., the ability to evaluate data intelligently. Consequently, each of us has to develop a discerning sense—an ability to use rational thought to interpret and understand the meaning of data. Quantitative literacy can help you make intelligent decisions, inferences, and generalizations; that is, it helps you *think critically* using statistics.

Statistical thinking involves applying rational thought and the science of statistics to critically assess data and inferences. Fundamental to the thought process is that variation exists in populations and process data.

To gain some insight into the role statistics plays in critical thinking, we present two examples of some misleading or faulty surveys.

Example 1.8
Biased Sample

Problem An article in the *New York Times* considered the question of whether motorcyclists should be required by law to wear helmets. In supporting his argument for no helmets, the editor of a magazine for Harley-Davidson bikers presented the results of one study that claimed "nine states without helmet laws had a lower fatality rate (3.05 deaths per 10,000 motorcycles) than those that mandated helmets (3.38)" and a survey that found "of 2,500 bikers at a rally, 98% of the respondents opposed such laws." Based on this information, do you think it is safer to ride a motorcycle without a helmet? What further statistical information would you like?

Solution You can use statistical thinking to help you critically evaluate the study. For example, before you can evaluate the validity of the 98% estimate, you would want to know how the data were collected. If a survey was, in fact, conducted, it's possible that the 2,500 bikers in the sample were not selected at random from the target population of all bikers, but rather were "self-selected." (Remember, they were all attending a rally—a rally likely for bikers who oppose the law.) If the respondents were likely to have strong opinions regarding the helmet law (e.g., strongly oppose the law), the resulting estimate is probably biased high. Also, if the biased sample was intentional, with the sole purpose to mislead the public, the researchers would be guilty of **unethical statistical practice.**

Ethics in Statistics

Intentionally selecting a biased sample in order to produce misleading statistics is considered *unethical statistical practice*.

You would also want more information about the study comparing the motorcycle fatality rate of the nine states without a helmet law to those states that mandate helmets. Were the data obtained from a published source? Where all 50 states included in the study, or were only certain states selected? That is, are you seeing sample data or population data? Furthermore, do the helmet laws vary among states? If so, can you really compare the fatality rates?

Look Back Questions such as these led a group of mathematics and statistics teachers attending an American Statistical Association course to discover a scientific and statistically sound study on helmets. The study reported a dramatic *decline* in motorcycle crash deaths after California passed its helmet law.

Example 1.9
Manipulative or Ambiguous Survey Questions

Problem Recently, Howard Stern moved his controversial radio program from free, over-the-air (AM/FM) radio to Sirius satellite radio (now called Sirius XM). The move was perceived in the industry to boost satellite radio subscriptions. This led American Media Services, a developer of AM/FM radio properties, to solicit a January 2006 nationwide random-digit dialing phone survey of 1,008 people. The purpose of the survey was to determine how much interest Americans really have in buying satellite radio service. After providing some background on Howard Stern's controversial radio program, one of the questions asked, "How likely are you to purchase a subscription to satellite radio after Howard Stern's move to Sirius?" The result: 86% of the respondents stated that they aren't likely to buy satellite radio because of Stern's move. Consequently, American Media Services concluded that "the Howard Stern Factor is overrated" and that "few Americans expect to purchase satellite radio"—claims that made the headlines of news reports and Web blogs. Do you agree?

Solution First, we need to recognize that American Media Services has a vested interest in the outcome of the survey—the company makes its money from over-the-air broadcast radio stations. Second, although the phone survey was conducted using random-digit dialing, there is no information provided on the response rate. It's possible that non-respondents (people who were not home or refused to answer the survey questions) tend to be people who use cell phones more than their landline phone, and, consequently, are more likely to use the latest in electronic technology, including satellite radio. Finally, the survey question itself is ambiguous. Do the respondents have negative feelings about satellite radio, Howard Stern, or both? If not for Howard Stern's program, would the respondents be more likely to buy satellite radio? To the critical thinker, it's unclear what the results of the survey imply.

Look Back Examining the survey results from the perspective of satellite radio providers, 14% of the respondents indicated that they would be likely to purchase satellite radio. Projecting the 14% back to the population of all American adults, this figure represents about 50 million people; what is interpreted as "few Americans" by American Media Services could be music to the ears of satellite radio providers.

As in both the motorcycle helmet study and satellite radio study, many statistical studies are based on survey data. Most of the problems with these surveys result from the use of *nonrandom* samples. These samples are subject to potential errors, such as *selection bias, nonresponse bias* (recall Example 1.6), and *measurement error*. Researchers who are aware of these problems and continue to use the sample data to make inferences are practicing unethical statistics.

Selection bias results when a subset of the experimental units in the population is excluded so that these units have no chance of being selected for the sample.

Nonresponse bias results when the researchers conducting a survey or study are unable to obtain data on all experimental units selected for the sample.

Measurement error refers to inaccuracies in the values of the data recorded. In surveys, the error may be due to ambiguous or leading questions and the interviewer's effect on the respondent.

Successful managers rely heavily on statistical thinking to help them make decisions. The role statistics can play in managerial decision making is displayed in the flow diagram in Figure 1.5. Every managerial decision-making problem begins with a real-world problem. This problem is then formulated in managerial terms and framed

Statistics IN Action Revisited — Critically Assessing the Ethics of a Statistical Study

The *New York Times* reported a strong link between a corporate executive's golf handicap and his/her company's stock performance. Thus, the newspaper inferred that the better the CEO is at golf, the better the company's stock performance will be. To critically assess this study, consider the following facts:

1. *Golf Digest* sent surveys to the CEOs at the 300 largest U.S. firms. Only 74 executives agreed to reveal their golf handicaps. Of these 74 CEOs, the *Times* collected data on stock performance for only 51 of the companies. (The other 23 CEOs were not in the stock performance database used by the newspaper.)
2. The *New York Times* researcher who performed the analysis of the data stated that "for all the different factors I've tested as possible links to predicting which CEOs are going to perform well or poorly, [golf handicap] is certainly one of the . . . strongest."
3. According to the *Times*, the researcher "scientifically sifted out a handful of CEOs because of their statistical extremes," in effect "removing seven CEOs from the final analysis because [their data] destroyed the trend lines."

These observations lead a critical thinker to doubt the validity of the inference made by the *New York Times* researcher. Consider first that the sample of CEOs analyzed was not randomly selected from all CEOs in the United States. In fact, they were self-selected—only those CEOs who chose to report their golf handicap were included in the study. (Not even all these "self-reporters" were included; some were eliminated because the newspaper did not have information on their company's stock performance in the database.) Thus, the potential for selection and/or nonresponse bias is extremely high.

Second, based on fact #2, it is likely that the researcher tested a multitude of factors and found only one (golf handicap) that had a link to stock performance. We will learn in subsequent chapters that when a plethora of irrelevant variables are tested statistically, by chance one or more of the variables will be found "statistically significant."

Finally, the researcher removed the data for seven CEOs based on their "statistical extremes." In the next chapter, we learn about statistical "outliers"—how to detect them and how to treat them when discovered. However, it can be shown (using the methods outlined in the text) that these seven data points are not outliers. If the data points are included in the analysis, the link between golf handicap and stock performance is found to be weak, at best.

Data Set: GOLFCEO

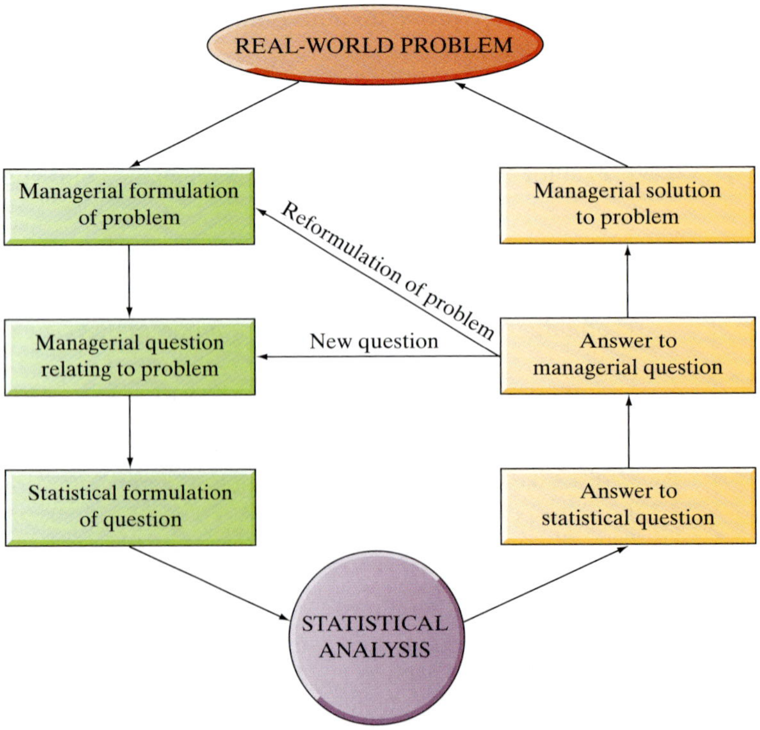

Figure 1.5
Flow diagram showing the role of statistics in managerial decision making *Source:* Chervany, N. L., Benson, G., and Iyer, R. K. (1980), "The Planning Stage in Statistical Reasoning," *The American Statistician.* © 1980 by the American Statistical Association. All rights reserved.

as a managerial question. The next sequence of steps (proceeding counterclockwise around the flow diagram) identifies the role that statistics can play in this process. The managerial question is translated into a statistical question, the sample data are collected and analyzed, and the statistical question is answered. The next step in the process is using the answer to the statistical question to reach an answer to the managerial question. The answer to the managerial question may suggest a reformulation of the original managerial problem, suggest a new managerial question, or lead to the solution of the managerial problem.

One of the most difficult steps in the decision-making process—one that requires a cooperative effort among managers and statisticians—is the translation of the managerial question into statistical terms (for example, into a question about a population). This statistical question must be formulated so that, when answered, it will provide the key to the answer to the managerial question. Thus, as in the game of chess, you must formulate the statistical question with the end result, the solution to the managerial question, in mind.

In the remaining chapters of the text, you'll become familiar with the tools essential for building a firm foundation in statistics and statistical thinking.

Activity 1.2 Identifying Misleading Statistics

In the *Statistics in Action* feature at the beginning of this chapter, several examples of false or misleading statistics were discussed. Claims such as *One in four American children under age 12 is hungry or at risk of hunger* are often used to persuade the public or the government to donate or allocate more money to charitable groups that feed the poor. Researchers sometimes claim a relationship exists between two seemingly unrelated quantities such as a CEO's golf handicap and the company's stock performance; such relationships are often weak at best and of little practical importance. Read the *Statistics in Action and Statistics in Action Revisited* features in this chapter before completing this activity.

1. Look for an article in a newspaper or on the Internet in which a large proportion or percentage of a population is purported to be "at risk" of some calamity, as in the childhood hunger example. Does the article cite a source or provide any information to support the proportion or percentage reported? Is the goal of the article to persuade some individual or group to take some action? If so, what action is being requested? Do you believe that the writer of the article may have some motive for exaggerating the problem? If so, give some possible motives.
2. Look for another article in which a relationship between two seemingly unrelated quantities is purported to exist, as in the CEO golf handicap and stock performance study. Select an article that contains some information on how the data were collected. Identify the target population and the data collection method. Based on what is presented in the article, do you believe that the data are representative of the population? Explain. Is the purported relationship of any practical interest? Explain.

CHAPTER NOTES

Key Terms

Note: Starred (*) terms are from the optional section in this chapter.

*Black box 11
Census 6
Data 3
Descriptive statistics 3
Designed experiment 14
Experimental unit 5
Inference 3
Inferential statistics 3
Measurement 6
Measurement error 19
Measure of reliability 9
Nonresponse bias 19
Observational study 15
Population 5
*Process 10
Published source 14
Qualitative data 13
Quantitative data 13
Quantitative literacy 18
Random sample 15
Reliability 8
Representative sample 15
Sample 6, 11
Selection bias 19
Statistical inference 7
Statistical thinking 18
Statistics 3
Survey 15
Unethical statistical practice 18
Variable 5

Key Ideas

Types of Statistical Applications

Descriptive

1. Identify **population** or **sample** (collection of **experimental units**)
2. Identify **variable(s)**
3. Collect **data**
4. **Describe** data

Inferential

1. Identify **population** (collection of *all* **experimental units**)
2. Identify **variable(s)**
3. Collect **sample** data (*subset* of population)
4. **Inference** about population based on sample
5. **Measure of reliability** for inference

Types of Data

1. **Quantitative** (numerical in nature)
2. **Qualitative** (categorical in nature)

Data-Collection Methods

1. **Observational**
2. **Published source**
3. **Survey**
4. **Designed experiment**

Problems with Nonrandom Samples

1. **Selection bias**
2. **Nonresponse bias**
3. **Measurement error**

Exercises 1.1–1.32

Note: Starred (*) exercises are from the optional section in this chapter.

Learning the Mechanics

1.1 What is statistics?

1.2 Explain the difference between descriptive and inferential statistics.

1.3 List and define the four elements of a descriptive statistics problem.

1.4 List and define the five elements of an inferential statistical analysis.

1.5 List the four major methods of collecting data and explain their differences.

1.6 Explain the difference between quantitative and qualitative data.

1.7 Explain how populations and variables differ.

1.8 Explain how populations and samples differ.

1.9 What is a representative sample? What is its value?

1.10 Why would a statistician consider an inference incomplete without an accompanying measure of its reliability?

*** 1.11** Explain the difference between a population and a process.

1.12 Define *statistical thinking*.

1.13 Suppose you're given a data set that classifies each sample unit into one of four categories: A, B, C, or D. You plan to create a computer database consisting of these data, and you decide to code the data as A = 1, B = 2, C = 3, and D = 4. Are the data consisting of the classifications A, B, C, and D qualitative or quantitative? After the data are input as 1, 2, 3, or 4, are they qualitative or quantitative? Explain your answers.

Applet Exercise 1.1

The *Random Numbers* applet generates a list of n random numbers from 1 to N, where n is the size of the sample and N is the size of the population. The list generated often contains repetitions of one or more numbers.

a. Using the applet *Random Numbers*, enter 1 for the minimum value, 10 for the maximum value, and 10 for the sample size. Then click on *Sample*. Look at the results and list any numbers that are repeated and the number of times each of these numbers occurs.

b. Repeat part **a,** changing the maximum value to 20 and keeping the size of the sample fixed at 10. If you still have repetitions, repeat the process, increasing the maximum value by 10 each time but keeping the size of the sample fixed. What is the smallest maximum value for which you had no repetitions?

c. Describe the relationship between the population size (maximum value) and the number of repetitions in the list of random numbers as the population size increases and the sample size remains the same. What can you conclude about using a random number generator to choose a relatively small sample from a large population?

Applet Exercise 1.2

The *Random Numbers* applet can be used to select a random sample from a population, but can it be used to simulate data? In parts **a** and **b,** you will use the applet to create data sets. Then you will explore whether those data sets are realistic.

a. In the activity *Keep the Change* on page 16, a data set called *Amounts Transferred* is described. Use the *Random Numbers* applet to simulate this data set by setting the minimum value equal to 0, the maximum value equal to 99, and the sample size equal to 30. Explain what the numbers in the list produced by the applet represent in the context of the activity. (You may need to read the activity.) Do the numbers produced by the applet seem reasonable? Explain.

b. Use the *Random Numbers* applet to simulate grades on a statistics test by setting the minimum value equal to 0, the maximum value equal to 100, and the sample size equal to 30. Explain what the numbers in the list produced by the applet represent in this context. Do the numbers produced by the applet seem reasonable? Explain.

c. Referring to parts **a** and **b,** why do the randomly generated data seem more reasonable in one situation than in the other? Comment on the usefulness of using a random number generator to produce data.

Applying the Concepts—Basic

1.14 NW **Credit card market shares.** Refer to the study of credit card purchases in the United States in Section 1.2 (Study 1, p. 4). Recall that the Nilson Report tracked all credit card purchases during 2007 and measured two variables: (1) the type of credit card used (VISA, MasterCard, American Express, or Discover), and (2) the amount (in dollars) of each purchase.

a. Identify the type (quantitative or qualitative) of each variable measured.

b. Does the data set collected represent a population or a sample? Explain.

1.15 NW **Opinion polls.** Pollsters regularly conduct opinion polls to determine the popularity rating of the current president. Suppose a poll is to be conducted tomorrow in which 2,000 individuals will be asked whether the president is doing a good or bad job. The 2,000 individuals will be selected by random-digit telephone dialing and asked the question over the phone.

a. What is the relevant population?
b. What is the variable of interest? Is it quantitative or qualitative?
c. What is the sample?
d. What is the inference of interest to the pollster?
e. What method of data collection is employed?
f. How likely is the sample to be representative?

1.16 NW **College application data.** Colleges and universities are requiring an increasing amount of information about applicants before making acceptance and financial aid decisions. Classify each of the following types of data required on a college application as quantitative or qualitative.

a. High school GPA
b. Honors, awards
c. Applicant's score on the SAT or ACT
d. Gender of applicant
e. Parents' income
f. Age of applicant

1.17 Microsoft survey of Windows. Windows Vista is a computer software product made by Microsoft Corporation. In designing Windows Vista, Microsoft telephoned thousands of users of an earlier version of Windows and asked them how the product could be improved. Assume customers were asked the following questions:

I. Are you the most frequent user of Windows in your household?
II. What is your age?
III. Are the tutorial instructions that accompany Windows helpful?
IV. When using a printer with Windows, do you most frequently use a laser printer or another type of printer?
V. If the speed of Windows could be changed, which one of the following would you prefer: slower, unchanged, or faster?
VI. How many people in your household have used Windows at least once?

Each of these questions defines a variable of interest to the company. Classify the data generated for each variable as quantitative or qualitative. Justify your classification.

1.18 Going online for health information. A *cyberchondriac* is defined as a person who regularly searches the Web for health care information. Each year a Harris Poll is conducted to determine the number of cyberchondriacs in the United States. In 2008, the Harris Poll surveyed 1,010 U.S. adults by telephone and asked the following questions:

1. Have you ever gone online to look for health care information?
2. How many times per month do you look for health care information online?
3. In the past year, have you ever discussed with your doctor the information you found online?

a. For each question, determine whether the type of data collected is quantitative or qualitative.
b. Do the data collected for the 1,010 adults represent a sample or a population? Explain.

1.19 The executive compensation scoreboard. Each year, *Forbes* publishes its Executive Compensation Scoreboard. (See Study #2, p. 4) For the 2008 scoreboard, data were collected for chief executive officers at the 500 largest U.S. companies and the following variables were measured for each CEO: (1) the industry type of the CEO's company (e.g., banking, retailing, etc.), (2) the CEO's total compensation ($ millions) for the year, (3) the CEO's total compensation ($ millions) over the previous 5 years, (4) the number of company stock shares (millions) held, (5) the CEO's age (years), and (6) the CEO's efficiency rating. (Data are saved in the file **CEOPAY2008.**)

a. Are the data for the 500 CEOs in the 2008 Executive Compensation Scoreboard a population or sample? Explain.
b. Identify the type (quantitative or qualitative) of each variable measured.

1.20 Annual survey of computer crimes. The Computer Security Institute (CSI) conducts an annual survey of computer crime at United States businesses. CSI sends survey questionnaires to computer security personnel at all U.S. corporations and government agencies. In 2006, 616 organizations responded to the CSI survey. Fifty-two percent of the respondents admitted unauthorized use of computer systems at their firms during the year. (*Computer Security Issues & Trends,* Spring 2006)

a. Identify the population of interest to CSI.
b. Identify the data-collection method used by CSI. Are there any potential biases in the method used?
c. Describe the variable measured in the CSI survey. Is it quantitative or qualitative?
d. What inference can be made from the study result?

Applying the Concepts—Intermediate

1.21 Satellite radio in cars. A recent survey conducted for the National Association of Broadcasters investigated satellite radio subscriber service and usage. The June 2007 survey, conducted by Wilson Research Strategies, consisted of a random sample of 501 satellite radio subscribers. One of the questions of interest was, "Do you have a satellite radio receiver in your car?" The survey found that 396 subscribers did, in fact, have a satellite receiver in their car.

a. Identify the population of interest to the National Association of Broadcasters.
b. Based on the survey question, what is the variable of interest?
c. Does the variable produce quantitative or qualitative data?
d. Describe the sample of interest. Is it representative of the population?
e. What inference can be made from the survey results?

1.22 New cancer screening method. According to the American Lung Association, lung cancer accounts for 28% of all cancer deaths in the United States. A new type of screening for lung cancer, computed tomography (CT), has been developed. Medical researchers believe CT scans are more sensitive than regular X-rays in pinpointing small tumors. The H. Lee Moffitt Cancer Center at the University of South Florida is currently conducting a clinical trial of 50,000 smokers nationwide to compare the effectiveness of CT scans with X-rays for detecting lung cancer (*Todays' Tomorrows,* Fall 2002). Each participating smoker is randomly assigned to one of two screening methods, CT or chest X-ray, and his or her progress is tracked over time. The age at which the scanning method first detects a tumor is the variable of interest.

a. Identify the data-collection method used by the cancer researchers.

b. Identify the experimental units of the study.
c. Identify the type (quantitative or qualitative) of the variable measured.
d. Identify the population and sample.
e. What is the inference that will ultimately be drawn from the clinical trial?

1.23 Inspection of highway bridges. All highway bridges in the United States are inspected periodically for structural deficiency by the Federal Highway Administration (FHWA). Data from the FHWA inspections are compiled into the National Bridge Inventory (NBI). Several of the nearly 100 variables maintained by the NBI are listed below. Classify each variable as quantitative or qualitative.
a. Length of maximum span (feet)
b. Number of vehicle lanes
c. Toll bridge (yes or no)
d. Average daily traffic
e. Condition of deck (good, fair, or poor)
f. Bypass or detour length (miles)
g. Route type (interstate, U.S., state, county, or city)

1.24 Structurally deficient highway bridges. Refer to Exercise 1.23. The 2007 NBI data were analyzed and the results made available at the FHWA Web site (www.fhwa.dot.gov). Using the FHWA inspection ratings, each of the 599,766 highway bridges in the United States was categorized as structurally deficient, functionally obsolete, or safe. About 12% of the bridges were found to be structurally deficient, while 13.5% were functionally obsolete.
a. What is the variable of interest to the researchers?
b. Is the variable of part **a** quantitative or qualitative?
c. Is the data set analyzed a population or a sample? Explain.
d. How did the NBI obtain the data for the study?

*__1.25 Monitoring product quality.__ The Wallace Company of Houston is a distributor of pipes, valves, and fittings to the refining, chemical, and petrochemical industries. The company was a recent winner of the Malcolm Baldrige National Quality Award. One of the steps the company takes to monitor the quality of its distribution process is to send out a survey twice a year to a subset of its current customers, asking the customers to rate the speed of deliveries, the accuracy of invoices, and the quality of the packaging of the products they have received from Wallace.
a. Describe the process studied.
b. Describe the variables of interest.
c. Describe the sample.
d. Describe the inferences of interest.
e. What are some of the factors that are likely to affect the reliability of the inferences?

1.26 Guilt in decision making. The effect of guilt emotion on how a decision maker focuses on the problem was investigated in the *Journal of Behavioral Decision Making* (January 2007). A total of 171 volunteer students participated in the experiment, where each was randomly assigned to one of three emotional states (guilt, anger, or neutral) through a reading/writing task. Immediately after the task, the students were presented with a decision problem (e.g., whether or not to spend money on repairing a very old car). The researchers found that a higher proportion of students in the guilty-state group chose to repair the car than those in the neutral-state and anger-state groups.
a. Identify the population, sample, and variables measured for this study.
b. Identify the data-collection method used.
c. What inference was made by the researcher?
d. In later chapters you will learn that the reliability of an inference is related to the size of the sample used. In addition to sample size, what factors might affect the reliability of the inference drawn in this study?

1.27 Accounting and Machiavellianism. *Behavioral Research in Accounting* (January 2008) published a study of Machiavellian traits in accountants. *Machiavellian* describes negative character traits that include manipulation, cunning, duplicity, deception, and bad faith. A questionnaire was administered to a random sample of 700 accounting alumni of a large southwestern university; however, due to nonresponse and incomplete answers, only 198 questionnaires could be analyzed. Several variables were measured, including age, gender, level of education, income, job satisfaction score, and Machiavellian ("Mach") rating score. The research findings suggest that Machiavellian behavior is not required to achieve success in the accounting profession.
a. What is the population of interest to the researcher?
b. What type of data (quantitative or qualitative) is produced by each of the variables measured?
c. Identify the sample.
d. Identify the data-collection method used.
e. What inference was made by the researcher?
f. How might the nonresponses impact the inference?

1.28 Success/failure of software reuse. The PROMISE Software Engineering Repository, hosted by the University of Ottawa, is a collection of publicly available data sets to serve researchers in building prediction software models. A PROMISE data set on software reuse, saved in the **SWREUSE** file, provides information on the success or failure of reusing previously developed software for each in a sample of 24 new software development projects. (Data source: *IEEE Transactions on Software Engineering*, Vol. 28, 2002.) Of the 24 projects, 9 were judged failures and 15 were successfully implemented.
a. Identify the experimental units for this study.
b. Describe the population from which the sample is selected.
c. What is the variable of interest in the study? Is it quantitative or qualitative?
d. Use the sample information to make an inference about the population.

Applying the Concepts—Advanced

1.29 Bank corporate mergers. *Corporate merger* is a means through which one firm (the bidder) acquires control of the assets of another firm (the target). During the late 1990s, there was a frenzy of bank mergers in the United States, as the banking industry consolidated into more efficient and more competitive units.
a. Construct a brief questionnaire (two or three questions) that could be used to query a sample of bank presidents

concerning their opinions of why the industry is consolidating and whether it will consolidate further.

b. Describe the population about which inferences could be made from the results of the survey.

c. Discuss the pros and cons of sending the questionnaire to all bank presidents versus a sample of 200.

*1.30 NW **Monitoring the production of soft drink cans.** The Wakefield plant of Coca-Cola and Schweppes Beverages Limited (CCSB) can produce 4,000 cans of soft drink per minute. The automated process consists of measuring and dispensing the raw ingredients into storage vessels to create the syrup, and then injecting the syrup, along with carbon dioxide, into the beverage cans. In order to monitor the subprocess that adds carbon dioxide to the cans, five filled cans are pulled off the line every 15 minutes, and the amount of carbon dioxide in each of these five cans is measured to determine whether the amounts are within prescribed limits.

a. Describe the process studied.

b. Describe the variable of interest.

c. Describe the sample.

d. Describe the inference of interest.

e. *Brix* is a unit for measuring sugar concentration. If a technician is assigned the task of estimating the average brix level of all 240,000 cans of beverage stored in a warehouse near Wakefield, will the technician be examining a process or a population? Explain.

1.31 Current population survey. The employment status (employed or unemployed) of each individual in the U.S. workforce is a set of data that is of interest to economists, businesspeople, and sociologists. To obtain information about the employment status of the workforce, the U.S. Bureau of the Census conducts what is known as the *Current Population Survey*. Each month interviewers visit about 59,000 of the 98 million households in the United States and question the occupants over 14 years of age about their employment status. Their responses enable the Bureau of the Census to *estimate* the percentage of people in the labor force who are unemployed (the *unemployment rate*).

a. Define the population of interest to the Census Bureau.

b. What variable is being measured? Is it quantitative or qualitative?

c. Is the problem of interest to the Census Bureau descriptive or inferential?

d. In order to monitor the rate of unemployment, it is essential to have a definition of *unemployed*. Different economists and even different countries define it in various ways. Develop your own definition of an "unemployed person." Your definition should answer such questions as: Are students on summer vacation unemployed? Are college professors who do not teach summer school unemployed? At what age are people considered to be eligible for the workforce? Are people who are out of work but not actively seeking a job unemployed?

Critical Thinking Challenge

1.32 *20/20* survey exposé. Refer to the "Statistics in Action" box of this chapter (p. 1). Recall that the popular prime-time ABC television program *20/20* presented several misleading (and possibly unethical) surveys in a segment titled "Facts or Fiction?—Exposés of So-Called Surveys." The information reported from four of these surveys are reproduced here (actual survey facts are provided in parentheses).

Quaker Oats study: Eating oat bran is a cheap and easy way to reduce your cholesterol count. (Fact: Diet must consist of nothing but oat bran to achieve a slightly lower cholesterol count.)

March of Dimes report: Domestic violence causes more birth defects than all medical issues combined. (Fact: No study—false report.)

American Association of University Women (AAUA) study: Only 29% of high school girls are happy with themselves, compared to 66% of elementary school girls. (Fact: Of 3,000 high school girls, 29% responded "Always true" to the statement "I am happy the way I am." Most answered "Sort of true" and "Sometimes true.")

Food Research and Action Center study: One in four American children under age 12 is hungry or at risk of hunger. (Fact: Based on responses to questions "Do you ever cut the size of meals?" and "Do you ever eat less than you feel you should?" and "Did you ever rely on limited numbers of foods to feed your children because you were running out of money to buy food for a meal?")

a. Refer to the Quaker Oats study relating oat bran to cholesterol levels. Discuss why it is unethical to report the results as stated.

b. Consider the false March of Dimes report on domestic violence and birth defects. Discuss the type of data required to investigate the impact of domestic violence on birth defects. What data-collection method would you recommend?

c. Refer to the AAUW study of self-esteem of high school girls. Explain why the results of the study are likely to be misleading. What data might be appropriate for assessing the self-esteem of high school girls?

d. Refer to the Food Research and Action Center study of hunger in America. Explain why the results of the study are likely to be misleading. What data would provide insight into the proportion of hungry American children?

References

Careers in Statistics, American Statistical Association, 2008 (www.amstat.org)

Cochran, W. G. *Sampling Techniques,* 3rd ed. New York: Wiley, 1977.

Deming, W. E. *Sample Design in Business Research.* New York: Wiley, 1963.

Ethical Guidelines for Statistical Practice, American Statistical Association, 1999. (www.amstat.org)

Hahn, G. J. and Doganaksoy, N. *The Role of Statistics in Business and Industry*. New York: Wiley, 2008.

Huff, D. *How to Lie with Statistics*. New York: Norton, 1982 (paperback 1993).

Hoerl, R. and Snee, R. *Statistical Thinking: Improving Business Performance*. Boston: Duxbury, 2002.

Kish, L. *Survey Sampling.* New York: Wiley, 1965 (paperback, 1995).

Peck, R., Casella, G., Cobb, G., Hoerl, R., Nolan, D., Starbuck, R., and Stern, H. *Statistics: A Guide to the Unknown,* 4th ed. Cengage Learning, 2005.

Scheaffer, R., Mendenhall, W., and Ott, R. L. *Elementary Survey Sampling,* 6th ed. Boston: Duxbury, 2005.

What Is a Survey? American Statistical Association (F. Scheuren, editor), 2nd ed., 2005 (www.amstat.org)

USING TECHNOLOGY

SPSS: Accessing and Listing Data

When you start an SPSS session, you will see a screen similar to Figure 1.S.1. The main portion of the screen is an empty spreadsheet, with columns representing variables and rows representing observations (or cases). The very top of the screen is the SPSS main menu bar, with buttons for the different functions and procedures available in SPSS. Once you have entered data into the spreadsheet, you can analyze the data by clicking the appropriate menu buttons.

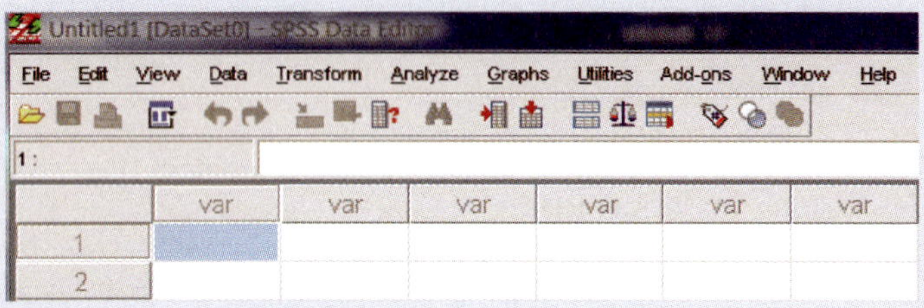

Figure 1.S.1 Initial screen viewed by the SPSS user

Entering Data

Step 1 To create an SPSS data file, enter data directly into the spreadsheet. See Figure 1.S.2, which shows data entered on a variable called "Fill."

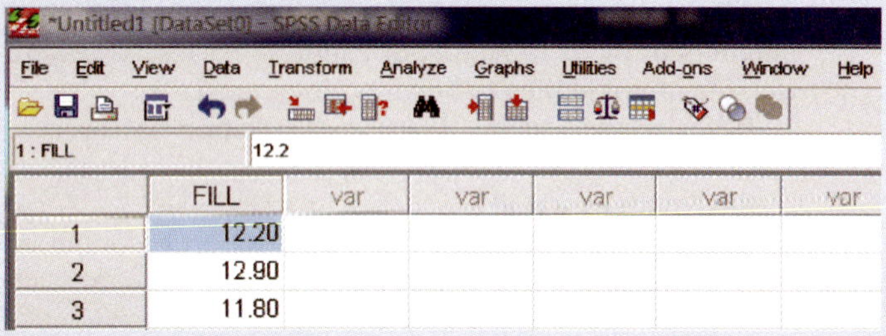

Figure 1.S.2 Data entered into the SPSS spreadsheet

Step 2 Name the variables (columns) by selecting the "Variable View" button at the bottom of the screen and typing in the name of each variable.

Accessing Data from a File

If the data are saved in an external data file, you can access the data using the options available in SPSS.

Step 1 Click the "File" button on the menu bar, and then click "Read Text Data," as shown in Figure 1.S.3. A dialog box similar to Figure 1.S.4 will appear.

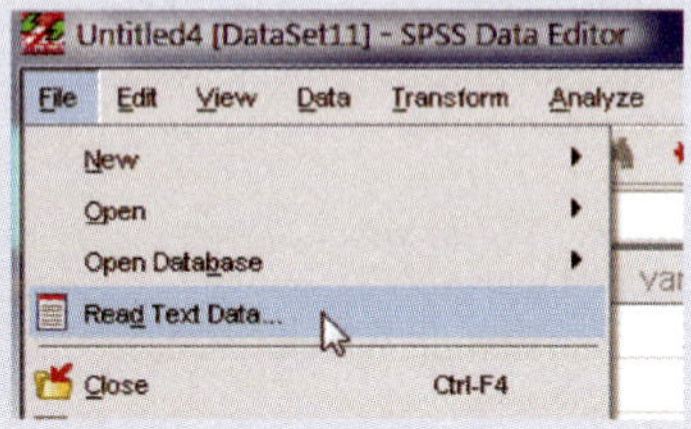

Figure 1.S.3 SPSS options for reading data from an external file

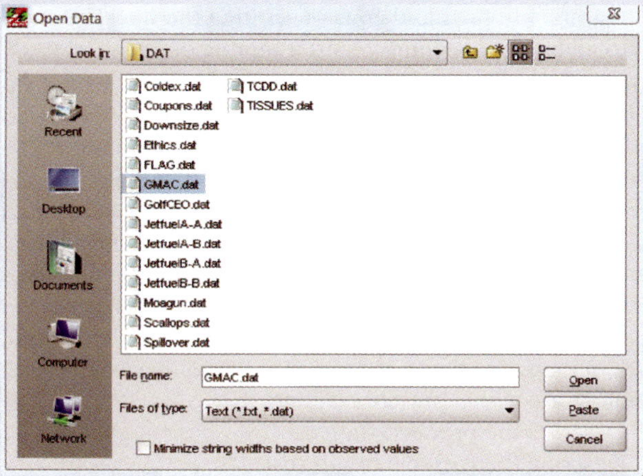

Figure 1.S.4 Selecting the external data file in SPSS

Step 2 Specify the disk drive and folder that contains the data file, click on the data file, and then click "Open," as shown in Figure 1.S.4. The SPSS Text Import Wizard Opens.

Step 3 The Text Import Wizard is a series of six screen menus. Make the appropriate selections on the screen, and click "Next" to go to the next screen.

Step 4 When finished, click "Finish." The SPSS spreadsheet will reappear with the data from the external data file, as shown in Figure 1.S.5.

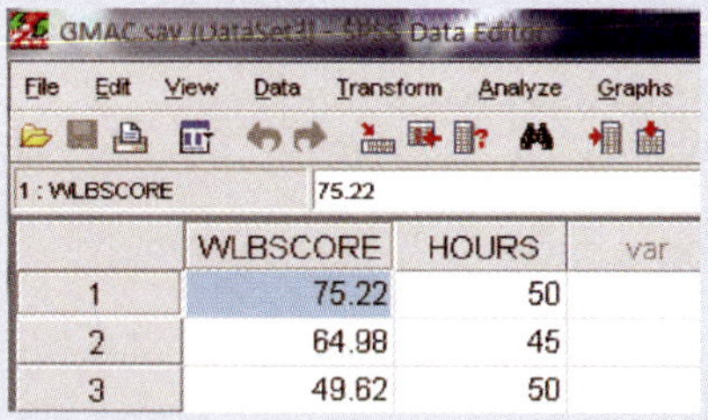

Figure 1.S.5 The SPSS spreadsheet with the imported data

Reminder: The variable (columns) can be named by selecting the "Variable View" button at the bottom of the spreadsheet screen and typing in the name of each variable.

To access a previously saved SPSS data file, click "File," then "Open," then select the SPSS file name.

Listing (Printing) Data

Step 1 Click on the "Analyze" button on the SPSS main menu bar, then click on "Reports," and then on "Report Summaries in Rows" (see Figure 1.S.6). The resulting menu, or dialog box, appears as in Figure 1.S.7.

Step 2 Enter the names of the variables you want to print in the "Data Columns" box (you can do this by simply clicking on the variables), check the "Display cases" box at the bottom left, and then click "OK." The printout will show up on your screen.

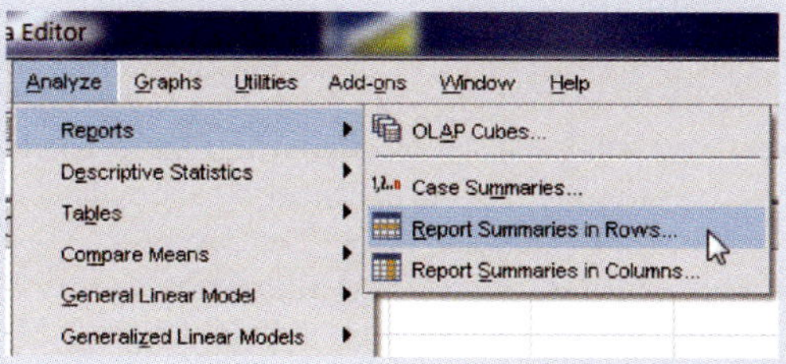

Figure 1.S.6 SPSS menu options for obtaining a data listing

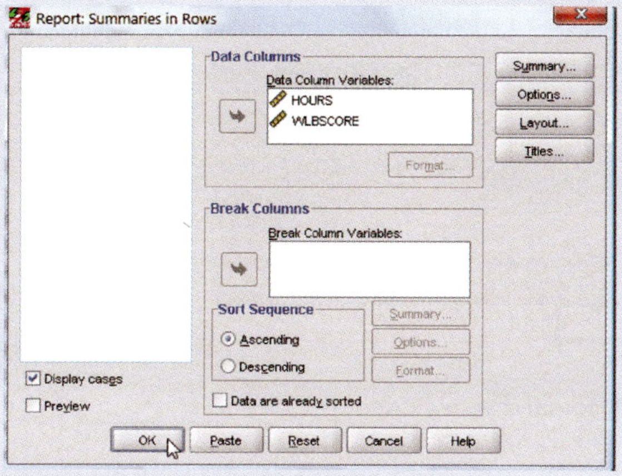

Figure 1.S.7 SPSS data report dialog box

Minitab: Accessing and Listing Data

When you start a Minitab session, you will see a screen similar to Figure 1.M.1. The bottom portion of the screen is an empty spreadsheet—called a Minitab worksheet—with columns representing variables and rows representing observations (or cases). The very top of the screen is the Minitab main menu bar, with buttons for the different functions and procedures available in Minitab. Once you have entered data into the spreadsheet, you can analyze the data by clicking the appropriate menu buttons. The results will appear in the Session window.

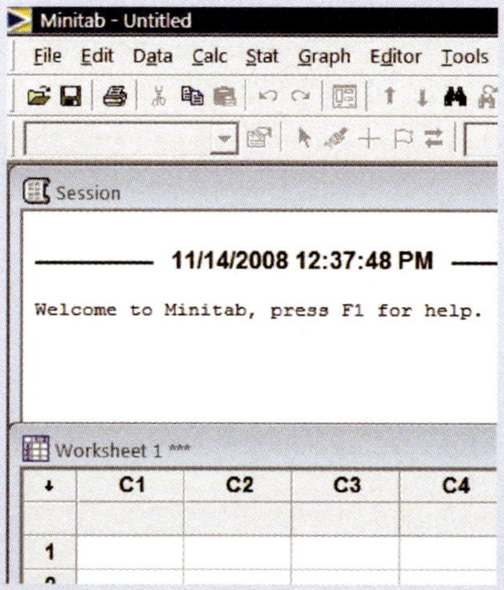

Figure 1.M.1 Initial screen viewed by the Minitab user

Entering Data

Create a Minitab data file by entering data directly into the worksheet. Figure 1.M.2 shows data entered for a variable called "FILL." Name the variables (columns) by typing in the name of each variable in the box below the column number.

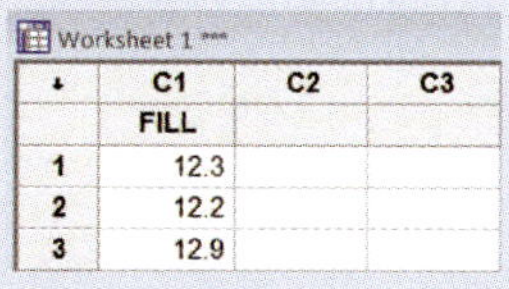

Figure 1.M.2 Data entered into the Minitab worksheet

Accessing External Data from a File

Step 1 Click the "File" button on the menu bar, and then click "Open Worksheet" as shown in Figure 1.M.3. A dialog box similar to Figure 1.M.4 will appear.

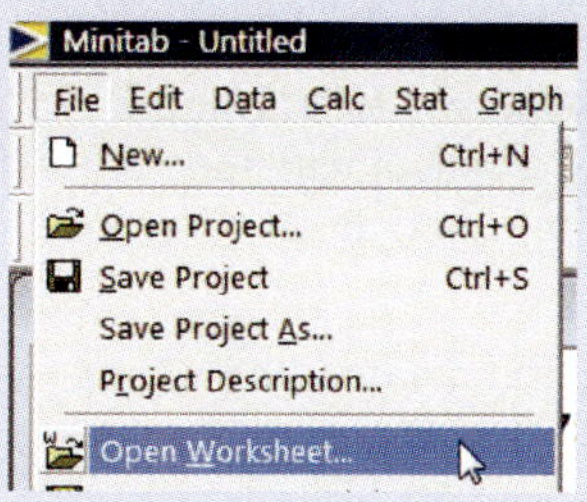

Figure 1.M.3 Minitab options for reading data from an external file

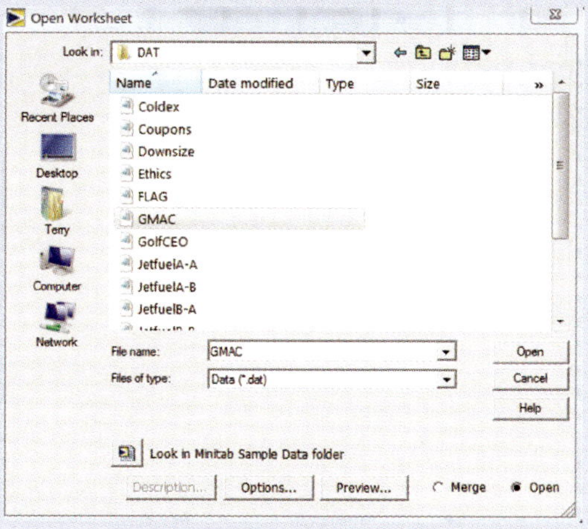

Figure 1.M.4 Selecting the external data file in Minitab

Step 2 Specify the disk drive and folder that contains the external data file and the file type, and then click on the file name, as shown in Figure 1.M.4.

Step 3 If the data set contains qualitative data or data with special characters, click on the "Options" button as shown in Figure 1.M.4. The Options dialog box, shown in Figure 1.M.5, will appear.

Step 4 Specify the appropriate options for the data set, and then click "OK" to return to the "Open Worksheet" dialog box (Figure 1.M.4).

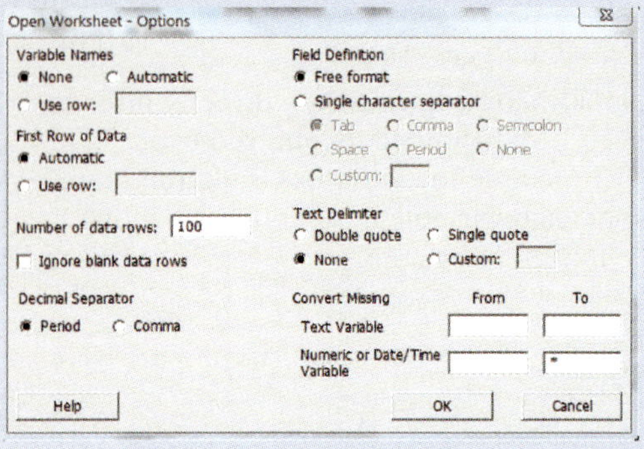

Figure 1.M.5 Selecting the Minitab data input options

Step 5 Click "Open" and the Minitab worksheet will appear with the data from the external data file, as shown in Figure 1.M.6.

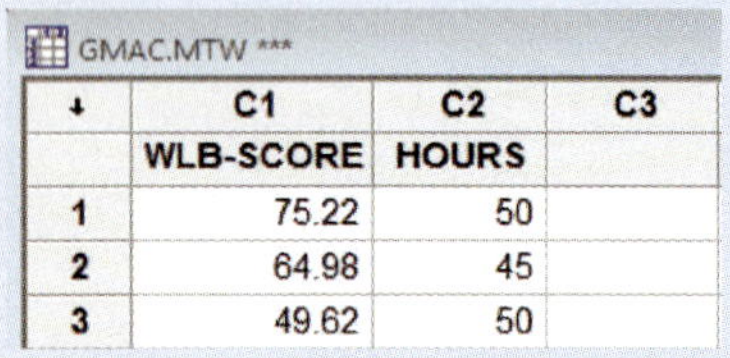

Figure 1.M.6 Minitab worksheet with the imported data

Reminder: The variables (columns) can be named by typing in the name of each variable in the box under the column number.

To access a previously saved Minitab worksheet, click "File," then "Open Worksheet," then select the Minitab file.

Listing (Printing) Data

Step 1 Click on the "Data" button on the Minitab main menu bar, and then click on "Display Data." The resulting menu, or dialog box, appears as in Figure 1.M.7.

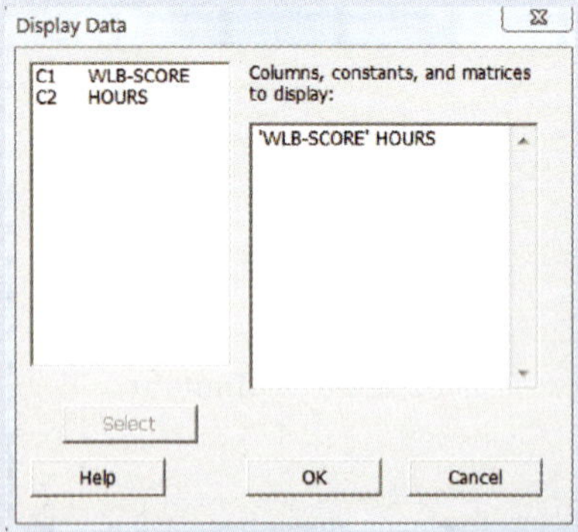

Figure 1.M.7 Minitab Display Data dialog box

Step 2 Enter the names of the variables you want to print in the "Columns, constants, and matrices to display" box (you can do this by simply double clicking on the variables), and then click "OK." The printout will show up on your Minitab session screen.

Excel: Accessing and Listing Data

When you open Excel, you will see a screen similar to Figure 1.E.1. The majority of the screen window is a spreadsheet—called an Excel workbook—with columns (labeled A, B, C, etc.) representing variables, and rows representing observations (or cases). The very top of the screen is the Excel main menu bar, with buttons for the different functions and procedures available in Excel. Once you have entered data into the spreadsheet, you can analyze the data by clicking the appropriate menu buttons. The results will appear in a new workbook.

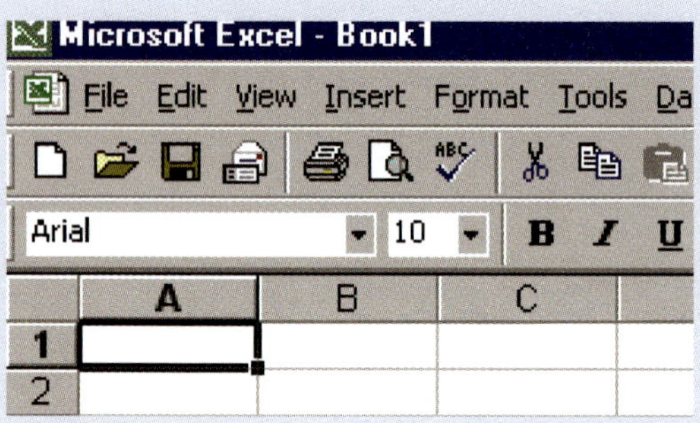

Figure 1.E.1 Initial screen viewed by Excel user

Entering Data

Enter data directly into the appropriate row and column of the spreadsheet. Figure 1.E.2 shows data entered in the first (A) column. Optionally, you can add names for the variables (columns) in the first row of the workbook.

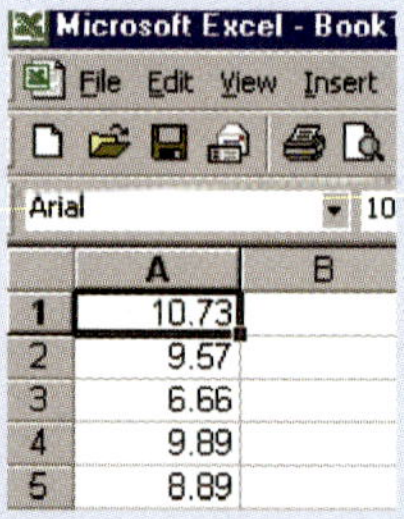

Figure 1.E.2 Data entered into the Excel workbook

Accessing Data from a File

Step 1 Click the "File" button on the menu bar, and then click "Open," as shown in Figure 1.E.3. A dialog box similar to Figure 1.E.4 will appear.

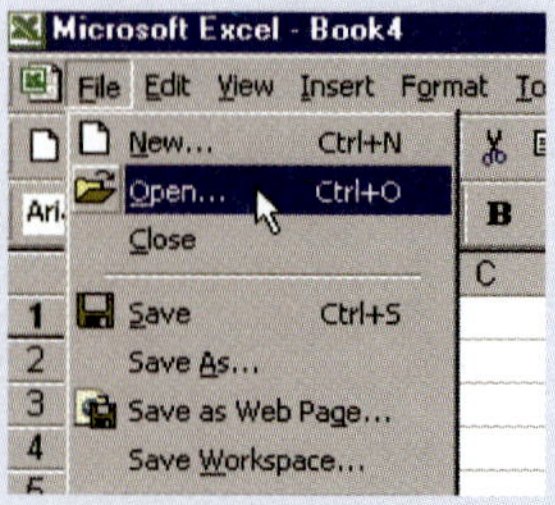

Figure 1.E.3 Excel options for reading data from an external file

Figure 1.E.4 Selecting the external data file in Excel

Step 2 Specify the disk drive and folder that contains the external data file and the file type, then click on the file name and click on "Open," as shown in Figure 1.E.4. The Excel Text Import Wizard opens (Figure 1.E.5.)

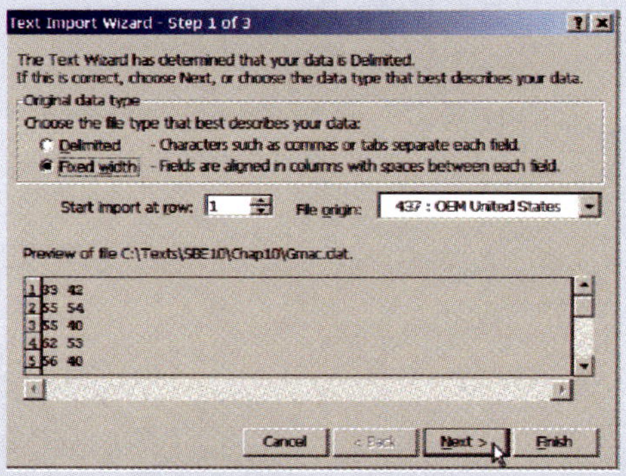

Figure 1.E.5 Excel Text Import Wizard, Screen 1

Step 3 Make the appropriate selections on the screen, and click "Next" to go to the next screen.

Step 4 When finished, click "Finish." The Excel workbook will reappear with the data from the external data file, as shown in Figure 1.E.6.

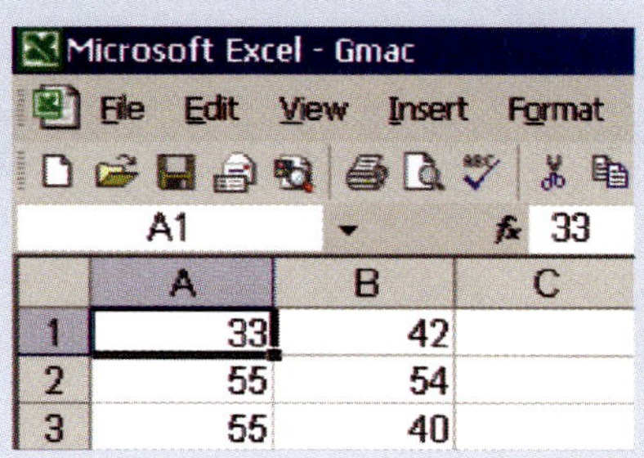

Figure 1.E.6 The Excel workbook with the imported data

Naming Variables

Step 1 Select "Insert" from the Excel main menu, and then select "Rows." A blank (empty) row will be added in the first row of the spreadsheet.

Step 2 Type the name of each variable in the first row under the appropriate column.

To access a previously created Excel workbook, click "File," then "Open," then select the Excel file.

Listing (Printing) Data

Step 1 Click on the "File" button on the Excel main menu bar.

Step 2 Click on "Print" (see Figure 1.E.7).

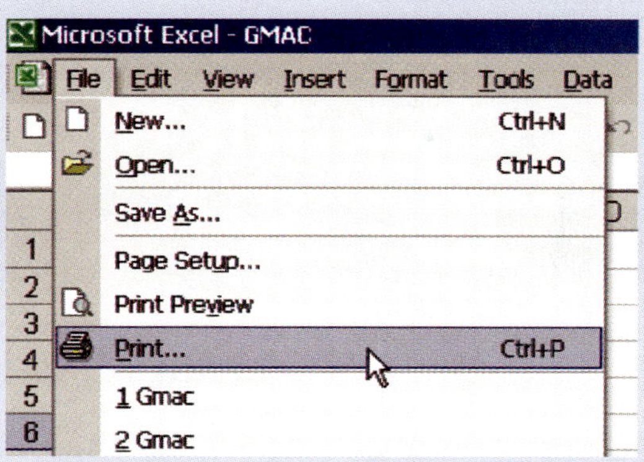

Figure 1.E.7 Excel menu options for obtaining a data listing

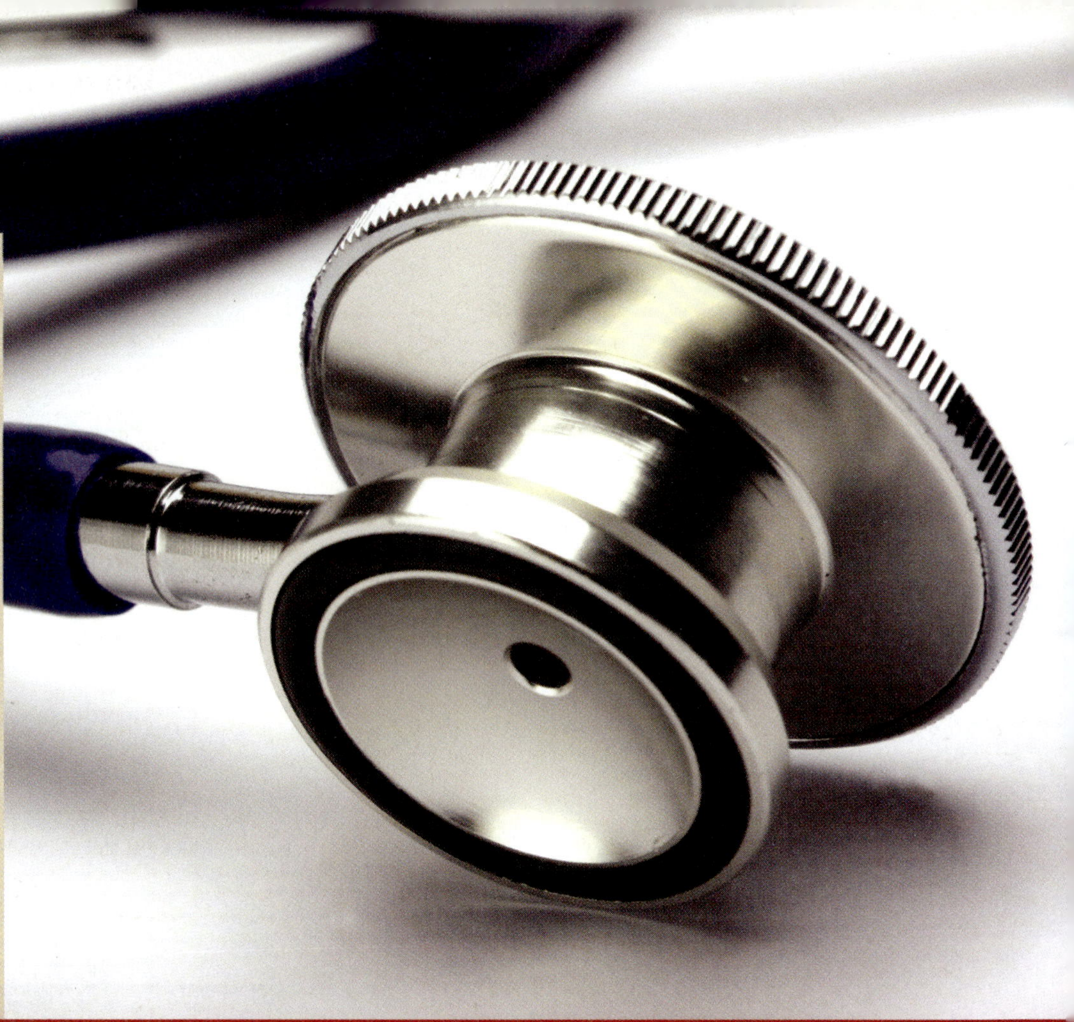

CONTENTS

2 Methods for Describing Sets of Data

Where We've Been

- Examined the difference between inferential and descriptive statistics
- Described the key elements of a statistical problem
- Learned about the two types of data—quantitative and qualitative
- Discussed the role of statistical thinking in managerial decision making

Where We're Going

- Describe data using graphs
- Describe data using numerical measures

Statistics IN Action Factors That Influence a Doctor to Refuse Ethics Consultation

Ethical dilemmas commonly arise in the course of a physician's clinical practice. These dilemmas include (but are not limited to) end-of-life issues, treatment of patients without health insurance, providing nonbeneficial treatment at the patient's request, obtaining informed consent, maintaining patient autonomy, truth-telling and confidentiality, involvement of children in research, designing of clinical trials, and termination of subject participation in research protocols. Empirical studies have found that an ethical issue will arise in one of every five clinical encounters.

Over the past 10 years, ethics consultation has evolved as a means of assisting physicians who are confused about how to best approach an ethical dilemma. About 80% of general hospitals in the United States now provide ethics consultation services to staff physicians. When faced with a difficult ethical decision, the doctor may request advice from a panel of ethics experts, with assurances that all communications will be anonymous and confidential. However, not all physicians take advantage of this service; in fact, some doctors refuse to use ethics consultation. *(continued)*

Medical researchers at University Community Hospital (UCH) in Tampa, Florida, undertook a study to determine the factors that might influence a physician's decision to request or to refuse ethics consultation.*

Survey questionnaires were distributed to all 746 physicians on staff at UCH; 118 of the questionnaires were returned, yielding a response rate of approximately 16%. The survey was designed to obtain data on the following variables for each physician:

1. *Level of previous ethics consultation use* **at UCH** ("used at least once" or "never used")
2. *Practitioner specialty* ("medical" or "surgical")
3. *Length of time in practice* (number of years)
4. *Amount of exposure to ethics in medical school* (number of hours)
5. *Would you ever consider using ethics consultation* **in the future** (yes or no)

The physicians were also asked to elicit opinions on the following statements about ethics consultants. (All responses were measured on a 5-point scale, where 1 = "strongly disagree," 2 = "somewhat disagree," 3 = "neither agree nor disagree," 4 = "somewhat disagree," or 5 = "strongly agree.")

6. *Ethics consultants have extensive training in ethics and ethics principles.*
7. *Ethics consultants participate in frequent ethics education.*
8. *Ethics consultants think they are "moral experts."*
9. *Ethics consultants cannot grasp the full picture from the "outside."*

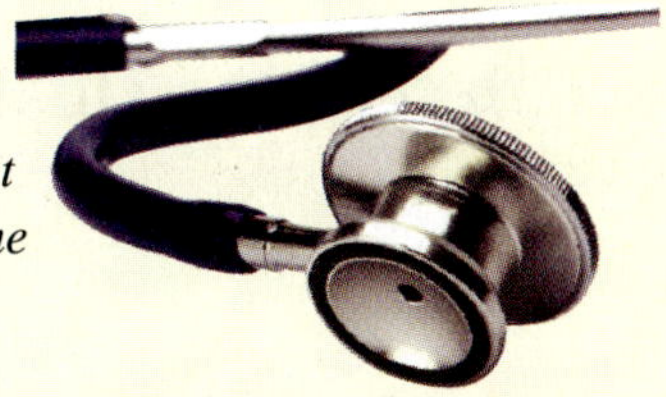

The UCH medical researchers wanted to use the survey results to develop insight into why certain physicians use ethics consultation and others do not. The researchers hypothesized that more experienced doctors and physicians who specialize in surgery would be less likely to use ethics consultation. The data for the study are stored in the file named **ETHICS** in SPSS, Minitab, and Excel.

In the following *Statistics in Action Revisited* sections, we apply the graphical and numerical descriptive techniques of this chapter to the **ETHICS** data to answer some of the researchers' questions.

Statistics IN Action Revisited

- Interpreting pie charts (p. 37)
- Interpreting histograms (p. 47)
- Interpreting numerical descriptive measures (p. 74)
- Detecting outliers (p. 88)
- Interpreting scatterplots (p. 92)

Data Set: ETHICS

Suppose you wish to evaluate the managerial capabilities of a class of 400 MBA students based on their Graduate Management Aptitude Test (GMAT) scores. How would you describe these 400 measurements? Characteristics of the data set include the typical or most frequent GMAT score, the variability in the scores, the highest and lowest scores, the "shape" of the data, and whether or not the data set contains any unusual scores. Extracting this information by "eye-balling" the data isn't easy. The 400 scores may provide too many bits of information for our minds to comprehend. Clearly, we need some formal methods for summarizing and characterizing the information in such a data set. Methods for describing data sets are also essential for statistical inference. Most populations are large data sets. Consequently, we need methods for describing a sample data set that let us make descriptive statements (inferences) about the population from which the sample was drawn.

Two methods for describing data are presented in this chapter, one *graphical* and the other *numerical*. Both play an important role in statistics. Section 2.1 presents both graphical and numerical methods for describing qualitative data. Graphical methods for describing quantitative data are presented in Sections 2.2, 2.8, and 2.10 and optional Section 2.9; numerical descriptive methods for quantitative data are presented in Sections 2.3–2.7. We end this chapter with a section on the *misuse* of descriptive techniques.

*Orlowski, J. P., Hein, S., Meinke, R., Christenson, J., and Sincich, T. "Why Physicians Use or Do Not Use Ethics Consultation," *Journal of Medical Ethics*, Vol. 32, 2006.

2.1 Describing Qualitative Data

Recall the "Executive Compensation Scoreboard" tabulated annually by *Forbes* (see Study 2 in Section 1.2). In addition to salary information, *Forbes* collects and reports personal data on the CEOs, including level of education. Do most CEOs have advanced degrees, such as masters degrees or doctorates? To answer this question, Table 2.1 gives the highest college degree obtained (bachelors, MBA, masters, law, PhD, or none) for each of the 40 best-paid CEOs in 2008.

For this study, the variable of interest, highest college degree obtained, is qualitative in nature. Qualitative data are nonnumerical; thus, the value of a qualitative variable can be classified only into categories called *classes*. The possible degree types—bachelors, MBA, masters, law, PhD, or none—represent the classes for this qualitative variable. We can summarize such data numerically in two ways: (1) by computing the *class frequency*—the number of observations in the data set that fall into each class;

Table 2.1 Data on 40 Best-Paid Executives

	CEO	Company	Salary ($ millions)	Age	Degree
1	Lawrence J Ellison	Oracle	192.92	63	None
2	Frederic M Poses	Trane	127.1	66	Bachelors
3	Aubrey K McClendon	Chesapeake Energy	116.89	47	Bachelors
4	Angelo R Mozilo	Countrywide Financial	102.84	69	Bachelors
5	Howard D Schultz	Starbucks	98.6	54	Bachelors
6	Nabeel Gareeb	MEMC Electronic Mats	79.56	43	None
7	Daniel P Amos	Aflac	75.16	56	Bachelors
8	Lloyd C Blankfein	Goldman Sachs Group	73.72	53	Law
9	Richard D Fairbank	Capital One Financial	73.17	57	MBA
10	Bob R Simpson	XTO Energy	72.27	59	MBA
11	Richard S Fuld Jr	Lehman Bros Holdings	71.9	62	MBA
12	Steven Roth	Vornado Realty	71.85	66	MBA
13	Marijn E Dekkers	Thermo Fisher	69	50	PhD
14	Steven A Burd	Safeway	67.17	58	Masters
15	Gregg L Engles	Dean Foods	66.08	50	Law
16	Nicholas D Chabraja	General Dynamics	60.26	65	Law
17	Leslie H Wexner	Limited Brands	56.06	70	Bachelors
18	David C Novak	Yum Brands	54.91	55	Bachelors
19	John T Chambers	Cisco Systems	54.77	58	MBA
20	William R Berkley	WR Berkley	54.6	62	MBA
21	Ray R Irani	Occidental Petroleum	54.4	73	PhD
22	Bradbury H Anderson	Best Buy	49.26	58	Bachelors
23	John W Rowe	Exelon	47.68	63	Law
24	Robert W Lane	Deere & Co	47.19	58	MBA
25	John B Hess	Hess	46.78	54	MBA
26	David E I Pyott	Allergan	46.19	55	Masters
27	Sol J Barer	Celgene	46.07	61	PhD
28	Albert L Lord	SLM	45.99	62	Bachelors
29	John H Hammergren	McKesson	44.91	49	MBA
30	Miles D White	Abbott Laboratories	44.76	53	MBA
31	J Willard Marriott Jr	Marriott International	44.09	76	Bachelors
32	Howard Solomon	Forest Labs	40.89	80	Law
33	Michael D Watford	Ultra Petroleum	40.64	54	MBA
34	Brian L Roberts	Comcast	38.98	48	Bachelors
35	James L Dolan	Cablevision	38.81	53	Bachelors
36	Richard C Adkerson	Freeport Copper	38.66	61	MBA
37	Paul J Evanson	Allegheny Energy	37.29	66	Law
38	Robert J Stevens	Lockheed Martin	36.56	56	Masters
39	Matthew K Rose	Burlington Santa Fe	36.52	49	Bachelors
40	Hugh Grant	Monsanto	35.78	50	Bachelors

Source: Data on 40 Best-Paid Executives, *Forbes*, April 30, 2008. Reprinted by permission of Forbes.com. © 2009 Forbes LLC.

Data Set: FORBES40

or (2) by computing the *class relative frequency*—the proportion of the total number of observations falling into each class.

A **class** is one of the categories into which qualitative data can be classified.

The **class frequency** is the number of observations in the data set falling into a particular class.

The **class relative frequency** is the class frequency divided by the total number of observations in the data set.

Examining Table 2.1, we observe that 2 of the 40 best-paid CEOs did not obtain a college degree, 14 obtained bachelors degrees, 12 MBAs, 3 masters degrees, 3 PhDs, and 6 law degrees. These numbers—2, 14, 12, 3, 3, and 6—represent the class frequencies for the six classes and are shown in the summary table, Figure 2.1, produced using SPSS.

The **class percentage** is the class relative frequency multiplied by 100.

Figure 2.1 also gives the relative frequency of each of the five degree classes. We know that we calculate the relative frequency by dividing the class frequency by the total number of observations in the data set. Thus, the relative frequencies for the five degree types are

$$\text{Bachelors: } \frac{14}{40} = .35$$

$$\text{Law: } \frac{6}{40} = .15$$

$$\text{Masters: } \frac{3}{40} = .075$$

$$\text{MBA: } \frac{12}{40} = .30$$

$$\text{None: } \frac{2}{40} = .05$$

$$\text{PhD: } \frac{3}{40} = .075$$

These values, expressed as a percentage, are shown in the "Percent" column in the SPSS summary table, Figure 2.1. If we sum the relative frequencies for MBA,

DEGREE

		Frequency	Percent	Valid Percent	Cumulative Percent
Valid	Bachelors	14	35.0	35.0	35.0
	Law	6	15.0	15.0	50.0
	Masters	3	7.5	7.5	57.5
	MBA	12	30.0	30.0	87.5
	None	2	5.0	5.0	92.5
	PhD	3	7.5	7.5	100.0
	Total	40	100.0	100.0	

Figure 2.1
SPSS summary table for degrees of 40 CEOs

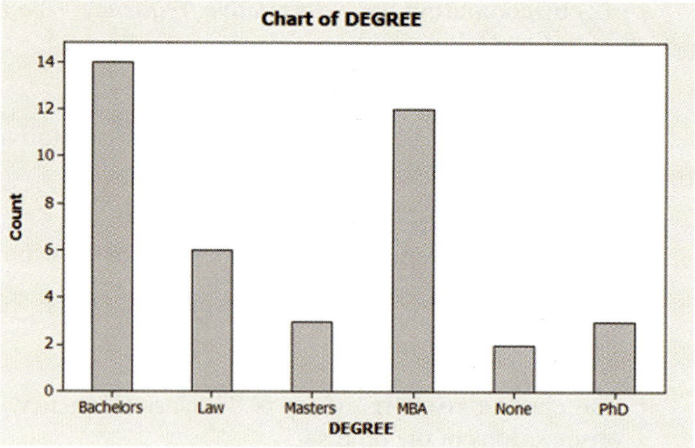

Figure 2.2
Minitab bar graph for degrees of 40 CEOs

masters, law, and PhD, we obtain $.30 + .075 + .15 + .075 = .60$. Therefore, 60% of the 40 best-paid CEOs obtained at least a masters degree (MBA, masters, law, or PhD).

Although the summary table in Figure 2.1 adequately describes the data in Table 2.1, we often want a graphical presentation as well. Figures 2.2 and 2.3 show two of the most widely used graphical methods for describing qualitative data—**bar graphs** and **pie charts.** Figure 2.2 is a bar graph for "highest degree obtained" produced with Minitab. Note that the height of the rectangle, or "bar," over each class is equal to the class frequency. (Optionally, the bar heights can be proportional to class relative frequencies.) In contrast, Figure 2.3 (also created using Minitab) shows the relative frequencies (expressed as a percentage) of the six degree types in a *pie chart.* Note that the pie is a circle (spanning 360°), and the size (angle) of the "pie slice" assigned to each class is proportional to the class relative frequency. For example, the slice assigned to the MBA degree is 30% of 360°, or $(.30)(360°) = 108°$.

Before leaving the data set in Table 2.1, consider the bar graph shown in Figure 2.4, produced using SPSS with annotations. Note that the bars for the CEO degree categories are arranged in descending order of height from left to right across the horizontal axis—that is, the tallest bar (bachelors) is positioned at the far left and the shortest bar is at the far right. This rearrangement of the bars in a bar graph is called a Pareto diagram. One goal of a Pareto diagram (named for the Italian economist, Vilfredo Pareto) is to make it easy to locate the "most important" categories—those with the largest frequencies. For the 40 best-paid CEOs in 2008, a bachelor's degree was the highest degree obtained by the most CEOs (35%).

BIOGRAPHY

VILFREDO PARETO (1843–1923)

The Pareto Principle

Born in Paris to an Italian aristocratic family, Vilfredo Pareto was educated at the University of Turin, where he studied engineering and mathematics. After the death of his parents, Pareto quit his job as an engineer and began writing and lecturing on the evils of the economic policies of the Italian government. While at the University of Lausanne in Switzerland in 1896, he published his first paper, "Cours d'economie politique." In the paper, Pareto derived a complicated mathematical formula to prove that the distribution of income and wealth in society is not random but that a consistent pattern appears throughout history in all societies. Essentially, Pareto showed that approximately 80% of the total wealth in a society lies with only 20% of the families. This famous law about the "vital few and the trivial many" is widely known as the Pareto principle in economics. ■

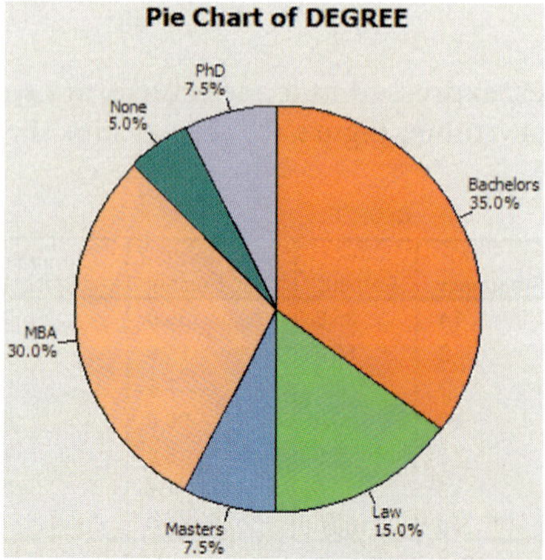

Figure 2.3
Minitab pie chart for degrees of 40 CEOs

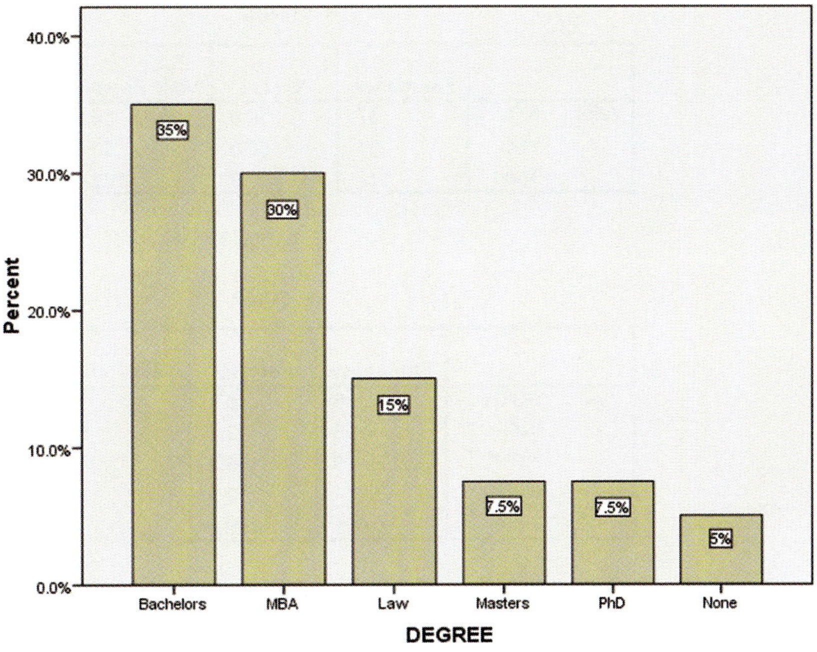

Figure 2.4
SPSS Pareto diagram for degrees of 40 CEOs

Summary of Graphical Descriptive Methods for Qualitative Data

Bar graph: The categories (classes) of the qualitative variable are represented by bars, where the height of each bar is either the class frequency, class relative frequency, or class percentage.

Pie chart: The categories (classes) of the qualitative variable are represented by slices of a pie (circle). The size of each slice is proportional to the class relative frequency.

Pareto diagram: A bar graph with the categories (classes) of the qualitative variable (i.e., the bars) arranged by height in descending order from left to right.

Let's look at a practical example that requires interpretation of the graphical results.

Example 2.1 Graphing and Summarizing Qualitative Data

Problem A group of cardiac physicians in southwest Florida have been studying a new drug designed to reduce blood loss in coronary artery bypass operations. Blood loss data for 114 coronary artery bypass patients (some who received a dosage of the drug and others who did not) are saved in the **BLOODLOSS** file. Although the drug shows promise in reducing blood loss, the physicians are concerned about possible side effects and complications. So their data set includes not only the qualitative variable, DRUG, which indicates whether or not the patient received the drug, but also the qualitative variable, COMP, which specifies the type (if any) of complication experienced by the patient. The four values of COMP recorded by the physicians are (1) redo surgery, (2) post-op infection, (3) both, or (4) none.

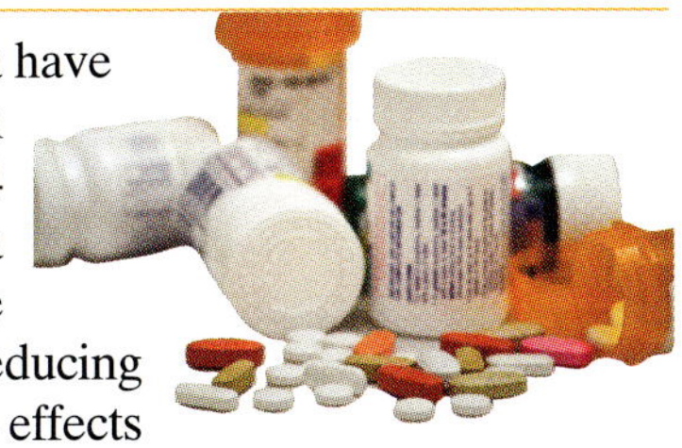

a. Figure 2.5, generated using SPSS, shows summary tables for the two qualitative variables, DRUG and COMP. Interpret the results.

b. Interpret the Minitab output shown in Figure 2.6 and the SPSS output shown in Figure 2.7.

Solution

a. The top table in Figure 2.5 is a summary frequency table for DRUG. Note that exactly half (57) of the 114 coronary artery bypass patients received the drug and half did not. The bottom table in Figure 2.5 is a summary frequency table for COMP. We see that about 69% of the 114 patients had no complications, leaving about 31% who experienced either a redo surgery, a post-op infection, or both.

DRUG

		Frequency	Percent	Valid Percent	Cumulative Percent
Valid	NO	57	50.0	50.0	50.0
	YES	57	50.0	50.0	100.0
	Total	114	100.0	100.0	

COMP

		Frequency	Percent	Valid Percent	Cumulative Percent
Valid	BOTH	6	5.3	5.3	5.3
	INFECT	15	13.2	13.2	18.4
	NONE	79	69.3	69.3	87.7
	REDO	14	12.3	12.3	100.0
	Total	114	100.0	100.0	

Figure 2.5
SPSS summary tables for DRUG and COMP

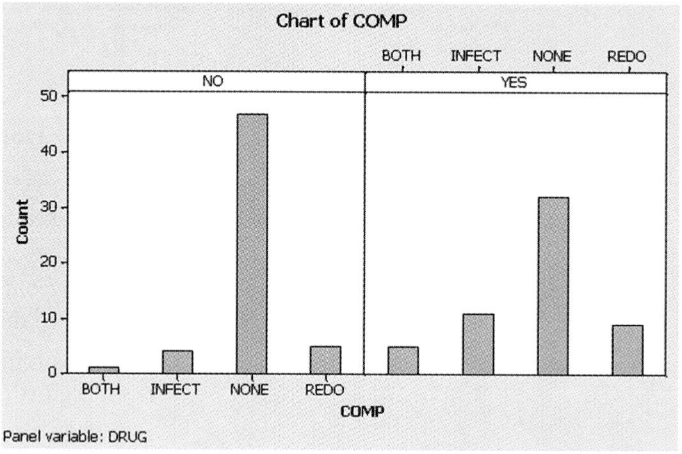

Figure 2.6
Minitab side-by-side bar graphs for COMP, by value of DRUG

COMP

DRUG			Frequency	Percent	Valid Percent	Cumulative Percent
NO	Valid	BOTH	1	1.8	1.8	1.8
		INFECT	4	7.0	7.0	8.8
		NONE	47	82.5	82.5	91.2
		REDO	5	8.8	8.8	100.0
		Total	57	100.0	100.0	
YES	Valid	BOTH	5	8.8	8.8	8.8
		INFECT	11	19.3	19.3	28.1
		NONE	32	56.1	56.1	84.2
		REDO	9	15.8	15.8	100.0
		Total	57	100.0	100.0	

Figure 2.7
SPSS summary tables for COMP by value of DRUG

b. Figure 2.6 is a Minitab side-by-side bar graph for the data. The four bars on the left represent the frequencies of COMP for the 57 patients who did not receive the drug; the four bars on the right represent the frequencies of COMP for the 57 patients who did receive a dosage of the drug. The graph clearly shows that patients who did not get the drug suffered fewer complications. The exact percentages are displayed in the SPSS summary tables of Figure 2.7. About 56% of the patients who got the drug had no complications, compared to about 83% for the patients who did not get the drug.

Look Back Although the drug may be effective in reducing blood loss, the results in Figures 2.6 and 2.7 also imply that patients on the drug may have a higher risk of complications. But before using this information to make a decision about the drug, the physicians will need to provide a measure of reliability for the inference—that is, the physicians will want to know whether the difference between the percentages of patients with complications observed in this sample of 114 patients is generalizable to the population of all coronary artery bypass patients.

Now Work Exercise 2.12

Statistics IN Action Revisited — Interpreting Pie Charts

In the survey of University Community Hospital physicians, the medical researchers measured three qualitative variables: *Level of previous ethics consultation use* ("never used" or "used"), *Practitioner specialty* ("medical" or "surgical"), and *Future use of ethics consultation* ("yes" or "no"). Pie charts and bar graphs can be used to summarize and describe the physicians' responses to these survey questions. Recall that the data are saved in the **ETHICS** file. These variables are named PREVUSE, SPEC, and FUTUREUSE in the data file. We created pie charts for these variables using both Minitab and Data Desk/XL, an Excel add-in (called DDXL).

Figure SIA2.1 is a pie chart for the PREVUSE variable. Clearly, a higher percentage of physicians (71.2%) has previously never used ethics consultation at the hospital than have (28.8%). The researchers want to know if this "previous use" pattern differs for the two practitioner specialties. Figure SIA2.2 shows side-by-side pie charts of the PREVUSE variable for each level of the SPEC variable. The left-side chart describes the pattern of previous use by medical specialists, and the right-side chart describes the pattern of previous use by surgeons. Figure SIA2.2 shows that slightly fewer surgeons (27.9%) have used ethics consultation in the past than have medical practitioners (29.3%).

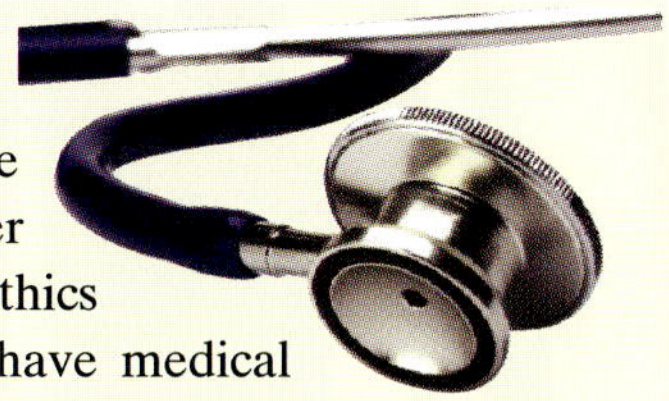

We produced a similar set of side-by-side pie charts to describe the qualitative variable FUTUREUSE in Figure SIA2.3. Apparently, the gap between surgeons and medical specialists has widened. These charts again show that fewer surgeons (76.7%) would consider using ethics consultation in the future than medical specialists (82.7%), but the difference in the percentages is greater than for previous use. The researchers' theory that surgical specialists at UCH are less likely to use ethics consultation than medical specialists is supported by the pie charts.

Data Set: ETHICS

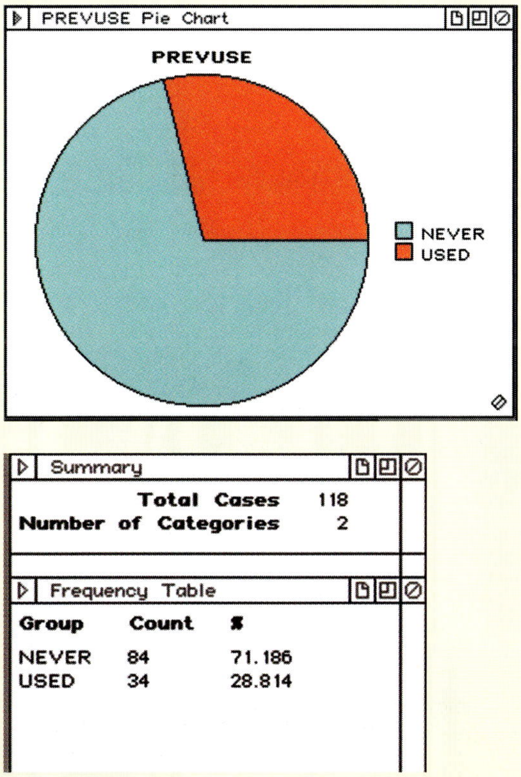

Figure SIA2.1
DDXL pie chart for previous use of ethics consultation

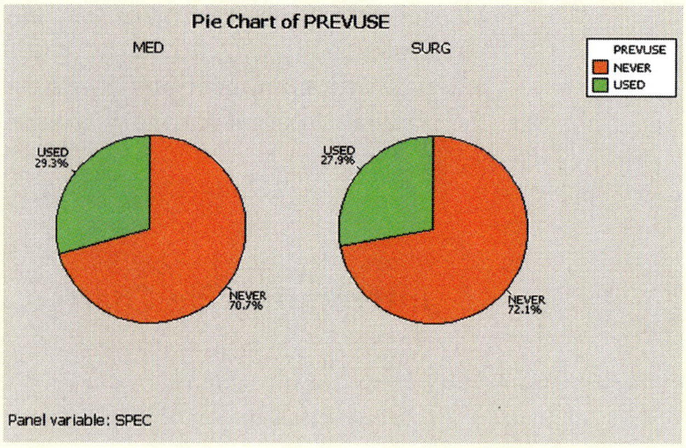

Figure SIA2.2
Minitab pie charts for previous use of ethics consultation—medical versus surgical specialty

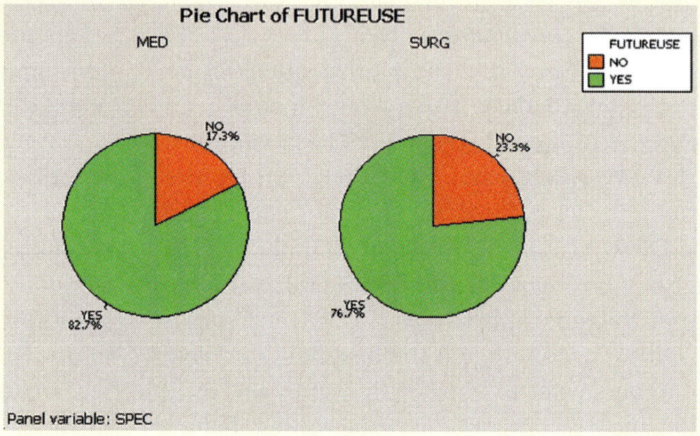

Figure SIA2.3
Minitab pie charts for future use of ethics consultation—medical versus surgical specialty

Exercises 2.1–2.15

Learning the Mechanics

2.1 Complete the following table.

Grade on Business Statistics Exam	Frequency	Relative Frequency
A: 90–100	—	.08
B: 80–89	36	—
C: 65–79	90	—
D: 50–64	30	—
F: Below 50	28	—
Total	200	1.00

2.2 A qualitative variable with three classes (X, Y, and Z) is measured for each of 20 units randomly sampled from a target population. The data (observed class for each unit) are listed below.

Y	X	X	Z	X	Y	Y	Y	X	X	Z	X
Y	Y	X	Z	Y	Y	Y	X				

a. Compute the frequency for each of the three classes.
b. Compute the relative frequency for each of the three classes.
c. Display the results, part **a,** in a frequency bar graph.
d. Display the results, part **b,** in a pie chart.

Applying the Concepts—Basic

2.3 Industrial robots. The Robotics Industries Association estimates that there are 184,000 industrial robots operating in North America. The graph on the next page shows the percentages of industrial robot units assigned to each of six task categories: (1) spot welding, (2) arc welding, (3) material removal, (4) material handling, (5) assembly, and (6) dispensing/coating.
a. What type of graph is used to describe the data?
b. Identify the variable measured for each of the 184,000 industrial robots.
c. Use the graph to identify the task that uses the highest percentage of industrial robots.
d. How many of the 184,000 industrial robots are used for spot welding?
e. What percentage of industrial robots are used for either spot welding or arc welding?

2.4 Who is to blame for rising health-care costs? Rising health-care costs are of major concern to Americans. A nationwide survey of 2,119 U.S. adults was conducted to elicit opinions on who is to blame for the rising costs (*The Harris Poll,* Oct. 28, 2008). The next table summarizes the responses to the

Graph for Exercise 2.3

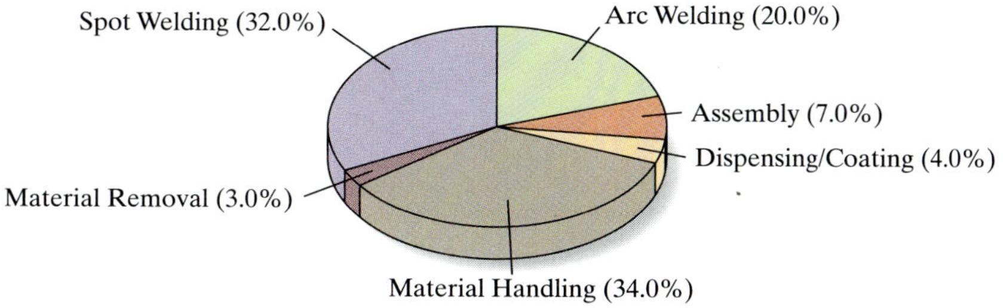

question "When you think of the rising costs of health care, who do you think is most responsible?"

a. Compute the relative frequencies in each response category.

b. Construct a relative frequency bar graph for the data.

c. Convert the relative frequency bar graph into a Pareto diagram. Interpret the graph.

Most Responsible for Rising Health-care Costs	Number Responding
Insurance companies	869
Pharmaceutical companies	339
Government	338
Hospitals	127
Physicians	85
Other	128
Not at all sure	233
Total	2,119

2.5 PIN pad shipments. Personal identification number (PIN) pads are devices that connect to point-of-sale electronic cash registers for debit and credit card purchases. The PIN pad allows the customer card to be accessed and the PIN encrypted before it is sent to the transaction manager. *The Nilson Report* (Oct. 2008) listed the volume of PIN pad shipments by manufacturers worldwide in 2007. For the 12 manufacturers listed in the table, a total of 334,039 PIN pads were shipped in 2007. (The data are saved in the **PINPADS** file.)

Manufacturer	Number Shipped (units)
Bitel	13,500
CyberNet	16,200
Fujian Landi	119,000
Glintt (ParaRede)	5,990
Intelligent	4,562
KwangWoo	42,000
Omron	20,000
Pax Tech.	10,072
ProvencoCadmus	20,000
SZZT Electronics	67,300
Toshiba TEC	12,415
Urmet	3,000

Source: The Nilson Report, No. 912, October 2008 (p. 9).

a. One of the 334,039 PIN pads is selected and the manufacturer of the pad is determined. What type of data (quantitative or qualitative) is measured?

b. Construct a frequency bar chart for the data summarized in the table.

c. Convert the frequency bar chart, part **b,** into a Pareto diagram. Interpret the results.

2.6 Management system failures. The U.S. Chemical Safety and Hazard Investigation Board (CSB) is responsible for determining the root cause of industrial accidents. Since its creation in 1998, the CSB has identified 83 incidents that were caused by management system failures (*Process Safety Progress,* Dec. 2004). The accompanying table gives a breakdown of the root causes of these 83 incidents.

Management System Cause Category	Number of Incidents
Engineering & Design	27
Procedures & Practices	24
Management & Oversight	22
Training & Communication	10
Total	83

Source: Blair, A. S. "Management system failures identified in incidents investigated by the U.S. Chemical Safety and Hazard Investigation Board," *Process Safety Progress,* Vol. 23, No. 4, Dec. 2004, pp. 232–236 (Table 1). Reprinted with permission of John Wiley & Sons, Inc.

a. Find the relative frequency of the number of incidents for each cause category.

b. Construct a Pareto diagram for the data.

c. From the Pareto diagram, identify the cause categories with the highest (and lowest) relative frequency of incidents.

2.7 Non-cash payments in the U.S. The "2007 Federal Reserve Payments Study," sponsored by the federal government, found that the number of non-cash payments in the United States reached 93.3 billion transactions during the year. (*CardTrak.com News,* Dec. 13, 2007.) The pie chart shown below describes the proportions of these non-cash transactions that were made with credit cards, debit cards, checks, electronic benefit transfer (EBT) cards for the Food Stamps program, and Automatic Clearing House (ACH) payments.

a. Describe the population of interest in the study.

b. Estimate the number of non-cash payments in 2007 that were made with credit cards.

c. What percentage of all non-cash payments in 2007 were made with either credit or debit cards?

d. Convert the pie chart into a Pareto diagram and interpret the figure.

Applying the Concepts—Intermediate

2.8 Products "Made in the USA." "Made in the USA" is a claim stated in many product advertisements or on product labels. Advertisers want consumers to believe that the product is manufactured with 100% U.S. labor and materials—which is often not the case. What does "Made in the USA" mean to the typical consumer? To answer this question, a group of marketing professors conducted an experiment at a shopping mall in Muncie, Indiana (*Journal of Global

Graph for Exercise 2.7

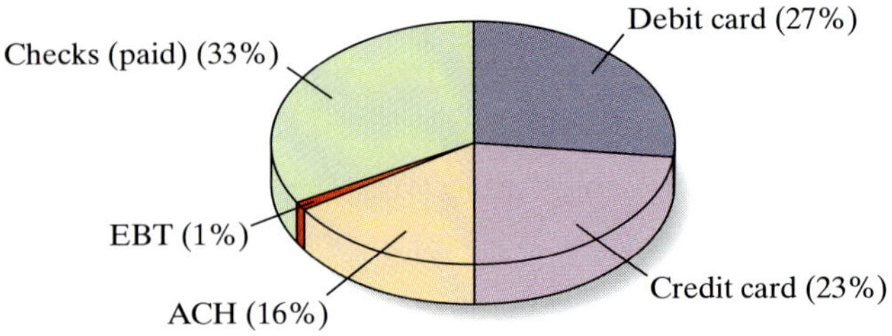

Business, Spring 2002). They asked every fourth adult entrant to the mall to participate in the study. A total of 106 shoppers agreed to answer the question, "'Made in the USA' means what percentage of U.S. labor and materials?" The responses of the 106 shoppers are summarized in the following table.

Response to "Made in the USA"	Number of Shoppers
100%	64
75 to 99%	20
50 to 74%	18
Less than 50%	4

Source: "'Made in the USA': Consumer Perceptions, Deception and Policy Alternatives," *Journal of Global Business,* Vol. 13, No. 24, Spring 2002 (Table 3).

a. What type of data-collection method was used?

b. What type of variable, quantitative or qualitative, is measured?

c. Present the data in the table in graphical form. Use the graph to make a statement about the percentage of consumers who believe "Made in the USA" means 100% U.S. labor and materials.

2.9 Blogs for Fortune 500 firms. Web site communication through blogs and forums is becoming a key marketing tool for companies. The *Journal of Relationship Marketing* (Vol. 7, 2008) investigated the prevalence of blogs and forums at Fortune 500 firms with both English and Chinese Web sites. Of the firms that provided blogs/forums as a marketing tool, the accompanying table gives a breakdown on the entity responsible for creating the blogs/forums. Use a graphical method to describe the data summarized in the table. Interpret the graph.

Blog/Forum	Percentage of Firms
Created by company	38.5
Created by employees	34.6
Created by third party	11.5
Creator not identified	15.4

Source: Mishra, K. & Li, C. "Relationship marketing in *Fortune* 500 U.S. and Chinese web sites," *Journal of Relationship Marketing,* Vol. 7, No. 1, 2008. Copyright © 2008 Routledge.

2.10 The Executive Compensation Scoreboard. Refer to the *Forbes* "Executive Compensation Scoreboard" for 2008, described in Chapter 1 (p. 4) and in Exercise 1.19 (p. 23). Recall that the industry type of the CEO's company (e.g., banking, retailing, etc.) was recorded for each of the 175 CEOs with the highest efficiency (pay-for-performance) ratings. (See Table 1.1, p. 4, for a list of the industries.) Access the **CEOPAYTOP175** file and use a graphical method to describe the frequency of occurrence of the industry types.

2.11 Color and clarity of diamonds. Diamonds are categorized according to the "four C's": carats, clarity, color, and cut. Each diamond stone that is sold on the open market is provided a certificate by an independent diamond assessor that lists these characteristics. Data for 308 diamonds were extracted from Singapore's *Business Times* and are saved in the **DIAMONDS** file (*Journal of Statistics Education,* Vol. 9, No. 1, 2001). Color is classified as D, E, F, G, H, or I, while clarity is classified as IF, VVS1, VVS2, VS1, or VS2. Use a graphical technique to summarize the color and clarity of the 308 diamond stones. What is the color and clarity that occurs most often? Least often?

2.12 History of corporate acquisitions. A corporate acquisition occurs when one corporation purchases all the stock shares of another, essentially taking over the other. The *Academy of Management Journal* (Aug. 2008) investigated the performance and timing of corporate acquisitions for a large sample of firms over the years 1980 to 2000. The accompanying data table (saved in the **ACQUISITIONS** file) gives the number of firms sampled and number that announced one or more acquisitions during the year for each year. Construct side-by-side bar charts to describe the firms with and without acquisitions in the years 1980 and 2000. Compare and contrast the bar charts.

Year	Number of Firms Sampled	Number with Acquisitions
1980	1,963	18
1981	2,044	115
1982	2,029	211
1983	2,187	273
1984	2,248	317
1985	2,238	182
1986	2,277	232
1987	2,344	258
1988	2,279	296
1989	2,231	350
1990	2,197	350
1991	2,261	370
1992	2,363	427
1993	2,582	532
1994	2,775	626
1995	2,890	652
1996	3,070	751
1997	3,099	799
1998	2,913	866
1999	2,799	750
2000	2,778	748
Total	51,567	9,123

Source: Iyer, D. N. & Miller, K. D. "Performance feedback, slack, and the timing of acquisitions," *Academy of Management Journal,* Vol. 51, No. 4, August 2008, pp. 808–822 (Table 1). Copyright 2008 by Academy of Management (NY). Reproduced with permission of Academy of Management (NY) in the format Textbook and electronic usage via Copyright Clearance Center.

Applying the Concepts—Advanced

2.13 Advertising with reader-response cards. "Reader-response cards" are used by marketers to advertise their product and obtain sales leads. These cards are placed in magazines and trade publications. Readers detach and mail in the cards to indicate their interest in the product, expecting literature or a phone call in return. How effective are these cards (called "bingo cards" in the industry) as a marketing tool? Performark, a Minneapolis business that helps companies close on sales leads, attempted to answer this question by responding to 17,000 card-advertisements placed by industrial marketers in a wide variety of trade publications over a 6-year period. Performark kept track of how long it took for each advertiser to respond. A summary of the response times, reported in *Inc.* magazine (July 1995), is given in the following table.

a. Describe the variable measured by Performark.

b. *Inc.* displayed the results in the form of a pie chart. Reconstruct the pie chart from the information given in the table.

c. How many of the 17,000 advertisers never responded to the sales lead?

d. Advertisers typically spend at least a million dollars on a reader-response card marketing campaign. Many industrial marketers feel these "bingo cards" are not worth their expense. Does the information in the pie chart, part **b,** support this contention? Explain why or why not. If not, what information can be gleaned from the pie chart to help potential "bingo card" campaigns?

Advertiser's Response Time	Percentage
Never responded	21
13–59 days	33
60–120 days	34
More than 120 days	12
Total	100

2.14 Stewardship at MBA programs. *Business Ethics* (Fall 2005) reported on a survey designed to rank master in business administration (MBA) programs worldwide on how well they prepare students for social and environmental stewardship. Each business school was ranked according to four criteria: *student exposure* (class time dedicated to social and environmental issues), *student opportunity* (courses with social and environmental content), *course content* (courses emphasize business as a force for positive social and environmental change), and *faculty research* (published articles that examine business in a social/environmental context). Each area was rated from 1 star (lowest rating) to 5 stars (highest rating). Overall, Stanford University received the top ranking, followed by ESADE (Spain), York University (Canada), Monterrey Technical Institute (Mexico), and the University of Notre Dame. A summary of the rankings (star ratings) for the top 30 MBA programs is shown in the table.

a. Illustrate the differences and similarities of the star-ranking distributions for the four different criteria.

b. Give a plausible reason why there were no 1-star ratings for the 30 MBA programs.

Criteria	5 Stars	4 Stars	3 Stars	2 Stars	1 Star	Total
Student Exposure	2	9	14	5	0	30
Student Opportunity	3	10	14	3	0	30
Course Content	3	9	17	1	0	30
Faculty Research	3	10	11	4	0	28

Source: Adapted from Stewardship at MBA programs – Biello, D. "MBA Programs for Social and Environmental Stewardship," *Business Ethics,* Fall 2005, p. 25. Used with permission of the Aspen Institute.

2.15 Groundwater contamination in wells. In New Hampshire, about half the counties mandate the use of reformulated gasoline. This has led to an increase in the contamination of groundwater with methyl *tert*-butyl ether (MTBE). *Environmental Science & Technology* (Jan. 2005) reported on the factors related to MTBE contamination in public and private New Hampshire wells. Data were collected for a sample of 223 wells. These data are saved in the **MTBE** file. Three of the variables are qualitative in nature: well class (public or private), aquifer (bedrock or unconsolidated), and detectible level of MTBE (below limit or detect). [Note: A detectible level of MTBE occurs if the MTBE value exceeds .2 micrograms per liter.] The data for 10 selected wells are shown in the accompanying table.

a. Use graphical methods to describe each of the three qualitative variables for all 223 wells.

b. Use side-by-side bar charts to compare the proportions of contaminated wells for private and public well classes.

c. Use side-by-side bar charts to compare the proportions of contaminated wells for bedrock and unconsolidated aquifers.

d. What inferences can be made from the bar charts, parts **a–c?**

Well Class	Aquifer	Detect MTBE
Private	Bedrock	Below Limit
Private	Bedrock	Below Limit
Public	Unconsolidated	Detect
Public	Unconsolidated	Below Limit
Public	Unconsolidated	Below Limit
Public	Unconsolidated	Below Limit
Public	Unconsolidated	Detect
Public	Unconsolidated	Below Limit
Public	Unconsolidated	Below Limit
Public	Bedrock	Detect
Public	Bedrock	Detect

Source: Ayotte, J. D., Argue, D. M., and McGarry, F. J. "Methyl *tert*-Butyl Ether Occurrence and Related Factors in Public and Private Wells in Southeast New Hampshire," *Environmental Science & Technology,* Vol. 39, No. 1, Jan. 2005, pp. 9–16. Copyright © 2005 American Chemical Society.

2.2 Graphical Methods for Describing Quantitative Data

Recall from Section 1.5 that quantitative data sets consist of data that are recorded on a meaningful numerical scale. For describing, summarizing, and detecting patterns in such data, we can use three graphical methods: **dot plots, stem-and-leaf displays,** and **histograms.** Because almost all statistical software packages can produce these graphs, we'll focus here on their interpretations rather than their construction.

For example, suppose a financial analyst is interested in the amount of resources spent by computer hardware and software companies on research and development

Table 2.2 **Percentage of Revenues Spent on Research and Development**

Company	Percentage	Company	Percentage	Company	Percentage	Company	Percentage
1	13.5	14	9.5	27	8.2	39	6.5
2	8.4	15	8.1	28	6.9	40	7.5
3	10.5	16	13.5	29	7.2	41	7.1
4	9.0	17	9.9	30	8.2	42	13.2
5	9.2	18	6.9	31	9.6	43	7.7
6	9.7	19	7.5	32	7.2	44	5.9
7	6.6	20	11.1	33	8.8	45	5.2
8	10.6	21	8.2	34	11.3	46	5.6
9	10.1	22	8.0	35	8.5	47	11.7
10	7.1	23	7.7	36	9.4	48	6.0
11	8.0	24	7.4	37	10.5	49	7.8
12	7.9	25	6.5	38	6.9	50	6.5
13	6.8	26	9.5				

Data Set: R&D

(R&D). She samples 50 of these high-technology firms and calculates the amount each spent last year on R&D as a percentage of their total revenue. The results are given in Table 2.2. As numerical measurements made on the sample of 50 units (the firms), these percentages represent quantitative data. The analyst's initial objective is to summarize and describe these data in order to extract relevant information.

A visual inspection of the data indicates some obvious facts. For example, the smallest R&D percentage is 5.2% (company 45) and the largest is 13.5% (companies 1 and 16). But it is difficult to provide much additional information on the 50 R&D percentages without resorting to some method of summarizing the data. One such method is a dot plot.

Dot Plots

A **dot plot** for the 50 R&D percentages, produced using SPSS software, is shown in Figure 2.8. The horizontal axis of Figure 2.8 is a scale for the quantitative variable, percent. The numerical value of each measurement in the data set is located on the horizontal scale by a dot. When data values repeat, the dots are placed above one another, forming a pile at that particular numerical location. As you can see, this dot plot shows that almost all of the R&D percentages are between 6% and 12%, with most falling between 7% and 9%.

Stem-and-Leaf Display

We used Minitab to generate another graphical representation of these same data, a **stem-and-leaf display,** in Figure 2.9. In this display the *stem* is the portion of the measurement (percentage) to the left of the decimal point, while the remaining portion to the right of the decimal point is the *leaf.*

The stems for the data set are listed in the second column of Figure 2.9 from the smallest (5) to the largest (13). Then the leaf for each observation is recorded in the row of the display corresponding to the observation's stem.* For example, the leaf 5 of the first observation (13.5) in Table 2.2 is placed in the row corresponding to the stem 13. Similarly, the leaf 4 for the second observation (8.4) in Table 2.2 is recorded in the row corresponding

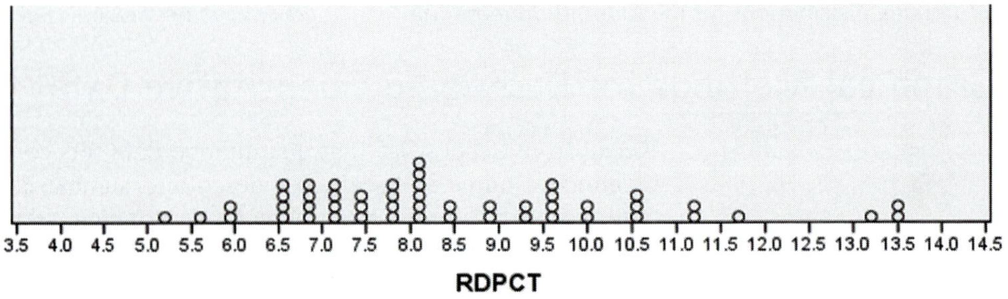

Figure 2.8
SPSS dot plot for 50 R&D percentages

*The first column of the Minitab stem-and-leaf display represents the cumulative number of measurements from the class interval to the nearest extreme class interval.

```
Stem-and-leaf of RDPct  N  = 50
Leaf Unit = 0.10


  3   5   269
 12   6   055568999
 23   7   11224557789
(9)   8   001222458
 18   9   02455679
 10  10   1556
  6  11   137
  3  12
  3  13   255
```

Figure 2.9
Minitab stem-and-leaf display for 50 R&D percentages

to the stem 8, while the leaf 5 for the third observation (10.5) is recorded in the row corresponding to the stem 10. (The leaves for these first three observations are shaded in Figure 2.9.) Typically, the leaves in each row are ordered as shown in Figure 2.9.

The stem-and-leaf display presents another compact picture of the data set. You can see at a glance that most of the sampled computer companies (37 of 50) spent between 6.0% and 9.9% of their revenues on R&D, and 11 of them spent between 7.0% and 7.9%. Relative to the rest of the sampled companies, three spent a high percentage of revenues on R&D—in excess of 13%.

The definitions of the stem and leaf can be modified to alter the graphical display. For example, suppose we had defined the stem as the tens digit for the R&D percentage data, rather than the ones and tens digits. With this definition, the stems and leaves corresponding to the measurements 13.5 and 8.4 would be as follows:

Stem	*Leaf*
1	3

Stem	*Leaf*
0	8

Note that the decimal portion of the numbers has been dropped. Generally, only one digit is displayed in the leaf.

If you look at the data, you'll see why we didn't define the stem this way. All the R&D measurements fall below 13.5, so all the leaves would fall into just two stem rows—1 and 0—in this display. The picture resulting from using only a few stems would not be nearly as informative as Figure 2.9.

Histograms

An **Excel/DDXL histogram** for these 50 R&D measurements is displayed in Figure 2.10. The horizontal axis for Figure 2.10, which gives the percentage amounts spent on R&D for each company, is divided into **class intervals** commencing with the interval (5.0–6.0) and proceeding in intervals of equal size to (13.0–14.0). The vertical axis gives the number (or *frequency*) of the 50 measurements that fall in each class interval. You can see that the class interval (7.0–8.0) (i.e., the class with the highest bar) contains the largest frequency of 11 R&D percentage measurements; the remaining class intervals tend to contain a smaller number of measurements as R&D percentage gets smaller or larger.

Histograms can be used to display either the *frequency* or *relative frequency* of the measurements falling into the class intervals. The class intervals, frequencies, and relative

BIOGRAPHY **JOHN TUKEY (1915–2000)**

The Picasso of Statistics

Like the legendary artist Pablo Picasso, who mastered and revolutionized a variety of art forms during his lifetime, John Tukey is recognized for his contributions to many subfields of statistics. Born in Massachusetts, Tukey was home-schooled, graduated with his bachelor's and master's degrees in chemistry from Brown University, and received his PhD in mathematics from Princeton University. While at Bell Telephone Laboratories in the 1960s and early 1970s, Tukey developed "exploratory data analysis," a set of graphical descriptive methods for summarizing and presenting huge amounts of data. Many of these tools, including the stem-and-leaf display and the box plot (see Section 2.8), are now standard features of modern statistical software packages. (In fact, it was Tukey himself who coined the term *software* for computer programs.) ■

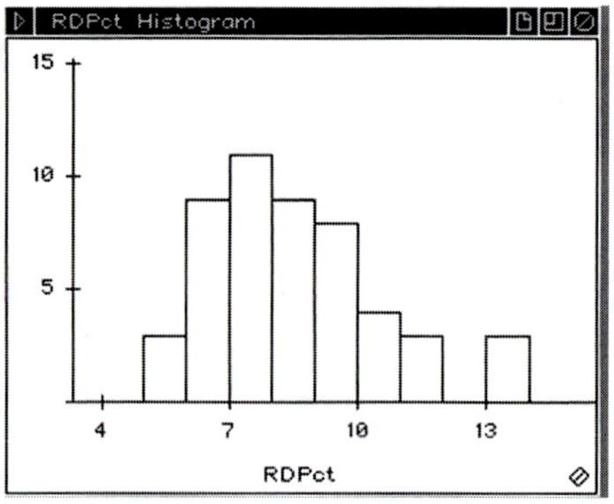

Figure 2.10
Excel/DDXL histogram for 50 R&D percentages

frequencies for the 50 R&D measurements are shown in Table 2.3.* By summing the relative frequencies in the intervals (6.0–7.0), (7.0–8.0), (8.0–9.0), (9.0–10.0), and (10.0–11.0), we find that .18 + .22 + .18 + .16 + .08 = .82, or 82%, of the R&D measurements are between 6.0 and 11.0. Similarly, summing the relative frequencies in the last two intervals, (12.0–13.0) and (13.0–14.0), we find that 6% of the companies spent over 12.0% of their revenues on R&D. Many other summary statements can be made by further study of the histogram.

When interpreting a histogram (say, the histogram in Figure 2.10), consider two important facts. First, the proportion of the total area under the histogram that falls above a particular interval of the horizontal axis is equal to the relative frequency of measurements falling in the interval. For example, the relative frequency for the class interval 7.0–8.0 is .22. Consequently, the rectangle above the interval contains 22% of the total area under the histogram.

Second, you can imagine the appearance of the relative frequency histogram for a very large set of data (say, a population). As the number of measurements in a data set is increased, you can obtain a better description of the data by decreasing the width of the class intervals. When the class intervals become small enough, a relative frequency histogram will (for all practical purposes) appear as a smooth curve (see Figure 2.11). Some recommendations for selecting the number of intervals in a histogram for smaller data sets are given in the box below Figure 2.11.

While histograms provide good visual descriptions of data sets—particularly very large ones—they do not let us identify individual measurements. In contrast, each of the original measurements is visible to some extent in a dot plot and clearly visible in a stem-and-leaf display. The stem-and-leaf display arranges the data in ascending order, so it's easy to locate the individual measurements. For example, in Figure 2.9, we can easily see that three of the R&D measurements are equal to 8.2, but we can't see that

Table 2.3 Class Intervals, Frequencies, and Relative Frequencies for the 50 R&D Measurements

Class	Class Interval	Class Frequency	Class Relative Frequency
1	5.0–6.0	3	3/50 = .06
2	6.0–7.0	9	9/50 = .18
3	7.0–8.0	11	11/50 = .22
4	8.0–9.0	9	9/50 = .18
5	9.0–10.0	8	8/50 = .16
6	10.0–11.0	4	4/50 = .08
7	11.0–12.0	3	3/50 = .06
8	12.0–13.0	0	0/50 = .00
9	13.0–14.0	3	3/50 = .06
Totals		50	1.00

*DDXL, like many statistical software packages, will classify an observation that falls on the borderline of a class interval into the next highest class interval. For example, the R&D measurement of 8.0, which falls on the border between the intervals (7.0–8.0) and (8.0–9.0), is classified into the (8.0–9.0) interval. The frequencies in Table 2.3 reflect this convention.

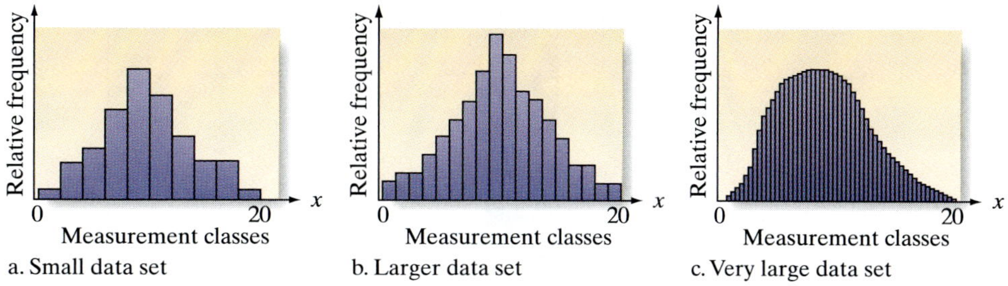

Figure 2.11
Effect of the size of a data set on the outline of a histogram

fact by inspecting the histogram in Figure 2.10. However, stem-and-leaf displays can become unwieldy for very large data sets. A very large number of stems and leaves causes the vertical and horizontal dimensions of the display to become cumbersome, diminishing the usefulness of the visual display.

Determining the Number of Classes in a Histogram

Number of Observations in Data Set	Number of Classes
Less than 25	5–6
25–50	7–14
More than 50	15–20

Example 2.2
Graphs for a Quantitative Variable

Problem A manufacturer of industrial wheels suspects that profitable orders are being lost because of the long time the firm takes to develop price quotes for potential customers. To investigate this possibility, 50 requests for price quotes were randomly selected from the set of all quotes made last year, and the processing time was determined for each quote. The processing times are displayed in Table 2.4, and each quote

Table 2.4 Price Quote Processing Time (Days)

Request Number	Processing Time	Lost?	Request Number	Processing Time	Lost?
1	2.36	No	26	3.34	No
2	5.73	No	27	6.00	No
3	6.60	No	28	5.92	No
4	10.05	Yes	29	7.28	Yes
5	5.13	No	30	1.25	No
6	1.88	No	31	4.01	No
7	2.52	No	32	7.59	No
8	2.00	No	33	13.42	Yes
9	4.69	No	34	3.24	No
10	1.91	No	35	3.37	No
11	6.75	Yes	36	14.06	Yes
12	3.92	No	37	5.10	No
13	3.46	No	38	6.44	No
14	2.64	No	39	7.76	No
15	3.63	No	40	4.40	No
16	3.44	No	41	5.48	No
17	9.49	Yes	42	7.51	No
18	4.90	No	43	6.18	No
19	7.45	No	44	8.22	Yes
20	20.23	Yes	45	4.37	No
21	3.91	No	46	2.93	No
22	1.70	No	47	9.95	Yes
23	16.29	Yes	48	4.46	No
24	5.52	No	49	14.32	Yes
25	1.44	No	50	9.01	No

Data Set: PRICEQUOTES

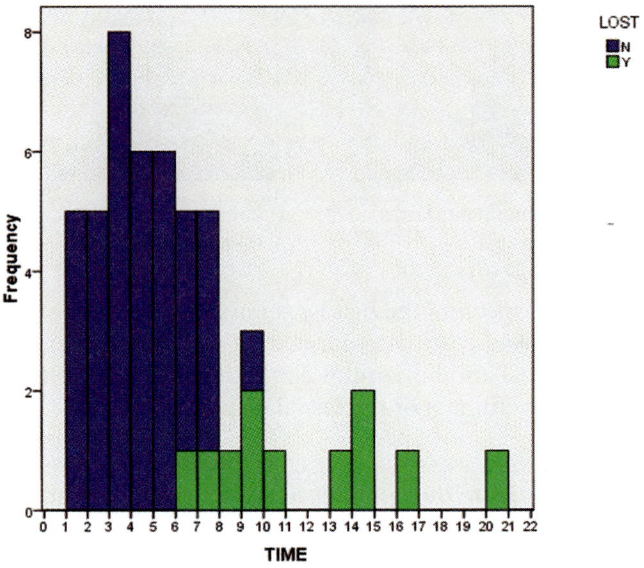

Figure 2.12
SPSS frequency histogram for price quote data

was classified according to whether the order was "lost" or not (i.e., whether or not the customer placed an order after receiving a price quote).

a. Use a statistical software package to create a frequency histogram for these data. Then shade the area under the histogram that corresponds to lost orders. Interpret the result.

b. Use a statistical software package to create a stem-and-leaf display for these data. Then shade each leaf of the display that corresponds to a lost order. Interpret the result.

Solution

a. We used SPSS to generate the frequency histogram in Figure 2.12. Note that 20 classes were formed by the SPSS program. The class intervals are (1.0–2.0), (2.0–3.0), . . . , (20.0–21.0). This histogram clearly shows the clustering of the measurements in the lower end of the distribution (between approximately 1 and 8 days), and the relatively few measurements in the upper end of the distribution (greater than 12 days). The shading of the area of the frequency histogram corresponding to lost orders (green bars) clearly indicates that they lie in the upper tail of the distribution.

b. We used Minitab to generate the stem-and-leaf display in Figure 2.13. Note that the stem (the second column of the printout) consists of the number of whole days (digits to the left of the decimal). The leaf (the third column of the printout) is the tenths digit (first digit after the decimal) of each measurement. Thus, the leaf 2 in the stem 20 (the last row of the printout) represents the time of 20.23 days. Like the histogram, the stem-and-leaf display shows the shaded "lost" orders in the upper tail of the distribution.

```
Stem-and-leaf of TIME  N  = 50
Leaf Unit = 0.10


 5   1  24789
10   2  03569
18   3  23344699
24   4  034469
(6)  5  114579
20   6  01467
15   7  24557
10   8  2
 9   9  049
 6  10  0
 5  11
 5  12
 5  13  4
 4  14  03
 2  15
 2  16  2
 1  17
 1  18
 1  19
 1  20  2
```

Figure 2.13
Minitab stem-and-leaf display for price quote data

Look Back As is usually the case for data sets that are not too large (say, fewer than 100 measurements), the stem-and-leaf display provides more detail than the histogram without being unwieldy. For instance, the stem-and-leaf display in Figure 2.13 clearly indicates that the lost orders are associated with high processing times (as does the histogram in Figure 2.12), and exactly which of the times correspond to lost orders. Histograms are most useful for displaying very large data sets, when the overall shape of the distribution of measurements is more important than the identification of individual measurements. Nevertheless, the message of both graphical displays is clear: Establishing processing time limits may well result in fewer lost orders.

Now Work Exercise 2.20

Most statistical software packages can be used to generate histograms, stem-and-leaf displays, and dot plots. All three are useful tools for graphically describing data sets. We recommend that you generate and compare the displays whenever you can. You'll find that histograms are generally more useful for very large data sets, while stem-and-leaf displays and dot plots provide useful detail for smaller data sets.

Summary of Graphical Descriptive Methods for Quantitative Data

Dot plot: The numerical value of each quantitative measurement in the data set is represented by a dot on a horizontal scale. When data values repeat, the dots are placed above one another vertically.

Stem-and-leaf display: The numerical value of the quantitative variable is partitioned into a "stem" and a "leaf." The possible stems are listed in order in a column. The leaf for each quantitative measurement in the data set is placed in the corresponding stem row. Leaves for observations with the same stem value are listed in increasing order horizontally.

Histogram: The possible numerical values of the quantitative variable are partitioned into class intervals, where each interval has the same width. These intervals form the scale of the horizontal axis. The frequency or relative frequency of observations in each class interval is determined. A vertical bar is placed over each class interval, with height equal to either the class frequency or class relative frequency.

Statistics IN Action Revisited — Interpreting Histograms

One of the quantitative variables measured in the ethics consultation survey of physicians was *Length of time in practice* (i.e., years of experience). Recall that the medical researchers hypothesize that older, more experienced physicians will be less likely to use ethics consultation in the future. To check the believability of this claim, we accessed the **ETHICS** data file in Minitab and created two frequency histograms for years of experience—one for physicians who indicated they would use ethics consultation in the future, and one for physicians who would not use ethics consultation. These side-by-side histograms are displayed in Figure SIA2.4.

From the histograms, you can see there is some support for the researchers' assertion. The histogram for the physicians who indicated they would use ethics consultation (the histogram on the right in Figure SIA2.4) shows that most of these physicians have been in practice between 10 and 20 years, while the histogram for nonusers (the histogram on the left in Figure SIA2.4) shows a tendency for these physicians to have more experience (over 20 years). However, the lack of data (only 21 observations) for the sample of physicians who would not use ethics consultation makes it difficult to reliably extend this inference to the population of physicians. In later chapters, we'll learn how to attach a measure of reliability to such an inference, even for small samples.

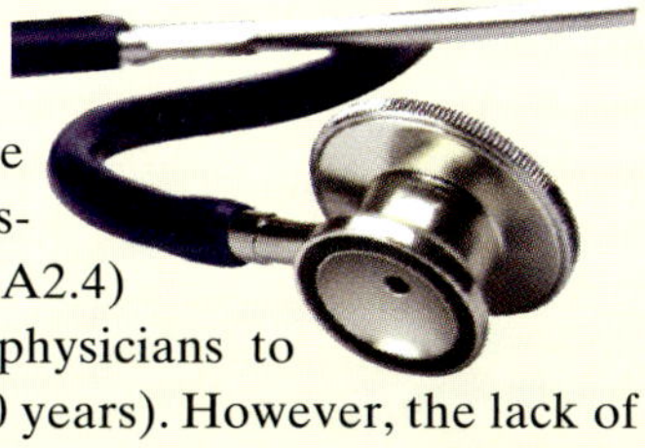

Data Set: ETHICS

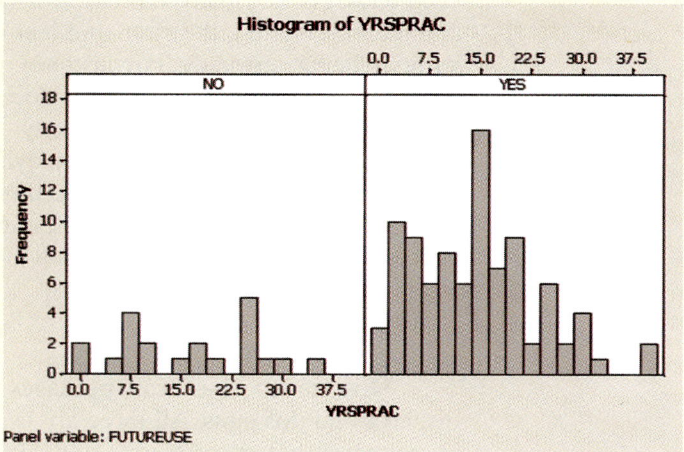

Figure SIA2.4
Minitab histograms for years of practice—ethics consultation users versus nonusers

Exercises 2.16–2.32

Learning the Mechanics

2.16 Graph the relative frequency histogram for the 500 measurements summarized in the accompanying relative frequency table.

Measurement Class	Relative Frequency
.5–2.5	.10
2.5–4.5	.15
4.5–6.5	.25
6.5–8.5	.20
8.5–10.5	.05
10.5–12.5	.10
12.5–14.5	.10
14.5–16.5	.05

2.17 Refer to Exercise 2.16. Calculate the number of the 500 measurements falling into each of the measurement classes. Then graph a frequency histogram for these data.

2.18 Consider the stem-and-leaf display shown here.

Stem	Leaf
5	1
4	457
3	00036
2	1134599
1	2248
0	012

a. How many observations were in the original data set?
b. In the bottom row of the stem-and-leaf display, identify the stem, the leaves, and the numbers in the original data set represented by this stem and its leaves.
c. Re-create all the numbers in the data set and construct a dot plot.

2.19 Minitab was used to generate the following histogram:

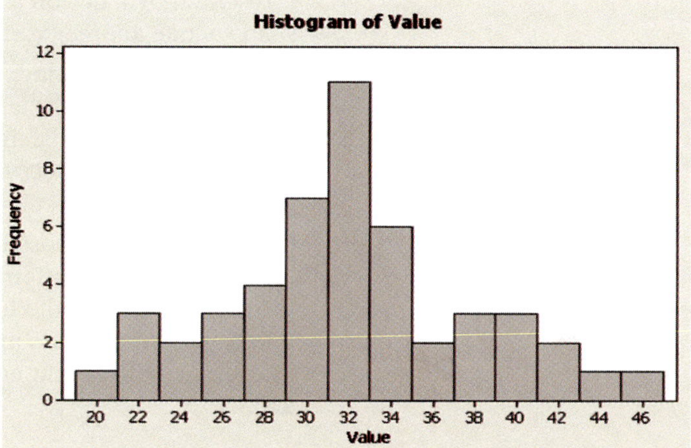

a. Is this a frequency histogram or a relative frequency histogram? Explain.
b. How many measurement classes were used in the construction of this histogram?
c. How many measurements are in the data set described by this histogram?

Applying the Concepts—Basic

2.20 NW **Computer security survey.** Refer to the 2006 CSI/FBI Computer Crime and Security Survey, Exercise 1.20 (p. 23). One of the survey questions asked respondents to indicate the percentage of computer security functions that their company outsources. Consequently, the quantitative variable of interest is measured as a percentage for each of 609 respondents in the 2006 survey. The following histogram summarizes the data.
a. Which measurement class contains the highest proportion of respondents?
b. What proportion of the 609 respondents indicate that they outsource between 20% and 40% of computer security functions?

c. What proportion of the 609 respondents outsource at least 40% of computer security functions?
d. How many of the 609 respondents outsource less than 20% of computer security functions?

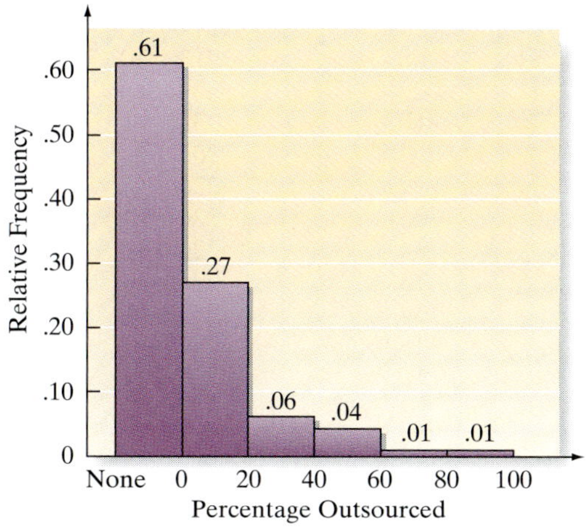

2.21 USGA golfing handicaps. The United States Golf Association (USGA) Handicap System is designed to allow golfers of differing abilities to enjoy fair competition. The handicap index is a measure of a player's potential scoring ability on an 18-hole golf course of standard difficulty. For example, on a par-72 course, a golfer with a handicap of 7 will typically have a score of 79 (seven strokes over par). Over 4.5 million golfers have an official USGA handicap index. The handicap indexes for both male and female golfers were obtained from the USGA and are summarized in the following two histograms.

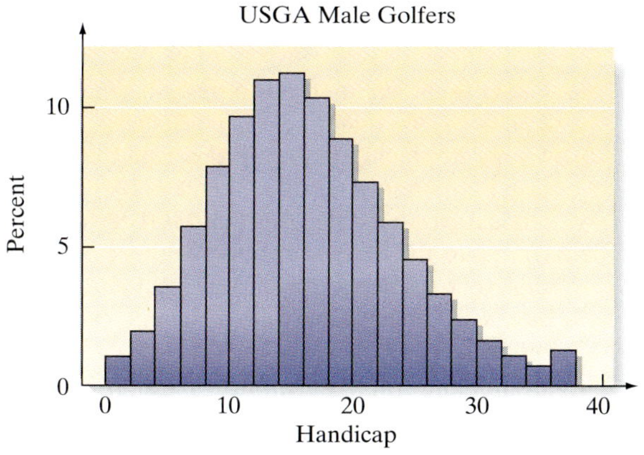

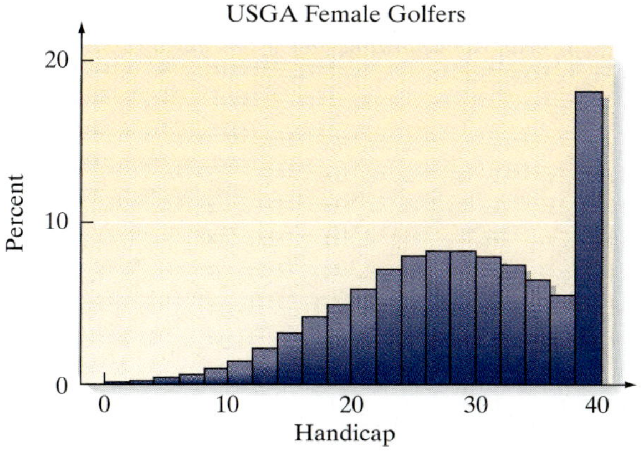

a. What percentage of male USGA golfers have a handicap greater than 20?
b. What percentage of female USGA golfers have a handicap greater than 20?

2.22 Sanitation inspection of cruise ships. To minimize the potential for gastrointestinal disease outbreaks, all passenger cruise ships arriving at U.S. ports are subject to unannounced sanitation inspections. Ships are rated on a 100-point scale by the Centers for Disease Control and Prevention. A score of 86 or higher indicates that the ship is providing an accepted standard of sanitation. The latest (as of Dec. 2008) sanitation scores for 183 cruise ships are saved in the **SHIPSANIT** file. The first five and last five observations in the data set are listed in the accompanying table.

Ship Name	Sanitation Score
Adventure of the Seas	99
Albatross	94
Amsterdam	98
AIDAaura	99
Amadea	84
⋮	⋮
Westerdam	98
Wind Spirit	88
Wind Surf	95
Zaandam	99
Zuiderdam	96

Source: U.S. Department of Health and Human Services, Centers for Disease Control and Prevention/National Center for Environmental Health and Agency for Toxic Substances and Disease Registry. May 24, 2006.

a. Generate both a stem-and-leaf display and histogram of the data.
b. Use the graphs to estimate the proportion of ships that have an accepted sanitation standard. Which graph did you use?
c. Locate the inspection score of 72 (*Legacy*) on the graph. Which graph did you use?

2.23 History of corporate acquisitions. Refer to the *Academy of Management Journal* (Aug. 2008) study of corporate acquisitions from 1980 to 2000, Exercise 2.12 (p. 40). The data are saved in the **ACQUISITIONS** file.
a. Construct either a dot plot or a stem-and-leaf display for the annual number of firms with at least one acquisition.
b. On the graph, part **a,** highlight (or circle) the values for the years 1996–2000. Do you detect a pattern? If so, what conclusion can you draw from the data?

Applying the Concepts—Intermediate

2.24 Most valuable NFL teams. Each year *Forbes* reports on the value of all teams in the National Football League. Although England's soccer team, Manchester United, is the most valuable team in the world ($1.8 billion), the NFL now has 19 teams worth at least $1 billion. For 2008, *Forbes* reports that the Dallas Cowboys are the most valuable team in the NFL, worth $1.6 billion. The current values (in $ millions) of all 32 NFL teams, as well as the percentage changes in values from 2007 to 2008, debt-to-value ratios, annual revenues, and operating incomes are listed in the accompanying table. (The data are saved in the **NFLTEAMVALUES** file.)
a. Use a graph to describe the distribution of current values for the 32 NFL teams.
b. Use a graph to describe the distribution of the 1-year change in current value for the 32 NFL teams.

Rank	Team	Current Value ($ mil)	1-Yr Value Change (%)	Debt/Value (%)	Revenue ($ mil)	Operating Income ($ mil)
1	Dallas Cowboys	1,612	7	39	269	30.6
2	Washington Redskins	1,538	5	16	327	58.1
3	New England Patriots	1,324	10	21	282	39.2
4	New York Giants	1,178	21	55	214	41.2
5	New York Jets	1,170	21	68	213	25.9
6	Houston Texans	1,125	7	27	239	43.9
7	Philadelphia Eagles	1,116	6	16	237	33.5
8	Indianapolis Colts	1,076	18	4	203	16.1
9	Chicago Bears	1,064	8	9	226	33.7
10	Baltimore Ravens	1,062	10	25	226	23
11	Denver Broncos	1,061	7	14	226	18.8
12	Tampa Bay Buccaneers	1,053	9	13	224	39.3
13	Miami Dolphins	1,044	11	34	232	36.1
14	Carolina Panthers	1,040	9	18	221	22.3
15	Cleveland Browns	1,035	7	10	220	19.3
16	Green Bay Packers	1,023	10	2	218	21.9
17	Kansas City Chiefs	1,016	6	13	214	11.9
18	Pittsburgh Steelers	1,015	9	10	216	14.4
19	Seattle Seahawks	1,010	10	12	215	8.9
20	Tennessee Titans	994	8	13	216	24.5
21	Cincinnati Bengals	941	3	11	205	22
22	New Orleans Saints	937	10	13	213	21.5
23	St. Louis Rams	929	2	6	206	26.4
24	Detroit Lions	917	5	38	204	−3.1
25	Arizona Cardinals	914	3	16	203	19.7
26	San Diego Chargers	888	7	11	207	19
27	Buffalo Bills	885	8	11	206	12.4
28	Jacksonville Jaguars	876	8	13	204	27.6
29	Atlanta Falcons	872	10	31	203	30.9
30	San Francisco 49ers	865	8	12	201	4.1
31	Oakland Raiders	861	6	6	205	27
32	Minnesota Vikings	839	7	38	195	18.9

Source: "Most Valuable Teams," *Forbes*, September 11, 2008. Reprinted by permission of Forbes.com. © 2009 Forbes LLC.

c. Use a graph to describe the distribution of debt-to-value ratios for the 32 NFL teams.

d. Use a graph to describe the distribution of the annual revenues for the 32 NFL teams.

e. Use a graph to describe the distribution of operating incomes for the 32 NFL teams.

f. Compare and contrast the graphs, parts **a–e**.

2.25 Color and clarity of diamonds. Refer to the *Journal of Statistics Education* study of diamonds, Exercise 2.11 (p. 40). In addition to color and clarity, the independent certification group (GIA, HRD, or IGI) and the number of carats were recorded for each of 308 diamonds for sale on the open market. Recall that the data are saved in the **DIAMONDS** file.

a. Use a graphical method to describe the carat distribution of all 308 diamonds.

b. Use a graphical method to describe the carat distribution of diamonds certified by the GIA group.

c. Repeat part **b** for the HRD and IGI certification groups.

d. Compare the three carat distributions, parts **b** and **c.** Is there one particular certification group that appears to be assessing diamonds with higher carats than the others?

2.26 Items arriving and departing a work center. In a manufacturing plant, a *work center* is a specific production facility that consists of one or more people and/or machines and is treated as one unit for the purposes of capacity requirements planning and job scheduling. If jobs arrive at

Number of Items Arriving at Work Center per Hour											
155	115	156	150	159	163	172	143	159	166	148	175
151	161	138	148	129	135	140	152	139			

Number of Items Departing Work Center per Hour											
156	109	127	148	135	119	140	127	115	122	99	106
171	123	135	125	107	152	111	137	161			

a particular work center at a faster rate than they depart, the work center impedes the overall production process and is referred to as a *bottleneck*. The data in the table on the bottom of page 50 (saved in the **WORKCTR** file) were collected by an operations manager for use in investigating a potential bottleneck work center. Construct dot plots for the two sets of data. Do the dot plots suggest that the work center may be a bottleneck? Explain.

2.27 Environmental failures of Arkansas companies. Any corporation doing business in the United States must be aware of and obey both federal and state environmental regulations. Failure to do so may result in irreparable damage to the environment and costly financial penalties to guilty corporations. Of the 55 civil actions filed against corporations within the state of Arkansas by the U.S. Department of Justice on behalf of the Environmental Protection Agency, 38 resulted in financial penalties. These penalties along with the laws that were violated are listed in the table below and in the data saved in the **CLEANAIR** file. (*Note:* Some companies were involved in more than one civil action.)

a. Construct a stem-and-leaf display for all 38 penalties.

b. Circle the individual leaves that are associated with penalties imposed for violations of the Clean Air Act (CAA).

c. What does the pattern of circles in part **b** suggest about the severity of the penalties imposed for CAA violations relative to the other types of violations reported in the table? Explain.

2.28 Is honey a cough remedy? Coughing at night is a common symptom of an upper respiratory tract infection, yet there is no accepted therapeutic cure. Does a teaspoon of honey before bed really calm a child's cough? To test the folk remedy, pediatric researchers at Pennsylvania State University carried out a designed study conducted over two nights (*Archives of Pediatrics and Adolescent Medicine,* Dec. 2007). A sample of 105 children who were ill with an upper respiratory tract infection and their parents participated in the study. On the first night, the parents rated their children's cough symptoms on a scale from 0 (no problems at all) to 6 (extremely severe) in five different areas. The total symptoms score (ranging from 0 to 30 points) was the variable of interest for the 105 patients. On the second night, the parents were instructed to give their

Honey Dosage:	12 11 15 11 10 13 10 4 15 16 9 14 10 6 10 8 11 12 12 8 9 5 12 12 9 11 15 10 15 9 13 8 12 10 8
DM Dosage:	4 6 9 4 7 7 7 9 12 10 11 6 3 4 9 12 7 6 8 12 12 4 12 10 15 9 13 7 10 13 9 4 4
No Dosage (Control):	5 8 6 1 0 8 12 8 7 7 1 6 7 7 12 7 9 7 9 5 11 9 5 1 4 3 6 8 8 6 7 10 9 4 8 7 3

Source: Paul, I. M., et al. "Effect of honey, dextromethorphan, and no treatment on nocturnal cough and sleep quality for coughing children and their parents," *Archives of Pediatrics and Adolescent Medicine,* Vol. 161, No. 12, Dec. 2007 (data simulated).

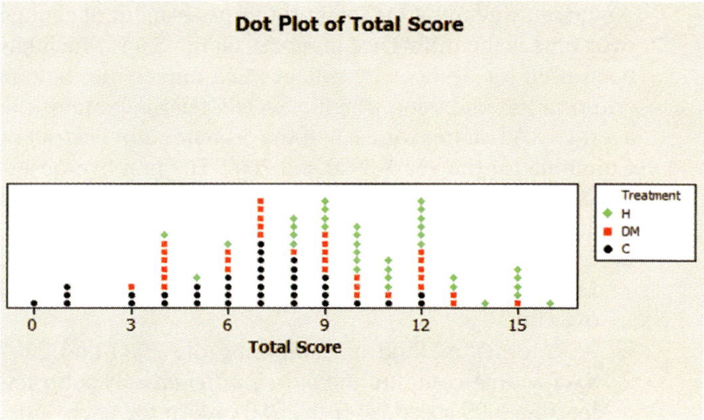

Company Identification Number	Penalty	Law*	Company Identification Number	Penalty	Law*
01	$ 930,000	CERCLA	17	20,000	CWA
02	10,000	CWA	18	40,000	CWA
03	90,600	CAA	19	20,000	CWA
04	123,549	CWA	20	40,000	CWA
05	37,500	CWA	21	850,000	CWA
06	137,500	CWA	22	35,000	CWA
07	2,500	SDWA	23	4,000	CAA
08	1,000,000	CWA	24	25,000	CWA
09	25,000	CAA	25	40,000	CWA
09	25,000	CAA	26	30,000	CAA
10	25,000	CWA	27	15,000	CWA
10	25,000	RCRA	28	15,000	CAA
11	19,100	CAA	29	105,000	CAA
12	100,000	CWA	30	20,000	CWA
12	30,000	CWA	31	400,000	CWA
13	35,000	CAA	32	85,000	CWA
13	43,000	CWA	33	300,000	CWA/
14	190,000	CWA			RCRA/
15	15,000	CWA			CERCLA
16	90,000	RCRA	34	30,000	CWA

*CAA: Clean Air Act; CERCLA: Comprehensive Environmental Response, Compensation, and Liability Act; RCRA: Resource Conservation and Recovery Act; SDWA: Safe Drinking Water Act.

Source: Tabor, R. H. and Stanwick, S. D. "Arkansas: An environmental perspective," *Arkansas Business and Economic Review,* Vol. 28, No. 2, Summer 1995, pp. 22–32 (Table 4).

sick child a dosage of liquid "medicine" prior to bedtime. Unknown to the parents, some were given a dosage of dextromethorphan (DM)—an over-the-counter cough medicine—while others were given a similar dose of honey. Also, a third group of parents (the control group) gave their sick children no dosage at all. Again, the parents rated their children's cough symptoms, and the improvement in total cough symptoms score was determined for each child. The data (improvement scores) for the study (saved in the **HONEYCOUGH** file) are shown in the table on the previous page, followed by a Minitab dot plot of the data. Notice that the green dots represent the children who received a dose of honey, the red dots represent those who got the DM dosage, and the black dots represent the children in the control group. What conclusions can pediatric researchers draw from the graph? Do you agree with the statement (extracted from the article), "honey may be a preferable treatment for the cough and sleep difficulty associated with childhood upper respiratory tract infection"?

Applying the Concepts—Advanced

2.29 State SAT scores. Educators are constantly evaluating the efficacy of public schools in the education and training of American students. One quantitative assessment of change over time is the difference in scores on the SAT, which has been used for decades by colleges and universities as one criterion for admission. The file **SATSCORES** contains the average SAT scores for each of the 50 states and District of Columbia for the years 2000 and 2007. The first five observations and last two observations in the data set are shown in the table.

a. Use graphs to display the two SAT score distributions. How have the distributions of state scores changed over time?

b. As another method of comparing the 2000 and 2007 SAT scores, compute the *paired difference* by subtracting the 2000 score from the 2007 score for each state. Summarize these differences with a graph.

c. Interpret the graph, part **b.** How do your conclusions compare to those of part **a?**

d. Based on the graph, part **b,** what is the largest improvement in SAT score between 2000 and 2007? Identify the state associated with this improvement.

State	2000	2007
Alabama	1114	1119
Alaska	1034	1036
Arizona	1044	1044
Arkansas	1117	1144
California	1015	1015
⋮	⋮	⋮
Wisconsin	1181	1185
Wyoming	1090	1136

Source: "College-Bound Seniors 2008." Copyright © 2008 the College Board. www.collegeboard.com. Reproduced with permission.

2.30 Time in bankruptcy. Financially distressed firms can gain protection from their creditors while they restructure by filing for protection under U.S. Bankruptcy Codes. In a *prepackaged bankruptcy,* a firm negotiates a reorganization plan with its creditors prior to filing for bankruptcy. This can result in a much quicker exit from bankruptcy than traditional bankruptcy filings. Brian Betker conducted a study of 49 prepackaged bankruptcies and reported the results in *Financial Management* (Spring 1995). The table below lists the time (in months) in bankruptcy for these 49 companies. The table also lists the results of a vote by each company's board of directors concerning their preferred reorganization plan. (*Note:* "Joint" = joint exchange offer with prepackaged bankruptcy solicitation; "Prepack" = prepackaged bankruptcy solicitation only;

Company	Prefiling Votes	Time in Bankruptcy (months)
AM International	None	3.9
Anglo Energy	Prepack	1.5
Arizona Biltmore*	Prepack	1.0
Astrex	None	10.1
Barry's Jewelers	None	4.1
Calton	Prepack	1.9
Cencor	Joint	1.4
Charter Medical*	Prepack	1.3
Cherokee*	Joint	1.2
Circle Express	Prepack	4.1
Cook Inlet Comm.	Prepack	1.1
Crystal Oil	None	3.0
Divi Hotels	None	3.2
Edgell Comm.*	Prepack	1.0
Endevco	Prepack	3.8
Gaylord Container	Joint	1.2
Great Amer. Comm.*	Prepack	1.0
Hadson	Prepack	1.5
In-Store Advertising	Prepack	1.0
JPS Textiles*	Prepack	1.4
Kendall*	Prepack	1.2
Kinder-Care	None	4.2
Kroy*	Prepack	3.0
Ladish*	Joint	1.5
LaSalle Energy*	Prepack	1.6
LIVE Entertainment	Joint	1.4
Mayflower Group*	Prepack	1.4
Memorex Telex*	Prepack	1.1
Munsingwear	None	2.9
Nat'l Environmental	Joint	5.2
Petrolane Gas	Prepack	1.2
Price Communications	None	2.4
Republic Health*	Joint	4.5
Resorts Int'l*	None	7.8
Restaurant Enterprises*	Prepack	1.5
Rymer Foods	Joint	2.1
SCI TV*	Prepack	2.1
Southland*	Joint	3.9
Specialty Equipment*	None	2.6
SPI Holdings*	Joint	1.4
Sprouse-Reitz	Prepack	1.4
Sunshine Metals	Joint	5.4
TIE/Communications	None	2.4
Trump Plaza	Prepack	1.7
Trump Taj Mahal	Prepack	1.4
Trump's Castle	Prepack	2.7
USG	Prepack	1.2
Vyquest	Prepack	4.1
West Point Acq.*	Prepack	2.9

*Leveraged buyout.

Source: Betker, B. L. "An empirical examination of prepackaged bankruptcy," *Financial Management,* Vol. 24, No. 1, Spring 1995, p. 6 (Table 2). Reprinted with permission of John Wiley & Sons, Inc.

"None" = no prefiling vote held.) (These data are saved in the **BANKRUPT** file.)

a. Construct a stem-and-leaf display for the length of time in bankruptcy for all 49 companies.

b. Summarize the information reflected in the stem-and-leaf display, part **a.** Make a general statement about the length of time in bankruptcy for firms using "prepacks."

c. Select a graphical technique that will permit a comparison of the time-in-bankruptcy distributions for the three types of "prepack" firms: those who held no prefiling vote; those who voted their preference for a joint solution; and those who voted their preference for a prepack.

d. The companies that were reorganized through a leveraged buyout are identified by an asterisk in the table. Identify these firms on the stem-and-leaf display, part **a,** by circling their bankruptcy times. Do you observe any pattern in the graph? Explain.

2.31 Phishing attacks to e-mail accounts. *Phishing* is the term used to describe an attempt to extract personal/financial information (e.g., PIN numbers, credit card information, bank account numbers) from unsuspecting people through fraudulent e-mail. An article in *Chance* (Summer 2007), demonstrates how statistics can help identify phishing attempts and make e-commerce safer. Data from an actual phishing attack against an organization were used to determine whether the attack may have been an "inside job" that originated within the company. The company set up a publicized e-mail account—called a "fraud box"—which enabled employees to notify them if they suspected an e-mail phishing attack. The interarrival times, i.e., the time differences (in seconds), for 267 fraud box e-mail notifications were recorded. *Chance* showed that if there is minimal or no collaboration or collusion from within the company, the interarrival times would have a frequency distribution similar to the one shown in the accompanying figure. The 267 interarrival times are saved in the **PHISHING** file. Construct a frequency histogram for the interarrival times. Give your opinion on whether the phishing attack against the organization was an "inside job."

2.32 Made-to-order delivery times. Production processes may be classified as *make-to-stock processes* or *make-to-order processes*. Make-to-stock processes are designed to produce a standardized product that can be sold to customers from the firm's inventory. Make-to-order processes are designed to produce products according to customer specifications (Schroeder, *Operations Management,* 2008). In general, performance of make-to-order processes is measured by delivery time—the time from receipt of an order until the product is delivered to the customer. The following data set (saved in the **DELTIMES** file) is a sample of delivery times (in days) for a particular make-to-order firm last year. The delivery times marked by an asterisk are associated with customers who subsequently placed additional orders with the firm.

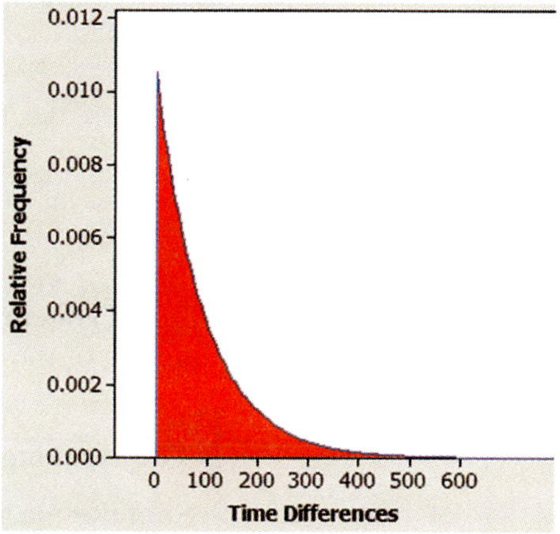

Concerned that they are losing potential repeat customers because of long delivery times, the management would like to establish a guideline for the maximum tolerable delivery time. Use a graphical method to help suggest a guideline. Explain your reasoning.

50*	64*	56*	43*	64*	82*	65*	49*	32*	63*	44*	71
54*	51*	102	49*	73*	50*	39*	86	33*	95	59*	51*
68											

2.3 Summation Notation

Now that we've examined some graphical techniques for summarizing and describing quantitative data sets, we turn to numerical methods for accomplishing this objective. Before giving the formulas for calculating numerical descriptive measures, let's look at some shorthand notation that will simplify our calculation instructions. Remember that such notation is used for one reason only—to avoid repeating the same verbal descriptions over and over. If you mentally substitute the verbal definition of a symbol each time you read it, you'll soon get used to it.

We denote the measurements of a quantitative data set as follows: $x_1, x_2, x_3, \ldots, x_n$, where x_1 is the first measurement in the data set, x_2 is the second measurement in the data set, x_3 is the third measurement in the data set, ..., and x_n is the nth (and last) measurement in the data set. Thus, if we have five measurements in a set of data, we will write x_1, x_2, x_3, x_4, x_5 to represent the measurements. If the actual numbers are 5, 3, 8, 5, and 4, we have $x_1 = 5$, $x_2 = 3$, $x_3 = 8$, $x_4 = 5$, and $x_5 = 4$.

Most of the formulas we use require a summation of numbers. For example, one sum we'll need to obtain is the sum of all the measurements in the data set, or $x_1 + x_2 + x_3 + \cdots + x_n$. To shorten the notation, we use the symbol Σ for the summation—that is, $x_1 + x_2 + x_3 + \dots + x_n = \sum_{i=1}^{n} x_i$. Verbally translate $\sum_{i=1}^{n} x_i$ as follows: "The sum of the measurements, whose typical member is x_i, beginning with the member x_1 and ending with the member x_n."

Suppose, as in our earlier example, that $x_1 = 5$, $x_2 = 3$, $x_3 = 8$, $x_4 = 5$, and $x_5 = 4$. Then the sum of the five measurements, denoted $\sum_{i=1}^{n} x_i$, is obtained as follows:

$$\sum_{i=1}^{5} x_i = x_1 + x_2 + x_3 + x_4 + x_5$$
$$= 5 + 3 + 8 + 5 + 4 = 25$$

Another important calculation requires that we square each measurement and then sum the squares. The notation for this sum is $\sum_{i=1}^{n} x_i^2$. For the preceding five measurements, we have

$$\sum_{i=1}^{5} x_i^2 = x_1^2 + x_2^2 + x_3^2 + x_4^2 + x_5^2$$
$$= 5^2 + 3^2 + 8^2 + 5^2 + 4^2$$
$$= 25 + 9 + 64 + 25 + 16 = 139$$

In general, the symbol following the summation sign represents the variable (or function of the variable) that is to be summed.

The Meaning of Summation Notation $\sum_{i=1}^{n} x_i$

Sum the measurements on the variable that appears to the right of the summation symbol, beginning with the 1st measurement and ending with the nth measurement.

Exercises 2.33–2.36

Learning the Mechanics

Note: In all exercises, Σ represents $\sum_{i=1}^{n}$.

2.33 A data set contains the observations 5, 1, 3, 2, 1. Find
- **a.** Σx
- **b.** Σx^2
- **c.** $\Sigma(x - 1)$
- **d.** $\Sigma(x - 1)^2$
- **e.** $(\Sigma x)^2$

2.34 Suppose a data set contains the observations 3, 8, 4, 5, 3, 4, 6. Find
- **a.** Σx
- **b.** Σx^2
- **c.** $\Sigma(x - 5)^2$
- **d.** $\Sigma(x - 2)^2$
- **e.** $(\Sigma x)^2$

2.35 Refer to Exercise 2.33. Find
- **a.** $\Sigma x^2 - \dfrac{(\Sigma x)^2}{5}$
- **b.** $\Sigma(x - 2)^2$
- **c.** $\Sigma x^2 - 10$

2.36 A data set contains the observations 6, 0, −2, −1, 3. Find
- **a.** Σx
- **b.** Σx^2
- **c.** $\Sigma x^2 - \dfrac{(\Sigma x)^2}{5}$

2.4 Numerical Measures of Central Tendency

When we speak of a data set, we refer to either a sample or a population. If statistical inference is our goal, we'll wish ultimately to use sample **numerical descriptive measures** to make inferences about the corresponding measures for the population.

As you'll see, a large number of numerical methods are available to describe quantitative data sets. Most of these methods measure one of two data characteristics:

1. The **central tendency** of the set of measurements—that is, the tendency of the data to cluster, or center, about certain numerical values (see Figure 2.14a).
2. The **variability** of the set of measurements—that is, the spread of the data (see Figure 2.14b).

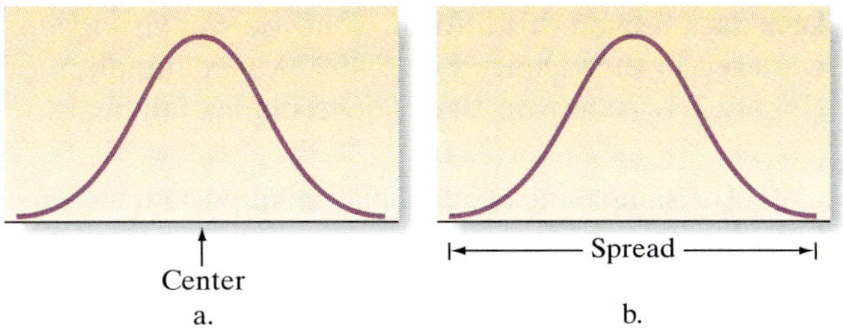

Figure 2.14
Numerical descriptive measures

In this section we concentrate on measures of central tendency. In the next section, we discuss measures of variability.

The most popular and best-understood measure of central tendency for quantitative data is the arithmetic mean (or simply the mean) of a data set.

> The **mean** of a set of quantitative data is the sum of the measurements divided by the number of measurements contained in the data set.

In everyday terms, the mean is the average value of the data set and is often used to represent a "typical" value. We denote the **mean of a sample** of measurements by $\overline{x}$ (read "x-bar") and represent the formula for its calculation as shown in the box below.

> **Formula for a Sample Mean**
>
> $$\overline{x} = \frac{\sum_{i=1}^{n} x_i}{n}$$

Example 2.3
Calculating the Sample Mean

Problem Calculate the mean of the following five sample measurements: 5, 3, 8, 5, 6.

Solution Using the definition of sample mean and the summation notation, we find

$$\overline{x} = \frac{\sum_{i=1}^{5} x_i}{5} = \frac{5 + 3 + 8 + 5 + 6}{5} = \frac{27}{5} = 5.4$$

Thus, the mean of this sample is 5.4.

Look Back There is no specific rule for rounding when calculating $\overline{x}$ because $\overline{x}$ is specifically defined to be the sum of all measurements divided by n—that is, it is a specific fraction. When $\overline{x}$ is used for descriptive purposes, it is often convenient to round the calculated value of $\overline{x}$ to the number of significant figures used for the original measurements. When $\overline{x}$ is to be used in other calculations, however, it may be necessary to retain more significant figures.

Now Work Exercise 2.38

Example 2.4
The Sample Mean on a Printout

Problem Calculate the sample mean for the R&D expenditure percentages of the 50 companies given in Table 2.2.

Solution The mean R&D percentage for the 50 companies is denoted

$$\overline{x} = \frac{\sum_{i=1}^{50} x_i}{50}$$

We employed Excel to compute the mean. The Excel printout is shown in Figure 2.15. The sample mean, highlighted on the printout, is $\overline{x} = 8.492$.

	A	B
1	*RDPct*	
2		
3	Mean	8.492
4	Standard Error	0.2801
5	Median	8.05
6	Mode	6.9
7	Standard Deviation	1.980604
8	Sample Variance	3.922792
9	Kurtosis	0.419288
10	Skewness	0.854601
11	Range	8.3
12	Minimum	5.2
13	Maximum	13.5
14	Sum	424.6
15	Count	50
16		

Figure 2.15
Excel numerical descriptive measures for 50 R&D percentages

Look Back Given this information, you can visualize a distribution of R&D percentages centered in the vicinity $\overline{x} = 8.492$. An examination of the relative frequency histogram (Figure 2.10) confirms that $\overline{x}$ does, in fact, fall near the center of the distribution.

The sample mean $\overline{x}$ will play an important role in accomplishing our objective of making inferences about populations based on sample information. For this reason, we need to use a different symbol for the *mean of a population*—the mean of the set of measurements on every unit in the population. We use the Greek letter μ(mu) for the population mean.

Symbols for the Sample and Population Mean

In this text, we adopt a general policy of using Greek letters to represent population numerical descriptive measures and Roman letters to represent corresponding descriptive measures for the sample. The symbols for the mean are

$$\overline{x} = \text{Sample mean}$$
$$\mu = \text{Population mean}$$

We'll often use the sample mean $\overline{x}$ to estimate (make an inference about) the population mean, μ. For example, the percentages of revenues spent on R&D by the population consisting of *all* U.S. companies has a mean equal to some value, μ. Our sample of 50 companies yielded percentages with a mean of $\overline{x} = 8.492$. If, as is usually the case, we don't have access to the measurements for the entire population, we could use $\overline{x}$ as an estimator or approximator for μ. Then we'd need to know something about the reliability of our inference—that is, we'd need to know how accurately we might expect $\overline{x}$ to estimate μ. In Chapter 5, we'll find that this accuracy depends on two factors:

1. The *size of the sample*. The larger the sample, the more accurate the estimate will tend to be.
2. The *variability,* or *spread, of the data*. All other factors remaining constant, the more variable the data, the less accurate the estimate.

Another important measure of central tendency is the *median*.

The **median** of a quantitative data set is the middle number when the measurements are arranged in ascending (or descending) order.

The median is of most value in describing large data sets. If the data set is characterized by a relative frequency histogram (Figure 2.16), the median is the point on the x-axis such that half the area under the histogram lies above the median and half lies below. [*Note:* In Section 2.2 we observed that the relative frequency associated with a particular interval on the horizontal axis is proportional to the amount of area under the histogram that lies above the interval.] We denote the *median of a sample* by m.

Calculating a Sample Median, m

Arrange the n measurements from smallest to largest.

1. If n is odd, m is the middle number.
2. If n is even, m is the mean of the middle two numbers.

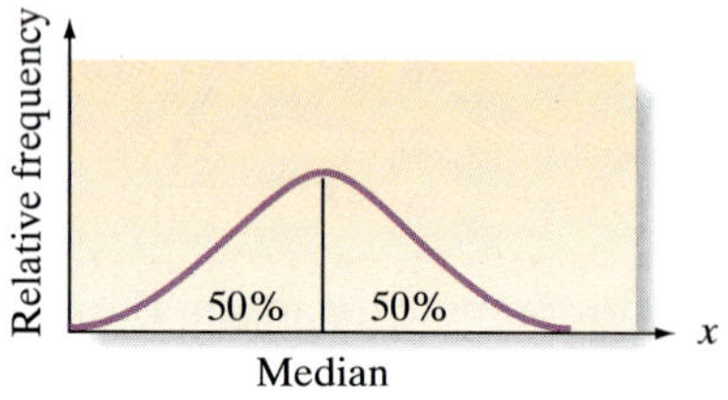

Figure 2.16
Location of the median

Example 2.5
Finding the Median

Problem Consider the following sample of $n = 7$ measurements: 5, 7, 4, 5, 20, 6, 2.

a. Calculate the median m of this sample.

b. Eliminate the last measurement (the 2) and calculate the median of the remaining $n = 6$ measurements.

Solution

a. The seven measurements in the sample are ranked in ascending order: 2, 4, 5, 5, 6, 7, 20. Because the number of measurements is odd, the median is the middle measurement. Thus, the median of this sample is $m = 5$ (the second 5 listed in the sequence).

b. After removing the 2 from the set of measurements, we rank the sample measurements in ascending order as follows: 4, 5, 5, 6, 7, 20. Now the number of measurements is even, so we average the middle two measurements. The median is $m = (5 + 6)/2 = 5.5$.

Look Back When the sample size n is even and the two middle numbers are different (as in part **b**), exactly half of the measurements will fall below the calculated median m. However, when n is odd (as in part **a**), the percentage of measurements that fall below m is approximately 50%. This approximation improves as n increases.

Now Work Exercise 2.37

In certain situations, the median may be a better measure of central tendency than the mean. In particular, the median is less sensitive than the mean to extremely large or small measurements. Note, for instance, that all but one of the measurements in part **a** of Example 2.5 center about $x = 5$. The single relatively large measurement, $x = 20$, does not affect the value of the median, 5, but it causes the mean, $\bar{x} = 7$, to lie to the right of most of the measurements.

As another example of data for which the central tendency is better described by the median than the mean, consider the salaries of professional athletes (e.g., National Basketball Association players). The presence of just a few athletes (e.g., LeBron James) with extremely high salaries will affect the mean more than the median. Thus, the median will provide a more accurate picture of the typical salary for the professional league. The mean could exceed the vast majority of the sample measurements (salaries), making it a misleading measure of central tendency.

Example 2.6
The Median on a Printout

Problem Calculate the median for the 50 R&D percentages given in Table 2.2. Compare the median to the mean found in Example 2.4.

Solution For this large data set, we again resort to a computer analysis. The median is highlighted on the Excel printout, Figure 2.15. You can see that the median is 8.05. This value implies that half of the 50 R&D percentages in the data set fall below 8.05 and half lie above 8.05.

Note that the mean (8.492) for these data is larger than the median. This fact indicates that the data are **skewed** to the right—that is, there are more extreme measurements in the right tail of the distribution than in the left tail (recall the histogram in Figure 2.10).

Look Back In general, extreme values (large or small) affect the mean more than the median because these values are used explicitly in the calculation of the mean. On the other hand, the median is not affected directly by extreme measurements because only the middle measurement (or two middle measurements) is explicitly used to calculate the median. Consequently, if measurements are pulled toward one end of the distribution (as with the R&D percentages), the mean will shift toward that tail more than the median.

A data set is said to be **skewed** if one tail of the distribution has more extreme observations than the other tail.

A comparison of the mean and the median gives us a general method for detecting skewness in data sets, as shown in the next box. With *rightward skewed* data, the right tail (high end) of the distribution has more extreme observations. These few, but large, measurements pull the mean away from the median toward the right; that is, rightward skewness indicates that the mean is greater than the median. Conversely, with *leftward skewed* data, the left tail (low end) of the distribution has more extreme observations. These few, but small, measurements also pull the mean away from the median but toward the left; consequently, leftward skewness implies that the mean is smaller than the median.

Detecting Skewness by Comparing the Mean and the Median

If the data set is skewed to the right, then the median is less than the mean.

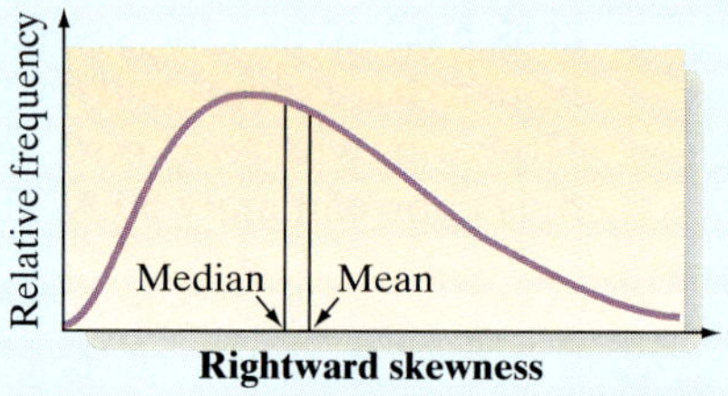

If the data set is symmetric, the mean equals the median.

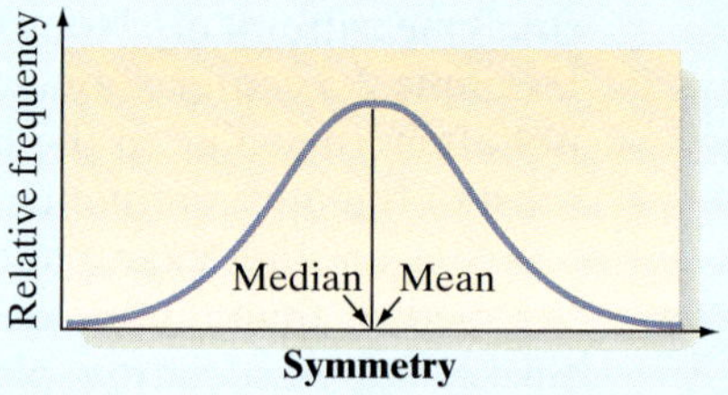

If the data set is skewed to the left, the mean is less than (to the left of) the median.

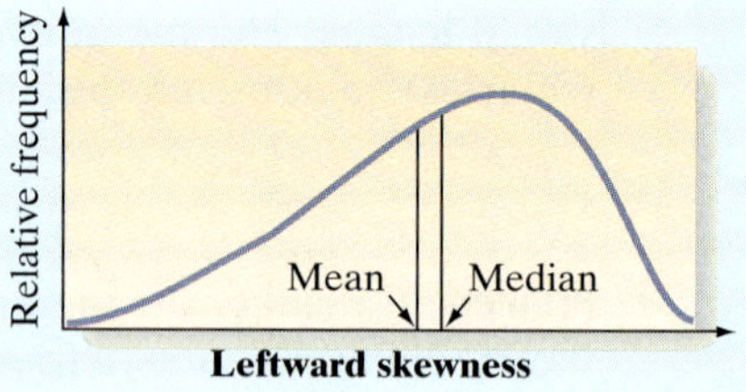

A third measure of central tendency is the *mode* of a set of measurements.

The **mode** is the measurement that occurs most frequently in the data set.

Example 2.7
Finding the Mode

Problem Each of 10 taste testers rated a new brand of barbecue sauce on a 10-point scale, where 1 = awful and 10 = excellent. Find the mode for the 10 ratings shown below.

8 7 9 6 8 10 9 9 5 7

Solution Because 9 occurs most often, the mode of the 10 taste-ratings is 9.

Look Back Note that the data are actually qualitative in nature (e.g., "awful," "excellent"). The mode is particularly useful for describing qualitative data. The modal category is simply the category (or class) that occurs most often.

Now Work Exercise 2.41

Because it emphasizes data concentration, the mode is used with quantitative data sets to locate the region in which much of the data are concentrated. A retailer of men's clothing would be interested in the modal neck size and sleeve length of potential customers. The modal income class of the laborers in the United States is of interest to the Labor Department.

For some quantitative data sets, the mode may not be very meaningful. For example, consider the percentages of revenues spent on R&D by 50 companies, Table 2.2. A reexamination of the data reveals that three of the measurements are repeated three times: 6.5%, 6.9%, and 8.2%. Thus, there are three modes in the sample, and none is particularly useful as a measure of central tendency.

A more meaningful measure can be obtained from a relative frequency histogram for quantitative data. The class interval containing the largest relative frequency is called the **modal class.** Several definitions exist for locating the position of the mode within a modal class, but the simplest is to define the mode as the midpoint of the modal class. For example, examine the relative frequency histogram for the price quote processing times in Figure 2.12. You can see that the modal class is the interval (3.0–4.0). The mode (the midpoint) is 3.5. This modal class (and the mode itself) identifies the area in which the data are most concentrated, and in that sense it is a measure of central tendency. However, for most applications involving quantitative data, the mean and median provide more descriptive information than the mode.

Example 2.8

Comparing the Mean, Median, and Mode of CEO Salaries

Problem Refer to *Forbes* magazine's "Executive Compensation Scoreboard," which lists the total annual pay for CEOs at the 500 largest U.S. firms. The data for the 2008 scoreboard, saved in the **CEOPAY2008** file, includes the quantitative variables total annual

Statistics

		PAY2008	AGE
N	Valid	497	500
	Missing	3	0
	Mean	12.87383	55.77
	Median	6.47000	56.00
	Mode	2.010[a]	53

a. Multiple modes exist. The smallest value is shown.

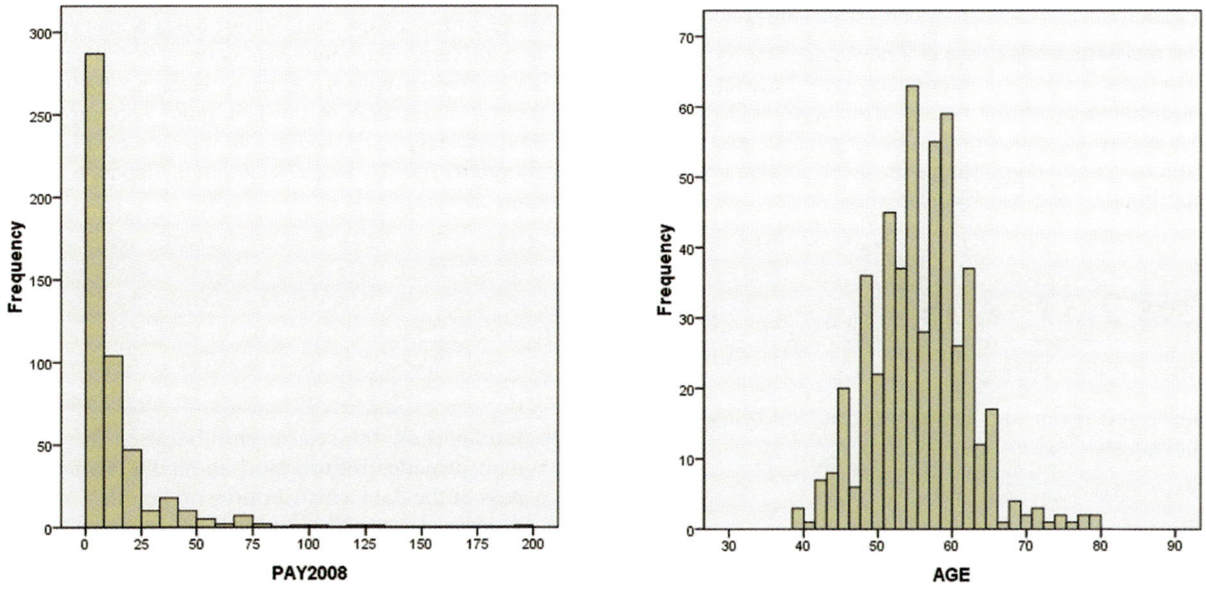

Figure 2.17
SPSS analysis of total 2008 pay and age for 500 CEOs

pay (in millions of dollars) and age. Find the mean, median, and mode for both of these variables. Which measure of central tendency is better for describing the distribution of total annual pay? Age?

Solution Measures of central tendency for the two variables were obtained using SPSS. The means, medians, and modes are displayed at the top of the SPSS printout, Figure 2.17.

For total annual pay, the mean, median, and mode are \$12.87 million, \$6.47 million, and \$2.01 million, respectively. Note that the mean is much greater than the median, indicating that the data are highly skewed right. This rightward skewness (graphically shown on the histogram for total pay in Figure 2.17) is due to several exceptionally high CEO salaries in 2008. Consequently, we would probably want to use the median, \$6.47 million, as the "typical" value for annual pay for CEOs at the 500 largest firms. The mode of \$2.01 million is the total pay value that occurs most often in the data set, but it is not very descriptive of the "center" of the total annual pay distribution.

For age, the mean, median, and mode are 55.77, 56, and 53 years, respectively. All three values are nearly the same, which is typical of symmetric distributions. From the age histogram to the right in Figure 2.17, you can see that the age distribution is nearly symmetric. Consequently, any of the three measures of central tendency could be used to describe the "middle" of the age distribution.

Look Back The choice of which measure of central tendency to use will depend on the properties of the data set analyzed and on the application. Consequently, it is vital that you understand how the mean, median, and mode are computed.

Now Work Exercise 2.46a, b

Activity 2.1: *Real Estate Sales:* Measures of Central Tendency

In recent years, the price of real estate in America's major metropolitan areas has skyrocketed. Newspapers usually report recent real estate sales data in their Saturday editions, both hard copies and online. These data usually include the actual prices paid for homes by geographical location during a certain time period, usually a one-week period six to eight weeks earlier, and some summary statistics, which might include comparisons to real estate sales data in other geographical locations or during other time periods.

1. Locate the real estate sales data in a newspaper for a major metropolitan area. From the information given, identify the time period during which the homes listed were sold. Then describe the way the sales prices are organized. Are they categorized by type of home (single family or condominium), by neighborhood or address, by sales price, etc.?
2. What summary statistics and comparisons are provided with the sales data? Describe several groups of people who might be interested in this data and how each of the summary statistics and comparisons would be helpful to them. Why are the measures of central tendency listed more useful in the real estate market than other measures of central tendency?
3. Based on the lowest and highest sales prices represented in the data, create ten intervals of equal size and use these intervals to create a relative frequency histogram for the sales data. Describe the shape of the histogram and explain how the summary statistics provided with the data are illustrated in the histogram. Based on the histogram, describe the "typical" home price.

Exercises 2.37–2.55

Learning the Mechanics

2.37 NW Calculate the mean and median of the following grade point averages:

3.2 2.5 2.1 3.7 2.8 2.0

2.38 NW Calculate the mean for samples where

a. $n = 10, \sum x = 85$
b. $n = 16, \sum x = 400$
c. $n = 45, \sum x = 35$
d. $n = 18, \sum x = 242$

2.39 Explain how the relationship between the mean and median provides information about the symmetry or skewness of the data's distribution.

2.40 Explain the difference between the calculation of the median for an odd and an even number of measurements. Construct one data set consisting of five measurements and another consisting of six measurements for which the medians are equal.

2.41 Calculate the mode, mean, and median of the following data:
NW

18 10 15 13 17 15 12 15 18 16 11

2.42 Calculate the mean, median, and mode for each of the following samples:

a. 7, − 2, 3, 3, 0, 4
b. 2, 3, 5, 3, 2, 3, 4, 3, 5, 1, 2, 3, 4
c. 51, 50, 47, 50, 48, 41, 59, 68, 45, 37

Applet Exercise 2.1

Use the applet entitled *Mean versus Median* to find the mean and median of each of the three data sets in Exercise 2.42. For each data set, set the lower limit to a number less than all of the data, set the upper limit to a number greater than all of the data, and then click on *Update*. Click on the approximate location of each data item on the number line. You can get rid of a point by dragging it to the trash can. To clear the graph between data sets, simply click on the trash can.

a. Compare the means and medians generated by the applet to those you calculated by hand in Exercise 2.42. If there are differences, explain why the applet might give values slightly different from the hand calculations.
b. Despite providing only approximate values of the mean and median of a data set, describe some advantages of using the applet to find these values.

2.43 Describe how the mean compares to the median for a distribution as follows:

a. Skewed to the left
b. Skewed to the right
c. Symmetric

Applet Exercise 2.2

Use the applet *Mean versus Median* to illustrate your descriptions in Exercise 2.43. For each part **a, b,** and **c,** create a data set with ten items that has the given property. Using the applet, verify that the mean and median have the relationship you described in Exercise 2.43.

Applet Exercise 2.3

Use the applet *Mean versus Median* to study the effect that an extreme value has on the difference between the mean and median. Begin by setting appropriate limits and plotting the given data on the number line provided in the applet.

0 6 7 7 8 8 8 9 9 10

a. Describe the shape of the distribution and record the value of the mean and median. Based on the shape of the distribution, do the mean and median have the relationship that you would expect?
b. Replace the extreme value of 0 with 2, then 4, and then 6. Record the mean and median each time. Describe what is happening to the mean as 0 is replaced by higher numbers. What is happening to the median? How is the difference between the mean and the median changing?
c. Now replace 0 with 8. What values does the applet give you for the mean and the median? Explain why the mean and the median should be the same.

Applying the Concepts—Basic

2.44 **Spending by banks following acquisitions.** Recently there has been a flurry of mergers of major U.S. banks, including Wells Fargo's acquisition of Wachovia, JPMorgan Chase's buyout of Washington Mutual, and Bank of America's takeover of Merrill Lynch. *The Nilson Report* (Oct. 2008) ranked U.S. banks with recent acquisitions by the amount charged to credit and debit cards issued by the banks. The following table gives the total amount charged in 2007 for the top ranked banks. (The data are saved in the **BANKCHARGE** file.)

U.S. Bank/ Card Issuer	Amount Charged ($ billions)
Bank of America	462.20
JPMorgan Chase	458.84
American Express	445.32
Citigroup	240.58
Wells Fargo	192.68

Source: The Nilson Report, No. 912, October 2008 (p. 8).

a. Find the mean amount charged for the top five banks and practically interpret this value.
b. Find the median amount charged for the top five banks and practically interpret this value.

2.45 **U.S. wine export markets.** The data in the next table, compiled by the Center for International Trade Development (CITD), provide a listing of the top 30 U.S. export markets for sparkling wines in 2007. (These data are saved in the **WINEEXPORTS** file.) Descriptive statistics for the amount exported (thousands of dollars) and three-year percentage change for the 30 countries are shown in the Minitab printout below.

a. Locate the mean amount exported on the printout and practically interpret its value.
b. Locate the median amount exported on the printout and practically interpret its value.
c. Locate the mean 3-year percentage change on the printout and practically interpret its value.
d. Locate the median 3-year percentage change on the printout and practically interpret its value.

Descriptive Statistics: EXPORT, CHANGE

Variable	N	N*	Mean	StDev	Minimum	Q1	Median	Q3	Maximum	IQR
EXPORT	30	0	653	1113	70	105	231	523	4952	418
CHANGE	28	2	481	1098	-49	21	156	499	5750	478

Top 30 U.S. Sparkling Wine Export Markets, 2007

Country	Export ($ Thousands)	3-Year Change (%)
Canada	4952	71.9
Japan	3714	−16.9
Mexico	2104	143.2
Cayman Islands	1576	280.7
United Kingdom	1041	465.8
Netherlands	807	550.8
Germany	645	658.8
Korea	482	20.5
France	449	−48.7
Russia	351	5750
Jamaica	350	21.5
China	339	539.6
Taiwan	309	505.9
Colombia	272	159
Hong Kong	232	114.8
Aruba	229	−31.6
Australia	225	1223.5
Haiti	191	478.8
Switzerland	181	70.8
Trinidad & Tobago	175	1490.9
Costa Rica	170	NA
Panama	126	447.8
Slovenia	106	NA
Neth. Antilles	103	−41.8
Bahamas	92	−20.7
New Zealand	91	193.5
Honduras	74	51
Philippines	72	242.9
Greece	71	153.6
Belgium	70	−16.7

Source: "Best Export Markets for U.S. Wines, 2008," *Center for International Trade Development* (Table II. A-3). Used with permission of Center for International Trade Development. citd.org.

2.46 Surface roughness of oil field pipe. Oil field pipes are internally coated in order to prevent corrosion. Researchers at the University of Louisiana, Lafayette, investigated the influence that coating may have on the surface roughness of oil field pipes (*Anti-corrosion Methods and Materials,* Vol. 50, 2003). A scanning probe instrument was used to measure the surface roughness of each in a sample of 20 sections of coated interior pipe. The data (in micrometers) are provided in the table and saved in the **ROUGHPIPE** file.

1.72	2.50	2.16	2.13	1.06	2.24	2.31	2.03	1.09	1.40
2.57	2.64	1.26	2.05	1.19	2.13	1.27	1.51	2.41	1.95

Source: Farshad, F., and Pesacreta, T. "Coated pipe interior surface roughness as measured by three scanning probe instruments." *Anti-corrosion Methods and Materials*, Vol. 50, No. 1, 2003 (Table III). Copyright © 2003 MCB UP Ltd.

NW **a.** Find and interpret the mean of the sample.

NW **b.** Find and interpret the median of the sample.

c. Which measure of central tendency—the mean or the median—best describes the surface roughness of the sampled pipe sections? Explain.

2.47 Is honey a cough remedy? Refer to the *Archives of Pediatrics and Adolescent Medicine* (Dec. 2007) study of honey as a remedy for coughing, Exercise 2.28 (p. 51). Recall that the 105 ill children in the sample were randomly divided into three groups: those who received a dosage of an over-the-counter cough medicine (DM), those who received a dosage of honey (H), and those who received no dosage (control group). The coughing improvement scores (as determined by the children's parents) for the patients, reproduced in the accompanying table, are saved in the **HONEYCOUGH** file.

a. Find the median improvement score for the honey dosage group.

b. Find the median improvement score for the DM dosage group.

c. Find the median improvement score for the control group.

d. Based on the results, parts **a–c,** what conclusions can pediatric researchers draw? (We show how to support these conclusions with a measure of reliability in subsequent chapters.)

Honey	12	11	15	11	10	13	10	4	15	16	9	14			
Dosage:	10	6	10	8	11	12	12	8	12	9	11	15			
	10	15	9	13	8	12	10	8	9	5	12				
DM	4	6	9	4	7	7	7	9	12	10	11	6	3	4	
Dosage:	9	12	7	6	8	12	12	4	12	13	7	10			
	13	9	4	4	10	15	9								
No Dosage	5	8	6	1	0	8	12	8	7	7	1	6	7	7	12
(Control):	7	9	7	9	5	11	9	5	6	8					
	8	6	7	10	9	4	8	7	3	1	4	3			

Source: Paul, I. M., et al. "Effect of honey, dextromethorphan, and no treatment on nocturnal cough and sleep quality for coughing children and their parents," *Archives of Pediatrics and Adolescent Medicine,* Vol. 161, No. 12, Dec. 2007 (data simulated).

Applying the Concepts—Intermediate

2.48 Most powerful business women in America. *Fortune* (Oct. 16, 2008) published a list of the 50 most powerful women in business in the United States. The data on age (in years) and title of each of these 50 women are stored in the **WPOWER50** file. The first five and last five observations of the data set are listed in the table on page 63.

a. Find the mean, median, and modal age of these 50 women.

b. What do the mean and median indicate about the skewness of the age distribution?

c. Construct a relative frequency histogram for the age data. What is the modal age class?

2.49 Size of diamonds sold at retail. Refer to Exercise 2.25 (p. 50) and the *Journal of Statistics Education* data on diamonds saved in the **DIAMONDS** file. Consider the quantitative variable, number of carats, recorded for each of the 308 diamonds for sale on the open market.

a. Find and interpret the mean of the data set.

b. Find and interpret the median of the data set.

c. Find and interpret the mode of the data set.

d. Which measure of central tendency best describes the 308 carat values? Explain.

Table for Exercise 2.48

Rank	Name	Age	Company	Title
1	Indra Nooyi	52	PepsiCo	CEO/Chairman
2	Irene Rosenfeld	55	Kraft Foods	CEO/Chairman
3	Pat Woertz	55	Archer Daniels Midland	CEO/Chairman
4	Anne Mulcahy	55	Xerox	CEO/Chairman
5	Angela Braley	47	Wellpoint	CEO/President
⋮	⋮	⋮	⋮	⋮
46	Lorrie Norrington	48	eBay	CEO
47	Terri Dial	58	Citigroup	CEO
48	Lynn Elsenhans	52	Sunoco	CEO/President
49	Cathie Black	64	Hearst Magazines	President
50	Marissa Mayer	33	Google	VP

Source: "50 most powerful women," *Fortune*, Oct. 16, 2008.

2.50 Ranking driving performance of professional golfers. A group of Northeastern University researchers developed a new method for ranking the total driving performance of golfers on the Professional Golf Association (PGA) tour. (*The Sport Journal,* Winter 2007.) The method requires knowing a golfer's average driving distance (yards) and driving accuracy (percent of drives that land in the fairway). The values of these two variables are used to compute a driving performance index. Data for the top 40 PGA golfers (as ranked by the new method) are saved in the **PGADRIVER** file. The first five and last five observations are listed in the accompanying table.

a. Find the mean, median, and mode for the 40 driving performance index values.

b. Interpret each of the measures of central tendency, part **a.**

c. Use the results, part **a,** to make a statement about the type of skewness in the distribution of driving performance indexes. Support your statement with a graph.

Rank	Player	Driving Distance (yards)	Driving Accuracy (%)	Driving Performance Index
1	Woods	316.1	54.6	3.58
2	Perry	304.7	63.4	3.48
3	Gutschewski	310.5	57.9	3.27
4	Wetterich	311.7	56.6	3.18
5	Hearn	295.2	68.5	2.82
⋮	⋮	⋮	⋮	⋮
36	Senden	291	66	1.31
37	Mickelson	300	58.7	1.30
38	Watney	298.9	59.4	1.26
39	Trahan	295.8	61.8	1.23
40	Pappas	309.4	50.6	1.17

Source: Wiseman, F., et al. "A New Method for Ranking Total Driving Performance on the PGA Tour," *The Sport Journal,* Vol. 10, No. 1, Winter 2007 (Table 2).

2.51 Semester hours taken by CPA candidates. In order to become a certified public accountant (CPA), you must pass the Uniform CPA Exam. Many states require a minimum of 150 semester hours of college education before a candidate can sit for the CPA exam. However, traditionally, colleges only require 128 semester hours for an undergraduate degree. A study of whether the "extra" 22 hours of college credit is warranted for CPA candidates was published in the *Journal of Accounting and Public Policy* (Spring 2002). For one aspect of the study, researchers sampled over 100,000 first-time candidates for the CPA exam and recorded the total semester hours of college credit for each candidate. The mean and median for the data set were 141.31 and 140 hours, respectively. Interpret these values. Make a statement about the type of skewness, if any, that exists in the distribution of total semester hours.

2.52 Symmetric or skewed? Would you expect the data sets described below to possess relative frequency distributions that are symmetric, skewed to the right, or skewed to the left? Explain.

a. The salaries of all persons employed by a large university

b. The grades on an easy test

c. The grades on a difficult test

d. The amounts of time students in your class studied last week

e. The ages of automobiles on a used-car lot

f. The amounts of time spent by students on a difficult examination (maximum time is 50 minutes)

Applying the Concepts—Advanced

2.53 Time in bankruptcy. Refer to the *Financial Management* (Spring 1995) study of prepackaged bankruptcy filings, Exercise 2.30 (p. 52). Recall that each of 49 firms that negotiated a reorganization plan with its creditors prior to filing for bankruptcy was classified in one of three categories: joint exchange offer with prepack, prepack solicitation only, and no prefiling vote held. Consider the quantitative variable length of time in bankruptcy (months) saved in the **BANKRUPT** file. Is it reasonable to use a single number (e.g., mean or median) to describe the center of the time-in-bankruptcy distributions? Or should three "centers" be calculated, one for each of the three categories of prepack firms? Explain.

2.54 Active nuclear power plants. The U.S. Energy Information Administration monitors all nuclear power plants operating in the United States. The table on the next page lists the number of active nuclear power plants operating in each of a sample of 20 states. (These data are saved in the **NUCLEAR** file.)

a. Find the mean, median, and mode of this data set.

b. Eliminate the largest value from the data set and repeat part **a.** What effect does dropping this measurement have on the measures of central tendency found in part **a?**

c. Arrange the 20 values in the table from lowest to highest. Next, eliminate the lowest two values and the highest two values from the data set and find the mean of the remaining data values. The result is called a *10% trimmed mean,* because it is calculated after removing the highest 10% and the lowest 10% of the data values. What advantages does a trimmed mean have over the regular arithmetic mean?

State	Number of Power Plants	State	Number of Power Plants
Alabama	5	New Hampshire	1
Arizona	3	New York	6
California	4	North Carolina	5
Florida	5	Ohio	3
Georgia	4	Pennsylvania	9
Illinois	11	South Carolina	7
Kansas	1	Tennessee	3
Louisiana	2	Texas	4
Massachusetts	1	Vermont	1
Mississippi	3	Wisconsin	3

Source: Statistical Abstract of the United States, 2005 (Table 906). U.S. Energy Information Administration, Electric Power Annual.

2.55 Professional athletes' salaries. The salaries of superstar professional athletes receive much attention in the media. The multimillion-dollar long-term contract is now commonplace among this elite group. Nevertheless, rarely does a season pass without negotiations between one or more of the players' associations and team owners for additional salary and fringe benefits for *all* players in their particular sports.

a. If a players' association wanted to support its argument for higher "average" salaries, which measure of central tendency do you think it should use? Why?

b. To refute the argument, which measure of central tendency should the owners apply to the players' salaries? Why?

2.5 Numerical Measures of Variability

Measures of central tendency provide only a partial description of a quantitative data set. The description is incomplete without a **measure of the variability,** or **spread,** of the data set. Knowledge of the data's variability along with its center can help us visualize the shape of a data set as well as its extreme values.

For example, suppose we are comparing the profit margin per construction job (as a percentage of the total bid price) for 100 construction jobs for each of two cost estimators working for a large construction company. The histograms for the two sets of 100 profit margin measurements are shown in Figure 2.18. If you examine the two histograms, you will notice that both data sets are symmetric with equal modes, medians, and means. However, cost estimator A (Figure 2.18a) has profit margins spread with almost equal relative frequency over the measurement classes, while cost estimator B (Figure 2.18b) has profit margins clustered about the center of the distribution. Thus, estimator B's profit margins are *less variable* than estimator A's. Consequently, you can see that we need a measure of variability as well as a measure of central tendency to describe a data set.

Perhaps the simplest measure of the variability of a quantitative data set is its *range*.

The **range** of a quantitative data set is equal to the largest measurement minus the smallest measurement.

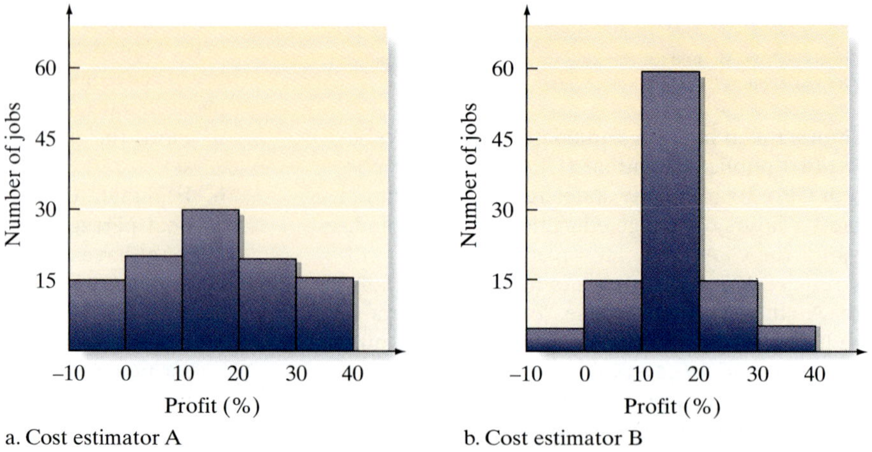

Figure 2.18
Profit margin histograms for two cost estimators

Table 2.5 Two Hypothetical Data Sets

	Sample 1	Sample 2
Measurements	1, 2, 3, 4, 5	2, 3, 3, 3, 4
Mean	$\bar{x} = \frac{1+2+3+4+5}{5} = \frac{15}{5} = 3$	$\bar{x} = \frac{2+3+3+3+4}{5} = \frac{15}{5} = 3$
Deviations of measurement values from $\bar{x}$	$(1-3), (2-3), (3-3), (4-3), (5-3)$, or $-2, -1, 0, 1, 2$	$(2-3), (3-3), (3-3), (3-3), (4-3)$, or $-1, 0, 0, 0, 1$

The range is easy to compute and easy to understand, but it is a rather insensitive measure of data variation when the data sets are large. This is because two data sets can have the same range and be vastly different with respect to data variation. This phenomenon is demonstrated in Figure 2.18. Although the ranges are equal and all central tendency measures are the same for these two symmetric data sets, there is an obvious difference between the two sets of measurements. The difference is that estimator B's profit margins tend to be more stable—that is, to pile up or to cluster about the center of the data set. In contrast, estimator A's profit margins are more spread out over the range, indicating a higher incidence of some high profit margins but also a greater risk of losses. Thus, even though the ranges are equal, the profit margin record of estimator A is more variable than that of estimator B, indicating a distinct difference in their cost-estimating characteristics.

Let's see if we can find a measure of data variation that is more sensitive than the range. Consider the two samples in Table 2.5: Each has five measurements. (We have ordered the numbers for convenience.)

Note that both samples have a mean of 3 and that we have also calculated the distance and direction, or *deviation,* between each measurement and the mean. What information do these deviations contain? If they tend to be large in magnitude, as in sample 1, the data are spread out, or highly variable, as shown in Figure 2.19a. If the deviations are mostly small, as in sample 2, the data are clustered around the mean, $\bar{x}$, and therefore do not exhibit much variability, as shown in Figure 2.19b. You can see that these deviations provide information about the variability of the sample measurements.

The next step is to condense the information in these deviations into a single numerical measure of variability. Averaging the deviations from $\bar{x}$ won't help because the negative and positive deviations cancel; that is, the sum of the deviations (and thus the average deviation) is always equal to zero.

Two methods come to mind for dealing with the fact that positive and negative deviations from the mean cancel. The first is to treat all the deviations as though they were positive, ignoring the sign of the negative deviations. We won't pursue this line of thought because the resulting measure of variability (the mean of the absolute values of the deviations) presents analytical difficulties beyond the scope of this text. A second method of eliminating the minus signs associated with the deviations is to square them. The quantity we can calculate from the squared deviations will provide a meaningful description of the variability of a data set and presents fewer analytical difficulties in inference making.

To use the squared deviations calculated from a data set, we first calculate the *sample variance*.

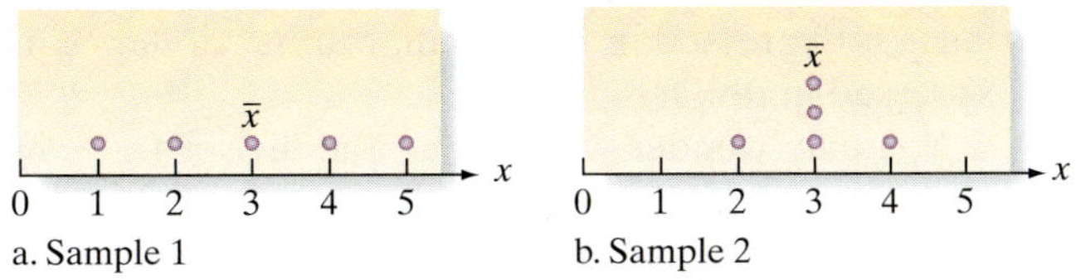

Figure 2.19
Dot plots for two data sets

The **sample variance** for a sample of n measurements is equal to the sum of the squared deviations from the mean divided by $(n - 1)$. In symbols, using s^2 to represent the sample variance,

$$s^2 = \frac{\sum_{i=1}^{n}(x_i - \bar{x})^2}{n - 1}$$

Note: A shortcut formula for calculating s^2 is

$$s^2 = \frac{\sum_{i=1}^{n} x_i^2 - \frac{\left(\sum_{i=1}^{n} x_i\right)^2}{n}}{n - 1}$$

Referring to the two samples in Table 2.5, you can calculate the variance for sample 1 as follows:

$$s^2 = \frac{(1-3)^2 + (2-3)^2 + (3-3)^2 + (4-3)^2 + (5-3)^2}{5-1}$$

$$= \frac{4 + 1 + 0 + 1 + 4}{4} = 2.5$$

The second step in finding a meaningful measure of data variability is to calculate the *standard deviation* of the data set.

The **sample standard deviation,** s, is defined as the positive square root of the sample variance, s^2. Thus, $s = \sqrt{s^2}$.

The population variance, denoted by the symbol σ^2 (sigma squared), is the average of the squared distances of the measurements on *all* units in the population from the mean, μ, and σ (sigma) is the square root of this quantity. Because we never really compute σ^2 or σ from the population (the object of sampling is to the avoid this costly procedure), we simply denote these two quantities by their respective symbols.

Symbols for Variance and Standard Deviation

s^2 = Sample variance

s = Sample standard deviation

σ^2 = Population variance

σ = Population standard deviation

Notice that, unlike the variance, the standard deviation is expressed in the original units of measurement. For example, if the original measurements are in dollars, the variance is expressed in the peculiar units "dollars squared," but the standard deviation is expressed in dollars.

You may wonder why we use the divisor $(n - 1)$ instead of n when calculating the sample variance. Wouldn't using n be more logical so that the sample variance would be the average squared deviation from the mean? The trouble is that using n tends to produce an underestimate of the population variance, σ^2, so we use $(n - 1)$ in the denominator to provide the appropriate correction for this tendency.* Because sample statistics such as s^2 are primarily used to estimate population parameters such as σ^2, $(n - 1)$ is preferred to n when defining the sample variance.

Example 2.9
Computing Measures of Variation

Problem Calculate the variance and standard deviation of the following sample: 2, 3, 3, 3, 4.

Solution If you calculate the values of s and s^2 using the formula in the box at the top of page 66, you first need to compute $\bar{x}$. From the table below, we see that $\Sigma x = 15$. Thus, $\bar{x} = \frac{\Sigma x}{n} = \frac{15}{5} = 3$. Now, for each measurement, find $(x - \bar{x})$ and $(x - \bar{x})^2$, as shown.

x	$(x - \bar{x})$	$(x - \bar{x})^2$
2	−1	1
3	0	0
3	0	0
3	0	0
4	1	1
$\Sigma x = 15$		$\Sigma(x - \bar{x})^2 = 2$

Then we use†

$$s^2 = \frac{\sum_{i=1}^{n}(x - \bar{x})^2}{n - 1} = \frac{2}{5 - 1} = \frac{2}{4} = .5$$

$$s = \sqrt{.5} = .71$$

Look Back As the sample size n increases, these calculations can become very tedious. As the next example shows, we can use the computer to find and s.

Now Work Exercise 2.57

Example 2.10
Measures of Variation on the Computer

Problem Use the computer to find the sample variance s^2 and the sample standard deviation s for the 50 companies' percentages of revenues spent on R&D.

Solution The Excel printout describing the R&D percentage data is reproduced in Figure 2.20. The variance and standard deviation, highlighted on the printout, are $s^2 = 3.922792$ and $s = 1.980604$.

	A	B
1	RDPct	
2		
3	Mean	8.492
4	Standard Error	0.2801
5	Median	8.05
6	Mode	6.9
7	Standard Deviation	1.980604
8	Sample Variance	3.922792
9	Kurtosis	0.419288
10	Skewness	0.854601
11	Range	8.3
12	Minimum	5.2
13	Maximum	13.5
14	Sum	424.6
15	Count	50
16		

Figure 2.20
Reproduction of Excel numerical descriptive measures for 50 R&D percentages

You now know that the standard deviation measures the variability of a set of data. The larger the standard deviation, the more variable the data. The smaller the standard deviation, the less variable the data. But how can we practically interpret the standard deviation and use it to make inferences? This is the topic of Section 2.6

*Appropriate here means that s^2 with the divisor $(n - 1)$ is an *unbiased estimator* of σ^2. We define and discuss *unbiasedness* of estimators in Chapter 4.

†When calculating s^2, how many decimal places should you carry? Although there are no rules for the rounding procedure, it's reasonable to retain twice as many decimal places in s^2 as you ultimately wish to have in s. If you wish to calculate s to the nearest hundredth (two decimal places), for example, you should calculate s^2 to the nearest ten-thousandth (four decimal places).

Activity 2.2: *Keep the Change:* Measures of Central Tendency and Variability

In this activity, we continue our study of the Bank of America *Keep the Change* savings program by looking at the measures of central tendency and variability for the three data sets collected in Activity 1.1 (p. 16).

1. Before performing any calculations, explain why you would expect greater variability in the data set *Purchase Totals* than in *Amounts Transferred.* Then find the mean and median of each of these two data sets. Are the mean and median essentially the same for either of these sets? If so, which one? Can you offer an explanation for these results?
2. Make a histogram for each of the data sets *Amounts Transferred* and *Bank Matching.* Describe any properties of the data that are evident in the histograms. Explain why it is more likely that *Bank Matching* is skewed to the right than *Amounts Transferred*. Based on your data and histogram, how concerned does Bank of America need to be about matching the maximum amount of $250 for its customers who are college students?
3. Form a fourth data set *Mean Amounts Transferred* by collecting the mean of the data set *Amounts Transferred* for each student in your class. Before performing any calculations, inspect the new data and describe any trends that you notice. Then find the mean and standard deviation of *Mean Amounts Transferred.* How close is the mean to $0.50? Without performing further calculations, determine whether the standard deviation of *Amounts Transferred* is less than or greater than the standard deviation of *Mean Amounts Transferred*. Explain.

Keep your results from this activity for use in other activities.

Exercises 2.56–2.70

Learning the Mechanics

2.56 Answer the following questions about variability of data sets:

a. What is the primary disadvantage of using the range to compare the variability of data sets?

b. Describe the sample variance using words rather than a formula. Do the same with the population variance.

c. Can the variance of a data set ever be negative? Explain. Can the variance ever be smaller than the standard deviation? Explain.

2.57 NW Calculate the range, variance, and standard deviation for the following samples:

a. 4, 2, 1, 0, 1

b. 1, 6, 2, 2, 3, 0, 3

c. 8, −2, 1, 3, 5, 4, 4, 1, 3, 3

d. 0.2, 0, 0, −1, 1, −2, 1, 0, −1, 1, −1, 0, −3, −2, −1, 0, 1

2.58 Calculate the variance and standard deviation for samples where

a. $n = 10, \Sigma x^2 = 84, \Sigma x = 20$

b. $n = 40, \Sigma x^2 = 380, \Sigma x = 100$

c. $n = 20, \Sigma x^2 = 18, \Sigma x = 17$

Applet Exercise 2.4

Use the applet entitled *Standard Deviation* to find the standard deviation of each of the four data sets in Exercise 2.57. For each data set, set the lower limit to a number less than all of the data, set the upper limit to a number greater than all of the data, and then click on *Update*. Click on the approximate location of each data item on the number line. You can get rid of a point by dragging it to the trash can. To clear the graph between data sets, simply click on the trash can.

a. Compare the standard deviations generated by the applet to those you calculated by hand in Exercise 2.57. If there are differences, explain why the applet might give values slightly different from the hand calculations.

b. Despite providing a slightly different value of the standard deviation of a data set, describe some advantages of using the applet.

2.59 Compute $\bar{x}$, s^2, and s for each of the following data sets. If appropriate, specify the units in which your answer is expressed.

a. 3, 1, 10, 10, 4

b. 8 feet, 10 feet, 32 feet, 5 feet

c. −1, −4, −3, 1, −4, −4

d. 1/5 ounce, 1/5 ounce, 1/5 ounce, 2/5 ounce, 1/5 ounce, 4/5 ounce,

2.60 Calculate the range, variance, and standard deviation for the following samples:

a. 39, 42, 40, 37, 41

b. 100, 4, 7, 96, 80, 3, 1, 10, 2

c. 100, 4, 7, 30, 80, 30, 42, 2

2.61 Using only integers between 0 and 10, construct two data sets with at least 10 observations each that have the same range but different means. Construct a dot plot for each of your data sets, and mark the mean of each data set on its dot diagram.

2.62 Using only integers between 0 and 10, construct two data sets with at least 10 observations each so that the two sets have the same mean but different variances. Construct dot plots for each of your data sets and mark the mean of each data set on its dot diagram.

2.63 Consider the following sample of five measurements: 2, 1, 1, 0, 3.

a. Calculate the range, s^2, and s.

b. Add 3 to each measurement and repeat part **a.**

c. Subtract 4 from each measurement and repeat part **a.**

d. Considering your answers to parts **a, b,** and **c,** what seems to be the effect on the variability of a data set by adding the same number to or subtracting the same number from each measurement?

Applet Exercise 2.5

Use the applet *Standard Deviation* to study the effect that multiplying or dividing each number in a data set by the same number has on the standard deviation. Begin by setting appropriate limits and plotting the given data on the number line provided in the applet.

0 1 1 1 2 2 3 4

a. Record the standard deviation. Then multiply each data item by 2, plot the new data items, and record the standard deviation. Repeat the process, first multiplying each of the original data items by 3 and then by 4. Describe what is happening to the standard deviation as the data items are multiplied by higher numbers. Divide each standard deviation by the standard deviation of the original data set. Do you see a pattern? Explain.

b. Divide each of the original data items by 2, plot the new data, and record the standard deviation. Repeat the process, first dividing each of the original data items by 3 and then by 4. Describe what is happening to the standard deviation as the data items are divided by higher numbers. Divide each standard deviation by the standard deviation of the original data set. Do you see a pattern? Explain.

c. Using your results from parts **a** and **b**, describe what happens to the standard deviation of a data set when each of the data items in the set is multiplied or divided by a fixed number n. Experiment by repeating parts **a** and **b** for other data sets if you need to.

Applet Exercise 2.6

Use the applet *Standard Deviation* to study the effect that an extreme value has on the standard deviation. Begin by setting appropriate limits and plotting the given data on the number line provided in the applet.

0 6 7 7 8 8 8 9 9 10

a. Record the standard deviation. Replace the extreme value of 0 with 2, then 4, and then 6. Record the standard deviation each time. Describe what is happening to the standard deviation as 0 is replaced by higher numbers.

b. How would the standard deviation of the data set compare to the original standard deviation if the 0 were replaced by 16? Explain.

Applying the Concepts—Basic

2.64 Spending by banks following acquisitions. Refer to *The Nilson Report* (Oct. 2008) study of the amount charged to credit and debit cards issued by U.S. banks with recent acquisitions, Exercise 2.44 (p. 61). The data for the five top-ranked banks, reproduced in the table, are saved in the **BANKCHARGE** file.

U.S. Bank/ Card Issuer	Amount Charged ($ billions)
Bank of America	462.20
JPMorgan Chase	458.84
American Express	445.32
Citigroup	240.58
Wells Fargo	192.68

Source: The Nilson Report, No. 912, October 2008 (p. 8).

a. Find the range of the data for the five top-ranked banks. Give the units of measurement for the range.

b. Find the variance of the data for the five top-ranked banks. If possible, give the units of measurement for the variance.

c. Find the standard deviation of the data for the five top-ranked banks. Give the units of measurement for the standard deviation.

2.65 U.S. wine export markets. Refer to the data on the top 30 U.S. export markets for sparkling wines, compiled by the Center for International Trade Development (CITD), given in Exercise 2.45 (p. 61). (These data are saved in the **WINEEXPORTS** file.) The Minitab descriptive statistics printout for the amount exported (thousands of dollars) and three-year percentage change for the 30 countries is reproduced below.

Descriptive Statistics: EXPORT, CHANGE

Variable	N	N*	Mean	StDev	Minimum	Q1	Median	Q3	Maximum	IQR
EXPORT	30	0	653	1113	70	105	231	523	4952	418
CHANGE	28	2	481	1098	-49	21	156	499	5750	478

a. Use the information on the printout to find the range of the amount exported.

b. Locate the standard deviation of the amount exported on the printout.

c. Use the result, part **b,** to find the variance of the amount exported.

2.66 Is honey a cough remedy? Refer to the *Archives of Pediatrics and Adolescent Medicine* (Dec. 2007) study of honey as a remedy for coughing, Exercises 2.28 and 2.47 (pp. 51, 62). The coughing improvement scores (as determined by the children's parents) for the patients in the over-the-counter cough medicine dosage (DM) group, honey dosage group, and control group are reproduced in the accompanying table. (These data are saved in the **HONEYCOUGH** file.)

a. Find the standard deviation of the improvement scores for the honey dosage group.

b. Find the standard deviation of the improvement scores for the DM dosage group.

c. Find the standard deviation of the improvement scores for the control group.

d. Based on the results, parts **a–c,** which group appears to have the most variability in coughing improvement scores? The least variability?

Honey Dosage:	12	11	15	11	10	13	10	4	15	16	9			
	14	10	6	10	8	11	12	12	8					
	12	9	11	15	10	15	9	13	8	12	10			
	8	9	5	12										
DM Dosage:	4	6	9	4	7	7	7	9	12	10	11	6		
	3	4	9	12	7	6	8	12	12	4	12			
	13	7	10	13	9	4	4	10	15	9				
No Dosage (Control):	5	8	6	1	0	8	12	8	7	7	1	6		
	7	7	12	7	9	7	9	5	11	9	5			
	6	8	8	6	7	10	9	4	8	7	3	1	4	3

Source: Paul, I. M., et al. "Effect of honey, dextromethorphan, and no treatment on nocturnal cough and sleep quality for coughing children and their parents," *Archives of Pediatrics and Adolescent Medicine,* Vol. 161, No. 12, Dec. 2007 (data simulated).

Applying the Concepts—Intermediate

2.67 Size of diamonds sold at retail. Refer to Exercise 2.25 (p. 50) and the *Journal of Statistics Education* data on diamonds saved in the **DIAMONDS** file. Consider the data on the number of carats for each of the 308 diamonds.

a. Find the range of the data set.

b. Find the variance of the data set.

c. Find the standard deviation of the data set.

d. Which measure of variation best describes the spread of the 308 carat values? Explain.

2.68 **Most powerful women in America.** Refer to Exercise 2.48 (p. 62) and *Fortune*'s (Oct. 16, 2008) list of the 50 most powerful women in America. The data are stored in the **WPOWER50** file.

a. Find the range of the ages for these 50 women.

b. Find the variance of the ages for these 50 women.

c. Find the standard deviation of the ages for these 50 women.

d. Suppose the standard deviation of the ages of the most powerful women in Europe is 10 years. For which location, the United States or Europe, is the age data more variable?

2.69 **Active nuclear power plants.** Refer to Exercise 2.54 (p. 63) and the U.S. Energy Information Administration's data on the number of nuclear power plants operating in each of 20 states. The data are saved in the **NUCLEAR** file.

a. Find the range, variance, and standard deviation of this data set.

b. Eliminate the largest value from the data set and repeat part **a.** What effect does dropping this measurement have on the measures of variation found in part **a**?

c. Eliminate the smallest and largest value from the data set and repeat part **a.** What effect does dropping both of these measurements have on the measures of variation found in part **a**?

Applying the Concepts—Advanced

2.70 **Estimating production time.** A widely used technique for estimating the length of time it takes workers to produce a product is the **time study.** In a time study, the task to be studied is divided into measurable parts, and each is timed with a stopwatch or filmed for later analysis. For each worker, this process is repeated many times for each subtask. Then the average and standard deviation of the time required to complete each subtask are computed for each worker. A worker's overall time to complete the task under study is then determined by adding his or her subtask-time averages (Gaither and Frazier, *Operations Management,* 2001). The data (in minutes) given in the table are the result of a time study of a production operation involving two subtasks. (These data are saved in the **TIMESTUDY** file.)

a. Find the overall time it took each worker to complete the manufacturing operation under study.

b. For each worker, find the standard deviation of the seven times for subtask 1.

c. In the context of this problem, what are the standard deviations you computed in part **b** measuring?

d. Repeat part **b** for subtask 2.

e. If you could choose workers similar to A or workers similar to B to perform subtasks 1 and 2, which type would you assign to each subtask? Explain your decisions on the basis of your answers to parts **a–d.**

	Worker A		Worker B	
Repetition	Subtask 1	Subtask 2	Subtask 1	Subtask 2
1	30	2	31	7
2	28	4	30	2
3	31	3	32	6
4	38	3	30	5
5	25	2	29	4
6	29	4	30	1
7	30	3	31	4

2.6 Interpreting the Standard Deviation

We've seen that if we are comparing the variability of two samples selected from a population, the sample with the larger standard deviation is the more variable of the two. Thus, we know how to interpret the standard deviation on a relative or comparative basis, but we haven't explained how it provides a measure of variability for a single sample.

To understand how the standard deviation provides a measure of variability of a data set, consider a specific data set and answer the following questions: How many measurements are within 1 standard deviation of the mean? How many measurements are within 2 standard deviations? For a specific data set, we can answer these questions by counting the number of measurements in each of the intervals. However, if we are interested in obtaining a general answer to these questions, the problem is more difficult.

Tables 2.6 and 2.7 give two sets of answers to the questions of how many measurements fall within 1, 2, and 3 standard deviations of the mean. The first, which applies to *any* set of data, is derived from a theorem proved by the Russian mathematician P. L. Chebyshev. The second, which applies to **mound-shaped,** symmetric distributions of data (where the mean, median, and mode are all about the same), is based upon empirical evidence that has accumulated over the years. However, the percentages given for the intervals in Table 2.7 provide remarkably good approximations even when the distribution of the data is slightly skewed or asymmetric. Note that both rules apply to either population data sets or sample data sets.

Table 2.6 Interpreting the Standard Deviation: Chebyshev's Rule

Chebyshev's Rule applies to any data set, regardless of the shape of the frequency distribution of the data.

a. No useful information is provided on the fraction of measurements that fall within 1 standard deviation of the mean [i.e., within the interval $(\bar{x} - s, \bar{x} + s)$ for samples and $(\mu - \sigma, \mu + \sigma)$ for populations].

b. At least $3/4$ will fall within 2 standard deviations of the mean [i.e., within the interval $(\bar{x} - 2s, \bar{x} + 2s)$ for samples and $(\mu - 2\sigma, \mu + 2\sigma)$ for populations].

c. At least $8/9$ of the measurements will fall within 3 standard deviations of the mean [i.e., within the interval $(\bar{x} - 3s, \bar{x} + 3s)$ for samples and $(\mu - 3\sigma, \mu + 3\sigma)$ for populations].

d. Generally, for any number k greater than 1, at least $(1 - 1/k^2)$ of the measurements will fall within k standard deviations of the mean [i.e., within the interval $(\bar{x} - ks, \bar{x} + ks)$ for samples and $(\mu - k\sigma, \mu + k\sigma)$ for populations].

Table 2.7 Interpreting the Standard Deviation: The Empirical Rule

The **Empirical Rule** is a rule of thumb that applies to data sets with frequency distributions that are mound-shaped and symmetric, as shown below.

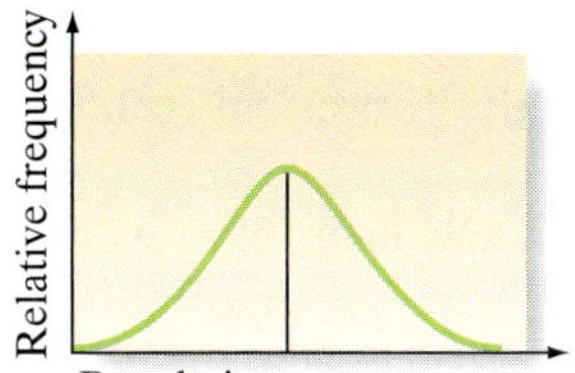

a. Approximately 68% of the measurements will fall within 1 standard deviation of the mean [i.e., within the interval $(\bar{x} - s, \bar{x} + s)$ for samples and $(\mu - \sigma, \mu + \sigma)$ for populations].

b. Approximately 95% of the measurements will fall within 2 standard deviations of the mean [i.e., within the interval $(\bar{x} - 2s, \bar{x} + 2s)$ for samples and $(\mu - 2\sigma, \mu + 2\sigma)$ for populations].

c. Approximately 99.7% (essentially all) of the measurements will fall within 3 standard deviations of the mean [i.e., within the interval $(\bar{x} - 3s, \bar{x} + 3s)$ for samples and $(\mu - 3\sigma, \mu + 3\sigma)$ for populations].

BIOGRAPHY PAFNUTY L. CHEBYSHEV (1821–1894)

The Splendid Russian Mathematician

P. L. Chebyshev was educated in mathematical science at Moscow University, eventually earning his master's degree. Following his graduation, Chebyshev joined St. Petersburg (Russia) University as a professor, becoming part of the well-known "Petersburg mathematical school." It was here that Chebyshev proved his famous theorem about the probability of a measurement being within *k* standard deviations of the mean (Table 2.6). His fluency in French allowed him to gain international recognition in probability theory. In fact, Chebyshev once objected to being described as a "splendid Russian mathematician," saying he surely was a "worldwide mathematician." One student remembered Chebyshev as "a wonderful lecturer" who "was always prompt for class," and "as soon as the bell sounded, he immediately dropped the chalk, and, limping, left the auditorium." ■

Example 2.11 Interpreting the Standard Deviation

Problem The 50 companies' percentages of revenues spent on R&D are repeated in Table 2.8. We have previously shown (see Figure 2.20, p. 67) that the mean and standard deviation of these data (rounded) are 8.49 and 1.98, respectively. Calculate the fraction of these measurements that lie within the intervals $\bar{x} \pm s$, $\bar{x} \pm 2s$, and $\bar{x} \pm 3s$ and compare the results with those predicted in Tables 2.6 and 2.7.

Solution We first form the interval

$$(\bar{x} - s, \bar{x} + s) = (8.49 - 1.98, 8.49 + 1.98) = (6.51, 10.47)$$

A check of the measurements reveals that 34 of the 50 measurements, or 68%, are within 1 standard deviation of the mean.

Table 2.8 R&D Percentages for 50 Companies

13.5	9.5	8.2	6.5	8.4	8.1	6.9	7.5	10.5	13.5
7.2	7.1	9.0	9.9	8.2	13.2	9.2	6.9	9.6	7.7
9.7	7.5	7.2	5.9	6.6	11.1	8.8	5.2	10.6	8.2
11.3	5.6	10.1	8.0	8.5	11.7	7.1	7.7	9.4	6.0
8.0	7.4	10.5	7.8	7.9	6.5	6.9	6.5	6.8	9.5

Data Set: R&D

The next interval of interest,

$$(\bar{x} - 2s, \bar{x} + 2s) = (8.49 - 3.96, 8.49 + 3.96) = (4.53, 12.45),$$

contains 47 of the 50 measurements, or 94%.

Finally, the 3-standard-deviation interval around $\bar{x}$,

$$(\bar{x} - 3s, \bar{x} + 3s) = (8.49 - 5.94, 8.49 + 5.94) = (2.55, 14.43),$$

contains all, or 100%, of the measurements.

In spite of the fact that the distribution of these data is skewed to the right (see Figure 2.10, p. 44), the percentages within 1, 2, and 3 standard deviations (68%, 94%, and 100%) agree very well with the approximations of 68%, 95%, and 99.7% given by the Empirical Rule (Table 2.7).

Look Back You will find that unless the distribution is extremely skewed, the mound-shaped approximations will be reasonably accurate. Of course, no matter what the shape of the distribution, Chebyshev's Rule (Table 2.6) assures that at least 75% and at least 89% of the measurements will lie within 2 and 3 standard deviations of the mean, respectively.

Now Work Exercise 2.74

Example 2.12
Check on the Calculation of *s*

Problem Chebyshev's Rule and the Empirical Rule are useful as a check on the calculation of the standard deviation. For example, suppose we calculated the standard deviation for the R&D percentages (Table 2.8) to be 3.92. Are there any "clues" in the data that enable us to judge whether this number is reasonable?

Solution The range of the R&D percentages in Table 2.8 is $13.5 - 5.2 = 8.3$. From Chebyshev's Rule and the Empirical Rule we know that most of the measurements (approximately 95% if the distribution is mound-shaped) will be within 2 standard deviations of the mean. And, regardless of the shape of the distribution and the number of measurements, almost all of them will fall within 3 standard deviations of the mean. Consequently, we would expect the range of the measurements to be between 4 (i.e., $\pm 2s$) and 6 (i.e., $\pm 3s$) standard deviations in length (see Figure 2.21).

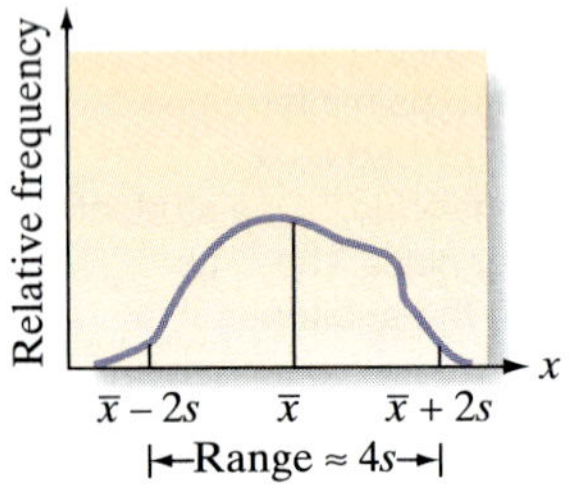

Figure 2.21
The relation between the range and the standard deviation

For the R&D data, this means that *s* should fall between

$$\frac{\text{Range}}{6} = \frac{8.3}{6} = 1.38 \quad \text{and} \quad \frac{\text{Range}}{4} = \frac{8.3}{4} = 2.08$$

In particular, the standard deviation should not be much larger than $1/4$ of the range, particularly for the data set with 50 measurements. Thus, we have reason to believe that the calculation of 3.92 is too large. A check of our work reveals that 3.92 is the variance s^2, not the standard deviation *s* (see Example 2.10). We "forgot" to take the square root (a common error); the correct value is $s = 1.98$. Note that this value is between $1/6$ and $1/4$ of the range.

Look Back In examples and exercises we'll sometimes use $s \approx \text{range}/4$ to obtain a crude, and usually conservatively large, approximation for *s*. However, we stress that this is no substitute for calculating the exact value of *s* when possible.

Now Work Exercise 2.75

In the next example, we use the concepts in Chebyshev's Rule and the Empirical Rule to build the foundation for statistical inference making.

Example 2.13
Making a Statistical Inference

Problem A manufacturer of automobile batteries claims that the average length of life for its grade A battery is 60 months. However, the guarantee on this brand is for just 36 months. Suppose the standard deviation of the life length is known to be 10 months, and the frequency distribution of the life-length data is known to be mound-shaped.

a. Approximately what percentage of the manufacturer's grade A batteries will last more than 50 months, assuming the manufacturer's claim is true?

b. Approximately what percentage of the manufacturer's batteries will last less than 40 months, assuming the manufacturer's claim is true?

c. Suppose your battery lasts 37 months. What could you infer about the manufacturer's claim?

Solution If the distribution of life length is assumed to be mound-shaped with a mean of 60 months and a standard deviation of 10 months, it would appear as shown in Figure 2.22. Note that we can take advantage of the fact that mound-shaped distributions are (approximately) symmetric about the mean, so that the percentages given by the Empirical Rule can be split equally between the halves of the distribution on each side of the mean.

For example, because approximately 68% of the measurements will fall within 1 standard deviation of the mean, the distribution's symmetry implies that approximately $^1/_2(68\%) = 34\%$ of the measurements will fall between the mean and 1 standard deviation on each side. This concept is illustrated in Figure 2.22. The figure also shows that 2.5% of the measurements lie beyond 2 standard deviations in each direction from the mean. This result follows from the fact that if approximately 95% of the measurements fall within 2 standard deviations of the mean, then about 5% fall outside 2 standard deviations; if the distribution is approximately symmetric, then about 2.5% of the measurements fall beyond 2 standard deviations on each side of the mean.

a. It is easy to see in Figure 2.22 that the percentage of batteries lasting more than 50 months is approximately 34% (between 50 and 60 months) plus 50% (greater than 60 months). Thus, approximately 84% of the batteries should have life length exceeding 50 months.

b. The percentage of batteries that last less than 40 months can also be easily determined from Figure 2.22. Approximately 2.5% of the batteries should fail prior to 40 months, assuming the manufacturer's claim is true.

c. If you are so unfortunate that your grade A battery fails at 37 months, you can make one of two inferences: either your battery was one of the approximately 2.5% that fail prior to 40 months, or something about the manufacturer's claim is not true. Because the chances are so small that a battery fails before 40 months, you would have good reason to have serious doubts about the manufacturer's claim. A mean

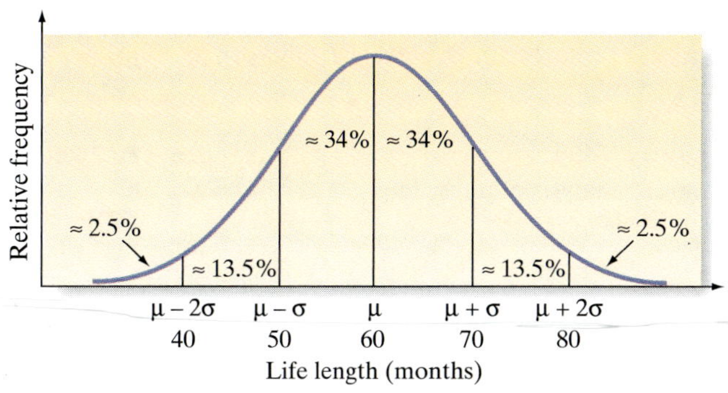

Figure 2.22
Battery life-length distribution: Manufacturer's claim assumed true

smaller than 60 months and/or a standard deviation longer than 10 months would both increase the likelihood of failure prior to 40 months.*

Look Back The approximations given in Figure 2.22 are more dependent on the assumption of a mound-shaped distribution than those given by the Empirical Rule (Table 2.7) because the approximations in Figure 2.22 depend on the (approximate) symmetry of the mound-shaped distribution. We saw in Example 2.11 that the Empirical Rule can yield good approximations even for skewed distributions. This will *not* be true of the approximations in Figure 2.22; the distribution *must* be mound-shaped and approximately symmetric.

Example 2.13 is our initial demonstration of the statistical inference-making process. At this point you should realize that we'll use sample information (in Example 2.13, your battery's failure at 37 months) to make inferences about the population (in Example 2.13, the manufacturer's claim about the life length for the population of all batteries). We'll build on this foundation as we proceed.

Statistics IN Action Revisited — Interpreting Numerical Descriptive Measures

We return to the analysis of length of time in practice for two groups of University Community Hospital physicians—those who indicate they are willing to use ethics consultation and those who would not use ethics consultation. Recall that the researchers propose that nonusers of ethics consultation will be more experienced than users. The Minitab descriptive statistics printout for the **ETHICS** data is displayed in Figure SIA2.5, with the means and standard deviations highlighted.

The sample mean for ethics consultation (EC) nonusers is 16.43, and the mean for EC users is 14.18. Our interpretation is that nonusers have slightly more experience (16.43 years, on average) than users (14.18 years, on average).

To interpret the standard deviation, we substitute into the formula, mean ± 2(standard deviation), to obtain the intervals:

EC Nonusers:

$$16.43 \pm 2(10.05) = 16.43 \pm 20.10 = (-3.67, 36.53)$$

EC Users:

$$14.18 \pm 2(8.95) = 14.18 \pm 17.90 = (-3.72, 32.08)$$

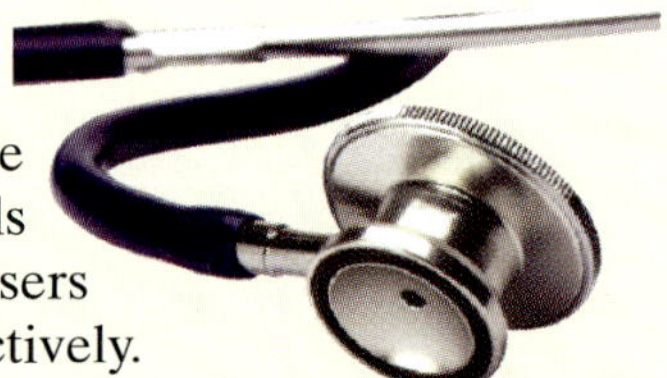

Because years of experience cannot take on a negative value, essentially the standard deviation intervals for EC nonusers and EC users are (0, 36.53) (0, 32.08), respectively.

From Chebyshev's Rule (Table 2.6), we know that at least 75% of the physicians who would not use ethics consultation will have anywhere between 0 and 36.5 years of experience. Similarly, we know that at least 75% of the EC users will have anywhere from 0 to 32.08 years of experience. Note that these ranges indicate that there is very little difference in the experience distributions of the two groups of physicians. However, if a physician on staff has 35 years of experience, it is very unlikely that the doctor would use ethics consultation because 35 years is above the mean ± 2(standard deviation) interval for EC users. Rather, the experience value of 35 is more likely to come from the distribution of years of experience for nonusers of ethics consultation.

Data Set: ETHICS

Descriptive Statistics: YRSPRAC

Variable	FUTUREUSE	N	N*	Mean	SE Mean	StDev	Variance	Minimum	Q1
YRSPRAC	NO	21	2	16.43	2.19	10.05	100.96	1.00	7.50
	YES	91	4	14.176	0.938	8.950	80.102	1.000	7.000

Variable	FUTUREUSE	Median	Q3	Maximum
YRSPRAC	NO	18.00	25.00	35.00
	YES	14.000	20.000	40.000

Figure SIA2.5
Minitab analysis of physicians' experience

*The assumption that the distribution is mound-shaped and symmetric may also be incorrect. However, if the distribution were skewed to the right, as life-length distributions often tend to be, the percentage of measurements more than 2 standard deviations *below* the mean would be even less than 2.5%.

Exercises 2.71–2.87

Learning the Mechanics

2.71 The output from a statistical software package indicates that the mean and standard deviation of a data set consisting of 200 measurements are \$1,500 and \$300, respectively.

a. What are the units of measurement of the variable of interest? Based on the units, what type of data is this: quantitative or qualitative?

b. What can be said about the number of measurements between \$900 and \$2,100? Between \$600 and \$2,400? Between \$1,200 and \$1,800? Between \$1,500 and \$2,100?

2.72 For any set of data, what can be said about the percentage of the measurements contained in each of the following intervals?

a. $\bar{x} - s$ to $\bar{x} + s$

b. $\bar{x} - 2s$ to $\bar{x} + 2s$

c. $\bar{x} - 3s$ to $\bar{x} + 3s$

2.73 For a set of data with a mound-shaped relative frequency distribution, what can be said about the percentage of the measurements contained in each of the intervals specified in Exercise 2.72?

2.74 NW The following (saved in the **LM2_74** file) is a sample of 25 measurements:

7	6	6	11	8	9	11	9	10	8	7	7	
5	9	10	7	7	7	7	9	12	10	10	8	6

a. Compute $\bar{x}$, s^2, and s for this sample.

b. Count the number of measurements in the intervals $\bar{x} \pm s$, $\bar{x} \pm 2s$, $\bar{x} \pm 3s$. Express each count as a percentage of the total number of measurements.

c. Compare the percentages found in part **b** to the percentages given by the Empirical Rule and Chebyshev's Rule.

d. Calculate the range and use it to obtain a rough approximation for s. Does the result compare favorably with the actual value for s found in part **a**?

2.75 NW Given a data set with a largest value of 760 and a smallest value of 135, what would you estimate the standard deviation to be? Explain the logic behind the procedure you used to estimate the standard deviation. Suppose the standard deviation is reported to be 25. Is this feasible? Explain.

Applying the Concepts—Basic

2.76 **Blogs for Fortune 500 firms.** Refer to the *Journal of Relationship Marketing* (Vol. 7, 2008) study of the prevalence of blogs and forums at Fortune 500 firms with both English and Chinese Web sites, Exercise 2.9 (p. 40). In a sample of firms that provide blogs and forums as marketing tools, the mean number of blogs/forums per site was 4.25, with a standard deviation of 12.02.

a. Provide an interval that is likely to contain the number of blogs/forums per site for at least 75% of the Fortune 500 firms in the sample.

b. Do you expect the distribution of the number of blogs/forums to be symmetric, skewed right, or skewed left? Explain.

2.77 **Semester hours taken by CPA candidates.** Refer to the *Journal of Accounting and Public Policy* (Spring 2002) study of 100,000 first-time candidates for the CPA exam, Exercise 2.51 (p. 63). Recall that the mean number of semester hours of college credit taken by the candidates was 141.31 hours. The standard deviation was reported to be 17.77 hours.

a. Compute the 2-standard deviation interval around the mean.

b. Make a statement about the proportion of first-time candidates for the CPA exam who have total college credit hours within the interval, part **a.**

c. For the statement, part **b,** to be true, what must be known about the shape of the distribution of total semester hours?

2.78 **Size of diamonds sold at retail.** Refer to the *Journal of Statistics Education* data on diamonds saved in the **DIAMONDS** file. In Exercise 2.49 (p. 62) you found the mean number of carats for the 308 diamonds in the data set, and in Exercise 2.67 (p. 69) you found the standard deviation. Use the mean and standard deviation to form an interval that will contain at least 75% of the carat values in the data set.

2.79 **Vehicle use of an intersection.** For each day of last year, the number of vehicles passing through a certain intersection was recorded by a city engineer. One objective of this study was to determine the percentage of days that more than 425 vehicles used the intersection. Suppose the mean for the data was 375 vehicles per day and the standard deviation was 25 vehicles.

a. What can you say about the percentage of days that more than 425 vehicles used the intersection? Assume you know nothing about the shape of the relative frequency distribution for the data.

b. What is your answer to part **a** if you know that the relative frequency distribution for the data is mound-shaped?

Applying the Concepts—Intermediate

2.80 **Sanitation inspection of cruise ships.** Refer to the Centers for Disease Control and Prevention listing of the Dec. 2008 sanitation scores for 183 cruise ships, Exercise 2.22 (p. 49). The data are saved in the **SHIPSANIT** file.

a. Find the mean and standard deviation of the sanitation scores.

b. Calculate the intervals $\bar{x} \pm s$, $\bar{x} \pm 2s$, $\bar{x} \pm 3s$.

c. Find the percentage of measurements in the data set that fall within each of the intervals, part **b.** Do these percentages agree with Chebyshev's Rule? The Empirical Rule?

2.81 **Bearing strength of concrete FRP strips.** Fiber reinforced polymer (FRP) composite materials are the standard for strengthening, retrofitting, and repairing concrete structures. Typically, FRP strips are fastened to the concrete with epoxy adhesive. Engineers at the University of Wisconsin-Madison have developed a new method of fastening the FRP strips using mechanical anchors (*Composites Fabrication Magazine,* Sep. 2004). To evaluate the new fastening method, 10 specimens of pultruded FRP strips mechanically fastened to highway bridges were tested for bearing strength. The strength measurements (recorded in mega pascal units, Mpa) are shown in the table and saved in the **FRP** file. Use the sample data to find an interval that is likely to contain the bearing strength of a pultruded FRP strip.

240.9	248.8	215.7	233.6	231.4	230.9	225.3	247.3	235.5	238.0

Source: Data are simulated from summary information provided in *Composites Fabrication Magazine,* Sep. 2004, p. 32 (Table 1).

2.82 Time in bankruptcy. Refer to the *Financial Management* (Spring 1995) study of 49 firms filing for prepackaged bankruptcy, Exercise 2.30 (p. 52). Data on the variable of interest, length of time (months) in bankruptcy for each firm, are saved in the **BANKRUPT** file.

a. Construct a histogram for the 49 bankruptcy times. Comment on whether the Empirical Rule is applicable for describing the bankruptcy time distribution for firms filing for prepackaged bankruptcy.

b. Find numerical descriptive statistics for the data set. Use this information to construct an interval that captures at least 75% of the bankruptcy times.

c. Count the number of the 49 bankruptcy times that fall within the interval, part **b,** and convert the result to a percentage. Does the result agree with Chebyshev's Rule? The Empirical Rule?

d. A firm is considering filing a prepackaged bankruptcy plan. Estimate the length of time the firm will be in bankruptcy.

2.83 Velocity of Winchester bullets. The *American Rifleman* (June 1993) reported on the velocity of ammunition fired from the FEG P9R pistol, a 9 mm gun manufactured in Hungary. Field tests revealed that Winchester bullets fired from the pistol had a mean velocity (at 15 feet) of 936 feet per second and a standard deviation of 10 feet per second. Tests were also conducted with Uzi and Black Hills ammunition.

a. Describe the velocity distribution of Winchester bullets fired from the FEG P9R pistol.

b. A bullet, brand unknown, is fired from the FEG P9R pistol. Suppose the velocity (at 15 feet) of the bullet is 1,000 feet per second. Is the bullet likely to be manufactured by Winchester? Explain.

2.84 Buy-side vs. sell-side analysts' earnings forecasts. Financial analysts are hired by investment companies to make forecasts of stock prices and recommendations about whether to buy, sell, or hold specific securities. These analysts can be categorized as either "buy-side" analysts or "sell-side" analysts based on a variety of factors, including scope of industry coverage, sources of information used, and target audience. A group of Harvard Business School professors compared earnings forecasts of buy-side and sell-side analysts (*Financial Analysts Journal,* Jul/Aug 2008). Data were collected on 3,526 forecasts made by buy-side analysts and 58,562 forecasts made by sell-side analysts, and the relative absolute forecast error was determined for each.

a. Frequency distributions for buy-side and sell-side analysts forecast errors (with the sell-side distribution superimposed over the buy-side distribution) are shown in the accompanying figure. Based on the figure, the researchers concluded "that absolute forecast errors for buy-side analysts have a higher mean and variance than those for the sell-side analysts." Do you agree? Explain.

b. The mean and standard deviation of forecast errors for both buy-side and sell-side analysts are given in the following table. For each type of analyst, provide an interval that will contain approximately 95% of the forecast errors. Compare these intervals. Which type of analyst is likely to have a relative forecast error of +2.00 or higher?

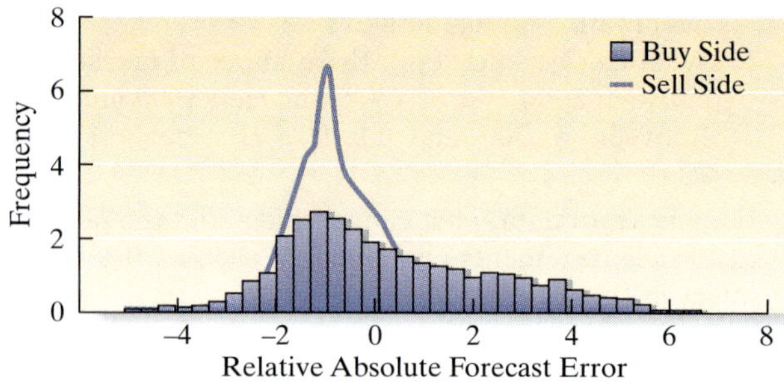

	Buy-Side Analysts	Sell-Side Analysts
Mean	0.85	−0.05
Standard Deviation	1.93	0.85

Source: Groysberg, B., Healy, P., & Chapman, C. *Financial Analysts Journal,* Vol. 64, No. 4, Jul/Aug. 2008 (Table 2).

Applying the Concepts—Advanced

2.85 Land purchase decision. A buyer for a lumber company must decide whether to buy a piece of land containing 5,000 pine trees. If 1,000 of the trees are at least 40 feet tall, the buyer will purchase the land; otherwise, he won't. The owner of the land reports that the height of the trees has a mean of 30 feet and a standard deviation of 3 feet. Based on this information, what is the buyer's decision?

2.86 Improving SAT scores. The National Education Longitudinal Survey (NELS) tracks a nationally representative sample of U.S. students from eighth grade through high school and college. Research published in *Chance* (Winter 2001) examined the Standardized Assessment Test (SAT) scores of 265 NELS students who paid a private tutor to help them improve their scores. The table summarizes the changes in both the SAT–Mathematics and SAT–Verbal scores for these students.

	SAT–Math	SAT–Verbal
Mean change in score	19	7
Standard deviation of score changes	65	49

a. Suppose one of the 265 students who paid a private tutor is selected at random. Give an interval that is likely to contain this student's change in the SAT–Math score.

b. Repeat part **a** for the SAT–Verbal score.

c. Suppose the selected student increased their score on one of the SAT tests by 140 points. Which test, the SAT–Math or SAT–Verbal, is the one most likely to have the 140-point increase? Explain.

2.87 Monitoring weights of flour bags. When it is working properly, a machine that fills 25-pound bags of flour dispenses an average of 25 pounds per fill; the standard deviation of the amount of fill is .1 pound. To monitor the performance of the machine, an inspector weighs the contents of a bag coming off the machine's conveyor belt every half hour during the day. If the contents of two consecutive bags fall more than 2 standard deviations from the mean (using the mean and standard deviation given above), the filling process is said to be out of control, and the machine is shut down briefly for adjustments. The data given in the following table (saved in the **FLOUR** file) are the weights measured by the inspector yesterday. Assume the machine is never shut down for more than 15 minutes at a time. At what times yesterday was the process shut down for adjustment? Justify your answer.

Time	Weight (pounds)
8:00 A.M.	25.10
8:30	25.15
9:00	24.81
9:30	24.75
10:00	25.00
10:30	25.05
11:00	25.23
11:30	25.25
12:00	25.01
12:30 P.M.	25.06
1:00	24.95
1:30	24.80
2:00	24.95
2:30	25.21
3:00	24.90
3:30	24.71
4:00	25.31
4:30	25.15
5:00	25.20

2.7 Numerical Measures of Relative Standing

We've seen that numerical measures of central tendency and variability describe the general nature of a quantitative data set (either a sample or a population). In addition, we may be interested in describing the *relative* quantitative location of a particular measurement within a data set. Descriptive measures of the relationship of a measurement to the rest of the data are called **measures of relative standing.**

One measure of the relative standing of a measurement is its **percentile ranking,** or **percentile score.** For example, if oil company A reports that its yearly sales are in the 90th percentile of all companies in the industry, the implication is that 90% of all oil companies have yearly sales less than company A's, and only 10% have yearly sales exceeding company A's. This is demonstrated in Figure 2.23. Similarly, if the oil company's yearly sales are in the 50th percentile (the median of the data set), 50% of all oil companies would have lower yearly sales and 50% would have higher yearly sales.

Percentile rankings are of practical value only for large data sets. Finding them involves a process similar to the one used in finding a median. The measurements are ranked in order, and a rule is selected to define the location of each percentile. Because we are primarily interested in interpreting the percentile rankings of measurements (rather than finding particular percentiles for a data set), we define the *pth percentile* of a data set.

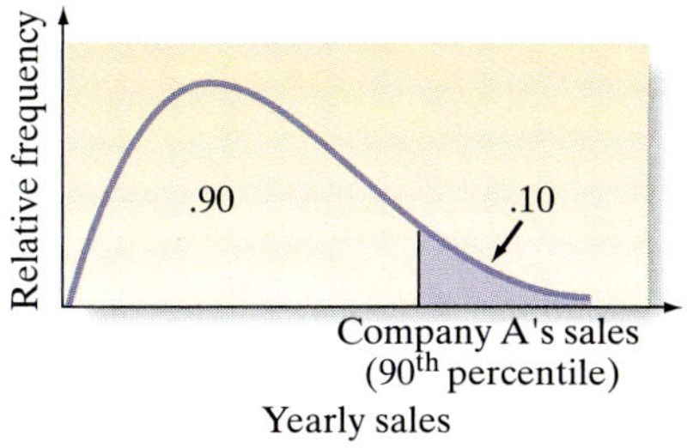

Figure 2.23
Location of 90th percentile for yearly sales of oil companies

For any set of n measurements (arranged in ascending or descending order), the ***pth* percentile** is a number such that p% of the measurements fall below the *pth* percentile and $(100 - p)$% fall above it.

Example 2.14
Finding and Interpreting Percentiles

Problem Refer to the percentages spent on R&D by the 50 high-technology firms listed in Table 2.8 (p. 72). A portion of the SPSS descriptive statistics printout is shown in Figure 2.24. Locate the 25th percentile and 95th percentile on the printout and interpret these values.

Solution Both the 25th percentile and 95th percentile are highlighted on the SPSS printout, Figure 2.24. These values are 7.05 and 13.335, respectively. Our interpretations are as follows: 25% of the 50 R&D percentages fall below 7.05 and 95% of the R&D percentages fall below 13.335.

Look Back The method for computing percentiles varies according to the software used. Some packages, like SPSS, give two different methods of computing percentiles. As the data set increases in size, these percentile values will converge to a single number.

Statistics

RDPCT

N	Valid	50
	Missing	0
Percentiles	5	5.765
	10	6.500
	25	7.050
	50	8.050
	75	9.625
	90	11.280
	95	13.335

Figure 2.24
SPSS percentiles for 50 R&D percentages

Now Work Exercise 2.89

Another measure of relative standing in popular use is the *z-score*. As you can see in the definition of z-score below, the z-score makes use of the mean and standard deviation of the data set in order to specify the relative location of a measurement. Note that the z-score is calculated by subtracting $\bar{x}$ (or μ) from the measurement x and then dividing the result by s (or σ). The final result, the z-score, represents the distance between a given measurement x and the mean, expressed in standard deviations.

The **sample *z*-score** for a measurement x is

$$z = \frac{x - \bar{x}}{s}$$

The **population *z*-score** for a measurement x is

$$z = \frac{x - \mu}{\sigma}$$

Example 2.15 Finding a *z*-Score

Problem Suppose 200 steelworkers are selected, and the annual income of each is determined. The mean and standard deviation are $\bar{x} = \$54{,}000$ and $s = \$2{,}000$. Suppose Joe Smith's annual income is \$52,000. What is his sample z-score?

Solution Joe Smith's annual income lies below the mean income of the 200 steelworkers (see Figure 2.25). We compute

$$z = \frac{x - \bar{x}}{s} = \frac{\$52{,}000 - \$54{,}000}{2{,}000} = -1.0$$

$$z = -1.0$$

which tells us that Joe Smith's annual income is 1.0 standard deviation *below* the sample mean, or, in short, his sample z-score is -1.0.

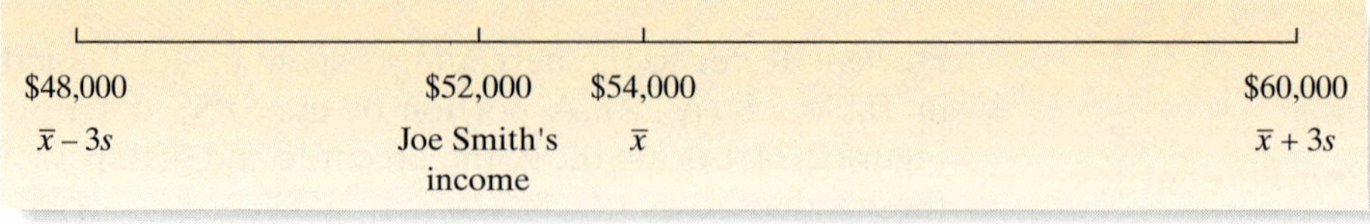

Figure 2.25
Annual income of steelworkers

Look Back The numerical value of the z-score reflects the relative standing of the measurement. A large positive z-score implies that the measurement is larger than almost all other measurements, whereas a large (in magnitude) negative z-score indicates that the measurement is smaller than almost every other measurement. If a z-score is 0 or near 0, the measurement is located at or near the mean of the sample or population.

Now Work Exercise 2.88

If we know that the frequency distribution of the measurements is mound-shaped, the following interpretation of the z-score can be given.

Interpretation of z-Scores for Mound-Shaped Distributions of Data

1. Approximately 68% of the measurements will have a z-score between −1 and 1.
2. Approximately 95% of the measurements will have a z-score between −2 and 2.
3. Approximately 99.7% (almost all) of the measurements will have a z-score between −3 and 3.

Note that this interpretation of z-scores is identical to that given by the Empirical Rule for mound-shaped distributions (Table 2.7). The statement that a measurement falls in the interval $(\mu - \sigma)$ to $(\mu + \sigma)$ is equivalent to the statement that a measurement has a population z-score between −1 and 1 because all measurements between $(\mu - \sigma)$ and $(\mu + \sigma)$ are within 1 standard deviation of μ. These z-scores are displayed in Figure 2.26.

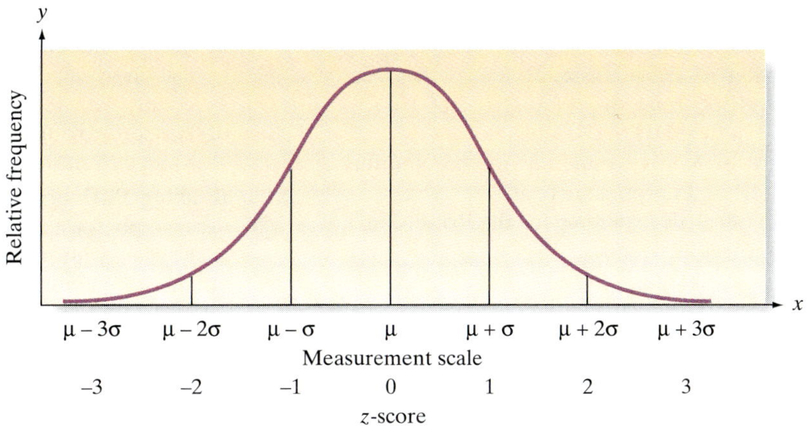

Figure 2.26
Population z-scores for a mound-shaped distribution

Exercises 2.88–2.102

Learning the Mechanics

2.88 NW Compute the z-score corresponding to each of the following values of x:

a. $x = 40, s = 5, \bar{x} = 30$
b. $x = 90, \mu = 89, \sigma = 2$
c. $\mu = 50, \sigma = 5, x = 50$
d. $s = 4, x = 20, \bar{x} = 30$
e. In parts **a–d,** state whether the z-score locates x within a sample or a population.
f. In parts **a–d,** state whether each value of x lies above or below the mean and by how many standard deviations.

2.89 NW Give the percentage of measurements in a data set that are above and below each of the following percentiles:

a. 75th percentile
b. 50th percentile
c. 20th percentile
d. 84th percentile

2.90 What is the 50th percentile of a quantitative data set called?

2.91 Compare the z-scores to decide which of the following x values lie the greatest distance above the mean and the greatest distance below the mean.

a. $x = 100, \mu = 50, \sigma = 25$
b. $x = 1, \mu = 4, \sigma = 1$
c. $x = 0, \mu = 200, \sigma = 100$
d. $x = 10, \mu = 5, \sigma = 3$

2.92 Suppose that 40 and 90 are two elements of a population data set and that their z-scores are −2 and 3, respectively. Using only this information, is it possible to determine the population's mean and standard deviation? If so, find them. If not, explain why it's not possible.

Applying the Concepts—Basic

2.93 Mathematics assessment test scores. According to the National Center for Education Statistics (2005), scores on a mathematics assessment test for United States eighth-graders have a mean of 279, a 10th percentile of 231, a 25th percentile of 255, a 75th percentile of 304, and a 90th percentile of 324. Interpret each of these numerical descriptive measures.

2.94 Drivers stopped by police. According to the Bureau of Justice Statistics (June 2006), 75% of all licensed drivers stopped by police are 25 years or older. Give a percentile ranking for the age of 25 years in the distribution of all ages of licensed drivers stopped by police.

2.95 Starting and mid-career salaries of bachelor's degree graduates. PayScale, Inc., an online provider of global compensation data, recently conducted a survey of 1.2 million bachelor's degree graduates with a minimum of ten years of work experience. Three of the many variables measured by PayScale were the graduate's starting salary, mid-career salary, and the college or university where they obtained their degree. A summary of the starting and mid-career salary data was reported in *The Wall Street Journal* (July 31, 2008). Descriptive statistics were provided for each of the over 300 colleges and universities that graduates attended. For example, graduates of the University of South Florida (USF) had a median starting salary of $41,100, a median mid-career salary of $71,100, and a mid-career 90th percentile salary of $131,000. Describe the salary distribution of USF bachelor's degree graduates by interpreting each of these summary statistics.

2.96 Sanitation inspection of cruise ships. Refer to the Dec. 2008 sanitation levels of cruise ships, Exercise 2.80 (p. 75). The data are saved in the **SHIPSANIT** file.

a. Give a measure of relative standing for the Nautilus Explorer score of 74. Interpret the result.

b. Give a measure of relative standing for the Rotterdam's score of 98. Interpret the result.

Applying the Concepts—Intermediate

2.97 Lead in drinking water. The U.S. Environmental Protection Agency (EPA) sets a limit on the amount of lead permitted in drinking water. The EPA *Action Level* for lead is .015 milligrams per liter (mg/L) of water. Under EPA guidelines, if 90% of a water system's study samples have a lead concentration less than .015 mg/L, the water is considered safe for drinking. I (coauthor Sincich) received a recent report on a study of lead levels in the drinking water of homes in my subdivision. The 90th percentile of the study sample had a lead concentration of .00372 mg/L. Are water customers in my subdivision at risk of drinking water with unhealthy lead levels? Explain.

2.98 Using *z*-scores for grades. At one university, the students are given *z*-scores at the end of each semester rather than the traditional GPAs. The mean and standard deviation of all students' cumulative GPAs, on which the *z*-scores are based, are 2.7 and .5, respectively.

a. Translate each of the following *z*-scores to corresponding GPA: $z = 2.0$, $z = -1.0$, $z = .5$, $z = -2.5$.

b. Students with *z*-scores below -1.6 are put on probation. What is the corresponding probationary GPA?

c. The president of the university wishes to graduate the top 16% of the students with *cum laude* honors and the top 2.5% with *summa cum laude* honors. Where (approximately) should the limits be set in terms of *z*-scores? In terms of GPAs? What assumption, if any, did you make about the distribution of the GPAs at the university?

2.99 Hazardous waste cleanup in Arkansas. The Superfund Act was passed by Congress to encourage state participation in the implementation of a law relating to the release and cleanup of hazardous substances. Hazardous waste sites financed by the Superfund Act are called *Superfund sites.* A total of 393 Superfund sites are operated by waste management companies in Arkansas (Tabor and Stanwick, *Arkansas Business and Economic Review,* Summer 1995). The number of these Superfund sites in each of Arkansas's 75 counties, shown in the next table, are saved in the **ARKFUND** file.

3	3	2	1	2	0	5	3	5	2	1	8	2
12	3	5	3	1	3	0	8	0	9	6	8	6
2	16	0	6	0	5	5	0	1	25	0	0	0
6	2	10	12	3	10	3	17	2	4	2	1	21
4	2	1	11	5	2	2	7	2	3	1	8	2
0	0	0	2	3	10	2	3	48	21			

Source: Tabor, R. H., and Stanwick, S. D. "Arkansas: An environmental perspective," *Arkansas Business and Economic Review,* Vol. 28, No. 2, Summer 1995, pp. 22–32 (Table 4).

a. Find the 10th percentile of the data set. Interpret the result.

b. Find the 95th percentile of the data set. Interpret the result.

c. Find the mean and standard deviation of the data; then use these values to calculate the *z*-score for an Arkansas county with 48 Superfund sites.

d. Based on your answer to part **c,** would you classify 48 as an extreme number of Superfund sites?

2.100 Ranking PhD programs in economics. Thousands of students apply for admission to graduate schools in economics each year with the intention of obtaining a PhD. The *Southern Economic Journal* (Apr. 2008) published a guide to graduate study in economics by ranking the PhD programs at 129 colleges and universities. Each program was evaluated according to the number of publications published by faculty teaching in the PhD program and by the quality of the publications. Data obtained from the Social Science Citation Index (SSCI) were used to calculate an overall productivity score for each PhD program. The mean and standard deviation of these 129 productivity scores were then used to compute a *z*-score for each economics program. Harvard University had the highest *z*-score ($z = 5.08$) and, hence, was the top-ranked school; Howard University was ranked last because it had the lowest *z*-score ($z = -0.81$). The data (*z*-scores) for all 129 economic programs are saved in the **ECOPHD** file.

a. Interpret the *z*-score for Harvard University.

b. Interpret the *z*-score for Howard University.

c. The authors of the *Southern Economic Journal* article note that "only 44 of the 129 schools have positive *z*-scores, indicating that the distribution of overall productivity is skewed to the right." Do you agree? (Check your answer by constructing a histogram for the *z*-scores in the **ECOPHD** file.)

Applying the Concepts—Advanced

2.101 Ranking PhD programs in economics (cont'd). Refer to the *Southern Economic Journal* (Apr. 2008) study of PhD programs in economics, Exercise 2.100. The authors also made the following observation: "A noticeable feature of this skewness is that distinction between schools diminishes as the rank declines. For example, the top-ranked school, Harvard, has a z-score of 5.08, and the fifth-ranked school, Yale, has a z-score of 2.18, a substantial difference. However, ..., the 70th-ranked school, the University of Massachusetts, has a z-score of –0.43, and the 80th-ranked school, the University of Delaware, has a z-score of –0.50, a very small difference. [Consequently] the ordinal rankings presented in much of the literature that ranks economics departments miss the fact that below a relatively small group of top programs, the differences in [overall] productivity become fairly small." Do you agree?

2.102 Blue versus red-colored exam study. In a study of how external clues influence performance, professors at the University of Alberta and Pennsylvania State University gave two different forms of a midterm examination to a large group of introductory students. The questions on the exam were identical and in the same order, but one exam was printed on blue paper and the other on red paper (*Teaching Psychology,* May 1998). Grading only the difficult questions on the exam, the researchers found that scores on the blue exam had a distribution with a mean of 53% and a standard deviation of 15%, while scores on the red exam had a distribution with a mean of 39% and a standard deviation of 12%. (Assume that both distributions are approximately mound-shaped and symmetric.)

a. Give an interpretation of the standard deviation for the students who took the blue exam.

b. Give an interpretation of the standard deviation for the students who took the red exam.

c. Suppose a student is selected at random from the group of students who participated in the study and the student's score on the difficult questions is 20%. Which exam form is the student more likely to have taken, the blue or the red exam? Explain.

Applet Exercise 2.7

Use the applet *Standard Deviation* to determine whether an item in a data set may be an outlier. Begin by setting appropriate limits and plotting the data below on the number line provided in the applet.

10	80	80	85	85	85	85	90	90	90	90	90	95	95	95	95	100	100

a. The green arrow shows the approximate location of the mean. Multiply the standard deviation given by the applet by 3. Is the data item 10 more than three standard deviations away from the green arrow (the mean)? Can you conclude that the 10 is an outlier?

b. Using the mean and standard deviation from part **a,** move the point at 10 on your plot to a point that appears to be about three standard deviations from the mean. Repeat the process in part **a** for the new plot and the new suspected outlier.

c. When you replaced the extreme value in part **a** with a number that appeared to be within three standard deviations of the mean, the standard deviation got smaller and the mean moved to the right, yielding a new data set where the extreme value was not within three standard deviations of the mean. Continue to replace the extreme value with higher numbers until the new value is within three standard deviations of the mean in the new data set. Use trial and error to estimate the smallest number that can replace the 10 in the original data set so that the replacement is not considered to be an outlier.

2.8 Methods for Detecting Outliers: Box Plots and z-Scores

Sometimes it is important to identify inconsistent or unusual measurements in a data set. An observation that is unusually large or small relative to the data values we want to describe is called an *outlier*.

Outliers are often attributable to one of several causes. First, the measurement associated with the outlier may be invalid. For example, the experimental procedure used to generate the measurement may have malfunctioned, the experimenter may have misrecorded the measurement, or the data might have been coded incorrectly in the computer. Second, the outlier may be the result of a misclassified measurement—that is, the measurement belongs to a population different from which the rest of the sample was drawn. Finally, the measurement associated with the outlier may be recorded correctly and from the same population as the rest of the sample but represents a rare (chance) event. Such outliers occur most often when the relative frequency distribution of the sample data is extremely skewed because such a distribution has a tendency to include extremely large or small observations relative to the others in the data set.

An observation (or measurement) that is unusually large or small relative to the other values in a data set is called an **outlier.** Outliers typically are attributable to one of the following causes:

1. The measurement is observed, recorded, or entered into the computer incorrectly.
2. The measurement comes from a different population.
3. The measurement is correct but represents a rare (chance) event.

Two useful methods for detecting outliers, one graphical and one numerical, are **box plots** and z-scores. The box plot is based on the *quartiles* of a data set. **Quartiles** are values that partition the data set into four groups, each containing 25% of the measurements. The *lower quartile* Q_L is the 25th percentile, the *middle quartile* is the median m (the 50th percentile), and the *upper quartile* Q_U is the 75th percentile (see Figure 2.27).

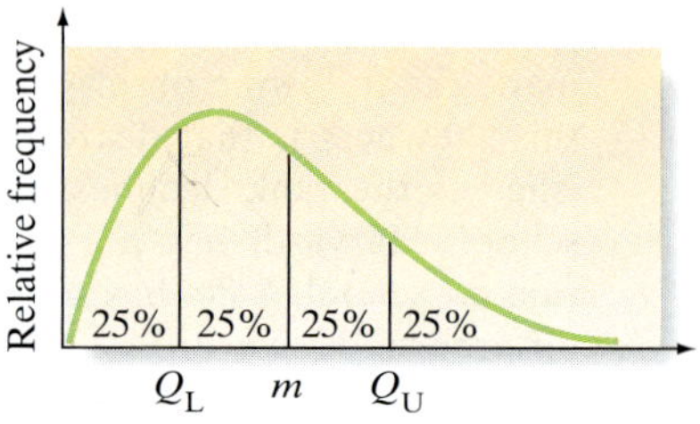

Figure 2.27
The quartiles for a data set

> The **lower quartile Q_L** is the 25th percentile of a data set. The **middle quartile m** is the median. The **upper quartile Q_U** is the 75th percentile.

A box plot is based on the *interquartile range* (IQR), the distance between the lower and upper quartiles:

$$\text{IQR} = Q_U - Q_L$$

> The **interquartile range (IQR)** is the distance between the lower and upper quartiles:
> $$\text{IQR} = Q_U - Q_L$$

An annotated Minitab box plot for the 50 companies' percentages of revenues spent on R&D (Table 2.2) is shown in Figure 2.28.* Note that a rectangle (the box) is drawn, with the top and bottom sides of the rectangle (the **hinges**) drawn at the quartiles Q_L and Q_U, respectively. By definition, then, the "middle" 50% of the observations—those between Q_L and Q_U—fall inside the box. For the R&D data, these quartiles are at 7.05 and 9.625 (see Figure 2.24, p. 78). Thus,

$$\text{IQR} = 9.625 - 7.05 = 2.575$$

The median is shown at 8.05 by a horizontal line within the box.

To guide the construction of the "tails" of the box plot, two sets of limits, called **inner fences** and **outer fences,** are used. Neither set of fences actually appears on the box plot. Inner fences are located at a distance of 1.5(IQR) from the hinges. Emanating from the hinges of the box are vertical lines called the **whiskers.** The two whiskers extend to the most extreme observation inside the inner fences. For example, the inner fence on the lower side (bottom) of the R&D percentage box plot is

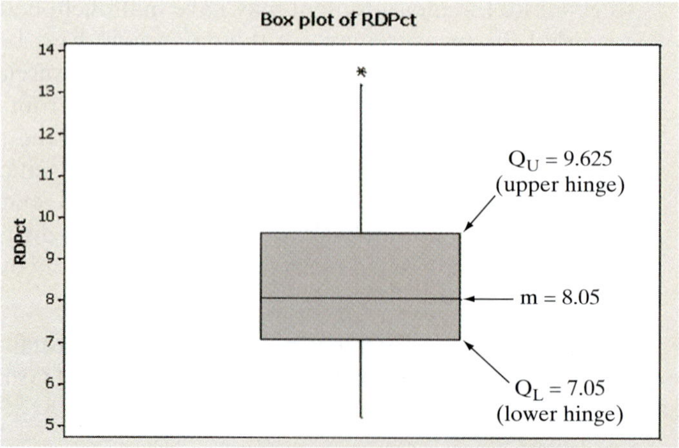

Figure 2.28
Annotated Minitab box plot for 50 R&D percentages

*Although box plots can be generated by hand, the amount of detail required makes them particularly well suited for computer generation. We use computer software to generate the box plots in this section.

$$\begin{aligned}\text{Lower inner fence} &= \text{Lower hinge} - 1.5(\text{IQR}) \\ &= 7.05 - 1.5(2.575) \\ &= 7.05 - 3.863 = 3.187\end{aligned}$$

The smallest measurement in the data set is 5.2, which is well inside this inner fence. Thus, the lower whisker extends to 5.2. Similarly, the upper whisker extends to the most extreme observation inside the upper inner fence, where

$$\begin{aligned}\text{Upper inner fence} &= \text{Upper hinge} + 1.5(\text{IQR}) \\ &= 9.625 + 1.5(2.575) \\ &= 9.625 + 3.863 = 13.488\end{aligned}$$

The largest measurement inside this fence is the third largest measurement, 13.2. Note that the longer upper whisker reveals the rightward skewness of the R&D distribution.

Values that are beyond the inner fences are deemed *potential outliers* because they are extreme values that represent relatively rare occurrences. In fact, for mound-shaped distributions, fewer than 1% of the observations are expected to fall outside the inner fences. Two of the 50 R&D measurements, both at 13.5, fall outside the upper inner fence. Each of these potential outliers is represented by the asterisk (*) at 13.5.

The other two imaginary fences, the outer fences, are defined at a distance 3(IQR) from each end of the box. Measurements that fall beyond the outer fence are represented by 0s (zeros) and are very extreme measurements that require special analysis. Because less than one-hundredth of 1% (.01% or .0001) of the measurements from mound-shaped distributions are expected to fall beyond the outer fence, these measurements are considered to be *outliers*. No measurement in the R&D percentage box plot (Figure 2.28) is represented by a 0; thus there are no outliers.

Recall that outliers are extreme measurements that stand out from the rest of the sample and may be faulty: They may be incorrectly recorded observations, members of a population different from the rest of the sample, or, at the least, very unusual measurements from the same population. For example, the two R&D measurements at 13.5 (identified by an asterisk) may be considered outliers. When we analyze these measurements, we find that they are correctly recorded. However, it turns out that both represent R&D expenditures of relatively young and fast-growing companies. Thus, the outlier analysis may have revealed important factors that relate to the R&D expenditures of high-tech companies: their age and rate of growth. Outlier analysis often reveals useful information of this kind and therefore plays an important role in the statistical inference-making process.

In addition to detecting outliers, box plots provide useful information on the variation in a data set. The elements (and nomenclature) of box plots are summarized in the next box. Some aids to the interpretation of box plots are also given.

Elements of a Box Plot

1. A rectangle (the **box**) is drawn with the ends (the **hinges**) drawn at the lower and upper quartiles (Q_L and Q_U). The median of the data is shown in the box, usually by a line or a symbol (such as "+").
2. The points at distances 1.5(IQR) from each hinge define the **inner fences** of the data set. Lines (the **whiskers**) are drawn from each hinge to the most extreme measurement inside the inner fence.
3. A second pair of fences, the **outer fences,** are defined at a distance of 3 interquartile ranges, 3(IQR) from the hinges. One symbol (usually "*") is used to represent measurements falling between the inner and outer fences, and another (usually "0") is used to represent measurements beyond the outer fences.
4. The symbols used to represent the median and the extreme data points (those beyond the fences) will vary depending on the software you use to construct the box plot. (You may use your own symbols if you are constructing a box plot by hand.) You should consult the program's documentation to determine exactly which symbols are used.

Aids to the Interpretation of Box Plots

1. Examine the length of the box. The IQR is a measure of the sample's variability and is especially useful for the comparison of two samples (see Example 2.17).
2. Visually compare the lengths of the whiskers. If one is clearly longer, the distribution of the data is probably skewed in the direction of the longer whisker.
3. Analyze any measurements that lie beyond the fences. Fewer than 5% should fall beyond the inner fences, even for very skewed distributions. Measurements beyond the outer fences are probably outliers, with one of the following explanations:
   a. The measurement is incorrect. It may have been observed, recorded, or entered into the computer incorrectly.
   b. The measurement belongs to a population different from the population that the rest of the sample was drawn from (see Example 2.17).
   c. The measurement is correct *and* from the same population as the rest. Generally, we accept this explanation only after carefully ruling out all others.

Example 2.16
Interpreting a Box Plot

Problem In Example 2.2 (p. 45) we analyzed 50 processing times (listed in Table 2.4) for the development of price quotes by the manufacturer of industrial wheels. The intent was to determine whether the success or failure in obtaining the order was related to the amount of time to process the price quotes. Each quote that corresponds to "lost" business was so classified. Use a statistical software package to draw a box plot for all 50 processing times. What does the box plot reveal about the data?

Solution The Minitab box plot printout for these data is shown in Figure 2.29. Note that the upper whisker is much longer than the lower whisker, indicating rightward skewness of the data. However, the most important feature of the data is made very obvious by the box plot: There are four measurements (indicated by asterisks) that are beyond the upper inner fence. Thus, the distribution is extremely skewed to the right, and several measurements—or outliers—need special attention in our analysis.

Look Back Before removing outliers from the data set, a good analyst will make a concerted effort to find the cause of the outliers. We offer an explanation for these processing time outliers in the next example.

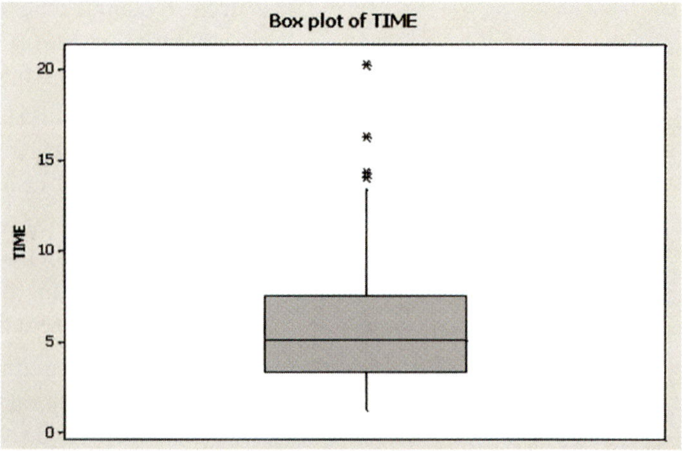

Figure 2.29
Minitab box plot for processing time data

Now Work Exercise 2.106

Example 2.17
Comparing Box Plots

Problem The box plot for the 50 processing times (Figure 2.29) does not explicitly reveal the differences, if any, between the set of times corresponding to the success and the set of times corresponding to the failure to obtain the business. Box plots corresponding to the 39 "won" and 11 "lost" bids were generated using SPSS and are shown in Figure 2.30. Interpret them.

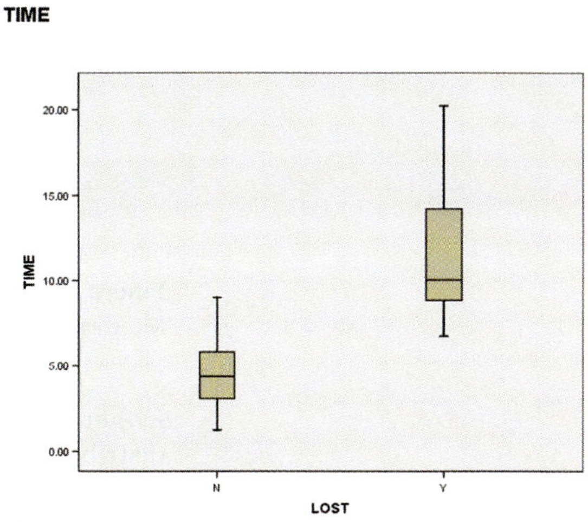

Figure 2.30
SPSS box plots of processing times for won and lost bids

Solution The division of the data set into two parts, corresponding to won and lost bids eliminates any observations that are beyond the inner fences. Furthermore, the skewness in the distributions has been reduced, as evidenced by the fact that the upper whiskers are only slightly longer than the lower. The box plots also reveal that the processing times corresponding to the lost bids tend to exceed those of the won bids. A plausible explanation for the outliers in the combined box plot (Figure 2.29) is that they are from a different population than the bulk of the times. In other words, there are two populations represented by the sample of processing times—one corresponding to lost bids and the other to won bids.

Look Back The box plots lend support to the conclusion that the price quote processing time and the success of acquiring the business are related. However, whether the visual differences between the box plots generalize to inferences about the populations corresponding to these two samples is a matter for inferential statistics, not graphical descriptions. We'll discuss how to use samples to compare two populations using inferential statistics in Chapter 7.

The following example illustrates how z-scores can be used to detect outliers and make inferences.

Example 2.18
Inference Using z-Scores

Problem Suppose a female bank employee believes that her salary is low as a result of sex discrimination. To substantiate her belief, she collects information on the salaries of her male counterparts in the banking business. She finds that their salaries have a mean of \$64,000 and a standard deviation of \$2,000. Her salary is \$57,000. Does this information support her claim of sex discrimination?

Solution The analysis might proceed as follows: First, we calculate the z-score for the woman's salary with respect to those of her male counterparts. Thus,

$$z = \frac{\$57{,}000 - \$64{,}000}{\$2{,}000} = -3.5$$

The implication is that the woman's salary is 3.5 standard deviations *below* the mean of the male salary distribution. Furthermore, if a check of the male salary data shows that the frequency distribution is mound-shaped, we can infer that very few salaries in this distribution should have a z-score less than -3, as shown in Figure 2.31. Clearly, a z-score of -3.5 represents an outlier.

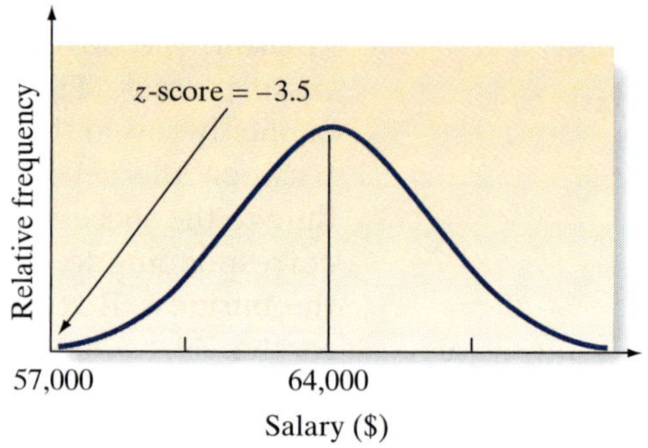

Figure 2.31
Male salary distribution

Either this female's salary is from a distribution different from the male salary distribution, or it is a very unusual (highly improbable) measurement from a distribution that is no different from the male salary distribution.

Look Back Which of the two situations do you think prevails? Statistical thinking would lead us to conclude that her salary does not come from the male salary distribution, lending support to the female bank employee's claim of sex discrimination. A careful investigator should aquire more information before inferring sex discrimination as the cause. We would want to know more about the data-collection technique the woman used and more about her competence at her job. Also, perhaps other factors such as length of employment should be considered in the analysis.

Now Work Exercise 2.103

Examples 2.17 and 2.18 exemplify an approach to statistical inference that might be called the **rare-event approach.** An experimenter hypothesizes a specific frequency distribution to describe a population of measurements. Then a sample of measurements is drawn from the population. If the experimenter finds it unlikely that the sample came from the hypothesized distribution, the hypothesis is concluded to be false. Thus, in Example 2.18, the woman believes her salary reflects discrimination. She hypothesizes that her salary should be just another measurement in the distribution of her male counterparts' salaries if no discrimination exists. However, it is so unlikely that the sample (in this case, her salary) came from the male frequency distribution that she rejects that hypothesis, concluding that the distribution from which her salary was drawn is different from the distribution for the men.

This rare-event approach to inference-making is discussed further in later chapters. Proper application of the approach requires a knowledge of probability, the subject of our next chapter.

We conclude this section with some rules of thumb for detecting outliers.

Rules of Thumb for Detecting Outliers*

Box Plots: Observations falling between the inner and outer fences are deemed *suspect outliers*. Observations falling beyond the outer fence are deemed *highly suspect outliers*.

***z*-scores:** Observations with z-scores greater than 3 in absolute value are considered outliers. (For some highly skewed data sets, observations with z-scores greater than 2 in absolute value may be outliers.)

*The z-score and box plot methods both establish rule-of-thumb limits outside of which a measurement is deemed to be an outlier. Usually, the two methods produce similar results. However, the presence of one or more outliers in a data set can inflate the computed value of s. Consequently, it will be less likely that an errant observation would have a z-score larger than 3 in absolute value. In contrast, the values of the quartiles used to calculate the intervals for a box plot are not affected by the presence of outliers.

Statistics IN Action Revisited — Detecting Outliers

In the ethics survey of University Community Hospital physicians, the medical researchers measured two quantitative variables: *Length of time in practice* (number of years) and *Amount of exposure to ethics in medical school* (number of hours). Are there any unusual values of these variables in the **ETHICS** data set? We will employ both the box plot and z-score methods to aid in identifying outliers in the data.

Descriptive statistics for these two variables, produced using Minitab, are shown in Figure SIA2.6. To employ the z-score method, we need the means and standard deviations. These values are highlighted on Figure SIA2.6. Then the 3-standard-deviation intervals are

$$\text{YRSPRAC:} \quad 14.6 \pm 3(9.2) = 14.6 \pm 27.6 = (-13.0, 42.2)$$

$$\text{EDHRS:} \quad 23.9 \pm 3(109.6) = 23.9 \pm 328.8 = (-304.9, 352.7)$$

[*Note:* Because neither of the variables can be negative, for practical purposes the intervals all begin at 0.]

In this application, we will focus on only the three largest values of the variables in the data set. For length of time in practice, these values are 35, 40, and 40 years. Note that all three values fall within the 3-standard-deviation interval—that is, they all have z-scores less than 3 in absolute value. Consequently, no outliers exist for the data on length of time in practice.

For ethics exposure, the three largest values are 75, 80, and 1,000 hours. Note that only one of these values, 1,000, falls beyond the 3-standard-deviation interval. Thus, the data for the physician who was exposed to 1,000 hours of ethics in medical school is considered an outlier using the z-score approach. However, notice that the standard deviation for the variable (109.6) is much larger than the mean (23.9). Typically, when S exceeds $\bar{x}$ (for a non-negative variable), a high degree of skewness exists. This skewness is due, in large part, to the extreme value of 1,000 hours. When such an extreme outlier occurs in the data, the standard deviation is inflated and the z-score method is less likely to detect unusual observations. (See the footnote on p. 86.) When this occurs, the box plot method for detecting outliers is preferred.

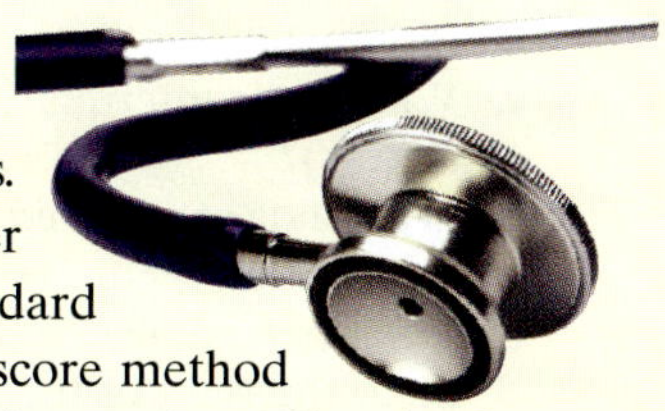

Rather than produce a box plot for the ethics exposure variable, we'll use the descriptive statistics in Figure SIA2.6 to find the inner and outer fences. From the Minitab printout, we see that $Q_L = 1$, $Q_U = 20$, and IQR $= 19$. Then the upper inner and outer fence boundaries of the box plot are

Upper inner fence:

$$Q_U + (1.5)\text{IQR} = 20 + (1.5)(19) = 48.5$$

Upper outer fence:

$$Q_U + (3)\text{IQR} = 20 + (3)(19) = 77.0$$

Now we see that the ethics exposure value of 80 hours is, indeed, a highly suspect outlier because it falls beyond the upper outer fence. Also, the exposure value of 75 hours is a suspect outlier because it falls beyond the upper inner fence. Thus, the box plot method detected an additional two outliers.

Before any type of inference is made concerning the population of ethics exposure values, we should consider whether these three outliers are legitimate observations (in which case they will remain in the data set) or are associated with physicians who are not members of the population of interest (in which case they will be removed from the data set).

Data Set: ETHICS

Descriptive Statistics: YRSPRAC, EDHRS

Variable	N	N*	Mean	StDev	Q1	Median	Q3	IQR
YRSPRAC	112	6	14.598	9.161	7.000	14.000	20.000	13.000
EDHRS	83	35	23.9	109.6	1.0	5.0	20.0	19.0

Figure SIA2.6
Minitab descriptive statistics for practice experience and ethics exposure

Exercises 2.103–2.116

Learning the Mechanics

2.103 NW A sample data set has a mean of 57 and a standard deviation of 11. Determine whether each of the following sample measurements are outliers.

a. 65
b. 21
c. 72
d. 98

2.104 Define the 25th, 50th, and 75th percentiles of a data set. Explain how they provide a description of the data.

2.105 Suppose a data set consisting of exam scores has a lower quartile $Q_L = 60$, a median $m = 75$, and an upper quartile $Q_U = 85$. The scores on the exam range from 18 to 100. Without having the actual scores available to you, construct as much of the box plot as possible.

2.106 Consider the horizontal box plot shown below.
NW
- **a.** What is the median of the data set (approximately)?
- **b.** What are the upper and lower quartiles of the data set (approximately)?
- **c.** What is the interquartile range of the data set (approximately)?
- **d.** Is the data set skewed to the left, skewed to the right, or symmetric?
- **e.** What percentage of the measurements in the data set lie to the right of the median? To the left of the upper quartile?
- **f.** Identify any outliers in the data.

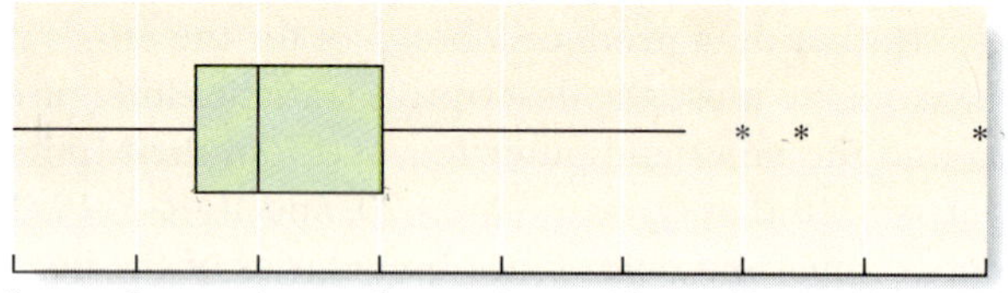

Applying the Concepts—Basic

2.107 Budget lapsing at army hospitals. Accountants use the term *budget lapsing* to describe the situation that occurs when unspent funds do not carry over from one budgeting period to the next. Due to budget lapsing, U.S. army hospitals tend to stockpile pharmaceuticals and other supplies toward the end of the fiscal year, leading to a spike in expenditures. This phenomenon was investigated in the *Journal of Management Accounting Research* (Vol. 19, 2007). Data on expenses per full-time equivalent employees for a sample of 1,751 army hospitals yielded the following summary statistics: $\bar{x} = \$6,563$, $m = \$6,232$, $s = \$2,484$, $Q_L = \$5,309$, and $Q_U = \$7,216$.
- **a.** Interpret, practically, the measures of relative standing.
- **b.** Compute the interquartile range, IQR, for the data.
- **c.** What proportion of the 1,751 army hospitals have expenses between \$5,309 and \$7,216?

2.108 Treating psoriasis with the "Doctorfish of Kangal." Psoriasis is a skin disorder with no known cure and no proven effective pharmacological treatment. An alternative treatment for psoriasis is ichthyotherapy, also known as therapy with the "Doctorfish of Kangal." Fish from the hot pools of Kangal, Turkey, feed on the skin scales of bathers, reportedly reducing the symptoms of psoriasis. In one study, 67 patients diagnosed with psoriasis underwent three weeks of ichthyotherapy (*Evidence-Based Research in Complementary and Alternative Medicine*, Dec. 2006). The Psoriasis Area Severity Index (PASI) of each patient was measured both before and after treatment. (The lower the PASI score, the better is the skin condition.) Box plots of the PASI scores, both before (baseline) and after three weeks of ichthyotherapy treatment, are shown in the accompanying diagram.
- **a.** Find the approximate 25th percentile, the median, and the 75th percentile for the PASI scores before treatment.
- **b.** Find the approximate 25th percentile, the median, and the 75th percentile for the PASI scores after treatment.
- **c.** Comment on the effectiveness of ichthyotherapy in treating psoriasis.

Output for Exercise 2.108

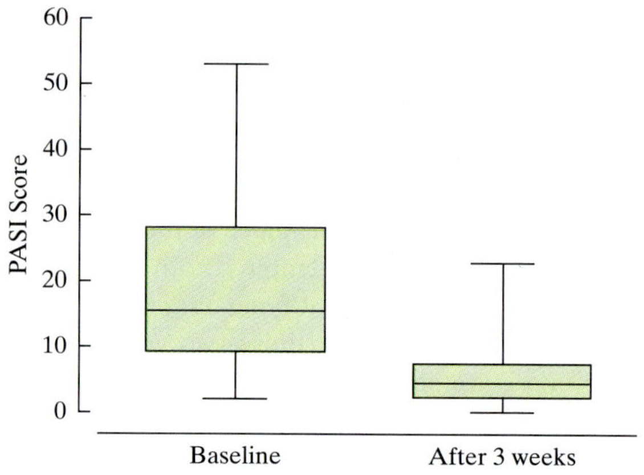

Source: Grassberger, M. and Hoch, W. "Ichthyotherapy as alternative treatment for patients with psoriasis: A pilot study." *Evidence-Based Research in Complementary and Alternative Medicine,* Vol. 3, No. 4, Dec. 2006, pp. 483–488 (Figure 3).

2.109 Semester hours taken by CPA candidates. Refer to the *Journal of Accounting and Public Policy* (Spring 2002) study of 100,000 first-time candidates for the CPA exam, Exercise 2.51 (p. 63). The number of semester hours of college credit earned by the candidates had a mean of 141.31 hours and a standard deviation of 17.77 hours.
- **a.** Find the z-score for a first-time candidate for the CPA exam who earned 160 semester hours of college credit. Is this observation considered an outlier?
- **b.** Give a value of number of semester hours that would, in fact, be considered an outlier in this data set.

2.110 Salary offers to MBAs. The **MBASAL** file contains the top salary offer (in thousands of dollars) received by each member of a sample of 50 MBA students who graduated from the Graduate School of Management at Rutgers, the state university of New Jersey. Descriptive statistics and a box plot for the data are shown on the accompanying DDXL printouts on p. 89.
- **a.** Find and interpret the z-score associated with the highest salary offer, the lowest salary offer, and the mean salary offer. Would you consider the highest offer to be unusually high? Why or why not?
- **b.** Based on the box plot for this data set, which salary offers (if any) are suspect or highly suspect outliers?

Applying the Concepts—Intermediate

2.111 Time in bankruptcy. Refer to the *Financial Management* (Spring 1995) study of 49 firms filing for prepackaged bankruptcies, Exercise 2.30 (p. 52). Recall that three types of "prepack" firms exist: (1) those who hold no prefiling vote, (2) those who vote their preference for a joint solution; and (3) those who vote their preference for a prepack. (These data are saved in the **BANKRUPT** file.)
- **a.** Construct a box plot for the time in bankruptcy (months) for each type of firm.
- **b.** Find the median bankruptcy times for the three types.
- **c.** How do the variabilities of the bankruptcy times compare for the three types?
- **d.** The standard deviations of the bankruptcy times are 2.47 for "none," 1.72 for "joint," and 0.96 for "prepack." Do the standard deviations agree with the interquartile

DDXL output for Exercise 2.110

Summary of SALARY

Count	50
Mean	52.334
Median	51
Std Dev	9.224
Variance	85.085
Range	40
Min	35
Max	75
IQR	13.7
25th%	45.3
75th%	59

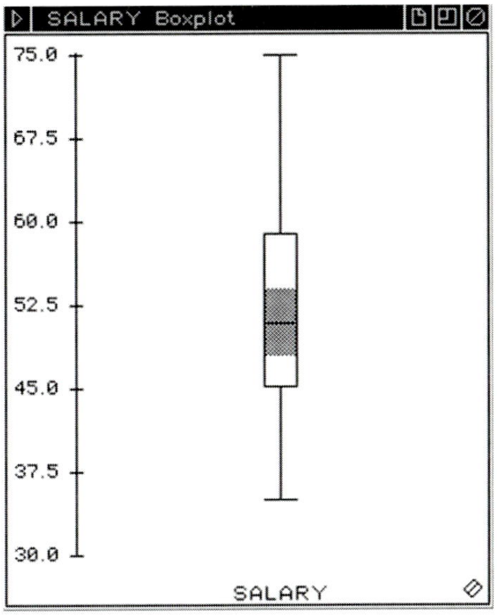

ranges with regard to the comparison of the variabilities of the bankruptcy times?

e. Is there evidence of outliers in any of the three distributions?

2.112 Most powerful women in America. Refer to the *Fortune* (Oct. 16, 2008) ranking of the 50 most powerful women in America, Exercise 2.48 (p. 62). The data are saved in the **WPOWER50** file. Use side-by-side box plots to compare the ages of the women in three groups based on their position within the firm: Group 1 (CEO/ chairman, CEO/president, COO/ president, EVP/president, or CFO/president); Group 2 (CEO, CFO, EVP/COO, CFO/EVP, CIO/EVP, chairman, COO, CRO, or president); Group 3 (EVP, executive, founder, SVP, treasurer, VP, director, or vice chair). Do you detect any outliers?

2.113 Sanitation inspection of cruise ships. Refer to Exercise 2.80 (p. 75) and the data on the sanitation levels of passenger cruise ships. The data are saved in the **SHIPSANIT** file.

a. Use the box plot method to detect any outliers in the data set.

b. Use the z-score method to detect any outliers in the data set.

c. Do the two methods agree? If not, explain why.

2.114 Hazardous waste cleanup in Arkansas. Refer to Exercise 2.99 (p. 80) and the data on the number of Superfund sites in each of 75 Arkansas counties. The data are saved in the **ARKFUND** file.

a. There is at least one outlier in the data. Use the methods of this chapter to detect the outliers.

b. Delete the outlier(s) found in part **a** from the data set and recalculate measures of central tendency and variation. Which measures are most affected by the removal of the outlier(s)?

2.115 Network server downtime. A manufacturer of network computer server systems is interested in improving its customer support services. As a first step, its marketing department has been charged with the responsibility of summarizing the extent of customer problems in terms of system downtime. The 40 most recent customers were surveyed to determine the amount of downtime (in hours) they had experienced during the previous month. These data are listed in the table and saved in the **DOWNTIME** file.

Customer Number	Downtime	Customer Number	Downtime
230	12	250	4
231	16	251	10
232	5	252	15
233	16	253	7
234	21	254	20
235	29	255	9
236	38	256	22
237	14	257	18
238	47	258	28
239	0	259	19
240	24	260	34
241	15	261	26
242	13	262	17
243	8	263	11
244	2	264	64
245	11	265	19
246	22	266	18
247	17	267	24
248	31	268	49
249	10	269	50

a. Construct a box plot for these data. Use the information reflected in the box plot to describe the frequency distribution of the data set. Your description should address central tendency, variation, and skewness.

b. Use your box plot to determine which customers are having unusually lengthy downtimes.

c. Find and interpret the z-scores associated with the customers you identified in part **b.**

Applying the Concepts—Advanced

2.116 Sensor motion of a robot. Researchers at Carnegie Mellon University developed an algorithm for estimating the sensor motion of a robotic arm by mounting a camera with inertia sensors on the arm (*International Journal of Robotics Research,* Dec. 2004). One variable of interest is the error of estimating arm translation (measured in centimeters). Data for 10 experiments are listed in the following table and saved in the **SENSOR**

Trial	Perturbed Intrinsics	Perturbed Projections	Translation Error (cm)
1	Yes	No	1.0
2	Yes	No	1.3
3	Yes	No	3.0
4	Yes	No	1.5
5	Yes	No	1.3
6	No	Yes	22.9
7	No	Yes	21.0
8	No	Yes	34.4
9	No	Yes	29.8
10	No	Yes	17.7

Source: Strelow, D. and Singh, S. "Motion estimation form image and inertial measurements," *International Journal of Robotics Research,* Vol. 23, No. 12, Dec. 2004 (Table 4). Copyright © 2004 SAGE Publications.

file. In each experiment, the perturbation of camera intrinsics and projections were varied. Suppose a trial resulted in a translation error of 4.5 cm. Is this value an outlier for trials with perturbed intrinsics but no perturbed projections? For trials with perturbed projections but no perturbed intrinsics? What type of camera perturbation most likely occurred for this trial?

2.9 Graphing Bivariate Relationships*

The claim is often made that the crime rate and the unemployment rate are "highly correlated." Another popular belief is that the gross domestic product (GDP) and the rate of inflation are "related." Some people even believe that the Dow Jones Industrial Average and the lengths of fashionable skirts are "associated." The words *correlated, related,* and *associated* imply a relationship between two variables—in the examples above, two *quantitative* variables.

One way to describe the relationship between two quantitative variables—called a **bivariate relationship**—is to plot the data in a **scattergram** (or **scatterplot**). A scattergram is a two-dimensional plot, with one variable's values plotted along the vertical axis and the other along the horizontal axis. For example, Figure 2.32 is a scattergram relating (1) the cost of mechanical work (heating, ventilating, and plumbing) to (2) the floor area of the building for a sample of 26 factory and warehouse buildings. Note that the scattergram suggests a general tendency for mechanical cost to increase as building floor area increases.

When an increase in one variable is generally associated with an increase in the second variable, we say that the two variables are "positively related" or "positively correlated."* Figure 2.32 implies that mechanical cost and floor area are positively correlated. Alternatively, if one variable has a tendency to decrease as the other increases, we say the

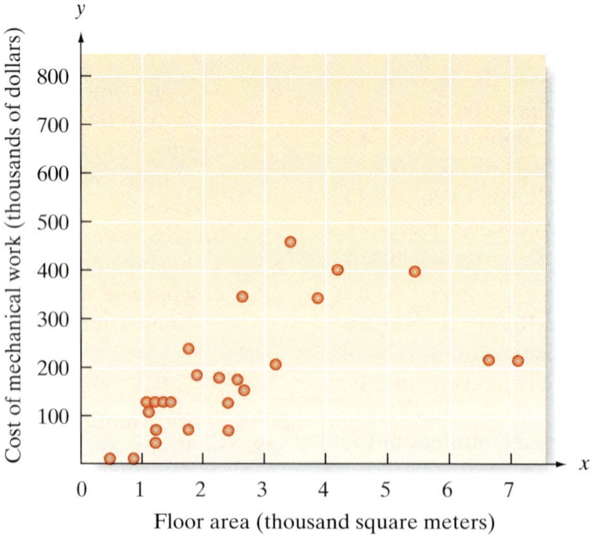

Figure 2.32
Scattergram of cost vs. floor area

*A formal definition of correlation is given in Chapter 10. We will learn that correlation measures the strength of the linear (or straight-line) relationship between two quantitative variables.

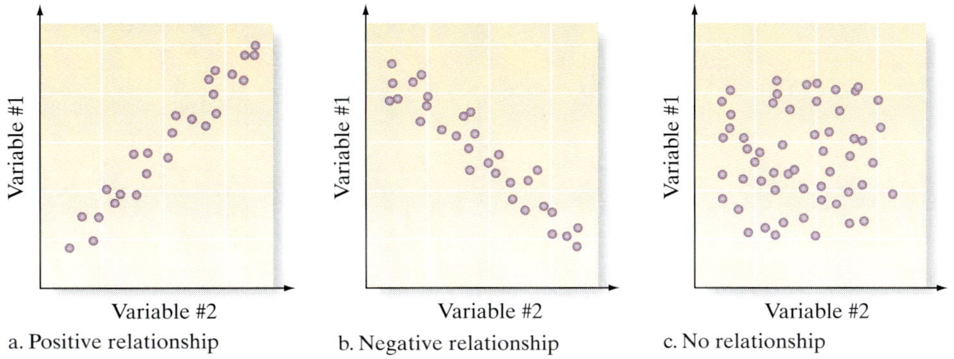

Figure 2.33
Hypothetical bivariate relationship

variables are "negatively correlated." Figure 2.33 shows several hypothetical scattergrams that portray a positive bivariate relationship (Figure 2.33a), a negative bivariate relationship (Figure 2.33b), and a situation where the two variables are unrelated (Figure 2.33c).

Example 2.19
Graphing Bivariate Data from a Hospital

Problem A medical item used to treat a hospital patient is called a *factor*. For example, factors can be intravenous (IV) tubing, IV fluid, needles, shave kits, bedpans, diapers, dressings, medications, and even code carts. The coronary care unit at Bayonet Point Hospital (St. Petersburg, Florida) recently investigated the relationship between the number of factors used per patient and the patient's length of stay (in days). Data on these two variables for a sample of 50 coronary care patients are given in Table 2.9. Use a scattergram to describe the relationship between the two variables of interest: number of factors and length of stay.

Solution Rather than construct the plot by hand, we resort to a statistical software package. The Excel plot of the data in Table 2.9, with length of stay (LOS) on the vertical axis and number of factors (FACTORS) on the horizontal axis, is shown in Figure 2.34.

Table 2.9 Data on Patient's Factors and Length of Stay

Number of Factors	Length of Stay (days)	Number of Factors	Length of Stay (days)
231	9	354	11
323	7	142	7
113	8	286	9
208	5	341	10
162	4	201	5
117	4	158	11
159	6	243	6
169	9	156	6
55	6	184	7
77	3	115	4
103	4	202	6
147	6	206	5
230	6	360	6
78	3	84	3
525	9	331	9
121	7	302	7
248	5	60	2
233	8	110	2
260	4	131	5
224	7	364	4
472	12	180	7
220	8	134	6
383	6	401	15
301	9	155	4
262	7	338	8

Source: Bayonet Point Hospital, Coronary Care Unit.

Data Set: MEDFACTORS

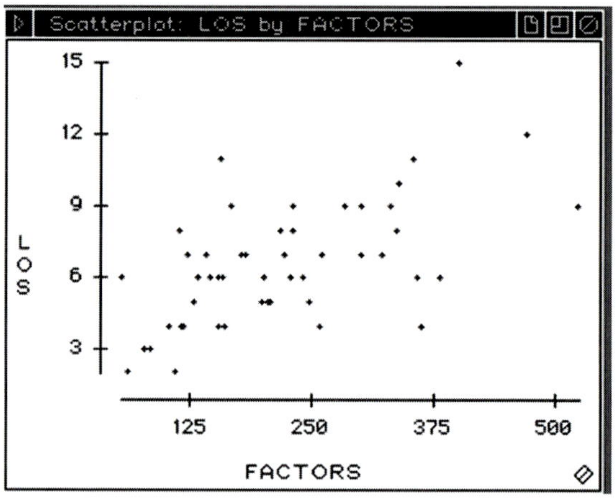

Figure 2.34
DDXL scatterplot of data in Table 2.9

Although the plotted points exhibit a fair amount of variation, the scattergram clearly shows an increasing trend. It appears that a patient's length of stay is positively correlated with the number of factors used in the patient's care.

Look Back If hospital administrators can be confident that the sample trend shown in Figure 2.34 accurately describes the trend in the population, then they may use this information to improve their forecasts of lengths of stay for future patients.

Now Work Exercise 2.120

The scattergram is a simple but powerful tool for describing a bivariate relationship. However, keep in mind that it is only a graph. No measure of reliability can be attached to inferences made about bivariate populations based on scattergrams of sample data. The statistical tools that enable us to make inferences about bivariate relationships are presented in Chapter 10.

Statistics IN Action Revisited — Interpreting Scatterplots

Consider the two quantitative variables, *Length of time in practice* (number of years) and *Amount of exposure to ethics in medical school* (number of hours), measured on a sample of University Community Hospital physicians. To investigate a possible relationship between these two variables, we created a scatterplot for the data in Figure SIA2.7 using Minitab.

At first glance, the graph appears to show almost no relationship between the variables. However, note the outlying data point to the far right of the scatterplot. This point corresponds to a physician

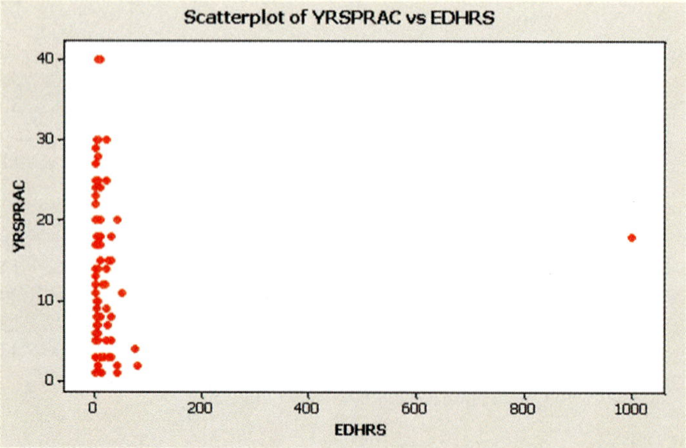

Figure SIA2.7
Minitab scatterplot of practice experience versus ethics exposure

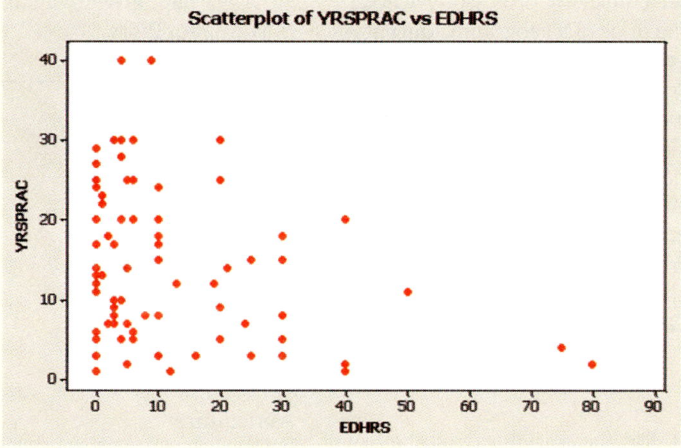

Figure SIA2.8
Minitab scatterplot of practice experience versus ethics exposure—outlier deleted

who reported 1,000 hours of exposure to ethics in medical school. Recall that we classified this data point as a highly suspect outlier in the previous SIA Revisited section (p. 89). If we remove this observation from the data set and rerun the scatterplot option of Minitab, the graph shown in Figure SIA2.8 is produced. Now the trend in the relationship is more apparent. For physicians with 20 or fewer hours of ethics exposure, there is little or no trend. However, for physicians with more than 20 hours of exposure to ethics, there appears to be a decreasing trend; that is, for physicians with high exposure to ethics, practice experience and exposure time are apparently negatively related.

Data Set: ETHICS

Exercises 2.117–2.126

Learning the Mechanics

2.117 Construct a scattergram for the data in the following table.

Variable 1:	5	1	1.5	2	2.5	3	3.5	4	4.5	5
Variable 2:	2	1	3	4	6	10	9	12	17	17

2.118 Construct a scattergram for the data in the following table.

Variable 1:	5	3	−1	2	7	6	4	0	8
Variable 2:	14	3	10	1	8	5	3	2	12

Applying the Concepts—Basic

2.119 Does elevation impact hitting performance in baseball? The Colorado Rockies play their major league home baseball games at Coors Field, Denver. Each year, the Rockies are among the leaders in team batting statistics (e.g., home runs, batting average, and slugging percentage). Many baseball experts attribute this phenomenon to the "thin air" of Denver—called the "mile-high" city due to its elevation. *Chance* (Winter 2006) investigated the effects of elevation on slugging percentage in Major League Baseball. Data were compiled on players' composite slugging percentage at each of 29 cities for the 2003 season, as well as each city's elevation (feet above sea level). The data are saved in the **MLBPARKS** file. (Selected observations are shown in the table.) Construct a scattergram for the data. Do you detect a trend?

City	Slug Pct.	Elevation
Anaheim	.480	160
Arlington	.605	616
Atlanta	.530	1,050
Baltimore	.505	130
Boston	.505	20
⋮	⋮	⋮
Denver	.625	5,277
⋮	⋮	⋮
Seattle	.550	350
San Francisco	.510	63
St. Louis	.570	465
Tampa	.500	10
Toronto	.535	566

Source: Schaffer, J. & Heiny, E.L. "The effects of elevation on slugging percentage in Major League Baseball." *Chance,* Vol. 19, No. 1, Winter 2006 (adapted from Figure 2, p. 30). Reprinted with permission from *Chance*. © 2006 by the American Statistical Association. All rights reserved.

2.120 State SAT scores. Refer to Exercise 2.29 (p. 52) and the data on state SAT scores saved in the **SATSCORES** file. NW Construct a scatterplot for the data, with 2000 SAT score on the horizontal axis and 2007 SAT score on the vertical axis. What type of trend do you detect?

2.121 Characteristics of diamonds sold at retail. Refer to the *Journal of Statistics Education* data on diamonds saved in the **DIAMONDS** file. In addition to the number of carats, the asking price for the each of the 308 diamonds for sale on the open market was recorded. Construct a scatterplot for the data, with number of carats on the horizontal axis and price on the vertical axis. What type of trend do you detect?

2.122 Are geography journals worth their cost? In *Geoforum* (Vol. 37, 2006), Simon Fraser University professor Nicholas Blomley assessed whether the price of a geography journal is correlated with quality. He collected pricing data (cost for a 1-year subscription in U.S. dollars) for a sample of 28 geography journals. In addition to cost, three other variables were measured: Journal Impact Factor (JIF), defined as the average number of times articles from the journal have been cited; number of citations for a journal over the past five years; and Relative Price Index (RPI), a measure developed by economists. [Note: A journal with an RPI less than 1.25 is considered a "good value."] The data for the 28 geography journals are saved in the **GEOJRNL** file and listed in the next table.

a. Construct a scatterplot for the variables JIF and cost. Do you detect a trend?

b. Construct a scatterplot for the variables number of cites and cost. Do you detect a trend?

c. Construct a scatterplot for the variables RPI and cost. Do you detect a trend?

Journal	Cost ($)	JIF	Cites	RPI
J. Econ. Geogr.	468	3.139	207	1.16
Prog. Hum. Geog.	624	2.943	544	0.77
T. I. Brit. Geogr.	499	2.388	249	1.11
Econ. Geogr.	90	2.325	173	0.30
A. A. A. Geogr.	698	2.115	377	0.93
Antipode	717	1.922	333	0.96
Reg. Stud.	1,312	1.652	445	1.49
Environ. Plan A	1,297	1.622	773	0.91
Environ. Plan D	479	1.591	297	0.99
Geoforum	1,118	1.560	298	1.58
Area	242	1.475	215	0.53
Polit. Geogr.	1,099	1.316	282	1.92
Int. J. Geogr. IS	1,733	1.234	240	3.42
Landscape. Ur. Plan	1,619	1.204	446	1.56
Sing. J. Trop. Geo.	197	1.029	62	1.13
Aust. Geogr.	345	0.947	72	1.83
Eur. Geogr. Eco.	499	0.780	62	1.99
Urban. Geogr.	530	0.667	135	1.81
J. Hist. Geogr.	388	0.661	96	1.44
Tij. Eco. Soc. Geo	343	0.518	99	1.00
Cult. Geogr.	538	0.500	24	NA
Pap. Reg. Sci.	397	0.481	89	1.36
Can. Geo. Can.	253	0.429	74	1.05
J. Geo. High. Ed.	1,115	0.413	79	4.33
Cartogr. J.	266	0.295	18	3.11
Geogr. Anal.	213	0.902	106	0.88
Geogr. J.	223	0.857	81	0.94
Appl. Geogr.	646	0.853	74	3.38

Source: Blomley, N. "Is this journal worth US$1118?," *Geoforum*, Vol. 37, 2006.

Applying the Concepts—Intermediate

2.123 Performance ratings of government agencies. The U.S. Office of Management and Budget (OMB) requires government agencies to produce annual performance and accounting reports (PARS) each year. A research team at George Mason University evaluated the quality of the PARS for 24 government agencies (*The Public Manager*, Summer 2008). Evaluation scores ranged from 12 (lowest) to 60 (highest). The PARS evaluation scores for 2007 and 2008, shown in the accompanying table, are saved in the **PARS** file.

a. Construct a scattergram for the data. Do you detect a trend in the data?

b. Based on the graph, identify one or two agencies that had greater than expected PARS evaluation scores for 2008.

Agency	Score07	Score08
Transportation	55	53
Labor	53	51
Veterans	51	51
NRC	39	34
Commerce	37	36
HHS	37	35
DHS	37	30
Justice	35	37
Treasury	35	35
GSA	34	40
Agriculture	33	35
EPA	33	36
Social Security	33	33
USAID	32	42
Education	32	36
Interior	32	31
NASA	32	32
Energy	31	34
HUD	31	30
NSF	31	31
State	31	50
OPM	27	28
SBA	22	31
Defense	17	32

Source: Ellig, J. & Wray, H. "Measuring performance reporting quality," *The Public Manager*, Vol. 37, No. 2, Summer 2008 (p. 66).

2.124 Spreading rate of spilled liquid. A contract engineer at DuPont Corp. studied the rate at which a spilled volatile liquid will spread across a surface (*Chemical Engineering*

Time (minutes)	Mass (pounds)
0	6.64
1	6.34
2	6.04
4	5.47
6	4.94
8	4.44
10	3.98
12	3.55
14	3.15
16	2.79
18	2.45
20	2.14
22	1.86
24	1.60
26	1.37
28	1.17
30	0.98
35	0.60
40	0.34
45	0.17
50	0.06
55	0.02
60	0.00

Source: Barry, J. "Estimating rates of spreading and evaporation of volatile liquids," *Chemical Engineering Progress*, Vol. 101, No. 1, Jan. 2005. Reproduced with permission.

Progress, Jan. 2005). Assume 50 gallons of methanol spills onto a level surface outdoors. The engineer used derived empirical formulas (assuming a state of turbulent-free convection) to calculate the mass (in pounds) of the spill after a period of time ranging from 0 to 60 minutes. The calculated mass values, given in the table on page 94, are saved in the **LIQUIDSPILL** file. Is there evidence to indicate that the mass of the spill tends to diminish as time increases? Support your answer with a scatterplot.

2.125 Most valuable NFL teams. Refer to the *Forbes* listing of the 2008 values of the 32 teams in the National Football League (NFL), Exercise 2.24 (p. 50). Access the data saved in the **NFLTEAMVALUES** file and construct a scattergram to investigate the relationship between 2008 value ($ millions) and operating income ($ millions). Would you recommend that an NFL executive use operating income to predict a team's current value? Explain.

Applying the Concepts–Advanced

2.126 Ranking driving performance of professional golfers. Refer to *The Sport Journal* (Winter 2007) analysis of a new method for ranking the total driving performance of golfers on the PGA tour, Exercise 2.50 (p. 63). Recall that the method uses both the average driving distance (yards) and driving accuracy (percent of drives that land in the fairway). Data on these two variables for the top 40 PGA golfers are saved in the **PGADRIVER** file. A professional golfer is practicing a new swing to increase his average driving distance. However, he is concerned that his driving accuracy will be lower. Is his concern a valid one? Explain.

2.10 The Time Series Plot

Each of the previous sections has been concerned with describing the information contained in a sample or population of data. Often these data are viewed as having been produced at essentially the same point in time. Thus, time has not been a factor in any of the graphical methods described so far.

Data of interest to managers are often produced and monitored over time. Examples include the daily closing price of their company's common stock, the company's weekly sales volume and quarterly profits, and characteristics—such as weight and length—of products produced by the company.

Data that are produced and monitored over time are called **time series data.**

Recall from Section 1.4 that a process is a series of actions or operations that generates output over time. Accordingly, measurements taken of a sequence of units produced by a process—such as a production process—are time series data. In general, any sequence of numbers produced over time can be thought of as being generated by a process.

When measurements are made over time, it is important to record both the numerical value and the time or the time period associated with each measurement. With this information, a **time series plot**—sometimes called a **run chart**—can be constructed to describe the time series data and to learn about the process that generated the data. A time series plot is simply a scatterplot with the measurements on the vertical axis and time or the order in which the measurements were made on the horizontal axis. The plotted points are usually connected by straight lines to make it easier to see the changes and movement in the measurements over time. For example, Figure 2.35 is a time series plot of a particular company's monthly sales (number of units sold per month). And Figure 2.36 is a time series plot of the weights of 30 one-gallon paint cans that were consecutively filled by the same filling head. Notice that the weights are plotted against the order in which the cans were filled rather than some unit of time. When monitoring production processes, it is often more convenient to record the order rather than the exact time at which each measurement was made.

Time series plots reveal the movement (trend) and changes (variation) in the variable being monitored. Notice how sales trend upward in the summer and how the variation in the weights of the paint cans increases over time. This kind of information would not be revealed by stem-and-leaf displays or histograms, as the following example illustrates.

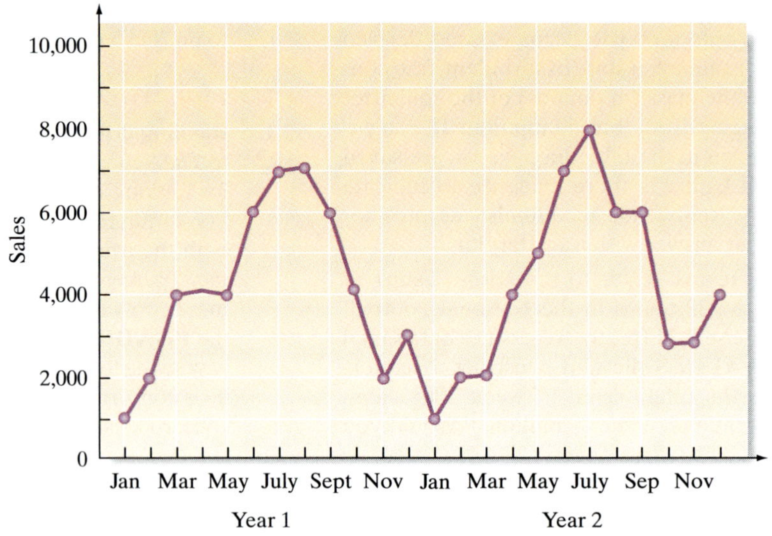

Figure 2.35
Time series plot of company sales

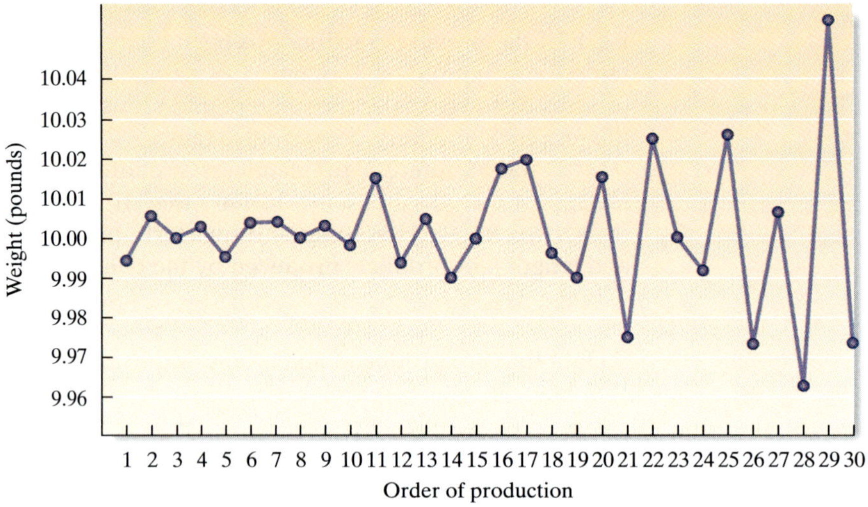

Figure 2.36
Time series plot of paint can weights

Example 2.20
Time Series Plot versus a Histogram

Problem W. Edwards Deming was one of America's most famous statisticians. He was best known for the role he played after World War II in teaching the Japanese how to improve the quality of their products by monitoring and continually improving their production processes. In his book *Out of the Crisis* (1986), Deming warned against the knee-jerk (i.e., automatic) use of histograms to display and extract information from data. As evidence, he offered the following example.

Fifty camera springs were tested in the order in which they were produced. The elongation of each spring was measured under the pull of 20 grams. Both a time series plot and a histogram were constructed from the measurements. They are shown in Figure 2.37, which has been reproduced from Deming's book. If you had to predict the elongation measurement of the next spring to be produced (i.e., spring 51) and could use only one of the two plots to guide your prediction, which would you use? Why?

Solution Only the time series plot describes the behavior *over time* of the process that produces the springs. The fact that the elongation measurements are decreasing over time can be gleaned only from the time series plot. Because the histogram does not reflect the order in which the springs were produced, it in effect represents all observations as having been produced simultaneously. Using the histogram to predict the elongation of the 51st spring would very likely lead to an overestimate.

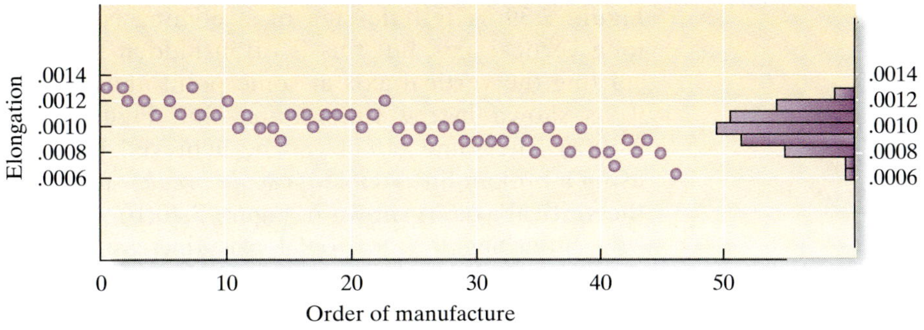

Figure 2.37
Deming's time series plot and histogram

Look Back The lesson from Deming's example is this: For displaying and analyzing data that have been generated over time by a process, the primary graphical tool is the time series plot, not the histogram.

We cover many other aspects of the statistical analysis of time series data in Chapter 13 (available on a CD that accompanies the text).

2.11 Distorting the Truth with Descriptive Techniques

A picture may be "worth a thousand words," but pictures can also color messages or distort them. In fact, the pictures in statistics (e.g., histograms, bar charts, time series plots, etc.) are susceptible to distortion, whether unintentional or as a result of unethical statistical practices. In this section, we will mention a few of the pitfalls to watch for when interpreting a chart, graph, or numerical descriptive measure.

Graphical Distortions

One common way to change the impression conveyed by a graph is to change the scale on the vertical axis, the horizontal axis, or both. For example, Figure 2.38 is a bar graph that shows the market share of sales for a company for each of the years 2004 to 2009. If you want to show that the change in firm A's market share over time is moderate, you should pack in a large number of units per inch on the vertical axis—that is, make the distance between successive units on the vertical scale small, as shown in Figure 2.38. You can see that a change in the firm's market share over time is barely apparent.

If you want to use the same data to make the changes in firm A's market share appear large, you should increase the distance between successive units on the vertical axis—that is, stretch the vertical axis by graphing only a few units per inch, as in

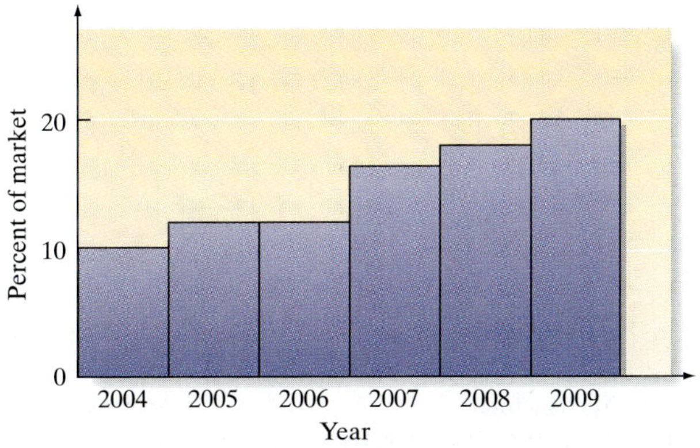

Figure 2.38
Firm A's market share from 2004 to 2009—packed vertical axis

Figure 2.39. A telltale sign of stretching is a long vertical axis, but this is often hidden by starting the vertical axis at some point above 0, as shown in the time series plot, Figure 2.40(a). The same effect can be achieved by using a broken line—called a *scale break*—for the vertical axis, as shown in Figure 2.40(b).

Stretching the horizontal axis (increasing the distance between successive units) may also lead you to incorrect conclusions. With bar graphs, a visual distortion can be achieved by making the width of the bars proportional to the height. For example, look at the bar chart in Figure 2.41(a), which depicts the percentage of a year's total automobile sales attributable to each of four major manufacturers. Now suppose we make both the width and the height grow as the market share grows. This change is shown in Figure 2.41(b). The reader may tend to equate the *area* of the bars with the relative market share of each manufacturer. But the fact is the true relative market share is proportional only to the *height* of the bars.

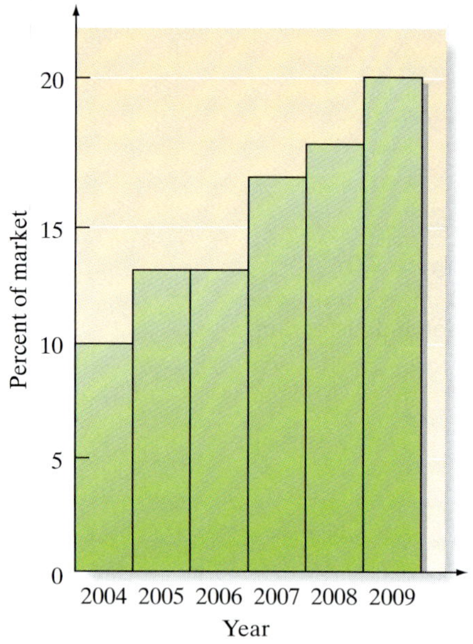

Figure 2.39
Firm A's market share from 2004 to 2009—stretched vertical axis

Ethics IN Statistics

Intentionally distorting a graph to portray a particular viewpoint is considered *unethical statistical practice*.

Sometimes we do not need to manipulate the graph to distort the impression it creates. Modifying the verbal description that accompanies the graph can change the interpretation that will be made by the viewer. Figure 2.42 provides good illustration of this ploy.

Although we've discussed only a few of the ways that graphs can be used to convey misleading pictures of phenomena, the lesson is clear. Look at all graphical descriptions of data with a critical eye. Particularly, check the axes and the size of the units on each axis. Ignore the visual changes and concentrate on the actual numerical changes indicated by the graph or chart.

Misleading Numerical Descriptive Statistics

The information in a data set can also be distorted by using numerical descriptive measures, as Example 2.21 indicates.

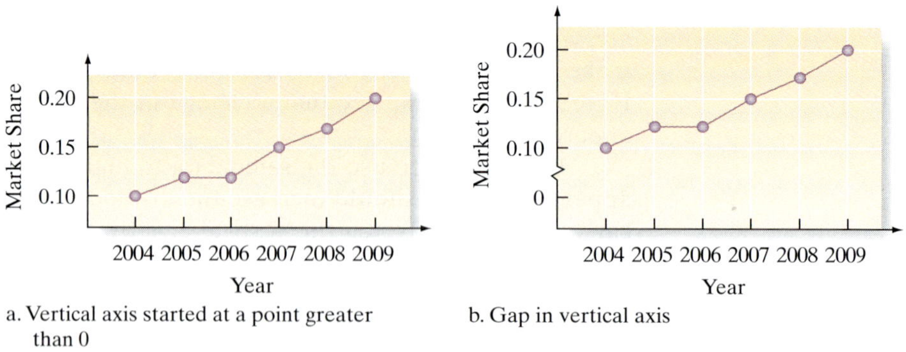

Figure 2.40
Firm A's market share from 2004 to 2009

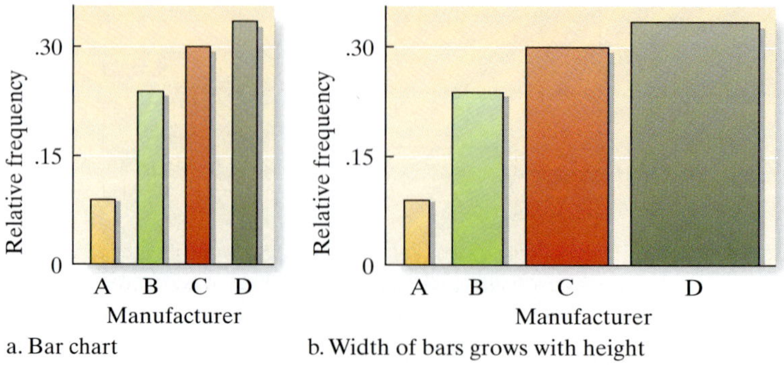

Figure 2.41
Relative share of the automobile market for each of four major manufacturers

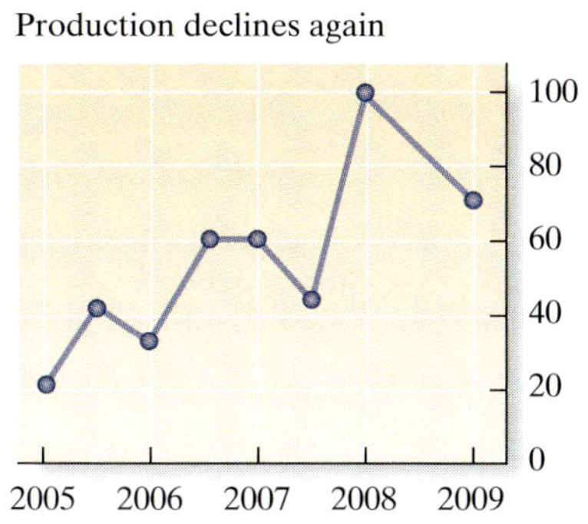

For our production, we need not even change the chart, so we can't be accused of fudging the data. Here we'll simply change the title so that for the Senate subcommittee, we'll indicate that we're not doing as well as in the past . . .

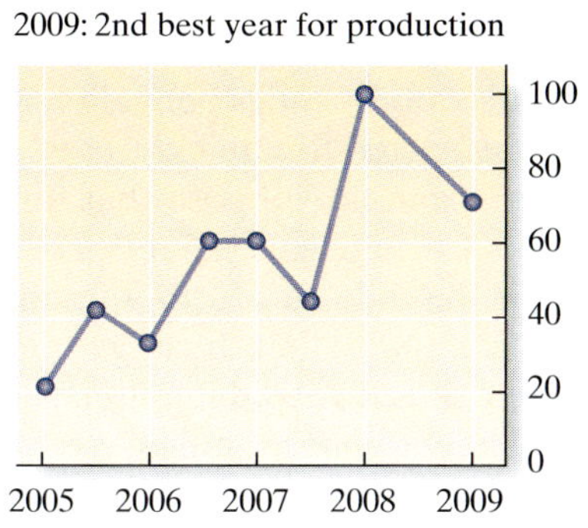

whereas for the general public, we'll tell them that we're still in the prime years.

Figure 2.42
Changing the verbal description to change a viewer's interpretation *Source:* Adapted from Selazny, G. "Grappling with Graphics," *Management Review,* Oct. 1975, p. 7.

Example 2.21
Misleading Descriptive Statistics

Problem Suppose you're considering working for a small law firm—one that currently has a senior member and three junior members. You inquire about the salary you could expect to earn if you join the firm. Unfortunately, you receive two answers:

Answer A: The senior member tells you that an "average employee" earns $107,500.

Answer B: One of the junior members later tells you that an "average employee" earns $95,000.

Which answer can you believe?

Solution The confusion exists because the phrase "average employee" has not been clearly defined. Suppose the four salaries paid are $95,000 for each of the three junior members and $145,000 for the senior member. Thus,

$$\text{Mean} = \frac{3(\$95{,}000) + \$145{,}000}{4} = \frac{\$430{,}000}{4} = \$107{,}500$$

$$\text{Median} = \$95{,}000$$

You can now see how the two answers were obtained. The senior member reported the mean of the four salaries, and the junior member reported the median. The information you received was distorted because neither person stated which measure of central tendency was being used.

Look Back Based on our earlier discussion of the mean and median, we would probably prefer the median as the measure that best describes the salary of the "average" employee.

Another distortion of information in a sample occurs when *only* a measure of central tendency is reported. Both a measure of central tendency and a measure of variability are needed to obtain an accurate mental image of a data set.

Suppose you want to buy a new car and are trying to decide which of two models to purchase. Because energy and economy are both important issues, you decide to purchase

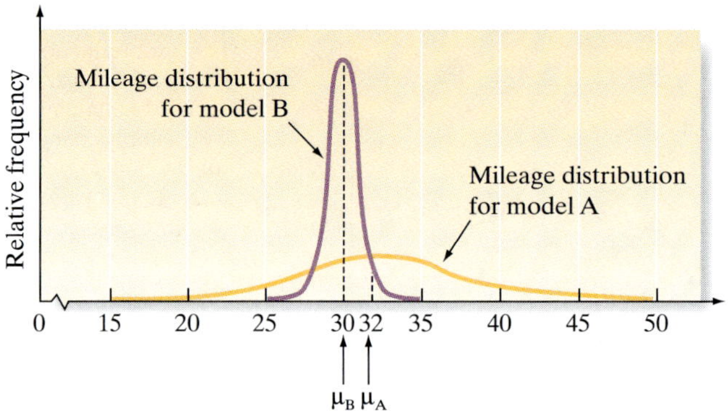

Figure 2.43 Mileage distributions for two car models

model A because its EPA mileage rating is 32 miles per gallon in the city, whereas the mileage rating for model B is only 30 miles per gallon in the city.

However, you may have acted too quickly. How much variability is associated with the ratings? As an extreme example, suppose that further investigation reveals that the standard deviation for model A mileages is 5 miles per gallon, whereas that for model B is only 1 mile per gallon. If the mileages form a mound-shaped distribution, they might appear as shown in Figure 2.43. Note that the larger amount of variability associated with model A implies that more risk is involved in purchasing model A—that is, the particular car you purchase is more likely to have a mileage rating that will greatly differ from the EPA rating of 32 miles per gallon if you purchase model A, while a model B car is not likely to vary from the 30-miles-per-gallon rating by more than 2 miles per gallon.

We conclude this section with another example on distorting the truth with numerical descriptive measures.

Example 2.22 More Misleading Descriptive Statistics

Problem *Children out of School in America* is a report on delinquency of school-age children prepared by the Children's Defense Fund (CDF), a government-sponsored organization. Consider the following three reported results of the CDF survey.

- Reported result 1: 25 percent of the 16- and 17-year-olds in the Portland, Maine, Bayside East Housing Project were out of school. Fact: *Only eight children were surveyed; two were found to be out of school.*
- Reported result 2: Of all the secondary-school students who had been suspended more than once in census tract 22 in Columbia, South Carolina, 33% had been suspended two times and 67% had been suspended three or more times. Fact: *CDF found only three children in that entire census tract who had been suspended; one child was suspended twice and the other two children, three or more times.*
- Reported result 3: In the Portland Bayside East Housing Project, 50% of all the secondary-school children who had been suspended more than once had been suspended three or more times. Fact: *The survey found two secondary-school children had been suspended in that area; one of them had been suspended three or more times.*

Identify the potential distortions in the results reported by the CDF.

Solution In each of these examples, the reporting of percentages (i.e., relative frequencies) instead of the numbers themselves is misleading. No inference we might draw from the cited examples would be reliable. (We'll see how to measure the reliability of estimated percentages in Chapter 5.) In short, either the report should state the numbers alone instead of percentages, or, better yet, it should state that the numbers were too small to report by region.

Look Back If several regions were combined, the numbers (and percentages) would be more meaningful.

Ethics IN Statistics

Purposeful reporting of numerical descriptive statistics in order to mislead the target audience is considered *unethical statistical practice*.

CHAPTER NOTES

Key Terms

Bar graph 35
Bivariate relationship 90
Box plots 82, 83
Central tendency 54
Chebyshev's Rule 71
Class 33
Class frequency 33
Class interval 43
Class percentage 33
Class relative frequency 33
Dot plot 42, 47
Empirical Rule 71
Hinges 82, 83
Histogram 43, 47
Inner fences 82, 83
Interquartile range 82
Lower quartile 82
Mean 55
Measures of central tendency 54
Measures of relative standing 77
Measures of variability or spread 64
Median 56
Middle quartile 82
Modal class 59
Mode 58
Mound-shaped distribution 70, 79
Numerical descriptive measures 54
Outer fences 82, 83
Outliers 81
p^{th} percentile 77
Pareto diagram 35
Percentile ranking/score 77
Pie chart 35
Population z-score 78
Quartiles 82
Range 64
Rare-event approach 86
Relative frequency histogram 43
Run chart 95
Sample standard deviation 66
Sample variance 66
Sample z-score 78
Scattergram 90
Scatterplot 90
Skewness 58
Spread 64
Standard deviation 66
Stem-and-leaf display 42, 47
Summation notation 54
Symmetric distribution 70
Time series data 95
Time series plot 95
Time study 70
Upper quartile 82
Variability 54
Variance 66
Whiskers 82, 83
z-score 78

Key Symbols

	Sample	*Population*
Mean:	$\bar{x}$	μ
Variance:	s^2	σ^2
Std. Dev.:	s	σ
Median:	m	
Lower Quartile:	Q_L	
Upper Quartile:	Q_U	
Interquartile Range:	IQR	

Key Ideas

Describing Qualitative Data

1. Identify **category classes**
2. Determine **class frequencies**
3. **Class relative frequency** = (class frequency)/n
4. **Graph** relative frequencies

Pie Chart:

Bar Graph:

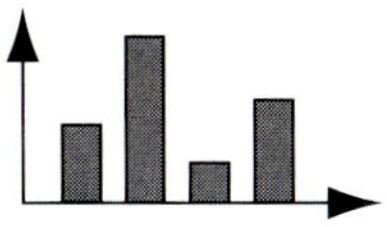

Pareto Diagram:

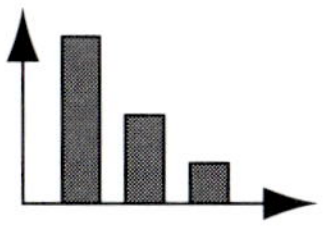

Graphing Quantitative Data

1 Variable

1. Identify class intervals
2. Determine **class interval frequencies**
3. **Class interval relative frequency** = (class interval frequency)/n
4. **Graph** class interval relative frequencies

Dot Plot:

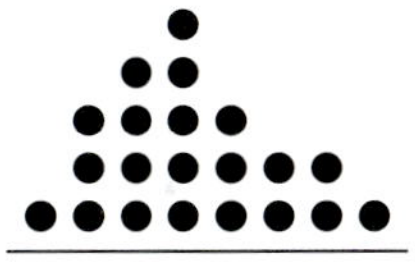

Stem-and-Leaf Display:

1	3
2	2489
3	126678
4	37
5	2

Histogram:

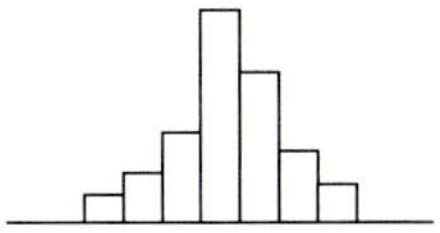

Box Plot:

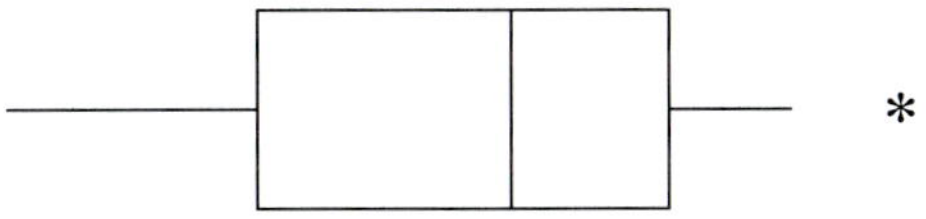

2 Variables

Scatterplot:

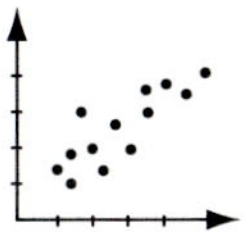

Numerical Description of Quantitative Data

Central Tendency

Mean: $\bar{x} = (\Sigma x_i)/n$

Median: Middle value when data ranked in order

Mode: Value that occurs most often

Variation

Range: Difference between largest and smallest value

Variance:

$$s^2 = \frac{\Sigma(x_i - \bar{x})^2}{n-1} = \frac{\Sigma x_i^2 - \frac{(\Sigma x_i)^2}{n}}{n-1}$$

Std Dev.: $s = \sqrt{s^2}$

Interquartile Range: $\text{IQR} = Q_U - Q_L$

Relative Standing

Percentile Score: Percentage of values that fall below x-score

z-score:

$$z = (x - \bar{x})/s$$

Rules for Detecting Quantitative Outliers

Interval	*Chebyshev's Rule*	*Empirical Rule*
$\bar{x} \pm s$	At least 0%	$\approx$ 68%
$\bar{x} \pm 2s$	At least 75%	$\approx$ 95%
$\bar{x} \pm 3s$	At least 89%	$\approx$ All

Rules for Detecting Quantitative Outliers

Method	*Suspect*	*Highly Suspect*
Box plot:	Values between inner & outer fences	Values beyond outer fences
z-score:	$2 < \lvert z \rvert < 3$	$\lvert z \rvert > 3$

Guide to Selecting the Data Description Method

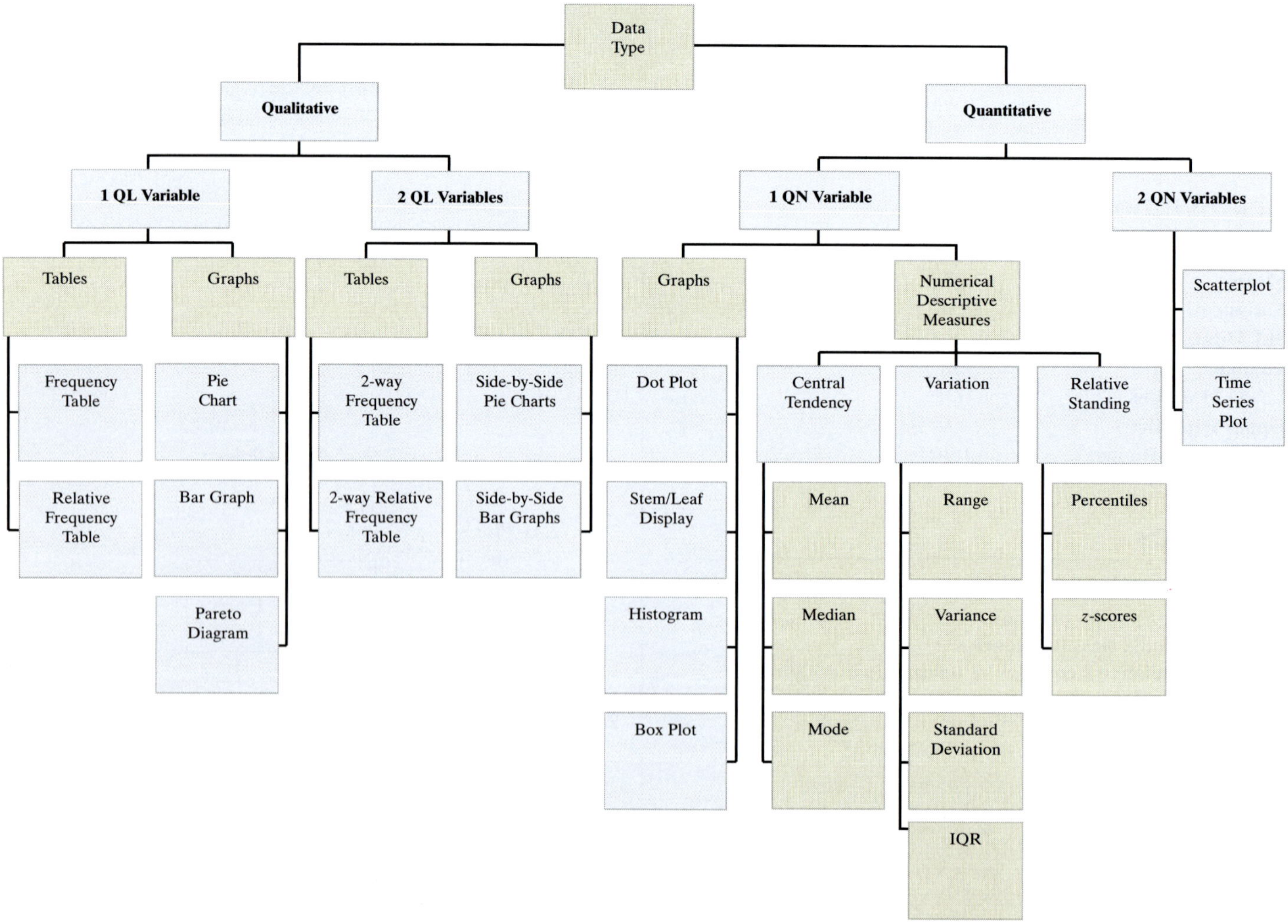

Supplementary Exercises 2.127–2.159

Learning the Mechanics

2.127 Construct a relative frequency histogram for the data summarized in the accompanying table.

Measurement Class	Relative Frequency
.00–.75	.02
.75–1.50	.01
1.50–2.25	.03
2.25–3.00	.05
3.00–3.75	.10
3.75–4.50	.14
4.50–5.25	.19
5.25–6.00	.15
6.00–6.75	.12
6.75–7.50	.09
7.50–8.25	.05
8.25–9.00	.04
9.00–9.75	.01

2.128 Discuss the conditions under which the median is preferred to the mean as a measure of central tendency.

2.129 Consider the following three measurements: 50, 70, 80. Find the z-score for each measurement if they are from a population with a mean and standard deviation equal to

a. $\mu = 60, \sigma = 10$ **b.** $\mu = 50, \sigma = 5$
c. $\mu = 40, \sigma = 10$ **d.** $\mu = 40, \sigma = 100$

2.130 Refer to Exercise 2.129. For parts **a–d,** determine whether the values 50, 70, and 80 are outliers.

2.131 For each of the following data sets, compute $\bar{x}$, s^2, and s:

a. 13, 1, 10, 3, 3
b. 13, 6, 6, 0
c. 1, 0, 1, 10, 11, 11, 15
d. 3, 3, 3, 3

2.132 For each of the following data sets, compute $\bar{x}$, s^2, and s. If appropriate, specify the units in which your answers are expressed.

a. 4, 6, 6, 5, 6, 7
b. −\$1, \$4, −\$3, \$0, −\$3, −\$6
c. $^3/_5$%, $^4/_5$%, $^2/_5$%, $^1/_5$%, $^1/_{16}$%
d. Calculate the range of each data set in parts **a–c.**

2.133 Explain why we generally prefer the standard deviation to the range as a measure of variability for quantitative data.

2.134 If the range of a set of data is 20, find a rough approximation to the standard deviation of the data set.

2.135 Construct a scattergram for the data in the following table.

Variable 1:	174	268	345	119	400	520	190	448	307	252
Variable 2:	8	10	15	7	22	31	15	20	11	9

Applying the Concepts—Basic

2.136 National firearms survey. In the journal *Injury Prevention* (Jan. 2007), researchers from the Harvard School of Public Health reported on the size and composition of privately held firearm stock in the United States. In a representative household telephone survey of 2,770 adults, 26% reported that they own at least one gun. The accompanying graphic summarizes the types of firearms owned.

a. What type of graph is shown?
b. Identify the qualitative variable described in the graph.
c. From the graph, identify the most common type of firearms.
d. Convert the graph into a Pareto diagram. Interpret the results.

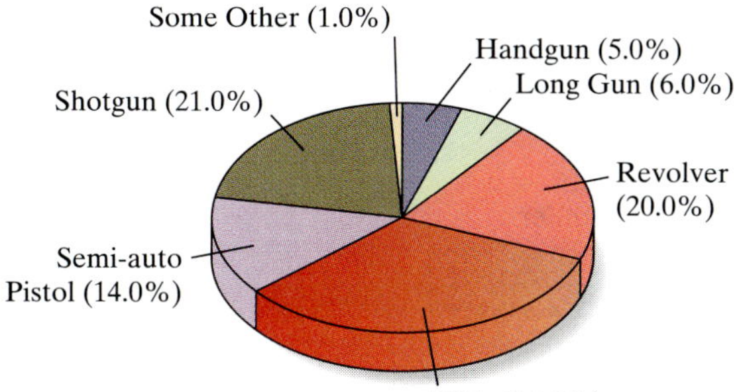

Source: Hepburn, M., Miller, D. A., and Hemenway, D. "The U.S. gun stock: Results from the 2004 national firearms survey," *Injury Prevention,* Vol. 13, No. 1, Jan. 2007 (Figure 1).

2.137 Business marketing publications. Business-to-business marketing describes the field of marketing between multiple business entities. The *Journal of Business-to-Business Marketing* (Vol. 15, 2008) produced a pie chart describing the number of business-to-business marketing articles published in all journals, by topical area, between 1971 and 2006. The data used to produce the pie chart are shown in the table and saved in the **MARPUB** file.

a. Compute the relative frequencies for the nine topical areas shown in the table. Interpret the relative frequency for Buyer Behavior.
b. Use the relative frequencies, part **a,** to construct a pie chart for the data. Why is the slice for Marketing Research smaller than the slice for Sales Management?

Area	Number
Global Marketing	235
Sales Management	494
Buyer Behavior	478
Relationships	498
Innovation	398
Marketing Strategy	280
Channels/Distribution	213
Marketing Research	131
Services	136
Total	2,863

Source: "Commentary on 'The essence of business marketing…' by Lichtenthal, Mummalaneni, and Wilson: The JBBM Comes of Age," by Peter J. LaPlaca, *Journal of Business-to-Business Marketing,* Vol. 15, No. 2, 2008 (Figure 4, p. 187). Copyright © 2008 Routledge.

2.138 A boom in U.S. bankruptcies. The American Bankruptcy Institute and the National Bankruptcy Research Center monitor the number of consumer bankruptcy filings each month. The table lists the number of bankruptcy filings for each of the first 10 months of 2008.

a. Explain why the data in the table represent time series data.

b. Construct a time series plot for the monthly number of bankruptcy filings.

c. Do you detect a trend in the time series plot? Explain.

Month in 2008	Number of bankruptcy filings
January	66,050
February	76,120
March	86,165
April	92,291
May	91,214
June	82,770
July	94,124
August	96,413
September	88,663
October	106,266

Source: CardData (www.carddata.com), Nov. 5, 2008.

2.139 Crash tests on new cars. The National Highway Traffic Safety Administration (NHTSA) crash-tests new car models to determine how well they protect the driver and front-seat passenger in a head-on collision. The NHTSA has developed a "star" scoring system for the frontal crash test, with results ranging from one star (*) to five stars (*****). The more stars in the rating, the better the level of crash protection in a head-on collision. The NHTSA crash test results for 98 cars (in a recent model year) are stored in the data file named **CRASH.** The driver-side star ratings for the 98 cars are summarized in the Minitab printout shown below. Use the information in the printout to form a pie chart. Interpret the graph.

Tally for Discrete Variables: DRIVSTAR

DRIVSTAR	Count	Percent
2	4	4.08
3	17	17.35
4	59	60.20
5	18	18.37
N=	98	

2.140 Crash tests on new cars (cont'd). Refer to Exercise 2.139. One quantitative variable recorded by the NHTSA is driver's severity of head injury (measured on a scale from 0 to 1,500). The mean and standard deviation for the 98 driver head-injury ratings in the **CRASH** file are displayed in the Minitab printout below. Use these values to find the z-score for a driver head-injury rating of 408. Interpret the result.

Descriptive Statistics: DRIVHEAD

Variable	N	Mean	StDev	Minimum	Q1	Median	Q3	Maximum
DRIVHEAD	98	603.7	185.4	216.0	475.0	605.0	724.3	1240.0

2.141 Defects in new automobiles. Consider the following data from the automobile industry (adapted from Kane 1989). All cars produced on a particular day were inspected for defects. The 145 defects found were categorized by type as shown in the accompanying table.

Defect Type	Number
Accessories	50
Body	70
Electrical	10
Engine	5
Transmission	10

a. Construct a Pareto diagram for the data. Use the graph to identify the most frequently observed type of defect.

b. All 70 car body defects were further classified as to type. The frequencies are provided in the following table. Form a Pareto diagram for type of body defect. (Adding this graph to the original Pareto diagram of part **a** is called *exploding the Pareto diagram.*) Interpret the result. What type of body defect should be targeted for special attention?

Body Defect	Number
Chrome	2
Dents	25
Paint	30
Upholstery	10
Windshield	3

2.142 Switching off air bags. Driver-side and passenger-side air bags are installed in all new cars to prevent serious or fatal injury in an automobile crash. However, air bags have been found to cause deaths in children and small people or people with handicaps in low-speed crashes. Consequently, in 1998, the federal government began allowing vehicle owners to request installation of an on–off switch for air bags. The table describes the reasons for requesting the installation of passenger-side on–off switches given by car owners over a two-year period.

Reason	Number of Requests
Infant	1,852
Child	17,148
Medical	8,377
Infant & medical	44
Child & medical	903
Infant & child	1,878
Infant & child & medical	135
Total	30,337

Source: National Highway Transportation Safety Administration, September 2000.

a. What type of variable, quantitative or qualitative, is summarized in the table? Give the values that the variable could assume.

b. Calculate the relative frequencies for each reason.

c. Display the information in the table in an appropriate graph.

d. What proportion of the car owners who requested on–off air bag switches gave medical as one of the reasons?

2.143 Orlando theme park prices. The entrance fees for adults to 14 theme parks and attractions in Orlando, Florida, are listed in the following table. (These data are saved in the **PARKFEES** file.) Find and interpret the mean, median, and mode of the entrance fees. Which measure of central tendency best describes the entrance-fee distribution of the data?

Theme Park/Attraction	Entrance Fee ($)
Blue Water Hot Air Balloons	185.00
Disney Animal Kingdom	56.70
Disney EPCOT Center	56.70
Disney Magic Kingdom	56.70
Disney MGM Studios	56.70
Disney Pleasure Island	20.95
Disney Wide World of Sports Complex	9.00
DisneyQuest	35.00
Gatorland	21.95
Holy Land Experience	29.99
SeaWorld Adventure Park	61.95
Universal CityWalk	11.95
Universal Islands of Adventure	63.00
Universal Studios	63.00

Source: American Automobile Association, 2006.

2.144 Collecting Beanie Babies. Beanie Babies are toy stuffed animals that have become valuable collector's items. *Beanie World Magazine* provided the age, retired status, and value of 50 Beanie Babies. The data are saved in the **BEANIE** file, with several of the observations shown in the table (in the next column).

a. Summarize the retired/current status of the 50 Beanie Babies with an appropriate graph. Interpret the graph.

b. Summarize the values of the 50 Beanie Babies with an appropriate graph. Interpret the graph.

Name	Age (Months)	Retired (R) Current (C)	Value ($)
1. Ally the Alligator	52	R	55.00
2. Batty the Bat	12	C	12.00
3. Bongo the Brown Monkey	28	R	40.00
4. Blackie the Bear	52	C	10.00
5. Bucky the Beaver	40	R	45.00
⋮	⋮	⋮	⋮
46. Stripes the Tiger(Gold/Black)	40	R	400.00
47. Teddy the 1997 Holiday Bear	12	R	50.00
48. Tuffy the Terrier	17	C	10.00
49. Tracker the Basset Hound	5	C	15.00
50. Zip the Black Cat	28	R	40.00

Source: Beanie World Magazine, Sept. 1998.

c. Use a graph to portray the relationship between a Beanie Baby's value and its age. Do you detect a trend?

d. According to Chebyshev's Rule, what percentage of the age measurements would you expect to find in the intervals $\bar{x} \pm .75s$, $\bar{x} \pm 2.5s$, $\bar{x} \pm 4s$?

e. What percentage of the age measurements actually fall in the intervals of part **d**? Compare your results with those of part **d.**

f. Repeat parts **d** and **e** for the value measurements.

2.145 Top Florida law firms. Data on the top-ranked law firms in Florida, obtained from *Florida Trend* magazine (April 2002), are provided in the table and saved in the **FLALAW** file.

a. Find the mean, median, and mode for the number of lawyers at the top-ranked Florida law firms. Interpret these values.

Rank	Firm	Headquarters	Number of Lawyers	Number of Offices
1	Holland & Knight	Tallahassee	529	11
2	Akerman Senterfit	Orlando	355	9
3	Greenberg Traurig	Miami	301	6
4	Carlton Fields	Tampa	207	6
5	Ruden McClosky Smit	Ft. Lauderdale	175	9
6	Fowler White Boggs	Tampa	175	7
7	Foley & Lardner	Orlando	159	5
8	Gray Harris	Orlando	158	6
9	Broad and Cassel	Orlando	150	7
10	Shutts & Bowen	Miami	144	5
11	Steel, Hector & Davis	Miami	141	5
12	Gunster Yoakley	W. Palm Beach	140	6
13	Adorno & Zeder	Miami	105	4
14	Becker & Poliakoff	Ft. Lauderdale	100	12
15	Lowndes Drosdick	Orlando	100	1
16	Conroy Simberg Ganon	Hollywood	91	6
17	Stearns Weaver	Miami	85	3
18	Wicker Smith O'Hara	Miami	85	6
19	Rogers Towers Bailey	Jacksonville	80	2
20	Butler Burnette	Tampa	77	3
21	Bilzin Sumberg Dunn	Miami	70	1
22	Morgan Colling	Orlando	70	4
23	White & Case	Miami	70	1
24	Fowler White Burnett	Miami	64	4
25	Rissman Weisberg	Orlando	63	3
26	Rumberger Kirk	Orlando	63	4

Source: Florida Trend magazine, April 2002, p. 105.

Data Set: FLALAW

b. Find the mean, median, and mode for the number of offices open by top-ranked Florida law firms. Interpret these values.

c. Construct a histogram for the number of lawyers. Which measure of central tendency best describes this distribution?

d. Construct a stem-and-leaf display for the number of offices. Which measure of central tendency best describes this distribution?

e. Construct a scatterplot for the data, with number of law offices on the horizontal axis and number of lawyers on the vertical axis. What type of trend do you detect?

Applying the Concepts—Intermediate

2.146 Survey of computer crime. Refer to the Computer Security Institute (CSI) annual survey of computer crime at U.S. businesses, Exercise 1.20 (p. 23). One question asked, "Did your business suffer unauthorized use of computer systems within the past year?" The responses are summarized in the next table for two survey years, 1999 and 2006. Compare the responses for the 2 years using side-by-side bar charts. What inference can be made from the charts?

Unauthorized Use of Computer Systems	Percentage in 1999	Percentage in 2006
Yes	62	52
No	17	38
Don't know	21	10
Totals	100	100

Source: "2006 CSI/FBI computer crime and security survey," *Computer Security Issues & Trends,* Spring 2006. Adapted from Fig. 12, p. 10.

2.147 Hull failures of oil tankers. Owing to several major ocean oil spills by tank vessels, Congress passed the 1990 Oil Pollution Act, which requires all tankers to be designed with thicker hulls. Further improvements in the structural design of a tank vessel have been proposed since then, each with the objective of reducing the likelihood of an oil spill and decreasing the amount of outflow in the event of a hull puncture. To aid in this development, *Marine Technology* (Jan. 1995) reported on the spillage amount (in thousands of metric tons) and cause of puncture for 42 recent major oil spills from tankers and carriers. [*Note:* Cause of puncture is classified as either collision (C), fire/explosion (FE), hull failure (HF), or grounding (G).] The data are saved in the **OILSPILL** file.

a. Use a graphical method to describe the cause of oil spillage for the 42 tankers. Does the graph suggest that any one cause is more likely to occur than any other? How is this information of value to the design engineers?

b. Find and interpret descriptive statistics for the 42 spillage amounts. Use this information to form an interval that can be used to predict the spillage amount of the next major oil spill.

2.148 Software defects. The Promise Software Engineering Repository is a collection of data sets available to serve businesses in building predictive software models. One such data set, saved in the **SWDEFECTS** file, contains information on 498 modules of software code. Each module was analyzed for defects and classified as "true" if it contained defective code and "false" if not. Access the data file and produce a bar graph or a pie chart for the defect variable. Use the graph to make a statement about the likelihood of defective software code.

2.149 Evaluating toothpaste brands. *Consumer Reports*, published by Consumers Union, is a magazine that contains ratings and reports for consumers on goods, services, health, and personal finances. Consumers Union reported on the testing of 46 brands of toothpaste (*Consumer Reports*, Sept. 1992). Each was rated on package design, flavor, cleaning ability, fluoride content, and cost per month (a cost estimate based on brushing with half-inch of toothpaste twice daily). The data in the next table (saved in the **TOOTHPASTE** file) are costs per month for the 46 brands. Costs marked by an asterisk represent those brands that carry the American Dental Association (ADA) seal verifying effective decay prevention.

a. Construct a stem-and-leaf display for the data.

b. Circle the individual leaves that represent those brands that carry the ADA seal.

c. What does the pattern of circles suggest about the costs of those brands approved by the ADA?

.58	.66	1.02	1.11	1.77	1.40	.73*	53*	.57*	1.34
1.29	.89*	.49	.53*	.52	3.90	4.73	1.26	.71*	.55*
.59*	.97	.44*	.74*	.51*	.68*	.67	1.22	.39	.55
.62	.66*	1.07	.64	1.32*	1.77*	.80*	.79	.89*	.64
.81*	.79*	.44*	1.09	1.04	1.12				

2.150 Time to develop price quotes. A manufacturer of industrial wheels is losing many profitable orders because of the long time it takes the firm's marketing, engineering, and accounting departments to develop price quotes for potential customers. To remedy this problem, the firm's management would like to set guidelines for the length of time each department should spend developing price quotes. To help develop these guidelines, 50 requests for price quotes were randomly selected from the set of price quotes made last year: the processing time (in days) was determined for each price quote for each department. These times are saved in the **LOSTQUOTES** file. Several observations are displayed in the table below. The price quotes are also classified by whether or not they were "lost" (i.e., whether or not the customer placed an order after receiving the price quote).

Request Number	Marketing	Engineering	Accounting	Lost?
1	7.0	6.2	.1	No
2	.4	5.2	.1	No
3	2.4	4.6	.6	No
4	6.2	13.0	.8	Yes
5	4.7	.9	.5	No
⋮	⋮	⋮	⋮	⋮
46	6.4	1.3	6.2	No
47	4.0	2.4	13.5	Yes
48	10.0	5.3	.1	No
49	8.0	14.4	1.9	Yes
50	7.0	10.0	2.0	No

a. Construct a stem-and-leaf display for the total processing time for each department. Shade the leaves that correspond to "lost" orders in each of the displays, and interpret each of the displays.

b. Using your results from part **a,** develop "maximum processing time" guidelines for each department that, if followed, will help the firm reduce the number of lost orders.

c. Generate summary statistics for the processing times. Interpret the results.

d. Calculate the z-score corresponding to the maximum processing time guideline you developed in part **b** for each department, and for the total processing time.

e. Calculate the maximum processing time corresponding to a z-score of 3 for each of the departments. What percentage of the orders exceed these guidelines? How does this agree with Chebyshev's Rule and the Empirical Rule?

f. Repeat part **e** using a z-score of 2.

g. Compare the percentage of "lost" quotes with corresponding times that exceed at least one of the guidelines in part **e** to the same percentage using the guidelines in part **f.** Which set of guidelines would you recommend be adopted? Why?

2.151 Misleading advertisement. A time series plot similar to the one shown next appeared in a recent advertisement for a well-known golf magazine. One person might interpret the plot's message as the longer you subscribe to the magazine, the better golfer you should become. Another person might interpret it as indicating that if you subscribe for 3 years, your game should improve dramatically.

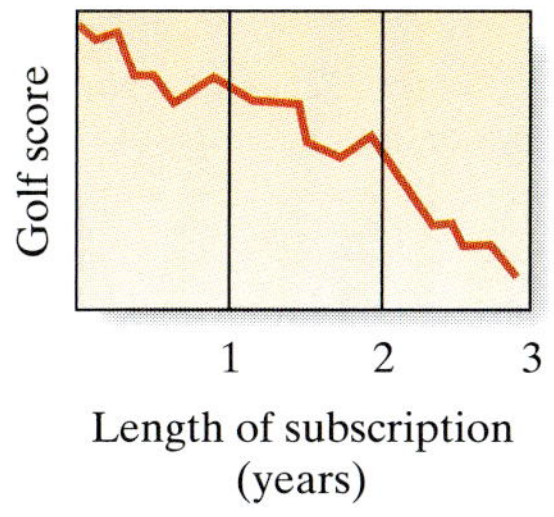

a. Explain why the plot can be interpreted in more than one way.

b. How could the plot be altered to rectify the current distortion?

2.152 History of corporate acquisitions. Refer to the *Academy of Management Journal* (Aug. 2008) study of corporate acquisitions from 1980 to 2000, Exercise 2.12 (p. 40). The data are saved in the **ACQUISITIONS** file.

a. Construct a time series plot of the number of firms with at least one acquisition.

b. For each year, compute the percentage of sampled firms with at least one acquisition. Then construct a time series plot of these percentages.

c. Which time series plot, part **a** or part **b,** is more informative about the history of corporate acquisitions over time? Explain.

2.153 Radiation levels in homes. In some locations, radiation levels in homes are measured at well above normal background levels in the environment. As a result, many architects and builders are making design changes to ensure adequate air exchange so that radiation will not be "trapped" in homes. In one such location, 50 homes levels were measured, and the mean level was 10 parts per billion (ppb), the median was 8 ppb, and the standard deviation was 3 ppb. Background levels in this location are at about 4 ppb.

a. Based on these results, is the distribution of the 50 homes' radiation levels symmetric, skewed to the left, or skewed to the right? Why?

b. Use both Chebyshev's Rule and the Empirical Rule to describe the distribution of radiation levels. Which do you think is most appropriate in this case? Why?

c. Use the results from part **b** to approximate the number of homes in this sample that have radiation levels above the background level.

d. Suppose another home is measured at a location 10 miles from the one sampled and has a level of 20 ppb. What is the z-score for this measurement relative to the 50 homes sampled in the other location? Is it likely that this new measurement comes from the same distribution of radiation levels as the other 50? Why? How would you go about confirming your conclusion?

2.154 Amount of zinc phosphide in commercial rat poison. A chemical company produces a substance composed of 98% cracked corn particles and 2% zinc phosphide for use in controlling rat populations in sugarcane fields. Production must be carefully controlled to maintain the 2% zinc phosphide because too much zinc phosphide will cause damage to the sugarcane and too little will be ineffective in controlling the rat population. Records from past production indicate that the distribution of the actual percentage of zinc phosphide present in the substance is approximately mound-shaped, with a mean of 2.0% and a standard deviation of .08%.

a. If the production line is operating correctly, approximately what proportion of batches from a day's production will contain less than 1.84% of zinc phosphide?

b. Suppose one batch chosen randomly actually contains 1.80% zinc phosphide. Does this indicate that there is too little zinc phosphide in today's production? Explain your reasoning.

2.155 U.S. peanut production. If not examined carefully, the graphical description of U.S. peanut production shown on the next page can be misleading.

a. Explain why the graph may mislead some readers.

b. Construct an undistorted graph of U.S. peanut production for the given years.

Applying the Concepts—Advanced

2.156 Investigating the claims of weight-loss clinics. The U.S. Federal Trade Commission assesses fines and other penalties against weight-loss clinics that make unsupported or misleading claims about the effectiveness of their programs. Brochures from two weight-loss clinics both advertise "statistical evidence" about the effectiveness of their programs. Clinic A claims that the *mean* weight loss during the first month is 15 pounds; Clinic B claims a *median* weight loss of 10 pounds.

a. Assuming the statistics are accurately calculated, which clinic would you recommend if you had no other information? Why?

b. Upon further research, the median and standard deviation for Clinic A are found to be 10 pounds and 20 pounds, respectively, while the mean and standard deviation for Clinic B are found to be 10 and 5 pounds,

Graph for Exercise 2.155

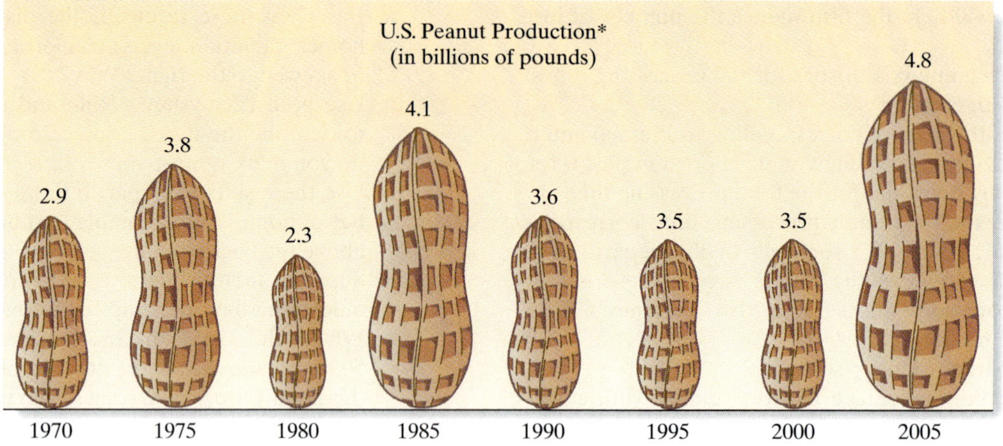

respectively. Both are based on samples of more than 100 clients. Describe the two clinics' weight-loss distributions as completely as possible given this additional information. What would you recommend to a prospective client now? Why?

c. Note that nothing has been said about how the sample of clients upon which the statistics are based was selected. What additional information would be important regarding the sampling techniques employed by the clinics?

2.157 **Age discrimination study.** The Age Discrimination in Employment Act mandates that workers 40 years of age or older be treated without regard to age in all phases of employment (hiring, promotions, firing, etc.). Age discrimination cases are of two types: *disparate treatment* and *disparate impact*. In the former, the issue is whether workers have been intentionally discriminated against. In the latter, the issue is whether employment practices adversely affect the protected class (i.e., workers 40 and over) even though no such effect was intended by the employer (Zabel 1989). A small computer manufacturer laid off 10 of its 20 software engineers. The ages of all engineers at the time of the layoff, shown below, are saved in the **LAYOFF** file. Analyze the data to determine whether the company may be vulnerable to a disparate impact claim.

Not laid off:	34	55	42	38	42	32	40	40	46	29
Laid off:	52	35	40	41	40	39	40	64	47	44

Critical Thinking Challenges

2.158 **No Child Left Behind Act.** According to the government, federal spending on K-12 education has increased dramatically over the past 20 years, but student performance has essentially stayed the same. Hence, in 2002, President George Bush signed into law the No Child Left Behind Act, a bill that promised improved student achievement for all U.S. children. *Chance* (Fall 2003) reported on a graphic obtained from the U.S. Department of Education Web site (www.ed.gov) that was designed to support the new legislation. The graphic is reproduced here. The bars in the graph represent annual federal spending on education, in billions of dollars (left-side vertical axis). The horizontal line represents the annual average fourth-grade children's reading ability score (right-side vertical axis). Critically assess the information portrayed in the graph. Does it, in fact, support the government's position that our children are not making classroom improvements despite federal spending on education? Use the following facts (divulged in the *Chance* article) to help you frame your answer: (1) The U.S. student population has also increased dramatically over the past 20 years, (2) fourth-grade reading test scores are designed to have an average of 250 with a standard deviation of 50, and (3) the reading test scores of seventh and twelfth grades and the mathematics scores of fourth graders did improve substantially over the past 20 years.

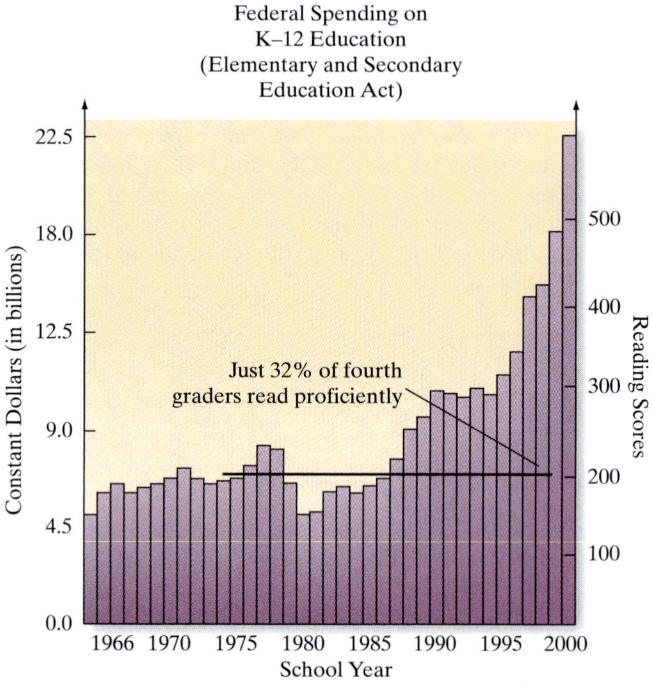

Source: U.S. Department of Education.

2.159 **Steel rod quality.** In his essay "Making Things Right," W. Edwards Deming considered the role of statistics in the quality control of industrial products.* In one example, Deming examined the quality-control process for a manufacturer of steel rods. Rods produced with diameters smaller than 1 centimeter fit too loosely in their bearings

*From Tanur, J., et al., eds. *Statistics: A Guide to the Unknown*. San Francisco: Holden-Day, 1978, pp. 279–81.

and ultimately must be rejected (thrown out). To determine whether the diameter setting of the machine that produces the rods is correct, 500 rods are selected from the day's production and their diameters are recorded. The distribution of the 500 diameters for one day's production is shown in the accompanying figure. Note that the symbol LSL in the figure represents the 1-centimeter lower specification limit of the steel rod diameters. There has been speculation that some of the inspectors are unaware of the trouble that an undersized rod diameter would cause later in the manufacturing process. Consequently, these inspectors may be passing rods with diameters that were barely below the lower specification limit and recording them in the interval centered at 1.000 centimeter. According to the figure, is there any evidence to support this claim? Explain.

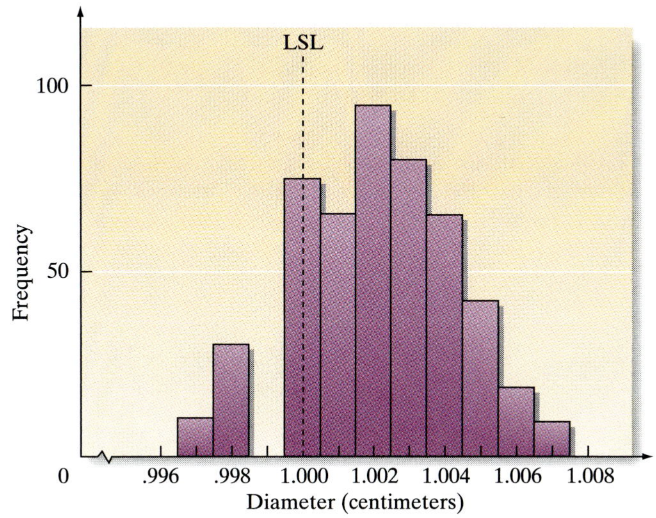

References

Deming, W. E. *Out of the Crisis*. Cambridge, Mass.: M.I.T. Center for Advanced Engineering Study, 1986.

Gitlow, H., Oppenheim, A., Oppenheim, R., and Levine, D. *Quality Management,* 3rd ed. Homewood, Ill.: Irwin, 2004.

Huff, D. *How to Lie with Statistics*. New York: Norton, 1954.

Ishikawa, K. *Guide to Quality Control,* 2nd ed. Asian Productivity Organization, 1986.

Juran, J. M. *Juran on Quality by Design: The New Steps for Planning Quality into Goods and Services*. New York: The Free Press, 1992.

Mendenhall, W., Beaver, R. J., and Beaver, B. M. *Introduction to Probability and Statistics,* 12th ed. North Scituate, Mass.: Duxbury, 2006.

Tufte, E. R. *Beautiful Evidence*. Cheshire Conn.: Graphics Press, 2006.

Tufte, E. R. *Envisioning Information*. Cheshire, Conn.: Graphics Press, 1990.

Tufte, E. R. *Visual Display of Quantiative Information*. Cheshire, Conn.: Graphics Press, 1983.

Tufte, E. R. *Visual Explanations*. Cheshire, Conn.: Graphics Press, 1997.

Tukey, J. *Exploratory Data Analysis*. Reading, Mass.: Addison-Wesley, 1977.

Zabel, S. L. "Statistical Proof of Employment Discrimination." *Statistics: A Guide to the Unknown*, 3rd ed. Pacific Grove, Calif.: Wadsworth, 1989.

USING TECHNOLOGY

SPSS: Describing Data

Graphing Data

Step 1 Click on the "Graphs" button on the SPSS menu bar and then select "Legacy Dialogs."

Step 2 Click on the graph of your choice (bar, pie, boxplot, scatter/dotplot, or histogram) to view the appropriate dialog box. The dialog box for a histogram is shown in Figure 2.S.1.

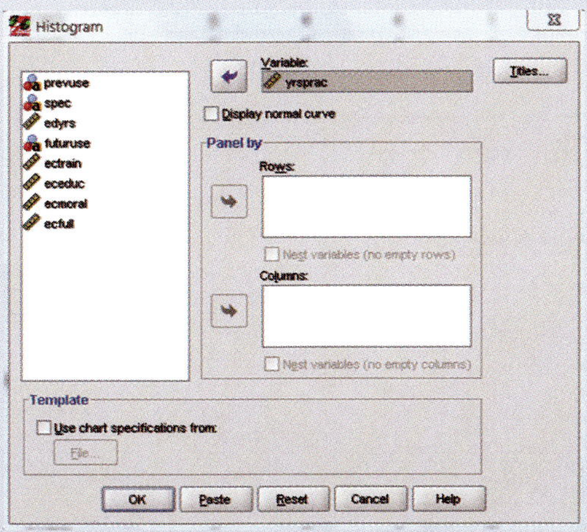

Figure 2.S.1 SPSS histogram dialog box

Step 3 Make the appropriate variable selections and click "OK" to view the graph.

Stem-and-Leaf Plots

Step 1 Select "Analyze" from the main SPSS menu, then "Descriptive Statistics," and then "Explore."

Step 2 In the "Explore" dialog box, select the variable to be analyzed in the "Dependent List" box, as shown in Figure 2.S.2.

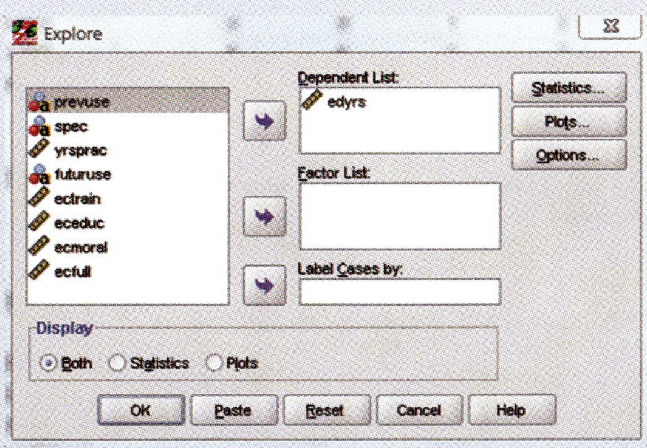

Figure 2.S.2 SPSS explore dialog box

Step 3 Click on either "Both" or "Plots" in the "Display" options and then click "OK" to display the stem-and-leaf graph.

Pareto Diagrams

Step 1 Select "Analyze" from the main SPSS menu, then "Quality Control," and then "Pareto Charts."

Step 2 Click "Define" on the resulting menu and then select the variable to be analyzed and move it to the "Category Axis" box, as shown in Figure 2.S.3.

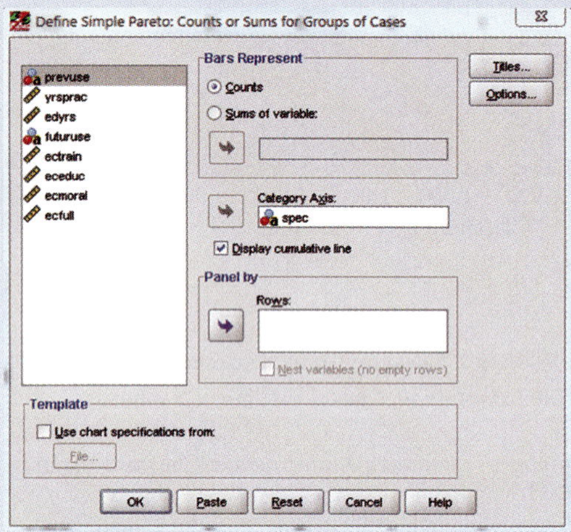

Figure 2.S.3 SPSS Pareto chart dialog box

Step 3 Click "OK" to display the Pareto diagram.

Numerical Descriptive Statistics

Step 1 Click on the "Analyze" button on the main menu bar and then click on "Descriptive Statistics."

Step 2 Select "Descriptives;" the dialog box is shown in Figure 2.S.4.

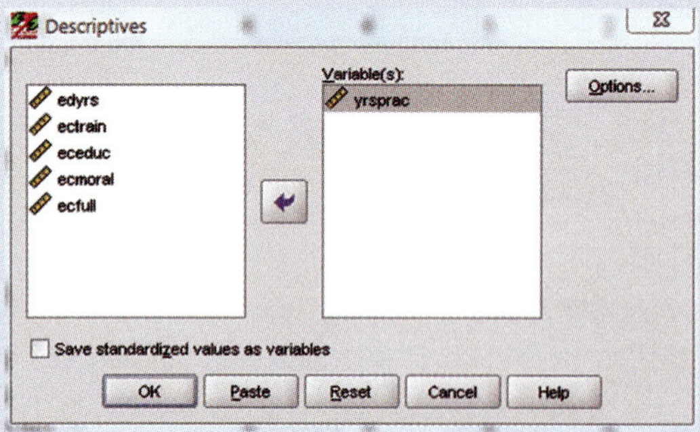

Figure 2.S.4 SPSS descriptive statistics dialog box

Step 3 Select the quantitative variables you want to analyze and place them in the "Variable(s)" box. You can control which descriptive statistics appear by clicking the "Options" button on the dialog box and making your selections.

Step 4 Click "OK" to view the descriptive statistics printout.

Percentiles

Step 1 Select "Explore" from the main SPSS menu.

Step 2 In the resulting dialog box (see Figure 2.5.2), select the "Statistics" button and check the "Percentiles" box.

Step 3 Return to the "Explore" dialog box and click "OK" to generate the descriptive statistics.

Minitab: Describing Data

Graphing Data

Step 1 Click on the "Graph" button on the Minitab menu bar.

Step 2 Click on the graph of your choice (bar, pie, scatterplot, histogram, dotplot, or stem-and-leaf) to view the appropriate dialog box. The dialog box for a histogram is shown in Figure 2.M.1.

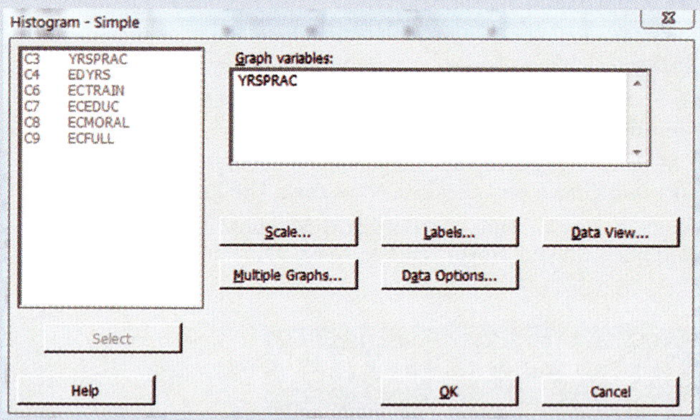

Figure 2.M.1 Minitab histogram dialog box

Step 3 Make the appropriate variable selections and then click "OK" to view the graph.

Numerical Descriptive Statistics

Step 1 Click on the "Stat" button on the main menu bar, click on "Basic Statistics," and then click on "Display Descriptive Statistics." The resulting dialog box appears in Figure 2.M.2.

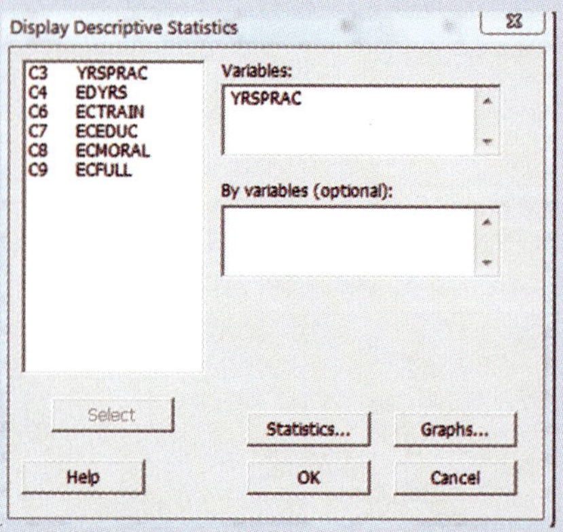

Figure 2.M.2 Minitab descriptive statistics dialog box

Step 2 Select the quantitative variables you want to analyze and place them in the "Variables" box. You can control which descriptive statistics appear by clicking the "Statistics" button on the dialog box and making your selections. (As an option, you can create histograms and dot plots for the data by clicking the "Graphs" button and making the appropriate selections.)

Step 3 Click "OK" to view the descriptive statistics printout.

Excel/DDXL: Describing Data

Graphing Data

Step 1 Highlight (select) the data columns you want to graph.

Step 2 Click the "Add-Ins" button on the Excel main menu bar and then click "DDXL."

Step 3 From the resulting menu select "Charts and Plots." The "Charts and Plots" dialog box will appear (as shown in Figure 2.E.1).

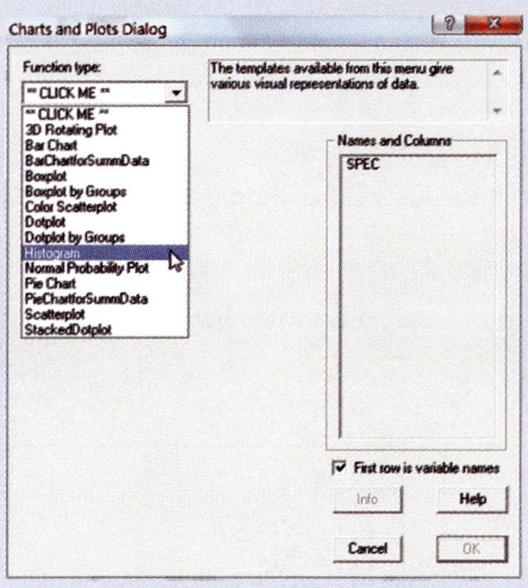

Figure 2.E.1 DDXL charts and plots menu options

Step 4 Click on the graph of your choice to view the appropriate dialog box. The dialog box for a histogram is shown in Figure 2.E.2.

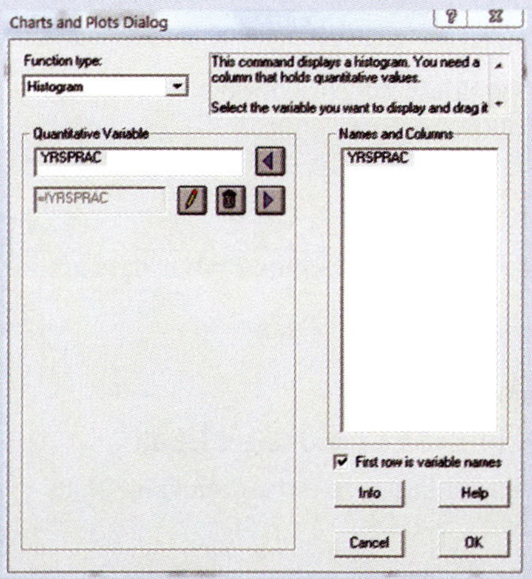

Figure 2.E.2 DDXL charts and plots menu for a histogram

Step 5 Move the variable of interest into the "Quantitative (or Qualitative)Variable" box and then click "OK" to view the graph.

Numerical Descriptive Statistics

Step 1 Highlight (select) the data columns you want to analyze, then click on "Add-Ins" in the main Excel menu bar and select "DDXL." On the resulting menu select "Summaries."

Step 2 Select "Summary of One Variable" to view the dialog box shown in Figure 2.E.3.

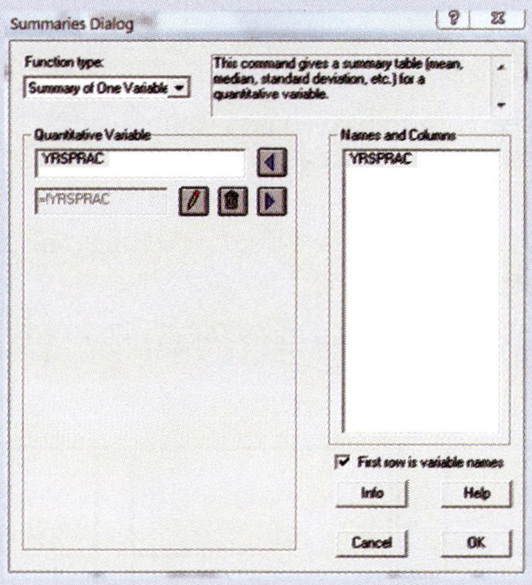

Figure 2.E.3 Excel/DDXL summary of one variable dialog box

Step 3 Move the variable of interest into the "Quantitative Variable" box and then click "OK" to view the summary statistics.

TI-84 Graphing Calculator: Describing Data

Histogram from Raw Data

Step 1 *Enter the data*

- Press **STAT** and select **1:Edit**

Note: If the list already contains data, clear the old data. Use the up arrow to highlight "**L1.**" Press **CLEAR ENTER.**

Use the arrow and **ENTER** keys to enter the data set into **L1.**

Step 2 *Set up the histogram plot*

- Press **2nd** and press **Y =** for **STAT PLOT**
- Press **1** for **Plot 1**
- Set the cursor so that **ON** is flashing
- For **Type,** use the arrow and Enter keys to highlight and select the histogram
- For **Xlist,** choose the column containing the data (in most cases, L1)

Note: Press **2nd 1** for **L1.**

Freq should be set to 1.

Step 3 *Select your window settings*

- Press **WINDOW** and adjust the settings as follows:

X min = lowest class boundary
X max = highest class boundary
X sel = class width
Y min = 0
Y max ≥ greatest class frequency
Y scl = 1
X res = 1

Step 4 *View the graph*

- Press **GRAPH**

Optional *Read class frequencies and class boundaries*

Step You can press **TRACE** to read the class frequencies and class boundaries. Use the arrow keys to move between bars.

Example The following figures show TI-84 window settings and histogram for the following sample data:

86, 70, 62, 98, 73, 56, 53, 92, 86, 37, 62, 83, 78, 49, 78, 37, 67, 79, 57

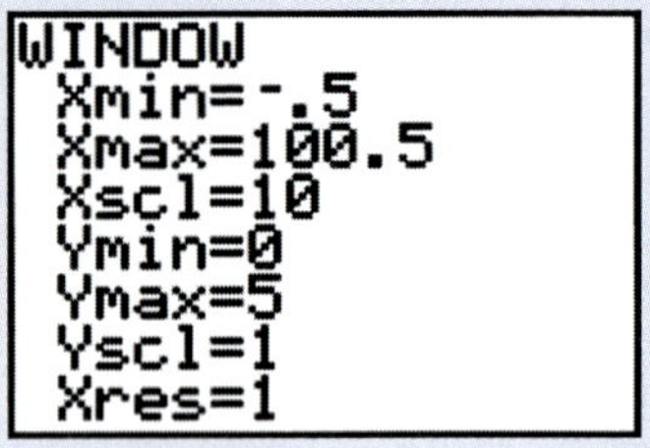

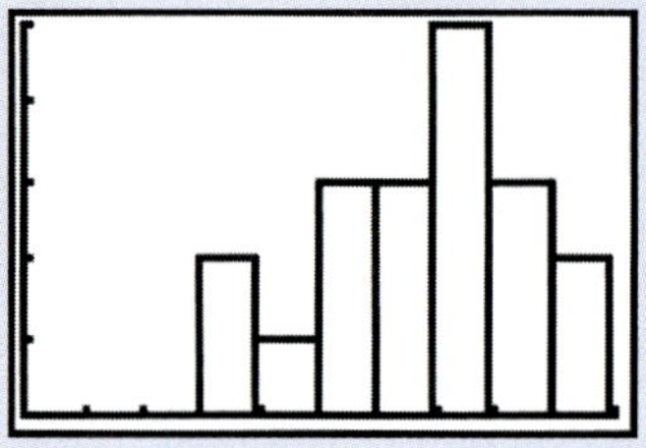

Histogram from a Frequency Table

Step 1 *Enter the data*

- Press **STAT** and select **1:Edit**

Note: If a list already contains data, clear the old data. Use the up arrow to highlight the list name, "**L1**" or "**L2.**"

- Press **CLEAR ENTER**
- Enter the midpoint of each class into **L1**
- Enter the class frequencies or relative frequencies into **L2**

Step 2 *Set up the histogram plot*

- Press **2nd** and **Y** = for **STAT PLOT**
- Press **1** for **Plot 1**
- Set the cursor so that **ON** is flashing
- For **Type,** use the arrow and Enter keys to highlight and select the histogram
- For **Xlist,** choose the column containing the midpoints
- For **Freq,** choose the column containing the frequencies or relative frequencies

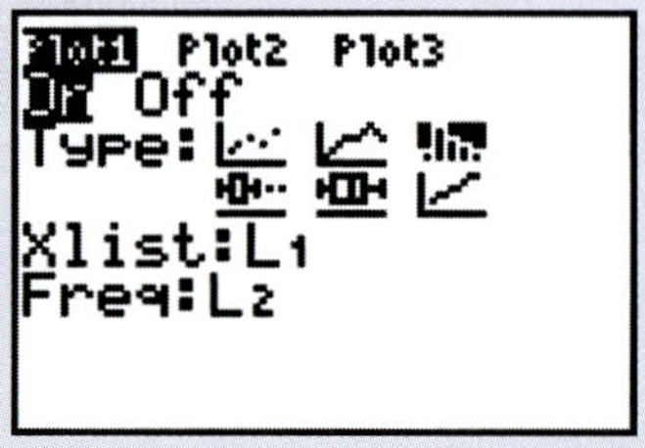

Step 3–4 *Follow steps 3–4 given above.*

Note: To set up the Window for relative frequencies, be sure to set **Ymax** to a value that is greater than or equal to the largest relative frequency.

One-Variable Descriptive Statistics

Step 1 *Enter the data*

- Press STAT and select **1:Edit**

Note: If the list already contains data, clear the old data. Use the up arrow to highlight "**L1.**" Press **CLEAR ENTER.**

- Use the arrow and **ENTER** keys to enter the data set into L1

Step 2 *Calculate descriptive statistics*

- Press **STAT**
- Press the right arrow key to highlight **CALC**
- Press **ENTER** for **1-Var Stats**
- Enter the name of the list containing your data
- Press **2nd 1** for **L1** (or **2nd 2** for **L2,** etc.)
- Press **ENTER**

You should see the statistics on your screen. Some of the statistics are off the bottom of the screen. Use the down arrow to scroll through to see the remaining statistics. Use the up arrow to scroll back up.

Example The descriptive statistics for the sample data set

86, 70, 62, 98, 73, 56, 53, 92, 86, 37, 62, 83, 78, 49, 78, 37, 67, 79, 57

The output screens for this example are shown below.

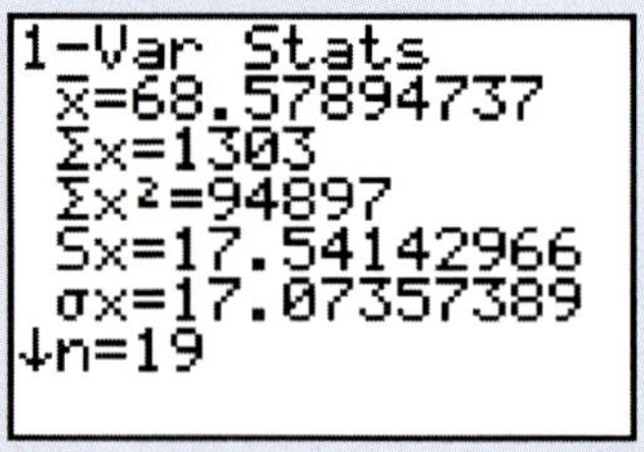

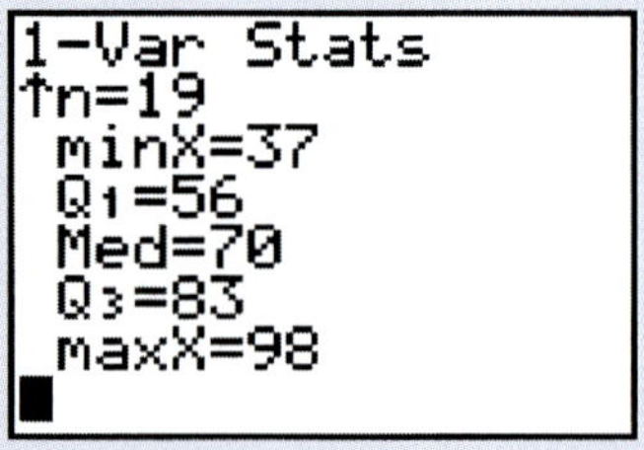

Sorting Data (to find the Mode)

The descriptive statistics do not include the mode. To find the mode, sort your data as follows:

- Press **STAT**
- Press **2** for **SORTA(**
- Enter the name of the list your data are in. If your data are in **L1,** press **2nd 1**
- Press **ENTER**
- The screen will say: **DONE**
- To see the sorted data, press **STAT** and select **1:Edit**
- Scroll down through the list and locate the data value that occurs most frequently

Box Plot

Step 1 *Enter the data*

- Press **STAT** and select **1:Edit**

Note: If the list already contains data, clear the old data. Use the up arrow to highlight "**L1.**" Press **CLEAR ENTER.**

- Use the arrow and **ENTER** keys to enter the data set into **L1**

Step 2 *Set up the box plot*

- Press **2nd Y** = for **STAT PLOT**
- Press **1** for **Plot 1**
- Set the cursor so that "**ON**" is flashing
- For **TYPE,** use the right arrow to scroll through the plot icons and select the box plot in the middle of the second row
- For **XLIST,** choose **L1**
- Set **FREQ** to **1**

Step 3 *View the graph*

- Press **ZOOM** and select **9:ZoomStat**

Optional *Read the five number summary*

- Press **TRACE**
- Use the left and right arrow keys to move between **minX, Q1, Med, Q3,** and **maxX**

Example Make a box plot for the given data:

86, 70, 62, 98, 73, 56, 53, 92, 86, 37, 62, 83, 78, 49, 78, 37, 67, 79, 57

The output screen for this example is shown below.

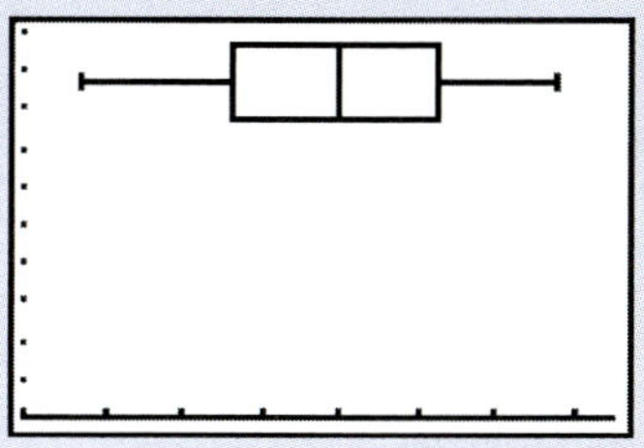

Scatterplots

Step 1 *Enter the data*

- Press **STAT** and select **1:Edit**

Note: If a list already contains data, clear the old data. Use the up arrow to highlight the list name, "**L1**" or "**L2.**"

- Press **CLEAR ENTER**
- Enter your x-data in **L1** and your y-data in **L2**

Step 2 *Set up the scatterplot*

- Press **2nd Y =** for **STAT PLOT**
- Press **1** for **Plot1**
- Set the cursor so that **ON** is flashing
- For **Type,** use the arrow and Enter keys to highlight and select the scatterplot (first icon in the first row)

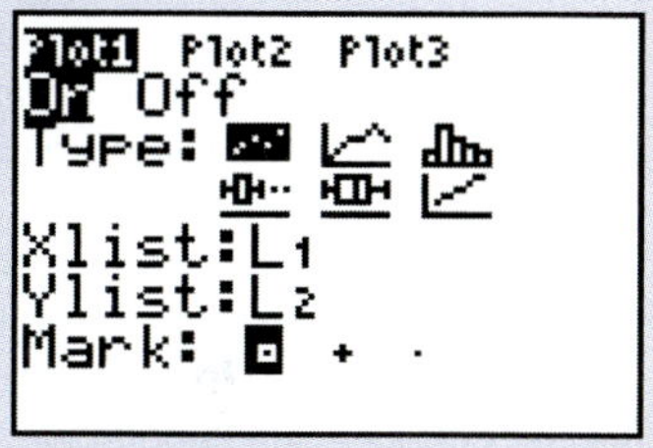

- For **Xlist,** choose the column containing the x-data
- For **Ylist,** choose the column containing the y-data

Step 3 *View the scatterplot*

- Press **ZOOM 9** for **ZoomStat**

Example The figures below show a table of data entered on the T1-84 and the scatterplot of the data obtained using the steps given above.

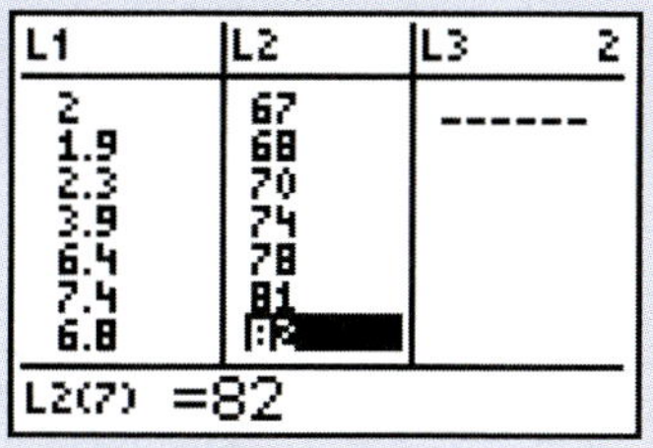

L1	L2	L3 2
2	67	------
1.9	68	
2.3	70	
3.9	74	
6.4	78	
7.4	81	
6.8	82	

L2(7) =82

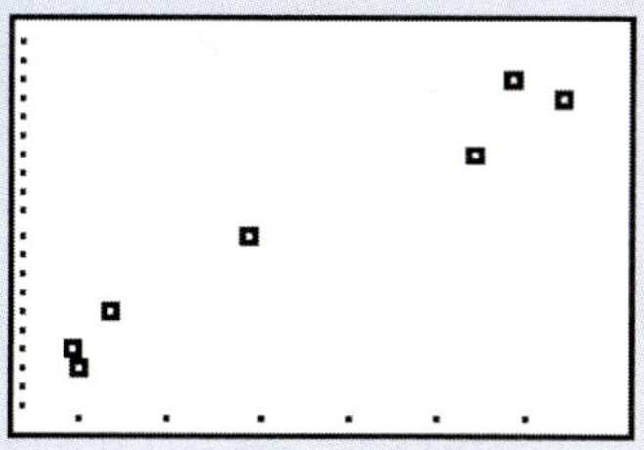

The Kentucky Milk Case—Part 1

Covers Chapters 1 and 2 – Many products and services are purchased by governments, cities, states, and businesses on the basis of scaled bids, and contracts are awarded to the lowest bidders. This process works extremely well in competitive markets, but it has the potential to increase the cost of purchasing if the markets are noncompetitive or if collusive practices are present. An investigation that began with a statistical analysis of bids in the Florida school milk market led to the recovery of more than $33,000,000 from dairies that had conspired to rig the bids there. The investigation spread quickly to other states, and to date, settlements and fines from dairies exceed $100.000,000 for school milk bidrigging in 20 other states. This case concerns a school milk bidrigging investigation in Kentucky.

Each year, the Commonwealth of Kentucky invites bids from dairies to supply half-pint containers of fluid milk products for its school districts. The products include whole white milk, low-fat white milk, and low-fat chocolate milk. In 13 school districts in northern Kentucky, the suppliers (dairies) were accused of "price-fixing" that is conspiring to allocate the districts so that the "winner" was predetermined. Because these districts are located in Boone, Campbell, and Kenton counties, the geographic market they represent is designated as the "tri-county" market. Over a 9-year period, two dairies—Meyer Dairy and Trauth Dairy—were the only bidders on the milk contracts in the school districts in the tri-county market. Consequently, these two companies were awarded all the milk contracts in the market. (In contrast, a large number of different dairies won the milk contracts for the school districts in the remainder of the northern Kentucky market—called the "surrounding" market.) The Commonwealth of Kentucky alleged that Meyer and Trauth conspired to allocate the districts in the tri-county market. To date, one of the dairies (Meyer) has admitted guilt, while the other (Trauth) steadfastly maintains its innocence.

The Commonwealth of Kentucky maintains a database on all bids received from the dairies competing for the milk contracts. Some of these data have been made available to you to analyze to determine whether there is empirical evidence of bid collusion in the tri-county market. The data, saved in the **MILK** file, are described in detail below. Some background information on the data and important economic theory regarding bid collusion is also provided. Use this information to guide your analysis. Prepare a professional document that presents the results of your analysis and gives your opinion regarding collusion.

Variable	Type	Description
YEAR	QN	Year in which milk contract awarded
MARKET	QL	Northern Kentucky Market (TRI-COUNTY or SURROUND)
WINNER	QL	Name of winning dairy
WWBID	QN	Winning bid price of whole white milk (dollars per half-pint)
WWQTY	QN	Quantity of whole white milk purchased (number of half-pints)
LFWBID	QN	Winning bid price of low-fat white milk (dollars per half-pints)
LFWQTY	QN	Quantity of low-fat white milk purchased (number of half-pints)
LFCBID	QN	Winning bid price of low-fat chocolate milk (dollars per half-pint)
LFCQTY	QN	Quantity of low-fat chocolate milk purchased (number of half-pints)
DISTRICT	QL	School district number
KYFMO	QN	FMO minimum raw cost of milk (dollars per half-pint)
MILESM	QN	Distance (miles) from Meyer processing plant to school district
MILEST	QN	Distance (miles) from Trauth processing plant to school district
LETDATE	QL	Date on which bidding on milk contract began (month/day/year)

(Number of observations: 392)

Data Set: MILK

Background Information

Collusive Market Environment

Certain economic features of a market create an environment in which collusion may be found. These basic features include the following:

1. *Few sellers and high concentration.* Only a few dairies control all or nearly all of the milk business in the market.
2. *Homogeneous products.* The products sold are essentially the same from the standpoint of the buyer (i.e., the school district).
3. *Inelastic demand.* Demand is relatively insensitive to price. (*Note:* The quantity of milk required by a school district is primarily determined by school enrollment, not price.)
4. *Similar costs.* The dairies bidding for the milk contracts face similar cost conditions. (*Note:* Approximately 60% of a dairy's production cost is raw milk, which is federally regulated. Meyer and Trauth are dairies of similar size, and both bought their raw milk from the same supplier.)

Although these market structure characteristics create an environment that makes collusive behavior easier, they do not necessarily indicate the existence of collusion. An analysis of the actual bid prices may provide additional information about the degree of competition in the market.

Collusive Bidding Patterns

The analyses of patterns in sealed bids reveal much about the level of competition, or lack thereof, among the vendors serving the market. Consider the following bid analyses:

1. *Market shares.* A market share for a dairy is the number of milk half-pints supplied by the dairy over a given school year, divided by the total number of half-pints supplied to the entire market.

One sign of potential collusive behavior is stable, nearly equal market shares over time for the dairies under investigation.

2. *Incumbency rates.* Market allocation is a common form of collusive behavior in bidrigging conspiracies. Typically, the same dairy controls the same school districts year after year. The incumbency rate for a market in a given school year is defined as the percentage of school districts that are won by the same vendor who won the previous year. An incumbency rate that exceeds 70% has been considered a sign of collusive behavior.
3. *Bid levels and dispersion.* In competitive sealed bid markets, vendors do not share information about their bids. Consequently, more dispersion or variability among the bids is observed than in collusive markets, where vendors communicate about their bids and have a tendency to submit bids in close proximity to one another in an attempt to make the bidding appear competitive. Furthermore, in competitive markets the bid dispersion tends to be directly proportional to the level of the bid: When bids are submitted at relatively high levels, there is more variability among the bids than when they are submitted at or near marginal cost, which will be approximately the same among dairies in the same geographic market.
4. *Price versus cost/distance.* In competitive markets, bid prices are expected to track costs over time. Thus, if the market is competitive, the bid price of milk should be highly correlated with the raw milk cost. Lack of such a relationship is another sign of collusion. Similarly, bid price should be correlated to the distance the product must travel from the processing plant to the school (due to delivery costs) in a competitive market.
5. *Bid sequence.* School milk bids are submitted over the spring and summer months, generally at the end of one school year and before the beginning of the next. When the bids are examined in sequence in competitive markets, the level of bidding is expected to fall as the bidding season progresses. (This phenomenon is attributable to the learning process that occurs during the season, with bids adjusted accordingly. Dairies may submit relatively high bids early in the season to "test the market," confident that volume can be picked up later if the early high bids lose. But, dairies who do not win much business early in the season are likely to become more aggressive in their bidding as the season progresses, driving price levels down.) Constant or slightly increasing price patterns of sequential bids in a market where a single dairy wins year after year is considered another indication of collusive behavior.
6. *Comparison of average winning bid prices.* Consider two similar markets, one in which bids are possibly rigged and the other in which bids are competitively determined. In theory, the mean winning price in the "rigged" market will be significantly higher than the mean price in the competitive market for each year in which collusion occurs.

CONTENTS

3 Probability

Where We've Been

- Identified the objective of inferential statistics: to make inferences about a population based on information in a sample.
- Introduced graphical and numerical descriptive measures for both quantitative and qualitative data.

Where We're Going

- Develop probability as a measure of uncertainty.
- Introduce basic rules for finding probabilities.
- Use probability as a measure of reliability for an inference.

Statistics IN Action Lotto Buster!

"Welcome to the Wonderful World of Lottery Bu$ters." So began the premier issue of Lottery Buster, *a monthly publication for players of the state lottery games.* Lottery Buster *provides interesting facts and figures on the 42 state lotteries and 2 multistate lotteries currently operating in the United States and, more importantly, tips on how to increase a player's odds of winning the lottery.*

New Hampshire, in 1963, was the first state in modern times to authorize a state lottery as an alternative to increasing taxes. (Prior to this time, beginning in 1895, lotteries were banned in America because of corruption.) Since then, lotteries have become immensely popular for two reasons: (1) They lure you with the opportunity to win millions of dollars with a $1 investment, and (2) when you lose, at least you believe your money is going to a good cause. Many state lotteries, such as Florida, designate a high percentage of lottery revenues to fund state education.

The popularity of the state lottery has brought with it an avalanche of "experts" and "mathematical wizards" (such as the editors of *Lottery Buster*) who provide advice on how to win the lottery—for a fee, of course! Many offer guaranteed *(continued)* "systems" of winning through computer software products with catchy names such as Lotto Wizard, Lottorobics, Win4d, and Lotto-luck.

For example, most knowledgeable lottery players would agree that the "golden rule" or "first rule" in winning lotteries is *game selection.* State lotteries generally offer three types of games: an Instant (scratch-off tickets or online) game, Daily Numbers (Pick-3 or Pick-4), and a weekly Pick-6 Lotto game.

One version of the Instant game involves scratching off the thin opaque covering on a ticket with the edge of a coin to determine whether you have won or lost. The cost of a ticket ranges from 50¢ to $5, and the amount won ranges from $1 to $100,000 in most states, and to as much as $1 million in others. *Lottery Buster* advises against playing the Instant game because it is "a pure chance play, and you can win only by dumb luck. No skill can be applied to this game."

The Daily Numbers game permits you to choose either a three-digit (Pick-3) or four-digit (Pick-4) number at a cost of $1 per ticket. Each night, the winning number is drawn. If your number matches the winning number, you win a large sum of money, usually $100,000. You do have some control over the Daily Numbers game (because you pick the numbers that you play), and, consequently, there are strategies available to increase your chances of winning. However, the Daily Numbers game, like the Instant game, is not available for out-of-state play.

To play Pick-6 Lotto, you select six numbers of your choice from a field of numbers ranging from 1 to *N,* where *N* depends on which state's game you are playing. For example, Florida's current Lotto game involves picking six numbers ranging from 1 to 53. (See Figure SIA3.1.) The cost of a ticket is $1, and the payoff, if your six numbers match the winning numbers drawn, is $7 million or more, depending on the number of tickets purchased. (To date, Florida has had the largest individual state weekly payoff of over $200 million.) In addition to the grand prize, you can win second-, third-, and fourth-prize payoffs by matching five, four, and three of the six numbers drawn, respectively. And you don't have to be a resident of the state to play the state's Lotto game.

Figure SIA3.1
Reproduction of Florida's 6/53 Lotto ticket

In this chapter, several Statistics in Action Revisited examples demonstrate how to use the basic concepts of probability to compute the odds of winning a state lottery game and to assess the validity of the strategies suggested by lottery "experts."

Statistics IN Action Revisited

- The Probability of Winning Lotto (p. 127)
- The Probability of Winning A Wheel System (p. 136)
- The Probability of Winning Cash 3 or Play 4 (p. 150)

Recall that one branch of statistics is concerned with decisions about a population based on sample information. You can see how this is accomplished more easily if you understand the relationship between population and sample—a relationship that becomes clearer if we reverse the statistical procedure of making inferences from sample to population. In this chapter, then, we assume that the population is known and calculate the chances of obtaining various samples from the population. Thus, we show that probability is the reverse of statistics: **In probability, we use the population information to infer the probable nature of the sample.**

Probability plays an important role in inference making. Suppose, for example, you have an opportunity to invest in an oil exploration company. Past records show that out of 10 previous oil drillings (a sample of the company's experiences), all 10 came up dry. What do you conclude? Do you think the chances are better than 50:50 that the company will hit a gusher? Should you invest in this company? Chances are, your answer to these questions will be an emphatic "no." If the company's exploratory prowess is sufficient to hit a producing well 50% of the time, a record of 10 dry wells out of 10 drilled is an event that is just too improbable.

Or suppose you're playing poker with what your opponents assure you is a well-shuffled deck of cards. In three consecutive five-card hands, the person on your right is dealt four aces. Based on this sample of three deals, do you think the cards are being adequately shuffled? Again, your answer is likely to be "no" because dealing three hands of four aces is just too improbable if the cards were properly shuffled.

Note that the decisions concerning the potential success of the oil drilling company and the adequacy of card shuffling both involve knowing the chance—or probability—of a certain sample result. Both situations were contrived so that you could easily conclude that the probabilities of the sample results were small. Unfortunately, the probabilities of many observed sample results are not so easy to evaluate intuitively. For these cases we will need the assistance of a theory of probability.

3.1 Events, Sample Spaces, and Probability

Let's begin our treatment of probability with simple examples that are easily described. With the aid of simple examples, we can introduce important definitions that will help us develop the notion of probability more easily.

Suppose a coin is tossed once and the up face is recorded. The result we see and record is called an *observation,* or *measurement,* and the process of making an observation is called an *experiment.* Notice that our definition of *experiment* is broader than the one used in the physical sciences, where you would picture test tubes, microscopes, and other laboratory equipment. Among other things, statistical experiments may include recording an Internet user's preference for a Web browser, recording a change in the Dow Jones Industrial Average from one day to the next, recording the weekly sales of a business firm, and counting the number of errors on a page of an accountant's ledger. The point is that a statistical experiment can be almost any act of observation as long as the outcome is uncertain.

> An **experiment** is an act or process of observation that leads to a single outcome that cannot be predicted with certainty.

Consider another simple experiment consisting of tossing a die and observing the number on the up face. The six basic possible outcomes to this experiment are as follows:

1. Observe a 1
2. Observe a 2
3. Observe a 3
4. Observe a 4
5. Observe a 5
6. Observe a 6

Note that if this experiment is conducted once, *you can observe one and only one of these six basic outcomes, and the outcome cannot be predicted with certainty.* Also, these possibilities cannot be decomposed into more basic outcomes. Because observing the outcome of an experiment is similar to selecting a sample from a population, the basic possible outcomes to an experiment are called *sample points.**

> A **sample point** is the most basic outcome of an experiment.

Example 3.1
Listing the Sample Points for a Coin-Tossing Experiment

Problem Two coins are tossed, and their up faces are recorded. List all the sample points for this experiment.

Solution Even for a seemingly trivial experiment, we must be careful when listing the sample points. At first glance, we might expect three basic outcomes: Observe two heads, Observe two tails, or Observe one head and one tail. However, further reflection reveals that the last of these, Observe one head and one tail, can be decomposed into two outcomes: Head on coin 1, Tail on coin 2; and Tail on coin 1, Head on coin 2.

A useful tool for illustrating this notion is a **tree diagram.** Figure 3.1 shows a tree diagram for this experiment. At the top of the "tree," there are two branches,

*Alternatively, the term *simple event* can be used.

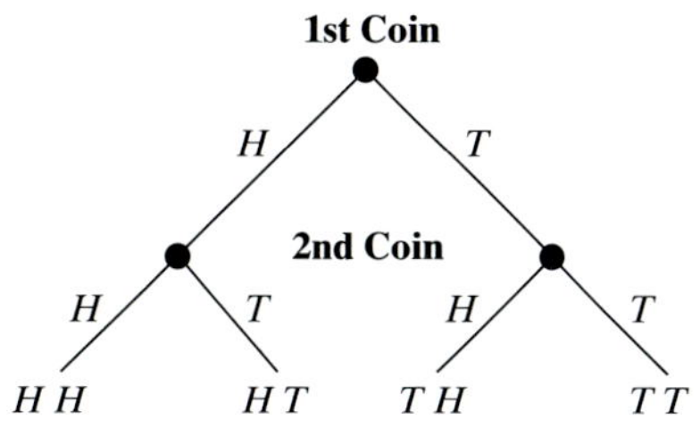

Figure 3.1
Tree diagram for the coin-tossing experiment

representing the two outcomes (*H* or *T*) for the first tossed coin. Each of these outcomes results in two more branches, representing the two outcomes (*H* or *T*) for the second tossed coin. Consequently, after tossing both coins, you can see that we have four sample points.

1. Observe *HH*
2. Observe *HT*
3. Observe *TH*
4. Observe *TT*

where *H* in the first position means "Head on coin 1," *H* in the second position means "Head on coin 2," and so on.

Look Back Even if the coins are identical in appearance, there are, in fact, two distinct coins. Thus, the sample points must account for this distinction.

Now Work Exercise 3.7a

We often wish to refer to the collection of all the sample points of an experiment. This collection is called the *sample space* of the experiment. For example, there are six sample points in the sample space associated with the die-toss experiment. The sample spaces for the experiments discussed thus far are shown in Table 3.1.

Table 3.1 Experiments and Their Sample Spaces

Experiment: Observe the up face on a coin.
Sample space: 1. Observe a head
2. Observe a tail

This sample space can be represented in set notation as a set containing two sample points:

$$S: \{H, T\}$$

where *H* represents the sample point Observe a head and *T* represents the sample point Observe a tail.

Experiment: Observe the up face on a die.
Sample space: 1. Observe a 1
2. Observe a 2
3. Observe a 3
4. Observe a 4
5. Observe a 5
6. Observe a 6

This sample space can be represented in set notation as a set of six sample points:

$$S: \{1, 2, 3, 4, 5, 6\}$$

Experiment: Observe the up faces on two coins.
Sample space: 1. Observe *HH*
2. Observe *HT*
3. Observe *TH*
4. Observe *TT*

This sample space can be represented in set notation as a set of four sample points:

$$S: \{HH, HT, TH, TT\}$$

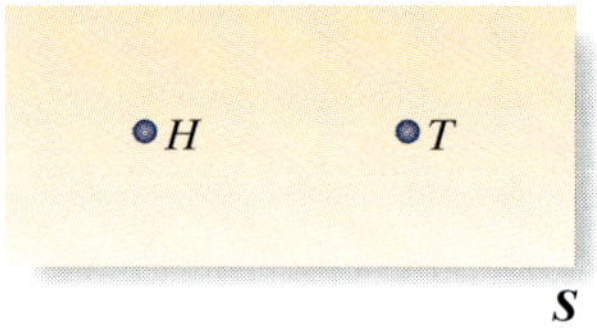

a. Experiment: Observe the up face on a coin

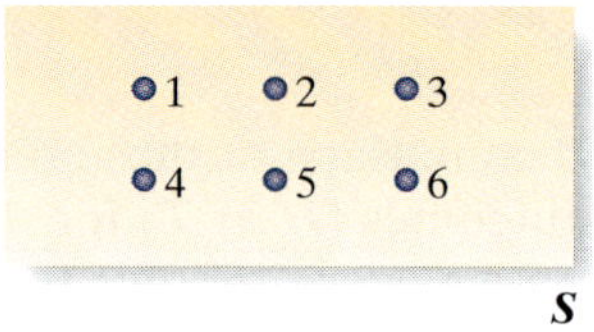

b. Experiment: Observe the up face on a die

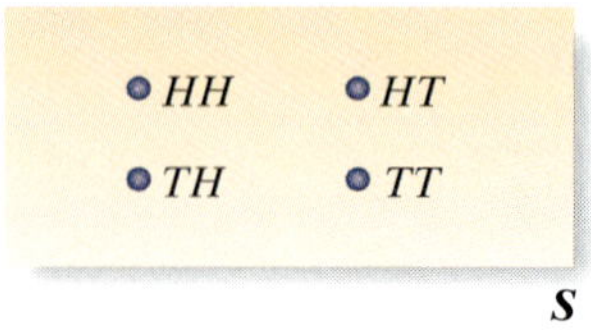

c. Experiment: Observe the up faces on two coins

Figure 3.2
Venn diagrams for the three experiments from Table 3.1

> The **sample space** of an experiment is the collection of all its sample points.

Just as graphs are useful in describing sets of data, a pictorial method for presenting the sample space will often be useful. Figure 3.2 shows such a representation for each of the experiments in Table 3.1. In each case, the sample space is shown as a closed figure, labeled *S*, containing all possible sample points. Each sample point is represented by a solid dot (i.e., a "point") and labeled accordingly. Such graphical representations are called **Venn diagrams.**

Now that we know that an experiment will result in *only one* basic outcome—called a *sample point*—and that the sample space is the collection of all possible sample points, we're ready to discuss the probabilities of the sample points. You've undoubtedly used the term *probability* and have some intuitive idea about its meaning. Probability is generally used synonymously with "chance," "odds," and similar concepts. For example, if a fair coin is tossed, we might reason that both the sample points, Observe a head and Observe a tail, have the same *chance* of occurring. Thus, we might state that "the probability of observing a head is 50%" or "the *odds* of seeing a head are 50:50." Both of these statements are based on an informal knowledge of probability. We'll begin our treatment of probability by using such informal concepts and then solidify what we mean later.

The probability of a sample point is a number between 0 and 1 inclusive that measures the likelihood that the outcome will occur when the experiment is performed. This number is usually taken to be the relative frequency of the occurrence of a sample point in a very long series of repetitions of an experiment.* For example, if we are assigning probabilities to the two sample points (Observe a head and Observe a tail) in the coin-toss experiment, we might reason that if we toss a balanced coin a very large number of times, the sample points Observe a head and Observe a tail will occur with the same relative frequency of .5.

Our reasoning is supported by Figure 3.3. The figure plots the relative frequency of the number of times that a head occurs when simulating (by computer) the toss of a coin N times, where N ranges from as few as 25 tosses to as many as 1,500 tosses of the coin. You can see that when N is large (i.e., $N = 1{,}500$), the relative frequency is converging to .5. Thus, the probability of each sample point in the coin-tossing experiment is .5.

For some experiments, we may have little or no information on the relative frequency of occurrence of the sample points; consequently, we must assign probabilities to the sample points based on general information about the experiment. For example, if the experiment is to invest in a business venture and to observe whether it succeeds or fails, the sample space would appear as in Figure 3.4.

We are unlikely to be able to assign probabilities to the sample points of this experiment based on a long series of repetitions because unique factors govern each performance of this kind of experiment. Instead, we may consider factors such as the personnel managing the venture, the general state of the economy at the time, the rate of success of similar ventures, and any other pertinent information. If we finally decide that the venture has an 80% chance of succeeding, we assign a probability of .8 to the sample point Success. This probability can be interpreted as a measure of our degree of belief in the outcome of the business venture; that is, it is a subjective probability. Notice, however, that such probabilities should be based on expert information that is carefully assessed. If not, we may be misled on any decisions based on these probabilities or based on any calculations in which they appear. [*Note:* For a text that deals in detail with the subjective evaluation of probabilities, see Winkler (1972) or Lindley (1985).]

BIOGRAPHY

JOHN VENN (1834–1923)

The English Logician

Born in Hull, England, John Venn is probably best known for his pictorial representation of unions and intersections (the Venn diagram). While lecturing in Moral Science at Cambridge University, Venn wrote *The Logic of Chance* and two other treatises on logic. In these works, Venn probably had his greatest contribution to the field of probability and statistics—the notion that the probability of an event is simply the long-run proportion of times the event occurs. Besides being a well-known mathematician, Venn was also a historian, philosopher, priest, and skilled machine builder. (His machine for bowling cricket balls once beat a top star on the Australian cricket team.) ■

*The result derives from an axiom in probability theory called the **Law of Large Numbers.** Phrased informally, this law states that the relative frequency of the number of times that an outcome occurs when an experiment is replicated over and over again (i.e., a large number of times) approaches the theoretical probability of the outcome.

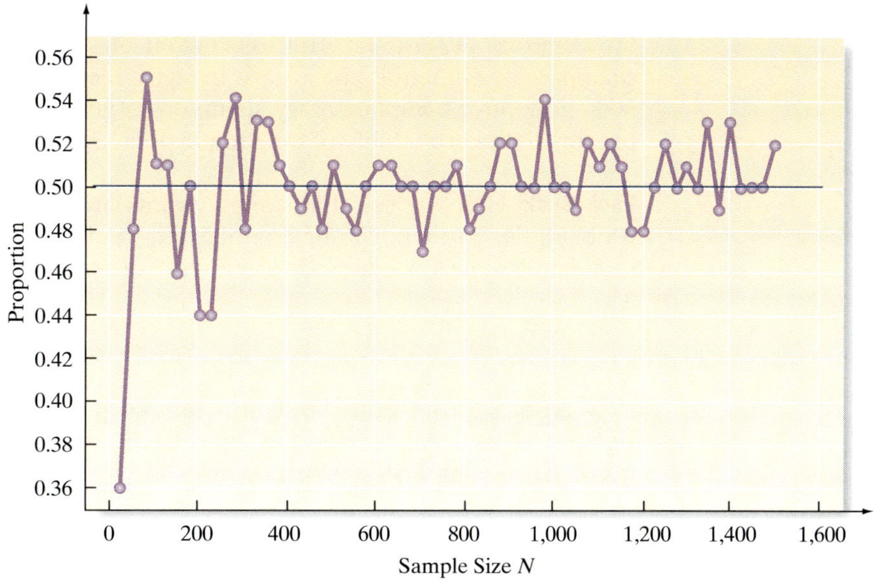

Figure 3.3
Proportion of heads in N coin tosses

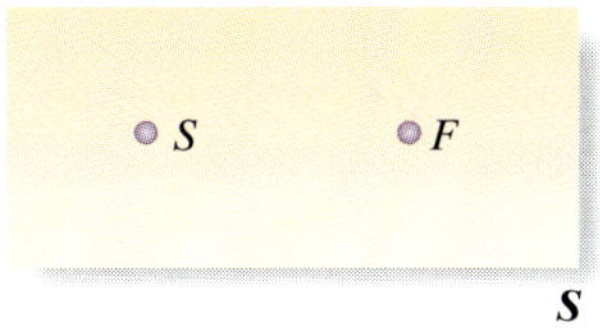

Figure 3.4
Experiment: Invest in a business venture and observe whether it succeeds (S) or fails (F)

No matter how you assign the probabilities to sample points, the probabilities assigned must obey two rules:

Probability Rules for Sample Points

Let p_i represent the probability of sample point i.

1. All sample point probabilities *must* lie between 0 and 1 (i.e., $0 \leq p_i \leq 1$).
2. The probabilities of all the sample points within a sample space *must* sum to 1 (i.e., $\sum p_i = 1$).

Assigning probabilities to sample points is easy for some experiments. For example, if the experiment is to toss a fair coin and observe the face, we would probably all agree to assign a probability of $\frac{1}{2}$ to the two sample points, Observe a head and Observe a tail. However, many experiments have sample points whose probabilities are more difficult to assign.

Example 3.2
Assigning Probabilities to the Sample Points for a Hotel Guest Room Survey

Problem Many American hotels offer complimentary shampoo in their guest rooms. Suppose you randomly select one hotel from a registry of all hotels in the United States and check whether the hotel offers complimentary shampoo. Show how this problem might be formulated in the framework of an experiment with sample points and a sample space. Indicate how probabilities might be assigned to the sample points.

Solution The experiment can be defined as the selection of an American hotel and the observation of whether complimentary shampoo is offered in the hotel's guest rooms. There are two sample points in the sample space corresponding to this experiment:

S: {The hotel offers complimentary shampoo.}
N: {No complimentary shampoo is offered by the hotel.}

The difference between this and the coin-toss experiment becomes apparent when we attempt to assign probabilities to the two sample points. What probability should we assign to the sample point S? If you answer .5, you are assuming that the events S and N

should occur with equal likelihood, just like the sample points Heads and Tails in the coin-toss experiment. But assignment of sample point probabilities for the hotel-shampoo experiment is not so easy. In fact, a recent survey of American hotels found that 80% now offer complimentary shampoo to guests. Then it might be reasonable to approximate the probability of the sample point *S* as .8 and that of the sample point *N* as .2.

Look Back Here we see that sample points are not always equally likely, so assigning probabilities to them can be complicated—particularly for experiments that represent real applications (as opposed to coin- and die-toss experiments).

Now Work Exercise 3.9

Although the probabilities of sample points are often of interest in their own right, it is usually probabilities of collections of sample points that are important. Example 3.3 demonstrates this point.

Example 3.3

Finding the Probability of a Collection of Sample Points from a Die-Tossing Experiment

Problem A fair die is tossed, and the up face is observed. If the face is even, you win \$1. Otherwise, you lose \$1. What is the probability that you win?

Solution Recall that the sample space for this experiment contains six sample points:

$$S: \{1, 2, 3, 4, 5, 6\}$$

Because the die is balanced, we assign a probability of $\frac{1}{6}$ to each of the sample points in this sample space. An even number will occur if one of the sample points, Observe a 2, Observe a 4, or Observe a 6, occurs. A collection of sample points such as this is called an *event*, which we denote by the letter *A*. Because the event *A* contains three sample points—each with probability $\frac{1}{6}$—and because no sample points can occur simultaneously, we reason that the probability of *A* is the sum of the probabilities of the sample points in *A*. Thus, the probability of *A* (i.e., the probability that you will win) is $\frac{1}{6} + \frac{1}{6} + \frac{1}{6} = \frac{1}{2}$.

Look Back Based on our notion of probability, $P(A) = \frac{1}{2}$ implies that, *in the long run*, you will win \$1 half the time and lose \$1 half the time.

Now Work Exercise 3.6

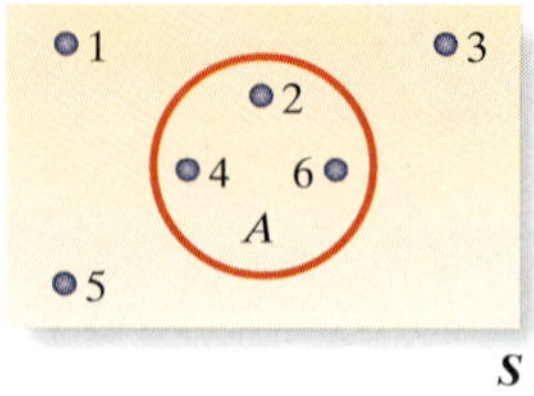

Figure 3.5
Die-toss experiment with event *A*: Observe an even number

Figure 3.5 is a Venn diagram depicting the sample space associated with a die-toss experiment and the event *A*, Observe an even number. The event *A* is represented by the closed figure inside the sample space *S*. This closed figure *A* contains all the sample points that comprise it.

To decide which sample points belong to the set associated with an event *A*, test each sample point in the sample space *S*. If event *A* occurs, then that sample point is in the event *A*. For example, the event *A*, Observe an even number, in the die-toss experiment will occur if the sample point Observe a 2 occurs. By the same reasoning, the sample points Observe a 4 and Observe a 6 are also in event *A*.

To summarize, we have demonstrated that an event can be defined in words or it can be defined as a specific set of sample points. This leads us to the following general definition of an event:

> An **event** is a specific collection of sample points. Further, a **simple event** contains only a single sample point, while a **compound event** contains two or more sample points.

Example 3.4

The Probability of an Event in a Coin-Tossing Experiment

Problem Consider the experiment of tossing two *unbalanced* coins. Because the coins are *not* balanced, their outcomes (H or T) are not equiprobable. Suppose the correct probabilities associated with the sample points are given in the table. [*Note:* The necessary properties for assigning probabilities to sample points are satisfied.]

Consider the events

A: {Observe exactly one head}

B: {Observe at least one head}

Calculate the probability of A and the probability of B.

Sample Point	Probability
HH	$4/9$
HT	$2/9$
TH	$2/9$
TT	$1/9$

Solution Event A contains the sample points HT and TH. Because two or more sample points cannot occur at the same time, we can easily calculate the probability of event A by summing the probabilities of the two sample points. Thus, the probability of observing exactly one head (event A), denoted by the symbol $P(A)$, is

$$P(A) = P(\text{Observe } HT) + P(\text{Observe } TH) = 2/9 + 2/9 = 4/9$$

Similarly, because B contains the sample points HH, HT, and TH,

$$P(B) = 4/9 + 2/9 + 2/9 = 8/9$$

Look Back Again, these probabilities should be interpreted *in the long run.* For example, $P(B) = 8/9 \approx .89$ implies that if we were to toss two coins an infinite number of times, we would observe at least 2 heads on about 89% of the tosses.

Now Work Exercise 3.3

The preceding example leads us to a general procedure for finding the probability of an event A:

Probability of an Event

The probability of an event A is calculated by summing the probabilities of the sample points in the sample space for A.

Thus, we can summarize the steps for calculating the probability of any event, as indicated in the next box.

Steps for Calculating Probabilities of Events

Step 1 Define the experiment; that is, describe the process used to make an observation and the type of observation that will be recorded.

Step 2 List the sample points.

Step 3 Assign probabilities to the sample points.

Step 4 Determine the collection of sample points contained in the event of interest.

Step 5 Sum the sample point probabilities to get the event probability.

Example 3.5

Applying the Five Steps to Find a Probability in Diversity Training

Problem Diversity training of employees is the latest trend in U.S. business. *USA Today* reported on the primary reasons businesses give for making diversity training part of their strategic planning process. The reasons are summarized in Table 3.2. Assume that one business is selected at random from all U.S. businesses that use diversity training, and the primary reason is determined.

Table 3.2 Primary Reasons for Diversity Training

Reason	Percentage
Comply with personnel policies (CPP)	7
Increase productivity (IP)	47
Stay competitive (SC)	38
Social responsibility (SR)	4
Other (O)	4
Total	100%

a. Define the experiment that generated the data in Table 3.2 and list the sample points.

b. Assign probabilities to the sample points.

c. What is the probability that the primary reason for diversity training is business related, that is, related to competition or productivity?

d. What is the probability that social responsibility is not a primary reason for diversity training?

Solution

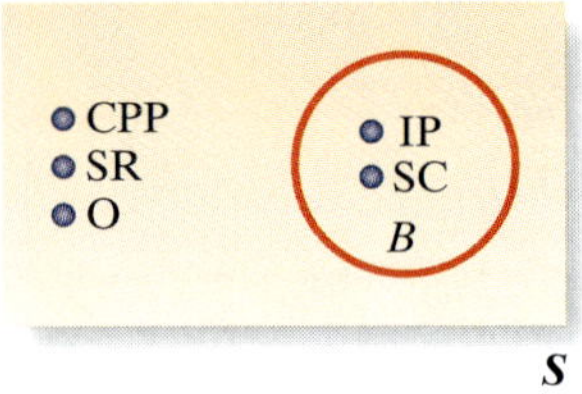

Figure 3.6
Venn diagram for diversity training survey

a. The experiment is the act of determining the primary reason for diversity training of employees at a U.S. business. The sample points, the simplest outcomes of the experiment, are the five response categories listed in Table 3.2. These sample points are shown in the Venn diagram in Figure 3.6.

b. If, as in Example 3.1, we were to assign equal probabilities in this case, each of the response categories would have a probability of one-fifth $\frac{1}{5}$, or .20. But, by examining Table 3.2 you can see that equal probabilities are not reasonable here because the response percentages were not even approximately the same in the five classifications. It is more reasonable to assign a probability equal to the response percentage in each class, as shown in Table 3.3.*

Table 3.3 Sample Point Probabilities for Diversity Training Survey

Sample Point	Probability
CPP	.07
IP	.47
SC	.38
SR	.04
O	.04

c. Let the symbol B represent the event that the primary reason for diversity training is business related. B is not a sample point because it consists of more than one of the response classifications (the sample points). In fact, as shown in Figure 3.6, B consists of two sample points, IP and SC. The probability of B is defined to be the sum of the probabilities of the sample points in B.

$$P(B) = P(\text{IP}) + P(\text{SC}) = .47 + .38 = .85$$

d. Let NSR represent the event that social responsibility is not a primary reason for diversity training. Then NSR consists of all sample points except SR, and the probability is the sum of the corresponding sample point probabilities:

$$P(NSR) = P(\text{CPP}) + P(\text{IP}) + P(\text{SC}) + P(\text{O})$$
$$= .07 + .47 + .38 + .04 = .96$$

Look Back The key to solving this problem is to follow the steps outlined in the box. We defined the experiment (Step 1) and listed the sample points (Step 2) in part **a.** The assignment of probabilities to the sample points (Step 3) was done in part **b.** For each probability in parts **c** and **d,** we identified the collection of points in the event (Step 4) and summed their probabilities (Step 5).

Now Work Exercise 3.17

*The response percentages were based on a sample of U.S. businesses; consequently, these assigned probabilities are estimates of the true population-response percentages. You'll learn how to measure the reliability of probability estimates in Chapter 5.

Example 3.6

The Probability of Investing in a Successful Venture

Problem You have the capital to invest in two of four ventures, each of which requires approximately the same amount of investment capital. Unknown to you, two of the investments will eventually fail and two will be successful. You research the four ventures because you think that your research will increase your probability of a successful choice over a purely random selection, and you eventually decide on two. If you used none of the information generated by your research and selected two ventures at random, what is the probability that you would select at least one successful venture?

Solution

Step 1: Denote the two successful enterprises as S_1 and S_2 and the two failing enterprises as F_1 and F_2. The experiment involves a random selection of two out of the four ventures, and each possible pair of ventures represents a sample point.

Step 2: The six sample points that make up the sample space are

1. (S_1, S_2)
2. (S_1, F_1)
3. (S_1, F_2)
4. (S_2, F_1)
5. (S_2, F_2)
6. (F_1, F_2)

Step 3: Next, we assign probabilities to the sample points. If we assume that the choice of any one pair is as likely as any other, then the probability of each sample point is $1/6$.

Step 4: The event of selecting at least one of the two successful ventures includes all the sample points except (F_1, F_2).

Step 5: Now, we find

$$\begin{aligned} P(\text{Select at least one success}) &= P(S_1,S_2) + P(S_1,F_1) + P(S_1,F_2) + \\ &\quad P(s_2, F_1) + P(S_2, F_2) \\ &= 1/6 + 1/6 + 1/6 + 1/6 + 1/6 = 5/6 \end{aligned}$$

Therefore, with a random selection, the probability of selecting at least one successful venture out of two is $5/6$.

The preceding examples have one thing in common: The number of sample points in each of the sample spaces was small; hence, the sample points were easy to identify and list. How can we manage this when the sample points run into the thousands or millions? For example, suppose you wish to select five business ventures from a group of 1,000. Then each different group of five ventures would represent a sample point. How can you determine the number of sample points associated with this experiment?

One method of determining the number of sample points for a complex experiment is to develop a counting system. Start by examining a simple version of the experiment. For example, see if you can develop a system for counting the number of ways to select two ventures from a total of four (this is exactly what was done in Example 3.6). If the ventures are represented by the symbols V_1, V_2, V_3, and V_4, the sample points could be listed in the following pattern:

$$\begin{array}{lll} (V_1, V_2) & (V_2, V_3) & (V_3, V_4) \\ (V_1, V_3) & (V_2, V_4) & \\ (V_1, V_4) & & \end{array}$$

Note the pattern and now try a more complex situation—say, sampling three ventures out of five. List the sample points and observe the pattern. Finally, see if you can deduce the pattern for the general case. Perhaps you can program a computer to produce the matching and counting for the number of samples of 5 selected from a total of 1,000.

A second method of determining the number of sample points for an experiment is to use **combinatorial mathematics.** This branch of mathematics is concerned with developing counting rules for given situations. For example, there is a simple rule for finding the number of different samples of five ventures selected from 1,000. This rule, called the **Combinations Rule,** is given in the box.

Combinations Rule

A sample of n elements is to be drawn from a set of N elements. Then, the number of different samples possible is denoted by $\binom{N}{n}$ and is equal to

$$\binom{N}{n} = \frac{N!}{n!(N-n)!}$$

where the factorial symbol (!) means that

$$n! = n(n-1)(n-2)\cdots(3)(2)(1)$$

For example, $5! = 5 \cdot 4 \cdot 3 \cdot 2 \cdot 1$. [*Note:* The quantity 0! is defined to be equal to 1.]

Example 3.7

Using the Combinations Rule to Determine the Number of Possible Investments: Selecting 2 from 4

Problem Refer to Example 3.6, where we selected two ventures from four in which to invest. Use the Combinations Rule to determine how many different selections can be made.

Solution For this example, $N = 4$, $n - 2$, and

$$\binom{4}{2} = \frac{4!}{2!2!} = \frac{4 \cdot 3 \cdot 2 \cdot 1}{(2 \cdot 1)(2 \cdot 1)} = \frac{4 \cdot 3}{2 \cdot 1} = \frac{12}{2} = 6$$

Look Back You can see that this agrees with the number of sample points obtained in Example 3.6.

Now Work Exercise 3.4

Example 3.8

Using the Combinations Rule to Determine the Number of Possible Investments: Selecting 5 from 20

Problem Suppose you plan to invest equal amounts of money in each of five business ventures. If you have 20 ventures from which to make the selection, how many different samples of five ventures can be selected from the 20?

Solution For this example, $N = 20$ and $n = 5$. Then the number of different samples of 5 that can be selected from the 20 ventures is

$$\binom{20}{5} = \frac{20!}{5!(20-5)!} = \frac{20!}{5!15!}$$

$$= \frac{20 \cdot 19 \cdot 18 \cdot \cdots \cdot 3 \cdot 2 \cdot 1}{(5 \cdot 4 \cdot 3 \cdot 2 \cdot 1)(15 \cdot 14 \cdot 13 \cdot \cdots \cdot 3 \cdot 2 \cdot 1)} = \frac{20 \cdot 19 \cdot 18 \cdot 17 \cdot 16}{5 \cdot 4 \cdot 3 \cdot 2 \cdot 1} = 15{,}504$$

Look Back You can see that attempting to list all the sample points for this experiment would be an extremely tedious and time consuming, if not practically impossible, task.

The Combinations Rule is just one of a large number of counting rules that have been developed by combinatorial mathematicians. This counting rule applies to situations in which the experiment calls for selecting n elements from a total of N elements, without replacing each element before the next is selected. If you are interested in learning other methods for counting sample points for various types of experiments, you will find a few of the basic counting rules in Appendix A. Others can be found in the chapter references.

Statistics IN Action Revisited The Probability of Winning Lotto

In Florida's state lottery game, called Pick-6 Lotto, you select six numbers of your choice from a set of numbers ranging from 1 to 53. We can apply the Combinations Rule to determine the total number of combinations of 6 numbers selected from 53 (i.e., the total number of sample points [or possible winning tickets]). Here, $N = 53$ and $n = 6$: therefore, we have

$$\binom{N}{n} = \frac{N!}{n!(N-n)!} = \frac{53!}{6!47!}$$
$$= \frac{(53)(52)(51)(50)(49)(48)(47!)}{(6)(5)(4)(3)(2)(1)(47!)}$$
$$= 22{,}957{,}480$$

Now, since the Lotto balls are selected at random, each of these 22,957,480 combinations is equally likely to occur. Therefore, the probability of winning Lotto is

$$P(\text{Win6/53 Lotto}) = 1/(22{,}957{,}480) = .00000004356$$

This probability is often stated as follows: The odds of winning the game with a single ticket are 1 in 22,957,480, or 1 in approximately 23 million. For all practical purposes, this probability is 0, implying that you have almost no chance of winning the lottery with a single ticket. Yet each week there is almost always a winner in the Florida Lotto. This apparent contradiction can be explained with the following analogy.

Suppose there is a line of Smart cars, front-to-back, from New York City to Los Angeles, California. Based on the distance between the two cities and the length of a standard minivan, there would be approximately 23 million Smart cars in line. Lottery officials will select, at random, one of the Smart cars and put a check for \$10 million dollars in the glove compartment. For a cost of \$1, you may roam the country and select one (and only one) Smart car and check the glove compartment. Do you think you will find \$10 million in the Smart car you choose? You can be almost certain that you won't. But now permit anyone to enter the lottery for \$1 and suppose that 50 million people do so. With such a large number of participants, it is very likely that someone will find the Smart car with the \$10 million—but it almost certainly won't be you! (This example illustrates an axiom in statistics called the Law of Large Numbers. See the footnote at the bottom of p. 120.)

Exercises 3.1–3.21

Learning the Mechanics

3.1 An experiment results in one of the following sample points: $E_1, E_2, E_3, E_4,$ or E_5.

a. Find $P(E_3)$ if $P(E_1) = .1$, $P(E_2) = .2$, $P(E_4) = .1$, and $P(E_5) = .1$.

b. Find $P(E_3)$ if $P(E_1) = P(E_3)$, $P(E_2) = .1$, $P(E_4) = .2$, and $P(E_5) = .1$.

c. Find $P(E_3)$ if $P(E_1) = P(E_2) = P(E_4) = P(E_5) = .1$.

3.2 The diagram below describes the sample space of a particular experiment and events A and B.

a. What is this type of diagram called?

b. Suppose the sample points are equally likely. Find $P(A)$ and $P(B)$.

c. Suppose $P(1) = P(2) = P(3) = P(4) = P(5) = {}^1/_{20}$ and $P(6) = P(7) = P(8) = P(9) = P(10) = {}^3/_{20}$. Find $P(A)$ and $P(B)$.

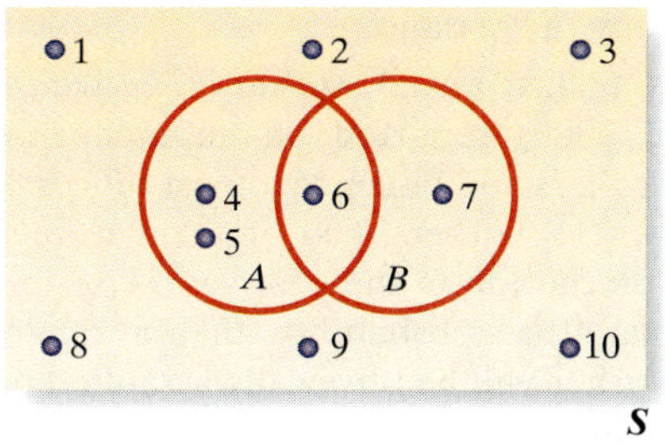

3.3 NW The sample space for an experiment contains five sample points with probabilities as shown in the table. Find the probability of each of the following events:

Sample Points	Probabilities
1	.05
2	.20
3	.30
4	.30
5	.15

A: {Either 1, 2, or 3 occurs}

B: {Either 1, 3, or 5 occurs}

C: {4 does not occur}

3.4 NW Compute each of the following:

a. $\binom{9}{4}$ **b.** $\binom{7}{2}$ **c.** $\binom{4}{4}$

d. $\binom{5}{0}$ **e.** $\binom{6}{5}$

3.5 Compute the number of ways you can select n elements from N elements for each of the following:

a. $n = 2, N = 5$

b. $n = 3, N = 6$

c. $n = 5, N = 20$

3.6 Two fair dice are tossed, and the face on each die is observed.
NW
a. Use a tree diagram to find the 36 sample points contained in the sample space.
b. Assign probabilities to the sample points in part **a.**
c. Find the probability of each of the following events:

$A = \{3 \text{ showing on each die}\}$

$B = \{\text{sum of two numbers showing is } 7\}$

$C = \{\text{sum of two numbers showing is even}\}$

3.7 Two marbles are drawn at random and without replacement from a box containing two blue marbles and three red marbles.
NW
a. List the sample points for this experiment.
b. Assign probabilities to the sample points.
c. Determine the probability of observing each of the following events:

A: {Two blue marbles are drawn.}

B: {A red and a blue marble are drawn.}

C: {Two red marbles are drawn.}

3.8 Simulate the experiment described in Exercise 3.7 using any five identically shaped objects, two of which are one color and three another color. Mix the objects, draw two, record the results, and then replace the objects. Repeat the experiment a large number of times (at least 100). Calculate the proportion of time events *A*, *B*, and *C* occur. How do these proportions compare with the probabilities you calculated in Exercise 3.7? Should these proportions equal the probabilities? Explain.

Applet Exercise 3.1

Use the applet *Simulating the Probability of Rolling a 6* to explore the relationship between the proportion of 6s rolled on several rolls of a die and the theoretical probability of rolling a 6 on a fair die.

a. To simulate rolling a die one time, click on the *Roll* button on the screen while $n = 1$. The outcome of the roll appears in the list at the right, and the cumulative proportion of 6s for one roll is shown above the graph and as a point on the graph corresponding to 1. Click *Reset* and repeat the process with $n = 1$ several times. What are the possible values of the cumulative proportion of 6s for one roll of a die? Can the cumulative proportion of 6s for one roll of a die equal the theoretical probability of rolling a 6 on a fair die? Explain.
b. Set $n = 10$ and click the *Roll* button. Repeat this several times, resetting after each time. Record the cumulative proportion of 6s for each roll. Compare the cumulative proportions for $n = 10$ to those for $n = 1$ in part **a.** Which tend to be closer to the theoretical probability of rolling a 6 on a fair die?
c. Repeat part **b** for $n = 1{,}000$, comparing the cumulative proportions for $n = 1{,}000$ to those for $n = 1$ in part **a** and for $n = 10$ in part **b.**
d. Based on your results for parts **a, b,** and **c,** do you believe that we could justifiably conclude that a die is unfair because we rolled it 10 times and didn't roll any 6s? Explain.

Applet Exercise 3.2

Use the applet *Simulating the Probability of a Head with a Fair Coin* to explore the relationship between the proportion of heads on several flips of a coin and the theoretical probability of getting heads on one flip of a fair coin.

a. Repeat parts **a–c** of Applet Exercise 3.1 for the experiment of flipping a coin and the event of getting heads.
b. Based on your results for part **a,** do you believe that we could justifiably conclude that a coin is unfair because we flipped it 10 times and didn't roll any heads? Explain.

Applying the Concepts—Basic

3.9 **Colors of M&M's candies.** In 1940, Forrest E. Mars Sr., formed the Mars Corporation to produce chocolate candies with a sugar shell that could be sold throughout the year and wouldn't melt during the summer. Originally, M&M's Plain Chocolate Candies came in only a brown color. Today, M&M's in standard bags come in six colors: brown, yellow, red, blue, orange, and green. According to Mars Corporation, today 24% of all M&M's produced are blue, 20% are orange, 16% are green, 14% are yellow, 13% are brown, and 13% are red. Suppose you purchase a randomly selected bag of M&M's Plain Chocolate Candies and randomly select one of the M&M's from the bag. The color of the selected M&M is of interest.
NW
a. Identify the outcomes (sample points) of this experiment.
b. Assign reasonable probabilities to the outcomes, part **a.**
c. What is the probability that the selected M&M is brown (the original color)?
d. In 1960, the colors red, green, and yellow were added to brown M&Ms. What is the probability that the selected M&M is either red, green, or yellow?
e. In 1995, based on voting by American consumers, the color blue was added to the M&M mix. What is the probability that the selected M&M is not blue?

3.10 **Workers' unscheduled absence survey.** Each year CCH, Inc., a firm that provides human resources and employment law information, conducts a survey on absenteeism in the workplace. The *2007 CCH Unscheduled Absence Survey* found that of all unscheduled work absences, 34% are due to "personal illness," 22% for "family issues," 18% for "personal needs," 13% for "entitlement mentality," and 13% due to "stress." Consider a randomly selected employee who has an unscheduled work absence.
a. List the sample points for this experiment.
b. Assign reasonable probabilities to the sample points.
c. What is the probability that the absence is due to something other than "personal illness"?

3.11 **Male nannies.** In a recent survey conducted by the International Nanny Association (INA) and reported at the INA Web site (www.nanny.org), 4,176 nannies were placed in a job in a given year. Only 24 of the nannies placed were men. Find the probability that a randomly selected nanny who was placed during the last year is a male nanny (a "mannie").

3.12 **Assaults against postal workers.** *The Wall Street Journal* (Sept. 1, 2000) reported on an independent study of postal workers and violence at post offices. In a sample of 12,000 postal workers, 600 of them were physically assaulted on the job in the past year. Use this information to estimate the probability that a randomly selected postal worker will be physically assaulted on the job during the year.

3.13 Going online for health information. A *cyberchondriac* is defined as a person who regularly searches the Web for health care information (see Exercise 1.18, p. 23). A 2008 Harris Poll surveyed 1,010 U.S. adults by telephone and asked each respondent how often (in the past month) he/she looked for health care information online. The results are summarized in the following table. Consider the response category of a randomly selected person who participated in the Harris Poll.

Response (# per Month)	Percentage of Respondents
None	25
1 or 2	31
3–5	25
6–9	5
10 or more	14
Total	100%

Source: The Harris Poll, July 29, 2008 (Table 3).

a. List the sample points for the experiment.
b. Assign reasonable probabilities to the sample points.
c. Find the probability that the respondent looks for health care information online more than two times per month.

3.14 Incorrectly filed and lost documents. Since the introduction of e-mail into organizations, the use of paper has increased dramatically. The amount of paperwork generated at large organizations can be overwhelming, leading to lack of storage space, incorrectly filed papers, and lost documents. An article in the *Communications of the ACM* (Vol. 43, 2000) reported that 3% of all documents produced at organizations are incorrectly filed, and 7.5% of all documents are lost forever. What is more likely to occur at a large organization, an incorrectly filed document or a document that is lost forever? Explain.

Applying the Concepts—Intermediate

3.15 USDA chicken inspection. The U.S. Department of Agriculture (USDA) reports that, under its standard inspection system, one in every 100 slaughtered chickens pass inspection with fecal contamination. (*Tampa Tribune,* Mar. 31, 2000.)

a. If a slaughtered chicken is selected at random, what is the probability that it passes inspection with fecal contamination?
b. The probability of part **a** was based on a USDA study that found that 306 of 32,075 chicken carcasses passed inspection with fecal contamination. Do you agree with the USDA's statement about the likelihood of a slaughtered chicken passing inspection with fecal contamination?

3.16 PIN pad shipments. Personal identification number (PIN) pads are devices that connect to point-of-sale electronic cash registers for debit and credit card purchases. Refer to *The Nilson Report* (Oct. 2008) listing of the volume of PIN pad shipments by manufacturers worldwide, Exercise 2.5 (p. 39). Recall that for the 12 manufacturers listed in the table, a total of 334,039 PIN pads were shipped in 2007. (The data are saved in the **PINPADS** file.) Suppose you randomly select one of these PIN pads and identify the manufacturer.

Manufacturer	Number Shipped (units)
Bitel	13,500
CyberNet	16,200
Fujian Landi	119,000
Glintt (ParaRede)	5,990
Intelligent	4,562
KwangWoo	42,000
Omron	20,000
Pax Tech.	10,072
ProvencoCadmus	20,000
SZZT Electronics	67,300
Toshiba TEC	12,415
Urmet	3,000

Source: The Nilson Report, No. 912, October 2008 (p. 9).

a. List the sample points for this experiment.
b. Explain why the probability of each sample point is not 1/12, then calculate the probability of each sample point.
c. Find the probability that the PIN pad is shipped by either Fujian Landi or SZZT Electronics.
d. Suppose that 1,000 of the PIN pads shipped in 2007 were found to be defective. Find the probability that the PIN pad selected is one of the defectives.

3.17 Management system failures. Refer to the *Process Safety Progress* (Dec. 2004) study of 83 industrial accidents caused by management system failures, Exercise 2.6 (p. 39). A summary of the root causes of these 83 incidents is reproduced in the following table.

Management System Cause Category	Number of Incidents
Engineering and Design	27
Procedures and Practices	24
Management and Oversight	22
Training and Communication	10
Total	83

Source: Blair, A. S. "Management system failures identified in incidents investigated by the U.S. Chemical Safety and Hazard Investigation Board," *Process Safety Progress,* Vol. 23, No. 4, Dec. 2004, pp. 232–236 (Table 1). Reprinted with permission of John Wiley & Sons, Inc.

a. Find and interpret the probability that an industrial accident is caused by faulty engineering and design.
b. Find and interpret the probability that an industrial accident is caused by something other than faulty procedures and practices.

3.18 Jai-alai bets. The Quinella bet at the paramutual game of jai-alai consists of picking the jai-alai players that will place first and second in a game *irrespective* of order. In jai-alai, eight players (numbered $1, 2, 3, \dots, 8$) compete in every game.

a. How many different Quinella bets are possible?
b. Suppose you bet the Quinella combination of 2–7. If the players are of equal ability, what is the probability that you win the bet?

3.19 Investing in stocks. From a list of 15 preferred stocks recommended by your broker, you will select three to invest in. How many different ways can you select the three stocks from the 15 recommended stocks?

3.20 Groundwater contamination in wells. Refer to the *Environmental Science & Technology* (Jan. 2005) study of methyl *tert*-butyl ether (MTBE) contamination in New Hampshire wells, Exercise 2.15 (p. 41). Data collected for a sample of 223 wells are saved in the **MTBE** file. Recall that each well was classified according to well class (public or private), aquifier (bedrock or unconsolidated), and detectable level of MTBE (below limit or detect).

a. Consider an experiment in which the well class, aquifier, and detectable MTBE level of a well are observed. List the sample points for this experiment. [Hint: One sample point is private/bedrock/below limit.]

b. Use statistical software to find the number of the 223 wells in each sample point outcome. Then use this information to compute probabilities for the sample points.

c. Find and interpret the probability that a well has a detectable level of MTBE.

3.21 Choosing portable grill displays. University of Maryland marketing professor R. W. Hamilton studied how people attempt to influence the choices of others by offering undesirable alternatives (*Journal of Consumer Research,* Mar. 2003). Such a phenomenon typically occurs when family members propose a vacation spot, friends recommend a restaurant for dinner, and realtors show the buyer potential homes. In one phase of the study, the researcher had each of 124 college students select showroom displays for portable grills. Five different displays (representing five different-sized grills) were available, but only three displays would be selected. The students were instructed to select the displays to maximize purchases of Grill #2 (a smaller-sized grill).

a. In how many possible ways can the three grill displays be selected from the five displays? List the possibilities.

b. The table shows the grill display combinations and the number of each selected by the 124 students. Use this information to assign reasonable probabilities to the different display combinations.

c. Find the probability that a student who participated in the study selected a display combination involving Grill #1.

Grill Display Combination	Number of Students
1-2-3	35
1-2-4	8
1-2-5	42
2-3-4	4
2-3-5	1
2-4-5	34

Source: Hamilton, R. W. "Why do people suggest what they do not want? Using context effects to influence others' choices," *Journal of Consumer Research,* Vol. 29, No. 4, Mar. 2003 (Table 1). Copyright © 2003 JCR, Inc.

3.22 Highest-rated new cars. *Consumer Reports* magazine annually asks readers to evaluate their experiences in buying a new car during the previous year. Analysis of the questionnaires for a recent year revealed that readers were most satisfied with the following three new cars (in no particular order): Infiniti M35, Toyota Prius, and Chevrolet Corvette (*Consumer Reports,* Apr. 2006).

a. List all possible sets of rankings for these top three cars.

b. Assuming that each set of rankings in part **a** is equally likely, what is the probability that readers ranked Toyota Prius first? That readers ranked Infiniti M35 third? That readers ranked Toyota Prius first and Chevrolet Corvette second (which is, in fact, what they did)?

Applying the Concepts—Advanced

3.23 Odds of winning a race. Handicappers for greyhound races express their belief about the probabilities that each greyhound will win a race in terms of **odds.** If the probability of event E is $P(E)$, then the *odds in favor of E* are $P(E)$ to $1 - P(E)$. Thus, if a handicapper assesses a probability of .25 that Oxford Shoes will win its next race, the odds in favor of Oxford Shoes are $^{25}/_{100}$ to $^{75}/_{100}$, or 1 to 3. It follows that the *odds against E* are $1 - P(E)$ to $P(E)$, or 3 to 1 against a win by Oxford Shoes. In general, if the odds in favor of event E are a to b, then $P(E) = a/(a + b)$.

a. A second handicapper assesses the probability of a win by Oxford Shoes to be $^1/_3$. According to the second handicapper, what are the odds in favor of Oxford Shoes winning?

b. A third handicapper assesses the odds in favor of Oxford Shoes to be 1 to 1. According to the third handicapper, what is the probability of Oxford Shoes winning?

c. A fourth handicapper assesses the odds against Oxford Shoes winning to be 3 to 2. Find this handicapper's assessment of the probability that Oxford Shoes will win.

3.24 Lead bullets as forensic evidence. *Chance* (Summer 2004) published an article on the use of lead bullets as forensic evidence in a federal criminal case. Typically, the Federal Bureau of Investigation (FBI) will use a laboratory method to match the lead in a bullet found at the crime scene with unexpended lead cartridges found in possession of the suspect. The value of this evidence depends on the chance of a *false positive* (i.e., the probability that the FBI finds a match given that the lead at the crime scene and the lead in possession of the suspect are actually from two different "melts," or sources). To estimate the false-positive rate, the FBI collected 1,837 bullets that they were confident all came from different melts. The FBI then examined every possible pair of bullets and counted the number of matches using its established criteria. According to *Chance,* the FBI found 693 matches. Use this information to compute the chance of a false positive. Is this probability small enough for you to have confidence in the FBI's forensic evidence?

3.25 Making your vote count. The recent Democratic and Republican presidential state primary elections were highlighted by the difference in the way winning candidates were awarded delegates. In Republican states, the winner is awarded all the state's delegates; conversely, the Democratic state winner is awarded delegates in proportion to the percentage of votes. This led to a *Chance* (Fall 2007) article on making your vote count. Consider the following scenario where you are one of five voters (for example, on a county commission where you are one of the five commissioners voting on an issue).

a. Determine the number of ways the five commissioners can vote, where each commissioner votes either for or against. (These outcomes represent the sample points for the experiment.)

b. Assume each commissioner is equally likely to vote for or against. Assign reasonable probabilities to the sample points, part **a.**

c. Your vote counts (i.e., is the decisive vote) only if the other four voters split, two in favor and two against.

Assuming you are commissioner number 5, how many of the sample points in part **a** result in a 2-2 split for the other four commissioners?

d. Use your answers to parts **a–c** to find the probability that your vote counts.

e. Now, suppose you convince two other commissioners to "vote in bloc," i.e., you all agree to vote among yourselves first, and whatever the majority decides is the way all three will vote. With only five total voters, this guarantees that the bloc vote will determine the outcome. In this scenario, your vote counts only if the other two commissioners in the bloc split votes, one in favor and one against. Find the probability that your vote counts.

3.2 Unions and Intersections

An event can often be viewed as a composition of two or more other events. Such events, which are called **compound events,** can be formed (composed) in two ways, as defined and illustrated here.

The **union** of two events A and B is the event that occurs if either A or B or both occur on a single performance of the experiment. We denote the union of events A and B by the symbol $A \cup B$. $A \cup B$ consists of all the sample points that belong to *A or B or both.* (See Figure 3.7a.)

The **intersection** of two events A and B is the event that occurs if both A and B occur on a single performance of the experiment. We write $A \cap B$ for the intersection of A and B. $A \cap B$ consists of all the sample points belonging to *both A and B.* (See Figure 3.7b.)

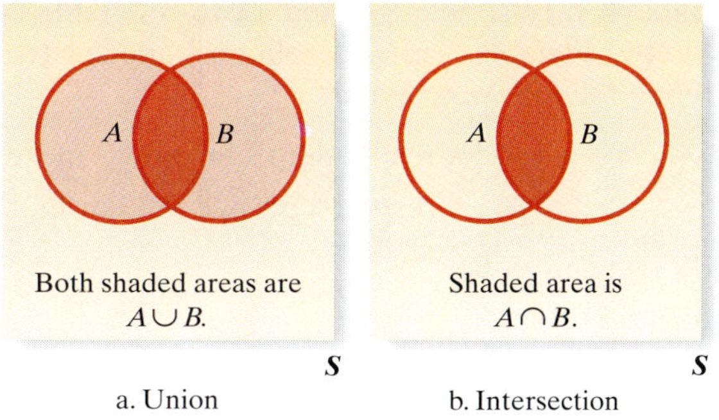

Figure 3.7
Venn diagrams for union and intersection

Example 3.9
Probabilities of Unions and Intersections in a Die-Toss Experiment

Problem Consider the die-toss experiment. Define the following events:

A: {Toss an even number}

B: {Toss a number less than or equal to 3}

a. Describe $A \cup B$ for this experiment.

b. Describe $A \cap B$ for this experiment.

c. Calculate $P(A \cup B)$ and $P(A \cap B)$ assuming the die is fair.

Solution Draw the Venn diagram as shown in Figure 3.8.

a. The union of A and B is the event that occurs if we observe either an even number, a number less than or equal to 3, or both on a single throw of the die. Consequently, the sample points in the event $A \cup B$ are those for which A occurs, B occurs, or both A and B occur. Checking the sample points in the entire sample space, we find that the collection of sample points in the union of A and B is

$$A \cup B = \{1, 2, 3, 4, 6\}$$

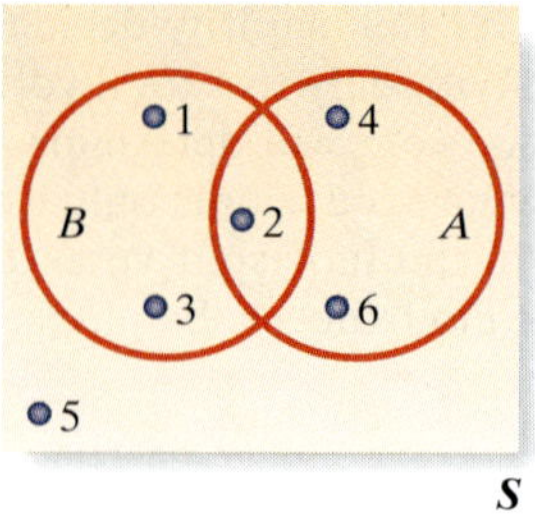

Figure 3.8
Venn diagram for die toss

b. The intersection of A and B is the event that occurs if we observe *both* an even number and a number less than or equal to 3 on a single throw of the die. Checking the sample points to see which imply the occurrence of *both* events A and B, we see that the intersection contains only one sample point:

$$A \cap B = \{2\}$$

In other words, the intersection of A and B is the sample point Observe a 2.

c. Recalling that the probability of an event is the sum of the probabilities of the sample points of which the event is composed, we have

$$\begin{aligned} P(A \cup B) &= P(1) + P(2) + P(3) + P(4) + P(6) \\ &= 1/6 + 1/6 + 1/6 + 1/6 + 1/6 = 5/6 \end{aligned}$$

and

$$P(A \cup B) = P(2) = 1/6$$

Now Work Exercise 3.25a–d

Example 3.10

Finding Probabilities in a Two-Way Table for a Mass Mailing: Income Versus Age

Problem Many firms undertake direct marketing campaigns to promote their products. The campaigns typically involve mailing information to millions of households. The response rates are carefully monitored to determine the demographic characteristics of respondents. By studying tendencies to respond, the firms can better target future mailings to those segments of the population most likely to purchase their products.

Suppose a distributor of mail-order tools is analyzing the results of a recent mailing. The probability of response is believed to be related to income and age. The percentages of the total number of respondents to the mailing are given by income and age classification in Table 3.4. This table is called a **two-way table** because responses are classified according to two variables, income (columns) and age (rows).

Define the following events:

A: {A respondent's income is more than \$50,000.}
B: {A respondents age is 30 or more.}

a. Find $P(A)$ and $P(B)$.
b. Find $P(A \cup B)$.
c. Find $P(A \cap B)$.

Solution Following the steps for calculating probabilities of events, we first note that the objective is to characterize the income and age distribution of respondents to the mailing. To accomplish this, we define the experiment to consist of selecting a respondent from the collection of all respondents and observing which income and age class he or she occupies. The sample points are the nine different age-income classifications:

E_1: {< 30 yrs, < \$25,000} E_4: {< 30 yrs, \$25,000–\$50,000} E_7: {< 30 yrs, > \$50,000}
E_2: {30–50 yrs, < \$25,000} E_5: {30–50 yrs, \$25,000–\$50,000} E_8: {30–50 yrs, > \$50,000}
E_3: {> 50 yrs, < \$25,000} E_6: {> 50 yrs, \$25,000–\$50,000} E_9: {> 50 yrs, > \$50,000}

Table 3.4 Two-Way Table with Percentage of Respondents in Age-Income Classes

	Income		
Age	< \$25,000	\$25,000–\$50,000	> \$50,000
< 30 yrs	5%	12%	10%
30–50 yrs	14%	22%	16%
> 50 yrs	8%	10%	3%

Next, we assign probabilities to the sample points. If we blindly select one of the respondents, the probability that he or she will occupy a particular age-income classification is the proportion, or relative frequency, of respondents in the classification. These proportions are given (as percentages) in Table 3.4. Thus,

$$P(E_1) = \text{Relative frequency of respondents in age-income class } \{< 30 \text{ yrs}, < \$25{,}000\} = .05$$
$$P(E_2) = .14$$
$$P(E_3) = .08$$
$$P(E_4) = .12$$
$$P(E_5) = .22$$
$$P(E_6) = .10$$
$$P(E_7) = .10$$
$$P(E_8) = .16$$
$$P(E_9) = .03$$

You may verify that the sample points probabilities add to 1.

a. To find $P(A)$, we first determine the collection of sample points contained in event A. Because A is defined as $\{> \$50{,}000\}$, we see from Table 3.4 that A contains the three sample points represented by the last column of the table. In other words, the event A consists of the income classification $\{> \$50{,}000\}$ in all three age classifications. The probability of A is the sum of the probabilities of the sample points in A:

$$P(A) = P(E_7) + P(E_8) + P(E_9) = .10 + .16 + .03 = .29$$

Similarly, $B = \{\geq 30 \text{ yrs}\}$ consists of the six sample points in the second and third rows of Table 3.4:

$$P(B) = P(E_2) + P(E_3) + P(E_5) + P(E_6) + P(E_8) + P(E_9)$$
$$= .14 + .08 + .22 + .10 + .16 + .03 = .73$$

b. The union of events A and B, $A \cup B$, consists of all the sample points in *either A or B or both*—that is, the union of A and B consists of all respondents whose income exceeds \$50,000 *or* whose age is 30 or more. In Table 3.4 this is any sample point found in the third column *or* the last two rows. Thus,

$$P(A \cup B) = .10 + .14 + .22 + .16 + .08 + .10 + .03 = .83$$

c. The intersection of events A and B, $A \cap B$, consists of all sample points in *both A and B*—that is, the intersection of A and B consists of all respondents whose income exceeds \$50,000 *and* whose age is 30 or more. In Table 3.4 this is any sample point found in the third column *and* the last two rows. Thus,

$$P(A \cap B) = .16 + .03 = .19$$

Look Back As with previous problems, the key to finding the probabilities of parts **b** and **c** is to identify the sample points that comprise the event of interest. In a two-way table like Table 3.4, the number of sample points will be equal to the number of rows times the number of columns.

Now Work Exercise 3.27f–g

3.3 Complementary Events

A very useful concept in the calculation of event probabilities is the notion of *complementary events:*

> The **complement** of an event A is the event that A does *not* occur—that is, the event consisting of all sample points that are not in event A. We denote the complement of A by A^c.

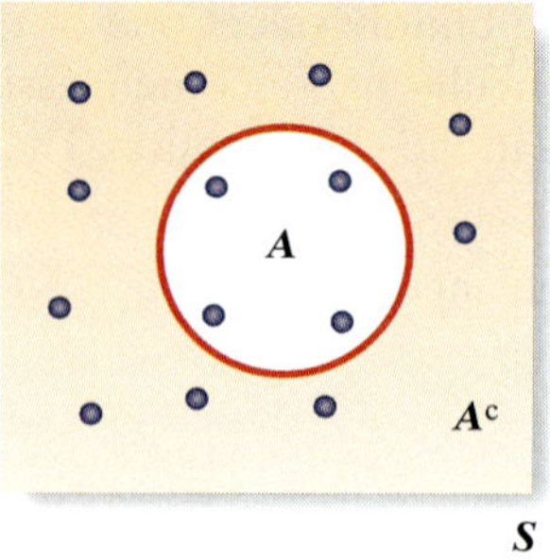

Figure 3.9
Venn diagram of complementary events

An event A is a collection of sample points, and the sample points included in A^c are those not in A. Figure 3.9 demonstrates this idea. Note from the figure that all sample points in S are included in *either* A or A^c and that *no* sample point is in both A and A^c. This leads us to conclude that the probabilities of an event and its complement *must sum to 1:*

Rule of Complements

The sum of the probabilities of complementary events equals 1: $P(A) + P(A^c) = 1$.

In many probability problems, calculating the probability of the complement of the event of interest is easier than calculating the event itself. Then, because

$$P(A) + P(A^c) = 1$$

we can calculate $P(A)$ by using the relationship

$$P(A) = 1 - P(A^c)$$

Example 3.11
Probabilities of Complementary Events in a Coin-Tossing Experiment

Problem Consider the experiment of tossing two fair coins. Use the complementary relationship to calculate the probability of event A: {Observing at least one head}.

Solution We know that the event A: {Observing at least one head} consists of the sample points

$$A: \{HH, HT, TH\}$$

The complement of A is defined as the event that occurs when A does not occur. Therefore,

$$A^c: \{\text{Observe no heads}\} = \{TT\}$$

This complementary relationship is shown in Figure 3.10. Assuming the coins are balanced,

$$P(A^c) = P(TT) = \tfrac{1}{4}$$

and

$$P(A) = 1 - P(A^c) = 1 - \tfrac{1}{4} = \tfrac{3}{4}$$

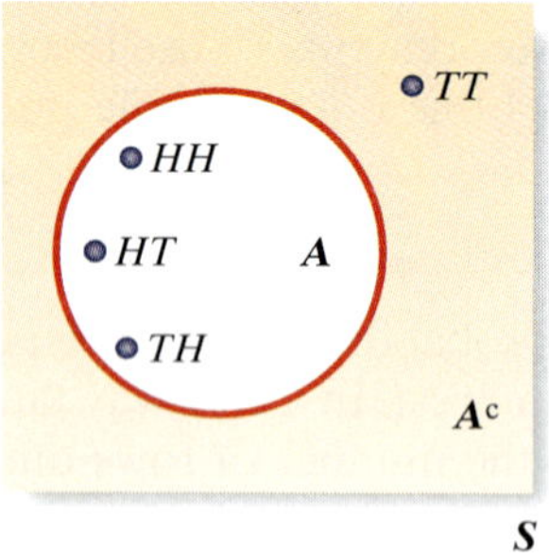

Figure 3.10
Complementary events in the toss of two coins

Look Back Note that we could have found $P(A)$ by summing the probabilities of the sample points HH, HT, and TH in A. Many times, it is easier to find the probability of A^c and use the rule of complements.

Now Work Exercise 3.25e–f

3.4 The Additive Rule and Mutually Exclusive Events

In Section 3.2 we saw how to determine which sample points are contained in a union and how to calculate the probability of the union by adding the probabilities of the sample points in the union. It is also possible to obtain the probability of the union of two events by using the **additive rule of probability.**

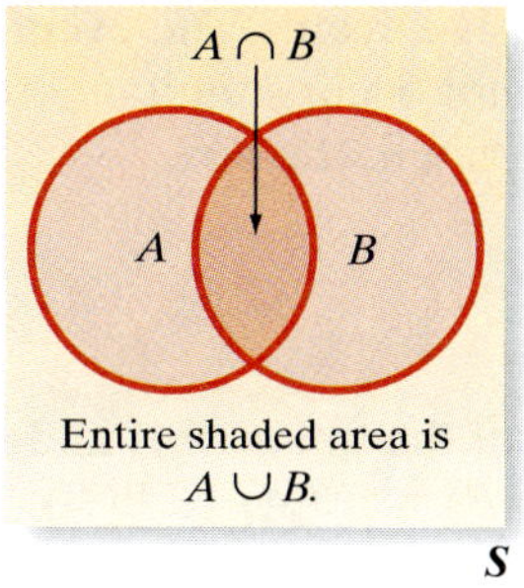

Figure 3.11
Venn diagram of union

The union of two events will often contain many sample points because the union occurs if either one or both of the events occur. By studying the Venn diagram in Figure 3.11, you can see that the probability of the union of two events, A and B, can be obtained by summing $P(A)$ and $P(B)$ and subtracting the probability corresponding to $A \cap B$. Therefore, the formula for calculating the probability of the union of two events is given in the next box.

Additive Rule of Probability

The probability of the union of events A and B is the sum of the probabilities of events A and B minus the probability of the intersection of events A and B—that is,

$$P(A \cup B) = P(A) + P(B) - P(A \cap B)$$

Example 3.12

Applying the Additive Rule to a Hospital Admission Study

Problem Hospital records show that 12% of all patients are admitted for surgical treatment, 16% are admitted for obstetrics, and 2% receive both obstetrics and surgical treatment. If a new patient is admitted to the hospital, what is the probability that the patient will be admitted either for surgery, obstetrics, or both? Use the additive rule of probability to arrive at the answer.

Solution Consider the following events:

A: {A patient admitted to the hospital receives surgical treatment.}

B: {A patient admitted to the hospital receives obstetrics treatment.}

Then, from the given information,

$$P(A) = .12$$
$$P(B) = .16$$

and the probability of the event that a patient receives both obstetrics and surgical treatment is

$$P(A \cap B) = .02$$

The event that a patient admitted to the hospital receives either surgical treatment, obstetrics treatment, or both is the union $A \cup B$. The probability of $A \cup B$ is given by the additive rule of probability:

$$P(A \cup B) = P(A) + P(B) - P(A \cap B) = .12 + .16 - .02 = .26$$

Thus, 26% of all patients admitted to the hospital receive either surgical treatment, obstetrics treatment, or both.

Look Back From the information given, it is not possible to list and assign probabilities to all the sample points. Consequently, we cannot proceed through the five-step process for finding the probability of an event and must use the additive rule.

Now Work Exercise 3.22

A very special relationship exists between events A and B when $A \cup B$ contains no sample points. In this case we call the events A and B *mutually exclusive events.*

Events A and B are **mutually exclusive** if $A \cap B$ contains no sample points—that is, if A and B have no sample points in common.

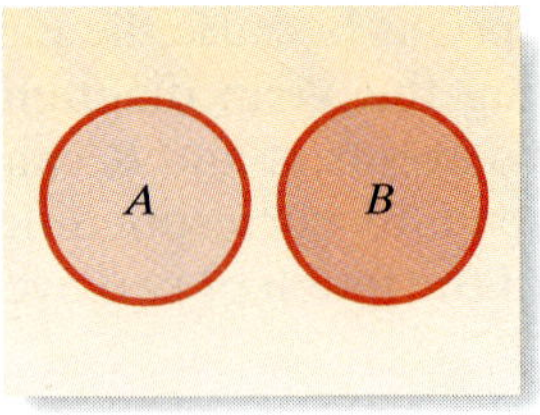

Figure 3.12
Venn diagram of mutually exclusive events

Figure 3.12 shows a Venn diagram of two mutually exclusive events. The events A and B have no sample points in common—that is, A and B cannot occur simultaneously, and $P(A \cap B) = 0$ Thus, we have the important relationship given in the box.

Probability of Union of Two Mutually Exclusive Events

If two events A and B are *mutually exclusive,* the probability of the union of A and B equals the sum of the probabilities of A and B; that is, $P(A \cup B) = P(A) + P(B)$.

CAUTION The formula just shown is *false* if the events are *not* mutually exclusive. For two nonmutually exclusive events, you must apply the general additive rule of probability. ▲

Example 3.13
The Union of Two Mutually Exclusive Events in a Coin-Tossing Experiment

Problem Consider the experiment of tossing two balanced coins. Find the probability of observing at *least* one head.

Solution Define the events

A: {Observe at least one head}
B: {Observe exactly one head}
C: {Observe exactly two heads}

Note that

$$A = B \cup C$$

and that $B \cap C$ contains no sample points (see Figure 3.13). Thus, B and C are mutually exclusive, so that

$$P(A) = B \cup C = P(B) + P(C) = {}^1/_2 + {}^1/_4 = {}^3/_4$$

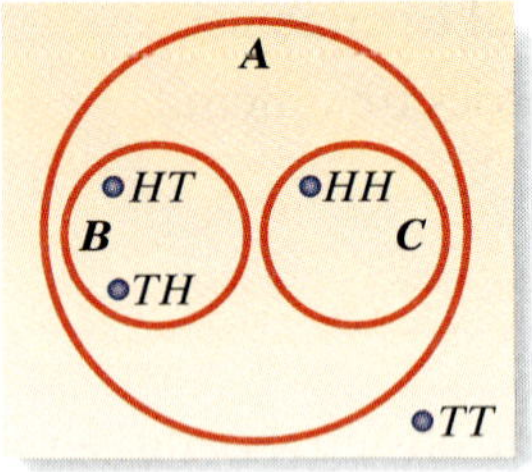

Figure 3.13
Venn diagram for coin-toss experiment

Look Back Although this example is very simple, it shows us that writing events with verbal descriptions that include the phrases *at least* or *at most* as unions of mutually exclusive events is very useful. This practice enables us to find the probability of the event by adding the probabilities of the mutually exclusive events.

Statistics IN Action Revisited — The Probability of Winning a Wheeling System

Refer to Florida's Pick-6 Lotto game in which you select six numbers of your choice from a field of numbers ranging from 1 to 53. In Section 3.1, we learned that the probability of winning Lotto on a single ticket is only 1 in approximately 23 million. The "experts" at Lotto Buster recommend many strategies for increasing the odds of winning the lottery. One strategy is to employ a wheeling system. In a complete wheeling system, you select more than six numbers, say, seven, and play every combination of six of those seven numbers.

Suppose you choose to "wheel" the following seven numbers: 2, 7, 18, 23, 30, 32, and 51. Every combination of six of these seven numbers is listed in Table SIA3.1. You can see that there are seven different possibilities. (Use the Combinations Rule with $N = 7$ and $n = 6$ to verify this.) Thus, we would purchase seven tickets (at a cost of $7) corresponding to these different combinations in a complete wheeling system.

To determine if this strategy does, in fact, increase our odds of winning, we need to find the probability that one of these seven combinations occurs during the 6/53 Lotto draw—that is, we need to find the probability that either Ticket 1 or Ticket 2 or Ticket 3 or Ticket 4 or Ticket 5 or Ticket 6 or Ticket 7 is the winning combination. Note that this probability is stated using the word *or,* implying a union of seven events. Letting T1 represent the event that Ticket 1 wins, and defining T2, T3, . . . , T7 in a similar fashion, we want to find

$$P(\text{T1 or T2 or T3 or T4 or T5 or T6 or T7})$$

Recall (Section 3.1, p. 127) that the 22,957,480 possible combinations in Pick-6 Lotto are mutually exclusive and equally likely to occur. Consequently, the probability of the union of the seven events is simply the sum of the probabilities of the individual events, where each event has probability of 1/(22,957,480):

$$P(\text{win Lotto with 7 wheeled numbers})$$
$$= P(\text{T1 or T2 or T3 or T4 or T5 or T6 or T7})$$
$$= 7/(22{,}957{,}480) = .0000003$$

In terms of odds, we now have 3 chances in 10 million of winning the Lotto with the complete wheeling system. The "experts" are correct—our odds of winning Lotto have increased (from 1 in 23 million). However, the probability of winning is so close to 0 we question whether the $7 spent on lottery tickets is worth the negligible increase in odds. In fact, it can be shown that to increase your chance of winning the 6/53 Lotto to 1 chance in 100 (i.e., .01) using a complete wheeling system, you would have to wheel 26 of your favorite numbers—a total of 230,230 combinations at a cost of $230,230!

Table SIA3.1 Wheeling the Six Numbers 2, 7, 18, 23, 30, 32, and 51

Ticket 1:	2	7	18	23	30	32
Ticket 2:	2	7	18	23	30	51
Ticket 3:	2	7	18	23	32	51
Ticket 4:	2	7	18	30	32	51
Ticket 5:	2	7	23	30	32	51
Ticket 6:	2	18	23	30	32	51
Ticket 7:	7	18	23	30	32	51

Exercises 3.26–3.46

Learning the Mechanics

3.26 NW Suppose $P(A) = .4$, $P(B) = .7$, and $P(A \cap B) = .3$. Find the following probabilities:

a. $P(B^c)$
b. $P(A^c)$
c. $P(A \cup B)$

3.27 A fair coin is tossed three times, and the events A and B are defined as follows:

A: {At least one head is observed.}
B: {The number of heads observed is odd.}

a. Identify the sample points in the events A, B, $A \cup B$, A^c, and $A \cap B$.
b. Find $P(A)$, $P(B)$, $P(A \cup B)$, $P(A^c)$, and $P(A \cap B)$ by summing the probabilities of the appropriate sample points.
c. Find $P(A \cup B)$ using the additive rule. Compare your answer to the one you obtained in part **b.**
d. Are the events A and B mutually exclusive? Why?

3.28 A pair of fair dice is tossed. Define the following events:

A: {You will roll a 7.} (i.e., The sum of the dots on the up faces of the two dice is equal to 7.)
B: {At least one of the two dice shows a 4.}

a. Identify the sample points in the events A, B, $A \cap B$, $A \cup B$, and A^c.
b. Find $P(A)$, $P(B)$, $P(A \cap B)$, $P(A \cup B)$, and $P(A^c)$ by summing the probabilities of the appropriate sample points.
c. Find $P(A \cup B)$ using the additive rule. Compare your answer to that for the same event in part **b.**
d. Are A and B mutually exclusive? Why?

3.29 NW Consider the Venn diagram below, where

$$P(E_1) = P(E_2) = P(E_3) = {}^1/_5,\ P(E_4) = P(E_5) = {}^1/_{20},$$
$$P(E_6) = {}^1/_{10}, \text{ and } P(E_7) = {}^1/_5.$$

Find each of the following probabilities:

a. $P(A)$
b. $P(B)$
c. $P(A \cup B)$
d. $P(A \cap B)$
e. $P(A^c)$
f. $P(B^c)$
g. $P(A \cup A^c)$
h. $P(A^c \cap B)$

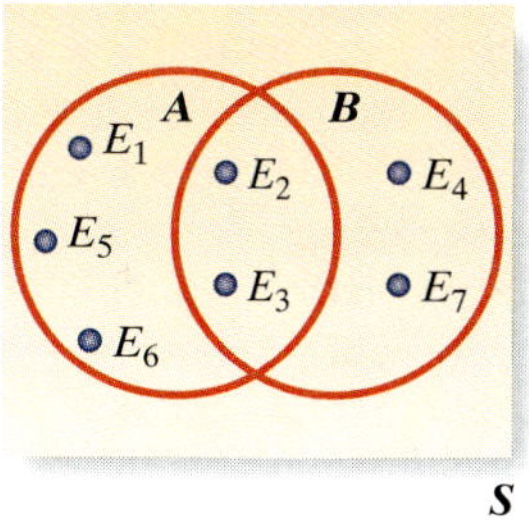

3.30 Consider the Venn diagram on the next page, where

$$P(E_1) = .10,\ P(E_2) = .05,\ P(E_3) = P(E_4) = .2,$$
$$P(E_5) = .06,\ P(E_6) = .3,\ P(E_7) = .06, \text{ and}$$
$$P(E_8) = .03.$$

Find the following probabilities:

a. $P(A^c)$
b. $P(B^c)$
c. $P(A^c \cap B)$
d. $P(A \cup B)$
e. $P(A \cap B)$
f. $P(A^c \cap B^c)$
g. Are events A and B mutually exclusive? Why?

Venn Diagram for Exercise 3.30

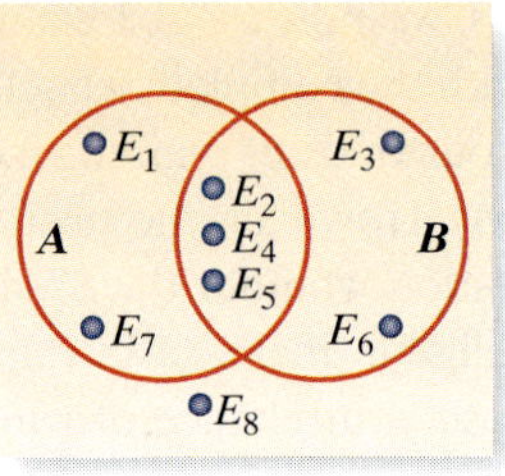

S

3.31 The outcomes of two variables are (Low, Medium, High) and (On, Off), respectively. An experiment is conducted in which the outcomes of each of the two variables are observed. The probabilities associated with each of the six possible outcome pairs are given in the accompanying two-way table.

	Low	Medium	High
On	.50	.10	.05
Off	.25	.07	.03

Consider the following events:

A: {On}
B: {Medium or On}
C: {Off and Low}
D: {High}

a. Find $P(A)$.
b. Find $P(B)$.
c. Find $P(C)$.
d. Find $P(D)$.
e. Find $P(A^c)$.
NW **f.** Find $P(A \cup B)$.
NW **g.** Find $P(A \cap C)$.
h. Consider each pair of events (*A* and *B*, *A* and *C*, *A* and *D*, *B* and *C*, *B* and *D*, *C* and *D*). List the pairs of events that are mutually exclusive. Justify your choices.

3.32 Refer to Exercise 3.31. Use the same event definitions to do the following exercises.
a. Write the event that the outcome is "On" and "High" as an intersection of two events.
b. Write the event that the outcome is "Low" or "Medium" as the complement of an event.

Applet Exercise 3.3

Use the applets *Simulating the Probability of Rolling a 6* and *Simulating the Probability of Rolling a 3 or 4* to explore the additive rule for probability.
a. Explain why the applet *Simulating the Probability of Rolling a 6* can also be used to simulate the probability of rolling a 3. Then use the applet with $n = 1{,}000$ to simulate the probability of rolling a 3. Record the cumulative proportion. Repeat the process to simulate the probability of rolling a 4.
b. Use the applet *Simulating the Probability of Rolling a 3 or 4* with $n = 1{,}000$ to simulate the probability of rolling a 3 or 4. Record the cumulative proportion.
c. Add the two cumulative proportions from part **a.** How does this sum compare to the cumulative proportion in part **b**? How does this illustrate the additive rule for probability?

Applet Exercise 3.4

Use the applets *Simulating the Probability of Rolling a 6* and *Simulating the Probability of Rolling a 3 or 4* to simulate the probability of the complement of an event.
a. Explain how the applet *Simulating the Probability of Rolling a 6* can also be used to simulate the probability of the event *rolling a 1, 2, 3, 4, or 5*. Then use the applet with $n = 1{,}000$ to simulate this probability.
b. Explain how the applet *Simulating the Probability of Rolling a 3 or 4* can also be used to simulate the probability of the event *rolling a 1, 2, 5, or 6*. Then use the applet with $n = 1{,}000$ to simulate this probability.
c. Which applet could be used to simulate the probability of the event *rolling a 1, 2, 3, or 4*? Explain.

Applying the Concepts—Basic

3.33 **Study of analysts' forecasts.** The *Journal of Accounting Research* (March 2008) published a study on relationship incentives and degree of optimism among analysts' forecasts. Participants were analysts at either a large or small brokerage firm who made their forecasts either early or late in the quarter. Also, some analysts were only concerned with making an accurate forecast, while others were also interested in their relationship with management. Suppose one of these analysts is randomly selected. Consider the following events:
A = {The analyst is concerned only with making an accurate forecast.}
B = {The analyst makes the forecast early in the quarter.}
C = {The analyst is from a small brokerage firm.}
Describe each of the following events in terms of unions, intersections, and complements (e.g., $A \cup B$, $A \cap B$, A^c, etc.).
a. The analyst makes an early forecast and is concerned only with accuracy.
b. The analyst is not concerned only with accuracy.
c. The analyst is from a small brokerage firm or makes an early forecast.
d. The analyst makes a late forecast and is not concerned only with accuracy.

3.34 **Problems at major companies.** The *Organization Development Journal* (Summer 2006) reported on the results of a survey of human resource officers (HROs) at major employers located in a southeastern city. The focus of the study was employee behavior, namely absenteeism, promptness to work, and turnover. The study found that 55% of the HROs had problems with employee absenteeism; also, 41% had problems with turnover. Suppose that 22% of the HROs had problems with both absenteeism and turnover. Use this information to find the probability that an HRO officer selected from the group surveyed had problems with either employee absenteeism or employee turnover.

3.35 **Use of cellular phone features.** Harris Interactive conducted a survey of cellular phone owners to estimate the likelihood of using extra phone features, such as the calendar, address book, instant messaging, e-mail, and so on. The results were reported online at *BusinessWeek.com* (Aug. 17, 2005). The following table shows the percentage of 18- to 34-year-olds who use the extra cell phone features. (Note that the percentages in the table do not sum to 100% because a cell phone owner may use more than one feature.)

Cell Phone Feature	Percentage of 18- to 34-Year-Olds
Calendar and address book	56%
Download and/or play games	54%
Download ringtones	47%
Instant messaging	43%
Take pictures	35%
Send and receive e-mail	28%
Access the Internet	34%
None of the above	16%

Source: BusinessWeek.com, Aug. 17, 2005.

a. Estimate the probability that an 18- to 34-year-old cell phone owner uses instant messaging.

b. Estimate the probability that an 18- to 34-year-old cell phone owner uses at least one of the features listed in the table.

3.36 Scanning errors at Wal-Mart. The National Institute for Standards and Technology (NIST) mandates that for every 100 items scanned through the electronic checkout scanner at a retail store, no more than 2 should have an inaccurate price. A recent study of the accuracy of checkout scanners at Wal-Mart stores in California was conducted (*Tampa Tribune,* Nov. 22, 2005). Of the 60 Wal-Mart stores investigated, 52 violated the NIST scanner accuracy standard. If one of the 60 Wal-Mart stores is randomly selected, what is the probability that the store does not violate the NIST scanner accuracy standard?

3.37 Inactive oil and gas structures. U.S. federal regulations require that operating companies clear all inactive offshore oil and gas structures within 1 year after production ceases. Researchers at the Louisiana State University Center for Energy Studies gathered data on both active and inactive oil and gas structures in the Gulf of Mexico (*Oil & Gas Journal,* Jan. 3, 2005). They discovered that the Gulf of Mexico had 2,175 active and 1,225 idle (inactive) structures at the end of 2003. The following table breaks down these structures by type (caisson, well protector, or fixed platform). Consider the structure type and active status of one of these oil/gas structures.

	Structure Type			
	Caisson	Well Protector	Fixed Platform	Totals
Active	503	225	1,447	2,175
Inactive	598	177	450	1,225

Source: Kaiser, M., and Mesyanzhinov, D. "Study tabulates idle Gulf of Mexico structures," *Oil & Gas Journal,* Vol. 103, No. 1, Jan. 3, 2005 (Table 2).

a. List the simple events for this experiment.

b. Assign reasonable probabilities to the simple events.

c. Find the probability that the structure is active.

d. Find the probability that the structure is a well protector.

e. Find the probability that the structure is an inactive caisson.

f. Find the probability that the structure is either inactive or a fixed platform.

g. Find the probability that the structure is not a caisson.

3.38 Social networking Web sites in the United Kingdom. In the United States, MySpace and FaceBook are considered the two most popular social networking Web sites. In the United Kingdom (UK), the competition for social networking is between MySpace and Bebo. According to Nielsen/NetRatings (April 2006), 4% of UK citizens visit MySpace, 3% visit Bebo, and 1% visit both MySpace and Bebo.

a. Draw a Venn diagram to illustrate the use of social networking sites in the United Kingdom.

b. Find the probability that a UK citizen visits either the MySpace or Bebo social networking sites.

c. Use your answer to part **b** to find the probability that a UK citizen does not visit either social networking site.

Applying the Concepts—Intermediate

3.39 Characteristics of a new product. The long-run success of a business depends on its ability to market products with superior characteristics that maximize consumer satisfaction and that give the firm a competitive advantage (Kotler & Keller, *Marketing Management,* 2006). Ten new products have been developed by a food-products firm. Market research has indicated that the 10 products have the characteristics described by the following Venn diagram:

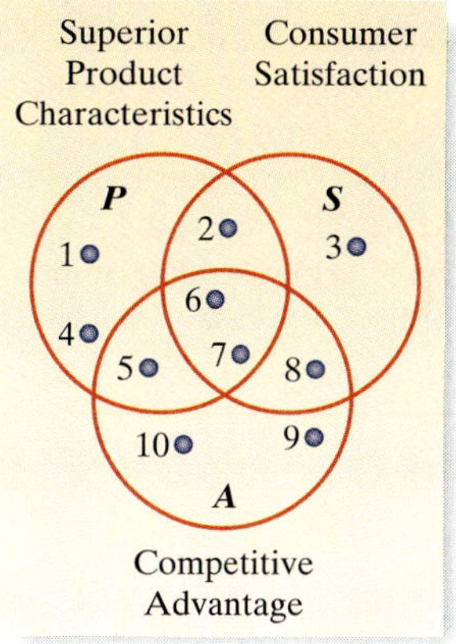

a. Write the event that a product possesses all the desired characteristics as an intersection of the events defined in the Venn diagram. Which products are contained in this intersection?

b. If one of the 10 products were selected at random to be marketed, what is the probability that it would possess all the desired characteristics?

c. Write the event that the randomly selected product would give the firm a competitive advantage or would satisfy consumers as a union of the events defined in the Venn diagram. Find the probability of this union.

d. Write the event that the randomly selected product would possess superior product characteristics and satisfy consumers. Find the probability of this intersection.

3.40 Guilt in decision making. The effect of guilt emotion on how a decision maker focuses on the problem was investigated in the Jan. 2007 issue of the *Journal of Behavioral Decision*

Making (see Exercise 1.26, p. 24). A total of 171 volunteer students participated in the experiment, where each was randomly assigned to one of three emotional states (guilt, anger, or neutral) through a reading/writing task. Immediately after the task, students were presented with a decision problem where the stated option has predominantly negative features (e.g., spending money on repairing a very old car). The results (number responding in each category) are summarized in the accompanying table. Suppose one of the 171 participants is selected at random.

Emotional State	Choose Stated Option	Do Not Choose Stated Option	Totals
Guilt	45	12	57
Anger	8	50	58
Neutral	7	49	56
Totals	60	111	171

Source: Gangemi, A., and Mancini, F. "Guilt and focusing in decision-making," *Journal of Behavioral Decision Making,* Vol. 20, Jan. 2007 (Table 2).

a. Find the probability that the respondent is assigned to the guilty state.
b. Find the probability that the respondent chooses the stated option (repair the car).
c. Find the probability that the respondent is assigned to the guilty state and chooses the stated option.
d. Find the probability that the respondent is assigned to the guilty state or chooses the stated option.

3.41 Appeals of federal civil trials. The *Journal of the American Law and Economics Association* (Vol. 3, 2001) published the results of a study of appeals of federal civil trials. The following table, extracted from the article, gives a breakdown of 2,143 civil cases that were appealed by either the plaintiff or the defendant. The outcome of the appeal, as well as the type of trial (judge or jury), was determined for each civil case. Suppose one of the 2,143 cases is selected at random and both the outcome of the appeal and type of trial are observed.

	Jury	Judge	Totals
Plaintiff trial win—reversed	194	71	265
Plaintiff trial win—affirmed/dismissed	429	240	669
Defendant trial win—reversed	111	68	179
Defendant trial win—affirmed/dismissed	731	299	1,030
Totals	1,465	678	2,143

a. Find $P(A)$, where $A = \{\text{jury trial}\}$.
b. Find $P(B)$, where $B = \{\text{plaintiff trial win is reversed}\}$.
c. Are A and B mutually exclusive events?

d. Find $P(A^c)$.
e. Find $P(A \cup B)$.
f. Find $P(A \cap B)$.

3.42 Likelihood of a tax return audit. At the beginning of each year, the Internal Revenue Service (IRS) releases information on the likelihood of a tax return being audited. In 2007, the IRS audited 1,384,563 individual tax returns from the total of 134,421,400 filed returns; also in 2007, the IRS audited 59,516 business returns from the total of 9,072,828 business returns filed (*IRS Fiscal Year 2007 Enforcement and Services Statistics*).

a. Suppose an individual tax return from 2007 is randomly selected. What is the probability that the return is audited by the IRS?
b. Refer to part **a.** Determine the probability that an individual return is not audited by the IRS.
c. Suppose a business tax return from 2007 is randomly selected. What is the probability that the return is audited by the IRS?
d. Refer to part **c.** Determine the probability that a business return is not audited by the IRS.

3.43 Online securities trading. E*TRADE Group Inc. was the first to provide online securities trading for its clients, offering an alternative to traditional investment firms. According to *Business Week,* online securities trading accounts for a significant share of the brokerage business. The table below reports the number of online and traditional accounts for five leading brokerages. Suppose a customer account is to be drawn at random from the population of accounts described in the table. Consider the following events:

A: {The account is with Merrill Lynch.}
B: {The account is an online account.}
C: {The account is with E*TRADE and is an online account.}
D: {The account is either with TD Waterhouse or E*TRADE and is an online account.}
E: { The account is with E*TRADE.}

a. Find the probability of each of the aforementioned events.
b. Find $P(A \cap B)$.
c. Find $P(A \cup B)$.
d. Find $P(B^c \cap E)$
e. Find $P(A \cup E)$.
f. Which pairs of events are mutually exclusive?

3.44 Home modifications for wheelchair users. The *American Journal of Public Health* (Jan. 2002) reported on a study of elderly wheelchair users who live at home. A sample of 306 wheelchair users, age 65 or older, were surveyed about whether they had an injurious fall during the year and whether their home features any one of five structural

Table for Exercise 3.43

Brokerage Firms	Online Accounts	Traditional Accounts	Total Accounts
Fidelity Investments	2.8 million	8.0 million	10.8 million
Merrill Lynch & Co.	0	8.0 million	8.0 million
Charles Schwab & Co.	2.8 million	3.5 million	6.3 million
TD Waterhouse Group Inc.	1.0 million	1.1 million	2.1 million
E*TRADE Group Inc.	1.24 million	0	1.24 million
Totals	7.84 million	20.6 million	28.44 million

Source: Business Week, Oct. 18, 1999, pp. 185–186.

modifications: bathroom modifications, widened doorways/hallways, kitchen modifications, installed railings, and easy-open doors. The responses are summarized in the accompanying table. Suppose we select, at random, one of the 306 surveyed wheelchair users.

Home Features	Injurious Fall(s)	No Falls	Totals
All 5	2	7	9
At least 1 but not all	26	162	188
None	20	89	109
Totals	48	258	306

Source: Berg, K., Hines, M., and Allen, S. "Wheelchair users at home: Few home modifications and many injurious falls," *American Journal of Public Health,* Vol. 92, No. 1, Jan. 2002 (Table 1).

a. Find the probability that the wheelchair user had an injurious fall.

b. Find the probability that the wheelchair user had all five features installed in the home.

c. Find the probability that the wheelchair user had no falls and none of the features installed in the home.

3.45 Reliability of gas station air pumps. Tire and automobile manufacturers and consumer safety experts all recommend that drivers maintain proper tire pressure in their cars. Consequently, many gas stations now provide air pumps and air gauges for their customers. In a *Research Note* (Nov. 2001), the National Highway Traffic Safety Administration studied the reliability of gas station air pumps. The next table gives the percentage of gas stations that provide air gauges that overreport the pressure level in the tire.

a. If the gas station air pressure gauge reads 35 psi, what is the probability that the pressure is overreported by 6 psi or more?

Station Gauge Pressure	Overreport by 4 psi or More (%)	Overreport by 6 psi or More (%)	Overreport by 8 psi or More (%)
25 psi	16	2	0
35 psi	19	9	0
45 psi	19	14	5
55 psi	20	15	9

b. If the gas station air pressure gauge reads 55 psi, what is the probability that the pressure is overreported by 8 psi or more?

c. If the gas station air pressure gauge reads 25 psi, what is the probability that the pressure is not overreported by 4 psi or more?

d. Are the events A = {overreport by 4 psi or more} and B = {overreport by 6 psi or more} mutually exclusive? Explain.

e. Based on your answer to part **d,** why do the probabilities in the table not sum to 1?

Applying the Concepts—Advanced

3.46 Galileo's Passedix game. Passedix is a game of chance played with three fair dice. Players bet whether the sum of the faces shown on the dice will be above or below 10. During the late sixteenth century, the astronomer and mathematician Galileo Galilei was asked by the Grand Duke of Tuscany to explain why "the chance of throwing a total of 9 with three fair dice was less than that of throwing a total of 10" (*Interstat,* Jan. 2004). The grand duke believed that the chance should be the same because "there are an equal number of partitions of the numbers 9 and 10." Find the flaw in the grand duke's reasoning and answer the question posed to Galileo.

3.5 Conditional Probability

The event probabilities we've been discussing give the relative frequencies of the occurrences of the events when the experiment is repeated a very large number of times. Such probabilities are often called **unconditional probabilities** because no special conditions are assumed, other than those that define the experiment.

Often, however, we have additional knowledge that might affect the likelihood of the outcome of an experiment, so we need to alter the probability of an event of interest. A probability that reflects such additional knowledge is called the **conditional probability** of the event. For example, we've seen that the probability of observing an even number (event A) on a toss of a fair die is $\frac{1}{2}$. But suppose we're given the information that on a particular throw of the die, the result was a number less than or equal to 3 (event B). Would the probability of observing an even number on that throw of the die still be equal to $\frac{1}{2}$? It can't be because making the assumption that B has occurred reduces the sample space from six sample points to three sample points (namely, those contained in event B). This reduced sample space is as shown in Figure 3.14.

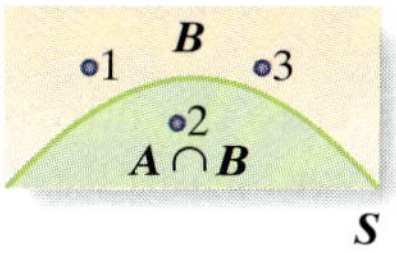

Figure 3.14
Reduced sample space for the die-toss experiment given that event B has occurred

Because the sample points for the die-toss experiment are equally likely, each of the three sample points in the reduced sample space is assigned an equal *conditional probability* of $\frac{1}{3}$. Because the only even number of the three in the reduced sample space B is the number 2 and the die is fair, we conclude that the probability that A occurs *given that B occurs* is $\frac{1}{3}$. We use the symbol $P(A|B)$ to represent the probability of event A given that event B occurs. For the die-toss example, $P(A|B) = \frac{1}{3}$.

To get the probability of event A given that event B occurs, we proceed as follows: We divide the probability of the part of A that falls within the reduced sample space B, namely $P(A \cap B)$, by the total probability of the reduced sample space, namely $P(B)$.

Thus, for the die-toss example with event *A:* {Observe an even number} and event *B:* {Observe a number less than or equal to 3}, we find

$$P(A|B) = \frac{P(A \cap B)}{P(B)} = \frac{P(2)}{P(1) + P(2) + P(3)} = \frac{\frac{1}{6}}{\frac{3}{6}} = \frac{1}{3}$$

The formula for $P(A|B)$ is true in general:

Conditional Probability Formula

To find the *conditional probability that event A occurs given that event B occurs,* divide the probability that *both A* and *B* occur by the probability that *B* occurs—that is,

$$P(A|B) = \frac{P(A \cap B)}{P(B)} \quad \text{[We assume that } P(B) \neq 0.]$$

This formula adjusts the probability of $A \cap B$ from its original value in the complete sample space *S* to a conditional probability in the reduced sample space *B*. If the sample points in the complete sample space are equally likely, then the formula will assign equal probabilities to the sample points in the reduced sample space, as in the die-toss experiment. If, on the other hand, the sample points have unequal probabilities, the formula will assign conditional probabilities proportional to the probabilities in the complete sample space. This is illustrated by the following practical examples.

Example 3.14
The Conditional Probability Formula Applied to Executives Who Cheat at Golf

Problem To develop programs for business travelers staying at convention hotels, a major hotel chain commissioned a study of executives who play golf. The study revealed that 55% of the executives admitted they had cheated at golf. Also, 20% of the executives admitted they had cheated at golf and had lied in business. Given an executive had cheated at golf, what is the probability that the executive also had lied in business?

Solution Let's define events *A* and *B* as follows:

$$A = \{\text{Executive who had cheated at golf}\}$$
$$B = \{\text{Executive who had lied in business}\}$$

From the study, we know that 55% of executives had cheated at golf, so $P(A) = .55$. Now, executives who both cheat at golf (event *A*) *and* lie in business (event *B*) represent the compound event $A \cap B$. From the study, $P(A \cap B) = .20$. We want to know the probability that an executive lied in business (event *B*), given he or she cheated at golf (event *A*)—that is, we want to know the conditional probability $P(A|B)$. Applying the preceding conditional probability formula, we have

$$P(B|A) = \frac{P(A \cap B)}{P(A)} = \frac{.20}{.55} = .364$$

Thus, given an executive had cheated at golf, the probability that the executive also had lied in business is .364.

Look Back One of the keys to correctly applying the formula is to write the information in the study in the form of probability statements involving the events of interest. The word *and* in the statement "cheat at golf *and* lie in business" implies an intersection of the two events, *A* and *B*. The word *given* in the phrase "*given* an executive cheats at golf" implies that event *A* is the given event.

Now Work Exercise 3.56

Example 3.15

Applying the Conditional Probability Formula to a Two-Way Table on Customer Desire to Buy

Figure 3.15
Sample space for contacting a sales prospect

Problem Suppose you are interested in the probability of the sale of a large piece of earthmoving equipment. A single prospect is contacted. Let F be the event that the buyer has sufficient money (or credit) to buy the product and let F^c denote the complement of F (the event that the prospect does not have the financial capability to buy the product). Similarly, let B be the event that the buyer wishes to buy the product and let B^c be the complement of that event. Then the four sample points associated with the experiment are shown in Figure 3.15, and their probabilities are given in Table 3.5. Use the sample point probabilities to find the probability that a single prospect will buy, given that the prospect is able to finance the purchase.

Table 3.5 Probabilities of Customer Desire to Buy and Ability to Finance

		Desire	
		To Buy, B	Not to Buy, B^c
Able to Finance	Yes, F	.2	.1
	No, F^c	.4	.3

Solution Suppose you consider the large collection of prospects for the sale of your product and randomly select one person from this collection. What is the probability that the person selected will buy the product? In order to buy the product, the customer must be financially able *and* have the desire to buy, so this probability would correspond to the entry in Table 3.5 {To buy, B} and next to {Yes, F}, or $P(B \cap F) = .2$. This is the unconditional probability of the event $B \cap F$.

In contrast, suppose you know that the prospect selected has the financial capability for purchasing the product. Now you are seeking the probability that the customer will buy given (the condition) that the customer has the financial ability to pay. This probability, the conditional probability of B given that F has occurred and denoted by the symbol $P(B|F)$, would be determined by considering only the sample points in the reduced sample space containing the sample points $B \cap F$ and $B^c \cap F$—that is, sample points that imply the prospect is financially able to buy. (This subspace is shaded in Figure 3.16.) From our definition of conditional probability,

$$P(B|F) = \frac{P(B \cap F)}{P(F)}$$

where $P(F)$ is the sum of the probabilities of the two sample points corresponding to $B \cap F$ and $B^c \cap F$ (given in Table 3.5). Then

$$P(F) = P(B \cap F) + P(B^c \cap F) = .2 + .1 = .3$$

and the conditional probability that a prospect buys, given that the prospect is financially able, is

$$P(B|F) = \frac{P(B \cap F)}{P(F)} = \frac{.2}{.3} = .667$$

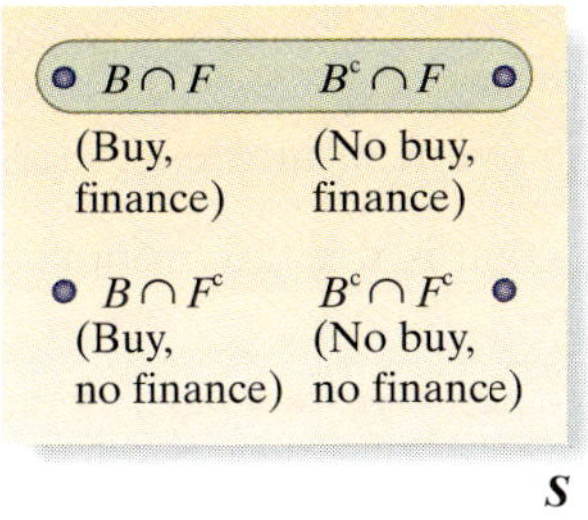

Figure 3.16
Subspace (shaded) containing sample points implying a financially able prospect

As we would expect, the probability that the prospect will buy, given that he or she is financially able, is higher than the unconditional probability of selecting a prospect who will buy.

Look Back Note that the conditional probability formula assigns a probability to the event $(B \cap F)$ in the reduced sample space that is proportional to the probability of the event in the complete sample space. To see this, note that the two sample points in the reduced sample space, $(B \cap F)$ and $(B^c \cap F)$, have probabilities of .2 and .1, respectively, in the complete sample space S. The formula assigns conditional probabilities $2/3$ and $1/3$ (use the formula to check the second one) to these sample points in the reduced sample space F so that the conditional probabilities retain the 2-to-1 proportionality of the original sample point probabilities.

Now Work Exercise 3.47a–b

Example 3.16

Applying the Conditional Probability Formula to a Two-Way Table on Customer Complaints

Problem The investigation of consumer product complaints by the Federal Trade Commission (FTC) has generated much interest by manufacturers in the quality of their products. A manufacturer of an electromechanical kitchen utensil conducted an analysis of a large number of consumer complaints and found that they fell into the six categories shown in Table 3.6. If a consumer complaint is received, what is the probability that the cause of the complaint was product appearance given that the complaint originated during the guarantee period?

Table 3.6 Distribution of Product Complaints

	Reason for Complaint			
Complaint Origin	Electrical	Mechanical	Appearance	Totals
During Guarantee Period	18%	13%	32%	63%
After Guarantee Period	12%	22%	3%	37%
Totals	30%	35%	35%	100%

Solution Let A represent the event that the cause of a particular complaint is product appearance and let B represent the event that the complaint occurred during the guarantee period. Checking Table 3.6, you can see that $(18 + 13 + 32)\% = 63\%$ of the complaints occur during the guarantee period. Hence, $P(B) = .63$. The percentage of complaints that were caused by appearance and occurred during the guarantee period (the event $A \cap B$) is 32%. Therefore, $P(A \cap B) = .32$.

Using these probability values, we can calculate the conditional probability $P(A|B)$ that the cause of a complaint is appearance, given that the complaint occurred during the guarantee time:

$$P(A|B) = \frac{P(A \cap B)}{P(B)} = \frac{.32}{.63} = .51$$

Consequently, we can see that slightly more than half the complaints that occurred during the guarantee period were due to scratches, dents, or other imperfections in the surface of the kitchen devices.

Look Back Note that the answer, $\frac{.32}{.63}$, is the proportion for the event of interest A (.32) divided by the row total proportion for the given event B (.63)—that is, it is the proportion of the time A occurs within the given event B.

Now Work Exercise 3.58

3.6 The Multiplicative Rule and Independent Events

The probability of an intersection of two events can be calculated using the multiplicative rule, which employs the conditional probabilities we defined in the previous section. Actually, we have already developed the formula in another context (Section 3.5). You will recall that the formula for calculating the conditional probability of A given B is

$$P(A|B) = \frac{P(A \cap B)}{P(B)}$$

Multiplying both sides of this equation by $P(B)$, we obtain a formula for the probability of the intersection of events A and B. This is often called the **multiplicative rule of probability.**

Multiplicative Rule of Probability

$$P(A \cap B) = P(A)P(B|A) \text{ or, equivalently, } P(A \cap B) = P(B)P(A|B)$$

Example 3.17

Applying the Multiplicative Rule to Find a Probability Associated with Wheat Futures

Problem An investor in wheat futures is concerned with the following events:

B: {U.S. production of wheat will be profitable next year.}

A: {A serious drought will occur next year.}

Based on available information, the investor believes that the probability is .01 that production of wheat will be profitable *assuming* a serious drought will occur in the same year and that the probability is .05 that a serious drought will occur—that is,

$$P(B|A) = .01 \text{ and } P(A) = .05$$

Based on the information provided, what is the probability that a serious drought will occur and that a profit will be made? That is, find $P(A \cap B)$, the probability of the intersection of events A and B.

Solution We want to calculate $P(A \cap B)$. Using the formula for the multiplicative rule, we obtain

$$P(A \cap B) = P(A)P(B|A) = (.05)(.01) = .0005$$

The probability that a serious drought occurs and the production of wheat is profitable is only .0005. As we might expect, this intersection is a very rare event.

Look Back The multiplicative rule can be expressed in two ways: $P(A \cap B) = P(A) \cdot P(B|A)$ or $P(A \cap B) = P(B) \cdot P(A|B)$. Select the formula that involves a given event for which you know the probability (e.g., event B in the example).

Now Work Exercise 3.60

Intersections often contain only a few sample points. In this case, the probability of an intersection is easy to calculate by summing the appropriate sample point probabilities. However, the formula for calculating intersection probabilities is invaluable when the intersection contains numerous sample points, as the next example illustrates.

Example 3.18

Applying the Multiplicative Rule to a Study of Welfare Workers

Problem A county welfare agency employs 10 welfare workers who interview prospective food stamp recipients. Periodically the supervisor selects, at random, the forms completed by 2 workers to audit for illegal deductions. Unknown to the supervisor, 3 of the workers have regularly been giving illegal deductions to applicants. What is the probability that both of the 2 workers chosen have been giving illegal deductions?

Solution Define the following two events:

A: {First worker selected gives illegal deductions}

B: {Second worker selected gives illegal deductions}

We want to find the probability of the event that both selected workers have been giving illegal deductions. This event can be restated as {First worker gives illegal deductions *and* second worker gives illegal deductions}. Thus, we want to find the probability of the intersection, $A \cap B$. Applying the multiplicative rule, we have

$$P(A \cap B) = P(A)P(B|A)$$

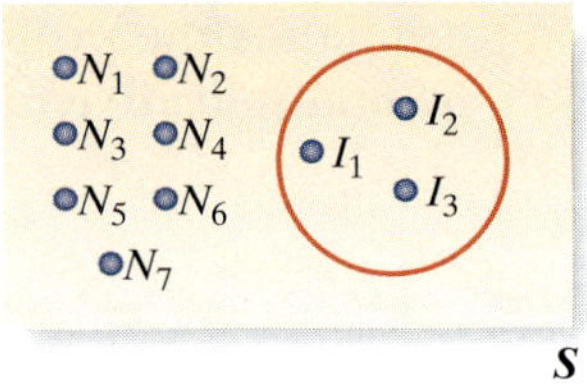

Figure 3.17
Venn diagram for finding $P(A)$

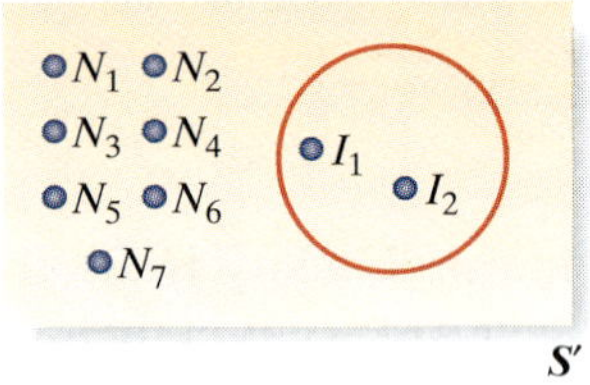

Figure 3.18
Venn diagram for finding $P(B|A)$

To find $P(A)$, it is helpful to consider the experiment as selecting 1 worker from the 10. Then the sample space for the experiment contains 10 sample points (representing the 10 welfare workers), where the 3 workers giving illegal deductions are denoted by the symbol I (I_1, I_2, I_3), and the 7 workers not giving illegal deductions are denoted by the symbol N ($N_1, \ldots, N_7$). The resulting Venn diagram is illustrated in Figure 3.17.

Because the first worker is selected at random from the 10, it is reasonable to assign equal probabilities to the 10 sample points. Thus, each sample point has a probability of $1/10$. The sample points in event A are $\{I_1, I_2, I_3,\}$—the 3 workers who are giving illegal deductions. Thus,

$$P(A) = P(I_1) + P(I_2) + P(I_3) = 1/10 + 1/10 + 1/10 = 3/10$$

To find the conditional probability, $P(B|A)$, we need to alter the sample space S. Because we know A has occurred [the first worker selected is giving illegal deductions (say I_3), only 2 of the 9 remaining workers in the sample space are giving illegal deductions]. The Venn diagram for this new sample space (S') is shown in Figure 3.18. Each of these nine sample points are equally likely, so each is assigned a probability of $1/9$. Because the event $(B|A)$ contains the sample points $\{I_1, I_2\}$, we have

$$P(B|A) = P(I_1) + P(I_2) = 1/9 + 1/9 = 2/9$$

Substituting $P(A) = 3/10$ and $P(B|A) = 2/9$ into the formula for the multiplicative rule, we find

$$P(A \cap B) = P(A)P(B|A) = (3/10)(2/9) = 6/90 = 1/15$$

Thus, there is a 1 in 15 chance that both workers chosen by the supervisor have been giving illegal deductions to food stamp recipients.

Look Back The key words *both* and *and* in the statement "both A and B occur" imply an intersection of two events, which in turn implies that we should *multiply* probabilities to obtain the probability of interest.

Now Work Exercise 3.51c

The sample space approach is only one way to solve the problem posed in Example 3.18. An alternative method employs the tree diagram (introduced in Example 3.1). Tree diagrams are helpful for calculating the probability of an intersection.

To illustrate, a tree diagram for Example 3.18 is displayed in Figure 3.19. The tree begins at the far left with two branches. These branches represent the two possible outcomes

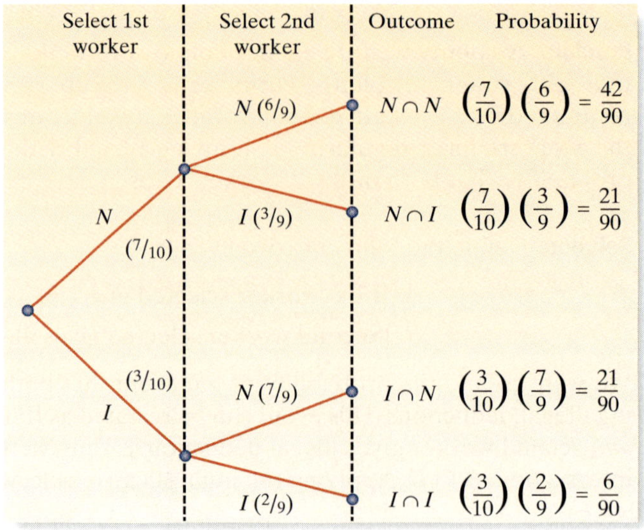

Figure 3.19
Tree diagram for Example 3.18

N (no illegal deductions) and I (illegal deductions) for the first worker selected. The unconditional probability of each outcome is given (in parentheses) on the appropriate branch—that is, for the first worker selected, $P(N) = 7/10$ and $P(I) = 3/10$. (These can be obtained by summing sample point probabilities as in Example 3.18.)

The next level of the tree diagram (moving to the right) represents the outcomes for the second worker selected. The probabilities shown here are conditional probabilities because the outcome for the first worker is assumed to be known. For example, if the first worker is giving illegal deductions (I), the probability that the second worker is also giving illegal deductions (I) is $2/9$ because of the 9 workers left to be selected, only 2 remain who are giving illegal deductions. This conditional probability, $2/9$, is shown in parentheses on the bottom branch of Figure 3.18.

Finally, the four possible outcomes of the experiment are shown at the end of each of the four tree branches. These events are intersections of two events (outcome of first worker and outcome of second worker). Consequently, the multiplicative rule is applied to calculate each probability, as shown in Figure 3.18. You can see that the intersection $\{I \cap I\}$ (i.e., the event that both workers selected are giving illegal deductions) has probability $6/90 = 1/15$—the same value obtained in Example 3.18.

In Section 3.5 we showed that the probability of an event A may be substantially altered by the knowledge that an event B has occurred. However, this will not always be the case. In some instances, the assumption that event B has occurred will *not* alter the probability of event A at all. When this is true, we say that the two events A and B are *independent events.*

Events A and B are **independent events** if the occurrence of B does not alter the probability that A has occurred; that is, events A and B are independent if

$$P(A|B) = P(A)$$

When events A and B are independent, it is also true that

$$P(B|A) = P(B)$$

Events that are not independent are said to be **dependent.**

Example 3.19

Checking for Independence in a Die-Tossing Experiment

Problem Consider the experiment of tossing a fair die and let

A: {Observe an even number.}

B: {Observe a number less than or equal to 4.}

Are events A and B independent?

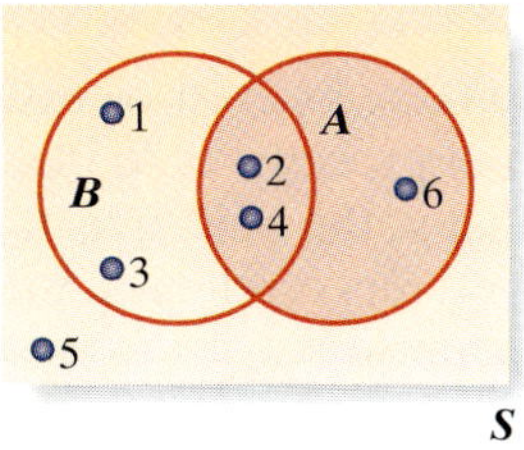

Figure 3.20
Venn diagram for die-toss experiment

Solution The Venn diagram for this experiment is shown in Figure 3.20. We first calculate

$$P(A) = P(2) + P(4) + P(6) = 1/2$$
$$P(B) = P(1) + P(2) + P(3) + P(4) = 2/3$$
$$P(A \cap B) = P(2) + P(4) = 1/3$$

Now assuming B has occurred, the conditional probability of A given B is

$$P(A|B) = \frac{P(A \cap B)}{P(B)} = \frac{1/3}{2/3} = \frac{1}{2} = P(A)$$

Thus, assuming that event B occurring does not alter the probability of observing an even number, $P(A)$ remains $1/2$. Therefore, the events A and B are independent.

Look Back Note that if we calculate the conditional probability of B given A, our conclusion is the same:

$$P(B|A) = \frac{P(A \cap B)}{P(A)} = \frac{1/3}{1/2} = \frac{2}{3} = P(B)$$

BIOGRAPHY **BLAISE PASCAL (1623–1662)**

Solver of Chevalier's Dilemma

As a precocious child growing up in France, Blaise Pascal showed an early inclination toward mathematics. Although his father would not permit Pascal to study mathematics before the age of 15 (removing all math texts from his house), at age 12 Blaise discovered on his own that the sum of the angles of a triangle are two right triangles.

Pascal went on to become a distinguished mathematician, as well as a physicist, theologian, and the inventor of the first digital calculator. Most historians attribute the beginning of the study of probability to the correspondence between Pascal and Pierre de Fermat in 1654. The two solved the Chevalier's dilemma—a gambling problem related to Pascal by his friend and Paris gambler the Chevalier de Mere. The problem involved determining the expected number of times one could roll two dice without throwing a double 6. (Pascal proved that the breakeven point was 25 rolls.) ■

Example 3.20 Checking for Independence in the Consumer Product Complaint Study

Problem Refer to the consumer product complaint study in Example 3.16. The percentages of complaints of various types during and after the guarantee period are shown in Table 3.6. Define the following events:

A: {Cause of complaint is product appearance.}

B: {Complaint occurred during the guarantee term.}

Are A and B independent events?

Solution Events A and B are independent if $P(A|B) = P(A)$. We calculated $P(A|B)$ in Example 3.16 to be .51, and from Table 3.6 we see that

$$P(A) = .32 + .03 = .35$$

Therefore, $P(A|B)$ is not equal to $P(A)$, and A and B are dependent events.

Now Work Exercise 3.47c

To gain an intuitive understanding of independence, think of situations in which the occurrence of one event does not alter the probability that a second event will occur. For example, suppose two small companies are being monitored by a financier for possible investment. If the businesses are in different industries and they are otherwise unrelated, then the success or failure of one company may be *independent* of the success or failure of the other—that is, the event that company A fails may not alter the probability that company B will fail.

As a second example, consider an election poll in which 1,000 registered voters are asked their preference between two candidates. Pollsters try to use procedures for selecting a sample of voters so that the responses are independent—that is, the objective of the pollster is to select the sample so the event that one polled voter prefers candidate A does not alter the probability that a second polled voter prefers candidate A.

We will make three final points about independence. The first is that the property of independence, unlike the mutually exclusive property, cannot be shown on or gleaned from a Venn diagram. This means *you can't trust your intuition.* In general, the only way to check for independence is by performing the calculations of the probabilities in the definition.

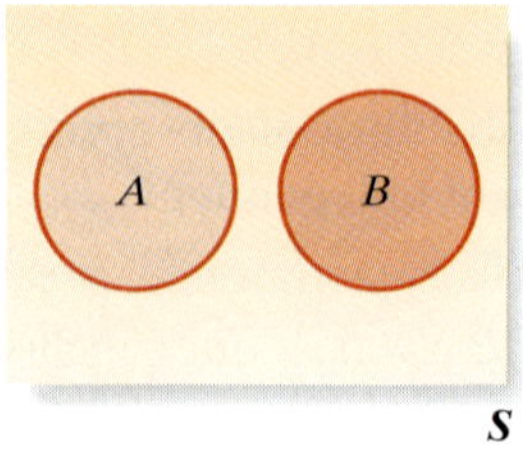

Figure 3.21
Mutually exclusive events are dependent events

The second point concerns the relationship between the mutually exclusive and independence properties. Suppose that events A and B are mutually exclusive, as shown in Figure 3.21, and both events have nonzero probabilities. Are these events independent or dependent? That is, does the assumption that B occurs alter the probability of the occurrence of A? It certainly does, because if we assume that B has occurred, it is impossible for A to have occurred simultaneously—that is, $P(A \mid B) = 0$. Thus, *mutually exclusive events are dependent events* because $P(A) \neq P(A \mid B)$.

The third point is that the probability of the intersection of independent events is very easy to calculate. Referring to the formula for calculating the probability of an intersection, we find

$$P(A \cap B) = P(A)P(B|A)$$

Thus, because $P(B|A) = P(B)$ when A and B are independent, we have the following useful rule:

Probability of Intersection of Two Independent Events

If events A and B are independent, the probability of the intersection of A and B equals the product of the probabilities of A and B; that is,

$$P(A \cap B) = P(A)P(B)$$

The converse is also true: If $P(A \cap B) = P(A)P(B)$, then events A and B are independent.

In the die-toss experiment, we showed in Example 3.19 that the events A: {Observe an even number} and B: {Observe a number less than or equal to 4} are independent if the die is fair. Thus,

$$P(A \cap B) = P(A)P(B) = (1/2)(2/3) = 1/3$$

This agrees with the result that we obtained in the example:

$$P(A \cap B) = P(2) + P(4) = 2/6 = 1/3$$

Example 3.21
Probability of Independent Events Occurring Simultaneously in the Diversity Training Study

Problem Refer to Example 3.5 (pp. 123–124). Recall that *USA Today* found that of all U.S. firms that use diversity training, 38% state that their primary reason for using it is to stay competitive.

a. What is the probability that in a sample of two firms that use diversity training, both primarily use it to stay competitive?

b. What is the probability that in a sample of 10 firms that use diversity training, all 10 primarily use it to stay competitive?

Solution

a. Let C_1 represent the event that firm 1 gives "stay competitive" as the primary reason for using diversity training. Define C_2 similarly for firm 2. The event that *both* firms give "stay competitive" as their primary reason is the intersection of the two events, $C_1 \cap C_2$. Based on the survey that found that 38% of U.S. firms use diversity training to stay competitive, we could reasonably conclude that $P(C_1) = .38$ and $P(C_2) = .38$. However, in order to compute the probability of $C_1 \cap C_2$ from the multiplicative rule, we must make the assumption that the two events are independent. Because the classification of any firm using diversity training is not likely to affect the classification of another firm, this assumption is reasonable. Assuming independence, we have

$$P(C_1 \cap C_2) = P(C_1)P(C_2) = (.38)(.38) = .1444$$

b. To see how to compute the probability that 10 of 10 firms will give "stay competitive" as their primary reason, first consider the event that 3 of 3 firms give "stay competitive" as the primary reason. Using the notation defined earlier, we want to compute the probability of the intersection $C_1 \cap C_2 \cap C_3$. Again assuming independence of the classifications, we have

$$P(C_1 \cap C_2 \cap C_3) = P(C_1)P(C_2)P(C_3) = (.38)(.38)(.38) = .054872$$

Similar reasoning leads us to the conclusion that the intersection of 10 such events can be calculated as follows:

$$P(C_1 \cap C_2 \cap C_3 \cap \ldots \cap C_{10}) = P(C_1)P(C_2)\ldots P(C_{10}) = (.38)^{10} = .0000628$$

Thus, the probability that 10 of 10 firms all give "stay competitive" as their primary reason for using diversity training is about 63 in 1 million, assuming the events (stated reasons for using diversity training) are independent.

Look Back The very small probability in part **b** makes it extremely unlikely that 10 of 10 firms would give "stay competitive" as their primary reason for diversity training. If this event should actually occur, we would need to reassess our estimate of the probability of .38 used in the calculation. If all 10 firms' reason is staying competitive, then the probability that any one firm gives staying competitive as their reason is much higher than .38. (This conclusion is another application of the rare event approach to statistical inference.)

Now Work Exercise 3.70a

Statistics IN Action Revisited — The Probability of Winning Cash 3 or Play 4

In addition to biweekly Lotto 6/53, the Florida Lottery runs several other games. Two popular daily games are "Cash 3" and "Play 4." In Cash 3, players pay \$1 to select three numbers in sequential order, where each number ranges from 0 to 9. If the three numbers selected (e.g., 2-8-4) match exactly the order of the three numbers drawn, the player wins \$500. Play 4 is similar to Cash 3, but players must match four numbers (each number ranging from 0 to 9). For a \$1 Play 4 ticket (e.g., 3-8-3-0), the player will win \$5,000 if the numbers match the order of the four numbers drawn.

During the official drawing for Cash 3, 10 Ping-Pong balls numbered 0, 1, 2, 3, 4, 5, 6, 7, 8, and 9 are placed into each of three chambers. The balls in the first chamber are colored pink, the balls in the second chamber are blue, and the balls in the third chamber are yellow. One ball of each color is randomly drawn, with the official order as pink-blue-yellow. In Play 4, a fourth chamber with orange balls is added, and the official order is pink-blue-yellow-orange. Because the draws of the colored balls are random and independent, we can apply an extension of the Probability Rule for the Intersection of Two Independent Events to find the odds of winning Cash 3 and Play 4. The probability of matching a numbered ball being drawn from a chamber is 1/10; therefore,

$$\begin{aligned} P(\text{Win Cash 3}) &= P(\text{match pink AND match blue) AND match yellow)} \\ &= P(\text{match pink}) \times P(\text{match blue})\, P(\text{match yellow}) \\ &= (1/10)(1/10)(1/10) = 1/1000 = .001 \end{aligned}$$

$$\begin{aligned} P(\text{Win Play 4}) &= P(\text{match pink AND match blue AND match yellow AND match orange}) \\ &= P(\text{match pink}) \times P(\text{match blue}) \times P(\text{match yellow}) \times P(\text{match orange}) \\ &= (1/10)(1/10)(1/10)(1/10) \\ &= 1/10{,}000 = .0001 \end{aligned}$$

Although the odds of winning one of these daily games is much better than the odds of winning Lotto 6/53, there is still only a 1 in 1,000 chance (for Cash 3) or 1 in 10,000 chance (for Play 4) of winning the daily game. And the payoffs (\$500 or \$5,000) are much smaller. In fact, it can be shown that you will lose an average of 50¢ every time you play either Cash 3 or Play 4!

Activity 3.1 *Exit Polls:* Conditional Probability

Exit polls are conducted in selected locations as voters leave their polling places after voting. In addition to being used to predict the outcome of elections before the votes are counted, these polls are used to gauge tendencies among voters. The results are usually stated in terms of conditional probabilities.

The table on the next page shows the results of exit polling that suggest men were almost evenly spit on voting for John McCain or Barack Obama, while women were more likely to vote for Obama in the 2008 presidential election. In addition, the table also suggests that more women than men voted in the election. The six percentages in the last three columns represent conditional probabilities where the given event is gender.

1. Find similar exit poll results where the voters are categorized by race, income, education, or some other criterion for a recent national, state, or local election. Choose two different examples and interpret the percentages given as probabilities, or conditional probabilities where appropriate.

2008 Presidential Election, Vote by Gender

	Obama	McCain	Other
Male (47%)	49%	48%	3%
Female (53%)	56%	43%	1%

Source: www.cnn.com

2. Use the multiplicative rule of probability to find the probabilities related to the percentages given. [For example, in the table above find *P*(Obama and Male) using the multiplicative rule.] Then interpret each of these probabilities and use them to determine the total percentage of the electorate who voted for each candidate.

3. Describe a situation where a business might use a form of exit polling to gauge customer reaction to a new product or service. Identify the type of business, the product or service, the criterion used to categorize customers, and how the customers' reactions will be determined. Then describe how the results will be summarized as conditional probabilities. How might the results of the poll benefit the business?

Exercises 3.47–3.74

Learning the Mechanics

3.47 For two events, A and B, $P(A) = .4$, $P(B) = .2$, and
NW $P(A \cap B) = .1$.

a. Find $P(A|B)$.
b. Find $P(B|A)$.
c. Are A and B independent events?

3.48 For two events, A and B, $P(A) = .4$, $P(B) = .2$, and $P(A|B) = .6$.

a. Find $P(A \cap B)$.
b. Find $P(B|A)$.

3.49 For two independent events, A and B, $P(A) = .4$ and $P(B) = .2$.

a. Find $P(A \cap B)$.
b. Find $P(A|B)$.
c. Find $P(A \cup B)$.

3.50 An experiment results in one of three mutually exclusive events, A, B, or C. It is known that $P(A) = .30$, $P(B) = .55$, and $P(C) = .15$. Find each of the following probabilities:

a. $P(A \cup B)$
b. $P(A \cap C)$
c. $P(A|B)$
d. $P(B \cup C)$
e. Are B and C independent events? Explain.

3.51 Consider the experiment depicted by the Venn diagram, with the sample space S containing five sample points. The sample points are assigned the following probabilities: $P(E_1) = .20$, $P(E_2) = .30$, $P(E_3) = .30$, $P(E_4) = .10$, $P(E_5) = .10$.

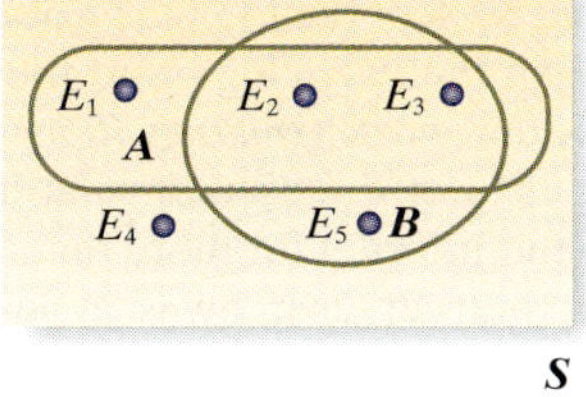

a. Calculate $P(A)$, $P(B)$, and $P(A \cap B)$.

b. Suppose we know that event A has occurred, so that the reduced sample space consists of the three sample points in A—namely, E_1, E_2, and E_3. Use the formula for conditional probability to adjust the probabilities of these three sample points for the knowledge that A has occurred [i.e., $P(E_i|A)$]. Verify that the conditional probabilities are in the same proportion to one another as the original sample point probabilities.

NW **c.** Calculate the conditional probability $P(B|A)$ in two ways: (1) Add the adjusted (conditional) probabilities of the sample points in the intersection $A \cap B$, as these represent the event that B occurs given that A has occurred; (2) use the formula for conditional probability:

$$P(B|A) = \frac{P(A \cap B)}{P(A)}$$

Verify that the two methods yield the same result.

d. Are events A and B independent? Why or why not?

3.52 Two fair coins are tossed, and the following events are defined:

A: {Observe at least one head}
B: {Observe exactly one head}

a. Draw a Venn diagram for the experiment, showing events A and B. Assign probabilities to the sample points.
b. Find $P(A)$, $P(B)$, and $P(A \cap B)$.
c. Use the formula for conditional probability to find $P(A|B)$ and $P(B|A)$. Verify your answer by inspecting the Venn diagram and using the concept of reduced sample spaces.

3.53 An experiment results in one of five sample points with the following probabilities: $P(E_1) = .22$, $P(E_2) = .31$, $P(E_3) = .15$, $P(E_4) = .22$, and $P(E_5) = .1$. The following events have been defined:

A: $\{E_1, E_3\}$
B: $\{E_2, E_3, E_4\}$
C: $\{E_1, E_5\}$

Find each of the following probabilities:

a. $P(A)$
b. $P(B)$
c. $P(A \cap B)$
d. $P(A|B)$
e. $P(B \cap C)$
f. $P(C|B)$
g. Consider each pair of events: A and B, A and C, and B and C. Are any of the pairs of events independent? Why?

3.54 Two fair dice are tossed, and the following events are defined:

A: {Sum of the numbers showing is odd}
B: {Sum of the numbers showing is 9, 11, or 12}

Are events A and B independent? Why?

3.55 A sample space contains six sample points and events *A*, *B*, and *C* as shown in the Venn diagram. The probabilities of the sample points are

$$P(1) = .20, P(2) = .05, P(3) = .30, P(4) = .10,$$
$$P(5) = .10, P(6) = .25.$$

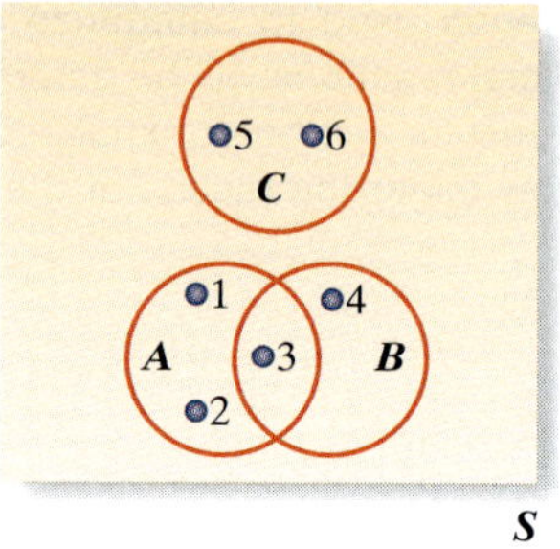

a. Which pairs of events, if any, are mutually exclusive? Why?
b. Which pairs of events, if any, are independent? Why?
c. Find $P(A \cup B)$ by adding the probabilities of the sample points and then by using the additive rule. Verify that the answers agree. Repeat for $P(A \cup C)$.

Applet Exercise 3.5

Use the applet *Simulating the Probability of Rolling a 6* to simulate conditional probabilities. Begin by running the applet twice with $n = 10$ without resetting between runs. The data on your screen represent 20 rolls of a die. The diagram above the *Roll* button shows the frequency of each of the six possible outcomes. Use this information to find each of the probabilities.

a. The probability of 6 given the outcome is 5 or 6
b. The probability of 6 given the outcome is even
c. The probability of 4 or 6 given the outcome is even
d. The probability of 4 or 6 given the outcome is odd

Applying the Concepts—Basic

3.56 NW **Speeding linked to fatal car crashes.** According to the National Highway Traffic and Safety Administration's National Center for Statistics and Analysis (NCSA), "speeding is one of the most prevalent factors contributing to fatal traffic crashes" (*NHTSA Technical Report,* Aug. 2005). The probability that speeding is a cause of a fatal crash is .3. Furthermore, the probability that speeding and missing a curve are causes of a fatal crash is .12. Given speeding is a cause of a fatal crash, what is the probability that the crash occurred on a curve?

3.57 **Largest nondomestic public companies.** *Forbes* (Mar. 30, 2006) conducted a survey of the 20 largest nondomestic public companies in the world. Of these 20 companies, 3 were banking companies based in the United Kingdom. A total of 5 UK companies were on the top 20 list. Suppose we select one of these 20 companies at random. Given the company is based in the UK, what is the probability that it is a banking company?

3.58 NW **Guilt in decision making.** Refer to the *Journal of Behavioral Decision Making* (Jan. 2007) study of the effect of guilt emotion on how a decision-maker focuses on the problem, Exercise 3.40 (p. 139). The results (number responding in each category) for the 171 study participants are reproduced in the table below. Suppose one of the 171 participants is selected at random.

Emotional State	Choose Stated Option	Do Not Choose Stated Option	Totals
Guilt	45	12	57
Anger	8	50	58
Neutral	7	49	56
Totals	60	111	171

Source: Gangemi, A., and Mancini, F. "Guilt and focusing in decision-making," *Journal of Behavioral Decision Making,* Vol. 20, Jan. 2007 (Table 2).

a. Given the respondent is assigned to the guilty state, what is the probability that the respondent chooses the stated option?
b. If the respondent does not choose to repair the car, what is the probability that the respondent is in the anger state?
c. Are the events {repair the car} and {guilty state} independent?

3.59 **National firearms survey.** Refer to the Harvard School of Public Health study of privately held firearm stock in the United States, Exercise 2.136 (p. 103). Recall that in a representative household telephone survey of 2,770 adults, 26% reported that they own at least one gun (*Injury Prevention,* Jan. 2007). Of those that own a gun 5% own a handgun. Suppose one of the 2,770 adults surveyed is randomly selected.

a. What is the probability that the adult owns at least one gun?
b. What is the probability that the adult owns at least one gun and the gun is a handgun?

3.60 NW **Monitoring quality of power equipment.** *Mechanical Engineering* (Feb. 2005) reported on the need for wireless networks to monitor the quality of industrial equipment. For example, consider Eaton Corp., a company that develops distribution products. Eaton estimates that 90% of the electrical switching devices it sells can monitor the quality of the power running through the device. Eaton further estimates that of the buyers of electrical switching devices capable of monitoring quality, 90% do not wire the equipment up for that purpose. Use this information to estimate the probability that an Eaton electrical switching device is capable of monitoring power quality and is wired up for that purpose.

3.61 **Internet user study.** In a study of Internet users, researchers discovered that 80% of the users own at least one computer and that 25% of the users log on to the Internet for more than 30 hours per week (*Internet Research,* Vol. 11, 2001). Suppose that 15% of the users own at least one computer and log on to the Internet for more than 30 hours per week.

a. Given an Internet user owns at least one computer, what is the probability that he or she logs on to the Internet for more than 30 hours per week?
b. Given an Internet user logs on to the Internet for more than 30 hours per week, what is the probability that he or she owns at least one computer?

Applying the Concepts—Intermediate

3.62 **Probability of winning a war.** Before a country enters into a war, a prudent government will assess the cost, utility, and probability of a victory. University of Georgia professor

P. L. Sullivan has developed a statistical model for determining the probability of winning a war based on a government's capabilities and resources (*Journal of Conflict Resolution,* Vol. 51, 2007). Now consider the current U.S.-Iraq conflict. One researcher used the model to estimate that the probability of a successful regime change in Iraq was .70 prior to the start of the war. Of course, we now know that the successful regime change was achieved. However, the model also estimates that given the mission is extended to support a weak Iraqi government, the probability of ultimate success is only .26. Assume these probabilities are accurate.

a. Prior to the start of the U.S.-Iraq war, what is the probability that a successful regime change is not achieved?

b. Given that the mission is extended to support a weak Iraqi government, what is the probability that a successful regime change is ultimately achieved?

c. Suppose the probability of the U.S. extending the mission to support a weak Iraqi government was .55. Find the probability that the mission is extended and results in a successful regime change.

3.63 Reversible errors in capital punishment cases. New York's Columbia University Law School released a report in February 2002 on all capital punishment (death penalty) cases that occurred over the past 25 years in the United States. Two of the findings are stated as follows: 68% of all capital punishment cases had a serious, reversible error. Of the cases that were overturned due to reversible error, 7% resulted in acquittal for the defendant during retrial.

a. Write each of these findings in the form of a probability.

b. Use the findings to find the probability that a capital case had a serious, reversible error and an innocent defendant.

3.64 Degrees of best-paid CEOs. Refer to the results of the *Forbes* (April 30, 2008) survey of the top 40 best-paid CEOs shown in Table 2.1 (p. 32). The data on highest degree obtained are summarized in the SPSS printout at the bottom of the page. Suppose you randomly select five of the CEOs (without replacement) and record the highest degree obtained by each.

a. What is the probability that the highest degree obtained by the first CEO you select is a bachelor's degree?

b. Suppose the highest degree obtained by each of the first four CEOs you select is a bachelor's degree. What is the probability that the highest degree obtained by the fifth CEO you select is a bachelor's degree?

3.65 Home modifications for wheelchair users. Refer to the *American Journal of Public Health* (Jan. 2002) study of elderly wheelchair users who live at home, Exercise 3.44 (p. 140). The table classifying a sample of 306 wheelchair users according to type of features installed in the home and whether or not they had an injurious fall follows. Suppose we select, at random, one of the 306 surveyed wheelchair users.

Home Features	Injurious Fall(s)	No Falls	Totals
All 5	2	7	9
At least 1	26	162	188
None	20	89	109
Totals	48	258	306

Source: Berg, K., Hines, M., and Allen, S. "Wheelchair users at home: Few home modifications and many injurious falls," *American Journal of Public Health,* Vol. 92, No. 1, Jan. 2002 (Table 1).

a. Given the wheelchair user had all five features installed, what is the probability that the user had an injurious fall?

b. Given the wheelchair user had none of the features installed, what is the probability that the user had an injurious fall?

3.66 Working mothers with children. The U.S. Census Bureau reports a decline in the percentage of mothers with infant children in the workforce. The following table gives a breakdown of the marital status and working status of the 3.9 million mothers with infant children in the year 2000. (The numbers in the table are reported in millions.) Consider the following events: $A = \{$Mom with infant works$\}$, $B = \{$Mom with infant is married and living with husband$\}$. Are A and B independent events?

	Working	Not Working
Married/living with husband	1,385	1,175
All other arrangements	786	588

Source: U.S. Census Bureau, *American Demographics,* Feb. 2002.

3.67 Intrusion detection systems. A computer intrusion detection system (IDS) is designed to provide an alarm whenever an intrusion (e.g., unauthorized access) is being attempted into a computer system. A probabilistic evaluation of a system with two independently operating intrusion detection systems (a double IDS) was published in the *Journal of Research of the National Institute of Standards and Technology* (November/December 2003). Consider a double IDS with system A and system B. If there is an intruder, system A sounds an alarm with probability .9, and system B sounds an alarm with probability .95. If there is no intruder, the probability that

SPSS Output for Exercise 3.64

DEGREE

		Frequency	Percent	Valid Percent	Cumulative Percent
Valid	Bachelors	14	35.0	35.0	35.0
	Law	6	15.0	15.0	50.0
	Masters	3	7.5	7.5	57.5
	MBA	12	30.0	30.0	87.5
	None	2	5.0	5.0	92.5
	PhD	3	7.5	7.5	100.0
	Total	40	100.0	100.0	

system A sounds an alarm (i.e., a false alarm) is .2, and the probability that system B sounds an alarm is .1. Assume that under a given condition (intruder or not), systems A and B operate independently.

a. Using symbols, express the four probabilities given in the example.

b. If there is an intruder, what is the probability that both systems sound an alarm?

c. If there is no intruder, what is the probability that both systems sound an alarm?

d. Given an intruder, what is the probability that at least one of the systems sounds an alarm?

3.68 **Detecting traces of TNT.** University of Florida researchers in the Department of Materials Science and Engineering have invented a technique to rapidly detect traces of TNT (*Today,* Spring 2005). The method, which involves shining a laser light on a potentially contaminated object, provides instantaneous results and gives no false positives. In this application, a false positive would occur if the laser light detects traces of TNT when, in fact, no TNT is actually present on the object. Let A be the event that the laser light detects traces of TNT. Let B be the event that the object contains no traces of TNT. The probability of a false positive is 0. Write this probability in terms of A and B using symbols such as $\cup$, $\cap$, and $|$.

3.69 **Are you really being served red snapper?** Red snapper is a rare and expensive reef fish served at upscale restaurants. Federal law prohibits restaurants from serving a cheaper look-alike variety of fish (e.g., vermillion snapper or lane snapper) to customers who order red snapper. Researchers at the University of North Carolina used DNA analysis to examine fish specimens labeled "red snapper" that were purchased from vendors across the country (*Nature,* July 15, 2004). The DNA tests revealed that 77% of the specimens were not red snapper but the cheaper look-alike variety of fish.

a. Assuming the results of the DNA analysis are valid, what is the probability that you are actually served red snapper the next time you order it at a restaurant?

b. If there are five customers at a restaurant, all who have ordered red snapper, what is the probability that at least one customer is actually served red snapper?

3.70 **Lie detector test.** A new type of lie detector—called the computerized voice stress analyzer (CVSA)—has been developed. The manufacturer claims that the CVSA is 98% accurate, and, unlike a polygraph machine, will not be thrown off by drugs and medical factors. However, laboratory studies by the U.S. Defense Department found that the CVSA had an accuracy rate of 49.8%—slightly less than pure chance (*Tampa Tribune,* Jan. 10, 1999). Suppose the CVSA is used to test the veracity of four suspects. Assume the suspects' responses are independent.

NW **a.** If the manufacturer's claim is true, what is the probability that the CVSA will correctly determine the veracity of all four suspects?

b. If the manufacturer's claim is true, what is the probability that the CVSA will yield an incorrect result for at least one of the four suspects?

c. Suppose that in a laboratory experiment conducted by the U.S. Defense Department on four suspects, the CVSA yielded incorrect results for two of the suspects. Use this result to make an inference about the true accuracy rate of the new lie detector.

Applying the Concepts—Advanced

3.71 **Patient medical instruction sheets.** Physicians and pharmacists sometimes fail to inform patients adequately about the proper application of prescription drugs and about the precautions to take in order to avoid potential side effects. One method of increasing patients' awareness of the problem is for physicians to provide patient medication instruction (PMI) sheets. The American Medical Association, however, has found that only 20% of the doctors who prescribe drugs frequently distribute PMI sheets to their patients. Assume that 20% of all patients receive the PMI sheet with their prescriptions and that 12% receive the PMI sheet and are hospitalized because of a drug-related problem. What is the probability that a person will be hospitalized for a drug-related problem given that the person has received the PMI sheet?

3.72 **Risk of a natural gas pipeline accident.** *Process Safety Progress* (Dec. 2004) published a risk analysis for a natural gas pipeline between Bolivia and Brazil. The most likely scenario for an accident would be natural gas leakage from a hole in the pipeline. The probability that the leak ignites immediately (causing a jet fire) is .01. If the leak does not immediately ignite, it may result in a delayed ignition of a gas cloud. Given no immediate ignition, the probability of delayed ignition (causing a flash fire) is .01. If there is no delayed ignition, the gas cloud will harmlessly disperse. Suppose a leak occurs in the natural gas pipeline. Find the probability that either a jet fire or a flash fire will occur. Illustrate with a tree diagram.

3.73 **Most likely coin-tossing sequence.** In *Parade Magazine*'s (Nov. 26, 2000) column "Ask Marilyn," the following question was posed: "I have just tossed a [balanced] coin 10 times, and I ask you to guess which of the following three sequences was the result. One (and only one) of the sequences is genuine."

(1) H H H H H H H H H H
(2) H H T T H T T H H H
(3) T T T T T T T T T T

a. Demonstrate that prior to actually tossing the coins, the three sequences are equally likely to occur.

b. Find the probability that the 10 coin tosses result in all heads or all tails.

c. Find the probability that the 10 coin tosses result in a mix of heads and tails.

d. Marilyn's answer to the question posed was "Though the chances of the three specific sequences occurring randomly are equal . . . it's reasonable for us to choose sequence (2) as the most likely genuine result." If you know that only one of the three sequences actually occurred, explain why Marilyn's answer is correct. [*Hint:* Compare the probabilities in parts **b** and **c.**]

3.74 **Software defects in NASA spacecraft instrument code.** Portions of computer software code that may contain undetected defects are called *blind spots.* The issue of blind spots in software code evaluation was addressed at the *8th IEEE International Symposium on High Assurance Software Engineering* (March 2004). The researchers developed guidelines for assessing methods of predicting software defects using data on 498 modules of software code written in "C" language for a NASA spacecraft instrument. One simple prediction algorithm is to count the lines of code in the module; any module with more than 50 lines of code is predicted to have a defect. The **SWDEFECTS** file contains the predicted and actual defect status of all 498 modules. A standard approach to evaluating a software defect prediction algorithm is to form a two-way summary table similar to the one shown here. In the table, *a, b, c,* and *d* represent the number of modules in each cell. Software engineers use these table entries to compute several probability measures, called *accuracy, detection rate, false alarm rate,* and *precision.*

		Module Has Defects	
		False	True
Algorithm	No	*a*	*b*
Predicts Defects	Yes	*c*	*d*

a. *Accuracy* is defined as the probability that the prediction algorithm is correct. Write a formula for *accuracy* as a function of the table values *a, b, c,* and *d.*

b. The *detection rate* is defined as the probability that the algorithm predicts a defect, given that the module actually is a defect. Write a formula for *detection rate* as a function of the table values *a, b, c,* and *d.*

c. The *false alarm rate* is defined as the probability that the algorithm predicts a defect, given that the module actually has no defect. Write a formula for *false alarm rate* as a function of the table values *a, b, c,* and *d.*

d. *Precision* is defined as the probability that the module has a defect, given that the algorithm predicts a defect. Write a formula for *precision* as a function of the table values *a, b, c,* and *d.*

e. Access the **SWDEFECTS** file and compute the values of accuracy, detection rate, false alarm rate, and precision. Interpret the results.

3.7 Random Sampling

How a sample is selected from a population is of vital importance in statistical inference because the probability of an observed sample will be used to infer the characteristics of the sampled population. To illustrate, suppose you deal yourself four cards from a deck of 52 cards and all four cards are aces. Do you conclude that your deck is an ordinary bridge deck, containing only four aces, or do you conclude that the deck is stacked with more than four aces? It depends on how the cards were drawn. If the four aces were always placed at the top of a standard bridge deck, drawing four aces is not unusual—it is certain. On the other hand, if the cards are thoroughly mixed, drawing four aces in a sample of four cards is highly improbable. The point, of course, is that in order to use the observed sample of four cards to draw inferences about the population (the deck of 52 cards), you need to know how the sample was selected from the deck.

One of the simplest and most frequently employed sampling procedures is implied in many of the previous examples and exercises. It produces what is known as a *random sample.* We learned in Section 1.6 (p. 15) that a random sample is likely to be *representative* of the population that it is selected from.

> If n elements are selected from a population in such a way that every set of n elements in the population has an equal probability of being selected, the n elements are said to be a **random sample.***

If a population is not too large and the elements can be numbered on slips of paper, poker chips, and so on, you can physically mix the slips of paper or chips and remove n elements from the total. The numbers that appear on the chips selected would indicate the population elements to be included in the sample. Becuase it is often difficult to achieve a thorough mix, such a procedure provides only an approximation to random sampling. Most researchers rely on **random number generators** to automatically generate the random sample. Random number generators are available in table form, and they are built into most statistical software packages.

Example 3.22
Selecting a Random Sample of Households

Problem Suppose you wish to randomly sample five households from a population of 100,000 households to participate in a study.

a. How many different samples can be selected?

b. Use a random number generator to select a random sample.

*Strictly speaking, this is a **simple random sample.** There are many different types of random samples. The simple random sample is the most common.

Solution

a. To determine the number of samples, we'll apply the Combinations Rule of Section 3.1. In this case, $N = 100{,}000$ and $n = 5$. Then

$$\binom{N}{n} = \binom{100{,}000}{5} = \frac{100{,}000!}{5!99{,}995!}$$

$$= \frac{100{,}000 \cdot 99{,}999 \cdot 99{,}998 \cdot 99{,}997 \cdot 99{,}996}{5 \cdot 4 \cdot 3 \cdot 2 \cdot 1}$$

$$= 8.33 \times 10^{22}$$

Thus, there are 83.3 billion trillion different samples of five households that can be selected from 100,000.

b. To ensure that each of the possible samples has an equal chance of being selected, as required for random sampling, we can employ a **random number table,** as provided in Table I in Appendix B. Random number tables are constructed in such a way that every number occurs with (approximately) equal probability. Furthermore, the occurrence of any one number in a position is independent of any of the other numbers that appear in the table. To use a table of random numbers, number the N elements in the population from 1 to N. Then turn to Table I and select a starting number in the table. Proceeding from this number either across the row or down the column, remove and record n numbers from the table.

To illustrate, first we number the households in the population from 1 to 100,000. Then, we turn to a page of Table I, say the first page. (A partial reproduction of the first page of Table I is shown in Table 3.7.) Now we arbitrarily select a starting number, say the random number appearing in the third row, second column. This number is 48,360. Then we proceed down the second column to obtain the remaining four random numbers. In this case we have selected five random numbers, which are shaded in Table 3.7. Using the first five digits to represent households from 1 to 99,999 and the number 00000 to represent household 100,000, we can see that the households numbered

48,360
93,093
39,975
6,907
72,905

should be included in our sample.

Note: Use only the necessary number of digits in each random number to identify the element to be included in the sample. If, in the course of recording the n numbers from the table, you select a number that has already been selected, simply discard the

Table 3.7 Partial Reproduction of Table I in Appendix B

Row \ Column	1	2	3	4	5	6
1	10480	15011	01536	02011	81647	91646
2	22368	46573	25595	85393	30995	89198
3	24130	48360	22527	97265	76393	64809
4	42167	93093	06243	61680	07856	16376
5	37570	39975	81837	16656	06121	91782
6	77921	06907	11008	42751	27756	53498
7	99562	72905	56420	69994	98872	31016
8	96301	91977	05463	07972	18876	20922
9	89579	14342	63661	10281	17453	18103
10	85475	36857	53342	53988	53060	59533
11	28918	69578	88231	33276	70997	79936
12	63553	40961	48235	03427	49626	69445
13	09429	93969	52636	92737	88974	33488

duplicate and select a replacement at the end of the sequence. Thus, you may have to record more than n numbers from the table to obtain a sample of n unique numbers.

Look Back Can we be perfectly sure that all 83.3 billion trillion samples have an equal chance of being selected? The fact is, we can't; but to the extent that the random number table contains truly random sequences of digits, the sample should be very close to random.

Now Work Exercise 3.76

Table I in Appendix B is just one example of a random number generator. For most scientific studies that require a large random sample, computers are used to generate the random sample. The EXCEL, Minitab, and SPSS statistical software packages all have easy-to-use random number generators.

For example, suppose we required a random sample of $n = 50$ households from the population of 100,000 households in Example 3.22. Here, we might employ the Minitab random number generator. Figure 3.22 shows a Minitab printout listing 50 random numbers (from a population of 100,000). The households with these identification numbers would be included in the random sample.

HOUSE50.MTW ***

↓	C1	C2	C3	C4	C5
	HouseID1	HouseID2	HouseID3	HouseID4	HouseID5
1	2036	22037	39884	62123	75750
2	3161	22055	44687	62421	80044
3	3930	22226	44689	63749	81857
4	9838	22773	45422	69079	81918
5	9961	22820	48104	69802	91680
6	11623	32989	50226	70895	94987
7	14114	33119	50443	71029	96110
8	17244	35045	50806	72879	97679
9	17598	35922	50813	73962	99232
10	18769	39272	61059	75517	99789
11					

Figure 3.22
Minitab worksheet with random sample of 50 households

ASPIRIN-STUDY.M

↓	C1	C2
	Physician	Treatment
1	1	3
2	2	11
3	3	10
4	4	14
5	5	2
6	6	7
7	7	6
8	8	16
9	9	17
10	10	9
11	11	
12	12	
13	13	
14	14	
15	15	
16	16	
17	17	
18	18	
19	19	
20	20	
21		

Figure 3.23
Minitab worksheet with random assignment of physicians

Recall our discussion of designed experiments in Section 1.6 (p. 14). The notion of random selection and randomization is key to conducting good research with a designed experiment. The next example illustrates a basic application.

Example 3.23
Randomization in a Designed Experiment

Problem A designed experiment in the medical field involving human subjects is referred to as a *clinical trial.* One recent clinical trial was designed to determine the potential of using aspirin in preventing heart attacks. Volunteer physicians were randomly divided into two groups—the *treatment* group and the *control* group. Each physician in the treatment group took one aspirin tablet a day for one year, while the physicians in the control group took an aspirin-free placebo made to look identical to an aspirin tablet. Because the physicians did not know which group, treatment or control, they were assigned to, the clinical trial is called a *blind study.* Assume 20 physicians volunteered for the study. Use a random number generator to randomly assign half of the physicians to the treatment group and half to the control group.

Solution Essentially, we want to select a random sample of 10 physicians from the 20. The first 10 selected will be assigned to the treatment group; the remaining 10 will be assigned to the control group. (Alternatively, we could randomly assign each physician, one by one, to either the treatment or control group. However, this would not guarantee exactly 10 physicians in each group.)

The Minitab random sample procedure was employed, producing the printout shown above in Figure 3.23. Numbering the physicians from 1 to 20, we see that physicians 3, 11, 10, 14, 2, 7, 6, 16, 17, and 9 are assigned to receive the aspirin (treatment). The remaining physicians are assigned the placebo (control).

Ethics IN Statistics

Intentionally selecting a nonrandom sample in an effort to support a particular viewpoint is considered *unethical statistical practice.*

Activity 3.2 *Keep the Change:* Independent Events

Once again we return to the Bank of America *Keep the Change* savings program, Activity 1.1 (p. 16). This time we look at whether certain events involving purchase totals and amounts transferred to savings are independent. Throughout this activity, the experiment consists of randomly selecting one purchase from a large group of purchases.

1. Define events *A* and *B* as follows:

   A: {Purchase total ends in $0.25.}
   B: {Amount transferred is less than $0.50.}

   Explain why events *A* and *B* are not independent. Are events *A* and *B* mutually exclusive? Use this example to explain the difference between independent events and mutually exclusive events.

2. Now define events *A* and *B* in this manner:

   A: {Purchase total is greater than $10.}
   B: {Amount transferred is less than $0.50.}

   Do you believe that these events are independent? Explain your reasoning.

3. To investigate numerically whether the events in Question 2 are independent, we will use the data collected in Activity 1.1, page 16. Pool your data with the data from other students or the entire class so that the combined data set represents at least 100 purchases. Complete the table by counting the number of purchases in each category.

Distribution of Purchases

	Purchase	Total	
Transfer Amount	≤$10	>$10	Totals
$0.00–$0.49			
$0.99–$0.99			
Totals			

Compute appropriate probabilities based on your completed table to test whether the events of Question 2 are independent. If you conclude that the events are not independent, can you explain your conclusion in terms of the original data?

Exercises 3.75–3.83

Learning the Mechanics

3.75 How are random samples related to representative samples?

3.76 NW Suppose you wish to sample $n = 2$ elements from a total of $N = 10$ elements.

a. Count the number of different samples that can be drawn, first by listing them, and then by using combinatorial mathematics (see Section 3.1).

b. If random sampling is to be employed, what is the probability that any particular sample will be selected?

c. Show how to use the random number table, Table I in Appendix B, to select a random sample of 2 elements from a population of 10 elements. Perform the sampling procedure 20 times. Do any two of the samples contain the same 2 elements? Given your answer to part **b**, did you expect repeated samples?

3.77 Suppose you wish to sample $n = 3$ elements from a total of $N = 600$ elements.

a. Count the number of different samples by using combinatorial mathematics (see Section 3.1).

b. If random sampling is to be employed, what is the probability that any particular sample will be selected?

c. Show how to use the random number table, Table I in Appendix B, to select a random sample of 3 elements from a population of 600 elements. Perform the sampling procedure 20 times. Do any two of the samples contain the same three elements? Given your answer to part **b**, did you expect repeated samples?

d. Use a computer to generate a random sample of 3 from the population of 600 elements.

3.78 Suppose that a population contains $N = 200{,}000$ elements. Use a computer or Table I in Appendix B to select a random sample of $n = 10$ elements from the population. Explain how you selected your sample.

Applying the Concepts—Basic

3.79 **Random-digit dialing.** To ascertain the effectiveness of their advertising campaigns, firms frequently conduct telephone interviews with consumers using *random-digit dialing.* With this approach, a random number generator mechanically creates the sample of phone numbers to be called.

a. Explain how the random number table (Table I in Appendix B) or a computer could be used to generate a sample of seven-digit telephone numbers.

b. Use the procedure you described in part **a** to generate a sample of 10 seven-digit telephone numbers.

c. Use the procedure you described in part **a** to generate five seven-digit telephone numbers whose first three digits are 373.

3.80 **Census sampling.** In addition to its decennial enumeration of the population, the U.S. Census Bureau regularly samples the population to estimate the level of and changes in a number of attributes, such as income, family size, employment, and marital status. Suppose the bureau plans to sample 1,000 households in a city that has a total of 534,322 households. Show how the bureau could use the random number table in Appendix B or a computer to generate the sample. Select the first 10 households to be included in the sample.

Applying the Concepts—Intermediate

3.81 **Auditing an accounting system.** In auditing a firm's financial statements, an auditor will (1) assess the capability of the firm's accounting system to accumulate, measure, and synthesize transactional data properly and (2) assess the operational effectiveness of the accounting system. In performing the second assessment, the auditor frequently relies on a random sample of actual transactions (Stickney and Weil, *Financial*

Accounting: An Introduction to Concepts, Methods, and Uses, 2002). A particular firm has 5,382 customer accounts that are numbered from 0001 to 5382.

a. One account is to be selected at random for audit. What is the probability that account number 3,241 is selected?

b. Draw a random sample of 10 accounts and explain in detail the procedure you used.

c. Refer to part **b.** The following are two possible random samples of size 10. Is one more likely to be selected than the other? Explain.

Sample Number 1				
5011	0082	0963	0772	3415
2663	1126	0008	0026	4189

Sample Number 2				
0001	0003	0005	0007	0009
0002	0004	0006	0008	0010

3.82 Sampling stocks on the NYSE. The results of the previous business day's transactions for stocks traded on the New York Stock Exchange (NYSE) and five regional exchanges—the Chicago, Pacific, Philadelphia, Boston, and Cincinnati stock exchanges—are summarized each business day in the NYSE–Composite Transactions table in the *Wall Street Journal.*

a. Examine the NYSE–Composite Transactions table in a recent issue of the *Wall Street Journal* and explain how to draw a random sample of stocks from the table.

b. Use the procedure you described in part **a** to draw a random sample of 20 stocks from a recent NYSE–Composite Transactions table. For each stock in the sample, list its name (i.e., the abbreviation given in the table), its sales volume, and its closing price.

Applying the Concepts—Advanced

3.83 Sampling TV markets for a court case. A recent court case involved a claim of satellite television subscribers obtaining illegal access to local TV stations. The defendant (the satellite TV company) wanted to sample TV markets nationwide and determine the percentage of its subscribers in each sampled market who have illegal access to local TV stations. To do this, defendant's expert witness drew a rectangular grid over the continental United States, with horizontal and vertical grid lines every .02 degrees of latitude and longitude, respectively. This created a total of 500 rows and 1,000 columns, or (500)(1,000) = 500,000 intersections. The plan was to randomly sample 900 intersection points and include the TV market at each intersection in the sample. Explain how you could use a random number generator to obtain a random sample of 900 intersections. Develop at least two plans: one that numbers the intersections from 1 to 500,000 prior to selection and another that selects the row and column of each sampled intersection (from the total of 500 rows and 1,000 columns).

3.8 Bayes's Rule

An early attempt to employ probability in making inferences is the basis for a branch of statistical methodology known as **Bayesian statistical methods.** The logic employed by the English philosopher Thomas Bayes in the mid-1700s involves converting an unknown conditional probability, say $P(B|A)$, to one involving a known conditional probability, say $P(A|B)$. The method is illustrated in the next example.

Example 3.24
Applying Bayes's Logic to an Intruder Detection System

Problem An unmanned monitoring system uses high-tech video equipment and microprocessors to detect intruders. A prototype system has been developed and is in use outdoors at a weapons munitions plant. The system is designed to detect intruders with a probability of .90. However, the design engineers expect this probability to vary with weather conditions. The system automatically records the weather condition each time an intruder is detected. Based on a series of controlled tests, in which an intruder was released at the plant under various weather conditions, the following information is available: Given the intruder was, in fact, detected by the system, the weather was clear 75% of the time, cloudy 20% of the time, and raining 5% of the time. When the system failed to detect the intruder, 60% of the days were clear, 30% cloudy, and 10% rainy. Use this information to find the probability of detecting an intruder, given rainy weather conditions. (Assume that an intruder has been released at the plant.)

Solution Define D to be the event that the intruder is detected by the system. Then D^c is the event that the system failed to detect the intruder. Our goal is to calculate the conditional probability, $P(D|\text{Rainy})$. From the statement of the problem, the following information is available:

$$P(D) = .90 \qquad P(D^c) = .10$$
$$P(\text{Clear}|D) = .75 \qquad P(\text{Clear}|D^c) = .60$$

$$P(\text{Cloudy}|D) = .20 \quad P(\text{Cloudy}|D^c) = .30$$
$$P(\text{Rainy}|D) = .05 \quad P(\text{Rainy}|D^c) = .10$$

Note that $P(D|\text{Rainy})$ is not one of the conditional probabilities that is known. However, we can find

$$P(\text{Rainy} \cap D) = P(D)P(\text{Rainy}|D) = (.90)(.05) = .045$$

and

$$P(\text{Rainy} \cap D^c) = P(D^c)P(\text{Rainy}|D^c) = (.10)(.10) = .01$$

using the multiplicative rule of probability. These two probabilities are highlighted on the tree diagram for the problem in Figure 3.24.

Now the event Rainy is the union of two mutually exclusive events, $(\text{Rainy} \cap D)$ and $(\text{Rainy} \cap D^c)$. Thus, applying the additive rule of probability, we have

$$P(\text{Rainy}) = P(\text{Rainy} \cap D) + P(\text{Rainy} \cap D^c) = .045 + .01 = .055$$

We now apply the formula for conditional probability to obtain

$$P(D|\text{Rainy}) = \frac{P(\text{Rainy} \cap D)}{P(\text{Rainy})} = \frac{P(\text{Rainy} \cap D)}{P(\text{Rainy} \cap D) + P(\text{Rainy} \cap D^c)}$$
$$= {}^{.045}/_{.055} = .818$$

Therefore, under rainy weather conditions, the prototype system can detect the intruder with a probability of .818—a value lower than the designed probability of .90.

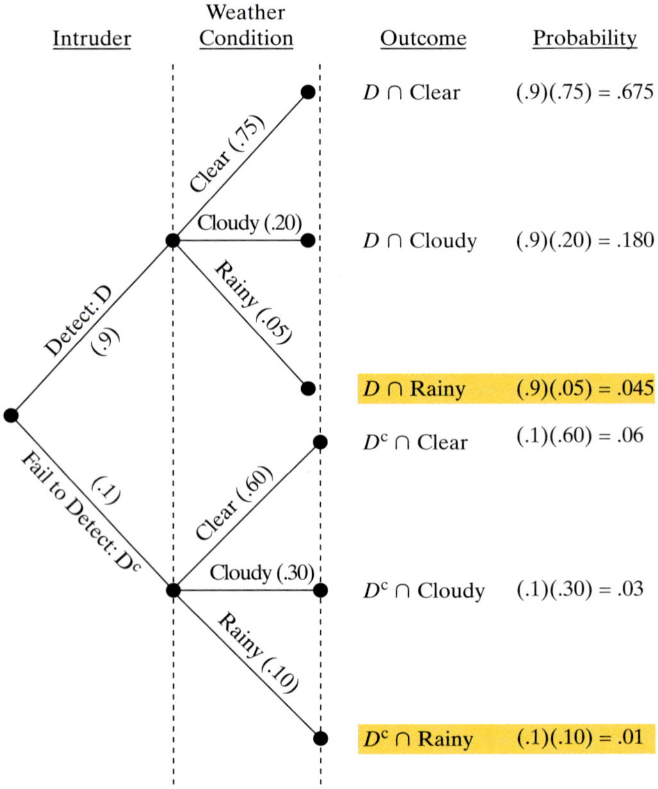

Figure 3.24
Tree diagram for Example 3.24

Now Work Exercise 3.84

The technique utilized in Example 3.24, called **Bayes's Rule,** can be applied when an observed event A occurs with any one of several mutually exclusive and exhaustive events, $B_1, B_2, \ldots, B_k$. The formula for finding the appropriate conditional probabilities is given in the box.

BIOGRAPHY

THOMAS BAYES (1702–1761)

The Inverse Probabilist

The Reverend Thomas Bayes was an ordained English Presbyterian minister who became a Fellow of the Royal Statistical Society without benefit of any formal training in mathematics or any published papers in science during his lifetime. His manipulation of the formula for conditional probability in 1761 is now known as Bayes's Theorem. At the time and for 200 years afterward, use of Bayes's Theorem (or inverse probability, as it was called) was controversial and, to some, considered an inappropriate practice. It was not until the 1960s that the power of the Bayesian approach to decision making began to be tapped. ■

Bayes's Rule

Given k mutually exclusive and exhaustive events, $B_1, B_2, \ldots, B_k$, such that $P(B_1) + P(B_2) + \cdots + P(B_k) = 1$, and an observed event A, then

$$P(B_i|A) = \frac{P(B_i \cap A)}{P(A)}$$

$$= \frac{P(B_i)P(A|B_i)}{P(B_1)P(A|B_1) + P(B_2)P(A|B_2) + \cdots + P(B_k)P(A|B_k)}$$

In applying Bayes's Rule to Example 3.24, the observed event $A = \{\text{Rainy}\}$ and the $k = 2$ mutually exclusive and exhaustive events are the complementary events $D = \{\text{intruder detected}\}$ and $D^c = \{\text{intruder not detected}\}$. Hence, the formula

$$P(D|\text{Rainy}) = \frac{P(D)P(\text{Rainy}|D)}{P(D)P(\text{Rainy}|D) + P(D^c)P(\text{Rainy}|D^c)}$$

$$= \frac{(.90)(.05)}{(.90)(.05) + (.10)(.10)} = .818$$

Exercises 3.84–3.94

Learning the Mechanics

3.84 NW Suppose the events B_1 and B_2 are mutually exclusive and complementary events, such that $P(B_1) = .75$ and $P(B_2) = .25$. Consider another event A such that $P(A|B_1) = .3$ and $P(A|B_2) = .5$.

a. Find $P(B_1 \cap A)$.
b. Find $P(B_2 \cap A)$.
c. Find $P(A)$ using the results in parts **a** and **b.**
d. Find $P(B_1|A)$.
e. Find $P(B_2|A)$.

3.85 Suppose the events B_1, B_2, and B_3 are mutually exclusive and complementary events, such that $P(B_1) = .2$, $P(B_2) = .15$, and $P(B_3) = .65$. Consider another event A such that $P(A|B_1) = .4$, $P(A|B_2) = .25$, and $P(A|B_3) = .6$. Use Bayes's Rule to find

a. $P(B_1|A)$
b. $P(B_2|A)$
c. $P(B_3|A)$

3.86 Suppose the events B_1, B_2, and B_3 are mutually exclusive and complementary events, such that $P(B_1) = .2$, $P(B_2) = .15$, and $P(B_3) = .65$. Consider another event A such that $P(A) = .4$. If A is independent of B_1, B_2, and B_3, use Bayes's Rule to show that $P(B_1|A) = P(B_1) = .2$.

Applying the Concepts—Basic

3.87 Tests for Down syndrome. Currently, there are three diagnostic tests available for chromosome abnormalities in a developing fetus: triple serum marker screening, ultrasound, and amniocentesis. The safest (to both the mother and fetus) and least expensive of the three is the ultrasound test. Two San Diego State University statisticians investigated the accuracy of using ultrasound to test for Down syndrome (*Chance,* Summer 2007). Let D denote that the fetus has a genetic marker for Down syndrome and N denote that the ultrasound test is normal (i.e., no indication of chromosome abnormalities). Then, the statisticians desire the probability $P(D|N)$. Use Bayes's Rule and the following probabilities (provided in the article) to find the desired probability: $P(D) = 1/80$, $P(D^c) = 79/80$, $P(N|D) = 1/2$, $P(N^c|D) = 1/2$, $P(N|D^c) = 1$, and $P(N^c|D^c) = 0$.

3.88 Errors in estimating job costs. A construction company employs three sales engineers. Engineers 1, 2, and 3 estimate the costs of 30%, 20%, and 50%, respectively, of all jobs bid by the company. For $i = 1, 2, 3$, define E_i to be the event that a job is estimated by engineer i. The following probabilities describe the rates at which the engineers make serious errors in estimating costs:

$$P(\text{error}\,|\,E_1) = .01,\ P(\text{error}\,|\,E_2) = .03, \text{ and}$$
$$P(\text{error}\,|\,E_3) = .02$$

a. If a particular bid results in a serious error in estimating job cost, what is the probability that the error was made by engineer 1?
b. If a particular bid results in a serious error in estimating job cost, what is the probability that the error was made by engineer 2?
c. If a particular bid results in a serious error in estimating job cost, what is the probability that the error was made by engineer 3?
d. Based on the probabilities, parts **a–c,** which engineer is most likely responsible for making the serious error?

3.89 Fish contaminated by a plant's toxic discharge. Refer to the U.S. Army Corps of Engineers' study on the DDT contamination of fish in the Tennessee River (Alabama), Example 1.5 (p. 13). Part of the investigation focused on how far upstream the contaminated fish have migrated. (A fish is considered to be contaminated if its measured DDT concentration is greater than 5.0 parts per million.)

a. Considering only the contaminated fish captured from the Tennessee River, the data (saved in the **DDT** file) reveal that 52% of the fish are found between 275 and 300 miles upstream, 39% are found 305 to 325 miles upstream, and 9% are found 330 to 350 miles upstream. Use these percentages to determine the probabilities, $P(275-300)$, $P(305-325)$, and $P(330-350)$.

b. Given that a contaminated fish is found a certain distance upstream, the probability that it is a channel catfish (CC) is determined from the data as $P(CC|275-300) = .775$, $P(CC|305-325) = .77$, and $P(CC|330-350) = .86$. If a contaminated channel catfish is captured from the Tennessee River, what is the probability that it was captured 275–300 miles upstream?

3.90 Drug testing in athletes. Due to inaccuracies in drug-testing procedures (e.g., false positives and false negatives) in the medical field, the results of a drug test represent only one factor in a physician's diagnosis. Yet, when Olympic athletes are tested for illegal drug use (i.e., doping), the results of a single test are used to ban the athlete from competition. In *Chance* (Spring 2004), University of Texas biostatisticians D. A. Berry and L. Chastain demonstrated the application of Bayes's Rule for making inferences about testosterone abuse among Olympic athletes. They used the following example: In a population of 1,000 athletes, suppose 100 are illegally using testosterone. Of the users, suppose 50 would test positive for testosterone. Of the nonusers, suppose 9 would test positive.

a. Given that the athlete is a user, find the probability that a drug test for testosterone will yield a positive result. (This probability represents the *sensitivity* of the drug test.)

b. Given the athlete is a nonuser, find the probability that a drug test for testosterone will yield a negative result. (This probability represents the *specificity* of the drug test.)

c. If an athlete tests positive for testosterone, use Bayes's Rule to find the probability that the athlete is really doping. (This probability represents the *positive predictive value* of the drug test.)

Applying the Concepts—Intermediate

3.91 Malfunctioning production lines. A manufacturing operation utilizes two production lines to assemble electronic fuses. Both lines produce fuses at the same rate and generally produce 2.5% defective fuses. However, production line 1 recently suffered mechanical difficulty and produced 6.0% defectives during a 3-week period. This situation was not known until several lots of electronic fuses produced in this period were shipped to customers. If one of the two fuses tested by a customer was found to be defective, what is the probability that the lot from which it came was produced on malfunctioning line 1? (Assume all the fuses in the lot were produced on the same line.)

3.92 Nondestructive evaluation. Nondestructive evaluation (NDE) describes methods that quantitatively characterize materials, tissues, and structures by noninvasive means, such as X-ray computed tomography, ultrasonics, and acoustic emission. Recently, NDE was used to detect defects in steel castings (*JOM,* May 2005). Assume that the probability that NDE detects a "hit" (i.e., predicts a defect in a steel casting) when, in fact, a defect exists is .97. (This is often called the *probability of detection.*) Also assume that the probability that NDE detects a hit when, in fact, no defect exists is .005. (This is called the *probability of a false call.*) Past experience has shown a defect occurs once in every 100 steel castings. If NDE detects a hit for a particular steel casting, what is the probability that an actual defect exists?

3.93 Purchasing microchips. An important component of your desktop or laptop personal computer (PC) is a microchip. The table gives the proportions of microchips that a certain PC manufacturer purchases from seven suppliers.

Supplier	Proportion
S_1	.15
S_2	.05
S_3	.10
S_4	.20
S_5	.12
S_6	.20
S_7	.18

a. It is known that the proportions of defective microchips produced by the seven suppliers are .001, .0003, .0007, .006, .0002, .0002, and .001, respectively. If a single PC microchip failure is observed, which supplier is most likely responsible?

b. Suppose the seven suppliers produce defective microchips at the same rate, .0005. If a single PC microchip failure is observed, which supplier is most likely responsible?

3.94 Intrusion detection systems. Refer to the *Journal of Research of the National Institute of Standards and Technology* (Nov.–Dec. 2003) study of a double intrusion detection system with independent systems, Exercise 3.67 (p. 153). Recall that if there is an intruder, system A sounds an alarm with probability .9 and system B sounds an alarm with probability .95. If there is no intruder, system A sounds an alarm with probability .2, and system B sounds an alarm with probability .1. Now assume that the probability of an intruder is .4. Also assume that under a given condition (intruder or not), systems A and B operate independently. If both systems sound an alarm, what is the probability that an intruder is detected?

CHAPTER NOTES

Key Terms

Sample point 118
Sample space 120
Simple event 122
Tree diagram 118
Two-way table 132
Unconditional probabilities 141
Union 131
Venn diagram 120

Key Symbols

S	**sample space** (collection of all sample points)
$A: \{1, 2\}$	set of **sample points in event** ***A***
$P(A)$	**probability** of event A
$A \cup B$	**union** of events A and B (either A or B can occur)
$A \cap B$	**intersection** of events A and B (both A and B occur)
A^c	**complement** of A (event A does not occur)
$A \mid B$	event A occurs, **given** event B occurs
$\binom{N}{n}$	number of **combinations** of N elements taken n at a time
$N!$	***N*** **factorial** $= N(N-1)(N-2)\ldots(2)(1)$

Key Ideas

Probability Rules for k Sample Points, $S_1, S_2, S_3, \ldots, S_k$

1. $0 \le P(S_i) \le 1$

2. $\Sigma P(S_i) = 1$

Random Sample

All possible such samples have equal probability of being selected.

Combinations Rule

Counting number of samples of n elements selected from N elements

$$\binom{N}{n} = \frac{N!}{n!(N-n)!} = \frac{N(N-1)(N-2)\cdots(N-n+1)}{n(n-1)(n-2)\cdots(2)(1)}$$

Bayes's Rule

$$P(S_i \mid A) = \frac{P(S_i)P(A \mid S_i)}{P(S_1)P(A \mid S_1) + P(S_2)P(A \mid S_2) + \cdots + P(S_k)P(A \mid S_k)}$$

Guide to Selecting Probability Rules

Type of Compound Event

- **Union (*A* or *B*)** — Addition Rule
  - **Mutually Exclusive**: $P(A \cup B) = P(A) + P(B)$
  - **Not Mutually Exclusive**: $P(A \cup B) = P(A) + P(B) - P(A \cap B)$
- **Intersection (*A* and *B*)** — Multiplication Rule
  - **Independent**: $P(A \cap B) = P(A) \cdot P(B)$
  - **Dependent**: $P(A \cap B) = P(A|B) \cdot P(B) = P(B|A) \cdot P(A)$
- **Complementary (not *A*)** — Rule of Complements
  - $P(A^c) = 1 - P(A)$
- **Conditional (*A given B*)** — Conditional Rule
  - **Independent**: $P(A|B) = P(A)$, $P(B|A) = P(B)$
  - **Dependent**: $P(A|B) = \frac{P(A \cap B)}{P(B)}$, $P(B|A) = \frac{P(A \cap B)}{P(A)}$

Supplementary Exercises 3.95–3.132

Learning the Mechanics

3.95 A sample space consists of four sample points, where $P(S_1) = .2$, $P(S_2) = .1$, $P(S_3) = .3$, and $P(S_4) = .4$.
a. Show that the sample points obey the two probability rules for a sample space.
b. If an event $A = \{S_1, S_4\}$, find $P(A)$.

3.96 For two events A and B, suppose $P(A) = .7$, $P(B) = .5$, and $P(A \cap B) = .4$. Find $P(A \cup B)$.

3.97 A and B are mutually exclusive events, with $P(A) = .2$ and $P(B) = .3$.
a. Find $P(A|B)$.
b. Are A and B independent events?

3.98 Which of the following pairs of events are mutually exclusive? Justify your response.
a. {The Dow Jones Industrial Average increases on Monday.}, {A large New York bank decreases its prime interest rate on Monday.}
b. {The next sale by a PC retailer is a laptop computer.}, {The next sale by a PC retailer is a desktop computer.}
c. {You reinvest all your dividend income in a limited partnership.}, {You reinvest all your dividend income in a money market fund.}

3.99 Given that $P(A \cap B) = .4$ and $P(A|B) = .8$, find $P(B)$.

3.100 Two events, A and B, are independent, with $P(A) = .3$ and $P(B) = .1$.
a. Are A and B mutually exclusive? Why?
b. Find $P(A|B)$and $P(B|A)$.
c. Find $P(A \cup B)$.

3.101 The Venn diagram below illustrates a sample space containing six sample points and three events, A, B, and C. The probabilities of the sample points are $P(1) = .3$, $P(2) = .2$, $P(3) = .1$, $P(4) = .1$, $P(5) = .1$, and $P(6) = .2$.

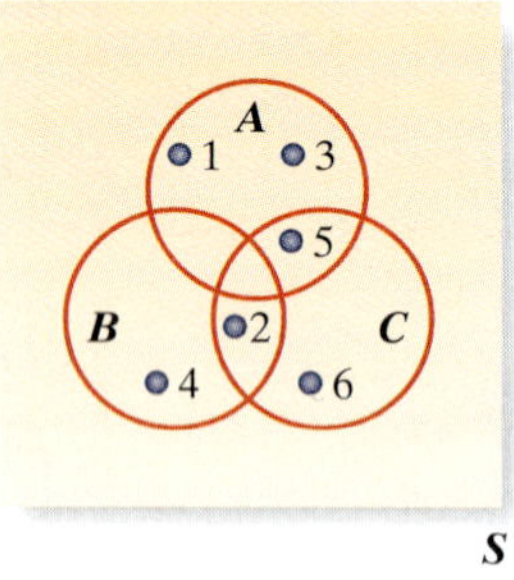

a. Find $P(A \cap B)$, $P(B \cap C)$, $P(A \cup C)$, $P(A \cup B \cup C)$, $P(B^c)$, $P(A^c \cap B)$, $P(B|C)$, and $P(B|A)$.
b. Are A and B independent? Mutually exclusive? Why?
c. Are B and C independent? Mutually exclusive? Why?

3.102 Find the numerical value of

a. $6!$ **b.** $\binom{10}{9}$ **c.** $\binom{10}{1}$ **d.** $\binom{6}{3}$ **e.** $0!$

3.103 A random sample of five graduate students is to be selected from 50 MBA majors for participation in a case competition.
a. In how many different ways can the sample be drawn?
b. Show how the random number table, Table I in Appendix B, can be used to select the sample of students.

Applet Exercise 3.6

Use the applet entitled *Random Numbers* to generate a list of 50 numbers between 1 and 100, inclusive. Use this list to find each of the probabilities.
a. The probability that a number chosen from the list is less than or equal to 50
b. The probability that a number chosen from the list is even
c. The probability that a number chosen from the list is less than or equal to 50 and even
d. The probability that a number chosen from the list is less than or equal to 50 given that the number is even
e. Do your results from parts **a–d** support the conclusion that the events *less than or equal to 50* and *even* are independent? Explain.

Applying the Concepts—Basic

3.104 Annual compensation and benefits report. Each year, Hudson Institute conducts a survey of 10,000 U.S. workers about their total compensation packages. One question in Hudson's *2006 Compensation and Benefits Report* focused on employees who received raises in the past year. Of these employees, 35% reported that their raise was based on job performance, 50% reported that it was based on a standard cost of living, and the remainder (15%) were unsure how their raises were determined. Suppose we select (at random) one of the U.S. workers surveyed who received a raise last year and inquire about how that worker's raise was determined.
a. List the sample points for this experiment.
b. Assign reasonable probabilities to the sample points.
c. Find the probability that the raise was based either on job performance or a standard cost of living.

3.105 Ownership of small businesses. According to the *Journal of Business Venturing* (Vol. 17, 2002), 27% of all small businesses owned by non-Hispanic whites nationwide are women-owned firms. If we select, at random, a small business owned by a non-Hispanic white, what is the probability that it is a male-owned firm?

3.106 Condition of public school facilities. The National Center for Education Statistics (NCES) conducted a survey on the condition of America's public school facilities. The survey revealed the following information. The probability that a public school building has inadequate plumbing is .25. Of the buildings with inadequate plumbing, the probability that the school has plans for repairing the building is .38. Find the probability that a public school building has inadequate plumbing and will be repaired.

3.107 New car crash tests. Refer to the National Highway Traffic Safety Administration (NHTSA) crash tests of new car models, Exercise 2.139 (p. 104). Recall that the NHTSA has developed a "star" scoring system, with results ranging from one star (*) to five stars (*****). The more stars in the rating, the better the level of crash protection in a head-on collision. A summary of the driver-side star ratings for 98 cars is reproduced in the accompanying Minitab printout. Assume that one of the 98 cars is selected at random. State whether each of the following is true or false.

a. The probability that the car has a rating of two stars is 4.

b. The probability that the car has a rating of four or five stars is .7857.

c. The probability that the car has a rating of one star is 0.

d. The car has a better chance of having a two-star rating than of having a five-star rating.

Tally for Discrete Variables: DRIVSTAR

DRIVSTAR	Count	Percent
2	4	4.08
3	17	17.35
4	59	60.20
5	18	18.37
N=	98	

3.108 ISO 9000 registered companies. The ISO 9000 is a series of standards for setting up and documenting quality systems, processes, and procedures. To gauge how managers view the standards or how the standards were achieved, a sample of 40 ISO 9000–registered companies in Colorado was selected, and the manager most responsible for ISO 9000 implementation was interviewed (*Quality Progress,* 1995). The following are some of the data obtained by the study:

Level of Top Management Involvement in the ISO 9000 Registration Process	Frequency
Very involved	9
Moderate involvement	16
Minimal involvement	12
Not involved	3

Length of Time to Achieve ISO 9000 Registration	Frequency
Less than 1 year	5
1–1.5 years	21
1.6–2 years	9
2.1–2.5 years	2
More than 2.5 years	3

Source: Weston, F. C. "What do managers really think of the ISO 9000 registration process?" *Quality Progress,* October 1995, pp. 68–69 (Tables 3 and 4). Reprinted with permission from *Quality Progress.* © 1995 American Society for Quality. No further distribution allowed without permission.

Suppose one of the 40 managers who were interviewed is to be randomly selected for additional questioning. Consider the events defined as follows:

A: {The manager was involved in the ISO 9000 registration.}

B: {The length of time to achieve ISO 9000 registration was more than 2 years.}

a. Find $P(A)$.

b. Find $P(B)$.

c. Explain why the preceding data are not sufficient to determine whether events A and B are independent.

3.109 Survey on energy conservation. A state energy agency mailed questionnaires on energy conservation to 1,000 homeowners in the state capital. Five hundred questionnaires were returned. Suppose an experiment consists of randomly selecting and reviewing one of the returned questionnaires. Consider the events:

A: {The home is constructed of brick.}

B: {The home is more than 30 years old.}

C: {The home is heated with oil.}

Describe each of the following events in terms of unions, intersections, and complements (i.e., $A \cup B$, $A \cap B$, A^c, etc.):

a. The home is more than 30 years old and is heated with oil.

b. The home is not constructed of brick.

c. The home is heated with oil or is more than 30 years old.

d. The home is constructed of brick and is not heated with oil.

3.110 Identifying urban counties. *Urban* and *rural* describe geographic areas for which land zoning regulations, school district policy, and public service policy are often set. However, the characteristics of urban/rural areas are not clearly defined. Researchers at the University of Nevada (Reno) asked a sample of county commissioners to give their perception of the single most important factor in identifying urban counties (*Professional Geographer,* Feb. 2000). In all, five factors were mentioned by the commissioners: total population, agricultural change, presence of industry, growth, and population concentration. The survey results are displayed in the pie chart below. Suppose one of the commissioners is selected at random and the most important factor specified by the commissioner is recorded.

a. List the sample points for this experiment.

b. Assign reasonable probabilities to the sample points.

c. Find the probability that the most important factor specified by the commissioner is population related.

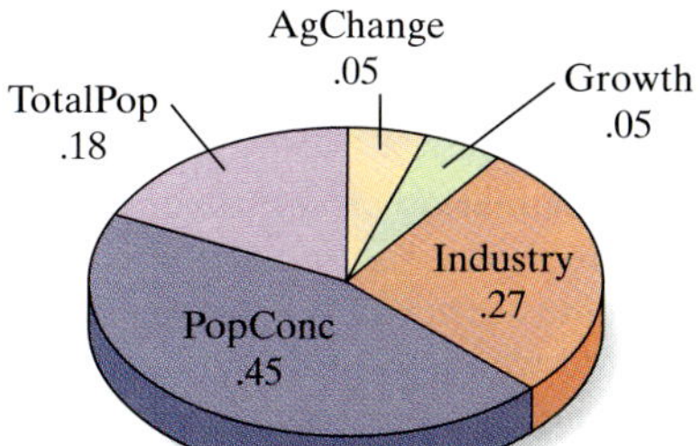

3.111 Use of country club facilities. A local country club has a membership of 600 and operates facilities that include an 18-hole championship golf course and 12 tennis courts. Before deciding whether to accept new members, the club president would like to know how many members regularly use each facility. A survey of the membership indicates that 70% regularly use the golf course, 50% regularly use the tennis courts, and 5% use neither of these facilities regularly.

a. Construct a Venn diagram to describe the results of the survey.

b. If one club member is chosen at random, what is the probability that the member uses either the golf course or the tennis courts or both?

c. If one member is chosen at random, what is the probability that the member uses both the golf and the tennis facilities?

d. A member is chosen at random from among those known to use the tennis courts regularly. What is the probability that the member also uses the golf course regularly?

3.112 Commodities shipped by rail. The following table, extracted from *Railway Age* (May 1999), lists the number of carloads of different types of commodities that were shipped by the major U.S. railroads during a week. Suppose the computer record for a carload shipped during the week is randomly selected from a masterfile of all carloads shipped that week and the commodity type shipped is identified.

Type of Commodity	Number of Carloads
Agricultural products	41,690
Chemicals	38,331
Coal	124,595
Forest products	21,929
Metallic ores and minerals	34,521
Motor vehicles and equipment	22,906
Nonmetallic minerals and products	37,416
Other carloads	14,382
Total	335,770

Source: Railway Age, May 1999, p. 1.

a. List or describe the sample points in this experiment.

b. Find the probability of each sample point.

c. What is the probability that the railcar was transporting automobiles? Nonagricultural products?

d. What is the probability that the railcar contained chemicals or coal?

e. One of the carloads shipped that week was in a boxcar with serial number 1003642. What is the probability that this particular carload would be the one randomly selected from the computer file? Justify your answer.

Applying the Concepts—Intermediate

3.113 Effect of stock market on consumer spending and saving. According to *Economic Inquiry* (Jan. 2002), 50% of all U.S. households own stock. A sample of these stockholders was asked what effect recent trends in the stock market have had on their saving and spending. The responses are summarized in the table. A U.S. household is selected at random.

Stock Market Effect on Spending/Saving	Percent of Stockholders
No effect	85.0
Spend more/save less	3.4
Spend less/save more	11.6
Total	100.0

Source: Starr-McCluer, M. "Stock market wealth and consumer spending," *Economic Inquiry,* Vol. 40, No. 1, Jan. 2002, pp. 69–79 (Table 3). Reprinted with permission of John Wiley & Sons, Inc.

a. What is the probability that the household does not own stock?

b. Given the household owns stock, what is the probability that the household spends more and saves less due to recent trends in the stock market?

c. What is the probability that the household owns stock but spends less and saves more due to recent trends in the stock market?

3.114 Assembling a panel of energy experts. The state legislature has appropriated $1 million to be distributed in the form of grants to individuals and organizations engaged in the research and development of alternative energy sources. You have been hired by the state's energy agency to assemble a panel of five energy experts whose task it will be to determine which individuals and organizations should receive the grant money. You have identified 11 equally qualified individuals who are willing to serve on the panel. How many different panels of five experts could be formed from these 11 individuals?

3.115 Testing a watch manufacturer's claim. A manufacturer of electronic digital watches claims that the probability of its watch running more than 1 minute slow or 1 minute fast after 1 year of use is .05. A consumer protection agency has purchased four of the manufacturer's watches with the intention of testing the claim.

a. Assuming that the manufacturer's claim is correct, what is the probability that none of the watches are as accurate as claimed?

b. Assuming that the manufacturer's claim is correct, what is the probability that exactly two of the four watches are as accurate as claimed?

c. Suppose that only one of the four tested watches is as accurate as claimed. What inference can be made about the manufacturer's claim? Explain.

d. Suppose that none of the watches tested are as accurate as claimed. Is it necessarily true that the manufacturer's claim is false? Explain.

3.116 Ranking razor blades. The corporations in the highly competitive razor blade industry do a tremendous amount of advertising each year. Corporation G gave a supply of three top-name brands, G, S, and W, to a consumer and asked her to use them and rank them in order of preference. The corporation was, of course, hoping the consumer would prefer its brand and rank it first, thereby giving them some material for a consumer interview advertising campaign. If the consumer did not prefer one blade over any other but was still required to rank the blades, what is the probability that

a. The consumer ranked brand G first?

b. The consumer ranked brand G last?

c. The consumer ranked brand G last and brand W second?

d. The consumer ranked brand W first, brand G second, and brand S third?

3.117 Link between cigar smoking and cancer. The *Journal of the National Cancer Institute* (Feb. 16, 2000) published the results of a study that investigated the association between cigar smoking and death from tobacco-related cancers. Data were obtained for a national sample of 137,243 American men. The results are summarized in the table on the next page. Each male in the study was classified according to his cigar-smoking status and whether or not he died from a tobacco-related cancer.

a. Find the probability that a man who never smoked cigars died from cancer.
b. Find the probability that a former cigar smoker died from cancer.
c. Find the probability that a current cigar smoker died from cancer.

	Died from Cancer	Did Not Die from Cancer	Totals
Never Smoked Cigars	782	120,747	121,529
Former Cigar Smoker	91	7,757	7,848
Current Cigar Smoker	141	7,725	7,866
Totals	1,014	136,229	137,243

Source: Shapiro, J. A., Jacobs, E. J., and Thun, M. J. "Cigar smoking in men and risk of death from tobacco-related cancers," *Journal of the National Cancer Institute,* Vol. 92, No. 4, Feb. 16, 2000, pp. 333–337 (Table 2). By permission of Oxford University Press.

3.118 Which events are independent? Use your intuitive understanding of independence to form an opinion about whether each of the following scenarios represent independent events.

a. The results of consecutive tosses of a coin
b. The opinions of randomly selected individuals in a preelection poll
c. A major league baseball player's results in two consecutive at-bats
d. The amount of gain or loss associated with investments in different stocks if these stocks are bought on the same day and sold on the same day one month later
e. The amount of gain or loss associated with investments in different stocks that are bought and sold in different time periods, 5 years apart
f. The prices bid by two different development firms in response to a building construction proposal

3.119 Employees' choices of health care plans. Most companies offer their employees a variety of health care plans to choose from—e.g., preferred provider organizations (PPOs) and health maintenance organizations (HMOs). A survey of 100 large, 100 medium, and 100 small companies that offer their employees HMOs, PPOs, and fee-for-service plans was conducted; each firm provided information on the plans chosen by their employees. These companies had a total employment of 833,303 people. A breakdown of the number of employees in each category by firm size and plan is provided in the table.

Company Size	Fee-for-Service	PPO	HMO	Totals
Small	1,808	1,757	1,456	5,021
Medium	8,953	6,491	6,938	22,382
Large	330,419	241,770	233,711	805,900
Totals	341,180	250,018	242,105	833,303

Source: Adapted from Bucci, M., and Grant, R. "Employer-sponsored health insurance: What's offered: What's chosen?" *Monthly Labor Review,* October 1995, Vol. 118, No. 10, pp. 38–43.

One employee from the 833,303 total employees is to be chosen at random for further analysis. Define the events A and B as follows:

A: {Observe an employee that chose fee-for-service}

B: {Observe an employee from a small company}

a. Find $P(B)$.
b. Find $P(A \cap B)$.
c. Find $P(A \cup B)$.
d. Find $P(A|B)$.
e. Are A and B independent? Justify your answer.

3.120 World Cup soccer match draws. Every four years the world's 32 best national soccer teams compete for the World Cup. The 2006 World Cup final between champion Italy and France was one of the most watched events in television history, with an estimated audience of 715 million people. Run by FIFA (Fédération Internationale de Football Association), national teams are placed into eight groups of four teams, with the group winners advancing to play for the World Cup. *Chance* (Spring 2007) investigated the fairness of the 2006 World Cup draw. Each of the top 8 seeded teams (teams ranked 1–8, called pot 1) were placed into one of the eight groups (named, Group A, B, C, D, E, F, G, and H). The remaining 24 teams were assigned to 3 pots of 8 teams each to achieve the best possible geographical distribution between the groups. The teams in pot 2 were assigned to groups as follows: The first team drawn was placed into Group A, the second team drawn was placed in Group B, etc. Teams in pots 3 and 4 were assigned to the groups in similar fashion. Because teams in pots 2–4 are not necessarily placed there based on their world ranking, this typically leads to a "group of death," i.e., a group involving at least two highly seeded teams where only one can advance.

a. In 2006, Germany (as the host country) was assigned as the top seed in Group A. What is the probability that Paraguay (with the highest ranking in pot 2) is assigned to Group A?
b. Many soccer experts view the South American teams (Ecuador and Paraguay) as the most dangerous teams in pot 2. What is the probability one of the South American teams is assigned to Group A?
c. In 2006, Group B was considered the "group of death," with England (world rank 2), Paraguay (highest rank in pot 2), Sweden (2nd highest rank in pot 3), and Trinidad and Tobago. What is the probability that Group B will include the team with the highest rank in pot 2 and the team with one of the top two ranks in pot 3?
d. In drawing teams from pot 2, there was a notable exception in 2006. If a South American team (either Ecuador or Paraguay) was drawn into a group with another South American team, it was automatically moved to the next group. This rule impacted Group C (Argentina as the top seed) and Group F (Brazil as the top seed), because they already have South American teams, and groups that follow these groups in the draw. Now Group D included the eventual champion Italy as its top seed. What is the probability that Group D was not assigned one of the dangerous South American teams in pot 2?

3.121 Chance of an Avon sale. The probability that an Avon salesperson sells beauty products to a prospective customer on the first visit to the customer is .4. If the salesperson fails to make the sale on the first visit, the probability that the sale will be made on the second visit is .65. The salesperson never visits a prospective customer more than twice. What is the probability that the salesperson will make a sale to a particular customer?

3.122 Repairing a computer network. The local area network (LAN) for the College of Business computing system at a large university is temporarily shut down for repairs. Previous shutdowns have been due to hardware failure, software failure, or power failure. Maintenance engineers have determined that the probabilities of hardware, software, and power problems are .01, .05, and .02, respectively. They have also determined that if the system experiences hardware problems, it shuts down 73% of the time. Similarly, if software problems occur, the system shuts down 12% of the time; and, if power failure occurs, the system shuts down 88% of the time. What is the probability that the current shutdown of the LAN is due to hardware failure? Software failure? Power failure?

3.123 Profile of a sustainable farmer. *Sustainable development* or *sustainable farming* means finding ways to live and work the Earth without jeopardizing the future. Studies were conducted in five Midwestern states to develop a profile of a sustainable farmer. The results revealed that farmers can be classified along a sustainability scale, depending on whether they are likely (L) or unlikely (U) to engage in the following practices: (1) Raise a broad mix of crops; (2) raise livestock; (3) use chemicals sparingly; and (4) use techniques for regenerating the soil, such as crop rotation.

a. List the different sets of classifications that are possible for the four practices (e.g., LUUL).

b. Suppose you are planning to interview farmers across the country to determine the frequency with which they fall into the classification sets you listed for part **a.** Because no information is yet available, assume initially that there is an equal chance of a farmer falling into any single classification set. Using that assumption, what is the probability that a farmer will be classified as unlikely on all four criteria (i.e., classified as a nonsustainable farmer)?

c. Using the same assumption as in part **b,** what is the probability that a farmer will be classified as likely on at least three of the criteria (i.e., classified as a near-sustainable farmer)?

3.124 Evaluating the performance of quality inspectors. The performance of quality inspectors affects both the quality of outgoing products and the cost of the products. A product that passes inspection is assumed to meet quality standards; a product that fails inspection may be reworked, scrapped, or reinspected. Quality engineers at Westinghouse Electric Corporation evaluated performances of inspectors in judging the quality of solder joints by comparing each inspector's classifications of a set of 153 joints with the consensus evaluation of a panel of experts. The results for a particular inspector are shown in the table. One of the 153 solder joints is to be selected at random.

	Inspector's Judgment	
Committee's Judgment	Joint Acceptable	Joint Rejectable
Joint acceptable	101	10
Joint rejectable	23	19

Source: Meagher, J., and Scazzero, J., "Measuring inspector variability," pp. 75–81. Reprinted with permission from Proceedings from the 39th Annual Quality Congress. © 1985 American Society for Quality. No further distribution allowed without permission.

a. What is the probability that the inspector judges the joint to be acceptable? That the committee judges the joint to be acceptable?

b. What is the probability that both the inspector and the committee judge the joint to be acceptable? That neither judge the joint to be acceptable?

c. What is the probability that the inspector and the committee disagree? Agree?

3.125 System components operating in series and parallel. Consider the two systems shown in the schematic below. System A operates properly only if all three components operate properly. (The three components are said to operate *in series*.) The probability of failure for system A components 1, 2, and 3 are .12, .09, and .11, respectively. Assume the components operate independently of each other. System B comprises two subsystems said to operate *in parallel.* Each subsystem has two components that operate in series. System B will operate properly as long as at least one of the subsystems functions properly. The probability of failure for each component in the system is .1. Assume the components operate independently of each other.

a. Find the probability that System A operates properly.

b. What is the probability that at least one of the components in system A will fail and therefore that the system will fail?

c. Find the probability that System B operates properly.

d. Find the probability that exactly one subsystem in system B fails.

e. Find the probability that system B fails to operate properly.

f. How many parallel subsystems like the two shown here would be required to guarantee that the system would operate properly at least 99% of the time?

Figures for Exercise 3.125

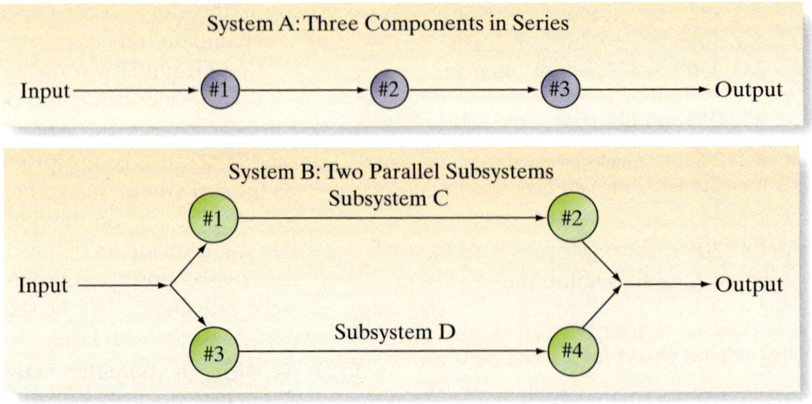

Applying the Concepts—Advanced

3.126 Rejected bottles of a brewery. A small brewery has two bottling machines. Machine A produces 75% of the bottles, and machine B produces 25%. 1 out of every 20 bottles filled by A is rejected for some reason, while 1 out of every 30 bottles from B is rejected. What proportion of bottles is rejected? What is the probability that a randomly selected bottle comes from machine A, given that it is accepted?

3.127 Scrap rate of machine parts. A press produces parts used in the manufacture of large-screen plasma televisions. If the press is correctly adjusted, it produces parts with a scrap rate of 5%. If it is not adjusted correctly, it produces scrap at a 50% rate. From past company records, the machine is known to be correctly adjusted 90% of the time. A quality control inspector randomly selects one part from those recently produced by the press and discovers it is defective. What is the probability that the machine is incorrectly adjusted?

3.128 Chance of winning at "craps." A version of the dice game "craps" is played in the following manner. A player starts by rolling two balanced dice. If the roll (the sum of the two numbers showing on the dice) results in a 7 or 11, the player wins. If the roll results in a 2 or a 3 (called *craps*), the player loses. For any other roll outcome, the player continues to throw the dice until the original roll outcome recurs (in which case the player wins) or until a 7 occurs (in which case the player loses).

a. What is the probability that a player wins the game on the first roll of the dice?

b. What is the probability that a player loses the game on the first roll of the dice?

c. If the player throws a total of 4 on the first roll, what is the probability that the game ends (win or lose) on the next roll?

3.129 Chance of winning blackjack. Blackjack, a favorite game of gamblers, is played by a dealer and at least one opponent (called a *player*). In one version of the game, 2 cards of a standard 52-card bridge deck are dealt to the player and 2 cards to the dealer. For this exercise, assume that drawing an ace and a face card is called *blackjack*. If the dealer does not draw a blackjack and the player does, the player wins. If both the dealer and player draw blackjack, a "push" (i.e., a tie) occurs.

a. What is the probability that the dealer will draw a blackjack?

b. What is the probability that the player wins with a blackjack?

3.130 Strategy in the game "Go." "Go" is one of the oldest and most popular strategic board games in the world, especially in Japan and Korea. The University of Virginia requires MBA students to learn Go to understand how the Japanese conduct business. This two-player game is played on a flat surface marked with 19 vertical and 19 horizontal lines. The objective is to control territory by placing pieces, called *stones,* on vacant points on the board. Players alternate placing their stones. The player using black stones goes first, followed by the player using white stones. *Chance* (Summer 1995) published an article that investigated the advantage of playing first (i.e., using the black stones) in Go. The results of 577 games recently played by professional Go players were analyzed.

a. In the 577 games, the player with the black stones, won 319 times, and the player with the white stones won 258 times. Use this information to assess the probability of winning when you play first in Go.

b. Professional Go players are classified by level. Group C includes the top-level players, followed by Group B (middle-level) and Group A (low-level) players. The following table describes the number of games won by the player with the black stones, categorized by level of the black player and level of the opponent. Assess the probability of winning when you play first in Go for each combination of player and opponent level.

c. If the player with the black stones is ranked higher than the player with the white stones, what is the probability that black wins?

d. Given the players are of the same level, what is the probability that the player with the black stones wins?

Black Player Level	Opponent Level	Number of Wins	Number of Games
C	A	34	34
C	B	69	79
C	C	66	118
B	A	40	54
B	B	52	95
B	C	27	79
A	A	15	28
A	B	11	51
A	C	5	39
Totals		319	577

Source: Kim, J., and Kim, H. J. "The advantage of playing first in Go," *Chance,* Vol. 8, No. 3, Summer 1995, p. 26 (Table 3). Reprinted with permission from *Chance.*

Critical Thinking Challenges

3.131 "Let's Make a Deal." Marilyn vos Savant, who is listed in *Guinness Book of World Records Hall of Fame* for "Highest IQ," writes a weekly column in the Sunday newspaper supplement, *Parade Magazine.* Her column, "Ask Marilyn," is devoted to games of skill, puzzles, and mind-bending riddles. In one issue (*Parade Magazine,* Feb. 24, 1991), vos Savant posed the following question:

Suppose you're on a game show, and you're given a choice of three doors. Behind one door is a car; behind the others, goats. You pick a door—say, #1—and the host, who knows what's behind the doors, opens another door—say #3—which has a goat. He then says to you, "Do you want to pick door #2?" Is it to your advantage to switch your choice?

Marilyn's answer: "Yes, you should switch. The first door has a $^1/_3$ chance of winning [the car], but the second has a $^2/_3$ chance [of winning the car]." Predictably, vos Savant's surprising answer elicited thousands of critical letters, many of them from PhD mathematicians, who disagreed with her. Who is correct, the PhDs or Marilyn?

3.132 Flawed Pentium computer chip. In October 1994, a flaw was discovered in the Pentium microchip installed in personal computers. The chip produced an incorrect result when dividing two numbers. Intel, the manufacturer of the Pentium chip, initially announced that such an error would occur once in 9 billion divisions, or "once in every 27,000 years" for a typical user; consequently, it did not immediately offer to replace the chip.

Depending on the procedure, statistical software packages (e.g., Minitab) may perform an extremely large number of divisions to produce the required output. For heavy users of the software 1 billion divisions over a short time frame is not unusual. Will the flawed chip be a problem for a heavy Minitab user? [*Note:* Two months after the flaw was discovered, Intel agreed to replace all Pentium chips free of charge.]

References

Bennett, D. J. *Randomness.* Cambridge, Mass.: Harvard University Press, 1998.

Epstein, R. A. *The Theory of Gambling and Statistical Logic,* rev. ed. New York: Academic Press, 1977.

Feller, W. *An Introduction to Probability Theory and Its Applications,* 3rd ed., Vol. 1. New York: Wiley, 1968.

Lindley, D. V. *Making Decisions,* 2nd ed. London: Wiley, 1985.

Parzen, E. *Modern Probability Theory and Its Applications.* New York: Wiley, 1960.

Wackerly, D., Mendenhall, W., and Scheaffer, R. L. *Mathematical Statistics with Applications,* 7th ed. Boston: Duxbury, 2008.

Williams, B. *A Sampler on Sampling.* New York: Wiley, 1978.

Winkler, R. L. *An Introduction to Bayesian Inference and Decision.* New York: Holt, Rinehart and Winston, 1972.

Wright, G., and Ayton, P., eds. *Subjective Probability.* New York: Wiley, 1994.

USING TECHNOLOGY

SPSS: Generating a Random Sample

Step 1 Click on the "Data" button on the SPSS menu bar and then click on "Select Cases," as shown in Figure 3.S.1. The resulting menu list appears as shown in Figure 3.S.2.

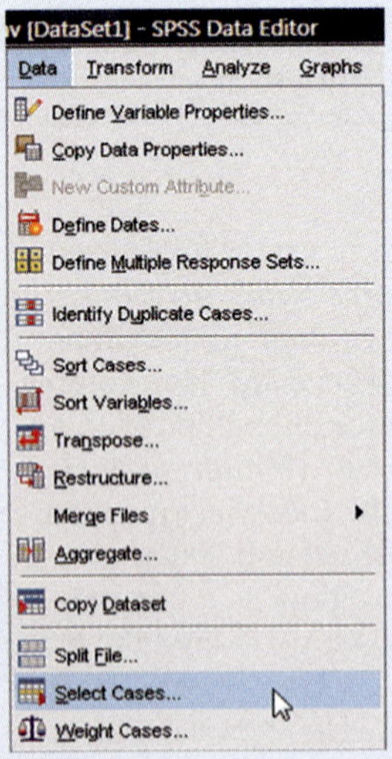

Figure 3.S.1 SPSS menu options for sampling from a data set

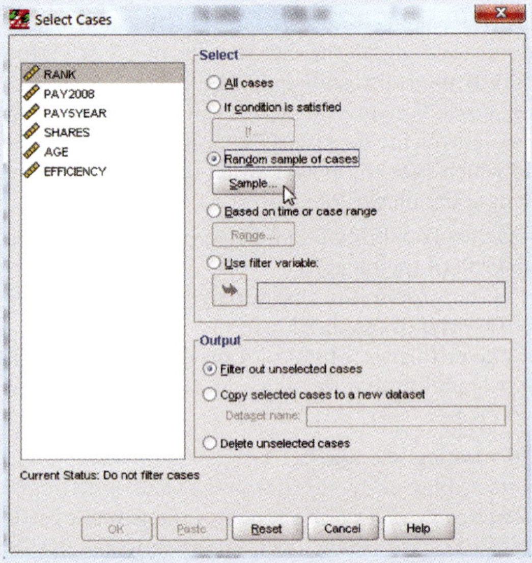

Figure 3.S.2 SPSS options for selecting a random sample

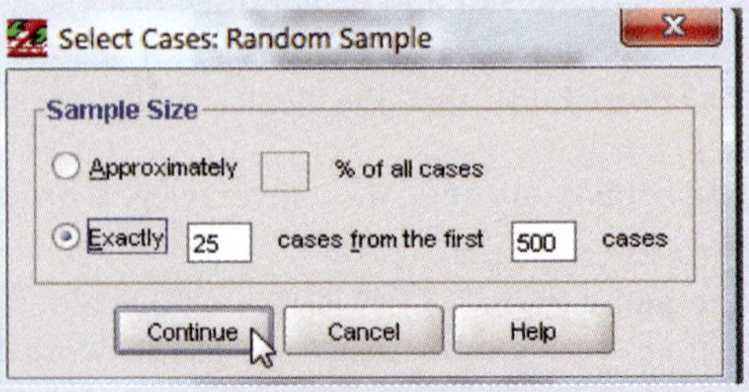

Figure 3.S.3 SPSS random sample dialog box

Step 2 Select "Random sample of cases" from the list and then click on the "Sample" button. The dialog box shown in Figure 3.S.3 will appear.

Step 3 Specify the sample size either as a percentage of cases or a raw number.

Step 4 Click "Continue" to return to the "Select Cases" dialog box (Figure 3.S.2) and then click "OK." The SPSS spreadsheet will reappear with the selected (sampled) cases.

Minitab: Generating a Random Sample

Step 1 Click on the "Calc" button on the Minitab menu bar and then click on "Random Data," and finally, click on "Sample From Columns," as shown in Figure 3.M.1. The resulting dialog box appears as shown in Figure 3.M.2.

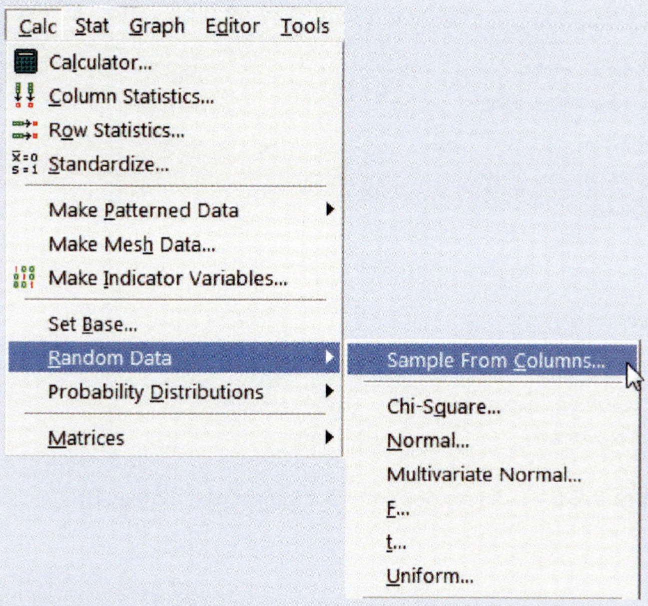

Figure 3.M.1 Minitab menu options for sampling from a data set

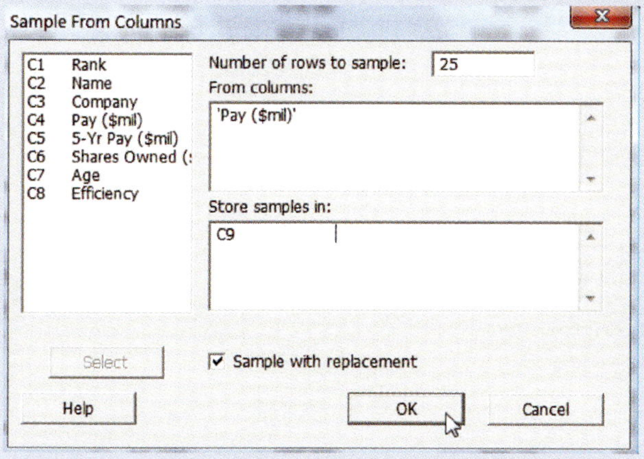

Figure 3.M.2 Minitab options for selecting a random sample from worksheet columns

Step 2 Specify the sample size (i.e., number of rows), the variable(s) to be sampled, and the column(s) where you want to save the sample.

Step 3 Click "OK" and the Minitab worksheet will reappear with the values of the variable for the selected (sampled) cases in the column specified.

In Minitab, you can also generate a sample of case numbers.

Step 1 From the Minitab menu, click on the "Calc" button and then click on "Random Data," and finally, click on the "Uniform" option (see Figure 3.M.1).

Step 2 In the resulting dialog box (shown in Figure 3.M.3), specify the number of cases (rows, i.e., the sample size), and the column where the case numbers selected will be stored.

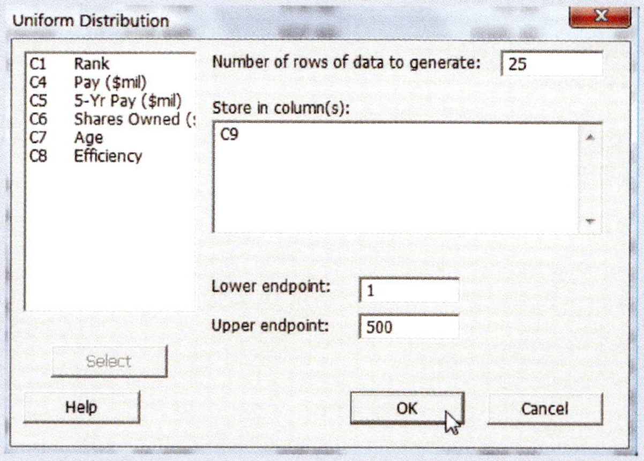

Figure 3.M.3 Minitab options for selecting a random sample of cases

Step 3 Click "OK" and the Minitab worksheet will reappear with the case numbers for the selected (sampled) cases in the column specified.

[*Note:* If you want the option of generating the same (identical) sample multiple times from the data set, then first click on the "Set Base" option shown in Figure 3.M.1. Specify an integer in the resulting dialog box. If you always select the same integer, Minitab will select the same sample when you choose the random sampling options.]

Excel: Generating a Random Sample

To obtain a random sample of numbers in EXCEL, perform the following:

Step 1 Select "Data," and then "Data Analysis" from the Excel main menu bar, as shown in Figure 3.E.1.

Step 2 Select "Random Number Generation" from the Data Analysis dialog box and then click "OK," as shown in Figure 3.E.2.

Step 3 In the resulting menu (shown in Figure 3.E.3), specify the number of variables and number of random numbers you want to generate and select "Uniform" for the distribution. Click "OK" to generate the random numbers.

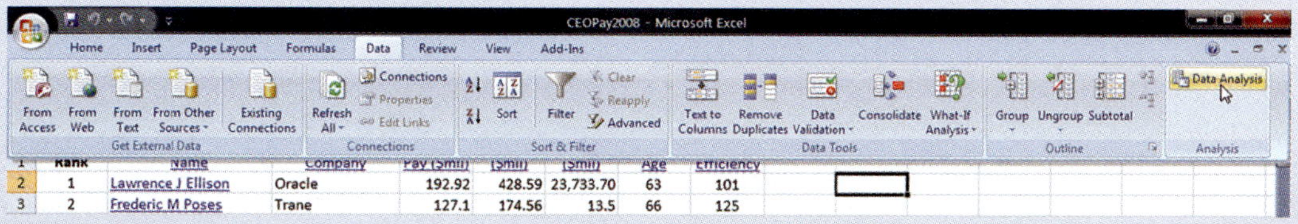

Figure 3.E.1 Excel menu options for generating a random sample of numbers

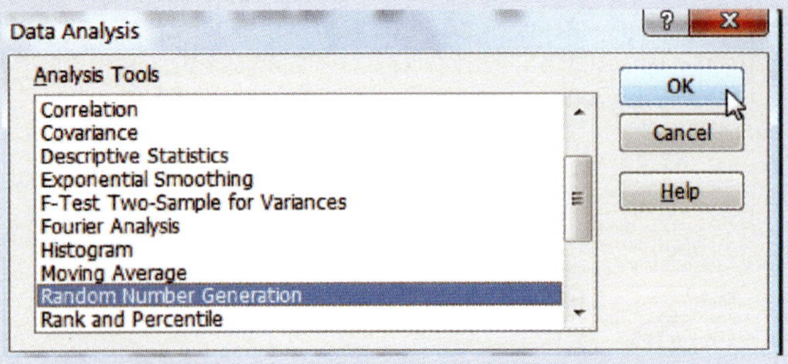

Figure 3.E.2 Excel data analysis tools options: Random number generation

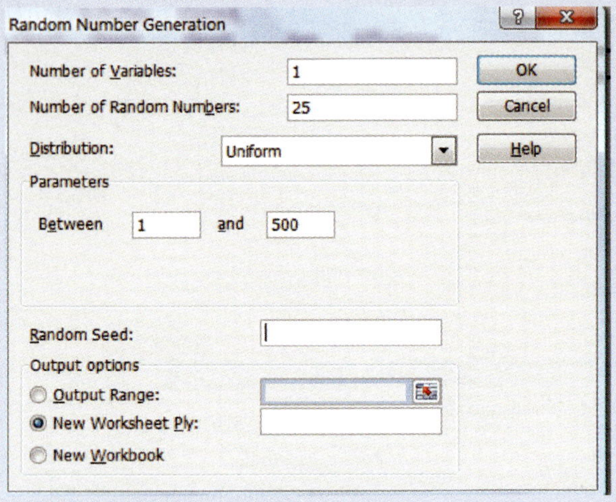

Figure 3.E.3 Excel random number generation dialog box

CONTENTS

4 Random Variables and Probability Distributions

Where We've Been

- Used probability to make an inference about a population from data in an observed sample
- Used probability to measure the reliability of the inference

Where We're Going

- Develop the notion of a random variable
- Learn that numerical data are observed values of either discrete or continuous random variables
- Study two important types of random variables and their probability models: the binomial and normal model
- Define a sampling distribution as the probability distribution of a sample statistic
- Learn that the sampling distribution of $\bar{x}$ follows a normal model

Statistics IN Action Super Weapons Development—Optimizing the Hit Ratio

The U.S. Army is working with a major defense contractor to develop a "super" weapon. The weapon is designed to fire a large number of sharp tungsten bullets—called flechettes*—with a single shot that will destroy a large number of enemy soldiers. Flechettes are about the size of an average nail, with small fins at one end to stabilize them in flight. Since World War I, when France dropped them in large quantities from aircraft on masses of ground troops, munitions experts have experimented with using flechettes in a variety of guns. The problem with using flechettes as ammunition is accuracy—current weapons that fire large quantities of flechettes have unsatisfactory hit ratios when fired at long distances.*

The defense contractor (not named here for both confidentiality and security reasons) has developed a prototype gun that fires 1,100 flechettes with a single round. In range tests, three 2-feet-wide targets were set up a distance of 500 meters (approximately 1,500 feet) from the weapon. Using a number line as a reference, the centers of the three targets were at 0, 5, and 10 feet, respectively, as shown in Figure SIA4.1. The prototype gun was aimed at the middle target (center at 5 feet) and fired once. The point where each *(continued)*

Statistics IN Action *(continued)*

of the 1,100 flechettes landed at the 500-meter distance was measured using a horizontal and vertical grid. For the purposes of this application, only the horizontal measurements are considered. These 1,100 measurements are saved in the **MOAGUN** file. (The data are simulated for confidentiality reasons.) For example, a flechette with a value of $x = 5.5$ hit the middle target, but a flechette with a value of $x = 2.0$ did not hit any of the three targets (see Figure SIA4.1).

The defense contractor is interested in the likelihood of any one of the targets being hit by a flechette and, in particular, wants to set the gun specifications to maximize the number of target hits. The weapon is designed to have a mean horizontal value equal to the aim point (e.g., $\mu = 5$ feet when aimed at the center target). By changing specifications, the contractor can vary the standard deviation, σ. The **MOAGUN** file contains flechette measurements for three different range tests—one with a standard deviation of $\sigma = 1$ foot, one with $\sigma = 2$ feet, and one with $\sigma = 4$ feet.

In this chapter, three Statistics in Action Revisited examples demonstrate how we can use one of the probability models discussed in this chapter—the normal probability distribution—to aid the defense contractor in developing its "super" weapon.

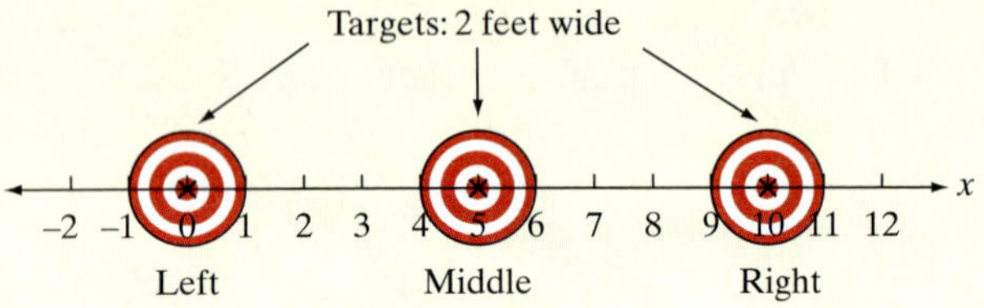

Figure SIA4.1
Target placement on gun range

Statistics IN Action Revisited

- Using the Normal Model (p. 214)
- Assessing Fit of the Normal Distribution (p. 221)
- Making an Inference about Mean Accuracy (p. 250)

Data Set: MOAGUN

You may have noticed that many of the examples of experiments in Chapter 3 generated quantitative (numerical) observations. The Consumer Price Index, the unemployment rate, the number of sales made in a week, and the yearly profit of a company are all examples of numerical measurements of some phenomenon. Thus, most experiments have sample points that correspond to values of some numerical variable.

To illustrate, consider the coin-tossing experiment of Chapter 3. Figure 4.1 is a Venn diagram showing the sample points when two coins are tossed and the up faces (heads or tails) of the coins are observed. One possible numerical outcome is the total number of heads observed. These values (0, 1, or 2) are shown in parentheses on the Venn diagram, one numerical value associated with each sample point. In the jargon of probability, the variable "total number of heads observed when two coins are tossed" is called a *random variable*.

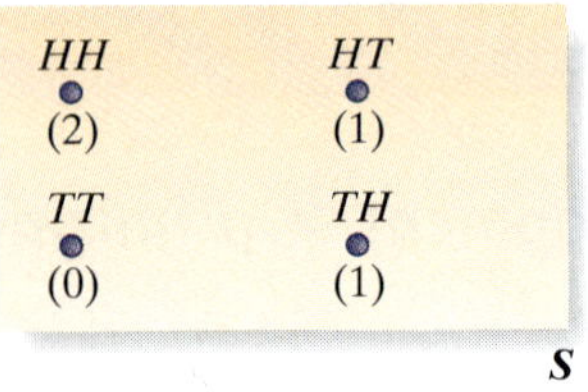

Figure 4.1
Venn diagram for coin-tossing experiment

> A **random variable** is a variable that assumes numerical values associated with the random outcomes of an experiment, where one (and only one) numerical value is assigned to each sample point.

The term *random variable* is more meaningful than the term *variable* because the adjective *random* indicates that the coin-tossing experiment may result in one of the several possible values of the variable—0, 1, and 2—according to the *random* outcome of the experiment, *HH, HT, TH,* and *TT*. Similarly, if the experiment is to count the number of customers who use the drive-up window of a bank each day, the random variable (the number of customers) will vary from day to day, partly because of random phenomena that influence whether customers use the drive-up window. Thus, the possible values of this random variable range from 0 to the maximum number of customers the window could possibly serve in a day.

We define two different types of random variables, *discrete* and *continuous,* in Section 4.1. Then we spend the remainder of this chapter discussing specific types of both discrete and continuous random variables and the aspects that make them important in business applications.

4.1 Two Types of Random Variables

Recall that the sample point probabilities corresponding to an experiment must sum to 1. Dividing one unit of probability among the sample points in a sample space and consequently assigning probabilities to the values of a random variable is not always as easy as the examples in Chapter 3 might lead you to believe. If the number of sample points can be completely listed, the job is straightforward. But if the experiment results in an infinite number of numerical sample points that are impossible to list, the task of assigning probabilities to the sample points is impossible without the aid of a probability model. The next three examples demonstrate the need for different probability models depending on the number of values that a random variable can assume.

Example 4.1

Values of a Discrete Random Variable—Wine Ratings

Problem A panel of 10 experts for the *Wine Spectator* (a national publication) is asked to taste a new white wine and assign a rating of 0, 1, 2, or 3. A score is then obtained by adding together the ratings of the 10 experts. How many values can this random variable assume?

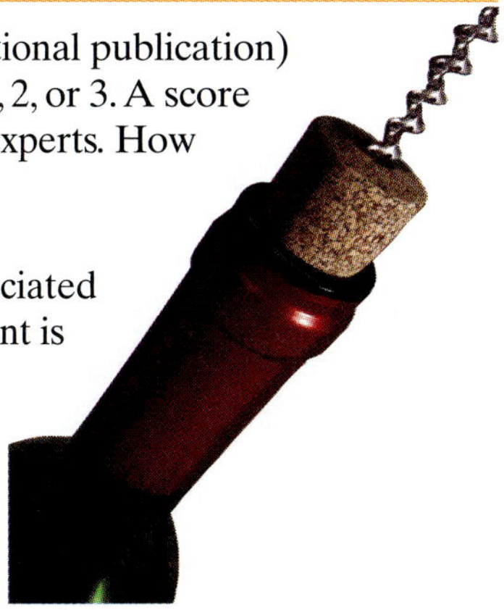

Solution A sample point is a sequence of 10 numbers associated with the rating of each expert. For example, one sample point is

$$\{1, 0, 0, 1, 2, 0, 0, 3, 1, 0\}$$

The random variable assigns a score to each one of these sample points by adding the 10 numbers together. Thus, the smallest score is 0 (if all 10 ratings are 0), and the largest score is 30 (if all 10 ratings are 3). Because every integer between 0 and 30 is a possible score, the random variable denoted by the symbol x can assume 31 values. Note that the value of the random variable for the sample point above is $x = 8$.*

Look Back This is an example of a *discrete random variable* because there is a finite number of distinct possible values. Whenever all the possible values a random variable can assume can be listed (or counted), the random variable is discrete.

Example 4.2

Values of a Discrete Random Variable—EPA Application

Problem Suppose the Environmental Protection Agency (EPA) takes readings once a month on the amount of pesticide in the discharge water of a chemical company. If the amount of pesticide exceeds the maximum level set by the EPA, the company is forced to take corrective action and may be subject to penalty. Consider the random variable, x, number of months before the company's discharge exceeds the EPA's maximum level. What values can x assume?

Solution The company's discharge of pesticide may exceed the maximum allowable level on the first month of testing, the second month of testing, and so on. It is possible that the company's discharge will *never* exceed the maximum level. Thus, the set of possible values for the number of months until the level is first exceeded is the set of all positive integers

$$1, 2, 3, 4, \dots$$

*The standard mathematical convention is to use a capital letter (e.g., X) to denote the theoretical random variable. The possible values (or realizations) of the random variable are typically denoted with a lowercase letter (e.g., x). Thus, in Example 4.1, the random variable X can take on the values $x = 0, 1, 2, \dots, 30$. Because this notation can be confusing for introductory statistics students, we simplify the notation by using the lowercase x to represent the random variable throughout.

Look Back If we can list the values of a random variable x, even though the list is never ending, we call the list **countable** and the corresponding random variable *discrete*. Thus, the number of months until the company's discharge first exceeds the limit is a *discrete random variable*.

Now Work Exercise 4.4

Example 4.3 Values of a Continuous Random Variable—Another EPA Application

Problem Refer to Example 4.2. A second random variable of interest is the amount x of pesticide (in milligrams per liter) found in the monthly sample of discharge waters from the chemical company. What values can this random variable assume?

Solution Unlike the *number* of months before the company's discharge exceeds the EPA's maximum level, the set of all possible values for the *amount* of discharge *cannot* be listed—that is, it is not countable. The possible values for the amounts of pesticide would correspond to the points on the interval between 0 and the largest possible value the amount of the discharge could attain, the maximum number of milligrams that could occupy 1 liter of volume. (Practically, the interval would be much smaller, say, between 0 and 500 milligrams per liter.)

Look Back When the values of a random variable are not countable but instead correspond to the points on some interval, we call it a *continuous random variable*. Thus, the *amount* of pesticide in the chemical plant's discharge waters is a *continuous random variable*.

Now Work Exercise 4.5

Random variables that can assume a *countable* number (finite or infinite) of values are called **discrete.**

Random variables that can assume values corresponding to any of the points contained in one or more intervals (i.e., values that are infinite and *uncountable*) are called **continuous.**

Several more examples of discrete random variables follow:

1. The number of sales made by a salesperson in a given week: $x = 0, 1, 2, \ldots$
2. The number of consumers in a sample of 500 who favor a particular product over all competitors: $x = 0, 1, 2, \ldots, 500$
3. The number of bids received in a bond offering: $x = 0, 1, 2, \ldots$
4. The number of errors on a page of an accountant's ledger: $x = 0, 1, 2, \ldots$
5. The number of customers waiting to be served in a restaurant at a particular time: $x = 0, 1, 2, \ldots$

Note that each of the examples of discrete random variables begins with the words "The number of ...." This wording is very common because the discrete random variables most frequently observed are counts.

We conclude this section with some more examples of continuous random variables:

1. The length of time between arrivals at a hospital clinic: $0 \leq x < \infty$ (infinity)
2. For a new apartment complex, the length of time from completion until a specified number of apartments are rented: $0 \leq x < \infty$
3. The amount of carbonated beverage loaded into a 12-ounce can in a can-filling operation: $0 \leq x \leq 12$
4. The depth at which a successful oil-drilling venture first strikes oil: $0 \leq x \leq c$, where c is the maximum depth obtainable

5. The weight of a food item bought in a supermarket: $0 \leq x \leq 500$ [*Note:* Theoretically, there is no upper limit on x, but it is unlikely that it would exceed 500 pounds.]

Discrete random variables and their probability distributions are discussed in Part I—Sections 4.2–4.4. Continuous random variables and their probability distributions are the topic of Part II—Sections 4.5–4.9.

Exercises 4.1–4.10

Applying the Concepts—Basic

4.1 Types of random variables. Which of the following describe continuous random variables? Which describe discrete random variables?

- **a.** The number of newspapers sold by the *New York Times* each month
- **b.** The amount of ink used in printing a Sunday edition of the *New York Times*
- **c.** The actual number of ounces in a one-gallon bottle of laundry detergent
- **d.** The number of defective parts in a shipment of nuts and bolts
- **e.** The number of people collecting unemployment insurance each month

4.2 Types of finance random variables. Security analysts are professionals who devote full-time efforts to evaluating the investment worth of a narrow list of stocks. The following variables are of interest to security analysts (Radcliffe, *Investments: Concepts, Analysis and Strategy*, 1997). Which are discrete and which are continuous random variables?

- **a.** The closing price of a particular stock on the New York Stock Exchange
- **b.** The number of shares of a particular stock that are traded each business day
- **c.** The quarterly earnings of a particular firm
- **d.** The percentage change in yearly earnings between 2008 and 2009 for a particular firm
- **e.** The number of new products introduced per year by a firm
- **f.** The time until a pharmaceutical company gains approval from the U.S. Food and Drug Administration to market a new drug

4.3 NHTSA crash tests. Refer to the National Highway Traffic Safety Administration (NHTSA) crash tests of new car models, Exercise 2.139 (p. 104). Recall that the NHTSA developed a driver-side "star" scoring system, with results ranging from one star (*) to five stars (*****). The more stars in the rating, the better the level of crash protection in a head-on collision. Suppose that a car is selected from the **CRASH** data and its driver-side star rating is determined. Let x equal the number of stars in the rating. Is x a discrete or continuous random variable?

4.4 NW **Customers in line at a Subway shop.** The number of customers, x, waiting in line to order sandwiches at a Subway shop at noon is of interest to the store manager. What values can x assume? Is x a discrete or continuous random variable?

4.5 NW **Executive Compensation Scoreboard.** Refer to *Forbes'* 2008 "Executive Compensation Scoreboard" (Table 1.1, p. 4). One variable saved in the **CEOPAY2008** file is the CEO's compensation (in $ millions) in 2008. Is x a discrete or continuous random variable?

Applying the Concepts—Intermediate

4.6 Banking. Give an example of a discrete random variable that would be of interest to a banker.

4.7 Economics. Give an example of a continuous random variable that would be of interest to an economist.

4.8 Hotel management. Give an example of a discrete random variable that would be of interest to the manager of a hotel.

4.9 Retailing. Give two examples of discrete random variables that would be of interest to the manager of a clothing store.

4.10 Stock market. Give an example of a continuous random variable that would be of interest to a stockbroker.

PART I: DISCRETE RANDOM VARIABLES

4.2 Probability Distributions for Discrete Random Variables

A complete description of a discrete random variable requires that we *specify the possible values the random variable can assume and the probability associated with each value.* To illustrate, consider Example 4.4.

Example 4.4
Finding a Probability Distribution for a Coin-Tossing Experiment

Problem Recall the experiment of tossing two coins (Section 4.1) and let x be the number of heads observed. Find the probability associated with each value of the random variable x, assuming the two coins are fair. Display these values in a table or graph.

Solution The sample space and sample points for this experiment are reproduced in Figure 4.2. Note that the random variable x can assume values 0, 1, and 2. Recall (from Chapter 3) that the probability associated with each of the four sample points is $\frac{1}{4}$.

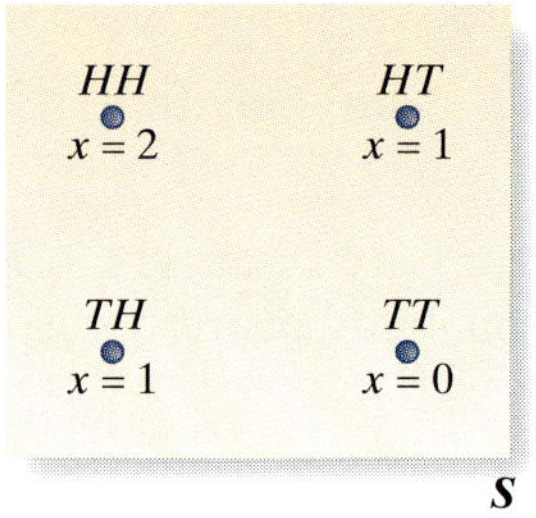

Figure 4.2
Venn diagram for the two-coin-toss experiment

Then, identifying the probabilities of the sample points associated with each of these values of x, we have

$$P(x = 0) = P(TT) = \tfrac{1}{4}$$
$$P(x = 1) = P(TH) + P(HT) = \tfrac{1}{4} + \tfrac{1}{4} = \tfrac{1}{2}$$
$$P(x = 2) = P(HH) = \tfrac{1}{4}$$

Thus, we now know the values the random variable can assume (0, 1, 2) and how the probability is *distributed over* these values (1/4, 1/2, 1/4). This completely describes the random variable and is referred to as the *probability distribution,* denoted by the symbol $p(x)$.* The probability distribution for the coin-toss example is shown in tabular form in Table 4.1 and in graphical form in Figure 4.3. Because the probability distribution for a discrete random variable is concentrated at specific points (values of x), the graph in Figure 4.3a represents the probabilities as the heights of vertical lines over the corresponding values of x. Although the representation of the probability distribution as a histogram, as in Figure 4.3b, is less precise (because the probability is spread over a unit interval), the histogram representation will prove useful when we approximate probabilities of certain discrete random variables in Section 4.4.

Table 4.1 Probability Distribution for Coin-Toss Experiment: Tabular Form

x	$p(x)$
0	1/4
1	1/2
2	1/4

Look Back We could also present the probability distribution for x as a formula, but this would unnecessarily complicate a very simple example. We give the formulas for the probability distributions of some common discrete random variables later in this chapter.

Now Work Exercise 4.15

The **probability distribution** of a **discrete random variable** is a graph, table, or formula that specifies the probability associated with each possible value the random variable can assume.

Two requirements must be satisfied by all probability distributions for discrete random variables.

Requirements for the Probability Distribution of a Discrete Random Variable, x

1. $p(x) \geq 0$ for all values of x
2. $\Sigma p(x) = 1$

where the summation of $p(x)$ is over all possible values of x.**

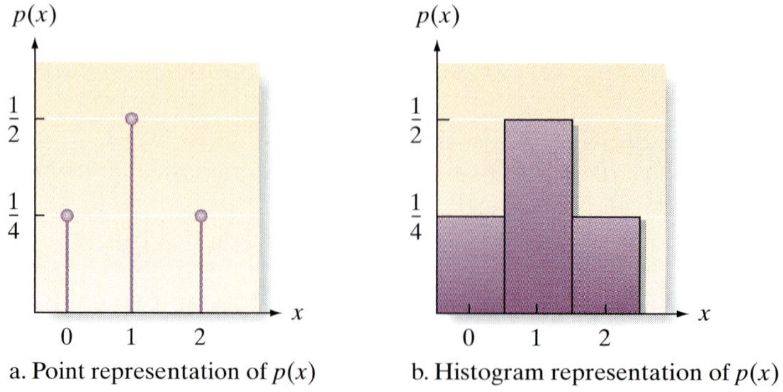

Figure 4.3
Probability distribution for coin-toss experiment: graphical form

*In standard mathematical notation, the probability that a random variable X takes on a value x is denoted $P(X = x) = p(x)$. Thus, $P(X = 0) = p(0)$, $P(X = 1) = p(1)$, etc. In this introductory text, we adopt the simpler $p(x)$ notation.
**Unless otherwise indicated, summations will always be over all possible values of x.

Example 4.4 illustrates how the probability distribution for a discrete random variable can be derived, but for many practical situations, the task is much more difficult. Fortunately, many experiments and associated discrete random variables observed in business possess identical characteristics. Thus, you might observe a random variable in a marketing experiment that would possess the same characteristics as a random variable observed in accounting, economics, or management. We classify random variables according to type of experiment, derive the probability distribution for each of the different types, and then use the appropriate probability distribution when a particular type of random variable is observed in a practical situation. The probability distributions for most commonly occurring discrete random variables have already been derived. This fact simplifies the problem of finding the appropriate probability distributions for the business analyst, as the next example illustrates.

Example 4.5
Probability Distribution for Texas Droughts Using a Formula

Problem A drought is a period of abnormal dry weather that causes serious problems in the farming industry of the region. University of Arizona researchers used historical annual data to study the severity of droughts in Texas (*Journal of Hydrologic Engineering*, Sept./Oct. 2003). The researchers showed that the distribution of x, the number of consecutive years that must be sampled until a dry (drought) year is observed, can be modeled using the formula

$$p(x) = (.3)(.7)^{x-1}, x = 1, 2, 3, \ldots$$

Find the probability that exactly 3 years must be sampled before a drought year occurs.

Solution We want to find the probability that $x = 3$. Using the formula, we have

$$p(3) = (.3)(.7)^{3-1} = (.3)(.7)^2 = (.3)(.49) = .147$$

Thus, there is about a 15% chance that exactly 3 years must be sampled before a drought year occurs in Texas.

Look Back The probability of interest can also be derived using the principles of probability developed in Chapter 3. The event of interest is $N_1N_2D_3$, where N_1 represents no drought occurs in the first sampled year, N_2 represents no drought occurs in the second sampled year, and D_3 represents a drought occurs in the third sampled year. The researchers discovered that the probability of a drought occurring in any sampled year is .3 (and, consequently, the probability of no drought occurring in any sampled year is .7). Using the multiplicative rule of probability for independent events, the probability of interest is $(.7)(.7)(.3) = .147$.

Because probability distributions are analogous to the relative frequency distributions of Chapter 2, it should be no surprise that the mean and standard deviation are useful descriptive measures.

If a discrete random variable x were observed a very large number of times, and the data generated were arranged in a relative frequency distribution, the relative frequency distribution would be indistinguishable from the probability distribution for the random variable. Thus, the probability distribution for a random variable is a theoretical model for the relative frequency distribution of a population. To the extent that the two distributions are equivalent (and we will assume they are), the probability distribution for x possesses a mean μ and a variance σ^2 that are identical to the corresponding descriptive measures for the population. How can you find the mean value for a random variable? We illustrate the procedure with an example.

Examine the probability distribution for x (the number of heads observed in the toss of two fair coins) in Figure 4.4. Try to locate the mean of the distribution intuitively. We may reason that the mean μ of this distribution is equal to 1 as follows: In a large number of experiments—say, 100,000—$^1/_4$ (or 25,000) should result in $x = 0$, $^1/_2$ (or 50,000) in $x = 1$, and $^1/_4$ (or 25,000) in $x = 2$ heads. Therefore, the average number of heads is

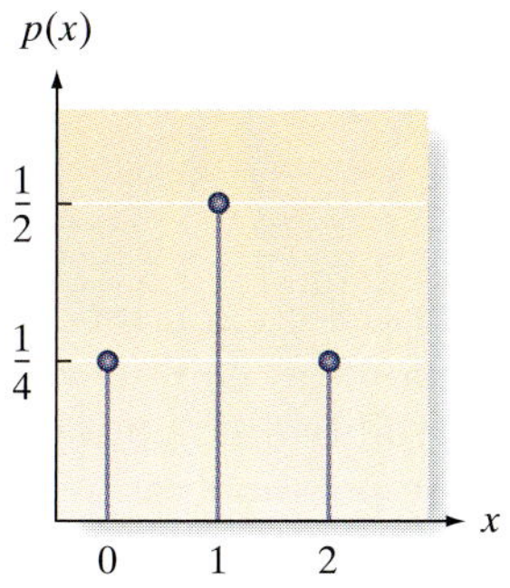

Figure 4.4
Probability distribution for a two-coin toss

$$\mu = \frac{0(25{,}000) + 1(50{,}000) + 2(25{,}000)}{100{,}000} = 0\left(\tfrac{1}{4}\right) + 1\left(\tfrac{1}{2}\right) + 2\left(\tfrac{1}{4}\right) = 0 + \tfrac{1}{2} + \tfrac{1}{2} = 1$$

Note that to get the population mean of the random variable x, we multiply each possible value of x by its probability $p(x)$ and then sum this product over all possible values of x. The *mean of x* is also referred to as the *expected value of x,* denoted $E(x)$.

> The **mean,** or **expected value,** of a discrete random variable x is
>
> $$\mu = E(x) = \sum xp(x)$$

Expected is a mathematical term and should not be interpreted as it is typically used. Specifically, a random variable might never be equal to its "expected value." Rather, the expected value is the mean of the probability distribution or a measure of its central tendency. You can think of μ as the mean value of x in a *very large* (actually, *infinite*) number of repetitions of the experiment, where the values of x occur in proportions equivalent to the probabilities of x.

Example 4.6
Finding an Expected Value—An Insurance Application

Problem Suppose you work for an insurance company, and you sell a $10,000 one-year term insurance policy at an annual premium of $290. Actuarial tables show that the probability of death during the next year for a person of your customer's age, sex, health, etc., is .001. What is the expected gain (amount of money made by the company) for a policy of this type?

Solution The experiment is to observe whether the customer survives the upcoming year. The probabilities associated with the two sample points, Live and Die, are .999 and .001, respectively. The random variable you are interested in is the gain x, which can assume the values shown in the following table.

Gain, x	Sample Point	Probability
$290	Customer lives	.999
−$9,710	Customer dies	.001

If the customer lives, the company gains the $290 premium as profit. If the customer dies, the gain is negative because the company must pay $10,000, for a net "gain" of $\$(290 - 10{,}000) = -\$9{,}710$. The expected gain is therefore

$$\mu = E(x) = \sum xp(x)$$
$$= (290)(.999) + (-9{,}710)(.001) = \$280$$

In other words, if the company were to sell a very large number of one-year $10,000 policies to customers possessing the characteristics previously described, it would (on the average) net $280 per sale in the next year.

Look Back Note that $E(x)$ need not equal a possible value of x—that is, the expected value is $280, but x will equal either $290 or −$9,710 each time the experiment is performed (a policy is sold and a year elapses). The expected value is a measure of central tendency—and in this case represents the average over a very large number of one-year policies—but is not a possible value of x.

Now Work Exercise 4.28a

We learned in Chapter 2 that the mean and other measures of central tendency tell only part of the story about a set of data. The same is true about probability distributions. We need to measure variability as well. Because a probability distribution can be viewed as a representation of a population, we will use the population variance to measure its variability.

The *population variance* σ^2 is defined as the average of the squared distance of x from the population mean μ. Because x is a random variable, the squared distance, $(x - \mu)^2$, is also a random variable. Using the same logic used to find the mean value of

x, we find the mean value of $(x - \mu)^2$ by multiplying all possible values of $(x - \mu)^2$ by $p(x)$ and then summing over all possible x values.* This quantity,

$$E[(x - \mu)^2] = \sum(x - \mu)^2 p(x)$$

is also called the *expected value of the squared distance from the mean;* that is, $\sigma^2 = E[(x - \mu)^2]$. The standard deviation of x is defined as the square root of the variance σ^2.

The **variance** of a **discrete random variable** x is

$$\sigma^2 = E[(x - \mu)^2] = \sum(x - \mu)^2 p(x)$$

The **standard deviation** of a **discrete random variable** is equal to the square root of the variance, i.e., $\sigma = \sqrt{\sigma^2}$.

Knowing the mean μ and standard deviation σ of the probability distribution of x, in conjunction with Chebyshev's Rule (Table 2.6, page 71) and the Empirical Rule (Table 2.7, page 71), we can make statements about the likelihood that values of x will fall within the intervals $\mu \pm \sigma$, $\mu \pm 2\sigma$, and $\mu \pm 3\sigma$. These probabilities are given in the box.

Probability Rules for a Discrete Random Variable

Let x be a discrete random variable with probability distribution $p(x)$, mean μ, and standard deviation σ. Then, depending on the shape of $p(x)$, the following probability statements can be made:

	Chebyshev's Rule	Empirical Rule
	Applies to any probability distribution (see Figure 4.5a)	Applies to probability distributions that are mound shaped and symmetric (see Figure 4.5b)
$P(\mu - \sigma < x < \mu + \sigma)$	≥ 0	$\approx .68$
$P(\mu - 2\sigma < x < \mu + 2\sigma)$	$\geq 3/4$	$\approx .95$
$P(\mu - 3\sigma < x < \mu + 3\sigma)$	$\geq 8/9$	≈ 1.00

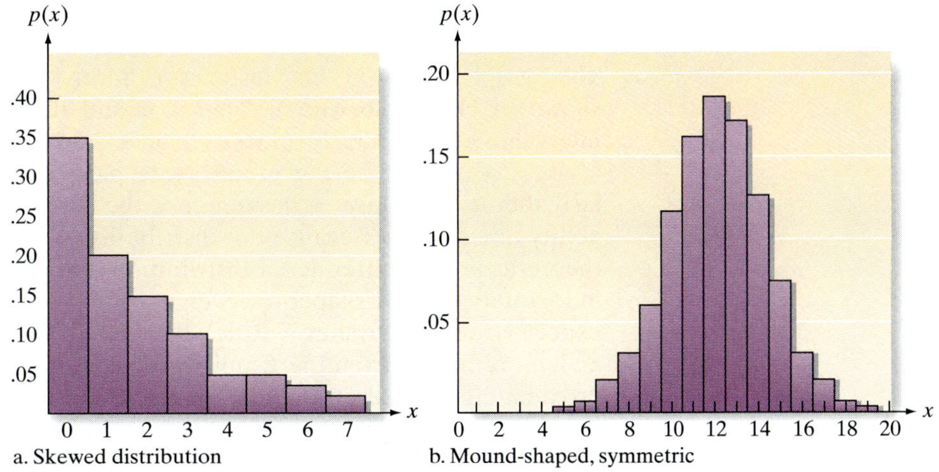

Figure 4.5
Shapes of two probability distributions for a discrete random variable x

*It can be shown that $E[(x - \mu^2)] = E(x^2) - \mu^2$, where $E(x^2) = \Sigma x^2 p(x)$. Note the similarity between this expression and the shortcut formula $\Sigma(x - \bar{x})^2 = \Sigma x^2 - (\Sigma x)^2/n$ given in Chapter 2.

Example 4.7

Finding μ and σ for an Internet Business Venture

Problem Suppose you invest a fixed sum of money in each of five Internet business ventures. Assume you know that 70% of such ventures are successful, the outcomes of the ventures are independent of one another, and the probability distribution for the number, x, of successful ventures out of five is

x	0	1	2	3	4	5
$p(x)$	.002	.029	.132	.309	.360	.168

a. Find $\mu = E(x)$. Interpret the result.

b. Find $\sigma = \sqrt{E[(x - \mu)^2]}$. Interpret the result.

c. Graph $p(x)$. Locate μ and the interval $\mu \pm 2\sigma$ on the graph. Use either Chebyshev's Rule or the Empirical Rule to approximate the probability that x falls in this interval. Compare this result with the actual probability.

d. Would you expect to observe fewer than two successful ventures out of five?

Solution

a. Applying the formula,

$$\mu = E(x) = \sum xp(x) = 0(.002) + 1(.029) + 2(.132) + 3(.309) + 4(.360) + 5(.168) = 3.50$$

On average, the number of successful ventures out of five will equal 3.5. Remember that this expected value has meaning only when the experiment—investing in five Internet business ventures—is repeated a large number of times.

b. Now we calculate the variance of x:

$$\sigma^2 = E[(x - \mu)^2] = \sum (x - \mu)^2 p(x)$$
$$= (0 - 3.5)^2(.002) + (1 - 3.5)^2(.029) + (2 - 3.5)^2(.132) + (3 - 3.5)^2(.309) + (4 - 3.5)^2(.360) + (5 - 3.5)^2(.168)$$
$$= 1.05$$

Thus, the standard deviation is

$$\sigma = \sqrt{\sigma^2} = \sqrt{1.05} = 1.02$$

This value measures the spread of the probability distribution of x, the number of successful ventures out of five. A more useful interpretation is obtained by answering parts **c** and **d.**

c. The graph of $p(x)$ in histogram form is shown in Figure 4.6 with the mean μ and the interval $\mu \pm 2\sigma = 3.50 \pm 2(1.02) = 3.50 \pm 2.04 = (1.46, 5.54)$ shown on the graph. Note particularly that $\mu = 3.5$ locates the center of the probability distribution. Because this distribution is a theoretical relative frequency distribution that is moderately mound shaped (see Figure 4.6), we expect (from Chebyshev's Rule) at least 75% and, more likely (from the Empirical Rule), approximately 95% of observed x values to fall in the interval $\mu \pm 2\sigma$—that is, between 1.46 and 5.54. You can see from Figure 4.6 that the actual probability that x falls in the interval $\mu \pm 2\sigma$ includes the sum of $p(x)$ for the values $x = 2$,

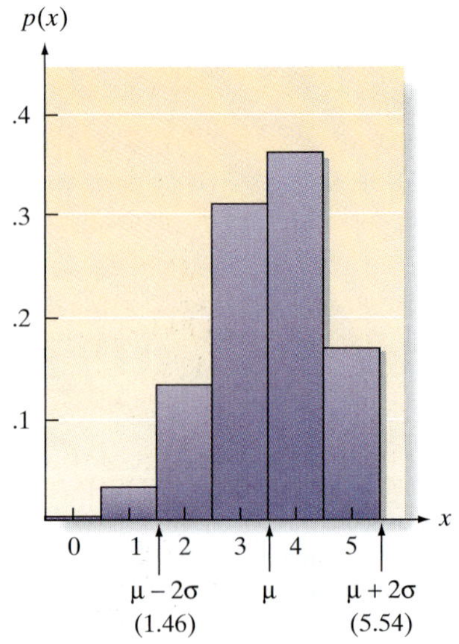

Figure 4.6
Graph of $p(x)$ for Example 4.6

$x = 3$, $x = 4$, and $x = 5$. This probability is $p(2) + p(3) + p(4) + p(5) = .132 + .309 + .360 + .168 = .969$. Therefore, 96.9% of the probability distribution lies within 2 standard deviations of the mean. This percentage is consistent with both Chebyshev's Rule and the Empirical Rule.

d. Fewer than two successful ventures out of five implies that $x = 0$ or $x = 1$. Because both these values of x lie outside the interval $\mu \pm 2\sigma$, we know from the Empirical Rule that such a result is unlikely (approximate probability of .05). The exact probability, $P(x \leq 1)$, is $p(0) + p(1) = .002 + .029 = .031$. Consequently, in a single experiment where we invest in five Internet business ventures, we would not expect to observe fewer than two successful ones.

Now Work Exercise 4.17

Exercises 4.11–4.35

Learning the Mechanics

4.11 A die is tossed. Let x be the number of spots observed on the upturned face of the die.

a. Find the probability distribution of x and display it in tabular form.

b. Display the probability distribution of x in graphical form.

4.12 The random variable x has the following discrete probability distribution:

x	1	3	5	7	9
$p(x)$	.1	.2	.4	.2	.1

a. List the values x may assume.

b. What value of x is most probable?

c. Display the probability distribution as a graph.

d. Find $P(x = 7)$.

e. Find $P(x \geq 5)$.

f. Find $P(x > 2)$.

4.13 A discrete random variable x can assume five possible values: 2, 3, 5, 8, and 10. Its probability distribution is shown here:

x	2	3	5	8	10
$p(x)$	.15	.10	–	.25	.25

a. What is $p(5)$?

b. What is the probability that x equals 2 or 10?

c. What is $P(x \leq 8)$?

4.14 Explain why each of the following is or is not a valid probability distribution for a discrete random variable x:

a.

x	0	1	2	3
$p(x)$	.1	.3	.3	.2

b.

x	−2	−1	0
$p(x)$	.25	.50	.25

c.

x	4	9	20
$p(x)$	−.3	.4	.3

d.

x	2	3	5	6
$p(x)$	.15	.15	.45	.35

4.15 NW Toss three fair coins and let x equal the number of heads observed.

a. Identify the sample points associated with this experiment and assign a value of x to each sample point.

b. Calculate $p(x)$ for each value of x.

c. Construct a graph for $p(x)$.

d. What is $P(x = 2 \text{ or } x = 3)$?

Applet Exercise 4.1

Use the applet *Random Numbers* to generate a list of 25 numbers between 1 and 3, inclusive. Let x represent a number chosen from this list.

a. What are the possible values of x?

b. Write a probability distribution for x in table form.

c. Use the probability distribution in part **b** to find the expected value of x.

d. Let y be a number randomly chosen from the set $\{1, 2, 3\}$. Write a probability distribution for y in table form and use it to find the expected value of y.

e. Compare the expected values of x and y in parts **c** and **d.** Why should these two numbers be approximately the same?

Applet Exercise 4.2

Run the applet *Simulating the Probability of a Head with a Fair Coin* ten times with $n = 2$, resetting between runs, to simulate flipping two coins ten times. Count and record the number of heads each time. Let x represent the number of heads on a single flip of the two coins.

a. What are the possible values of x?

b. Use the results of the simulation to write a probability distribution for x in table form and then use it to find the expected value of x.

c. Explain why the expected value in part **b** should be close to 1.

4.16 Consider the probability distribution for the random variable x shown here:

x	10	20	30	40	50	60
$p(x)$	.05	.20	.30	.25	.10	.10

a. Find μ, σ^2, and σ.
b. Graph $p(x)$.
c. Locate μ and the interval $\mu \pm 2\sigma$ on your graph. What is the probability that x will fall within the interval $\mu \pm 2\sigma$?

4.17 NW Consider the probability distribution shown here:

x	−4	−3	−2	−1	0	1	2	3	4
$p(x)$	.02	.07	.10	.15	.30	.18	.10	.06	.02

a. Calculate μ, σ^2, and σ.
b. Graph $p(x)$. Locate μ, $\mu - 2\sigma$, and $\mu + 2\sigma$ on the graph.
c. What is the probability that x is in the interval $\mu \pm 2\sigma$?

4.18 Consider the probability distributions shown here:

x	0	1	2
$p(x)$	.3	.4	.3

y	0	1	2
$p(y)$	.1	.8	.1

a. Use your intuition to find the mean for each distribution. How did you arrive at your choice?
b. Which distribution appears to be more variable? Why?
c. Calculate μ and σ^2 for each distribution. Compare these answers to your answers in parts **a** and **b**.

Applying the Concepts—Basic

4.19 **NHTSA crash tests.** Refer to the NHTSA crash tests of new car models, Exercise 4.3 (p. 177). A summary of the driver-side star ratings for the 98 cars in the **CRASH** file is reproduced in the accompanying Minitab printout. Assume that one of the 98 cars is selected at random and let x equal the number of stars in the car's driver-side star rating.

Tally for Discrete Variables: DRIVSTAR

```
DRIVSTAR  Count  Percent
       2      4     4.08
       3     17    17.35
       4     59    60.20
       5     18    18.37
      N=     98
```

a. Use the information in the printout to find the probability distribution for x.
b. Find $P(x = 5)$.
c. Find $P(x \le 2)$.
d. Find $\mu = E(x)$ and practically interpret the result.

4.20 **Ages of "dot-com" employees.** The age distribution for the employees of a highly successful "dot-com" company headquartered in Atlanta is shown in the next table. An employee is to be randomly selected from this population.

Age	20	21	22	23	24	25	26	27	28	29	30	31	32	33
Proportion	.02	.04	.05	.07	.04	.02	.07	.02	.11	.07	.09	.13	.15	.12

Source: Personal communication from P. George Benson.

a. Can the relative frequency distribution in the table be interpreted as a probability distribution? Explain.
b. Graph the probability distribution.
c. What is the probability that the randomly selected employee is over 30 years of age? Over 40 years of age? Under 30 years of age?
d. What is the probability that the randomly selected employee will be 25 or 26 years old?

4.21 **Customer arrivals at Wendy's.** The probability distribution for the number of customer arrivals per 15-minute period at a Wendy's in New Jersey is shown below.

x	5	6	7	8	9	10	11	12	13	14	15
$p(x)$	.01	.02	.03	.05	.08	.09	.11	.13	.12	.10	.08

x	16	17	18	19	20	21
$p(x)$	.06	.05	.03	.02	.01	.01

Source: Ford, R., Roberts, D., and Saxton. P. *Queuing Models*. Graduate School of Management, Rutgers University, 1992.

a. Does this distribution meet the two requirements for the probability distribution of a discrete random variable? Justify your answer.
b. What is the probability that exactly 16 customers enter the restaurant in the next 15 minutes?
c. Find $p(x \le 10)$.
d. Find $p(5 \le x \le 15)$.

4.22 **Choosing portable grill displays.** Refer to the *Journal of Consumer Research* (Mar. 2003) marketing study of influencing consumer choices by offering undesirable alternatives, Exercise 3.21 (p. 130). Recall that each of 124 college students selected showroom displays for portable grills. Five different displays (representing five different-sized grills) were available, but the students were instructed to select only three displays in order to maximize purchases of Grill #2 (a smaller-sized grill). The table shows the grill display combinations and the number of times each was selected by the 124 students. Suppose one of the 124 students is selected at random. Let x represent the sum of the grill numbers selected by this student. (This value is an indicator of the size of the grills selected.)
a. Find the probability distribution for x.
b. What is the probability that x exceeds 10?

Grill Display Combination	Number of Students
1-2-3	35
1-2-4	8
1-2-5	42
2-3-4	4
2-3-5	1
2-4-5	34

Source: Hamilton, R. W. "Why do people suggest what they do not want? Using context effects to influence others' choices," *Journal of Consumer Research*, Vol. 29, Mar. 2003 (Table 1). Copyright © 2003 JCR, Inc.

4.23 Solar energy cells. According to *Wired* (June 2008), 35% of the world's solar energy cells are manufactured in China. Consider a random sample of five solar energy cells and let x represent the number in the sample that are manufactured in China. In the next section, we show that the probability distribution for x is given by the formula,

$$p(x) = \frac{(5!)(.35)^x(.65)^{5-x}}{(x!)(5-x)!}, \text{ where}$$

$$n! = (n)(n-1)(n-2)\cdots(2)(1)$$

a. Explain why x is a discrete random variable.
b. Find $p(x)$ for $x = 0, 1, 2, 3, 4$, and 5.
c. Show that the properties for a discrete probability distribution are satisfied.
d. Find the probability that at least four of the five solar energy cells in the sample are manufactured in China.

Applying the Concepts—Intermediate

4.24 USDA chicken inspection. In Exercise 3.15 (p. 129) you learned that one in every 100 slaughtered chickens passes USDA inspection with fecal contamination. Consider a random sample of three slaughtered chickens that all pass USDA inspection. Let x equal the number of chickens in the sample that have fecal contamination.
a. Find $p(x)$ for $x = 0, 1, 2, 3$.
b. Graph $p(x)$.
c. Find $P(x \leq 1)$.

4.25 Contaminated gun cartridges. A weapons manufacturer uses a liquid propellant to produce gun cartridges. During the manufacturing process, the propellant can get mixed with another liquid to produce a contaminated cartridge. A University of South Florida statistician, hired by the company to investigate the level of contamination in the stored cartridges, found that 23% of the cartridges in a particular lot were contaminated. Suppose you randomly sample (without replacement) gun cartridges from this lot until you find a contaminated one. Let x be the number of cartridges sampled until a contaminated one is found. It is known that the probability distribution for x is given by the formula

$$p(x) = (.23)(.77)^{x-1}, x = 1, 2, 3, \ldots$$

a. Find $p(1)$. Interpret this result.
b. Find $p(5)$. Interpret this result.
c. Find $P(x \geq 2)$. Interpret this result.

4.26 Appeals of federal civil trials. Refer to the *Journal of the American Law and Economics Association* (Vol. 3, 2001) study of appeals of federal civil trials, Exercise 3.41 (p. 140). A breakdown of the 678 civil cases that were originally tried in front of a judge (rather than a jury) and appealed by either the plaintiff or defendant is reproduced in the next table. Suppose each civil case is awarded points (positive or negative) based on the outcome of the appeal for the purpose of evaluating federal judges. If the appeal is affirmed or dismissed, +5 points are awarded. If the appeal of a plaintiff trial win is reversed, −1 point is awarded. If the appeal of a defendant trial win is reversed, −3 points are awarded. Suppose one of the 678 cases is selected at random and the number, x, of points awarded is determined. Find and graph the probability distribution for x.

Outcome of Appeal	Number of Cases
Plaintiff trial win—reversed	71
Plaintiff trial win—affirmed/dismissed	240
Defendant trial win—reversed	68
Defendant trial win—affirmed/dismissed	299
Total	678

4.27 Mailrooms contaminated with anthrax. During autumn 2001, there was a highly publicized outbreak of anthrax cases among U.S. Postal Service workers. In *Chance* (Spring 2002), research statisticians discussed the problem of sampling mailrooms for the presence of anthrax spores. Let x equal the number of mailrooms contaminated with anthrax spores in a random sample of n mailrooms selected from a population of N mailrooms. The researchers showed that the probability distribution for x is given by the formula

$$p(x) = \frac{\binom{k}{x}\binom{N-k}{n-x}}{\binom{N}{n}}$$

where k is the number of contaminated mailrooms in the population. (In Section 4.4 we identify this probability distribution as the *hypergeometric distribution.*) Suppose $N = 100$, $n = 3$, and $k = 20$.
a. Find $p(0)$.
b. Find $p(1)$.
c. Find $p(2)$.
d. Find $p(3)$.

4.28 Investment risk analysis. The risk of a portfolio of financial assets is sometimes called *investment risk* (Radcliffe, 1997). In general, investment risk is typically measured by computing the variance or standard deviation of the probability distribution that describes the decision maker's potential outcomes (gains or losses). The greater the variation in potential outcomes, the greater the uncertainty faced by the decision maker; the smaller the variation in potential outcomes, the more predictable the decision maker's gains or losses. The two discrete probability distributions given in the next table were developed from historical data. They describe the potential total physical damage losses next year to the fleets of delivery trucks of two different firms.

Firm A		Firm B	
Loss Next Year	Probability	Loss Next Year	Probability
$ 0	.01	$ 0	.00
500	.01	200	.01
1,000	.01	700	.02
1,500	.02	1,200	.02
2,000	.35	1,700	.15
2,500	.30	2,200	.30
3,000	.25	2,700	.30
3,500	.02	3,200	.15
4,000	.01	3,700	.02
4,500	.01	4,200	.02
5,000	.01	4,700	.01

NW **a.** Verify that both firms have the same expected total physical damage loss.
b. Compute the standard deviation of each probability distribution and determine which firm faces the greater risk of physical damage to its fleet next year.

4.29 Reliability of a bridge network. The journal *Networks* periodically publishes studies on the reliability of flow networks. For example, *Networks* (Sep. 2007) gave applications in mobile ad hoc and sensor networks. Another network examined in a May issue, illustrated below, is a bridge network with arcs a_1, a_2, a_3, a_4, a_5, and a_6. The probability distribution of the capacity x for each of the six arcs is provided in the following table.

Arc	Capacity (x)	p(x)	Arc	Capacity (x)	p(x)
a_1	3	.60	a_4	1	.90
	2	.25		0	.10
	1	.10			
	0	.05			
a_2	2	.60	a_5	1	.90
	1	.30		0	.10
	0	.10			
a_3	1	.90	a_6	2	.70
	0	.10		1	.25
				0	.05

Source: Lin, J., et al. "On reliability evaluation of capacitated-flow network in terms of minimal pathsets," *Networks,* Vol. 25, No. 3, May 1995, p. 135 (Table 1), 1995, John Wiley and Sons.

a. Verify that the properties of discrete probability distributions are satisfied for each arc capacity distribution.
b. Find the probability that the capacity for arc a_1 will exceed 1.
c. Repeat part **b** for each of the remaining five arcs.
d. Compute the mean capacity for each arc and interpret its value.
e. Compute σ for each arc and interpret the results.

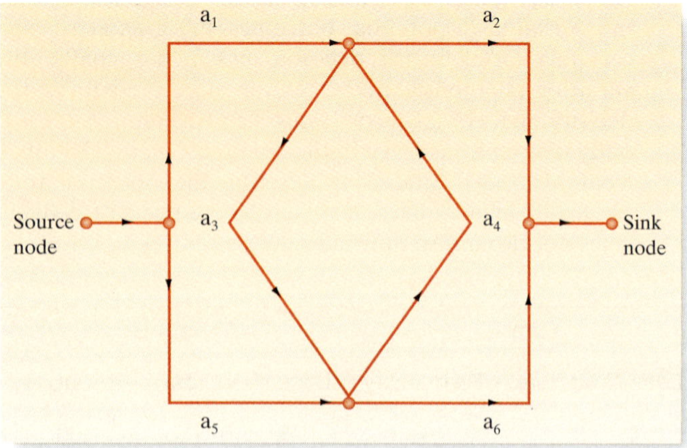

4.30 Checkout lanes at a supermarket. A team of consultants working for a large national supermarket chain based in the New York metropolitan area developed a statistical model for predicting the annual sales of potential new store locations. Part of their analysis involved identifying variables that influence store sales, such as the size of the store (in square feet), the size of the surrounding population, and the number of checkout lanes. They surveyed 52 supermarkets in a particular region of the country and constructed the relative frequency distribution shown below to describe the number of checkout lanes per store, x.

x	1	2	3	4	5	6	7	8	9	10
Relative Frequency	.01	.04	.04	.08	.10	.15	.25	.20	.08	.05

Source: Adapted from Chow, W. et. al. "A model for predicting a supermarket's annual sales per square foot," Graduate School of Management, Rutgers University, 1994.

a. Why do the relative frequencies in the table represent the approximate probabilities of a randomly selected supermarket having x number of checkout lanes?
b. Find $E(x)$ and interpret its value in the context of the problem.
c. Find the standard deviation of x.
d. According to Chebyshev's Rule (Chapter 2, p. 71), what percentage of supermarkets would be expected to fall within $\mu \pm \sigma$? Within $\mu \pm 2\sigma$?
e. What is the actual number of supermarkets that fall within $\mu \pm \sigma$? $\mu \pm 2\sigma$? Compare your answers to those of part **d.** Are the answers consistent?

4.31 Expected loss due to flood damage. The National Weather Service issues precipitation forecasts that indicate the likelihood of measurable precipitation ($\geq .01$ inch) at a specific point (the official rain gauge) during a given time period. Suppose that if a measurable amount of rain falls during the next 24 hours, a river will reach flood stage and a business will incur damages of \$300,000. The National Weather Service has indicated that there is a 30% chance of a measurable amount of rain during the next 24 hours.
a. Construct the probability distribution that describes the potential flood damages.
b. Find the firm's expected loss due to flood damage.

4.32 Expected lotto winnings. Most states offer weekly lotteries to generate revenue for the state. Despite the long odds of winning, residents continue to gamble on the lottery each week. In SIA, Chapter 3 (p. 116), you learned that the chance of winning Florida's Pick-6 Lotto game is 1 in approximately 23 million. Suppose you buy a \$1 Lotto ticket in anticipation of winning the \$7 million grand prize. Calculate your expected net winnings. Interpret the result.

4.33 The showcase showdown. On the popular television game show *The Price Is Right*, contestants can play "The Showcase Showdown." The game involves a large wheel with 20 nickel values, 5, 10, 15, 20, . . . , 95, 100, marked on it. Contestants spin the wheel once or twice, with the objective of obtaining the highest total score *without going over a dollar (100)*. [According to the *American Statistician* (Aug. 1995), the optimal strategy for the first spinner in a three-player game is to spin a second time only if the value of the initial spin is 65 or less.] Let x represent the total score for a single contestant playing "The Showcase Showdown." Assume a "fair" wheel (i.e., a wheel with equally likely outcomes). If the total of the player's spins exceeds 100, the total score is set to 0.
a. If the player is permitted only one spin of the wheel, find the probability distribution for x.
b. Refer to part **a.** Find $E(x)$ and interpret this value.
c. Refer to part **a.** Give a range of values within which x is likely to fall.

d. Suppose the player will spin the wheel twice, no matter what the outcome of the first spin. Find the probability distribution for x.

e. What assumption did you make to obtain the probability distribution, part **d**? Is it a reasonable assumption?

f. Find μ and σ for the probability distribution, part **d**, and interpret the results.

g. Refer to part **d.** What is the probability that in two spins the player's total score exceeds a dollar (i.e., is set to 0)?

h. Suppose the player obtains a 20 on the first spin and decides to spin again. Find the probability distribution for x.

i. Refer to part **h,** What is the probability that the player's total score exceeds a dollar?

j. Given the player obtains a 65 on the first spin and decides to spin again, find the probability that the player's total score exceeds a dollar.

k. Repeat part **j** for different first-spin outcomes. Use this information to suggest a strategy for the one-player game.

Applying the Concepts—Advanced

4.34 Robot-sensor system configuration. Engineers at Broadcom Corp. and Simon Fraser University collaborated on research involving a robot-sensor system in an unknown environment (*International Journal of Robotics Research,* Dec. 2004). As an example, the engineers presented the three-point, single-link robotic system shown in the accompanying figure. Each point (A, B, or C) in the physical space of the system has either an "obstacle" status or a "free" status. There are two single links in the system: A ↔ B and B ↔ C. A link has a "free" status if and only if both points in the link are "free." Otherwise, the link has an "obstacle" status. Of interest is the random variable Y, the total number of links in the system that are "free."

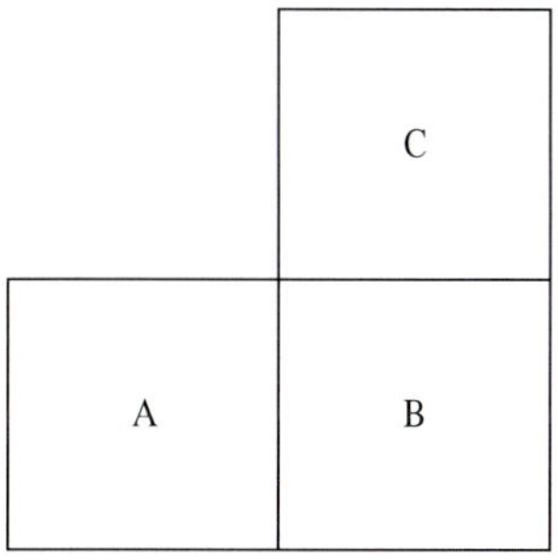

a. List the possible values of Y for the system.

b. The researchers stated that the probability of any point in the system having a "free" status is .5. Assuming the three points in the system operate independently, find the probability distribution for Y.

4.35 Parlay card betting. Odds makers try to predict which professional and college football teams will win and by how much (the *spread*). If the odds makers do this accurately, adding the spread to the underdog's score should make the final score a tie. Suppose a bookie will give you \$6 for every \$1 you risk if you pick the winners in three games (adjusted by the spread) on a "parlay" card. Thus, for every \$1 bet, you will either lose \$1 or gain \$5. What is the bookie's expected earnings per dollar wagered?

4.3 The Binomial Distribution

Many experiments result in *dichotomous* responses—that is, responses for which there exist two possible alternatives, such as Yes-No, Pass-Fail, Defective-Nondefective, or Male-Female. A simple example of such an experiment is the coin-toss experiment. A coin is tossed a number of times, say 10. Each toss results in one of two outcomes, Head or Tail. Ultimately, we are interested in the probability distribution of x, the number of heads observed. Many other experiments are equivalent to tossing a coin (either balanced or unbalanced) a fixed number n of times and observing the number x of times that one of the two possible outcomes occurs. Random variables that possess these characteristics are called **binomial random variables.**

Public opinion and consumer preference polls (e.g., the CNN, Gallup, and Harris polls) frequently yield observations on binomial random variables. For example, suppose a sample of 100 current customers is selected from a firm's database and each person is asked whether he or she prefers the firm's product (a Head) or prefers a competitor's product (a Tail). Suppose we are interested in x, the number of customers in the sample who prefer the firm's product. Sampling 100 customers is analogous to tossing the coin 100 times. Thus, you can see that consumer preference polls like the one described here are real-life equivalents of coin-toss experiments. We have been describing a **binomial experiment;** it is identified by the following characteristics.

Characteristics of a Binomial Experiment

1. The experiment consists of n identical trials.
2. There are only two possible outcomes on each trial. We will denote one outcome by S (for success) and the other by F (for failure).

3. The probability of S remains the same from trial to trial. This probability is denoted by p, and the probability of F is denoted by q. Note that $q = 1 - p$.
4. The trials are independent.
5. The binomial random variable x is the number of S's in n trials.

Example 4.8
Assessing Whether x Is Binomial

Problem For the following examples, decide whether x is a binomial random variable.

a. You randomly select 3 bonds out of a possible 10 for an investment portfolio. Unknown to you, 8 of the 10 will maintain their present value, and the other 2 will lose value due to a change in their ratings. Let x be the number of the 3 bonds you select that lose value.

b. Before marketing a new product on a large scale, many companies will conduct a consumer preference survey to determine whether the product is likely to be successful. Suppose a company develops a new diet soda and then conducts a taste preference survey in which 100 randomly chosen consumers state their preferences among the new soda and the two leading sellers. Let x be the number of the 100 who choose the new brand over the two others.

c. Some surveys are conducted by using a method of sampling other than simple random sampling (defined in Chapter 3). For example, suppose a television cable company plans to conduct a survey to determine the fraction of households in the city that would use the cable television service. The sampling method is to choose a city block at random and then survey every household on that block. This sampling technique is called *cluster sampling*. Suppose 10 blocks are so sampled, producing a total of 124 household responses. Let x be the number of the 124 households that would use the television cable service.

Solution

a. In checking the binomial characteristics in the box, a problem arises with both characteristic 3 (probabilities remaining the same from trial to trial) and characteristic 4 (independence). The probability that the first bond you pick loses value is clearly $^2/_{10}$. Now suppose the first bond you picked was 1 of the 2 that will lose value. This reduces the chance that the second bond you pick will lose value to $^1/_9$ because now only 1 of the 9 remaining bonds are in that category. Thus, the choices you make are dependent, and therefore x, the number of 3 bonds you select that lose value, is *not* a binomial random variable.

b. Surveys that produce dichotomous responses and use random sampling techniques are classic examples of binomial experiments. In our example, each randomly selected consumer either states a preference for the new diet soda or does not. The sample of 100 consumers is a very small proportion of the totality of potential consumers, so the response of one would be, for all practical purposes, independent of another.* Thus, x is a binomial random variable.

c. This example is a survey with dichotomous responses (Yes or No to the cable service), but the sampling method is not simple random sampling. Again, the binomial characteristic of independent trials would probably not be satisfied. The responses of households within a particular block would be dependent because the households within a block tend to be similar with respect to income, level of education, and general interests. Thus, the binomial model would not be satisfactory for x if the cluster sampling technique were employed.

Look Back Nonbinomial random variables with two outcomes on every trial typically occur because they do not satisfy characteristics 3 or 4 of a binomial distribution.

Now Work Exercise 4.43a

BIOGRAPHY

JACOB BERNOULLI (1654–1705)

The Bernoulli Distribution

Son of a magistrate and spice maker in Basel, Switzerland, Jacob Bernoulli completed a degree in theology at the University of Basel. While at the university, however, he studied mathematics secretly and against the will of his father. Jacob taught mathematics to his younger brother Johan, and they both went on to become distinguished European mathematicians. At first the brothers collaborated on the problems of the time (e.g., calculus); unfortunately, they later became bitter mathematical rivals. Jacob applied his philosophical training and mathematical intuition to probability and the theory of games of chance, where he developed the Law of Large Numbers. In his book *Ars Conjectandi*, published in 1713 (eight years after his death), the binomial distribution was first proposed. Jacob showed that the binomial distribution was a sum of independent 0–1 variables, now known as "Bernoulli" random variables. ■

*In most real-life applications of the binomial distribution, the population of interest has a finite number of elements (trials), denoted N. When N is large and the sample size n is small relative to N, say $n/N \le .05$, the sampling procedure, for all practical purposes, satisfies the conditions of a binomial experiment.

Example 4.9

Deriving the Binomial Probability Distribution in a PC Purchase Application

Problem A computer retailer sells both desktop and laptop personal computers (PCs) online. Assume that 80% of the PCs that the retailer sells online are desktops and 20% are laptops.

a. Use the steps given in Chapter 3 (box on page 123) to find the probability that all of the next four online PC purchases are laptops.

b. Find the probability that three of the next four online PC purchases are laptops.

c. Let x represent the number of the next four online PC purchases that are laptops. Explain why x is a binomial random variable.

d. Use the answers to parts **a** and **b** to derive a formula for $p(x)$, the probability distribution of the binomial random variable x.

Solution

a. **1.** The first step is to define the experiment. Here we are interested in observing the type of PC purchased online by each of the next four (buying) customers: desktop (D) or laptop (L).

2. Next, we list the sample points associated with the experiment. Each sample point consists of the purchase decisions made by the four online customers. For example, $DDDD$ represents the sample point that all four purchase desktop PCs, while $LDDD$ represents the sample point that customer 1 purchases a laptop, while customers 2, 3, and 4 purchase desktops. The 16 sample points are listed in Table 4.2.

3. We now assign probabilities to the sample points. Note that each sample point can be viewed as the intersection of four customers' decisions and, assuming the decisions are made independently, the probability of each sample point can be obtained using the multiplicative rule, as follows:

$$\begin{aligned} P(DDDD) &= P[(\text{customer 1 chooses desktop}) \cap (\text{customer 2 chooses desktop}) \\ &\qquad \cap (\text{customer 3 chooses desktop}) \cap (\text{customer 4 chooses desktop})] \\ &= P(\text{customer 1 chooses desktop}) \times P(\text{customer 2 chooses} \\ &\qquad \text{desktop}) \times P(\text{customer 3 chooses desktop}) \\ &\qquad \times P(\text{customer 4 chooses desktop}) \\ &= (.8)(.8)(.8)(.8) = (.8)^4 = .4096 \end{aligned}$$

All other sample point probabilities are calculated using similar reasoning. For example,

$$P(LDDD) = (.2)(.8)(.8)(.8) = (.2)(.8)^3 = .1024$$

You can check that this reasoning results in sample point probabilities that add to 1 over the 16 points in the sample space.

4. Finally, we add the appropriate sample point probabilities to obtain the desired event probability. The event of interest is that all four online customers purchase laptops. In Table 4.2 we find only one sample point, LLLL,

Table 4.2 Sample Points for PC Experiment of Example 4.9

$DDDD$	$LDDD$	$LLDD$	$DLLL$	LLLL
	$DLDD$	$LDLD$	$LDLL$	
	$DDLD$	$LDDL$	$LLDL$	
	$DDDL$	$DLLD$	$LLLD$	
		$DLDL$		
		$DDLL$		

contained in this event. All other sample points imply that at least one desktop is purchased. Thus,

$$P(\text{All four purchase laptops}) = P(LLLL) = (.2)^4 = .0016$$

That is, the probability is only 16 in 10,000 that all four customers purchase laptop PCs.

b. The event that three of the next four online buyers purchase laptops consists of the four sample points in the fourth column of Table 4.2: *DLLL, LDLL, LLDL,* and *LLLD.* To obtain the event probability, we add the sample point probabilities:

$$\begin{aligned} &P(\text{3 of next 4 customers purchase laptops}) \\ &\quad = P(DLLL) + P(LDLL) + P(LLDL) + P(LLLD) \\ &\quad = (.2)^3(.8) + (.2)^3(.8) + (.2)^3(.8) + (.2)^3(.8) \\ &\quad = 4(.2)^3(.8) = .0256 \end{aligned}$$

Note that each of the four sample point probabilities is the same because each sample point consists of three L's and one D; the order does not affect the probability because the customers' decisions are (assumed) independent.

c. We can characterize the experiment as consisting of four identical trials—the four customers' purchase decisions. There are two possible outcomes to each trial, D or L, and the probability of L, $p = .2$, is the same for each trial. Finally, we are assuming that each customer's purchase decision is independent of all others, so that the four trials are independent. Then it follows that x, the number of the next four purchases that are laptops, is a binomial random variable.

d. The event probabilities in parts **a** and **b** provide insight into the formula for the probability distribution $p(x)$. First, consider the event that three purchases are laptops (part **b**). We found that

$$\begin{aligned} P(x = 3) &= (\text{Number of sample points for which } x = 3) \\ &\quad \times (.2)^{\text{Number of laptops purchased}} \times (.8)^{\text{Number of desktops purchased}} \\ &= 4(.2)^3(.8)^1 \end{aligned}$$

In general, we can use combinatorial mathematics to count the number of sample points. For example,

Number of sample points for which $x = 3$

$$\begin{aligned} &= \text{Number of different ways of selecting 3 of the 4 trials for } L \text{ purchases} \\ &= \binom{4}{3} = \frac{4!}{3!(4-3)!} = \frac{4 \cdot 3 \cdot 2 \cdot 1}{(3 \cdot 2 \cdot 1) \cdot 1} = 4 \end{aligned}$$

The formula that works for any value of x can be deduced as follows. Because

$$P(x = 3) = \binom{4}{3}(.2)^3(.8)^1,$$

then $p(x) = \binom{4}{x}(.2)^x(.8)^{4-x}$

The component $\binom{4}{x}$ counts the number of sample points with x laptops and the component $(.2)^x(.8)^{4-x}$ is the probability associated with each sample point having x laptops. For the general binomial experiment, with n trials and probability of success p on each trial, the probability of x successes is

$$p(x) = \binom{n}{x} \cdot p^x(1-p)^{n-x}$$

Number of sample points with x S's — $\binom{n}{x}$

Probability of x S's and $(n - x)$ F's in any sample point — $p^x(1-p)^{n-x}$

Look Back In theory, you could always resort to the principles developed in this example to calculate binomial probabilities; list the sample points and sum their probabilities. However, as the number of trials (n) increases, the number of sample points grows very rapidly (the number of sample points is 2^n). Thus, we prefer the formula for calculating binomial probabilities because its use avoids listing sample points.

The binomial distribution* is summarized in the box.

The Binomial Probability Distribution

$$p(x) = \binom{n}{x} p^x q^{n-x} \ (x = 0, 1, 2, \dots, n)$$

where

$$\begin{aligned} p &= \text{Probability of a success on a single trial} \\ q &= 1 - p \\ n &= \text{Number of trials} \\ x &= \text{Number of successes in } n \text{ trials} \\ n - x &= \text{Number of failures in } n \text{ trials} \\ \binom{n}{x} &= \frac{n!}{x!(n-x)!} \end{aligned}$$

As noted in Chapter 3, the symbol 5! means $5 \cdot 4 \cdot 3 \cdot 2 \cdot 1 = 120$. Similarly, $n! = n(n-1)(n-2) \cdots 3 \cdot 2 \cdot 1$; remember, $0! = 1$.

Example 4.10
Applying the Binomial Distribution to the Manufacture of Automobiles

Problem A machine that produces stampings for automobile engines is malfunctioning and producing 10% defectives. The defective and nondefective stampings proceed from the machine in a random manner. If the next five stampings are tested, find the probability that three of them are defective.

Solution Let x equal the number of defectives in $n = 5$ trials. Then x is a binomial random variable with p, the probability that a single stamping will be defective, equal to .1, and $q = 1 - p = 1 - .1 = .9$. The probability distribution for x is given by the expression

$$\begin{aligned} p(x) &= \binom{n}{x} p^x q^{n-x} = \binom{5}{x} (.1)^x (.9)^{5-x} \\ &= \frac{5!}{x!(5-x)!} (.1)^x (.9)^{5-x} \qquad (x = 0, 1, 2, 3, 4, 5) \end{aligned}$$

To find the probability of observing $x = 3$ defectives in a sample of $n = 5$, substitute $x = 3$ into the formula for $p(x)$ to obtain

$$\begin{aligned} p(3) &= \frac{5!}{3!(5-3)!} (.1)^3 (.9)^{5-3} = \frac{5!}{3!2!} (.1)^3 (.9)^2 \\ &= \frac{5 \cdot 4 \cdot 3 \cdot 2 \cdot 1}{(3 \cdot 2 \cdot 1)(2 \cdot 1)} (.1)^3 (.9)^2 = 10(.1)^3 (.9)^2 \\ &= .0081 \end{aligned}$$

Look Back Note that the binomial formula tells us that there are 10 sample points having 3 defectives (check this by listing them), each with probability $(.1)^3(.9)^2$.

Now Work Exercise 4.39

*The binomial distribution is so named because the probabilities, $p(x)$, $x = 0, 1, \dots, n$, are terms of the binomial expansion $(q + p)^n$.

The mean, variance, and standard deviation for the binomial random variable x are shown in the box.

Mean, Variance, and Standard Deviation for a Binomial Random Variable

Mean: $\mu = np$
Variance: $\sigma^2 = npq$
Standard deviation: $\sigma = \sqrt{npq}$

As we demonstrated in Chapter 2, the mean and standard deviation provide measures of the central tendency and variability, respectively, of a distribution. Thus, we can use μ and σ to obtain a rough visualization of the probability distribution for x when the calculation of the probabilities is too tedious. The next example illustrates this idea.

Example 4.11

Finding μ and σ in the Automobile Manufacturing Application

Problem Refer to Example 4.10 and find the values of $p(0)$, $p(1)$, $p(2)$, $p(4)$, and $p(5)$. Graph $p(x)$. Calculate the mean μ and standard deviation σ. Locate μ and the interval $\mu - 2\sigma$ to $\mu + 2\sigma$ on the graph. If the experiment were to be repeated many times, what proportion of the x observations would fall within the interval $\mu - 2\sigma$ to $\mu + 2\sigma$?

Solution Again, $n = 5$, $p = .1$, and $q = .9$. Then, substituting into the formula for $p(x)$:

$$p(0) = \frac{5!}{0!(5-0)!}(.1)^0(.9)^{5-0} = \frac{5\cdot4\cdot3\cdot2\cdot1}{(1)(5\cdot4\cdot3\cdot2\cdot1)}(1)(.9)^4 = .59049$$

$$p(1) = \frac{5!}{1!(5-1)!}(.1)^1(.9)^{5-1} = 5(.1)(.9)^4 = .32805$$

$$p(2) = \frac{5!}{2!(5-2)!}(.1)^2(.9)^{5-2} = (10)(.1)^2(.9)^3 = .07290$$

$$p(3) = .0081 \text{ (from Example 4.10)}$$

$$p(4) = \frac{5!}{4!(5-4)!}(.1)^4(.9)^{5-4} = (5)(.1)^4(.9) = .00045$$

$$p(5) = \frac{5!}{5!(5-5)!}(.1)^5(.9)^{5-5} = (.1)^5 = .00001$$

The graph of $p(x)$ is shown as a probability histogram in Figure 4.7.

To calculate the values of μ and σ, substitute $n = 5$ and $p = .1$ into the following formulas:

$$\mu = np = (5)(.1) = .5$$

$$\sigma = \sqrt{npq} = \sqrt{(5)(.1)(.9)} = \sqrt{.45} = .67$$

To find the interval $\mu - 2\sigma$ to $\mu + 2\sigma$, we calculate

$$\mu - 2\sigma = .5 - 2(.67) = -.84$$

$$\mu + 2\sigma = .5 + 2(.67) = 1.84$$

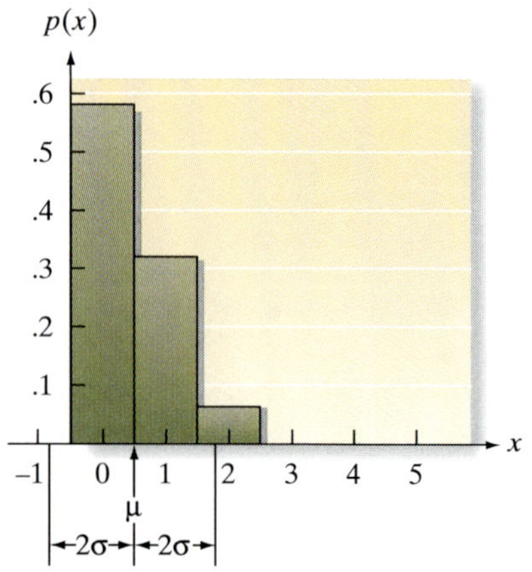

Figure 4.7
The binomial distribution: $n = 5$, $p = .1$

If the experiment was repeated a large number of times, what proportion of the x observations would fall within the interval $\mu - 2\sigma$ to $\mu + 2\sigma$? You can see from Figure 4.7 that all observations equal to 0 or 1 will fall within the interval. The probabilities corresponding to these values are .5905 and .3280, respectively. Consequently, you would expect $.5905 + .3280 = .9185$, or approximately 91.9%, of the observations to fall within the interval $\mu - 2\sigma$ to $\mu + 2\sigma$.

Look Back This result again emphasizes that for most probability distributions, observations rarely fall more than 2 standard deviations from μ.

Table 4.3 **Reproduction of Part of Table II in Appendix B: Binomial Probabilities for $n = 10$**

k \ p	.01	.05	.10	.20	.30	.40	.50	.60	.70	.80	.90	.95	.99
0	.904	.599	.349	.107	.028	.006	.001	.000	.000	.000	.000	.000	.000
1	.996	.914	.736	.376	.149	.046	.011	.002	.000	.000	.000	.000	.000
2	1.000	.988	.930	.678	.383	.167	.055	.012	.002	.000	.000	.000	.000
3	1.000	.999	.987	.879	.650	.382	.172	.055	.011	.001	.000	.000	.000
4	1.000	1.000	.998	.967	.850	.633	.377	.166	.047	.006	.000	.000	.000
5	1.000	1.000	1.000	.994	.953	.834	.623	.367	.150	.033	.002	.000	.000
6	1.000	1.000	1.000	.999	.989	.945	.828	.618	.350	.121	.013	.001	.000
7	1.000	1.000	1.000	1.000	.998	.988	.945	.833	.617	.322	.070	.012	.000
8	1.000	1.000	1.000	1.000	1.000	.998	.989	.954	.851	.624	.264	.086	.004
9	1.000	1.000	1.000	1.000	1.000	1.000	.999	.994	.972	.893	.651	.401	.096

Using Binomial Tables

Calculating binomial probabilities becomes tedious when n is large. For some values of n and p, the binomial probabilities have been tabulated in Table II in Appendix B. Part of Table II is shown in Table 4.3; a graph of the binomial probability distribution for $n = 10$ and $p = .10$ is shown in Figure 4.8. Table II actually contains a total of nine tables, labeled (a) through (i), each one corresponding to $n = 5, 6, 7, 8, 9, 10, 15, 20$, and 25. In each of these tables, the columns correspond to values of p, and the rows correspond to values (k) of the random variable x. The entries in the table represent **cumulative binomial probabilities,** $p(x \le k)$. Thus, for example, the entry in the column corresponding to $p = .10$ and the row corresponding to $k = 2$ is .930 (shaded), and its interpretation is

$$P(x \le 2) = P(x = 0) + P(x = 1) + P(x = 2) = .930$$

This probability is also shaded in the graphical representation of the binomial distribution with $n = 10$ and $p = .10$ in Figure 4.8.

You can also use Table II to find the probability that x equals a specific value. For example, suppose you want to find the probability that $x = 2$ in the binomial distribution with $n = 10$ and $p = .10$. This is found by subtraction as follows:

$$\begin{aligned} P(x = 2) &= [P(x = 0) + P(x = 1) + P(x = 2)] - [P(x = 0) + P(x = 1)] \\ &= P(x \le 2) - P(x \le 1) = .930 - .736 = .194 \end{aligned}$$

The probability that a binomial random variable exceeds a specified value can be found using Table II and the notion of complementary events. For example, to find the probability that x exceeds 2 when $n = 10$ and $p = .10$, we use

$$P(x > 2) = 1 - P(x \le 2) = .930 = .070$$

Note that this probability is represented by the unshaded portion of the graph in Figure 4.8.

All probabilities in Table II are rounded to three decimal places. Thus, although none of the binomial probabilities in the table is exactly zero, some are small enough (less than .0005) to round to .000. For example, using the formula to find $P(x = 0)$ when $n = 10$ and $p = .6$, we obtain

$$P(x = 0) = \binom{10}{0}(.6)^0(.4)^{10-0} = .4^{10} = .00010486$$

but this is rounded to .000 in Table II in Appendix B (see Table 4.3).

Similarly, none of the table entries is exactly 1.0, but when the cumulative probabilities exceed .9995, they are rounded to 1.000. The row corresponding to the largest possible value for x, $x = n$, is omitted because all the cumulative probabilities in that row are equal to 1.0 (exactly). For example, in Table 4.3, with $n = 10$, $P(x \le 10) = 1.0$, no matter what the value of p.

The following example further illustrates the use of Table II.

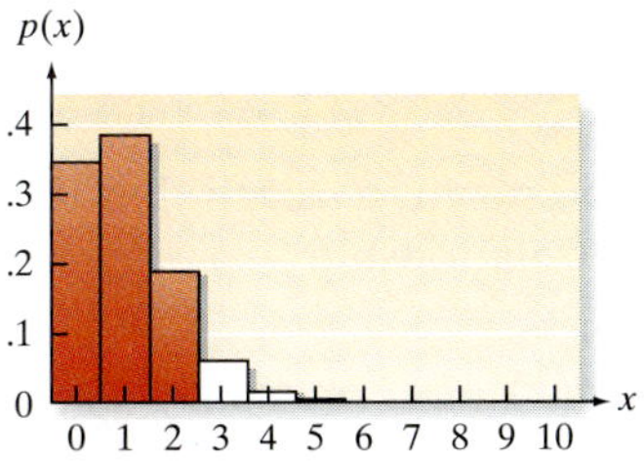

Figure 4.8
Binomial probability distribution for $n = 10$ and $p = .10$; $P(x \le 2)$ shaded

Example 4.12

Using the Binomial Table to Find Probabilities Associated with Worker Unionization

Problem Suppose a poll of 20 employees is taken in a large company. The purpose is to determine x, the number who favor unionization. Suppose that 60% of all the company's employees favor unionization.

a. Find the mean and standard deviation of x.

b. Use Table II in Appendix B to find the probability that $x \leq 10$.

c. Use Table II to find the probability that $x > 12$.

d. Use Table II to find the probability that $x = 11$.

e. Graph the probability distribution of x and locate the interval $\mu \pm 2\sigma$ on the graph.

Solution

a. The number of employees polled is presumably small compared with the total number of employees in this company. Thus, we may treat x, the number of the 20 who favor unionization, as a binomial random variable. The value of p is the fraction of the total employees who favor unionization; that is, $p = .6$. Therefore, we calculate the mean and variance:

$$\mu = np = 20(.6) = 12$$
$$\sigma^2 = npq = 20(.6)(.4) = 4.8$$
$$\sigma = \sqrt{4.8} = 2.19$$

b. Looking in the $k = 10$ row and the $p = .6$ column of Table II (Appendix B) for $n = 20$, we find the value of .245. Thus,

$$P(x \leq 10) = .245$$

c. To find the probability

$$P(x > 12) = \sum_{x=13}^{20} p(x)$$

we use the fact that for all probability distributions, $\sum_{\text{All } x} p(x) = 1$. Therefore,

$$P(x > 12) = 1 - P(x \leq 12) = 1 - \sum_{x=0}^{12} p(x)$$

Consulting Table II, we find the entry in row $k = 12$, column $p = .6$ to be .584. Thus,

$$P(x > 12) = 1 - .584 = .416$$

d. To find the probability that exactly 11 employees favor unionization, recall that the entries in Table II are cumulative probabilities and use the relationship

$$\begin{aligned} P(x = 11) &= [p(0) + p(1) + \cdots + p(11)] - [p(0) + p(1) + \cdots + p(10)] \\ &= P(x \leq 11) - P(x \leq 10) \end{aligned}$$

Then

$$P(x = 11) = .404 - .245 = .159$$

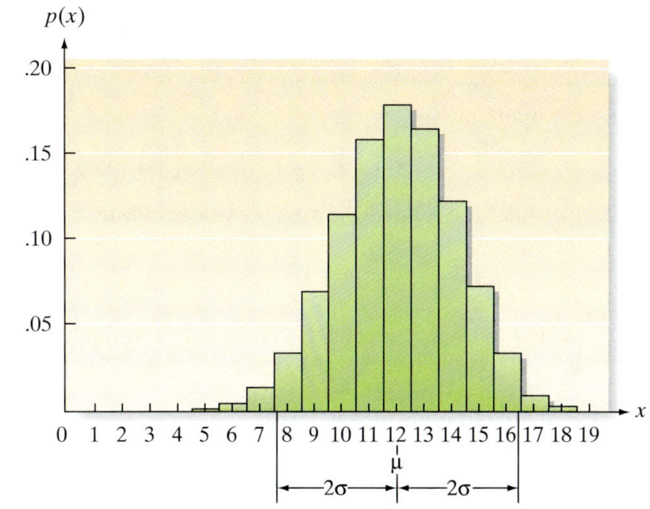

Figure 4.9
The binomial probability distribution for x in Example 4.12: $n = 20$, $p = .6$

e. The probability distribution for x in this example is shown in Figure 4.9. Note that

$$\mu - 2\sigma = 12 - 2(2.2) = 7.6$$
$$\mu + 2\sigma = 12 + 2(2.2) = 16.4$$

The interval (7.6, 16.4) is also shown in Figure 4.9. The probability that x falls in this interval is $P(x = 8, 9, 10, \ldots, 16) = P(x \le 16) - P(x \le 7) = .984 - .021 = .963$. This probability is very close to the .95 given by the Empirical Rule. Thus, we expect the number of employees in the sample of 20 who favor unionization to be between 8 and 16.

Activity 4.1 *Warehouse Club Memberships:* Exploring a Binomial Random Variable

Warehouse clubs are retailers that offer lower prices than traditional retailers, but they sell only to customers who have purchased memberships and often merchandise must be purchased in large quantities. A warehouse club may offer more than one type of membership, such as a regular membership R for a low annual fee and an upgraded membership U for a higher annual fee. The upgraded membership has additional benefits that might include extended shopping hours, additional discounts on certain products, or cash back on purchases.

A local warehouse club has determined that 20% of its customer base has the upgraded membership.

1. What is the probability $P(U)$ that a randomly chosen customer entering the store has an upgraded membership? What is the probability $P(R)$ that a randomly chosen customer entering the store has a regular (not upgraded) membership?

   In an effort to sell more upgraded memberships, sales associates are placed at the entrance to the store to explain the benefits of the upgraded membership to customers as they enter. Suppose that in a given time period five customers enter the store.

2. Given that 20% of the store's customers have an upgraded membership, how many of the five customers would you expect to have an upgraded membership?

3. Because there are five customers and each customer either has an upgraded membership (U) or does not (R), there are $2^5 = 32$ different possible combinations of membership types among the five customers. List these 32 possibilities.

4. Find the probability of each of the 32 outcomes. Assume that each of the five customers' membership type is independent of each of the other customers' membership type and use your probabilities $P(U)$ and $P(R)$ with the multiplicative rule. For example, $P(RRUUR) = P(R)P(R)P(U)P(U)P(R)$.

5. Notice that $P(URRRR) = P(RURRR) = P(RRURR) = P(RRRUR) = P(RRRRU) = P(U)^1P(R)^4$ so that P(exactly one U) $= P(URRRR) + P(RURRR) + P(RRURR) + P(RRRUR) + P(RRRRU) = 5P(U)^1P(R)^4$ where 5 is the number of ways that exactly one U can occur. Find P(exactly one U). Use similar reasoning to establish that P(no U's) $= 1P(U)^0P(R)^5$, P(exactly two U's) $= 10P(U)^2P(R)^3$, P(exactly three U's) $= 10P(U)^3P(R)^2$, P(exactly four U's) $= 5P(U)^4P(R)^1$, and P(five U's) $= 1P(U)^5P(R)^0$.

6. Let x be the number of upgraded memberships U in a sample of five customers. Use the results of Part **5** to write a probability distribution for the random variable x in table form. Find the mean and standard deviation of the distribution using the formulas in Section 4.2.

7. Calculate np and $\sqrt{npq}$ where $n = 5$, $p = P(U)$, and $q = P(R)$. How do these numbers compare to the mean and standard deviation of the random variable x in Part **6** and to the expected number of customers with upgraded memberships in Part **1**?

8. Explain how the characteristics of a binomial random variable are illustrated in this activity.

Exercises 4.36–4.56

Learning the Mechanics

4.36 Compute the following:

a. $\frac{6!}{2!(6-2)!}$ **b.** $\binom{5}{2}$ **c.** $\binom{7}{0}$ **d.** $\binom{6}{6}$ **e.** $\binom{4}{3}$

4.37 Consider the following probability distribution:

$$p(x) = \binom{5}{x}(.7)^x(.3)^{5-x} \quad (x = 0, 1, 2, \ldots, 5)$$

a. Is x a discrete or a continuous random variable?
b. What is the name of this probability distribution?
c. Graph the probability distribution.
d. Find the mean and standard deviation of x.
e. Show the mean and the 2-standard-deviation interval on each side of the mean on the graph you drew in part **c.**

4.38 Suppose x is a binomial random variable with $n = 3$ and $p = .3$.
a. Calculate the value of $p(x)$, $x = 0, 1, 2, 3$, using the formula for a binomial probability distribution.
b. Using your answers to part **a,** give the probability distribution for x in tabular form.

4.39 NW If x is a binomial random variable, compute $p(x)$ for each of the following cases:
a. $n = 5, x = 1, p = .2$
b. $n = 4, x = 2, q = .4$
c. $n = 3, x = 0, p = .7$
d. $n = 5, x = 3, p = .1$
e. $n = 4, x = 2, q = .6$
f. $n = 3, x = 1, p = .9$

4.40 If x is a binomial random variable, use Table II in Appendix B to find the following probabilities:
a. $P(x = 2)$ for $n = 10, p = .4$
b. $P(x \le 5)$ for $n = 15, p = .6$
c. $P(x > 1)$ for $n = 5, p = .1$
d. $P(x < 10)$ for $n = 25, p = .7$
e. $P(x \ge 10)$ for $n = 15, p = .9$
f. $P(x = 2)$ for $n = 20, p = .2$

4.41 If x is a binomial random variable, calculate μ, σ^2, and σ for each of the following:
a. $n = 25, p = .5$
b. $n = 80, p = .2$
c. $n = 100, p = .6$
d. $n = 70, p = .9$
e. $n = 60, p = .8$
f. $n = 1{,}000, p = .04$

4.42 The binomial probability distribution is a family of probability distributions with each single distribution depending on the values of n and p. Assume that x is a binomial random variable with $n = 4$.
a. Determine a value of p such that the probability distribution of x is symmetric.
b. Determine a value of p such that the probability distribution of x is skewed to the right.
c. Determine a value of p such that the probability distribution of x is skewed to the left.
d. Graph each of the binomial distributions you obtained in parts **a, b,** and **c.** Locate the mean for each distribution on its graph.
e. In general, for what values of p will a binomial distribution be symmetric? Skewed to the right? Skewed to the left?

Applet Exercise 4.3

Use the applets *Simulating the Probability of a Head with an Unfair Coin (P(H) = .2),* and *Simulating the Probability of a Head with an Unfair Coin (P(H) = .8)* to study the mean μ of a binomial distribution.

a. Run each applet mentioned above once with $n = 1{,}000$ and record the cumulative proportions. How does the cumulative proportion for each applet compare to the value of $P(H)$ given for the applet?
b. Using the cumulative proportion from each applet as p, compute $\mu = np$ for each applet where $n = 1{,}000$. What does the value of μ represent in terms of the results from running each applet in part **a**?
c. In your own words, describe what the mean μ of a binomial distribution represents.

Applet Exercise 4.4

Open the applet *Sample from a Population.* On the pull-down menu to the right of the top graph, select *Binary*. Set $n = 10$ as the sample size and repeatedly choose samples from the population. For each sample, record the number of 1s in the sample. Let x be the number of 1s in a sample of size 10. Explain why x is a binomial random variable.

Applet Exercise 4.5

Use the applet *Simulating the Stock Market* to estimate the probability that the stock market will go up each of the next two days. Repeatedly run the applet for $n = 2$, recording the number of up's each time. Use the proportion of 2s among your results as the estimate of the probability. Compare to the binomial probability where $x = 2, n = 2,$ and $p = 0.5$.

Applying the Concepts—Basic

4.43 **Using wireless Internet at home.** According to a survey conducted by the *Pew Internet & American Life Project* (Feb. 2007), 20% of Internet users go online at home using a wireless network. In a random sample of 100 Internet users, let x be the number who go online at home using a wireless network.
NW **a.** Explain why x is a binomial random variable (to a reasonable degree of approximation).
b. What is the value of p? Interpret this value.
c. What is the expected value of x? Interpret this value.

4.44 **Analysis of bottled water.** Is the bottled water you're drinking really purified water? A four-year study of bottled water brands conducted by the Natural Resources Defense Council found that 25% of bottled water is just tap water packaged in a bottle (*Scientific American,* July 2003). Consider a sample of five bottled water brands and let x equal the number of these brands that use tap water.
a. Explain why x is (approximately) a binomial random variable.
b. Give the probability distribution for x as a formula.
c. Find $P(x = 2)$.
d. Find $P(x \le 1)$.

4.45 **Women-owned businesses.** According to the *Journal of Business Venturing* (Vol. 17, 2002), 27% of all small businesses owned by non-Hispanic whites nationwide are women-owned firms.

a. In a sample of 200 small businesses owned by non-Hispanic whites, how many would you expect to be female owned?

b. If eight small businesses owned by non-Hispanic whites were randomly selected, what is the probability that none are female owned? That half are female owned?

4.46 Physicians' opinions on a career in medicine. Many primary care doctors feel overworked and burdened by potential lawsuits. In fact, the Physicians' Foundation reported that 60% of all general practice physicians in the United States do not recommend medicine as a career (*Reuters,* Nov. 18, 2008). Let x represent the number of sampled general practice physicians who do not recommend medicine as a career.

a. Explain why x is approximately a binomial random variable.

b. Use the Physicians' Foundation report to estimate p for the binomial random variable.

c. Consider a random sample of 25 general practice physicians. Use p from part **b** to find the mean and standard deviation of x, the number who do not recommend medicine as a career.

d. For the sample of part **c,** find the probability that at least one general practice physician does not recommend medicine as a career.

4.47 Tracking missiles with satellite imagery. The U.S. government has devoted considerable funding to missile defense research over the past 20 years. The latest development is the Space-Based Infrared System (SBIRS), which uses satellite imagery to detect and track missiles (*Chance,* Summer 2005). The probability that an intruding object (e.g., a missile) will be detected on a flight track by SBIRS is .8. Consider a sample of 20 simulated tracks, each with an intruding object. Let x equal the number of these tracks where SBIRS detects the object.

a. Demonstrate that x is (approximately) a binomial random variable.

b. Give the values of p and n for the binomial distribution.

c. Find $P(x = 15)$, the probability that SBIRS will detect the object on exactly 15 tracks.

d. Find $P(x \geq 15)$, the probability that SBIRS will detect the object on at least 15 tracks.

e. Find $E(x)$ and interpret the result.

Applying the Concepts—Intermediate

4.48 Making your vote count. Refer to the *Chance* (Fall 2007) study on making your vote count, Exercise 3.25 (p. 130). Recall the scenario where you are one of five county commissioners voting on an issue, and each commissioner is equally likely to vote for or against.

a. Your vote counts (i.e., is the decisive vote) only if the other four voters split, two in favor and two against. Use the binomial distribution to find the probability that your vote counts.

b. If you convince two other commissioners to "vote in bloc" (i.e., you all agree to vote among yourselves first, and whatever the majority decides is the way all three will vote, guaranteeing that the issue is decided by the bloc), your vote counts only if these two commissioners split their bloc votes, one in favor and one against. Again, use the binomial distribution to find the probability that your vote counts.

4.49 Bridge inspection ratings. According to the National Bridge Inspection Standard (NBIS), public bridges over 20 feet in length must be inspected and rated every 2 years. The NBIS rating scale ranges from 0 (poorest rating) to 9 (highest rating). University of Colorado engineers used a probabilistic model to forecast the inspection ratings of all major bridges in Denver (*Journal of Performance of Constructed Facilities,* Feb. 2005). For the year 2020, the engineers forecast that 9% of all major Denver bridges will have ratings of 4 or below.

a. Use the forecast to find the probability that in a random sample of 10 major Denver bridges, at least 3 will have an inspection rating of 4 or below in 2020.

b. Suppose that you actually observe 3 or more of the sample of 10 bridges with inspection ratings of 4 or below in 2020. What inference can you make? Why?

4.50 Detecting a computer virus attack. *Chance* (Winter 2004) presented basic methods for detecting virus attacks (e.g., Trojan programs or worms) on a network computer that are sent from a remote host. These viruses reach the network through requests for communication (e.g., e-mail, Web chat, or remote log-in) that are identified as "packets." For example, the "SYN flood" virus ties up the network computer by "flooding" the network with multiple packets. Cybersecurity experts can detect this type of virus attack if at least one packet is observed by a network sensor. Assume that the probability of observing a single packet sent from a new virus is only .001. If the virus actually sends 150 packets to a network computer, what is the probability that the virus is detected by the sensor?

4.51 Tax returns audited by the IRS. According to the Internal Revenue Service (IRS), the chances of your tax return being audited are about 1 in 100 if your income is less than $1 million and 9 in 100 if your income is $1 million or more (*IRS Fiscal Year 2007 Enforcement and Services Statistics*).

a. What is the probability that a taxpayer with income less than $1 million will be audited by the IRS? With income $1 million or more?

b. If five taxpayers with incomes under $1 million are randomly selected, what is the probability that exactly one will be audited? That more than one will be audited?

c. Repeat part **b** assuming that five taxpayers with incomes of $1 million or more are randomly selected.

d. If two taxpayers with incomes under $1 million are randomly selected and two with incomes more than $1 million are randomly selected, what is the probability that none of these taxpayers will be audited by the IRS?

e. What assumptions did you have to make in order to answer these questions using the methodology presented in this section?

4.52 Student gambling on sports. A study of gambling activity at the University of West Georgia (UWG) discovered that 60% of the male students wagered on sports in the past year (*The Sport Journal,* Fall 2006). Consider a random sample of 50 UWG male students. How many of these students would you expect to have gambled on sports in the past year? Give a range that is likely to include the number of male students who have gambled on sports.

4.53 FDA report on pesticides in food. Every quarter, the Food and Drug Administration (FDA) produces a report called the *Total Diet Study*. The FDA's report covers a variety of food items, each of which is analyzed for potentially harmful chemical compounds. A recent *Total Diet Study* reported that no pesticides at all were found in 70% of the domestically produced food samples (*FDA Pesticide Program: Residue Monitoring,* 2006). Consider a random sample of 800 food items analyzed for the presence of pesticides.

a. Compute μ and σ for the random variable x, the number of food items found that showed no trace of pesticide.

b. Based on a sample of 800 food items, is it likely you would observe less than half without any traces of pesticide? Explain.

Applying the Concepts—Advanced

4.54 Purchasing decision. Suppose you are a purchasing officer for a large company. You have purchased 5 million electrical switches, and your supplier has guaranteed that the shipment will contain no more than .1% defectives. To check the shipment, you randomly sample 500 switches, test them, and find that four are defective. Based on this evidence, do you think the supplier has complied with the guarantee? Explain.

4.55 USGA golf ball specifications. According to the U.S. Golf Association (USGA), "The weight of the [golf] ball shall not be greater than 1.620 ounces avoirdupois (45.93 grams). … The diameter of the ball shall not be less than 1.680 inches. … The velocity of the ball shall not be greater than 250 feet per second" (USGA 2002). The USGA periodically checks the specifications of golf balls sold in the United States by randomly sampling balls from pro shops around the country. Two dozen of each kind are sampled, and if more than three do not meet size and/or velocity requirements, that kind of ball is removed from the USGA's approved-ball list.

a. What assumptions must be made and what information must be known in order to use the binomial probability distribution to calculate the probability that the USGA will remove a particular kind of golf ball from its approved-ball list?

b. Suppose 10% of all balls produced by a particular manufacturer are less than 1.680 inches in diameter, and assume that the number of such balls, x, in a sample of two dozen balls can be adequately characterized by a binomial probability distribution. Find the mean and standard deviation of the binomial distribution.

c. Refer to part **b.** If x has a binomial distribution, then so does the number, y, of balls in the sample that meet the USGA's minimum diameter. [*Note*: $x + y = 24$.] Describe the distribution of y. In particular, what are p, q, and n? Also, find $E(y)$ and the standard deviation of y.

4.56 Reliability of a "one-shot" device. A "one-shot" device can be used only once; after use, the device (e.g., a nuclear weapon, space shuttle, automobile air bag) is either destroyed or must be rebuilt. The destructive nature of a one-shot device makes repeated testing either impractical or too costly. Hence, the reliability of such a device must be determined with minimal testing. Consider a one-shot device that has some probability, p, of failure. Of course, the true value of p is unknown, so designers will specify a value of p that is the largest defective rate they are willing to accept. Designers will conduct n tests of the device and determine the success or failure of each test. If the number of observed failures, x, is less than or equal to some specified value, K, then the device is considered to have the desired failure rate. Consequently, the designers want to know the minimum sample size n needed so that observing K or fewer defectives in the sample will demonstrate that the true probability of failure for the one-shot device is no greater than p.

a. Suppose the desired failure rate for a one-shot device is $p = .10$. Also, suppose designers will conduct $n = 20$ tests of the device and conclude that the device is performing to specifications if $K = 1$ (i.e., if 1 or no failures are observed in the sample). Find $P(x \leq 1)$.

b. In reliability analysis, $1 - P(x \leq K)$ is often called the *level of confidence* for concluding that the true failure rate is less than or equal to p. Find the level of confidence for the one-shot device described in part **a.** In your opinion, is this an acceptable level? Explain.

c. Demonstrate that the confidence level can be increased by either (1) increasing the sample size n or (2) decreasing the number K of failures allowed in the sample.

d. Typically, designers want a confidence level of .90, .95, or .99. Find the values of n and K to use so that the designers can conclude (with at least 95% confidence) that the failure rate for the one-shot device of part **a** is no greater than $p = .10$.

[*Note:* The U.S. Department of Defense Reliability Analysis Center (DoD RAC) provides designers with free access to tables and toolboxes that give the minimum sample size n required to obtain a desired confidence level for a specified number of observed failures in the sample.]

4.4 Other Discrete Distributions: Poisson and Hypergeometric

Poisson Random Variable

A type of discrete probability distribution that is often useful in describing the number of rare events that will occur in a specific period of time or in a specific area or volume is the **Poisson distribution** (named after the 18th-century physicist and mathematician, Siméon Poisson). Typical examples of random variables for which the Poisson probability distribution provides a good model are as follows:

1. The number of industrial accidents per month at a manufacturing plant
2. The number of noticeable surface defects (scratches, dents, etc.) found by quality inspectors on a new automobile

3. The parts per million of some toxin found in the water or air emission from a manufacturing plant
4. The number of customer arrivals per unit of time at a supermarket checkout counter
5. The number of death claims received per day by an insurance company
6. The number of errors per 100 invoices in the accounting records of a company

Characteristics of a Poisson Random Variable

1. The experiment consists of counting the number of times a certain event occurs during a given unit of time or in a given area or volume (or weight, distance, or any other unit of measurement).
2. The probability that an event occurs in a given unit of time, area, or volume is the same for all the units.
3. The number of events that occur in one unit of time, area, or volume is independent of the number that occur in any other mutually exclusive unit.
4. The mean (or expected) number of events in each unit is denoted by the Greek letter lambda, λ.

BIOGRAPHY

SIMÉON D. POISSON (1781–1840)

A Lifetime Mathematician

Growing up in France during the French Revolution, Siméon Denis Poisson was sent away by his father to become an apprentice surgeon, but he lacked the manual dexterity to perform the delicate procedures required and returned home. He eventually enrolled in Ecole Polytechnique University to study mathematics. In his final year of study, Poisson wrote a paper on the theory of equations that was of such quality that he was allowed to graduate without taking the final examination. Two years later, Poisson was named a professor at the university. During his illustrious career, Poisson published between 300 and 400 mathematics papers. He is most known for his 1837 paper where the distribution of a rare event—the Poisson distribution—first appears. (In fact, the distribution was actually described years earlier by one of the Bernoulli brothers.) Poisson dedicated his life to mathematics, once stating that "life is good for only two things: to study mathematics and to teach it." ■

The characteristics of the Poisson random variable are usually difficult to verify for practical examples. The examples given satisfy them well enough that the Poisson distribution provides a good model in many instances. As with all probability models, the real test of the adequacy of the Poisson model is in whether it provides a reasonable approximation to reality—that is, whether empirical data support it.

The probability distribution, mean, and variance for a Poisson random variable are shown in the next box.

Probability Distribution, Mean, and Variance for a Poisson Random Variable*

$$p(x) = \frac{\lambda^x e^{-\lambda}}{x!} \quad (x = 0, 1, 2, \ldots)$$

$$\mu = \lambda$$

$$\sigma^2 = \lambda$$

where

λ = Mean number of events during given unit of time, area, volume, etc.

$e = 2.71828\ldots$

The calculation of Poisson probabilities is made easier by the use of Table III in Appendix B, which gives the cumulative probabilities $P(x \leq k)$ for various values of λ. The use of Table III is illustrated in Example 4.13.

Example 4.13 Finding Poisson Probabilities Associated with Worker Absenteeism

Problem Suppose the number, x, of a company's employees who are absent on Mondays has (approximately) a Poisson probability distribution. Furthermore, assume that the average number of Monday absentees is 2.6.

a. Find the mean and standard deviation of x, the number of employees absent on Monday.

b. Use Table III to find the probability that fewer than two employees are absent on a given Monday.

*The Poisson probability distribution also provides a good approximation to a binomial distribution with mean $\lambda = np$ when n is large and p is small (say, $np \leq 7$).

c. Use Table III to find the probability that more than five employees are absent on a given Monday.

d. Use Table III to find the probability that exactly five employees are absent on a given Monday.

Solution

a. The mean and variance of a Poisson random variable are both equal to λ. Thus, for this example,

$$\mu = \lambda = 2.6$$
$$\sigma^2 = \lambda = 2.6$$

Then the standard deviation of x is

$$\sigma = \sqrt{2.6} = 1.61$$

Remember that the mean measures the central tendency of the distribution and does not necessarily equal a possible value of x. In this example, the mean is 2.6 absences, and although there cannot be 2.6 absences on a given Monday, the average number of Monday absences is 2.6. Similarly, the standard deviation of 1.61 measures the variability of the number of absences per week. Perhaps a more helpful measure is the interval $\mu \pm 2\sigma$, which in this case stretches from $-.62$ to 5.82. We expect the number of absences to fall in this interval most of the time—with at least 75% relative frequency (according to Chebyshev's Rule) and probably with approximately 95% relative frequency (the Empirical Rule). The mean and the 2-standard-deviation interval around it are shown in Figure 4.10.

b. A partial reproduction of Table III is shown in Table 4.4. The rows of the table correspond to different values of λ, and the columns correspond to different values (k) of the Poisson random variable x. The entries in the table (like the binomial probabilities in Table II) give the cumulative probability $P(x \leq k)$. To find the probability that fewer than two employees are absent on Monday, we first note that

$$P(x < 2) = P(x \leq 1)$$

This probability is cumulative and therefore is the entry in Table III in the row corresponding to $\lambda = 2.6$ and the column corresponding to $k = 1$. The entry is .267, shown shaded in Table 4.4. This probability corresponds to the shaded area in Figure 4.10 and may be interpreted as meaning that there is a 26.7% chance that fewer than two employees will be absent on a given Monday.

c. To find the probability that more than five employees are absent on a given Monday, we consider the complementary event

$$P(x > 5) = 1 - P(x \leq 5) = 1 - .951 = .049$$

where .951 is the entry in Table III corresponding to $\lambda = 2.6$ and $k = 5$ (see Table 4.4). Note from Figure 4.10 that this is the area in the interval $\mu \pm 2\sigma$, or $-.62$ to 5.82. Then the number of absences should exceed 5—or, equivalently, should be more than 2 standard deviations from the mean—during only about 4.9% of all Mondays. Note that this percentage agrees remarkably well with that given by the Empirical Rule for

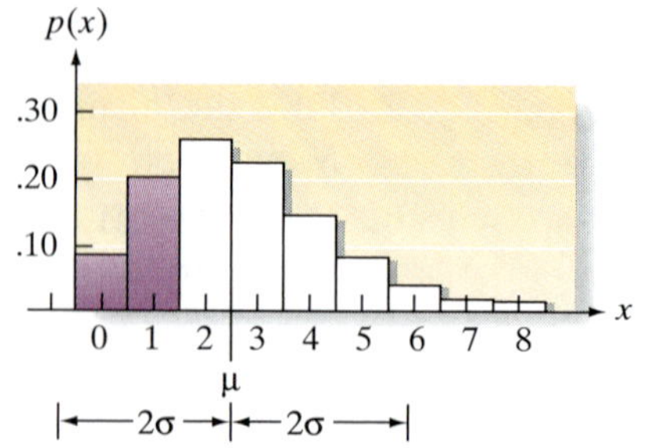

Figure 4.10
Probability distribution for number of Monday absences

Table 4.4 Reproduction of Part of Table III in Appendix B

λ \ k	0	1	2	3	4	5	6	7	8	9
2.2	.111	.355	.623	.819	.928	.975	.993	.998	1.000	1.000
2.4	.091	.308	.570	.779	.904	.964	.988	.997	.999	1.000
2.6	.074	.267	.518	.736	.877	.951	.983	.995	.999	1.000
2.8	.061	.231	.469	.692	.848	.935	.976	.992	.998	.999
3.0	.050	.199	.423	.647	.815	.916	.966	.988	.996	.999
3.2	.041	.171	.380	.603	.781	.895	.955	.983	.994	.998
3.4	.033	.147	.340	.558	.744	.871	.942	.977	.992	.997
3.6	.027	.126	.303	.515	.706	.844	.927	.969	.988	.996
3.8	.022	.107	.269	.473	.668	.816	.909	.960	.984	.994
4.0	.018	.092	.238	.433	.629	.785	.889	.949	.979	.992
4.2	.015	.078	.210	.395	.590	.753	.867	.936	.972	.989
4.4	.012	.066	.185	.359	.551	.720	.844	.921	.964	.985
4.6	.010	.056	.163	.326	.513	.686	.818	.905	.955	.980
4.8	.008	.048	.143	.294	.476	.651	.791	.887	.944	.975
5.0	.007	.040	.125	.265	.440	.616	.762	.867	.932	.968
5.2	.006	.034	.109	.238	.406	.581	.732	.845	.918	.960
5.4	.005	.029	.095	.213	.373	.546	.702	.822	.903	.951
5.6	.004	.024	.082	.191	.342	.512	.670	.797	.886	.941
5.8	.003	.021	.072	.170	.313	.478	.638	.771	.867	.929
6.0	.002	.017	.062	.151	.285	.446	.606	.744	.847	.916

mound-shaped distributions, which tells us to expect approximately 5% of the measurements (values of the random variable) to lie farther than 2 standard deviations from the mean.

d. To use Table III to find the probability that *exactly* five employees are absent on a Monday, we must write the probability as the difference between two cumulative probabilities:

$$P(x = 5) = P(x \leq 5) - P(x \leq 4) = .951 - .877 = .074$$

Now Work Exercise 4.58

Note that the probabilities in Table III are all rounded to three decimal places. Thus, although in theory a Poisson random variable can assume infinitely large values, the values of k in Table III are extended only until the cumulative probability is 1.000. This does not mean that x *cannot* assume larger values, but only that the likelihood is less than .001 (in fact, less than .0005) that it will do so.

Finally, you may need to calculate Poisson probabilities for values of λ not found in Table III. You may be able to obtain an adequate approximation by interpolation, but if not, consult more extensive tables for the Poisson distribution.

Hypergeometric Random Variable

The **hypergeometric probability distribution** provides a realistic model for some types of enumerative (countable) data. The characteristics of the hypergeometric distribution are listed in the following box:

Characteristics of a Hypergeometric Random Variable

1. The experiment consists of randomly drawing n elements without replacement from a set of N elements, r of which are S's (for success) and $(N - r)$ of which are F's (for failure).
2. The hypergeometric random variable x is the number of S's in the draw of n elements.

Note that both the hypergeometric and binomial characteristics stipulate that each draw, or trial, results in one of two outcomes. The basic difference between these random variables is that the hypergeometric trials are dependent, while the binomial trials are independent. The draws are dependent because the probability of drawing an *S* (or an *F*) is dependent on what occurred on preceding draws.

To illustrate the dependence between trials, we note that the probability of drawing an *S* on the first draw is r/N. Then the probability of drawing an *S* on the second draw depends on the outcome of the first. It will be either $(r-1)/(N-1)$ or $r/(N-1)$, depending on whether the first draw was an *S* or an *F*. Consequently, the results of the draws represent dependent events.

For example, suppose we define x as the number of women hired in a random selection of three applicants from a total of six men and four women. This random variable satisfies the characteristics of a **hypergeometric random variable** with $N = 10$ and $n = 3$. The possible outcomes on each trial are either the selection of a female (*S*) or the selection of a male (*F*). Another example of a hypergeometric random variable is the number x of defective large-screen plasma televisions in a random selection of $n = 4$ from a shipment of $N = 8$. Finally, as a third example, suppose $n = 5$ stocks are randomly selected from a list of $N = 15$ stocks. Then the number x of the five companies selected that pay regular dividends to stockholders is a hypergeometric random variable.

The hypergeometric probability distribution is summarized in the following box:

Probability Distribution, Mean, and Variance of the Hypergeometric Random Variable

$$p(x) = \frac{\binom{r}{x}\binom{N-r}{n-x}}{\binom{N}{n}} \quad [x = \text{Maximum}\ [0, n-(N-r)], \ldots, \text{Minimum}(r, n)]$$

$$\mu = \frac{nr}{N} \qquad \sigma^2 = \frac{r(N-r)n(N-n)}{N^2(N-1)}$$

where

N = Total number of elements

r = Number of *S*'s in the N elements

n = Number of elements drawn

x = Number of *S*'s drawn in the n elements

Example 4.14 Applying the Hypergeometric Distribution—Selecting Teaching Assistants

Problem Suppose a marketing professor randomly selects three new teaching assistants from a total of ten applicants—six male and four female students. Let x be the number of females who are hired.

a. Find the mean and standard deviation of x.

b. Find the probability that no females are hired.

Solution

a. Because x is a hypergeometric random variable with $N = 10$, $n = 3$, and $r = 4$, the mean and variance are

$$\mu = \frac{nr}{N} = \frac{(3)(4)}{10} = 1.2$$

$$\sigma^2 = \frac{r(N-r)n(N-n)}{N^2(N-1)} = \frac{4(10-4)3(10-3)}{(10)^2(10-1)}$$

$$= \frac{(4)(6)(3)(7)}{(100)(9)} = .56$$

The standard deviation is

$$\sigma = \sqrt{.56} = .75$$

b. The probability that no female students are hired by the professor, assuming that the selection is truly random, is

$$P(x = 0) = p(0) = \frac{\binom{4}{0}\binom{10-4}{3-0}}{\binom{10}{3}}$$

$$= \frac{\dfrac{4!}{0!(4-0)!}\dfrac{6!}{3!(6-3)!}}{\dfrac{10!}{3!(10-3)!}} = \frac{(1)(20)}{120} = \frac{1}{6}$$

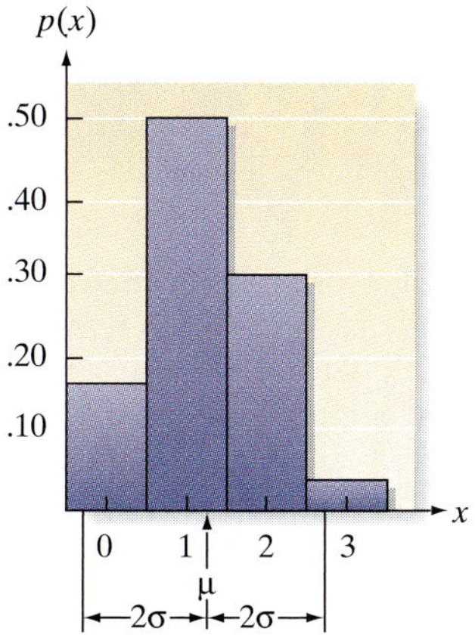

Figure 4.11
Probability distribution for x in Example 4.14

Look Back The entire probability distribution for x is shown in Figure 4.11. The mean $\mu = 1.2$ and the interval $\mu \pm 2\sigma = (-.3, 2.7)$ are indicated. You can see that if this random variable were to be observed over and over again a large number of times, most of the values of x would fall within the interval $\mu \pm 2\sigma$.

Now Work Exercise 4.73

Exercises 4.57–4.78

Learning the Mechanics

4.57 Consider the probability distribution shown here:

$$p(x) = \frac{3^x e^{-3}}{x!} \quad (x = 0, 1, 2, \dots)$$

a. Is x a discrete or continuous random variable? Explain.
b. What is the name of this probability distribution?
c. Graph the probability distribution.
d. Find the mean and standard deviation of x.

4.58 NW Assume that x is a random variable having a Poisson probability distribution with a mean of 1.5. Use Table III to find the following probabilities:
a. $P(x \le 3)$ **b.** $P(x \ge 3)$
c. $P(x = 3)$ **d.** $P(x = 0)$
e. $P(x > 0)$ **f.** $P(x > 6)$

4.59 Given that x is a hypergeometric random variable, compute $p(x)$ for each of the following cases:
a. $N = 5, n = 3, r = 3, x = 1$
b. $N = 9, n = 5, r = 3, x = 3$
c. $N = 4, n = 2, r = 2, x = 2$
d. $N = 4, n = 2, r = 2, x = 0$

4.60 Given that x is a hypergeometric random variable with $N = 8, n = 3$, and $r = 5$, compute the following:
a. $P(x = 1)$ **b.** $P(x = 0)$
c. $P(x = 3)$ **d.** $P(x \ge 4)$

4.61 Given that x is a random variable for which a Poisson probability distribution provides a good approximation, use Table III to compute the following:
a. $P(x \le 2)$ when $\lambda = 1$
b. $P(x \le 2)$ when $\lambda = 2$
c. $P(x \le 2)$ when $\lambda = 3$
d. What happens to the probability of the event $\{x \le 2\}$ as λ increases from 1 to 3? Is this intuitively reasonable?

4.62 Suppose x is a random variable for which a Poisson probability distribution with $\lambda = 5$ provides a good characterization.
a. Graph $p(x)$ for $x = 0, 1, 2, \dots, 15$.
b. Find μ and σ for x and locate μ and the interval $\mu \pm 2\sigma$ on the graph.
c. What is the probability that x will fall within the interval $\mu \pm 2\sigma$?

4.63 Suppose you plan to sample 10 items from a population of 100 items and would like to determine the probability of observing 4 defective items in the sample. Which probability distribution should you use to compute this probability under the conditions listed here? Justify your answers.
a. The sample is drawn without replacement.
b. The sample is drawn with replacement.

4.64 Given that x is a hypergeometric random variable with $N = 10, n = 5$, and $r = 7$:
a. Display the probability distribution for x in tabular form.
b. Compute the mean and variance of x.
c. Graph $p(x)$ and locate μ and the interval $\mu \pm 2\sigma$ on the graph.
d. What is the probability that x will fall within the interval $\mu \pm 2\sigma$?

Applying the Concepts—Basic

4.65 **On-site treatment of hazardous waste.** The Resource Conservation and Recovery Act mandates the tracking and disposal of hazardous waste produced at U.S. facilities. *Professional Geographer* (Feb. 2000) reported the hazardous-waste generation and disposal characteristics of 209 facilities.

Only 8 of these facilities treated hazardous waste on-site. Use the hypergeometric distribution to answer the following:

a. In a random sample of 10 of the 209 facilities, what is the expected number in the sample that treat hazardous waste on-site? Interpret this result.

b. Find the probability that 4 of the 10 selected facilities treat hazardous waste on-site.

4.66 FDIC bank failures. The Federal Deposit Insurance Corporation (FDIC) normally insures deposits of up to \$100,000 in banks that are members of the Federal Reserve System against losses due to bank failure or theft. Over the last 5 years, the average number of bank failures per year among insured banks was 3.8 (*FDIC Stats at a Glance,* Sep. 2008). Assume that x, the number of bank failures per year among insured banks, can be adequately characterized by a Poisson probability distribution with mean 4.

a. Find the expected value and standard deviation of x.

b. In 2005, no insured bank failed. How far (in standard deviations) does $x = 0$ lie below the mean of the Poisson distribution? That is, find the z-score for $x = 0$.

c. In 2002, 10 insured banks failed. Find $P(x \leq 10)$.

4.67 Airline fatalities. U.S. airlines average about 4.5 fatalities per month (*Statistical Abstract of the United States: 2008*). Assume the probability distribution for x, the number of fatalities per month, can be approximated by a Poisson probability distribution.

a. What is the probability that no fatalities will occur during any given month?

b. What is the probability that one fatality will occur during a month?

c. Find $E(x)$ and the standard deviation of x.

4.68 Male nannies. According to the International Nanny Association (INA), 4,176 nannies were placed in a job last year (www.nanny.org, 2007). Of these, only 24 were men. In Exercise 3.11 (p. 128) you found the probability that a randomly selected nanny who was placed last year is a man. Now use the hypergeometric distribution to find the probability that in a random sample of 10 nannies who were placed last year, at least 1 is a man.

4.69 Contaminated gun cartridges. Refer to the investigation of contaminated gun cartridges at a weapons manufacturer, presented in Exercise 4.25 (p. 185). In a sample of 158 cartridges from a certain lot, 36 were found to be contaminated and 122 were "clean." If you randomly select 5 of these 158 cartridges, what is the probability that all 5 will be "clean"?

4.70 Spare line replacement units. The DoD RAC publishes Selected Topics in Assurance Related Technologies (START) sheets to help improve the quality of manufactured components and systems. One START sheet, titled "Application of the Poisson Distribution" (Vol. 9, No. 1, 2002), focuses on a spare line replacement unit (LRU). Failures of LRUs are assumed to follow a Poisson distribution with a mean failure rate of 1.2 failures per 10,000 hours.

a. Find the probability that there are no LRU failures during the next 10,000 hours of operation.

b. Find the probability that there are at least two LRU failures during the next 10,000 hours of operation.

Applying the Concepts—Intermediate

4.71 LAN video conferencing. A network administrator is installing a video conferencing module to a local area network (LAN) computer system. Of interest is the capacity of the LAN to handle users who attempt to call in for video conferencing during peak hours. Calls are blocked if the user finds all LAN lines are "busy." The capacity is directly related to the rate at which calls are blocked ("Traffic Engineering Model for LAN Video Conferencing," Intel, 2005). Let x equal the number of calls blocked during the peak hour (busy) video conferencing call time. The network administrator believes that x has a Poisson distribution with mean $\lambda = 5$.

a. Find the probability that fewer than 3 calls are blocked during the peak hour.

b. Find $E(x)$ and interpret its value.

4.72 Lot inspection sampling. Imagine that you are purchasing small lots of a manufactured product. If it is very costly to test a single item, it may be desirable to test a sample of items from the lot instead of testing every item in the lot. Suppose each lot contains 10 items. You decide to sample 4 items per lot and reject the lot if you observe 1 or more defectives.

a. If the lot contains 1 defective item, what is the probability that you will accept the lot?

b. What is the probability that you will accept the lot if it contains 2 defective items?

4.73 Testing for spoiled wine. Suppose that you are purchasing cases of wine (12 bottles per case) and that, periodically, you select a test case to determine the adequacy of the bottles' seals. To do this, you randomly select and test 3 bottles in the case. If a case contains 1 spoiled bottle of wine, what is the probability that this bottle will turn up in your sample?
NW

4.74 Customer arrivals at a bakery. As part of a project targeted at improving the services of a local bakery, a management consultant (L. Lei of Rutgers University) monitored customer arrivals for several Saturdays and Sundays. Using the arrival data, she estimated the average number of customer arrivals per 10-minute period on Saturdays to be 6.2. She assumed that arrivals per 10-minute interval followed the Poisson distribution (some of whose values are missing) shown in the table below.

a. Compute the missing probabilities.

b. Plot the distribution.

c. Find μ and σ and plot the intervals $\mu \pm \sigma$, $\mu \pm 2\sigma$, and $\mu \pm 3\sigma$ on your plot of part **b.**

Probability Distribution for Exercise 4.74

x	0	1	2	3	4	5	6	7	8	9	10	11	12	13
$p(x)$	.002	.013	—	.081	.125	.155	—	.142	.110	.076	—	.026	.014	.007

Source: Lei, L. *Dorsi's Bakery: Modeling Service Operations*. Graduate School of Management, Rutgers University, 1993.

d. The owner of the bakery claims that more than 75 customers per hour enter the store on Saturdays. Based on the consultant's data, is this likely? Explain.

4.75 Making high-stakes insurance decisions. The *Journal of Economic Psychology* (Sep. 2008) published the results of a high stakes experiment where subjects were asked how much they would pay for insuring a valuable painting. The painting was threatened by fire and theft, hence, the need for insurance. To make the risk realistic, the subjects were informed that if it rained on exactly 24 days in July, the painting was considered to be stolen; if it rained on exactly 23 days in August, the painting was considered to be destroyed by fire. Although the probability of these two events, "fire" and "theft," was ambiguous for the subjects, the researchers estimated their probabilities of occurrence at .0001. Rain frequencies for the months of July and August were shown to follow a Poisson distribution with a mean of 10 days per month.

a. Find the probability that it will rain on exactly 24 days in July.

b. Find the probability that it will rain on exactly 23 days in August.

c. Are the probabilities, parts **a** and **b,** good approximations to the probabilities of "fire" and "theft"?

4.76 Reliability of incandescent lightbulbs. A large manufacturing plant has 3,200 incandescent lightbulbs illuminating the manufacturing floor. The rate at which the bulbs fail follows a Poisson distribution with a mean of 3 bulbs per hour. What is the probability that no bulbs fail in an 8-hour shift?

Applying the Concepts—Advanced

4.77 Gender discrimination suit. The *Journal of Business & Economic Statistics* (July 2000) presented a case in which a charge of gender discrimination was filed against the U.S. Postal Service. At the time, there were 302 U.S. Postal Service employees (229 men and 73 women) who applied for promotion. Of the 72 employees who were awarded promotion, 5 were female. Make an inference about whether or not females at the U.S. Postal Service were promoted fairly.

4.78 Waiting for a car wash. An automatic car wash takes exactly 5 minutes to wash a car. On average, 10 cars per hour arrive at the car wash. Suppose that 30 minutes before closing time, 5 cars are in line. If the car wash is in continuous use until closing time, is it likely anyone will be in line at closing time?

PART II: CONTINUOUS RANDOM VARIABLES

4.5 Probability Distributions for Continuous Random Variables

The graphical form of the probability distribution for a **continuous random variable** x is a smooth curve that might appear as shown in Figure 4.12. This curve, a function of x, is denoted by the symbol $f(x)$ and is variously called a **probability density function (pdf),** a **frequency function,** or a **probability distribution.**

The areas under a probability distribution correspond to probabilities for x. For example, the area A beneath the curve between the two points a and b, as shown in Figure 4.12, is the probability that x assumes a value between a and b $(a < x < b)$. Because there is no area over a point, say $x = a$, it follows that (according to our model) the probability associated with a particular value of x is equal to 0; that is, $P(x = a) = 0$ and hence $P(a < x < b) = P(a \leq x \leq b)$. In other words, the probability is the same whether or not you include the endpoints of the interval. Also, because areas over intervals represent probabilities, it follows that the total area under a probability distribution, the probability assigned to all values of x, should equal 1. Note that probability distributions for continuous random variables possess different shapes depending on the relative frequency distributions of real data that the probability distributions are supposed to model.

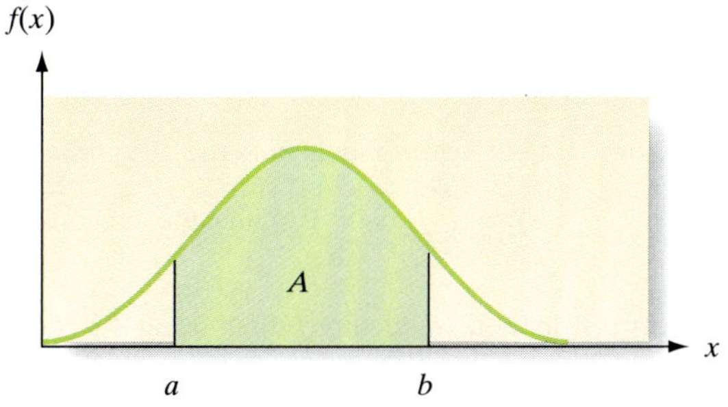

Figure 4.12
A probability distribution $f(x)$ for a continuous random variable x

The areas under most probability distributions are obtained by using calculus or numerical methods.* Because these methods often involve difficult procedures, we will give the areas for some of the most common probability distributions in tabular form in Appendix B. Then, to find the area between two values of x, say $x = a$ and $x = b$, you simply have to consult the appropriate table.

For each of the continuous random variables presented in this chapter, we will give the formula for the probability distribution along with its mean μ and standard deviation σ. These two numbers will enable you to make some approximate probability statements about a random variable even when you do not have access to a table of areas under the probability distribution.

4.6 The Normal Distribution

One of the most commonly observed continuous random variables has a **bell-shaped** probability distribution (or **bell curve**), as shown in Figure 4.13. It is known as a **normal random variable,** and its probability distribution is called a **normal distribution.**

The normal distribution plays a very important role in the science of statistical inference. Moreover, many business phenomena generate random variables with probability distributions that are very well approximated by a normal distribution. For example, the monthly rate of return for a particular stock is approximately a normal random variable, and the probability distribution for the weekly sales of a corporation might be approximated by a normal probability distribution. The normal distribution might also provide an accurate model for the distribution of scores on an employment aptitude test. You can determine the adequacy of the normal approximation to an existing population by comparing the relative frequency distribution of a large sample of the data to the normal probability distribution. Methods to detect disagreement between a set of data and the assumption of normality are presented in Section 4.7.

The normal distribution is perfectly symmetric about its mean μ, as can be seen in the examples in Figure 4.14. Its spread is determined by the value of its standard deviation σ. The formula for the normal probability distribution is showsn in the box. When plotted, this formula yields a curve like that shown in Figure 4.13.

Note that the mean μ and standard deviation σ appear in this formula, so no separate formulas for μ and σ are necessary. To graph the normal curve, we have to know the numerical values of μ and σ. Computing the area over intervals under the normal

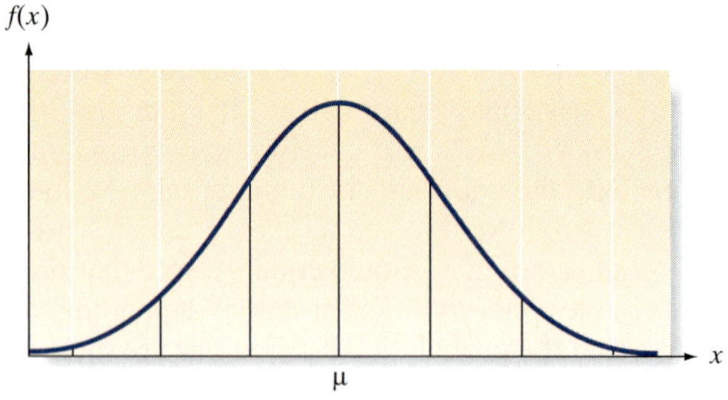

Figure 4.13
A normal probability distribution

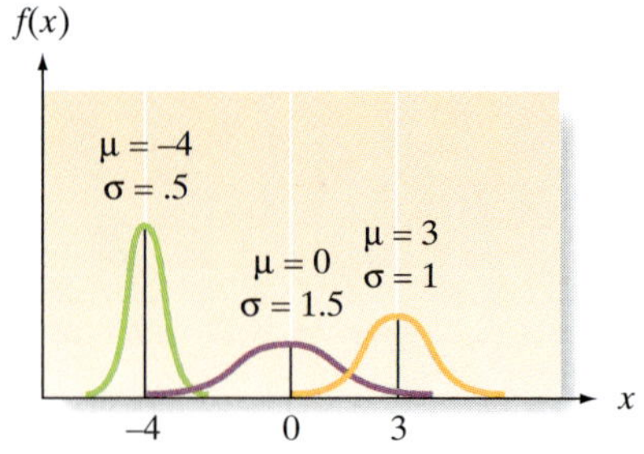

Figure 4.14
Several normal distributions with different means and standard deviations

*Students with knowledge of calculus should note that the probability that x assumes a value in the interval $a < x < b$ is $P(a < x < b) = \int_a^b f(x)dx$, assuming the integral exists. Similar to the requirement for a discrete probability distribution, we require $f(x) \geq 0$ and $\int_{-\infty}^{\infty} f(x)dx = 1$.

BIOGRAPHY

THE GAUSSIAN DISTRIBUTION (1777–1855)

Carl F. Gauss

The normal distribution began in the 18th century as a theoretical distribution for errors in disciplines where fluctuations in nature were believed to behave randomly. Although he may not have been the first to discover the formula, the normal distribution was named the Gaussian distribution after Carl Friedrich Gauss. A well-known and respected German mathematician, physicist, and astronomer, Gauss applied the normal distribution while studying the motion of planets and stars. Gauss's prowess as a mathematician was exemplified by one of his most important discoveries. At the young age of 22, Gauss constructed a regular 17-gon by ruler and compasses—a feat that was the most major advance in mathematics since the time of the ancient Greeks. In addition to publishing close to 200 scientific papers, Gauss invented the heliograph as well as a primitive telegraph device. ■

Probability Distribution for a Normal Random Variable x

Probability density function: $f(x) = \dfrac{1}{\sigma\sqrt{2\pi}} e^{-(1/2)[(x-\mu)/\sigma]^2}$

where

μ = Mean of the normal random variable x

σ = Standard deviation

$\pi = 3.1415\ldots$

$e = 2.71828\ldots$

$P(x < a)$ is obtained from a table of normal probabilities

probability distribution is a difficult task.* Consequently, we will use the computed areas listed in Table IV in Appendix B. Although there are an infinitely large number of normal curves—one for each pair of values for μ and σ—we have formed a single table that will apply to any normal curve.

Table IV is based on a normal distribution with mean $\mu = 0$ and standard deviation $\sigma = 1$, called a *standard normal distribution*. A random variable with a standard normal distribution is typically denoted by the symbol z. The formula for the probability distribution of z is given by

$$f(z) = \frac{1}{\sqrt{2\pi}} e^{-(1/2)z^2}$$

Figure 4.15 shows the graph of a standard normal distribution.

The **standard normal distribution** is a normal distribution with $\mu = 0$ and $\sigma = 1$. A random variable with a standard normal distribution, denoted by the symbol z, is called a *standard normal random variable*.

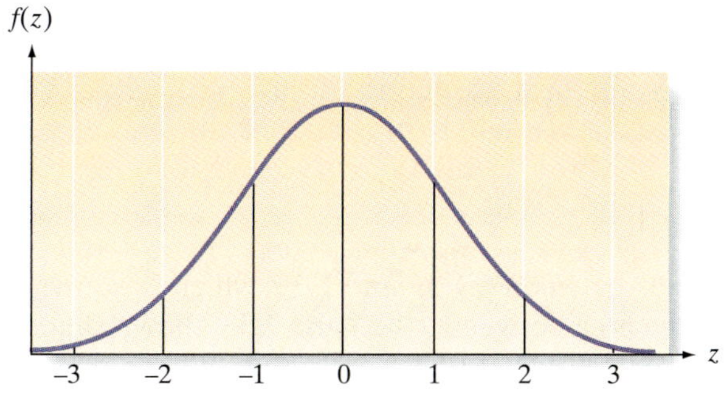

Figure 4.15
Standard normal distribution: $\mu = 0, \sigma = 1$

*The student with knowledge of calculus should note that there is not a closed-form expression for $P(a < x < b) = \int_a^b f(x)\,dx$ for the normal probability distribution. The value of this definite integral can be obtained to any desired degree of accuracy by numerical approximation procedures. For this reason, it is tabulated for the user.

Table 4.5 **Reproduction of Part of Table IV in Appendix B**

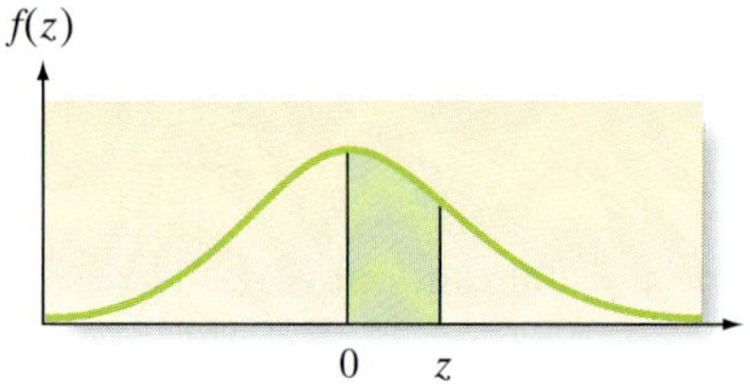

z	.00	.01	.02	.03	.04	.05	.06	.07	.08	.09
.0	.0000	.0040	.0080	.0120	.0160	.0199	.0239	.0279	.0319	.0359
.1	.0398	.0438	.0478	.0517	.0557	.0596	.0636	.0675	.0714	.0753
.2	.0793	.0832	.0871	.0910	.0948	.0987	.1026	.1064	.1103	.1141
.3	.1179	.1217	.1255	.1293	.1331	.1368	.1406	.1443	.1480	.1517
.4	.1554	.1591	.1628	.1664	.1700	.1736	.1772	.1808	.1844	.1879
.5	.1915	.1950	.1985	.2019	.2054	.2088	.2123	.2157	.2190	.2224
.6	.2257	.2291	.2324	.2357	.2389	.2422	.2454	.2486	.2517	.2549
.7	.2580	.2611	.2642	.2673	.2704	.2734	.2764	.2794	.2823	.2852
.8	.2881	.2910	.2939	.2967	.2995	.3023	.3051	.3078	.3106	.3133
.9	.3159	.3186	.3212	.3238	.3264	.3289	.3315	.3340	.3365	.3389
1.0	.3413	.3438	.3461	.3485	.3508	.3531	.3554	.3577	.3599	.3621
1.1	.3643	.3665	.3686	.3708	.3729	.3749	.3770	.3790	.3810	.3830
1.2	.3849	.3869	.3888	.3907	.3925	.3944	.3962	.3980	.3997	.4015
1.3	.4032	.4049	.4066	.4082	.4099	.4115	.4131	.4147	.4162	.4177
1.4	.4192	.4207	.4222	.4236	.4251	.4265	.4279	.4292	.4306	.4319
1.5	.4332	.4345	.4357	.4370	.4382	.4394	.4406	.4418	.4429	.4441

Because we will ultimately convert all normal random variables to standard normal in order to use Table IV to find probabilities, it is important that you learn to use Table IV well. A partial reproduction of Table IV is shown in Table 4.5. Note that the values of the standard normal random variable z are listed in the left-hand column. The entries in the body of the table give the area (probability) between 0 and z. Examples 4.15–4.18 illustrate the use of the table.

Example 4.15

Using the Standard Normal Table to Find $P(-z_0 < z < z_0)$

Problem Find the probability that the standard normal random variable z falls between -1.33 and 1.33.

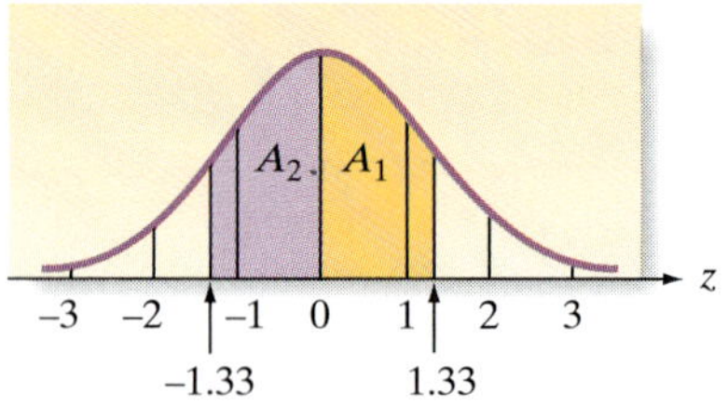

Figure 4.16
Areas under the standard normal curve for Example 4.15

Solution The standard normal distribution is shown again in Figure 4.16. Because all probabilities associated with standard normal random variables can be depicted as areas under the standard normal curve, you should always draw the curve and then equate the desired probability to an area.

In this example, we want to find the probability that z falls between -1.33 and 1.33, which is equivalent to the area between -1.33 and 1.33, shown shaded in Figure 4.16. Table IV provides the area between $z = 0$ and any value of z, so that if we look up $z = 1.33$, we find that the area between $z = 0$ and $z = 1.33$ is .4082. (The value of 1.33 and the area of .4082 are both highlighted in Table 4.5.) This is the area labeled A_1 in Figure 4.16. To find the area A_2 located between $z = 0$ and $z = -1.33$, we note that the symmetry of the normal distribution implies that the area between $z = 0$ and any point to the left is equal to the area between $z = 0$ and the point equidistant to the right. Thus, in this example, the area between $z = 0$ and $z = -1.33$ is equal to the area between $z = 0$ and $z = 1.33$. That is,

$$A_1 = A_2 = .4082$$

The probability that z falls between -1.33 and 1.33 is the sum of the areas of A_1 and A_2. We summarize in probabilistic notation:

$$P(-1.33 < z < 1.33) = P(-1.33 < z < 0) + P(0 < z < 1.33)$$
$$= A_1 + A_2 = .4082 + .4082 = .8164$$

Look Back Remember that "<" and "≤" are equivalent in events involving z because the inclusion (or exclusion) of a single point does not alter the probability of an event involving a continuous random variable.

Now Work Exercise 4.81

Example 4.16

Using the Standard Normal Table to Find $P(z > z_0)$

Problem Find the probability that a standard normal random variable exceeds 1.64; that is, find $P(z > 1.64)$.

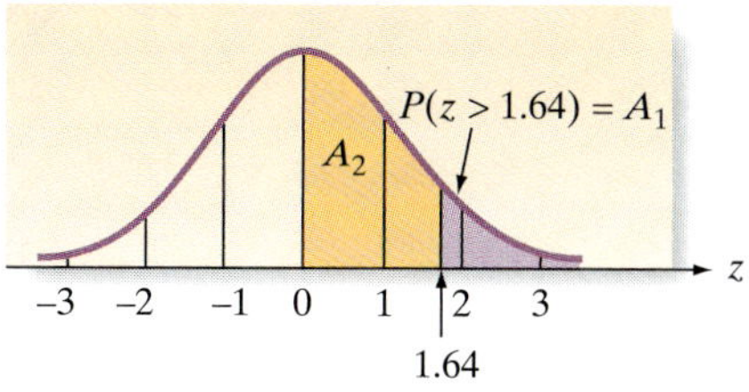

Figure 4.17
Areas under the standard normal curve for Example 4.16

Solution The area under the standard normal distribution to the right of 1.64 is the shaded area labeled A_1 in Figure 4.17. This area represents the desired probability that z exceeds 1.64. However, when we look up $z = 1.64$ in Table IV, we must remember that the probability given in the table corresponds to the area between $z = 0$ and $z = 1.64$ (the area labeled A_2 in Figure 4.17). From Table IV, we find that $A_2 = .4495$. To find the area A_1 to the right of 1.64, we make use of two facts:

1. The standard normal distribution is symmetric about its mean, $z = 0$.
2. The total area under the standard normal probability distribution equals 1.

Taken together, these two facts imply that the areas on either side of the mean $z = 0$ equal .5; thus, the area to the right of $z = 0$ in Figure 4.17 is $A_1 + A_2 = .5$. Then

$$P(z > 1.64) = A_1 = .5 - A_2 = .5 - .4495 = .0505$$

Look Back To attach some practical significance to this probability, note that the implication is that the chance of a standard normal random variable exceeding 1.64 is approximately .05.

Now Work Exercise 4.79a

Example 4.17

Using the Standard Normal Table to Find $P(z < z_0)$

Problem Find the probability that a standard normal random variable lies to the left of .67.

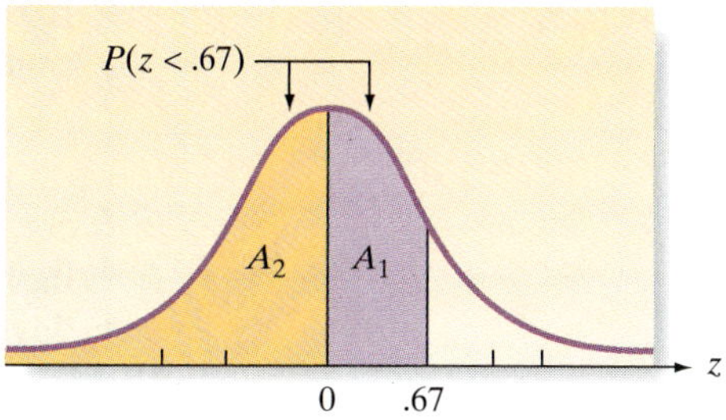

Figure 4.18
Areas under the standard normal curve for Example 4.17

Solution The event is shown as the highlighted area in Figure 4.18. We want to find $P(z < .67)$. We divide the highlighted area into two parts: the area A_1 between $z = 0$ and $z = .67$ and the area A_2 to the left of $z = 0$. We must always make such a division when the desired area lies on both sides of the mean ($z = 0$) because Table IV contains areas between $z = 0$ and the point you look up. We look up $z = .67$ in Table IV to find that $A_1 = .2486$. The symmetry of the standard normal distribution also implies that half the distribution lies on each side of the mean, so the area A_2 to the left of $z = 0$ is .5. Then

$$P(z < .67) = A_1 + A_2 = .2486 + .5 = .7486$$

Look Back Note that this probability is approximately .75. Thus, about 75% of the time the standard normal random variable z will fall below .67. This implies that $z = .67$ represents the approximate 75th percentile for the distribution.

Now Work Exercise 4.82f

Example 4.18

Using the Standard Normal Table to Find $P(|z| > z_0)$

Problem Find the probability that a standard normal random variable exceeds 1.96 in absolute value.

Solution We want to find

$$P(|z| > 1.96) = P(z < -1.96 \text{ or } z > 1.96)$$

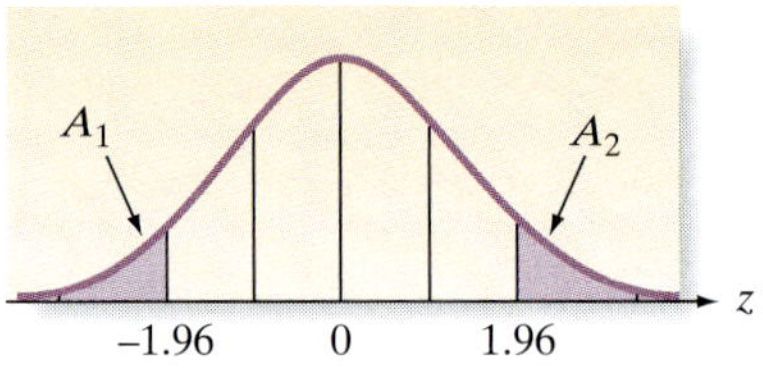

Figure 4.19 Areas under the standard normal curve for Example 4.18

This probability is the shaded area in Figure 4.19. Note that the total shaded area is the sum of two areas, A_1 and A_2—areas that are equal because of the symmetry of the normal distribution.

We look up $z = 1.96$ and find the area between $z = 0$ and $z = 1.96$ to be .4750. Then the area to the right of 1.96, A_2, is $.5 - .4750 = .0250$, so that

$$P(|z| > 1.96) = A_1 + A_2 = .0250 + .0250 = .05$$

Look Back We emphasize, again, the importance of drawing the standard normal curve when finding normal probabilities.

To apply Table IV to a normal random variable x with any mean μ and any standard deviation σ, we must first convert the value of x to a z-score. The population z-score for a measurement was defined (in Section 2.7) as the *distance* between the measurement and the population mean, divided by the population standard deviation. Thus, the z-score gives the distance between a measurement and the mean in units equal to the standard deviation. In symbolic form, the z-score for the measurement x is

$$z = \frac{x - \mu}{\sigma}$$

Note that when $x = \mu$, we obtain $z = 0$.

An important property of the normal distribution is that if x is normally distributed with any mean and any standard deviation, z is always normally distributed with mean 0 and standard deviation 1. That is, z is a standard normal random variable.

Property of Normal Distributions

If x is a normal random variable with mean μ and standard deviation σ, then the random variable z, defined by the formula

$$z = \frac{x - \mu}{\sigma}$$

has a standard normal distribution. The value z describes the number of standard deviations between x and μ.

Recall from Example 4.18 that $P(|z| > 1.96) = .05$. This probability coupled with our interpretation of z implies that any normal random variable lies more than 1.96 standard deviations from its mean only 5% of the time. Compare this to the Empirical Rule (Chapter 2), which tells us that about 5% of the measurements in mound-shaped distributions will lie beyond 2 standard deviations from the mean. The normal distribution actually provides the model on which the Empirical Rule is based, along with much "empirical" experience with real data that often approximately obey the rule, whether drawn from a normal distribution or not.

Example 4.19

Finding the Probability of a Normal Random Variable—Cell Phone Application

Problem Assume that the length of time, x, between charges of a cellular phone is normally distributed with a mean of 10 hours and a standard deviation of 1.5 hours. Find the probability that the cell phone will last between 8 and 12 hours between charges.

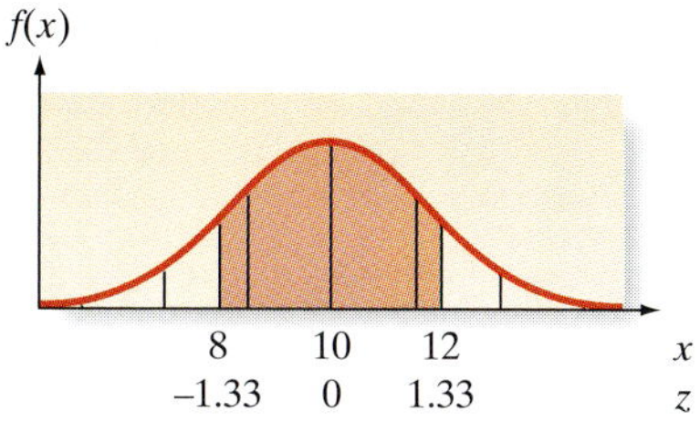

Figure 4.20
Areas under the normal curve for Example 4.19

Solution The normal distribution with mean $\mu = 10$ and $\sigma = 1.5$ is shown in Figure 4.20. The desired probability that the charge lasts between 8 and 12 hours is shaded. In order to find the probability, we must first convert the distribution to standard normal, which we do by calculating the z-score:

$$z = \frac{x - \mu}{\sigma}$$

The z-scores corresponding to the important values of x are shown beneath the x values on the horizontal axis in Figure 4.20. Note that $z = 0$ corresponds to the mean of $\mu = 10$ hours, whereas the x values 8 and 12 yield z-scores of -1.33 and 1.33, respectively. Thus, the event that the cell phone charge lasts between 8 and 12 hours is equivalent to the event that a standard normal random variable lies between -1.33 and 1.33. We found this probability in Example 4.15 (see Figure 4.16) by doubling the area corresponding to $z = 1.33$ in Table IV. That is,

$$P(8 \le x \le 12) = P(-1.33 \le z \le 1.33) = 2(.4082) = .8164$$

Now Work Exercise 4.92a–c

The steps to follow when calculating a probability corresponding to a normal random variable are shown in the box.

Steps for Finding a Probability Corresponding to a Normal Random Variable

1. Sketch the normal distribution and indicate the mean of the random variable x. Then shade the area corresponding to the probability you want to find.
2. Convert the boundaries of the shaded area from x values to standard normal random variable z values using the formula

$$z = \frac{x - \mu}{\sigma}$$

   Show the z values under the corresponding x values on your sketch.
3. Use Table IV in Appendix B to find the areas corresponding to the z values. If necessary, use the symmetry of the normal distribution to find areas corresponding to negative z values and the fact that the total area on each side of the mean equals .5 to convert the areas from Table IV to the probabilities of the event you have shaded.

Example 4.20

Using Normal Probabilities to Make an Inference about Advertised Gas Mileage

Problem Suppose an automobile manufacturer introduces a new model that has an advertised mean in-city mileage of 27 miles per gallon. Although such advertisements seldom report any measure of variability, suppose you write the manufacturer for the details of the tests, and you find that the standard deviation is 3 miles per gallon. This information leads you to formulate a probability model for the random variable x, the in-city mileage for this car model. You believe that the probability distribution of x can be approximated by a normal distribution with a mean of 27 and a standard deviation of 3.

a. If you were to buy this model of automobile, what is the probability that you would purchase one that averages less than 20 miles per gallon for in-city driving? In other words, find $P(x < 20)$.

b. Suppose you purchase one of these new models and it does get less than 20 miles per gallon for in-city driving. Should you conclude that your probability model is incorrect?

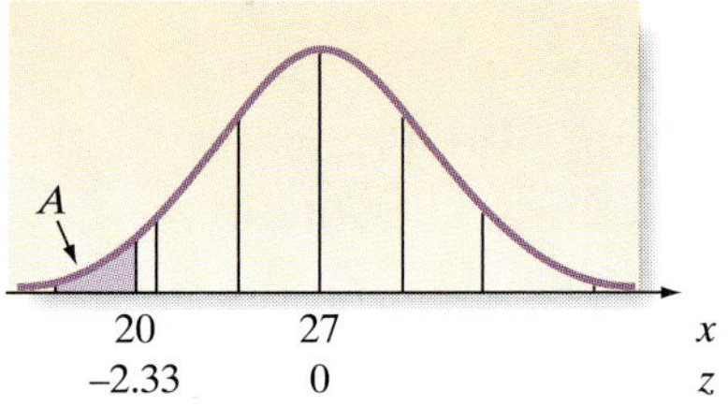

Figure 4.21
Areas under the normal curve for Example 4.20

Solution

a. The probability model proposed for x, the in-city mileage, is shown in Figure 4.21. We are interested in finding the area A to the left of 20 because this area corresponds to the probability that a measurement chosen from this distribution falls below 20. In other words, if this model is correct, the area A represents the fraction of cars that can be expected to get less than 20 miles per gallon for in-city driving. To find A, we first calculate the z-value corresponding to $x = 20$. That is,

$$z = \frac{x - \mu}{\sigma} = \frac{20 - 27}{3} = -\frac{7}{3} = -2.33$$

Then

$$P(x < 20) = P(z < -2.33)$$

as indicated by the shaded area in Figure 4.21. Because Table IV gives areas only to the right of the mean (and because the normal distribution is symmetric about its mean), we look up 2.33 in Table IV and find that the corresponding area is .4901. This is equal to the area between $z = 0$ and $z = -2.33$, so we find

$$P(x < 20) = A = .5 - .4901 = .0099 \approx .01$$

According to this probability model, you should have only about a 1% chance of purchasing a car of this make with an in-city mileage under 20 miles per gallon.

b. Now you are asked to make an inference based on a sample—the car you purchased. You are getting less than 20 miles per gallon for in-city driving. What do you infer? We think you will agree that one of two possibilities is true:

1. The probability model is correct. You simply were unfortunate to have purchased one of the cars in the 1% that get less than 20 miles per gallon in the city.
2. The probability model is incorrect. Perhaps the assumption of a normal distribution is unwarranted or the mean of 27 is an overestimate, or the standard deviation of 3 is an underestimate, or some combination of these errors was made. At any rate, the form of the actual probability model certainly merits further investigation.

You have no way of knowing with certainty which possibility is correct, but the evidence points to the second one. We are again relying on the rare-event approach to statistical inference that we introduced earlier. The sample (one measurement in this case) was so unlikely to have been drawn from the proposed probability model that it casts serious doubt on the model. We would be inclined to believe that the model is somehow in error.

Look Back When applying the rare event approach, the calculated probability must be small (say, less than or equal to .05) in order to infer that the observed event is, indeed, unlikely.

Now Work Exercise 4.92d

Occasionally you will be given a probability and will want to find the values of the normal random variable that correspond to the probability. For example, suppose the scores on a college entrance examination are known to be normally distributed, and a certain prestigious university will consider for admission only those applicants whose scores exceed the 90th percentile of the test score distribution. To determine the minimum score for admission consideration, you will need to be able to use Table IV in reverse, as demonstrated in the following example.

Example 4.21
Using the Normal Table in Reverse

Problem Find the value of z, call it z_0, in the standard normal distribution that will be exceeded only 10% of the time—that is, find z_0 such that $P(z \geq z_0) = .10$.

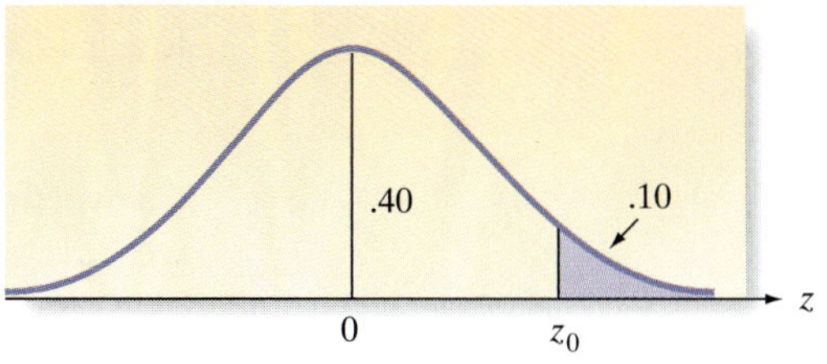

Figure 4.22
Area under the standard normal curve for Example 4.21

Solution In this case we are given a probability, or an area, and asked to find the value of the standard normal random variable that corresponds to the area. Specifically, we want to find the value z_0 such that only 10% of the standard normal distribution exceeds z_0 (see Figure 4.22).

We know that the total area to the right of the mean $z = 0$ is .5, which implies that z_0 must lie to the right of (above) 0. To pinpoint the value, we use the fact that the area to the right of z_0 is .10, which implies that the area between $z = 0$ and z_0 is $.5 - .1 = .4$. But areas between $z = 0$ and some other z-value are exactly the types given in Table IV. Therefore, we look up the area .4000 in the body of Table IV and find that the corresponding z-value is (to the closest approximation) $z_0 = 1.28$. The implication is that the point 1.28 standard deviations above the mean is the 90th percentile of a normal distribution.

Look Back As with earlier problems, it is critical to draw correctly the normal probability of interest on the normal curve. Placement of z_0 to the left or right of 0 is the key. Be sure to shade the probability (area) involving z_0. If it does not agree with the probability of interest (i.e., the shaded area is greater than .5 and the probability of interest is smaller than .5), then you need to place z_0 on the opposite side of 0.

Now Work Exercise 4.85

Example 4.22
Using the Normal Table in Reverse

Problem Find the value of z_0 such that 95% of the standard normal z values lie between $-z_0$ and z_0; that is, $P(-z_0 \leq z \leq z_0) = .95$.

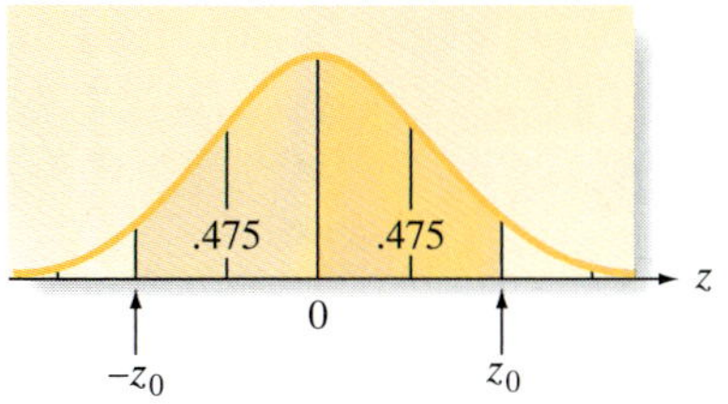

Figure 4.23
Areas under the standard normal curve for Example 4.22

Solution Here we wish to move an equal distance z_0 in the positive and negative directions from the mean $z = 0$ until 95% of the standard normal distribution is enclosed. This means that the area on each side of the mean will be equal to $\frac{1}{2}(.95) = .475$, as shown in Figure 4.23. Because the area between $z = 0$ and z_0 is .475, we look up .475 in the body of Table IV to find the value $z_0 = 1.96$. Thus, as we found in the reverse order in Example 4.21, 95% of a normal distribution lies between -1.96 and 1.96 standard deviations of the mean.

Now Work Exercise 4.88

Now that you have learned to use Table IV to find a standard normal z-value that corresponds to a specified probability, we demonstrate a practical application in Example 4.23.

Example 4.23
Using the Normal Table in Reverse—Paint Manufacturing Application

Problem Suppose a paint manufacturer has a daily production, x, that is normally distributed with a mean of 100,000 gallons and a standard deviation of 10,000 gallons. Management wants to create an incentive bonus for the production crew when the daily production exceeds the 90th percentile of the distribution, in hopes that the crew will, in turn, become more productive. At what level of production should management pay the incentive bonus?

Solution In this example, we want to find a production level, x_0, such that 90% of the daily levels (x values) in the distribution fall below x_0 and only 10% fall above x_0—that is,

$$P(x \leq x_0) = .90$$

Converting x to a standard normal random variable, where $\mu = 100{,}000$ and $\sigma = 10{,}000$, we have

$$P(x \leq x_0) = P\left(z \leq \frac{x_0 - \mu}{\sigma}\right)$$
$$= P\left(z \leq \frac{x_0 - 100{,}000}{10{,}000}\right) = .90$$

In Example 4.21 (see Figure 4.22) we found the 90th percentile of the standard normal distribution to be $z_0 = 1.28$—that is, we found $P(z \leq 1.28) = .90$. Consequently, we know the production level x_0 at which the incentive bonus is paid corresponds to a z-score of 1.28; that is,

$$\frac{x_0 - 100{,}000}{10{,}000} = 1.28$$

If we solve this equation for x_0, we find

$$x_0 = 100{,}000 + 1.28(10{,}000) = 100{,}000 + 12{,}800 = 112{,}800$$

This x value is shown in Figure 4.24. Thus, the 90th percentile of the production distribution is 112,800 gallons. Management should pay an incentive bonus when a day's production exceeds this level if its objective is to pay only when production is in the top 10% of the current daily production distribution.

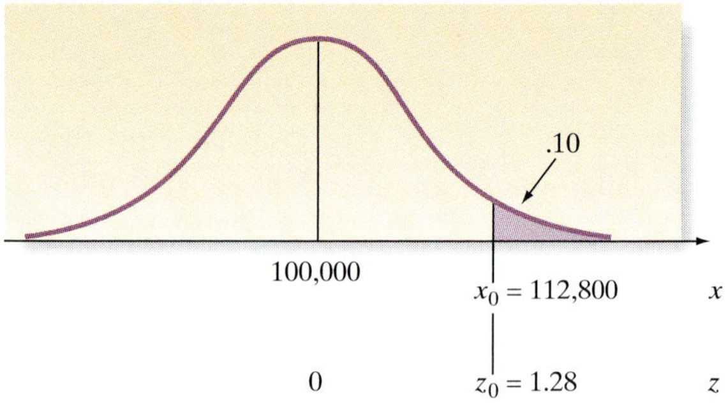

Figure 4.24
Area under the normal curve for Example 4.23

Look Back As this example shows, in practical applications of the normal table in reverse, first find the value of z_0, then convert the value to the units of x using the z-score formula in reverse.

Statistics in Action Revisited Using the Normal Model

Recall that a defense contractor has developed a prototype gun for the U.S. Army that fires 1,100 flechettes with a single round. The specifications of the weapon are set so that when the gun is aimed at a target 500 meters away, the mean horizontal grid value of the flechettes is equal to the aim point. In the range test, the weapon was aimed at the center target in Figure SIA4.1 (p. 174); thus, $\mu = 5$ feet. For three different tests, the standard deviation was set at $\sigma = 1$ feet, $\sigma = 2$ feet, and $\sigma = 4$ feet. From past experience, the defense contractor has found that the distribution of the horizontal hit point measurements is closely approximated by a normal distribution. Therefore, we can use the normal distribution to find the probability that a single flechette shot from the weapon will hit any one of the three targets. Recall from Figure SIA4.1 (p. 174) that the three targets range from -1 to 1, 4 to 6, and 9 to 11 feet on the horizontal grid.

Consider, first, the middle target. Letting x represent the horizontal measurement for a flechette shot from the gun, the flechette will hit the target if $4 \le x \le 6$. The probability that this flechette will hit the target when $\mu = 5$ and $\sigma = 1$ is, using the normal probability table (Table IV, Appendix B),

$$\underset{\sigma = 1}{\textit{Middle}:}\ P(4 \le x \le 6) = P\left(\frac{4-5}{1} < z < \frac{6-5}{1}\right)$$
$$= P(-1 < z < 1)$$
$$= 2(.3413) = .6826$$

Similarly, we find the probability that the flechette hits the left and right targets shown in Figure SIA4.1.

$$\underset{\sigma = 1}{\textit{Left}:}\ P(-1 \le x \le 1) = P\left(\frac{-1-5}{1} < z < \frac{1-5}{1}\right)$$
$$= P(-6 < z < -4) \approx 0$$
$$\underset{\sigma = 1}{\textit{Right}:}\ P(9 \le x \le 11) = P\left(\frac{9-5}{1} < z < \frac{11-5}{1}\right)$$
$$= P(4 < z < 6) \approx 0$$

You can see that there is about a 68% chance that a flechette will hit the middle target but virtually no chance that one will hit the left and right targets when the standard deviation is set at 1 foot.

To find these three probabilities for $\sigma = 2$ and $\sigma = 4$, we use the normal probability function in Minitab. Figure SIA4.2 is a Minitab worksheet giving the cumulative probabilities of a normal random variable (with $\mu = 5, \sigma = 1$) falling below the x values in the first column. The cumulative probabilities for $\sigma = 2$ and $\sigma = 4$ are given in the columns named "sigma2" and "sigma4," respectively.

Using the cumulative probabilities in the figure to find the three probabilities when $\sigma = 2$, we have

$$\underset{\sigma = 2}{\textit{Middle}:}\ P(4 \le x \le 6) = P(x \le 6) - P(x \le 4)$$
$$= .6915 - .3085 = .3830$$

↓	C1	C2	C3	C4
	x	sigma1	sigma2	sigma4
1	-1	0.00000	0.001350	0.066807
2	1	0.00003	0.022750	0.158655
3	4	0.15866	0.308538	0.401294
4	6	0.84134	0.691462	0.598706
5	9	0.99997	0.977250	0.841345
6	11	1.00000	0.998650	0.933193

Figure SIA4.2
Minitab worksheet with cumulative normal probabilities

$$\underset{\sigma = 2}{\textit{Left}:}\ P(-1 \le x \le 1) = P(x \le 1) - P(x \le -1)$$
$$= .0228 - .0014 = .0214$$

$$\underset{\sigma = 2}{\textit{Right}:}\ P(9 \le x \le 11) = P(x \le 11) - P(x \le 9)$$
$$= .9987 - .9773 = .0214$$

Thus, when $\sigma = 2$, there is about a 38% chance that a flechette will hit the middle target, a 2% chance that one will hit the left target, and a 2% chance that one will hit the right target. The probability that a flechette will hit either the middle or left or right target is simply the sum of these three probabilities (an application of the additive rule of probability). This sum is $.3830 + .0214 + .0214 = .4258$; consequently, there is about a 42% chance of hitting any one of the three targets when specifications are set so that $\sigma = 2$.

Now, we use the cumulative probabilities in Figure SIA4.2 to find the three hit probabilities when $\sigma = 4$:

$$\underset{\sigma = 4}{\textit{Middle}}\ P(4 \le x \le 6) = P(x \le 6) - P(x \le 4)$$
$$= .5987 - .4013 = .1974$$

$$\underset{\sigma = 4}{\textit{Left}:}\ P(-1 \le x \le 1) = P(x \le 1) - P(x \le -1)$$
$$= .1587 - .0668 = .0919$$

$$\underset{\sigma = 9}{\textit{Right}:}\ P(9 \le x \le 11) = P(x \le 11) - P(x \le 9)$$
$$= .9332 - .8413 = .0919$$

Thus, when $\sigma = 4$, there is about a 20% chance that a flechette will hit the middle target, a 9% chance that one will hit the left target, and a 9% chance that one will hit the right target. The probability that a flechette will hit any one of the three targets is $.1974 + .0919 + .0919 = .3812$.

These probability calculations reveal a few patterns. First, the probability of hitting the middle target (the target where the gun is aimed) is reduced as the standard deviation is increased. Obviously, if the U.S. Army wants to maximize the chance of hitting the target that the prototype gun is aimed at, it will want specifications set with a small value of σ. But if the Army wants to hit multiple targets with a single shot of the weapon, σ should be increased. With a larger σ, not as many of the flechettes will hit the target aimed at, but more will hit peripheral targets. Whether σ should be set at 4 or 6 (or some other value) depends on how high of a hit rate is required for the peripheral targets.

Data Set: MOAGUN

Exercises 4.79–4.104

Learning the Mechanics

4.79 Find the following probabilities for the standard normal random variable z:

NW **a.** $P(z > 1.46)$ **b.** $P(z < -1.56)$
c. $P(.67 \leq z < 2.41)$ **d.** $P(-1.96 \leq z < -.33)$
e. $P(z \geq 0)$ **f.** $P(-2.33 < z < 1.50)$

4.80 Find the area under the standard normal probability distribution between the following pairs of z-scores:
a. $z = 0$ and $z = 2.00$ **b.** $z = 0$ and $z = 3$
c. $z = 0$ and $z = 1.5$ **d.** $z = 0$ and $z = .80$

4.81 Find each of the following probabilities for the standard
NW normal random variable z:
a. $P(-1 \leq z \leq 1)$ **b.** $P(-1.96 \leq z \leq 1.96)$
c. $P(-1.645 \leq z \leq 1.645)$ **d.** $P(-2 \leq z \leq 2)$

4.82 Find the following probabilities for the standard normal random variable z:
a. $P(-1 < z < 1)$ **b.** $P(-2 < z < 2)$
c. $P(-2.16 \leq z \leq .55)$ **d.** $P(-.42 < z < 1.96)$
e. $P(z \geq -2.33)$ NW **f.** $P(z < 2.33)$

4.83 Find each of the following probabilities for a standard normal random variable z:
a. $P(z = 1)$ **b.** $P(z \leq 1)$
c. $P(z < 1)$ **d.** $P(z > 1)$

4.84 Find a value of the standard normal random variable z, call it z_0, such that
a. $P(z \geq z_0) = .05$ **b.** $P(z \geq z_0) = .025$
c. $P(z \leq z_0) = .025$ **d.** $P(z \geq z_0) = .10$
e. $P(z > z_0) = .10$

4.85 Find a value of the standard normal random variable z,
NW call it z_0, such that
a. $P(z \leq z_0) = .2090$
b. $P(z \leq z_0) = .7090$
c. $P(-z_0 \leq z < z_0) = .8472$
d. $P(-z_0 \leq z \leq z_0) = .1664$
e. $P(z_0 \leq z \leq 0) = .4798$
f. $P(-1 < z < z_0) = .5328$

4.86 Give the z-score for a measurement from a normal distribution for the following:
a. 1 standard deviation above the mean
b. 1 standard deviation below the mean
c. Equal to the mean
d. 2.5 standard deviations below the mean
e. 3 standard deviations above the mean

4.87 Suppose the random variable x is best described by a normal distribution with $\mu = 30$ and $\sigma = 4$. Find the z-score that corresponds to each of the following x values:
a. $x = 20$ **b.** $x = 30$
c. $x = 2.75$ **d.** $x = 15$
e. $x = 35$ **f.** $x = 25$

4.88 The random variable x has a normal distribution with
NW $\mu = 1{,}000$ and $\sigma = 10$.
a. Find the probability that x assumes a value more than 2 standard deviations from its mean. More than 3 standard deviations from μ.
b. Find the probability that x assumes a value within 1 standard deviation of its mean. Within 2 standard deviations of μ.
c. Find the value of x that represents the 80th percentile of this distribution. The 10th percentile.

4.89 Suppose x is a normally distributed random variable with $\mu = 11$ and $\sigma = 2$. Find each of the following:
a. $P(10 \leq x \leq 12)$ **b.** $P(6 \leq x \leq 10)$
c. $P(13 \leq x \leq 16)$ **d.** $P(7.8 \leq x \leq 12.6)$
e. $P(x \geq 13.24)$ **f.** $P(x \geq 7.62)$

4.90 Suppose x is a normally distributed random variable with $\mu = 50$ and $\sigma = 3$. Find a value of the random variable, call it x_0, such that
a. $P(x \leq x_0) = .8413$
b. $P(x > x_0) = .025$
c. $P(x > x_0) = .95$
d. $P(41 \leq x < x_0) = .8630$
e. 10% of the values of x are less than x_0.
f. 1% of the values of x are greater than x_0.

Applet Exercise 4.6

Open the applet *Sample from a Population*. On the pull-down menu to the right of the top graph, select *Bell shaped*. The box to the left of the top graph displays the population mean, median, and standard deviation.
a. Run the applet for each available value of n on the pull-down menu for the sample size. Go from the smallest to the largest value of n. For each value of n, observe the shape of the graph of the sample data and record the mean, median, and standard deviation of the sample.
b. Describe what happens to the shape of the graph and the mean, median, and standard deviation of the sample as the sample size increases.

Applying the Concepts—Basic

4.91 **Most powerful business women in America.** Refer to the *Fortune* (Oct. 16, 2008) list of the 50 most powerful women in America, Exercise 2.48 (p. 62). Recall that the data on age (in years) of each woman is stored in the **WPOWER50** file. The ages in the data set can be shown to be approximately normally distributed with a mean of 50 years and a standard deviation of 6 years. A powerful woman is randomly selected from the data, and her age is observed.
a. Find the probability that her age will fall between 55 and 60 years.
b. Find the probability that her age will fall between 48 and 52 years.
c. Find the probability that her age will be less than 35 years.
d. Find the probability that her age will exceed 40 years.

4.92 **The business of casino gaming.** Casino gaming yields over
NW \$35 billion in revenue each year in the United States. In *Chance* (Spring 2005), University of Denver statistician R. C. Hannum discussed the business of casino gaming and its reliance on the laws of probability. Casino games of pure chance (e.g., craps, roulette, baccarat, and keno) always yield a "house advantage." For example, in the game of double-zero roulette, the expected casino win percentage is 5.26% on bets made on whether the outcome will be either black or red. (This implies that for every \$5 bet on black or red, the casino will earn a net of about 25 cents.) It can be shown that in 100 roulette plays on black/red, the average casino win percentage is normally distributed with mean 5.26% and standard deviation 10%. Let x represent the average

casino win percentage after 100 bets on black/red in double-zero roulette.

a. Find $P(x > 0)$. (This is the probability that the casino wins money.)
b. Find $P(5 < x < 15)$.
c. Find $P(x < 1)$.
d. If you observed an average casino win percentage of −25% after 100 roulette bets on black/red, what would you conclude?

4.93 Buy-side vs. sell-side analysts' earnings forecasts. Financial analysts who make forecasts of stock prices are categorized as either "buy-side" analysts or "sell-side" analysts. Refer to the *Financial Analysts Journal* (Jul./Aug. 2008) comparison of earnings forecasts of buy-side and sell-side analysts, Exercise 2.84 (p. 76). The mean and standard deviation of forecast errors for both types of analysts are reproduced in the table. Assume that the distribution of forecast errors are approximately normally distributed.

a. Find the probability that a buy-side analyst has a forecast error of +2.00 or higher.
b. Find the probability that a sell-side analyst has a forecast error of +2.00 or higher.

	Buy-Side Analysts	Sell-Side Analysts
Mean	0.85	−0.05
Standard Deviation	1.93	0.85

Source: Groysberg, B., Healy, P., and Chapman, C. *Financial Analysts Journal,* Vol. 64, No. 4, Jul./Aug. 2008.

4.94 NHTSA crash safety tests. Refer to Exercise 4.19 (p. 184) and the NHTSA crash test data for new cars. One of the variables saved in the **CRASH** file is the severity of a driver's head injury when the car is in a head-on collision with a fixed barrier while traveling at 35 miles per hour. The more points assigned to the head injury rating, the more severe the injury. The head injury ratings can be shown to be approximately normally distributed with a mean of 605 points and a standard deviation of 185 points. One of the crash-tested cars is randomly selected from the data, and the driver's head injury rating is observed.

a. Find the probability that the rating will fall between 500 and 700 points.
b. Find the probability that the rating will fall between 400 and 500 points.
c. Find the probability that the rating will be less than 850 points.
d. Find the probability that the rating will exceed 1,000 points.

4.95 Transmission delays in wireless technology. Resource reservation protocol (RSVP) was originally designed to establish signaling links for stationary networks. In *Mobile Networks and Applications* (Dec. 2003), RSVP was applied to mobile wireless technology (e.g., a PC notebook with wireless LAN card for Internet access). A simulation study revealed that the transmission delay (measured in milliseconds) of an RSVP linked wireless device has an approximate normal distribution with mean $\mu = 48.5$ milliseconds and $\sigma = 8.5$ milliseconds.

a. What is the probability that the transmission delay is less than 57 milliseconds?
b. What is the probability that the transmission delay is between 40 and 60 milliseconds?

Applying the Concepts—Intermediate

4.96 Mean shifts on a production line. *Six Sigma* is a comprehensive approach to quality goal setting that involves statistics. An article in *Aircraft Engineering and Aerospace Technology* (Vol. 76, No. 6, 2004) demonstrated the use of the normal distribution in Six Sigma goal setting at Motorola Corporation. Motorola discovered that the average defect rate for parts produced on an assembly line varies from run-to-run and is approximately normally distributed with a mean equal to 3 defects per million. Assume that the goal at Motorola is for the average defect rate to vary no more than 1.5 standard deviations above or below the mean of 3. How likely is it that the goal will be met?

4.97 Rating employee performance. Almost all companies utilize some type of year-end performance review for their employees. Human Resources (HR) at the University of Texas Health Science Center provides guidelines for supervisors rating their subordinates. For example, raters are advised to examine their ratings for a tendency to be either too lenient or too harsh. According to HR, "if you have this tendency, consider using a normal distribution—10% of employees (rated) exemplary, 20% distinguished, 40% competent, 20% marginal, and 10% unacceptable." Suppose you are rating an employee's performance on a scale of 1 (lowest) to 100 (highest). Also, assume the ratings follow a normal distribution with a mean of 50 and a standard deviation of 15.

a. What is the lowest rating you should give to an "exemplary" employee if you follow the University of Texas HR guidelines?
b. What is the lowest rating you should give to a "competent" employee if you follow the University of Texas HR guidelines?

4.98 Cotton crop yield study. The crop yield for a particular farm in a particular year is typically measured as the amount of the crop produced per acre. For example, cotton is measured in pounds per acre. It has been demonstrated that the normal distribution can be used to characterize crop yields over time (*American Journal of Agricultural Economics,* May 1999). Historical data indicate that next summer's cotton yield for a particular Georgia farmer can be characterized by a normal distribution with mean 1,500 pounds per acre and standard deviation 250. The farm in question will be profitable if it produces at least 1,600 pounds per acre.

a. What is the probability that the farm will lose money next summer?
b. Assume the same normal distribution is appropriate for describing cotton yield in each of the next two summers. Also assume that the two yields are statistically independent. What is the probability that the farm will lose money for two straight years?
c. What is the probability that the cotton yield falls within 2 standard deviations of 1,500 pounds per acre next summer?

4.99 Personnel dexterity tests. Personnel tests are designed to test a job applicant's cognitive and/or physical abilities. The Wonderlic IQ test is an example of the former; the Purdue Pegboard speed test involving the arrangement of pegs on a peg board is an example of the latter. A particular dexterity test is administered nationwide by a private testing service. It is known that for all tests administered last year, the distribution of scores was approximately normal with mean 75 and standard deviation 7.5.

a. A particular employer requires job candidates to score at least 80 on the dexterity test. Approximately what percentage of the test scores during the past year exceeded 80?

b. The testing service reported to a particular employer that one of its job candidate's scores fell at the 98th percentile of the distribution (i.e., approximately 98% of the scores were lower than the candidate's, and only 2% were higher). What was the candidate's score?

4.100 Manufacturing hourly pay rate. Government data indicate that the mean hourly wage for manufacturing workers in the United States is $17 (*Statistical Abstract of the United States: 2008*). Suppose the distribution of manufacturing wage rates nationwide can be approximated by a normal distribution with standard deviation $1.25 per hour. The first manufacturing firm contacted by a particular worker seeking a new job pays $18.30 per hour.

a. If the worker were to undertake a nationwide job search, approximately what proportion of the wage rates would be greater than $18.30 per hour?

b. If the worker were to randomly select a U.S. manufacturing firm, what is the probability the firm would pay more than $18.30 per hour?

c. The population median, call it η, of a continuous random variable x is the value such that $P(x \geq \eta) = P(x \leq \eta) = .5$—that is, the median is the value η such that half the area under the probability distribution lies above η and half lies below it. Find the median of the random variable corresponding to the wage rate and compare it to the mean wage rate.

4.101 Model for long-term construction cost. Before negotiating a long-term construction contract, building contractors must carefully estimate the total cost of completing the project. Benzion Barlev of New York University proposed a model for total cost of a long-term contract based on the normal distribution (*Journal of Business Finance and Accounting*, July 1995). For one particular construction contract, Barlev assumed total cost, x, to be normally distributed with mean $850,000 and standard deviation $170,000. The revenue, R, promised to the contractor is $1,000,000.

a. The contract will be profitable if revenue exceeds total cost. What is the probability that the contract will be profitable for the contractor?

b. What is the probability that the project will result in a loss for the contractor?

c. Suppose the contractor has the opportunity to renegotiate the contract. What value of R should the contractor strive for in order to have a .99 probability of making a profit?

Applying the Concepts—Advanced

4.102 Industrial filling process. The characteristics of an industrial filling process in which an expensive liquid is injected into a container was investigated in the *Journal of Quality Technology* (July 1999). The quantity injected per container is approximately normally distributed with mean 10 units and standard deviation .2 units. Each unit of fill costs $20 per unit. If a container contains less than 10 units (i.e., is underfilled), it must be reprocessed at a cost of $10. A properly filled container sells for $230.

a. Find the probability that a container is underfilled. Not underfilled.

b. A container is initially underfilled and must be reprocessed. Upon refilling, it contains 10.60 units. How much profit will the company make on this container?

c. The operations manager adjusts the mean of the filling process upward to 10.10 units in order to make the probability of underfilling approximately zero. Under these conditions, what is the expected profit per container?

4.103 Load on frame structures. In the *Journal of the International Association for Shell and Spatial Structures* (Apr. 2004), Japanese environmental researchers studied the performance of truss and frame structures subjected to uncertain loads. The load was assumed to have a normal distribution with a mean of 20,000 pounds. Also, the probability that the load is between 10,000 and 30,000 pounds is .95. Based on this information, find the standard deviation of the load distribution.

4.104 Box plots and the standard normal distribution. What relationship exists between the standard normal distribution and the box-plot methodology (Section 2.8) for describing distributions of data using quartiles? The answer depends on the true underlying probability distribution of the data. Assume for the remainder of this exercise that the distribution is normal.

a. Calculate the values of the standard normal random variable z, call them z_L and z_U, that correspond to the hinges of the box plot—that is, the lower and upper quartiles, Q_L and Q_U—of the probability distribution.

b. Calculate the z values that correspond to the inner fences of the box plot for a normal probability distribution.

c. Calculate the z values that correspond to the outer fences of the box plot for a normal probability distribution.

d. What is the probability that an observation lies beyond the inner fences of a normal probability distribution? The outer fences?

e. Can you better understand why the inner and outer fences of a box plot are used to detect outliers in a distribution? Explain.

4.7 Descriptive Methods for Assessing Normality

In the chapters that follow, we learn how to make inferences about the population based on information in the sample. Several of these techniques are based on the assumption that the population is approximately normally distributed. Consequently, it will be important to determine whether the sample data come from a normal population before we can properly apply these techniques.

Several descriptive methods can be used to check for normality. In this section, we consider the four methods summarized in the box.

Determining Whether the Data Are from an Approximately Normal Distribution

1. Construct either a histogram or stem-and-leaf display for the data and note the shape of the graph. If the data are approximately normal, the shape of the histogram or stem-and-leaf display will be similar to the normal curve, Figure 4.13 (i.e., mound shaped and symmetric about the mean).
2. Compute the intervals $\bar{x} \pm s$, $\bar{x} \pm 2s$, and $\bar{x} \pm 3s$, and determine the percentage of measurements falling in each. If the data are approximately normal, the percentages will be approximately equal to 68%, 95%, and 100%, respectively.
3. Find the interquartile range, IQR, and standard deviation, s, for the sample, then calculate the ratio IQR/s. If the data are approximately normal, then IQR/$s \approx 1.3$.
4. Examine a *normal probability plot* for the data. If the data are approximately normal, the points will fall (approximately) on a straight line.

The first two methods come directly from the properties of a normal distribution established in Section 4.6. Method 3 is based on the fact that for normal distributions, the z values corresponding to the 25th and 75th percentiles are $-.67$ and $.67$, respectively (see Example 4.17). Because $\sigma = 1$ for a standard normal distribution,

$$\frac{\text{IQR}}{\sigma} = \frac{Q_U - Q_L}{\sigma} = \frac{.67 - (-.67)}{1} = 1.34$$

The final descriptive method for checking normality is based on a *normal probability plot*. In such a plot, the observations in a data set are ordered from smallest to largest and then plotted against the expected z-scores of observations calculated under the assumption that the data come from a normal distribution. When the data are, in fact, normally distributed, a linear (straight-line) trend will result. A nonlinear trend in the plot suggests that the data are nonnormal.

A **normal probability plot** for a data set is a scatterplot with the ranked data values on one axis and their corresponding expected z-scores from a standard normal distribution on the other axis. [*Note:* Computation of the expected standard normal z-scores are beyond the scope of this text. Therefore, we will rely on available statistical software packages to generate a normal probability plot.]

Example 4.24

Checking for Normal Data—EPA Estimated Gas Mileages

Problem The Environmental Protection Agency (EPA) performs extensive tests on all new car models to determine their mileage ratings. The results of 100 EPA tests on a certain new car model are displayed in Table 4.6. Numerical and graphical descriptive measures for the data are shown on the Minitab and SPSS printouts, Figures 4.25a–c. Determine whether the EPA mileage ratings are from an approximate normal distribution.

Table 4.6 EPA Gas Mileage Ratings for 100 Cars (miles per gallon)

36.3	41.0	36.9	37.1	44.9	36.8	30.0	37.2	42.1	36.7
32.7	37.3	41.2	36.6	32.9	36.5	33.2	37.4	37.5	33.6
40.5	36.5	37.6	33.9	40.2	36.4	37.7	37.7	40.0	34.2
36.2	37.9	36.0	37.9	35.9	38.2	38.3	35.7	35.6	35.1
38.5	39.0	35.5	34.8	38.6	39.4	35.3	34.4	38.8	39.7
36.3	36.8	32.5	36.4	40.5	36.6	36.1	38.2	38.4	39.3
41.0	31.8	37.3	33.1	37.0	37.6	37.0	38.7	39.0	35.8
37.0	37.2	40.7	37.4	37.1	37.8	35.9	35.6	36.7	34.5
37.1	40.3	36.7	37.0	33.9	40.1	38.0	35.2	34.8	39.5
39.9	36.9	32.9	33.8	39.8	34.0	36.8	35.0	38.1	36.9

Data Set: EPAGAS

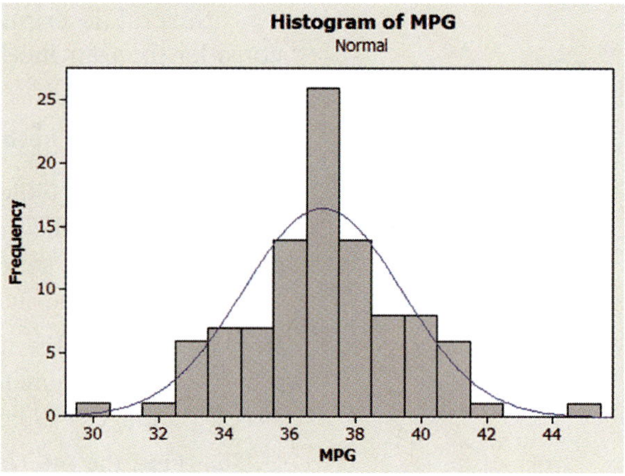

Figure 4.25a
Minitab histogram for gas mileage data

Descriptive Statistics: MPG

Variable	N	Mean	StDev	Minimum	Q1	Median	Q3	Maximum
MPG	100	36.994	2.418	30.000	35.625	37.000	38.375	44.900

Figure 4.25b
Minitab descriptive statistics for gas mileage data

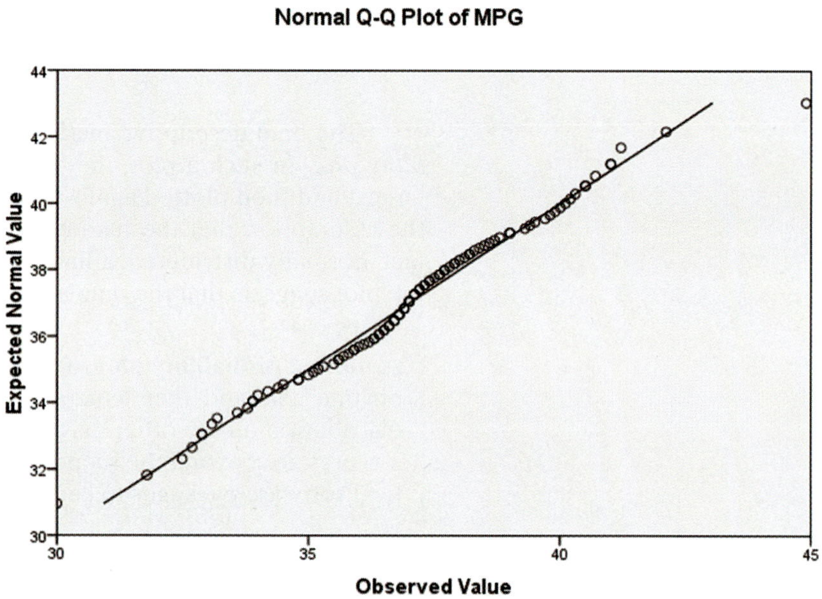

Figure 4.25c
SPSS normal probability plot for gas mileage data

Solution As a first check, we examine the Minitab histogram of the data shown in Figure 4.25a. Clearly, the mileages fall in an approximately mound-shaped, symmetric distribution centered around the mean of approximately 37 mpg. Note that a normal curve is superimposed on the figure. Therefore, using check #1 in the box, the data appear to be approximately normal.

To apply check #2, we obtain $\bar{x} = 37$ and $s = 2.4$ from the Minitab printout, Figure 4.25b. The intervals $\bar{x} \pm s$, $\bar{x} \pm 3s$, and $\bar{x} \pm 3s$ are shown in Table 4.7, as well as the percentage of mileage ratings that fall in each interval. These percentages agree almost exactly with those from a normal distribution.

Table 4.7 Describing the 100 EPA Mileage Ratings

Interval	Percentage in Interval
$\bar{x} \pm s = (34.6, 39.4)$	68
$\bar{x} \pm 2s = (32.2, 41.8)$	96
$\bar{x} \pm 3s = (29.8, 44.2)$	99

Check #3 in the box requires that we find the ratio IQR/s. From Figure 4.25(b), the 25th percentile (called Q_1 by Minitab) is $Q_L = 35.625$, and the 75th percentile (labeled Q_3 by Minitab) is $Q_U = 38.375$. Then, $\text{IQR} = Q_U - Q_L = 2.75$, and the ratio is

$$\frac{\text{IQR}}{s} = \frac{2.75}{2.4} = 1.15$$

Because this value is approximately equal to 1.3, we have further confirmation that the data are approximately normal.

A fourth descriptive method is to interpret a normal probability plot. An SPSS normal probability plot for the mileage data is shown in Figure 4.25c. Notice that the ordered mileage values (shown on the horizontal axis) fall reasonably close to a straight line when plotted against the expected z-scores from a normal distribution. Thus, check #4 also suggests that the EPA mileage data are likely to be approximately normally distributed.

Look Back The checks for normality given in the box are simple, yet powerful, techniques to apply, but they are only descriptive in nature. It is possible (although unlikely) that the data are nonnormal even when the checks are reasonably satisfied. Thus, we should be careful not to claim that the 100 EPA mileage ratings of Example 4.24 are, in fact, normally distributed. We can only state that it is reasonable to believe that the data are from a normal distribution.*

Now Work Exercise 4.108

As we will learn in the next chapter, several inferential methods of analysis require the data to be approximately normal. If the data are clearly nonnormal, inferences derived from the method may be invalid. Therefore, it is advisable to check the normality of the data prior to conducting the analysis.

Statistics IN Action Revisited — Assessing Fit of the Normal Distribution

In Statistics in Action Revisited, Section 4.6, we used the normal distribution to find the probability that a single flechette from a super weapon that shoots 1,100 flechettes at once hits one of three targets at 500 meters. Recall that for three range tests, the weapon was always aimed at the center target (i.e., the specification mean was set at $\mu = 5$ feet), but the specification standard deviation was varied at $\sigma = 1$ foot, $\sigma = 2$ feet, and $\sigma = 4$ feet. Table SIA4.1 shows the calculated normal probabilities of hitting the three targets for the different values of σ, as well as the actual results of the three range tests. (Recall that the actual data are saved in the **MOAGUN** file.) You can see that the proportion of the 1,100

Table SIA4.1 Summary of Normal Probability Calculations and Actual Range Test Results

Target	Specification	Normal Probability	Actual Number of Hits	Hit Ratio (Hits/1,100)
Left (–1 to 1)	$\sigma = 1$	.0000	0	.000
	$\sigma = 2$	.0214	30	.027
	$\sigma = 4$	.0919	73	.066
Middle (4 to 6)	$\sigma = 1$	.6826	764	.695
	$\sigma = 2$	.3820	409	.372
	$\sigma = 4$	.1974	242	.220
Right (9 to 11)	$\sigma = 1$	.0000	0	.000
	$\sigma = 2$	.0214	23	.021
	$\sigma = 4$	.0919	93	.085

Data Set: MOAGUN

*Statistical tests of normality that provide a measure of reliability for the inference are available. However, these tests tend to be very sensitive to slight departures from normality (i.e., they tend to reject the hypothesis of normality for any distribution that is not perfectly symmetrical and mound shaped). Consult the references (see Ramsey & Ramsey, 1990) if you want to learn more about these tests.

flechettes that actually hit each target—called the *hit ratio*—agrees very well with the estimated probability of a hit using the normal distribution.

Consequently, it appears that our assumption that the horizontal hit measurements are approximately normally distributed is reasonably satisfied. Further evidence of this is provided by the Minitab histograms of the horizontal hit measurements shown in Figures SIA4.3a–c. The normal curves superimposed on the histograms fit the data very well.

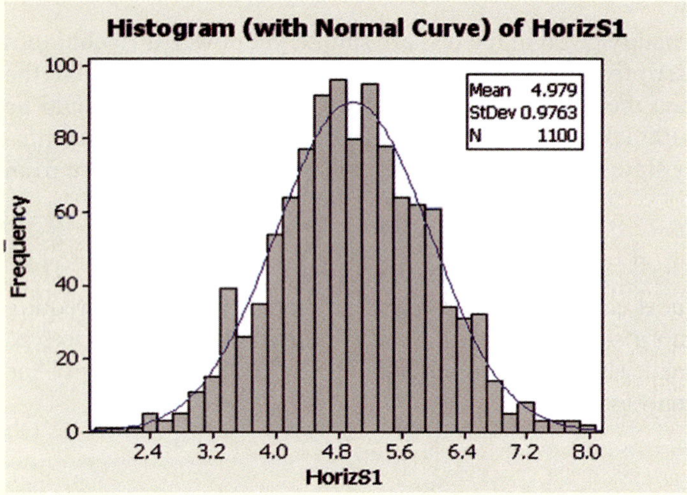

Figure SIA4.3a
Minitab histogram for the horizontal hit measurements when $\sigma = 1$

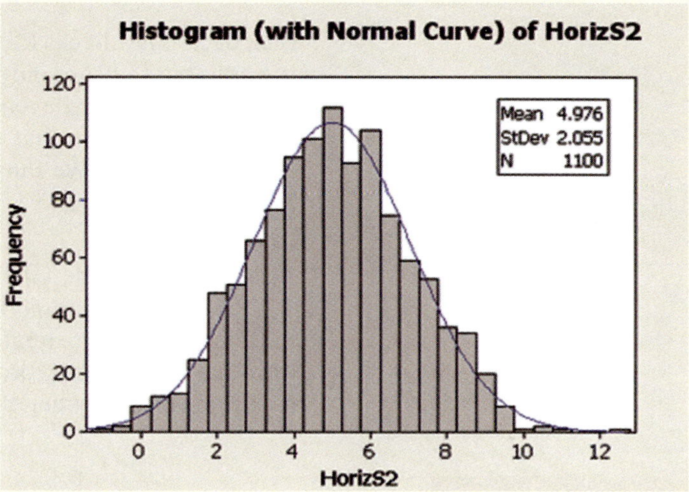

Figure SIA4.3b
Minitab histogram for the horizontal hit measurements when $\sigma = 2$

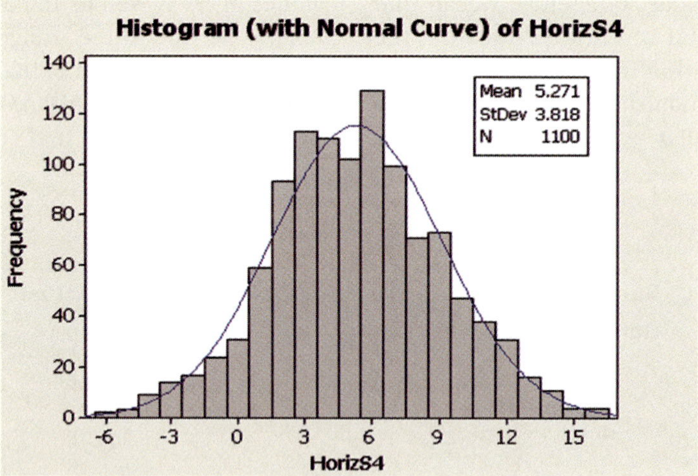

Figure SIA4.3c
Minitab histogram for the horizontal hit measurements when $\sigma = 4$

Exercises 4.105–4.117

Learning the Mechanics

4.105 If a population data set is normally distributed, what is the proportion of measurements you would expect to fall within the following intervals?

a. $\mu \pm \sigma$
b. $\mu \pm 2\sigma$
c. $\mu \pm 3\sigma$

4.106 Consider a sample data set with the following summary statistics: $s = 95$, $Q_L = 72$, $Q_U = 195$.

a. Calculate IQR.
b. Calculate IQR/s.
c. Is the value of IQR/s approximately equal to 1.3? What does this imply?

Normal probability plots for Exercise 4.107

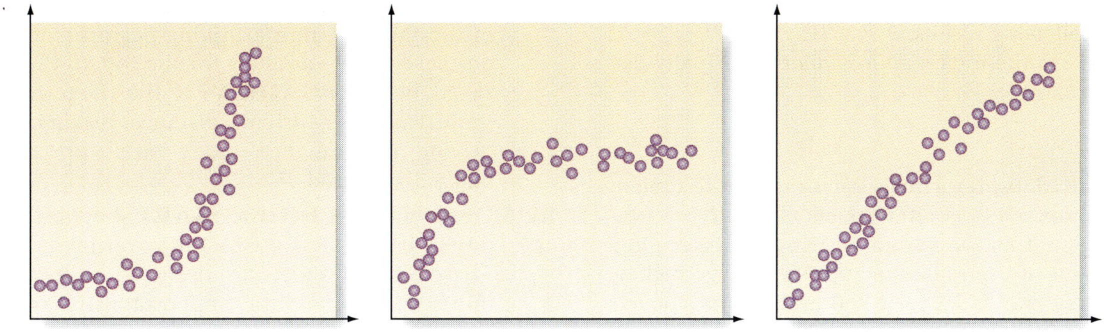

a. b. c.

4.107 Normal probability plots for three data sets are shown above. Which plot indicates that the data are approximately normally distributed?

4.108 Examine the sample data saved in the **LM4_18** file.

NW

5.9	5.3	1.6	7.4	8.6	1.2	2.1
4.0	7.3	8.4	8.9	6.7	4.5	6.3
7.6	9.7	3.5	1.1	4.3	3.3	8.4
1.6	8.2	6.5	1.1	5.0	9.4	6.4

a. Construct a stem-and-leaf plot to assess whether the data are from an approximately normal distribution.

b. Compute s for the sample data.

c. Find the values of Q_L and Q_U and the value of s from part **b** to assess whether the data come from an approximately normal distribution.

d. Generate a normal probability plot for the data and use it to assess whether the data are approximately normal.

Applying the Concepts—Basic

4.109 Most powerful business women in America. Refer to the *Fortune* (Oct. 16, 2008) list of the 50 most powerful women in America. In Exercise 4.91 (p. 216), you assumed that the ages (in years) of these women are approximately normally distributed. An Excel/DDXL printout with summary statistics for the age variable saved in the **WPOWER50** file is reproduced below.

a. Use the relevant statistics on the printout to compute the IQR. Does the result agree with the value shown on the printout?

Summary of AGE

Count	50
Mean	50.02
Median	51
Std Dev	6.444
Variance	41.53
Range	36
Min	28
Max	64
IQR	7
25th%	47
75th%	54

b. Locate the value of the standard deviation, s, on the printout.

c. Use the results, parts **a** and **b,** to demonstrate that the age distribution is approximately normal.

d. In Exercise 2.48 (p. 62) you constructed a relative frequency histogram for the age data. Use this graph to support your assumption of normality.

4.110 Ranking PhD programs in economics. Refer to the *Southern Economic Journal* (Apr. 2008) rankings of PhD programs in economics at 129 colleges and universities, Exercise 2.100 (p. 80). Recall that the number of publications published by faculty teaching in the PhD program and the quality of the publications were used to calculate an overall productivity score for each program. The mean and standard deviation of these 129 productivity scores were then used to compute a z-score for each economics program. The data (z-scores) for all 129 economic programs are saved in the **ECOPHD** file. A Minitab normal probability plot for the z-scores are shown below. Use the graph to assess whether the data are approximately normal.

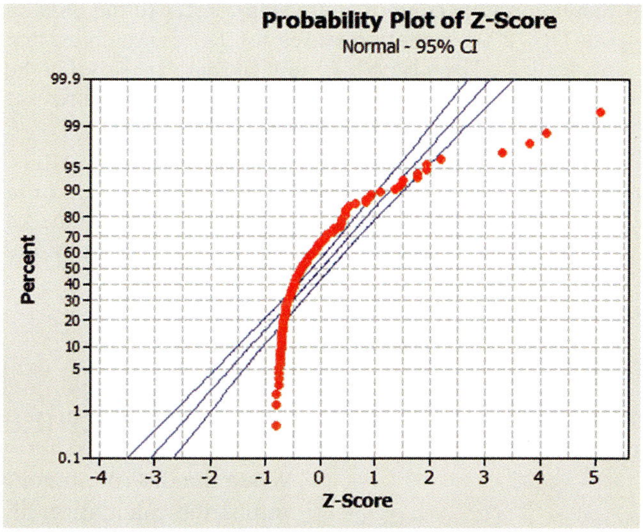

4.111 Software file updates. Software configuration management was used to monitor a software engineering team's performance at Motorola, Inc. (*Software Quality Professional,* Nov. 2004). One of the variables of interest was the number of updates to a file changed because of a problem report. Summary statistics for $n = 421$ files yielded the following results: $\bar{x} = 4.71$, $s = 6.09$, $Q_L = 1$, and $Q_U = 6$. Are these data approximately normally distributed? Explain.

Applying the Concepts—Intermediate

4.112 Wear-out of used display panels. Wear-out failure time of electronic components is often assumed to have a normal distribution. Can the normal distribution be applied to the wear-out of used manufactured products, such as colored display panels? A lot of 50 used display panels was purchased by an outlet store. Each panel displays 12 to 18 color characters. Prior to acquisition, the panels had been used for about one-third of their expected lifetimes. The data in the accompanying table (saved in the **PANELFAIL** file) give the failure times (in years) of the 50 used panels. Use the techniques of this section to determine whether the used panel wear-out times are approximately normally distributed.

0.01	1.21	1.71	2.30	2.96	0.19	1.22	1.75	2.30	2.98	0.51
1.24	1.77	2.41	3.19	0.57	1.48	1.79	2.44	3.25	0.70	1.54
1.88	2.57	3.31	0.73	1.59	1.90	2.61	1.19	0.75	1.61	1.93
2.62	3.50	0.75	1.61	2.01	2.72	3.50	1.11	1.62	2.16	2.76
3.50	1.16	1.62	2.18	2.84	3.50					

Source: Irony, T. Z., Lauretto, M., Pereira, C., and Stern, J. M. "A Weibull wearout test: Full Bayesian approach," paper presented at *Mathematical Sciences Colloquium,* Binghamton University, Binghamton, UK, December 2001.

4.113 NHTSA crash tests. Refer to the NHTSA crash test data for new cars. In Exercise 4.94 (p. 217), you assumed that the driver's head-injury rating is approximately normally distributed. Apply the methods of this chapter to the data saved in the **CRASH** file to support this assumption.

4.114 Sanitation inspection of cruise ships. Refer to the data on the Dec. 2008 sanitation scores for 183 cruise ships, first presented in Exercise 2.22 (p. 49). The data are saved in the **SHIPSANIT** file. Assess whether the sanitation scores are approximately normally distributed.

4.115 Ranking driving performance of professional golfers. Refer to *The Sport Journal* (Winter 2007) article on a new method for ranking the driving performance of PGA golfers, Exercise 2.50 (p. 63). Recall that the method incorporates a golfer's average driving distance (yards) and driving accuracy (percent of drives that land in the fairway) into a driving performance index. Data on these three variables for the top 40 PGA golfers are saved in the **PGADRIVER** file. Determine which of these variables—driving distance, driving accuracy, and driving performance index—are approximately normally distributed.

4.116 Department of Defense START sheets. The DoD RAC publishes START sheets for government and industry. These START sheets are designed to improve the reliability, maintainability, supportability, and quality of manufactured components and systems. In Volume 11 (2004), the DoD analyzed the following data set on tensile strength measurements taken at two different temperatures. (The data are saved in the **TENSILE** file.) In the START sheet, the DoD demonstrated that the data for the 11 tensile strength measurements were sampled from an approximately normal distribution. Do you agree?

Temperature	Strength
75	328.2
75	334.7
75	347.8
75	346.3
75	338.7
75	340.8
−67	343.6
−67	334.2
−67	348.7
−67	356.3
−67	344.1

Applying the Concepts—Advanced

4.117 Semester hours taken by CPA candidates. Refer to the *Journal of Accounting and Public Policy* (Spring 2002) study of first-time candidates for the CPA exam, Exercise 2.51 (p. 63). The variable of interest is the total semester hours of college credit for each candidate. Recall that the mean and median for the data set were 141.31 and 140 hours, respectively, and the standard deviation was 17.77 hours. Demonstrate why the probability distribution for the variable, total semester hours, is unlikely to be normally distributed.

4.8 Approximating a Binomial Distribution with a Normal Distribution

When the discrete binomial random variable (Section 4.3) can assume a large number of values, the calculation of its probabilities may become very tedious. To contend with this problem, we provide tables in Appendix B to give the probabilities for some values of n and p, but these tables are by necessity incomplete. Recall that the binomial probability table (Table II) can be used only for $n = 5, 6, 7, 8, 9, 10, 15, 20,$ or 25. To deal with this limitation, we seek approximation procedures for calculating the probabilities associated with a binomial probability distribution.

When n is large, a normal probability distribution may be used to provide a good approximation to the probability distribution of a binomial random variable. To show how this approximation works, we refer to Example 4.12, in which we used the binomial distribution to model the number x of 20 employees who favor unionization. We assumed that 60% of all the company's employees favored unionization. The mean and standard deviation of x were found to be $\mu = 12$ and $\sigma = 2.2$. The binomial distribution for $n = 20$ and $p = .6$ is shown in Figure 4.26, and the approximating normal distribution with mean $\mu = 12$ and standard deviation $\sigma = 2.2$ is superimposed.

As part of Example 4.12, we used Table II to find the probability that $x \leq 10$. This probability, which is equal to the sum of the areas contained in the rectangles (shown in Figure 4.26) that correspond to $p(0), p(1), p(2), \ldots, p(10)$, was found to equal .245. The portion of the approximating normal curve that would be used to approximate the area $p(0) + p(1) + p(2) + \cdots + p(10)$ is shaded in Figure 4.26. Note that this shaded area lies to the left of 10.5 (not 10), so we may include all of the probability in the rectangle corresponding to $p(10)$. Because we are approximating a discrete distribution (the binomial) with a continuous distribution (the normal), we call the use of 10.5 (instead of 10 or 11) a **correction for continuity,** that is, we are correcting the discrete distribution so that it can be approximated by the continuous one. The use of the correction for continuity leads to the calculation of the following standard normal z-value:

$$z = \frac{x - \mu}{\sigma} = \frac{10.5 - 12}{2.2} = -.68$$

Using Table IV, we find the area between $z = 0$ and $z = .68$ to be .2517. Then the probability that x is less than or equal to 10 is approximated by the area under the normal distribution to the left of 10.5, shown shaded in Figure 4.26—that is,

$$P(x \leq 10) \approx P(z \leq -.68) = .5 - P(-.68 < z \leq 0) = .5 - .2517 = .2483$$

The approximation differs only slightly from the exact binomial probability, .245. Of course, when tables of exact binomial probabilities are available, we will use the exact value rather than a normal approximation.

Use of the normal distribution will not always provide a good approximation for binomial probabilities. The following is a useful rule of thumb to determine when n is large enough for the approximation to be effective: *The interval $\mu \pm 3\sigma$ should lie within the range of the binomial random variable* x *(i.e., 0 to* n*) in order for the normal approximation to be adequate.* The rule works well because almost all of the normal distribution falls within 3 standard deviations of the mean, so if this interval is contained within the range of x values, there is "room" for the normal approximation to work.

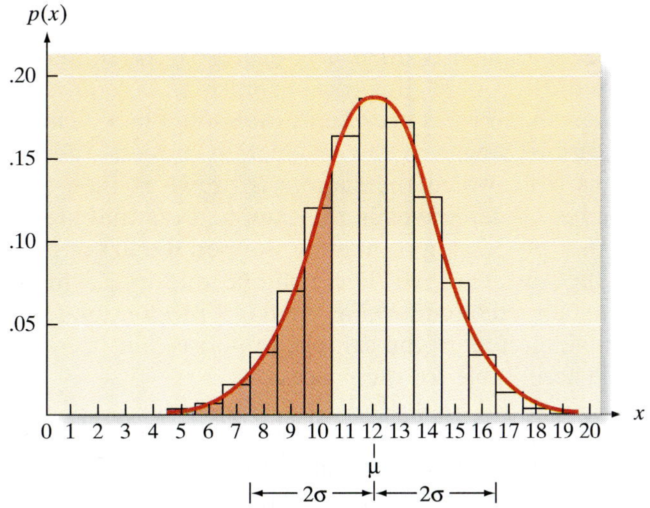

Figure 4.26
Binomial distribution for $n = 20$, $p = .6$ and normal distribution with $\mu = 12$, $\sigma = 2.2$

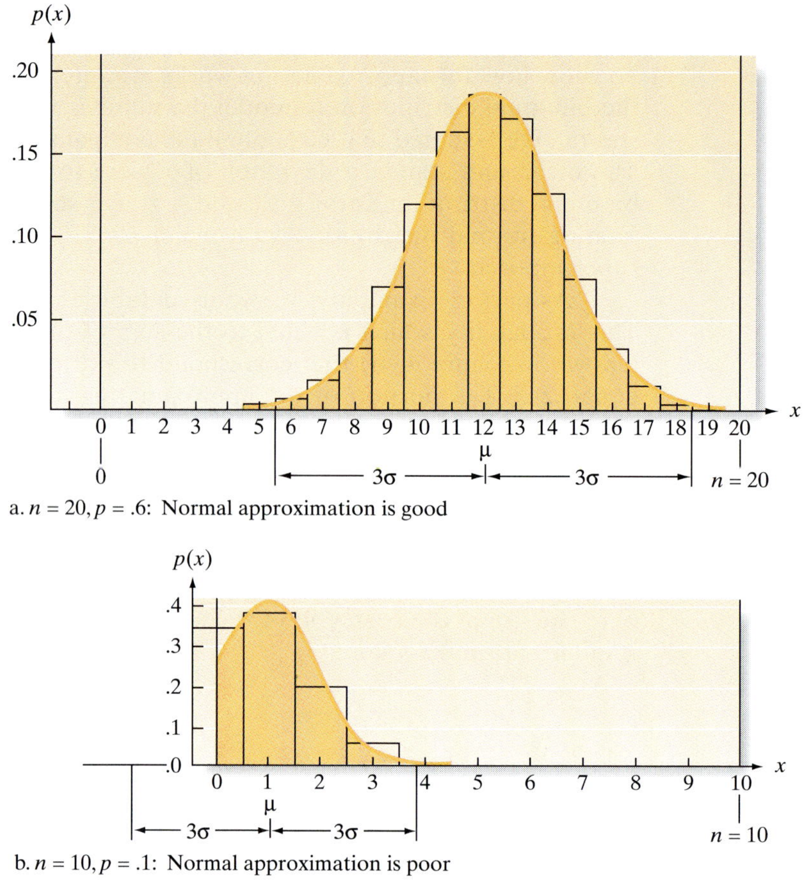

Figure 4.27
Rule of thumb for normal approximation to binomial probabilities

BIOGRAPHY

ABRAHAM DE MOIVRE (1667–1754)

Advisor to Gamblers

French-born mathematician Abraham de Moivre moved to London when he was 21 years old to escape religious persecution. In England, he earned a living first as a traveling teacher of mathematics and then as an advisor to gamblers, underwriters, and annuity brokers. De Moivre's major contributions to probability theory are contained in two of his books: *The Doctrine of Chances* (1718) and *Miscellanea Analytica* (1730). In these works he defines statistical independence, develops the formula for the normal probability distribution, and derives the normal curve as an approximation to the binomial distribution. Despite his eminence as a mathematician, de Moivre died in poverty. He is famous for predicting the day of his own death using an arithmetic progression. ■

As shown in Figure 4.27a for the preceding example with $n = 20$ and $p = .6$, the interval $\mu \pm 3\sigma = 12 \pm 3(2.2) = (5.4, 18.6)$ lies within the range 0 to 20. However, if we were to try to use the normal approximation with $n = 10$ and $p = .1$, the interval $\mu \pm 3\sigma$ is $1 \pm 3(.95)$, or $(-1.85, 3.85)$. As shown in Figure 4.27b, this interval is not contained within the range of x because $x = 0$ is the lower bound for a binomial random variable. Note in Figure 4.27b that the normal distribution will not "fit" in the range of x, and therefore it will not provide a good approximation to the binomial probabilities.

Example 4.25

Applying the Normal Approximation to a Binomial Probability—Lot Acceptance Sampling

Problem One problem with any product (e.g., a graphing calculator) that is mass produced is quality control. The process must somehow be monitored or audited to be sure the output of the process conforms to requirements. One method of dealing with this problem is *lot acceptance sampling*, in which items being produced are sampled at various stages of the production process and are carefully inspected. The lot of items from which the sample is drawn is then accepted or rejected, based on the number of defectives in the sample. Lots that are accepted may be sent forward for further processing or may be shipped to customers; lots that are rejected may be reworked or scrapped. For example, suppose a manufacturer of calculators chooses 200 stamped circuits from the day's production and determines x, the number of defective circuits in the sample. Suppose that up to a 6% rate of defectives is considered acceptable for the process.

a. Find the mean and standard deviation of x, assuming the defective rate is 6%.

b. Use the normal approximation to determine the probability that 20 or more defectives are observed in the sample of 200 circuits (i.e., find the approximate probability that $x \geq 20$).

Solution

a. The random variable x is binomial with $n = 200$ and the fraction defective $p = .06$. Thus,

$$\mu = np = 200(.06) = 12$$
$$\sigma = \sqrt{npq} = \sqrt{200(.06)(.94)} = \sqrt{11.28} = 3.36$$

We first note that

$$\mu \pm 3\sigma = 12 \pm 3(3.36) = 12 \pm 10.08 = (1.92, 22.08)$$

lies completely within the range from 0 to 200. Therefore, a normal probability distribution should provide an adequate approximation to this binomial distribution.

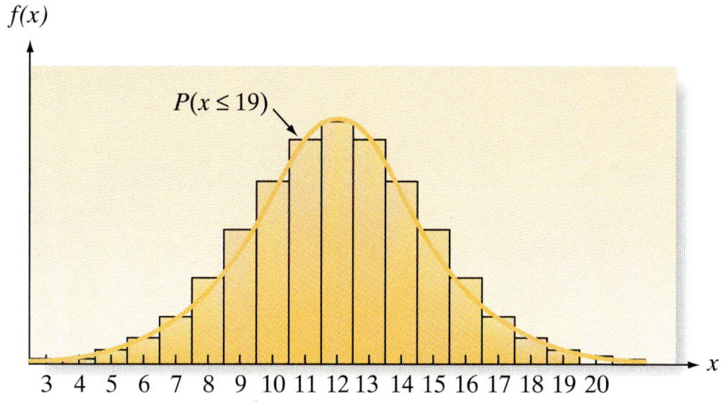

Figure 4.28
Normal approximation to the binomial distribution with $n = 200$, $p = .06$

b. Using the rule of complements, $P(x \geq 20) = 1 - P(x \leq 19)$. To find the approximating area corresponding to $x \leq 19$, refer to Figure 4.28. Note that we want to include all the binomial probability histogram from 0 to 19, inclusive. Because the event is of the form $x \leq a$, the proper correction for continuity is $a + .5 = 19 + .5 = 19.5$. Thus, the z-value of interest is

$$z = \frac{(a + .5) - \mu}{\sigma} = \frac{19.5 - 12}{3.36} = \frac{7.5}{3.36} = 2.23$$

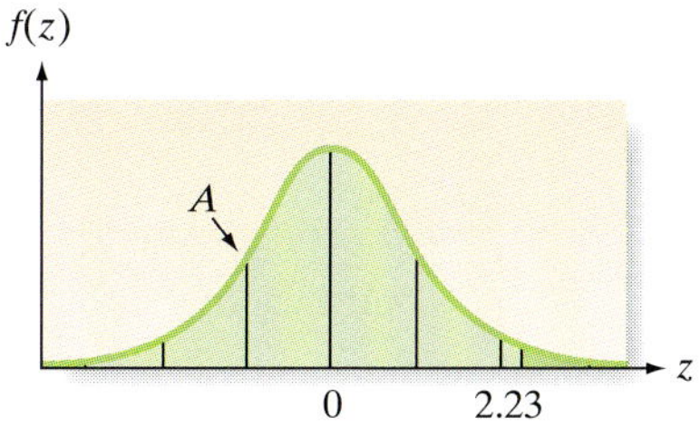

Figure 4.29
Standard normal distribution

Referring to Table IV in Appendix B, we find that the area to the right of the mean 0 corresponding to $z = 2.23$ (see Figure 4.29) is .4871. So the area $A = P(z \leq 2.23)$ is

$$A = .5 + .4871 = .9871$$

Thus, the normal approximation to the binomial probability we seek is

$$P(x \geq 20) = 1 - P(x \leq 19) \approx 1 - .9871 = .0129$$

In other words, the probability is extremely small that 20 or more defectives will be observed in a sample of 200 circuits—*if in fact the true fraction of defectives is .06.*

Look Back If the manufacturer observes $x \geq 20$, the likely reason is that the process is producing more than the acceptable 6% defectives. The lot acceptance sampling procedure is another example of using the rare-event approach to make inferences.

Now Work Exercise 4.118

The steps for approximating a binomial probability by a normal probability are given in the next box.

Using a Normal Distribution to Approximate Binomial Probabilities

1. After you have determined n and p for the binomial distribution, calculate the interval

$$\mu \pm 3\sigma = np \pm 3\sqrt{npq}$$

If the interval lies in the range 0 to n, the normal distribution will provide a reasonable approximation to the probabilities of most binomial events.

2. Express the binomial probability to be approximated in the form $P(x \leq a)$ or $P(x \leq b) - P(x \leq a)$. For example,

$$P(x < 3) = P(x \leq 2)$$
$$P(x \geq 5) = 1 - P(x \leq 4)$$
$$P(7 \leq x \leq 10) = P(x \leq 10) - P(x \leq 6)$$

3. For each value of interest a, the correction for continuity is $(a + .5)$, and the corresponding standard normal z-value is

$$z = \frac{(a + .5) - \mu}{\sigma} \quad \text{(See Figure 4.30)}$$

4. Sketch the approximating normal distribution and shade the area corresponding to the probability of the event of interest, as in Figure 4.30. Verify that the rectangles you have included in the shaded area correspond to the event probability you wish to approximate. Using Table IV and the z-value(s) you calculated in step **3,** find the shaded area. This is the approximate probability of the binomial event.

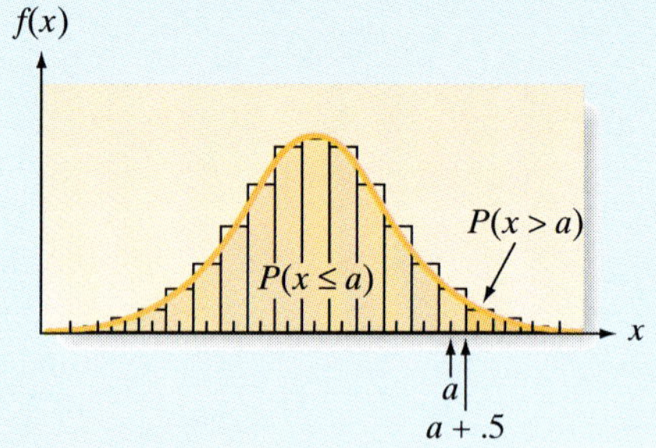

Figure 4.30
Approximating binomial probabilities by normal probabilities

Exercises 4.118–4.132

Learning the Mechanics

4.118 Suppose x is a binomial random variable with $p = .4$ and $n = 25$.
NW

a. Would it be appropriate to approximate the probability distribution of x with a normal distribution? Explain.

b. Assuming that a normal distribution provides an adequate approximation to the distribution of x, what are the mean and variance of the approximating normal distribution?

c. Use Table II in Appendix B to find the exact value of $P(x \geq 9)$.

d. Use the normal approximation to find $P(x \geq 9)$.

4.119 Assume that x is a binomial random variable with n and p as specified in parts **a–f.** For which cases would it be appropriate to use a normal distribution to approximate the binomial distribution?

a. $n = 100, p = .01$ **b.** $n = 20, p = .6$
c. $n = 10, p = .4$ **d.** $n = 1{,}000, p = .05$
e. $n = 100, p = .8$ **f.** $n = 35, p = .7$

4.120 Assume that x is a binomial random variable with $n = 1{,}000$ and $p = .50$. Use a normal approximation to find each of the following probabilities:

a. $P(x > 500)$
b. $P(490 \leq x < 500)$
c. $P(x > 550)$

4.121 Assume that x is a binomial random variable with $n = 100$ and $p = .40$. Use a normal approximation to find the following:

a. $P(x \leq 35)$
b. $P(40 \leq x \leq 50)$
c. $P(x \geq 38)$

Applying the Concepts—Basic

4.122 Using wireless Internet at home. In Exercise 4.43 (p. 196), you learned that 20% of Internet users go online at home using a wireless network (*Pew Internet & American Life Project,* Feb. 2007). In a random sample of 100 Internet users, let x be the number who go online at home using a wireless network.

a. Find the mean of x. (This value should agree with your answer to Exercise 4.43c.)
b. Find the standard deviation of x.
c. Find the z-score for the value $x = 27.5$.
d. Find the approximate probability that the number of Internet users who go online at home using a wireless network in a sample of 100 is less than or equal to 25.

4.123 LASIK surgery complications. According to recent studies, 1% of all patients who undergo laser surgery (i.e., LASIK) to correct their vision have serious postlaser vision problems (*All About Vision,* 2006). In a sample of 100,000 patients, what is the approximate probability that fewer than 950 will experience serious postlaser vision problems?

4.124 Women-owned businesses. In Exercise 4.45 (p. 196), you learned that 27% of all small businesses owned by non-Hispanic whites nationwide are women-owned firms (*Journal of Business Venturing,* Vol. 17, 2002). In a random sample of 350 small businesses owned by non-Hispanic whites, let x be the number that are owned by a woman.
a. Find the mean of x.
b. Find the standard deviation of x.
c. Find the z-score for the value $x = 99.5$.
d. Find the approximate probability that the number of small businesses in the sample of 350 that are owned by a woman is 100 or more.

4.125 Using the Internet for banking. Security First Network Bank (SFNB), founded in 1995 in Atlanta, Georgia, is considered to be the first ever Internet bank. Today, almost all financial institutions offer services via the Internet. According to *Consumer Interests Annual* (2007), the proportion of U.S. households that do some or all of their banking on the Internet is .28. Consider a random sample of 1,000 U.S. households.
a. How many of these households would you expect to use Internet banking?
b. Find the approximate probability that more than 750 of these households use Internet banking.

Applying the Concepts—Intermediate

4.126 Defects in semiconductor wafers. The computer chips in today's notebook and laptop computers are produced from semiconductor wafers. Certain semiconductor wafers are exposed to an environment that generates up to 100 possible defects per wafer. The number of defects per wafer, x, was found to follow a binomial distribution if the manufacturing process is stable and generates defects that are randomly distributed on the wafers (*IEEE Transactions on Semiconductor Manufacturing,* May 1995). Let p represent the probability that a defect occurs at any one of the 100 points of the wafer. For each of the following cases, determine whether the normal approximation can be used to characterize x.
a. $p = .01$
b. $p = .50$
c. $p = .90$

4.127 Analysis of bottled water. Refer to the *Scientific American* (July 2003) report on whether bottled water is really purified water, Exercise 4.44 (p. 196). Recall that the Natural Resources Defense Council found that 25% of bottled water brands fill their bottles with just tap water. In a random sample of 65 bottled water brands, is it likely that 20 or more brands will contain tap water? Explain.

4.128 Credit card market shares. The following table reports the U.S. credit card industry's market share data for 2007. A random sample of 100 credit card users is to be questioned regarding their satisfaction with their credit card company. For simplification, assume that each credit card user carries just one credit card and that the market share percentages are the percentages of all credit card customers that carry each brand.

Credit Card	Market Share %
Visa	46
MasterCard	36
American Express	12
Discover	6

Source: Nilson Report, May 2008.

a. Propose a procedure for randomly selecting the 100 credit card users.
b. For random samples of 100 credit card users, what is the expected number of customers who carry Visa? Discover?
c. What is the probability that half or more of the sample of credit card users carry Visa? American Express?
d. Justify the use of the normal approximation to the binomial in answering the question in part **c**.

4.129 Testing a manufacturer's claim. A manufacturer of CD-ROMs claims that 99.4% of its CDs are defect-free. A large software company that buys and uses a large number of the CDs wants to verify this claim, so it selects 1,600 CDs to be tested. The tests reveal 12 CDs to be defective. Assuming that the manufacturer's claim is correct, what is the probability of finding 12 or more defective CDs in a sample of 1,600? Does your answer cast doubt on the manufacturer's claim? Explain.

4.130 Financial aid for college students. The *Chronicle of Higher Education Almanac* reports that the percentage of undergraduates in the United States receiving federal financial aid is 45% at public four-year institutions and 52% at private four-year institutions. The U.S. Department of Education is interested in questioning a random sample of 100 U.S. undergraduate students to assess their satisfaction with federal financial aid procedures and policies.
a. Explain the difficulties of obtaining the desired random sample.
b. Assume the appropriate percentage above applies to your institution. If a random sample of 100 students from your institution were contacted, what is the approximate probability that 50 or more receive financial aid? Less than 25?
c. What assumptions must be made in order to answer part **b** using the normal approximation to the binomial?

Applying the Concepts—Advanced

4.131 Waiting time at an emergency room. According to *Health Affairs* (Oct. 28, 2004), the median time a patient waits to see a doctor in a typical U.S. emergency room is 30 minutes. Consider a U.S. emergency room on a day when 150 patients visit. What is the probability that

a. more than half will wait more than 30 minutes?
b. more than 85 will wait more than 30 minutes?
c. more than 60, but fewer than 90, will wait more than 30 minutes?

4.132 Luggage inspection at Newark airport. *New Jersey Business* reports that Newark Liberty International Airport's new terminal handles an average of 3,000 international passengers an hour but is capable of handling twice that number. Also, after scanning all luggage, 20% of arriving international passengers are detained for intrusive luggage inspection. The inspection facility can handle 600 passengers an hour without unreasonable delays for the travelers.

a. When international passengers arrive at the rate of 1,500 per hour, what is the expected number of passengers who will be detained for luggage inspection?
b. In the future, it is expected that as many as 4,000 international passengers will arrive per hour. When that occurs, what is the expected number of passengers who will be detained for luggage inspection?
c. Refer to part **b.** Find the approximate probability that more than 600 international passengers will be detained for luggage inspection. (This is also the probability that travelers will experience unreasonable luggage inspection delays.)

4.9 Other Continuous Distributions: Uniform and Exponential

Uniform Random Variable

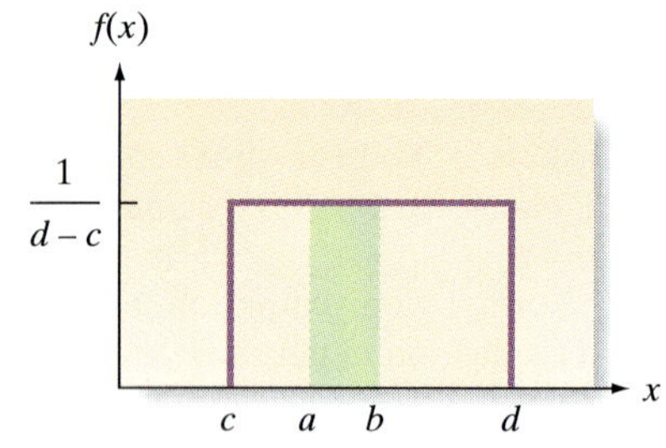

Figure 4.31
The uniform probability distribution

Continuous random variables that appear to have equally likely outcomes over their range of possible values possess a **uniform probability distribution.** For example, if a short exists in a 5-meter stretch of electrical wire, it may have an equal probability of being in any particular 1-centimeter segment along the line. Or if a safety inspector plans to choose a time at random during the four afternoon work hours to pay a surprise visit to a certain area of a plant, then each 1-minute time interval in this 4-work-hour period will have an equally likely chance of being selected for the visit.

Suppose the random variable x can assume values only in an interval $c \leq x \leq d$. Then the **uniform frequency function** has a rectangular shape, as shown in Figure 4.31. Note that the possible values of x consist of all points in the interval between point c and point d. The height of $f(x)$ is constant in that interval and equals $1/(d - c)$. Therefore, the total area under $f(x)$ is given by

$$\text{Total area of rectangle} = (\text{Base})(\text{Height}) = (d - c)\left(\frac{1}{d - c}\right) = 1$$

The uniform probability distribution provides a model for continuous random variables that are *evenly distributed* over a certain interval—that is, a uniform random variable is one that is just as likely to assume a value in one interval as it is to assume a value in any other interval of equal size. There is no clustering of values around any value; instead, there is an even spread over the entire region of possible values.

The uniform distribution is sometimes referred to as the **randomness distribution** because one way of generating a uniform random variable is to perform an experiment in which a point is *randomly selected* on the horizontal axis between the points c and d. If we were to repeat this experiment infinitely often, we would create a uniform probability distribution like that shown in Figure 4.31. The random selection of points in an interval can also be used to generate random numbers such as those in Table I in Appendix B. Recall that random numbers are selected in such a way that every number would have an equal probability of selection. Therefore, random numbers are realizations of a uniform random variable. (Random numbers were used to draw random samples in Section 3.7.) The formulas for the uniform probability distribution, its mean, and standard deviation are shown in the next box.

Suppose the interval $a < x < b$ lies within the domain of x; that is, it falls within the larger interval $c \leq x \leq d$. Then the probability that x assumes a value within the

Probability Distribution for a Uniform Random Variable *x*

Probability density function: $f(x) = \frac{1}{d - c} \quad c \le x \le d$

Mean: $\mu = \frac{c + d}{2}$ Standard deviation: $\sigma = \frac{d - c}{\sqrt{12}}$

$P(a < x < b) = (b - a)/(d - c), c \le a < b \le d$

interval $a < x < b$ is equal to the area of the rectangle over the interval, namely, $(b - a)/(d - c)$.* (See the shaded area in Figure 4.31.)

Example 4.26

Applying the Uniform Distribution to Steel Manufacturing

Problem Suppose the research department of a steel manufacturer believes that one of the company's rolling machines is producing sheets of steel of varying thickness. The thickness is a uniform random variable with values between 150 and 200 millimeters. Any sheets less than 160 millimeters must be scrapped because they are unacceptable to buyers.

a. Calculate and interpret the mean and standard deviation of *x*, the thickness of the sheets produced by this machine.

b. Graph the probability distribution of *x*, and show the mean on the horizontal axis. Also show 1- and 2-standard-deviation intervals around the mean.

c. Calculate the fraction of steel sheets produced by this machine that have to be scrapped.

Solution

a. To calculate the mean and standard deviation for *x*, we substitute 150 and 200 millimeters for *c* and *d*, respectively, in the formulas for uniform random variables. Thus,

$$\mu = \frac{c + d}{2} = \frac{150 + 200}{2} = 175 \text{ millimeters}$$

and

$$\sigma = \frac{d - c}{\sqrt{12}} = \frac{200 - 150}{\sqrt{12}} = \frac{50}{3.464} = 14.43 \text{ millimeters}$$

Our interpretations follow:

The average thickness of all manufactured steel sheets is $\mu = 175$ millimeters. From Chebyshev's Rule (Table 2.6, p. 71), we know that at least 75% of the thickness values, *x*, in the distribution will fall in the interval

$$\mu \pm 2\sigma = 175 \pm 2(14.43)$$
$$= 175 \pm 28.86$$

or between 146.14 and 203.86 millimeters. (This demonstrates, once again, the conservativeness of Chebyshev's Rule because we know that all values of *x* fall between 150 and 200 millimeters.)

b. The uniform probability distribution is

$$f(x) = \frac{1}{d - c} = \frac{1}{200 - 150} = \frac{1}{50} \quad (150 \le x \le 200)$$

*The student with knowledge of calculus should note that

$$P(a < x < b) \int_a^b f(x)d(x) = \int_a^b 1/(d - c)dx = (b - a)/(d - c)$$

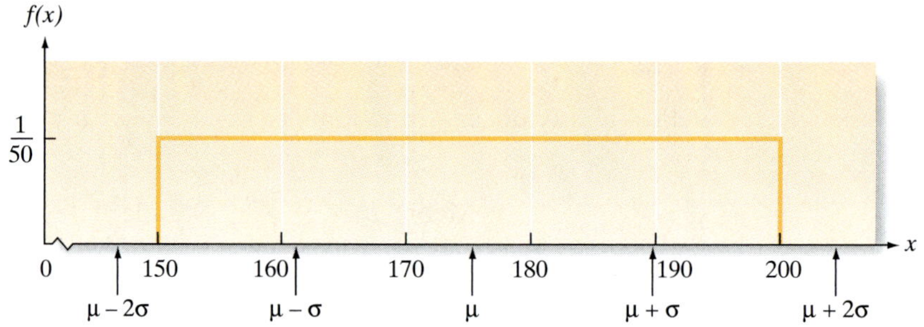

Figure 4.32
Distribution for x in Example 4.26

The graph of this function is shown in Figure 4.32. The mean and 1- and 2-standard-deviation intervals around the mean are shown on the horizontal axis.

c. To find the fraction of steel sheets produced by the machine that have to be scrapped, we must find the probability that x, the thickness, is less than 160 millimeters. As indicated in Figure 4.33, we need to calculate the area under the frequency function $f(x)$ between the points $x = 150$ and $x = 160$. Therefore, in this case, $a = 150$ and $b = 160$. Applying the formula in the box, we have

$$P(x < 160) = P(150 < x < 160)$$
$$= \frac{b - a}{d - c} = \frac{160 - 150}{200 - 150} = \frac{10}{50} = \frac{1}{5}$$

That is, 20% of all the sheets made by this machine must be scrapped.

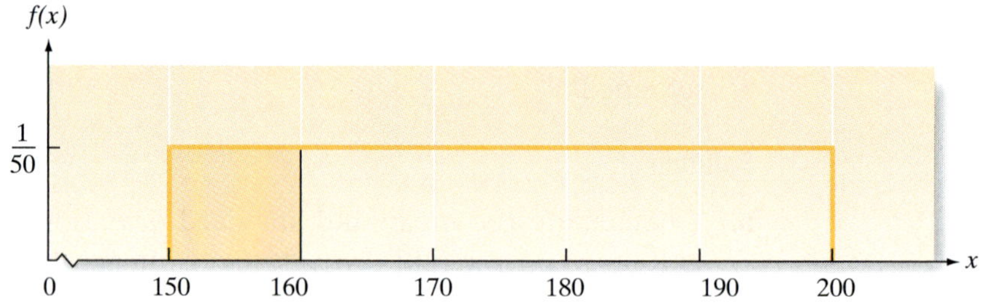

Figure 4.33
Probability that sheet thickness, x, is between 150 and 160 millimeters

Look Back The calculated probability in part **c** is the area of a rectangle with base $160 - 150 = 10$ and height $1/50$. Alternatively, we can find the fraction that has to be scrapped as

$$P(x < 160) = (\text{Base})(\text{Height}) = (10)\left(\frac{1}{50}\right) = \frac{1}{5}$$

Now Work Exercise 4.142

Exponential Random Variable

The length of time between emergency arrivals at a hospital, the length of time between breakdowns of manufacturing equipment, and the length of time between catastrophic events (e.g., a stockmarket crash), are all continuous random phenomena that we might want to describe probabilistically. The length of time or the distance between occurrences of random events like these can often be described by the **exponential probability distribution.** For this reason, the exponential distribution is sometimes called the **waiting-time distribution.** The formula for the exponential probability distribution is shown in the following box, along with its mean and standard deviation.

Probability Distribution for an Exponential Random Variable x

Probability density function: $f(x) = \frac{1}{\theta} e^{-x/\theta} (x > 0)$

Mean: $\mu = \theta$

Standard deviation: $\sigma = \theta$

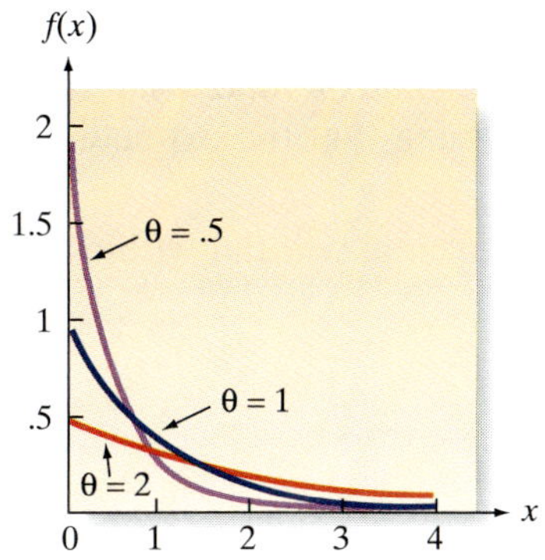

Figure 4.34
Exponential distributions

Unlike the normal distribution, which has a shape and location determined by the values of the two quantities μ and σ, the shape of the exponential distribution is governed by a single quantity: θ. Further, it is a probability distribution with the property that its mean equals its standard deviation. Exponential distributions corresponding to $\theta = .5$, 1, and 2 are shown in Figure 4.34.

To calculate probabilities of **exponential random variables,** we need to be able to find areas under the exponential probability distribution. Suppose we want to find the area A to the right of some number a, as shown in Figure 4.35. This area can be calculated by means of the formula shown in the box that follows. Use Table V in Appendix B or a calculator with an exponential function to find the value of $e^{-a/\theta}$ after substituting the appropriate numerical values for θ and a.

Finding the Area *A* to the Right of a Number *a* for an Exponential Distribution*

$$A = P(x \geq a) = e^{-a/\theta}$$

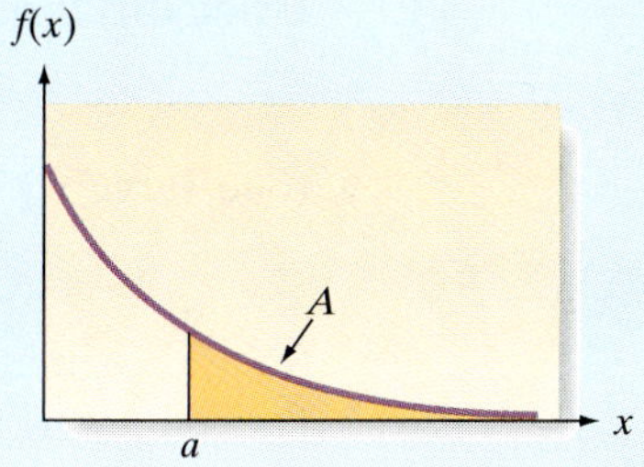

Figure 4.35
The area A to the right of a number a for an exponential distribution

Example 4.27
Finding an Exponential Probability—Hospital Emergency Arrivals

Problem Suppose the length of time (in hours) between emergency arrivals at a certain hospital is modeled as an exponential distribution with $\theta = 2$. What is the probability that more than 5 hours pass without an emergency arrival?

Solution The probability we want is the area A to the right of $a = 5$ in Figure 4.36. To find this probability, we use the area formula:

$$A = e^{-a/\theta} = e^{-(5/2)} = e^{-2.5}$$

Referring to Table V, we find that

$$A = e^{-2.5} = .082085$$

Our exponential model indicates that the probability that more than 5 hours pass between emergency arrivals is about .08 for this hospital.

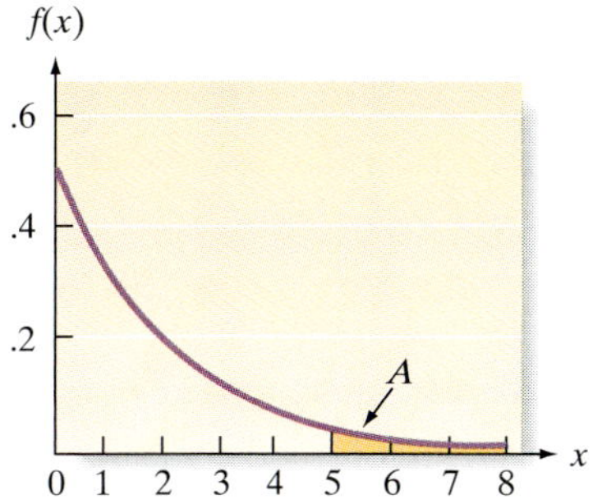

Figure 4.36
Area to the right of $a = 5$ for Example 4.27

Look Back The value $e^{-2.5}$ can also be found with a standard calculator. Or you can find the desired probability with a statistical software package.

Now Work Exercise 4.137

Example 4.28
The Mean and Variance of an Exponential Random Variable—Length of Life of a Microwave Oven

Problem A manufacturer of microwave ovens is trying to determine the length of warranty period it should attach to its magnetron tube, the most critical component in the oven. Preliminary testing has shown that the length of life (in years), x, of a magnetron tube has an exponential probability distribution with $\theta = 6.25$.

*For students with a knowledge of calculus, the shaded area in Figure 4.35 corresponds to the integral

$$\int_a^b \frac{1}{\theta} e^{-x/\theta} dx = -e^{-x/\theta} \Big]_a^{\infty} = e^{-a/\theta}.$$

a. Find the mean and standard deviation of x.

b. Suppose a warranty period of five years is attached to the magnetron tube. What fraction of tubes must the manufacturer plan to replace, assuming that the exponential model with $\theta = 6.25$ is correct?

c. Find the probability that the length of life of a magnetron tube will fall within the interval $\mu - 2\sigma$ to $\mu + 2\sigma$.

Solution

a. Because $\theta = \mu = \sigma$, both μ and σ equal 6.25.

b. To find the fraction of tubes that will have to be replaced before the five-year warranty period expires, we need to find the area between 0 and 5 under the distribution. This area, A, is shown in Figure 4.37. To find the required probability, we recall the formula

$$P(x > a) = e^{-a/\theta}$$

Using this formula, we find that

$$P(x > 5) = e^{-a/\theta} = e^{-5/6.25} = e^{-.80} = .449329$$

(See Table V.) To find the area A, we use the complementary relationship:

$$P(x \leq 5) = 1 - P(x > 5) = 1 - .449329 = .550671$$

So approximately 55% of the magnetron tubes will have to be replaced during the five-year warranty period.

c. We would expect the probability that the life of a magnetron tube, x, falls within the interval $\mu - 2\sigma$ to $\mu + 2\sigma$ to be quite large. A graph of the exponential distribution showing the interval from $\mu - 2\sigma$ to $\mu + 2\sigma$ is given in Figure 4.38. Because the point $\mu - 2\sigma$ lies below $x = 0$, we need to find only the area between $x = 0$ and $x = \mu + 2\sigma = 6.25 + 2(6.25) = 18.75$.

This area, P, which is shaded in Figure 4.38, is

$$P = 1 - P(x > 18.75) = 1 - e^{-18.75/\theta} = 1 - e^{-18.75/6.25} = 1 - e^{-3}$$

Using Table V or a calculator, we find that $e^{-3} = .049787$. Therefore, the probability that the life x of a magnetron tube will fall within the interval $\mu - 2\sigma$ to $\mu + 2\sigma$ is

$$P = 1 - e^{-3} = 1 - .049787 = .950213$$

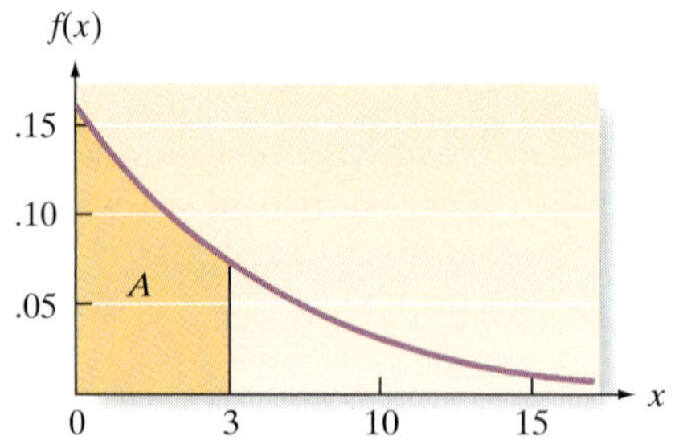

Figure 4.37
Area to the left of $a = 5$ for Example 4.28

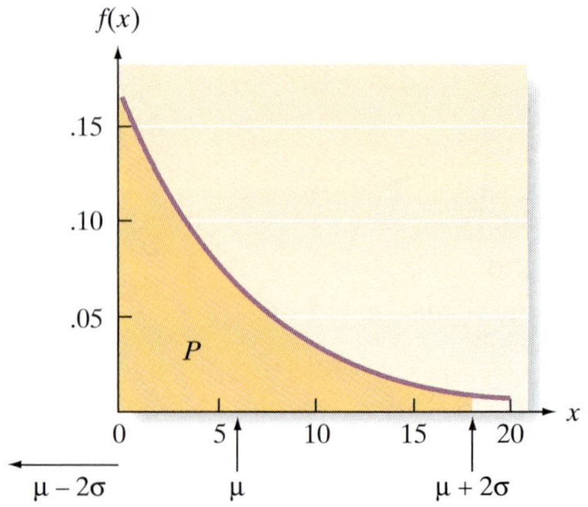

Figure 4.38
Area in the interval $\mu \pm 2\sigma$ for Example 4.28

Look Back You can see that this probability agrees well with the interpretation of a standard deviation given by the Empirical Rule (Table 2.7, p. 71), even though the probability distribution that we were given is not mound shaped. (It is strongly skewed to the right.)

Exercises 4.133–4.155

Learning the Mechanics

4.133 Suppose x is a random variable best described by a uniform probability distribution with $c = 20$ and $d = 45$.
   a. Find $f(x)$.
   b. Find the mean and standard deviation of x.
   c. Graph $f(x)$ and locate μ and the interval $\mu \pm 2\sigma$ on the graph. Note that the probability that x assumes a value within the interval $\mu \pm 2\sigma$ is equal to 1.

4.134 Refer to Exercise 4.133. Find the following probabilities:
   a. $P(20 \le x \le 30)$ **b.** $P(20 < x \le 30)$
   c. $P(x \ge 30)$ **d.** $P(x \ge 45)$
   e. $P(x \le 40)$ **f.** $P(x < 40)$
   g. $P(15 \le x \le 35)$ **h.** $P(21.5 \le x \le 31.5)$

4.135 Suppose x is a random variable best described by a uniform probability distribution with $c = 3$ and $d = 7$.
   a. Find $f(x)$.
   b. Find the mean and standard deviation of x.
   c. Find $P(\mu - \sigma \le x \le \mu + \sigma)$.

4.136 Use Table V in Appendix B to determine the value of $e^{-a/\theta}$ for each of the following cases:
   a. $\theta = 1, a = 1$ **b.** $\theta = 1, a = 2.5$
   c. $\theta = .4, a = 3$ **d.** $\theta = .2, a = .3$

4.137 NW Suppose x has an exponential distribution with $\theta = 1$. Find the following probabilities:
   a. $P(x > 1)$ **b.** $P(x \le 3)$
   c. $P(x > 1.5)$ **d.** $P(x \le 5)$

4.138 Suppose x has an exponential distribution with $\theta = 2.5$. Find the following probabilities:
   a. $P(x \le 4)$ **b.** $P(x > 5)$
   c. $P(x \le 2)$ **d.** $P(x > 3)$

4.139 The random variable x is best described by a uniform probability distribution with $c = 100$ and $d = 200$. Find the probability that x assumes a value
   a. More than 2 standard deviations from μ.
   b. Less than 3 standard deviations from μ.
   c. Within 2 standard deviations of μ.

4.140 The random variable x can be adequately approximated by an exponential probability distribution with $\theta = 2$. Find the probability that x assumes a value
   a. More than 3 standard deviations from μ.
   b. Less than 2 standard deviations from μ.
   c. Within half a standard deviation of μ.

Applet Exercise 4.7

Open the applet *Sample from a Population*. On the pull-down menu to the right of the top graph, select *Uniform*. The box to the left of the top graph displays the population mean, median, and standard deviation.
   a. Run the applet for each available value of n on the pull-down menu for the sample size. Go from the smallest to the largest value of n. For each value of n, observe the shape of the graph of the sample data and record the mean, median, and standard deviation of the sample.
   b. Describe what happens to the shape of the graph and the mean, median, and standard deviation of the sample as the sample size increases.

Applet Exercise 4.8

Suppose we set the *Random Numbers* applet to generate one number between 1 and 100, inclusive. We let the value of the random variable x be the number generated when the *Sample* button is clicked. Explain why the distribution of x is approximately uniform even though x is a discrete rather than continuous random variable.

Applying the Concepts—Basic

4.141 **Maintaining pipe wall temperature.** Maintaining a constant pipe wall temperature in some hot-process applications is critical. A new technique that utilizes bolt-on trace elements to maintain temperature was presented in the *Journal of Heat Transfer* (Nov. 2000). Without bolt-on trace elements, the pipe wall temperature of a switch condenser used to produce plastic has a uniform distribution ranging from 260° to 290°F. When several bolt-on trace elements are attached to the piping, the wall temperature is uniform from 278° to 285°F.
   a. Ideally, the pipe wall temperature should range between 280° and 284°F. What is the probability that the temperature will fall in this ideal range when no bolt-on trace elements are used? When bolt-on trace elements are attached to the pipe?
   b. When the temperature is 268°F or lower, the hot liquid plastic hardens (or plates), causing a buildup in the piping. What is the probability of plastic plating when no bolt-on trace elements are used? When bolt-on trace elements are attached to the pipe?

4.142 **New method for detecting anthrax.** Researchers at the University of South Florida Center for Biological Defense have developed a safe method for rapidly detecting anthrax spores in powders and on surfaces (*USF Magazine,* Summer 2002). The method has been found to work well even when there are very few anthrax spores in a powder specimen. Consider a powder specimen that has exactly 10 anthrax spores. Suppose that the number of anthrax spores in the sample detected by the new method follows an approximate uniform distribution between 0 and 10.
   a. Find the probability that 8 or fewer anthrax spores are detected in the powder specimen.
   b. Find the probability that between 2 and 5 anthrax spores are detected in the powder specimen.

4.143 **Lead in metal shredder residue.** On the basis of data collected from metal shredders across the nation, the amount x of extractable lead in metal shredder residue has an approximate exponential distribution with mean $\theta = 2.5$ milligrams per liter (Florida Shredder's Association).
   a. Find the probability that x is greater than 2 milligrams per liter.
   b. Find the probability that x is less than 5 milligrams per liter.

4.144 **Critical-part failures in NASCAR vehicles.** In NASCAR races such as the Daytona 500, 43 drivers start the race; however, about 10% of the cars do not finish due to the failure of critical parts. University of Portland professors conducted a study of critical-part failures from 36 NASCAR races (*The Sport Journal,* Winter 2007). The researchers discovered that the time (in hours) until the first critical-part failure is exponentially distributed with a mean of .10 hour.

a. Find the probability that the time until the first critical-part failure is 1 hour or more.
b. Find the probability that the time until the first critical-part failure is less than 30 minutes.

4.145 Random numbers. The data set listed in the table was created using the Minitab random number generator and is saved in the **RANUNI** file. Construct a relative frequency histogram for the data. Except for the expected variation in relative frequencies among the class intervals, does your histogram suggest that the data are observations on a uniform random variable with $c = 0$ and $d = 100$? Explain.

38.8759	98.0716	64.5788	60.8422	.8413
88.3734	31.8792	32.9847	.7434	93.3017
12.4337	11.7828	87.4506	94.1727	23.0892
47.0121	43.3629	50.7119	88.2612	69.2875
62.6626	55.6267	78.3936	28.6777	71.6829
44.0466	57.8870	71.8318	28.9622	23.0278
35.6438	38.6584	46.7404	11.2159	96.1009
95.3660	21.5478	87.7819	12.0605	75.1015

Applying the Concepts—Intermediate

4.146 Marine losses for an oil company. The frequency distribution shown in the next table depicts the property and marine losses incurred by a large oil company over the last two years. This distribution can be used by the company to predict future losses and to help determine an appropriate level of insurance coverage. In analyzing the losses within an interval of the distribution, for simplification, analysts may treat the interval as a uniform probability distribution (*Research Review,* Summer 1998). In the insurance business, intervals like these are often called *layers*.

Layer	Property and Marine Losses (millions of $)	Frequency
1	0.00–0.01	668
2	0.01–0.05	38
3	0.05–0.10	7
4	0.10–0.25	4
5	0.25–0.50	2
6	0.50–1.00	1
7	1.00–2.50	0

Source: Cozzolino, J. M., and Mikolaj, P. J. "Applications of the piecewise constant pareto distribution," *Research Review,* Summer 1998.

a. Use a uniform distribution to model the loss amount in layer 2. Graph the distribution. Calculate and interpret its mean and variance.
b. Repeat part **a** for layer 6.
c. If a loss occurs in layer 2, what is the probability that it exceeds $10,000? That it is under $25,000?
d. If a layer-6 loss occurs, what is the probability that it is between $750,000 and $1,000,000? That it exceeds $900,000? That it is exactly $900,000?

4.147 Soft-drink dispenser. The manager of a local soft-drink bottling company believes that when a new beverage-dispensing machine is set to dispense 7 ounces, it in fact dispenses an amount x at random anywhere between 6.5 and 7.5 ounces inclusive. Suppose x has a uniform probability distribution.

a. Is the amount dispensed by the beverage machine a discrete or a continuous random variable? Explain.
b. Graph the frequency function for x, the amount of beverage the manager believes is dispensed by the new machine when it is set to dispense 7 ounces.
c. Find the mean and standard deviation for the distribution graphed in part **b** and locate the mean and the interval $\mu \pm 2\sigma$ on the graph.
d. Find $P(x \geq 7)$.
e. Find $P(x < 6)$.
f. Find $P(6.5 \leq x \leq 7.25)$.
g. What is the probability that each of the next six bottles filled by the new machine will contain more than 7.25 ounces of beverage? Assume that the amount of beverage dispensed in one bottle is independent of the amount dispensed in another bottle.

4.148 NHL overtime games. In the National Hockey League (NHL), games that are tied at the end of three periods are sent into "sudden-death" overtime. In overtime, the team to score the first goal wins. An analysis of NHL overtime games showed that the length of time elapsed before the winning goal is scored has an exponential distribution with mean 9.15 minutes (*Chance,* Winter 1995).

a. For a randomly selected overtime NHL game, find the probability that the winning goal is scored in 3 minutes or less.
b. In the NHL, each period (including overtime) lasts 20 minutes. If neither team scores a goal in overtime, the game is considered a tie. What is the probability of an NHL game ending in a tie?

4.149 Ship-to-shore transfer times. Lack of port facilities or shallow water may require cargo on a large ship to be transferred to a pier in smaller craft. The smaller craft may have to cycle back and forth from ship to shore many times. Researchers G. Horne (Center for Naval Analysis) and T. Irony (George Washington University) developed models of this transfer process that provide estimates of ship-to-shore transfer times (*Naval Research Logistics*, Vol. 41, 1994). They used an exponential distribution to model the time between arrivals of the smaller craft at the pier.

a. Assume that the mean time between arrivals at the pier is 17 minutes. Give the value of θ for this exponential distribution. Graph the distribution.
b. Suppose there is only one unloading zone at the pier available for the small craft to use. If the first craft docks at 10:00 A.M. and doesn't finish unloading until 10:15 A.M., what is the probability that the second craft will arrive at the unloading zone and have to wait before docking?

4.150 Cycle availability of a system. In the jargon of system maintenance, *cycle availability* is defined as the probability that the system is functioning at any point in time. The DoD developed a series of performance measures for assessing system cycle availability (START, Vol. 11, 2004). Under certain assumptions about the failure time and maintenance time of a system, cycle availability is shown to be uniformly distributed between 0 and 1. Find the following parameters for cycle availability: mean, standard deviation, 10th percentile, lower quartile, and upper quartile. Interpret the results.

4.151 Product failure behavior. An article in *Hotwire* (Dec. 2002) discussed the length of time till failure of a product produced at Hewlett-Packard. At the end of the product's lifetime, the time till failure is modeled using an exponential distribution with mean 500 thousand hours. In reliability jargon this is known as the "wear-out" distribution for the product. During its normal (useful) life, assume the product's time till failure is uniformly distributed over the range 100 thousand to 1 million hours.

a. At the end of the product's lifetime, find the probability that the product fails before 700 thousand hours.

b. During its normal (useful) life, find the probability that the product fails before 700 thousand hours.

c. Show that the probability of the product failing before 830 thousand hours is approximately the same for both the normal (useful) life distribution and the wear-out distribution.

4.152 Reliability of CD-ROMs. In *Reliability Ques* (March 2004), the exponential distribution was used to model the lengths of life of CD-ROM drives in a two-drive system. The two CD-ROM drives operate independently, and at least one drive must be operating for the system to operate successfully. Both drives have a mean length of life of 25,000 hours.

a. The reliability $R(t)$ of a single CD-ROM drive is the probability that the life of the drive exceeds t hours. Give a formula for $R(t)$.

b. Use the result from part **a** to find the probability that the life of the single CD-ROM drive exceeds 8,760 hours (the number of hours of operation in a year).

c. The reliability $S(t)$ of the two CD-ROM drive system is the probability that the life of at least one drive exceeds t hours. Give a formula for $S(t)$. [*Hint:* Use the rule of complements and the fact that the two drives operate independently.]

d. Use the result from part **c** to find the probability that the two-drive CD-ROM system has a life whose length exceeds 8,760 hours.

e. Compare the probabilities you found in parts **b** and **d.**

Applying the Concepts—Advanced

4.153 Gouges on a spindle. A tool-and-die machine shop produces extremely high-tolerance spindles. The spindles are 18-inch slender rods used in a variety of military equipment. A piece of equipment used in the manufacture of the spindles malfunctions on occasion and places a single gouge somewhere on the spindle. However, if the spindle can be cut so that it has 14 consecutive inches without a gouge, then the spindle can be salvaged for other purposes. Assuming that the location of the gouge along the spindle is random, what is the probability that a defective spindle can be salvaged?

4.154 Reliability of a robotic device. The *reliability* of a piece of equipment is frequently defined to be the probability, p, that the equipment performs its intended function successfully for a given period of time under specific conditions (Render and Heizer, *Principles of Operations Management,* 1995). Because p varies from one point in time to another, some reliability analysts treat p as if it were a random variable. Suppose an analyst characterizes the uncertainty about the reliability of a particular robotic device used in an automobile assembly line using the following distribution:

$$f(p) = \begin{cases} 1 & 0 \le p \le 1 \\ 0 & \text{otherwise} \end{cases}$$

a. Graph the analyst's probability distribution for p.

b. Find the mean and variance of p.

c. According to the analyst's probability distribution for p, what is the probability that p is greater than .95? Less than .95?

d. Suppose the analyst receives the additional information that p is definitely between .90 and .95, but that there is complete uncertainty about where it lies between these values. Describe the probability distribution the analyst should now use to describe p.

4.155 Length of life of a halogen bulb. For a certain type of halogen light bulb, an old bulb that has been in use for a while tends to have a longer life than a new bulb. Let x represent the life (in hours) of a new halogen light bulb and assume that x has an exponential distribution with mean $\theta = 250$ hours. According to *Microelectronics and Reliability* (Jan. 1986), the "life" distribution of x is considered *new better than used* (NBU) if

$$P(x > a + b) \le P(x > a)P(x > b)$$

Alternatively, a "life" distribution is considered *new worse than used* (NWU) if

$$P(x > a + b) \ge P(x > a)P(x > b)$$

a. Show that when $a = 300$ and $b = 200$, the exponential distribution is both NBU and NWU.

b. Choose any two positive numbers a and b and repeat part **a.**

c. Show that, in general, for any positive a and b, the exponential distribution with mean θ is both NBU and NWU. Such a "life" distribution is said to be *new same as used,* or *memoryless.* Explain why.

PART III: RANDOM VARIABLES BASED ON SAMPLING DISTRIBUTIONS

4.10 Sampling Distributions

In the previous sections, we assumed that we knew the probability distribution of a random variable, and using this knowledge we were able to compute the mean, variance, and probabilities associated with the random variable. However, in most practical applications, this information is not available. In reality, the true mean and standard deviation are unknown quantities that would have to be estimated. Numerical quantities that describe probability distributions are called *parameters*. Thus, p, the probability of a success in a binomial experiment, and μ and σ, the mean and standard deviation of a normal distribution, are examples of parameters.

> A **parameter** is a numerical descriptive measure of a population. Because it is based on all the observations in the population, its value is almost always unknown.

We have also discussed the sample mean $\bar{x}$, sample variance s^2, sample standard deviation s, etc., which are numerical descriptive measures calculated from the sample. We will often use the information contained in these *sample statistics* to make inferences about the parameters of a population.

> A **sample statistic** is a numerical descriptive measure of a sample. It is calculated from the observations in the sample.

Note that the term *statistic* refers to a *sample* quantity, and the term *parameter* refers to a *population* quantity, as shown in Table 4.8.

Table 4.8 List of Population Parameters and Corresponding Sample Statistics

	Population Parameter	Sample Statistic
Mean	μ	$\bar{x}$
Variance	σ^2	s^2
Standard deviation	σ	s
Binomial proportion	p	$\hat{p}$

Before we can show you how to use sample statistics to make inferences about population parameters, we need to be able to evaluate their properties. Does one sample statistic contain more information than another about a population parameter? On what basis should we choose the "best" statistic for making inferences about a parameter? For example, if we want to estimate a parameter of a population—say, the population mean μ—we could use a number of sample statistics for our estimate. Two possibilities are the sample mean $\bar{x}$ and the sample median m. Which of these do you think will provide a better estimate of μ?

Before answering this question, consider the following example: Toss a fair die and let x equal the number of dots showing on the up face. Suppose the die is tossed three times, producing the sample measurements 2, 2, 6. The sample mean is $\bar{x} = 3.33$ and the sample median is $m = 2$. Because the population mean of x is $\mu = 3.5$, you can see that for this sample of three measurements, the sample mean $\bar{x}$ provides an estimate that falls closer to μ than does the sample median (see Figure 4.39a). Now suppose we toss the die three more times and obtain the sample measurements 3, 4, 6. The mean and median of this sample are $\bar{x} = 4.33$ and $m = 4$, respectively. This time m is closer to μ (see Figure 4.39b).

This simple example illustrates an important point: Neither the sample mean nor the sample median will *always* fall closer to the population mean. Consequently, we cannot compare these two sample statistics, or, in general, any

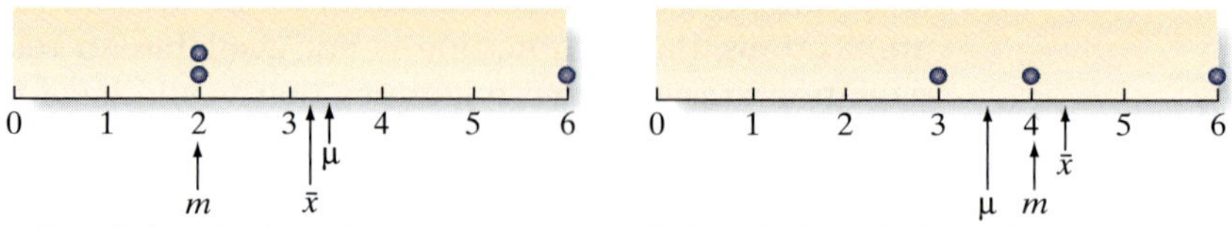

a. Sample 1: $\bar{x}$ is closer than m to μ

b. Sample 2: m is closer than $\bar{x}$ to μ

Figure 4.39
Comparing the sample mean ($\bar{x}$) and sample median (m) as estimators of the population mean (μ)

two sample statistics, on the basis of their performance for a single sample. Instead, we need to recognize that sample statistics are themselves random variables because different samples can lead to different values for the sample statistics. As random variables, sample statistics must be judged and compared on the basis of their probability distributions (i.e., the *collection* of values and associated probabilities of each statistic that would be obtained if the sampling experiment was repeated a *very large number of times*). We will illustrate this concept with another example.

Suppose it is known that the connector module manufactured for a certain brand of pacemaker has a mean length of $\mu = .3$ inch and a standard deviation of .005 inch. Consider an experiment consisting of randomly selecting 25 recently manufactured connector modules, measuring the length of each, and calculating the sample mean length $\bar{x}$. If this experiment was repeated a very large number of times, the value of $\bar{x}$ would vary from sample to sample. For example, the first sample of 25 length measurements might have a mean $\bar{x} = .301$, the second sample a mean $\bar{x} = .298$, the third sample a mean $\bar{x} = .303$, etc. If the sampling experiment was repeated a very large number of times, the resulting histogram of sample means would be approximately the probability distribution of $\bar{x}$. If $\bar{x}$ is a good estimator of μ, we would expect the values of $\bar{x}$ to cluster around μ, as shown in Figure 4.40. This probability distribution is called a *sampling distribution* because it is generated by repeating a sampling experiment a very large number of times.

The **sampling distribution** of a sample statistic calculated from a sample of n measurements is the probability distribution of the statistic.

In actual practice, the sampling distribution of a statistic is obtained mathematically or (at least approximately) by simulating the sample on a computer using a procedure similar to that just described.

If $\bar{x}$ has been calculated from a sample of $n = 25$ measurements selected from a population with mean $\mu = .3$ and standard deviation $\sigma = .005$, the sampling distribution (Figure 4.40) provides information about the behavior of $\bar{x}$ in repeated sampling. For example, the probability that you will draw a sample of 25 length measurements and obtain a value of $\bar{x}$ in the interval $.299 \leq \bar{x} \leq .3$ will be the area under the sampling distribution over that interval.

Because the properties of a statistic are typified by its sampling distribution, it follows that to compare two sample statistics, you compare their sampling distributions. For example, if you have two statistics, A and B, for estimating the same parameter (for purposes of illustration, suppose the parameter is the population variance σ^2) and if their sampling distributions are as shown in Figure 4.41, you would choose statistic A in preference to statistic B. You would make this choice because the sampling distribution for statistic A centers over σ^2 and has less spread (variation) than the sampling distribution for statistic B. When you draw a single sample in a practical sampling situation, the probability is higher that statistic A will fall nearer σ^2.

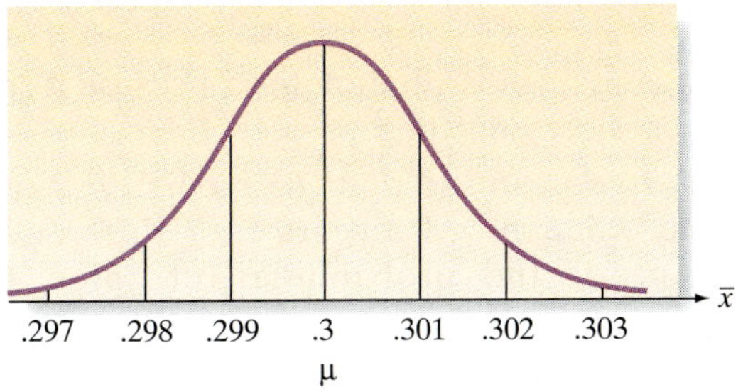

Figure 4.40
Sampling distribution for $\bar{x}$ based on a sample of $n = 25$ length measurements

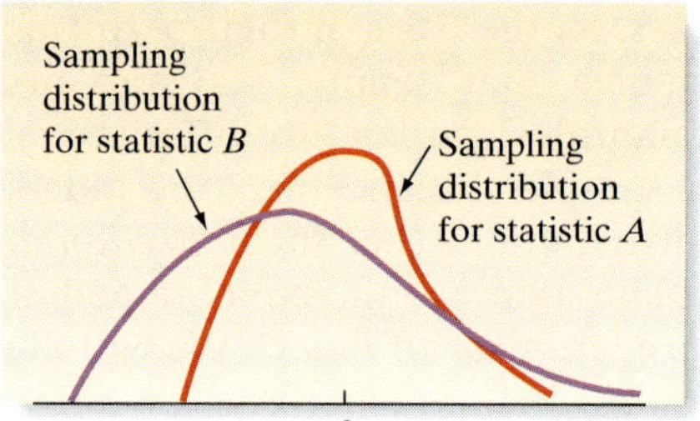

Figure 4.41
Two sampling distributions for estimating the population variance, σ^2

Remember that in practice we will not know the numerical value of the unknown parameter σ^2, so we will not know whether statistic A or statistic B is closer to σ^2 for a sample. We have to rely on our knowledge of the theoretical sampling distributions to choose the best sample statistic and then use it sample after sample. The procedure for finding the sampling distribution for a statistic is demonstrated in Example 4.29.

Example 4.29
Finding a Sampling Distribution

Problem Consider a game played with a standard 52-card bridge deck in which you can score 0, 3, or 12 points per hand. Suppose the population of points per hand is described by the probability distribution shown here. A random sample of $n = 3$ hands is selected from the population.

Points, x	0	3	12
$p(x)$	$1/2$	$1/4$	$1/4$

a. Find the sampling distribution of the sample mean number of points, $\bar{x}$.

b. Find the sampling distribution of the sample median number of points, m.

Solution Points for every possible sample of $n = 3$ hands listed in Table 4.9, along with the sample mean and median. The probability of each sample is obtained using the multiplicative rule. For example, the probability of the sample (0, 0, 3) is $p(0) \cdot p(0) \cdot p(3) = (1/2)(1/2)(1/4) = 1/16$. The probability for each sample is also listed in Table 4.9. Note that the sum of these probabilities is equal to 1.

a. From Table 4.9 you can see that $\bar{x}$ can assume the values 0, 1, 2, 3, 4, 5, 6, 8, 9, and 12. Because $\bar{x} = 0$ occurs in only the first listed sample, $P(\bar{x} = 0) = 8/64$. Similarly, $\bar{x} = 1$ occurs in 3 samples: (0, 0, 3), (0, 3, 0), and (3, 0, 0). Therefore, $P(\bar{x} = 1)$ is the sum of the probabilities for these 3 samples, i.e., $P(\bar{x} = 1) = 4/64 + 4/64 + 4/64 = 12/64$. Calculating the probabilities of the remaining values of $\bar{x}$ and arranging them in a table, we obtain the probability distribution shown below.

$\bar{x}$	0	1	2	3	4	5	6	8	9	12
$p(\bar{x})$	$8/64$	$12/64$	$6/64$	$1/64$	$12/64$	$12/64$	$3/64$	$6/64$	$3/64$	$1/64$

This is the sampling distribution for $\bar{x}$ because it specifies the probability associated with each possible value of $\bar{x}$.

b. In Table 4.9 you can see that the median m can assume one of the values 0, 3, or 12. The value $m = 0$ occurs in 7 different samples. Therefore, $P(m = 0)$ is the sum of the probabilities of these seven samples, i.e., $P(m = 0) = 8/64 + 4/64 + 4/64 + 4/64 + 4/64 + 4/64 + 4/64 = 32/64$. Similarly, $m = 3$ occurs in

Table 4.9 All Possible Samples of $n = 3$ Hands of a Card Game, Example 4.29

Possible Samples	$\bar{x}$	m	Probability
0, 0, 0	0	0	$(1/2)(1/2)(1/2) = 1/8 = 8/64$
0, 0, 3	1	0	$(1/2)(1/2)(1/4) = 1/16 = 4/64$
0, 0, 12	4	0	$(1/2)(1/2)(1/4) = 1/16 = 4/64$
0, 3, 0	1	0	$(1/2)(1/4)(1/2) = 1/16 = 4/64$
0, 3, 3	2	3	$(1/2)(1/4)(1/4) = 1/32 = 2/64$
0, 3, 12	5	3	$(1/2)(1/4)(1/4) = 1/32 = 2/64$
0, 12, 0	4	0	$(1/2)(1/4)(1/2) = 1/16 = 4/64$
0, 12, 3	5	3	$(1/2)(1/4)(1/4) = 1/32 = 2/64$
0, 12, 12	8	12	$(1/2)(1/4)(1/4) = 1/32 = 2/64$
3, 0, 0	1	0	$(1/4)(1/2)(1/2) = 1/16 = 4/64$
3, 0, 3	2	3	$(1/4)(1/2)(1/4) = 1/32 = 2/64$
3, 0, 12	5	3	$(1/4)(1/2)(1/4) = 1/32 = 2/64$
3, 3, 0	2	3	$(1/4)(1/4)(1/2) = 1/32 = 2/64$
3, 3, 3	3	3	$(1/4)(1/4)(1/4) = 1/64$
3, 3, 12	6	3	$(1/4)(1/4)(1/4) = 1/64$
3, 12, 0	5	3	$(1/4)(1/4)(1/2) = 1/32 = 2/64$
3, 12, 3	6	3	$(1/4)(1/4)(1/4) = 1/64$
3, 12, 12	9	12	$(1/4)(1/4)(1/4) = 1/64$
12, 0, 0	4	0	$(1/4)(1/2)(1/2) = 1/16 = 4/64$
12, 0, 3	5	3	$(1/4)(1/2)(1/4) = 1/32 = 2/64$
12, 0, 12	8	12	$(1/4)(1/2)(1/4) = 1/32 = 2/64$
12, 3, 0	5	3	$(1/4)(1/4)(1/2) = 1/32 = 2/64$
12, 3, 3	6	3	$(1/4)(1/4)(1/4) = 1/64$
12, 3, 12	9	12	$(1/4)(1/4)(1/4) = 1/64$
12, 12, 0	8	12	$(1/4)(1/4)(1/2) = 1/32 = 2/64$
12, 12, 3	9	12	$(1/4)(1/4)(1/4) = 1/64$
12, 12, 12	12	12	$(1/4)(1/4)(1/4) = 1/64$
			$Sum = 64/64 = 1$

13 samples and $m = 12$ occurs in 7 samples, and these probabilities are obtained by summing the probabilities of their respective sample points. Therefore, the probability distribution (i.e., the sampling distribution) for the median m is as shown below.

m	0	3	12
$p(m)$	$32/64$	$22/64$	$10/64$

Look Back The sampling distributions of parts **a** and **b** are found by first listing all possible distinct values of the statistic and then calculating the probability of each value. Note that if the values of x were equally likely, then the 27 sample points in Table 4.9 would all have the same probability of occurring, namely $1/27$.

Now Work Exercise 4.156

Example 4.29 demonstrates the procedure for finding the exact sampling distribution of a statistic when the number of different samples that could be selected from the population is relatively small. In the real world, populations often consist of a large number of different values, making samples difficult (or impossible) to enumerate. When this situation occurs, we may choose to obtain the approximate sampling distribution for a statistic by simulating the sampling over and over again and recording the proportion of times different values of the statistic occur. Example 4.30 illustrates this procedure.

Example 4.30

Simulating a Sampling Distribution

Problem Refer to Example 4.26 (p. 231). Recall that the thickness of a steel sheet follows a uniform distribution with values between 150 and 200 millimeters. Suppose we perform the following experiment over and over again: Randomly sample 11 steel sheets from the production line and record the thickness, x, of each. Calculate the two sample statistics

$$\bar{x} = \text{Sample mean} = \frac{\Sigma x}{11}$$

$$m = \text{Median} = \text{Sixth sample measurement when the 11 thicknesses are arranged in ascending order}$$

Obtain approximations to the sampling distributions of $\bar{x}$ and m.

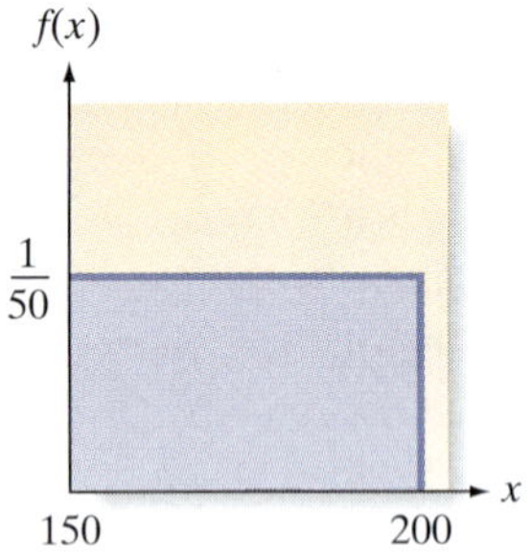

Figure 4.42
Uniform distribution for thickness of steel sheets

Solution Recall from Section 4.9 that the population of thicknesses follows the uniform distribution shown in Figure 4.42. We used Minitab to generate 1,000 samples from this population, each with $n = 11$ observations. Then we compute $\bar{x}$ and m for each sample. Our goal is to obtain approximations to the sampling distributions of $\bar{x}$ and m to find out which sample statistic ($\bar{x}$ or m) contains more information about μ. [*Note:* In this particular example, we *know* the population mean is $\mu = 175$ mm. (See Section 4.9.)] The first 10 of the 1,000 samples generated are presented in Table 4.10. For instance, the first computer-generated sample from the uniform distribution (arranged in ascending order) contained the following thickness measurements: 151, 157, 162, 169, 171, 173, 181, 182, 187, 188, and 193 millimeters. The sample mean $\bar{x}$ and median m computed for this sample are

$$\bar{x} = \frac{151 + 157 + \cdots + 193}{11} = 174.0$$

$$m = \text{Sixth ordered measurement} = 173$$

The Minitab relative frequency histograms for $\bar{x}$ and m for the 1,000 samples of size $n = 11$ are shown in Figure 4.43. These histograms represent approximations to the true sampling distributions of $\bar{x}$ and m.

Look Back You can see that the values of $\bar{x}$ tend to cluster around μ to a greater extent than do the values of m. Thus, on the basis of the observed sampling distributions, we conclude that $\bar{x}$ contains more information about μ than m does—at least for samples of $n = 11$ measurements from the uniform distribution.

Table 4.10 First 10 Samples of $n = 11$ Thickness Measurements from Uniform Distribution

Sample	Thickness Measurements											Mean	Median
1	173	171	187	151	188	181	182	157	162	169	193	174.00	173
2	181	190	182	171	187	177	162	172	188	200	193	182.09	182
3	192	195	187	187	172	164	164	189	179	182	173	180.36	182
4	173	157	150	154	168	174	171	182	200	181	187	172.45	173
5	169	160	167	170	197	159	174	174	161	173	160	169.46	169
6	179	170	167	174	173	178	173	170	173	198	187	176.55	173
7	166	177	162	171	154	177	154	179	175	185	193	172.09	175
8	164	199	152	153	163	156	184	151	198	167	180	169.73	164
9	181	193	151	166	180	199	180	184	182	181	175	179.27	181
10	155	199	199	171	172	157	173	187	190	185	150	176.18	173

Data Set: SIMUNI

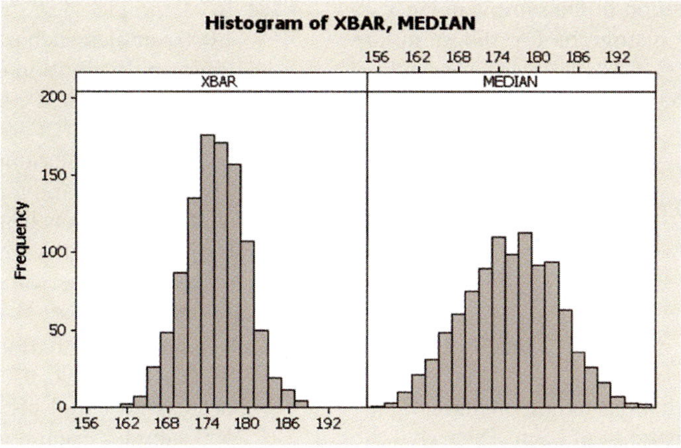

Figure 4.43
Minitab histograms for sample mean and sample median, Example 4.30

Now Work Exercise 4.161

As noted earlier, many sampling distributions can be derived mathematically, but the theory necessary to do this is beyond the scope of this text. Consequently, when we need to know the properties of a statistic, we will present its sampling distribution and simply describe its properties. The important properties of the sampling distribution of the sample mean are discussed in the next section.

Exercises 4.156–4.162

Learning the Mechanics

4.156 NW The probability distribution shown here describes a population of measurements that can assume values of 0, 2, 4, and 6, each of which occurs with the same relative frequency:

x	0	2	4	6
$p(x)$	$1/4$	$1/4$	$1/4$	$1/4$

a. List all the different samples of $n = 2$ measurements that can be selected from this population.

b. Calculate the mean of each different sample listed in part **a.**

c. If a sample of $n = 2$ measurements is randomly selected from the population, what is the probability that a specific sample will be selected?

d. Assume that a random sample of $n = 2$ measurements is selected from the population. List the different values of $\bar{x}$ found in part **b** and find the probability of each. Then give the sampling distribution of the sample mean $\bar{x}$ in tabular form.

e. Construct a probability histogram for the sampling distribution of $\bar{x}$.

4.157 Simulate sampling from the population described in Exercise 4.156 by marking the values of x, one on each of four identical coins (or poker chips, etc.). Place the coins (marked 0, 2, 4, and 6) into a bag, randomly select one, and observe its value. Replace this coin, draw a second coin, and observe its value. Finally, calculate the mean $\bar{x}$ for this sample of $n = 2$ observations randomly selected from the population (Exercise 4.156, part **b**). Replace the coins, mix, and using the same procedure, select a sample of $n = 2$ observations from the population. Record the numbers and calculate $\bar{x}$ for this sample. Repeat this sampling process until you acquire 100 values of $\bar{x}$. Construct a relative frequency distribution for these 100 sample means. Compare this distribution to the exact sampling distribution of $\bar{x}$ found in part **e** of Exercise 4.156. [*Note:* The distribution obtained in this exercise is an approximation to the exact sampling distribution. But, if you were to repeat the sampling procedure, drawing two coins not 100 times but 10,000 times, the relative frequency distribution for the 10,000 sample means would be almost identical to the sampling distribution of $\bar{x}$ found in Exercise 4.156, part **e.**]

4.158 Consider the population described by the probability distribution shown below.

x	1	2	3	4	5
$p(x)$	.2	.3	.2	.2	.1

The random variable x is observed twice. If these observations are independent, verify that the different samples of size 2 and their probabilities are as shown below.

Sample	Probability	Sample	Probability
1, 1	.04	3, 4	.04
1, 2	.06	3, 5	.02
1, 3	.04	4, 1	.04
1, 4	.04	4, 2	.06
1, 5	.02	4, 3	.04
2, 1	.06	4, 4	.04
2, 2	.09	4, 5	.02
2, 3	.06	5, 1	.02
2, 4	.06	5, 2	.03
2, 5	.03	5, 3	.02
3, 1	.04	5, 4	.02
3, 2	.06	5, 5	.01
3, 3	.04		

a. Find the sampling distribution of the sample mean $\overline{x}$.
b. Construct a probability histogram for the sampling distribution of $\overline{x}$.
c. What is the probability that $\overline{x}$ is 4.5 or larger?
d. Would you expect to observe a value of $\overline{x}$ equal to 4.5 or larger? Explain.

4.159 Refer to Exercise 4.158 and find $E(x) = \mu$. Then use the sampling distribution of $\overline{x}$ found in Exercise 4.158 to find the expected value of $\overline{x}$. Note that $E(\overline{x}) = \mu$.

4.160 Refer to Exercise 4.158. Assume that a random sample of $n = 2$ measurements is randomly selected from the population.
a. List the different values that the sample median m may assume and find the probability of each. Then give the sampling distribution of the sample median.
b. Construct a probability histogram for the sampling distribution of the sample median and compare it with the probability histogram for the sample mean (Exercise 4.158, part **b**).

4.161 NW In Example 4.30 we used a computer to generate 1,000 samples, each containing $n = 11$ observations, from a uniform distribution over the interval from 150 to 200. For this exercise, use a computer to generate 500 samples, each containing $n = 15$ observations, from this population.
a. Calculate the sample mean for each sample. To approximate the sampling distribution of $\overline{x}$, construct a relative frequency histogram for the 500 values of $\overline{x}$.
b. Repeat part **a** for the sample median. Compare this approximate sampling distribution with the approximate sampling distribution of $\overline{x}$ found in part **a.**

4.162 Consider a population that contains values of x equal to 00, 01, 02, 03, . . . , 96, 97, 98, 99. Assume that these values of x occur with equal probability. Use a computer to generate 500 samples, each containing $n = 25$ measurements, from this population. Calculate the sample mean $\overline{x}$ and sample variance s^2 for each of the 500 samples.
a. To approximate the sampling distribution of $\overline{x}$, construct a relative frequency histogram for the 500 values of $\overline{x}$.
b. Repeat part **a** for the 500 values of s^2.

4.11 The Sampling Distribution of a Sample Mean and the Central Limit Theorem

Estimating the mean useful life of automobiles, the mean monthly sales for all computer dealers in a large city, and the mean breaking strength of new plastic are practical problems with something in common. In each case, we are interested in making an inference about the mean μ of some population. As we mentioned in Chapter 2, the sample mean $\overline{x}$ is, in general, a good estimator of μ. We now develop pertinent information about the sampling distribution for this useful statistic.

Example 4.31
Describing the Sampling Distribution of $\overline{x}$

Problem Suppose a population has the uniform probability distribution given in Figure 4.44. The mean and standard deviation of this probability distribution are $\mu = 175$ and $\sigma = 14.43$. (See Section 4.9 for the formulas for μ and σ.) Now suppose a sample of 11 measurements is selected from this population. Describe the sampling distribution of the sample mean $\overline{x}$ based on the 1,000 sampling experiments discussed in Example 4.30.

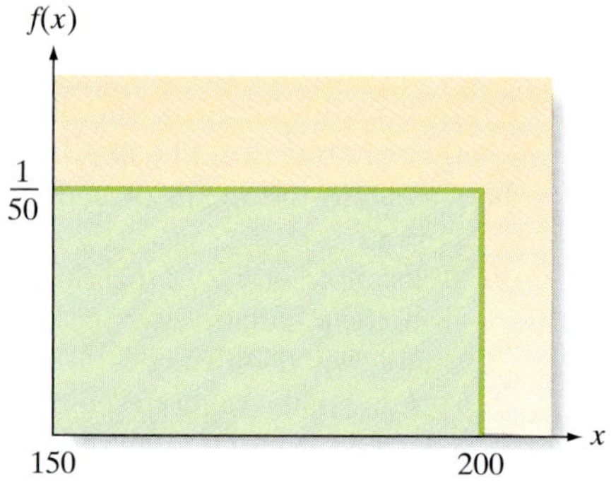

Figure 4.44
Sampled uniform population

Solution You will recall that in Example 4.30 we generated 1,000 samples of $n = 11$ measurements each. The Minitab histogram for the 1,000 sample means is shown in Figure 4.45 with a normal probability distribution superimposed. You can see that this normal probability distribution approximates the computer-generated sampling distribution very well.

To fully describe a normal probability distribution, it is necessary to know its mean and standard deviation. Minitab gives these statistics for the 1,000 $\overline{x}$'s in the upper right corner of the histogram, Figure 4.45. You can see that the mean is 175.2, and the standard deviation is 4.383.

To summarize our findings based on 1,000 samples, each consisting of 11 measurements from a uniform population, the sampling distribution of $\overline{x}$ appears to be approximately normal with a mean of about 175 and a standard deviation of about 4.38.

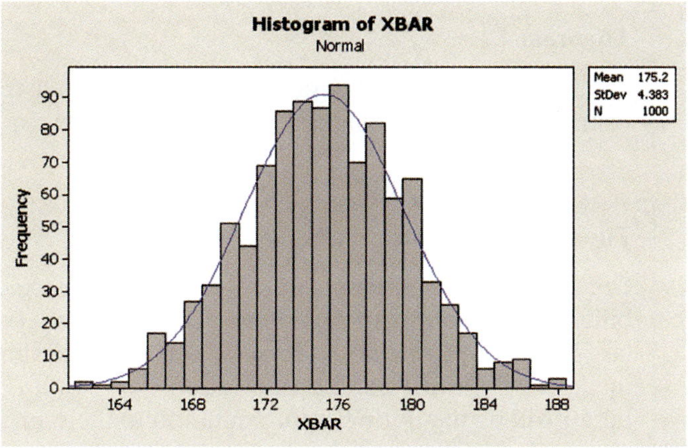

Figure 4.45
Minitab histogram for sample mean in 1,000 samples

Look Back Note that the simulated value $\mu_{\bar{x}} = 175.2$ is very close to $\mu = 175$ for the uniform distribution—that is, the simulated sampling distribution of $\bar{x}$ appears to provide an accurate estimate of μ.

The true sampling distribution of $\bar{x}$ has the properties given in the next box, assuming only that a random sample of n observations has been selected from *any* population.

Properties of the Sampling Distribution of $\bar{x}$

1. Mean of sampling distribution equals mean of sampled population*, that is, $\mu_{\bar{x}} = E(\bar{x}) = \mu$.
2. Standard deviation of sampling distribution equals

$$\frac{\text{Standard deviation of sampled population}}{\text{Square root of sample size}}$$

That is, $\sigma_{\bar{x}} = \sigma/\sqrt{n}$.†

The standard deviation $\sigma_{\bar{x}}$ is often referred to as the **standard error of the mean.**

You can see that our approximation to $\mu_{\bar{x}}$ in Example 4.31 was precise because property 1 assures us that the mean is the same as that of the sampled population: 175 mm. Property 2 tells us how to calculate the standard deviation of the sampling distribution of $\bar{x}$. Substituting $\sigma = 14.43$, the standard deviation of the sampled uniform distribution, and the sample size $n = 11$ into the formula for $\sigma_{\bar{x}}$, we find

$$\sigma_{\bar{x}} = \frac{\sigma}{\sqrt{n}} = \frac{14.43}{\sqrt{11}} = 4.35$$

Thus, the approximation we obtained in Example 4.31, $\sigma_{\bar{x}} = 4.38$, is very close to the exact value, $\sigma_{\bar{x}} = 4.35$.**

What can be said about the shape of the sampling distribution of $\bar{x}$? Two important theorems provide this information.

*When this property holds, we say that $\bar{x}$ is an *unbiased* estimate of μ.

†If the sample size, n, is large relative to the number, N, of elements in the population (e.g., 5% or more), $\sigma/\sqrt{n}$ must be multiplied by a finite population correction factor, $\sqrt{(N - n)/(N - 1)}$. For most sampling situations, this correction factor will be close to 1 and can be ignored.

**It can be shown (proof omitted) that the value of $\sigma^2_{\bar{x}}$ is the smallest variance among all unbiased estimators of μ—thus, $\bar{x}$ is the MVUE (minimum variance unbiased estimator) for μ.

Theorem 4.1

If a random sample of n observations is selected from a population with a normal distribution, the sampling distribution of $\bar{x}$ will be a normal distribution.

Theorem 4.2 (Central Limit Theorem)

Consider a random sample of n observations selected from a population (*any* probability distribution) with mean μ and standard deviation σ. Then, when n is sufficiently large, the sampling distribution of $\bar{x}$ will be approximately a normal distribution with mean $\mu_{\bar{x}} = \mu$ and standard deviation $\sigma_{\bar{x}} = \sigma/\sqrt{n}$. The larger the sample size, the better will be the normal approximation to the sampling distribution of $\bar{x}$.*

BIOGRAPHY

PIERRE-SIMON LAPLACE (1749–1827)

The CLT Originator

As a boy growing up in Normandy, France, Pierre-Simon Laplace attended a Benedictine priory school. Upon graduation, he entered Caen University to study theology. During his two years there, he discovered his mathematical talents and began his career as an eminent mathematician. In fact, he considered himself the best mathematician in France. Laplace's contributions to mathematics ranged from introducing new methods of solving differential equations to complex analyses of motions of astronomical bodies. While studying the angles of inclination of comet orbits in 1778, Laplace showed that the sum of the angles were normally distributed. Consequently, he is considered to be the originator of the Central Limit Theorem. (A rigorous proof of the theorem, however, was not provided until the early 1930s by another French mathematician, Paul Levy.) Laplace also discovered Bayes's theorem and established Bayesian statistical analysis as a valid approach to many practical problems of his time. ■

Thus, for sufficiently large samples, the sampling distribution of $\bar{x}$ is approximately normal. How large must the sample size n be so that the normal distribution provides a good approximation for the sampling distribution of $\bar{x}$? The answer depends on the shape of the distribution of the sampled population, as shown by Figure 4.46. Generally speaking, the greater the skewness of the sampled population distribution, the larger the sample size must be before the normal distribution is an adequate approximation for the sampling distribution of $\bar{x}$. For most sampled populations, sample sizes of $n \geq 30$ will suffice for the normal approximation to be reasonable. We will use the normal approximation for the sampling distribution of $\bar{x}$ when the sample size is at least 30.

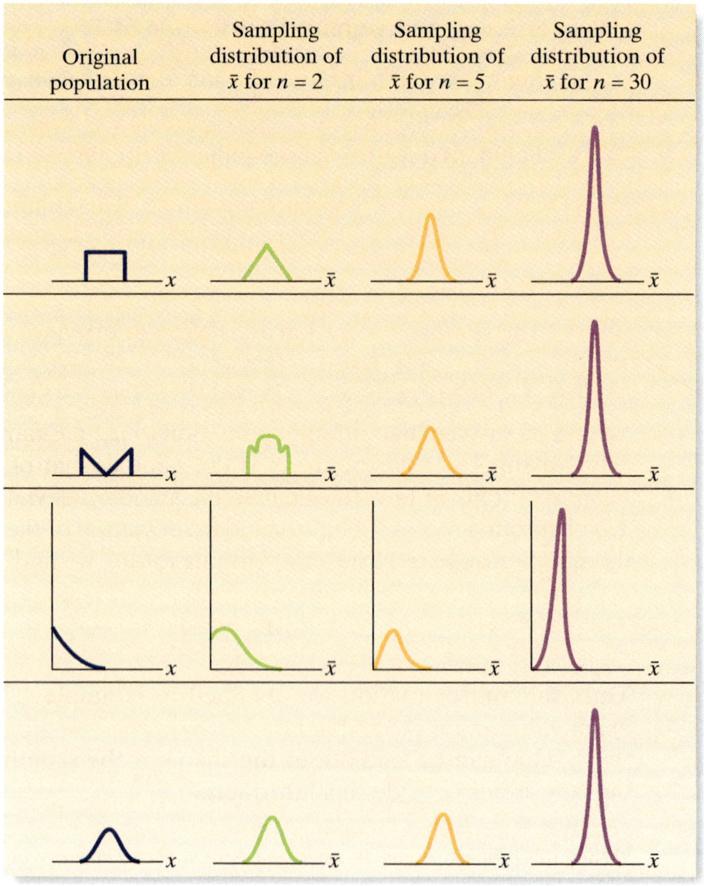

Figure 4.46
Sampling distributions of $\bar{x}$ for different populations and different sample sizes

*Moreover, because of the Central Limit Theorem, the sum of a random sample on n observations, Σx, will possess a sampling distribution that is approximately normal for large samples. This distribution will have a mean equal to $n\mu$ and a variance equal to $n\sigma^2$. Proof of the Central Limit Theorem is beyond the scope of this book, but it can be found in many mathematical statistics texts.

Example 4.32

Using the Central Limit Theorem to Find a Probability

Problem Suppose we have selected a random sample of $n = 36$ observations from a population with mean equal to 80 and standard deviation equal to 6. It is known that the population is not extremely skewed.

a. Sketch the relative frequency distributions for the population and for the sampling distribution of the sample mean, $\bar{x}$.

b. Find the probability that $\bar{x}$ will be larger than 82.

Solution

a. We do not know the exact shape of the population relative frequency distribution, but we do know that it should be centered about $\mu = 80$, its spread should be measured by $\sigma = 6$, and it is not highly skewed. One possibility is shown in Figure 4.47a. From the Central Limit Theorem, we know that the sampling distribution of $\bar{x}$ will be approximately normal because the sampled population distribution is not extremely skewed. We also know that the sampling distribution will have mean and standard deviation

$$\mu_{\bar{x}} = \mu = 80 \quad \text{and} \quad \sigma_{\bar{x}} = \frac{\sigma}{\sqrt{n}} = \frac{6}{\sqrt{36}} = 1$$

The sampling distribution of $\bar{x}$ is shown in Figure 4.47b.

b. The probability that $\bar{x}$ will exceed 82 is equal to the darker shaded area in Figure 4.48. To find this area, we need to find the z-value corresponding to $\bar{x} = 82$. Recall that the standard normal random variable z is the difference between any normally distributed random variable and its mean, expressed in units of its standard deviation. Because $\bar{x}$ is approximately a normally distributed random variable with mean $\mu_{\bar{x}} = \mu$ and $\sigma_{\bar{x}} = \sigma/\sqrt{n}$, it follows that the standard normal z-value corresponding to the sample mean, $\bar{x}$, is

$$z = \frac{(\text{Normal random variable}) - (\text{Mean})}{\text{Standard deviation}} = \frac{\bar{x} - \mu_{\bar{x}}}{\sigma_{\bar{x}}}$$

Therefore, for $\bar{x} = 82$, we have

$$z = \frac{\bar{x} - \mu_{\bar{x}}}{\sigma_{\bar{x}}} = \frac{82 - 80}{1} = 2$$

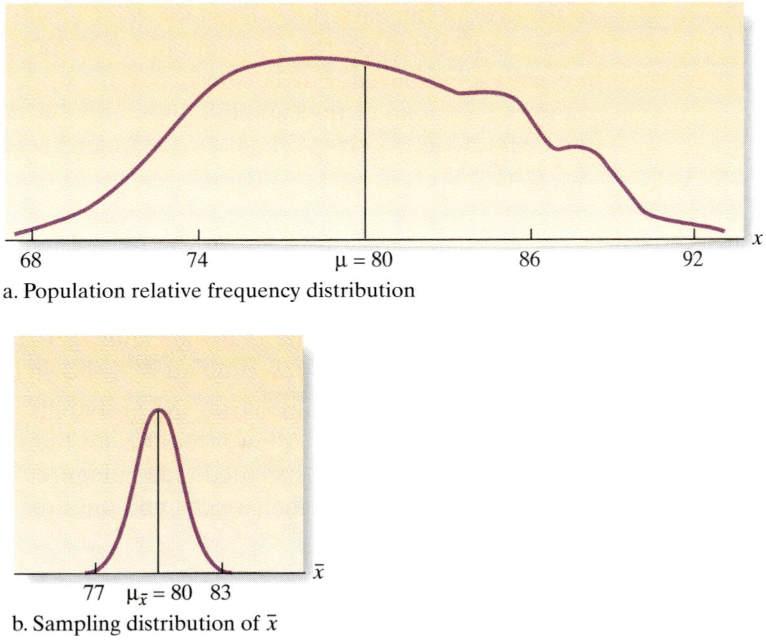

Figure 4.47
A population relative frequency distribution and the sampling distribution for $\bar{x}$

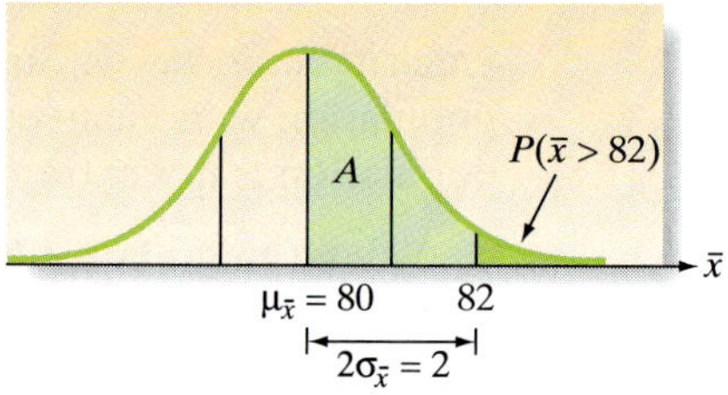

Figure 4.48
The sampling distribution of $\bar{x}$

The area A in Figure 4.48 corresponding to $z = 2$ is given in the table of areas under the normal curve (see Table IV in Appendix B) as .4772. Therefore, the tail area corresponding to the probability that $\bar{x}$ exceeds 82 is

$$P(\bar{x} > 82) = P(z > 2) = .5 - .4772 = .0228$$

Look Back The key to finding the probability, part **b,** is to recognize that the distribution of $\bar{x}$ is normal with mean $\mu_{\bar{x}} = \mu$ and $\sigma_{\bar{x}} = \sigma/\sqrt{n}$.

Now Work Exercises 4.166–4.167

Example 4.33
A Practical Application of the Central Limit Theorem to Test a Manufacturer's Claim

Problem A manufacturer of automobile batteries claims that the distribution of the lengths of life of its best battery has a mean of 54 months and a standard deviation of 6 months. Recently, the manufacturer has received a rash of complaints from unsatisfied customers whose batteries have died earlier than expected. Suppose a consumer group decides to check the manufacturer's claim by purchasing a sample of 50 of these batteries and subjecting them to tests that determine battery life.

a. Assuming that the manufacturer's claim is true, describe the sampling distribution of the mean lifetime of a sample of 50 batteries.

b. Assuming that the manufacturer's claim is true, what is the probability that the consumer group's sample has a mean life of 52 or fewer months?

Solution

a. Even though we have no information about the shape of the probability distribution of the lives of the batteries, we can use the Central Limit Theorem to deduce that the sampling distribution for a sample mean lifetime of 50 batteries is approximately normally distributed. Furthermore, the mean of this sampling distribution is the same as the mean of the sampled population, which is $\mu = 54$ months according to the manufacturer's claim. Finally, the standard deviation of the sampling distribution is given by

$$\sigma_{\bar{x}} = \frac{\sigma}{\sqrt{n}} = \frac{6}{\sqrt{50}} = .85 \text{ month}$$

Note that we used the claimed standard deviation of the sampled population, $\sigma = 6$ months. Thus, if we assume that the claim is true, the sampling distribution for the mean life of the 50 batteries sampled is as shown in Figure 4.49.

b. If the manufacturer's claim is true, the probability that the consumer group observes a mean battery life of 52 or fewer months for their sample of 50 batteries, $P(\bar{x} \leq 52)$, is equivalent to the darker shaded area in Figure 4.49. Because the sampling distribution is approximately normal, we can find this area by computing the standard normal z-value:

$$z = \frac{\bar{x} - \mu_{\bar{x}}}{\sigma_{\bar{x}}} = \frac{52 - 54}{.85} = -2.35$$

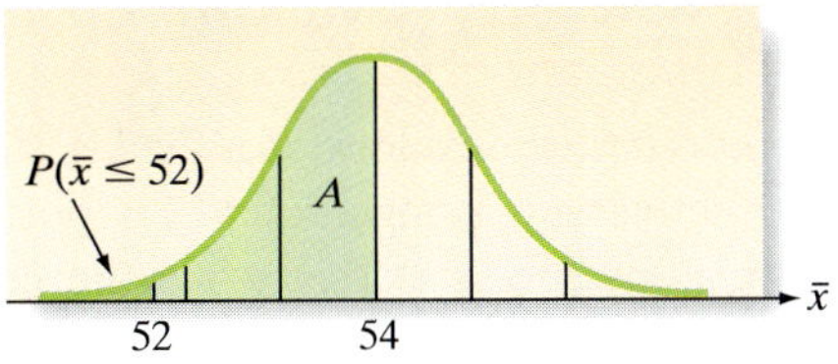

Figure 4.49
Sampling distribution of $\bar{x}$ in Example 4.33 for $n = 50$

where $\mu_{\bar{x}}$, the mean of the sampling distribution of $\bar{x}$, is equal to μ, the mean of the lives of the sampled population, and $\sigma_{\bar{x}}$ is the standard deviation of the sampling distribution of $\bar{x}$. Note that z is the familiar standardized distance (z-score) of Section 2.7 and because $\bar{x}$ is approximately normally distributed, it will possess (approximately) the standard normal distribution of Section 4.6.

The area A shown in Figure 4.49 between $\bar{x} = 52$ and $\bar{x} = 54$ (corresponding to $z = -2.35$) is found in Table IV in Appendix B to be .4906. Therefore, the area to the left of $\bar{x} = 52$ is

$$P(\bar{x} \leq 52) = .5 - A = .5 - .4906 = .0094$$

Thus, the probability that the consumer group will observe a sample mean of 52 or less is only .0094 if the manufacturer's claim is true.

Look Back If the 50 tested batteries do exhibit a mean of 52 or fewer months, the consumer group will have strong evidence that the manufacturer's claim is untrue because such an event is very unlikely to occur if the claim is true. (This is still another application of the *rare-event approach to statistical inference*.)

Now Work Exercise 4.172

We conclude this section with two final comments on the sampling distribution of $\bar{x}$. First, from the formula $\sigma_{\bar{x}} = \sigma/\sqrt{n}$, we see that the standard deviation of the sampling distribution of $\bar{x}$ gets smaller as the sample size n gets larger. For example, we computed $\sigma_{\bar{x}} = .85$ when $n = 50$ in Example 4.33. However, for $n = 100$, we obtain $\sigma_{\bar{x}} = \sigma/\sqrt{n} = 6/\sqrt{100} = .60$. This relationship will hold true for most of the sample statistics encountered in this text—that is, *the standard deviation of the sampling distribution decreases as the sample size increases*. Consequently, the larger the sample size, the more accurate the sample statistic (e.g., $\bar{x}$) is in estimating a population parameter (e.g., μ). We will use this result in Chapter 5 to help us determine the sample size needed to obtain a specified accuracy of estimation.

Our second comment concerns the Central Limit Theorem. In addition to providing a very useful approximation for the sampling distribution of a sample mean, the Central Limit Theorem offers an explanation for the fact that many relative frequency distributions of data possess mound-shaped distributions. Many of the measurements we take in business are really means or sums of a large number of small phenomena. For example, a company's sales for 1 year are the total of the many individual sales the company made during the year. Similarly, we can view the length of time a construction company takes to build a house as the total of the times taken to complete a multitude of distinct jobs, and we can regard the monthly demand for blood at a hospital as the total of the many individual patients' needs. Whether or not the observations entering into these sums satisfy the assumptions basic to the Central Limit Theorem is open to question. However, it is a fact that many distributions of data in nature are mound shaped and possess the appearance of normal distributions.

Statistics IN Action Revisited — Making an Inference about Mean Accuracy

Consider, again, the prototype gun developed for the U.S. Army that fires over 1,000 flechettes with a single round. Suppose the specifications of the weapon are set so that when the gun is aimed at the center target (see Figure SIA4.1, page 174) from 500 meters away, the mean horizontal grid value is $\mu = 5$ feet (i.e., the aim point) and the standard deviation is $\sigma = 1$ foot. After firing one round of the super weapon, the U.S. Army randomly samples 100 of the discharged flechettes and records x, the horizontal measurement, for each. The sample mean was determined to be $\bar{x} = 5.3$. Is this result consistent with the gun specifications or is the true mean μ set higher than 5?

To answer this question, we need to examine the likelihood of obtaining the sample result, given that the gun specifications are met—that is, how likely is it to observe a sample mean of $\bar{x} = 5.3$ or higher for a sample of $n = 100$ flechettes when the mean and standard deviation of the population of horizontal grid measurements are set at $\mu = 5$ and $\sigma = 1$? Thus, the probability of interest is $P(\bar{x} > 5.3)$.

To find this probability, we invoke the Central Limit Theorem. According to the theorem, the sampling distribution of $\bar{x}$ is approximately normally distributed with the following mean and standard deviation:

$$\mu_{\bar{x}} = \mu = 5 \qquad \sigma_{\bar{x}} = \sigma/\sqrt{n} = 1/\sqrt{100} = .1$$

[Note that because we've already established (see SIA on pp. 214–215) that the population of horizontal grid measurements are normally distributed, Theorem 4.1 also guarantees that the distribution of $\bar{x}$ will be normally distributed.] Therefore, we find the desired probability (using the standard normal table) as follows:

$$P(\bar{x} > 5.3) = P\left(z > \frac{5.3 - \mu_{\bar{x}}}{\sigma_{\bar{x}}}\right)$$

$$= P\left(z > \frac{5.3 - 5}{.1}\right) = P(z > 3) = .0013$$

In other words, the probability that we observe a sample mean at least as large as $\bar{x} = 5.3$ feet *if the mean and standard deviation of the horizontal grid measurements are* $\mu = 5$ *and* $\sigma = 1$ *(i.e., if the desired specifications are met) is almost 0.*

This small probability leads us to make one of two conclusions: Either the gun specifications are met and the U.S. Army has observed an extremely rare event (one with almost no chance of happening), or one or both of the gun specifications ($\mu = 5$ and $\sigma = 1$) are not being met. The *rare event approach* to making statistical inferences, of course, would favor the second conclusion. Possibly the gun specifications were set so that the true mean μ is higher than 5 feet, or the standard deviation was set at some number other than $\sigma = 1$. [In fact, you can show that if the specifications are set at $\sigma = 4$, $P(\bar{x} > 5.3) = .2266$. Thus, the observed sample mean has a reasonable chance of occurring if the standard deviation is inadvertently set at $\sigma = 4$.]

Activity 4.2 *Keep the Change:* Simulating a Sampling Distribution

In this activity, we will once again be working with the data set *Amounts Transferred* from Activity 1.1 on page 16. This activity is designed for small groups or the entire class.

1. Pool your *Amounts Transferred* data with other class members or the entire class so that the pooled data set has at least 100 data items. Have someone in the group calculate the mean and the standard deviation of the pooled data set.
2. Devise a convenient way to choose random samples from the pooled data set. You can assign each data item a number beginning with 1 and use a random number generator to select a sample, or you can write each data item on a small card and then draw cards without looking.
3. Choose a random sample of size $n = 30$ from the pooled data and find the mean of the sample, called a *sample mean*. Group members should repeat the process of choosing a sample of size $n = 30$ from the pooled data and finding the sample mean until the group has accumulated at least 25 sample means. Call this new data set *Sample Means*.
4. Find the mean and standard deviation of the data set *Sample Means*. Explain how the Central Limit Theorem is illustrated in this activity.

Keep the data set from this activity for use in future activities.

Exercises 4.163–4.181

Learning the Mechanics

4.163 Will the sampling distribution of $\bar{x}$ always be approximately normally distributed? Explain.

4.164 Suppose a random sample of $n = 25$ measurements is selected from a population with mean μ and standard deviation σ. For each of the following values of μ and σ, give the values of $\mu_{\bar{x}}$ and $\sigma_{\bar{x}}$.

a. $\mu = 10, \sigma = 3$ **b.** $\mu = 100, \sigma = 25$
c. $\mu = 20, \sigma = 40$ **d.** $\mu = 10, \sigma = 100$

4.165 Suppose a random sample of n measurements is selected from a population with mean $\mu = 100$ and variance $\sigma^2 = 100$. For each of the following values of n, give the mean and standard deviation of the sampling distribution of the sample mean $\bar{x}$.

a. $n = 4$ **b.** $n = 25$
c. $n = 100$ **d.** $n = 50$
e. $n = 500$ **f.** $n = 1{,}000$

4.166 NW A random sample of $n = 64$ observations is drawn from a population with a mean equal to 20 and a standard deviation equal to 16.

a. Give the mean and standard deviation of the (repeated) sampling distribution of $\bar{x}$.
b. Describe the shape of the sampling distribution of $\bar{x}$. Does your answer depend on the sample size?
c. Calculate the standard normal z-score corresponding to a value of $\bar{x} = 15.5$.
d. Calculate the standard normal z-score corresponding to $\bar{x} = 23$.

4.167 NW Refer to Exercise 4.166. Find the probability that

a. $\bar{x}$ is less than 16.
b. $\bar{x}$ is greater than 23.
c. $\bar{x}$ is greater than 25.
d. $\bar{x}$ falls between 16 and 22.
e. $\bar{x}$ is less than 14.

4.168 A random sample of $n = 900$ observations is selected from a population with $\mu = 100$ and $\sigma = 10$.

a. What are the largest and smallest values of $\bar{x}$ that you would expect to see?
b. How far, at the most, would you expect $\bar{x}$ to deviate from μ?
c. Did you have to know μ to answer part **b**? Explain.

4.169 A random sample of $n = 100$ observations is selected from a population with $\mu = 30$ and $\sigma = 16$. Approximate the following probabilities:

a. $P(\bar{x} \geq 28)$ **b.** $P(22.1 \leq \bar{x} \leq 26.8)$
c. $P(\bar{x} \leq 28.2)$ **d.** $P(\bar{x} \geq 27.0)$

4.170 Consider a population that contains values of x equal to 0, 1, 2, ..., 97, 98, 99. Assume that the values of x are equally likely. For each of the following values of n, use a computer to generate 500 random samples and calculate $\bar{x}$ for each sample. For each sample size, construct a relative frequency histogram of the 500 values of $\bar{x}$. What changes occur in the histograms as the value of n increases? What similarities exist? Use $n = 2, n = 5, n = 10, n = 30$, and $n = 50$.

Applet Exercise 4.9

Open the applet *Sampling Distributions*. On the pull-down menu to the right of the top graph, select *Binary*.

a. Run the applet for the sample size $n = 10$ and the number of samples $N = 1{,}000$. Observe the shape of the graph of the sample proportions and record the mean, median, and standard deviation of the sample proportions.
b. How does the mean of the sample proportions compare to the mean $\mu = 0.5$ of the original distribution?
c. Compute the standard deviation of the original distribution using the formula $\sigma = \sqrt{np(1-p)}$ where $n = 1$ and $p = 0.5$. Divide the result by $\sqrt{10}$, the square root of the sample size used in the sampling distribution. How does this result compare to the standard deviation of the sample proportions?
d. Explain how the graph of the distribution of sample proportions suggests that the distribution may be approximately normal.
e. Explain how the results of parts **b–d** illustrate the Central Limit Theorem.

Applet Exercise 4.10

Open the applet *Sampling Distributions*. On the pull-down menu to the right of the top graph, select *Uniform*. The box to the left of the top graph displays the population mean, median, and standard deviation of the original distribution.

a. Run the applet for the sample size $n = 30$ and the number of samples $N = 1{,}000$. Observe the shape of the graph of the sample means and record the mean, median, and standard deviation of the sample means.
b. How does the mean of the sample means compare to the mean of the original distribution?
c. Divide the standard deviation of the original distribution by $\sqrt{30}$, the square root of the sample size used in the sampling distribution. How does this result compare to the standard deviation of the sample proportions?
d. Explain how the graph of the distribution of sample means suggests that the distribution may be approximately normal.
e. Explain how the results of parts **b–d** illustrate the Central Limit Theorem.

Applying the Concepts—Basic

4.171 Semester hours taken by CPA candidates. Refer to the *Journal of Accounting and Public Policy* (Spring 2002) study of first-time candidates for the CPA exam, Exercise 2.51 (p. 63). The number of semester hours of college credit taken by candidates has a distribution with a mean of 141 hours and a standard deviation of 18 hours. Consider a random sample of 100 first-time candidates for the CPA exam and let $\bar{x}$ represent the mean number of hours of college credit taken for the sample.

a. What is $\mu_{\bar{x}}$?
b. What is $\sigma_{\bar{x}}$?
c. Describe the shape of the sampling distribution of $\bar{x}$.
d. Find the z-score for the value $\bar{x} = 142$ hours.
e. Find $P(\bar{x} > 142)$

4.172 NW **Salary of a travel management professional.** According to a National Business Travel Association (NBTA) 2008 survey, the average salary of a travel management professional is \$97,300. Assume that the standard deviation of such salaries is \$30,000. Consider a random sample of 50 travel management professionals and let $\bar{x}$ represent the mean salary for the sample.

a. What is $\mu_{\bar{x}}$?
b. What is $\sigma_{\bar{x}}$?
c. Describe the shape of the sampling distribution of $\bar{x}$.
d. Find the z-score for the value $\bar{x} = 89{,}500$.
e. Find $P(\bar{x} > 89{,}500)$.

4.173 Improving SAT scores. Refer to the *Chance* (Winter 2001) examination of Scholastic Assessment Test (SAT) scores of students who pay a private tutor to help them improve their results, Exercise 2.86 (p. 76). On the SAT–Mathematics test, these students had a mean score change of +19 points, with a standard deviation of 65 points. In a random sample of

100 students who pay a private tutor to help them improve their results, what is the likelihood that the sample mean score change is less than 10 points?

4.174 Critical part failures in NASCAR vehicles. Refer to *The Sport Journal* (Winter 2007) analysis of critical part failures at NASCAR races, Exercise 4.144 (p. 236). Recall that researchers found that the time x (in hours) until the first critical part failure is exponentially distributed with $\mu = .10$ and $\sigma = .10$. Now consider a random sample of $n = 50$ NASCAR races and let $\bar{x}$ represent the sample mean time until the first critical part failure.

a. Find $E(\bar{x})$ and $\text{Var}(\bar{x})$.

b. Although x has an exponential distribution, the sampling distribution of $\bar{x}$ is approximately normal. Why?

c. Find the probability that the sample mean time until the first critical part failure exceeds .13 hour.

Applying the Concepts—Intermediate

4.175 Levelness of concrete slabs. Geotechnical engineers use water-level "manometer" surveys to assess the levelness of newly constructed concrete slabs. Elevations are typically measured at eight points on the slab; of interest is the maximum differential between elevations. The *Journal of Performance of Constructed Facilities* (Feb. 2005) published an article on the levelness of slabs in California residential developments. Elevation data collected for over 1,300 concrete slabs *before tensioning* revealed that maximum differential, x, has a mean of $\mu = .53$ inch and a standard deviation of $\sigma = .193$ inch. Consider a sample of $n = 50$ slabs selected from those surveyed and let $\bar{x}$ represent the mean of the sample.

a. Fully describe the sampling distribution of $\bar{x}$.

b. Find $P(\bar{x} > .58)$.

c. The study also revealed that the mean maximum differential of concrete slabs measured *after tensioning and loading* is $\mu = .58$ inch. Suppose the sample data yield $\bar{x} = .59$ inch. Comment on whether the sample measurements were obtained before tensioning or after tensioning and loading.

4.176 Surface roughness of pipe. Refer to the *Anti-Corrosion Methods and Materials* (Vol. 50, 2003) study of the surface roughness of oil field pipes, Exercise 2.46 (p. 62). Recall that a scanning probe instrument was used to measure the surface roughness x (in micrometers) of 20 sampled sections of coated interior pipe. The data are saved in the **ROUGHPIPE** file. Consider the sample mean, $\bar{x}$.

a. Assume that the surface roughness distribution has a mean of $\mu = 1.8$ micrometers and a standard deviation of $\sigma = .5$ micrometer. Use this information to find the probability that $\bar{x}$ exceeds 1.85 micrometers.

b. The sample data are reproduced in the following table. Compute $\bar{x}$.

c. Based on the result, part **b**, comment on the validity of the assumptions made in part **a**.

1.72	2.50	2.16	2.13	1.06	2.24	2.31	2.03	1.09	1.40
2.57	2.64	1.26	2.05	1.19	2.13	1.27	1.51	2.41	1.95

Source: Farshad, F., and Pesacreta, T. "Coated pipe interior surface roughness as measured by three scanning probe instruments," *Anti-Corrosion Methods and Materials,* Vol. 50, No. 1, 2003 (Table III). Copyright © 2003 MCB UP Ltd.

4.177 Is exposure to a chemical in Teflon-coated cookware hazardous? Perfluorooctanoic acid (PFOA) is a chemical used in Teflon-coated cookware to prevent food from sticking. The EPA is investigating the potential risk of PFOA as a cancer-causing agent (*Science News Online,* August 27, 2005). It is known that the blood concentration of PFOA in people in the general population has a mean of $\mu = 6$ parts per billion (ppb) and a standard deviation of $\sigma = 10$ ppb. *Science News Online* reported on tests for PFOA exposure conducted on a sample of 326 people who live near DuPont's Teflon-making Washington (West Virginia) Works facility.

a. What is the probability that the average blood concentration of PFOA in the sample is greater than 7.5 ppb?

b. The actual study resulted in $\bar{x} = 300$ ppb. Use this information to make an inference about the true mean μ PFOA concentration for the population of people who live near DuPont's Teflon facility.

4.178 Rental car fleet evaluation. National Car Rental Systems, Inc., commissioned the U.S. Automobile Club (USAC) to conduct a survey of the general condition of the cars rented to the public by Hertz, Avis, National, and Budget Rent-a-Car.* USAC officials evaluate each company's cars using a demerit point system. Each car starts with a perfect score of 0 points and incurs demerit points for each discrepancy noted by the inspectors. One measure of the overall condition of a company's cars is the mean of all scores received by the company (i.e., the company's *fleet mean score*). To estimate the fleet mean score of each rental car company, 10 major airports were randomly selected, and 10 cars from each company were randomly rented for inspection from each airport by USAC officials (i.e., a sample of size $n = 100$ cars from each company's fleet was drawn and inspected).

a. Describe the sampling distribution of $\bar{x}$, the mean score of a sample of $n = 100$ rental cars.

b. Interpret the mean of $\bar{x}$ in the context of this problem.

c. Assume $\mu = 30$ and $\sigma = 60$ for one rental car company. For this company, find $P(\bar{x} \geq 45)$.

d. Refer to part **c.** The company claims that their true fleet mean score "couldn't possibly be as high as 30." The sample mean score tabulated by USAC for this company was Does this result tend to support or refute the claim? Explain.

4.179 Analysis of supplier lead time. In determining when to place orders to replenish depleted product inventories, a retailer should take into consideration the lead times for the products. *Lead time* is the time between placing the order and having the product available to satisfy customer demand. It includes time for placing the order, receiving the shipment from the supplier, inspecting the units received, and placing them in inventory (Clauss, *Applied Management Science and Spreadsheet Modeling,* 1996). Interested in average lead time, μ, for a particular supplier of men's apparel, the

*Information by personal communication with Rajiv Tandon, corporate vice president and general manager of the Car Rental Division, National Car Rental Systems, Inc., Minneapolis, Minnesota.

purchasing department of a national department store chain randomly sampled 50 of the supplier's lead times and found $\bar{x} = 44$ days.

a. Describe the shape of the sampling distribution of $\bar{x}$.

b. If μ and σ are really 40 and 12, respectively, what is the probability that a second random sample of size 50 would yield $\bar{x}$ greater than or equal to 44?

c. Using the values for μ and σ in part **b,** what is the probability that a sample of size 50 would yield a sample mean within the interval $\mu \pm 2\sigma/\sqrt{n}$?

Applying the Concepts—Advanced

4.180 Plastic fill process. University of Louisville researchers J. Usher, S. Alexander, and D. Duggins examined the process of filling plastic pouches of dry blended biscuit mix (*Quality Engineering,* Vol. 91, 1996). The current fill mean of the process is set at $\mu = 406$ grams, and the process fill standard deviation is $\sigma = 10.1$ grams. (According to the researchers, "The high level of variation is due to the fact that the product has poor flow properties and is, therefore, difficult to fill consistently from pouch to pouch.") Operators monitor the process by randomly sampling 36 pouches each day and measuring the amount of biscuit mix in each. Consider $\bar{x}$, the mean fill amount of the sample of 36 products. Suppose that on one particular day, the operators observe $\bar{x} = 400.8$. One of the operators believes that this indicates that the true process fill mean μ for that day is less than 406 grams. Another operator argues that $\mu = 406$, and the small value of $\bar{x}$ observed is due to random variation in the fill process. Which operator do you agree with? Why?

4.181 Handwashing versus handrubbing. The *British Medical Journal* (Aug. 17, 2002) published the results of a study to compare the effectiveness of handwashing with soap and handrubbing with alcohol. Health care workers who used handrubbing had a mean bacterial count of 35 per hand with a standard deviation of 59. Health care workers who used handwashing had a mean bacterial count of 69 per hand with a standard deviation of 106. In a random sample of 50 health care workers, all using the same method of cleaning their hands, the mean bacterial count per hand, $\bar{x}$, is less than 30. Give your opinion on whether this sample of workers used handrubbing with alcohol or handwashing with soap.

CHAPTER NOTES

Key Terms

Key Symbols

$p(x)$	Probability distribution for discrete random variable, x
$f(x)$	Probability distribution for continuous random variable, x
S	Outcome of binomial trial denoted "success"
F	Outcome of binomial trial denoted "failure"
p	$P(S)$ in binomial trial
q	$P(F)$ in binomial trial $= 1 - p$
e	Constant used in normal and Poisson probability distributions, $e = 2.71828\ldots$
π	Constant used in normal probability distributions, $\pi = 3.1415\ldots$
$\mu_{\bar{x}}$	Mean of the population sampling distribution of $\bar{x}$
$\sigma_{\bar{x}}$	Standard deviation of the population sampling distribution of $\bar{x}$

Key Formulas

Random Variable	Probability Distribution	Mean	Variance
General Discrete:	Table, formula, or graph for $p(x)$	$\sum_{\text{all } x} x \cdot p(x)$	$\sum_{\text{all } x} (x - \mu)^2 \cdot p(x)$
Binomial:	$p(x) = \binom{n}{x} p^x q^{n-x}$ $x = 0, 1, 2, \ldots, n$	np	npq
Poisson:	$p(x) = \dfrac{\lambda^x e^{-\lambda}}{x!}$ $x = 0, 1, 2, \ldots$	λ	λ
Hypergeometric	$p(x) = \dfrac{\binom{r}{x}\binom{N-r}{n-x}}{\binom{N}{n}}$	$\dfrac{nr}{N}$	$\dfrac{r(N-r)n(N-n)}{N^2(N-1)}$
Uniform:	$f(x) = 1/(d - c)$ $(c \le x \le d)$	$(c + d)/2$	$(d - c)^2/12$
Normal:	$f(x) = \dfrac{1}{\sigma\sqrt{2\pi}} e^{-1/2[(x-\mu)/\sigma]^2}$	μ	σ^2
Exponential	$f(x) = \dfrac{1}{\theta} e^{-x/\theta}$	θ	θ
Standard Normal:	$f(z) = \dfrac{1}{\sqrt{2\pi}} e^{-1/2(z)^2}$ $z = (x - \mu)/\sigma$	$\mu = 0$	$\sigma^2 = 1$
***Sample Mean:* (large *n*)**	$f(\bar{x}) = \dfrac{1}{\sigma_{\bar{x}}\sqrt{2\pi}} e^{-1/2[(\bar{x}-\mu)/\sigma_{\bar{x}}]^2}$	$\mu_{\bar{x}} = \mu$	$\sigma_{\bar{x}}^2 = \sigma^2/n$

Key Ideas

Properties of Probability Distributions

Discrete Distributions

1. $p(x) \ge 0$
2. $\sum_{\text{all } x} p(x) = 1$

Continuous Distributions

1. $P(x = a) = 0$
2. $P(a < x < b)$ is area under curve between a and b

Normal Approximation to Binomial

x is binomial (n, p)
$P(x \le a) \approx P\{z < (a + .5) - \mu\}$

Methods for Assessing Normality

1. *Histogram*

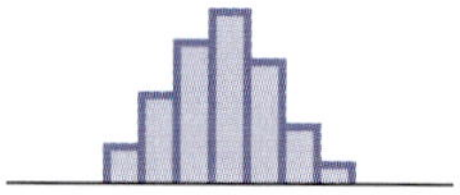

2. *Stem-and-leaf display*

1	7
2	3389
3	245677
4	19
5	2

3. $(IQR)/S \approx 1.3$
4. *Normal probability plot*

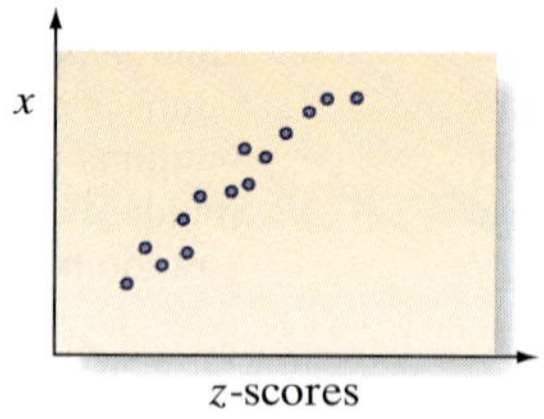

Generating the Sampling Distribution of $\bar{x}$

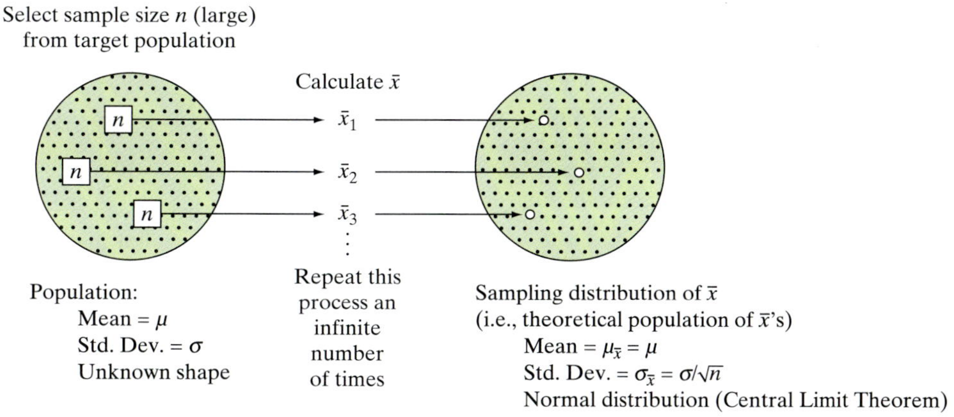

Guide to Selecting a Probability Distribution

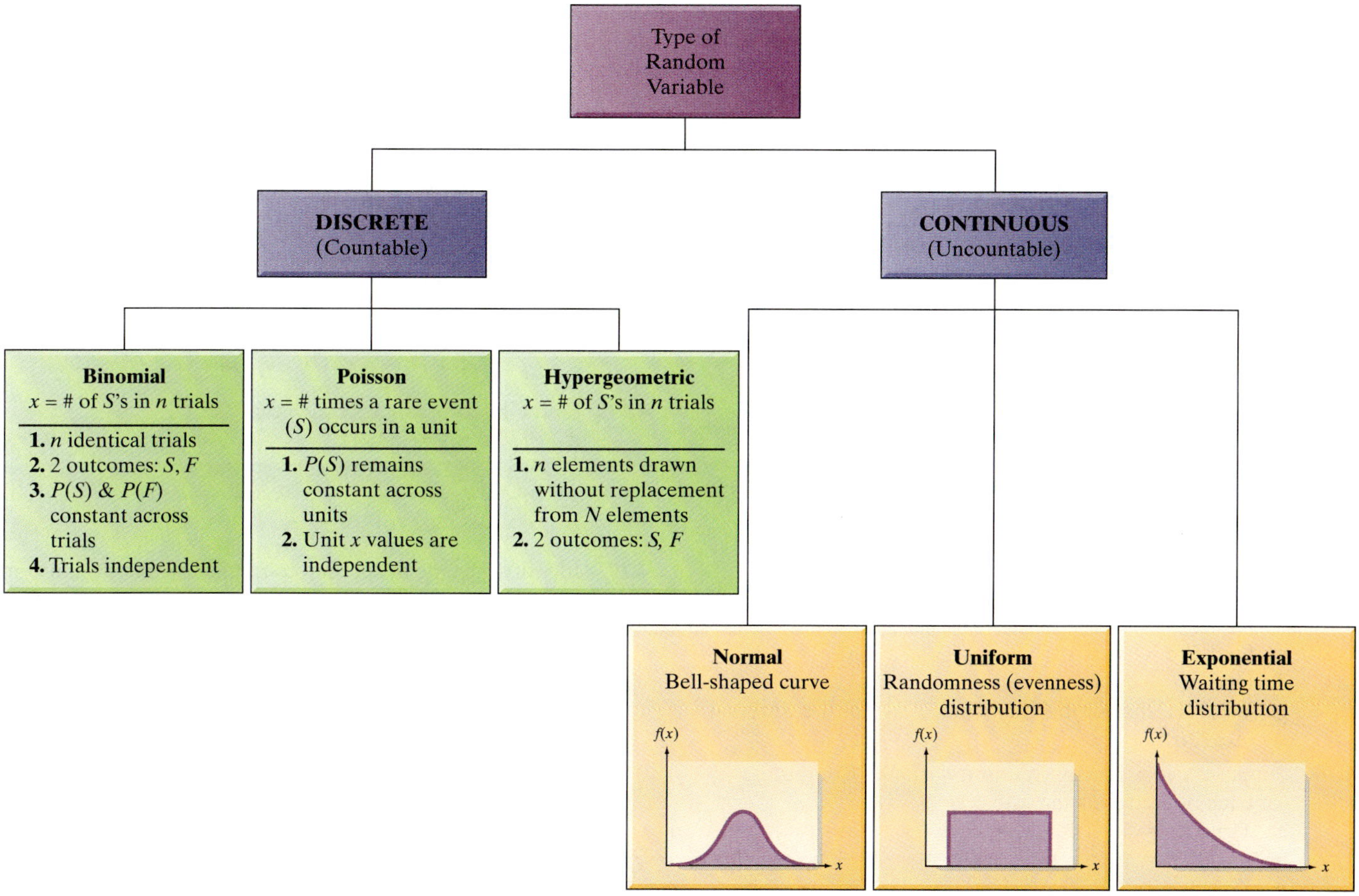

Supplementary Exercises 4.182–4.227

Learning the Mechanics

4.182 For each of the following examples, decide whether x is a binomial random variable and explain your decision:

a. A manufacturer of computer chips randomly selects 100 chips from each hour's production in order to estimate the proportion defective. Let x represent the number of defectives in the 100 sampled chips.

b. Of five applicants for a job, two will be selected. Although all applicants appear to be equally qualified, only three have the ability to fulfill the expectations of the company. Suppose that the two selections are made at random from the five applicants and let x be the number of qualified applicants selected.

c. A software developer establishes a support hotline for customers to call in with questions regarding use of the software. Let x represent the number of calls received on the support hotline during a specified workday.

d. Florida is one of a minority of states with no state income tax. A poll of 1,000 registered voters is conducted to determine how many would favor a state income tax in light of the state's current fiscal condition. Let x be the number in the sample who would favor the tax.

4.183 Given that x is a binomial random variable, compute $p(x)$ for each of the following cases:

a. $n = 7, x = 3, p = .5$

b. $n = 4, x = 3, p = .8$

c. $n = 15, x = 1, p = .1$

4.184 Consider the discrete probability distribution shown here.

x	10	12	18	20
$p(x)$	.2	.3	.1	.4

a. Calculate μ, σ^2, and σ.

b. What is $P(x < 15)$?

c. Calculate $\mu \pm 2\sigma$.

d. What is the probability that x is in the interval $\mu \pm 2\sigma$?

4.185 Suppose x is a binomial random variable with $n = 20$ and $p = .7$.

a. Find $P(x = 14)$.

b. Find $P(x \leq 12)$.

c. Find $P(x > 12)$.

d. Find $P(9 \leq x \leq 18)$.

e. Find $P(8 < x < 18)$.

f. Find μ, σ^2, and σ.

g. What is the probability that x is in the interval $\mu \pm 2\sigma$?

4.186 Suppose x is a Poisson random variable. Compute $p(x)$ for each of the following cases:

a. $\lambda = 2, x = 3$

b. $\lambda = 1, x = 4$

c. $\lambda = .5, x = 2$

4.187 Identify the type of random variable—binomial, Poisson, or hypergeometric—described by each of the following probability distributions:

a. $p(x) = \dfrac{.5^x e^{-5}}{x!} \quad (x = 0, 1, 2, \ldots)$

b. $p(x) = \binom{6}{x}(.2)^x(.8)^{6-x} \quad (x = 0,1,2,\ldots, 6)$

c. $p(x) = \dfrac{10!}{x!(10-x)!}(.9)^x(.1)^{10-x}$ $(x = 0, 1, 2,\ldots, 10)$

4.188 Given that x is a hypergeometric random variable, compute $p(x)$ for each of the following cases:

a. $N = 8, n = 5, r = 3, x = 2$

b. $N = 6, n = 2, r = 2, x = 2$

c. $N = 5, n = 4, r = 4, x = 3$

4.189 Which of the following describe discrete random variables, and which describe continuous random variables?

a. The number of damaged inventory items

b. The average monthly sales revenue generated by a salesperson over the past year

c. Square feet of warehouse space a company rents

d. The length of time a firm must wait before its copying machine is fixed

4.190 Assume that x is a random variable best described by a uniform distribution with $c = 10$ and $d = 90$.

a. Find $f(x)$.

b. Find the mean and standard deviation of x.

c. Graph the probability distribution for x and locate its mean and the interval $\mu \pm 2\sigma$ on the graph.

d. Find $P(x \leq 60)$.

e. Find $P(x \geq 90)$.

f. Find $P(x \leq 80)$.

g. Find $P(\mu - \sigma \leq x \leq \mu + \sigma)$.

h. Find $P(x > 75)$.

4.191 Find the following probabilities for the standard normal random variable z:

a. $P(z \leq 2.1)$ **b.** $P(z \geq 2.1)$

c. $P(z \geq -1.65)$ **d.** $P(-2.13 \leq z \leq -.41)$

e. $P(-1.45 \leq z \leq 2.15)$ **f.** $P(z \leq -1.43)$

4.192 Find a z-score, call it z_0, such that

a. $P(z \leq z_0) = .5080$

b. $P(z \geq z_0) = .5517$

c. $P(z \geq z_0) = .1492$

d. $P(z_0 \leq z \leq .59) = .4773$

4.193 Identify the type of continuous random variable—uniform, normal, or exponential—described by each of the following probability density functions:

a. $f(x) = (e^{-x/7})/7, x > 0$

b. $f(x) = 1/20, 5 < x < 25$

c. $f(x) = \dfrac{e^{-.5[(x-10)/5]^2}}{5\sqrt{2\pi}}$

4.194 Assume that x has an exponential distribution with $\theta = 3$. Find

a. $P(x \leq 1)$ **b.** $P(x > 1)$

c. $P(x = 1)$ **d.** $P(x \leq 6)$

e. $P(2 \leq x \leq 10)$

4.195 The random variable x has a normal distribution with $\mu = 75$ and $\sigma = 10$. Find the following probabilities:

a. $P(x \leq 80)$ **b.** $P(x \geq 85)$

c. $P(70 \leq x \leq 75)$ **d.** $P(x > 80)$

e. $P(x = 78)$ **f.** $P(x \leq 110)$

4.196 Assume that x is a binomial random variable with $n = 100$ and $p = .5$. Use the normal probability distribution to approximate the following probabilities:

a. $P(x \leq 48)$ **b.** $P(50 \leq x \leq 65)$
c. $P(x \geq 70)$ **d.** $P(55 \leq x \leq 58)$
e. $P(x = 62)$ **f.** $P(x \leq 49 \text{ or } x \geq 72)$

4.197 The random variable x has a normal distribution with $\mu = 40$ and $\sigma^2 = 36$. Find a value of x, call it x_0, such that

a. $P(x \geq x_0) = .10$ **b.** $P(\mu \leq x < x_0) = .40$
c. $P(x < x_0) = .05$ **d.** $P(x \geq x_0) = .40$
e. $P(x_0 \leq x < \mu) = .45$

4.198 A random sample of 40 observations is to be drawn from a large population of measurements. It is known that 30% of the measurements in the population are 1s, 20% are 2s, 20% are 3s, and 30% are 4s.

a. Give the mean and standard deviation of the (repeated) sampling distribution of $\bar{x}$, the sample mean of the 40 observations.
b. Describe the shape of the sampling distribution of $\bar{x}$. Does your answer depend on the sample size?

4.199 A random sample of $n = 68$ observations is selected from a population with $\mu = 19.6$ and $\sigma = 3.2$. Approximate each of the following probabilities.

a. $P(\bar{x} \leq 19.6)$ **b.** $P(\bar{x} \leq 19)$
c. $P(\bar{x} \geq 20.1)$ **d.** $P(19.2 \leq \bar{x} \leq 20.6)$

4.200 Use a statistical software package to generate 100 random samples of size $n = 2$ from a population characterized by a uniform probability distribution (Section 4.9) with $c = 0$ and $d = 10$. Compute $\bar{x}$ for each sample and plot a frequency distribution for the 100 $\bar{x}$ values. Repeat this process for $n = 5, 10, 30$, and 50. Explain how your plots illustrate the Central Limit Theorem.

4.201 A random sample of size n is to be drawn from a large population with mean 100 and standard deviation 10, and the sample mean $\bar{x}$ is to be calculated. To see the effect of different sample sizes on the standard deviation of the sampling distribution of $\bar{x}$, plot $\sigma/\sqrt{n}$ against n for $n = 1$, 5, 10, 20, 30, 40, and 50.

Applying the Concepts—Basic

4.202 **Back injuries at work.** A poll by the Gallup Organization sponsored by Philadelphia-based CIGNA Integrated Care found that about 40% of employees have missed work due to a musculoskeletal (back) injury of some kind (*National Underwriter,* Apr. 5, 1999). Let x be the number of sampled workers who have missed work due to a back injury.

a. Explain why x is approximately a binomial random variable.
b. Use the Gallup poll data to estimate p for the binomial random variable of part **a.**
c. A random sample of 10 workers is to be drawn from a particular manufacturing plant. Use the p from part **b** to find the mean and standard deviation of x, the number of workers who missed work due to back injuries.
d. For the sample in part **c,** find the probability that exactly one worker missed work due to a back injury. That more than one worker missed work due to a back injury.

4.203 **Dentists' use of laughing gas.** According to the American Dental Association, 60% of all dentists use nitrous oxide in their practice. If x equals the number of dentists in a random sample of five dentists who use laughing gas in practice, then the probability distribution of x is

x	0	1	2	3	4	5
$p(x)$	.0102	.0768	.2304	.3456	.2592	.0778

a. Verify that the probabilities for x sum to 1.
b. Find $P(x = 4)$.
c. Find $P(x < 2)$.
d. Find $P(x \geq 3)$.
e. Find $\mu = E(x)$ and practically interpret this value.

4.204 **Analysis of "no-hitters" in baseball.** In baseball, a *no-hitter* is a regulation nine-inning game in which the pitcher yields no hits to the opposing batters. *Chance* (Summer 1994) reported on a study of no-hitters in Major League Baseball (MLB). The initial analysis focused on the total number of hits yielded per game per team for nine-inning MLB games. The distribution of hits/nine-innings is approximately normal with mean 8.72 and standard deviation 1.10.

a. What percentage of nine-inning MLB games result in fewer than six hits?
b. Demonstrate, statistically, why a no-hitter is considered an extremely rare occurrence in MLB.

4.205 **Trajectory of an electrical circuit.** Researchers at the University of California–Berkeley have designed, built, and tested a switched-capacitor circuit for generating random signals (*International Journal of Circuit Theory and Applications,* May–June 1990). The circuit's trajectory was shown to be uniformly distributed on the interval (0, 1).

a. Give the mean and variance of the circuit's trajectory.
b. Compute the probability that the trajectory falls between .2 and .4.
c. Would you expect to observe a trajectory that exceeds .995? Explain.

4.206 **Deep-draft vessel casualties.** University of New Mexico economists Gawande and Wheeler modeled the number of casualties experienced by a deep-draft marine vessel over a 3-year period as a Poisson random variable, x. They estimated $E(x)$ to be .03 (*Management Science,* January 1999).

a. Find the variance x.
b. Discuss the conditions that would make the researchers' Poisson assumption plausible.
c. What is the probability that a deep-draft U.S. flag vessel will have exactly one casualty in a 3-year time period? No casualties in a 3-year period?

4.207 **Ocean quahog harvesting.** The ocean quahog is a type of clam found in the coastal waters of New England and the mid-Atlantic states. A federal survey of offshore ocean quahog harvesting in New Jersey, conducted from 1980 to 1992, revealed an average catch per unit effort (CPUE) of 89.34 clams. The CPUE standard deviation was 7.74 (*Journal of Shellfish Research,* June 1995). Let represent the mean CPUE for a sample of 35 attempts to catch ocean quahogs off the New Jersey shore.

a. Compute $\mu_{\bar{x}}$ and $\sigma_{\bar{x}}$. Interpret their values.
b. Sketch the sampling distribution of $\bar{x}$.

c. Find $P(\bar{x} > 88)$.
d. Find $P(\bar{x} < 87)$.

4.208 Dutch elm disease. A nursery advertises that it has 10 elm trees for sale. Unknown to the nursery, 3 of the trees have already been infected with Dutch elm disease and will die within a year.
a. If a buyer purchases 2 trees, what is the probability that both trees will be healthy?
b. Refer to part **a.** What is the probability that at least 1 of the trees is infected?

4.209 Hospital patient interarrival times. The length of time between arrivals at a hospital clinic has an approximately exponential probability distribution. Suppose the mean time between arrivals for patients at a clinic is 4 minutes.
a. What is the probability that a particular interarrival time (the time between the arrival of two patients) is less than 1 minute?
b. What is the probability that the next four interarrival times are all less than 1 minute?
c. What is the probability that an interarrival time will exceed 10 minutes?

4.210 Switching banks after a merger. Banks that merge with others to form "mega-banks" sometimes leave customers dissatisfied with the impersonal service. A poll by the Gallup Organization found 20% of retail customers switched banks after their banks merged with another (*Bank Marketing,* Feb. 1999). One year after the acquisition of First Fidelity by First Union, a random sample of 25 retail customers who had banked with First Fidelity were questioned. Let x be the number of those customers who switched their business from First Union to a different bank.
a. What assumptions must hold in order for x to be a binomial random variable? In the remainder of this exercise, use the data from the Gallop Poll to estimate p.
b. What is the probability that $x \leq 10$?
c. Find $E(x)$ and the standard deviation of x.
d. Calculate the interval $\mu \pm 2\sigma$.
e. If samples of size 25 were drawn repeatedly a large number of times and x determined for each sample, what proportion of the x values would fall within the interval you calculated in part **d**?

4.211 Machine repair times. An article in *IEEE Transactions* (Mar. 1990) gave an example of a flexible manufacturing system with four machines operating independently. The repair rates for the machines (i.e., the time, in hours, it takes to repair a failed machine) are exponentially distributed with means $\mu_1 = 1$, $\mu_2 = 2$, $\mu_3 = .5$, and $\mu_4 = .5$, respectively.
a. Find the probability that the repair time for machine 1 exceeds 1 hour.
b. Repeat part **a** for machine 2.
c. Repeat part **a** for machines 3 and 4.
d. If all four machines fail simultaneously, find the probability that the repair time for the entire system exceeds 1 hour.

4.212 Estimating demand for white bread. A bakery has determined that the number of loaves of its white bread demanded daily has a normal distribution with mean 7,200 loaves and standard deviation 300 loaves. Based on cost considerations, the company has decided that its best strategy is to produce a sufficient number of loaves so that it will fully supply demand on 94% of all days.
a. How many loaves of bread should the company produce?
b. Based on the production in part **a,** on what percentage of days will the company be left with more than 500 loaves of unsold bread?

4.213 Millions of suburban commuters are finding railroads to be a convenient, time-saving, less stressful alternative to the automobile. While generally perceived as a safe mode of transportation, the average number of deaths per week due to railroad accidents is a surprisingly high 20 (U.S. National Center for Health Statistics, *Vital Statistics of the United States, 2001*).
a. Construct arguments both for and against the use of the Poisson distribution to characterize the number of deaths per week due to railroad accidents.
b. For the remainder of this exercise, assume the Poisson distribution is an adequate approximation for x, the number of deaths per week due to railroad accidents. Find $E(x)$ and the standard deviation of x.
c. Based strictly on your answers to part **b,** is it likely that only four or fewer deaths occur next week? Explain.
d. Find $P(x \leq 4)$. Is this probability consistent with your answer to part **c**? Explain.

4.214 Environmental failures of Arkansas companies. Refer to the study of 38 Arkansas corporations that were penalized for violating one or more environmental laws, Exercise 2.27 (p. 51). The financial penalties assessed to the companies are saved in the **CLEANAIR** file. Determine whether the financial penalties are approximately normally distributed.

4.215 Time in bankruptcy. Refer to the *Financial Management* (Spring 1995) study of 49 companies that filed for a prepackaged bankruptcy, Exercise 2.30 (p. 52). The time in bankruptcy (measured in months) for each company is saved in the **BANKRUPT** file. Determine whether the bankruptcy times are approximately normally distributed.

4.216 Length of time working for an employer. Workers are much less likely to remain with one employer for many years than their parents were a generation before (*Georgia Trend,* December 1999). Do today's college students understand that the workplace they are about to enter is vastly different than the one their parents entered? To help answer this question, researchers at the Terry College of Business at the University of Georgia sampled 344 business students and asked them this question: Over the course of your lifetime, what is the maximum number of years you expect to work for any one employer? The resulting sample had $\bar{x} = 19.1$ years and $s = 6$ years. Assume the sample of students was randomly selected from the 5,800 undergraduate students in the Terry College.
a. Describe the sampling distribution of $\bar{x}$.
b. If the population mean was 18.5 years, what is $P(\bar{x} \geq 19.1 \text{ years})$?
c. If the population mean was 19.5 years, what is $P(\bar{x} \geq 19.1 \text{ years})$?
d. If $P(\bar{x} \geq 19.1) = .5$, what is the population mean?
e. If $P(\bar{x} \geq 19.1) = .2$, is the population mean greater or less than 19.1 years? Justify your answer.

4.217 Errors in measuring truck weights. To help highway planners anticipate the need for road repairs and design future construction projects, data are collected on the estimated volume and weight of truck traffic on specific roadways (*Transportation Planning Handbook,* 2008) using specialized "weigh-in-motion" equipment. In an experiment performed by the Minnesota Department of Transportation involving repeated weighing of a 27,907-pound truck, it was found that the weights recorded by the weigh-in-motion equipment were approximately normally distributed with a mean of 27,315 and a standard deviation of 628 pounds (Minnesota Department of Transportation). It follows that the difference between the actual weight and recorded weight, the error of measurement, is normally distributed with a mean of 592 pounds and a standard deviation of 628 pounds.

a. What is the probability that the weigh-in-motion equipment understates the actual weight of the truck?

b. If a 27,907-pound truck was driven over the weigh-in-motion equipment 100 times, approximately how many times would the equipment overstate the truck's weight?

c. What is the probability that the error in the weight recorded by the weigh-in-motion equipment for a 27,907-pound truck exceeds 400 pounds?

d. It is possible to adjust (or *calibrate*) the weigh-in-motion equipment to control the mean error of measurement. At what level should the mean error be set so the equipment will understate the weight of a 27,907-pound truck 50% of the time? Only 40% of the time?

4.218 Errors in filling prescriptions A large number of preventable errors (e.g., overdoses, botched operations, misdiagnoses) are being made by doctors and nurses in U.S. hospitals (*New York Times,* July 18, 1995). A study of a major metropolitan hospital revealed that of every 100 medications prescribed or dispensed, 1 was in error, but only 1 in 500 resulted in an error that caused significant problems for the patient. It is known that the hospital prescribes and dispenses 60,000 medications per year.

a. What is the expected number of errors per year at this hospital? The expected number of significant errors per year?

b. Within what limits would you expect the number of significant errors per year to fall?

c. What assumptions did you need to make in order to answer these questions?

4.219 Whistle-blowing among federal employees. *Whistle-blowing* refers to an employee's reporting of wrongdoing by coworkers. A survey found that about 5% of employees contacted had reported wrongdoing during the past 12 months. Assume that a sample of 25 employees in one agency are contacted and let x be the number who have observed and reported wrongdoing in the last 12 months. Assume that the probability of whistle-blowing is .05 for any federal employee over the past 12 months.

a. Find the mean and standard deviation of x. Can x be equal to its expected value? Explain.

b. Write the event that at least 5 of the employees are whistle-blowers in terms of x. Find the probability of the event.

c. If 5 of the 25 contacted have been whistle-blowers over the past 12 months, what would you conclude about the applicability of the 5% assumption to this agency? Use your answer to part **b** to justify your conclusion.

4.220 Piercing rating of fencing safety jackets. A manufacturer produces safety jackets for competitive fencers. These jackets are rated by the minimum force, in newtons, that will allow a weapon to pierce the jacket. When this process is operating correctly, it produces jackets that have ratings with an average of 840 newtons and a standard deviation of 15 newtons. FIE, the international governing body for fencing, requires jackets to be rated at a minimum of 800 newtons. To check whether the process is operating correctly, a manager takes a sample of 50 jackets from the process, rates them, and calculates $\bar{x}$, the mean rating for jackets in the sample. She assumes that the standard deviation of the process is fixed but is worried that the mean rating of the process may have changed.

a. What is the sampling distribution of $\bar{x}$ if the process is still operating correctly?

b. Suppose the manager's sample has a mean rating of 830 newtons. What is the probability of getting an $\bar{x}$ of 830 newtons or lower if the process is operating correctly?

c. Given the manager's assumption that the standard deviation of the process is fixed, what does your answer to part **b** suggest about the current state of the process (i.e., does it appear that the mean jacket rating is still 840 newtons)?

d. Now suppose that the mean of the process has not changed, but the standard deviation of the process has increased from 15 newtons to 45 newtons. What is the sampling distribution of $\bar{x}$ in this case? What is the probability of getting an $\bar{x}$ of 830 newtons or lower when $\bar{x}$ has this distribution?

4.221 Awarding of home improvement grants. A curious event was described in the *Minneapolis Star and Tribune*. The Minneapolis Community Development Agency (MCDA) makes home improvement grants each year to homeowners in depressed city neighborhoods. Of the $708,000 granted one year, $233,000 was awarded by the city council via a "random selection" of 140 homeowners' applications from among a total of 743 applications: 601 from the north side, and 142 from the south side, of Minneapolis. Oddly, all 140 grants awarded were from the north side—clearly a highly improbable outcome if, in fact, the 140 winners were randomly selected from among the 743 applicants.

a. Suppose the 140 winning applications were randomly selected from among the total of 743 and let x equal the number in the sample from the north side. Find the mean and standard deviation of x.

b. Use the results of part **a** to support a contention that the grant winners were not randomly selected.

4.222 Time delays at a bus stop. A bus is scheduled to stop at a certain bus stop every half hour on the hour and the half hour. At the end of the day, buses still stop after every 30 minutes, but because delays often occur earlier in the day, the bus is never early and likely to be late. The director of the bus line claims that the length of time a bus is late is uniformly distributed, and the maximum time that a bus is late is 20 minutes.

a. If the director's claim is true, what is the expected number of minutes a bus will be late?

b. If the director's claim is true, what is the probability that the last bus on a given day will be more than 19 minutes late?

c. If you arrive at the bus stop at the end of a day at exactly half past the hour and must wait more than 19 minutes for the bus, what would you conclude about the director's claim? Why?

Applying the Concepts—Advanced

4.223 **How many questionnaires to mail?** The probability that a consumer responds to a marketing department's mailed questionnaire is .4. How many questionnaires should be mailed if you want to be reasonably certain that at least 100 will be returned?

4.224 **Establishing tolerance limits.** The *tolerance limits* for a particular quality characteristic (e.g., length, weight, or strength) of a product are the minimum and/or maximum values at which the product will operate properly. Tolerance limits are set by the engineering design function of the manufacturing operation (Moss, *Applying TQM to Product Design and Development,* 1996). The tensile strength of a particular metal part can be characterized as being normally distributed with a mean of 25 pounds and a standard deviation of 2 pounds. The upper and lower tolerance limits for the part are 30 pounds and 21 pounds, respectively. A part that falls within the tolerance limits results in a profit of $10. A part that falls below the lower tolerance limit costs the company $2; a part that falls above the upper tolerance limit costs the company $1. Find the company's expected profit per metal part produced.

4.225 **Purchasing decision.** A building contractor has decided to purchase a load of factory-reject aluminum siding as long as the average number of flaws per piece of siding in a sample of size 35 from the factory's reject pile is 2.1 or less. If it is known that the number of flaws per piece of siding in the factory's reject pile has a Poisson probability distribution with a mean of 2.5, find the approximate probability that the contractor will not purchase a load of siding.

Critical Thinking Challenges

4.226 **Reverse cocaine sting.** *The American Statistician* (May 1991) described an interesting application of a probability distribution in a case involving illegal drugs. During a drug bust, police seized 496 foil packets of a white, powdery substance that appeared to be cocaine. The police laboratory randomly selected 4 packets and found that all 4 tested positive for cocaine. This finding led to the conviction of the drug traffickers. Following the conviction, the police used 2 of the remaining 492 foil packets (i.e., those not tested) in a reverse sting operation. The 2 randomly selected packets were sold by undercover officers to a buyer who disposed of the evidence before being arrested. Is there beyond a reasonable doubt that the 2 packets contained cocaine? To solve the dilemma, assume that of the 496 original packets confiscated, 331 contained genuine cocaine and 165 contained an inert (legal) powder. (A statistician, hired as an expert witness on the case, showed that the chance of the defendant being found not guilty is maximized when 331 packets contain cocaine and 165 do not.) [Hint: First, find the probability that 4 packets randomly selected from the original 496 will test positive for cocaine. Next find the probability that the 2 packets sold in the reverse sting did not contain cocaine. Finally, find the probability that both events occur (i.e., that the first 4 packets selected test positive for cocaine but that the next 2 packets selected do not). In each of these probability calculations, apply the binomial probability distribution to approximate the probabilities.]

4.227 **The insomnia pill.** A research report published in the *Proceedings of the National Academy of Sciences* (Mar. 1994) brought encouraging news to insomniacs and international business travelers who suffer from jet lag. Neuroscientists at the Massachusetts Institute of Technology (MIT) have been experimenting with melatonin—a hormone secreted by the pineal gland in the brain—as a sleep-inducing hormone. Because the hormone is naturally produced, it is nonaddictive. The researchers believe melatonin may be effective in treating jet lag. In the MIT study, young male volunteers were given various doses of melatonin or a placebo (a dummy medication containing no melatonin). Then they were placed in a dark room at midday and told to close their eyes for 30 minutes. The variable of interest was the time (in minutes) elapsed before each volunteer fell asleep.

Now, consider a random sample of 40 young males, each of whom is given a dosage of the sleep-inducing hormone, melatonin. The times (in minutes) to fall asleep for these 40 males are listed in the table and saved in the **INSOMNIA** file. The researchers know that with the placebo (i.e., no hormone), the mean time to fall asleep is $\mu = 15$ minutes and the standard deviation is $\sigma = 10$ minutes. They want to use the data to make an inference about the true value of μ for those taking the melatonin. Is melatonin an effective drug against insomnia?

7.6	2.1	1.4	1.5	3.6	17.0	14.9	4.4	4.7	20.1
7.7	2.4	8.1	1.5	10.3	1.3	3.7	2.5	3.4	10.7
2.9	1.5	15.9	3.0	1.9	8.5	6.1	4.5	2.2	2.6
7.0	6.4	2.8	2.8	22.8	1.5	4.6	2.0	6.3	3.2

[Note: These data are simulated sleep times based on summary information provided in the MIT study.]

References

Deming, W. E. *Out of the Crisis*. Cambridge, Mass.: MIT Center for Advanced Engineering Study, 1986.

Hogg, R. V., McKean, J. W., and Craig, A. T. *Introduction to Mathematical Statistics,* 6th ed. Upper Saddle River, N.J.: Prentice Hall, 2005.

Larsen, R. J., and Marx, M. L. *An Introduction to Mathematical Statistics and Its Applications,* 4th ed. Upper Saddle River, N.J.: Prentice Hall, 2005.

Lindgren, B. W. *Statistical Theory,* 4th ed. New York: Chapman & Hall, 1993.

Ramsey, P. P., and Ramsey, P. H. "Simple tests of normality in small samples," *Journal of Quality Technology,* Vol. 22, 1990.

Wackerly, D., Mendenhall, W., and Scheaffer, R, *Mathematical Statistics with Applications,* 7th ed. North Scituate, Mass.: Duxbury, 2008.

USING TECHNOLOGY

SPSS: Discrete Probabilities, Continuous Probabilities, Normal Probability Plots, and Simulated Sampling Distributions

Discrete and Continuous Probabilities

Step 1 Select "Transform" on the SPSS menu bar and then click on "Compute Variable," as shown in Figure 4.S.1. The resulting dialog box appears as shown in Figure 4.S.2.

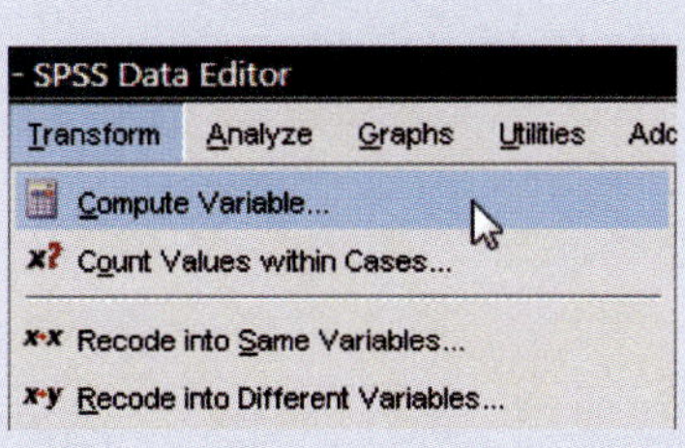

Figure 4.S.1 SPSS menu options for obtaining discrete and continuous probabilities

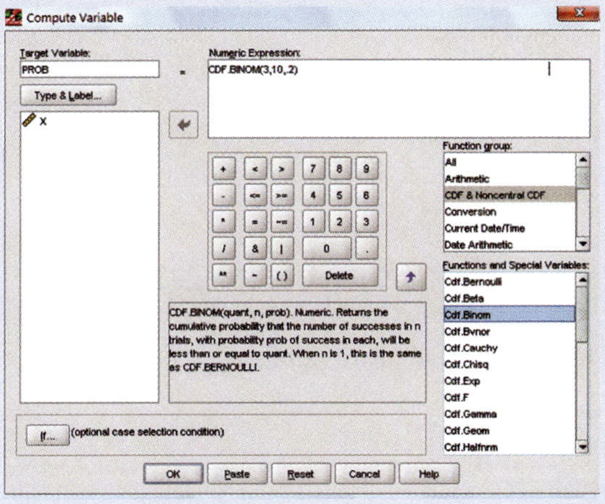

Figure 4.S.2 SPSS compute cumulative binomial probabilities dialog box

Step 2 Specify a name for the "Target Variable."

Step 3 Select either "CDF" (for cumulative probabilities) or "PDF" (for exact probabilities) in the "Function Group" box.

Step 4 Select the appropriate probability distribution in the "Functions and Special Variables" box. (For example, for cumulative probabilities of the binomial, use the function CDF. BINOM. For exact binomial probabilities, use the PDF.BINOM function.)

Step 5 Enter the parameters of the distribution in the "Numeric Expression" box. For example, Figure 4.S.2 shows the cumulative binomial function with parameters of $x = 3$ (the first number in the function), $n = 10$ (the second number), and $p = .2$ (the third number).

Step 6 Click "OK;" SPSS will compute the requested probability (in this example, the cumulative binomial probability that x is less than or equal to 3) and display it on the SPSS spreadsheet. Similarly, for example, Figure 4.S.3 shows the cumulative normal function with parameters of $x = 3.7$ (the first number in the function), $\mu = 5$ (the second number), and $\sigma = 2$ (the third number). When you click "OK," SPSS will compute the requested probability (in this example, the cumulative normal probability that x is less than 3.7) and display it on the SPSS spreadsheet.

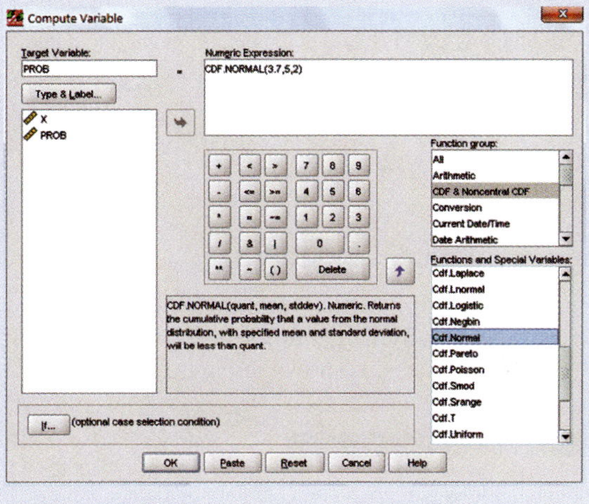

Figure 4.S.3 SPSS compute cumulative normal probabilities dialog box

Normal Probability Plot

Step 1 Select "Analyze" on the SPSS menu bar, and then select "Descriptive Statistics" and "Q-Q Plots," as shown in Figure 4.S.4. The resulting dialog box appears as shown in Figure 4.S.5.

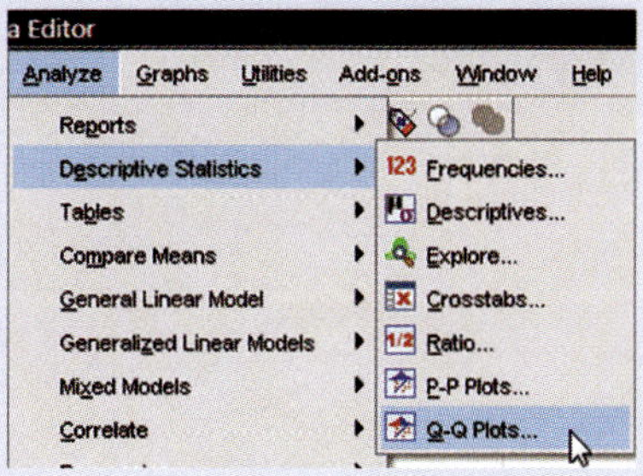

Figure 4.S.4 SPSS options for a normal probability plot

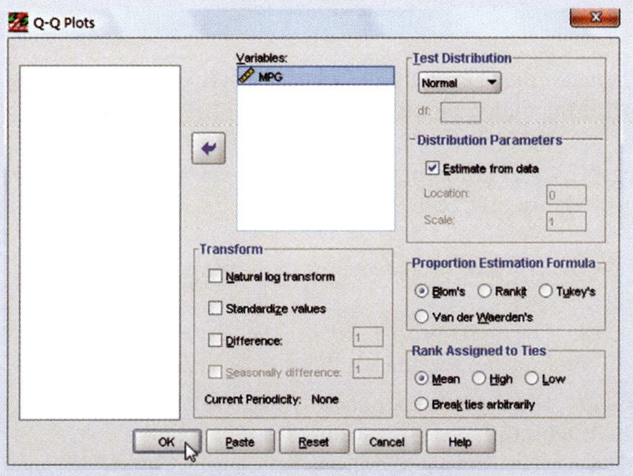

Figure 4.S.5 SPSS normal probability plot dialog box

Step 2 Specify the variable of interest in the "Variables" box and select "Normal" in the "Test Distribution" box.

Step 3 Click "OK" to generate the normal probability plot.

Sampling Distributions

Generating sampling distributions with SPSS is not a simple process. Consult the SPSS user's guide for help with this feature.

Minitab: Discrete Probabilities, Continuous Probabilities, Normal Probability Plots, and Simulated Sampling Distributions

Discrete and Continuous Probabilities

Step 1 Select the "Calc" button on the Minitab menu bar, click on "Probability Distributions" and then finally select the distribution of your choice (e.g., "Binomial"), as shown in Figure 4.M.1.

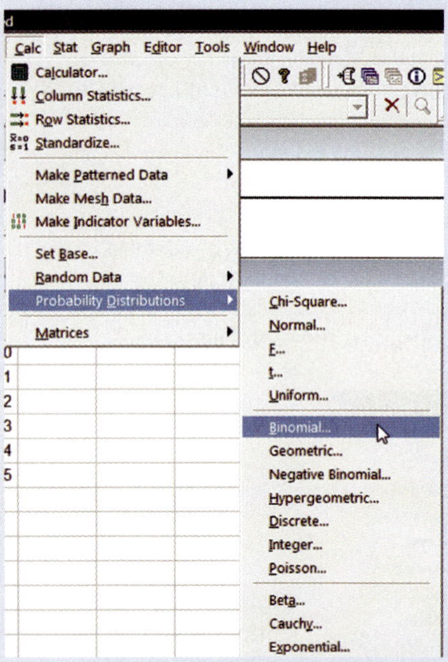

Figure 4.M.1 Minitab menu options for obtaining discrete and continuous probabilities

Step 2 Select either "Probability" or "Cumulative probability" on the resulting dialog box.

Step 3 Specify the parameters of the distribution (e.g., sample size n, probability of success p, μ or σ.)

Step 4 Specify the value of x of interest in the "Input constant" box.

Step 5 Click "OK." The probability for the value of x will appear on the Minitab session window.

[*Note:* Figure 4.M.2 gives the specifications for finding $P(x = 1)$ in a binomial distribution with $n = 5$ and $P = 2$. Figure 4.M.3 gives the specifications for finding $P(x \leq 20)$ in a normal distribution with $\mu = 24.5$ and $\sigma = 1$.]

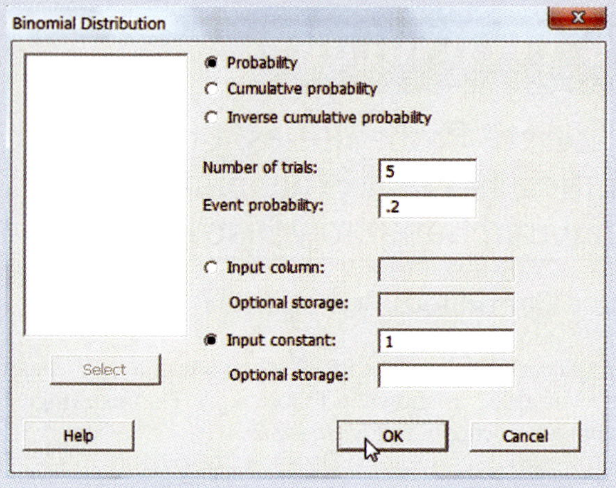

Figure 4.M.2 Minitab binomial distribution dialog box

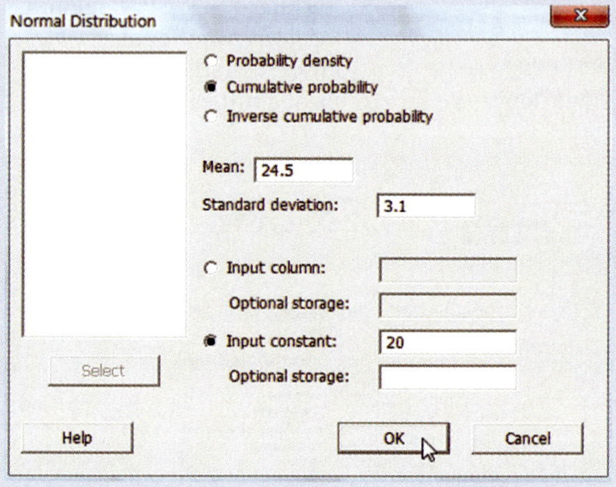

Figure 4.M.3 Minitab normal distribution dialog box

Normal Probability Plot

Step 1 Select "Graph" on the Minitab menu bar and then click on "Probability Plot," as shown in Figure 4.M.4.

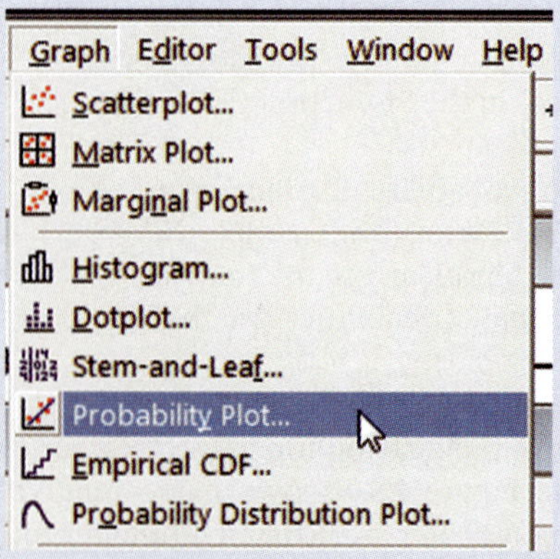

Figure 4.M.4 Minitab options for a normal probability plot

Step 2 Select "Single" (for one variable) on the next box, and the dialog box will appear as shown in Figure 4.M.5.

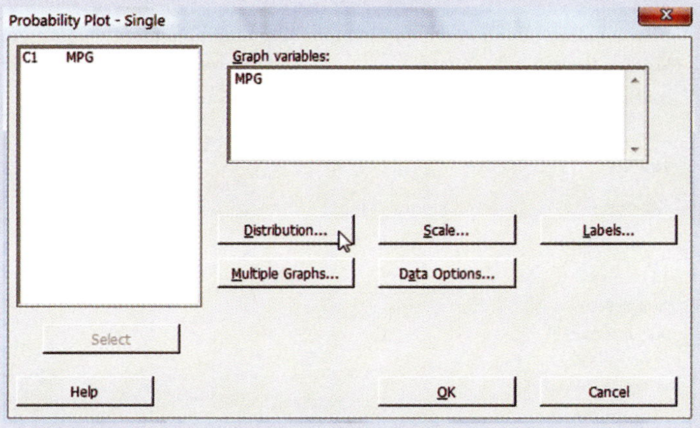

Figure 4.M.5 Minitab normal probability plot dialog box

Step 3 Specify the variable of interest in the "Graph variables" box and then click the "Distribution" button and select the "Normal" option. Click "OK" to return to the Probability Plot dialog box.

Step 4 Click "OK" to generate the normal probability plot.

Sampling Distributions

Step 1 Select "Calc" on the Minitab menu bar and then click on "Random Data" (see Figure 4.M.1).

Step 2 On the resulting menu list, click on the distribution of your choice (e.g., "Uniform"). A dialog box similar to the one (the Uniform Distribution) shown in Figure 4.M.6 will appear.

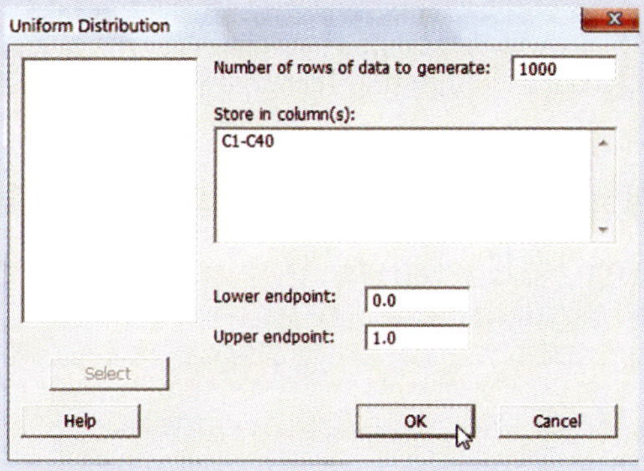

Figure 4.M.6 Minitab dialog box for simulating the uniform distribution

Step 3 Specify the number of samples (e.g., 1,000) to generate in the "Number of rows to generate" box and the columns where the data will be stored in the "Store in columns" box. (The number of columns will be equal to the sample size, e.g., 40.) Finally, specify the parameters of the distribution (e.g., the lower and upper range of the uniform distribution). When you click "OK," the simulated data will appear on the Minitab worksheet.

Step 4 Calculate the value of the sample statistic of interest for each sample. To do this, click on the "Calc" button on the Minitab menu bar and then click on "Row Statistics," as shown in Figure 4.M.7. The resulting dialog box appears in Figure 4.M.8.

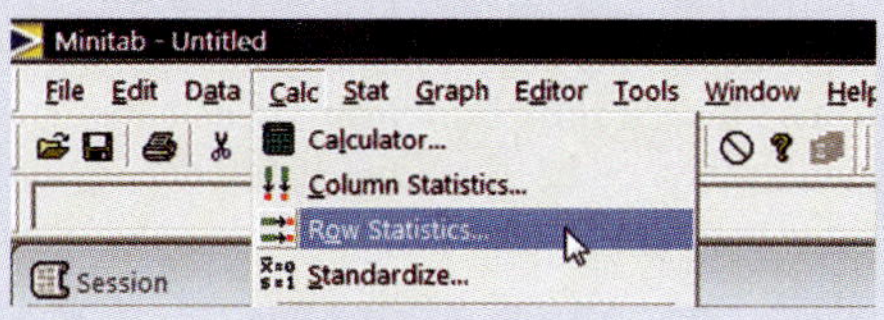

Figure 4.M.7 Minitab selections for generating sample statistics for the simulated data

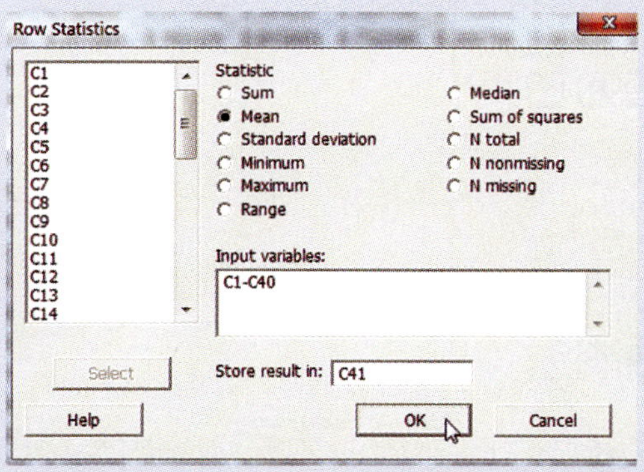

Figure 4.M.8 Minitab row statistics dialog box

Step 5 Check the sample statistic (e.g., the mean) you want to calculate, specify the "Input variables" (or columns), and the column where you want the value of the sample statistic to be saved. Click "OK" and the value of the statistic for each sample will appear on the Minitab worksheet.

[*Note:* Use the Minitab menu choices provided in the Using Technology section in Chapter 2, p. 110 to generate a histogram of the sampling distribution of the statistic or to find the mean and variance of the sampling distribution.]

Excel/DDXL: Normal Probability Plot and Simulated Sampling Distributions

[*Note:* Discrete and continuous probability functions are unavailable in Excel and DDXL.]

Normal Probability Plot

Step 1 Highlight the cells (columns) with the data for the variable to be analyzed on the Excel spreadsheet.

Step 2 Click on the "Add-ins" button on the Excel menu bar, and then select "DDXL."

Step 3 Select "Charts and Plots" on the resulting menu, as shown in Figure 4.E.1.

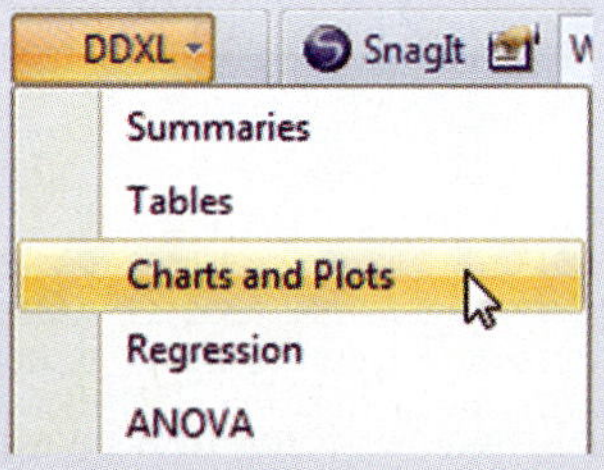

Figure 4.E.1 Excel/DDXL options for a normal probability plot

Step 4 On the Charts and Plots dialog box (see Figure 4.E.2), specify "Normal Probability Plot" in the "Function Type" box, and select the variable to be analyzed in the "Quantitative Variable" box.

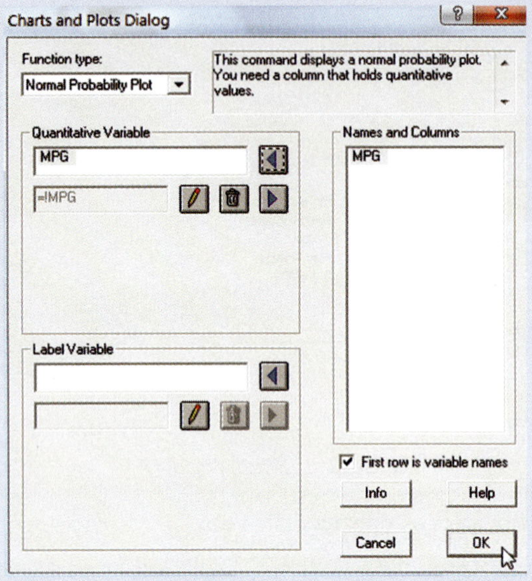

Figure 4.E.2 Excel/DDXL charts and plots dialog box

Step 5 Click "OK" to view the normal probability plot.

Sampling Distributions

Step 1 Select "Data" on the main menu bar and then click on "Data Analysis." The resulting menu is shown in Figure 4.E.3.

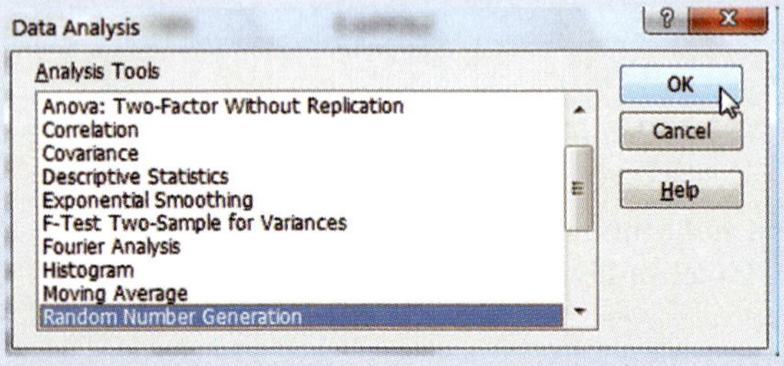

Figure 4.E.3 Excel data analysis menu options

Step 2 Select "Random Number Generation" from the Data Analysis menu and then click "OK." The dialog box in Figure 4.E.4 will appear.

Step 3 Specify the number of samples (e.g., 10) to generate in the "Number of Variables" box and the sample size of each (e.g., 100) in the "Number of Random Numbers" box.

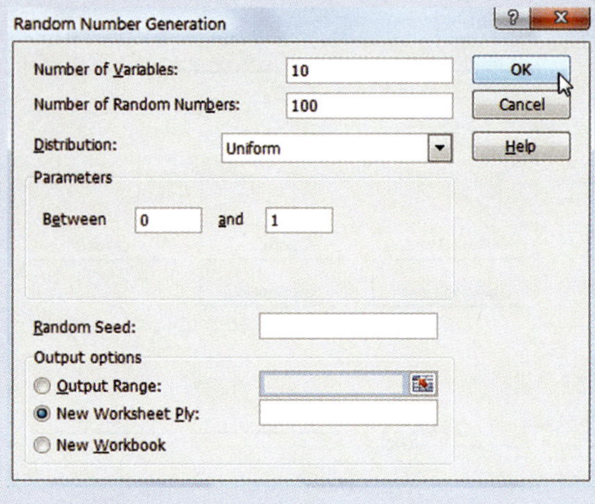

Figure 4.E.4 Excel dialog box for generating uniform random numbers

Step 4 Specify the distribution (e.g., uniform) that the samples will be selected from and the parameters of the distribution.

Step 5 Click "OK;" the random numbers will appear on a new Excel spreadsheet (Figure 4.E.5).

Step 6 Specify a formula for the sample mean on the resulting Excel spreadsheet. For example, the formula "=AVERAGE (A2:J2)" in column "L" computes the mean of the 10 random numbers in each row of the spreadsheet shown in Figure 4.E.5.

Step 7 Once the means of the random samples are computed, you can generate a histogram or summary statistics for the mean by following the steps outlined in Using Technology, Chapter 2.

TI-84 Graphing Calculator: Discrete and Continuous Random Variables and Normal Probability Plots

Calculating the Mean and Standard Deviation of a Discrete Random Variable

Step 1 *Enter the data*

- Press **STAT** and select **1:Edit**

Note: If the lists already contain data, clear the old data. Use the up **ARROW** to highlight '**L1**'

- Press **CLEAR ENTER**
- Use the up **ARROW** to highlight '**L2**'

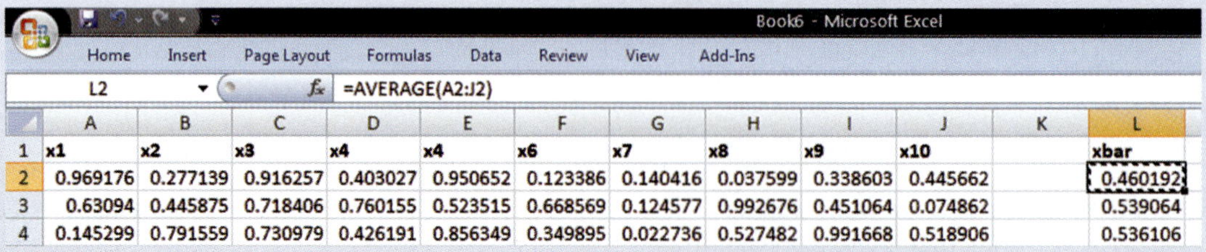

	A	B	C	D	E	F	G	H	I	J	K	L
1	x1	x2	x3	x4	x4	x6	x7	x8	x9	x10		xbar
2	0.969176	0.277139	0.916257	0.403027	0.950652	0.123386	0.140416	0.037599	0.338603	0.445662		0.460192
3	0.63094	0.445875	0.718406	0.760155	0.523515	0.668569	0.124577	0.992676	0.451064	0.074862		0.539064
4	0.145299	0.791559	0.730979	0.426191	0.856349	0.349895	0.022736	0.527482	0.991668	0.518906		0.536106

Figure 4.E.5 Excel formula for computing sample means from random numbers

- Press **CLEAR ENTER**
- Use the **ARROW** and **ENTER** keys to enter the x-values of the variable into **L1**
- Use the **ARROW** and **ENTER** keys to enter the probabilities, $P(x)$, into **L2**

Step 2 *Access the Calc Menu*

- Press **STAT**
- Arrow right to **CALC**
- Select **1-Var Stats**
- Press **ENTER**
- Press **2nd 1** for **L1**
- Press **COMMA**
- Select **2nd 2** for **L2**
- Press **ENTER**

The mean and standard deviation will be displayed on the screen, as well as the quartiles, min, and max.

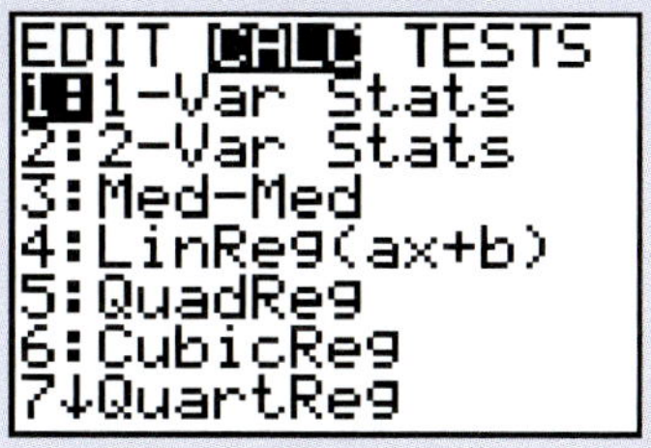

Calculating Binomial Probabilities

I. $P(x = k)$

To compute the probability of k successes in n trials, where p is the probability of success for each trial, use the **binompdf(** command. *Binompdf* stands for "binomial probability density function." This command is under the **DISTR**ibution menu and has the format **binompdf(*n, p, k*).**

Example Compute the probability of 5 successes in 8 trials, where the probability of success for a single trial is 40%. In this example, $n = 8$, $p = .4$, and $k = 5$.

Step 1 *Enter the binomial parameters*

- Press **2nd VARS** for **DISTR**
- Press the down **ARROW** key until **binompdf** is highlighted
- Press **ENTER**
- After **binompdf(**, type **8, .4, 5)** (*Note:* be sure to use the **COMMA** key between each parameter)
- Press **ENTER**
- You should see

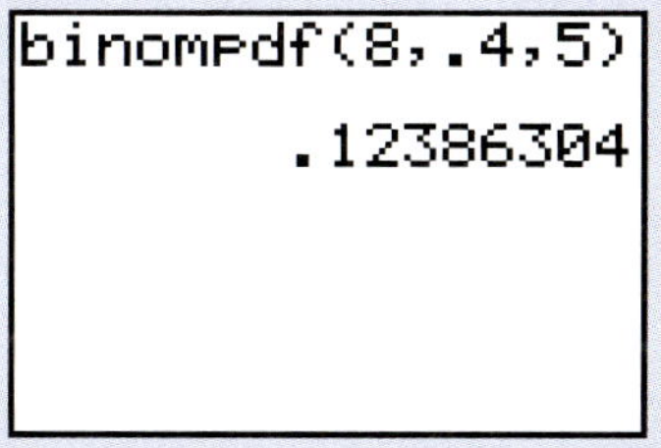

Thus, $P(x = 5)$ is about 12.4%.

II. $P(x \leq k)$

To compute the probability of k or fewer successes in n trials, where p is the probability of success for each trial, use the **binomcdf(** command. *Binomcdf* stands for "binomial ***cumulative*** probability density function." This command is under the **DISTR**ibution menu and has the format **binomcdf(*n, p, k*).**

Example Compute the probability of 5 or fewer successes in 8 trials, where the probability of success for a single trial is 40%. In this example, $n = 8$, $p = .4$, and $k = 5$.

Step 2 *Enter the binomial parameters*

- Press **2nd VARS** for **DISTR**
- Press down the **ARROW** key until **binomcdf** is highlighted
- Press **ENTER**
- After **binomcdf(**, type **8, .4, 5)**
- Press **ENTER**
- You should see

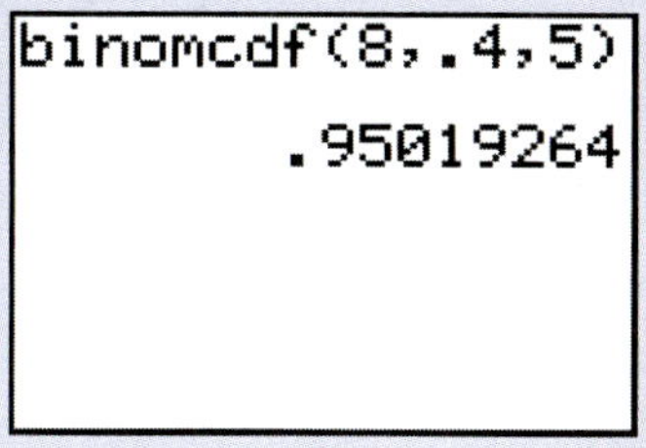

Thus, $P(x \leq k)$ is about 95%.

III. $P(x < k)$, $P(x > k)$, $P(x \geq k)$

To find the probability of less than k successes $P(x < k)$, more than k successes $P(x > k)$, or at least k successes $P(x \geq k)$, variations of the **binomcdf(** command must be used as shown below.

- $P(x < k)$ use **binomcdf(*n, p, k*** **− 1)**
- $P(x > k)$ use **1 − binomcdf(*n, p, k*)**
- $P(x \geq k)$ use **1 − binomcdf(*n, p, k*** **− 1)**

Calculating Poisson Probabilities

I. $P(x = k)$

To compute $P(x = k)$, the probability of exactly k successes in a specified interval where λ is the mean number of successes in the interval, use the **poissonpdf(** command. *Poissonpdf* stands for "Poisson probability density function." This command is under the **DISTR**ibution menu and has the format **poissonpdf( ).**

Example Suppose that the number, x, of reported sightings per week of blue whales is recorded. Assume that x has approximately a Poisson probability distribution and that the average number of weekly sightings is 2.6. Compute the probability that exactly five sightings are made during a given week. In this example, $\lambda = 2.6$ and $k = 5$.

Step 1 *Enter the poisson parameters*

- Press **2nd VARS** for **DISTR**
- Press the down **ARROW** key until **poissonpdf** is highlighted
- Press **ENTER**
- After **poissonpdf(**, type **2.6, 5)** (*Note:* be sure to use the **COMMA** key between each parameter)
- Press **ENTER**
- You should see

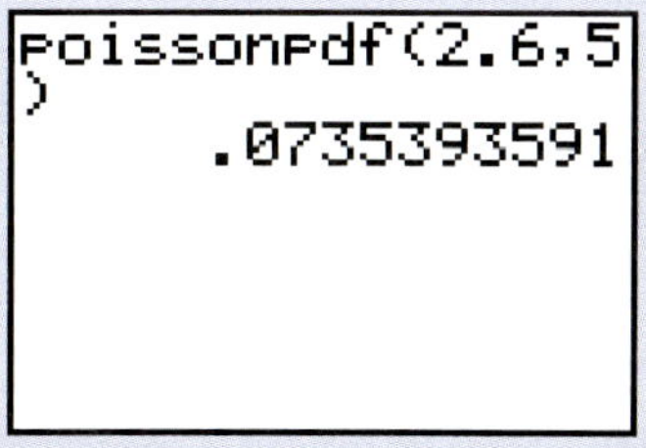

Thus, the $P(x = 5)$ is about 7.4%.

II. $P(x \leq k)$

To compute the probability of k or fewer successes in a specified interval, where λ is the mean number of successes in the interval, use the **poissoncdf(** command. *Poissoncdf* stands for "Poisson *cumulative* probability density function." This command is under the **DISTR**ibution menu and has the format **poissoncdf(**.

Example In the preceding example, compute the probability that five or fewer sightings are made during a given week. In this example, $\lambda = 2.6$ and $k = 5$.

Step 1 *Enter the poisson parameters*

- Press **2nd VARS** for **DISTR**
- Press the down **ARROW** key until **poissoncdf** is highlighted
- Press **ENTER**
- After **poissoncdf(**, type **2.6, 5)**
- Press **ENTER**
- You should see:

```
poissoncdf(2.6,5
)
          .9509628481
```

Thus, the $P(x = 5)$ is about 95.1%.

III. $P(x < k)$, $P(x > k)$, $P(x \geq k)$

To find the probability of less than k successes, more than k successes, or at least k successes, variations of the **poissoncdf(** command must be used as shown below.

- $P(x < k)$ use **poissoncdf(** λ**,** $k - 1$**)**
- $P(x > k)$ use 1 – **poissoncdf(** λ**,** k**)**
- $P(x \geq k)$ use 1 – **poissoncdf(** λ**,** $k - 1$**)**

Graphing the Area under the Standard Normal Curve

Step 1 *Turn off all plots*

- Press **2nd PRGM** and select **1:ClrDraw**
- Press **ENTER** and "Done" will appear on the screen
- Press **2nd** Y= and select **4:PlotsOff**
- Press **ENTER** and "Done" will appear on the screen

Step 2 *Set the viewing window (Recall that almost all of the area under the standard normal curve falls between −5 and 5. A height of 0.5 is a good choice for Ymax.)*

Note: When entering a negative number, be sure to use the negative sign (−), not the minus sign.

- Set **Xmin** = **−5**
- **Xmax** = **5**
- **Xscl** = **1**
- **Ymin** = **0**
- **Ymax** = **.5**
- **Yscl** = **0**
- **Xres** = **1**

Step 3 *View graph*

- Press **2nd VARS**
- Arrow right to **DRAW**
- Press **ENTER** to select **1:ShadeNorm(**
- Enter your lower limit (e.g., −5)
- Press **COMMA**
- Enter your upper limit (e.g., 1.5)
- Press **)**
- Press **ENTER**

```
ShadeNorm(-5,1.5
)█
```

The graph will be displayed along with the area, lower limit, and upper limit.

Thus, $P(z < 1.5) = .9332$.

Nonstandard Normal Probabilities

I. Finding Normal Probabilities without a Graph

To compute probabilities for a normal distribution, use the **normalcdf(** command. *Normalcdf* stands for "normal cumulative density function." This command is under the **DISTR**ibution menu and has the format **normalcdf(*lower limit, upper limit, mean, standard deviation*).**

Step 1 *Find probability*

- Press **2nd VARS** for **DISTR** and select **Normalcdf(**
- After **Normalcdf(,** type in the lower limit
- Press **COMMA**
- Enter the upper limit
- Press **COMMA**
- Enter the mean
- Press **COMMA**
- Enter the standard deviation
- Press **)**
- Press **ENTER**

The probability will be displayed on the screen.

Example What is $P(x < 115)$ for a normal distribution with $\mu = 100$ and $\sigma = 10$? In this example, the lower limit is $-\infty$, the upper limit is 115, the mean is 100, and the standard deviation is 10. To represent $-\infty$ on the calculator, enter **(−) 1,** press **2nd** and press the **COMMA** key for **EE,** and then press **99.** The screen appears as follows:

```
normalcdf(-1E99,
115,100,10)
          .9331927713
```

Thus, $P(x < 115)$ is .9332.

II. Finding Normal Probabilities with a Graph

Step 1 *Turn off all plots*

- Press **Y=** and **CLEAR** all functions from the Y registers
- Press **2nd Y=** and select **4:PlotsOff**
- Press **ENTER ENTER,** and 'Done' will appear on the screen

Step 2 *Set the viewing window (These values depend on the mean and standard deviation of the data.) Note:* When entering a negative number, be sure to use the negative sign (−), not the minus sign.

- Press **WINDOW**
- Set **Xmin** = $\mu - 5\sigma$

- **Xmax** = $\mu + 5\sigma$
- **Xscl** = σ
- **Ymin** = $-.125/\sigma$
- **Ymax** = $.5/\sigma$
- **Yscl** = 1
- **Xres** = 1

Step 3 *View graph*

- Press **2nd VARS**
- **ARROW** right to **DRAW**
- Press **ENTER** to select **1:ShadeNorm(**
- Enter the lower limit
- Press **COMMA**
- Enter the upper limit
- Press **COMMA**
- Enter the mean
- Press **COMMA**
- Enter the standard deviation
- Press)
- Press **ENTER**

The graph will be displayed along with the area, lower limit, and upper limit.

Example What is $P(x < 115)$ for a normal distribution with $\mu = 100$ and $\sigma = 10$? In this example, the lower limit is $-\infty$, the upper limit is 115, the mean is 100, and the standard deviation is 10. To represent $-\infty$ on the calculator, enter **(−) 1,** press **2nd** and press the **comma** key for **EE,** and then press **99**. The screens appear as follows:

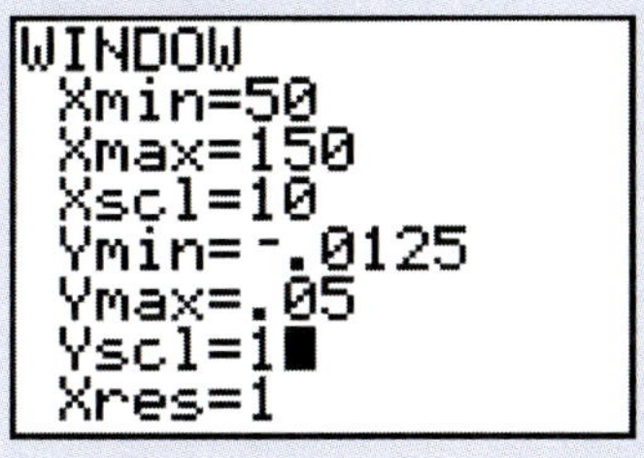

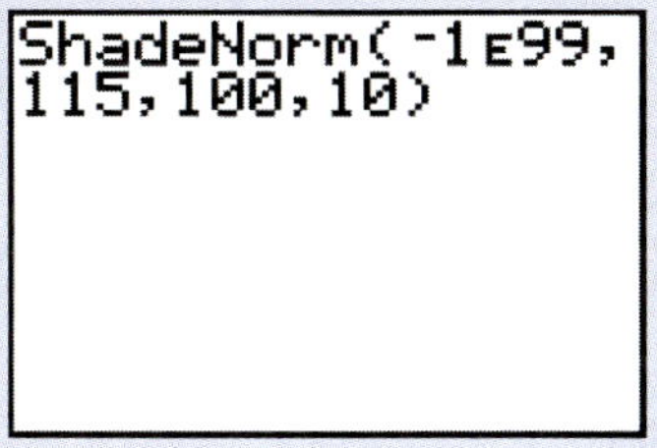

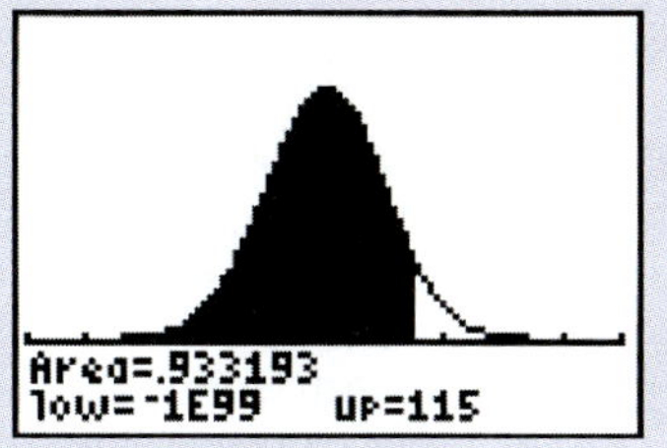

Graphing a Normal Probability Plot

Step 1 *Enter the data*

- Press **STAT** and select **1:Edit**

Note: If the list already contains data, clear the old data. Use the up arrow to highlight **L1**.

- Press **CLEAR ENTER**
- Use the **ARROW** and **ENTER** keys to enter the data set into **L1**

Step 2 *Set up the normal probability plot*

- Press **Y=** and **CLEAR** all functions from the Y registers
- Press **2nd** and press **Y=** for **STAT PLOT**
- Press **1** for **Plot 1**
- Set the cursor so that **ON** is flashing
- For **Type**, use the **ARROW** and **ENTER** keys to highlight and select the last graph in the bottom row
- For **Data List**, choose the column containing the data (in most cases, L1) (Note: Press **2nd** 1 for **L1**)
- For **Data Axis,** choose **X** and press **ENTER**

Step 3 *View plot*

- Press **ZOOM 9**

Your data will be displayed against the expected *z*-scores from a normal distribution. If you see a "generally" linear relationship, your data set is approximately normal.

The Furniture Fire Case

Covers Chapters 3 and 4 — A wholesale furniture retailer stores in-stock items at a large warehouse located in Tampa, Florida. Early in the year, a fire destroyed the warehouse and all the furniture in it. After determining the fire was an accident, the retailer sought to recover costs by submitting a claim to its insurance company.

As is typical in a fire insurance policy of this type, the furniture retailer must provide the insurance company with an estimate of "lost" profit for the destroyed items. Retailers calculate profit margin in percentage form using the Gross Profit Factor (GPF). By definition, the GPF for a single sold item is the ratio of the profit to the item's selling price measured as a percentage, that is,

$$\text{Item GPF} = (\text{Profit/Sales price}) \times 100\%$$

Of interest to both the retailer and the insurance company is the average GPF for all of the items in the warehouse. Because these furniture pieces were all destroyed, their eventual selling prices and profit values are obviously unknown. Consequently, the average GPF for all the warehouse items is unknown.

One way to estimate the mean GPF of the destroyed items is to use the mean GPF of similar, recently sold items. The retailer sold 3,005 furniture items in the year prior to the fire and kept paper invoices on all sales. Rather than calculate the mean GPF for all 3,005 items (the data were not computerized), the retailer sampled a total of 253 of the invoices and computed the mean GPF for these items. The 253 items were obtained by first selecting a sample of 134 items and then augmenting this sample with a second sample of 119 items. The mean GPFs for the two subsamples were calculated to be 50.6% and 51.0%, respectively, yielding an overall average GPF of 50.8%. This average GPF can be applied to the costs of the furniture items destroyed in the fire to obtain an estimate of the "lost" profit.

According to experienced claims adjusters at the insurance company, the GPF for sale items of the type destroyed in the fire rarely exceeds 48%. Consequently, the estimate of 50.8% appeared to be unusually high. (A 1% increase in GPF for items of this type equates to, approximately, an additional $16,000 in profit.) When the insurance company questioned the retailer on this issue, the retailer responded "Our estimate was based on selecting two independent, random samples from the population of 3,005 invoices. Because the samples were selected randomly and the total sample size is large, the mean GPF estimate of 50.8% is valid."

A dispute arose between the furniture retailer and the insurance company, and a lawsuit was filed. In one portion of the suit, the insurance company accused the retailer of fraudulently representing their sampling methodology. Rather than selecting the samples randomly, the retailer was accused of selecting an unusual number of "high-profit" items from the population in order to increase the average GPF of the overall sample.

To support their claim of fraud, the insurance company hired a CPA firm to independently assess the retailer's Gross Profit Factor. Through the discovery process, the CPA firm legally obtained the paper invoices for the entire population of 3,005 items sold and input the information into a computer. The selling price, profit, profit margin, and month sold for these 3,005 furniture items are stored in the **FIRE** file, described below.

Your objective in this case is to use these data to determine the likelihood of fraud. Is it likely that a random sample of 253 items selected from the population of 3,005 items would yield a mean GPF of at least 50.8%? Or, is it likely that two independent, random samples of sizes 134 and 119 will yield mean GPFs of at least 50.6% and 51.0%, respectively? (These were the questions posed to a statistician retained by the CPA firm.) Use the ideas of probability and sampling distributions to guide your analysis.

Prepare a professional document that presents the results of your analysis and gives your opinion regarding fraud. Be sure to describe the assumptions and methodology used to arrive at your findings.

Variable	Type	Description
MONTH	QL	Month in which item was sold in 1991
INVOICE	QN	Invoice number
SALES	QN	Sales price of item in dollars
PROFIT	QN	Profit amount of item in dollars
MARGIN	QN	Profit margin of item = (Profit/Sales) × 100%

Data Set: FIRE

CONTENTS

5 Inferences Based on a Single Sample *Estimation with Confidence Intervals*

Where We've Been

- Learned that populations are characterized by numerical descriptive measures called *parameters*
- Found that decisions about population parameters are based on *statistics* computed from the sample
- Discovered that *inferences* about parameters are subject to uncertainty and that this uncertainty is reflected in the *sampling distribution* of a statistic

Where We're Going

- Estimate a population parameter (means or proportion) based on a large sample selected from the population
- Use the sampling distribution of a statistic to form a confidence interval for the population parameter
- Show how to select the proper sample size for estimating a population parameter

Statistics IN Action Medicare Fraud Investigations

United States Department of Justice (USDOJ) press release (May 8, 2008): *Eleven people have been indicted in . . . a targeted criminal, civil and administrative effort against individuals and health care companies that fraudulently bill the Medicare program. The indictments in the Central District of California resulted from the creation of a multi-agency team of federal, state and local investigators designed specifically to combat Medicare fraud through the use of real-time analysis of Medicare billing data.*

USDOJ press release (Jun 27, 2008): *The owners of four Miami-based healthcare corporations were sentenced and remanded to prison yesterday for their roles in schemes to defraud the Medicare program. Collectively, the three defendants through their companies collected more than $14 million from the Medicare program for unnecessary medicine, durable medical equipment (DME) and home health care services.*

USDOJ press release (Oct. 7, 2008): *Eight Miami-Dade County residents have been charged in a 16-count indictment for their alleged roles in a Medicare fraud scheme involving fake HIV infusion treatments.* (continued)

Statistics IN Action *(continued)*

As the above press releases imply, the U.S. Department of Justice (USDOJ) and its Medicare Fraud Strike Force conducts investigations into suspected fraud and abuse of the Medicare system by health care providers. According to published reports, the Strike Force was responsible for almost 25% of the Medicare fraud charges brought nationwide in 2007.

One way in which Medicare fraud occurs is through the use of "upcoding," which refers to the practice of providers coding Medicare claims at a higher level of care than was actually provided to the patient. For example, suppose a particular kind of claim can be coded at three levels, where Level 1 is a routine office visit, Level 2 is a thorough examination involving advanced diagnostic tests, and Level 3 involves performing minor surgery. The amount of Medicare payment is higher for each increased level of claim. Thus, upcoding would occur if Level 1 services were billed at Level 2 or Level 3 payments, or if Level 2 services were billed at Level 3 payment.

The USDOJ relies on sound statistical methods to help identify Medicare fraud. Once the USDOJ has determined that possible upcoding has occurred, it next seeks to further investigate whether it is the result of legitimate practice (perhaps the provider is a specialist giving higher levels of care) or the result of fraudulent action on the part of the provider. To further its investigation, the USDOJ will next ask a statistician to select a sample of the provider's claims. For example, the statistician might determine that a random sample of 52 claims from the 1,000 claims in question will provide a sufficient sample to estimate the overcharge reliably. The USDOJ then asks a health care expert to audit each of the medical files corresponding to the sampled claims and determine whether the level of care matches the level billed by the provider, and, if not, to determine what level should have been billed. Once the audit has been completed, the USDOJ will calculate the overcharge.

In this chapter, we present a recent Medicare fraud case investigated by the USDOJ. Results for the audit of 52 sampled claims, with the amount paid for each claim, the amount disallowed by the auditor, and the amount that should have been paid for each claim, are saved in the **MCFRAUD** file.* Knowing that a total of $103,500 was paid for the 1,000 claims, the USDOJ wants to use the sample results to extrapolate the overpayment amount to the entire population of 1,000 claims.

Statistics IN Action Revisited

- Estimating the Mean Overpayment (p. 286)
- Estimating the Coding Error Rate (p. 294)
- Determining Sample Size (p. 301)

Data Set: MCFRAUD

5.1 Identifying and Estimating the Target Parameter

In this chapter, our goal is to estimate the value of an unknown population parameter, such as a population mean or a proportion from a binomial population. For example, we might want to know the mean gas mileage for a new car model, the average expected life of a flat-screen computer monitor, or the proportion of dot-com companies that fail within a year of start-up.

You'll see that different techniques are used for estimating a mean or proportion, depending on whether a sample contains a large or small number of measurements. Nevertheless, our objectives remain the same. We want to use the sample information to estimate the population parameter of interest (called the **target parameter**) and assess the reliability of the estimate.

> The unknown population parameter (e.g., mean or proportion) that we are interested in estimating is called the **target parameter.**

Often, there are one or more key words in the statement of the problem that indicate the appropriate target parameter. Some key words associated with the two parameters covered in this section are listed in the table on the next page.

*Data provided (with permission) from Info Tech, Inc., Gainesville, Florida.

Determining the Target Parameter

Parameter	Key Words or Phrases	Type of Data
μ	Mean; average	Quantitative
p	Proportion; percentage; fraction; rate	Qualitative

For the examples given above, the words *mean* in *mean gas mileage* and *average* in *average life expectancy* imply that the target parameter is the population mean, μ. The word *proportion* in *proportion of dot-com companies that fail within one year of start-up* indicates that the target parameter is the binomial proportion, p.

In addition to key words and phrases, the type of data (quantitative or qualitative) collected is indicative of the target parameter. With quantitative data, you are likely to be estimating the mean of the data. With qualitative data with two outcomes (success or failure), the binomial proportion of successes is likely to be the parameter of interest.

A single number calculated from the sample that estimates a target population parameter is called a **point estimator.** For example, we'll use the sample mean, $\bar{x}$, to estimate the population mean μ. Consequently, $\bar{x}$ is a point estimator. Similarly, we'll learn that the sample proportion of successes, denoted $\hat{p}$, is a point estimator for the binomial proportion p. Also, we will attach a measure of reliability to our estimate by obtaining an **interval estimator**—a range of numbers that contain the target parameter with a high degree of confidence. For this reason the interval estimate is also called a **confidence interval.**

A **point estimator** of a population parameter is a rule or formula that tells us how to use the sample data to calculate a *single* number that can be used as an *estimate* of the target parameter.

An **interval estimator** (or **confidence interval**) is a formula that tells us how to use the sample data to calculate an *interval* that *estimates* the target parameter.

We consider confidence intervals for estimating a population mean using a large sample in Section 5.2 and a small sample in Section 5.3. Confidence intervals for a population proportion are presented in Section 5.4. In Section 5.5, we show how to determine the sample sizes necessary for reliable estimates of the target parameters based on simple random sampling. Finally, in Sections 5.6 and 5.7 we introduce more complex sample survey designs and their application to business.

5.2 Confidence Interval for a Population Mean: Normal (z) Statistic

Suppose a large bank wants to estimate the average amount of money owed by its delinquent debtors (i.e., debtors who are more than 2 months behind in payment). To accomplish this objective, the bank plans to randomly sample 100 of its delinquent accounts and use the sample mean, $\bar{x}$, of the amounts overdue to estimate μ, the mean for *all* delinquent accounts. Because the sample mean $\bar{x}$ represents a single number estimator, it is the *point estimator* of the target parameter μ. How can we assess the accuracy of this point estimator?

According to the Central Limit Theorem, the sampling distribution of the sample mean is approximately normal for large samples, as shown in Figure 5.1. Let us calculate the interval estimator:

$$\bar{x} \pm 1.96\sigma_{\bar{x}} = \bar{x} \pm \frac{1.96\sigma}{\sqrt{n}}$$

That is, we form an interval from 1.96 standard deviations below the sample mean to 1.96 standard deviations above the mean. *Prior to drawing the sample,* what are the chances that this interval will enclose μ, the population mean?

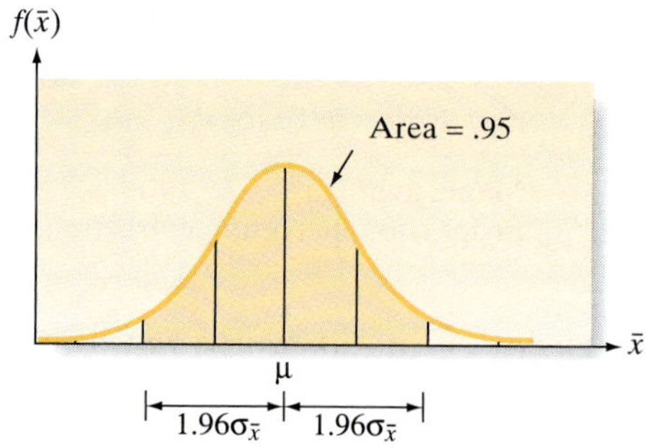

Figure 5.1
Sampling distribution of $\overline{x}$

To answer this question, refer to Figure 5.1. If the 100 measurements yield a value of $\overline{x}$ that falls between the two lines on either side of μ (i.e., within 1.96 standard deviations of μ), then the interval $\overline{x} \pm 1.96\sigma_{\overline{x}}$ will contain μ; if $\overline{x}$ falls outside these boundaries, the interval $\overline{x} \pm 1.96\sigma_{\overline{x}}$ will not contain μ. From Section 4.7, we know that the area under the normal curve (the sampling distribution of $\overline{x}$) between these boundaries is exactly .95. Thus, the probability that a randomly selected interval, $\overline{x} \pm 1.96\sigma_{\overline{x}}$, will contain μ is equal to .95.

Example 5.1
Estimating the Mean Overdue Amount, σ Known

Problem Consider the large bank that wants to estimate the average amount of money owed by its delinquent debtors, μ. The bank randomly samples $n = 100$ of its delinquent accounts and finds that the sample mean amount owed is $\overline{x} = \$230$. Also, suppose it is known that the standard deviation of the amount owed for all delinquent accounts is $\sigma = \$90$. Use the interval estimator $\overline{x} \pm 1.96\sigma_{\overline{x}}$ to calculate a confidence interval for the target parameter, μ.

Solution Substituting $\overline{x} = 230$ and $\sigma = 90$ into the interval estimator formula, we obtain:

$$\overline{x} \pm 1.96\sigma_{\overline{x}} = \overline{x} \pm (1.96)\sigma/\sqrt{n} = 230 \pm (1.96)(90/\sqrt{100}) = 230 \pm 17.64$$

Or, (212.36, 247.64)

Look Back Because we know the probability that the interval $\overline{x} \pm 1.96\sigma_{\overline{x}}$ will contain μ is .95, we call the interval estimator a *95% confidence interval* for μ.

Now Work Exercise 5.3a

The interval $\overline{x} \pm 1.96\sigma_{\overline{x}}$ in Example 5.1 is called a *large-sample* 95% confidence interval for the population mean μ. The term *large-sample* refers to the sample being of sufficiently large size that we can apply the Central Limit Theorem and the normal (z) statistic to determine the form of the sampling distribution of $\overline{x}$. Empirical research suggests that a sample size n exceeding a value between 20 and 30 will usually yield a sampling distribution of $\overline{x}$ that is approximately normal. This result led many practitioners to adopt the rule of thumb that a sample size of $n \geq 30$ is required to use large-sample confidence interval procedures. Keep in mind, though, that 30 is not a magical number and, in fact, is quite arbitrary.

Also, note that the large-sample interval estimator requires knowing the value of the population standard deviation, σ. In most (if not nearly all) practical business applications, however, the value of σ will be unknown. For large samples, the fact that σ is unknown poses only a minor problem because the sample standard deviation s provides a very good approximation to σ. The next example illustrates the more realistic large-sample confidence interval procedure.

Example 5.2
Estimating the Mean Overdue Amount, σ Unknown

Problem Refer to Example 5.1 and the problem of estimating μ, the average amount of money owed by a bank's delinquent debtors. The overdue amounts for the $n = 100$ delinquent accounts are shown in Table 5.1. Use the data to find a 95% confidence interval for μ and interpret the result.

Table 5.1 Overdue Amounts (in Dollars) for 100 Delinquent Accounts

195	243	132	133	209	400	142	312	221	289
221	162	134	275	355	293	242	458	378	148
278	222	236	178	202	222	334	208	194	135
363	221	449	265	146	215	113	229	221	243
512	193	134	138	209	207	206	310	293	310
237	135	252	365	371	238	232	271	121	134
203	178	180	148	162	160	86	234	244	266
119	259	108	289	328	331	330	227	162	354
304	141	158	240	82	17	357	187	364	268
368	274	278	190	344	157	219	77	171	280

Data Set: OVERDUE

Solution The large bank almost surely does not know the true standard deviation, σ, of the population of overdue amounts. However, because the sample size is large, we will use the sample standard deviation, s, as an estimate for σ in the confidence interval formula. An Excel printout of summary statistics for the sample of 100 overdue amounts is shown in Figure 5.2a. From the shaded portion of the printout, we find $\bar{x} = 233.28$ and $s = 90.34$. Substituting these values into the interval estimator formula, we obtain:

$$\bar{x} \pm (1.96)\sigma/\sqrt{n} \approx \bar{x} \pm (1.96)s/\sqrt{n} = 233.28 \pm (1.96)(90.34)/\sqrt{100} = 233.28 \pm 17.71$$

Or, (215.57, 250.99). That is, we estimate the mean amount of delinquency for all accounts to fall within the interval \$215.57 to \$250.99.

	A	B
1	AMOUNT	
2		
3	Mean	233.28
4	Standard Error	9.033988347
5	Median	222
6	Mode	221
7	Standard Deviation	90.33988347
8	Sample Variance	8161.294545
9	Kurtosis	0.254810234
10	Skewness	0.476799829
11	Range	495
12	Minimum	17
13	Maximum	512
14	Sum	23328
15	Count	100

Figure 5.2a
Excel summary statistics for overdue amounts

Look Back The confidence interval is also shown at the bottom of the Excel/DDXL printout, Figure 5.2b. Note that the end-points of the interval vary slightly from those computed in the example. This is due to the fact that when σ is unknown and n is large,

AMOUNT Confidence Interval

Summary Statistics

Count	Mean	Std Dev	df
100	233.28	90.34	99

Interval Results

Confidence Interval

With 95% Confidence, 215.355 < μ < 251.205

Figure 5.2b
Excel/DDXL 95% confidence interval for mean overdue amount

the sampling distribution of $\bar{x}$ will deviate slightly from the normal (z) distribution. In practicc, these differences can be ignored.

Now Work Exercise 5.10

Can we be sure that μ, the true mean, is in the interval (215.57, 250.99) in Example 5.2? We cannot be certain, but we can be reasonably confident that it is. This confidence is derived from the knowledge that if we were to draw repeated random samples of 100 measurements from this population and form the interval $\bar{x} \pm 1.96\sigma_{\bar{x}}$ each time, 95% of the intervals would contain μ. We have no way of knowing (without looking at all the delinquent accounts) whether our sample interval is one of the 95% that contains μ or one of the 5% that does not, but the odds certainly favor its containing μ. The probability, .95, that measures the confidence we can place in the interval estimate is called a *confidence coefficient*. The percentage, 95%, is called the *confidence level* for the interval estimate.

> The **confidence coefficient** is the probability that a randomly selected confidence interval encloses the population parameter—that is, the relative frequency with which similarly constructed intervals enclose the population parameter when the estimator is used repeatedly a very large number of times. The **confidence level** is the confidence coefficient expressed as a percentage.

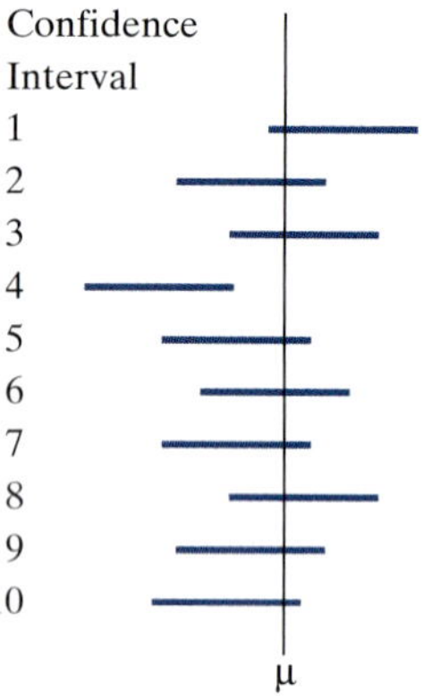

Figure 5.3
Confidence intervals for μ: 10 samples

Now we have seen how an interval can be used to estimate a population mean. When we use an interval estimator, we can usually calculate the probability that the estimation *process* will result in an interval that contains the true value of the population mean—that is, the probability that the interval contains the parameter in repeated usage is usually known. Figure 5.3 shows what happens when 10 different samples are drawn from a population, and a confidence interval for μ is calculated from each. The location of μ is indicated by the vertical line in the figure. Ten confidence intervals, each based on one of 10 samples, are shown as horizontal line segments. Note that the confidence intervals move from sample to sample—sometimes containing μ and other times missing μ. *If our confidence level is 95%, then in the long run, 95% of our confidence intervals will contain μ and 5% will not.*

Suppose you wish to choose a confidence coefficient other than .95. Notice in Figure 5.1 that the confidence coefficient .95 is equal to the total area under the sampling distribution, less .05 of the area, which is divided equally between the two tails. Using this idea, we can construct a confidence interval with any desired confidence coefficient by increasing or decreasing the area (call it α) assigned to the tails of the sampling distribution (see Figure 5.4). For example, if we place area $\alpha/2$ in each tail and if $z_{\alpha/2}$ is the z-value such that the area $\alpha/2$ lies to its right, then the confidence interval with confidence coefficient $(1 - \alpha)$ is

$$\bar{x} \pm (z_{\alpha/2})\sigma_{\bar{x}}$$

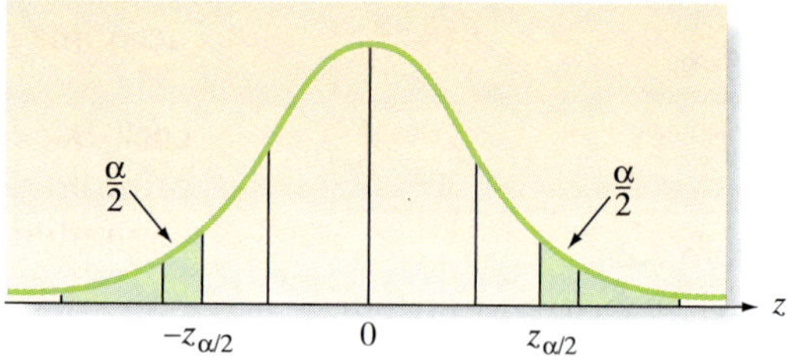

Figure 5.4
Locating $z_{\alpha/2}$ on the standard normal curve

> **Conditions Required for a Valid Large-Sample Confidence Interval for μ**
>
> 1. A random sample is selected from the target population.
> 2. The sample size n is large (i.e., $n \geq 30$). Due to the Central Limit Theorem, this condition guarantees that the sampling distribution of $\bar{x}$ is approximately normal. Also, for large n, s will be a good estimator of σ.)

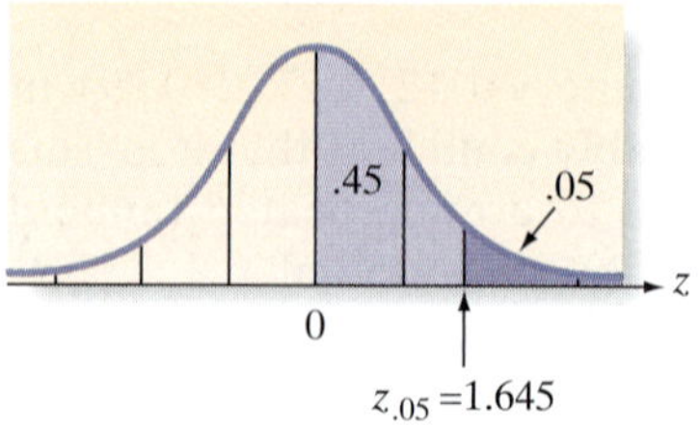

Figure 5.5
The z-value ($z_{.05}$) corresponding to an area equal to .05 in the upper tail of the z-distribution

To illustrate, for a confidence coefficient of .90, we have $(1 - \alpha) = .90$, $\alpha = .10$, and $\alpha/2 = .05$; $z_{.05}$ is the z-value that locates area .05 in the upper tail of the sampling distribution. Recall that Table IV in Appendix B gives the areas between the mean and a specified z-value. Because the total area to the right of the mean is .5, we find that $z_{.05}$ will be the z-value corresponding to an area of $.5 - .05 = .45$ to the right of the mean (see Figure 5.5). This z-value is $z_{.05} = 1.645$.

Confidence coefficients used in practice usually range from .90 to .99. The most commonly used confidence coefficients with corresponding values of α and $z_{\alpha/2}$ are shown in Table 5.2.

BIOGRAPHY JERZEY NEYMAN (1894–1981)

Speaking Statistics with a Polish Accent

Polish-born Jerzey Neyman was educated at the University of Kharkov (Russia) in elementary mathematics but taught himself graduate mathematics by studying journal articles on the subject. After receiving his doctorate in 1924 from the University of Warsaw (Poland), Neyman accepted a position at University College (London). There, he developed a friendship with Egon Pearson; Neyman and Pearson together developed the theory of hypothesis testing (Chapter 6). In 1934, in a talk to the Royal Statistical Society, Neyman first proposed the idea of interval estimation, which he called *confidence intervals.* (It is interesting that Neyman rarely receives credit in textbooks as the originator of the confidence interval procedure.) In 1938, he emigrated to the United States and the University of California at Berkeley. At Berkeley, he built one of the strongest statistics departments in the country. Jerzey Neyman is considered one of the great founders of modern statistics. He was a superb teacher and innovative researcher who loved his students, always sharing his ideas with them. Neyman's influence on those he met is best expressed by a quote from prominent statistician David Salsburg: "We have all learned to speak statistics with a Polish accent." ■

Table 5.2 Commonly Used Values of $z_{\alpha/2}$

Confidence Level $100(1 - \alpha)$	α	$\alpha/2$	$z_{\alpha/2}$
90%	.10	.05	1.645
95%	.05	.025	1.96
99%	.01	.005	2.575

Large-Sample $(1 - \alpha)$% Confidence Interval for μ

$$\bar{x} \pm z_{\alpha/2}\sigma_{\bar{x}} = \bar{x} \pm z_{\alpha/2}\left(\frac{\sigma}{\sqrt{n}}\right)$$

where $z_{\alpha/2}$ is the z-value with an area $\alpha/2$ to its right (see Figure 5.4) and $\sigma_{\bar{x}} = \sigma/\sqrt{n}$. The parameter σ is the standard deviation of the sampled population, and n is the sample size.

Note: When σ is unknown (as is almost always the case) and n is large (say, $n \geq 30$), the confidence interval is approximately equal to

$$\bar{x} \pm z_{\alpha/2}\left(\frac{s}{\sqrt{n}}\right)$$

where s is the sample standard deviation.

Example 5.3

Large-Sample Confidence Interval for the Mean Number of Unoccupied Seats per Flight, μ

Problem Unoccupied seats on flights cause airlines to lose revenue. Suppose a large airline wants to estimate its average number of unoccupied seats per flight over the past year. To accomplish this, the records of 225 flights are randomly selected, and the number of unoccupied seats is noted for each of the sampled flights. (The data are saved in the **AIRNOSHOWS** file.) Descriptive statistics for the data are displayed in the Minitab printout, Figure 5.6.

```
Variable    N      Mean    StDev  SE Mean          90% CI
NOSHOWS   225   11.5956   4.1026   0.2735  (11.1438, 12.0473)
```

Figure 5.6

Minitab confidence interval for mean, Example 5.3

Estimate μ, the mean number of unoccupied seats per flight during the past year, using a 90% confidence interval.

Solution The general form of the 90% confidence interval for a population mean is

$$\bar{x} \pm z_{\alpha/2}\sigma_{\bar{x}} = \bar{x} \pm z_{.05}\sigma_{\bar{x}} = \bar{x} \pm 1.645\left(\frac{\sigma}{\sqrt{n}}\right)$$

From Figure 5.6, we find (after rounding) $\bar{x} = 11.6$. Because we do not know the value of σ (the standard deviation of the number of unoccupied seats per flight for all flights of the year), we use our best approximation—the sample standard deviation s. Then the 90% confidence interval is, approximately,

$$11.6 \pm 1.645\left(\frac{4.1}{\sqrt{225}}\right) = 11.6 \pm .45$$

or from 11.15 to 12.05—that is, at the 90% confidence level, we estimate the mean number of unoccupied seats per flight to be between 11.15 and 12.05 during the sampled year. This result is verified (except for rounding) on the right side of the Minitab printout in Figure 5.6.

Look Back We stress that the confidence level for this example, 90%, refers to the procedure used. If we were to apply this procedure repeatedly to different samples, approximately 90% of the intervals would contain μ. Although we do not know whether this particular interval (11.15, 12.05) is one of the 90% that contain μ or one of the 10% that do not, our knowledge of probability gives us "confidence" that the interval contains μ.

Now Work Exercise 5.12

The interpretation of confidence intervals for a population mean is summarized in the next box.

Interpretation of a Confidence Interval for a Population Mean

When we form a $100(1 - \alpha)\%$ confidence interval for μ, we usually express our confidence in the interval with a statement such as, "We can be $100(1 - \alpha)\%$ confident that μ lies between the lower and upper bounds of the confidence interval," where for a particular application, we substitute the appropriate numerical values for the confidence and for the lower and upper bounds. *The statement reflects our confidence in the estimation process rather than in the particular interval that is calculated from the sample data.* We know that repeated application of the same procedure will result in different lower and upper bounds on the interval. Furthermore, we know that $100(1 - \alpha)\%$ of the resulting intervals will contain μ. There is (usually) no way to determine whether any particular interval is one of those that contain μ, or one that does not. However, unlike point estimators, confidence intervals have some measure of reliability, the confidence coefficient, associated with them. For that reason they are generally preferred to point estimators.

Sometimes, the estimation procedure yields a confidence interval that is too wide for our purposes. In this case, we will want to reduce the width of the interval to obtain a more precise estimate of μ. One way to accomplish this is to decrease the confidence coefficient, $1 - \alpha$. For example, reconsider the problem of estimating the mean amount owed, μ, for all delinquent accounts. Recall that for a sample of 100 accounts, $\bar{x} = \$233.28$ and $s = \$90.34$. A 90% confidence interval for μ is

$$\bar{x} \pm 1.645\sigma/\sqrt{n} \approx 233.28 \pm (1.645)(90.34/\sqrt{100}) = 233.28 \pm 14.86$$

or (\$218.42, \$248.14). You can see that this interval is narrower than the previously calculated 95% confidence interval (\$215.57, \$250.99). Unfortunately, we also have "less confidence" in the 90% confidence interval. An alternative method used to decrease the width of an interval without sacrificing "confidence" is to increase the sample size n. We demonstrate this method in Section 5.5.

Exercises 5.1–5.20

Learning the Mechanics

5.1 Find $z_{\alpha/2}$ for each of the following:
NW
a. $\alpha = .10$ **b.** $\alpha = .01$
c. $\alpha = .05$ **d.** $\alpha = .20$

5.2 What is the confidence level of each of the following confidence intervals for μ?

a. $\bar{x} \pm 1.96\left(\frac{\sigma}{\sqrt{n}}\right)$ **b.** $\bar{x} \pm 1.645\left(\frac{\sigma}{\sqrt{n}}\right)$

c. $\bar{x} \pm 2.575\left(\frac{\sigma}{\sqrt{n}}\right)$ **d.** $\bar{x} \pm 1.282\left(\frac{\sigma}{\sqrt{n}}\right)$

e. $\bar{x} \pm .99\left(\frac{\sigma}{\sqrt{n}}\right)$

5.3 A random sample of n measurements was selected from a population with unknown mean μ and known standard deviation σ. Calculate a 95% confidence interval for μ for each of the following situations:
NW
a. $n = 75, \bar{x} = 28, \sigma^2 = 12$
b. $n = 200, \bar{x} = 102, \sigma^2 = 22$
c. $n = 100, \bar{x} = 15, \sigma = .3$
d. $n = 100, \bar{x} = 4.05, \sigma = .83$
e. Is the assumption that the underlying population of measurements is normally distributed necessary to ensure the validity of the confidence intervals in parts **a–d**? Explain.

5.4 A random sample of 90 observations produced a mean $\bar{x} = 25.9$ and a standard deviation $s = 2.7$.
a. Find an approximate 95% confidence interval for the population mean μ.
b. Find an approximate 90% confidence interval for μ.
c. Find an approximate 99% confidence interval for μ.

5.5 A random sample of 70 observations from a normally distributed population possesses a sample mean equal to 26.2 and a sample standard deviation equal to 4.1.
a. Find an approximate 95% confidence interval for μ.
b. What do you mean when you say that a confidence coefficient is .95?
c. Find an approximate 99% confidence interval for μ.
d. What happens to the width of a confidence interval as the value of the confidence coefficient is increased while the sample size is held fixed?
e. Would your confidence intervals of parts **a** and **c** be valid if the distribution of the original population was not normal? Explain.

Applet Exercise 5.1

Use the applet *Confidence Intervals for a Mean (the impact of confidence level)* to investigate the situation in Exercise 5.5 further. For this exercise, assume that $\mu = 2.62$ is the population mean and $\sigma = 4.1$ is the population standard deviation.
a. Using $n = 70$ and the normal distribution with the mean and standard deviation above, run the applet one time. How many of the 95% confidence intervals contain the mean? How many would you expect to contain the mean? How many of the 99% confidence intervals contain the mean? How many would you expect to contain the mean?
b. Which confidence level has a greater frequency of intervals that contain the mean? Is this result what you would expect? Explain.
c. Without clearing, run the applet several more times. What happens to the proportion of 95% confidence intervals that contain the mean as you run the applet more and more? What happens to the proportion of 99% confidence intervals that contain the mean as you run the applet more and more? Interpret these results in terms of the meanings of the 95% confidence interval and the 99% confidence interval.
d. Change the distribution to *right skewed*, clear, and run the applet several more times. Do you get the same results as in part **c**? Would you change your answer to part **e** of Exercise 5.5? Explain.

Applet Exercise 5.2

Use the applet *Confidence Intervals for a Mean (the impact of confidence level)* to investigate the effect of the sample size on the proportion of confidence intervals that contain the mean when the underlying distribution is skewed. Set the distribution to *right skewed*, the mean to 10, and the standard deviation to 1.
a. Using $n = 30$, run the applet several times without clearing. What happens to the proportion of 95% confidence intervals that contain the mean as you run the applet more and more? What happens to the proportion of 99% confidence intervals that contain the mean as you run the applet more and more? Do the proportions seem to be approaching the values that you would expect?
b. Clear and run the applet several times using $n = 100$. What happens to the proportions of 95% confidence intervals and 99% confidence intervals that contain the mean this time? How do these results compare to your results in part **a**?
c. Clear and run the applet several times using $n = 1{,}000$. How do the results compare to your results in parts **a** and **b**?
d. Describe the effect of sample size on the likelihood that a confidence interval contains the mean for a skewed distribution.

5.6 Explain what is meant by the statement, "We are 95% confident that an interval estimate contains μ."

5.7 Explain the difference between an interval estimator and a point estimator for μ.

5.8 The mean and standard deviation of a random sample of n measurements are equal to 33.9 and 3.3, respectively.

a. Find a 95% confidence interval for μ if $n = 100$.

b. Find a 95% confidence interval for μ if $n = 400$.

c. Find the widths of the confidence intervals found in parts **a** and **b.** What is the effect on the width of a confidence interval of quadrupling the sample size while holding the confidence coefficient fixed?

5.9 Will a large-sample confidence interval be valid if the population from which the sample is taken is not normally distributed? Explain.

Applying the Concepts—Basic

5.10 NW **Latex allergy in health care workers.** Health care workers who use latex gloves with glove powder on a daily basis are particularly susceptible to developing a latex allergy. Symptoms of a latex allergy include conjunctivitis, hand eczema, nasal congestion, skin rash, and shortness of breath. Each in a sample of 46 hospital employees who were diagnosed with latex allergy based on a skin-prick test reported on their exposure to latex gloves (*Current Allergy & Clinical Immunology*, Mar. 2004). Summary statistics for the number of latex gloves used per week are $\bar{x} = 19.3$, $s = 11.9$.

a. Give a point estimate for the average number of latex gloves used per week by all health care workers with a latex allergy.

b. Form a 95% confidence interval for the average number of latex gloves used per week by all health care workers with a latex allergy.

c. Give a practical interpretation of the interval, part **b.**

d. Give the conditions required for the interval, part **b,** to be valid.

5.11 **Budget lapsing at army hospitals.** Budget lapsing occurs when unspent funds do not carry over from one budgeting period to the next. Refer to the *Journal of Management Accounting Research* (Vol. 19, 2007) study on budget lapsing at U.S. army hospitals, Exercise 2.107 (p. 87). Because budget lapsing often leads to a spike in expenditures at the end of the fiscal year, the researchers recorded expenses per full-time equivalent employee for each in a sample of 1,751 army hospitals. The sample yielded the following summary statistics: $\bar{x} = \$6,563$ and $s = \$2,484$. Estimate the mean expenses per full-time equivalent employee of all U.S. army hospitals using a 90% confidence interval. Interpret the result.

5.12 NW **Tax-exempt charities.** Donations to tax-exempt organizations such as the Red Cross, the Salvation Army, the YMCA, and the American Cancer Society not only go to the stated charitable purpose but are also used to cover fund-raising expenses and overhead. The **CHARITY** file (see next table) contains the 2008 charitable commitments (i.e., the percentange of expenses that go toward the charitable purpose) for a sample of 30 charities. A Minitab analysis of the data is shown in the next column.

a. Give a point estimate for the mean charitable commitment of tax-exempt organizations.

b. Locate a 95% confidence interval for the true mean charitable commitment of tax-exempt organizations on the printout. Interpret the result.

c. Why is the confidence interval of part **b** a better estimator of the mean charitable commitment than the point estimator of part **a?** Explain.

```
Variable   N   Mean   StDev  SE Mean        95% CI
COMMIT    30  79.67   10.25     1.87  (75.84, 83.49)
```

Organization	Charitable Commitment
American Cancer Society	71%
American National Red Cross	89
Big Brothers Big Sisters of America	78
Boy Scouts of America	87
Boys & Girls Clubs of America	80
CARE USA	90
Covenant House	67
Disabled American Veterans	69
Ducks Unlimited	86
Feed the Children	84
Girl Scouts of the USA	82
Goodwill Industries International	88
Habitat for Humanity International	83
Mayo Clinic	97
March of Dimes Foundation	77
Multiple Sclerosis Society	78
Museum of Modern Art	80
Nature Conservancy	80
Paralyzed Veterans of America	62
Planned Parenthood Federation	82
Salvation Army	82
Shriners Hospital for Children	92
Smithsonian Institution	82
Special Olympics	74
Trust for Public Land	89
United States Olympic Committee	71
United Way	86
WGBH Educational Foundation	79
World Trade Center Memorial Foundation	44
YMCA of the USA	81

Source: "America's 200 largest charities," *Forbes,* Nov. 19, 2008. Reprinted by permission of Forbes.com. © 2009 Forbes LLC.

5.13 **Blogs for Fortune 500 firms.** Refer to the *Journal of Relationship Marketing* (Vol. 7, 2008) study of the prevalence of blogs and forums at Fortune 500 firms with both English and Chinese Web sites, Exercise 2.76 (p. 75). In a sample of 56 firms that provide blogs and forums as marketing tools, the mean number of blogs/forums per site was 4.25, with a standard deviation of 12.02.

a. Find a 99% confidence interval for the mean number of blogs/forums per site of all Fortune 500 firms that provide blogs and forums as marketing tools.

b. Give a practical interpretation of the interval, part **a.**

c. The number of blogs/forums per site is known to have a highly skewed distribution. Does this impact the validity of the inference, part **b**?

5.14 **Wear-out of used display panels.** Refer to Exercise 4.112 (p. 224) and the study of the wear-out failure time of used colored display panels purchased by an outlet store. Recall that prior to acquisition, the panels had been used for about one-third of their expected lifetimes. The failure times (in years) for a sample of 50 used panels (saved in the **PANELFAIL** file) are reproduced in the next table, followed by an SPSS printout of the analysis of the data.

SPSS Output for Exercise 5.14

Descriptives

			Statistic	Std. Error
FAILTIME	Mean		1.9350	.13133
	95% Confidence Interval for Mean	Lower Bound	1.6711	
		Upper Bound	2.1989	
	5% Trimmed Mean		1.9454	
	Median		1.8350	
	Variance		.862	
	Std. Deviation		.92865	
	Minimum		.01	
	Maximum		3.50	
	Range		3.49	
	Interquartile Range		1.43	
	Skewness		-.008	.337
	Kurtosis		-.755	.662

a. Locate a 95% confidence interval for the true mean failure time of used colored display panels on the printout.
b. Give a practical interpretation of the interval, part **a.**
c. In repeated sampling of the population of used colored display panels, where a 95% confidence interval for the mean failure time is computed for each sample, what proportion of all the confidence intervals generated will capture the true mean failure time?

0.01	1.21	1.71	2.30	2.96	0.19	1.22	1.75	2.30	2.98	0.51
1.24	1.77	2.41	3.19	0.57	1.48	1.79	2.44	3.25	0.70	1.54
1.88	2.57	3.31	0.73	1.59	1.90	2.61	1.19	0.75	1.61	1.93
2.62	3.50	0.75	1.61	2.01	2.72	3.50	1.11	1.62	2.16	2.76
3.50	1.16	1.62	2.18	2.84	3.50					

Source: Based on Irony, T. Z., Lauretto, M., Pereira, C., and Stern, J. M. "A Weibull wearout Test: Full Bayesian approach," paper presented at *Mathematical Sciences Colloquium*, Binghamton University, Binghamton, UK, December 2001.

Applying the Concepts—Intermediate

5.15 Executive Compensation Scoreboard. Refer to *Forbes*' 2008 "Executive Compensation Scoreboard." Recall that the **CEOPAY08** file contains the 2008 salaries (in $ millions) of the 500 CEOs that participated in the *Forbes*' survey. Suppose you are interested in estimating the mean 2008 salary for these 500 CEOs.

a. What is the target parameter?
b. Obtain a random sample of 50 salaries from the data set.
c. Find the mean of the 50 salaries, part **b.**
d. Verify that the standard deviation for the population of 500 salaries is $\sigma = \$18.5$ million.
e. Use the information, parts **c** and **d,** to form a 99% confidence interval for the true mean 2008 salary of the 500 CEOs in the *Forbes*' survey.
f. Give a practical interpretation of the interval, part **e.**
g. Find the true mean salary of the 500 CEOs and check to see if this value falls within the 99% confidence interval, part **e.**

5.16 401(k) Participation rates. Named for the section of the 1978 Internal Revenue Code that authorized them, 401(k) plans permit employees to shift part of their before-tax salaries into investments such as mutual funds. Employers typically match 50% of the employee's contribution up to about 6% of salary. One company, concerned with what it believed was a low employee participation rate in its 401(k) plan, sampled 30 other companies with similar plans and asked for their 401(k) participation rates. The following rates (in percentages) were obtained and are saved in the **RATE401K** file.

80	76	81	77	82	80	85	60	80	79	82	70
88	85	80	79	83	75	87	78	80	84	72	75
90	84	82	77	75	86						

a. Construct a 90% confidence interval for the mean participation rate for all companies that have 401(k) plans.
b. Interpret the interval in the context of this problem.
c. What assumption is necessary to ensure the validity of this confidence interval?
d. If the company that conducted the sample has a 71% participation rate, can it safely conclude that its rate is below the population mean rate for all companies with 401(k) plans? Explain.
e. If in the data set the 60% had been 80%, how would the center and width of the confidence interval you constructed in part **a** be affected?

5.17 Accounting and Machiavellianism. Refer to the *Behavioral Research in Accounting* (Jan. 2008) study of Machiavellian traits in accountants, Exercise 1.27 (p. 24). Recall that *Machiavellian* describes negative character traits that include manipulation, cunning, duplicity, deception, and bad faith. A Machiavellian ("Mach") rating score was determined for each in a sample of accounting alumni of a large southwestern university. Scores range from a low of 40 to a high of 160, with the theoretical neutral Mach rating score of 100. The 122 purchasing managers in the sample had a mean Mach rating score of 99.6, with a standard deviation of 12.6.

a. From the sample, estimate the true mean Mach rating score of all purchasing managers.
b. Form a 95% confidence interval for the estimate, part **b.**

c. Give a practical interpretation of the interval, part **c.**

d. A director of purchasing at a major firm claims that the true mean Mach rating score of all purchasing managers is 85. Is there evidence to dispute this claim?

5.18 **Size of diamonds sold at retail.** Refer to Exercise 2.49 (p. 62) and the *Journal of Statistics Education* data on diamonds saved in the **DIAMONDS** file. Consider the quantitative variable, number of carats, recorded for each of the 308 diamonds for sale on the open market.

a. Select a random sample of 30 diamonds from the 308 diamonds.

b. Find the mean and standard deviation of the number of carats per diamond for the sample.

c. Use the sample information, part **b,** to construct a 95% confidence interval for the mean number of carats in the population of 308 diamonds.

d. Interpret the phrase *95% confidence* when applied to the interval, part **c.**

e. Refer to the mean of all 308 diamonds you calculated in Exercise 2.49. Does the "population" mean fall within the confidence interval of part **c**?

Applying the Concepts—Advanced

5.19 **Improving SAT scores.** Refer to the *Chance* (Winter 2001) and National Education Longitudinal Survey (NELS) study of 265 students who paid a private tutor to help them improve their SAT scores, Exercise 2.86 (p. 76). The changes in both the SAT–Mathematics and SAT–Verbal scores for these students are reproduced in the table. Suppose the true population mean change in score on one of the SAT tests for all students who paid a private tutor is 15. Which of the two tests, SAT–Mathematics or SAT–Verbal, is most likely to have this mean change? Explain.

	SAT–Math	SAT–Verbal
Mean change in score	19	7
Standard deviation of score changes	65	49

5.20 **The "Raid" test kitchen.** According to scientists, the cockroach has had 300 million years to develop a resistance to destruction. In a study conducted by researchers for S.C. Johnson & Son, Inc. (manufacturers of Raid and Off), 5,000 roaches (the expected number in a roach-infested house) were released in the Raid test kitchen. One week later, the kitchen was fumigated, and 16,298 dead roaches were counted, a gain of 11,298 roaches for the 1-week period. Assume that none of the original roaches died during the 1-week period and that the standard deviation of x, the number of roaches produced per roach in a 1-week period, is 1.5. Use the number of roaches produced by the sample of 5,000 roaches to find a 95% confidence interval for the mean number of roaches produced per week for each roach in a typical roach-infested house.

5.3 Confidence Interval for a Population Mean: Student's t-Statistic

Federal legislation requires pharmaceutical companies to perform extensive tests on new drugs before they can be marketed. Initially, a new drug is tested on animals. If the drug is deemed safe after this first phase of testing, the pharmaceutical company is then permitted to begin human testing on a limited basis. During this second phase, inferences must be made about the safety of the drug based on information in very small samples.

Suppose a pharmaceutical company must estimate the average increase in blood pressure of patients who take a certain new drug. Assume that only six patients (randomly selected from the population of all patients) can be used in the initial phase of human testing. The use of a *small sample* in making an inference about μ presents two immediate problems when we attempt to use the standard normal z as a test statistic.

BIOGRAPHY **WILLIAM S. GOSSET (1876–1937)**

Student's t-Distribution

At the age of 23, William Gosset earned a degree in chemistry and mathematics at prestigious Oxford University. He was immediately hired by the Guinness Brewing Company in Dublin, Ireland, for his expertise in chemistry. However, Gosset's mathematical skills allowed him to solve numerous practical problems associated with brewing beer. For example, Gosset applied the Poisson distribution to model the number of yeast cells per unit volume in the fermentation process. His most important discovery was that of the t-distribution in 1908. Because most applied researchers worked with small samples, Gosset was interested in the behavior of the mean in the small sample case. He tediously took numerous small sets of numbers, calculated the mean and standard deviation, obtained their t-ratio, and plotted the results on graph paper. The shape of the distribution was always the same—the t-distribution. Under company policy, employees were forbidden to publish their research results, so Gosset used the pen name *Student* to publish a paper on the subject. Hence, the distribution has been called Student's t-distribution. ■

Problem 1 The shape of the sampling distribution of the sample mean $\bar{x}$ (and the z-statistic) now depends on the shape of the population that is sampled. We can no longer assume that the sampling distribution of $\bar{x}$ is approximately normal because the Central Limit Theorem ensures normality only for samples that are sufficiently large.

Solution to Problem 1 According to Theorem 4.1, the sampling distribution of $\bar{x}$ (and z) is exactly normal, even for small samples, if the sampled population is normal. It is approximately normal if the sampled population is approximately normal.

Problem 2 The population standard deviation σ is almost always unknown. Although it is still true that $\sigma_{\bar{x}} = \sigma/\sqrt{n}$, the sample standard deviation s may provide a poor approximation for σ when the sample size is small.

Solution to Problem 2 Instead of using the standard normal statistic

$$z = \frac{\bar{x} - \mu}{\sigma_{\bar{x}}} = \frac{\bar{x} - \mu}{\sigma/\sqrt{n}}$$

which requires knowledge of or a good approximation to σ, we define and use the statistic

$$t = \frac{\bar{x} - \mu}{s/\sqrt{n}}$$

in which the sample standard deviation, s, replaces the population standard deviation, σ.

If we are sampling from a normal distribution, the ***t*-statistic** has a sampling distribution very much like that of the z-statistic: mound-shaped, symmetric, with mean 0. The primary difference between the sampling distributions of t and z is that the t-statistic is more variable than the z, which follows intuitively when you realize that t contains two random quantities ($\bar{x}$ and s), whereas z contains only one ($\bar{x}$).

The actual amount of variability in the sampling distribution of t depends on the sample size n. A convenient way of expressing this dependence is to say that the t-statistic has $(n - 1)$ **degrees of freedom (df).** Recall that the quantity $(n - 1)$ is the divisor that appears in the formula for s^2. This number plays a key role in the sampling distribution of s^2 and appears in discussions of other statistics in later chapters. In particular, the smaller the number of degrees of freedom associated with the t-statistic, the more variable will be its sampling distribution.

In Figure 5.7 we show both the sampling distribution of z and the sampling distribution of a t-statistic with 4 df. You can see that the increased variability of the t-statistic means that the t-value, t_α, that locates an area α in the upper tail of the t-distribution is larger than the corresponding value z_α. For any given value of α, the t-value t_α increases as the df decreases. Values of t that will be used in forming small-sample confidence intervals of μ are given in Table V in Appendix B. A partial reproduction of this table is shown in Table 5.3.

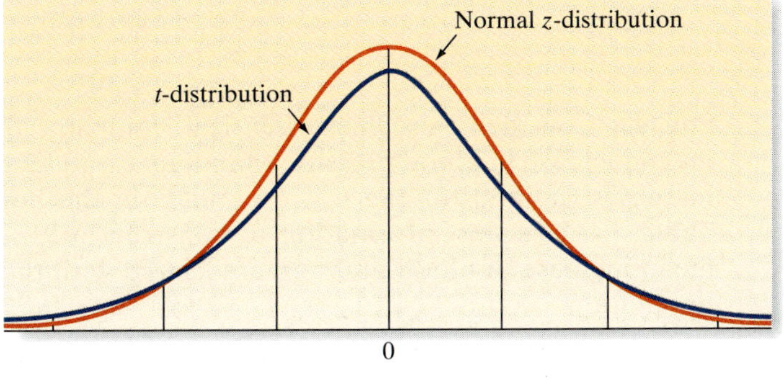

Figure 5.7
Standard normal (z) distribution and t-distribution with 4 df

Table 5.3 Reproduction of Part of Table V in Appendix B

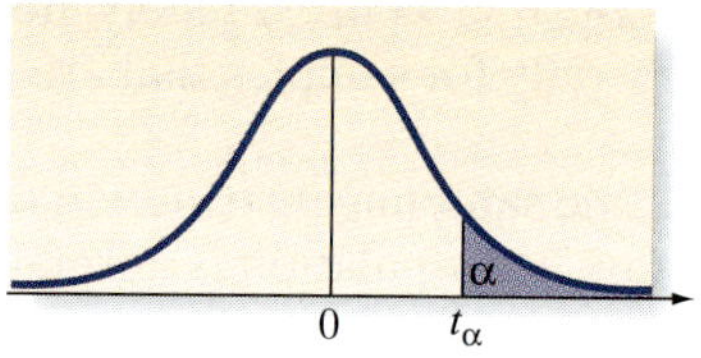

Degrees of Freedom	$t_{.100}$	$t_{.050}$	$t_{.025}$	$t_{.010}$	$t_{.005}$	$t_{.001}$	$t_{.0005}$
1	3.078	6.314	12.706	31.821	63.657	318.13	636.62
2	1.886	2.920	4.303	6.965	9.925	22.326	21.598
3	1.638	2.353	3.182	4.541	5.841	10.213	12.924
4	1.533	2.132	2.776	3.747	4.604	7.173	8.610
5	1.476	2.015	2.571	3.365	4.032	5.893	6.869
6	1.440	1.943	2.447	3.132	3.707	5.208	5.959
7	1.415	1.895	2.365	2.998	3.499	4.785	5.408
8	1.397	1.860	2.306	2.896	3.355	4.501	5.041
9	1.383	1.833	2.262	2.821	3.250	4.297	4.781
10	1.372	1.812	2.228	2.764	3.169	4.144	4.587
11	1.363	1.796	2.201	2.718	3.106	4.025	4.437
12	1.356	1.782	2.179	2.681	3.055	3.930	4.318
13	1.350	1.771	2.160	2.650	3.012	3.852	4.221
14	1.345	1.761	2.145	2.624	2.977	3.787	4.140
15	1.341	1.753	2.131	2.602	2.947	3.733	4.073
⋮	⋮	⋮	⋮	⋮	⋮	⋮	⋮
∞	1.282	1.645	1.960	2.326	2.576	3.090	3.291

Note that t_α values are listed for various degrees of freedom, where α refers to the tail area under the t-distribution to the right of t_α. For example, if we want the t-value with an area of .025 to its right and 4 df, we look in the table under the column $t_{.025}$ for the entry in the row corresponding to 4 df. This entry is $t_{.025} = 2.776$, as shown in Figure 5.8. The corresponding standard normal z-score is $z_{.025} = 1.96$.

Note that the last row of Table V, where df $= \infty$ (infinity), contains the standard normal z-values. This follows from the fact that as the sample size n grows very large, s becomes closer to σ and thus t becomes closer in distribution to z. In fact, when df $= 29$, there is little difference between corresponding tabulated values of z and t. Thus, researchers often choose the arbitrary cutoff of $n = 30$ (df $= 29$) to distinguish between the large-sample and small-sample inferential techniques.

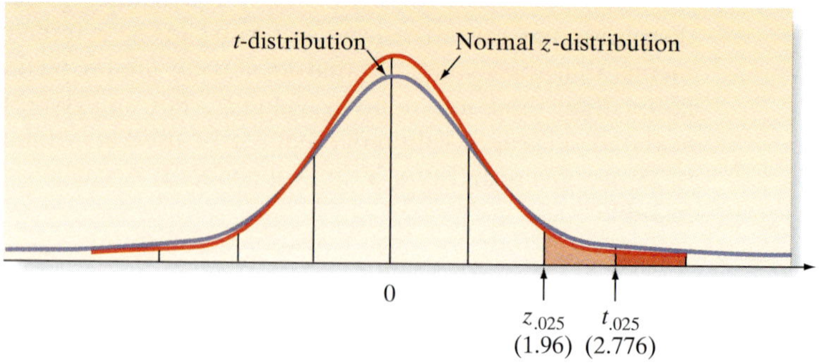

Figure 5.8
The $t_{.025}$ value in a t-distribution with 4 df and the corresponding $z_{.025}$ value

Example 5.4
A Confidence Interval for Mean Blood Pressure Increase Using the *t*-statistic

Problem Consider the pharmaceutical company that desires an estimate of the mean increase in blood pressure of patients who take a new drug. The blood pressure increases (points) for the $n = 6$ patients in the human testing phase are shown in Table 5.4. Use this information to construct a 95% confidence interval for μ, the mean increase in blood pressure associated with the new drug for all patients in the population.

Table 5.4 **Blood Pressure Increases for $n = 6$ Patients**

1.7	3.0	.8	3.4	2.7	2.1

Data Set: BPINCR

Solution First, note that we are dealing with a sample too small to assume that the sample mean $\bar{x}$ is approximately normally distributed by the Central Limit Theorem—that is, we do not get the normal distribution of $\bar{x}$ "automatically" from the Central Limit Theorem when the sample size is small. Instead, the measured variable, in this case the increase in blood pressure, must be normally distributed in order for the distribution of $\bar{x}$ to be normal.

Second, unless we are fortunate enough to know the population standard deviation σ, which in this case represents the standard deviation of *all* the patients' increases in blood pressure when they take the new drug, we cannot use the standard normal z-statistic to form our confidence interval for μ. Instead, we must use the t-distribution, with $(n - 1)$ degrees of freedom.

In this case, $n - 1 = 5$ df, and the t-value is found in Table 5.3 to be $t_{.025} = 2.571$ with 5 df. Recall that the large-sample confidence interval would have been of the form

$$\bar{x} \pm z_{\alpha/2}\sigma_{\bar{x}} = \bar{x} \pm z_{\alpha/2}\frac{\sigma}{\sqrt{n}} = \bar{x} \pm z_{.025}\frac{\sigma}{\sqrt{n}}$$

where 95% is the desired confidence level. To form the interval for a small sample from *a normal distribution, we simply substitute t for z and s for σ* in the preceding formula:

$$\bar{x} \pm t_{\alpha/2}\frac{s}{\sqrt{n}}$$

An SPSS printout showing descriptive statistics for the six blood pressure increases is displayed in Figure 5.9. Note that $\bar{x} = 2.283$ and $s = .950$. Substituting these numerical values into the confidence interval formula, we get

$$2.283 \pm (2.571)\left(\frac{.950}{\sqrt{6}}\right) = 2.283 \pm .997$$

or 1.286 to 3.280 points. Note that this interval agrees (except for rounding) with the confidence interval generated by SPSS in Figure 5.9.

We interpret the interval as follows: We can be 95% confident that the mean increase in blood pressure associated with taking this new drug is between 1.286 and 3.28 points. As with our large-sample interval estimates, our confidence is in the process, not in this particular interval. We know that if we were to repeatedly use this estimation procedure, 95% of the confidence intervals produced would contain the true mean μ, *assuming that the probability distribution of changes in blood pressure from which our sample was selected is normal.* The latter assumption is necessary for the small-sample interval to be valid.

Descriptives

			Statistic	Std. Error
BPINCR	Mean		2.283	.3877
	95% Confidence Interval for Mean	Lower Bound	1.287	
		Upper Bound	3.280	
	5% Trimmed Mean		2.304	
	Median		2.400	
	Variance		.902	
	Std. Deviation		.9496	
	Minimum		.8	
	Maximum		3.4	
	Range		2.6	
	Interquartile Range		1.625	
	Skewness		-.573	.845
	Kurtosis		-.389	1.741

Figure 5.9
SPSS confidence interval for mean blood pressure increase

Look Back What price did we pay for having to use a small sample to make the inference? First, we had to assume the underlying population is normally distributed, and if the assumption is invalid, our interval might also be invalid.* Second, we had to form the interval using a t-value of 2.571 rather than a z-value of 1.96, resulting in a wider interval to achieve the same 95% level of confidence. If the interval from 1.286 to 3.28 is too wide to be of use, then we know how to remedy the situation: increase the number of patients sampled to decrease the interval width (on average).

Now Work Exercise 5.22

The procedure for forming a small-sample confidence interval is summarized in the accompanying boxes.

Small-Sample Confidence Interval† for μ

$$\bar{x} \pm t_{\alpha/2}\left(\frac{s}{\sqrt{n}}\right)$$

where $t_{\alpha/2}$ is based on $(n - 1)$ degrees of freedom.

Conditions Required for a Valid Small-Sample Confidence Interval for μ

1. A random sample is selected from the target population.
2. The population has a relative frequency distribution that is approximately normal.

Example 5.5

A Small-Sample Confidence Interval for μ in Destructive Sampling

Problem Some quality-control experiments require *destructive sampling* (i.e., the test to determine whether the item is defective destroys the item) in order to measure some particular characteristic of the product. The cost of destructive sampling often dictates small samples. For example, suppose a manufacturer of printers for personal computers wishes to estimate the mean number of characters printed before the printhead fails. Suppose the printer manufacturer tests $n = 15$ randomly selected printheads and records the number of characters printed until failure for each. These 15 measurements (in millions of characters) are listed in Table 5.5, followed by an Excel/DDXL summary statistics printout in Figure 5.10.

a. Form a 99% confidence interval for the mean number of characters printed before the printhead fails. Interpret the result.

b. What assumption is required for the interval, part **a,** to be valid? Is it reasonably satisfied?

Table 5.5 Number of Characters (in Millions) for $n = 15$ Printhead Tests

1.13	1.55	1.43	.92	1.25
1.36	1.32	.85	1.07	1.48
1.20	1.33	1.18	1.22	1.29

Data Set: PRINTHEAD

*By *invalid,* we mean that the probability that the procedure will yield an interval that contains μ is not equal to $(1 - \alpha)$. Generally, if the underlying population is approximately normal, then the confidence coefficient will approximate the probability that a randomly selected interval contains μ.

†The procedure given in the box assumes that the population standard deviation σ is unknown, which is almost always the case. If σ is known, we can form the small-sample confidence interval just as we would a large-sample confidence interval using a standard normal z-value instead of t. However, we must still assume that the underlying population is approximately normal.

Solution

a. For this small sample ($n = 15$), we use the t-statistic to form the confidence interval. We use a confidence coefficient of .99 and $n - 1 = 14$ degrees of freedom to find $t_{\alpha/2}$ in Table V:

$$t_{\alpha/2} = t_{.005} = 2.977$$

[*Note:* The small sample forces us to extend the interval almost 3 standard deviations (of $\bar{x}$) on each side of the sample mean in order to form the 99% confidence interval.] From the Excel/DDXL printout, Figure 5.10, we find $\bar{x} = 1.239$ and $s = .193$. Substituting these values into the confidence interval formula, we obtain

$$\bar{x} \pm t_{.005}\left(\frac{s}{\sqrt{n}}\right) = 1.239 \pm 2.977\left(\frac{.193}{\sqrt{15}}\right)$$

$$= 1.239 \pm .148 \text{ or } (1.091, 1.387)$$

NUMBER Confidence Interval

Summary Statistics

Count	Mean	Std Dev	df
15	1.239	0.193	14

Interval Results

Confidence Interval

With 99% Confidence, 1.09 < μ < 1.387

Figure 5.10
Excel/DDXL summary statistics and confidence interval for data in Table 5.5

This interval is shown at the bottom of the printout, Figure 5.10.

Our interpretation is as follows: The manufacturer can be 99% confident that the printhead has a mean life of between 1.091 and 1.387 million characters. If the manufacturer were to advertise that the mean life of its printheads is (at least) 1 million characters, the interval would support such a claim. Our confidence is derived from the fact that 99% of the intervals formed in repeated applications of this procedure would contain μ.

b. Because n is small, we must assume that the number of characters printed before printhead failure is a random variable from a normal distribution—that is, we assume that the population from which the sample of 15 measurements is selected is distributed normally. One way to check this assumption is to graph the distribution of data in Table 5.5. If the sample data are approximately normal, then the population from which the sample is selected is very likely to be normal. A Minitab stem-and-leaf plot for the sample data is displayed in Figure 5.11. The distribution is mound shaped and nearly symmetric. Therefore, the assumption of normality appears to be reasonably satisfied.

Stem-and-Leaf Display: NUMBER

```
Stem-and-leaf of NUMBER  N  = 15
Leaf Unit = 0.010

  1    8   5
  2    9   2
  3   10   7
  5   11   38
 (4)  12   0259
  6   13   236
  3   14   38
  1   15   5
```

Figure 5.11
Minitab stem-and-leaf display of data in Table 5.5

Look Back Other checks for normality, such as a normal probability plot and the ratio IQR/S, may also be used to verify the normality condition.

Now Work Exercise 5.31

We have emphasized throughout this section that an assumption that the population is approximately normally distributed is necessary for making small-sample inferences about μ when using the t-statistic. Although many phenomena do have approximately normal distributions, it is also true that many random phenomena have distributions that are not normal or even mound shaped. Empirical evidence acquired over the years has shown that the t-distribution is rather insensitive to moderate departures from normality—that is, use of the t-statistic when sampling from slightly or moderately skewed mound shaped populations generally produces credible results; however, for cases in which the distribution is distinctly nonnormal, we must either take a large sample or use a *nonparametric method* (the topic of Chapter 14).

What Do You Do When the Population Relative Frequency Distribution Departs Greatly from Normality?

Answer: Use the nonparametric statistical methods of Chapter 14.

Statistics IN Action Revisited — Estimating the Mean Overpayment

Refer to the Medicare fraud investigation described in the Statistics in Action (pp. 269–270). Recall that the United States Department of Justice (USDOJ) obtained a random sample of 52 claims from a population of 1,000 Medicare claims. For each claim, the amount paid, the amount disallowed (denied) by the auditor, and the amount that should have been paid (allowed) were recorded and saved in the **MCFRAUD** file. The USDOJ wants to use these data to calculate an estimate of the overpayment for all 1,000 claims in the population.

One way to do this is to first use the sample data to estimate the mean overpayment per claim for the population, then use the estimated mean to extrapolate the overpayment amount to the population of all 1,000 claims. The difference between the amount paid and the amount allowed by the auditor represents the overpayment for each claim. This value is recorded as the amount denied in the **MCFRAUD** file. These overpayment amounts are listed in the accompanying table.

Minitab software is used to find a 95% confidence interval for μ, the mean overpayment amount. The Minitab printout is displayed in Figure SIA5.1. The 95% confidence interval for μ (highlighted on the printout) is (16.51, 27.13). Thus, the USDOJ can be 95% confident that the mean overpayment amount for the population of 1,000 claims is between \$16.51 and \$27.13.

Now, let x_i represent the overpayment amount for the ith claim. If the true mean μ were known, then the total overpayment amount for all 1,000 claims would be equal

$$\sum_{i=1}^{1,000} x_i = (1,000)\left[\sum_{i=1}^{1,000} x_i\right]/(1,000) = (1,000)\mu$$

Consequently, to estimate the total overpayment amount for the 1,000 claims, the USDOJ will simply multiply the end points of the interval by 1,000. This yields the 95% confidence interval (\$16,510, \$27,130).* Typically, the USDOJ is willing to give the Medicare provider in question the benefit of the doubt by demanding a repayment equal to the lower 95% confidence bound—in this case, \$16,510.

Table SIA5.1 Overpayment Amounts for Sample of 52 Claims

\$0.00	\$31.00	\$0.00	\$37.20	\$37.20	\$0.00	\$43.40	\$0.00
\$37.20	\$43.40	\$0.00	\$37.20	\$0.00	\$24.80	\$0.00	\$0.00
\$37.20	\$0.00	\$37.20	\$0.00	\$37.20	\$37.20	\$37.20	\$0.00
\$37.20	\$37.20	\$0.00	\$0.00	\$37.20	\$0.00	\$43.40	\$37.20
\$0.00	\$37.20	\$0.00	\$37.20	\$37.20	\$0.00	\$37.20	\$37.20
\$0.00	\$37.20	\$43.40	\$0.00	\$37.20	\$37.20	\$37.20	\$0.00
\$43.40	\$0.00	\$43.40	\$0.00				

Data Set: MCFRAUD

```
Variable        N   Mean   StDev  SE Mean         95% CI
Denied_Amount  52  21.82   19.08     2.65   (16.51, 27.13)
```

Figure SIA5.1
Minitab confidence interval for mean overpayment

*This interval represents an approximation to the true 95% confidence interval for the total amount of overpayment. The precise interval involves use of a continuity correction for population size (see Section 5.6).

Exercises 5.21–5.35

Learning the Mechanics

5.21 Suppose you have selected a random sample of $n = 5$ measurements from a normal distribution. Compare the standard normal z-values with the corresponding t-values if you were forming the following confidence intervals.

a. 80% confidence interval
b. 90% confidence interval
c. 95% confidence interval
d. 98% confidence interval
e. 99% confidence interval
f. Use the table values you obtained in parts **a–e** to sketch the z- and t-distributions. What are the similarities and differences?

Applet Exercise 5.3

Use the applet *Confidence Intervals for a Mean (the impact of not knowing the standard deviation)* to compare proportions of z-intervals and t-intervals that contain the mean for a population that is normally distributed.

a. Using $n = 5$ and the normal distribution with mean 50 and standard deviation 10, run the applet several times. How do the proportions of z-intervals and t-intervals that contain the mean compare?
b. Repeat part **a** first for $n = 10$ and then for $n = 20$. Compare your results to part **a.**
c. Describe any patterns you observe between the proportion of z-intervals that contain the mean and the proportion of t-intervals that contain the mean as the sample size increases.

Applet Exercise 5.4

Use the applet *Confidence Intervals for a Mean (the impact of not knowing the standard deviation)* to compare proportions of z-intervals and t-intervals that contain the mean for a population with a skewed distribution.

a. Using $n = 5$ and the right skewed distribution with mean 50 and standard deviation 10, run the applet several times. How do the proportions of z-intervals and t-intervals that contain the mean compare?
b. Repeat part **a** first for $n = 10$ and then for $n = 20$. Compare your results to part **a.**
c. Describe any patterns you observe between the proportion of z-intervals that contain the mean and the proportion of t-intervals that contain the mean as the sample size increases.
d. How does skewedness of the underlying distribution affect the proportions of z-intervals and t-intervals that contain the mean?

5.22 NW Explain the differences in the sampling distributions of $\bar{x}$ for large and small samples under the following assumptions.

a. The variable of interest, x, is normally distributed.
b. Nothing is known about the distribution of the variable x.

5.23 Let t_0 be a particular value of t. Use Table V in Appendix B to find t_0 values such that the following statements are true.

a. $P(-t_0 < t < t_0) = .95$ where df $= 10$
b. $P(t \leq -t_0 \text{ or } t \geq t_0) = .05$ where df $= 10$
c. $P(t \leq t_0) = .05$ where df $= 10$
d. $P(t \leq -t_0 \text{ or } t \geq t_0) = .10$ where df $= 20$
e. $P(t \leq -t_0 \text{ or } t \geq t_0) = .01$ where df $= 5$

5.24 Let t_0 be a specific value of t. Use Table V in Appendix B to find t_0 values such that the following statements are true.

a. $P(t \geq t_0) = .025$ where df $= 11$
b. $P(t \geq t_0) = .01$ where df $= 9$
c. $P(t \leq t_0) = .005$ where df $= 6$
d. $P(t \leq t_0) = .05$ where df $= 18$

5.25 The following random sample was selected from a normal distribution: 4, 6, 3, 5, 9, 3.

a. Construct a 90% confidence interval for the population mean μ.
b. Construct a 95% confidence interval for the population mean μ.
c. Construct a 99% confidence interval for the population mean μ.
d. Assume that the sample mean $\bar{x}$ and sample standard deviation s remain exactly the same as those you just calculated but are based on a sample of $n = 25$ observations rather than $n = 6$ observations. Repeat parts **a–c.** What is the effect of increasing the sample size on the width of the confidence intervals?

5.26 The following sample of 16 measurements (saved in the **LM5_26** file) was selected from a population that is approximately normally distributed:

91	80	99	110	95	106	78	121	106	100	97	82
100	83	115	104								

a. Construct an 80% confidence interval for the population mean.
b. Construct a 95% confidence interval for the population mean and compare the width of this interval with that of part **a.**
c. Carefully interpret each of the confidence intervals and explain why the 80% confidence interval is narrower.

Applying the Concepts—Basic

5.27 **Assessing the bending strength of a wooden roof.** The white wood material used for the roof of an ancient Japanese temple is imported from Northern Europe. The wooden roof must withstand as much as 100 centimeters of snow in the winter. Architects at Tohoku University (Japan) conducted a study to estimate the mean bending strength of the white wood roof (*Journal of the International Association for Shell and Spatial Structures,* Aug. 2004). A sample of 25 pieces of the imported wood were tested and yielded the following statistics on breaking strength (MPa): $\bar{x} = 75.4$, $s = 10.9$. Estimate the true mean breaking strength of the white wood with a 90% confidence interval. Interpret the result.

5.28 **Hospital length of stay.** Health insurers and the federal government are both putting pressure on hospitals to shorten the average length of stay (LOS) of their patients. The average LOS for men in the United States is 5.2 days, and the average for women is 4.5 days (*Statistical Abstract of the United States: 2008*). A random sample of 20 hospitals in one state had a mean LOS for women of 3.8 days and a standard deviation of 1.2 days.

a. Use a 90% confidence interval to estimate the population mean LOS for women for the state's hospitals.
b. Interpret the interval in terms of this application.
c. What is meant by the phrase "90% confidence interval"?

Minitab output for Exercise 5.29

One-Sample T: GrwthRate

```
Variable    N  Mean  StDev  SE Mean        95% CI
GrwthRate  12  2801   4912     1418  (-320, 5922)
```

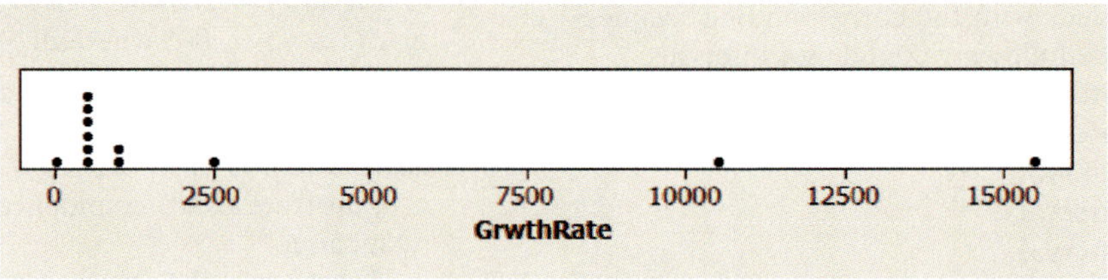

5.29 Technology Fast 500. Deloitte & Touche rank the 500 fastest-growing technology companies in the United States based on percentage growth over a 5-year period. Consider a random sample of 12 companies similar to those ranked by D&T. The rankings and growth rates are shown in the table and saved in the **FAST500** file. A Minitab analysis of the data appears above.

Rank	5-year Revenue Growth Rate
19	15,554%
30	10,361
87	2,551
161	1,176
195	945
246	706
279	613
326	435
351	393
398	343
441	291
496	245

a. Locate a 95% confidence interval for the true mean 5-year revenue growth rate on the printout. Interpret the result.

b. In order to estimate the mean described in part **a** with a small-sample confidence interval, what characteristic must the population possess?

c. Explain why the required population characteristics may not hold in this case.

5.30 Surface roughness of pipe. Refer to the *Anti-corrosion Methods and Materials* (Vol. 50, 2003) study of the surface roughness of coated interior pipe used in oil fields, Exercise 2.46 (p. 62). The data (in micrometers) for 20 sampled pipe sections (saved in the **ROUGHPIPE** file) are reproduced in the next table; a Minitab analysis of the data appears below.

a. Locate a 95% confidence interval for the mean surface roughness of coated interior pipe on the accompanying Minitab printout.

b. Would you expect the average surface roughness to be as high as 2.5 micrometers? Explain.

1.72	2.50	2.16	2.13	1.06	2.24	2.31	2.03	1.09	1.40
2.57	2.64	1.26	2.05	1.19	2.13	1.27	1.51	2.41	1.95

Source: Farshad, F., and Pesacreta, T. "Coated pipe interior surface roughness as measured by three scanning probe instruments," *Anti-corrosion Methods and Materials*, Vol. 50, No. 1, 2003 (Table III). Copyright © 2003 MCB UP Ltd.

One-Sample T: ROUGH

```
Variable   N     Mean    StDev  SE Mean            95% CI
ROUGH     20  1.88100  0.52391  0.11715  (1.63580, 2.12620)
```

Applying the Concepts—Intermediate

5.31 Minimizing tractor skidding distance. When planning for a new forest road to be used for tree harvesting, planners must select the location to minimize tractor skidding distance. In the *Journal of Forest Engineering* (July 1999), researchers wanted to estimate the true mean skidding distance along a new road in a European forest. The skidding distances (in meters) were measured at 20 randomly selected road sites. These values are given in the accompanying table and saved in the **SKIDDING** file.

a. Estimate the true mean skidding distance for the road with a 95% confidence interval.

b. Give a practical interpretation of the interval, part **a.**

c. What conditions are required for the inference, part **b,** to be valid? Are these conditions reasonably satisfied?

d. A logger working on the road claims the mean skidding distance is at least 425 meters. Do you agree?

488	350	457	199	285	409	435	574	439	546
385	295	184	261	273	400	311	312	141	425

Source: Extracted from Tujek, J., and Pacola, E. "Algorithms for skidding distance modeling on a raster digital terrain model" *Journal of Forest Engineering*, Vol. 10, No. 1, July 1999 (Table 1).

5.32 Contamination of New Jersey wells. Methyl t-butyl ether (MTBE) is an organic water contaminant that often results from gasoline spills. The level of MTBE (in parts per billion) was measured for a sample of 12 well sites located near a gasoline service station in New Jersey (*Environmental Science & Technology,* Jan. 2005). The data are listed in the accompanying table and saved in the **NJGAS** file.

150	367	38	12	11	134
12	251	63	8	13	107

Source: Reprinted with permission from Kuder, T., et al. "Enrichment of stable carbon and hydrogen isotopes during anaerobic biodegradation of MTBE: Microcosm and field evidence," *Environmental Science & Technology,* Vol. 39, No. 1, Jan. 2005, pp. 213–220 (Table 1). © 2005 American Chemical Society

a. Give a point estimate for μ, the true mean MTBE level for all well sites located near the New Jersey gasoline service station.

b. Calculate and interpret a 99% confidence interval for μ.

c. What assumptions are required for the interval, part **b,** to be valid? Are these assumptions reasonably satisfied?

5.33 Al Qaeda attacks on the United States. *Studies in Conflict & Terrorism* (Vol. 29, 2006) published an empirical analysis of recent incidents involving suicide terrorist attacks. The data in the table (saved in the **ALQAEDA** file) are the number of individual suicide bombings and/or attacks for each in a sample of 21 incidents involving a transgression against the United States by the Al Qaeda terrorist group. (For example, the infamous 9/11/2001 Al Qaeda attack involved 4 separate hijacked airplane attacks—two planes crashed into the World Trade Center towers, one crashed into the Pentagon, and a fourth plane crashed in a field in Pennsylvania.)

a. Find the mean and standard deviation of the sample data.

b. Describe the population from which the sample is selected.

c. Use the information, part **a,** to find a 90% confidence interval for the mean, μ, of the population.

d. Give a practical interpretation of the result, part **c.**

e. In repeated sampling, where intervals similar to the one computed in part **c** are generated, what proportion of the intervals will enclose the true value of μ?

1	1	2	1	2	4	1	1	1	1	2	3	4	5	1	1	1	2	2	2	1

Source: Moghadam, A. "Suicide terrorism, occupation, and the globalization of martyrdom: A critique of *Dying to Win,*" *Studies in Conflict & Terrorism,* Vol. 29, No. 8, 2006, pp. 707–729 (Table 3). Reprinted by permission of the publisher (Taylor & Francis Group www.informaworld.com).

5.34 Overbooking policies for major airlines. Airlines overbook flights in order to reduce the odds of flying with unused seats. An article in *Transportation Research* (Vol. 38, 2002) investigated the optimal overbooking policies for 10 major U.S. airlines. One of the variables measured for each airline was the compensation (in dollars) per bumped passenger required to maximize future revenue. These *threshold* levels of compensation for the 10 airlines are listed in the next table and saved in the **OVERBOOK** file. Suppose the data represent a sample of threshold compensation levels selected from all major airlines. Estimate the true mean threshold compensation level for all major airlines using a 90% confidence interval. Interpret the result.

Airline	Threshold Compensation per Bumped Passenger
Alaska	$2,112
America West	752
American	1,312
Continental	1,264
Delta	2,048
Northwest	1,104
Southwest	720
Trans World	1,456
United	1,520
US Air	1,392

Source: Reprinted from Suzuki, Y. "An empirical analysis of the optimal overbooking policies for US major airlines," *Transportation Research, Part E,* Vol. 38, 2002, pp. 135–149 (Table 4). © 2002 with permission from Elsevier.

5.35 Largest private companies. IPOs—initial public offerings of stock—create billions of dollars of new wealth for owners, managers, and employees of companies that were previously privately owned. Nevertheless, hundreds of large and thousands of small companies remain privately owned. The revenues of a random sample of 15 firms from *Forbes* 441 Largest Private Companies list (saved in the **PRIVCOMP** file) is given in the table below.

Company	Revenue (in billions)
Enterprise Rent-A-Car	$13.10
Flying J	14.32
Tenaska Energy	11.60
Wawa	5.05
Ergon	4.49
Brookshire Grocery	2.20
BrightStar	3.66
Bose	2.18
Mary Kay	2.40
LL Bean	1.62
Rooms to Go	1.75
SAS Institute	2.15
Blue Tee	1.10
Printpack	1.36
US Oil	2.19

Source: "America's Largest Private Companies," *Forbes,* Nov. 3, 2008. Reprinted by permission of Forbes.com. © 2009 Forbes LLC.

a. Describe the population from which the random sample was drawn.

b. Use a 98% confidence interval to estimate the mean revenue of the population of companies in question.

c. Interpret your confidence interval in the context of the problem.

d. What characteristic must the population possess to ensure the appropriateness of the estimation procedure used in part **b**?

e. Suppose *Forbes* reports that the true mean revenue of the 441 companies on the list is $5.0 billion. Is the claim believable?

5.4 Large-Sample Confidence Interval for a Population Proportion

The number of public opinion polls has grown at an astounding rate in recent years. Almost daily, the news media report the results of some poll. Pollsters regularly determine the percentage of people who approve of the president's on-the-job performance, the fraction of voters in favor of a certain candidate, the fraction of customers who prefer a particular product, and the proportion of households that watch a particular TV program. In each case, we are interested in estimating the percentage (or proportion) of some group with a certain characteristic. In this section, we consider methods for making inferences about population proportions when the sample is large.

Example 5.6
Estimating a Population Proportion—Preference for Breakfast Cereal

Problem A food-products company conducted a market study by randomly sampling and interviewing 1,000 consumers to determine which brand of breakfast cereal they prefer. Suppose 313 consumers were found to prefer the company's brand. How would you estimate the true fraction of *all* consumers who prefer the company's cereal brand?

Solution In this study, consumers are asked which brand of breakfast cereal they prefer. Note that "brand" is a qualitative variable and what we are asking is how you would estimate the probability p of success in a binomial experiment, where p is the probability that a chosen consumer prefers the company's brand. One logical method of estimating p for the population is to use the proportion of successes in the sample—that is, we can estimate p by calculating

$$\hat{p} = \frac{\text{Number of consumers sampled who prefer the company's brand}}{\text{Number of consumers sampled}}$$

where $\hat{p}$ is read "p hat." Thus, in this case,

$$\hat{p} = \frac{313}{1{,}000} = .313$$

Look Back To determine the reliability of the estimator $\hat{p}$, we need to know its sampling distribution—that is, if we were to draw samples of 1,000 consumers over and over again, each time calculating a new estimate $\hat{p}$, what would be the frequency distribution of all the $\hat{p}$ values? The answer lies in viewing $\hat{p}$ as the average, or mean, number of successes per trial over the n trials. If each success is assigned a value equal to 1 and a failure is assigned a value of 0, then the sum of all n sample observations is x, the total number of successes, and $\hat{p} = x/n$ is the average, or mean, number of successes per trial in the n trials. The Central Limit Theorem tells us that the relative frequency distribution of the sample mean for any population is approximately normal for sufficiently large samples.

Now Work Exercise 5.45a

The repeated sampling distribution of $\hat{p}$ has the characteristics listed in the box on page 291 and shown in Figure 5.12.

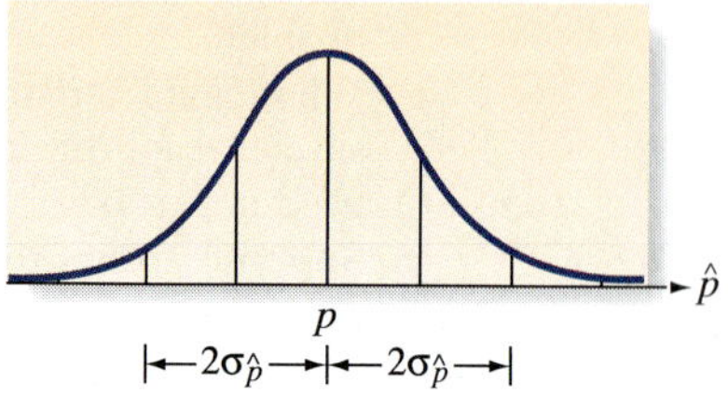

Figure 5.12
Sampling distribution of $\hat{p}$

Sampling Distribution of $\hat{p}$

1. The mean of the sampling distribution of $\hat{p}$ is p; that is, $\hat{p}$ is an unbiased estimator of p.
2. The standard deviation of the sampling distribution of $\hat{p}$ is $\sqrt{pq/n}$; that is, $\sigma\hat{p} = \sqrt{pq/n}$, where $q = 1 - p$.
3. For large samples, the sampling distribution of $\hat{p}$ is approximately normal. A sample size is considered large if both $n\hat{p} \geq 15$ and $n\hat{q} \geq 15$.

The fact that $\hat{p}$ is a "sample mean number of successes per trial" allows us to form confidence intervals about p in a manner that is completely analogous to that used for large-sample estimation of μ.

Large-Sample Confidence Interval for $\hat{p}$

$$\hat{p} \pm z_{\alpha/2}\sigma_{\hat{p}} = \hat{p} \pm z_{\alpha/2}\sqrt{\frac{pq}{n}} \approx \hat{p} \pm z_{a/2}\sqrt{\frac{\hat{p}\hat{q}}{n}}$$

where $\hat{p} = \dfrac{x}{n}$ and $\hat{q} = 1 - \hat{p}$

Note: When n is large, $\hat{p}$ can approximate the value of p in the formula for $\sigma_{\hat{p}}$.

Conditions Required for a Valid Large-Sample Confidence Interval for p

1. A random sample is selected from the target population.
2. The sample size n is large. (This condition will be satisfied if both $n\hat{p} \geq 15$ and $n\hat{q} \geq 15$. Note that $n\hat{p}$ and $n\hat{q}$ are simply the number of successes and number of failures, respectively, in the sample.)

Thus, if 313 of 1,000 consumers prefer the company's cereal brand, a 95% confidence interval for the proportion of *all* consumers who prefer the company's brand is

$$\hat{p} \pm z_{\alpha/2}\sigma_{\hat{p}} = .313 \pm 1.96\sqrt{\frac{pq}{1{,}000}}$$

where $q = 1 - p$. Just as we needed an approximation for σ in calculating a large-sample confidence interval for μ, we now need an approximation for p. As Table 5.7 shows, the approximation for p does not have to be especially accurate because the value of $\sqrt{pq}$ needed for the confidence interval is relatively insensitive to changes in p. Therefore, we can use $\hat{p}$ to approximate p. Keeping in mind that $\hat{q} = 1 - \hat{p}$, we substitute these values into the formula for the confidence interval:

$$\hat{p} \pm 1.96\sqrt{\frac{pq}{1{,}000}} \approx \hat{p} \pm 1.96\sqrt{\frac{\hat{p}\hat{q}}{1{,}000}}$$
$$= .313 \pm 1.96\sqrt{\frac{(.313)(.687)}{1{,}000}}$$
$$= .313 \pm .029$$
$$= (.284, .342)$$

Table 5.7 Values of pq for Several Different Values of p

p	pq	$\sqrt{pq}$
.5	.25	.50
.6 or .4	.24	.49
.7 or .3	.21	.46
.8 or .2	.16	.40
.9 or .1	.09	.30

The company can be 95% confident that the interval from 28.4% to 34.2% contains the true percentage of *all* consumers who prefer its brand—that is, in repeated construction of confidence intervals, approximately 95% of all samples would produce confidence intervals that enclose p. Note that the guidelines for interpreting a confidence interval about μ also apply to interpreting a confidence interval for p because p is the "population fraction of successes" in a binomial experiment.

Example 5.7

Finding a Large-Sample Confidence Interval for *p*, the Proportion Optimistic about the Economy

Problem Many public polling agencies conduct surveys to determine the current consumer sentiment concerning the state of the economy. For example, the Bureau of Economic and Business Research (BEBR) at the University of Florida conducts quarterly surveys to gauge consumer sentiment in the Sunshine State. Suppose that the BEBR randomly samples 484 consumers and finds that 257 are optimistic about the state of the economy. Use a 90% confidence interval to estimate the proportion of all consumers in Florida who are optimistic about the state of the economy. Based on the confidence interval, can BEBR infer that the majority of Florida consumers are optimistic about the economy?

Solution The number, x, of the 484 sampled consumers who are optimistic about the Florida economy is a binomial random variable if we can assume that the sample was randomly selected from the population of Florida consumers and that the poll was conducted identically for each sampled consumer.

The point estimate of the proportion of Florida consumers who are optimistic about the economy is

$$\hat{p} = \frac{x}{n} = \frac{257}{484} = .531$$

We first check to be sure that the sample size is sufficiently large that the normal distribution provides a reasonable approximation for the sampling distribution of $\hat{p}$. We require both $n\hat{p}$ and $n\hat{q}$ to be at least 15. Now

$$n\hat{p} = 484(.531) = 257$$

and

$$n\hat{q} = 484(1 - .531) = 227$$

Because both of these values are at least 15, we may conclude that the normal approximation is reasonable.

We now proceed to form the 90% confidence interval for p, the true proportion of Florida consumers who are optimistic about the state of the economy:

$$\hat{p} \pm z_{\alpha/2}\sigma_{\hat{p}} = \hat{p} \pm z_{\alpha/2}\sqrt{\frac{pq}{n}} \approx \hat{p} \pm z_{\alpha/2}\sqrt{\frac{\hat{p}\hat{q}}{n}}$$
$$= .531 \pm 1.645\sqrt{\frac{(.531)(.469)}{484}} = .531 \pm .037 = (.494, .568)$$

(This interval is also shown on the Minitab printout, Figure 5.13.) Thus, we can be 90% confident that the proportion of all Florida consumers who are confident about the economy is between .494 and .568. As always, our confidence stems from the fact that 90% of all similarly formed intervals will contain the true proportion p and not from any knowledge about whether this particular interval does.

Can we conclude that the majority of Florida consumers are optimistic about the economy based on this interval? If we wished to use this interval to infer that a majority is optimistic, the interval would have to support the inference that p exceeds .5—that is, that more than 50% of the Florida consumers are optimistic about the economy. Note that the interval contains some values below .5 (as low as .494) as well as some above .5 (as high as .568). Therefore, we cannot conclude that the true value of p exceeds .5 based on this 90% confidence interval.

Sample	X	N	Sample p	90% CI
1	257	484	0.530992	(0.493681, 0.568303)

Figure 5.13
Portion of Minitab printout with 90% confidence interval for *p*

Look Back If the entire confidence interval fell above .5 (e.g., an interval from .52 to .54), then we could conclude (with 90% confidence) that the true proportion of consumers who are optimistic exceeds .5.

Now Work Exercise 5.45b,c

We conclude this section with a warning and an illustrative example.

CAUTION Unless n is extremely large, the large-sample procedure presented in this section performs poorly when p is near 0 or near 1. ▲

The problem stated in the above warning can be illustrated as follows. Suppose you want to estimate the proportion of executives who die from a work-related injury using a sample size of $n = 100$. This proportion is likely to be near 0, say $p \approx .001$. If so, then $np \approx 100(.001) = .1$ is less than the recommended value of 15 (see Conditions in the box on p. 291). Consequently, a confidence interval for p based on a sample of $n = 100$ will probably be misleading.

To overcome this potential problem, an *extremely* large sample size is required. Because the value of n required to satisfy "extremely large" is difficult to determine, statisticians (see Agresti & Coull, 1998) have proposed an alternative method, based on the Wilson (1927) point estimator of p. The procedure is outlined in the box below. Researchers have shown that this confidence interval works well for any p, even when the sample size n is very small.

Adjusted $(1 - \alpha)100\%$ Confidence Interval for a Population Proportion, p

$$\tilde{p} \pm z_{\alpha/2}\sqrt{\frac{\tilde{p}(1-\tilde{p})}{n+4}}$$

where $\tilde{p} = \frac{x+2}{n+4}$ is the adjusted sample proportion of observations with the characteristic of interest, x is the number of successes in the sample, and n is the sample size.

Example 5.8

Applying the Adjusted Confidence Interval Procedure for p, Proportion Who Are Victims of a Violent Crime

Problem According to *True Odds: How Risk Affects Your Everyday Life* (Walsh, 1997), the probability of being the victim of a violent crime is less than .01. Suppose that in a random sample of 200 Americans, 3 were victims of a violent crime. Estimate the true proportion of Americans who were victims of a violent crime using a 95% confidence interval.

Solution Let p represent the true proportion of Americans who were victims of a violent crime. Because p is near 0, an "extremely large" sample is required to estimate its value using the usual large-sample method. Note that the number of "successes," 3, is less than 15. Thus, we doubt whether the sample size of 200 is large enough to apply the large-sample method. Alternatively, we will apply the adjustment outlined in the box.

Because the number of "successes" (i.e., number of violent crime victims) in the sample is $x = 3$, the adjusted sample proportion is

$$\tilde{p} = \frac{x+2}{n+4} = \frac{3+2}{200+4} = \frac{5}{204} = .025$$

Note that this adjusted sample proportion is obtained by adding a total of four observations—two "successes" and two "failures"—to the sample data. Substituting $\tilde{p} = .025$ into the equation for a 95% confidence interval, we obtain

$$\tilde{p} \pm 1.96\sqrt{\frac{\tilde{p}(1-\tilde{p})}{n+4}} = .025 \pm 1.96\sqrt{\frac{(.025)(.975)}{204}}$$
$$= .025 \pm .021$$

or (.004, .046). Consequently, we are 95% confident that the true proportion of Americans who are victims of a violent crime falls between .004 and .046.

Look Back If we apply the standard large sample confidence interval formula, where $\hat{p} = 3/200 = .015$, we obtain

$$\hat{p} \pm 1.96\sqrt{\frac{\hat{p}\hat{q}}{200}} = .015 \pm 1.96\sqrt{\frac{(.015)(.985)}{200}}$$
$$= .015 \pm .017 \text{ or } (-.002, .032)$$

Note that the interval contains negative (nonsensical) values for the true proportion. Such a result is typical when the large-sample method is misapplied.

Now Work Exercise 5.52

Statistics IN Action Revisited — Estimating the Coding Error Rate

In the previous Statistics in Action Revisited (p. 286), we showed how to estimate the mean overpayment amount for claims in a Medicare fraud study. In addition to estimating overcharges, the USDOJ also is interested in estimating the *coding error rate* of a Medicare provider. The coding error rate is defined as the proportion of Medicare claims that are coded incorrectly. Thus, for this inference, the USDOJ is interested in estimating a population proportion, p. Typically, the USDOJ finds that about 50% of the claims in a Medicare fraud case are incorrectly coded.

If you examine the sample data in Table SIA5.1, you can verify that of the 52 audited claims, 31 were determined to be coded incorrectly, resulting in an overcharge. These are the claims with a disallowed amount greater than \$0. Therefore, an estimate of the coding error rate, p, for this Medicare provider is

$$\hat{p} = 31/52 = .596$$

A 95% confidence interval for p can be obtained with the use of a confidence interval formula or with statistical software.

The **MCFRAUD** file includes a qualitative variable (called "Coding Error") at two levels, where "Yes" represents that the claim is coded incorrectly and "No" represents that the claim is coded correctly. Thus, the USDOJ desires an estimate of the proportion of "Yes" values for the "Coding Error" variable. A confidence interval for the proportion of incorrectly coded claims is highlighted on the accompanying Minitab printout, Figure SIA5.2. The interval, (.45, .73), implies that the true coding error rate for the 1,000 claims in the population falls between .45 and .73, with 95% confidence. Note that the value .5—the proportion of incorrectly coded claims expected by the USDOJ—falls within the 95% confidence interval.

Data Set: MCFRAUD

```
Event = Yes


Variable        X   N  Sample p          95% CI
Coding_Error   31  52  0.596154  (0.451016, 0.729940)
```

Figure SIA5.2
Minitab confidence interval for coding error rate

Exercises 5.36–5.55

Learning the Mechanics

5.36 Describe the sampling distribution of $\hat{p}$ based on large samples of size n—that is, give the mean, the standard deviation, and the (approximate) shape of the distribution of $\hat{p}$ when large samples of size n are (repeatedly) selected from the binomial distribution with probability of success p.

Applet Exercise 5.5

Use the applet *Confidence Intervals for a Proportion* to investigate the effect of the value of p on the number of confidence intervals that contain the population proportion p for a fixed sample size. For this exercise, use sample size $n = 10$.

a. Run the applet several times without clearing for $p = .1$. What proportion of the 95% confidence intervals contain

p? What proportion of the 99% confidence intervals contain p? Do the results surprise you? Explain.

b. Repeat part **a** for each value of p: $p = .2, p = .3, p = .4, p = .5, p = .6, p = .7, p = .8$, and $p = .9$.

c. Which value of p yields the greatest proportion of each type of interval that contain p?

d. Based on your results, what values of p will yield more reliable confidence intervals for a fixed sample size n? Explain.

Applet Exercise 5.6

Use the applet *Confidence Intervals for a Proportion* to investigate the effect of the sample size on the number of confidence intervals that contain the population proportion p for a value of p close to 0 or 1.

a. Run the applet several times without clearing for $p = .5$ and $n = 50$. Record the proportion of the 99% confidence intervals containing p.

b. Now set $p = .1$ and run the applet several times without clearing for $n = 50$. How does the proportion of the 99% confidence intervals containing p compare to that in part **a**?

c. Repeat part **b** keeping $p = .1$ and increasing the sample size by 50 until you find a sample size that yields a similar proportion of the 99% confidence intervals containing p as that in part **a.**

d. Based on your results, describe how the value of p affects the sample size needed to guarantee a certain level of confidence.

5.37 For the binomial sample information summarized in each part, indicate whether the sample size is large enough to use the methods of this chapter to construct a confidence interval for p.

a. $n = 400, \hat{p} = .10$ **b.** $n = 50, \hat{p} = .10$
c. $n = 20, \hat{p} = .5$ **d.** $n = 20, \hat{p} = .3$

5.38 A random sample of size $n = 121$ yielded $\hat{p} = .88$.

a. Is the sample size large enough to use the methods of this section to construct a confidence interval for p? Explain.

b. Construct a 90% confidence interval for p.

c. What assumption is necessary to ensure the validity of this confidence interval?

5.39 A random sample of size $n = 225$ yielded $\hat{p} = .46$.

a. Is the sample size large enough to use the methods of this section to construct a confidence interval for p? Explain.

b. Construct a 95% confidence interval for p.

c. Interpret the 95% confidence interval.

d. Explain what is meant by the phrase "95% confidence interval."

5.40 A random sample of 50 consumers taste-tested a new snack food. Their responses were coded (0: do not like; 1: like; 2: indifferent), recorded, and saved in the **SNACK** file as follows:

1	0	0	1	2	0	1	1	0	0
0	1	0	2	0	2	2	0	0	1
1	0	0	0	0	1	0	2	0	0
0	1	0	0	1	0	0	1	0	1
0	2	0	0	1	1	0	0	0	1

a. Use an 80% confidence interval to estimate the proportion of consumers who like the snack food.

b. Provide a statistical interpretation for the confidence interval you constructed in part **a.**

Applying the Concepts—Basic

5.41 **Is Starbucks coffee overpriced?** The *Minneapolis Star Tribune* (August 12, 2008) reported that 73% of Americans say that Starbucks coffee is overpriced. The source of this information was a national telephone survey of 1,000 American adults conducted by Rasmussen Reports.

a. Identify the population of interest in this study.

b. Identify the sample for the study.

c. Identify the parameter of interest in the study.

d. Find and interpret a 95% confidence interval for the parameter of interest.

5.42 **Satellite radio in cars.** Refer to the June 2007 survey on satellite radio subscriber service and usage conducted for the National Association of Broadcasters, Exercise 1.21 (p. 23). Recall that a random sample of 501 satellite radio subscribers were asked, "Do you have a satellite radio receiver in your car?" The survey found that 396 subscribers did, in fact, have a satellite receiver in their car.

a. From the sample, calculate an estimate of the true proportion of satellite radio subscribers who have a satellite radio receiver in their car.

b. Form a 90% confidence interval for the estimate, part **a.**

c. Give a practical interpretation of the interval, part **b.**

5.43 **Contractors with a company Web site.** Each year, construction contractors and equipment distributors from across the United States participate in a survey called the CIT Construction Industry Forecast. Recently, 900 contractors were interviewed for the survey. Of these, 414 indicated that they either already have a company Web site or plan to have a company Web site by the end of the year (*Contractor Magazine,* Jan. 2005).

a. Use the survey information to find a point estimate for the true proportion of contractors in the United States who have a company Web site or who will have one by the end of the year.

b. Find an interval estimate for the proportion, part **a.** Use a 90% confidence interval.

c. Give a practical interpretation of the interval, part **b.** Your answer should begin with "We are 90% confident…."

d. Explain the meaning of the phrase, "90% confident."

5.44 **Material safety data sheets.** For over 20 years, the Occupational Safety & Health Administration has required companies that handle hazardous chemicals to complete material safety data sheets (MSDSs). These MSDSs have been criticized for being too hard to understand and complete by workers. Although improvements were implemented in 1990, a more recent study of 150 MSDSs revealed that only 11% were satisfactorily completed (*Chemical & Engineering News,* Feb. 7, 2005).

a. Give a point estimate of p, the true proportion of MSDSs that are satisfactorily completed.

b. Find a 95% confidence interval for p.

c. Give a practical interpretation of the interval, part **b.**

5.45 NW **Cell phone use by drivers.** In a July 2001 research note, the U.S. Department of Transportation reported the results of the *National Occupant Protection Use Survey*. One focus of the survey was to determine the level of cell phone use by drivers while they are in the act of driving a motor passenger vehicle. Data collected by observers at randomly selected intersections across the country revealed that in a sample of 1,165 drivers, 35 were using their cell phone.

a. Give a point estimate of p, the true driver cell phone use rate (i.e., the true proportion of drivers who are using a cell phone while driving).

b. Compute a 95% confidence interval for p.

c. Give a practical interpretation of the interval, part **b.**

5.46 **Products "Made in the USA."** Refer to Exercise 2.8 (p. 39) and the *Journal of Global Business* (Spring 2002) survey to determine what "Made in the USA" means to consumers. Recall that 106 shoppers at a shopping mall in Muncie, Indiana, responded to the question, "'Made in the USA' means what percentage of U.S. labor and materials?" Sixty-four shoppers answered "100%."

a. Define the population of interest in the survey.

b. What is the characteristic of interest in the population?

c. Estimate the true proportion of consumers who believe "Made in the USA" means 100% U.S. labor and materials using a 90% confidence interval.

d. Give a practical interpretation of the interval, part **c.**

e. Explain what the phrase "90% confidence" means for this interval.

Applying the Concepts—Intermediate

5.47 **Adoption of high-speed Internet connection at home.** The Pew Internet & American Life Project reports periodically on the growth of home broadband adoption in the United States. A recent (May 2006) random-digit-dialing telephone survey of 4,000 American adults found that 42% have access to a high-speed Internet connection at home.

a. Give a point estimate of the true percentage of all American adults who have access to a high-speed Internet connection at home.

b. Construct a 95% confidence interval for the estimate, part **a,** and interpret the result.

c. In 2005, the Pew Internet & American Life Project reported that 30% of all American adults had access to a high-speed Internet connection at home. Is there evidence to conclude that the percentage has increased since 2005? Explain.

5.48 **Interviewing candidates for a job.** The costs associated with conducting interviews for a job opening have skyrocketed over the years. According to a Harris Interactive survey, 211 of 502 senior human resources executives at U.S. companies believe that their hiring managers are interviewing too many people to find qualified candidates for the job (*Business Wire,* June 8, 2006).

a. Describe the population of interest in this study.

b. Identify the population parameter of interest, p.

c. Is the sample size large enough to provide a reliable estimate of p?

d. Find and interpret an interval estimate for the true proportion of senior human resources executives who believe that their hiring managers interview too many candidates during a job search. Use a confidence level of 98%.

e. If you had constructed a 90% confidence interval, would it be wider or narrower?

5.49 **History of corporate acquisitions.** Refer to the *Academy of Management Journal* (Aug. 2008) investigation of the performance and timing of corporate acquisitions, Exercise 2.12 (p. 40). Recall that a corporate acquisition occurs when one corporation purchases all the stock shares of another. The investigation discovered that in a sample of 2,778 firms, 748 announced one or more acquisitions during the year 2000. Estimate the true percentage of all firms that announced one or more acquisitions during the year 2000 using a 90% confidence interval. Interpret the result.

5.50 **Diamonds sold on the open market.** Refer to the sample of 308 diamond stones that were listed for sale on the open market in Singapore's *Business Times*. The data are saved in the **DIAMONDS** file. Recall that the color of each diamond is classified as D, E, F, G, H, or I, while the clarity of each is classified as VVS1, VVS2, VS1, or VS2.

a. Find a 99% confidence interval for the proportion of all diamonds for sale on the open market that are classified as "D" color. Interpret the result.

b. Find a 99% confidence interval for the proportion of all diamonds for sale on the open market that are classified as "VS1" clarity. Interpret the result.

5.51 **Accuracy of price scanners at Wal-Mart.** The National Institute for Standards and Technology (NIST) mandates that for every 100 items scanned through the electronic checkout scanner at a retail store, no more than 2 should have an inaccurate price. A recent study of the accuracy of checkout scanners at Wal-Mart stores in California was conducted (*Tampa Tribune,* Nov. 22, 2005). At each of 60 randomly selected Wal-Mart stores, 100 random items were scanned. The researchers found that 52 of the 60 stores had more than 2 items that were inaccurately priced.

a. Give an estimate of p, the proportion of Wal-Mart stores in California that have more than 2 inaccurately priced items per 100 items scanned.

b. Construct a 95% confidence interval for p.

c. Give a practical interpretation of the interval, part **b.**

d. Suppose a Wal-Mart spokesperson claims that 99% of California Wal-Mart stores are in compliance with the NIST mandate on accuracy of price scanners. Comment on the believability of this claim.

e. Are the conditions required for a valid large-sample confidence interval for p satisfied in this application? If not, comment on the validity of the inference in part **d.**

5.52 NW **Male nannies.** The International Nanny Association reports that in a sample of 4,176 nannies who were placed in a job in 2007, only 24 were men (www.nanny.org). Use Wilson's adjustment to find a 95% confidence interval for the true proportion of all working nannies who are men. Interpret the resulting interval.

5.53 **Are you really being served red snapper?** Refer to the *Nature* (July 15, 2004) study of fish specimens labeled "red snapper," Exercise 3.69 (p. 154). Recall that federal law prohibits restaurants from serving a cheaper look-alike variety of fish (e.g., vermillion snapper or lane snapper) to customers who order red snapper. A team of University of North Carolina (UNC) researchers analyzed the meat from each in a sample of 22 "red snapper" fish fillets purchased from vendors across the United States in an effort to estimate the true proportion of fillets that are really red snapper. DNA tests revealed that 17 of the 22 fillets (or 77%) were not red snapper but the cheaper look-alike variety of fish.

a. Identify the parameter of interest to the UNC researchers.

b. Explain why a large-sample confidence interval is inappropriate to apply in this study.

c. Construct a 95% confidence interval for the parameter of interest using Wilson's adjustment.

d. Give a practical interpretation of the confidence interval.

Applying the Concepts—Advanced

5.54 **Latex allergy in health care workers.** Refer to the *Current Allergy & Clinical Immunology* (Mar. 2004) study of health care workers who use latex gloves, Exercise 5.10 (p. 278). In addition to the 46 hospital employees who were diagnosed with a latex allergy based on a skin-prick test, another 37 health care workers were diagnosed with the allergy using a latex-specific serum test. Of these 83 workers with confirmed latex allergy, only 36 suspected that they had the allergy when asked on a questionnaire. Make a statement about the likelihood that a health care worker with latex allergy suspects he or she actually has the allergy. Attach a measure of reliability to your inference.

5.55 **U.S. Postal Service's performance.** The accounting firm of Price Waterhouse annually monitors the U.S. Postal Service's performance. One parameter of interest is the percentage of mail delivered on time. In a sample of 332,000 items mailed between Dec. 10 and Mar. 3—the most difficult delivery season due to bad weather and holidays—Price Waterhouse determined that 282,200 items were delivered on time (*Tampa Tribune,* Mar. 26, 1995). Use this information to make a statement about the likelihood of an item being delivered on time by the U.S. Postal Service.

5.5 Determining the Sample Size

Recall (Section 1.6) that one way to collect the relevant data for a study used to make inferences about the population is to implement a designed (planned) experiment. Perhaps the most important design decision faced by the analyst is to determine the size of the sample. We show in this section that the appropriate sample size for making an inference about a population mean or proportion depends on the desired reliability.

Estimating a Population Mean

Consider the example from Section 5.2 in which we estimated the mean overdue amount for all delinquent accounts in a large credit corporation. A sample of 100 delinquent accounts produced the 95% confidence interval $\bar{x} \pm 1.96\sigma_{\bar{x}} \approx 233.28 \pm 17.71$. Consequently, our estimate $\bar{x}$ was within \$17.71 of the true mean amount due, μ, for all the delinquent accounts at the 95% confidence level—that is, the 95% confidence interval for μ was $2(17.71) = \$35.42$ wide when 100 accounts were sampled. This is illustrated in Figure 5.14(a).

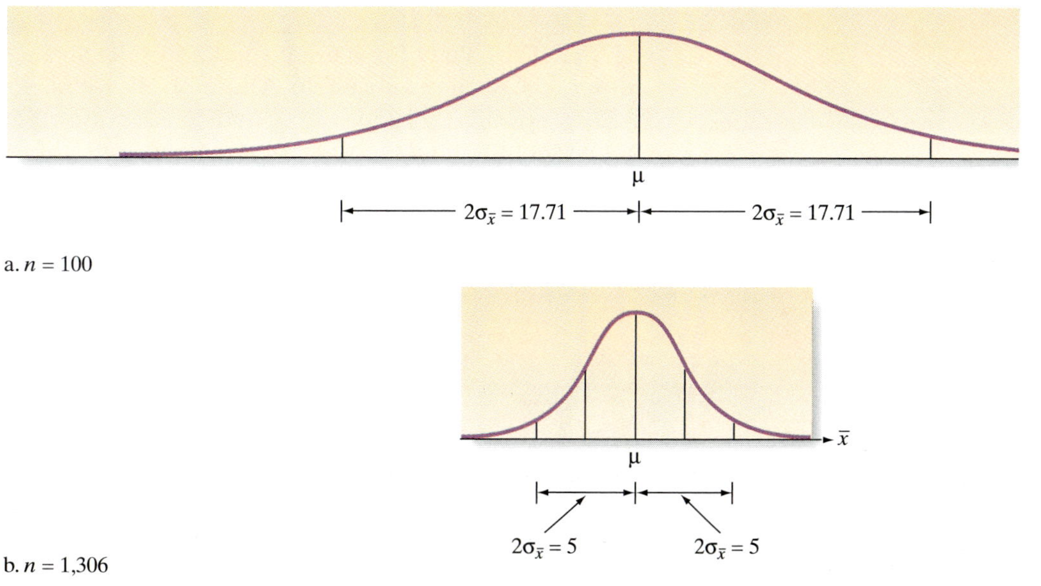

Figure 5.14
Relationship between sample size and width of confidence interval: delinquent debtors example

Now suppose we want to estimate μ to within \$5 with 95% confidence—that is, we want to narrow the width of the confidence interval from \$35.42 to \$10, as shown in Figure 5.14(b). How much will the sample size have to be increased to accomplish this? If we want the estimator $\bar{x}$ to be within \$5 of μ, we must have

$$2\sigma_{\bar{x}} = 5 \qquad \text{or, equivalently,} \qquad 2\left(\frac{\sigma}{\sqrt{n}}\right) = 5$$

The necessary sample size is obtained by solving this equation for n. To do this, we need an approximation for σ. We have an approximation from the initial sample of 100 accounts—namely, the sample standard deviation, $s = 90.34$. Thus,

$$2\left(\frac{\sigma}{\sqrt{n}}\right) \approx 2\left(\frac{s}{\sqrt{n}}\right) = 2\left(\frac{90.34}{\sqrt{n}}\right) = 5$$

$$\sqrt{n} = \frac{2(90.34)}{5} = 36.136$$

$$n = (36.136)^2 = 1{,}305.81 \approx 1{,}306$$

Approximately 1,306 accounts will have to be randomly sampled to estimate the mean overdue amount μ to within \$5 with (approximately) 95% confidence. The confidence interval resulting from a sample of this size will be approximately \$10 wide (see Figure 5.14b).

In general, we express the reliability associated with a confidence interval for the population mean μ by specifying the **sampling error,** within which we want to estimate μ with $100(1 - \alpha)\%$ confidence. The sampling error (denoted SE), then, is equal to the half-width of the confidence interval, as shown in Figure 5.15.

Sample Size Determination for $100(1 - \alpha)\%$ Confidence Interval for μ

In order to estimate μ with a sampling error and with $100(1 - \alpha)\%$ confidence, the required sample size is found as follows:

$$z_{\alpha/2}\left(\frac{\sigma}{\sqrt{n}}\right) = \text{SE}$$

The solution for n is given by the equation

$$n = \frac{(z_{\alpha/2})^2\sigma^2}{(\text{SE})^2}$$

Note: The value of σ is usually unknown. It can be estimated by the standard deviation, s, from a prior sample. Alternatively, we may approximate the range R of observations in the population, and (conservatively) estimate $\sigma \approx R/4$. In any case, you should round the value of n obtained *upward* to ensure that the sample size will be sufficient to achieve the specified reliability.

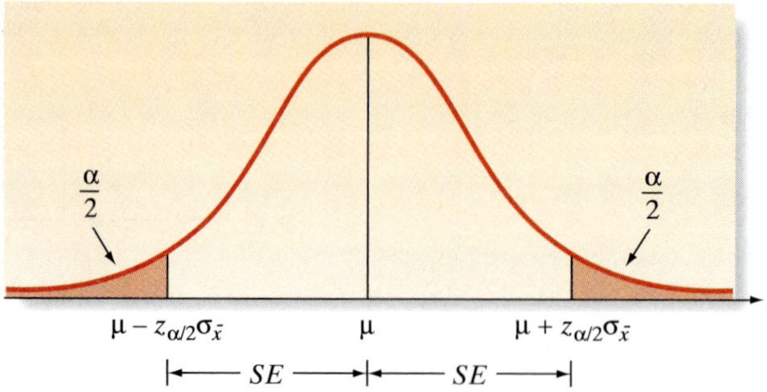

Figure 5.15
Specifying the sampling error as the half-width of a confidence interval

Example 5.9

Finding the Sample Size for Estimating μ, the Mean Inflation Pressure of Footballs

Problem The manufacturer of official NFL footballs uses a machine to inflate its new balls to a pressure of 13.5 pounds. When the machine is properly calibrated, the mean inflation pressure is 13.5 pounds, but uncontrollable factors cause the pressures of individual footballs to vary randomly from about 13.3 to 13.7 pounds. For quality-control purposes, the manufacturer wishes to estimate the mean inflation pressure to within .025 pound of its true value with a 99% confidence interval. What sample size should be used?

Solution We desire a 99% confidence interval that estimates μ with a sampling error of SE = .025 pound. For a 99% confidence interval, we have $z_{\alpha/2} = z_{.005} = 2.575$. No previous estimate of s is available; however, we are given that the range of observations is $R = 13.7 - 13.3 = .4$. A conservative estimate (based on Chebychev's Rule) is $\sigma \approx R/4 = .1$. Now we use the formula derived in the box to find the sample size n:

$$n = \frac{(z_{\alpha/2})^2\sigma^2}{(\text{SE})^2} \approx \frac{(2.575)^2(.1)^2}{(.025)^2} = 106.09$$

We round this up to $n = 107$. Realizing that σ was approximated by $R/4$, we might even advise that the sample size be specified as $n = 110$ to be more certain of attaining the objective of a 99% confidence interval with a sampling error of .025 pound or less.

Look Back To determine the value of the sampling error, look for the value that follows the key words "estimate μ to within. . . ."

Now Work Exercise 5.64

Sometimes the formula will yield a small sample size (say, $n < 30$). Unfortunately, this solution is invalid because the procedures and assumptions for small samples differ from those for large samples, as we discovered in Section 5.3. Therefore, if the formulas yield a small sample size, one simple strategy is to select a sample size $n = 30$.

Estimating a Population Proportion

The method outlined above is easily applied to a population proportion p. For example, in Section 5.4, a company used a sample of 1,000 consumers to calculate a 95% confidence interval for the proportion of consumers who preferred its cereal brand, obtaining the interval $.313 \pm .029$. Suppose the company wishes to estimate its market share more precisely, say to within .015 with a 95% confidence interval.

The company wants a confidence interval with a sampling error for the estimate of p of SE = .015. The sample size required to generate such an interval is found by solving the following equation for n (see Figure 5.16):

$$z_{\alpha/2}\sigma_{\hat{p}} = \text{SE} \qquad \text{or} \qquad z_{\alpha/2}\sqrt{\frac{pq}{n}} = .015$$

Because a 95% confidence interval is desired, the appropriate z-value is $z_{\alpha/2} = z_{.025} = 1.96 \approx 2$. We must approximate the value of the product pq before we can solve the

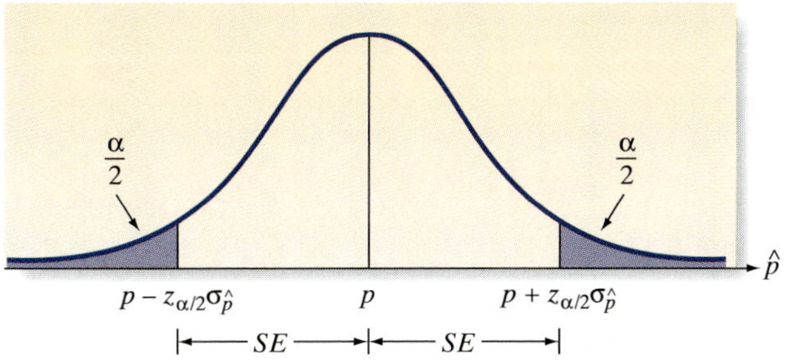

Figure 5.16
Specifying the sampling error of a confidence interval for a population proportion p

equation for n. As shown in Table 5.6, the closer the values of p and q to .5, the larger the product pq. Thus, to find a conservatively large sample size that will generate a confidence interval with the specified reliability, we generally choose an approximation of p close to .5. In the case of the food-products company, however, we have an initial sample estimate of $\hat{p} = .313$. A conservatively large estimate of pq can therefore be obtained by using, say, $p = .35$. We now substitute into the equation and solve for n:

$$2\sqrt{\frac{(.35)(.65)}{n}} = .015$$

$$n = \frac{(2)^2(.35)(.65)}{(.015)^2}$$

$$= 4{,}044.44 \approx 4{,}045$$

The company must sample about 4,045 consumers to estimate the percentage who prefer its brand to within .015 with a 95% confidence interval.

The procedure for finding the sample size necessary to estimate a population proportion p with a specified sampling error SE is given in the box.

Sample Size Determination for $100(1 - \alpha)\%$ Confidence Interval for p

In order to estimate a binomial probability p with sampling error SE and with $100(1 - \alpha)$ % confidence, the required sample size is found by solving the following equation for n:

$$z_{\alpha/2}\sqrt{\frac{pq}{n}} = \text{SE}$$

The solution for n can be written as follows:

$$n = \frac{(z_{\alpha/2})^2(pq)}{(\text{SE})^2}$$

Note: Because the value of the product pq is unknown, it can be estimated by using the sample fraction of successes, $\hat{p}$, from a prior sample. Remember (Table 5.6) that the value of pq is at its maximum when p equals .5, so you can obtain conservatively large values of n by approximating p by .5 or values close to .5. In any case, you should round the value of n obtained *upward* to ensure that the sample size will be sufficient to achieve the specified reliability.

Example 5.10

Finding the Sample Size for Estimating p, the Fraction of Defective Cell Phones

Problem A cellular telephone manufacturer that entered the postregulation market too quickly has an initial problem with excessive customer complaints and consequent returns of the cell phones for repair or replacement. The manufacturer wants to determine the magnitude of the problem in order to estimate its warranty liability. How many cellular telephones should the company randomly sample from its warehouse and check in order to estimate the fraction defective, p, to within .01 with 90% confidence?

Solution In order to estimate p to within .01 of its true value, we set the half-width of the confidence interval equal to SE = .01, as shown in Figure 5.17.

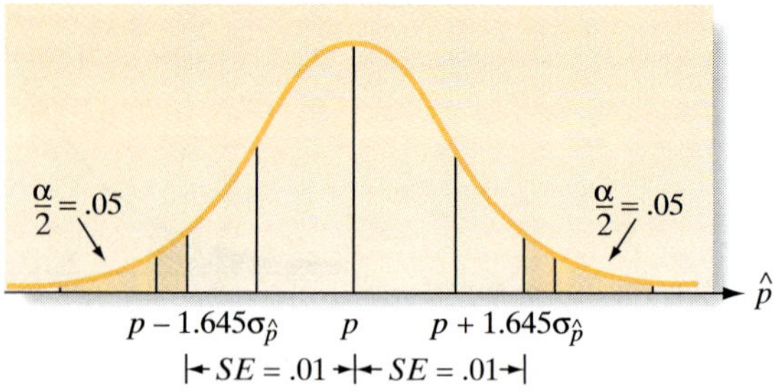

Figure 5.17
Specified reliability for estimate of fraction defective in Example 5.10

The equation for the sample size n requires an estimate of the product pq. We could most conservatively estimate $pq = .25$ (i.e., use $p = .5$), but this may be overly conservative when estimating a fraction defective. A value of .1, corresponding to 10% defective, will probably be conservatively large for this application. The solution is therefore

$$n = \frac{(z_{\alpha/2})^2(pq)}{(\text{SE})^2} = \frac{(1.645)^2(.1)(.9)}{(.01)^2} = 2{,}435.4 \approx 2{,}436$$

Thus, the manufacturer should sample 2,436 cellular telephones in order to estimate the fraction defective, p, to within .01 with 90% confidence.

Look Back Remember that this answer depends on our approximation for pq, where we used .09. If the fraction defective is closer to .05 than to .10, we can use a sample of 1,286 cell phones (check this) to estimate p to within .01 with 90% confidence.

Now Work Exercise 5.65

The cost of sampling will also play an important role in the final determination of the sample size to be selected to estimate either μ or p. Although more complex formulas can be derived to balance the reliability and cost considerations, we will solve for the necessary sample size and note that the sampling budget may be a limiting factor. Consult the references for a more complete treatment of this problem.

Statistics IN Action Revisited Determining Sample Size

In the previous Statistics in Action applications in this chapter, we used confidence intervals (1) to estimate μ, the mean overpayment amount for claims in a Medicare fraud study, and (2) to estimate p, the coding error rate (i.e., proportion of claims that are incorrectly coded) of a Medicare provider. Both of these confidence intervals were based on selecting a random sample of 52 claims from the population of claims handled by the Medicare provider. How does the USDOJ determine how many claims to sample for auditing?

Consider the problem of estimating the coding error rate, p. As stated in a previous Statistics in Action Revisited, the USDOJ typically finds that about 50% of the claims in a Medicare fraud case are incorrectly coded. Suppose the USDOJ wants to estimate the true coding error rate of a Medicare provider to within .1 with 95% confidence. How many claims should be randomly sampled for audit in order to attain the desired estimate?

Here, the USDOJ desires a sampling error of SE = .1, a confidence level of $1 - \alpha = .95$ (for which $z_{\alpha/2} = 1.96$), and uses an estimate $p \approx .50$. Substituting these values into the sample size formula (p. 300), we obtain:

$$\begin{aligned} n &= (z_{\alpha/2})^2(pq)/(\text{SE})^2 \\ &= (1.96)^2(.5)(.5)/(.1)^2 \\ &= 96.04 \end{aligned}$$

Consequently, the USDOJ should audit about 97 randomly selected claims to attain a 95% confidence interval for p with a sampling error of .10.

[*Note:* You may wonder why the sample actually used in the fraud analysis included only 52 claims. The sampling strategy employed involved more than selecting a simple random sample; rather, it used a more sophisticated sampling scheme, called stratified random sampling. The 52 claims represented the sample for just one of the strata.]

Exercises 5.56–5.72

Learning the Mechanics

5.56 If you wish to estimate a population mean with a sampling error of SE = .3 using a 95% confidence interval, and you know from prior sampling that σ^2 is approximately equal to 7.2, how many observations would have to be included in your sample?

5.57 Suppose you wish to estimate a population mean correct to within .20 with probability equal to .90. You do not know σ^2, but you know that the observations will range in value between 30 and 34.

a. Find the approximate sample size that will produce the desired accuracy of the estimate. You wish to be conservative to ensure that the sample size will be ample to achieve the desired accuracy of the estimate. [*Hint:* Using your knowledge of data variation from Section 2.6, assume that the range of the observations will equal 4σ.]

b. Calculate the approximate sample size, making the less conservative assumption that the range of the observations is equal to 6σ.

5.58 In each case, find the approximate sample size required to construct a 95% confidence interval for p that has sampling error of SE = .08.

a. Assume p is near .2.

b. Assume you have no prior knowledge about p, but you wish to be certain that your sample is large enough to achieve the specified accuracy for the estimate.

5.59 The following is a 90% confidence interval for p: (.26, .54). How large was the sample used to construct this interval?

5.60 It costs you \$10 to draw a sample of size $n = 1$ and measure the attribute of interest. You have a budget of \$1,500.

a. Do you have sufficient funds to estimate the population mean for the attribute of interest with a 95% confidence interval 5 units in width? Assume $\sigma = 14$.

b. If you used a 90% confidence level, would your answer to part **a** change? Explain.

5.61 Suppose you wish to estimate the mean of a normal population using a 95% confidence interval, and you know from prior information that $\sigma^2 \approx 1$.

a. To see the effect of the sample size on the width of the confidence interval, calculate the width of the confidence interval for $n = 16, 25, 49, 100$, and 400.

b. Plot the width as a function of sample size n on graph paper. Connect the points by a smooth curve and note how the width decreases as n increases.

5.62 If nothing is known about p, .5 can be substituted for p in the sample-size formula for a population proportion. But when this is done, the resulting sample size may be larger than needed. Under what circumstances will using $p = .5$ in the sample-size formula yield a sample size larger than needed to construct a confidence interval for p with a specified bound and a specified confidence level?

Applying the Concepts—Basic

5.63 Assessing the bending strength of a wooden roof. Refer to the *Journal of the International Association for Shell and Spatial Structures* (Aug. 2004) study to estimate the mean bending strength of imported white wood used on the roof of an ancient Japanese temple, Exercise 5.27 (p. 287). Suppose you want to estimate the true mean breaking strength of the white wood to within 4 MPa using a 90% confidence interval. How many pieces of the imported wood need to be tested? Recall that the sample standard deviation of the breaking strengths from the study was 10.9 MPa.

5.64 NW **Accounting and Machiavellianism.** Refer to the *Behavioral Research in Accounting* (Jan. 2008) study of Machiavellian traits in accountants, Exercise 5.17 (p. 279), where a Mach rating score was determined for each in a sample of accounting alumni who work as purchasing managers. Suppose you want to reduce the width of the 95% confidence interval for the true mean Mach rating score of all purchasing managers you obtained in Exercise 5.17b. How many purchasing managers should be included in the sample if you desire a sampling error of only 1.5 Mach rating points? Use $\sigma \approx 12$ in your calculations.

5.65 NW **Aluminum cans contaminated by fire.** A gigantic warehouse located in Tampa, Florida, stores approximately 60 million empty aluminum beer and soda cans. Recently, a fire occurred at the warehouse. The smoke from the fire contaminated many of the cans with blackspot, rendering them unusable. A University of South Florida statistician was hired by the insurance company to estimate p, the true proportion of cans in the warehouse that were contaminated by the fire. How many aluminum cans should be randomly sampled to estimate p to within .02 with 90% confidence?

5.66 "Made in the USA" survey. Refer to Exercise 5.46 (p. 296) and the *Journal of Global Business* (Spring 2002) survey to determine what "Made in the USA" means to consumers. Recall that 64 of 106 shoppers at a shopping mall in Muncie, Indiana, believe that "Made in the USA" implies all labor and materials are produced in the United States. Suppose the researchers want to increase the sample size in order to estimate the true proportion, p, to within .05 of its true value using a 90% confidence interval.

a. What is the confidence level desired by the researchers?

b. What is the sampling error desired by the researchers?

c. Compute the sample size necessary to obtain the desired estimate.

Applying the Concepts—Intermediate

5.67 Bacteria in bottled water. Is the bottled water you drink safe? The Natural Resources Defense Council warns that the bottled water you are drinking may contain more bacteria and other potentially carcinogenic chemicals than allowed by state and federal regulations. Of the more than 1,000 bottles studied, nearly one-third exceeded government levels (www.nrdc.org). Suppose that the Natural Resources Defense Council wants an updated estimate of the population proportion of bottled water that violates at least one government standard. Determine the sample size (number of bottles) needed to estimate this proportion to within ± 0.01 with 99% confidence.

5.68 Scanning errors at Wal-Mart. Refer to the National Institute for Standards and Technology (NIST) study of the accuracy of checkout scanners at Wal-Mart stores in California, Exercise 5.51 (p. 296). Recall that in a sample of 60 Wal-Mart stores, 52 violated the NIST scanner accuracy standard (*Tampa Tribune,* Nov. 22, 2005). You demonstrated in Exercise 5.51e that the conditions for a valid large-sample confidence interval for the true proportion of Wal-Mart stores in California that violate the NIST standard were not met. Determine the number of Wal-Mart stores that must be sampled in order to estimate the true proportion to within .05 with 90% confidence using the large-sample method.

5.69 Monitoring phone calls to a toll-free number. A large food-products company receives about 100,000 phone calls a year from consumers on its toll-free number. A computer monitors and records how many rings it takes for an operator to answer, how much time each caller spends "on hold," and other data. However, the reliability of the monitoring system has been called into question by the operators and their labor union. As a check on the computer system, approximately how many calls should be manually monitored during the next year to estimate the true mean time that callers spend on hold to within 3 seconds with 95% confidence? Answer this question for the following values of the standard deviation of waiting times (in seconds): 10, 20, and 30.

5.70 USGA golf ball tests. The United States Golf Association USGA tests all new brands of golf balls to ensure that they meet USGA specifications. One test conducted is intended to measure the average distance traveled when the ball is hit by a machine called "Iron Byron," a name inspired by the swing of the famous golfer Byron Nelson. Suppose the USGA wishes to estimate the mean distance for a new brand to within 1 yard with 90% confidence. Assume that past tests have indicated that the standard deviation of the distances Iron Byron hits golf balls is approximately 10 yards. How many golf balls should be hit by Iron Byron to achieve the desired accuracy in estimating the mean?

Applying the Concepts—Advanced

5.71 Is caffeine addictive? Does the caffeine in coffee, tea, and cola induce an addiction similar to that induced by alcohol, tobacco, heroine, and cocaine? In an attempt to answer this question, researchers at Johns Hopkins University examined 27 caffeine drinkers and found 25 who displayed some type of withdrawal symptoms when abstaining from caffeine. [*Note*: The 27 caffeine drinkers volunteered for the study.] Furthermore, of 11 caffeine drinkers who were diagnosed as caffeine dependent, 8 displayed dramatic withdrawal symptoms (including impairment in normal functioning) when they consumed a caffeine-free diet in a controlled setting. The National Coffee Association claimed, however, that the study group was too small to draw conclusions. Is the sample large enough to estimate the true proportion of caffeine drinkers who are caffeine dependent to within .05 of the true value with 99% confidence? Explain.

5.72 Preventing production of defective items. It costs more to produce defective items—because they must be scrapped or reworked—than it does to produce nondefective items. This simple fact suggests that manufacturers should ensure the quality of their products by perfecting their production processes rather than through inspection of finished products (Deming, 1986). In order to better understand a particular metal-stamping process, a manufacturer wishes to estimate the mean length of items produced by the process during the past 24 hours.

a. How many parts should be sampled in order to estimate the population mean to within .1 millimeter (mm) with 90% confidence? Previous studies of this machine have indicated that the standard deviation of lengths produced by the stamping operation is about 2 mm.

b. Time permits the use of a sample size no larger than 100. If a 90% confidence interval for μ is constructed using $n = 100$, will it be wider or narrower than would have been obtained using the sample size determined in part **a**? Explain.

c. If management requires that μ be estimated to within .1 mm and that a sample size of no more than 100 be used, what is (approximately) the maximum confidence level that could be attained for a confidence interval that meets management's specifications?

5.6 Finite Population Correction for Simple Random Sampling

The large-sample confidence intervals for a population mean μ and a population proportion p presented in the previous sections are based on a simple random sample selected from the target population. Although we did not state it, the procedure also assumes that the number N of measurements (i.e., sampling units) in the population is large relative to the sample size n.

In some sampling situations, the sample size n may represent 5% or perhaps 10% of the total number N of sampling units in the population. When the sample size is large relative to the number of measurements in the population (see the next box), the standard errors of the estimators of μ and p given in Sections 5.2 and 5.4, respectively, should be multiplied by a **finite population correction factor.**

The form of the finite population correction factor depends on how the population variance σ^2 is defined. In order to simplify the formulas of the standard errors, it is common to define σ^2 as division of the sum of squares of deviations by $N - 1$ rather than by N (analogous to the way we defined the sample variance). If we adopt this convention, the finite population correction factor becomes $\sqrt{(N - n)/N}$. Then the estimated standard errors of $\bar{x}$ (the estimator of μ) and $\hat{p}$ (the estimator of p) are as shown in the box on page 304.*

Rule of Thumb for Finite Population Correction Factor

Use the finite population correction factor (shown in the next box) when $n/N > .05$.

*For most surveys and opinion polls, the finite population correction factor is approximately equal to 1 and, if desired, can be safely ignored. However, if $n/N > .05$, the finite population correction factor should be included in the calculation of the standard error.

Simple Random Sampling with Finite Population of Size N

Estimation of the Population Mean

Estimated standard error:

$$\hat{\sigma}_{\bar{x}} = \frac{s}{\sqrt{n}}\sqrt{\frac{N-n}{N}}$$

Approximate 95% confidence interval: $\bar{x} \pm 2\hat{\sigma}_{\bar{x}}$

Estimation of the Population Proportion

Estimated standard error:

$$\hat{\sigma}_{\hat{p}} = \sqrt{\frac{\hat{p}(1-\hat{p})}{n}}\sqrt{\frac{N-n}{N}}$$

Approximate 95% confidence interval: $\hat{p} \pm 2\,\hat{\sigma}_{\hat{p}}$

Note: The confidence intervals are "approximate" because we are using 2 to approximate the value $z_{.025} = 1.96$.

Example 5.11

Applying the Finite Population Correction Factor in the Manufacture of Sheet Aluminum Foil

Problem A specialty manufacturer wants to purchase remnants of sheet aluminum foil. The foil, all of which is the same thickness, is stored on 1,462 rolls, each containing a varying amount of foil. To obtain an estimate of the total number of square feet of foil on all the rolls, the manufacturer randomly sampled 100 rolls and measured the number of square feet on each roll. The sample mean was 47.4, and the sample standard deviation was 12.4.

a. Find an approximate 95% confidence interval for the mean amount of foil on the 1,462 rolls.

b. Estimate the total number of square feet of foil on all the rolls by multiplying the confidence interval, part **a,** by 1,462. Interpret the result.

Solution

a. Each roll of foil is a sampling unit, and there are $N = 1{,}462$ units in the population, and the sample size is $n = 100$. Because $n/N = 100/1{,}462 = .068$ exceeds .05, we need to apply the finite population correction factor. We have $n = 100$, $\bar{x} = 47.4$, and $s = 12.4$. Substituting these quantities, we obtain the approximate 95% confidence interval:

$$\bar{x} \pm 2\frac{s}{\sqrt{n}}\sqrt{\frac{(N-n)}{N}} = (47.4) \pm 2\frac{12.4}{\sqrt{100}}\sqrt{\frac{(1{,}462 - 100)}{1{,}462}}$$
$$= 47.4 \pm 2.39$$

or, (45.01, 49.79).

b. For finite populations of size N, the sum of all measurements in the population—called a *population total*—is

$$\sum_{i=1}^{N} x_i = N\mu$$

Because the confidence interval, part **a,** estimates μ, an estimate of the population total is obtained by multiplying the endpoints of the interval by N. For we have

$$\text{Lower Limit} = N(45.01) = 1{,}462(45.01) = 65{,}804.6$$
$$\text{Upper Limit} = N(49.79) = 1{,}462(49.79) = 72{,}793.0$$

Consequently, the manufacturer estimates the total amount of foil to be in the interval of 65,805 square feet to 72,793 square feet with 95% confidence.

Look Back If the manufacturer wants to adopt a conservative approach, the bid for the foil will be based on the lower confidence limit, 65,805 square feet of foil.

Now Work Exercise 5.80a

Exercises 5.73–5.85

Learning the Mechanics

5.73 Calculate the percentage of the population sampled and the finite population correction factor for each of the following situations.

a. $n = 1{,}000, N = 2{,}500$
b. $n = 1{,}000, N = 5{,}000$
c. $n = 1{,}000, N = 10{,}000$
d. $n = 1{,}000, N = 100{,}000$

5.74 Suppose the standard deviation of the population is known to be $\sigma = 200$. Calculate the standard error of $\bar{x}$ for each of the situations described in Exercise 5.73.

5.75 Suppose $N = 10{,}000$, $n = 2{,}000$, and $s = 50$.

a. Compute the standard error of $\bar{x}$ using the finite population correction factor.
b. Repeat part **a** assuming $n = 4{,}000$.
c. Repeat part **a** assuming $n = 10{,}000$.
d. Compare parts **a, b,** and **c** and describe what happens to the standard error of $\bar{x}$ as n increased.
e. The answer to part **c** is 0. This indicates that there is no sampling error in this case. Explain.

5.76 Suppose $N = 5{,}000$, $n = 64$, and $s = 24$.

a. Compare the size of the standard error of $\bar{x}$ computed with and without the finite population correction factor.
b. Repeat part **a,** but this time assume $n = 400$.
c. Theoretically, when sampling from a finite population, the finite population correction factor should always be used in computing the standard error of $\bar{x}$. However, when n is small relative to N, the finite population correction factor is close to 1 and can safely be ignored. Explain how parts **a** and **b** illustrate this point.

5.77 Suppose you want to estimate a population proportion, p, and $\hat{p} = .42$, $N = 6{,}000$, and $n = 1{,}600$. Find an approximate 95% confidence interval for p.

5.78 Suppose you want to estimate a population mean, μ, and $\bar{x} = 422$, $s = 14$, $N = 375$, and $n = 40$. Find an approximate 95% confidence interval for μ.

5.79 A random sample of size $n = 30$ was drawn from a population of size $N = 300$. The following measurements were obtained and are saved in the **LM5_79** file.

21	33	19	29	22	38	58	29	52	36	37	30
53	37	29	18	35	42	36	41	35	36	33	38
29	38	39	54	42	42						

a. Estimate μ with an approximate 95% confidence interval.
b. Estimate p, the proportion of measurements in the population that are greater than 30, with an approximate 95% confidence interval.

Applying the Concepts—Basic

5.80 **Magazine subscriber salaries.** Each year, the trade magazine *Quality Progress* publishes a study of subscribers' salaries. One year, the 223 vice presidents sampled had a mean salary of \$116,754 and a standard deviation of \$39,185. Suppose the goal of the study is to estimate the true mean salary of all vice presidents who subscribe to *Quality Progress*.

NW **a.** If 2,193 vice presidents subscribe to *Quality Progress*, estimate the mean with an approximate 95% confidence interval.
b. Interpret the result.

5.81 **NFL player survey.** Researchers at the University of Pennsylvania's Wharton Sports Business Initiative collaborated with the National Football League Players Association (NFLPA) to produce the first NFL Player Survey in 2007. Of the 1,696 active NFL players, 1,355 (almost 80%) responded to the survey. One of the survey questions asked, "Who is the coach—professional, college, or high school—that has been the most influential in your career?" Of the 1,355 respondents, 759 selected an NFL (professional) coach.

a. Construct a 95% confidence interval for the true proportion of active NFL players who select a professional coach as the most influential in their careers.
b. Why is it necessary to use the continuity correction factor in the construction of the interval, part **a**?
c. Give a practical interpretation of the interval, part **a.**

Applying the Concepts—Intermediate

5.82 **Furniture brand familiarity.** A brand name that consumers recognize is a highly valued commodity in any industry. To assess brand familiarity in the furniture industry, NPD (a market research firm) surveyed 1,333 women who head U.S. households that have incomes of \$25,000 or more. The sample was drawn from a database of 25,000 households that match the criteria listed above. Of the 10 furniture brands evaluated, La-Z-Boy was the most recognized brand; 70.8% of the respondents indicated they were "very familiar" with La-Z-Boy (*HFN,* Oct. 11, 1999).

a. Describe the population being investigated by NPD.
b. In constructing a confidence interval to estimate the proportion of households that are very familiar with the La-Z-Boy brand, is it necessary to use the finite population correction factor? Explain
c. What estimate of the standard error of $\hat{p}$ should be used in constructing the confidence interval of part **b**?
d. Construct a 90% confidence interval for the true proportion and interpret it in the context of the problem.

5.83 **Auditing sampling methods.** Since the early 1950s, auditors have relied to a great extent on sampling techniques, rather than 100% audits, to help them test and evaluate the financial records of a client firm. When sampling is used to obtain an estimate of the total dollar value of an account—the account balance—the examination is known as *a substantive test* (*Audit Sampling—AICPA Audit Guide,* 2008). In order to evaluate the reasonableness of a firm's stated total value of its parts inventory, an auditor randomly samples 100 of the total of 500 parts in stock, prices each part, and reports the results shown in the next table. These data are saved in the **AUDPARTS** file.

Part Number	Part Price	Sample Size
002	$ 108	3
101	55	2
832	500	1
077	73	10
688	300	1
910	54	4
839	92	6
121	833	5
271	50	9
399	125	12
761	1,000	2
093	62	8
505	205	7
597	88	11
830	100	19

a. Give a point estimate of the mean value of the parts inventory.
b. Find the estimated standard error of the point estimate of part **a.**
c. Construct an approximate 95% confidence interval for the mean value of the parts inventory.
d. The firm reported a mean parts inventory value of $300. What does your confidence interval of part **c** suggest about the reasonableness of the firm's reported figure? Explain.

5.84 Invoice errors in a billing system. In a study of invoice errors in a company's new billing system, an auditor randomly sampled 35 invoices produced by the new system and recorded actual amount (A), invoice amount (I), and the difference (or error), $x = (A - I)$. The results were $\bar{x} = \$1$ and $s = \$124$. At the time that the sample was drawn, the new system had produced 1,500 invoices. Use this information to find an approximate 95% confidence interval for the true mean error per invoice of the new system. Interpret the result.

Applying the Concepts—Advanced

5.85 Pesticide residue in corn products. The U.S. Environmental Protection Agency (EPA) bans use of the cancer-causing pesticide ethylene dibromide (EDB) as a fumigant for grain- and flour-milling equipment. EDB was once used to protect against infestation by microscopic roundworms called *nematodes*. The EPA sets maximum safe levels for EDB presence in raw grain, flour, cake mixes, cereals, bread, and other grain products on supermarket shelves and in warehouses. Of the 3,000 corn-related products sold in one state, tests indicated that 15 of a random sample of 175 had EDB residues above the safe level. Will more than 7% of the corn-related products in this state have to be removed from shelves and warehouses? Explain.

5.7 Sample Survey Designs*

The confidence interval methodology developed in Sections 5.2–5.5 is based on simple random sampling (Chapter 3). A (simple) random sample is just one of several different sampling designs used in *sample surveys*.

The term *sample survey* is used in conjunction with the sampling of populations (i.e., collections of people, households, businesses, etc.). A consumer preference poll is an example of a sample survey. Samplings conducted to estimate the general level of business inventories or to estimate the proportion of households that watched a particular television program are also examples of sample surveys.

Sample surveys cost time and money, and sometimes they are almost impossible to conduct. For example, suppose we want to obtain an estimate of the proportion of U.S. households that plan to purchase new television sets next year, and we plan to base our estimate on the intentions of a random sample of 3,000 households. What are the problems associated with collecting these data? In order to use a random number generator (Chapter 3) to select the sample, we would need a list of all the households in the United States. Obtaining such a list would be a monumental obstacle. After we obtain a list of households, we need to contact each of the 3,000 selected for the sample. Will all be at home when the surveyor reaches the household? And will all answer the surveyor's question? You can see that collecting a random sample is easier said than done.

The large body of knowledge underlying **survey sampling** or **sample survey design** was developed to help solve some of the problems we have noted. It includes sample survey designs that help reduce the cost and time involved in conducting a sample survey, and it includes the confidence interval procedures associated with those designs. Because survey sampling is a course in itself, we will present only a few of the most widely used sample survey designs and address only a few of the problems you might encounter in this optional section. Further information on this important subject can be found in the references at the end of the text.

One of the most common sampling designs (besides random sampling) is called *stratified random sampling*. **Stratified random sampling** is used when the sampling units (i.e., the units that are sampled) associated with the population can be physically separated into two or more groups of sampling units (called **strata**) where the within-stratum response variation is less than the variation within the entire population. For example, suppose we want to estimate the average amount of rent paid for a two-bedroom apartment in New York City. Because the variation in rent paid within New York City is likely to be large, we may want to divide the city into regions (strata) where the rents within each stratum are relatively homogeneous. Then we would estimate the population mean by selecting random samples from within each stratum and combining the stratum estimates.

Stratified random sampling often produces estimators with smaller standard errors than those achieved using simple random sampling. Furthermore, by sampling from each stratum, we are more likely to obtain a sample representative of the entire population. In addition, the administrative and labor costs of selecting the stratum samples are often less than those for simple random sampling.

Sometimes it is difficult or too costly to select random samples. For example, it would be easier to obtain a sample of student opinions at a large university by systematically selecting every hundredth name from the student directory. This type of sample design is called a **systematic sample.** Although systematic samples are usually easier to select than other types of samples, one difficulty is the possibility of a systematic sampling bias. For example, if every fifth item in an assembly line is selected for quality-control inspection, and if five different machines are sequentially producing the items, all the items sampled may have been manufactured by the same machine. If we use systematic sampling, we must be certain that no cycles (like every fifth item manufactured by the same machine) exist in the list of the sampling units.

A third alternative to the simple random sampling design is *randomized response sampling*. **Randomized response sampling** is particularly useful when the questions of the pollsters are likely to elicit false answers. For example, suppose each person in a sample of wage earners is asked whether he or she cheated on an income tax return. A person who has not cheated most likely would give an honest answer to this question. A cheater might lie, thus biasing an estimate of what proportion of persons cheats on their income tax returns.

One method of coping with the false responses produced by sensitive questions is randomized response sampling. Each person is presented *two* questions; one question is the object of the survey, and the other is an innocuous question to which the interviewee will give an honest answer. For example, each person might be asked these two questions:

1. Did you cheat on your income tax return?

2. Did you drink coffee this morning?

Then a procedure is used to select randomly which of the two questions the person is to answer. For example, the interviewee might be asked to flip a coin. If the coin shows a head, the interviewee answers the sensitive question, 1. If the coin shows a tail, the interviewee answers the innocuous question, 2. Because the interviewer never has the opportunity to see the coin, the interviewee can answer the question and feel assured that his or her guilt (if guilty) will not be exposed. Consequently, the random response procedure can elicit an honest response to a sensitive question. Sophisticated methodology is then used to derive an estimate of the percentage of "yes" responses to the sensitive question.

As mentioned earlier, in any sample survey design, cost (either in time, labor, or money) may be an issue. Two methods for reducing the cost of random sampling are to use a telephone survey or a mailed survey. Although this type of sampling eliminates transportation costs and reduces labor costs, it introduces a serious difficulty—the problem of **nonresponse.** By this, we mean that sampling units contained in a sample do not produce sample observations. For example, an individual may not be at home when telephoned or may refuse to complete and mail back a questionnaire.

Nonresponse is a serious problem because it may lead to very biased results. There may be a high association between the type of response and whether or not a person responds. For example, most citizens in a community might have an opinion on a school bond issue, but the respondents in a mail survey might very well be those with vested interests in the outcome of the survey—say, parents with children of school age, school teachers, or those whose taxes might be substantially affected. Others with no vested interests might have opinions on the issue but might not take the time to respond. For this example, the absence of the nonrespondents' data could lead to a larger estimate of the percentage in favor of the issue than was actually the case. In other words, the absence of the nonrespondents' data could lead to a biased estimate.

Nonresponse is a very important sampling problem. If your sampling plan calls for a specific collection of sampling units, failure to acquire the responses from those units may violate your sampling plan and lead to biased estimates. If you intend to select a random sample and you cannot obtain responses from some of the sampling units, then your sampling procedure is *no longer random,* and the methodology based on it (e.g., confidence intervals) and the product of the methodology (e.g., inferences) are suspect.

Ethics IN Statistics

In survey sampling, intentional omission of a group of respondents in order to bias the results toward a particular view or outcome is considered *unethical statistical practice*.

There are ways for coping with nonresponse. Most involve tracking down and questioning all or part of the nonrespondents and using the additional information to adjust for the missing nonrespondent data. For mailed surveys, however, it has been found that the inclusion of a monetary incentive with the questionnaire—even as little as 25¢—will substantially increase the response rate of the survey.

There are many sampling designs available to a sample surveyor; some are variations on simple random sampling and stratified random sampling, and others are completely different. In addition, different types of estimators can be used with these designs. In this brief introduction to survey sampling, our intent was to present only a few of the most important sample survey designs and some of their inherent problems. Thorough presentations of the different sample survey designs are given in textbooks devoted to this topic (see the references on page 314).

CHAPTER NOTES

Key Terms

Adjusted $(1 - \alpha)100\%$ confidence interval for a population parameter, p 293
Confidence coefficient 274
Confidence interval 271
Confidence level 274
Degrees of freedom 281
Finite population correction factor 303
Interval estimator 271
Large-sample confidence interval for μ 291
Nonresponse 307
Point estimator 271
Randomized response sampling 307
Sample survey design 306
Sampling error 298
Small-sample confidence interval for μ 284
Sample size determination for $100(1 - \alpha)\%$ confidence interval for μ 298
Sample size determination for $100(1 - \alpha)\%$ confidence interval for p 300
Strata 307
Stratified random sampling 306
Survey sampling 306
Systematic sampling 307
Target parameter 270
t-statistic 281

Key Symbols

μ	Population mean
σ	Population standard deviation
p	Population proportion; P(Success) in binomial trial
q	$1 - p$
$\bar{x}$	Sample mean (estimator of μ)
$\hat{p}$	Sample proportion (estimator of p)
$\mu_{\bar{x}}$	Mean of the population sampling distribution of $\bar{x}$
$\sigma_{\bar{x}}$	Standard deviation of the sampling distribution of $\bar{x}$
$\sigma_{\hat{p}}$	Standard deviation of the sampling distribution of $\hat{p}$
SE	Sampling error in estimation
α	$(1 - \alpha)$ represents the confidence coefficient
$z_{\alpha/2}$	z-value used in a $100(1 - \alpha)\%$ large-sample confidence interval
$t_{\alpha/2}$	Student's t-value used in a $100(1 - \alpha)\%$ small-sample confidence interval
N	Number of observations in the target population

Key Ideas

Population Parameters, Estimators, and Standard Errors

Parameter (θ)	Estimator ($\hat{\theta}$)	Standard Error of Estimator ($\sigma_{\hat{\theta}}$)	Estimated Std. Error ($\hat{\sigma}_{\hat{\theta}}$)
Mean, μ	$\bar{x}$	$\sigma/\sqrt{n}$	$s/\sqrt{n}$
Proportion, p	$\hat{p}$	$\sqrt{pq/n}$	$\sqrt{\hat{p}\hat{q}/n}$

Confidence Interval: An interval that encloses an unknown population parameter with a certain level of confidence $(1 - \alpha)$

Confidence Coefficient: The probability $(1 - \alpha)$ that a randomly selected confidence interval encloses the true value of the population parameter.

Key Words for Identifying the Target Parameter

μ—Mean, Average
p—Proportion, Fraction, Percentage, Rate, Probability

Sample Survey Designs

1. *Simple random sampling*
2. *Stratified random sampling*
3. *Systematic sampling*
4. *Random response sampling*

Commonly Used z-Values for a Large-Sample Confidence Interval

90% CI: $(1 - \alpha) = .10$ $z_{.05} = 1.645$
95% CI: $(1 - \alpha) = .05$ $z_{.025} = 1.96$
99% CI: $(1 - \alpha) = .01$ $z_{.005} = 2.575$

Determining the Sample Size n

Estimating μ: $n = (z_{\alpha/2})^2(\sigma^2)/(\text{SE})^2$
Estimating p: $n = (z_{\alpha/2})^2(pq)/(\text{SE})^2$

Finite Population Correction Factor

Required when $n/N > .05$

Illustrating the Notion of "95% Confidence"

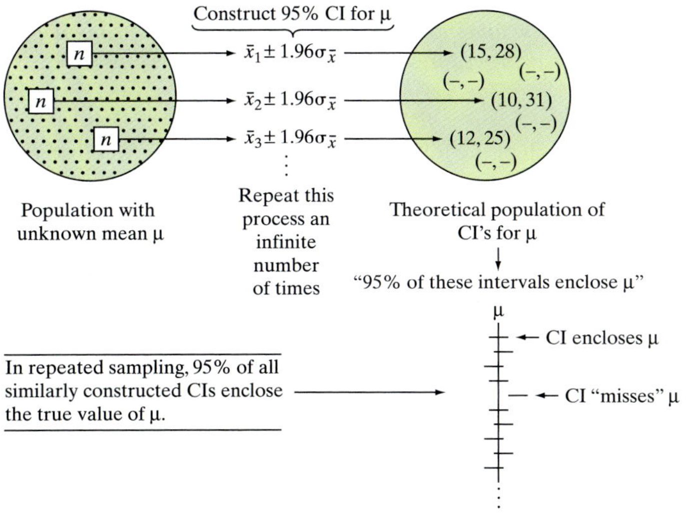

Guide to Forming a Confidence Interval

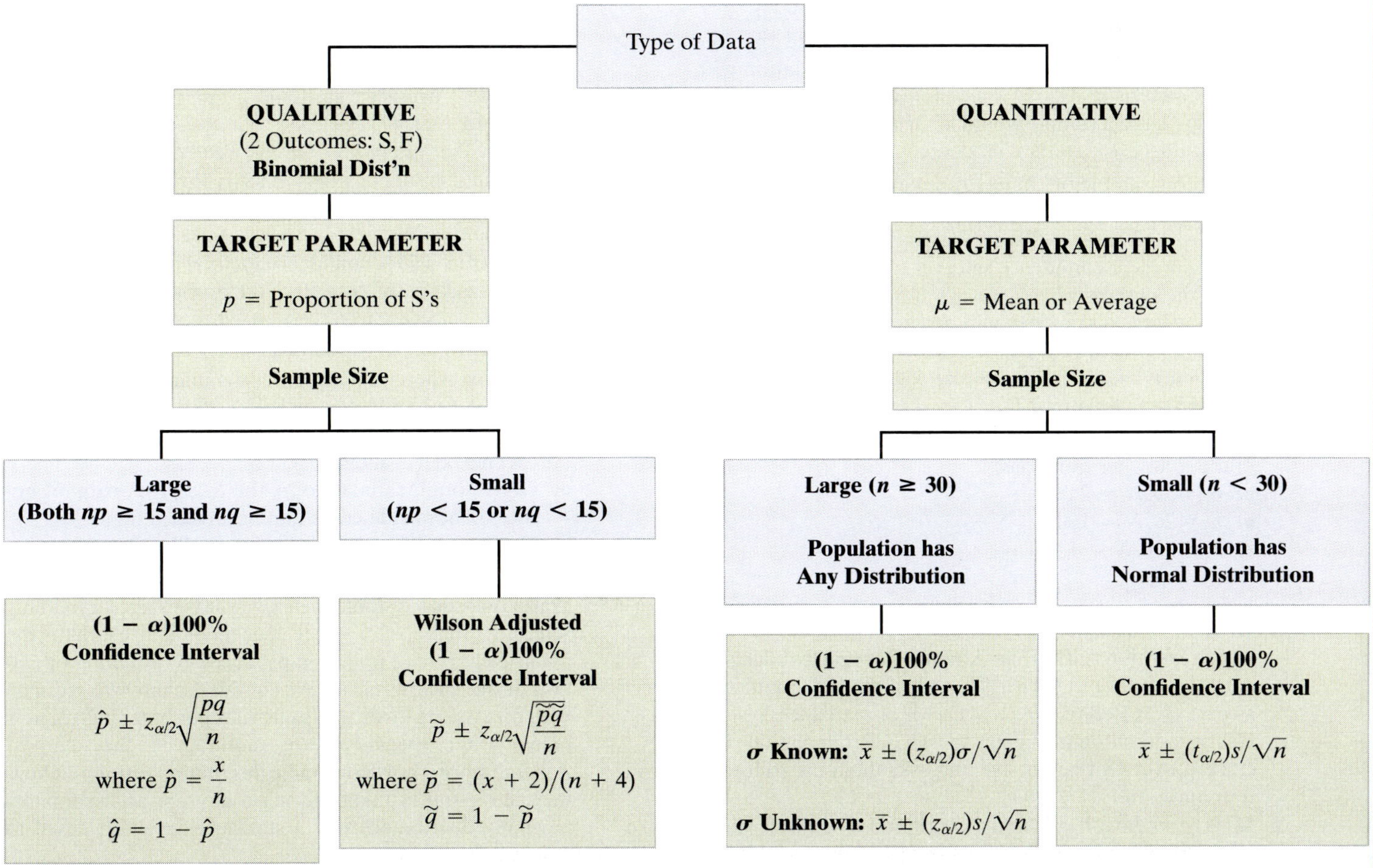

Supplementary Exercises 5.86–5.115

Note: List the assumptions necessary for the valid implementation of the statistical procedures you use in solving all these exercises.

Learning the Mechanics

5.86 In each of the following instances, determine whether you would use a z- or t-statistic (or neither) to form a 95% confidence interval and then look up the appropriate z- or t-value.

a. Random sample of size $n = 23$ from a normal distribution with unknown mean μ and standard deviation σ

b. Random sample of size $n = 135$ from a normal distribution with unknown mean μ and standard deviation σ

c. Random sample of size $n = 10$ from a normal distribution with unknown mean μ and standard deviation $\sigma = 5$

d. Random sample of size $n = 73$ from a distribution about which nothing is known

e. Random sample of size $n = 12$ from a distribution about which nothing is known

5.87 Let t_0 represent a particular value of t from Table V in Appendix B. Find the table values such that the following statements are true.

a. $P(t \leq t_0) = .05$ where df $= 20$

b. $P(t \geq t_0) = .005$ where df $= 9$

c. $P(t \leq -t_0 \text{ or } t \geq t_0) = .10$ where df $= 8$

d. $P(t \leq -t_0 \text{ or } t \geq t_0) = .01$ where df $= 17$

5.88 In a random sample of 400 measurements, 227 of the measurements possess the characteristic of interest, A.

a. Use a 95% confidence interval to estimate the true proportion p of measurements in the population with characteristic A.

b. How large a sample would be needed to estimate p to within .02 with 95% confidence?

5.89 A random sample of 225 measurements is selected from a population, and the sample mean and standard deviation are $\bar{x} = 32.5$ and $s = 30.0$, respectively.

a. Use a 99% confidence interval to estimate the mean of the population, μ.

b. How large a sample would be needed to estimate μ to within .5 with 99% confidence?

c. What is meant by the phrase *99% confidence* as it is used in this exercise?

5.90 Calculate the finite population correction factor for each of the following situations:

a. $n = 50, N = 2{,}000$

b. $n = 20, N = 100$

c. $n = 300, N = 1{,}500$

Applying the Concepts—Basic

5.91 General health survey. The Centers for Disease Control and Prevention (CDCP) in Atlanta, Georgia, conducts an annual survey of the general health of the U.S. population as part of its Behavioral Risk Factor Surveillance System. Using random-digit dialing, the CDCP telephones U.S. citizens over 18 years of age and asks them the following four questions:

1. Is your health generally excellent, very good, good, fair, or poor?
2. How many days during the previous 30 days was your physical health not good because of injury or illness?
3. How many days during the previous 30 days was your mental health not good because of stress, depression, or emotional problems?
4. How many days during the previous 30 days did your physical or mental health prevent you from performing your usual activities?

Identify the parameter of interest for each question.

5.92 Semester hours taken by CPA candidates. Refer to the *Journal of Accounting and Public Policy* (Spring 2002) study of 100,000 first-time candidates for the CPA exam, Exercise 2.51 (p. 63). Recall that the mean number of semester hours of college credit taken by the candidates was 141.31 hours. The standard deviation was reported to be 17.77 hours.

a. Compute a 99% confidence interval for the mean number of semester hours taken by all first-time candidates for the CPA exam.

b. Give a practical interpretation of the interval, part **a.**

c. For the interpretation, part **b,** to be valid, what conditions must hold?

5.93 Security of information submitted over the Internet. As Internet usage proliferates, so do questions of security and confidentiality of personal information, including such things as social security and credit card numbers. NCR Corporation surveyed 1,000 U.S. adults and asked them under what circumstances they would give personal information to a company. Twenty-nine percent said they would never give personal data to a company, while 51% said they would if the company had strict privacy guidelines in place (*Precision Marketing,* Oct. 4, 1999).

a. Verify that the sample size is large enough to construct a valid confidence interval for p, the proportion of all U.S. adults who would never give personal information to a company.

b. Construct a 95% confidence interval for p and interpret your result in the context of the problem.

5.94 Homicide on the job. In a study of homicide on the job, University of North Carolina researchers collected data on workplaces where an employee was murdered (*American Journal of Epidemiology,* Vol. 154, 2001). In a sample of 105 cases, 67 of the homicides occurred during night hours (between 9:00 P.M. and 6:00 A.M.).

a. Give a point estimate of p, the true proportion of on-the-job homicide cases that occurred at night.

b. Compute a 95% confidence interval for p.

c. Give a practical interpretation of the interval, part **b.**

5.95 Water pollution testing. The EPA wants to test a randomly selected sample of n water specimens and estimate the mean daily rate of pollution produced by a mining operation. If the EPA wants a 95% confidence interval estimate with a sampling error of 1 milligram per liter (mg/L), how many water specimens are required in the sample? Assume prior knowledge indicates that pollution readings in water samples taken during a day are approximately normally distributed with a standard deviation equal to 5 mg/L.

5.96 Lead and copper in drinking water. Periodically, the Hillsborough County (Florida) Water Department tests the drinking water of homeowners for contaminants such as lead and copper. The lead and copper levels in water specimens collected for a sample of 10 residents of the Crystal Lakes Manors subdivision (saved in the **LEADCOPP** file) are shown below, followed by a Minitab printout analyzing the data at the bottom of this page.

Lead (μg/L)	Copper (mg/L)
1.32	.508
0	.279
13.1	.320
.919	.904
.657	.221
3.0	.283
1.32	.475
4.09	.130
4.45	.220
0	.743

Source: Hillsborough County Water Department Environmental Laboratory, Tampa, Florida.

a. Locate a 90% confidence interval for the mean lead level in water specimens from Crystal Lakes Manors on the printout.

b. Locate a 90% confidence interval for the mean copper level in water specimens from Crystal Lakes Manors on the printout.

c. Interpret the intervals, parts **a** and **b,** in the words of the problem.

d. Discuss the meaning of the phrase, "90% confident."

5.97 Employees with substance abuse problems. According to the New Jersey *Governor's Council for a Drug-Free Workplace Report,* 50 of the 72 sampled businesses that are members of the council admitted that they had employees with substance abuse problems. At the time of the survey, 251 New Jersey businesses were members of the Governor's Council. Use the finite population correction factor to find a 95% confidence interval for the proportion of all New Jersey Governor's Council business members who have employees with substance abuse problems. Interpret the resulting interval.

5.98 Sick leave taken by employees. A company is interested in estimating μ, the mean number of days of sick leave taken by all its employees. The firm's statistician selects at random 100 personnel files and notes the number of sick days taken by each employee. The following sample statistics are computed: $\bar{x} = 12.2$ days, $s = 10$ days.

a. Estimate μ using a 90% confidence interval. Interpret the result.

b. How many personnel files would the statistician have to select in order to estimate μ to within 2 days with a 99% confidence interval?

5.99 Fish contaminated by a plant's discharge. Refer (Ex 1.5 p. 13)to the U.S. Army Corps of Engineers data on a sample of 144 contaminated fish collected from the river adjacent to a chemical plant. The data are saved in the **DDT** file. Estimate the proportion of contaminated fish that are of the channel catfish species. Use a 90% confidence interval and interpret the result.

Applying the Concepts—Intermediate

5.100 Improving the productivity of chickens. Farmers have discovered that the more domestic chickens peck at objects placed in their environment, the healthier and more productive the chickens seem to be. White string has been found to be a particularly attractive pecking stimulus. In one experiment, 72 chickens were exposed to a string stimulus. Instead of white string, blue-colored string was used. The number of pecks each chicken took at the blue string over a specified time interval was recorded. Summary statistics for the 72 chickens were $\bar{x} = 1.13$ pecks, $s = 2.21$ pecks (*Applied Animal Behaviour Science,* Oct. 2000).

a. Estimate the population mean number of pecks made by chickens pecking at blue string using a 99% confidence interval. Interpret the result.

b. Previous research has shown that $\mu = 7.5$ pecks if chickens are exposed to white string. Based on the results, part **a,** is there evidence that chickens are more apt to peck at white string than blue string? Explain.

5.101 Petroleum waste contamination. Accidental spillage and misguided disposal of petroleum wastes have resulted in extensive contamination of soils across the country. A common hazardous compound found in the contaminated soil is benzo(a)pyrene [B(a)p]. An experiment was conducted to determine the effectiveness of a method designed to remove B(a)p from soil (*Journal of Hazardous Materials,* June 1995). Three soil specimens contaminated with a known amount of B(a)p were treated with a toxin that inhibits microbial growth. After 95 days of incubation, the percentage of B(a)p removed from each soil specimen was measured. The experiment produced the following summary statistics: $\bar{x} = 49.3$ and $s = 1.5$.

a. Use a 99% confidence interval to estimate the mean percentage of B(a)p removed from a soil specimen in which the toxin was used.

b. Interpret the interval in terms of this application.

c. What assumption is necessary to ensure the validity of this confidence interval?

d. Comment on whether the true mean percent removed could be as high as 50%.

5.102 IRS answers to gift tax questions. According to estimates made by the General Accounting Office, the Internal Revenue Service (IRS) answered 18.3 million telephone inquiries during a recent tax season, and 17% of the IRS

Minitab output for Exercise 5.96

```
Variable   N    Mean   StDev  SE Mean        90% CI
LEAD      10    2.89    3.92     1.24  ( 0.61,     5.16)
COPPER    10  0.4083  0.2495   0.0789  (0.2637, 0.5529)
```

offices provided wrong answers. These estimates were based on data collected from sample calls to numerous IRS offices. How many IRS offices should be randomly selected and contacted in order to estimate the proportion of IRS offices that fail to correctly answer questions about gift taxes with a 90% confidence interval of width .06?

5.103 IQ comparison of older vs. younger workers. The 1967 Age Discrimination in Employment Act (ADEA) made it illegal to discriminate against workers 40 years of age and older. Opponents of the law argue that there are sound economic reasons why employers would not want to hire and train workers who are very close to retirement. They also argue that people's abilities tend to deteriorate with age. In fact, *Forbes* (Dec. 13, 1999) reported that 25-year-olds did significantly better than 60-year-olds on the Wechsler Adult Intelligence Scale, the most popular IQ test. The following data (saved in the **IQ25** and **IQ60** files) are raw test scores (i.e., not the familiar normalized IQ scores) for a sample of thirty-six 25-year-olds and thirty-six 60-year-olds:

a. Estimate the mean raw test score for all 25-year-olds using a 99% confidence interval. Give a practical interpretation of the confidence interval.

b. What assumption(s) must hold for the method of estimation used in part **a** to be appropriate?

c. Find a 95% confidence interval for the mean raw score of all 60-year-olds and interpret your result.

25-Year-Olds					
54	61	80	92	41	63
59	68	66	76	82	80
82	47	81	77	88	94
49	86	55	82	45	51
70	72	63	50	52	67
75	60	58	49	63	68

60-Year-Olds					
42	54	38	22	58	37
60	49	51	60	45	42
73	28	65	65	60	34
34	33	40	28	36	60
45	61	47	30	45	45
45	37	27	40	37	58

Source: Adapted from "The case for age discrimination," *Forbes*, Dec. 13, 1999, p. 13. Reprinted by permission of Forbes.com. © 2009 Forbes LLC.

5.104 Air bags pose danger for children. By law, all new cars must be equipped with both driver-side and passenger-side safety air bags. There is concern, however, over whether air bags pose a danger for children sitting on the passenger side. In a National Highway Traffic Safety Administration (NHTSA) study of 55 people killed by the explosive force of air bags, 35 were children seated on the front-passenger side (www.nhtsa.org). This study led some car owners with children to disconnect the passenger-side air bag. Consider all fatal automobile accidents in which it is determined that air bags were the cause of death. Let p represent the true proportion of these accidents involving children seated on the front-passenger side.

a. Use the data from the NHTSA study to estimate p.

b. Construct a 99% confidence interval for p.

c. Interpret the interval, part **b,** in the words of the problem.

d. NHTSA investigators determined that 24 of 35 children killed by the air bags were not wearing seat belts or were improperly restrained. How does this information impact your assessment of the risk of an air bag fatality?

5.105 Air bags pose danger for children (cont'd). Refer to Exercise 5.104. How many fatal accidents should the NHTSA sample in order to estimate the proportion to within .1 of its true value using a 99% confidence interval?

5.106 Laying off cancer patients. According to the U.S. Bureau of Labor Statistics, 1 of every 80 American workers (i.e., 1.3%) is fired or laid off. Are employees with cancer fired or laid off at the same rate? To answer this question, *Working Women* magazine and Amgen—a company that makes drugs to lessen chemotherapy side effects—conducted a telephone survey of 100 cancer survivors who worked while undergoing treatment. Of these 100 cancer patients, 7 were fired or laid off due to their illness.

a. Construct a 90% confidence interval for the true percentage of all cancer patients who are fired or laid off due to their illness.

b. Give a practical interpretation of the interval, part **a.**

c. Are employees with cancer fired or laid off at the same rate as all U.S. workers? Explain.

5.107 Vacation times at major companies. The primary determinant of the amount of vacation time U.S. employees receive is their length of service. According to data released by Hewitt Associates (*Management Review,* Nov. 1995), more than 8 of 10 employers provide 2 weeks of vacation after the first year. After 5 years, 75% of employers provide 3 weeks, and after 15 years most provide 4-week vacations. To more accurately estimate p, the proportion of U.S. employers who provide only 2 weeks of vacation to new hires, a random sample of 24 major U.S. companies was contacted. The following vacation times reported (in days) are saved in the **VACTIMES** file.

10	12	10	10	10	10
15	10	10	10	10	10
10	10	10	10	10	15
10	10	15	10	10	10

a. Construct a 95% confidence interval for p.

b. Is the sample size large enough to ensure that the normal distribution provides a reasonable approximation to the sampling distribution of $\hat{p}$? Justify your answer.

c. How large a sample would be required to estimate p to within .02 with 95% confidence?

5.108 Performance appraisal process. The relationship between an employee's participation in the performance appraisal process and subsequent subordinate reactions toward the appraisal was investigated in the *Journal of Applied Psychology* (Aug. 1998). In Chapter 10 we will discuss a quantitative measure of the relationship between two variables, called the *coefficient* of *correlation* r. The researchers obtained r for a sample of 34 studies that examined the relationship between appraisal participation and a subordinate's satisfaction with the appraisal. These correlations are listed in the table and saved in the **CORR34** file. (Values of r near $+1$ reflect a strong positive relationship between the variables.) Find a 95% confidence interval for the mean of the data and interpret it in the words of the problem.

.50	.58	.71	.46	.63	.66	.31	.35	.51	.06	.35	.19
.40	.63	.43	.16	−.08	.51	.59	.43	.30	.69	.25	.20
.39	.20	.51	.68	.74	.65	.34	.45	.31	.27		

Source: Cawley, B. D., Keeping, L. M., and Levy, P. E. "Participation in the performance appraisal process and employee reactions: A meta-analytic review of field investigations," *Journal of Applied Psychology,* Vol. 83, No. 4, Aug. 1998, pp. 632–633 (Appendix). Copyright © 1998 by the American Psychological Association. Reproduced with permission. The use of APA information does not imply endorsement by APA.

5.109 Interpreting sampling error. When a poll reports, for example, that 61% of the public supports a program of national health insurance, it usually also reports the sampling error. For example, a poll might report that the estimate is accurate to within plus or minus 3%. A classic essay in *Time* magazine ("How not to read polls," Apr. 28, 1980) points out the following:

> *Readers consistently misinterpret the meaning of this "warning label." ... [The sampling error warning] says nothing about errors that might be caused by a sloppily worded question or a biased one or a single question that evokes complex feelings. Example: "Are you satisfied with your job?" Most important of all, warning labels about sampling error say nothing about whether or not the public is conflict-ridden or has given a subject much thought. This is the most serious source of opinion poll misinterpretation.*

Carefully explain the difference between sampling error and nonsampling error, both in general and in the context of the above quotation.

5.110 Salmonella poisoning from eating an ice cream bar. Recently, a case of salmonella (bacterial) poisoning was traced to a particular brand of ice cream bar, and the manufacturer removed the bars from the market. Despite this response, many consumers refused to purchase *any* brand of ice cream bars for some period of time after the event (McClave, personal consulting). One manufacturer conducted a survey of consumers 6 months after the outbreak. A sample of 244 ice cream bar consumers was contacted, and 23 respondents indicated that they would not purchase ice cream bars because of the potential for food poisoning.

a. What is the point estimate of the true fraction of the entire market who refuse to purchase bars 6 months after the outbreak?

b. Is the sample size large enough to use the normal approximation for the sampling distribution of the estimator of the binomial probability? Justify your response.

c. Construct a 95% confidence interval for the true proportion of the market who still refuse to purchase ice cream bars 6 months after the event.

d. Interpret both the point estimate and confidence interval in terms of this application.

5.111 Salmonella poisoning from eating an ice cream bar (cont'd). Refer to Exercise 5.110. Suppose it is now 1 year after the outbreak of food poisoning was traced to ice cream bars. The manufacturer wishes to estimate the proportion who still will not purchase bars to within .02 using a 95% confidence interval. How many consumers should be sampled?

Applying the Concepts—Advanced

5.112 Accountants salary survey. Each year, *Management Accounting* reports the results of a salary survey of the members of the Institute of Management Accountants (IMA). One year, the 2,112 members responding had a salary distribution with a 20th percentile of $35,100; a median of $50,000; and an 80th percentile of $73,000.

a. Use this information to determine the minimum sample size that could be used in next year's survey to estimate the mean salary of IMA members to within $2,000 with 98% confidence. [*Hint:* To estimate *s*, first apply Chebyshev's Theorem to find k such that at least 60% of the data fall within k standard deviations of μ. Then find $s \approx$ (80th percentile – 20th percentile)/$2k$.]

b. Explain how you estimated the standard deviation required for the sample size calculation.

c. List any assumptions you make.

5.113 Internal auditing of invoices. A firm's president, vice presidents, department managers, and others use financial data generated by the firm's accounting system to help them make decisions regarding such things as pricing, budgeting, and plant expansion. To provide reasonable certainty that the system provides reliable data, internal auditors periodically perform various checks of the system (Horngren, Foster, and Datar, *Cost Accounting: A Managerial Emphasis,* 2005). Suppose an internal auditor is interested in determining the proportion of sales invoices in a population of 5,000 sales invoices for which the "total sales" figure is in error. She plans to estimate the true proportion of invoices in error based on a random sample of size 100.

a. Assume that the population of invoices is numbered from 1 to 5,000 and that every invoice ending with a 0 is in error (i.e., 10% are in error). Use a random number generator to draw a random sample of 100 invoices from the population of 5,000 invoices. For example, random number 456 stands for invoice number 456. List the invoice numbers in your sample and indicate which of your sampled invoices are in error (i.e., those ending in a 0).

b. Use the results of your sample of part **a** to construct a 90% confidence interval for the true proportion of invoices in error.

c. Recall that the true population proportion of invoices in error is equal to .1. Compare the true proportion with the estimate of the true proportion you developed in part **b.** Does your confidence interval include the true proportion?

Critical Thinking Challenge

5.114 "Out of control" production process. When companies employ control charts to monitor the quality of their products, a series of small samples is typically used to determine if the process is "in control" during the period of time in which each sample is selected. (We cover quality-control charts in Chapter 12.) Suppose a concrete-block manufacturer samples nine blocks per hour and tests the breaking strength of each. During 1 hour's test, the mean and standard deviation are 985.6 pounds per square inch (psi) and 22.9 psi, respectively. The process is to be considered "out of control" if the true mean strength differs from 1,000 psi. The manufacturer wants to be reasonably certain that the process is really out of control before shutting down the process and trying to determine the problem. What is your recommendation?

5.115 **Scallops, sampling, and the law.** In *Interfaces* (March–April 1995), the case involved a ship that fishes for scallops off the coast of New England. In order to protect baby scallops from being harvested, the U.S. Fisheries and Wildlife Service requires that "the average meat per scallop weigh at least $\frac{1}{36}$ of a pound." The ship was accused of violating this weight standard. Bennett lays out the scenario:

> *The vessel arrived at a Massachusetts port with 11,000 bags of scallops, from which the harbormaster randomly selected 18 bags for weighing. From each such bag, his agents took a large scoopful of scallops; then, to estimate the bag's average meat per scallop, they divided the total weight of meat in the scoopful by the number of scallops it contained. Based on the 18 [numbers] thus generated, the harbormaster estimated that each of the ship's scallops possessed an average of $\frac{1}{39}$ of a pound of meat (that is, they were about seven percent lighter than the minimum requirement). Viewing this outcome as conclusive evidence that the weight standard had been violated, federal authorities at once confiscated 95 percent of the catch (which they then sold at auction). The fishing voyage was thus transformed into a financial catastrophe for its participants.*

The actual scallop weight measurements for each of the 18 sampled bags are listed in the accompanying table and are saved in the **SCALLOPS** file. For ease of exposition, Bennett expressed each number as a multiple of $\frac{1}{36}$ of a pound, the minimum permissible average weight per scallop. Consequently, numbers below 1 indicate individual bags that do not meet the standard.

The ship's owner filed a lawsuit against the federal government, declaring that his vessel had fully complied with the weight standard. A Boston law firm was hired to represent the owner in legal proceedings, and Bennett was retained by the firm to provide statistical litigation support and, if necessary, expert witness testimony.

a. Recall that the harbormaster sampled only 18 of the ship's 11,000 bags of scallops. One of the questions the lawyers asked Bennett was, "Can a reliable estimate of the mean weight of all the scallops be obtained from a sample of size 18?" Give your opinion on this issue.

b. As stated in the article, the government's decision rule is to confiscate a catch if the sample mean weight of the scallops is less than $\frac{1}{36}$ of a pound. Do you see any flaws in this rule?

c. Develop your own procedure for determining whether a ship is in violation of the minimum-weight restriction. Apply your rule to the data. Draw a conclusion about the ship in question.

.93	.88	.85	.91	.91	.84	.90	.98	.88
.89	.98	.87	.91	.92	.99	1.14	1.06	.93

Source: Barnett, A. "Misapplications review: Jail terms," *Interfaces,* Vol. 25, No. 2, Mar.–Apr. 1995, p. 20.

References

Agresti, A., and Coull, B. A. "Approximate is better than 'exact' for interval estimation of binomial proportions" *The American Statistician,* Vol. 52, No. 2, May 1998, pp. 119–126.

Arkin, H. *Sampling Methods for the Auditor*. New York: McGraw-Hill, 1982.

Cochran, W. G. *Sampling Techniques,* 3rd ed. New York: Wiley, 1977.

Freedman, D., Pisani, R., and Purves, R. *Statistics*. New York: Norton, 1978.

Kish, L. *Survey Sampling*. New York: Wiley, 1965.

Mendenhall, W., Beaver, R. J., and Beaver, B. *Introduction to Probability and Statistics,* 13th ed. Belmont, CA: Brooks/Cole, 2009.

Wilson, E. G. "Probable inference, the law of succession, and statistical inference," *Journal of the American Statistical Association,* Vol. 22, 1927, pp. 209–212.

USING TECHNOLOGY

SPSS: Confidence Intervals

SPSS can be used to obtain one-sample confidence intervals for a population mean but cannot currently produce confidence intervals for a population proportion. To generate a confidence interval for the mean perform the following:

Step 1 Access the SPSS spreadsheet file that contains the sample data.

Step 2 Click on the "Analyze" button on the SPSS menu bar and then click on "Descriptive Statistics" and "Explore," as shown in Figure 5.S.1.

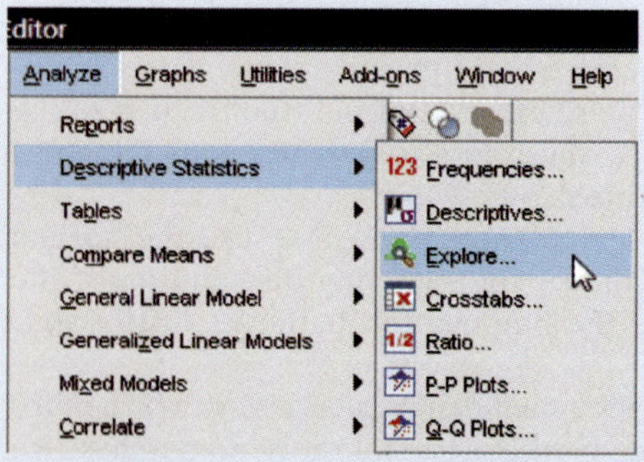

Figure 5.S.1 SPSS menu options—confidence interval for the mean

Step 3 On the resulting dialog box (shown in Figure 5.S.2.), specify the quantitative variable of interest in the "Dependent List" and then click on the "Statistics" button.

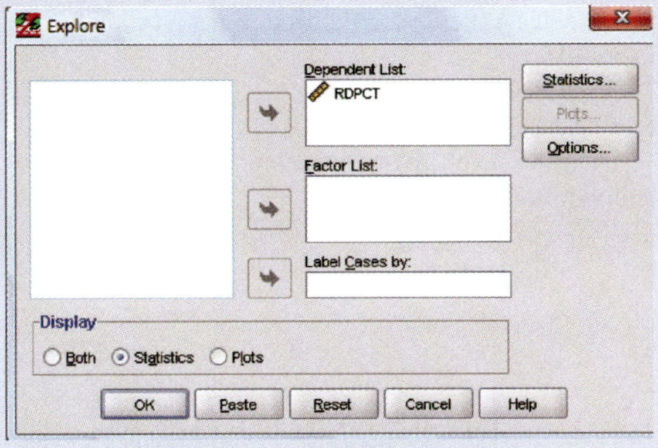

Figure 5.S.2 SPSS explore dialog box

Step 4 Specify the confidence level in the resulting dialog box, as shown in Figure 5.S.3.

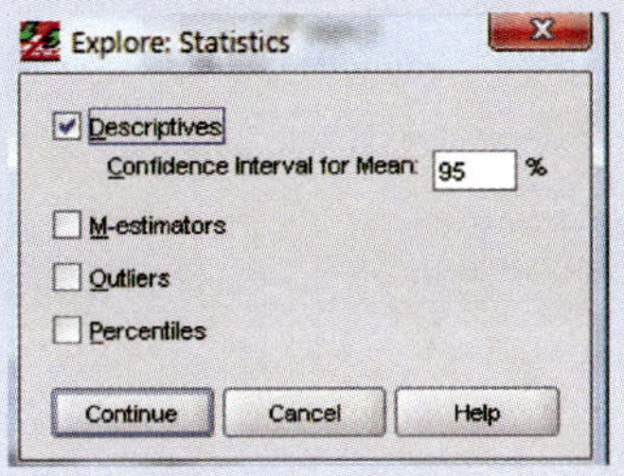

Figure 5.S.3 SPSS explore statistics options

Step 5 Click "Continue" to return to the "Explore" dialog box and then click "OK" to produce the confidence interval.

Minitab: Confidence Intervals

Minitab can be used to obtain one-sample confidence intervals for both a population mean and a population proportion.

Confidence Interval for a Mean

Step 1 Access the Minitab data worksheet that contains the quantitative variable of interest.

Step 2 Click on the "Stat" button on the Minitab menu bar and then click on "Basic Statistics" and "1-Sample t," as shown in Figure 5.M.1.

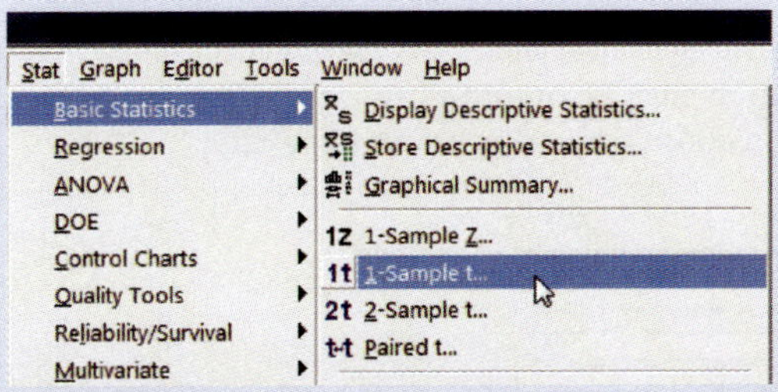

Figure 5.M.1 Minitab menu options—confidence interval for the mean

Step 3 On the resulting dialog box (shown in Figure 5.M.2.), click on "Samples in Columns," and then specify the quantitative variable of interest in the open box.

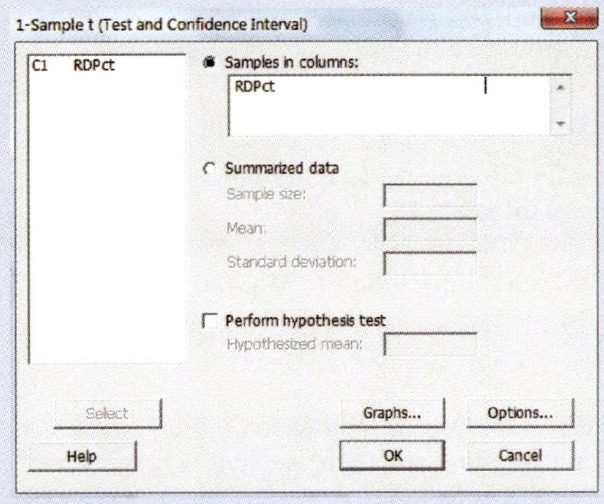

Figure 5.M.2 Minitab 1-sample t dialog box

Step 4 Click on the "Options" button at the bottom of the dialog box and specify the confidence level in the resulting dialog box, as shown in Figure 5.M.3.

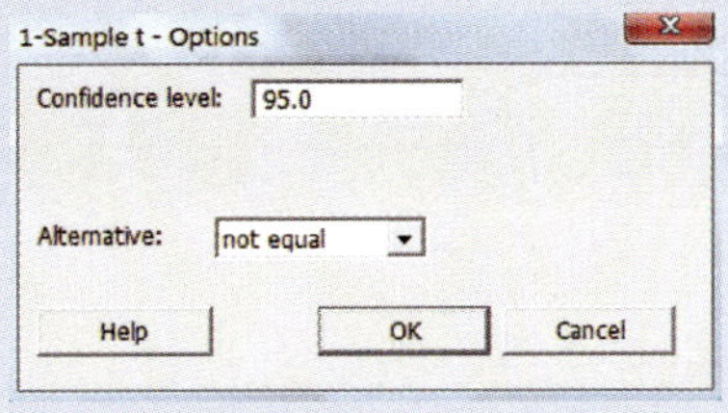

Figure 5.M.3 Minitab 1-sample t options

Step 5 Click "OK" to return to the "1-Sample t" dialog box, and then click "OK" again to produce the confidence interval.

Note: If you want to produce a confidence interval for the mean from summary information (e.g., the sample mean, sample standard deviation, and sample size), click on "Summarized data" in the "1-Sample t" dialog box, as shown in Figure 5.M.4. Enter the values of the summary statistics and then click "OK."

Important: The Minitab 1-sample t procedure uses the t-statistic to generate the confidence interval. When the sample size n is small, this is the appropriate method. When the sample size n is large, the t-value will be approximately equal to the large-sample z-value and the resulting interval will still be valid. If you have a large sample and you know the value of the population standard deviation σ (which is rarely the case),

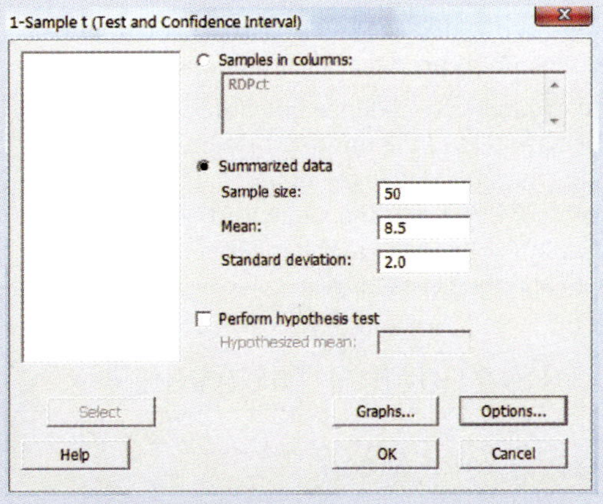

Figure 5.M.4 Minitab 1-sample t dialog box with summary statistics options

select "1-sample Z" from the "Basic Statistics" menu options (see Figure 5.M.1) and make the appropriate selections.

Confidence Interval for a Proportion

Step 1 Access the Minitab data worksheet that contains the qualitative variable of interest.

Step 2 Click on the "Stat" button on the Minitab menu bar and then click on "Basic Statistics" and "1 Proportion" (see Figure 5.M.1).

Step 3 On the resulting dialog box (shown in Figure 5.M.5.), click on "Samples in Columns" and then specify the qualitative variable of interest in the open box.

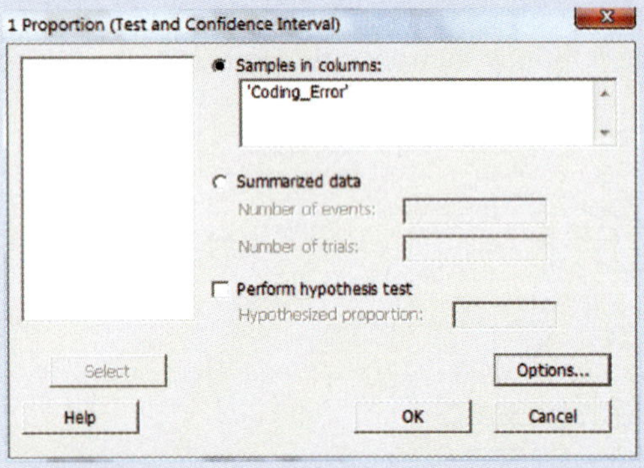

Figure 5.M.5 Minitab 1-proportion dialog box

Step 4 Click on the "Options" button at the bottom of the dialog box and specify the confidence level in the resulting dialog box, as shown in Figure 5.M.6. Also, check the "Use test and interval based on normal distribution" box at the bottom.

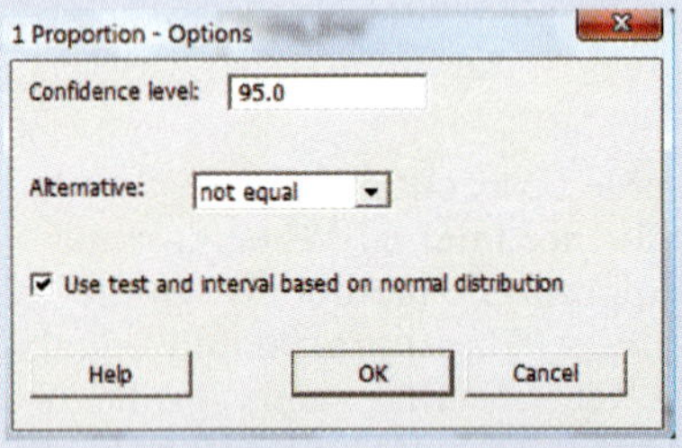

Figure 5.M.6 Minitab 1-proportion dialog box options

Step 5 Click "OK" to return to the "1- Proportion" dialog box and then click "OK" again to produce the confidence interval.

Note: If you want to produce a confidence interval for a proportion from summary information (e.g., the number of successes and the sample size), click on "Summarized data" in the "1-Proportion" dialog box (see Figure 5.M.5). Enter the value for the number of trials (i.e., the sample size) and the number of events (i.e., the number of successes) and then click "OK."

Excel/DDXL: Confidence Intervals

Confidence Interval for a Mean

To obtain a confidence interval for the mean of a quantitative variable, perform the following:

Step 1 Highlight (select) the data column you want to analyze on the Excel spreadsheet.

Step 2 Click on "Add-Ins" in the main Excel menu bar and select "DDXL." On the resulting menu, select "Confidence Intervals," as shown in Figure 5.E.1.

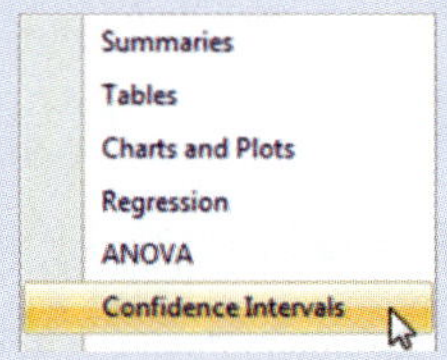

Figure 5.E.1 Excel/DDXL menu options—confidence intervals

Step 3 On the resulting menu, select "1 Var t Interval" in the Function Type box, as shown in Figure 5.E.2.

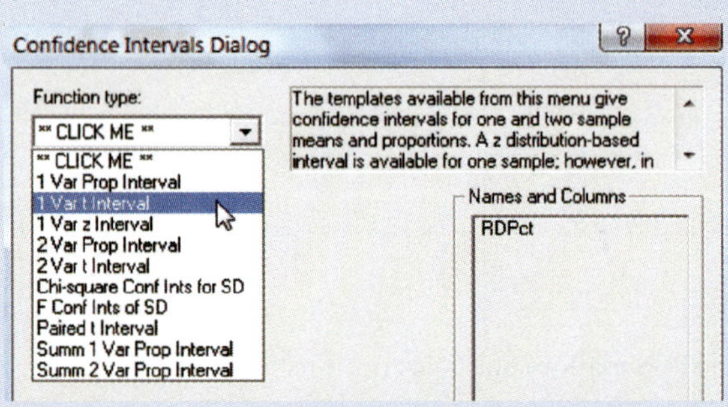

Figure 5.E.2 Excel/DDXL confidence intervals dialog box

Step 4 Move the variable of interest into the "Quantitative Variable" box and then click "OK," as shown in Figure 5.E.3.

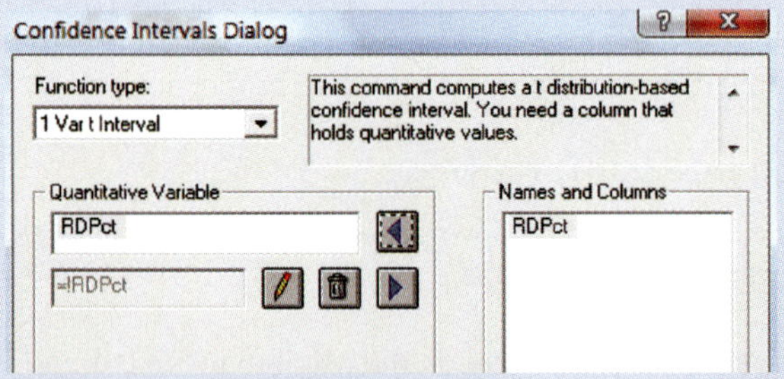

Figure 5.E.3 Excel/DDXL confidence interval for mean options

Step 5 On the resulting menu, click the desired confidence level (e.g., 95%), then click "Compute Interval," as shown in Figure 5.E.4. The confidence interval will appear.

Figure 5.E.4 Excel/DDXL confidence interval for mean setup

Confidence Interval for a Proportion

To obtain a confidence interval for the proportion of successes for a qualitative variable, perform the following:

Step 1 Highlight (select) the data column you want to analyze on the Excel spreadsheet.

Step 2 Click on "Add-Ins" in the main Excel menu bar and select "DDXL." On the resulting menu select "Confidence Intervals," as shown in Figure 5.E.1.

Step 3 On the resulting menu, select "1 Var Prop Interval" in the Function Type box and move the variable of interest into the "Proportions Variable" box, as shown in Figure 5.E.5. Click "OK."

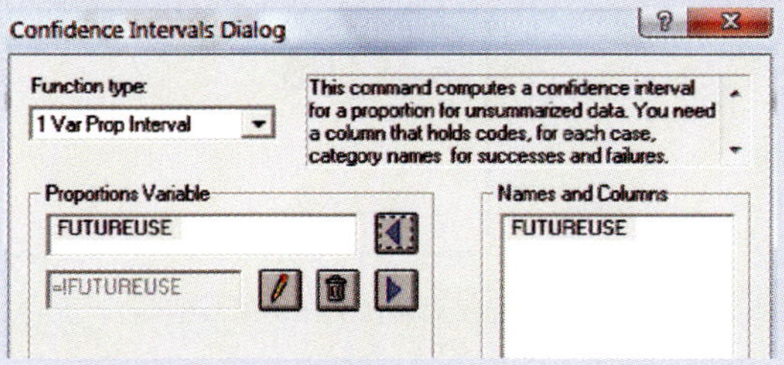

Figure 5.E.5 Excel/DDXL confidence interval for proportion options

Step 4 On the resulting menu, select the value of the qualitative variable that represents a success, click the desired confidence level (e.g., 95%), and click "Compute Interval," as shown in Figure 5.E.6. The confidence interval will appear.

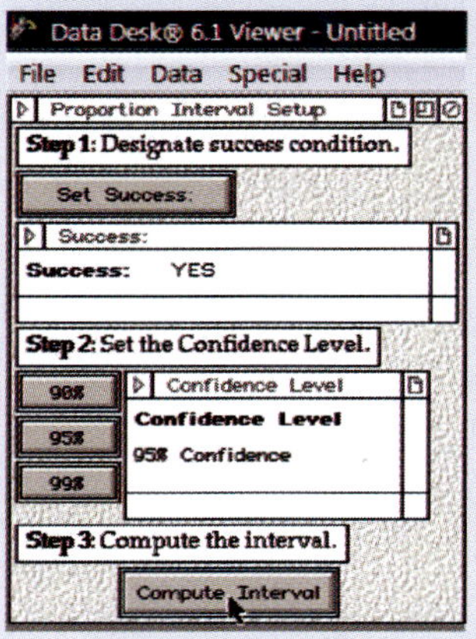

Figure 5.E.6 Excel/DDXL confidence interval for proportion setup

TI-84 Graphing Calculator: Confidence Intervals

Creating a Confidence Interval for a Population Mean (Known σ or $n \geq 30$)

Step 1 *Enter the data (Skip to Step 2 if you have summary statistics, not raw data)*

- Press **STAT** and select **1:Edit**

Note: If the list already contains data, clear the old data. Use the up **ARROW** to highlight "**L1.**"

- Press **CLEAR ENTER**
- Use the **ARROW** and **ENTER** keys to enter the data set into **L1**

Step 2 *Access the Statistical Tests Menu*

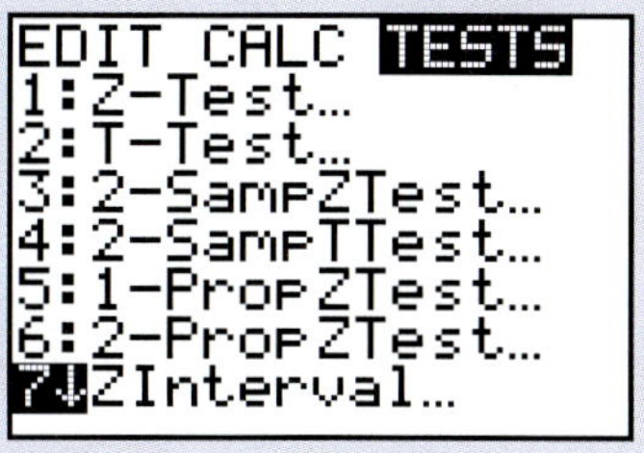

- Press **STAT**
- Arrow right to **TESTS**
- Arrow down to **ZInterval**
- Press **ENTER**

Step 3 *Choose "**Data,**" or "**Stats.**" ("Data" is selected when you have entered the raw data into a List. "Stats" is selected when you are given only the mean, standard deviation, and sample size.)*

- Press **ENTER**

If you selected "Data," enter a value for σ (the best approximation is *s*, the sample standard deviation)

- Set **List** to **L1**
- Set **Freq** to **1**
- Set **C-Level** to the confidence level
- Arrow down to "**Calculate**"
- Press **ENTER**

If you selected "Stats," enter a value for σ (the best approximation is *s*, the sample standard deviation)

- Enter the sample mean and sample size
- Set **C-Level** to the confidence level
- Arrow down to "**Calculate**"
- Press **ENTER**

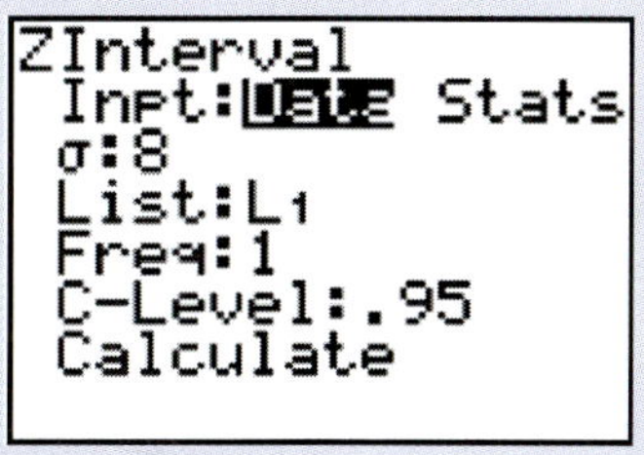

(The screen at right is set up for an example with a standard deviation of 20, a mean of 200, and a sample size of 40.) The confidence interval will be displayed along with the sample mean and the sample size.

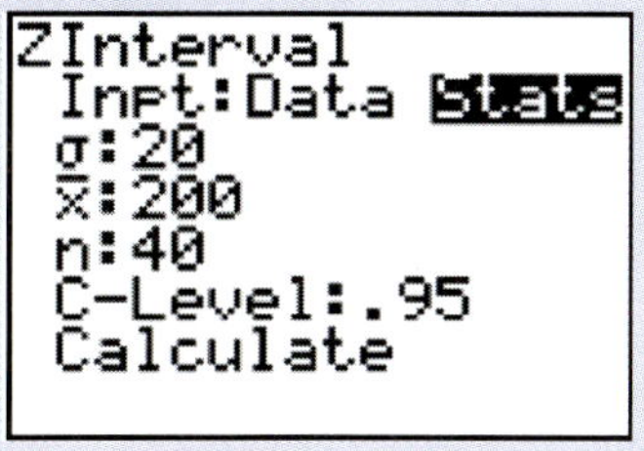

Creating a Confidence Interval for a Population Mean ($n < 30$)

Step 1 *Enter the data (skip to Step 2 if you have summary statistics, not raw data)*

- Press **STAT** and select **1:Edit**

Note: If the list already contains data, clear the old data. Use the up **ARROW** to highlight "**L1.**"

- Press **CLEAR ENTER**
- Use the **ARROW** and **ENTER** keys to enter the data set into **L1**

Step 2 *Access the Statistical Tests Menu*

- Press **STAT**
- Arrow right to **TESTS**
- Arrow down to **TInterval**
- Press **ENTER**

Step 3 *Choose "**Data**" or "**Stats.**" ("Data" is selected when you have entered the raw data into a List. "Stats" is selected* *when you are given only the mean, standard deviation, and sample size.)*

- Press **ENTER**
- If you selected "Data," set **List** to **L1**
- Set **Freq** to **1**
- Set **C-Level** to the confidence level
- Arrow down to "**Calculate**"
- Press **ENTER**

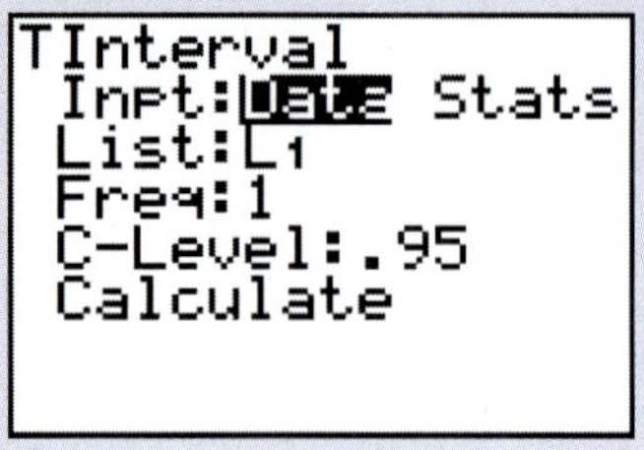

If you selected "Stats," enter the mean, standard deviation, and sample size.

- Set **C-Level** to the confidence level
- Arrow down to "**Calculate**"
- Press **ENTER**

(The screen here is set up for an example with a mean of 100 and a standard deviation of 10.)

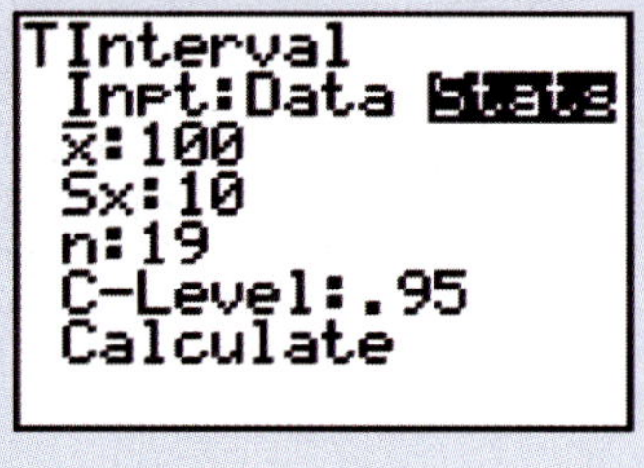

The confidence interval will be displayed with the mean, standard deviation, and sample size.

Creating a Confidence Interval for a Population Proportion (Large Samples)

Step 1 *Access the Statistical Tests Menu*

- Press **STAT**
- Arrow right to **TESTS**
- Arrow down to **1-PropZInt**
- Press **ENTER**

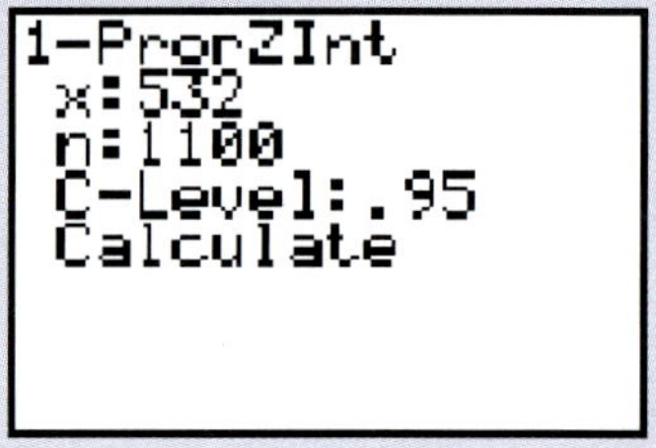

Step 2 *Enter the values for **x, n,** and **C-Level***

- where **x** = number of successes
- **n** = sample size
- **C-Level** = level of confidence
- Arrow down to "**Calculate**"
- Press **ENTER**

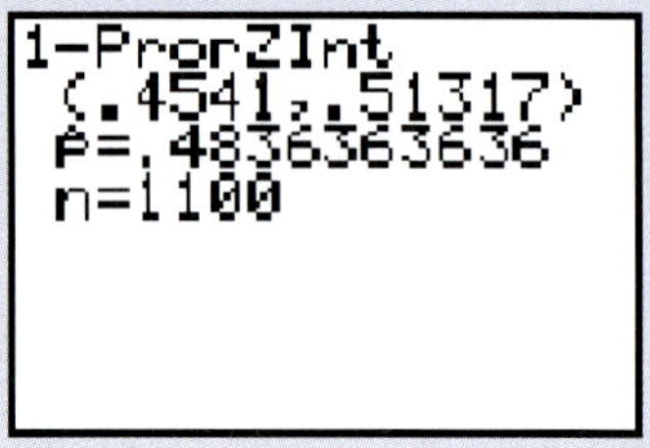

(The screens at the right are set up for an example with $x = 532$, $n = 1{,}100$, and confidence level of .95.)

CONTENTS

6 Inferences Based on a Single Sample *Tests of Hypotheses*

Where We've Been

- Used sample information to provide a *point estimate* of a population parameter
- Used the sampling distribution of a statistic to assess the reliability of an estimate through a *confidence interval*

Where We're Going

- Test a specific value of a population parameter (mean or proportion), called a *test of hypothesis*
- Provide a measure of reliability for the hypothesis test, called the *significance level* of the test

Statistics IN Action Diary of a Kleenex® User

In 1924, Kimberly-Clark Corporation invented a facial tissue for removing cold cream and began marketing it as Kleenex® brand tissues. Today, Kleenex® is recognized as the top-selling brand of tissue in the world. A wide variety of Kleenex® products are available, ranging from extra-large tissues to tissues with lotion. Over the past 80 years, Kimberly-Clark Corporation has packaged the tissues in boxes of different sizes and shapes and varied the number of tissues packaged in each box. For example, currently a family-size box contains 144 two-ply tissues, a cold-care box contains 70 tissues (coated with lotion), and a convenience pocket pack contains 15 miniature tissues.

How does Kimberly-Clark Corp. decide how many tissues to put in each box? According to the *Wall Street Journal,* marketing experts at the company use the results of a survey of Kleenex® customers to help determine how many tissues are packed in a box. In the mid-1980s, when Kimberly-Clark Corp. developed the cold-care box designed especially for people who have a cold, the company conducted their initial survey of customers for this purpose. Hundreds of customers were asked to keep *(continued)*

Statistics IN Action *(continued)*

count of their Kleenex® use in diaries. According to the *Wall Street Journal* report, the survey results left "little doubt that the company should put 60 tissues in each box." The number 60 was "the average number of times people blow their nose during a cold." (*Note:* In 2000, the company increased the number of tissues packaged in a cold-care box to 70.)

From summary information provided in the *Wall Street Journal* article, we constructed a data set that represents the results of a survey similar to the one described above. In the data file named **TISSUES,** we recorded the number of tissues used by each of 250 consumers during a period when they had a cold. We apply the hypothesis testing methodology presented in this chapter to this data set in several Statistics in Action Revisited examples.

Statistics IN Action Revisited

- Key Elements of a Hypothesis Test (p. 329)
- Testing a Population Mean (p. 340)
- Testing a Population Proportion (p. 353)

Data Set: TISSUES

Suppose you wanted to determine whether the mean waiting time in the drive-through line of a fast-food restaurant is less than 5 minutes, or whether the majority of consumers are optimistic about the economy. In both cases you are interested in making an inference about how the value of a parameter relates to a specific numerical value. Is it less than, equal to, or greater than the specified number? This type of inference, called a **test of hypothesis,** is the subject of this chapter.

We introduce the elements of a test of hypothesis in Sections 6.1 and 6.2. We then show how to conduct a large-sample test of hypothesis about a population mean in Sections 6.3 and 6.4. In Section 6.5 we use small samples to conduct tests about means. Large-sample tests about binomial probabilities are the subject of Section 6.6, and some advanced methods for determining the reliability of a test are covered in Section 6.7. Finally, we show how to conduct a test about a population variance in Section 6.8.

6.1 The Elements of a Test of Hypothesis

Suppose building specifications in a certain city require that the average breaking strength of residential sewer pipe be more than 2,400 pounds per foot of length (i.e., per linear foot). Each manufacturer who wants to sell pipe in this city must demonstrate that its product meets the specification. Note that we are interested in making an inference about the mean μ of a population. However, in this example we are less interested in estimating the value of μ than we are in testing a **hypothesis** about its value—that is, *we want to decide whether the mean breaking strength of the pipe exceeds 2,400 pounds per linear foot.*

A statistical **hypothesis** is a statement about the numerical value of a population parameter.

The method used to reach a decision is based on the rare-event concept explained in earlier chapters. We define two hypotheses: (1) The **null hypothesis** is that which represents the status quo to the party performing the sampling experiment—the hypothesis that will be accepted unless the data provide convincing evidence that it is false. (2) The **alternative,** or **research, hypothesis** is that which will be accepted only if the data provide convincing evidence of its truth. From the point of view of the city conducting the tests, the null hypothesis is that the manufacturer's pipe does *not* meet specifications unless the tests provide convincing evidence otherwise. The null and alternative hypotheses are therefore

Null hypothesis (H_0): $\mu \leq 2{,}400$
(i.e., the manufacturer's pipe does not meet specifications)
Alternative (*research*) *hypothesis* (H_a): $\mu > 2{,}400$
(i.e., the manufacturer's pipe meets specifications)

The **null hypothesis,** denoted H_0, represents the hypothesis that will be accepted unless the data provide convincing evidence that it is false. This usually represents the "status quo" or some claim about the population parameter that the researcher wants to test.

The **alternative (research) hypothesis,** denoted H_a, represents the hypothesis that will be accepted only if the data provide convincing evidence of its truth. This usually represents the values of a population parameter for which the researcher wants to gather evidence to support.

How can the city decide when enough evidence exists to conclude that the manufacturer's pipe meets specifications? Because the hypotheses concern the value of the population mean μ, it is reasonable to use the sample mean $\bar{x}$ to make the inference, just as we did when forming confidence intervals for μ in Sections 5.2 and 5.3. The city will conclude that the pipe meets specifications only when the sample mean $\bar{x}$ convincingly indicates that the population mean exceeds 2,400 pounds per linear foot.

"Convincing" evidence in favor of the alternative hypothesis will exist when the value of $\bar{x}$ exceeds 2,400 by an amount that cannot be readily attributed to sampling variability. To decide, we compute a **test statistic,** i.e., a numerical value computed from the sample. Here, the test statistic is the z-value that measures the distance between the value of $\bar{x}$ and the value of μ specified in the null hypothesis. When the null hypothesis contains more than one value of μ, as in this case (H_0: $\mu \leq 2{,}400$), we use the value of μ closest to the values specified in the alternative hypothesis. The idea is that if the hypothesis that μ *equals* 2,400 can be rejected in favor of $\mu > 2{,}400$, then μ *less than or equal to* 2,400 can certainly be rejected. Thus, the test statistic is

$$z = \frac{\bar{x} - 2{,}400}{\sigma_{\bar{x}}} = \frac{\bar{x} - 2{,}400}{\sigma/\sqrt{n}}$$

Note that a value of $z = 1$ means that $\bar{x}$ is 1 standard deviation above $\mu = 2{,}400$; a value of $z = 1.5$ means that $\bar{x}$ is 1.5 standard deviations above $\mu = 2{,}400$, and so on. How large must z be before the city can be convinced that the null hypothesis can be rejected in favor of the alternative and conclude that the pipe meets specifications?

The **test statistic** is a sample statistic, computed from information provided in the sample, that the researcher uses to decide between the null and alternative hypotheses.

If you examine Figure 6.1, you will note that the chance of observing $\bar{x}$ more than 1.645 standard deviations above 2,400 is only .05—*if in fact the true mean μ is 2,400.*

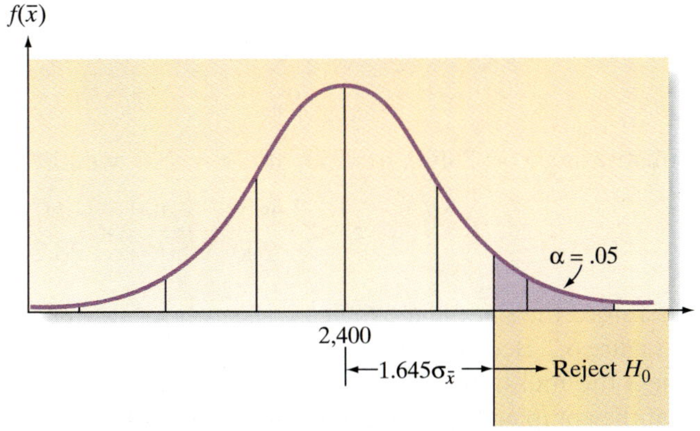

Figure 6.1
The sampling distribution of $\bar{x}$, assuming $\mu = 2{,}400$

Thus, if the sample mean is more than 1.645 standard deviations above 2,400, either H_0 is true and a relatively rare event has occurred (.05 probability), or H_a is true and the population mean exceeds 2,400. Because we would most likely reject the notion that a rare event has occurred, we would reject the null hypothesis ($\mu \leq 2{,}400$) and conclude that the alternative hypothesis ($\mu > 2{,}400$) is true. What is the probability that this procedure will lead us to an incorrect decision?

Such an incorrect decision—deciding that the null hypothesis is false when in fact it is true—is called a **Type I error.** As indicated in Figure 6.1, the risk of making a Type I error is denoted by the symbol α—that is,

$$\alpha = P(\text{Type I error})$$
$$= P(\text{Rejecting the null hypothesis when in fact the null hypothesis is true})$$

A **Type I error** occurs if the researcher rejects the null hypothesis in favor of the alternative hypothesis when, in fact, H_0 is true. The probability of committing a Type I error is denoted by α.

In our example,

$$\alpha = P(z > 1.645 \text{ when in fact } \mu = 2{,}400) = .05$$

We now summarize the elements of the test:

H_0: $\mu \leq 2{,}400$ (Pipe does not meet specifications)

H_a: $\mu > 2{,}400$ (Pipe meets specifications)

Test statistic: $z = \dfrac{\bar{x} - 2{,}400}{\sigma_{\bar{x}}}$

Rejection region: $z > 1.645$, which corresponds to $\alpha = .05$

Note that the **rejection region** refers to the values of the test statistic for which we will *reject the null hypothesis.*

The **rejection region** of a statistical test is the set of possible values of the test statistic for which the researcher will reject H_0 in favor of H_a.

To illustrate the use of the test, suppose we test 50 sections of sewer pipe and find the mean and standard deviation for these 50 measurements to be

$$\bar{x} = 2{,}460 \text{ pounds per linear foot}$$
$$s = 200 \text{ pounds per linear foot}$$

As in the case of estimation, we can use s to approximate σ when s is calculated from a large set of sample measurements.

The test statistic is

$$z = \frac{\bar{x} - 2{,}400}{\sigma_{\bar{x}}} = \frac{\bar{x} - 2{,}400}{\sigma/\sqrt{n}} \approx \frac{\bar{x} - 2{,}400}{s/\sqrt{n}}$$

Substituting $\bar{x} = 2{,}460$, $n = 50$, and $s = 200$, we have

$$z \approx \frac{2{,}460 - 2{,}400}{200/\sqrt{50}} = \frac{60}{28.28} = 2.12$$

Therefore, the sample mean lies $2.12\sigma_{\bar{x}}$ above the hypothesized value of μ, 2,400, as shown in Figure 6.2. Because this value of z exceeds 1.645, it falls in the rejection region. That is, we reject the null hypothesis that $\mu = 2{,}400$ and conclude that $\mu > 2{,}400$. Thus, it appears that the company's pipe has a mean strength that exceeds 2,400 pounds per linear foot.

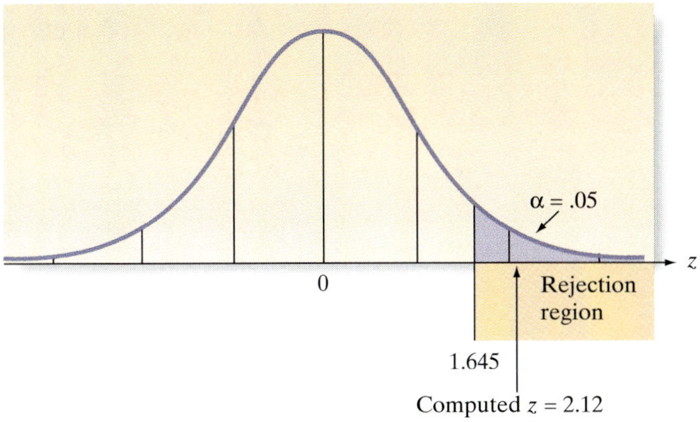

Figure 6.2
Location of the test statistic for a test of the hypothesis H_0: $\mu = 2{,}400$

BIOGRAPHY

EGON S. PEARSON (1895–1980)

The Neyman-Pearson Lemma

Egon Pearson was the only son of noteworthy British statistician Karl Pearson (see Biography, p. 524). As you might expect, Egon developed an interest in the statistical methods developed by his father and, upon completing graduate school, accepted a position to work for Karl in the Department of Applied Statistics at University College, London. Egon is best known for his collaboration with Jerzey Neyman (see Biography, p. 275) on the development of the theory of hypothesis testing. One of the basic concepts in the Neyman-Pearson approach was that of the "null" and "alternative" hypotheses. Their famous Neyman-Pearson lemma was published in *Biometrika* in 1928. Egon Pearson had numerous other contributions to statistics and was known as an excellent teacher and lecturer. In his last major work, Egon fulfilled a promise made to his father by publishing an annotated version of Karl Pearson's lectures on the early history of statistics. ■

How much faith can be placed in this conclusion? What is the probability that our statistical test could lead us to reject the null hypothesis (and conclude that the company's pipe meets the city's specifications) when in fact the null hypothesis is true? The answer is $\alpha = .05$—that is, we selected the level of risk, α, of making a Type I error when we constructed the test. Thus, the chance is only 1 in 20 that our test would lead us to conclude the manufacturer's pipe satisfies the city's specifications when in fact the pipe does *not* meet specifications.

Now, suppose the sample mean breaking strength for the 50 sections of sewer pipe turned out to be $\bar{x} = 2{,}430$ pounds per linear foot. Assuming that the sample standard deviation is still $s = 200$, the test statistic is

$$z = \frac{2{,}430 - 2{,}400}{200/\sqrt{50}} = \frac{30}{28.28} = 1.06$$

Therefore, the sample mean $\bar{x} = 2{,}430$ is only 1.06 standard deviations above the null hypothesized value of $\mu = 2{,}400$. As shown in Figure 6.3, this value does not fall into the rejection region ($z > 1.645$). Therefore, we know that we cannot reject H_0 using $\alpha = .05$. Even though the sample mean exceeds the city's specification of 2,400 by 30 pounds per linear foot, it does not exceed the specification by enough to provide *convincing* evidence that the *population mean* exceeds 2,400.

Should we accept the null hypothesis H_0: $\mu \leq 2{,}400$ and conclude that the manufacturer's pipe does not meet specifications? To do so would be to risk a **Type II error**—that of concluding that the null hypothesis is true (the pipe does not meet specifications) when in fact it is false (the pipe does meet specifications). We denote the probability of committing a Type II error by β, and we show in Section 6.7 that β is often difficult to determine

> A **Type II error** occurs if the researcher accepts the null hypothesis when, in fact, H_0 is false. The probability of committing a Type II error is denoted by β.

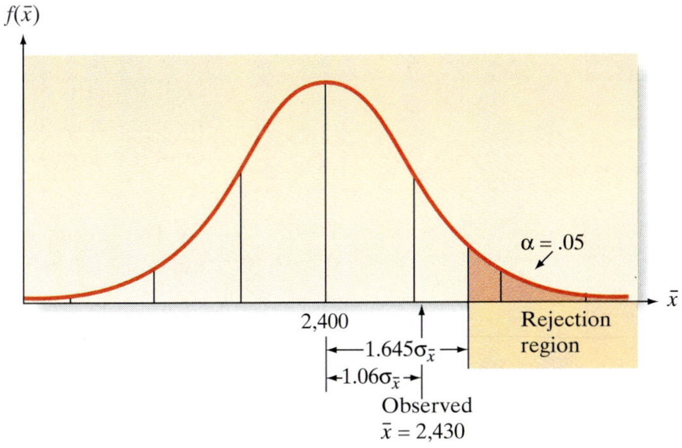

Figure 6.3
Location of test statistic when $\bar{x} = 2{,}430$

Table 6.1 Conclusions and Consequences for a Test of Hypothesis

Conclusion	True State of Nature	
	H_0 True	H_a True
Accept H_0 (Assume H_0 True)	Correct decision	Type II error (probability β)
Reject H_0 (Assume H_a True)	Type I error (probability α)	Correct decision

precisely. Rather than make a decision (accept H_0) for which the probability of error (β) is unknown, we avoid the potential Type II error by avoiding the conclusion that the null hypothesis is true. Instead, we will simply state that *the sample evidence is insufficient to reject H_0 at $\alpha = .05$*. Because the null hypothesis is the "status-quo" hypothesis, the effect of not rejecting H_0 is to maintain the status quo. In our pipe-testing example, the effect of having insufficient evidence to reject the null hypothesis that the pipe does not meet specifications is probably to prohibit the use of the manufacturer's pipe unless and until there is sufficient evidence that the pipe does meet specifications—that is, until the data indicate convincingly that the null hypothesis is false, we usually maintain the status quo implied by its truth.

Table 6.1 summarizes the four possible outcomes of a test of hypothesis. The "true state of nature" columns in Table 6.1 refer to the fact that either the null hypothesis H_0 is true or the alternative hypothesis H_a is true. Note that the true state of nature is unknown to the researcher conducting the test. The "decision" rows in Table 6.1 refer to the action of the researcher, assuming that he or she will either conclude that H_0 is true or that H_a is true, based on the results of the sampling experiment. Note that a Type I error can be made *only* when the null hypothesis is rejected in favor of the alternative hypothesis, and a Type II error can be made *only* when the null hypothesis is accepted. Our policy will be to make a decision only when we know the probability of making the error that corresponds to that decision. Because α is usually specified by the analyst, we will generally be able to reject H_0 (accept H_a) when the sample evidence supports that decision. However, because β is usually not specified, *we will generally avoid the decision to accept H_0, preferring instead to state that the sample evidence is insufficient to reject H_0 when the test statistic is not in the rejection region.*

CAUTION Be careful not to "accept H_0" when conducting a test of hypothesis because the measure of reliability, $\beta = P(\text{Type II error})$, is almost always unknown. If the test statistic does not fall into the rejection region, it is better to state the conclusion as "insufficient evidence to reject H_0."* ▲

The elements of a test of hypothesis are summarized in the following box. Note that the first four elements are all specified *before* the sampling experiment is performed. In no case will the results of the sample be used to determine the hypotheses; the data are collected to test the predetermined hypotheses, not to formulate them.

Elements of a Test of Hypothesis

1. Null hypothesis (H_0): A theory about the specific values of one or more population parameters. The theory generally represents the status quo, which we adopt until it is proven false. The theory is always stated as H_0: parameter = value.
2. *Alternative (research) hypothesis*(H_a): A theory that contradicts the null hypothesis. The theory generally represents that which we will adopt only when sufficient evidence exists to establish its truth.

*In many practical business applications of hypothesis testing, nonrejection leads management to behave as if the null hypothesis were accepted. Accordingly, the distinction between acceptance and nonrejection is frequently blurred in practice. We discuss the issues connected with the acceptance of the null hypothesis and the calculation of β in more detail in Section 6.7.

3. *Test statistic:* A sample statistic used to decide whether to reject the null hypothesis.
4. *Rejection region:* The numerical values of the test statistic for which the null hypothesis will be rejected. The rejection region is chosen so that the probability is α that it will contain the test statistic when the null hypothesis is true, thereby leading to a Type I error. The value of α is usually chosen to be small (e.g., .01, .05, or .10) and is referred to as the **level of significance** of the test.
5. *Assumptions:* Clear statement(s) of any assumptions made about the population(s) being sampled.
6. *Experiment and calculation of test statistic:* Performance of the sampling experiment and determination of the numerical value of the test statistic.
7. *Conclusion:*
   a. If the numerical value of the test statistic falls in the rejection region, we reject the null hypothesis and conclude that the alternative hypothesis is true. We know that the hypothesis-testing process will lead to this conclusion incorrectly (Type I error) only $100\alpha\%$ of the time when H_0 is true.
   b. If the test statistic does not fall in the rejection region, we do not reject H_0. Thus, we reserve judgment about which hypothesis is true. We do not conclude that the null hypothesis is true because we do not (in general) know the probability β that our test procedure will lead to an incorrect acceptance of H_0 (Type II error).

As with confidence intervals, the methodology for testing hypotheses varies depending on the target population parameter. In this chapter, we develop methods for testing a population mean, a population proportion, and a population variance. Some key words and the type of data associated with these target parameters are listed in the accompanying box.

Determining the Target Parameter

Parameter	Key Words or Phrases	Type of Data
μ	Mean; average	Quantitative
p	Proportion; percentage; fraction; rate	Qualitative
σ^2	Variance; variability; spread	Quantitative

6.2 Formulating Hypotheses and Setting Up the Rejection Region

In Section 6.1 we learned that the null and alternative hypotheses form the basis for inference using a test of hypothesis. The null and alternative hypotheses may take one of several forms. In the sewer pipe example, we tested the null hypothesis that the population mean strength of the pipe is less than or equal to 2,400 pounds per linear foot against the alternative hypothesis that the mean strength exceeds 2,400—that is, we tested

$$H_0: \mu \leq 2{,}400 \quad \text{(Pipe does not meet specifications)}$$

$$H_a: \mu > 2{,}400 \quad \text{(Pipe meets specifications)}$$

This is a **one-tailed** (or **one-sided**) **statistical test** because the alternative hypothesis specifies that the population parameter (the population mean μ in this example) is strictly greater than a specified value (2,400 in this example). If the null hypothesis had been H_0: $\mu \geq 2{,}400$ and the alternative hypothesis had been H_a: $\mu < 2{,}400$, the test would still be one-sided because the parameter is still specified to be on "one side" of the null hypothesis value. Some statistical investigations seek to show that the population parameter is *either larger or smaller* than some specified value. Such an alternative hypothesis is called a **two-tailed** (or **two-sided**) **hypothesis.**

While alternative hypotheses are always specified as strict inequalities, such as $\mu < 2{,}400$, $\mu > 2{,}400$, or $\mu \neq 2{,}400$, null hypotheses are usually specified as equalities, such as $\mu = 2{,}400$. Even when the null hypothesis is an inequality, such as $\mu \leq 2{,}400$, we specify H_0: $\mu = 2{,}400$, reasoning that if sufficient evidence exists to show that H_a: $\mu > 2{,}400$ is true when tested against H_0: $\mu = 2{,}400$, then surely sufficient evidence exists to reject $\mu < 2{,}400$ as well. Therefore, the null hypothesis is specified as the value of μ closest to a one-sided alternative hypothesis and as the only value *not* specified in a two-tailed alternative hypothesis. The steps for selecting the null and alternative hypotheses are summarized in the following box.

Steps for Selecting the Null and Alternative Hypotheses

1. Select the *alternative hypothesis* as that which the sampling experiment is intended to establish. The alternative hypothesis will assume one of three forms:
   a. One-tailed, upper-tailed (e.g., H_a: $\mu > 2{,}400$)
   b. One-tailed, lower-tailed (e.g., H_a: $\mu < 2{,}400$)
   c. Two-tailed (e.g., H_a: $\mu \neq 2{,}400$)
2. Select the *null hypothesis* as the status quo, that which will be presumed true unless the sampling experiment conclusively establishes the alternative hypothesis. The null hypothesis will be specified as that parameter value closest to the alternative in one-tailed tests and as the complementary (or only unspecified) value in two-tailed tests.

   (e.g., H_0: $\mu = 2{,}400$)

A **one-tailed test** of hypothesis is one in which the alternative hypothesis is directional and includes the symbol "<" or ">."

A **two-tailed test** of hypothesis is one in which the alternative hypothesis does not specify departure from H_0 in a particular direction and is written with the symbol " $\neq$."

Example 6.1
Formulating H_0 and H_a for a Test of a Population Mean

Problem A metal lathe is checked periodically by quality control inspectors to determine whether it is producing machine bearings with a mean diameter of .5 inch. If the mean diameter of the bearings is larger or smaller than .5 inch, then the process is out of control and must be adjusted. Formulate the null and alternative hypotheses for a test to determine whether the bearing production process is out of control.

Solution The hypotheses must be stated in terms of a population parameter. Here, we define μ as the true mean diameter (in inches) of all bearings produced by the metal lathe. If either $\mu > .5$ or $\mu < .5$, then the lathe's production process is out of control. Because the inspectors want to be able to detect either possibility (indicating that the process is in need of adjustment), these values of μ represent the alternative (or research) hypothesis. Alternatively, because $\mu = .5$ represents an in-control process (the status quo), this represents the null hypothesis. Therefore, we want to conduct the two-tailed test:

H_0: $\mu = .5$ (i.e., the process is in control)
H_a: $\mu \neq .5$ (i.e., the process is out of control)

Look Back Here, the alternative hypothesis is not necessarily the hypothesis that the quality control inspectors desire to support. However, they will make adjustments to the metal lathe settings only if there is strong evidence to indicate that the process is out of control. Consequently, $\mu \neq .5$ must be stated as the alternative hypothesis.

Now Work Exercise 6.10a

Example 6.2

Formulating H_0 and H_a for a Test of a Population Proportion

Problem Cigarette advertisements are required by federal law to carry the following statement: "Warning: The surgeon general has determined that cigarette smoking is dangerous to your health." However, this warning is often located in inconspicuous corners of the advertisements and printed in small type. Suppose the Federal Trade Commission (FTC) claims that 80% of cigarette consumers fail to see the warning. A marketer for a large tobacco firm wants to gather evidence to show that the FTC's claim is too high, i.e., that fewer than 80% of cigarette consumers fail to see the warning. Specify the null and alternative hypotheses for a test of the FTC's claim.

Solution The marketer wants to make an inference about p, the true proportion of all cigarette consumers who fail to see the surgeon general's warning. In particular, the marketer wants to collect data to show that fewer than 80% of cigarette consumers fail to see the warning, i.e., $p < .80$. Consequently, $p < .80$. represents the alternative hypothesis and $p = .80$ (the claim made by the FTC) represents the null hypothesis. That is, the marketer desires the one-tailed (lower-tailed) test:

$$H_0\text{: } p = .80 \text{ (i.e., the FTC's claim is true)}$$
$$H_a\text{: } p < .80 \text{ (i.e., the FTC's claim is false)}$$

Look Back Whenever a claim is made about the value of a particular population parameter and the researcher wants to test the claim, believing that it is false, the claimed value will represent the null hypothesis.

Now Work Exercise 6.11

The rejection region for a two-tailed test differs from that for a one-tailed test. When we are trying to detect departure from the null hypothesis in *either* direction, we must establish a rejection region in both tails of the sampling distribution of the test statistic. Figures 6.4a and 6.4b show the one-tailed rejection regions for lower- and upper-tailed tests, respectively. The two-tailed rejection region is illustrated in Figure 6.4c. Note that a rejection region is established in each tail of the sampling distribution for a two-tailed test.

The rejection regions corresponding to typical values selected for α are shown in Table 6.2 for one- and two-tailed tests. Note that the smaller α you select, the more evidence (the larger z) you will need before you can reject H_0.

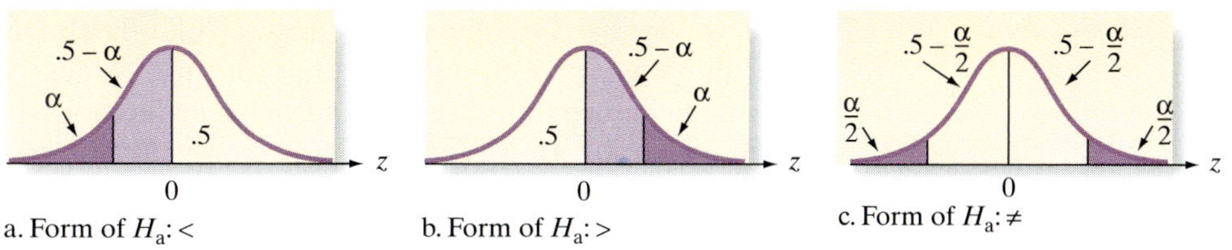

Figure 6.4
Rejection regions corresponding to one- and two-tailed tests

Table 6.2 Rejection Regions for Common Values of α

	Alternative Hypotheses		
	Lower-Tailed	Upper-Tailed	Two-Tailed
$\alpha = .10$	$z < -1.28$	$z > 1.28$	$z < -1.645$ or $z > 1.645$
$\alpha = .05$	$z < -1.645$	$z > 1.645$	$z < -1.96$ or $z > 1.96$
$\alpha = .01$	$z < -2.33$	$z > 2.33$	$z < -2.575$ or $z > 2.575$

Example 6.3

Setting Up a Hypothesis Test for μ, the Mean Amount of Cereal in a Box

Problem A manufacturer of cereal wants to test the performance of one of its filling machines. The machine is designed to discharge a mean amount of 12 ounces per box, and the manufacturer wants to detect any departure from this setting. This quality study calls for randomly sampling 100 boxes from today's production run and determining whether the mean fill for the run is 12 ounces per box. Set up a test of hypothesis for this study, using $\alpha = .01$.

Solution The key word *mean* in the statement of the problem implies that the target parameter is μ, the mean amount of cereal discharged into the box. Because the manufacturer wishes to detect a departure from the setting of $\mu = 12$ in either direction, $\mu < 12$ or $\mu > 12$, we conduct a two-tailed statistical test. Following the procedure for selecting the null and alternative hypotheses, we specify as the alternative hypothesis that the mean differs from 12 ounces because detecting the machine's departure from specifications is the purpose of the quality control study. The null hypothesis is the presumption that the filling machine is operating properly unless the sample data indicate otherwise. Thus,

H_0: $\mu = 12$ (Population mean fill amount is 12 ounces)

H_a: $\mu \neq 12$ (i.e., $\mu < 12$ or $\mu > 12$; machine is under- or overfilling the box)

The test statistic measures the number of standard deviations between the observed value of $\bar{x}$ and the null hypothesized value $\mu = 12$:

$$\textit{Test statistic: } z = \frac{\bar{x} - 12}{\sigma_{\bar{x}}}$$

The rejection region must be designated to detect a departure from $\mu = 12$ in *either* direction, so we will reject H_0 for values of z that are either too small (negative) or too large (positive). To determine the precise values of z that comprise the rejection region, we first select α, the probability that the test will lead to incorrect rejection of the null hypothesis. Then we divide α equally between the lower and upper tails of the distribution of z, as shown in Figure 6.5. In this example, $\alpha = .01$, so $\alpha/2 = .005$ is placed in each tail. The areas in the tails correspond to $z = -2.575$ and $z = 2.575$, respectively (from Table 6.2):

$$\textit{Rejection region: } z < -2.575 \text{ or } z > 2.575 \text{ (see Figure 6.5)}$$

Assumptions: Because the sample size of the experiment is large enough ($n > 30$), the Central Limit Theorem will apply, and no assumptions need be made about the population of fill measurements. The sampling distribution of the sample mean fill of 100 boxes will be approximately normal regardless of the distribution of the individual boxes' fills.

Look Back Note that the test is set up *before* the sampling experiment is conducted. The data are not used to develop the test. Evidently, the manufacturer does not want to disrupt the filling process to adjust the machine, unless the sample data provide very convincing evidence that it is not meeting specifications, because the value of α has been set quite low at .01. If the sample evidence results in the rejection of H_0, the manufacturer will confidently conclude that the machine needs adjustment because there is only a .01 probability of a Type I error.

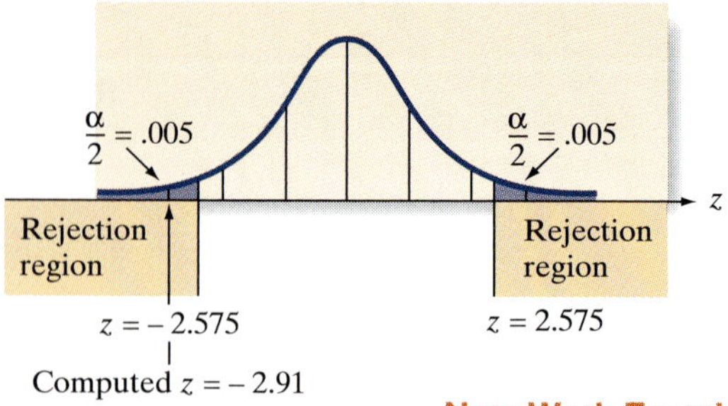

Figure 6.5
Two-tailed rejection region: $\alpha = .01$

Now Work Exercise 6.10b, c

Once the test is set up, the manufacturer is ready to perform the sampling experiment and conduct the test. The test is performed in the next section.

Statistics IN Action Revisited Key Elements of a Hypothesis Test

In Kimberly-Clark Corp.'s survey of people with colds, each of 250 customers was asked to keep count of his or her use of Kleenex® tissues in diaries. One goal of the company was to determine how many tissues to package in a cold-care box of Kleenex®; consequently, the total number of tissues used was recorded for each person surveyed. Because number of tissues is a quantitative variable, the parameter of interest is μ, the mean number of tissues used by all customers with colds.

Now, according to a *Wall Street Journal* report, there was "little doubt that the company should put 60 tissues" in a cold-care box of Kleenex® tissues. This statement was based on a claim made by marketing experts that 60 is the average number of times a person will blow his or her nose during a cold. Consequently, the marketers are claiming that $\mu = 60$. Suppose we disbelieve the claim that $\mu = 60$, believing instead that the population mean is smaller than 60 tissues. In order to test the claim against our belief, we set up the following null and alternative hypotheses:

$$H_0: \mu = 60 \qquad H_a: \mu < 60$$

We'll conduct this test in the next Statistics in Action Revisited on p. 340.

Activity 6.1 *Challenging a Company's Claim:* Tests of Hypotheses

Use the Internet or a newspaper or magazine to find an example of a claim made by a company about the reliability or efficiency of one of its products. In this activity, you represent a consumer group that believes the claim may be false.

1. In your example, what kinds of evidence might exist that would cause one to suspect that the claim might be false and therefore worthy of a statistical study? Be specific. If the claim were false, how would consumers be hurt?
2. Explain the steps necessary to reject the company's claim at level α. State the null and alternative hypotheses. If you reject the claim, does it mean that the claim is false?
3. If you reject the claim when the claim is actually true, what type of error has occurred? What is the probability of this error occurring?
4. If you were to file a lawsuit against the company based on your rejection of its claim, how might the company use your results to defend itself?

Exercises 6.1–6.18

Learning the Mechanics

6.1 Which hypothesis, the null or the alternative, is the status-quo hypothesis? Which is the research hypothesis?

6.2 Which element of a test of hypothesis is used to decide whether to reject the null hypothesis in favor of the alternative hypothesis?

6.3 What is the level of significance of a test of hypothesis?

6.4 What is the difference between Type I and Type II errors in hypothesis testing? How do α and β relate to Type I and Type II errors?

6.5 List the four possible results of the combinations of decisions and true states of nature for a test of hypothesis.

6.6 We (generally) reject the null hypothesis when the test statistic falls in the rejection region, but we do not accept the null hypothesis when the test statistic does not fall in the rejection region. Why?

6.7 If you test a hypothesis and reject the null hypothesis in favor of the alternative hypothesis, does your test prove that the alternative hypothesis is correct? Explain.

6.8 For each of the following rejection regions, sketch the sampling distribution for z and indicate the location of the rejection region.

a. $z > 1.96$
b. $z > 1.645$
c. $z > 2.575$
d. $z < -1.28$
e. $z < -1.645$ or $z > 1.645$
f. $z < -2.575$ or $z > 2.575$
g. For each of the rejection regions specified in parts **a–f,** what is the probability that a Type I error will be made?

Applet Exercise 6.1

Use the applet *Hypothesis Test for a Mean* to investigate the frequency of Type I and Type II errors. For this exercise, use $n = 100$ and the normal distribution with mean 50 and standard deviation 10.

a. Set the null mean equal to 50 and the alternative to *not equal*. Run the applet one time. How many times was the null hypothesis rejected at level .05? In this case, the null hypothesis is true. Which type of error occurred each time the true null hypothesis was rejected? What is the probability of rejecting a true null hypothesis at level .05? How does the proportion of times the null hypothesis was rejected compare to this probability?

b. Clear the applet, then set the null mean equal to 47 and keep the alternative at *not equal*. Run the applet one time. How many times was the null hypothesis *not* rejected at level .05? In this case, the null hypothesis is false. Which type of error occurred each time the null hypothesis was *not* rejected? Run the applet several more times without clearing. Based on your results, what can you conclude about the probability of failing to reject the null hypothesis for the given conditions?

Applying the Concepts—Basic

6.9 Effectiveness of online courses. The Sloan Survey of Online Learning, "Making the Grade: Online Education in the United States," reported that 60% of college presidents believe that their online education courses are as good as or superior to courses that use traditional face-to-face instruction (*Inside Higher Ed,* Nov. 2006).

a. Give the null hypothesis for testing the claim made by the Sloan Survey.

b. Give the rejection region for a two-tailed test conducted at $\alpha = .01$.

6.10 NW **Play Golf America program.** The Professional Golf Association (PGA) and *Golf Digest* have developed the Play Golf America program, in which teaching professionals at participating golf clubs provide a free 10-minute lesson to new customers. According to *Golf Digest* (July 2008), golf facilities that participate in the program gain, on average, $2,400 in green fees, lessons, or equipment expenditures. A teaching professional at a golf club believes that the average gain in greens fees, lessons, or equipment expenditures for participating golf facilities exceeds $2,400.

a. In order to support the claim made by the teaching professional, what null and alternative hypotheses should you test?

b. Suppose you select $\alpha = .05$. Interpret this value in the words of the problem.

c. For $\alpha = .05$, specify the rejection region of a large-sample test.

6.11 NW **Student loan default rate.** The national student loan default rate has dropped steadily over the last decade. Five years ago, the Department of Education reported the default rate (i.e., the proportion of college students who default on their loans) at .045. Set up the null and alternative hypotheses if you want to determine if the student loan default rate this year is less than .045.

6.12 Work travel policy. American Express Consulting reported in *USA Today* (June 15, 2001) that 80% of U.S. companies have formal, written travel and entertainment policies for their employees. Give the null hypothesis for testing the claim made by American Express Consulting.

6.13 Calories in school lunches. A University of Florida economist conducted a study of Virginia elementary school lunch menus. During the state-mandated testing period, school lunches averaged 863 calories (National Bureau of Economic Research, Nov. 2002). The economist claims that after the testing period ends, the average caloric content of Virginia school lunches drops significantly. Set up the null and alternative hypotheses to test the economist's claim.

6.14 Libor interest rate. The interest rate at which London banks lend money to one another is called the *London interbank offered rate,* or *Libor.* The British Bankers Association regularly surveys international banks for the Libor rate. One recent report (Bankrate.com, Aug. 23, 2006) had the average Libor rate at 5.40% for 3-month loans—a value considered high by many Western banks. Set up the null and alternative hypotheses for testing the reported value.

Applying the Concepts—Intermediate

6.15 FDA certification of new drugs. According to *Chemical Marketing Reporter,* pharmaceutical companies spend $15 billion per year on research and development of new drugs. The pharmaceutical company must subject each new drug to lengthy and involved testing before receiving the necessary permission from the Food and Drug Administration (FDA) to market the drug. The FDA's policy is that the pharmaceutical company must provide substantial evidence that a new drug is safe prior to receiving FDA approval, so that the FDA can confidently certify the safety of the drug to potential consumers.

a. If the new drug testing were to be placed in a test of hypothesis framework, would the null hypothesis be that the drug is safe or unsafe? The alternative hypothesis?

b. Given the choice of null and alternative hypotheses in part **a,** describe Type I and Type II errors in terms of this application. Define α and β in terms of this application.

c. If the FDA wants to be very confident that the drug is safe before permitting it to be marketed, is it more important that α or β be small? Explain.

6.16 Authorizing computer users. One of the most pressing problems in high-technology industries is computer security. Computer security is typically achieved by using a *password*—a collection of symbols (usually letters and numbers) that must be supplied by the user before the computer permits access to the account. The problem is that persistent hackers can create programs that enter millions of combinations of symbols into a target system until the correct password is found. The newest systems solve this problem by requiring authorized users to identify themselves by unique body characteristics. For example, a system developed by Palmguard, Inc. tests the hypothesis

H_0: The proposed user is authorized
H_a: The proposed user is unauthorized

by checking characteristics of the proposed user's palm against those stored in the authorized users' data bank (*Omni,* 1984).

a. Define a Type I error and Type II error for this test. Which is the more serious error? Why?

b. Palmguard reports that the Type I error rate for its system is less than 1%, whereas the Type II error rate is .00025%. Interpret these error rates.

c. Another successful security system, the EyeDentifyer, "spots authorized computer users by reading the one-of-a-kind patterns formed by the network of minute blood vessels across the retina at the back of the eye." The EyeDentifyer reports Type I and II error rates of .01% (1 in 10,000) and .005% (5 in 100,000), respectively. Interpret these rates.

Applying the Concepts—Advanced

6.17 Jury trial outcomes. Sometimes, the outcome of a jury trial defies the "common sense" expectations of the general public (e.g., the O. J. Simpson verdict in the "Trial of the Century"). Such a verdict is more acceptable if we understand that the jury trial of an accused murderer is analogous to the statistical hypothesis-testing process. The null hypothesis in a jury trial is that the accused is innocent. (The status-quo hypothesis in the U.S. system of justice is innocence, which is assumed to be true until proven *beyond a reasonable doubt.*) The alternative hypothesis is guilt, which is accepted only when sufficient evidence exists to establish its truth. If the vote of the jury is unanimous in favor of guilt, the null hypothesis of innocence is rejected, and the court concludes that the accused murderer is guilty. Any vote other than a unanimous one for guilt results in a "not guilty" verdict. The court never accepts the null hypothesis; that is, the court never declares the accused "innocent." A "not guilty" verdict (as in the O. J. Simpson case) implies that the court could not find the defendant guilty *beyond a reasonable doubt.*

a. Define Type I and Type II errors in a murder trial.

b. Which of the two errors is the more serious? Explain.

c. The court does not, in general, know the values of α and β; but ideally, both should be small. One of these probabilities is assumed to be smaller than the other in a jury trial. Which one, and why?

d. The court system relies on the belief that the value of α is made very small by requiring a unanimous vote before guilt is concluded. Explain why this is so.

e. For a jury prejudiced against a guilty verdict as the trial begins, will the value of α increase or decrease? Explain.

f. For a jury prejudiced against a guilty verdict as the trial begins, will the value of β increase or decrease? Explain.

6.18 Intrusion detection systems. Refer to the *Journal of Research of the National Institute of Standards and Technology* (Nov.–Dec. 2003) study of a computer intrusion detection system (IDS), Exercise 3.67 (p. 153). Recall that an IDS is designed to provide an alarm whenever unauthorized access (e.g., an intrusion) to a computer system occurs. The probability of the system giving a false alarm (i.e., providing a warning when no intrusion occurs) is defined by the symbol α, while the probability of a missed detection (i.e., no warning given when an intrusion occurs) is defined by the symbol β. These symbols are used to represent Type I and Type II error rates, respectively, in a hypothesis testing scenario.

a. What is the null hypothesis, H_0?

b. What is the alternative hypothesis, H_a?

c. According to actual data on the EMERALD system collected by the Massachusetts Institute of Technology Lincoln Laboratory, only 1 in 1,000 computer sessions with no intrusions resulted in a false alarm. For the same system, the laboratory found that only 500 of 1,000 intrusions were actually detected. Use this information to estimate the values of α and β.

6.3 Test of Hypothesis about a Population Mean: Normal (z) Statistic

When testing a hypothesis about a population mean μ, the test statistic we use will depend on whether the sample size n is large (say, $n \geq 30$) or small, and whether or not we know the value of the population standard deviation, σ. In this section, we consider the large-sample case.

Because the sample size is large, the Central Limit Theorem guarantees that the sampling distribution of $\bar{x}$ is approximately normal. Consequently, the test statistic for a test based on large samples will be based on the normal z-statistic. Although the z-statistic requires that we know the true population standard deviation σ, this is rarely, if ever, the case. However, we established in Chapter 5 that when n is large, the sample standard deviation s provides a good approximation to σ, and the z-statistic can be approximated as follows:

$$z = \frac{\bar{x} - \mu_0}{\sigma_{\bar{x}}} = \frac{\bar{x} - \mu_0}{\sigma/\sqrt{n}} \approx \frac{\bar{x} - \mu_0}{s/\sqrt{n}}$$

where μ_0 represents the value of μ specified in the null hypothesis.

The setup of a large-sample test of hypothesis about a population mean is summarized in the following boxes. Both the one- and two-tailed tests are shown.

Large-Sample Test of Hypothesis about μ

One-Tailed Test	Two-Tailed Test
$H_0: \mu = \mu_0$ $H_a: \mu < \mu_0$ (or $H_a: \mu > \mu_0$)	$H_0: \mu = \mu_0$ $H_a: \mu \neq \mu_0$

Test statistic: $z = \dfrac{\bar{x} - \mu_0}{\sigma_{\bar{x}}} \approx \dfrac{\bar{x} - \mu_0}{s/\sqrt{n}}$	*Test statistic:* $z = \dfrac{\bar{x} - \mu_0}{\sigma_{\bar{x}}} \approx \dfrac{\bar{x} - \mu_0}{s/\sqrt{n}}$
Rejection region: $z < -z_{\alpha}$ (or $z > z_{\alpha}$ when H_a: $\mu > \mu_0$) where z_{α} is chosen so that $P(z > z_{\alpha}) = \alpha$	*Rejection region:* $\lvert z \rvert > z_{\alpha/2}$ where $z_{\alpha/2}$ is chosen so that $P(\lvert z \rvert > z_{\alpha/2}) = \alpha/2$

Note: μ_0 is the symbol for the numerical value assigned to μ under the null hypothesis.

Conditions Required for a Valid Large-Sample Hypothesis Test for μ

1. A random sample is selected from the target population.
2. The sample size n is large (i.e., $n \geq 30$). (Due to the Central Limit Theorem, this condition guarantees that the test statistic will be approximately normal regardless of the shape of the underlying probability distribution of the population.)

Once the test has been set up, the sampling experiment is performed and the test statistic calculated. The next box contains possible conclusions for a test of hypothesis, depending on the result of the sampling experiment.

Possible Conclusions for a Test of Hypothesis

1. If the calculated test statistic falls in the rejection region, reject H_0 and conclude that the alternative hypothesis H_a is true. State that you are rejecting H_0 at the α level of significance. Remember that the confidence is in the testing *process,* not the particular result of a single test.
2. If the test statistic does not fall in the rejection region, conclude that the sampling experiment does not provide sufficient evidence to reject H_0 at the α level of significance. [Generally, we will not "accept" the null hypothesis unless the probability β of a Type II error has been calculated (see Section 6.7).]

Example 6.4

Carrying Out a Hypothesis Test for μ, the Mean Amount of Cereal in a Box

Problem Refer to the quality control test set up in Example 6.3 (p. 328). Recall that a machine is designed to discharge a mean of 12 ounces of cereal per box. A sample of 100 boxes yielded the fill amounts (in ounces) shown in Table 6.3. Use these data to conduct the test.

$$H_0: \mu = 12 \quad \text{(Population mean fill amount is 12 ounces)}$$
$$H_a: \mu \neq 12 \quad \text{(Machine is under- or overfilling the box)}$$

Table 6.3 Fill Amounts from Quality Control Tests

12.3	12.2	12.9	11.8	12.1	11.7	11.8	11.3	12.0	11.7
11.0	12.7	11.2	11.8	11.4	11.3	11.5	12.1	12.5	11.7
12.3	11.7	11.6	11.6	11.1	12.1	12.4	11.4	11.6	11.4
10.9	11.0	11.5	11.6	11.6	11.4	11.9	11.1	11.7	12.1
12.2	11.7	11.6	11.4	12.4	11.0	11.8	12.9	13.2	11.5
11.5	12.0	11.9	11.8	12.5	11.8	12.4	12.0	12.2	12.4
11.8	12.6	11.8	11.8	11.5	12.0	12.7	11.5	11.0	11.8
11.2	12.6	12.0	12.6	12.0	12.0	12.5	12.0	12.8	11.8
12.6	12.4	10.9	12.0	11.9	11.6	11.3	12.1	11.8	12.2
12.2	11.5	12.7	11.5	11.0	11.7	12.5	11.6	11.3	11.1

Data Set: CEREAL

Descriptive Statistics: FILL

Variable	N	Mean	StDev	Minimum	Q1	Median	Q3	Maximum
FILL	100	11.851	0.512	10.900	11.500	11.800	12.200	13.200

Figure 6.6
Minitab descriptive statistics for fill amounts, Example 6.4

Solution To carry out the test, we need to find the values of $\bar{x}$ and s. These values, $\bar{x} = 11.851$ and $s = .512$, are shown (highlighted) on the Minitab printout, Figure 6.6. Now, we substitute these sample statistics into the test statistic and obtain:

$$z = \frac{\bar{x} - 12}{\sigma_{\bar{x}}} = \frac{\bar{x} - 12}{\sigma/\sqrt{n}} = \frac{11.851 - 12}{\sigma/\sqrt{100}}$$

$$\approx \frac{11.851 - 12}{s/10} = \frac{-.149}{.512/10} = -2.91$$

The implication is that the sample mean, 11.851, is (approximately) 3 standard deviations below the null hypothesized value of 12.0 in the sampling distribution of $\bar{x}$. You can see in Figure 6.5 that this value of z is in the lower-tail rejection region, which consists of all values of $z < -2.575$. These sample data provide sufficient evidence to reject H_0 and conclude, at the $\alpha = .01$ level of significance, that the mean fill differs from the specification of $\mu = 12$ ounces. It appears that the machine is, on average, underfilling the boxes.

Look Back Three points about the test of hypothesis in this example apply to all statistical tests:

1. Because z is less than -2.575, it is tempting to state our conclusion at a significance level lower than $\alpha = .01$. We resist the temptation because the level of α is determined before the sampling experiment is performed. If we decide that we are willing to tolerate a 1% Type I error rate, the result of the sampling experiment should have no effect on that decision. In general, *the same data should not be used both to set up and to conduct the test.*
2. When we state our conclusion at the .01 level of significance, we are referring to the failure rate of the procedure, not the result of this particular test. We know that the test procedure will lead to the rejection of the null hypothesis only 1% of the time when in fact $\mu = 12$. Therefore, *when the test statistic falls in the rejection region, we infer that the alternative $\mu \neq 12$ is true and express our confidence in the procedure by quoting the α level of significance, or the $100(1 - \alpha)\%$ confidence level.*
3. Although a test may lead to a "statistically significant" result (i.e., rejecting H_0 at significance level α, as in the test above), it may not be "practically significant." For example, suppose the quality control study tested $n = 100{,}000$ cereal boxes, resulting in $\bar{x} = 11.995$ and $s = .5$. Now, a two-tailed hypothesis test of H_0: $\mu = 12$ results in a test statistic of $z = \dfrac{(11.995 - 12)}{.5/\sqrt{100{,}000}} = -3.16.$

   This result at $\alpha = .01$ leads us to "reject H_0" and conclude that the mean, μ, is "statistically different" from 12. However, for all practical purposes, the sample mean, $\bar{x} = 11.995$, and hypothesized mean, $\mu = 12$, are the same. Because the result is not "practically significant," the company is not likely to spend money fixing a machine that, for all practical purposes, is dispensing an average of 12 ounces of cereal into the boxes. Consequently, *not all "statistically significant" results are "practically significant."*

Now Work Exercise 6.23

Exercises 6.19–6.33

Learning the Mechanics

6.19 Suppose you are interested in conducting the statistical test of H_0: $\mu = 255$ against H_a: $\mu > 255$, and you have decided to use the following decision rule: Reject H_0 if the sample mean of a random sample of 81 items is more than 270. Assume that the standard deviation of the population is 63.

a. Express the decision rule in terms of z.

b. Find α, the probability of making a Type I error, by using this decision rule.

6.20 A random sample of 100 observations from a population with standard deviation 60 yielded a sample mean of 110.

a. Test the null hypothesis that $\mu = 100$ against the alternative hypothesis that $\mu > 100$ using $\alpha = .05$. Interpret the results of the test.

b. Test the null hypothesis that $\mu = 100$ against the alternative hypothesis that $\mu \neq 100$ using $\alpha = .05$. Interpret the results of the test.

c. Compare the results of the two tests you conducted. Explain why the results differ.

6.21 A random sample of 64 observations produced the following summary statistics: $\bar{x} = .323$ and $s^2 = .034$.

a. Test the null hypothesis that $\mu = .36$ against the alternative hypothesis that $\mu < .36$ using $\alpha = .10$.

b. Test the null hypothesis that $\mu = .36$ against the alternative hypothesis that $\mu \neq .36$ using $\alpha = .10$. Interpret the result.

Applet Exercise 6.2

Use the applet *Hypotheses Test for a Mean* to investigate the effect of the underlying distribution on the proportion of Type I errors. For this exercise use $n = 100$, mean $= 50$, standard deviation $= 10$, null mean $= 50$, and alternative $<$.

a. Select the normal distribution and run the applet several times without clearing. What happens to the proportion of times the null hypothesis is rejected at the .05 level as the applet is run more and more times?

b. Clear the applet and then repeat part **a** using the right skewed distribution. Do you get similar results? Explain.

c. Describe the effect that the underlying distribution has on the probability of making a Type I error.

Applet Exercise 6.3

Use the applet *Hypotheses Test for a Mean* to investigate the effect of the underlying distribution on the proportion of Type II errors. For this exercise use $n = 100$, mean $= 50$, standard deviation $= 10$, null mean $= 52$, and alternative $<$.

a. Select the normal distribution and run the applet several times without clearing. What happens to the proportion of times the null hypothesis is rejected at the .01 level as the applet is run more and more times? Is this what you would expect? Explain.

b. Clear the applet and then repeat part **a** using the right skewed distribution. Do you get similar results? Explain.

c. Describe the effect that the underlying distribution has on the probability of making a Type II error.

Applet Exercise 6.4

Use the applet *Hypotheses Test for a Mean* to investigate the effect of the null mean on the probability of making a Type II error. For this exercise use $n = 100$, mean $= 50$, standard deviation $= 10$, and alternative $<$ with the normal distribution. Set the null mean to 55 and run the applet several times without clearing. Record the proportion of Type II errors that occurred at the .01 level. Clear the applet and repeat for null means of 54, 53, 52, and 51. What can you conclude about the probability of a Type II error as the null mean gets closer to the actual mean? Can you offer a reasonable explanation for this?

Applying the Concepts—Basic

6.22 Accounting and Machiavellianism. Refer to the *Behavioral Research in Accounting* (Jan. 2008) study of Machiavellian traits in accountants, Exercise 5.17 (p. 279). A Mach rating score was determined for each in a random sample of 122 purchasing managers, with the following results: $\bar{x} = 99.6$, $s = 12.6$. Recall that a director of purchasing at a major firm claims that the true mean Mach rating score of all purchasing managers is 85.

a. Suppose you want to test the director's claim. Specify the null and alternative hypothesis for the test.

b. Give the rejection region for the test using $\alpha = .10$.

c. Find the value of the test statistic.

d. Use the result, part **c,** to make the appropriate conclusion.

6.23 NW **Latex allergy in health care workers.** Refer to the *Current Allergy & Clinical Immunology* (Mar. 2004) study of $n = 46$ hospital employees who were diagnosed with a latex allergy from exposure to the powder on latex gloves, Exercise 5.10 (p. 278). The number of latex gloves used per week by the sampled workers is summarized as follows: $\bar{x} = 19.3$ and $s = 11.9$. Let μ represent the mean number of latex gloves used per week by all hospital employees. Consider testing H_0: $\mu = 20$ against H_a: $\mu < 20$.

a. Give the rejection region for the test at a significance level of $\alpha = .01$.

b. Calculate the value of the test statistic.

c. Use the results, parts **a** and **b,** to make the appropriate conclusion.

6.24 Prices of hybrid cars. *BusinessWeek.com* provides consumers with retail prices of new cars at dealers from across the country. The July 2006 prices for the hybrid Toyota Prius were obtained from a sample of 160 dealers. These 160 prices are saved in the **HYBRIDCARS** file.

a. Give the null and alternative hypotheses for testing whether the mean July 2006 dealer price of the Toyota Prius, μ, differs from \$25,000.

b. Access the **HYBRIDCARS** file and find the sample mean and sample standard deviation.

c. Use the information in part **b** to find the test statistic for the hypothesis test.

d. Give the rejection region for the hypothesis test, using $\alpha = .05$.

e. State the appropriate conclusion for the hypothesis test.

Table for Exercise 6.28

14622	13196	11948	11289	11964	10526	10387	10592	10460	10086
14628	13396	11726	11252	12449	11030	10787	10603	10144	11674
11510	10946	10508	10604	10270	10529	10360	14796	12913	12270
11842	10656	11360	11136	10814	13523	11289	11183	10951	9722
10481	9812	9669	9643	9115	9115	11588	10888	9738	9295
9421	9105	10233	10186	9918	9209	9532	9933	9152	9295
16243	14628	12766	8714	9469	11948	12414			

6.25 Identifying type of urban land cover. For planning purposes, urban land-cover must be identified as either grassland, commercial, or residential. This is typically done using remote sensing data from satellite pictures. In *Geographical Analysis* (Oct. 2006), researchers from Arizona State, Florida State, and Louisiana State universities collaborated on a new method for analyzing remote sensing data. A satellite photograph of an urban area was divided into 4x4 meter areas (called pixels). Of interest is a numerical measure of the distribution of gaps or hole sizes in the pixel, called *lacunarity*. The mean and standard deviation of the lacunarity measurements for a sample of 100 pixels randomly selected from a specific urban area are 225 and 20, respectively. It is known that the mean lacunarity measurement for all grassland pixels is 220. Do the data suggest that the area sampled is grassland? Test at $\alpha = .01$.

6.26 Size of diamonds sold at retail. Refer to the *Journal of Statistics Education* data on diamonds saved in the **DIAMONDS** file. In Exercise 5.18 (p. 280) you selected a random sample of 30 diamonds from the 308 diamonds and found the mean and standard deviation of the number of carats per diamond for the sample. Let μ represent the mean number of carats in the population of 308 diamonds. Suppose you want to test H_0: $\mu = .6$ against H_a: $\mu \neq .6$.

a. In the words of the problem, define a Type I error and a Type II error.

b. Use the sample information to conduct the test at a significance level of $\alpha = .05$.

c. Conduct the test, part **b,** using $\alpha = .10$.

d. What do the results suggest about the choice of α in a test of hypothesis?

6.27 Producer's and consumer's risk. In quality control applications of hypothesis testing, the null and alternative hypotheses are frequently specified as

H_0: The production process is performing satisfactorily.

H_a: The process is performing in an unsatisfactory manner.

Accordingly, α is sometimes referred to as the *producer's risk,* while β is called the *consumer's risk* (Stevenson, *Operations Management,* 2008).

An injection molder produces plastic golf tees. The process is designed to produce tees with a mean weight of .250 ounce. To investigate whether the injection molder is operating satisfactorily, 40 tees were randomly sampled from the last hour's production. Their weights (in ounces) are listed in the following table and saved in the **TEES** file.

a. Write H_0 and H_a in terms of the true mean weight of the golf tees, μ.

b. Do the data provide sufficient evidence to conclude that the process is not operating satisfactorily? Test using $\alpha = .01$.

c. In the context of this problem, explain why it makes sense to call α the producer's risk and β the consumer's risk.

.247	.251	.254	.253	.253	.248	.253	.255	.256	.252
.253	.252	.253	.256	.254	.256	.252	.251	.253	.251
.253	.253	.248	.251	.253	.256	.254	.250	.254	.255
.249	.250	.254	.251	.251	.255	.251	.253	.252	.253

Applying the Concepts—Intermediate

6.28 Cooling method for gas turbines. During periods of high electricity demand, especially during the hot summer months, the power output from a gas turbine engine can drop dramatically. One way to counter this drop in power is by cooling the inlet air to the gas turbine. An increasingly popular cooling method uses high pressure inlet fogging. The performance of a sample of 67 gas turbines augmented with high-pressure inlet fogging was investigated in the *Journal of Engineering for Gas Turbines and Power* (Jan. 2005). One measure of performance is heat rate (kilojoules per kilowatt per hour). Heat rates for the 67 gas turbines, saved in the **GASTURBINE** file, are listed in the table above. Suppose that a standard gas turbine has, on average, a heat rate of 10,000 kJ/kWh. Conduct a test to determine if the mean heat rate of gas turbines augmented with high-pressure inlet fogging exceeds 10,000 kJ/kWh. Use $\alpha = .05$.

6.29 Point spreads of NFL games. During the National Football League (NFL) season, Las Vegas oddsmakers establish a point spread on each game for betting purposes. For example, the champion Pittsburgh Steelers were established as 7-point favorites over the Arizona Cardinals in the 2009 Super Bowl. The final scores of NFL games were compared against the final point spreads established by the oddsmakers in *Chance* (Fall 1998). The difference between the game outcome and point spread (called a *point-spread error*) was calculated for 240 NFL games. The mean and standard deviation of the point-spread errors are $\bar{x} = -1.6$ and $s = 13.3$. Use this information to test the hypothesis that the true mean point-spread error for all NFL games differs from 0. Conduct the test at $\alpha = .01$ and interpret the result.

6.30 Revenue for a full-service funeral. According to the National Funeral Directors Association (NFDA), the nation's 22,000 funeral homes collected an average of \$6,500 per full-service funeral in 2005 (*NFDA Fact Sheet,* 2009). A random sample of 36 funeral homes reported revenue data for 2009. Among other measures, each reported its average fee for a full-service funeral. These data (in thousands of dollars) are shown in the following table and saved in the **FUNERAL** file.

7.4	9.4	5.3	8.4	7.5	6.5	6.2	8.3	6.7
11.6	6.3	5.9	6.7	5.8	5.2	6.4	6.0	7.4
7.2	6.6	6.3	5.3	6.6	5.6	8.4	7.2	7.4
5.8	6.3	6.1	7.0	7.2	6.1	5.4	7.4	6.6

a. What are the appropriate null and alternative hypotheses to test whether the average full-service fee of U. S. funeral homes in 2009 exceeds $6,500?

b. Conduct the test at $\alpha = .05$. Do the sample data provide sufficient evidence to conclude that the average fee in 2009 was higher than in 2005?

c. In conducting the test, was it necessary to assume that the population of average full-service fees was normally distributed? Justify your answer.

6.31 Salaries of postgraduates. The *Economics of Education Review* (Vol. 21, 2002) published a paper on the relationship between education level and earnings. The data for the research were obtained from the National Adult Literacy Survey of over 25,000 respondents. The survey revealed that males with a postgraduate degree had a mean salary of $61,340 (with standard error $s_{\bar{x}} = \$2{,}185$), while females with a postgraduate degree had a mean of $32,227 (with standard error $s_{\bar{x}} = \$932$).

a. The article reports that a 95% confidence interval for μ_M, the population mean salary of all males with postgraduate degrees, is ($57,050, $65,631). Based on this interval, is there evidence to say that μ_M differs from $60,000? Explain.

b. Use the summary information to test the hypothesis that the true mean salary of males with postgraduate degrees differs from $60,000. Use $\alpha = .05$. (*Note:* $s_{\bar{x}} = s/\sqrt{n}$.)

c. Explain why the inferences in parts **a** and **b** agree.

d. The article reports that a 95% confidence interval for μ_F, the population mean salary of all females with postgraduate degrees, is ($30,396, $34,058). Based on this interval, is there evidence to say that μ_F differs from $33,000? Explain.

e. Use the summary information to test the hypothesis that the true mean salary of females with postgraduate degrees differs from $33,000. Use $\alpha = .05$. (*Note:* $s_{\bar{x}} = s/\sqrt{n}$.)

f. Explain why the inferences in parts **d** and **e** agree.

6.32 Solder-joint inspections. Current technology uses high resolution X-rays and lasers for inspection of solder-joint defects on printed circuit boards (PCBs) (*Global SMT & Packaging,* April 2008). A particular manufacturer of laser-based inspection equipment claims that its product can inspect on average at least 10 solder joints per second when the joints are spaced .1 inch apart. The equipment was tested by a potential buyer on 48 different PCBs. In each case, the equipment was operated for exactly 1 second. The number of solder joints inspected on each run follows (data are saved in the **PCB** file):

10	9	10	10	11	9	12	8	8	9	6	10
7	10	11	9	9	13	9	10	11	10	12	8
9	9	9	7	12	6	9	10	10	8	7	9
11	12	10	0	10	11	12	9	7	9	9	10

a. The potential buyer wants to know whether the sample data refute the manufacturer's claim. Specify the null and alternative hypotheses that the buyer should test.

b. In the context of this exercise, what is a Type I error? A Type II error?

c. Conduct the hypothesis test you described in part **a** and interpret the test's results in the context of this exercise. Use $\alpha = .05$.

Applying the Concepts—Advanced

6.33 Why do small firms export? What motivates small firms to export their products? To answer this question, California State University professor Ralph Pope conducted a survey of 137 exporting firms listed in the *California International Trade Register* (*Journal of Small Business Management,* Vol. 40, 2002.) Firm CEOs were asked to respond to the statement "Management believes that the firm can achieve economies of scale by exporting" on a scale of 1 (strongly disagree) to 5 (strongly agree). Summary statistics for the $n = 137$ scale scores were reported as $\bar{x} = 3.85$ and $s = 1.5$. In the journal article, the researcher hypothesized that if the true mean scale score exceeds 3.5, then CEOs at all California small firms generally agree with the statement.

a. Conduct the appropriate test using $\alpha = .05$. State your conclusion in the words of the problem.

b. Explain why the results of the study, although "statistically significant," may not be practically significant.

c. The scale scores for the sample of 137 small firms are unlikely to be normally distributed. Does this invalidate the inference you made in part **a**? Explain.

6.4 Observed Significance Levels: *p*-Values

According to the statistical test procedure described in Section 6.2, the rejection region and, correspondingly, the value of α are selected prior to conducting the test, and the conclusions are stated in terms of rejecting or not rejecting the null hypothesis. A second method of presenting the results of a statistical test is one that reports the extent to which the test statistic disagrees with the null hypothesis and leaves to the reader the task of deciding whether to reject the null hypothesis. This measure of disagreement is called the *observed significance level* (or *p-value*) for the test.

The **observed significance level,** or *p*-value, for a specific statistical test is the probability (assuming H_0 is true) of observing a value of the test statistic that is at least as contradictory to the null hypothesis, and supportive of the alternative hypothesis, as the actual one computed from the sample data.

For example, the value of the test statistic computed for the sample of $n = 50$ sections of sewer pipe was $z = 2.12$. Because the test is one-tailed—that is, the alternative (research) hypothesis of interest is H_a: $\mu > 2{,}400$—values of the test statistic even more contradictory to H_0 than the one observed would be values larger than $z = 2.12$. Therefore, the observed significance level (p-value) for this test is

$$p\text{-value} = P(z > 2.12)$$

or, equivalently, the area under the standard normal curve to the right of $z = 2.12$ (see Figure 6.7).

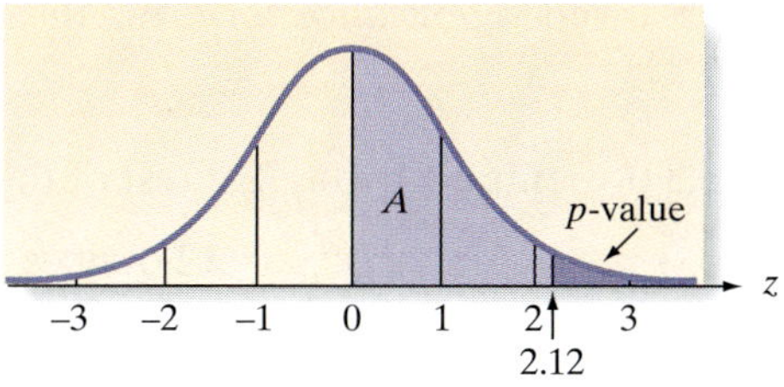

Figure 6.7
Finding the p-value for an upper-tailed test when $z = 2.12$

The area A in Figure 6.7 is given in Table IV in Appendix B as .4830. Therefore, the upper-tail area corresponding to $z = 2.12$ is

$$p\text{-value} = .5 - .4830 = .0170$$

Consequently, we say that these test results are "very significant" (i.e., they disagree rather strongly with the null hypothesis, H_0: $\mu = 2{,}400$, and favor H_a: $\mu > 2{,}400$). The probability of observing a z-value as large as 2.12 is only .0170, if in fact the true value of μ is 2,400.

If you are inclined to select $\alpha = .05$ for this test, then you would reject the null hypothesis because the p-value for the test, .0170, is less than .05. In contrast, if you choose $\alpha = .01$, you would not reject the null hypothesis because the p-value for the test is larger than .01. Thus, the use of the observed significance level is identical to the test procedure described in the preceding sections except that the choice of α is left to you.

The steps for calculating the p-value corresponding to a test statistic for a population mean are given in the next box.

Steps for Calculating the p-value for a Test of Hypothesis

1. Determine the value of the test statistic z corresponding to the result of the sampling experiment.
2. **a.** If the test is one-tailed, the p-value is equal to the tail area beyond z in the same direction as the alternative hypothesis. Thus, if the alternative hypothesis is of the form $>$, the p-value is the area to the right of, or above, the observed z-value. Conversely, if the alternative is of the form $<$, the p-value is the area to the left of, or below, the observed z-value. (See Figure 6.8.)

   b. If the test is two-tailed, the p-value is equal to twice the tail area beyond the observed z-value in the direction of the sign of z—that is, if z is positive, the p-value is twice the area to the right of, or above, the observed z-value. Conversely, if z is negative, the p-value is twice the area to the left of, or below, the observed z-value. (See Figure 6.9.)

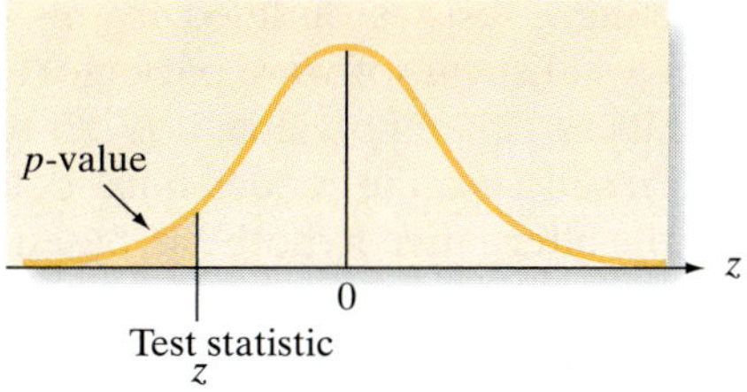

a. Lower-tailed test, H_a: $\mu < \mu_0$

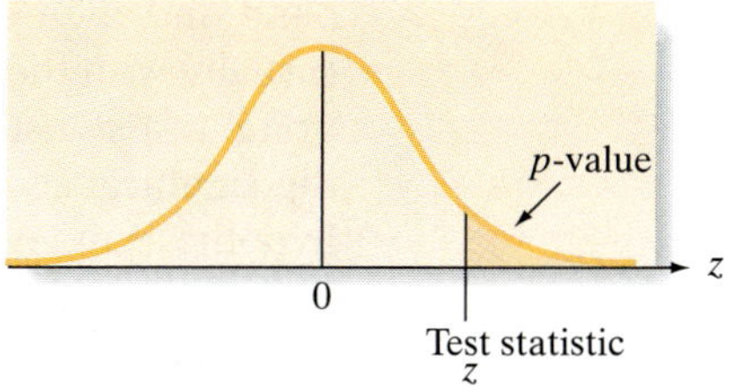

b. Upper-tailed test, H_a: $\mu > \mu_0$

Figure 6.8
Finding the p-value for a one-tailed test

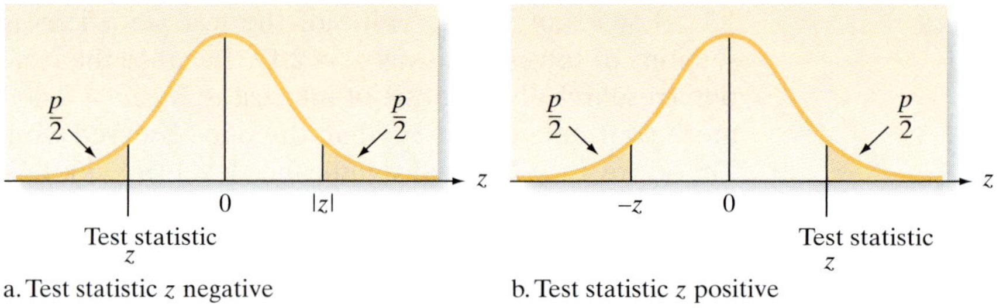

Figure 6.9
Finding the p-value for a two-tailed test: p-value $= 2(p/2)$

Example 6.5
Computing a *p*-Value for the Test of Mean Filling Weight

Problem Find the observed significance level for the test of the mean filling weight in Examples 6.3 and 6.4.

Solution Example 6.3 presented a two-tailed test of the hypothesis

$$H_0: \mu = 12 \text{ ounces}$$

against the alternative hypothesis

$$H_a: \mu \neq 12 \text{ ounces}$$

The observed value of the test statistic in Example 6.4 was $z = -2.91$, and any value of z less than -2.91 or greater than 2.91 (because this is a two-tailed test) would be even more contradictory to H_0. Therefore, the observed significance level for the test is

$$p\text{-value} = P(z < -2.91 \text{ or } z > 2.91) = P(|z| > 2.91)$$

Thus, we calculate the area below the observed z-value, $z = -2.91$, and double it. Consulting Table IV in Appendix B, we find that $P(z < -2.91) = .5 - .4982 = .0018$. Therefore, the p-value for this two-tailed test is

$$2P(z < -2.91) = 2(.0018) = .0036$$

This p-value can also be obtained using statistical software. The rounded p-value is shown (highlighted) on the Minitab printout, Figure 6.10.

```
Test of mu = 12 vs not = 12

Variable    N      Mean   StDev  SE Mean            95% CI             T      P
FILL      100   11.8510  0.5118   0.0512  (11.7495, 11.9525)      -2.91  0.004
```

Figure 6.10
Minitab test of mean fill amount, Example 6.5

Look Back We can interpret this p-value as a strong indication that the machine is not filling the boxes according to specifications because we would observe a test statistic this extreme or more extreme only 36 in 10,000 times if the machine were meeting specifications ($\mu = 12$). The extent to which the mean differs from 12 could be better determined by calculating a confidence interval for μ.

Now Work Exercise 6.39

When publishing the results of a statistical test of hypothesis in journals, case studies, reports, and so on, many researchers make use of p-values. Instead of selecting α beforehand and then conducting a test, as outlined in this chapter, the researcher computes (usually with the aid of a statistical software package) and reports the value of the appropriate test statistic and its associated p-value. It is left to the reader of the report to judge the significance of the result (i.e., the reader must determine whether to reject the null hypothesis in favor of the alternative hypothesis, based on the reported p-value). Usually, the null hypothesis is rejected if the observed significance level is *less than* the fixed significance level, α, chosen by the reader. The inherent advantage of reporting test results in this manner are twofold: (1) Readers are permitted to select the maximum value of α that they would be willing to tolerate if they actually carried out a standard test of hypothesis in the manner outlined in this chapter, and (2) a measure of the degree of significance of the result (i.e., the *p*-value) is provided.

Ethics IN Statistics

Selecting the value of α *after* computing the observed significance level (*p*-value) in order to guarantee a preferred conclusion is considered *unethical statistical practice*.

Reporting Test Results as *p*-Values: How to Decide Whether to Reject H_0

1. Choose the maximum value of α that you are willing to tolerate.
2. If the observed significance level (*p*-value) of the test is less than the chosen value of α, reject the null hypothesis. Otherwise, do not reject the null hypothesis.

Example 6.6

Using *p*-Values to Test Mean Hospital Length of Stay

Problem Knowledge of the amount of time a patient occupies a hospital bed—called *length of stay* (LOS)—is important for allocating resources. At one hospital, the mean LOS was determined to be 5 days. A hospital administrator believes that the mean LOS may now be less than 5 days due to a newly adopted managed care system. To check this, the LOSs (in days) for 100 randomly selected hospital patients were recorded; these are listed in Table 6.4. Test the hypothesis that the true mean LOS at the hospital is less than 5 days; that is,

$$H_0: \mu = 5 \quad \text{(Mean LOS is 5 days.)}$$
$$H_a: \mu < 5 \quad \text{(Mean LOS is less than 5 days.)}$$

Use the data in the table to conduct the test at $\alpha = .05$.

Solution The data were entered into a computer, and Excel/DDXL was used to conduct the analysis. The DDXL printout for the lower-tailed test is displayed in Figure 6.11. Both the test statistic, $z \approx -1.28$, and *p*-value of the test, $p = .102$, are highlighted

Table 6.4 Lengths of Stay for 100 Hospital Patients

2	3	8	6	4	4	6	4	2	5
8	10	4	4	4	2	1	3	2	10
1	3	2	3	4	3	5	2	4	1
2	9	1	7	17	9	9	9	4	4
1	1	1	3	1	6	3	3	2	5
1	3	3	14	2	3	9	6	6	3
5	1	4	6	11	22	1	9	6	5
2	2	5	4	3	6	1	5	1	6
17	1	2	4	5	4	4	3	2	3
3	5	2	3	3	2	10	2	4	2

Data Set: HOSPLOS

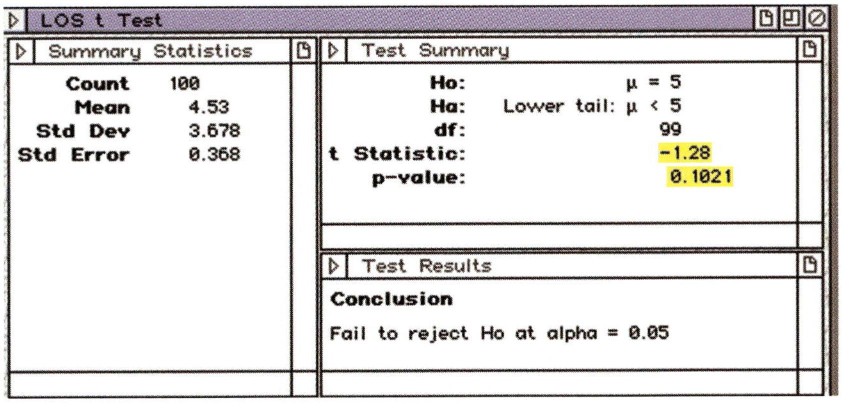

Figure 6.11
Excel/DDXL printout for the hypothesis test of Example 6.6

on the printout. Because the p-value exceeds our selected α value, $\alpha = .05$, we cannot reject the null hypothesis. Hence, there is insufficient evidence (at $\alpha = .05$) to conclude that the true mean LOS at the hospital is less than 5 days.

Now Work Exercise 6.42

Note: Some statistical software packages (e.g., SPSS) will conduct only two-tailed tests of hypothesis. For these packages, you obtain the p-value for a one-tailed test as shown in the box:

Converting a Two-Tailed p-Value from a Printout to a One-Tailed p-value

$$p = \frac{\text{Reported } p\text{-value}}{2} \quad \text{if} \begin{cases} H_a \text{ is of form } > \text{ and } z \text{ is positive} \\ H_a \text{ is of form } < \text{ and } z \text{ is negative} \end{cases}$$

$$p = 1 - \left(\frac{\text{Reported } p\text{-value}}{2}\right) \quad \text{if} \begin{cases} H_a \text{ is of form } > \text{ and } z \text{ is negative} \\ H_a \text{ is of form } < \text{ and } z \text{ is positive} \end{cases}$$

Statistics IN Action Revisited — Testing a Population Mean

Refer to Kimberly-Clark Corp.'s survey of 250 people who kept a count of their use of Kleenex® tissues in diaries (pp. 319–320). We want to test the claim made by marketing experts that $\mu = 60$ is the average number of tissues used by people with colds against our belief that the population mean is smaller than 60 tissues—that is, we want to test

$$H_0: \mu = 60 \quad H_a: \mu < 60$$

We will select $\alpha = .05$ as the level of significance for the test.

The survey results for the 250 sampled Kleenex® users are stored in the **TISSUES** data file. A Minitab analysis of the data yielded the printout displayed in Figure SIA6.1.

The observed significance level of the test, highlighted on the printout, is p-value $= .018$. Because this p-value is less than $\alpha = .05$, we have sufficient evidence to reject H_0; therefore, we conclude that the mean number of tissues used by a person with a cold is less than 60 tissues.

[*Note:* If we conduct the same test using $\alpha = .01$ as the level of significance, we would have insufficient evidence to reject H_0 because the p-value $= .018$ is greater than $\alpha = .01$. Thus, at $\alpha = .01$, there is not sufficient evidence to support our alternative that the population mean is less than 60.]

Data Set: TISSUES

Test of mu = 60 vs < 60

Variable	N	Mean	StDev	SE Mean	95% Upper Bound	T	P
NUMUSED	250	56.6760	25.0343	1.5833	59.2900	-2.10	0.018

Figure SIA6.1
Minitab test of $\mu = 60$ for Kleenex® survey

Exercises 6.34–6.48

Learning the Mechanics

6.34 If a hypothesis test were conducted using $\alpha = .05$, for which of the following p-values would the null hypothesis be rejected?

a. .06 **b.** .10
c. .01 **d.** .001
e. .251 **f.** .042

6.35 For each α and observed significance level (p-value) pair, indicate whether the null hypothesis would be rejected.

a. $\alpha = .05$, p-value $= .10$
b. $\alpha = .10$, p-value $= .05$
c. $\alpha = .01$, p-value $= .001$
d. $\alpha = .025$, p-value $= .05$
e. $\alpha = .10$, p-value $= .45$

6.36 In a test of the hypothesis $H_0: \mu = 50$ versus $H_a: \mu > 50$, a sample of $n = 100$ observations possessed mean $\bar{x} = 49.4$ and standard deviation $s = 4.1$. Find and interpret the p-value for this test.

6.37 In a test of $H_0: \mu = 100$ against $H_a: \mu > 100$, the sample data yielded the test statistic $z = 2.17$. Find and interpret the p-value for the test.

6.38 In a test of the hypothesis H_0: $\mu = 10$ versus H_a: $\mu \neq 10$, a sample of $n = 50$ observations possessed mean $\bar{x} = 10.7$ and standard deviation $s = 3.1$. Find and interpret the p-value for this test.

6.39 NW In a test of H_0: $\mu = 100$ against H_a: $\mu \neq 100$, the sample data yielded the test statistic $z = 2.17$. Find the p-value for the test.

6.40 In a test of H_0: $\mu = 75$ performed using the computer, SPSS reports a two-tailed p-value of .1032. Make the appropriate conclusion for each of the following situations:

a. H_a: $\mu < 75$, $z = -1.63$, $\alpha = .05$
b. H_a: $\mu < 75$, $z = 1.63$, $\alpha = .10$
c. H_a: $\mu > 75$, $z = 1.63$, $\alpha = .10$
d. H_a: $\mu \neq 75$, $z = -1.63$, $\alpha = .01$

6.41 An analyst tested the null hypothesis $\mu \geq 20$ against the alternative hypothesis that $\mu < 20$. The analyst reported a p-value of .06. What is the smallest value of α for which the null hypothesis would be rejected?

Applying the Concepts—Basic

6.42 NW **Prices of hybrid cars.** Refer to the *BusinessWeek.com* listing of the July 2006 dealer prices for the hybrid Toyota Prius, Exercise 6.24 (p. 334). Recall that the data are saved in the **HYBRIDCARS** file. The results of the test of H_0: $\mu = 25{,}000$ versus H_a: $\mu \neq 25.000$ is shown in the Minitab printout below. Find and interpret the p-value of the test.

6.43 **Size of diamonds sold at retail.** Refer to the carat data for 308 diamonds saved in the **DIAMONDS** file. In Exercise 6.26 (p. 335), you tested H_0: $\mu = .6$ against H_a: $\mu \neq .6$ based on a random sample of 30 diamonds.

a. Use a statistical software package to find the p-value of the test.
b. Compare the p-value to $\alpha = .05$ and make the appropriate conclusion.

6.44 **Testing mean weight of golf tees.** In Exercise 6.27 (p. 335), you tested H_0: $\mu = .250$ versus H_a: $\mu \neq .250$, where μ is the population mean weight of plastic golf tees. Recall that the data are saved in the **TEES** file. An SPSS printout for the hypothesis test is shown below. Locate the p-value on the printout and interpret its value.

6.45 **Revenue for a full-service funeral.** Refer to the National Funeral Directors Association study of the average fee charged for a full-service funeral, Exercise 6.30 (p. 335). Recall that a test was conducted to determine if the true mean fee charged exceeds \$6,500. The data (saved in the **FUNERAL** file) for the sample of 36 funeral homes were analyzed using Excel/DDXL. The resulting printout of the test of hypothesis is shown below.

a. Locate the p-value for this upper-tailed test of hypothesis.
b. Use the p-value to make a decision regarding the null hypothesis tested. Does the decision agree with your decision in Exercise 6.30?

Minitab Output for Exercise 6.42

```
Test of mu = 25000 vs not = 25000

Variable    N     Mean    StDev  SE Mean          95% CI              T      P
PRICE     160  25476.7   2429.8    192.1   (25097.3, 25856.1)   2.48  0.014
```

SPSS Output for Exercise 6.44

One-Sample Test

	Test Value = .250					
					95% Confidence Interval of the Difference	
	t	df	Sig. (2-tailed)	Mean Difference	Lower	Upper
WEIGHT	7.019	39	.000	.002475	.00176	.00319

Excel/DDXL Output for Exercise 6.45

Revenue06 t Test

Summary Statistics

Count	36
Mean	6.819
Std Dev	1.265
Std Error	0.211

Test Summary

Ho:	μ = 6.5
Ha:	Upper tail: μ > 6.5
df:	35
t Statistic:	1.52
p-value:	0.0693

Test Results

Conclusion

Fail to reject Ho at alpha = 0.05

Applying the Concepts—Intermediate

6.46 Feminized faces in TV commercials. Television commercials most often employ females, or "feminized" males, to pitch a company's product. Research published in *Nature* (Aug. 27, 1998) revealed that people are, in fact, more attracted to "feminized" faces, regardless of gender. In one experiment, 50 human subjects viewed both a Japanese female face and a Caucasian male face on a computer. Using special computer graphics, each subject could morph the faces (by making them more feminine or more masculine) until they attained the "most attractive" face. The level of feminization x (measured as a percentage) was measured.

a. For the Japanese female face, $\bar{x} = 10.2\%$ and $s = 31.3\%$. The researchers used this sample information to test the null hypothesis of a mean level of feminization equal to 0%. Verify that the test statistic is equal to 2.3.

b. Refer to part **a.** The researchers reported the p-value of the test as $p = .021$. Verify and interpret this result.

c. For the Caucasian male face, $\bar{x} = 15.0\%$ and $s = 25.1\%$. The researchers reported the test statistic (for the test of the null hypothesis stated in part **a**) as 4.23 with an associated p-value of approximately 0. Verify and interpret these results.

6.47 Buy-side vs. sell-side analysts' earnings forecasts. Refer to the *Financial Analysts Journal* (Jul./Aug. 2008) study of earnings forecasts of buy-side and sell-side analysts, Exercise 2.84 (p. 76). Buy-side analysts differ from sell-side analysts on a variety of factors, including scope of industry coverage, sources of information used, and target audience. Recall that data were collected on 3,526 forecasts made by buy-side analysts and 58,562 forecasts made by sell-side analysts, and the relative absolute forecast error was determined for each. A positive forecast error indicates that the analyst is overestimating earnings, while a negative forecast error implies that the analyst is underestimating earnings. Summary statistics for the forecast errors in the two samples are reproduced in the next table.

	Buy-Side Analysts	Sell-Side Analysts
Mean	0.85	−0.05
Standard Deviation	1.93	0.85

Source: Groysberg, B., Healy, P., and Chapman, C. "Buy-side vs. sell-side analysts' earnings forecast," *Financial Analysts Journal,* Vol. 64, No. 4, Jul./Aug. 2008.

a. Conduct a test (at $\alpha = .01$) to determine if the true mean forecast error for buy-side analysts is positive. Use the observed significance level (p-value) of the test to make your decision and state your conclusion in the words of the problem.

b. Conduct a test (at $\alpha = .01$) to determine if the true mean forecast error for sell-side analysts is negative. Use the observed significance level (p-value) of the test to make your decision and state your conclusion in the words of the problem.

Applying the Concepts—Advanced

6.48 Ages of cable TV shoppers. In a paper presented at the 2000 Conference of the International Association for Time Use Research, Professor Margaret Sanik of Ohio State University reported the results of her study on American cable TV viewers who purchase items from one of the home shopping channels. She found that the average age of these cable TV shoppers was 51 years. Suppose you want to test the null hypothesis, H_0: $\mu = 51$, using a sample of $n = 50$ cable TV shoppers.

a. Find the p-value of a two-tailed test if $\bar{x} = 52.3$ and $s = 7.1$.

b. Find the p-value of an upper-tailed test if $\bar{x} = 52.3$ and $s = 7.1$.

c. Find the p-value of a two-tailed test if $\bar{x} = 52.3$ and $s = 10.4$,

d. For each of the tests, parts **a–c,** give a value of α that will lead to a rejection of the null hypothesis.

e. If $\bar{x} = 52.3$, give a value of s that will yield a two-tailed p-value of .01 or less.

6.5 Test of Hypothesis about a Population Mean: Student's t-Statistic

A manufacturing operation consists of a single-machine-tool system that produces an average of 15.5 transformer parts every hour. After undergoing a complete overhaul, the system was monitored by observing the number of parts produced in each of 17 randomly selected 1-hour periods. The mean and standard deviation for the 17 production runs are

$$\bar{x} = 15.42 \qquad s = .16$$

Does this sample provide sufficient evidence to conclude that the true mean number of parts produced every hour by the overhauled system differs from 15.5?

This inference can be placed in a test of hypothesis framework. We establish the preoverhaul mean as the null hypothesized value and use a two-tailed alternative that the true mean of the overhauled system differs from the preoverhaul mean:

H_0: $\mu = 15.5$ (Mean of overhauled system equals 15.5 parts per hour.)

H_a: $\mu \neq 15.5$ (Mean of overhauled system differs from 15.5 parts per hour.)

Recall from Section 5.3 that when we are faced with making inferences about a population mean using the information in a small sample, two problems emerge:

1. The normality of the sampling distribution for $\bar{x}$ does not follow from the Central Limit Theorem when the sample size is small. We must assume that the distribution of measurements from which the sample was selected is

approximately normally distributed in order to ensure the approximate normality of the sampling distribution of $\bar{x}$.

2. If the population standard deviation σ is unknown, as is usually the case, then we cannot assume that s will provide a good approximation for σ when the sample size is small. Instead, we must use the t-distribution rather than the standard normal z-distribution to make inferences about the population mean μ.

Therefore, as the test statistic of a small-sample test of a population mean, we use the t-statistic:

$$\textit{Test statistic: } t = \frac{\bar{x} - \mu_0}{s/\sqrt{n}} = \frac{\bar{x} - 15.5}{s/\sqrt{n}}$$

where μ_0 is the null hypothesized value of the population mean μ. In our example, $\mu_0 = 15.5$.

To find the rejection region, we must specify the value of α, the probability that the test will lead to rejection of the null hypothesis when it is true, and then consult the t-table (Table V in Appendix B). Using $\alpha = .05$, the two-tailed rejection region is

Rejection region: $t_{\alpha/2} = t_{.025} = 2.120$ with $n - 1 = 16$ df

Reject H_0 if $t < -2.120$ or $t > 2.120$

The rejection region is shown in Figure 6.12.

We are now prepared to calculate the test statistic and reach a conclusion:

$$t = \frac{\bar{x} - \mu_0}{s/\sqrt{n}} = \frac{15.42 - 15.50}{.16/\sqrt{n}} = \frac{-.08}{.0388} = -2.06$$

Because the calculated value of t does not fall in the rejection region (Figure 6.12), we cannot reject H_0 at the $\alpha = .05$ level of significance. Based on the sample evidence, we should not conclude that the mean number of parts produced per hour by the overhauled system differs from 15.5.

It is interesting to note that the calculated t-value, -2.06, is *less than* $-z_{.05} = -1.96$. The implication is that if we had *incorrectly* used a z-statistic for this test, we would have rejected the null hypothesis at $\alpha = .05$, concluding that the mean production per hour of the overhauled system differs from 15.5 parts. The important point is that the statistical procedure to be used must always be closely scrutinized, with all the assumptions understood. Many statistical distortions are the result of misapplications of otherwise valid procedures.

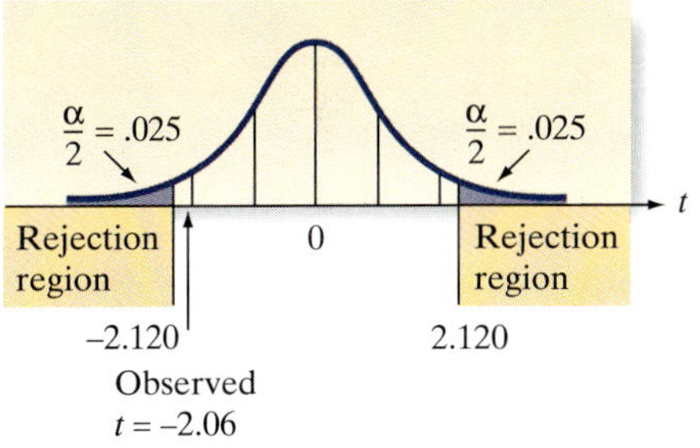

Figure 6.12
Two-tailed rejection region for small-sample t-test

The technique for conducting a small-sample test of hypothesis about a population mean is summarized in the following box.

Small-Sample Test of Hypothesis about μ

One-Tailed Test	Two-Tailed Test
$H_0: \mu = \mu_0$	$H_0: \mu = \mu_0$
$H_a: \mu < \mu_0$ (or $H_a: \mu > \mu_0$)	$H_a: \mu \neq \mu_0$
Test statistic: $t = \frac{\bar{x} - \mu_0}{s/\sqrt{n}}$	*Test statistic:* $t = \frac{\bar{x} - \mu_0}{s/\sqrt{n}}$
Rejection region: $t < -t_\alpha$ (or $t > t_\alpha$ when $H_a: \mu > \mu_0$)	*Rejection region:* $\lvert t \rvert > t_{\alpha/2}$

where t_α and $t_{\alpha/2}$ are based on $(n - 1)$ degrees of freedom

Conditions Required for a Valid Small-Sample Hypothesis Test for μ

1. A random sample is selected from the target population.
2. The population from which the sample is selected has a distribution that is approximately normal.

Example 6.7

Conducting a Small-Sample Test for μ—Does a New Engine Meet Air Pollution Standards?

Problem A major car manufacturer wants to test a new engine to determine whether it meets new air pollution standards. The mean emission μ of all engines of this type must be less than 20 parts per million of carbon. Ten engines are manufactured for testing purposes, and the emission level of each is determined. The data (in parts per million) are listed in Table 6.5.

Do the data supply sufficient evidence to allow the manufacturer to conclude that this type of engine meets the pollution standard? Assume that the production process is stable and the manufacturer is willing to risk a Type I error with probability $\alpha = .01$.

Table 6.5 **Emission Levels for Ten Engines**

15.6	16.2	22.5	20.5	16.4	19.4	19.6	17.9	12.7	14.9

Data Set: EMISSIONS

Solution The manufacturer wants to support the research hypothesis that the mean emission level μ for all engines of this type is less than 20 parts per million. The elements of this small-sample one-tailed test are

H_0: $\mu = 20$ (Mean emission level equals 20 ppm.)

H_a: $\mu < 20$ (Mean emission level is less than 20 ppm—i.e., engine meets pollution standard.)

Test statistic: $$t = \frac{\bar{x} - 20}{s/\sqrt{n}}$$

Rejection region: For $\alpha = .01$ and df $= n - 1 = 9$, the one-tailed rejection region (see Figure 6.13) is $t < -t_{.01} = -2.821$.

Assumption: The relative frequency distribution of the population of emission levels for all engines of this type is approximately normal. Based on the shape of the Minitab stem-and-leaf display of the data shown in Figure 6.14, this assumption appears to be reasonably satisfied.

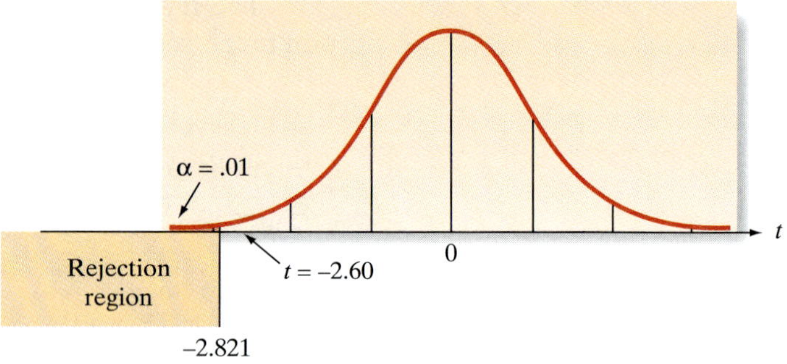

Figure 6.13
A *t*-distribution with 9 df and the rejection region for Example 6.7

To calculate the test statistic, we entered the data into a computer and analyzed it using Minitab. The Minitab printout is shown in Figure 6.14. From the printout, we obtain $\bar{x} = 17.57$, $s = 2.95$. Substituting these values into the test statistic formula, we get

$$t = \frac{\bar{x} - 20}{s/\sqrt{n}} = \frac{17.57 - 20}{2.95/\sqrt{10}} = -2.60$$

Stem-and-Leaf Display: E-LEVEL

```
Stem-and-leaf of E-LEVEL  N  = 10
Leaf Unit = 1.0


 1   1  2
 3   1  45
(3)  1  667
 4   1  99
 2   2  0
 1   2  2
```

One-Sample T: E-LEVEL

```
Test of mu = 20 vs < 20
```

Variable	N	Mean	StDev	SE Mean	95% Upper Bound	T	P
E-LEVEL	10	17.5700	2.9522	0.9336	19.2814	-2.60	0.014

Figure 6.14
Minitab analysis of 10 emission levels, Example 6.7

Because the calculated t falls outside the rejection region (see Figure 6.13), the manufacturer cannot reject H_0. There is insufficient evidence to conclude that $\mu < 20$ parts per million. Consequently, we cannot conclude that the new engine type meets the pollution standard.

Look Back Are you satisfied with the reliability associated with this inference? The probability is only $\alpha = .01$ that the test would support the research hypothesis if, in fact, it were false.

Now Work Exercise 6.52a,b

Example 6.8
The p-Value for a Small-Sample Test of μ

Problem Find the observed significance level for the test described in Example 6.7. Interpret the result.

Solution The test of Example 6.7 was a lower-tailed test: H_0: $\mu = 20$ versus H_a: $\mu < 20$. Because the value of t computed from the sample data was $t = -2.60$, the observed significance level (or p-value) for the test is equal to the probability that t would assume a value less than or equal to -2.60 if, in fact, H_0 were true. This is equal to the area in the lower tail of the t-distribution (shaded in Figure 6.15).

One way to find this area (i.e., the p-value for the test) is to consult the t-table (Table V in Appendix B). Unlike the table of areas under the normal curve, Table V gives only the t-values corresponding to the areas .100, .050, .025, .010, .005, .001, and .0005. Therefore, we can only approximate the p-value for the test. Because the observed t-value was based on 9 degrees of freedom, we use the df = 9 row in Table V and move across the row until we reach the t-values that are closest to the observed $t = -2.60$. [*Note:* We ignore the minus sign.] The t-values corresponding to p-values of .010 and .025 are 2.821 and 2.262, respectively. Because the observed t-value falls between $t_{.010}$ and $t_{.025}$, the p-value for the test lies between .010 and .025. In other words, $.010 < p\text{-value} < .025$. Thus, we would reject the null hypothesis, H_0: $\mu = 20$ parts per million, for any value of α larger than .025 (the upper bound of the p-value).

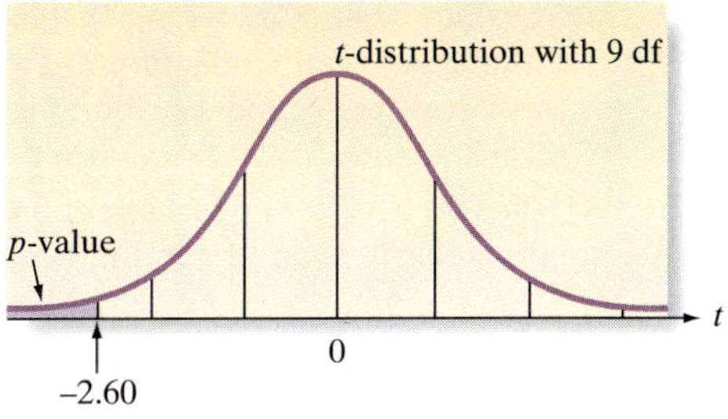

Figure 6.15
The observed significance level for the test of Example 6.7

A second, more accurate, way to obtain the p-value is to use a statistical software package to conduct the test of hypothesis. The Minitab printout shown in Figure 6.14 gives both the test statistic (−2.60) and p-value (.014).

You can see that the actual p-value of the test falls within the bounds obtained from Table V. Based on the actual p-value, we will reject H_0: $\mu = 20$ in favor of H_a: $\mu < 20$ for any α level larger than .014.

Now Work Exercise 6.52c

Small-sample inferences typically require more assumptions and provide less information about the population parameter than do large-sample inferences. Nevertheless, the t-test is a method of testing a hypothesis about a population mean of a normal distribution when only a small number of observations are available. What can be done if you know that the population relative frequency distribution is decidedly nonnormal, say highly skewed?

What Can Be Done If the Population Relative Frequency Distribution Departs Greatly from Normal?

Answer: Use one of the nonparametric statistical methods of Chapter 14.

Activity 6.2 *Keep the Change:* Tests of Hypotheses

In this activity, we will test claims that the mean amount transferred for any single purchase is \$0.50 and that the mean amount that Bank of America matches for a customer during the first 90 days of enrollment is at least \$25. We will be working with data sets from Activity 1.1, p. 16 and Activity 4.2, p. 250.

1. Based on the assumption that all transfer amounts between \$0.00 and \$0.99 seem to be equally likely to occur, one may conclude that the mean of the amounts transferred is about \$0.50. Explain how someone who doesn't believe this conclusion would use a test of hypothesis to argue that the conclusion is false.
2. Suppose that your original data set *Amounts Transferred* from Activity 1.1 represents a random sample of amounts transferred for all Bank of America customers' purchases. Does your sample meet the requirements for performing either a large-sample or a small-sample test of hypothesis about the population mean? Explain. If your data meet the criteria for one of the tests, perform that test at $\alpha = .05$.
3. Use the pooled data set of *Amounts Transferred* from Activity 4.2 to represent a random sample of amounts transferred for all Bank of America customers' purchases. Explain how the conditions for the large-sample test of hypothesis about a population mean are met. Then perform the test at $\alpha = .05$. Do your results suggest that the mean may be something other than \$0.50? Explain.
4. A friend suggests to you that the mean amount the bank matches for a customer during the first 90 days is at least \$25. Explain how you could use a test of hypothesis to argue that your friend is wrong.
5. Suppose that your data set *Bank Matching* from Activity 1.1 represents a random sample of all Bank of America customers' bank matching. Perform an appropriate test of hypotheses at $\alpha = .05$ against your friend's claim. Assume that the underlying distribution is normal, if necessary. Does the test provide evidence that your friend's claim is false?

Keep the results from this activity for use in other activities.

Exercises 6.49–6.63

Learning the Mechanics

6.49 **a.** Consider testing H_0: $\mu = 80$. Under what conditions should you use the t-distribution to conduct the test?

b. In what ways are the distributions of the z-statistic and t-test statistic alike? How do they differ?

6.50 For each of the following rejection regions, sketch the sampling distribution of t and indicate the location of the rejection region on your sketch:

a. $t > 1.440$ where df $= 6$

b. $t < -1.782$ where df $= 12$

c. $t < -2.060$ or $t > 2.060$ where df $= 25$

d. For each of parts **a–c**, what is the probability that a Type I error will be made?

6.51 A random sample of n observations is selected from a normal population to test the null hypothesis that $\mu = 10$. Specify the rejection region for each of the following combinations of H_a, α, and n:

a. H_a: $\mu \neq 10$; $\alpha = .05$; $n = 14$

b. H_a: $\mu > 10$; $\alpha = .01$; $n = 24$

c. H_a: $\mu > 10$; $\alpha = .10$; $n = 9$

d. H_a: $\mu < 10$; $\alpha = .01$; $n = 12$

e. H_a: $\mu \neq 10$; $\alpha = .10$; $n = 20$

f. H_a: $\mu < 10$; $\alpha = .05$; $n = 4$

6.52 NW A sample of five measurements, randomly selected from a normally distributed population, resulted in the following summary statistics: $\bar{x} = 4.8$, $s = 1.3$.

a. Test the null hypothesis that the mean of the population is 6 against the alternative hypothesis, $\mu < 6$. Use $\alpha = .05$.

b. Test the null hypothesis that the mean of the population is 6 against the alternative hypothesis, $\mu \neq 6$. Use $\alpha = .05$.

c. Find the observed significance level for each test.

6.53 Suppose you conduct a *t*-test for the null hypothesis H_0: $\mu = 1{,}000$ versus the alternative hypothesis H_a: $\mu > 1{,}000$ based on a sample of 17 observations. The test results are $t = 1.89$ and p-value $= .038$.

a. What assumptions are necessary for the validity of this procedure?

b. Interpret the results of the test.

c. Suppose the alternative hypothesis had been the two-tailed H_a: $\mu \neq 1{,}000$. If the *t*-statistic were unchanged, then what would the *p*-value be for this test? Interpret the *p*-value for the two-tailed test.

Applying the Concepts—Basic

6.54 A new dental bonding agent. When bonding teeth, orthodontists must maintain a dry field. A new bonding adhesive (called *Smartbond*) has been developed to eliminate the necessity of a dry field. However, there is concern that the new bonding adhesive is not as strong as the current standard, a composite adhesive (*Trends in Biomaterials & Artificial Organs,* Jan. 2003). Tests on a sample of 10 extracted teeth bonded with the new adhesive resulted in a mean breaking strength (after 24 hours) of $\bar{x} = 5.07$ Mpa and a standard deviation of $s = .46$ Mpa. Orthodontists want to know if the true mean breaking strength of the new bonding adhesive is less than 5.70 Mpa, the mean breaking strength of the composite adhesive.

a. Set up the null and alternative hypotheses for the test.

b. Find the rejection region for the test using $\alpha = .01$.

c. Compute the test statistic.

d. Give the appropriate conclusion for the test.

e. What conditions are required for the test results to be valid?

6.55 Surface roughness of pipe. Refer to the *Anti-corrosion Methods and Materials* (Vol. 50, 2003) study of the surface roughness of coated interior pipe used in oil fields, Exercise 5.30 (p. 288). The data (in micrometers) for 20 sampled pipe sections are reproduced in the next table and saved in the **ROUGHPIPE** file.

a. Give the null and alternative hypotheses for testing whether the mean surface roughness of coated interior pipe, μ, differs from 2 micrometers.

b. Find the test statistic for the hypothesis test.

c. Give the rejection region for the hypothesis test, using $\alpha = .05$.

d. State the appropriate conclusion for the hypothesis test.

e. A Minitab printout giving the test results is shown at the bottom of the page. Find and interpret the *p*-value of the test.

f. In Exercise 5.30 you found a 95% confidence interval for μ. Explain why the confidence interval and test lead to the same conclusion about μ.

1.72	2.50	2.16	2.13	1.06	2.24	2.31	2.03	1.09	1.40
2.57	2.64	1.26	2.05	1.19	2.13	1.27	1.51	2.41	1.95

Source: Farshad, F., and Pesacreta, T. "Coated pipe interior surface roughness as measured by three scanning probe instruments," *Anti-corrosion Methods and Materials,* Vol. 50, No. 1, 2003 (Table III). Copyright © 2003 MCB UP Ltd.

6.56 Product usage index. Information Resources, Inc., a Chicago-based research organization, tracks supermarket sales in 28 metropolitan markets in the United States. They convert their data for specific products to an index that measures product usage relative to the national average usage. For example, Green Bay, Wisconsin's, ketchup index is 143, the highest in the nation. This means that Green Bay residents consume 43% more ketchup, on average, than the mean national consumption rate. The table lists the salad dressings index for each in a sample of seven Southeastern cities. These data are saved in the **SALAD** file.

Salad Dressings Index (U.S. mean = 100)	
Charlotte, N.C.	124
Birmingham, Al.	99
Raleigh, N.C.	124
Knoxville, Tenn.	99
Memphis, Tenn.	90
Atlanta, Ga.	111
Nashville, Tenn.	89

Source: Wall Street Journal Interactive Edition, Jan. 5, 2000. Copyright 2000 by Dow Jones & Company Inc. Reproduced with permission of Dow Jones & Company, Inc. in the format Textbook via Copyright Clearance Center.

a. Specify the appropriate null and alternative hypotheses for testing whether the true mean consumption rate of salad dressings in the Southeastern United States is different than the mean national consumption rate of 100.

b. What assumptions about the sample and population must hold in order for it to be appropriate to use a *t*-statistic in conducting the hypothesis test?

Minitab Output for Exercise 6.55

One-Sample T: ROUGH

```
Test of mu = 2 vs not = 2

Variable   N     Mean    StDev  SE Mean           95% CI              T      P
ROUGH     20  1.88100  0.52391  0.11715  (1.63580, 2.12620)  -1.02  0.322
```

Excel/DDXL Output for Exercise 6.57

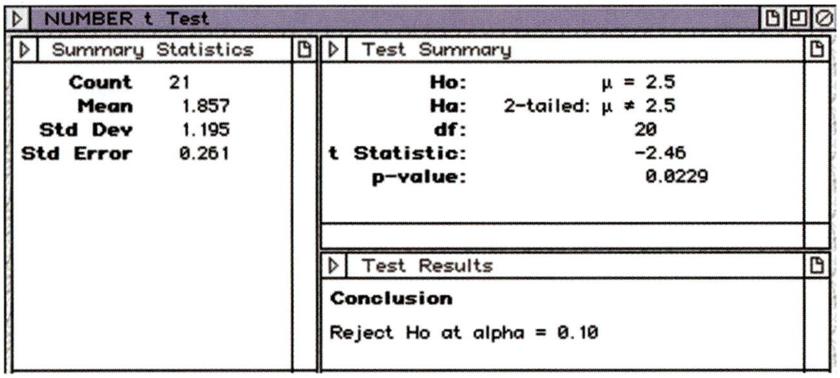

c. Conduct the hypothesis test using $\alpha = .05$.
d. Is the observed significance level of the test greater or less than .05? Justify your answer.

6.57 Al Qaeda attacks on the United States. Refer to the *Studies in Conflict & Terrorism* (Vol. 29, 2006) analysis of recent incidents involving suicide terrorist attacks, Exercise 5.33 (p. 289). Data on the number of individual suicide bombings that occurred in each of 21 sampled Al Qaeda attacks against the United States is reproduced in the table. (Recall the data are saved in the **ALQAEDA** file).

a. Do the data indicate that the true mean number of suicide bombings for all Al Qaeda attacks against the United States differs from 2.5? Use $\alpha = .10$ and the accompanying Excel/DDXL printout to answer the question.
b. In Exercise 5.33 you found a 90% confidence interval for the mean, μ, of the population. Answer the question in part **a** based on the 90% confidence interval.
c. Do the inferences derived from the test (part **a**) and confidence interval (part **b**) agree? Explain why or why not.
d. What assumption about the data must be true for the inferences to be valid?
e. Use a graph to check whether the assumption, part **d**, is reasonably satisfied. Comment on the validity of the inference.

1	1	2	1	2	4	1	1	1	1	2	3	4	5	1	1	1	2	2	2	1

Source: Moghadam, A. "Suicide terrorism, occupation, and the globalization of martyrdom: A critique of *Dying to Win*," *Studies in Conflict & Terrorism*, Vol. 29, No. 8, 2006, pp. 707–729 (Table 3). Reprinted by permission of the publisher (Taylor & Francis Group, www.informaworld.com).

Applying the Concepts—Intermediate

6.58 Spending by banks following acquisitions. Refer to *The Nilson Report* (Oct. 2008) study of U.S. banks with recent acquisitions and mergers, Exercise 2.44 (p. 61). The amount charged to credit and debit cards issued by the banks was determined. The next table gives the total amount charged in 2007 for a sample of five banks. These data are saved in the **BANKCHARGE** file.

a. Give the null and alternative hypothesis for determining whether the mean amount charged to credit and debit cards in 2007 by all banks with recent acquisitions differs from $200 billion.
b. Use the results shown on the SPSS printout at the top of page 349 to conduct the test at $\alpha = .05$.
c. Repeat part **b**, but conduct the test at $\alpha = .10$.
d. You may recall from Exercise 2.44 that the five banks in the sample represent the top ranked banks in the United States with recent acquisitions. Explain how this sample may impact the validity of your conclusions in parts **b** and **c.**

U.S. Bank/Card Issuer	Amount Charged ($ billions)
Bank of America	462.20
JPMorgan Chase	458.84
American Express	445.32
Citigroup	240.58
Wells Fargo	192.68

Source: The Nilson Report, No. 912, October 2008 (p. 8).

6.59 Minimizing tractor skidding distance. Refer to the *Journal of Forest Engineering* (July 1999) study of minimizing tractor skidding distances along a new road in a European forest, Exercise 5.31 (p. 288). The skidding distances (in meters) were measured at 20 randomly selected road sites. The data (saved in the **SKIDDING** file) are repeated below. Recall that a logger working on the road claims the mean skidding distance is at least 425 meters. Is there sufficient evidence to refute this claim? Use $\alpha = .10$.

488	350	457	199	285	409	435	574	439	546
385	295	184	261	273	400	311	312	141	425

Source: Extracted from Tujek, J., and Pacola, E. "Algorithms for skidding distance modeling on a raster digital terrain model," *Journal of Forest Engineering*, Vol. 10, No. 1, July 1999 (Table 1).

SPSS Output for Exercise 6.58

One-Sample Statistics

	N	Mean	Std. Deviation	Std. Error Mean
AMOUNT	5	359.9240	132.05186	59.05539

One-Sample Test

	Test Value = 200					
					95% Confidence Interval of the Difference	
	t	df	Sig. (2-tailed)	Mean Difference	Lower	Upper
AMOUNT	2.708	4	.054	159.92400	-4.0400	323.8880

6.60 Testing the effectiveness of a mosquito repellent. A study was conducted to evaluate the effectiveness of a mosquito repellent designed by the U.S. Army to be applied as camouflage face paint (*Journal of the Mosquito Control Association,* June 1995). The repellent was applied to the forearms of five volunteers, and then the arms were exposed to 15 active mosquitoes for a 10-hour period. The percentage of the forearm surface area protected from bites (called *percent repellency*) was calculated for each of the five volunteers. For one color of paint (loam), the following summary statistics were obtained:

$$\bar{x} = 83\% \qquad s = 15\%$$

a. The new repellent is considered effective if it provides a percent repellency of at least 95. Conduct a test to determine whether the mean repellency percentage of the new mosquito repellent is less than 95. Test using $\alpha = .10$

b. What assumptions are required for the hypothesis test in part **a** to be valid?

6.61 Crack intensity of paved highways. The Mississippi Department of Transportation collected data on the number of cracks (called *crack intensity*) in an undivided two-lane highway using van-mounted, state-of-the-art video technology (*Journal of Infrastructure Systems,* Mar. 1995). The mean number of cracks found in a sample of eight 50-meter sections of the highway was $\bar{x} = .210$, with a variance of $s^2 = .011$. Suppose the American Association of State Highway and Transportation Officials (AASHTO) recommends a maximum mean crack intensity of .100 for safety purposes. Is there evidence to say that the true mean crack intensity of the Mississippi highway exceeds the AASHTO recommended maximum? Use $\alpha = .01$ in the test.

6.62 Active nuclear power plants. Refer to the U.S. Energy Information Administration's list of active nuclear power plants operating in each of a sample of 20 states, Exercise 2.54 (p. 63). The data, saved in the **NUCLEAR** file, are reproduced in the next table.

a. Is there sufficient evidence to claim that the mean number of active nuclear power plants operating in all states exceeds 3? Test using $\alpha = .10$.

b. Are the conditions required for a valid small-sample test reasonably satisfied? Explain.

c. Eliminate the lowest two values and the highest two values from the data set, then conduct the test of part **a** on the smaller data set. What impact does this have on the test results?

d. Why is it dangerous to eliminate data points in order to satisfy an assumption for a test of hypothesis?

State	Number of Power Plants	State	Number of Power Plants
Alabama	5	New Hampshire	1
Arizona	3	New York	6
California	4	North Carolina	5
Florida	5	Ohio	3
Georgia	4	Pennsylvania	9
Illinois	11	South Carolina	7
Kansas	1	Tennessee	3
Louisiana	2	Texas	4
Massachusetts	1	Vermont	1
Mississippi	3	Wisconsin	3

Source: Statistical Abstract of the United States, 2005 (Table 906). U.S. Energy Information Administration, Electric Power Annual.

Applying the Concepts—Advanced

6.63 Arsenic in smelters. The Occupational Safety and Health Act (OSHA) allows issuance of engineering standards to ensure safe workplaces for all Americans. The maximum allowable mean level of arsenic in smelters, herbicide production facilities, and other places where arsenic is used is .004 milligrams per cubic meter of air. Suppose smelters at two plants are being investigated to determine whether they are meeting OSHA standards. Two analyses of the air are made at each plant, and the results (in milligrams per cubic meter of air) are shown in the table. The data are saved in the **ARSENIC** file. A claim is made that the OSHA standard is violated at Plant 2 but not at Plant 1. Do you agree?

Plant 1		Plant 2	
Observation	Arsenic Level	Observation	Arsenic Level
1	.01	1	.05
2	.005	2	.09

6.6 Large-Sample Test of Hypothesis about a Population Proportion

Inferences about population proportions (or percentages) are often made in the context of the probability, p, of "success" for a binomial distribution. We saw how to use large samples from binomial distributions to form confidence intervals for p in Section 5.4. We now consider tests of hypotheses about p.

For example, consider the problem of *insider trading* in the stock market. Insider trading is the buying and selling of stock by an individual privy to inside information in a company, usually a high-level executive in the firm. The Securities and Exchange Commission (SEC) imposes strict guidelines about insider trading so that all investors can have equal access to information that may affect the stock's price. An investor wishing to test the effectiveness of the SEC guidelines monitors the market for a period of a year and records the number of times a stock price increases the day following a significant purchase of stock by an insider. For a total of 576 such transactions, the stock increased the following day 327 times. Does this sample provide evidence that the stock price may be affected by insider trading?

We first view this as a binomial experiment, with the 576 transactions as the trials, with success representing an increase in the stock's price the following day. Let p represent the probability that the stock price will increase following a large insider purchase. If the insider purchase has no effect on the stock price (that is, if the information available to the insider is identical to that available to the general market), then the investor expects the probability of a stock increase to be the same as that of a decrease, or $p = .5$. On the other hand, if insider trading affects the stock price (indicating that the market has not fully accounted for the information known to the insiders), then the investor expects the stock either to decrease or to increase more than half the time following significant insider transactions; that is, $p \neq .5$.

We can now place the problem in the context of a test of hypothesis:

H_0: $p = .5$ (Probability of stock increase equals .5–
i.e., insider purchase has no effect on stock price.)

H_a: $p \neq .5$ (Probability of stock increase differs from .5–
i.e., insider trading effects stock price.)

Recall that the sample proportion, $\hat{p}$, is really just the sample mean of the outcomes of the individual binomial trials and, as such, is approximately normally distributed (for large samples) according to the Central Limit Theorem. Thus, for large samples, we can use the standard normal z as the test statistic:

$$\textit{Test statistic: } z = \frac{\text{Sample proportion} - \text{Null hypothesized proportion}}{\text{Standard deviation of sample proportion}} = \frac{\hat{p} - p_0}{\sigma_{\hat{p}}}$$

where we use the symbol p_0 to represent the null hypothesized value of p.

Rejection region: We use the standard normal distribution to find the appropriate rejection region for the specified value of α. Using $\alpha = .05$, the two-tailed rejection region is

$$z < -z_{\alpha/2} = -z_{.025} = -1.96 \text{ or } z > z_{\alpha/2} = z_{.025} = 1.96$$

See Figure 6.16.

We are now prepared to calculate the value of the test statistic. Before doing so, we want to be sure that the sample size is large enough to ensure that the normal approximation for the sampling distribution of $\hat{p}$ is reasonable. Recall from Section 5.4 that we require both np and nq to be at least 15. Because the null hypothesized value, p_0, is assumed to be the true value of p until our test procedure indicates otherwise, then we check to see if $np_0 \geq 15$ and $nq_0 \geq 15$ (where $q_0 = 1 - p_0$). Now, $np_0 = (576)(.5) = 288$ and $nq_0 = (576)(.5) = 288$;

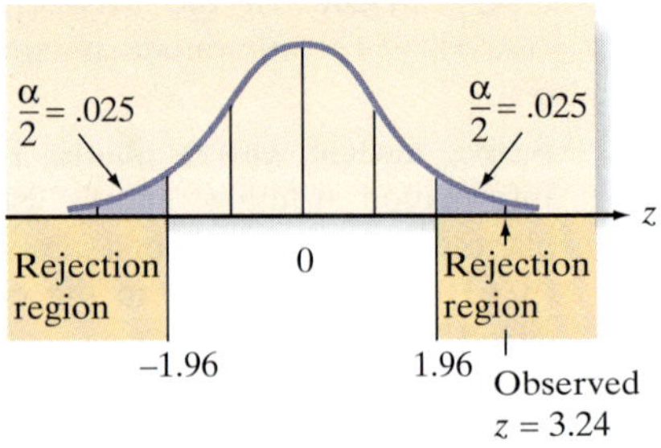

Figure 6.16
Rejection region for insider trading example

therefore, the normal distribution will provide a reasonable approximation for the sampling distribution of $\hat{p}$.

Returning to the hypothesis test at hand, the proportion of the sampled transactions that resulted in a stock increase is

$$\hat{p} = \frac{327}{576} = .568$$

Finally, we calculate the number of standard deviations (the z-value) between the sampled and hypothesized value of the binomial proportion:

$$z = \frac{\hat{p} - p_0}{\sigma_{\hat{p}}} = \frac{\hat{p} - p_0}{\sqrt{p_0 q_0 / n}} = \frac{.568 - .5}{.021} = \frac{.068}{.021} = 3.24$$

The implication is that the observed sample proportion is (approximately) 3.24 standard deviations above the null hypothesized proportion .5 (Figure 6.16). Therefore, we reject the null hypothesis, concluding at the .05 level of significance that the true probability of an increase or decrease in a stock's price differs from .5 the day following insider purchase of the stock. It appears that an insider purchase significantly increases the probability that the stock price will increase the following day. (To estimate the magnitude of the probability of an increase, a confidence interval can be constructed.)

The test of hypothesis about a population proportion p is summarized in the next box. Note that the procedure is entirely analogous to that used for conducting large-sample tests about a population mean.

Large-Sample Test of Hypothesis about *p*

One-Tailed Test	Two-Tailed Test
H_0: $p = p_0$	H_0: $p = p_0$
H_a: $p < p_0$ (or H_a: $p > p_0$)	H_a: $p \neq p_0$
Test statistic: $z = \frac{\hat{p} - p_0}{\sigma_{\hat{p}}}$	*Test statistic:* $z = \frac{\hat{p} - p_0}{\sigma_{\hat{p}}}$

where, according to H_0, $\sigma_{\hat{p}} = \sqrt{p_0 q_0 / n}$ and $q_0 = 1 - p_0$

Rejection region: $z < -z_\alpha$ (or $z > z_\alpha$ when H_a: $p > p_0$)	*Rejection region:* $\lvert z \rvert > z_{\alpha/2}$

Note: p_0 is the symbol for the numerical value of p assigned in the null hypothesis.

Conditions Required for a Valid Large-Sample Hypothesis Test for *p*

1. A random sample is selected from a binomial population.
2. The sample size n is large. (This condition will be satisfied if both $np_0 \geq 15$ and $nq_0 \geq 15$.)

Example 6.9

Conducting a Hypothesis Test for *p*, the Proportion of Defective Batteries

Problem The reputations (and hence sales) of many businesses can be severely damaged by shipments of manufactured items that contain a large percentage of defectives. For example, a manufacturer of alkaline batteries may want to be reasonably certain that fewer than 5% of its batteries are defective. Suppose 300 batteries are randomly selected from a very large shipment; each is tested, and 10 defective batteries are found. Does this provide sufficient evidence for the manufacturer to conclude that the fraction defective in the entire shipment is less than .05? Use $\alpha = .01$.

Solution The objective of the sampling is to determine whether there is sufficient evidence to indicate that the fraction defective, p, is less than .05. Consequently, we will test the null hypothesis that $p = .05$ against the alternative hypothesis that $p < .05$. The elements of the test are

H_0: $p = .05$ (Fraction of defective batteries equals .05.)

H_a: $p < .05$ (Fraction of defective batteries is less than .05.)

Test statistic: $z = \dfrac{\hat{p} - p_0}{\sigma_{\hat{p}}}$

Rejection region: $z < -z_{.01} = -2.33$ (see Figure 6.17)

Before conducting the test, we check to determine whether the sample size is large enough to use the normal approximation to the sampling distribution of $\hat{p}$. Because $np_0 = (300)(.05) = 15$ and $nq_0 = (300)(.95) = 285$ are both at least 15, the normal approximation will be adequate.

We now calculate the test statistic:

$$z = \frac{\hat{p} - .05}{\sigma_{\hat{p}}} = \frac{(10/300) - .05}{\sqrt{p_0 q_0/n}} = \frac{.03333 - .05}{\sqrt{p_0 q_0/300}}$$

Notice that we use p_0 to calculate $\sigma_{\hat{p}}$ because, in contrast to calculating $\sigma_{\hat{p}}$ for a confidence interval, the test statistic is computed on the assumption that the null hypothesis is true—that is, $p = p_0$. Therefore, substituting the values for $\hat{p}$ and p_0 into the z-statistic, we obtain

$$z \approx \frac{-.01667}{\sqrt{(.05)(.95)/300}} = \frac{-.01667}{.0126} = -1.32$$

As shown in Figure 6.17, the calculated z-value does not fall in the rejection region. Therefore, there is insufficient evidence at the .01 level of significance to indicate that the shipment contains fewer than 5% defective batteries.

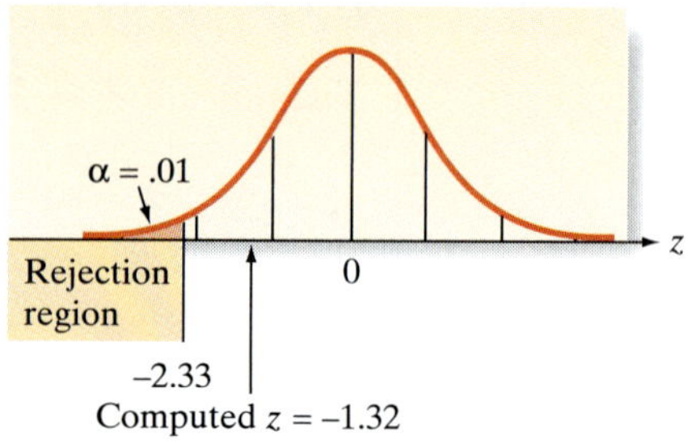

Figure 6.17
Rejection region for Example 6.9

Now Work Exercise 6.64a, b

Example 6.10
The p-Value for a Test about p

Problem In Example 6.9 we found that we did not have sufficient evidence, at the $\alpha = .01$ level of significance, to indicate that the fraction defective p of alkaline batteries was less than $p = .05$. How strong was the weight of evidence favoring the alternative hypothesis (H_a: $p < .05$)? Find the observed significance level for the test.

Solution The computed value of the test statistic z was $z = -1.32$. Therefore, for this lower-tailed test, the observed significance level is

$$\text{Observed significance level} = P(z \leq -1.32)$$

This lower-tail area is shown in Figure 6.18. The area between $z = 0$ and $z = 1.32$ is given in Table IV in Appendix B as .4066. Therefore, the observed significance level is $.5 - .4066 = .0934$.

Note: The observed significance level can also be obtained with statistical software. The Minitab printout shown in Figure 6.19 gives the p-value (highlighted).

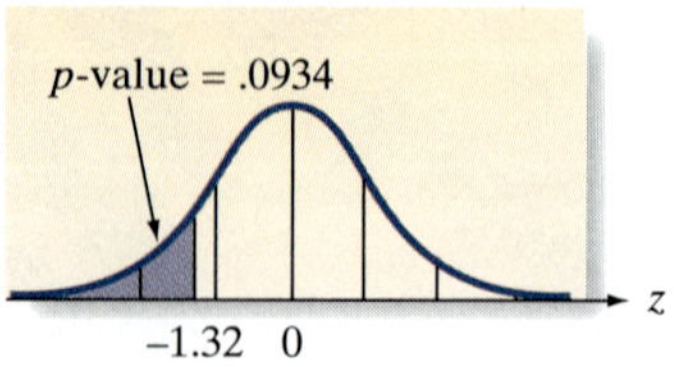

Figure 6.18
The observed significance level for Example 6.10

Look Back Although we did not reject H_0: $p = .05$ at $\alpha = .01$, the probability of observing a z-value as small as or smaller than -1.32 is only .0934 if, in fact, H_0 is true.

Test and CI for One Proportion

```
Test of p = 0.05 vs p < 0.05

                                  95% Upper
Sample    X    N  Sample p          Bound  Z-Value  P-Value
1        10  300  0.033333       0.050380    -1.32    0.093

Using the normal approximation.
```

Figure 6.19
Minitab lower-tailed test of *p*, Example 6.10

Therefore, we would reject H_0 if we choose $\alpha = .10$ (because the observed significance level is less than .10), and we would not reject H_0 (the conclusion of Example 6.9) if we choose $\alpha = .05$ or $\alpha = .01$.

Now Work Exercise 6.64c

Small-sample test procedures are also available for *p*, although most surveys use samples that are large enough to employ the large-sample tests presented in this section. A test of proportions that can be applied to small samples is discussed in Chapter 9.

Statistics IN Action Revisited — Testing a Population Proportion

In the previous "Statistics in Action Revisited" (p. 340), we investigated Kimberly-Clark Corp.'s assertion that the company should put 60 tissues in a cold-care box of Kleenex® tissues. We did this by testing the claim that the mean number of tissues used by a person with a cold is $\mu = 60$, using data collected from a survey of 250 Kleenex® users. Another approach to the problem is to consider the proportion of Kleenex® users who use fewer than 60 tissues when they have a cold. Now the population parameter of interest is *p*, the proportion of all Kleenex® users who use fewer than 60 tissues when they have a cold.

Kimberly-Clark Corp.'s belief that the company should put 60 tissues in a cold-care box will be supported if the *median* number of tissues used is 60. Now, if the true median is, in fact, 60, then half of the Kleenex® users will use less than 60 tissues and half will use more than 60 tissues (i.e., $p = .5$). Is there evidence to indicate that the population proportion differs from .5? To answer this question, we set up the following null and alternative hypotheses:

$$H_0{:}p = .5 \qquad H_a{:}p \neq .5$$

Recall that the survey results for the 250 sampled Kleenex® users are stored in the **TISSUES** data file. In addition to the number of tissues used by each person, the file contains a qualitative variable—called USED60—representing whether the person used fewer or more than 60 tissues. (The values of USED60 in the data set are "BELOW" or "ABOVE.") A Minitab analysis of this variable yielded the printout displayed in Figure SIA6.2.

On the Minitab printout, X represents the number of the 250 people with colds who used less than 60 tissues. Note that X = 143. This value is used to compute the test statistic, $z = 2.28$, highlighted on the printout. The *p*-value of the test, also highlighted on the printout, is .023. Because this value is less than $\alpha = .05$, there is sufficient evidence (at $\alpha = .05$) to reject H_0; we conclude that the proportion of all Kleenex® users who use fewer than 60 tissues when they have a cold differs from .5. However, if we test at $\alpha = .01$, there is insufficient evidence to reject H_0. Consequently, our choice of α (as in the previous "Statistics in the Action Revisited") is critical to our decision.

Data Set: TISSUES

```
Test of p = 0.5 vs p not = 0.5

Event = BELOW

Variable    X    N  Sample p         95% CI          Z-Value  P-Value
USED60    143  250  0.572000  (0.510666, 0.633334)     2.28    0.023
```

Figure SIA6.2
Minitab test of $p = .5$ for Kleenex® survey

Exercises 6.64–6.79

Learning the Mechanics

6.64 NW Suppose a random sample of 100 observations from a binomial population gives a value of $\hat{p} = .63$ and you wish to test the null hypothesis that the population parameter p is equal to .70 against the alternative hypothesis that p is less than .70.

a. Noting that $\hat{p} = .63$, what does your intuition tell you? Does the value of $\hat{p}$ appear to contradict the null hypothesis?

b. Use the large-sample z-test to test H_0: $p = .70$ against the alternative hypothesis, H_a: $p < .70$. Use $\alpha = .05$. How do the test results compare with your intuitive decision from part **a**?

c. Find and interpret the observed significance level of the test you conducted in part **b.**

6.65 Suppose the sample in Exercise 6.64 has produced $\hat{p} = .83$ and we wish to test H_0: $p = .9$ against the alternative H_a: $p < .9$.

a. Calculate the value of the z-statistic for this test.

b. Note that the numerator of the z-statistic ($\hat{p} - p_0 = .83 - .90 = -.07$) is the same as for Exercise 6.64. Considering this, why is the absolute value of z for this exercise larger than that calculated in Exercise 6.64?

c. Complete the test using $\alpha = .05$ and interpret the result.

d. Find the observed significance level for the test and interpret its value.

6.66 A statistics student used a computer program to test the null hypothesis H_0: $p = .5$ against the one-tailed alternative, H_a: $p > .5$. A sample of 500 observations are input into SPSS, which returns the following results: $z = .44$, two-tailed p-value = .33.

a. The student concludes, based on the p-value, that there is a 33% chance that the alternative hypothesis is true. Do you agree? If not, correct the interpretation.

b. How would the p-value change if the alternative hypothesis was two-tailed, H_a: $p \neq .5$? Interpret this p-value.

6.67 Refer to Exercise 5.40 (p. 295), in which 50 consumers taste-tested a new snack food. Their responses (where 0 = do not like; 1 = like; 2 = indifferent) are reproduced below and saved in the **SNACK** file.

a. Test H_0: $p = .5$ against H_a: $p > .5$, where p is the proportion of customers who do not like the snack food. Use $\alpha = .10$.

b. Find the observed significance level of your test.

1	0	0	1	2	0	1	1	0	0	0	1
0	2	0	2	2	0	0	1	1	0	0	0
0	1	0	2	0	0	0	1	0	0	1	0
0	1	0	1	0	2	0	0	1	1	0	0
0	1										

6.68 For the binomial sample sizes and null hypothesized values of p in each part, determine whether the sample size is large enough to use the normal approximation methodology presented in this section to conduct a test of the null hypothesis H_0: $p = p_0$.

a. $n = 900$, $p_0 = .975$
b. $n = 125$, $p_0 = .01$
c. $n = 40$, $p_0 = .75$
d. $n = 15$, $p_0 = .75$
e. $n = 12$, $p_0 = .62$

Applet Exercise 6.5

Use the applet *Hypotheses Test for a Proportion* to investigate the relationships between the probabilities of Type I and Type II errors occurring at level .05 and .01. For this exercise, use $n = 100$, true $p = 0.5$, and alternative *not equal.*

a. Set null $p = .5$. What happens to the proportion of times the null hypothesis is rejected at the .05 level and at the .01 level as the applet is run more and more times? What type of error has occurred when the null hypothesis is rejected in this situation? Based on your results, is this type of error more likely to occur at level .05 or at level .01? Explain.

b. Set null $p = .6$. What happens to the proportion of times the null hypothesis is *not* rejected at the .05 level and at the .01 level as the applet is run more and more times? What type of error has occurred when the null hypothesis is *not* rejected in this situation? Based on your results, is this type of error more likely to occur at level .05 or at level .01? Explain.

c. Use your results from parts **a** and **b** to make a general statement about the probabilities of Type I and Type II errors at levels .05 and .01.

Applet Exercise 6.6

Use the applet *Hypotheses Test for a Proportion* to investigate the effect of the true population proportion p on the probability of a Type I error occurring. For this exercise, use $n = 100$, and alternative *not equal.*

a. Set true $p = .5$ and null $p = .5$. Run the applet several times and record the proportion of times the null hypothesis is rejected at the .01 level.

b. Clear the applet and repeat part **a** for true $p = .1$ and null $p = .1$. Then repeat one more time for true $p = .01$ and null $p = .01$.

c. Based on your results from parts **a** and **b,** what can you conclude about the probability of a Type I error occurring as the true population proportion gets closer to 0?

Applying the Concepts—Basic

6.69 **Adoption of high-speed Internet connection at home.** Refer to the Pew Internet & American Life Project report on the growth of home broadband adoption in the United States, Exercise 5.47 (p. 296). Recall that in a May 2006 random-digit-dialing telephone survey of 4,000 American adults, 42% of the sample had access to a high-speed Internet connection at home. Let p represent the true proportion of all American adults who had access to a high-speed Internet connection at home in 2006.

a. Compute a point estimate of p.

b. Recall that in 2005, the Pew Internet & American Life Project reported that 30% of all American adults had access to a high-speed Internet connection at home. Set up the null and alternative hypothesis to test whether p has increased since 2005.

c. Compute the test statistic for the test, part **b.**

d. Locate the rejection region for the test if $\alpha = .01$.

e. Use the results, parts **c** and **d,** to make the appropriate conclusion.

f. Find the observed significance level of the test and confirm the conclusion stated in part **e.**

6.70 "Made in the USA" survey. Refer to the *Journal of Global Business* (Spring 2002) study of what "Made in the USA" means to consumers, Exercise 2.8 (p. 39). Recall that 64 of 106 randomly selected shoppers believed "Made in the USA" means 100% of labor and materials are from the United States. Let p represent the true proportion of consumers who believe "Made in the USA" means 100% of labor and materials are from the United States.

a. Calculate a point estimate for p.

b. A claim is made that $p = .70$. Set up the null and alternative hypotheses to test this claim.

c. Calculate the test statistic for the test, part **b.**

d. Find the rejection region for the test if $\alpha = .01$.

e. Use the results, parts **c** and **d,** to make the appropriate conclusion.

6.71 History of corporate acquisitions. Refer to the *Academy of Management Journal* (Aug. 2008) investigation of the performance and timing of corporate acquisitions, Exercise 5.49 (p. 296). Recall that the investigation discovered that in a random sample of 2,778 firms, 748 announced one or more acquisitions during the year 2000. Does the sample provide sufficient evidence to indicate that the true percentage of all firms that announced one or more acquisitions during the year 2000 is less than 30%? Use $\alpha = .05$ to make your decision.

6.72 Accuracy of price scanners at Wal-Mart. Refer to Exercise 5.51 (p. 296) and the study of the accuracy of checkout scanners at Wal-Mart stores in California. Recall that the National Institute for Standards and Technology (NIST) mandates that for every 100 items scanned through the electronic checkout scanner at a retail store, no more than two should have an inaccurate price. A study of random items purchased at California Wal-Mart stores found that 8.3% had the wrong price (*Tampa Tribune,* Nov. 22, 2005). Assume that the study included 1,000 randomly selected items.

a. Identify the population parameter of interest in the study.

b. Set up H_0 and H_a for a test to determine if the true proportion of items scanned at California Wal-Mart stores exceeds the 2% NIST standard.

c. Find the test statistic and rejection region (at $\alpha = .05$) for the test.

d. Give a practical interpretation of the test.

e. What conditions are required for the inference, part **d,** to be valid? Are these conditions met?

6.73 Toothpaste brands with the ADA seal. *Consumer Reports* evaluated and rated 46 brands of toothpaste. One attribute examined in the study was whether or not a toothpaste brand carries an American Dental Association (ADA) seal verifying effective decay prevention. The data for the 46 brands (coded 1 = ADA seal, 0 = no ADA seal) are listed here and saved in the **ADA** file.

0	0	0	0	0	0	1	1	1	0	0	1
0	1	0	0	0	0	1	1	1	0	1	1
1	1	0	0	0	0	0	1	0	0	1	1
1	0	1	0	1	1	1	0	0	0		

a. Give the null and alternative hypotheses for testing whether the true proportion of toothpaste brands with the ADA seal verifying effective decay prevention is less than .5.

b. Locate the p-value on the Minitab printout below.

c. Make the appropriate conclusion using $\alpha = .10$.

Minitab Output for Exercise 6.73

```
Test of p = 0.5 vs p < 0.5

Event = 1

                                        95%
                                      Upper     Exact
Variable    X   N  Sample p           Bound   P-Value
ADASEAL    20  46  0.434783        0.566289     0.231
```

6.74 Minority vacation-home owners. The *National Association of Realtors* (NAR) reported the results of a May 2006 survey of vacation-home owners. In a sample of 416 households that owned one or more vacation homes, 46 were minorities. Prior to 2003, 6% of vacation-home owners were minorities.

a. Do the survey results allow the NAR to conclude (at $\alpha = .01$) that the percentage of vacation-home owners in 2006 who are minorities is larger than 6%?

b. The NAR sent the survey questionnaire to a nationwide sample of 45,000 households that owned vacation homes. Only 416 of the households responded to the survey. How might this bias the survey results?

Applying the Concepts—Intermediate

6.75 Effectiveness of skin cream. Pond's Age-Defying Complex, a cream with alpha hydroxy acid, advertises that it can reduce wrinkles and improve the skin. In a study published in *Archives of Dermatology* (June 1996), 33 middle-aged women used a cream with alpha hydroxy acid for 22 weeks. At the end of the study period, a dermatologist judged whether each woman exhibited skin improvement. The results for the 33 women (where I = improved skin and N = no improvement) are listed in the next table and saved in the **SKINCREAM** file. [Note: Pond's recently discontinued the production of this cream product, replacing it with Age-Defying Towlettes.]

a. Do the data provide sufficient evidence to conclude that the cream will improve the skin of more than 60% of middle-aged women? Test using $\alpha = .05$.

b. Find and interpret the p-value of the test.

I	I	N	I	N	N	I	I	I	I	I	I
N	I	I	I	N	I	I	I	N	I	N	I
I	I	I	I	I	N	I	I	N			

6.76 Reporting company changes to the SEC. The SEC requires a company to file Form 8-K to report material changes in its financial condition or operation. Common material events are changes in directors, auditors, and assets disposition. Up until 2004, companies had 15 business days to submit Form 8-K. Currently, the SEC requires that the form be submitted within 4 business days of the material event (*Sarbanes-Oxley Act,* Section 409). Previous research has indicated that as high as 10% of firms were in violation of the old, 15-day rule. Old Dominion University Accounting Professor Rob Pinsker conducted a study to determine if firms with material changes were able to comply with the new "4-day" rule (Information Systems Mid-Year Meeting, 2004). In a sample of 462 firms with material events, only 23 were in violation of the new 4-day requirement for filing. Are you able to conclude that the true percentage of firms in violation of the new 4-day rule for reporting material changes is less than 10%? Make your inference at a significance level of $\alpha = .01$.

6.77 Graduation rates of student-athletes. Are student-athletes at Division I universities poorer students than nonathletes? The National Collegiate Athletic Association (NCAA) measures the academic outcomes of student-athletes with the Graduation Success Rate (GSR). The GSR is measured as the percentage of eligible athletes who graduate within six years of entering college. According to the NCAA, the GSR for all scholarship athletes at Division I institutions is 63% (*Inside Higher Ed,* Nov. 10, 2006). It is well known that the GSR for all students at Division I colleges is 60%.

a. Suppose the NCAA report was based on a sample of 500 student-athletes, of which 315 graduated within six years. Is this sufficient information to conclude that the GSR for all scholarship athletes at Division I institutions differs from 60%? Test using $\alpha = .01$.

b. The GSR statistics were also broken down by gender and sport. For example, men's Division I college basketball players had a GSR of 42% (compared to a known GSR of 58% for all male college students). Suppose this statistic was based on a sample of 200 male basketball players, of which 84 graduated within six years. Is this sufficient information to conclude that the GSR for all male basketball players at Division I institutions differs from 58%? Test using $\alpha = .01$.

Applying the Concepts—Advanced

6.78 Choosing portable grill displays. Refer to the *Journal of Consumer Research* (Mar. 2003) experiment on influencing the choices of others by offering undesirable alternatives, Exercise 3.21 (p. 130). Recall that each of 124 college students selected three portable grills from five to display on the showroom floor. The students were instructed to include Grill #2 (a smaller-sized grill) and select the remaining two grills in the display to maximize purchases of Grill #2. If the six possible grill display combinations (1-2-3, 1-2-4, 1-2-5, 2-3-4, 2-3-5, and 2-4-5) are selected at random, then the proportion of students selecting any display will be 1/6 = .167. One theory tested by the researcher is that the students will tend to choose the three-grill display so that Grill #2 is a compromise between a more desirable and a less desirable grill (i.e., display 1-2-3, 1-2-4, or 1-2-5). Of the 124 students, 85 selected a three-grill display that was consistent with this theory. Use this information to test the theory proposed by the researcher at $\alpha = .05$.

6.79 The Pepsi Challenge. "Take the Pepsi Challenge" was a famous marketing campaign used by the Pepsi-Cola Company. Coca-Cola drinkers participated in a blind taste test where they were asked to taste unmarked cups of Pepsi and Coke and were asked to select their favorite. In one Pepsi television commercial, an announcer states that "in recent blind taste tests, more than half the Diet Coke drinkers surveyed said they preferred the taste of Diet Pepsi." Suppose 100 Diet Coke drinkers took the Pepsi Challenge and 56 preferred the taste of Diet Pepsi. Determine if more than half of all Diet Coke drinkers will select Diet Pepsi in the blind taste test. Select α to minimize the probability of a Type I error. What are the consequences of the test results from Coca-Cola's perspective?

6.7 Calculating Type II Error Probabilities: More about β*

In our introduction to hypothesis testing in Section 6.1, we showed that the probability of committing a Type I error, α, can be controlled by the selection of the rejection region for the test. Thus, when the test statistic falls in the rejection region and we make the decision to reject the null hypothesis, we do so knowing the error rate for incorrect rejections of H_0. The situation corresponding to accepting the null hypothesis, and thereby risking a Type II error, is not generally as controllable. For that reason, we adopted a policy of nonrejection of H_0 when the test statistic does not fall in the rejection region, rather than risking an error of unknown magnitude.

To see how β, the probability of a Type II error, can be calculated for a test of hypothesis, recall the example in Section 6.1 in which a city tests a manufacturer's pipe to see whether it meets the requirement that the mean strength exceeds 2,400 pounds per linear foot. The setup for the test is as follows:

$$H_0: \mu = 2{,}400$$
$$H_a: \mu > 2{,}400$$

Test statistic: $z = \dfrac{\bar{x} - 2{,}400}{\sigma/\sqrt{n}}$

Rejection region: $z > 1.645$ for $\alpha = .05$

Figure 6.20a shows the rejection region for the **null distribution**—that is, the distribution of the test statistic assuming the null hypothesis is true. The area in the rejection region is .05, and this area represents α, the probability that the test statistic leads to rejection of H_0 when in fact H_0 is true.

The Type II error probability β is calculated assuming that the null hypothesis is false because it is defined as the *probability of accepting H_0 when it is false.* Because H_0 is false for any value of μ exceeding 2,400, one value of β exists for each possible value of μ greater than 2,400 (an infinite number of possibilities). Figures 6.20b–d show three of the possibilities, corresponding to alternative hypothesis values of μ equal to 2,425, 2,450, and 2,475, respectively. Note that β is the area in the *nonrejection* (or *acceptance*) *region* in each of these distributions and that β decreases as the true value of μ moves farther from the null hypothesized value of $\mu = 2{,}400$. This is sensible because the probability of incorrectly accepting the null hypothesis should decrease as the distance between the null and alternative values of μ increases.

In order to calculate the value of β for a specific value of μ in H_a, we proceed as follows:

1. Calculate the value of $\bar{x}$ that corresponds to the border between the acceptance and rejection regions. For the sewer pipe example, this is the value of $\bar{x}$ that lies 1.645 standard deviations above $\mu = 2{,}400$ in the sampling distribution of $\bar{x}$. Denoting this value by $\bar{x}_0$, corresponding to the largest value of $\bar{x}$ that supports the null hypothesis, we find (recalling that $s = 200$ and $n = 50$)

$$\bar{x}_0 = \mu_0 + 1.645\sigma_{\bar{x}} = 2{,}400 + 1.645\left(\frac{\sigma}{\sqrt{n}}\right)$$
$$\approx 2{,}400 + 1.645\left(\frac{s}{\sqrt{n}}\right) = 2{,}400 + 1.645\left(\frac{200}{\sqrt{50}}\right)$$
$$= 2{,}400 + 1.645(28.28) = 2{,}446.5$$

2. For a particular alternative distribution corresponding to a value of μ, denoted by μ_a, we calculate the z-value corresponding to $\bar{x}_0$, the border between the rejection and acceptance regions. We then use this z-value and Table IV in Appendix B to determine the area in the *acceptance region* under the alternative distribution. This area is the value of β corresponding to the particular alternative μ_a. For example, for the alternative $\mu_a = 2{,}425$, we calculate

$$z = \frac{\bar{x}_0 - 2{,}425}{\sigma_{\bar{x}}} = \frac{\bar{x}_0 - 2{,}425}{\sigma/\sqrt{n}}$$
$$\approx \frac{\bar{x}_0 - 2{,}425}{\sigma_{\bar{x}}} = \frac{2{,}446.5 - 2{,}425}{28.28} = .76$$

Note in Figure 6.20b that the area in the acceptance region is the area to the left of $z = .76$. This area is

$$\beta = .5 + .2764 = .7764$$

Thus, the probability that the test procedure will lead to an incorrect acceptance of the null hypothesis $\mu = 2{,}400$ when in fact $\mu = 2{,}425$ is about .78. As the average strength of the pipe increases to 2,450, the value of β decreases to .4522 (Figure 6.20c). If the mean strength is further increased to 2,475, the value of β is

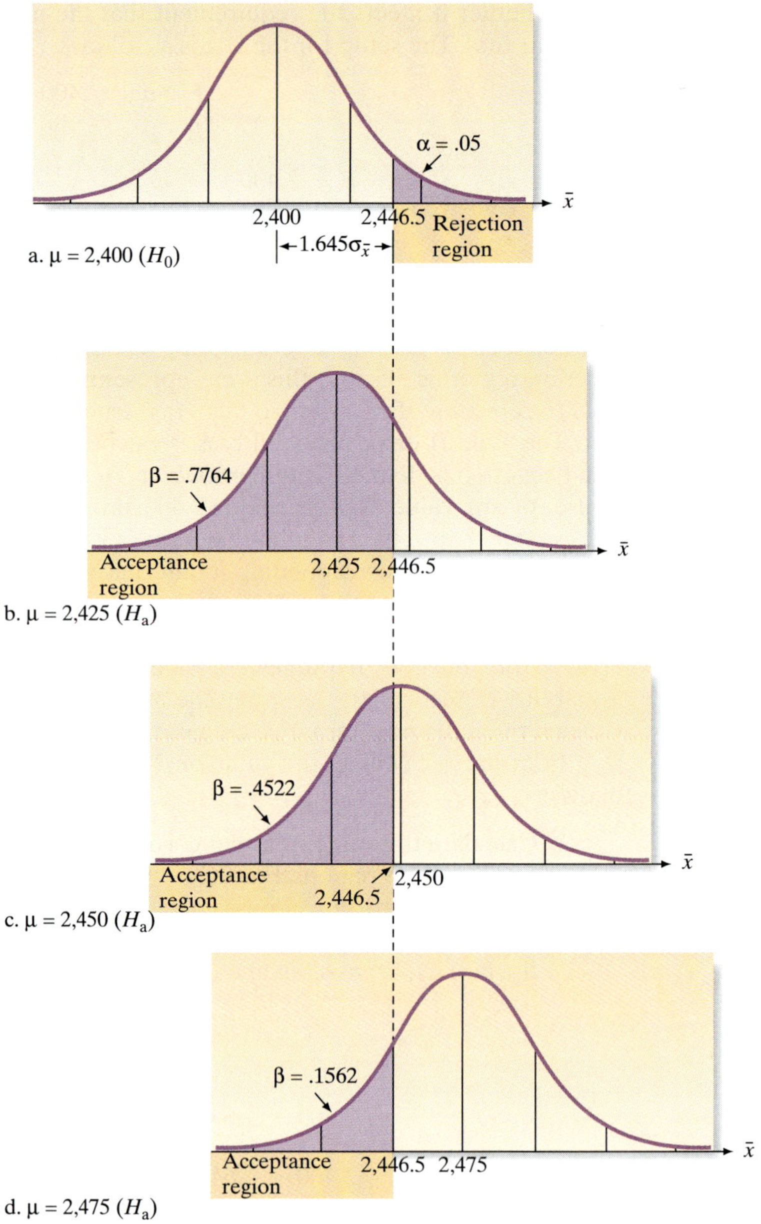

Figure 6.20
Values of α and β for various values of μ

further decreased to .1562 (Figure 6.20d). Thus, even if the true mean strength of the pipe exceeds the minimum specification by 75 pounds per linear foot, the test procedure will lead to an incorrect acceptance of the null hypothesis (rejection of the pipe) approximately 16% of the time. The upshot is that the pipe must be manufactured so that the mean strength well exceeds the minimum requirement if the manufacturer wants the probability of its acceptance by the city to be large (i.e., β to be small).

The steps for calculating β for a large-sample test about a population mean are summarized in the next box.

Steps for Calculating β for a Large-Sample Test about μ

1. Calculate the value(s) of $\bar{x}$ corresponding to the border(s) of the rejection region. There will be one border value for a one-tailed test and two for a two-tailed test. The formula is one of the following, corresponding to a test with level of significance α:

Upper-tailed test: $\bar{x}_0 = \mu_0 + z_\alpha \sigma_{\bar{x}} \approx \mu_0 + z_\alpha\left(\frac{s}{\sqrt{n}}\right)$

Lower-tailed test: $\bar{x}_0 = \mu_0 - z_\alpha \sigma_{\bar{x}} \approx \mu_0 - z_\alpha\left(\frac{s}{\sqrt{n}}\right)$

Two-tailed test: $\bar{x}_{0,\text{L}} = \mu_0 - z_{\alpha/2}\sigma_{\bar{x}} \approx \mu_0 - z_{\alpha/2}\left(\frac{s}{\sqrt{n}}\right)$

$\bar{x}_{0,\text{U}} = \mu_0 + z_{\alpha/2}\sigma_{\bar{x}} \approx \mu_0 + z_{\alpha/2}\left(\frac{s}{\sqrt{n}}\right)$

2. Specify the value of μ_a in the alternative hypothesis for which the value of β is to be calculated. Then convert the border value(s) of $\bar{x}_0$ to z-value(s) using the alternative distribution with mean μ_a. The general formula for the z-value is

$$z = \frac{\bar{x}_0 - \mu_a}{\sigma_{\bar{x}}}$$

Sketch the alternative distribution (centered at μ_a) and shade the area in the acceptance (nonrejection) region. Use the z-statistic(s) and Table IV in Appendix B to find the shaded area, which is β.

Following the calculation of β for a particular value of μ_a, you should interpret the value in the context of the hypothesis testing application. It is often useful to interpret the value of $1 - \beta$, which is known as the *power of the test* corresponding to a particular alternative, μ_a. Because β is the probability of accepting the null hypothesis when the alternative hypothesis is true with $\mu = \mu_a$, $1 - \beta$ is the probability of the complementary event, or the probability of rejecting the null hypothesis when the alternative H_a: $\mu = \mu_a$ is true—that is, the power $1 - \beta$ measures the likelihood that the test procedure will lead to the correct decision (reject H_0) for a particular value of the mean in the alternative hypothesis.

The **power of a test** is the probability that the test will correctly lead to the rejection of the null hypothesis for a particular value of μ in the alternative hypothesis. The power is equal to $1 - \beta$ for the particular alternative considered.

For example, in the sewer pipe example we found that $\beta = .7764$ when $\mu = 2{,}425$. This is the probability that the test leads to the (incorrect) acceptance of the null hypothesis when $\mu = 2{,}425$. Or, equivalently, the power of the test is $1 - .7764 = .2236$, which means that the test will lead to the (correct) rejection of the null hypothesis only 22% of the time when the pipe exceeds specifications by 25 pounds per linear foot. When the manufacturer's pipe has a mean strength of 2,475 (that is, 75 pounds per linear foot in excess of specifications), the power of the test increases to $1 - .1562 = .8438$—that is, the test will lead to the acceptance of the manufacturer's pipe 84% of the time if $\mu = 2{,}475$.

Example 6.11 Finding the Power of the Test

Problem Recall the quality control study in Examples 6.3 and 6.4, in which we tested to determine whether a cereal box filling machine was deviating from the specified mean fill of $\mu = 12$ ounces. The test setup is repeated here:

H_0: $\mu = 12$

H_a: $\mu \neq 12$ (i.e., $\mu < 12$ or $\mu > 12$)

Test statistic: $z = \dfrac{\bar{x} - 12}{\sigma_{\bar{x}}}$

Rejection region: $z < -1.96$ or $z > 1.96$ for $\alpha = .05$

$z < -2.575$ or $z > 2.575$ for $\alpha = .01$

Note that two rejection regions have been specified corresponding to values of $\alpha = .05$ and $\alpha = .01$, respectively. Assume that $n = 100$ and $s = .5$.

a. Suppose the machine is underfilling the boxes by an average of .1 ounce (i.e., $\mu = 11.9$). Calculate the values of β corresponding to the two rejection regions. Discuss the relationship between the values of α and β.

b. Calculate the power of the test for each of the rejection regions when $\mu = 11.9$.

Solution

a. We first consider the rejection region corresponding to $\alpha = .05$. The first step is to calculate the border values of $\bar{x}$ corresponding to the two-tailed rejection region, $z < -1.96$ or $z > 1.96$:

$$\bar{x}_{0.L} = \mu_0 - 1.96\sigma_{\bar{x}} \approx \mu_0 - 1.96\left(\frac{s}{\sqrt{n}}\right) = 12 - 1.96\left(\frac{.5}{10}\right) = 11.902$$

$$\bar{x}_{0.U} = \mu_0 + 1.96\sigma_{\bar{x}} \approx \mu_0 + 1.96\left(\frac{s}{\sqrt{n}}\right) = 12 + 1.96\left(\frac{.5}{10}\right) = 12.098$$

These border values are shown in Figure 6.21.

Next, we convert these values to z-values in the alternative distribution with $\mu_a = 11.9$:

$$z_L = \frac{x_{0.L} - \mu_a}{\sigma_{\bar{x}}} \approx \frac{11.902 - 11.9}{.05} = .04$$

$$z_U = \frac{x_{0.U} - \mu_a}{\sigma_{\bar{x}}} \approx \frac{12.098 - 11.9}{.05} = 3.96$$

These z-values are shown in Figure 6.21b: You can see that the acceptance (or nonrejection) region is the area between them. Using Table IV in Appendix B, we find that the area between $z = 0$ and $z = .04$ is .0160, and the area between $z = 0$ and $z = 3.96$ is (approximately) .5 (because $z = 3.96$ is off the scale of Table IV). Then the area between $z = .04$ and $z = 3.96$ is, approximately,

$$\beta = .5 - .0160 = .4840$$

Thus, the test with $\alpha = .05$ will lead to a Type II error about 48% of the time when the machine is underfilling, on average, by .1 ounce.

For the rejection region corresponding to $\alpha = .01$, $z < -2.575$, or $z > 2.575$, we find

$$\bar{x}_{0.L} = 12 - 2.575\left(\frac{.5}{10}\right) = 11.871$$

$$\bar{x}_{0.U} = 12 + 2.575\left(\frac{.5}{10}\right) = 12.129$$

These border values of the rejection region are shown in Figure 6.21c.

Converting these to z-values in the alternative distribution with $\mu_a = 11.9$, we find $z_L = -.58$ and $z_U = 4.58$. The area between these values is, approximately,

$$\beta = .2190 + .5 = .7190$$

Thus, the chance that the test procedure with $\alpha = .01$ will lead to an incorrect acceptance of H_0 is about 72%.

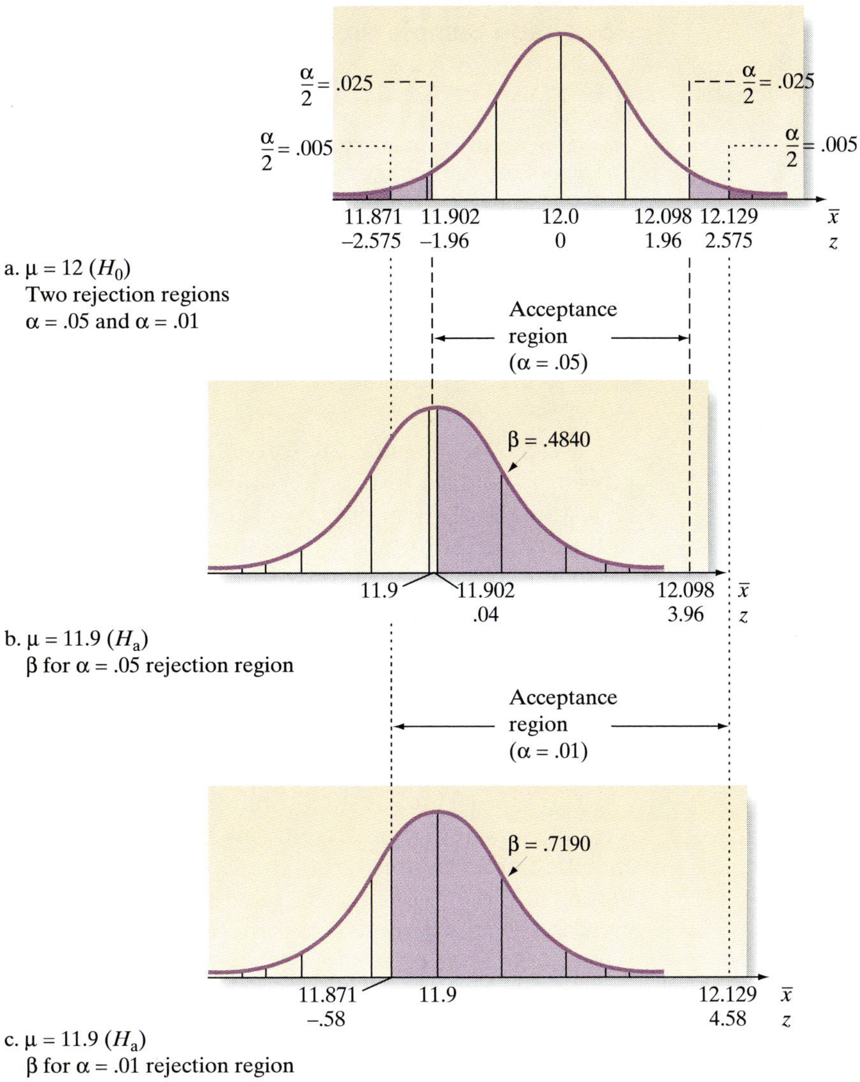

Figure 6.21
Calculation of β for filling machine example

Note that the value of β increases from .4840 to .7190 when we decrease the value of α from .05 to .01. This is a general property of the relationship between α and β: *as α is decreased (increased), β is increased (decreased).*

b. The power is defined to be the probability of (correctly) rejecting the null hypothesis when the alternative is true. When $\mu = 11.9$ and $\alpha = .05$, we find

$$\text{Power} = 1 - \beta = 1 - .4840 = .5160$$

When $\mu = 11.9$ and $\alpha = .01$, we find

$$\text{Power} = 1 - \beta = 1 - .7190 = .2810$$

You can see that the power of the test is decreased as the level of α is decreased. This means that as the probability of incorrectly rejecting the null hypothesis is decreased, the probability of correctly accepting the null hypothesis for a given alternative is also decreased.

Look Back A key point of this example is that the value of α must be selected carefully, with the realization that a test is made less capable of detecting departures from the null hypothesis when the value of α is decreased.

Figure 6.22
Minitab power analysis for Example 6.11

```
Power and Sample Size

1-Sample Z Test

Testing mean = null (versus not = null)
Calculating power for mean = null + difference
Alpha = 0.05  Assumed standard deviation = 0.5

                Sample
Difference        Size      Power
       0.1         100   0.516005
```

Note: Most statistical software packages now have options for computing the power of standard tests of hypothesis. Usually you will need to specify the type of test (z-test or t-test), form of H_a ($<$, $>$, or $\neq$), standard deviation, sample size, and the value of the parameter in H_a (or the difference between the value in H_0 and the value in H_a). The Minitab power analysis for Example 6.11 when $\alpha = .05$ is displayed in Figure 6.22. The power of the test (.516) is highlighted on the printout.

We have shown that the probability of committing a Type II error, β, is inversely related to α (Example 6.11) and that the value of β decreases as the value of μ_a moves farther from the null hypothesis value (sewer pipe example). The sample size n also affects β. Remember that the standard deviation of the sampling distribution of $\bar{x}$ is inversely proportional to the square root of the sample size ($\sigma_{\bar{x}} = \sigma/\sqrt{n}$) Thus, as illustrated in Figure 6.23, the variability of both the null and alternative sampling distributions is decreased as n is increased. If the value of α is specified and remains fixed, the value of β decreases as n increases, as illustrated in Figure 6.23. Conversely, the power of the test for a given alternative hypothesis is increased as the sample size is increased. The properties of β and power are summarized in the box.

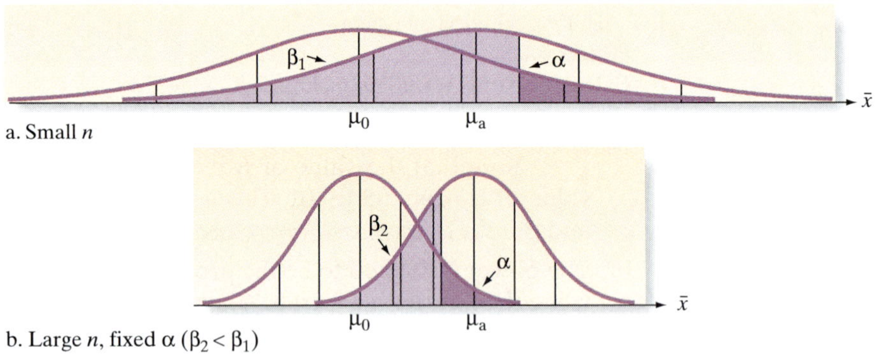

Figure 6.23
Relationship between α, β, and n

Properties of β and Power

1. For fixed n and α, the value of β decreases, and the power increases as the distance between the specified null value μ_0 and the specified alternative value μ_a increases (see Figure 6.20).
2. For fixed n and values of μ_0 and μ_a, the value of β increases, and the power decreases as the value of α is decreased (see Figure 6.21).
3. For fixed α and values of μ_0 and μ_a, the value of β decreases, and the power increases as the sample size n is increased (see Figure 6.23).

Exercises 6.80–6.89

Learning the Mechanics

6.80 **a.** List three factors that will increase the power of a test.
b. What is the relationship between β, the probability of committing a Type II error, and the power of a test?

6.81 Suppose you want to test H_0: $\mu = 500$ against H_a: $\mu > 500$ using $\alpha = .05$. The population in question is normally distributed with standard deviation 100. A random sample of size $n = 25$ will be used.
a. Sketch the sampling distribution of $\bar{x}$ assuming that H_0 is true.
b. Find the value of $\bar{x}_0$, that value of $\bar{x}$ above which the null hypothesis will be rejected. Indicate the rejection region on your graph of part **a.** Shade the area above the rejection region and label it α.
c. On your graph of part **a,** sketch the sampling distribution of $\bar{x}$ if $\mu = 550$. Shade the area under this distribution that corresponds to the probability that $\bar{x}$ falls in the nonrejection region when $\mu = 550$. Label this area β.
d. Find β.
e. Compute the power of this test for detecting the alternative H_a: $\mu = 550$.

6.82 Refer to Exercise 6.81.
a. If $\mu = 575$ instead of 550, what is the probability that the hypothesis test will incorrectly fail to reject H_0? That is, what is β?
b. If $\mu = 575$, what is the probability that the test will correctly reject the null hypothesis? That is, what is the power of the test?
c. Compare β and the power of the test when $\mu = 575$ to the values you obtained in Exercise 6.81 for $\mu = 550$. Explain the differences.

6.83 It is desired to test H_0: $\mu = 75$ against H_a: $\mu < 75$ using $\alpha = .10$. The population in question is uniformly distributed with standard deviation 15. A random sample of size 49 will be drawn from the population.
a. Describe the (approximate) sampling distribution of $\bar{x}$ under the assumption that H_0 is true.
b. Describe the (approximate) sampling distribution of $\bar{x}$ under the assumption that the population mean is 70.
c. If μ were really equal to 70, what is the probability that the hypothesis test would lead the investigator to commit a Type II error?
d. What is the power of this test for detecting the alternative H_a: $\mu = 70$?

6.84 Refer to Exercise 6.83.
a. Find β for each of the following values of the population mean: 74, 72, 70, 68, and 66.
b. Plot each value of β you obtained in part **a** against its associated population mean. Show β on the vertical axis and μ on the horizontal axis. Draw a curve through the five points on your graph.
c. Use your graph of part **b** to find the approximate probability that the hypothesis test will lead to a Type II error when $\mu = 73$.
d. Convert each of the β values you calculated in part **a** to the power of the test at the specified value of μ. Plot the power on the vertical axis against μ on the horizontal axis. Compare the graph of part **b** to the *power curve* of this part.
e. Examine the graphs of parts **b** and **d.** Explain what they reveal about the relationships among the distance between the true mean μ and the null hypothesized mean μ_0, the value of β, and the power.

6.85 Suppose you want to conduct the two-tailed test of H_0: $\mu = 30$ against H_a: $\mu \neq 30$ using $\alpha = .05$. A random sample of size 121 will be drawn from the population in question. Assume the population has a standard deviation equal to 1.2.
a. Describe the sampling distribution of $\bar{x}$ under the assumption that H_0 is true.
b. Describe the sampling distribution of $\bar{x}$ under the assumption that $\mu = 29.8$.
c. If μ were really equal to 29.8, find the value of β associated with the test.
d. Find the value of β for the alternative H_a: $\mu = 30.4$.

Applying the Concepts—Intermediate

6.86 **Square-footage of new California homes.** The average size of single-family homes built in the United States is 2,230 square feet, an increase of over 200 square feet a decade earlier (*Wall Street Journal Interactive Edition,* Jan. 7, 2000). A random sample of 100 new homes sold in California yielded the following size information: $\bar{x} = 2{,}347$ square feet and $s = 257$ square feet.
a. Assume the average size of U.S. homes is known with certainty. Do the sample data provide sufficient evidence to conclude that the mean size of California homes built exceeds the national average? Test using $\alpha = .01$.
b. Suppose the actual mean size of new California homes was 2,330 square feet. What is the power of the test in part **a** to detect this 100-square-foot difference?
c. If the California mean were actually 2,280 square feet, what is the power of the test in part **a** to detect this 50-square-foot difference?

6.87 **Manufacturers that practice sole sourcing.** If a manufacturer (the vendee) buys all items of a particular type from a particular vendor, the manufacturer is practicing *sole sourcing* (Schonberger and Knod, *Operations Management,* 2001). As part of a sole-sourcing arrangement, a vendor agrees to periodically supply its vendee with sample data from its production process. The vendee uses the data to investigate whether the mean length of rods produced by the vendor's production process is truly 5.0 millimeters (mm) or more, as claimed by the vendor and desired by the vendee.
a. If the production process has a standard deviation of .01 mm, the vendor supplies $n = 100$ items to the vendee, and the vendee uses $\alpha = .05$ in testing H_0: $\mu = 5.0$ mm against H_a: $\mu < 5.0$ mm, what is the probability that the vendee's test will fail to reject the null hypothesis when in fact $\mu = 4.9975$ mm? What is the name given to this type of error?

b. Refer to part **a.** What is the probability that the vendee's test will reject the null hypothesis when in fact $\mu = 5.0$? What is the name given to this type of error?

c. What is the power of the test to detect a departure of .0025 mm below the specified mean rod length of 5.0 mm?

6.88 Fuel economy of the Honda Civic. According to the Environmental Protection Agency (EPA) *Fuel Economy Guide,* the 2009 Honda Civic automobile obtains a mean of 36 miles per gallon (mpg) on the highway. Suppose Honda claims that the EPA has underestimated the Civic's mileage. To support its assertion, the company selects $n = 50$ model 2009 Civic cars and records the mileage obtained for each car over a driving course similar to the one used by the EPA. The following data resulted: $\bar{x} = 38.3$ mpg, $s = 6.4$ mpg.

a. If Honda wishes to show that the mean mpg for 2009 Civic autos is greater than 36 mpg, what should the alternative hypothesis be? The null hypothesis?

b. Do the data provide sufficient evidence to support the auto manufacturer's claim? Test using $\alpha = .05$. List any assumptions you make in conducting the test.

c. Calculate the power of the test for the mean values of 36.5, 37.0, 37.5, 38.0, and 38.5, assuming $s = 6.4$ is a good estimate of σ.

d. Plot the power of the test on the vertical axis against the mean on the horizontal axis. Draw a curve through the points.

e. Use the power curve of part **d** to estimate the power for the mean value $\mu = 37.75$. Calculate the power for this value of μ and compare it to your approximation.

f. Use the power curve to approximate the power of the test when $\mu = 41$. If the true value of the mean mpg for this model is really 41 what (approximately) are the chances that the test will fail to reject the null hypothesis that the mean is 36?

6.89 Solder-joint inspections. Refer to Exercise 6.32 (p. 336), in which the performance of a particular type of laser-based inspection equipment was investigated. Assume that the standard deviation of the number of solder joints inspected on each run is 1.2. If $\alpha = .05$ is used in conducting the hypothesis test of interest using a sample of 48 circuit boards, and if the true mean number of solder joints that can be inspected is really equal to 9.5, what is the probability that the test will result in a Type II error?

6.8 Test of Hypothesis about a Population Variance

Although many practical problems involve inferences about a population mean (or proportion), it is sometimes of interest to make an inference about a population variance, σ^2. To illustrate, a quality control supervisor in a cannery knows that the exact amount each can contains will vary because there are certain uncontrollable factors that affect the amount of fill. The mean fill per can is important, but equally important is the variation of fill. If σ^2, the variance of the fill, is large, some cans will contain too little and others too much. Suppose regulatory agencies specify that the standard deviation of the amount of fill should be less than .1 ounce. To determine whether the process is meeting this specification, the supervisor randomly selects 10 cans and weighs the contents of each. The results are given in Table 6.6.

Table 6.6 Fill Weights (in Ounces) of 10 Cans

16.00	16.06	15.95	16.04	16.10	16.05	16.02	16.03	15.99	16.02

Data Set: FILLAMOUNTS

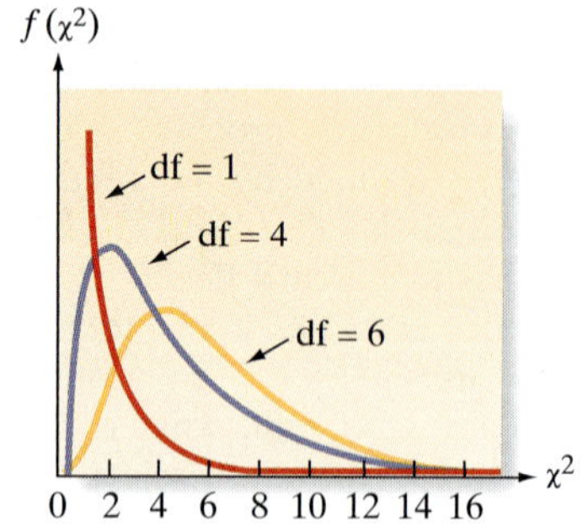

Figure 6.24
Several χ^2 probability distributions

Do these data provide sufficient evidence to indicate that the variability is as small as desired? To answer this question, we need a procedure for testing a hypothesis about σ^2.

Intuitively, it seems that we should compare the sample variance σ^2 to the hypothesized value of σ^2 (or s to σ) in order to make a decision about the population's variability. The quantity

$$\frac{(n - 1)s^2}{\sigma^2}$$

has been shown to have a sampling distribution called a **chi-square (χ^2) distribution** when the population from which the sample is taken is *normally distributed.* Several chi-square distributions are shown in Figure 6.24.

BIOGRAPHY

FRIEDRICH R. HELMERT (1843–1917)

Helmert Transformations

German Friedrich Helmert studied engineering sciences and mathematics at Dresden University, where he earned his PhD, then accepted a position as a professor of geodesy—the scientific study of the earth's size and shape—at the technical school in Aachen. Helmert's mathematical solutions to geodesy problems led him to several statistics-related discoveries. His greatest statistical contribution occurred in 1876, when he was the first to prove that the sampling distribution of the sample variance, s^2, is a chi-square distribution. Helmert used a series of mathematical transformations to obtain the distribution of s^2—transformations that have since been named "Helmert transformations" in his honor. Later in life, Helmert was appointed professor of advanced geodesy at the prestigious University of Berlin and director of the Prussian Geodesic Institute. ■

The upper-tail areas for this distribution have been tabulated and are given in Table VI in Appendix B, a portion of which is reproduced in Table 6.7. The table gives the values of χ^2, denoted as χ^2_α, that locate an area of α in the upper tail of the chi-square distribution; that is, $P(\chi^2 > \chi^2_\alpha) = \alpha$. In this case, as with the t-statistic, the shape of the chi-square distribution depends on the degrees of freedom associated with s^2, namely $(n - 1)$. Thus, for $n = 10$ and an upper-tail value $\alpha = .05$, you will have $n - 1 = 9$ df and $\chi^2_{.05} = 16.9190$ (highlighted in Table 6.7). To further illustrate the use of Table VI, we return to the can-filling example.

Table 6.7 Reproduction of Part of Table VI in Appendix B: Critical Values of Chi Square

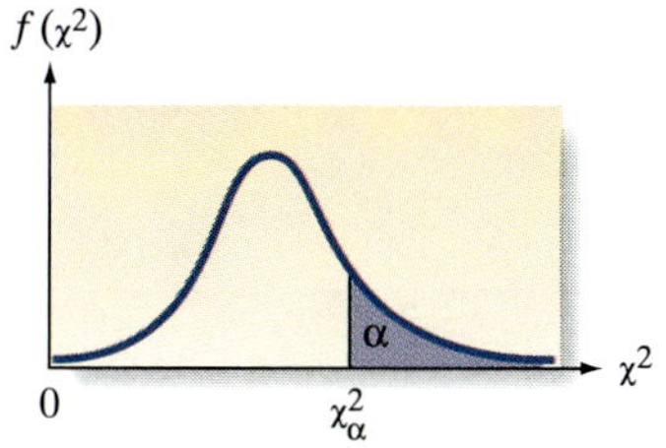

Degrees of Freedom	$\chi^2_{.100}$	$\chi^2_{.050}$	$\chi^2_{.025}$	$\chi^2_{.010}$	$\chi^2_{.005}$
1	2.70554	3.84146	5.02389	6.63490	7.87944
2	4.60517	5.99147	7.37776	9.21034	10.5966
3	6.25139	7.81473	9.34840	11.3449	12.8381
4	7.77944	9.48773	11.1433	13.2767	14.8602
5	9.23635	11.0705	12.8325	15.0863	16.7496
6	10.6446	12.5916	14.4494	16.8119	18.5476
7	12.0170	14.0671	16.0128	18.4753	20.2777
8	13.3616	15.5073	17.5346	20.0902	21.9550
9	14.6837	16.9190	19.0228	21.6660	23.5893
10	15.9871	18.3070	20.4831	23.2093	25.1882
11	17.2750	19.6751	21.9200	24.7250	25.7569
12	18.5494	21.0261	23.3367	26.2170	28.2995
13	19.8119	22.3621	24.7356	27.6883	29.8194
14	21.0642	23.6848	26.1190	29.1413	31.3193
15	22.3072	24.9958	27.4884	30.5779	32.8013
16	23.5418	26.2862	28.8454	31.9999	34.2672
17	24.7690	27.5871	30.1910	33.4087	35.7185
18	25.9894	28.8693	31.5264	34.8053	37.1564
19	27.2036	30.1435	32.8523	36.1908	38.5822

Example 6.12

Conducting a Test for the Variance, σ^2, of Fill Weights

Problem Refer to the fill weights for the sample of ten 16-ounce cans in Table 6.6. Do the data provide sufficient evidence to indicate that the true standard deviation σ of the fill measurements of all 16-ounce cans is less than .1 ounce?

Solution Here, we want to test whether $\sigma < .1$. Because the null and alternative hypotheses must be stated in terms of σ^2 rather than σ, we want to test the null hypothesis that $\sigma^2 = (.1)^2 = .01$ against the alternative that $\sigma^2 < .01$. Therefore, the elements of the test are

H_0: $\sigma^2 = .01$ (Fill variance equals .01—i.e., process specifications are not met.)

H_a: $\sigma^2 < .01$ (Fill variance is less than .01—i.e., process specifications are met.)

$$\textit{Test statistic: } \chi^2 = \frac{(n-1)s^2}{\sigma^2}$$

Assumption: The distribution of the amounts of fill is approximately normal.

Rejection region: The smaller the value of s^2 we observe, the stronger the evidence in favor of H_a. Thus, we reject H_0 for "small values" of the test statistic. With $\alpha = .05$ and 9 df, the χ^2 value for rejection is found in Table VI and pictured in Figure 6.25. We will reject H_0 if $\chi^2 < 3.32511$.

(Remember that the area given in Table VI is the area to the *right* of the numerical value in the table. Thus, to determine the lower-tail value, which has $\alpha = .05$ to its *left,* we used the $\chi^2_{.95}$ column in Table VI.)

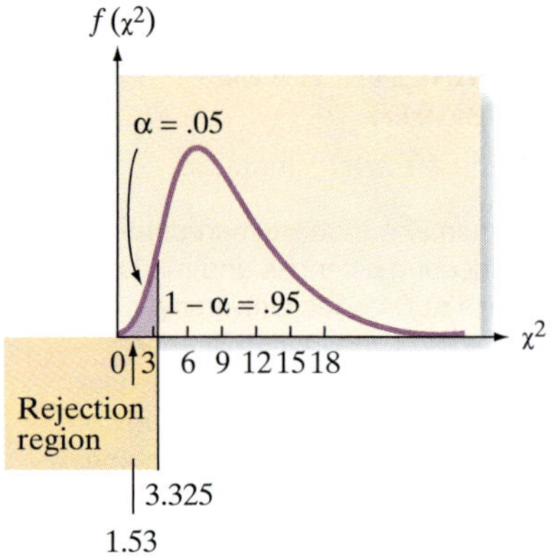

Figure 6.25
Rejection region for Example 6.12

An Excel/DDXL analysis of the data in Table 6.6 is displayed in Figure 6.26. The value of s (highlighted) on the printout is $s = .0412$. Substituting into the formula for the test statistic, we have

$$\chi^2 = \frac{(n-1)s^2}{\sigma^2} = \frac{9(.0412)^2}{.01} = 1.53$$

Because the test statistic falls into the rejection region, we reject H_0 in favor of H_a—that is, the supervisor can conclude that the variance σ^2 of the population of all amounts of fill is less than .01 ($\sigma < .1$) with probability of a Type I error equal to $\alpha = .05$. If this procedure is repeatedly used, it will incorrectly reject H_0 only 5% of the time. Thus, the quality control supervisor is confident in the decision that the cannery is operating within the desired limits of variability.

Look Back Note that both the test statistic (rounded) and the lower-tailed p-value of the test (.003) are highlighted at the bottom of the printout, Figure 6.26. Because $\alpha = .05$ exceeds the p-value, our decision to reject H_0 is confirmed.

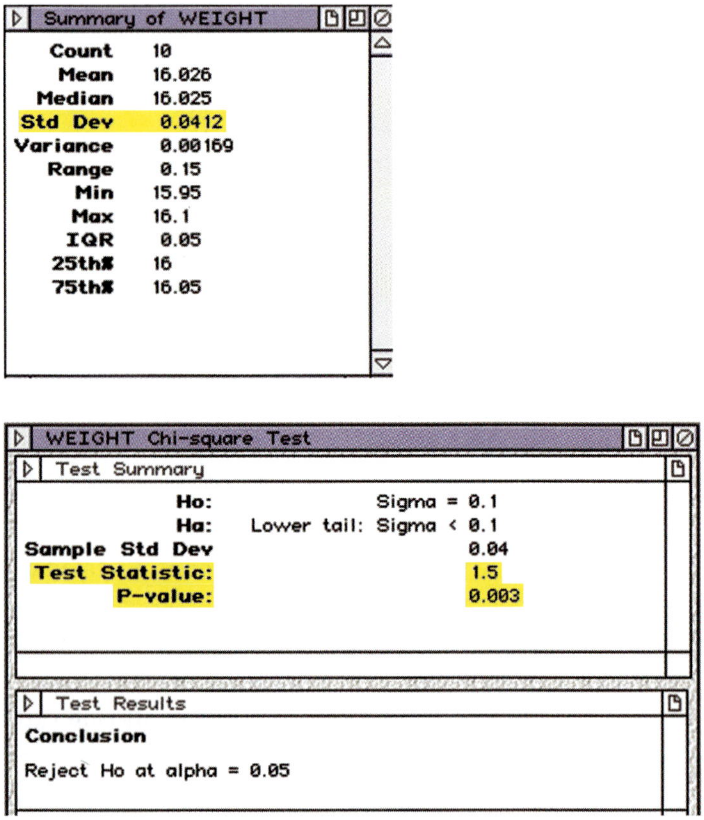

Figure 6.26
Excel/DDXL analysis of fill weights, Example 6.12

Now Work Exercise 6.98

One-tailed and two-tailed tests of hypothesis for σ^2 are given in the following box.*

Test of a Hypothesis about σ^2

One-Tailed Test	Two-Tailed Test
H_0: $\sigma^2 = \sigma_0^2$	H_0: $\sigma^2 = \sigma_0^2$
H_a: $\sigma^2 < \sigma_0^2$ (or H_a: $\sigma^2 > \sigma_0^2$)	H_a: $\sigma^2 \neq \sigma_0^2$
Test statistic: $\chi^2 = \dfrac{(n-1)s^2}{\sigma_0^2}$	*Test statistic:* $\chi^2 = \dfrac{(n-1)s^2}{\sigma_0^2}$
Rejection region: $\chi^2 < \chi^2_{(1-\alpha)}$ (or $\chi^2 > \chi^2_{\alpha}$ when H_a: $\sigma^2 > \sigma_0^2$)	*Rejection region:* $\chi^2 < \chi^2_{(1-\alpha/2)}$ or $\chi^2 > \chi^2_{(\alpha/2)}$

where σ_0^2 is the hypothesized variance and the distribution of χ^2 is based on $(n-1)$ degrees of freedom.

Conditions Required for a Valid Hypothesis Test for s^2

1. A random sample is selected from the target population.
2. The population from which the sample is selected has a distribution that is approximately normal.

CAUTION The procedure for conducting a hypothesis test for σ^2 in the above examples requires an assumption regardless of whether the sample size n is large or small. We must assume that the population from which the sample is selected has an approximate normal distribution. Unlike small sample tests for μ based on the t-statistic, *slight to moderate departures from normality will render the chi-square test invalid.* ▲

Exercises 6.90–6.101

Learning the Mechanics

6.90 Let χ_0^2 be a particular value of χ^2. Find the value of χ_0^2 such that

a. $P(\chi^2 > \chi_0^2) = .10$ for $n = 12$
b. $P(\chi^2 > \chi_0^2) = .05$ for $n = 9$
c. $P(\chi^2 > \chi_0^2) = .025$ for $n = 5$

6.91 A random sample of n observations is selected from a normal population to test the null hypothesis that $\sigma^2 = 25$. Specify the rejection region for each of the following combinations of H_a, α, and n:

a. H_a: $\sigma^2 \neq 25$; $\alpha = .05$; $n = 16$
b. H_a: $\sigma^2 > 25$; $\alpha = .01$; $n = 23$
c. H_a: $\sigma^2 > 25$; $\alpha = .10$; $n = 15$
d. H_a: $\sigma^2 < 25$; $\alpha = .01$; $n = 13$
e. H_a: $\sigma^2 \neq 25$; $\alpha = .10$; $n = 7$
f. H_a: $\sigma^2 < 25$; $\alpha = .05$; $n = 25$

6.92 A random sample of seven measurements gave $\bar{x} = 9.4$ and $s^2 = 4.84$.

a. What assumptions must you make concerning the population in order to test a hypothesis about σ^2?
b. Suppose the assumptions in part **a** are satisfied. Test the null hypothesis, $\sigma^2 = 1$, against the alternative hypothesis, $\sigma^2 > 1$. Use $\alpha = .05$.
c. Test the null hypothesis that $\sigma^2 = 1$ against the alternative hypothesis that $\sigma^2 \neq 1$. Use $\alpha = .05$.

6.93 Refer to Exercise 6.92. Suppose we had $n = 100$, $\bar{x} = 9.4$, and $s^2 = 4.84$.

a. Test the null hypothesis, H_0: $\sigma^2 = 1$, against the alternative hypothesis, H_a: $\sigma^2 > 1$.
b. Compare your test result with that of Exercise 6.92.

6.94 A random sample of $n = 7$ observations from a normal population produced the following measurements: 4, 0, 6,

*A confidence interval for σ^2 can also be formed using the chi-square distribution with $(n-1)$ degrees of freedom. A $(1-a)$ 100% confidence interval is

$$\frac{(n-1)s^2}{\chi^2_{a/2}} < \sigma^2 < \frac{(n-1)s^2}{\chi^2_{(1-a/2)}}$$

3, 3, 5, 9. Do the data provide sufficient evidence to indicate that $\sigma^2 < 1$? Test using $\alpha = .05$.

Applying the Concepts—Basic

6.95 Latex allergy in health care workers. Refer to the *Current Allergy & Clinical Immunology* (Mar. 2004) study of $n = 46$ hospital employees who were diagnosed with a latex allergy from exposure to the powder on latex gloves, Exercise 6.23 (p. 334). Recall that the number of latex gloves used per week by the sampled workers is summarized as follows: $\bar{x} = 19.3$ and $s = 11.9$. Let σ^2 represent the variance in the number of latex gloves used per week by all hospital employees. Consider testing H_0: $\sigma^2 = 100$ against H_a: $\sigma^2 \neq 100$.

a. Give the rejection region for the test at a significance level of $\alpha = .01$.

b. Calculate the value of the test statistic.

c. Use the results, parts **a** and **b,** to make the appropriate conclusion.

6.96 A new dental bonding agent. Refer to the *Trends in Biomaterials & Artificial Organs* (Jan. 2003) study of a new dental bonding adhesive (called *Smartbond*), Exercise 6.54 (p. 347). Recall that tests on a sample of 10 extracted teeth bonded with the new adhesive resulted in a mean breaking strength (after 24 hours) of $\bar{x} = 5.07$ Mpa and a standard deviation of $s = .46$ Mpa. The manufacturer must demonstrate that the breaking strength variance of the new adhesive is less than the variance of the standard composite adhesive, $\sigma^2 = .25$.

a. Set up the null and alternative hypotheses for the test.

b. Find the rejection region for the test using $\alpha = .01$.

c. Compute the test statistic.

d. Give the appropriate conclusion for the test.

e. What conditions are required for the test results to be valid?

6.97 Golf tees produced from an injection mold. Refer to Exercise 6.27 (p. 335) and the weights of tees produced by an injection mold process. If operating correctly, the process will produce tees with a weight variance of .000004 (ounces)^2. If the weight variance differs from .000004, the injection molder is out of control.

a. Set up the null and alternative hypotheses for testing whether the injection mold process is out of control.

b. Use the data saved in the **TEES** file to conduct the test, part **a.** Use $\alpha = .01$.

c. What conditions are required for inferences derived from the test to be valid? Are they reasonably satisfied?

Applying the Concepts—Intermediate

6.98 NW **Identifying type of land cover.** Refer to the *Geographical Analysis* (Oct. 2006) study of a new method for analyzing remote sensing data from satellite pixels, Exercise 6.25 (p. 335). Recall that the method uses a numerical measure of the distribution of gaps or hole sizes in the pixel, called lacunarity. Summary statistics for the lacunarity measurements in a sample of 100 grassland pixels are $\bar{x} = 225$ and $s = 20$. As stated in Exercise 6.25, it is known that the mean lacunarity measurement for all grassland pixels is 220. The method will be effective in identifying land cover if the standard deviation of the measurements is 10% (or less) of the true mean, i.e., if the standard deviation is less than 22.

a. Give the null and alternative hypotheses for a test to determine if, in fact, the standard deviation of all grassland pixels is less than 22.

b. A Minitab analysis of the data is shown at the bottom of the page. Locate and interpret the p-value of the test. Use $\alpha = .10$.

6.99 Do ball bearings conform to specifications? It is essential in the manufacture of machinery to use parts that conform to specifications. In the past, diameters of the ball bearings produced by a certain manufacturer had a variance of .00156. To cut costs, the manufacturer instituted a less expensive production method. The variance of the diameters of 100 randomly sampled bearings produced by the new process was .00211. Do the data provide sufficient evidence to indicate that diameters of ball-bearings produced by the new process are more variable than those produced by the old process?

6.100 Cooling method for gas turbines. Refer to the *Journal of Engineering for Gas Turbines and Power* (Jan. 2005) study of the performance of augmented gas turbine engines, Exercise 6.28 (p. 335). Recall that performance for each in a sample of 67 gas turbines was measured by heat rate

Minitab output for Exercise 6.98

Test for One Standard Deviation

```
Method
Null hypothesis         Sigma = 22
Alternative hypothesis  Sigma = < 22

The standard method is only for the normal distribution.


Statistics
N   StDev  Variance
100   20.0       400


Tests
Method    Chi-Square  DF   P-Value
Standard       81.82  99     0.105
```

(kilojoules per kilowatt per hour). The data are saved in the **GASTURBINE** file. Suppose that standard gas turbines have heat rates with a standard deviation of 1,500 kJ/kWh. Is there sufficient evidence to indicate that the heat rates of the augmented gas turbine engine are more variable than the heat rates of the standard gas turbine engine? Test using $\alpha = .05$.

Applying the Concepts—Advanced

6.101 Why do small firms export? Refer to the *Journal of Small Business Management* (Vol. 40, 2002) study of what motivates small firms to export, Exercise 6.33 (p. 336). Recall that in a survey of 137 exporting firms, each CEO was asked to respond to the statement "Management believes that the firm can achieve economies of scale by exporting" on a scale of 1 (strongly disagree) to 5 (strongly agree). Summary statistics for the $n = 137$ scale scores were reported as $\bar{x} = 3.85$ and $s = 1.5$.

a. Explain why the researcher will be unable to conclude that the true mean scale score exceeds 3.5 (as in Exercise 6.33) if the standard deviation of the scale scores is too large.

b. Give the largest value of the true standard deviation, σ, for which you will reject the null hypothesis H_0: $\mu = 3.5$ in favor of the alternative hypothesis H_a: $\mu > 3.5$ using $\alpha = .01$.

c. Based on the study results, is there evidence (at $\alpha = .01$) to indicate that σ is smaller than the value you determined in part **b**?

CHAPTER NOTES

Key Terms

Key Symbols

μ	Population mean
p	Population proportion, P(Success), in binomial trial
σ^2	Population variance
$\bar{x}$	Sample mean (estimator of μ)
$\hat{p}$	Sample proportion (estimator of p)
s^2	Sample variance (estimator of σ^2)
H_0	Null hypothesis
H_a	Alternative hypothesis
α	Probability of a Type I error
β	Probability of a Type II error
χ^2	Chi-square (sampling distribution of s^2 for normal data)

Key Ideas

Key Words for Identifying the Target Parameter

μ—Mean, Average

p—Proportion, Fraction, Percentage, Rate, Probability

σ^2—Variance, Variability, Spread

Elements of a Hypothesis Test

1. *Null hypothesis* (H_0)
2. *Alternative hypothesis* (H_a)
3. *Test statistic* (z, t, *or* χ^2)
4. *Significance level* (α)
5. *p-value*
6. *Conclusion*

Errors in Hypothesis Testing

Type I Error = Reject H_0 when H_0 is true (occurs with probabilitys α)

Type II Error = Accept H_0 when H_0 is false (occurs with probability β)

Power of a Test = P(Reject H_0 when H_0 is false) $= 1 - \beta$

Forms of Alternative Hypothesis

Lower-tailed: H_a: $\mu < 50$

Upper-tailed: H_a: $\mu > 50$

Two-tailed: H_a: $\mu \neq 50$

Using *p*-values to Decide

1. Choose significance level (α)
2. Obtain p-value of the test
3. If $\alpha > p$-value, reject H_0

Guide to Selecting a One-Sample Hypothesis Test

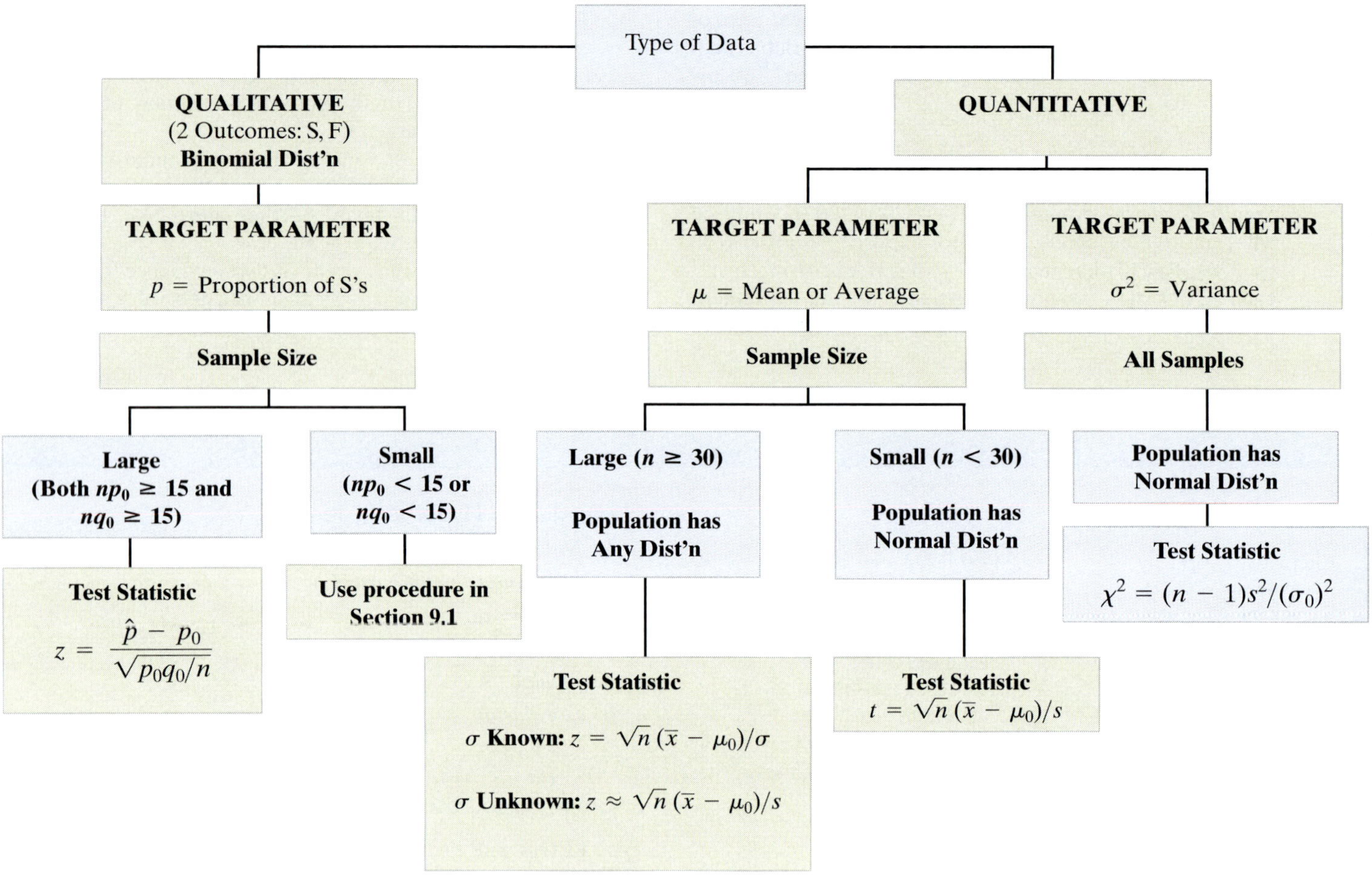

Supplementary Exercises 6.102–6.135

Note: List the assumptions necessary for the valid implementation of the statistical procedures you use in solving all these exercises.

Learning the Mechanics

6.102 Specify the differences between a large-sample and small-sample test of hypothesis about a population mean μ. Focus on the assumptions and test statistics.

6.103 *Complete the following statement:* The smaller the p-value associated with a test of hypothesis, the stronger the support for the _____ hypothesis. Explain your answer.

6.104 Which of the elements of a test of hypothesis can and should be specified *prior* to analyzing the data that are to be used to conduct the test?

6.105 If you select a very small value for α when conducting a hypothesis test, will β tend to be big or small? Explain.

6.106 If the rejection of the null hypothesis of a particular test would cause your firm to go out of business, would you want α to be small or large? Explain.

6.107 A random sample of 20 observations selected from a normal population produced $\bar{x} = 72.6$ and $s^2 = 19.4$.

a. Test H_0: $\mu = 80$ against H_a: $\mu < 80$. Use $\alpha = .05$.
b. Test H_0: $\mu = 80$ against H_a: $\mu \neq 80$. Use $\alpha = .01$.

6.108 A random sample of 175 measurements possessed a mean $\bar{x} = 8.2$ and a standard deviation $s = .79$.

a. Test H_0: $\mu = 8.3$ against H_a: $\mu \neq 8.3$. Use $\alpha = .05$.
b. Test H_0: $\mu = 8.4$ against H_a: $\mu \neq 8.4$. Use $\alpha = .05$.
c. Test H_0: $\sigma = 1$ against H_a: $\sigma \neq 1$. Use $\alpha = .05$.
d. Find the power of the test, part **a**, if $\mu_a = 8.5$.

6.109 A random sample of $n = 200$ observations from a binomial population yields $\hat{p} = .29$.

a. Test H_0: $p = .35$ against H_a: $p < .35$. Use $\alpha = .05$.
b. Test H_0: $p = .35$ against H_a: $p \neq .35$. Use $\alpha = .05$.

6.110 A t-test is conducted for the null hypothesis H_0: $\mu = 10$ versus the alternative H_a: $\mu > 10$ for a random sample of $n = 17$ observations. The test results are $t = 1.174$, p-value $= .1288$.

a. Interpret the p-value.
b. What assumptions are necessary for the validity of this test?
c. Calculate and interpret the p-value assuming the alternative hypothesis was instead H_a: $\mu \neq 10$.

6.111 A random sample of 41 observations from a normal population possessed a mean $\bar{x} = 88$ and a standard deviation $s = 6.9$.

a. Test H_0: $\sigma^2 = 30$ against H_a: $\sigma^2 > 30$. Use $\alpha = .05$.
b. Test H_0: $\sigma^2 = 30$ against H_a: $\sigma^2 \neq 30$. Use $\alpha = .05$.

Applying the Concepts—Basic

6.112 Semester hours taken by CPA candidates. Refer to the *Journal of Accounting and Public Policy* (Spring 2002) study of $n = 100{,}000$ first-time candidates for the CPA exam, Exercise 5.92 (p. 310). The number of semester hours of college credit taken by the sampled candidates is summarized as follows: $\bar{x} = 141.31$ hours and $s = 17.77$ hours. Let μ represent the mean number of semester hours taken by all first-time candidates for the CPA exam. Consider testing H_0: $\mu = 140$ against H_a: $\mu > 140$.

a. Give the rejection region for the test at a significance level of $\alpha = .01$.

b. Calculate the value of the test statistic.

c. Use the results, parts **a** and **b,** to make the appropriate conclusion.

6.113 Unauthorized computer use. Refer to the Computer Security Institute (CSI) survey of computer crime, Exercise 2.146 (p. 106). Recall that in 1999, 62% of businesses suffered unauthorized use of computer systems. In a survey taken 7 years later, 320 of a sample of 616 businesses reported unauthorized use of computer systems. (*Computer Security Issues & Trends,* Spring 2006). Let p represent the true proportion of businesses that reported unauthorized use of computer systems in 2006.

a. Calculate a point estimate for p.

b. Set up the null and alternative hypothesis to test whether the value of p has changed since 1999.

c. Calculate the test statistic for the test, part **b.**

d. Find the rejection region for the test if $\alpha = .05$.

e. Use the results, parts **c** and **d,** to make the appropriate conclusion.

f. Find the p-value of the test and confirm that the conclusion based on the p-value agrees with the conclusion in part **e.**

6.114 Quality of cable TV news. In a survey of 500 television viewers with access to cable TV, each was asked whether they agreed with the statement, "Overall, I find the quality of news on cable networks (such as CNN, FOXNews, CNBC, and MSNBC) to be better than news on the ABC, CBS, and NBC networks." A total of 248 viewers agreed with the statement (*Cabletelevision Advertising Bureau,* May 2002). [*Note:* The survey respondents were contacted via e-mail on the Internet.]

a. Identify the population parameter of interest in the survey.

b. Give a point estimate of the population parameter.

c. Set up H_0 and H_a for testing whether the true percentage of TV viewers who find cable news to be better quality than network news differs from 50%.

d. Conduct the test, part **c,** using $\alpha = .10$ Make the appropriate conclusion in the words of the problem.

e. What conditions are required for the inference, part **d,** to be valid? Do they appear to be satisfied?

6.115 Beta-value of a stock. The "beta coefficient" of a stock is a measure of the stock's volatility (or risk) relative to the market as a whole. Stocks with beta coefficients greater than 1 generally bear greater risk (more volatility) than the market, whereas stocks with beta coefficients less than 1 are less risky (less volatile) than the overall market (Alexander, Sharpe, and Bailey, *Fundamentals of Investments,* 2000). A random sample of 15 high-technology stocks was selected at the end of 2009, and the mean and standard deviation of the beta coefficients were calculated: $\bar{x} = 1.23$, $s = .37$.

a. Set up the appropriate null and alternative hypotheses to test whether the average high-technology stock is riskier than the market as a whole.

b. Establish the appropriate test statistic and rejection region for the test. Use $\alpha = .10$.

c. What assumptions are necessary to ensure the validity of the test?

d. Calculate the test statistic and state your conclusion.

e. What is the approximate p-value associated with this test? Interpret it.

f. Conduct a test to determine if the variance of the stock beta values differs from .15. Use $\alpha = .05$.

6.116 A camera that detects liars. According to *New Scientist* (Jan. 2, 2002), a new thermal imaging camera that detects small temperature changes is now being used as a polygraph device. The U.S. Department of Defense Polygraph Institute (DDPI) claims the camera can correctly detect liars 75% of the time by monitoring the temperatures of their faces.

a. Give the null hypothesis for testing the claim made by the DDPI.

b. What is a Type I error for this problem? A Type II error?

6.117 Pouring temperature of an iron automotive casting. The Cleveland Casting Plant produces iron automotive castings for Ford Motor Company (*Quality Engineering,* Vol. 7, 1995). The pouring temperatures (in degrees Fahrenheit) for a sample of 10 crankshafts produced at the plant are listed on the next page. These data are saved in the **IRONTEMP** file. When the process is stable, the target pouring temperature of the molton iron is 2,550 degrees. Conduct a test to determine whether the true mean pouring temperature differs from the target setting. Test using $\alpha = .01$. Use the SPSS printout below to carry out the test.

SPSS Output for Exercise 6.117

One-Sample Test

	Test Value = 2550					
					95% Confidence Interval of the Difference	
	t	df	Sig. (2-tailed)	Mean Difference	Lower	Upper
POURTEMP	1.210	9	.257	8.70	-7.57	24.97

Data for Exercise 6.117

2,543	2,541	2,544	2,620	2,560	2,559	2,562
2,553	2,552	2,553				

Source: Price, B., and Barth, B. "A structural model relating process inputs and final product characteristics," *Quality Engineering,* Vol. 7, No. 4, 1995, p. 696 (Table 2). Reprinted by permission of the publisher (Taylor & Francis Group, www.informaworld.com).

Applying the Concepts—Intermediate

6.118 Consumers' use of discount coupons. In 1894, druggist Asa Candler began distributing handwritten tickets to his customers for free glasses of Coca-Cola at his soda fountain. That was the genesis of the discount coupon. In 1975, it was estimated that 65% of U.S. consumers regularly used discount coupons when shopping. In a more recent consumer survey, 72% said they regularly redeem coupons (Prospectiv 2008 Consumer Coupon Poll). Assume the recent survey consisted of a random sample of 1,000 shoppers.

a. Does the survey provide sufficient evidence that the percentage of shoppers using cents-off coupons exceeds 65%? Test using $\alpha = .05$.

b. Is the sample size large enough to use the inferential procedures presented in this section? Explain.

c. Find the observed significance level for the test you conducted in part **a** and interpret its value.

6.119 Errors in medical tests. Medical tests have been developed to detect many serious diseases. A medical test is designed to minimize the probability that it will produce a "false positive" or a "false negative." A *false positive* refers to a positive test result for an individual who does not have the disease, whereas a false negative is a negative test result for an individual who does have the disease.

a. If we treat a medical test for a disease as a statistical test of hypothesis, what are the null and alternative hypotheses for the medical test?

b. What are the Type I and Type II errors for the test? Relate each to false positives and false negatives.

c. Which of the errors has graver consequences? Considering this error, is it more important to minimize α or β? Explain.

6.120 Drivers' use of the Lincoln Tunnel. The Lincoln Tunnel (under the Hudson River) connects suburban New Jersey to midtown Manhattan. On Mondays at 8:30 A.M., the mean number of cars waiting in line to pay the Lincoln Tunnel toll is 1,220. Because of the substantial wait during rush hour, the Port Authority of New York and New Jersey is considering raising the amount of the toll between 7:30 and 8:30 A.M. to encourage more drivers to use the tunnel at an earlier or later time. Suppose the Port Authority experiments with peak-hour pricing for 6 months, increasing the toll from $4 to $7 during the rush hour peak. On 10 different workdays at 8:30 A.M. aerial photographs of the tunnel queues are taken and the number of vehicles counted. The results (saved in the **TUNNEL** file) follow:

1,260	1,052	1,201	942	1,062	999	931	849	867	735

Analyze the data for the purpose of determining whether peak-hour pricing succeeded in reducing the average number of vehicles attempting to use the Lincoln Tunnel during the peak rush hour.

6.121 Point spreads of NFL games. Refer to the *Chance* (Fall 1998) study of point-spread errors in NFL games, Exercise 6.29 (p. 335). Recall that the difference between the actual game outcome and the point spread established by oddsmakers—the point-spread error—was calculated for 240 NFL games. The results are summarized as follows: $\bar{x} = -1.6$, $s = 13.3$. Suppose the researcher wants to know whether the true standard deviation of the point-spread errors exceeds 15. Conduct the analysis using $\alpha = .10$.

6.122 Improving the productivity of chickens. Refer to the *Applied Animal Behaviour Science* (Oct. 2000) study of the color of string preferred by pecking domestic chickens, Exercise 5.100 (p. 311). Recall that $n = 72$ chickens were exposed to blue string and the number of pecks each chicken took at the string over a specified time interval had a mean of $\bar{x} = 1.13$ pecks and a standard deviation of $s = 2.21$ pecks. Also recall that previous research has shown that $\mu = 7.5$ pecks if chickens are exposed to white string.

a. Conduct a test (at $\alpha = .01$) to determine if the true mean number of pecks at blue string is less than $\mu = 7.5$ pecks.

b. In Exercise 5.100, you used a 99% confidence interval as evidence that chickens are more apt to peck at white string than blue string. Do the test results, part **a,** support this conclusion? Explain.

6.123 Are manufacturers satisfied with trade promotions? Sales promotions that are used by manufacturers to entice retailers to carry, feature, or push the manufacturer's products are called *trade promotions.* A survey of 132 manufacturers conducted by Nielsen, found that 36% of the manufacturers were satisfied with their spending for trade promotions (*Survey of Trade Promotion Practices,* 2004). Is this sufficient evidence to reject a previous claim by the American Marketing Association that no more than half of all manufacturers are dissatisfied with their trade promotion spending?

a. Conduct the appropriate hypothesis test at $\alpha = .02$. Begin your analysis by determining whether the sample size is large enough to apply the testing methodology presented in this chapter.

b. Report the observed significance level of the test and interpret its meaning in the context of the problem.

c. Calculate β, the probability of a Type II error, if in fact 55% of all manufacturers are dissatisfied with their trade promotion spending.

6.124 Arresting shoplifters. Shoplifting in the United States costs retailers about $35 million a day. Despite the seriousness of the problem, the National Association of Shoplifting Prevention (NASP) claims that only 50% of all shoplifting are turned over to police (www.shopliftingprevention.org, 2009). A random sample of 40 U.S. retailers were questioned concerning the disposition of the most recent shoplifter they apprehended. A total of 24 were turned over to police. Do these data provide sufficient evidence to contradict the NASP?

a. Conduct a hypothesis test to answer the question of interest. Use $\alpha = .05$.

b. Is the sample size large enough to use the inferential procedure of part **a**?

c. Find the observed significance level of the hypothesis test in part **a.** Interpret the value.

d. For what values of α would the observed significance level be sufficient to reject the null hypothesis of the test you conducted in part **b**?

6.125 Arresting shoplifters (cont'd). Refer to Exercise 6.124.

a. Describe a Type II error in terms of this application.

b. Calculate the probability β of a Type II error for this test assuming that the true fraction of shoplifters turned over to the police is $p = .55$.

c. Suppose the number of retailers sampled is increased from 40 to 100. How does this affect the probability of a Type II error for $p = .55$?

6.126 Frequency marketing programs by restaurants. To instill customer loyalty, airlines, hotels, rental car companies, and credit card companies (among others) have initiated *frequency marketing programs* that reward their regular customers. More than 80 million people are members of the frequent flier programs of the airline industry (www.frequentflier.com). A large fast-food restaurant chain wished to explore the profitability of such a program. They randomly selected 12 of their 1,200 restaurants nationwide and instituted a frequency program that rewarded customers with a \$5.00 gift certificate after every 10 meals purchased at full price. They ran the trial program for 3 months. The restaurants not in the sample had an average increase in profits of \$1,047.34 over the previous 3 months, whereas the restaurants in the sample had the following changes in profit. These data are saved in the **PROFIT** file.

\$2,232.90	\$ 545.47	\$3,440.70	\$1,809.10
\$6,552.70	\$4,798.70	\$2,965.00	\$2,610.70
\$3,381.30	\$1,591.40	\$2,376.20	−\$2,191.00

Note that the last number is negative, representing a decrease in profits.

a. Specify the appropriate null and alternative hypotheses for determining whether the mean profit change for restaurants with frequency programs is significantly greater (in a statistical sense) than \$1,047.34.

b. Conduct the test of part **b** using $\alpha = .05$. Does it appear that the frequency program would be profitable for the company if adopted nationwide?

6.127 EPA limits on vinyl chloride. The EPA sets an airborne limit of 5 parts per million (ppm) on vinyl chloride, a colorless gas used to make plastics, adhesives, and other chemicals. It is both a carcinogen and a mutagen (New Jersey Department of Health, *Hazardous Substance Fact Sheet,* 2005). A major plastics manufacturer, attempting to control the amount of vinyl chloride its workers are exposed to, has given instructions to halt production if the mean amount of vinyl chloride in the air exceeds 3.0 ppm. A random sample of 50 air specimens produced the following statistics: $\bar{x} = 3.1$ ppm, $s = .5$ ppm.

a. Do these statistics provide sufficient evidence to halt the production process? Use $\alpha = .01$.

b. If you were the plant manager, would you want to use a large or a small value for α for the test in part **a**? Explain.

c. Find the p-value for the test and interpret its value.

6.128 EPA limits vinyl chloride (cont'd). Refer to Exercise 6.127.

a. In the context of the problem, define a Type II error.

b. Calculate β for the test described in part **a** of Exercise 6.127, assuming that the true mean is $\mu = 3.1$ ppm.

c. What is the power of the test to detect a departure from the manufacturer's 3.0 ppm limit when the mean is 3.1 ppm?

d. Repeat parts **b** and **c** assuming that the true mean is 3.2 ppm. What happens to the power of the test as the plant's mean vinyl chloride level departs further from the limit?

6.129 EPA limits on vinyl chloride (cont'd). Refer to Exercises 6.127 and 6.128.

a. Suppose an α value of .05 is used to conduct the test. Does this change favor halting production? Explain.

b. Determine the value of β and the power for the test when $\alpha = .05$ and $\mu = 3.1$.

c. What happens to the power of the test when α is increased?

6.130 Evaluating a measuring instrument. One way of evaluating a measuring instrument is to repeatedly measure the same item and compare the average of these measurements to the item's known measured value. The difference is used to assess the instrument's accuracy (American Society for Quality). To evaluate a particular Metlar scale, an item whose weight is known to be 16.01 ounces is weighed five times by the same operator. The measurements (in ounces) are saved in the **METLAR** file and follow:

15.99	16.00	15.97	16.01	15.96

a. In a statistical sense, does the average measurement differ from 16.01? Conduct the appropriate hypothesis test at $\alpha = .05$. What does your analysis suggest about the accuracy of the instrument?

b. List any assumptions you make in conducting the hypothesis test, part **a.**

c. Evaluate the instrument's precision by testing whether the standard deviation of the weight measurements is greater than .01. Use $\alpha = .05$.

6.131 Testing the validity of a TV advertisement. The manufacturer of an over-the-counter analgesic claims that its product brings pain relief to headache sufferers in less than 3.5 minutes, on average. In order to be able to make this claim in its television advertisements, the manufacturer was required by a particular television network to present statistical evidence in support of the claim. The manufacturer reported that for a random sample of 50 headache sufferers, the mean time to relief was 3.3 minutes and the standard deviation was 1.1 minutes.

a. Do these data support the manufacturer's claim? Test using $\alpha = .05$.

b. Report the p-value of the test.

c. In general, do large p-values or small p-values support the manufacturer's claim? Explain.

Applying the Concepts—Advanced

6.132 NCAA "March Madness." For three weeks each March, the NCAA holds its annual men's basketball championship tournament. The 64 best college basketball teams in the nation play a single-elimination tournament—a total of 63 games—to determine the NCAA champion. Tournament followers, from hardcore gamblers to the casual fan who enters the office betting pool, have a strong interest in handicapping the games. To provide insight into this phenomenon, statisticians Hal Stern and Barbara Mock analyzed data from 13 previous NCAA tournaments and published their results in *Chance* (Winter 1998). The results of first-round games are summarized in the next table.

Summary of First-Round NCAA Tournament Games, 1985–1997

			Margin of Victory (Points)	
Matchup (Seeds)	Number of Games	Number Won by Favorite (Higher Seed)	Mean	Standard Deviation
1 vs 16	52	52	22.9	12.4
2 vs 15	52	49	17.2	11.4
3 vs 14	52	41	10.6	12.0
4 vs 13	52	42	10.0	12.5
5 vs 12	52	37	5.3	10.4
6 vs 11	52	36	4.3	10.7
7 vs 10	52	35	3.2	10.5
8 vs 9	52	22	−2.1	11.0

Source: Stern, H. S., and Mock, B. "College basketball upsets: Will a 16-seed ever beat a 1-seed?" *Chance*, Vol. 11, No. 1, Winter 1998, p. 29 (Table 3). Reprinted with permission from *Chance*.

a. A common perception among fans, media, and gamblers is that the higher seeded team has a better than 50-50 chance of winning a first-round game. Is there evidence to support this perception? Conduct the appropriate test for each matchup. What trends do you observe?

b. Is there evidence to support the claim that a 1-, 2-, 3-, or 4-seeded team will win by an average of more than 10 points in first-round games? Conduct the appropriate test for each matchup.

c. Is there evidence to support the claim that a 5-, 6-, 7-, or 8-seeded team will win by an average of less than 5 points in first-round games? Conduct the appropriate test for each matchup.

d. For each matchup, test the null hypothesis that the standard deviation of the victory margin is 11 points.

e. The researchers also calculated the difference between the game outcome (victory margin, in points) and point spread established by Las Vegas oddsmakers for a sample of 360 recent NCAA tournament games. The mean difference is .7, and the standard deviation of the difference is 11.3. If the true mean difference is 0, then the point spread can be considered a good predictor of the game outcome. Use this sample information to test the hypothesis that the point spread, on average, is a good predictor of the victory margin in NCAA tournament games.

6.133 Factors that inhibit learning in marketing. What factors inhibit the learning process in the classroom? To answer this question, researchers at Murray State University surveyed 40 students from a senior-level marketing class (*Marketing Education Review,* Fall 1994). Each student was given a list of factors and asked to rate the extent to which each factor inhibited the learning process in courses offered in their department. A 7-point rating scale was used, where 1 = "not at all" and 7 = "to a great extent." The factor with the highest rating was instructor related: "Professors who place too much emphasis on a single right answer rather than overall thinking and creative ideas." Summary statistics for the student ratings of this factor are $\bar{x} = 4.70$, $s = 1.62$.

a. Conduct a test to determine if the true mean rating for this instructor-related factor exceeds 4. Use $\alpha = .05$. Interpret the test results.

b. Examine the results of the study from a practical view, then discuss why "statistically significant" does not always imply "practically significant."

c. Because the variable of interest, rating, is measured on a 7-point scale, it is unlikely that the population of ratings will be normally distributed. Consequently, some analysts may perceive the test, part **a,** to be invalid and search for alternative methods of analysis. Defend or refute this argument.

6.134 Testing the placebo effect. *The placebo effect* describes the phenomenon of improvement in the condition of a patient taking a placebo—a pill that looks and tastes real but contains no medically active chemicals. Physicians at a clinic in La Jolla, California, gave what they thought were drugs to 7,000 asthma, ulcer, and herpes patients. Although the doctors later learned that the drugs were really placebos, 70% of the patients reported an improved condition (*Forbes,* May 22, 1995). Use this information to test (at $\alpha = .05$) the placebo effect at the clinic. Assume that if the placebo is ineffective, the probability of a patient's condition improving is .5.

Critical Thinking Challenge

6.135 The hot tamale caper. "Hot tamales" are chewy, cinnamon-flavored candies. A bulk vending machine is known to dispense, on average, 15 hot tamales per bag with a standard deviation of 3 per bag. *Chance* (Fall 2000) published an article on a classroom project in which students were required to purchase bags of hot tamales from the machine and count the number of candies per bag. One student group claimed they purchased five bags that had the following candy counts: 25, 23, 21, 21, and 20. These data are saved in the **TAMALES** file. There was some question as to whether the students had fabricated the data. Use a hypothesis test to gain insight into whether or not the data collected by the students are fabricated. Use a level of significance that gives the benefit of the doubt to the students.

References

Snedecor, G. W., and Cochran, W. G. *Statistical Methods,* 7th ed. Ames: Iowa State University Press, 1980.

Wackerly, D., Mendenhall, W., and Scheaffer, R. *Mathematical Statistics with Applications,* 7th ed. Belmont, CA: Thomson, Brooks/Cole, 2008.

USING TECHNOLOGY

SPSS: Tests of Hypotheses

Note: SPSS cannot currently conduct a test for a population variance.

Testing μ

Step 1 Access the SPSS spreadsheet file that contains the sample data.

Step 2 Click on the "Analyze" button on the SPSS menu bar and then click on "Compare Means" and "One-Sample T Test," as shown in Figure 6.S.1.

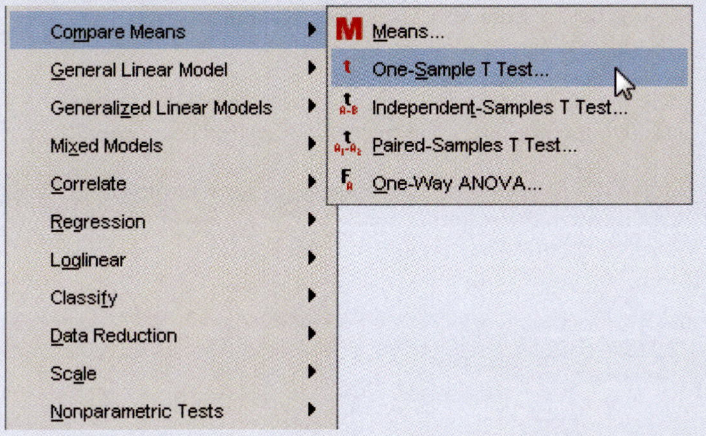

Figure 6.S.1 SPSS menu options for a test on a mean

Step 3 On the resulting dialog box (shown in Figure 6.S.2), specify the quantitative variable of interest in the "Test Variable(s)" box and the value of μ_0 for the null hypothesis in the "Test Value" box.

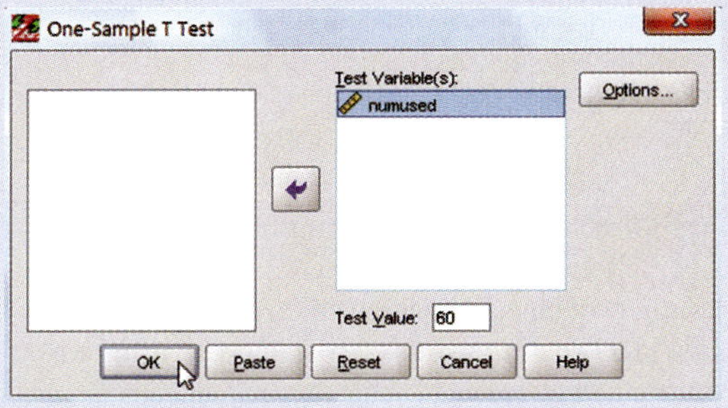

Figure 6.S.2 SPSS 1-sample t test for mean dialog box

Step 4 Click "OK." SPSS will automatically conduct a two-tailed test of hypothesis.

Important Note: The SPSS one-sample t-procedure uses the t-statistic to conduct the test of hypothesis. When the sample size n is small, this is the appropriate method. When the sample size n is large, the t-value will be approximately equal to the large-sample z-value and the resulting test will still be valid.

Testing p

Step 1 Access the SPSS spreadsheet file that contains the sample data.

Step 2 Click on the "Analyze" button on the SPSS menu bar and then click on "Nonparametric Tests" (see Figure 6.S.1).

Step 3 Select "Binomial" from the next menu.

Step 4 On the resulting dialog box (shown in Figure 6.S.3), specify the binomial variable of interest in the "Test Variable List" box and the value of p_0 for the null hypothesis in the "Test Proportion" box.

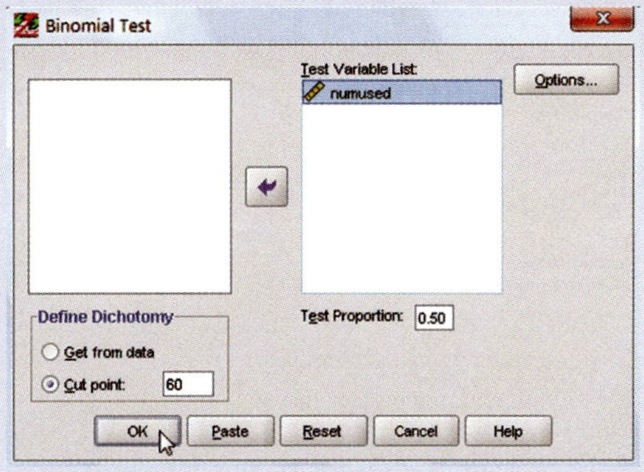

Figure 6.S.3 SPSS binomial test for proportion dialog box

Step 5 Click "OK" to generate the two-tailed test of hypothesis.

Important: SPSS requires the binomial variable to be entered as a quantitative variable. Typically, this is accomplished by entering two numerical values (e.g., 0 and 1) for the two outcomes of the variable. If the data have been entered in this fashion, select the "Get from data" option in the "Define Dichotomy" area of the dialog box.

You can also create the two outcome values for a quantitative variable by selecting the "Cut point" option in the "Define Dichotomy" area and specifying a numerical value. All values of the variables less than or equal to the cut point value are assigned to one group (success), and all other values are assigned to the other group (failure).

Minitab: Tests of Hypotheses

Testing μ

Step 1 Access the Minitab data worksheet that contains the sample data.

Step 2 Click on the "Stat" button on the Minitab menu bar and then click on "Basic Statistics" and "1-Sample t," as shown in Figure 6.M.1.

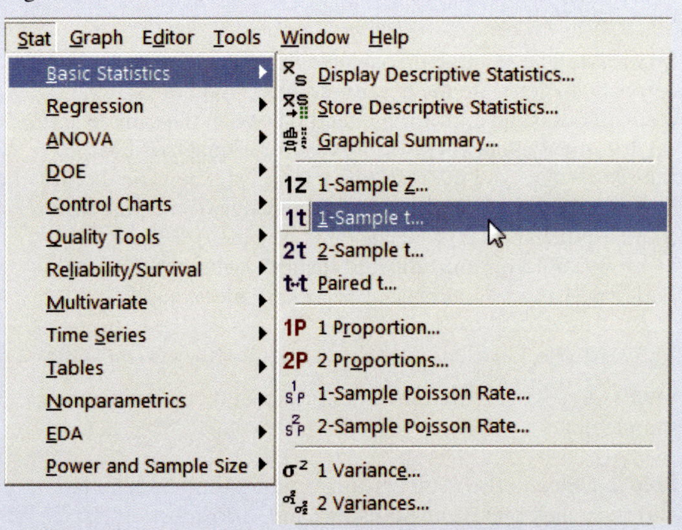

Figure 6.M.1 Minitab menu options for a test on a mean

Step 3 On the resulting dialog box (shown in Figure 6.M.2), click on "Samples in Columns" and then specify the quantitative variable of interest in the open box.

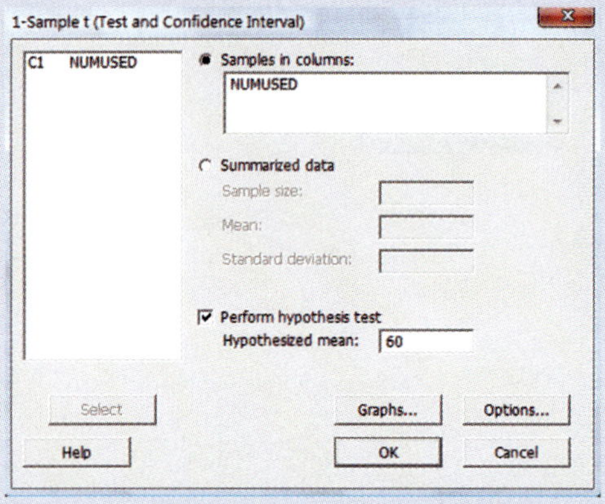

Figure 6.M.2 Minitab 1-sample t test for mean dialog box

Step 4 Check "Perform hypothesis test" and then specify the value of μ_0 for the null hypothesis in the "Hypothesized mean" box.

Step 5 Click on the "Options" button at the bottom of the dialog box and specify the form of the alternative hypothesis, as shown in Figure 6.M.3.

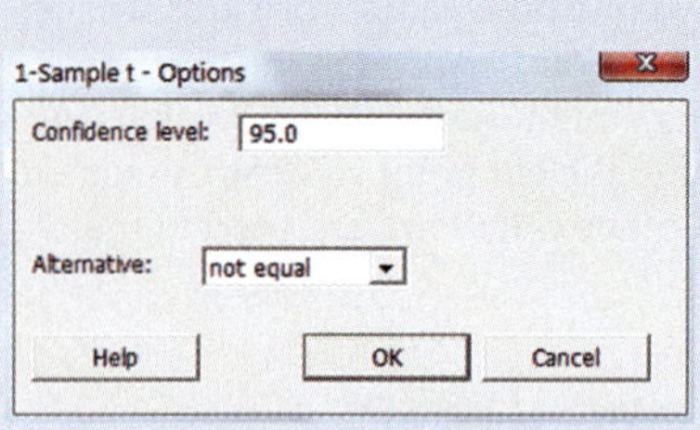

Figure 6.M.3 Minitab 1-sample t test options

Step 6 Click "OK" to return to the "1-Sample t" dialog box and then click "OK" again to produce the hypothesis test.

Note: If you want to produce a test for the mean from summary information (e.g., the sample mean, sample standard deviation, and sample size), click on "Summarized data" in the "1-Sample t" dialog box, enter the values of the summary statistics and μ_0, and then click "OK."

Important: The Minitab one-sample t-procedure uses the t-statistic to generate the hypothesis test. When the sample size n is small, this is the appropriate method. When the sample size n is large, the t-value will be approximately equal to the large-sample z-value, and the resulting test will still be valid. If you have a large sample and you know the value of the population standard deviation σ (which is rarely the case), select "1-sample Z" from the "Basic Statistics" menu options (see Figure 6.M.1) and make the appropriate selections.

Testing p

Step 1 Access the Minitab data worksheet that contains the sample data.

Step 2 Click on the "Stat" button on the Minitab menu bar and then click on "Basic Statistics" and "1 Proportion" (see Figure 6.M.1).

Step 3 On the resulting dialog box (shown in Figure 6.M.4), click on "Samples in Columns," and then specify the qualitative variable of interest in the open box.

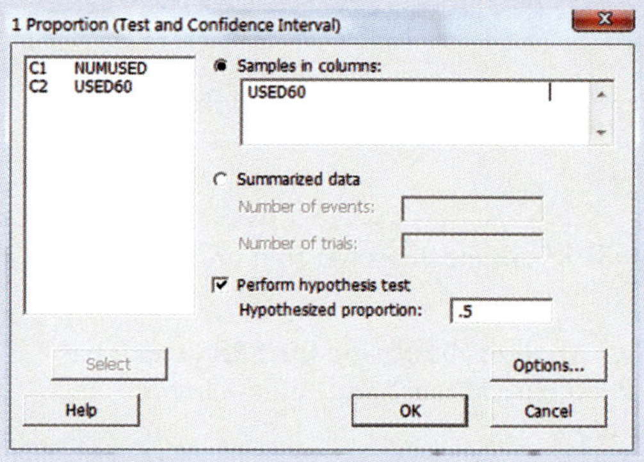

Figure 6.M.4 Minitab 1 proportion test dialog box

Step 4 Check "Perform hypothesis test" and then specify the null hypothesis value p_0 in the "Hypothesized proportion" box.

Step 5 Click "Options," then specify the form of the alternative hypothesis in the resulting dialog box, as shown in Figure 6.M.5. Also, check the "Use test and interval based on normal distribution" box at the bottom.

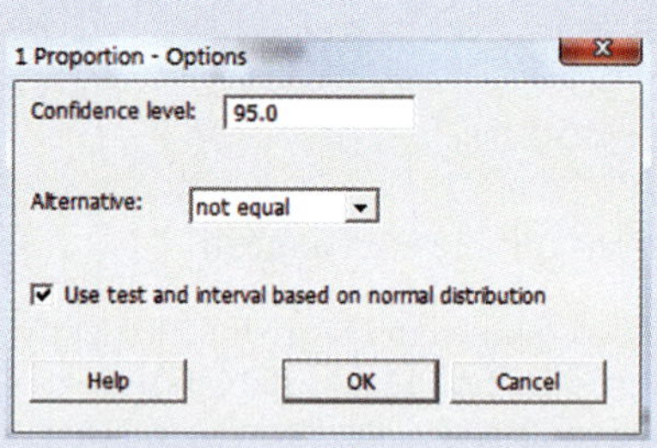

Figure 6.M.5 Minitab 1 proportion test options

Step 6 Click "OK" to return to the "1-Proportion" dialog box and then click "OK" again to produce the test results.

Note: If you want to produce a confidence interval for a proportion from summary information (e.g., the number of successes and the sample size), click on "Summarized data" in the "1 Proportion" dialog box (see Figure 6.M.4). Enter the value for the number of trials (i.e., the sample size) and the number of events (i.e., the number of successes), and then click "OK."

Testing σ^2

Step 1 Access the Minitab data worksheet that contains the sample data set.

Step 2 Click on the "Stat" button on the Minitab menu bar and and then click on "Basic Statistics" and "1 Variance" (see Figure 6.M.1).

Step 3 Once the resulting dialog box appears (see Figure 6.M.6), click on "Samples in Columns" and then specify the quantitative variable of interest in the open box.

Step 4 Check "Perform hypothesis test" and specify the null hypothesis value of the standard deviation σ_0 in the open box.

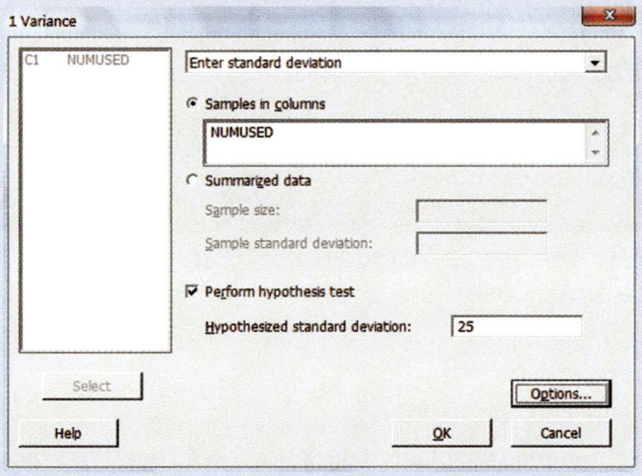

Figure 6.M.6 Minitab 1 variance test dialog box

Step 5 Click on the "Options" button at the bottom of the dialog box and specify the form of the alternative hypothesis (similar to Figure 6.M.3).

Step 6 Click "OK" twice to produce the hypothesis test.

Note: If you want to produce a test for the variance from summary information (e.g., the sample standard deviation and sample size), click on "Summarized data" in the "1 Variance" dialog box (Figure 6.M.6) and enter the values of the summary statistics.

Excel/DDXL: Tests of Hypotheses

Testing μ

Step 1 Highlight (select) the data column you want to analyze on the Excel spreadsheet.

Step 2 Click on "Add-Ins" in the main Excel menu bar and select "DDXL." On the resulting menu, select "Hypothesis Tests," as shown in Figure 6.E.1.

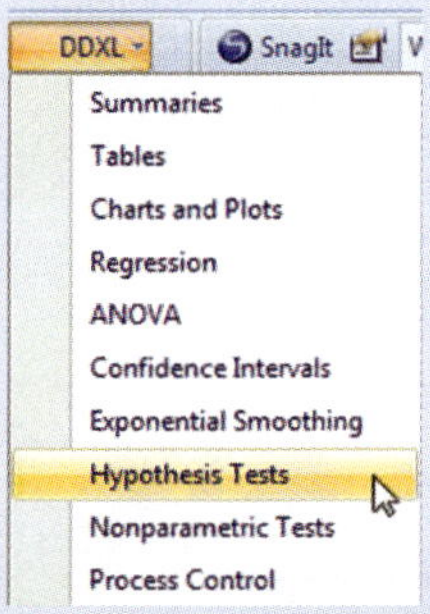

Figure 6.E.1 Excel/DDXL menu options for hypothesis tests

Step 3 In the resulting window, select "1 Var t Test" in the Function Type box, as shown in Figure 6.E.2.

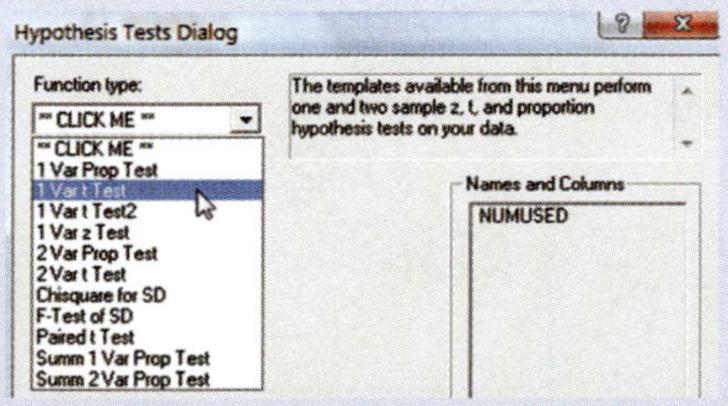

Figure 6.E.2 DDXL hypothesis tests dialog box

Step 4 Move the variable of interest into the "Quantitative Variable" box and then click "OK," as shown in Figure 6.E.3.

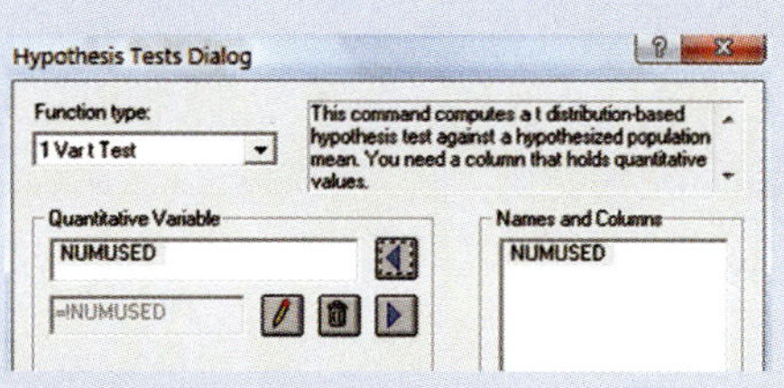

Figure 6.E.3 DDXL menu selections for test on a mean

Step 5 In the resulting window, specify the hypothesized value of μ, α, and the form of the alternative hypothesis and then click "Compute," as shown in Figure 6.E.4. The test results will appear.

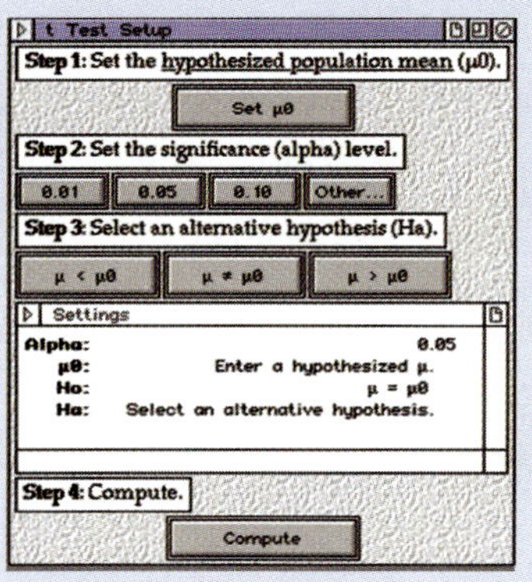

Figure 6.E.4 DDXL t test on a mean setup options

Testing p

Step 1 Highlight (select) the data column you want to analyze on the Excel spreadsheet.

Step 2 Click on "Add-Ins" in the main Excel menu bar and select "DDXL." On the resulting menu, select "Hypothesis Tests," as shown in Figure 6.E.1.

Step 3 In the resulting window, select "1 Var Prop Test" in the Function Type box (see Figure 6.E.2).

Step 4 Move the variable of interest into the "Proportions Variable" box and then click "OK," as shown in Figure 6.E.5.

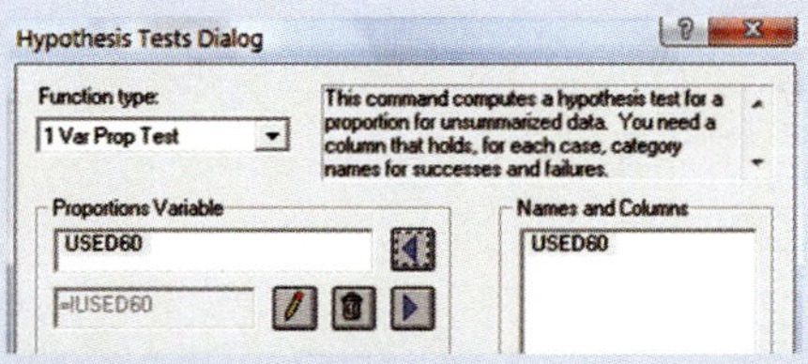

Figure 6.E.5 DDXL menu selections for test of a proportion

Step 5 In the resulting window, specify the hypothesized value of p, α, and the form of the alternative hypothesis and then click "Compute" as shown in Figure 6.E.6. The test results will appear.

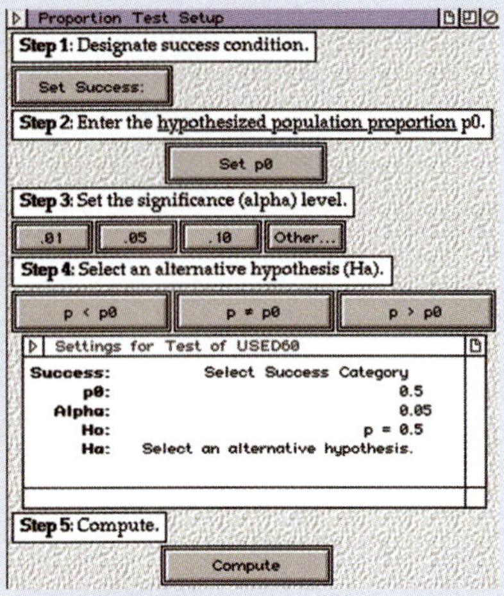

Figure 6.E.6 DDXL test on a proportion setup options

Testing σ^2

Step 1 Highlight (select) the data column you want to analyze on the Excel spreadsheet.

Step 2 Click on "Add-Ins" in the main Excel menu bar and select "DDXL." In the resulting window, select "Hypothesis Tests," as shown in Figure 6.E.1.

Step 3 In the resulting window, select "Chisquare for SD" in the Function Type box (see Figure 6.E.2).

Step 4 Move the variable of interest into the "Quantitative Variable" box and then click "OK," as shown in Figure 6.E.7.

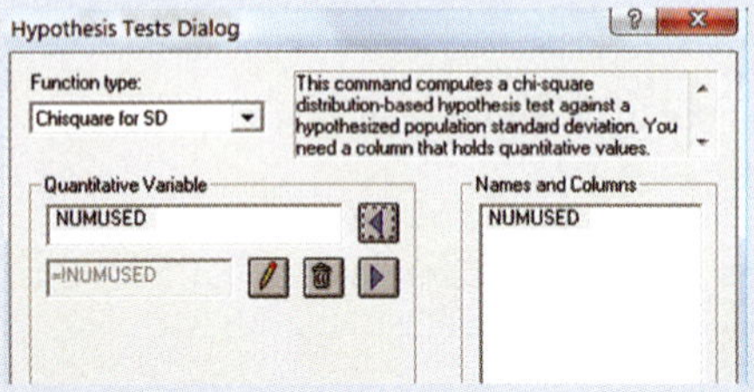

Figure 6.E.7 DDXL menu selections for test of a variance

Step 5 In the resulting window, specify the hypothesized value of σ, α, and the form of the alternative hypothesis and then click "Compute" as shown in Figure 6.E.8. The test results will appear.

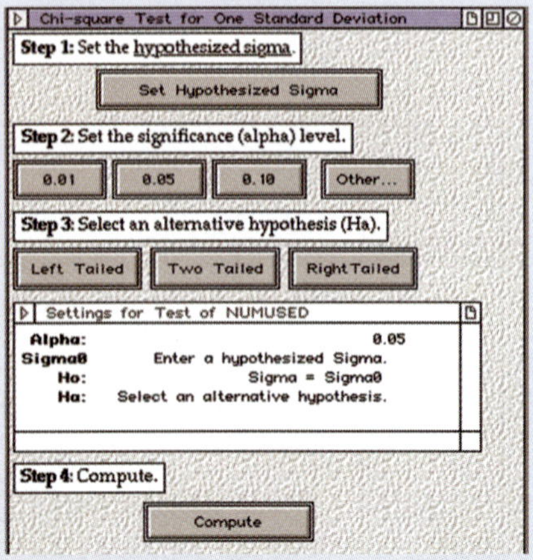

Figure 6.E.8 DDXL test on a variance setup options

TI-84 Graphing Calculator: Tests of Hypotheses

Note: The TI-84 graphing calculator cannot currently conduct a test for a population variance.

Hypothesis Test for a Population Mean (Large Sample Case)

Step 1 *Enter the Data (Skip to Step 2 if you have summary statistics, not raw data)*

- Press **STAT** and select **1:Edit**

Note: If the list already contains data, clear the old data. Use the up **ARROW** to highlight "**L1.**"

- Press **CLEAR ENTER**
- Use the **ARROW** and **ENTER** keys to enter the data set into **L1**

Step 2 *Access the Statistical Tests Menu*

- Press **STAT**
- Arrow right to **TESTS**
- Press **ENTER** to select either **Z-Test** (if large sample and σ known) or **T-Test** (if σ unknown)

Step 3 *Choose "**Data**" or "**Stats**." ("Data" is selected when you have entered the raw data into a List. "Stats" is selected when you are given only the mean, standard deviation, and sample size.)*

- Press **ENTER**

If you selected "Data," enter the values for the hypothesis test where μ_0 = the value for μ in the null hypothesis, σ = assumed value of the population standard deviation.

- Set **List** to **L1**
- Set **Freq** to **1**
- Use the **ARROW** to highlight the appropriate alternative hypothesis
- Press **ENTER**
- Arrow down to "**Calculate**"
- Press **ENTER**

If you selected "Stats," enter the values for the hypothesis test where μ_0 = the value for μ in the null hypothesis, σ = assumed value of the population standard deviation.

- Enter the sample mean, sample standard deviation, and sample size
- Use the **ARROW** to highlight the appropriate alternative hypothesis
- Press **ENTER**
- Arrow down to "**Calculate**"
- Press **ENTER**

The chosen test will be displayed as well as the z (or t) test statistic, the p-value, the sample mean, and the sample size.

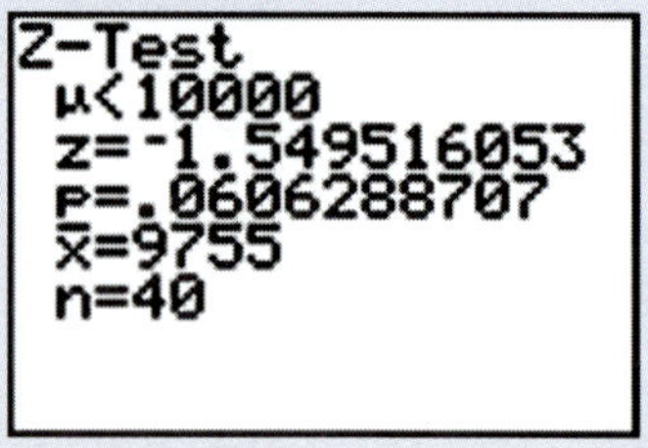

Testing p

Step 1 *Enter the Data (Skip to Step 2 if you have summary statistics, not raw data)*

- Press **STAT** and select **1:Edit**

Note: If the list already contains data, clear the old data. Use the up **ARROW** to highlight "**L1.**"

- Press **CLEAR ENTER**
- Use the **ARROW** and **ENTER** keys to enter the data set into **L1**

Step 2 *Access the Statistical Tests Menu*

- Press **STAT**
- Arrow right to **TESTS**
- Press **ENTER** after selecting **1-Prop Z Test**

Step 3 *Enter the hypothesized proportion p_0, the number of success x, and the sample size n*

- Use the **ARROW** to highlight the appropriate alternative hypothesis
- Press **ENTER**
- Arrow down to "**Calculate**"
- Press **ENTER**

The chosen test will be displayed as well as the z-test statistic and the p-value.

CONTENTS

7 Inferences Based on Two Samples

Confidence Intervals and Tests of Hypotheses

Where We've Been

- Explored two methods for making statistical inferences: *confidence intervals* and *tests of hypotheses*
- Studied confidence intervals and tests for a single population mean μ, a single population proportion p, and a single population variance σ^2
- Learned how to select the sample size necessary to estimate a population parameter with a specified margin of error

Where We're Going

- Learn how to compare two populations using confidence intervals and tests of hypotheses
- Apply these inferential methods to problems where we want to compare two population means, two population proportions, or two population variances
- Determine the sizes of the samples necessary to estimate the difference between two population parameters with a specified margin of error

Statistics IN Action *ZixIt Corp. v. Visa USA Inc.*—A Libel Case

The National Law Journal *(Aug. 26–Sep. 2, 2002) reported on an interesting court case involving ZixIt Corp., a start-up Internet credit card clearing center. ZixIt claimed that its new online credit card processing system would allow Internet shoppers to make purchases without revealing their credit card numbers. This claim violated the established protocols of most major credit card companies, including Visa. Without the company's knowledge, a Visa vice president for technology research and development began writing e-mails and Web site postings on a Yahoo! message board for ZixIt investors, challenging ZixIt's claim and urging investors to sell their ZixIt stock. The Visa executive posted over 400 e-mails and notes before he was caught. Once it was discovered that a Visa executive was responsible for the postings, ZixIt filed a lawsuit against Visa Corp., alleging that Visa—using the executive as its agent—had engaged in a "malicious two-part scheme to disparage and interfere with ZixIt" and its efforts to market the new online credit card processing system. In the libel case ZixIt asked for $699 million in damages.*

Dallas lawyers Jeff Tillotson and Mike Lynn, of the law firm Lynn Tillotson & Pinker, were hired to defend Visa in the lawsuit. The lawyers, in turn, hired *(continued)*

Dr. James McClave (coauthor of this text) as their expert statistician. McClave testified in court on an "event study" he did matching the Visa executive's e-mail postings with movement of ZixIt's stock price the next business day. McClave's testimony, showing that there was an equal number of days when the stock went up as went down after a posting, helped the lawyers representing Visa to prevail in the case. *The National Law Journal* reported that, after two-and-a-half days of deliberation, "the jurors found [the Visa executive] was not acting in the scope of his employment and that Visa had not defamed ZixIt or interfered with its business."

In this chapter, we demonstrate several of the statistical analyses McClave used to infer that the Visa executive's postings had no effect on ZixIt's stock price. The daily ZixIt stock prices as well as the timing of the Visa executive's postings are saved in the **ZIXITVISA** file.* We apply the statistical methodology presented in this chapter to this data set in two Statistics in Action Revisited examples.

Statistics IN Action Revisited

- Comparing Mean Price Changes (p. 391)
- Comparing Proportions (p. 412)

Data Set: ZIXITVISA

7.1 Identifying the Target Parameter

Many experiments involve a comparison of two populations. For instance, a realtor may want to estimate the difference in mean sales price between city and suburban homes. A consumer group might test whether two major brands of food freezers differ in the average amount of electricity they use. A television market researcher wants to estimate the difference in the proportions of younger and older viewers who regularly watch a popular TV program. A golf ball supplier may be interested in comparing the variability in the distance that two competing brands of golf balls travel when struck with the same club. In this chapter, we consider techniques for using two samples to compare the populations from which they were selected.

The same procedures that are used to estimate and test hypotheses about a single population can be modified to make inferences about two populations. As in Chapters 5 and 6, the methodology used will depend on the sizes of the samples and the parameter of interest (i.e., the *target parameter*). Some key words and the type of data associated with the parameters covered in this chapter are listed in the box.

Determining the Target Parameter

Parameter	Key Words or Phrases	Type of Data
$\mu_1 - \mu_2$	Mean difference; difference in averages	Quantitative
$p_1 - p_2$	Difference between proportions, percentages, fractions, or rates; compare proportions	Qualitative
$(\sigma_1)^2/(\sigma_2)^2$	Ratio of variances; difference in variability or spread; compare variation	Quantitative

You can see that the key words *difference* and *compare* help identify the fact that two populations are to be compared. For the examples given above, the words *mean* in *mean sales price* and *average* in *average amount of electricity* imply that the target parameter is the difference in population means, $\mu_1 - \mu_2$. The word *proportions* in *proportions of younger and older adults* indicates that the target parameter is the difference in proportions, $p_1 - p_2$. Finally, the key word *variability* in *variability in the distance* identifies the ratio of population variances, $(\sigma_1)^2/(\sigma_2)^2$, as the target parameter.

As with inferences about a single population, the type of data (quantitative or qualitative) collected on the two samples is also indicative of the target parameter. With

*Data provided (with permission) from Info Tech, Inc., Gainesville, Florida.

quantitative data, you are likely to be interested in comparing the means or variances of the data. With qualitative data with two outcomes (success or failure), a comparison of the proportions of successes is likely to be of interest.

We consider methods for comparing two population means in Sections 7.2 and 7.3. A comparison of population proportions is presented in Section 7.4 and population variances in Section 7.6. We show how to determine the sample sizes necessary for reliable estimates of the target parameters in Section 7.5.

7.2 Comparing Two Population Means: Independent Sampling

In this section we develop both large-sample and small-sample methodologies for comparing two population means. In the large-sample case, we use the z-statistic (where $z \approx t$ when the population variances are unknown), while in the small-sample case we use the t-statistic.

Large Samples

Example 7.1

Comparing Mean Prices of Japanese and U.S. Cars—A Large Sample Confidence Interval for $\mu_1 - \mu_2$

Problem In recent years, the United States and Japan have engaged in intense negotiations regarding restrictions on trade between the two countries. One of the claims made repeatedly by U.S. officials is that many Japanese manufacturers price their goods higher in Japan than in the United States, in effect subsidizing low prices in the United States by extremely high prices in Japan. According to the U.S. argument, Japan accomplishes this by keeping competitive U.S. goods from reaching the Japanese marketplace.

An economist decided to test the hypothesis that higher retail prices are being charged for Japanese automobiles in Japan than in the United States. She obtained independent random samples of 50 retail sales in the United States and 50 retail sales in Japan over the same time period and for the same model of automobile and converted the Japanese sales prices from yen to dollars using current conversion rates. The data, saved in the **AUTOSTUDY** file, are listed in Table 7.1. Form a 95% confidence interval for the difference between the population mean retail prices of this automobile model for the two countries. Interpret the result.

Solution Recall that the general form of a large-sample confidence interval for a single mean μ is $\bar{x} \pm z_{\alpha/2}\sigma_{\bar{x}}$—that is, we add and subtract $z_{\alpha/2}$ standard deviations of the sample estimate, $\bar{x}$, to the value of the estimate. We employ a similar procedure to form the confidence interval for the difference between two population means.

Let μ_1 represent the mean of the population of retail sales prices for this car model sold in the United States. Let μ_2 be similarly defined for retail sales in Japan. We wish to form a confidence interval for $(\mu_1 - \mu_2)$. An intuitively appealing estimator for

Table 7.1 Automobile Retail Prices (Thousands of Dollars)

USA Sales:	18.2	16.2	17.2	18.7	18.4	16.6	14.9	16.8	12.1	10.8
	18.5	15.5	16.2	16.3	18.2	19.5	13.2	16.8	12.9	17.2
	18.2	16.3	16.8	16.4	18.6	15.6	17.1	18.1	18.9	19.0
	17.3	18.8	14.9	16.7	20.3	17.1	14.6	17.2	13.0	18.4
	16.9	13.3	16.3	15.9	16.6	17.6	16.0	17.1	14.6	18.0
Japan Sales:	18.5	14.0	18.2	21.1	13.9	18.7	14.9	16.4	16.3	18.0
	16.8	19.8	17.3	16.6	14.9	16.3	16.5	15.4	17.6	20.1
	16.4	18.0	17.5	18.4	19.8	14.8	18.2	16.7	20.2	16.2
	20.4	17.9	15.5	15.4	17.7	17.1	17.9	17.4	18.2	16.2
	18.5	16.9	17.6	14.4	21.6	18.6	16.2	14.3	12.5	20.0

Data Set: AUTOSTUDY

Group Statistics

	COUNTRY	N	Mean	Std. Deviation	Std. Error Mean
PRICE	USA	50	16596.00	1981.440	280.218
	JAPAN	50	17236.00	1974.093	279.179

Independent Samples Test

		Levene's Test for Equality of Variances		t-test for Equality of Means						
									95% Confidence Interval of the Difference	
		F	Sig.	t	df	Sig. (2-tailed)	Mean Difference	Std. Error Difference	Lower	Upper
PRICE	Equal variances assumed	.118	.732	-1.618	98	.109	-640.000	395.554	-1424.964	144.964
	Equal variances not assumed			-1.618	97.999	.109	-640.000	395.554	-1424.964	144.964

Figure 7.1
SPSS summary statistics and confidence interval for automobile price study

$(\mu_1 - \mu_2)$ is the difference between the sample means, $(\bar{x}_1 - \bar{x}_2)$. Thus, we will form the confidence interval of interest by

$$(\bar{x}_1 - \bar{x}_2) \pm z_{\alpha/2}\sigma_{(\bar{x}_1 - \bar{x}_2)}$$

Assuming the two samples are independent, the standard deviation of the difference between the sample means is

$$\sigma_{(\bar{x}_1 - \bar{x}_2)} = \sqrt{\frac{\sigma_1^2}{n_1} + \frac{\sigma_2^2}{n_2}} \approx \sqrt{\frac{s_1^2}{n_1} + \frac{s_2^2}{n_2}}$$

Note that we have substituted s_1^2 and s_2^2 for the usually unknown values of σ_1^2 and σ_2^2, respectively. With large samples, this will be a good approximation.

Summary statistics for the car sales data are displayed in the SPSS printout, Figure 7.1. Note that $\bar{x}_1 = \$16{,}596$, $\bar{x}_2 = \$17{,}236$, $s_1 = \$1{,}981$, and $s_2 = \$1{,}974$. Using these values and noting that $\alpha = .05$ and $z_{.025} = 1.96$, we find that the 95% confidence interval is, approximately,

$$(16{,}596 - 17{,}236) \pm 1.96\sqrt{\frac{(1{,}981)^2}{50} + \frac{(1{,}974)^2}{50}} = -640 \pm (1.96)(396)$$

$$= -640 \pm 776$$

or $(-1416, 136)$. This interval is also given at the bottom of Figure 7.1. (Differences in the results are due to rounding and normal approximation.)

Using this estimation procedure over and over again for different samples, we know that approximately 95% of the confidence intervals formed in this manner will enclose the difference in population means $(\mu_1 - \mu_2)$. Therefore, we are highly confident that the difference in mean retail prices in the United States and Japan is between $-\$1{,}416$ and \$136. Because 0 falls in this interval, it is possible for the difference to be 0 (i.e., for $\mu_1 = \mu_2$); thus, the economist cannot conclude that a significant difference exists between the mean retail prices in the two countries.

Look Back If the confidence interval for $(\mu_1 - \mu_2)$ contains all positive numbers [e.g., (527, 991)], then we would conclude that the difference between the means is positive and that $\mu_1 > \mu_2$. Alternatively, if the interval contains all negative numbers [e.g., $(-722, -145)$], then we would conclude that the difference between the means is negative and that $\mu_1 < \mu_2$.

Now Work Exercise 7.3a

The justification for the procedure used in Example 7.1 to estimate $(\mu_1 - \mu_2)$ relies on the properties of the sampling distribution of $(\bar{x}_1 - \bar{x}_2)$. The performance of the estimator in repeated sampling is pictured in Figure 7.2, and its properties are summarized in the box on the next page.

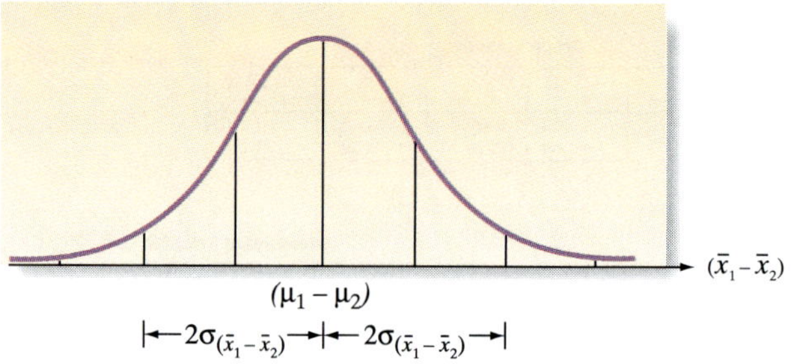

Figure 7.2
Sampling distribution of $(\bar{x}_1 - \bar{x}_2)$

Properties of the Sampling Distribution of $(\bar{x}_1 - \bar{x}_2)$

1. The mean of the sampling distribution $(\bar{x}_1 - \bar{x}_2)$ is $(\mu_1 - \mu_2)$.
2. If the two samples are independent, the standard deviation of the sampling distribution is

$$\sigma_{(\bar{x}_1 - \bar{x}_2)} = \sqrt{\frac{\sigma_1^2}{n_1} + \frac{\sigma_2^2}{n_2}}$$

   where σ_1^2 and σ_2^2 are the variances of the two populations being sampled, and n_1 and n_2 are the respective sample sizes. We also refer to $\sigma_{(\bar{x}_1 - \bar{x}_2)}$ as the **standard error** of the statistic $(\bar{x}_1 - \bar{x}_2)$.
3. The sampling distribution of $(\bar{x}_1 - \bar{x}_2)$ is approximately normal for *large samples* by the Central Limit Theorem.

In Example 7.1, we noted the similarity in the procedures for forming a large-sample confidence interval for one population mean and a large-sample confidence interval for the difference between two population means. When we are testing hypotheses, the procedures are again very similar. The general large-sample procedures for forming confidence intervals and testing hypotheses about $(\mu_1 - \mu_2)$ are summarized in the following boxes.

Large Sample Confidence Interval for $(\mu_1 - \mu_2)$

$$(\bar{x}_1 - \bar{x}_2) \pm z_{\alpha/2}\sigma_{(\bar{x}_1 - \bar{x}_2)} = (\bar{x}_1 - \bar{x}_2) \pm z_{\alpha/2}\sqrt{\frac{\sigma_1^2}{n_1} + \frac{\sigma_2^2}{n_2}}$$

Large-Sample Test of Hypothesis for $(\mu_1 - \mu_2)$

One-Tailed Test	Two-Tailed Test
H_0: $(\mu_1 - \mu_2) = D_0$ H_a: $(\mu_1 - \mu_2) < D_0$ [or H_a: $(\mu_1 - \mu_2) > D_0$]	H_0: $(\mu_1 - \mu_2) = D_0$ H_a: $(\mu_1 - \mu_2) \neq D_0$

where D_0 = Hypothesized difference between the means (this difference is often hypothesized to be equal to 0)

Test statistic:

$$z = \frac{(\bar{x}_1 - \bar{x}_2) - D_0}{\sigma_{(\bar{x}_1 - \bar{x}_2)}} \quad \text{where} \quad \sigma_{(\bar{x}_1 - \bar{x}_2)} = \sqrt{\frac{\sigma_1^2}{n_1} + \frac{\sigma_2^2}{n_2}} \approx \sqrt{\frac{s_1^2}{n_1} + \frac{s_2^2}{n_2}}$$

One-Tailed Test	Two-Tailed Test
Rejection region: $z < -z_\alpha$ [or $z > z_\alpha$ when H_a: $(\mu_1 - \mu_2) > D_0$]	*Rejection region:* $\lvert z \rvert > z_{\alpha/2}$

Conditions Required for Valid Large-Sample Inferences about $(\mu_1 - \mu_2)$

1. The two samples are randomly selected in an independent manner from the two target populations.
2. The sample sizes, n_1 and n_2, are both large (i.e., $n_1 \geq 30$ and $n_2 \geq 30$). [Due to the Central Limit Theorem, this condition guarantees that the sampling distribution of $(\bar{x}_1 - \bar{x}_2)$ will be approximately normal regardless of the shapes of the underlying probability distributions of the populations. Also, s_1^2 and s_2^2 will provide good approximations to σ_1^2 and σ_2^2 when the samples are both large.]

Example 7.2

Comparing Mean Prices of Japanese and U.S. Cars with a Large-Sample Test for $\mu_1 - \mu_2$

Problem Refer to the study of retail prices of an automobile sold in the United States and Japan, Example 7.1. Another way to compare the mean retail prices for the two countries is to conduct a test of hypothesis. Use the information on the SPSS printout, Figure 7.1, to conduct the test. Use $\alpha = .05$.

Solution Again, we let μ_1 and μ_2 represent the population mean retail sales prices in the United States and Japan, respectively. If the claim made by the U.S. government is true, then the mean retail price in Japan will exceed the mean in the United States [i.e., $\mu_1 < \mu_2$ or $(\mu_1 - \mu_2) < 0$]. Thus, the elements of the test are as follows:

H_0: $(\mu_1 - \mu_2) = 0$ (i.e., $\mu_1 = \mu_2$; note that $D_0 = 0$ for this hypothesis test)

H_a: $(\mu_1 - \mu_2) < 0$ (i.e., $\mu_1 < \mu_2$)

Test statistic: $z = \dfrac{(\bar{x}_1 - \bar{x}_2) - D_0}{\sigma_{(\bar{x}_1 - \bar{x}_2)}} = \dfrac{(\bar{x}_1 - \bar{x}_2) - 0}{\sigma_{(\bar{x}_1 - \bar{x}_2)}}$

Rejection region: $z < -z_{.05} = -1.645$ (see Figure 7.3)

Substituting the summary statistics given in Figure 7.1 into the test statistic, we obtain

$$z = \frac{(\bar{x}_1 - \bar{x}_2) - 0}{\sigma_{(\bar{x}_1 - \bar{x}_2)}} = \frac{(16{,}596 - 17{,}236)}{\sqrt{\dfrac{\sigma_1^2}{n_1} + \dfrac{\sigma_2^2}{n_2}}}$$

$$\approx \frac{-640}{\sqrt{\dfrac{s_1^2}{n_1} + \dfrac{s_2^2}{n_2}}} = \frac{-640}{\sqrt{\dfrac{(1{,}981)^2}{50} + \dfrac{(1{,}974)^2}{50}}} = \frac{-640}{396} = -1.62$$

[Note: This value of the test statistic is shown (highlighted) at the bottom of the SPSS printout, Figure 7.1 (p. 383)]

As you can see in Figure 7.3, the calculated z-value does not fall in the rejection region. Therefore, the samples do not provide sufficient evidence, at $\alpha = .05$, for the economist to conclude that the mean retail price in Japan exceeds that in the United States.

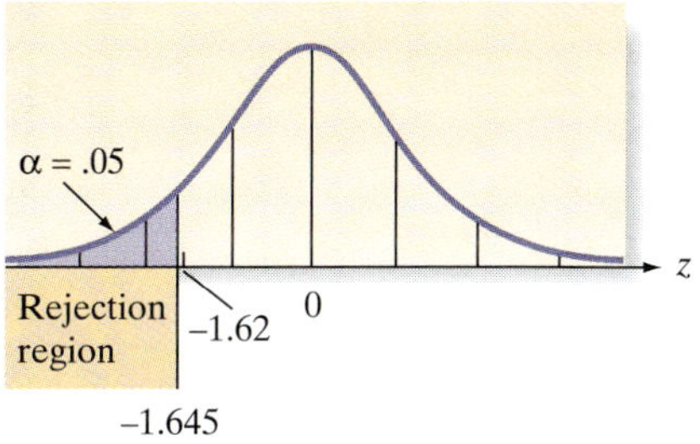

Figure 7.3
Rejection region for Example 7.2

Look Back First, note that this conclusion agrees with the inference drawn from the 95% confidence interval in Example 7.1. Generally, however, a confidence interval will provide more information on the difference in means than a test. A test can detect only whether or not a difference between the means exists, while the confidence interval provides information on the magnitude of the difference. Second, a one-tailed hypothesis test and a confidence interval (which is two-tailed) may not always agree. However, a two-tailed test of hypothesis and a confidence interval will *always* give the same inference about the target parameter, as long as the value of α is the same for both.

Example 7.3
The p-Value of a Test for $\mu_1 - \mu_2$

Problem Find the observed significance level for the test in Example 7.2. Interpret the result.

Solution The alternative hypothesis in Example 7.2, H_a: $(\mu_1 - \mu_2) < 0$, required a lower one-tailed test using

$$z = \frac{\bar{x}_1 - \bar{x}_2}{\sigma_{(\bar{x}_1 - \bar{x}_2)}}$$

as a test statistic. Because the approximate z-value calculated from the sample data was -1.62, the observed significance level (p-value) for the lower-tailed test is the probability of observing a value of z more contradictory to the null hypothesis as $z = -1.62$; that is,

$$p\text{-value} = P(z < -1.62)$$

This probability is computed assuming H_0 is true and is equal to the shaded area shown in Figure 7.4.

The tabulated area corresponding to $z = 1.62$ in Table IV in Appendix B is .4474. Therefore, the observed significance level of the test is

$$p\text{-value} \approx .5 - .4474 = .0526$$

Because our selected α value, .05, is less than this p-value, we have insufficient evidence to reject H_0: $(\mu_1 - \mu_2) = 0$ in favor of H_a: $(\mu_1 - \mu_2) < 0$.

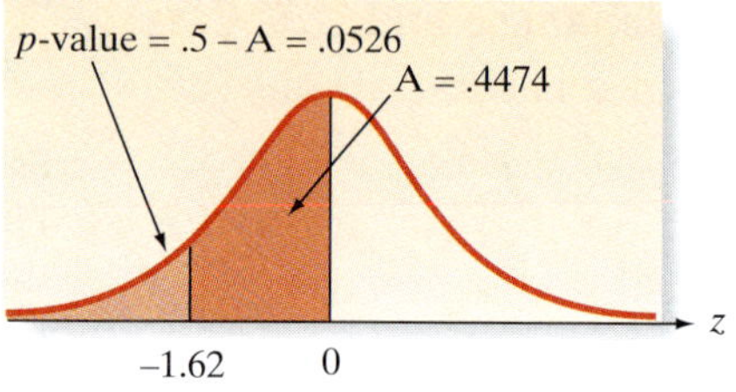

Figure 7.4
The observed significance level for Example 7.2

Look Back The p-value of the test is more easily obtained from a statistical software package. A Minitab printout for the hypothesis test is displayed in Figure 7.5. The one-tailed p-value, highlighted on the printout, is .054, which agrees (except for rounding) with our approximated p-value.

Two-Sample T-Test and CI: USA, JAPAN

```
Two-sample T for USA vs JAPAN

        N   Mean  StDev  SE Mean
USA    50  16596   1981      280
JAPAN  50  17236   1974      279


Difference = mu (USA) - mu (JAPAN)
Estimate for difference:  -640.000
95% upper bound for difference:  16.838
T-Test of difference = 0 (vs <): T-Value = -1.62  P-Value = 0.054  DF = 98
Both use Pooled StDev = 1977.7703
```

Figure 7.5
Minitab analysis for comparing U.S. and Japan mean auto prices

Now Work Exercise 7.3b

Small Samples

When comparing two population means with small samples (say, $n_1 < 30$ and $n_2 < 30$), the methodology of the previous three examples is invalid. The reason? When the sample sizes are small, estimates of σ_1^2 and σ_2^2 are unreliable, and the Central Limit Theorem (which guarantees that the z-statistic is normal) can no longer be applied. But as in the case of a single mean (Section 6.5), we use the familiar Student's t-distribution described in Chapter 5.

To use the t*-distribution, both sampled populations must be approximately normally distributed with equal population variances, and the random samples must be selected independently of each other.* The normality and equal variances assumptions imply relative frequency distributions for the populations that would appear as shown in Figure 7.6.

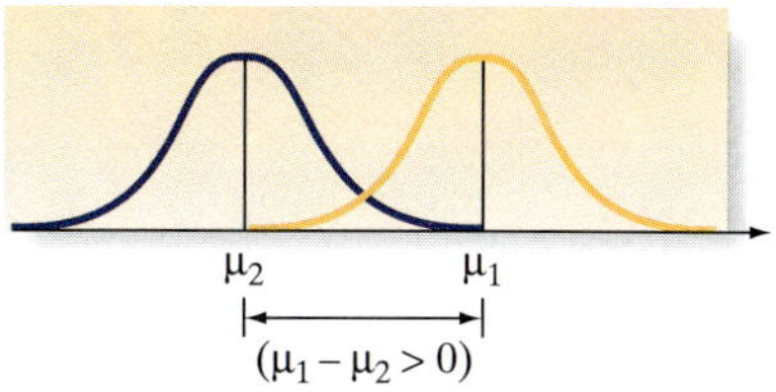

Figure 7.6
Assumptions for the two-sample *t*: (1) normal populations, (2) equal variances

Because we assume the two populations have equal variances ($\sigma_1^2 = \sigma_2^2 = \sigma^2$), it is reasonable to use the information contained in both samples to construct a **pooled sample estimator of σ^2** for use in confidence intervals and test statistics. Thus, if s_1^2 and s_2^2 are the two sample variances (both estimating the variance σ^2 common to both populations), the pooled estimator of σ^2, denoted as s_p^2, is

$$s_p^2 = \frac{(n_1 - 1)s_1^2 + (n_2 - 1)s_2^2}{(n_1 - 1) + (n_2 - 1)} = \frac{(n_1 - 1)s_1^2 + (n_2 - 1)s_2^2}{n_1 + n_2 - 2}$$

or

$$s_p^2 = \frac{\overbrace{\sum(x_1 - \bar{x}_1)^2}^{\textbf{From sample 1}} + \overbrace{\sum(x_2 - \bar{x}_2)^2}^{\textbf{From sample 2}}}{n_1 + n_2 - 2}$$

where x_1 represents a measurement from sample 1 and x_2 represents a measurement from sample 2. Recall that the term *degrees of freedom* was defined in Section 5.2 as 1 less than the sample size. Thus, in this case, we have $(n_1 - 1)$ degrees of freedom for sample 1 and $(n_2 - 1)$ degrees of freedom for sample 2. Because we are pooling the information on σ^2 obtained from both samples, the degrees of freedom associated with the pooled variance s_p^2 is equal to the sum of the degrees of freedom for the two samples, namely, the denominator of s_p^2; that is, $(n_1 - 1) + (n_2 - 1) = n_1 + n_2 - 2$.

Note that the second formula given for s_p^2 shows that the pooled variance is simply a *weighted average* of the two sample variances, s_1^2 and s_2^2. The weight given each variance is proportional to its degrees of freedom. If the two variances have the same number of degrees of freedom (i.e., if the sample sizes are equal), then the pooled variance is a simple average of the two sample variances. The result is an average or "pooled" variance that is a better estimate of σ^2 than either s_1^2 or s_2^2 alone.

Both the confidence interval and the test of hypothesis procedures for comparing two population means with small samples are summarized in the following boxes.

Small-Sample Confidence Interval for $(\mu_1 - \mu_2)$ (Independent Samples)

$$(\bar{x}_1 - \bar{x}_2) \pm t_{\alpha/2}\sqrt{s_p^2\left(\frac{1}{n_1} + \frac{1}{n_2}\right)}$$

where $s_p^2 = \dfrac{(n_1 - 1)s_1^2 + (n_2 - 1)s_2^2}{n_1 + n_2 - 2}$

and $t_{\alpha/2}$ is based on $(n_1 + n_2 - 2)$ degrees of freedom.

BIOGRAPHY

BRADLEY EFRON (1938–PRESENT)

The Bootstrap Method

Bradley Efron was raised in St. Paul, Minnesota, the son of a truck driver who was the amateur statistician for his bowling and baseball leagues. Efron received a BS in mathematics from the California Institute of Technology in 1960, but, by his own admission, had no talent for modern abstract math. His interest in the science of statistics developed after reading a book, cover to cover, by Harold Cramer. Efron went to Stanford University to study statistics and earned his PhD there in 1964. He has been a faculty member in Stanford's Department of Statistics since 1966. Over his career, Efron has received numerous awards and prizes for his contributions to modern statistics, including the MacArthur Prize Fellow (1983), the American Statistical Association Wilks Medal (1990), and the Parzen Prize for Statistical Innovation (1998). In 1979, Efron invented a method—called the *bootstrap*—of estimating and testing population parameters in situations where either the sampling distribution is unknown or the assumptions are violated. The method involves repeatedly taking samples of size *n* (with replacement) from the original sample and calculating the value of the point estimate. Efron showed that the sampling distribution of the estimator is simply the frequency distribution of the bootstrap estimates. ■

Small-Sample Test of Hypothesis for $(\mu_1 - \mu_2)$ (Independent Samples)

One-Tailed Test	Two-Tailed Test
$H_0: (\mu_1 - \mu_2) = D_0$	$H_0: (\mu_1 - \mu_2) = D_0$
$H_a: (\mu_1 - \mu_2) < D_0$	$H_a: (\mu_1 - \mu_2) \neq D_0$
[or $H_a: (\mu_1 - \mu_2) > D_0$]	

$$\textit{Test statistic: } t = \frac{(\bar{x}_1 - \bar{x}_2) - D_0}{\sqrt{s_p^2\left(\frac{1}{n_1} + \frac{1}{n_2}\right)}}$$

Rejection region: $t < -t_\alpha$	*Rejection region:* $\lvert t\rvert > t_{\alpha/2}$
[or $t > t_\alpha$ when $H_a: (\mu_1 - \mu_2) > D_0$]	

where t_α and $t_{\alpha/2}$ are based on $(n_1 + n_2 - 2)$ degrees of freedom.

Conditions Required for Valid Small-Sample Inferences about $(\mu_1 - \mu_2)$

1. The two samples are randomly selected in an independent manner from the two target populations.
2. Both sampled populations have distributions that are approximately normal.
3. The population variances are equal (i.e., $\sigma_1^2 = \sigma_2^2$).

Example 7.4

Comparing Manager Performance with a Small-Sample Confidence Interval for $(\mu_1 - \mu_2)$

Problem Behavioral researchers have developed an index designed to measure managerial success. The index (measured on a 100-point scale) is based on the manager's length of time in the organization and his or her level within the firm; the higher the index, the more successful the manager. Suppose a researcher wants to compare the average success index for two groups of managers at a large manufacturing plant. Managers in group 1 engage in a high volume of interactions with people outside the manager's work unit. (Such interactions include phone and face-to-face meetings with customers and suppliers, outside meetings, and public relations work.) Managers in group 2 rarely interact with people outside their work unit. Independent random samples of 12 and 15 managers are selected from groups 1 and 2, respectively, and the success index of each is recorded. The results of the study are given in Table 7.2.

a. Use the data in the table to estimate the true mean difference between the success indexes of managers in the two groups. Use a 95% confidence interval.

b. Interpret the interval, part **a.**

c. What assumptions must be made in order that the estimate be valid? Are they reasonably satisfied?

Table 7.2 Managerial Success Indexes for Two Groups of Managers

Group 1						Group 2					
Interaction with Outsiders						Few Interactions					
65	58	78	60	68	69	62	53	36	34	56	50
66	70	53	71	63	63	42	57	46	68	48	42
						52	53	43			

Data Set: MANSUCCESS

```
Two-sample T for SUCCESS

GROUP    N    Mean   StDev   SE Mean
1       12   65.33    6.61       1.9
2       15   49.47    9.33       2.4


Difference = mu (1) - mu (2)
Estimate for difference:  15.87
95% CI for difference:   (9.29, 22.45)
T-Test of difference = 0 (vs not =): T-Value = 4.97  P-Value = 0.000  DF = 25
Both use Pooled StDev = 8.2472
```

Figure 7.7

Minitab printout for Example 7.4

Solution

a. For this experiment, let μ_1 and μ_2 represent the mean success index of group 1 and group 2 managers, respectively. Then, the objective is to obtain a 95% confidence interval for $(\mu_1 - \mu_2)$.

The first step in constructing the confidence interval is to obtain summary statistics (e.g., $\bar{x}$ and s) on the success index for each group of managers. The data of Table 7.2 were entered into a computer, and Minitab was used to obtain these descriptive statistics. The Minitab printout appears in Figure 7.7. Note that $\bar{x}_1 = 65.33$, $s_1 = 6.61$, $\bar{x}_2 = 49.47$, and $s_2 = 9.33$.

Next, we calculate the pooled estimate of variance:

$$s_p^2 = \frac{(n_1 - 1)s_1^2 + (n_2 - 1)s_2^2}{n_1 + n_2 - 2}$$

$$= \frac{(12 - 1)(6.61)^2 + (15 - 1)(9.33)^2}{12 + 15 - 2} = 67.97$$

where s_p^2 is based on $(n_1 + n_2 - 2) = (12 + 15 - 2) = 25$ degrees of freedom. Also, we find $t_{\alpha/2} = t_{.025} = 2.06$ (based on 25 degrees of freedom) from Table V in Appendix B.

Finally, the 95% confidence interval for $(\mu_1 - \mu_2)$, the difference between mean managerial success indexes for the two groups, is

$$(\bar{x}_1 - \bar{x}_2) \pm t_{\alpha/2}\sqrt{s_p^2\left(\frac{1}{n_1} + \frac{1}{n_2}\right)} = 65.33 - 49.47 \pm t_{.025}\sqrt{67.97\left(\frac{1}{12} + \frac{1}{15}\right)}$$

$$= 15.86 \pm (2.06)(3.19)$$

$$= 15.86 \pm 6.58$$

or (9.28, 22.44). This interval agrees (except for rounding) with the one shown at the bottom of the Minitab printout, Figure 7.7.

b. Notice that the confidence interval includes positive differences only. Consequently, we are 95% confident that $(\mu_1 - \mu_2)$ exceeds 0. In fact, we estimate the mean success index, μ_1, for managers with a high volume of outsider interaction (group 1) to be anywhere between 9.28 and 22.44 points higher than the mean success index, μ_2, of managers with few interactions (group 2).

c. To properly use the small-sample confidence interval, the following assumptions must be satisfied:

1. The samples of managers are randomly and independently selected from the populations of group 1 and group 2 managers.
2. The success indexes are normally distributed for both groups of managers.
3. The variance of the success indexes are the same for the two populations, (i.e., $\sigma_1^2 = \sigma_2^2$).

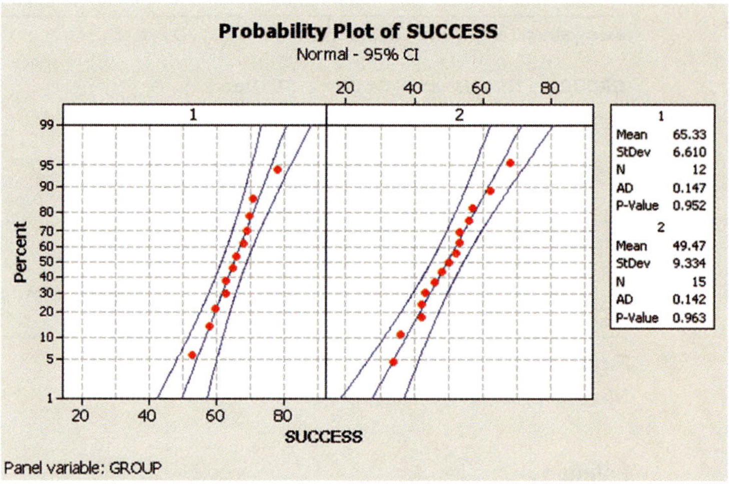

Figure 7.8
Minitab normal probability plots for manager success index

The first assumption is satisfied, based on the information provided about the sampling procedure in the problem description. To check the plausibility of the remaining two assumptions, we resort to graphical methods. Figure 7.8 is a Minitab printout that displays normal probability plots for the success indexes of the two samples of managers. The near straight-line trends on both plots indicate that the success index distributions are approximately mound shaped and symmetric. Consequently, each sample data set appears to come from a population that is approximately normal.

One way to check assumption #3 is to test the null hypothesis H_0: $\sigma_1^2 = \sigma_2^2$. This test is covered in Section 7.6. Another approach is to examine box plots for the sample data. Figure 7.9 is a Minitab printout that shows side-by-side vertical box plots for the success indexes in the two samples. Recall from Section 2.9 that the box plot represents the "spread" of a data set. The two box plots appear to have about the same spread; thus, the samples appear to come from populations with approximately the same variance.

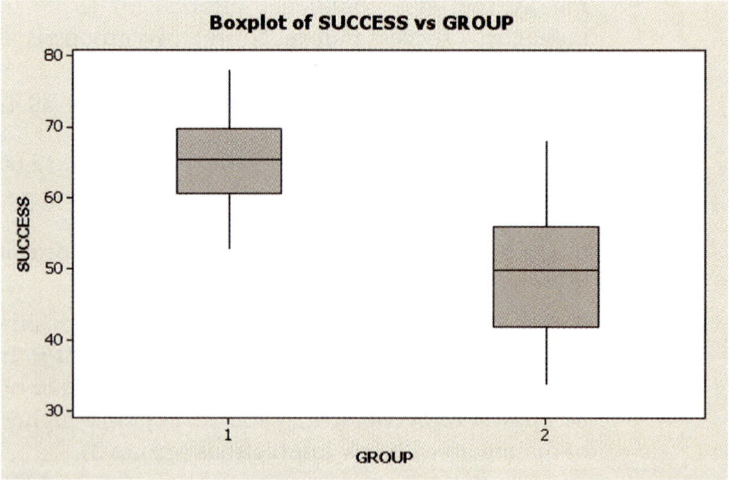

Figure 7.9
Minitab box plots for manager success index

Look Back All three assumptions appear to be reasonably satisfied for this application of the small-sample confidence interval.

Now Work Exercise 7.7

The two-sample t-statistic is a powerful tool for comparing population means when the assumptions are satisfied. It has also been shown to retain its usefulness when the sampled populations are only approximately normally distributed. And when the sample sizes are equal, the assumption of equal population variances can be relaxed—that is, if $n_1 = n_2$, then σ_1^2 and σ_2^2 can be quite different, and the test statistic will still possess, approximately, a Student's t-distribution. In the case where $\sigma_1^2 \neq \sigma_2^2$ and $n_1 \neq n_2$, an approximate small-sample confidence interval or test can be obtained by modifying the degrees of freedom associated with the t-distribution.

The next box gives the approximate small-sample procedures to use when the assumption of equal variances is violated. The test for the "unequal sample sizes" case is based on Satterthwaite's (1946) approximation.

Approximate Small-Sample Procedures when $\sigma_1^2 \neq \sigma_2^2$

1. **Equal sample sizes** ($n_1 = n_2 = n$)

   Confidence interval: $(\bar{x}_1 - \bar{x}_2) \pm t_{\alpha/2}\sqrt{(s_1^2 + s_2^2)/n}$

   Test statistic for H_0: $(\mu_1 - \mu_2) = 0$: $\quad t = (\bar{x}_1 - \bar{x}_2)/\sqrt{(s_1^2 + s_2^2)/n}$

   where t is based on $\nu = n_1 + n_2 - 2 = 2(n - 1)$ degrees of freedom.

2. **Unequal sample sizes** ($n_1 \neq n_2$)

   Confidence interval: $(\bar{x}_1 - \bar{x}_2) \pm t_{\alpha/2}\sqrt{(s_1^2/n_1) + (s_2^2/n_2)}$

   Test statistic for H_0: $(\mu_1 - \mu_2) = 0$: $t = (\bar{x}_1 - \bar{x}_2)/\sqrt{(s_1^2/n_1) + (s_2^2/n_2)}$ where t is based on degrees of freedom equal to

$$\nu = \frac{(s_1^2/n_1 + s_2^2/n_2)^2}{\dfrac{(s_1^2/n_1)^2}{n_1 - 1} + \dfrac{(s_2^2/n_2)^2}{n_2 - 1}}$$

Note: The value of ν will generally not be an integer. Round ν down to the nearest integer to use the t-table.

When the assumptions are clearly not satisfied, you can select larger samples from the populations or you can use other available statistical tests (nonparametric statistical tests, which are described in Chapter 14).

What Should You Do if the Assumptions Are Not Satisfied?

Answer: If you are concerned that the assumptions are not satisfied, use the Wilcoxon rank sum test for independent samples to test for a shift in population distributions. See Chapter 14.

Statistics IN Action Revisited — Comparing Mean Price Changes

Refer to the *ZixIt v. Visa* court case described in the Statistics in Action (pp. 380–381). Recall that a Visa executive wrote e-mails and made Web site postings in an effort to undermine a new online credit card processing system developed by ZixIt. ZixIt sued Visa for libel, asking for $699 million in damages. An expert statistician, hired by the defendants (Visa), performed an "event study" in which he matched the Visa executive's e-mail postings with movement of ZixIt's stock price the next business day. The data were collected daily from September 1 to December 30, 1999 (an 83-day period) and are available in the **ZIXITVISA** file. In addition to daily closing price (dollars) of ZixIt stock, the file contains a variable for whether or not the Visa executive posted an e-mail and the change in price of the stock the following business day. During the 83-day period, the executive posted e-mails on 43 days and had no postings on 40 days.

If the daily postings by the Visa executive had a negative impact on ZixIt stock, then the average price change following nonposting days should exceed the average price change

Two-Sample T-Test and CI: PriceChange, Posting

```
Two-sample T for PriceChange

Posting   N   Mean  StDev  SE Mean
NO       40  -0.13   3.46     0.55
POST     43   0.06   2.20     0.34


Difference = mu (NO) - mu (POST)
Estimate for difference:  -0.188
95% CI for difference:  (-1.470, 1.093)
T-Test of difference = 0 (vs not =): T-Value = -0.29  P-Value = 0.770  DF = 65
```

Figure SIA7.1
Minitab comparison of two price change means

following posting days. Consequently, one way to analyze the data is to conduct a comparison of two population means through either a confidence interval or a test of hypothesis. Here, we let μ_1 represent the mean price change of ZixIt stock following all nonposting days and μ_2 represent the mean price change of ZixIt stock following posting days. If, in fact, the charges made by ZixIt are true, then μ_1 will exceed μ_2. However, if the data do not support ZixIt's claim, then we will not be able to reject the null hypothesis H_0: $(\mu_1 - \mu_2) = 0$ in favor of H_a: $(\mu_1 - \mu_2) > 0$. Similarly, if a confidence interval for $(\mu_1 - \mu_2)$ contains the value 0, then there will be no evidence to support ZixIt's claim.

Because both sample sizes ($n_1 = 40$ and $n_2 = 43$) are large, we can apply the large-sample z-test or large-sample confidence interval procedure for independent samples. A Minitab printout for this analysis is shown in Figure SIA7.1. Both the 95% confidence interval and p-value for a two-tailed test of hypothesis are highlighted on the printout. Note that the 95% confidence interval, (−\$1.47, \$1.09), includes the value \$0, and the p-value for the two-tailed hypothesis test (.770) implies that the two population means are not significantly different. Also, interestingly, the sample mean price change after posting days ($\bar{x}_1 = \$.06$) is small and positive, while the sample mean price change after nonposting days ($\bar{x}_2 = -\$.13$) is small and negative, totally contradicting ZixIt's claim.

The statistical expert for the defense presented these results to the jury, arguing that the "average price change following posting days is small and similar to the average price change following nonposting days" and "the difference in the means is not statistically significant."

Note: The statistician also compared the mean ZixIt trading volume (number of ZixIt stock shares traded) after posting days to the mean trading volume after nonposting days. These results are shown in Figure SIA7.2. You can see that the 95% confidence interval for the difference in mean trading volume (highlighted) includes 0, and the p-value for a two-tailed test of hypothesis for a difference in means (also highlighted) is not statistically significant. These results were also presented to the jury in defense of Visa.

Data Set: ZIXITVISA

```
Two-sample T for VolumeAfter

Posting   N    Mean   StDev  SE Mean
NO       40  719645  430837    68121
POST     43  578665  333921    50922


Difference = mu (NO) - mu (POST)
Estimate for difference:  140980
95% CI for difference:  (-28526, 310485)
T-Test of difference = 0 (vs not =): T-Value = 1.66  P-Value = 0.102  DF = 73
```

Figure SIA7.2
Minitab comparison of two trading volume means analysis

Exercises 7.1–7.24

Learning the Mechanics

7.1 The purpose of this exercise is to compare the variability of $\bar{x}_1$ and $\bar{x}_2$ with the variability of $(\bar{x}_1 - \bar{x}_2)$.

a. Suppose the first sample is selected from a population with mean $\mu_1 = 150$ and variance $\sigma_1^2 = 900$. Within what range should the sample mean vary about 95% of the time in repeated samples of 100 measurements from this distribution? That is, construct an interval extending 2 standard deviations of $\bar{x}_1$ on each side of μ_1.

b. Suppose the second sample is selected independently of the first from a second population with mean $\mu_2 = 150$ and variance $\sigma_2^2 = 1{,}600$. Within what range should the sample mean vary about 95% of the time in repeated samples of 100 measurements from this distribution? That is, construct an interval extending 2 standard deviations of $\bar{x}_2$ on each side of μ_2.

c. Now consider the difference between the two sample means $(\bar{x}_1 - \bar{x}_2)$. What are the mean and standard deviation of the sampling distribution of $(\bar{x}_1 - \bar{x}_2)$?

d. Within what range should the difference in sample means vary about 95% of the time in repeated independent samples of 100 measurements each from the two populations?

e. What, in general, can be said about the variability of the difference between independent sample means relative to the variability of the individual sample means?

7.2 Independent random samples of 64 observations each are chosen from two normal populations with the following means and standard deviations:

Population 1	Population 2
$\mu_1 = 12$	$\mu_2 = 10$
$\sigma_1 = 4$	$\sigma_2 = 3$

Let $\bar{x}_1$ and $\bar{x}_2$ denote the two sample means.

a. Give the mean and standard deviation of the sampling distribution of $\bar{x}_1$.

b. Give the mean and standard deviation of the sampling distribution of $\bar{x}_2$.

c. Suppose you were to calculate the difference between the sample means $(\bar{x}_1 - \bar{x}_2)$. Find the mean and standard deviation of the sampling distribution of $(\bar{x}_1 - \bar{x}_2)$.

d. Will the statistic $(\bar{x}_1 - \bar{x}_2)$ be normally distributed? Explain.

7.3 NW In order to compare the means of two populations, independent random samples of 400 observations are selected from each population, with the following results:

Sample 1	Sample 2
$\bar{x}_1 = 5{,}275$	$\bar{x}_2 = 5{,}240$
$s_1 = 150$	$s_2 = 200$

a. Use a 95% confidence interval to estimate the difference between the population means $(\mu_1 - \mu_2)$. Interpret the confidence interval.

b. Test the null hypothesis H_0: $(\mu_1 - \mu_2) = 0$ versus the alternative hypothesis H_a: $(\mu_1 - \mu_2) \neq 0$. Give the significance level of the test and interpret the result.

c. Suppose the test in part **b** was conducted with the alternative hypothesis H_a: $(\mu_1 - \mu_2) > 0$. How would your answer to part **b** change?

d. Test the null hypothesis H_0: $(\mu_1 - \mu_2) = 25$ versus H_a: $(\mu_1 - \mu_2) \neq 25$. Give the significance level and interpret the result. Compare your answer to the test conducted in part **b.**

e. What assumptions are necessary to ensure the validity of the inferential procedures applied in parts **a–d**?

7.4 To use the t-statistic to test for a difference between the means of two populations, what assumptions must be made about the two populations? About the two samples?

7.5 Two populations are described in each of the following cases. In which cases would it be appropriate to apply the small-sample t-test to investigate the difference between the population means?

a. Population 1: Normal distribution with variance σ_1^2. Population 2: Skewed to the right with variance $\sigma_2^2 = \sigma_1^2$.

b. Population 1: Normal distribution with variance σ_1^2. Population 2: Normal distribution with variance $\sigma_2^2 \neq \sigma_1^2$.

c. Population 1: Skewed to the left with variance σ_1^2. Population 2: Skewed to the left with variance $\sigma_2^2 = \sigma_1^2$.

d. Population 1: Normal distribution with variance σ_1^2. Population 2: Normal distribution with variance $\sigma_2^2 = \sigma_1^2$.

e. Population 1: Uniform distribution with variance σ_1^2. Population 2: Uniform distribution with variance $\sigma_2^2 = \sigma_1^2$.

7.6 Assume that $\sigma_1^2 = \sigma_2^2 = \sigma^2$. Calculate the pooled estimator of σ^2 for each of the following cases:

a. $s_1^2 = 120$, $s_2^2 = 100$, $n_1 = n_2 = 25$

b. $s_1^2 = 12$, $s_2^2 = 20$, $n_1 = 20$, $n_2 = 10$

c. $s_1^2 = .15$, $s_2^2 = .20$, $n_1 = 6$, $n_2 = 10$

d. $s_1^2 = 3{,}000$, $s_2^2 = 2{,}500$, $n_1 = 16$, $n_2 = 17$

Note that the pooled estimate is a weighted average of the sample variances. To which of the variances does the pooled estimate fall nearer in each of the above cases?

7.7 NW Independent random samples from normal populations produced the results shown in the next table. The data are saved in the **LM7_7** file.

Sample 1	Sample 2
1.2	4.2
3.1	2.7
1.7	3.6
2.8	3.9
3.0	

a. Calculate the pooled estimate of σ^2.

b. Do the data provide sufficient evidence to indicate that $\mu_2 > \mu_1$? Test using $\alpha = .10$.

c. Find a 90% confidence interval for $(\mu_1 - \mu_2)$.

d. Which of the two inferential procedures, the test of hypothesis in part **b** or the confidence interval in part **c**, provides more information about $(\mu_1 - \mu_2)$?

7.8 Two independent random samples have been selected—100 observations from population 1 and 100 from population 2. Sample means $\bar{x}_1 = 15.5$ and $\bar{x}_2 = 26.6$ were obtained. From previous experience with these populations, it is known that the variances are $\sigma_1^2 = 9$ and $\sigma_2^2 = 16$.

a. Find $\sigma_{(\bar{x}_1 - \bar{x}_2)}$.

b. Sketch the approximate sampling distribution for $(\bar{x}_1 - \bar{x}_2)$ assuming $(\mu_1 - \mu_2) = 10$.

c. Locate the observed value of $(\bar{x}_1 - \bar{x}_2)$ on the graph you drew in part **b.** Does it appear that this value contradicts the null hypothesis H_0: $(\mu_1 - \mu_2) = 10$?

d. Use the z-table on the inside of the front cover to determine the rejection region for the test of H_0: $(\mu_1 - \mu_2) = 10$ against H_0: $(\mu_1 - \mu_2) \neq 10$. Use $\alpha = .05$.

e. Conduct the hypothesis test of part **d** and interpret your result.

f. Construct a 95% confidence interval for $(\mu_1 - \mu_2)$. Interpret the interval.

g. Which inference provides more information about the value of $(\mu_1 - \mu_2)$—the test of hypothesis in part **e** or the confidence interval in part **f**?

7.9 Independent random samples of $n_1 = 233$ and $n_2 = 312$ are selected from two populations and used to test the hypothesis H_0: $(\mu_1 - \mu_2) = 0$ against the alternative H_a: $(\mu_1 - \mu_2) \neq 0$.

a. The two-tailed p-value of the test is .1150. Interpret this result.

b. If the alternative hypothesis had been H_a: $(\mu_1 - \mu_2) < 0$, how would the p-value change? Interpret the p-value for this one-tailed test.

7.10 Independent random samples from approximately normal populations produced the results shown below. (The data are saved in the **LM7_10** file.)

Sample 1				Sample 2			
52	33	42	44	52	43	47	56
41	50	44	51	62	53	61	50
45	38	37	40	56	52	53	60
44	50	43		50	48	60	55

a. Do the data provide sufficient evidence to conclude that $(\mu_2 - \mu_1) > 10$? Test using $\alpha = .01$.

b. Construct a 98% confidence interval for $(\mu_2 - \mu_1)$. Interpret your result.

7.11 Independent random samples selected from two normal populations produced the sample means and standard deviations shown below.

Sample 1	Sample 2
$n_1 = 17$	$n_2 = 12$
$\bar{x}_1 = 5.4$	$\bar{x}_2 = 7.9$
$s_1 = 3.4$	$s_2 = 4.8$

a. Conduct the test H_0: $(\mu_1 - \mu_2) = 0$ against H_a: $(\mu_1 - \mu_2) \neq 0$. Interpret the results.

b. Estimate $(\mu_1 - \mu_2)$ using a 95% confidence interval.

Applying the Concepts—Basic

7.12 Effectiveness of teaching software. Educational software—ranging from video-game-like programs played on Sony PlayStations to rigorous drilling exercises used on computers—has become very popular in school districts across the country. The U.S. Department of Education (DOE) recently conducted a national study of the effectiveness of educational software. In one phase of the study, a sample of 1,516 first-grade students in classrooms that used educational software was compared to a sample of 1,103 first-grade students in classrooms that did not use the technology. In its *Report to Congress* (March 2007), the DOE concluded that "[mean] test scores [of students on the SAT reading test] were not significantly higher in classrooms using reading . . . software products" than in classrooms that did not use educational software.

a. Identify the parameter of interest to the DOE.

b. Specify the null and alternative hypotheses for the test conducted by the DOE.

c. The p-value for the test was reported as .62. Based on this value, do you agree with the conclusion of the DOE? Explain.

7.13 Children's recall of TV ads. Marketing professors at Robert Morris and Kent State universities examined children's recall and recognition of television advertisements (*Journal of Advertising,* Spring 2006). Two groups of children were shown a 60-second commercial for Sunkist FunFruit Rock-n-Roll Shapes. One group (the A/V group) was shown the ad with both audio and video; the second group (the video-only group) was shown only the video portion of the commercial. Following the viewing, the children were asked to recall 10 specific items from the ad. The number of the 10 items recalled correctly by each child is summarized in the table. The researchers theorized that "children who receive an audiovisual presentation will have the same level of mean recall of ad information as those who receive only the visual aspects of the ad."

Video Only Group	A/V Group
$n_1 = 20$	$n_2 = 20$
$\bar{x}_1 = 3.70$	$\bar{x}_2 = 3.30$
$s_1 = 1.98$	$s_2 = 2.13$

Source: Maher, J. K., Hu, M. Y., and Kolbe, R. H. "Children's recall of television ad elements," *Journal of Advertising,* Vol. 35, No. 1, Spring 2006, pp. 23–33 (Table 1).

a. Set up the appropriate null and alternative hypotheses to test the researchers' theory.

b. Find the value of the test statistic.

c. Give the rejection region for $\alpha = .10$.

d. Make the appropriate inference. What can you say about the researchers' theory?

e. The researchers reported the p-value of the test as p-value $= .542$. Interpret this result.

f. What conditions are required for the inference to be valid?

7.14 Rating service at five-star hotels. A study published in *The Journal of American Academy of Business, Cambridge* (March 2002) examined whether the perception of service quality at five-star hotels in Jamaica differed by gender. Hotel guests were randomly selected from the lobby and restaurant areas and asked to rate 10 service-related items (e.g., "The personal attention you received from our employees"). Each item was rated on a 5-point scale (1 = "much worse than I expected," 5 = "much better than I expected"), and the sum of the items for each guest was determined. A summary of the guest scores is provided in the next table.

Gender	Sample Size	Mean Score	Standard Deviation
Males	127	39.08	6.73
Females	114	38.79	6.94

a. Construct a 90% confidence interval for the difference between the population mean service-rating scores given by male and female guests at Jamaican five-star hotels.

b. Use the interval, part **a,** to make an inference about whether the perception of service quality at five-star hotels in Jamaica differs by gender.

7.15 Buy-side vs. sell-side analysts' earnings forecasts. Refer to the *Financial Analysts Journal* (Jul./Aug. 2008) study of financial analysts' forecast earnings, Exercise 2.84 (p. 76). Recall that data were collected on 3,526 forecasts made by buy-side analysts and 58,562 forecasts made by sell-side analysts, and the relative absolute forecast error was determined for each. The mean and standard deviation of forecast errors for both types of analysts are given in the table.

	Buy-Side Analysts	Sell-Side Analysts
Sample Size	3,526	58,562
Mean	0.85	−0.05
Standard Deviation	1.93	0.85

Source: Groysberg, B., Healy, P., and Chapman, C. "Buy-side vs. sell-side analysts' earnings forecasts," *Financial Analysts Journal,* Vol. 64, No. 4, Jul./Aug. 2008.

a. Construct a 95% confidence interval for the difference between the mean forecast error of buy-side analysts and the mean forecast error of sell-side analysts.

b. Based on the interval, part **a,** which type of analysis has the greater mean forecast error? Explain.

c. What assumptions about the underlying populations of forecast errors (if any) are necessary for the validity of the inference, part **b**?

7.16 Homework assistance for accounting students. How much assistance should accounting professors provide students for completing homework? Is too much assistance counterproductive? These were some of the questions of interest in a *Journal of Accounting Education* (Vol. 25, 2007) article. A total of 75 junior-level accounting majors who were enrolled in Intermediate Financial Accounting participated in an experiment. All students took a pretest on a topic not covered in class; then each was given a homework problem to solve on the same topic. However, the students were randomly assigned different levels of assistance on the homework. Some (20 students) were given the completed solution, some (25 students) were given check figures at various steps of the solution, and the rest (30 students) were given no help. After finishing the homework, each student was given a posttest on the subject. One of the variables of interest to the researchers was the knowledge gain (or, test score improvement), measured as the difference between the posttest and pretest scores. The sample mean knowledge gains for the three groups of students are provided in the table.

	No Solutions	Check Figures	Completed Solutions
Sample Size	30	25	20
Sample Mean	2.43	2.72	1.95

Source: Lindquist, T. M., and Olsen, L. M. "How much help, is too much help? An experimental investigation of the use of check figures and completed solutions in teaching intermediate accounting," *Journal of Accounting Education,* Vol. 25, No. 3, 2007, pp. 103–117 (Table 1, Panel B). © 2007 with permission of Elsevier.

a. The researchers theorize that as the level of homework assistance increases, the test score improvement from pretest to posttest will decrease. Do the sample means reported in the table support this theory?

b. What is the problem with using only the sample means to make inferences about the population mean knowledge gains for the three groups of students?

c. The researchers conducted a statistical test of hypothesis to compare the mean knowledge gain of students in the "no solutions" group to the mean knowledge gain of students in the "check figures" group. Based on the theory, part **a,** set up the null and alternative hypothesis for the test.

d. The observed significance level of the t-test of part **c** was reported as .8248. Using $\alpha = .05$, interpret this result.

e. The researchers conducted a statistical test of hypothesis to compare the mean knowledge gain of students in the "completed solutions" group to the mean knowledge gain of students in the "check figures" group. Based on the theory, part **a,** set up the null and alternative hypothesis for the test.

f. The observed significance level of the t-test of part **e** was reported as .1849. Using $\alpha = .05$, interpret this result.

g. The researchers conducted a statistical test of hypothesis to compare the mean knowledge gain of students in the "no solutions" group to the mean knowledge gain of students in the "completed solutions" group. Based on the theory, part **a,** set up the null and alternative hypothesis for the test.

h. The observed significance level of the t-test of part **c** was reported as .2726. Using $\alpha = .05$, interpret this result.

Applying the Concepts—Intermediate

7.17 Patent infringement case. *Chance* (Fall 2002) described a lawsuit where Intel Corp. was charged with infringing on a patent for an invention used in the automatic manufacture of computer chips. In response, Intel accused the inventor of adding material to his patent notebook after the patent was witnessed and granted. The case rested on whether a patent witness' signature was written on top of key text in the notebook or under the key text. Intel hired a physicist who used an X-ray beam to measure the relative concentration of certain elements (e.g., nickel, zinc, potassium) at several spots on the notebook page. The zinc measurements for three notebook locations—on a text line, on a witness line, and on the intersection of the witness and text line—are provided in the table and saved in the **PATENT** file.

Text Line	.335	.374	.440			
Witness Line	.210	.262	.188	.329	.439	.397
Intersection	.393	.353	.285	.295	.319	

a. Use a test or a confidence interval (at $\alpha = .05$) to compare the mean zinc measurement for the text line with the mean for the intersection.

b. Use a test or a confidence interval (at $\alpha = .05$) to compare the mean zinc measurement for the witness line with the mean for the intersection.

c. From the results, parts **a** and **b**, what can you infer about the mean zinc measurements at the three notebook locations?

d. What assumptions are required for the inferences to be valid? Are they reasonably satisfied?

7.18 Computer-mediated communication study. Computer-mediated communication (CMC) is a form of interaction that heavily involves technology (e.g., instant messaging, e-mail). A study was conducted to compare relational intimacy in people interacting via CMC to people meeting face-to-face (FTF) (*Journal of Computer-Mediated Communication,* April 2004). Participants were 48 undergraduate students, of which half were randomly assigned to the CMC group and half assigned to the FTF group. Each group was given a task that required communication with the group members. Those in the CMC group used the "chat" mode of instant-messaging software; those in the FTF group met in a conference room. The variable of interest, relational intimacy score, was measured (on a 7-point scale) for each participant after each of three different meeting sessions and are saved in the **INTIMACY** file. Scores for the first meeting session are given below. The researchers hypothesized that, after the first meeting, the mean relational intimacy score for participants in the CMC group will be lower than the mean relational intimacy score for participants in the FTF group. Test the researchers' hypothesis using $\alpha = .10$.

CMC	4	3	3	4	3	3	3	3	4	4	3	4
	3	3	2	4	2	4	5	4	4	4	5	3
FTF	5	4	4	4	3	3	3	4	3	3	3	3
	4	4	4	4	4	3	3	3	4	4	2	4

Note: Data simulated from descriptive statistics provided in article.

7.19 Does rudeness really matter in the workplace? Studies have established that rudeness in the workplace can lead to retaliatory and counterproductive behavior. However, there has been little research on how rude behaviors influence a victim's task performance. Such a study was conducted, with the results published in the *Academy of Management Journal* (Oct. 2007). College students enrolled in a management course were randomly assigned to one of two experimental conditions: rudeness condition (45 students) and control group (53 students). Each student was asked to write down as many uses for a brick as possible in five minutes. For those students in the rudeness condition, the facilitator displayed rudeness by generally berating students for being irresponsible and unprofessional (due to a late-arriving confederate). No comments were made about the late-arriving confederate for students in the control group. The number of different uses for a brick was recorded for each of the 98 students, and the data were saved in the **RUDE** file, shown at the bottom of the page. Conduct a statistical analysis (at $\alpha = .01$) to determine if the true mean performance level for students in the rudeness condition is lower than the true mean performance level for students in the control group.

7.20 Should you purchase the lecture notes? Some college professors make bound lecture notes available to their classes in an effort to improve teaching effectiveness. *Marketing Educational Review* (Fall 1994) published a study of business students' opinions of lecture notes. Two groups of students were surveyed—86 students enrolled in a promotional strategy class that required the purchase of lecture notes and 35 students enrolled in a sales/retailing elective that did not offer lecture notes. At the end of the semester, students were asked to respond to the statement, "Having a copy of the lecture notes was [would be] helpful in understanding the material." Responses were measured on a 9-point semantic difference scale, where 1 = strongly disagree and 9 = strongly agree. A summary of the results is reported in the table.

Purchased Lecture Notes	Did Not Purchase Lecture Notes
$n_1 = 86$	$n_2 = 35$
$\bar{x}_1 = 8.48$	$\bar{x}_2 = 7.80$
$s_1^2 = 0.94$	$s_2^2 = 2.99$

Source: Gray, J. I., and Abernathy, A. M. "Pros and cons of lecture notes and handout packages: Faculty and student opinions," *Marketing Education Review,* Vol. 4, No. 3, Fall 1984, p. 25 (Table 4). Reproduced with permission of American Marketing Association in the format Textbook and electronic usage via Copyright Clearance Center.

a. Describe the two populations involved in the comparison.

b. Do the samples provide sufficient evidence to conclude that there is a difference in the mean responses of the two groups of students? Test using $\alpha = .01$.

c. Construct a 99% confidence interval for $(\mu_1 - \mu_2)$. Interpret the result.

d. Would a 95% confidence interval for $(\mu_1 - \mu_2)$ be narrower or wider than the one you found in part **c**? Why?

Control Group:

1 24 5 16 21 7 20 1 9 20 19 10 23 16 0 4 9 13 17 13 0 2 12 11 7 3 11

1 19 9 12 18 5 21 30 15 4 2 12 11 10 13 11 3 6 10 13 16 12 28 19 12 20

Rudeness Condition:

4 11 18 11 9 6 5 11 9 12 7 5 7 3 11 1 9 11 10 7 8 9 10 7

11 4 13 5 4 7 8 3 8 15 9 16 10 0 7 15 13 9 2 13 10

7.21 Is honey a cough remedy? Refer to the *Archives of Pediatrics and Adolescent Medicine* (Dec. 2007) study of honey as a children's cough remedy, Exercise 2.28 (p. 51). Children who were ill with an upper respiratory tract infection and their parents participated in the study. Parents were instructed to give their sick child a dosage of liquid "medicine" prior to bedtime. Unknown to the parents, some were given a dosage of dextromethorphan (DM)—an over-the-counter cough medicine—while others were given a similar dose of honey. (*Note:* A third group gave their children no medicine.) Parents then rated their children's cough symptoms, and the improvement in total cough symptoms score was determined for each child. The data (improvement scores) for the 35 children in the DM dosage group and the 35 children in the honey dosage group (saved in the **HONEYCOUGH** file) are reproduced in the table below. Do you agree with the statement (extracted from the article), "honey may be a preferable treatment for the cough and sleep difficulty associated with childhood upper respiratory tract infection"? Use the comparison of two means methodology presented in this section to answer the question.

Honey Dosage:
12 11 15 11 10 13 10 4 15 16 9 14 10 6 10 8 11
12 12 8 12 9 11 15 10 15 9 13 8 12 10 8 9 5 12

DM Dosage:
4 6 9 4 7 7 7 9 12 10 11 6 3 4 9 7 8
12 12 4 12 13 7 10 13 9 4 4 10 15 9 12 6

Source: Paul, I. M., et al. "Effect of honey, dextromethorphan, and no treatment on nocturnal cough and sleep quality for coughing children and their parents," *Archives of Pediatrics and Adolescent Medicine,* Vol. 161, No. 12, Dec. 2007 (data simulated).

7.22 Bacteria counts in a plant's liquid waste discharge. Suppose you manage a plant that purifies its liquid waste and discharges the water into a local river. An EPA inspector has collected water specimens of the discharge of your plant and also water specimens in the river upstream from your plant. Each water specimen is divided into five parts, the bacteria count is read on each, and the median count for each specimen is reported. The bacteria counts for each of six specimens are reported in the following table for the two locations; the data are saved in the **BACTERIA** file.

Plant Discharge			Upstream		
30.1	36.2	33.4	29.7	30.3	26.4
28.2	29.8	34.9	27.3	31.7	32.3

a. Why might the bacteria counts shown here tend to be approximately normally distributed?
b. What are the appropriate null and alternative hypotheses to test whether the mean bacteria count for the plant discharge exceeds that for the upstream location? Be sure to define any symbols you use.
c. Conduct the test, part **b.** Carefully interpret the results.
d. What assumptions are necessary to ensure the validity of this test?

7.23 Cooling method for gas turbines. Refer to the *Journal of Engineering for Gas Turbines and Power* (Jan. 2005) study of gas turbines augmented with high-pressure inlet fogging, Exercise 6.28 (p. 335). The researchers classified gas turbines into three categories: traditional, advanced, and aeroderivative. Summary statistics on heat rate (kilojoules per kilowatt per hour) for each of the three types of gas turbines and saved in the **GASTURBINE** file are shown in the Minitab printout at the bottom of the page.

a. Is there sufficient evidence of a difference between the mean heat rates of traditional augmented gas turbines and aeroderivative augmented gas turbines? Test using $\alpha = .05$.
b. Is there sufficient evidence of a difference between the mean heat rates of advanced augmented gas turbines and aeroderivative augmented gas turbines? Test using $\alpha = .05$.

Applying the Concepts—Advanced

7.24 CareerBank.com annual salary survey. CareerBank.com conducts an annual salary survey of accounting, finance, and banking professionals. For the 2004 survey, data were collected for 2,800 responses submitted online by professionals across the country who voluntarily responded to CareerBank.com's Web-based survey. Salary comparisons were made by gender, education, and marital status. Some of the results are shown in the accompanying table.

	Males	Females
Mean Salary	\$69,848	\$52,012
Number of Respondents	1,400	1,400

a. Suppose you want to make an inference about the difference between the mean salaries of male and female accounting/finance/banking professionals at a 95% level of confidence. Why is this impossible to do using the information in the table?
b. Give values of the missing standard deviations that would lead you to conclude that the mean salary for males is significantly higher than the mean salary for females at a 95% level of confidence.
c. In your opinion, are the sample standard deviations, part **b,** reasonable values for the salary data? Explain.
d. How does the data-collection method impact any inferences derived from the data?

Minitab Output for Exercise 7.23

Descriptive Statistics: HEATRATE

Variable	ENGINE	N	Mean	StDev	Minimum	Maximum
HEATRATE	Advanced	21	9764	639	9105	11588
	Aeroderiv	7	12312	2652	8714	16243
	Traditional	39	11544	1279	10086	14796

7.3 Comparing Two Population Means: Paired Difference Experiments

Suppose you want to compare the mean daily sales of two restaurants located in the same city. If you were to record the restaurants' total sales for each of 12 randomly selected days during a 6-month period, the results might appear as shown in Table 7.3. Do these data provide evidence of a difference between the mean daily sales of the two restaurants?

Table 7.3 Daily Sales for Two Restaurants

Day	Restaurant 1 x_1	Restaurant 2 x_1
1 (Wednesday)	$1,005	$ 918
2 (Saturday)	2,073	1,971
3 (Tuesday)	873	825
4 (Wednesday)	1,074	999
5 (Friday)	1,932	1,827
6 (Thursday)	1,338	1,281
7 (Thursday)	1,449	1,302
8 (Monday)	759	678
9 (Friday)	1,905	1,782
10 (Monday)	693	639
11 (Saturday)	2,106	2,049
12 (Tuesday)	981	933

Data Set: RESTSALES

We want to test the null hypothesis that the mean daily sales, μ_1 and μ_2, for the two restaurants are equal against the alternative hypothesis that they differ; that is,

$$H_0: (\mu_1 - \mu_2) = 0$$
$$H_a: (\mu_1 - \mu_2) \neq 0$$

Many researchers mistakenly use the t-statistic for two independent samples (Section 7.2) to conduct this test. The analysis is shown on the Excel/DDXL printout, Figure 7.10. The test statistic, $t = .38$, is highlighted on the printout, as well as the p-value of the test, p-value $= .7047$. At $\alpha = .10$, the p-value exceeds α. Thus, from *this* analysis we might conclude that insufficient evidence exists to infer that there is a difference in mean daily sales for the two restaurants.

If you carefully examine the data in Table 7.3, however, you will find this conclusion difficult to accept. The sales of restaurant 1 exceed those of restaurant 2 *for every one of the randomly selected 12 days.* This, in itself, is strong evidence to indicate that μ_1 differs from μ_2, and we will subsequently confirm this fact. Why, then, was the t-test unable to detect this difference? The answer is, *the independent samples* t*-test is not a valid procedure to use with this set of data.*

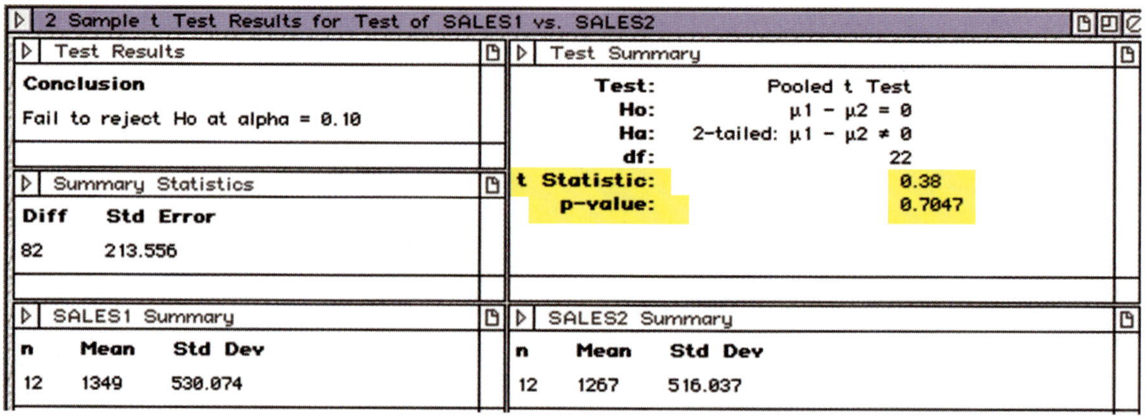

Figure 7.10
Excel/DDXL analysis of daily restaurant sales

The t-test is inappropriate because the assumption of independent samples is invalid. We have randomly chosen *days;* thus, once we have chosen the sample of days for restaurant 1, we have *not* independently chosen the sample of days for restaurant 2. The dependence between observations within days can be seen by examining the pairs of daily sales, which tend to rise and fall together as we go from day to day. This pattern provides strong visual evidence of a violation of the assumption of independence required for the two-sample t-test of Section 7.2. Also, substituting $s_1^2 = 530.07$ and $s_2^2 = 516.04$ (obtained from the printout, Figure 7.10) into the formula for s_p^2, we obtain

$$s_p^2 = \frac{(n_1 - 1)s_1^2 + (n_2 - 1)s_2^2}{n_1 + n_2 - 2}$$

$$= \frac{(12 - 1)(530.07)^2 + (12 - 1)(516.04)^2}{12 + 12 - 2} = 273{,}635.7$$

Thus, there is a *large variation within samples* (reflected by the large value of s_p^2) in comparison to the relatively *small difference between the sample means.* Because s_p^2 is so large, the t-test of Section 7.2 is unable to detect a possible difference between μ_1 and μ_2.

Table 7.4 Daily Sales and Differences for Two Restaurants

Day	Restaurant 1 x_1	Restaurant 2 x_2	Difference $d = x_1 - x_2$
1 (Wednesday)	\$1,005	\$ 918	\$ 87
2 (Saturday)	2,073	1,971	102
3 (Tuesday)	873	825	48
4 (Wednesday)	1,074	999	75
5 (Friday)	1,932	1,827	105
6 (Thursday)	1,338	1,281	57
7 (Thursday)	1,449	1,302	147
8 (Monday)	759	678	81
9 (Friday)	1,905	1,782	123
10 (Monday)	693	639	54
11 (Saturday)	2,106	2,049	57
12 (Tuesday)	981	933	48

We now consider a valid method of analyzing the data of Table 7.3. In Table 7.4, we add the column of differences between the daily sales of the two restaurants, $d = x_1 - x_2$. We can regard these daily differences in sales as a random sample of all daily differences, past and present. Then we can use this sample to make inferences about the mean of the population of differences, μ_d, which is equal to the difference $(\mu_1 - \mu_2)$—that is, the mean of the population (and sample) of differences equals the difference between the population (and sample) means. Thus, our test becomes

$$H_0\text{: } \mu_d = 0 \text{ [i.e., } (\mu_1 - \mu_2) = 0]$$
$$H_a\text{: } \mu_d \neq 0 \text{ [i.e., } (\mu_1 - \mu_2) \neq 0]$$

The test statistic is a one-sample t (Section 6.5) because we are now analyzing a single sample of differences for small n:

$$\textit{Test statistic: } t = \frac{\bar{d} - 0}{s_d/\sqrt{n_d}}$$

where $\bar{d}$ = Sample mean difference
s_d = Sample standard deviation of differences
n_d = Number of differences = number of pairs

Assumptions: The population of differences in daily sales is approximately normally distributed. The sample differences are randomly selected from the population differences. [*Note:* We do not need to make the assumption that $\sigma_1^2 = \sigma_2^2$.]

Rejection region: At significance level $\alpha = .05$, we will reject H_0 if $|t| > t_{.05}$, where $t_{.05}$ is based on $(n_d - 1)$ degrees of freedom.

Referring to Table V in Appendix B, we find the t-value corresponding to $\alpha = .025$ and $n_d - 1 = 12 - 1 = 11$ df to be $t_{.025} = 2.201$. Then we will reject the null hypothesis if $|t| > 2.201$ (see Figure 7.11). Note that the number of degrees of freedom has decreased from $n_1 + n_2 - 2 = 22$ to 11 when we use the paired difference experiment rather than the two independent random samples design.

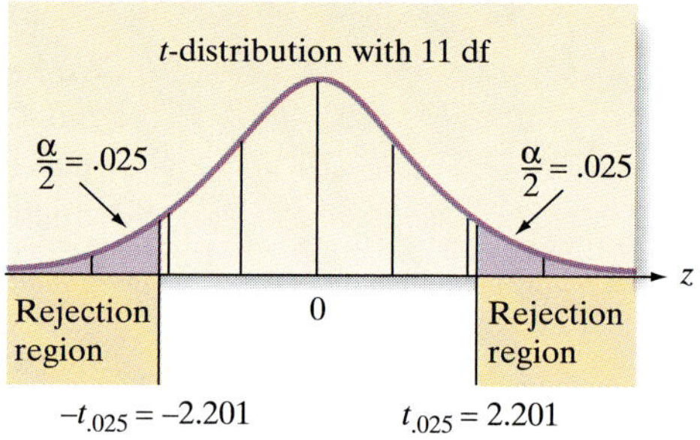

Figure 7.11
Rejection region for restaurant sales example

Summary statistics for the $n = 12$ differences are shown on the Minitab printout, Figure 7.12. Note that $\bar{d} = 82.0$ and $s_d = 32.0$ (rounded). Substituting these values into the formula for the test statistic, we have

$$t = \frac{\bar{d} - 0}{s_d/\sqrt{n_d}} = \frac{82}{32/\sqrt{12}} = 8.88$$

Because this value of t falls in the rejection region, we conclude (at $\alpha = .05$) that the difference in population mean daily sales for the two restaurants differs from 0. We can reach the same conclusion by noting that the p-value of the test, highlighted in Figure 7.12, is approximately 0. The fact that $(\bar{x}_1 - \bar{x}_2) = \bar{d} = \$82.00$ strongly suggests that the mean daily sales for rcstaurant 1 cxcccds thc mcan daily sales for restaurant 2.

This kind of experiment, in which observations are paired and the differences are analyzed, is called a **paired difference experiment.** In many cases, a paired difference experiment can provide more information about the difference between population means than an independent samples experiment. The idea is to compare population means by comparing the differences between pairs of experimental units (objects, people, etc.) that were very similar prior to the experiment. The differencing removes sources of variation that tend to inflate σ^2. For instance, in the restaurant example, the day-to-day variability in daily sales is removed by analyzing the differences between the restaurants' daily sales. Making comparisons within groups of similar experimental units is called **blocking,** and the paired difference experiment is an example of a **randomized block experiment.** In our example, the days represent the blocks.

Some other examples for which the paired difference experiment might be appropriate are the following:

1. Suppose you want to estimate the difference $(\mu_1 - \mu_2)$ in mean price per gallon between two major brands of premium gasoline. If you choose two independent random samples of stations for each brand, the variability in price due to geographic location may be large. To eliminate this source of variability, you could choose pairs of stations of similar size, one station for each brand, in close geographic proximity and use the sample of differences between the prices of the brands to make an inference about $(\mu_1 - \mu_2)$.

```
Paired T for SALES1 - SALES2

              N    Mean   StDev   SE Mean
SALES1       12    1349     530       153
SALES2       12    1267     516       149
Difference   12   82.00   31.99      9.23

95% CI for mean difference: (61.68, 102.32)
T-Test of mean difference = 0 (vs not = 0): T-Value = 8.88  P-Value = 0.000
```

Figure 7.12
Minitab analysis of differences in daily restaurant sales

2. A college placement center wants to estimate the difference ($\mu_1 - \mu_2$) in mean starting salaries for men and women graduates who seek jobs through the center. If it independently samples men and women, the starting salaries may vary because of their different college majors and differences in grade point averages. To eliminate these sources of variability, the placement center could match male and female job seekers according to their majors and grade point averages. Then the differences between the starting salaries of each pair in the sample could be used to make an inference about ($\mu_1 - \mu_2$).

3. To compare the performance of two automobile salespeople, we might test a hypothesis about the difference ($\mu_1 - \mu_2$) in their respective mean monthly sales. If we randomly choose n_1 months of salesperson 1's sales and independently choose n_2 months of salesperson 2's sales, the month-to-month variability caused by the seasonal nature of new car sales might inflate s_p^2 and prevent the two-sample t-statistic from detecting a difference between μ_1 and μ_2, if such a difference actually exists. However, by taking the difference in monthly sales for the two salespeople for each of n months, we eliminate the month-to-month variability (seasonal variation) in sales, and the probability of detecting a difference between μ_1 and μ_2, if a difference exists, is increased.

The hypothesis-testing procedures and the method of forming confidence intervals for the difference between two means using a paired difference experiment are summarized in the following boxes for both large and small n.

Paired Difference Test of Hypothesis for $\mu_d = (\mu_1 - \mu_2)$

One-Tailed Test	Two-Tailed Test
H_0: $\mu_d = D_0$	H_0: $\mu_d = D_0$
H_a: $\mu_d < D_0$ [or H_a: $\mu_d > D_0$]	H_a: $\mu_d \neq D_0$

Large Sample

Test statistic: $z = \dfrac{\bar{d} - D_0}{\sigma_d/\sqrt{n_d}} \approx \dfrac{\bar{d} - D_0}{s_d/\sqrt{n_d}}$

Rejection region: $z < -z_\alpha$ [or $z > z_\alpha$ when H_a: $\mu_d > D_0$]	*Rejection region:* $\lvert z\rvert > z_{\alpha/2}$

Small Sample

Test statistic: $t = \dfrac{\bar{d} - D_0}{s_d/\sqrt{n_d}}$

Rejection region: $t < -t_\alpha$ [or $t > t_\alpha$ when H_a: $\mu_d > D_0$]	*Rejection region:* $\lvert t\rvert > t_{\alpha/2}$

where t_α and $t_{\alpha/2}$ are based on $(n_d - 1)$ degrees of freedom

Paired Difference Confidence Interval for $\mu_d = (\mu_1 - \mu_2)$

Large Sample

$$\bar{d} \pm z_{\alpha/2}\frac{\sigma_d}{\sqrt{n_d}} \approx \bar{d} \pm z_{\alpha/2}\frac{s_d}{\sqrt{n_d}}$$

Small Sample

$$\bar{d} \pm t_{\alpha/2}\frac{s_d}{\sqrt{n_d}}$$

where $t_{\alpha/2}$ is based on $(n_d - 1)$ degrees of freedom.

Conditions Required for Valid Large-Sample Inferences about μ_d

1. A random sample of differences is selected from the target population of differences.
2. The sample size n_d is large (i.e., $n_d \geq 30$); due to the Central Limit Theorem, this condition guarantees that the test statistic will be approximately normal regardless of the shape of the underlying probability distribution of the population).

Conditions Required for Valid Small-Sample Inferences about μ_d

1. A random sample of differences is selected from the target population of differences.
2. The population of differences has a distribution that is approximately normal.

Example 7.5

Using a Confidence Interval for μ_d to Compare Mean Salaries of Males and Females

Problem An experiment is conducted to compare the starting salaries of male and female college graduates who find jobs. Pairs are formed by choosing a male and a female with the same major and similar grade point averages (GPAs). Suppose a random sample of 10 pairs is formed in this manner and the starting annual salary of each person is recorded. The results are shown in Table 7.5. Compare the mean starting salary, μ_1, for males to the mean starting salary, μ_2, for females using a 95% confidence interval. Interpret the results.

Table 7.5 Data on Annual Salaries for Matched Pairs of College Graduates

Pair	Male x_1	Female x_2	Difference $d = x_1 - x_2$
1	\$29,300	\$28,800	\$500
2	41,500	41,600	−100
3	40,400	39,800	600
4	38,500	38,500	0
5	43,500	42,600	900
6	37,800	38,000	−200
7	69,500	69,200	300
8	41,200	40,100	1,100
9	38,400	38,200	200
10	59,200	58,500	700

Data Set: GRADPAIRS

Solution Because the data on annual salary are collected in pairs of males and females matched on GPA and major, a paired difference experiment is performed. To conduct the analysis, we first compute the differences between the salaries, as shown in Table 7.5. Summary statistics for these $n = 10$ differences are displayed in the Minitab printout, Figure 7.13.

The 95% confidence interval for $\mu_d = (\mu_1 - \mu_2)$ for this small sample is

$$\bar{d} \pm t_{\alpha/2}\frac{s_d}{\sqrt{n_d}}$$

where $t_{\alpha/2} = t_{.025} = 2.262$ (obtained from Table V Appendix B) is based on $n_d - 1 = 9$ degrees of freedom. Substituting the values of $\bar{d} = 400$ and $s_d = 435$ shown on the printout, we obtain

$$\bar{d} \pm t_{.025}\frac{s_d}{\sqrt{n_d}} = 400 \pm 2.262\left(\frac{435}{\sqrt{10}}\right)$$
$$= 400 \pm 311 = (\$89, \$711)$$

```
Paired T for MALE - FEMALE

             N   Mean  StDev  SE Mean
MALE        10  43930  11665     3689
FEMALE      10  43530  11617     3674
Difference  10    400    435      137


95% CI for mean difference: (89, 711)
T-Test of mean difference = 0 (vs not = 0): T-Value = 2.91  P-Value = 0.017
```

Figure 7.13
Minitab analysis of salary differences

[*Note:* This interval is also shown on the Minitab printout, Figure 7.13.] Our interpretation is that the true mean difference between the starting salaries of males and females falls between \$89 and \$711, with 95% confidence. Because the interval falls above 0, we infer that $\mu_1 - \mu_2 > 0$; that is, that the mean salary for males exceeds the mean salary for females.

Look Back Remember that $\mu_d = \mu_1 - \mu_2$. So, if $\mu_d > 0$, then $\mu_1 > \mu_2$. Alternatively, if $\mu_d < 0$, then $\mu_1 < \mu_2$.

Now Work Exercise 7.37

To measure the amount of information about $(\mu_1 - \mu_2)$ gained by using a paired difference experiment in Example 7.5 rather than an independent samples experiment, we can compare the relative widths of the confidence intervals obtained by the two methods. A 95% confidence interval for $(\mu_1 - \mu_2)$ using the paired difference experiment is, from Example 7.5, (\$89, \$711). If we analyzed the same data as though this were an independent samples experiment,* we would first obtain the descriptive statistics shown in the SPSS printout, Figure 7.14.

Then we substitute the sample means and standard deviations shown on the printout into the formula for a 95% confidence interval for $(\mu_1 - \mu_2)$ using independent samples:

$$(\bar{x}_1 - \bar{x}_2) \pm t_{.025}\sqrt{s_p^2\left(\frac{1}{n_1} + \frac{1}{n_2}\right)}$$

where

$$s_p^2 = \frac{(n_1 - 1)s_1^2 + (n_2 - 1)s_2^2}{n_1 + n_2 - 2}$$

Group Statistics

	GENDER	N	Mean	Std. Deviation	Std. Error Mean
SALARY	M	10	43930.00	11665.148	3688.844
	F	10	43530.00	11616.946	3673.601

Independent Samples Test

		Levene's Test for Equality of Variances		t-test for Equality of Means						
									95% Confidence Interval of the Difference	
		F	Sig.	t	df	Sig. (2-tailed)	Mean Difference	Std. Error Difference	Lower	Upper
SALARY	Equal variances assumed	.000	.991	.077	18	.940	400.00	5206.046	-10537.5	11337.50
	Equal variances not assumed			.077	18.000	.940	400.00	5206.046	-10537.5	11337.51

Figure 7.14
SPSS analysis of salaries, assuming independent samples

*This is done only to provide a measure of the increase in the amount of information obtained by a paired design in comparison to an unpaired design. Actually, if an experiment is designed using pairing, an unpaired analysis would be invalid because the assumption of independent samples would not be satisfied.

SPSS performed these calculations and obtained the interval (−\$10,537.50, \$11,337.50). This interval is highlighted in Figure 7.14.

Notice that the independent samples interval includes 0. Consequently, if we were to use this interval to make an inference about $(\mu_1 - \mu_2)$, we would incorrectly conclude that the mean starting salaries of males and females do not differ! You can see that the confidence interval for the independent sampling experiment is about five times wider than for the corresponding paired difference confidence interval. Blocking out the variability due to differences in majors and grade point averages significantly increases the information about the difference in male and female mean starting salaries by providing a much more accurate (smaller confidence interval for the same confidence coefficient) estimate of $(\mu_1 - \mu_2)$.

You may wonder whether conducting a paired difference experiment is always superior to an independent samples experiment. The answer is—most of the time but not always. We sacrifice half the degrees of freedom in the t-statistic when a paired difference design is used instead of an independent samples design. This is a loss of information, and unless this loss is more than compensated for by the reduction in variability obtained by blocking (pairing), the paired difference experiment will result in a net loss of information about $(\mu_1 - \mu_2)$. Thus, we should be convinced that the pairing will significantly reduce variability before performing the paired difference experiment. Most of the time this will happen.

One final note: The pairing of the observations is determined before the experiment is performed (that is, by the *design* of the experiment). A paired difference experiment is *never* obtained by pairing the sample observations after the measurements have been acquired.

Ethics IN Statistics

In a two-group analysis, intentionally pairing observations after the data have been collected in order to produce a desired result is considered *unethical statistical practice.*

What Do You Do When the Assumption of a Normal Distribution for the Population of Differences Is Not Satisfied?

Answer: Use the Wilcoxon signed rank test for the paired difference design (Chapter 14).

Activity 7.1 *Box Office Receipts:* Comparing Population Means

Use the Internet to find the daily box office receipts for two different hit movies during the first eight weeks after their releases. In this activity, you will compare the mean daily box office receipts of these movies in two different ways.

1. Independently select random samples of size $n = 30$ from the data sets for each of the movies' daily box office receipts. Find the mean and standard deviation of each sample. Then find a confidence interval for the difference of the means.
2. Now pair the data for the two movies by day, that is, the box office receipts for the day of release for each movie are paired, the box office receipts for each movie's second day are paired, and so forth. Calculate the difference in box office receipts for each day and select a random sample of size $n = 30$ from the daily differences. Then find a confidence interval for the sample mean.
3. Compare the confidence intervals from Exercises 1 and 2. Explain how the sampling for the paired difference experiment is different from the independent sampling. How might this sampling technique yield a better comparison of the two means in the box office example?
4. Compute the actual means for the daily box office receipts for each of the movies and then find the difference of the means. Does the difference of the means lie in both confidence intervals you found? Is the exact difference remarkably closer to one of the estimates? Explain.

Exercises 7.25–7.40

Learning the Mechanics

7.25 A paired difference experiment yielded n_d pairs of observations. In each case, what is the rejection region for testing $H_0: \mu_d > 2$?

a. $n_d = 12, \alpha = .05$

b. $n_d = 24, \alpha = .10$

c. $n_d = 4, \alpha = .025$

d. $n_d = 80, \alpha = .01$

7.26 NW The data for a random sample of six paired observations are shown in the next table and saved in the **LM7_26** file.

a. Calculate the difference between each pair of observations by subtracting observation 2 from observation 1. Use the differences to calculate $\bar{d}$ and s_d^2.

b. If μ_1 and μ_2 are the means of populations 1 and 2, respectively, express μ_d in terms of μ_1 and μ_2.

c. Form a 95% confidence interval for μ_d.

Pair	Sample from Population 1 (Observation 1)	Sample from Population 2 (Observation 2)
1	7	4
2	3	1
3	9	7
4	6	2
5	4	4
6	8	7

d. Test the null hypothesis $H_0: \mu_d = 0$ against the alternative hypothesis $H_a: \mu_d \neq 0$. Use $\alpha = .05$.

7.27 The data for a random sample of 10 paired observations are shown in the table below and saved in the **LM7_27** file.

Pair	Sample from Population 1	Sample from Population 2
1	19	24
2	25	27
3	31	36
4	52	53
5	49	55
6	34	34
7	59	66
8	47	51
9	17	20
10	51	55

a. If you wish to test whether these data are sufficient to indicate that the mean for population 2 is larger than that for population 1, what are the appropriate null and alternative hypotheses? Define any symbols you use.

b. Conduct the test, part **a,** using $\alpha = .10$.

c. Find a 90% confidence interval for μ_d. Interpret this result.

d. What assumptions are necessary to ensure the validity of this analysis?

7.28 A paired difference experiment produced the following results:

$$n_d = 38 \quad \bar{x}_1 = 92 \quad \bar{x}_2 = 95.5 \quad \bar{d} = -3.5 \quad s_d^2 = 21$$

a. Determine the values of z for which the null hypothesis, $\mu_1 - \mu_2 = 0$, would be rejected in favor of the alternative hypothesis, $\mu_1 - \mu_2 < 0$. Use $\alpha = .10$.

b. Conduct the paired difference test described in part **a.** Draw the appropriate conclusions.

c. What assumptions are necessary so that the paired difference test will be valid?

d. Find a 90% confidence interval for the mean difference μ_d.

e. Which of the two inferential procedures, the confidence interval of part **d** or the test of hypothesis of part **b,** provides more information about the differences between the population means?

7.29 A paired difference experiment yielded the following results:

$$n_d = 40 \quad \Sigma d = 468 \quad \Sigma d^2 = 6{,}880$$

a. Test $H_0: \mu_d = 10$ against $H_a: \mu_d \neq 10$, where $\mu_d = (\mu_1 - \mu_2)$. Use $\alpha = .05$.

b. Report the p-value for the test you conducted in part **a.** Interpret the p-value.

c. Do you need to assume that the population of differences is normally distributed? Explain.

Applying the Concepts—Basic

7.30 **Performance ratings of government agencies.** The U.S. Office of Management and Budget (OMB) requires government agencies to produce annual performance and accounting reports (PARS) each year. A research team at George Mason University evaluated the quality of the PARS for 24 government agencies (*The Public Manager,* Summer 2008), where evaluation scores ranged from 12 (lowest) to 60 (highest). The **PARS** file contains the 2007 and 2008 evaluation scores for all 24 agencies. (See Exercise 2.123, p. 94.) Data for a random sample of five of these agencies are shown in the accompanying table. Suppose you want to conduct a paired-difference test to determine whether the true mean evaluation score of government agencies in 2008 exceeds the true mean evaluation score in 2007.

Agency	Score07	Score08
GSA	34	40
Agriculture	33	35
Social Security	33	33
USAID	32	42
Defense	17	32

Source: Ellig, J., and Wray, H. "Measuring performance reporting quality," *The Public Manager,* Vol. 37, No. 2, Summer 2008 (p. 66). Reproduced with permission of *The Public Manager* in the format electronic usage via Copyright Clearance Center.

a. Explain why the data should be analyzed using a paired-difference test.

b. Compute the difference between the 2008 score and the 2007 score for each sampled agency.

c. Find the mean and standard deviation of the differences, part **b.**

d. Use the summary statistics, part **c,** to find the test statistic.

e. Give the rejection region for the test using $\alpha = .10$.

f. Make the appropriate conclusion in the words of the problem.

7.31 **Salaries of technology professionals.** The data in the table on the next page, obtained from *Business Week's* (June, 22, 2006) technology section, represents typical salaries of technology professionals in 13 metropolitan areas for 2003 and 2005. These data are saved in the **TECHPRO** file. Suppose you want to determine if the mean salary of technology professionals at all U.S. metropolitan areas has increased between 2003 and 2005.

a. Set up the null and alternative hypothesis for the test.

b. Calculate the difference between the 2003 and 2005 salaries for each metropolitan area.

c. Find the mean and standard deviation of the differences, part **b.**

d. Use the results, part **c,** to calculate the test statistic.

e. Find the rejection region for the test at $\alpha = .10$.

f. Make the appropriate conclusion.

g. What conditions are required for the inference in part **f** to be valid? Are these conditions reasonably satisfied?

Metro Area	2003 Salary ($ Thousands)	2005 Salary ($ Thousands)
Silicon Valley	87.7	85.9
New York	78.6	80.3
Washington, D.C.	71.4	77.4
Los Angeles	70.8	77.1
Denver	73.0	77.1
Boston	76.3	80.1
Atlanta	73.6	73.2
Chicago	71.1	73.0
Philadelphia	69.5	69.8
San Diego	69.0	77.1
Seattle	71.0	66.9
Dallas-Ft. Worth	73.0	71.0
Detroit	62.3	64.1

Source: Dice, Inc.; www.businessweek.com, June 22, 2006. Dice Salary Survey—Metro Area Salaries. Reprinted from June 22, 2006 issue of *Business Week* by special permission. Copyright © 2006 by the McGraw-Hill Companies, Inc.

7.32 Life expectancy of Oscar winners. Movie actors who win an Oscar usually can command a greater fee for their next motion picture. Does winning an Academy of Motion Picture Arts and Sciences award (aka, an Oscar) lead to long-term mortality for movie actors? In an article in the *Annals of Internal Medicine* (May 15, 2001), researchers sampled 762 Academy Award winners and matched each one with another actor of the same sex who was in the same winning film and was born in the same era. The life expectancy (age) of each pair of actors was compared.

a. Explain why the data should be analyzed as a paired difference experiment.

b. Set up the null hypothesis for a test to compare the mean life expectancies of Academy Award winners and nonwinners.

c. The sample mean life expectancies of Academy Award winners and nonwinners were reported as 79.7 years and 75.8 years, respectively. The p-value for comparing the two population means was reported as $p = .003$. Interpret this value in the context of the problem.

7.33 "I am not selling anything" surveys. To improve response rates in telephone surveys, interviewers are often instructed by the polling company to state "I am not selling anything" at the outset of the call. The effectiveness of the "I am not selling anything" strategy was investigated in the *International Journal of Public Opinion Research* (Winter 2004). The sample consisted of 29 different telephone surveys. However, in each survey about half the people were contacted by interviewers using the "I am not selling anything" introduction and the other half were contacted by interviewers using the standard (no mention of "not selling") introduction. Thus, for each of the 29 surveys, both the "not selling" and standard interviewing techniques were employed. Summary statistics on response rates (proportion of people called who actually respond to the survey questions) are given in the accompanying table. The goal of the researchers was to compare the mean response rates of the two interviewing methods with the specific purpose to determine if the mean response rate for "not selling" is higher than that for the standard.

a. Explain why the data should be analyzed as a paired-difference experiment.

b. Analyze the data in the table using the independent-samples t-test. Do you detect a significant difference between the mean response rates of the two methods using $\alpha = .05$?

c. The researchers applied the paired-difference t procedure and obtained an observed significance level of p-value $= .001$. Interpret this result if $\alpha = 0.5$.

d. Compare the inferences you made in parts **b** and **c.**

	"Not selling" Introduction	Standard Introduction
Number of Surveys	29	29
Mean Response Rate	.262	.246
Standard Deviation	.12	.11

Source: De Leeuw, E. D. and Hox, J. J. "I am not selling anything: 29 experiments in telephone introductions," *International Journal of Public Opinion Research*, Vol. 16, No. 4, Winter 2004, pp. 464–473 (Table 1). By permission of Oxford University Press.

7.34 NHTSA new car crash tests. Refer to the National Highway Traffic Safety Administration (NHTSA) crash test data for new cars saved in the **CRASH** file. Crash-test dummies were placed in the driver's seat and front passenger's seat of a new car model, and the car was steered by remote control into a head-on collision with a fixed barrier while traveling at 35 miles per hour. Two of the variables measured for each of the 98 new cars in the data set are (1) the severity of the driver's chest injury and (2) the severity of the passenger's chest injury. (The more points assigned to the chest injury rating, the more severe the injury.) Suppose the NHTSA wants to determine whether the true mean driver chest injury rating exceeds the true mean passenger chest injury rating and, if so, by how much.

a. State the parameter of interest to the NHTSA.

b. Explain why the data should be analyzed as matched pairs.

c. Find a 99% confidence interval for the true difference between the mean chest injury ratings of drivers and front-seat passengers.

d. Interpret the interval, part **c.** Does the true mean driver chest injury rating exceed the true mean passenger chest injury rating? If so, by how much?

e. What conditions are required for the analysis to be valid? Do these conditions hold for these data?

Applying the Concepts—Intermediate

7.35 Taking "power naps" during work breaks. Lack of sleep costs companies about $18 billion a year in lost productivity, according to the National Sleep Foundation. Companies are waking up to the problem, however. Some even have quiet rooms available for study or sleep. "Power naps" are in vogue (*Athens Daily News*, Jan. 9, 2000). A major airline recently began encouraging reservation agents to nap during their breaks. The table on the next page (saved in the **POWERNAP** file) lists the number of complaints received about each of a sample of 10 reservation agents during the 6 months before naps were encouraged and during the 6 months after the policy change.

a. Do the data present sufficient evidence to conclude that the new napping policy reduced the mean number of customer complaints about reservation agents? Test using $\alpha = .05$.

b. What assumptions must hold to ensure the validity of the test?

c. What variables, not controlled in the study, could lead to an invalid conclusion?

Operator	Before Policy	After Policy
1	10	5
2	3	0
3	16	7
4	11	4
5	8	6
6	2	4
7	1	2
8	14	3
9	5	5
10	6	1

7.36 Computer-mediated communication study. Refer to the *Journal of Computer-Mediated Communication* (Apr. 2004) study to compare relational intimacy in people interacting via CMC to people meeting FTF, Exercise 7.18 (p. 396). Recall that a relational intimacy score was measured (on a 7-point scale) for each participant after each of three different meeting sessions. The researchers also hypothesized that the mean relational intimacy score for participants in the CMC group will significantly increase between the first and third meetings, but the difference between the first and third meetings will not significantly change for participants in the FTF group.

a. For the CMC group comparison, give the null and alternative hypotheses of interest.

b. The researchers made the comparison, part **a,** using a paired t-test. Explain why the data should be analyzed as matched pairs.

c. For the CMC group comparison, the reported test statistic was $t = 3.04$ with p-value $= .003$. Interpret these results. Is the researchers' hypothesis supported?

d. For the FTF group comparison, give the null and alternative hypotheses of interest.

e. For the FTF group comparison, the reported test statistic was $t = .39$ with p-value $= .70$. Interpret these results. Is the researchers' hypothesis supported?

7.37 NW **Testing electronic circuits.** Japanese researchers have developed a compression/depression method of testing electronic circuits based on Huffman coding (*IEICE Transactions on Information & Systems,* Jan. 2005). The new method is designed to reduce the time required for input decompression and output compression—called the *compression ratio.* Experimental results were obtained by testing a sample of 11 benchmark circuits (all of different sizes) from a SUN Blade 1000 workstation. Each circuit was tested using the standard compression/depression method and the new Huffman-based coding method, and the compression ratio was recorded. The data (saved in the **CIRCUITS** file) are given in the accompanying table. Compare the two methods with a 95% confidence interval. Which method has the smaller mean compression ratio?

Circuit	Standard Method	Huffman-Coding Method
1	.80	.78
2	.80	.80
3	.83	.86
4	.53	.53
5	.50	.51
6	.96	.68
7	.99	.82
8	.98	.72
9	.81	.45
10	.95	.79
11	.99	.77

Source: Ichihara, H., Shintani, M., and Inoue, T. "Huffman-based test response coding," *IEICE Transactions on Information & Systems,* Vol. E88-D, No. 1, Jan. 2005 (Table 3). Copyright 2005 by Oxford University Press – Journals. Reproduced with permission of Oxford University Press – Journals in the format Textbook and electronic usage via Copyright Clearance Center.

7.38 Impact of red light cameras on car crashes. To combat red-light-running crashes—the phenomenon of a motorist entering an intersection after the traffic signal turns red and causing a crash—many states are adopting photo-red enforcement programs. In these programs, red light cameras installed at dangerous intersections photograph the license plates of vehicles that run the red light. How effective are photo-red enforcement programs in reducing red-light-running crash incidents at intersections? The Virginia Department of Transportation (VDOT) conducted a comprehensive study of its newly adopted photo-red enforcement program and published the results in a June 2007 report. In one portion of the study, the VDOT provided crash data both before and after installation of red light cameras at several intersections. The data (measured as the number of crashes caused by red light running per intersection per year) for 13 intersections in Fairfax County, Virginia, are given in the table and saved in the **REDLIGHT** file. Analyze the data for the VDOT. What do you conclude?

Intersection	Before Camera	After Camera
1	3.60	1.36
2	0.27	0
3	0.29	0
4	4.55	1.79
5	2.60	2.04
6	2.29	3.14
7	2.40	2.72
8	0.73	0.24
9	3.15	1.57
10	3.21	0.43
11	0.88	0.28
12	1.35	1.09
13	7.35	4.92

Source: Virginia Transportation Research Council, "Research report: The impact of red light cameras (photo-red enforcement) on crashes in Virginia," June 2007.

7.39 Evaluating a new drug. Merck Research Labs conducted an experiment to evaluate the effect of a new drug using the single-T swim maze. Nineteen impregnated dam rats were captured and allocated a dosage of 12.5 milligrams of the drug. One male and one female rat pup were randomly selected from each resulting litter to perform in the swim maze. Each pup was placed in the water at one end of the maze and allowed to swim until it escaped at the opposite end. If the pup failed to escape after a certain period of time, it was placed at the beginning of the maze and given another chance. The experiment was repeated until each pup accomplished three successful escapes. The table on the next page (saved in the **RATPUPS** file) reports the number of swims required by each pup to perform three successful escapes. Is there sufficient evidence of a difference between the mean number of swims required by male and female pups? Conduct the test (at $\alpha = .10$). Comment on the assumptions required for the test to be valid.

Litter	Male	Female	Litter	Male	Female
1	8	5	11	6	5
2	8	4	12	6	3
3	6	7	13	12	5
4	6	3	14	3	8
5	6	5	15	3	4
6	6	3	16	8	12
7	3	8	17	3	6
8	5	10	18	6	4
9	4	4	19	9	5
10	4	4			

Source: Thomas E. Bradstreet, Merck Research Labs, BL 3–2, West Point, PA 19486.

Applying the Concepts—Advanced

7.40 Alcoholic fermentation in wines. Determining alcoholic fermentation in wine is critical to the wine-making process. Must/wine density is a good indicator of the fermentation point because the density value decreases as sugars are converted into alcohol. For decades, winemakers have measured must/wine density with a hydrometer. Although accurate, the hydrometer employs a manual process that is very time-consuming. Consequently, large wineries are searching for more rapid measures of density measurement. An alternative method uses the hydrostatic balance instrument (similar to the hydrometer but digital). A winery in Portugal collected the must/wine density measurements for white wine samples randomly selected from the fermentation process for a recent harvest. For each sample, the density of the wine at 20°C was measured with both the hydrometer and the hydrostat balance. The densities for 40 wine samples are saved in the **WINE40** file. The first five and last five observations are shown in the table. The winery will use the alternative method of measuring wine density only if it can be demonstrated that the mean difference between the density measurements of the two methods does not exceed .002. Perform the analysis for the winery. Provide the winery with a written report of your conclusions.

Sample	Hydrometer	Hydrostatic
1	1.08655	1.09103
2	1.00270	1.00272
3	1.01393	1.01274
4	1.09467	1.09634
5	1.10263	1.10518
⋮	⋮	⋮
36	1.08084	1.08097
37	1.09452	1.09431
38	0.99479	0.99498
39	1.00968	1.01063
40	1.00684	1.00526

Source: Cooperative Cellar of Borba (*Adega Cooperative de Borba*), Portugal.

7.4 Comparing Two Population Proportions: Independent Sampling

Suppose a personal water craft (PWC) manufacturer wants to compare the potential market for its products in the northeastern United States to the market in the southeastern United States. Such a comparison would help the manufacturer decide where to concentrate sales efforts. Using telephone directories, the company randomly chooses 1,000 households in the southeast (SE) and 1,000 households in the northeast (NE) and determines whether each household plans to buy a PWC within the next 5 years. The objective is to use this sample information to make an inference about the difference $(p_1 - p_2)$ between the proportion p_1 of *all* households in the SE and the proportion p_2 of *all* households in the NE that plan to purchase a PWC within 5 years.

The two samples represent independent binomial experiments. (See Section 4.3 for the characteristics of binomial experiments.) The binomial random variables are the numbers x_1 and x_2 of the 1,000 sampled households in each area that indicate they will purchase a PWC within 5 years. The results are summarized in Table 7.6.

Table 7.6 Results of Telephone Survey

SE	NE
$n_1 = 1{,}000$	$n_2 = 1{,}000$
$x_1 = 42$	$x_2 = 24$

We can now calculate the sample proportions $\hat{p}_1$ and $\hat{p}_2$ of the households in the SE and NE, respectively, that are prospective buyers:

$$\hat{p}_1 = \frac{x_1}{n_1} = \frac{42}{1{,}000} = .042$$

$$\hat{p}_2 = \frac{x_2}{n_2} = \frac{24}{1{,}000} = .024$$

The difference between the sample proportions $(\hat{p}_1 - \hat{p}_2)$ makes an intuitively appealing point estimator of the difference between the population parameters $(p_1 - p_2)$. For our example, the estimate is

$$(\hat{p}_1 - \hat{p}_2) = .042 - .024 = .018$$

To judge the reliability of the estimator $(\hat{p}_1 - \hat{p}_2)$, we must observe its performance in repeated sampling from the two populations—that is, we need to know the sampling distribution of $(\hat{p}_1 - \hat{p}_2)$. The properties of the sampling distribution are given in the next box. Remember that $\hat{p}_1$ and $\hat{p}_2$ can be viewed as means of the number of successes per trial in the respective samples, so the Central Limit Theorem applies when the sample sizes are large.

Properties of the Sampling Distribution of $(p_1 - p_2)$

1. The mean of the sampling distribution of $(\hat{p}_1 - \hat{p}_2)$ is $(p_1 - p_2)$; that is,

$$E(\hat{p}_1 - \hat{p}_2) = p_1 - p_2$$

Thus, $(\hat{p}_1 - \hat{p}_2)$ is an unbiased estimator of $(p_1 - p_2)$.

2. The standard deviation of the sampling distribution of $(\hat{p}_1 - \hat{p}_2)$ is

$$\sigma_{(\hat{p}_1 - \hat{p}_2)} = \sqrt{\frac{p_1q_1}{n_1} + \frac{p_2q_2}{n_2}}$$

3. If the sample sizes n_1 and n_2 are large (see Section 5.4 for a guideline), the sampling distribution of $(\hat{p}_1 - \hat{p}_2)$ is approximately normal.

Because the distribution of $(\hat{p}_1 - \hat{p}_2)$ in repeated sampling is approximately normal, we can use the z-statistic to derive confidence intervals for $(p_1 - p_2)$ or test a hypothesis about $(p_1 - p_2)$.

For the PWC example, a 95% confidence interval for the difference $(p_1 - p_2)$ is

$$(\hat{p}_1 - \hat{p}_2) \pm 1.96\sigma_{(\hat{p}_1 - \hat{p}_2)} \quad \text{or} \quad (\hat{p}_1 - \hat{p}_2) \pm 1.96\sqrt{\frac{p_1q_1}{n_1} + \frac{p_2q_2}{n_2}}$$

The quantities p_1q_1 and p_2q_2 must be estimated to complete the calculation of the standard deviation $\sigma_{(\hat{p}_1 - \hat{p}_2)}$ and hence the calculation of the confidence interval. In Section 5.4 we showed that the value of pq is relatively insensitive to the value chosen to approximate p. Therefore, $\hat{p}_1\hat{q}_1$ and $\hat{p}_2\hat{q}_2$ will provide satisfactory estimates to approximate p_1q_1 and p_2q_2, respectively. Then

$$\sqrt{\frac{p_1q_1}{n_1} + \frac{p_2q_2}{n_2}} \approx \sqrt{\frac{\hat{p}_1\hat{q}_1}{n_1} + \frac{\hat{p}_2\hat{q}_2}{n_2}}$$

and we will approximate the 95% confidence interval by

$$(\hat{p}_1 - \hat{p}_2) \pm 1.96\sqrt{\frac{\hat{p}_1\hat{q}_1}{n_1} + \frac{\hat{p}_2\hat{q}_2}{n_2}}$$

Substituting the sample quantities yields

$$(.042 - .024) \pm 1.96\sqrt{\frac{(.042)(.958)}{1{,}000} + \frac{(.024)(.976)}{1{,}000}}$$

or, $.018 \pm .016$. Thus, we are 95% confident that the interval from .002 to .034 contains $(p_1 - p_2)$.

We infer that there are between .2% and 3.4% more households in the southeast than in the northeast that plan to purchase PWCs in the next 5 years.

The general form of a confidence interval for the difference $(p_1 - p_2)$ between population proportions is given in the box below.

Large-Sample $(1 - \alpha)$% Confidence Interval for $(p_1 - p_2)$

$$(\hat{p}_1 - \hat{p}_2) \pm z_{\alpha/2}\sigma_{(\hat{p}_1 - \hat{p}_2)} = (\hat{p}_1 - \hat{p}_2) \pm z_{\alpha/2}\sqrt{\frac{p_1q_1}{n_1} + \frac{p_2q_2}{n_2}}$$

$$\approx (\hat{p}_1 - \hat{p}_2) \pm z_{\alpha/2}\sqrt{\frac{\hat{p}_1\hat{q}_1}{n_1} + \frac{\hat{p}_2\hat{q}_2}{n_2}}$$

Conditions Required for Valid Large-Sample Inferences about $(p_1 - p_2)$

1. The two samples are randomly selected in an independent manner from the two target populations.
2. The sample sizes, n_1 and n_2, are both large so that the sampling distribution of $(\hat{p}_1 - \hat{p}_2)$ will be approximately normal. (This condition will be satisfied if both $n_1\hat{p}_1 \geq 15$, $n_1\hat{q}_1 \geq 15$, and $n_2\hat{p}_2 \geq 15$, $n_2\hat{q}_2 \geq 15$.)

The z-statistic,

$$z = \frac{(\hat{p}_1 - \hat{p}_2) - (p_1 - p_2)}{\sigma_{(\hat{p}_1 - \hat{p}_2)}}$$

is used to test the null hypothesis that $(p_1 - p_2)$ equals some specified difference, say D_0. For the special case where $D_0 = 0$—that is, where we want to test the null hypothesis $H_0: (p_1 - p_2) = 0$ (or, equivalently, $H_0: p_1 = p_2$)—the best estimate of $p_1 = p_2 = p$ is obtained by dividing the total number of successes $(x_1 + x_2)$ for the two samples by the total number of observations $(n_1 + n_2)$; that is,

$$\hat{p} = \frac{x_1 + x_2}{n_1 + n_2} \quad \text{or} \quad \hat{p} = \frac{n_1\hat{p}_1 + n_2\hat{p}_2}{n_1 + n_2}$$

The second equation shows that $\hat{p}$ is a weighted average of $\hat{p}_1$ and $\hat{p}_2$, with the larger sample receiving more weight. If the sample sizes are equal, then $\hat{p}$ is a simple average of the two sample proportions of successes.

We now substitute the weighted average $\hat{p}$ for both p_1 and p_2 in the formula for the standard deviation of $(\hat{p}_1 - \hat{p}_2)$:

$$\sigma_{(\hat{p}_1 - \hat{p}_2)} = \sqrt{\frac{p_1q_1}{n_1} + \frac{p_2q_2}{n_2}} \approx \sqrt{\frac{\hat{p}\hat{q}}{n_1} + \frac{\hat{p}\hat{q}}{n_2}} = \sqrt{\hat{p}\hat{q}\left(\frac{1}{n_1} + \frac{1}{n_2}\right)}$$

The test is summarized in the next box.

Large-Sample Test of Hypothesis about $(p_1 - p_2)$

One-Tailed Test	Two-Tailed Test
$H_0: (p_1 - p_2) = 0$*	$H_0: (p_1 - p_2) = 0$
$H_a: (p_1 - p_2) < 0$	$H_a: (p_1 - p_2) \neq 0$
[or $H_a: (p_1 - p_2) > 0$]	

Test statistic: $z = \dfrac{(\hat{p}_1 - \hat{p}_2)}{\sigma_{(\hat{p}_1 - \hat{p}_2)}}$

One-Tailed Test	Two-Tailed Test
Rejection region: $z < -z_\alpha$	*Rejection region:* $\lvert z \rvert > z_{\alpha/2}$
[or $z > z_\alpha$ when $H_a: (p_1 - p_2) > 0$]	

Note: $\sigma_{(\hat{p}_1 - \hat{p}_2)} = \sqrt{\dfrac{p_1q_1}{n_1} + \dfrac{p_2q_2}{n_2}} \approx \sqrt{\hat{p}\hat{q}\left(\dfrac{1}{n_1} + \dfrac{1}{n_2}\right)}$, where $\hat{p} = \dfrac{x_1 + x_2}{n_1 + n_2}$.

*The test can be adapted to test for a difference $D_0 \neq 0$. Because most applications call for a comparison of p_1 and p_2, implying $D_0 = 0$, we will confine our attention to this case.

Example 7.6

Comparing Repair Rates of Two Car Models Using a Large-Sample Test about $p_1 - p_2$

Problem A consumer advocacy group wants to determine whether there is a difference between the proportions of the two leading automobile models that need major repairs (more than \$500) within 2 years of their purchase. A sample of 400 two-year owners of model 1 is contacted, and a sample of 500 two-year owners of model 2 is contacted. The numbers x_1 and x_2 of owners who report that their cars needed major repairs within the first 2 years are 53 and 78, respectively. Test the null hypothesis that no difference exists between the proportions in populations 1 and 2 needing major repairs against the alternative that a difference does exist. Use $\alpha = .10$.

Solution If we define p_1 and p_2 as the true proportions of model 1 and model 2 owners, respectively, whose cars need major repairs within 2 years, the elements of the test are

$$H_0: (p_1 - p_2) = 0$$

$$H_a: (p_1 - p_2) \neq 0$$

$$\textit{Test statistic: } z = \frac{(\hat{p}_1 - \hat{p}_2) - 0}{\sigma_{(\hat{p}_1 - \hat{p}_2)}}$$

$$\textit{Rejection region } (\alpha = .10): |z| > z_{\alpha/2} = z_{.05} = 1.645$$

(see Figure 7.15)

We now calculate the sample proportions of owners who need major car repairs,

$$\hat{p}_1 = \frac{x_1}{n_1} = \frac{53}{400} = .1325$$

$$\hat{p}_2 = \frac{x_2}{n_2} = \frac{78}{500} = .1560$$

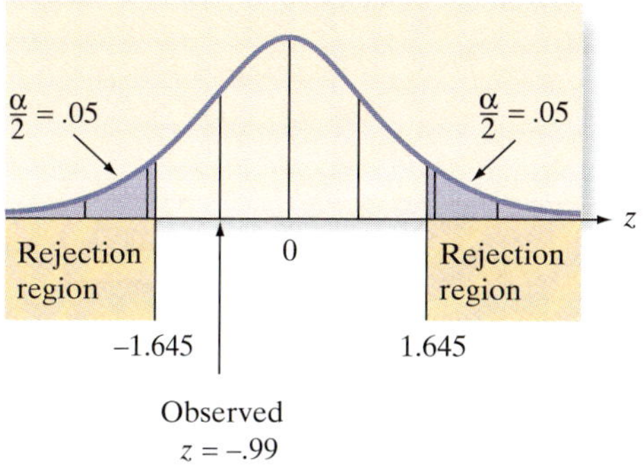

Figure 7.15
Rejection region for Example 7.6

Then

$$z = \frac{(\hat{p}_1 - \hat{p}_2) - 0}{\sigma_{(\hat{p}_1 - \hat{p}_2)}} \approx \frac{(\hat{p}_1 - \hat{p}_2)}{\sqrt{\hat{p}\hat{q}\left(\frac{1}{n_1} + \frac{1}{n_2}\right)}}$$

where

$$\hat{p} = \frac{x_1 + x_2}{n_1 + n_2} = \frac{53 + 78}{400 + 500} = .1456$$

Note that $\hat{p}$ is a weighted average of $\hat{p}_1$ and $\hat{p}_2$, with more weight given to the larger sample of model 2 owners.

Thus, the computed value of the test statistic is

$$z = \frac{.1325 - .1560}{\sqrt{(.1456)(.8544)\left(\frac{1}{400} + \frac{1}{500}\right)}} = \frac{-.0235}{.0237} = -.99$$

The samples provide insufficient evidence at $\alpha = .10$ to detect a difference between the proportions of the two models that need repairs within 2 years. Even though 2.35% more sampled owners of model 2 found they needed major repairs, this difference is less than 1 standard deviation ($z = -.99$) from the hypothesized zero difference between the true proportions.

Example 7.7

Finding the Observed Significance Level of a Test for $p_1 - p_2$

Problem Use a statistical software package to conduct the test in Example 7.6. Find and interpret the p-value of the test.

Solution We entered the sample sizes (n_1 and n_2) and number of successes (x_1 and x_2) into Excel/DDXL and obtained the printout shown in Figure 7.16. The test statistic for this two-tailed test is shaded on the printout, as well as the observed significance level (p-value). Note that p-value = .3205 exceeds α = .10. Consequently, there is no evidence of a difference between the true population proportions.

```
Proportion Test for the Difference Between S1 and N1

Summary Statistics                 Test Summary
          n1     400                    p:                      0
      p-hat1      0.133                Ho:              p1 - p2 = 0
          n2     500                   Ha:   2-tailed:  p1 - p2 ≠ 0
      p-hat2      0.156       z Statistic:                  -0.99
  Difference     -0.0235         p-value:                  0.3205
    pooled n     900
pooled p-hat      0.146         Test Results
     Std Err      0.0237        Conclusion
                                Fail to reject Ho at alpha =  0.10
```

Figure 7.16
Excel/DDXL printout for test of two proportions

Statistics IN Action Revisited Comparing Proportions

In the first Statistics in Action Revisited in this chapter (pp. 391–392), we demonstrated how the expert statistician used a comparison of two means to defend Visa in a libel case. Recall that ZixIt claims that a Visa executive's e-mail postings had a negative impact on ZixIt's attempt to develop a new online credit card processing system. Here, we demonstrate another way to analyze the data, one successfully presented in court by the statistician.

In addition to daily closing price and trading volume of ZixIt stock, the **ZIXITVISA** file also contains a qualitative variable that indicates whether the stock price increased or not (decreased or stayed the same) on the following day. This variable was created by the statistician to compare the proportion of days on which ZixIt stock went up for posting and nonposting days. Let p_1 represent the proportion of days where the ZixIt stock price increased following all nonposting days and p_2 represent the proportion of days where the ZixIt stock price increased following posting days. Then, if the charges made by ZixIt are true (i.e., that postings will have a negative impact on ZixIt stock), p_1 will exceed p_2. Thus, a comparison of two population proportions is appropriate. Recall that during the 83-day period of interest, the executive posted e-mails on 43 days and had no postings on 40 days. Again, both sample sizes (n_1 = 40 and n_2 = 43) are large, so we can apply the large-sample z-test or large-sample confidence interval procedure for independent samples. (Can you demonstrate this?) A Minitab printout for this analysis is shown in Figure SIA7.3.

From the printout you can see that following the 40 nonposting days, the price increased on 20 days; following the 43 posting days, the stock price increased on 18 days. Thus, the sample proportions are p_1 = 20/40 = .5 and p_2 = 18/43 = .42. Are these sample proportions different enough for us to conclude that the population proportions are different and that ZixIt's claim is true? Not according to the statistical analysis. Note that the 95% confidence interval for $(p_1 - p_2)$, $(-.133, .295)$, includes the value 0, and the p-value for the two-tailed test of H_0: $(p_1 - p_2) = 0$, p-value = .456, exceeds, say α = .05. Both imply that the two population proportions are not significantly different. Also, neither sample proportion is significantly different from .5. (Can you demonstrate this?) Consequently, in courtroom testimony, the statistical expert used these results to conclude that "the direction of ZixIt's stock price movement following days with postings is random, just like days with no postings."

Data Set: ZIXITVISA

Test and CI for Two Proportions: Up/Down, Posting

```
Event = UP

Posting   X   N  Sample p
NO       20  40  0.500000
POST     18  43  0.418605

Difference = p (NO) - p (POST)
Estimate for difference:  0.0813953
95% CI for difference:  (-0.132500, 0.295291)
Test for difference = 0 (vs not = 0):  Z = 0.75  P-Value = 0.456

Fisher's exact test: P-Value = 0.513
```

Figure SIA7.3
Minitab comparison of two proportions analysis

Activity 7.2 *Keep the Change:* Inferences Based on Two Samples

In this activity, you will compare the mean amounts transferred for two different Bank of America customers as well as design some studies that might help the marketing department determine where to allocate more of its advertising budget. You will be working with data sets from Activity 1.1, *Keep the Change: Collecting Data* on page 16.

1. You will need to work with another student in the class on this exercise. Each of you should use your data set *Amounts Transferred* from the Chapter 1 Activity as a random sample from a theoretically larger set of all your amounts ever transferred. Then the means and standard deviations of your data sets will be the sample means and standard deviations. Write a confidence interval for the difference of the two means at the 95% level. Does the interval contain 0? Are your mean amounts transferred significantly different? Explain.
2. Design a study to determine whether there is a significant difference in the mean amount of Bank of America matches for customers in California enrolled in the program and the mean amount of Bank of America matches for customers in Florida enrolled in the program. Be specific about sample sizes, tests used, and how a conclusion will be reached. How might the results of this study help Bank of America estimate costs in the program?
3. Design a study to determine whether there is a significant difference in the percentage of Bank of America customers in California enrolled in the program and the percentage of Bank of America customers in Florida enrolled in the program. Be specific about sample sizes, tests used, and how a conclusion will be reached. How might the results of this study help Bank of America's marketing department?

Keep the results from this activity for use in other activities.

Exercises 7.41–7.58

Learning the Mechanics

7.41 Consider making an inference about $p_1 - p_2$, where there are x_1 successes in n_1 binomial trials and x_2 successes in n_2 binomial trials.

a. Describe the distributions of x_1 and x_2.

b. Explain why the Central Limit Theorem is important in finding an approximate distribution for $(\hat{p}_1 - \hat{p}_2)$.

7.42 For each of the following values of α, find the values of z for which $H_0: (p_1 - p_2) = 0$ would be rejected in favor of $H_a: (p_1 - p_2) < 0$.

a. $\alpha = .01$ **b.** $\alpha = .025$
c. $\alpha = .05$ **d.** $\alpha = .10$

7.43 In each case, determine whether the sample sizes are large enough to conclude that the sampling distribution of $(\hat{p}_1 - \hat{p}_2)$ is approximately normal.

a. $n_1 = 12, n_2 = 14, \hat{p}_1 = .42, \hat{p}_2 = .57$
b. $n_1 = 12, n_2 = 14, \hat{p}_1 = .92, \hat{p}_2 = .86$
c. $n_1 = n_2 = 30, \hat{p}_1 = .70, \hat{p}_2 = .73$
d. $n_1 = 100, n_2 = 250, \hat{p}_1 = .93, \hat{p}_2 = .97$
e. $n_1 = 125, n_2 = 200, \hat{p}_1 = .08, \hat{p}_2 = .12$

7.44 Construct a 95% confidence interval for $(p_1 - p_2)$ in each of the following situations:

a. $n_1 = 400, \hat{p}_1 = .65; n_2 = 400, \hat{p}_2 = .58$
b. $n_1 = 180, \hat{p}_1 = .31; n_2 = 250, \hat{p}_2 = .25$
c. $n_1 = 100, \hat{p}_1 = .46; n_2 = 120, \hat{p}_2 = .61$

7.45 Independent random samples, each containing 800 observations, were selected from two binomial populations. The samples from populations 1 and 2 produced 320 and 400 successes, respectively.

a. Test $H_0: (p_1 - p_2) = 0$ against $H_a: (p_1 - p_2) > 0$. Use $\alpha = .05$.

b. Test $H_0: (p_1 - p_2) = 0$ against $H_a: (p_1 - p_2) \neq 0$. Use $\alpha = .01$.

c. Test $H_0: (p_1 - p_2) = 0$ against $H_a: (p_1 - p_2) < 0$. Use $\alpha = .01$.

d. Form a 90% confidence interval for $(p_1 - p_2)$.

7.46 Random samples of size $n_1 = 55$ and $n_2 = 65$ were drawn from populations 1 and 2, respectively. The samples yielded $\hat{p}_1 = .7$ and $\hat{p}_2 = .6$. Test $H_0: (p_1 - p_2) = 0$ against $H_a: (p_1 - p_2) > 0$ using $\alpha = .05$.

Applying the Concepts—Basic

7.47 NW **The "winner's curse" in auction bidding.** In auction bidding, the "winner's curse" is the phenomenon of the winning (or highest) bid price being above the expected value of the item being auctioned. *The Review of Economics and Statistics* (Aug. 2001) published a study on whether bid experience impacts the likelihood of the winner's curse occurring. Two groups of bidders in a sealed-bid auction were compared: (1) super-experienced bidders and (2) less-experienced bidders. In the super-experienced group, 29 of 189 winning bids were above the item's expected value; in the less-experienced group, 32 of 149 winning bids were above the item's expected value.

a. Find an estimate of p_1, the true proportion of super-experienced bidders who fall prey to the winner's curse.

b. Find an estimate of p_2, the true proportion of less-experienced bidders who fall prey to the winner's curse.

c. Construct a 90% confidence interval for $p_1 - p_2$.

d. Give a practical interpretation of the confidence interval, part **c.** Make a statement about whether bid experience impacts the likelihood of the winner's curse occurring.

7.48 **Is steak your favorite barbeque food?** July is National Grilling Month in the United States. On July 1, 2008, *The Harris Poll #70* reported on a survey of Americans' grilling preferences. When asked about their favorite food prepared on a barbeque, 662 of 1,250 randomly sampled Democrats preferred steak, as compared to 586 of 930 randomly sampled Republicans.

a. Give a point estimate for the proportion of all Democrats who prefer steak as their favorite barbeque food.

b. Give a point estimate for the proportion of all Republicans who prefer steak as their favorite barbeque food.

c. Give a point estimate for the difference between the proportions of all Democrats and all Republicans who prefer steak as their favorite barbeque food.

d. Construct a 95% confidence interval for the difference between the proportions of all Democrats and all Republicans who prefer steak as their favorite barbeque food.

e. Give a practical interpretation of the interval, part **d.**

f. Explain the meaning of the phrase *95% confident* in your answer to part **e.**

7.49 **Likelihood of getting a routine medical checkup.** Who is more likely to get a routine medical checkup—employed or unemployed people? To answer this question, a team of physicians and public health professors collected data on a sample of over 2,200 individuals (*American Journal of Public Health,* Jan. 2002). Of the 1,140 individuals who were employed, 642 visited a physician for a routine checkup within the past year. In contrast, 740 of the 1,106 unemployed individuals had a routine medical checkup within the past year.

a. Specify the parameter of interest to the research team.

b. Set up the null and alternative hypotheses for testing whether there is a difference between the percentages of employed and unemployed people who had a recent routine medical checkup.

c. Compute the test statistic for the test.

d. Give the rejection region for the test using $\alpha = .01$.

e. The research team reported the p-value for the test as $p\text{-value} \approx 0$. Do you agree?

f. Make the appropriate conclusion.

7.50 **Planning habits survey.** *American Demographics* (Jan. 2002) reported the results of a survey on the planning habits of men and women. In response to the question, "What is your preferred method of planning and keeping track of meetings, appointments, and deadlines?" 56% of the men and 46% of the women answered, "Keep them in my head." A nationally representative sample of 1,000 adults participated in the survey; therefore, assume that 500 were men and 500 were women.

a. Set up the null and alternative hypotheses for testing whether the percentage of men who prefer keeping track of appointments in their head is larger than the corresponding percentage of women.

b. Compute the test statistic for the test.

c. Give the rejection region for the test using $\alpha = .01$.

d. Find the p-value for the test.

e. Make the appropriate conclusion.

7.51 **Cell phone usage differs by gender.** The role of the cell phone in modern life was investigated by a Pew Internet & American Life Project (April 2006) survey. A total of 1,286 cell phone users were interviewed in the sample. One of the objectives was to compare male and female cell phone users. For example, 32% of men admitted they sometimes don't drive safely while talking or texting on a cell phone compared to 25% of women. Also, 71% of men used their cell phone in an emergency compared to 77% of women. Assume that half (643) of the cell phone users in the sample were men and half (643) were women.

a. Describe the two populations of interest in the survey.

b. Give an estimate of the proportion of men and the proportion of women who sometimes do not drive safely while talking or texting on a cell phone.

c. Find a 90% confidence interval for the difference between the proportions of men and women who sometimes do not drive safely while talking or texting on a cell phone.

d. From your answer to part **c,** can you conclude that men are more likely than women to sometimes not drive safely while talking or texting on a cell phone? Explain.

e. Give an estimate of the proportion of men and the proportion of women who used their cell phone in an emergency.

f. Conduct a test to determine whether the proportions of men and women who used their cell phone in an emergency differ. Use $\alpha = .10$.

Applying the Concepts—Intermediate

7.52 **Electronic versus printed surveys.** The rapid evolution of computer hardware and software has made it easy for businesses to conduct computer-based and Web-based (i.e., electronic) surveys. Professors at Michigan State and DePaul universities collaborated on a study designed to compare the response rates of electronic surveys and traditional print surveys (Decision Sciences Institute, *Decision Line,* July 2001). The two surveys were developed for customers who had purchased products over the Internet from a leading retailer of office supplies. Of the 631 customers mailed the printed survey, 261 returned usable responses. Of the 414 customers who were sent a computer disk with the electronic survey, 155 returned usable responses.

a. Estimate the difference between the response rates of the two survey types using a 90% confidence interval. Interpret the result.

b. If the difference in response rates is 5% or less, the researchers will infer that there is no "practical" difference in response rates for the two surveys. Are the researchers able to make this inference? Explain.

7.53 Racial profiling by the LAPD. *Racial profiling* is a term used to describe any police action that relies on ethnicity rather than behavior to target suspects engaged in criminal activities. Does the Los Angeles Police Department (LAPD) invoke racial profiling in stops and searches of LA drivers? This question was addressed in *Chance* (Spring 2006).

Race	Number Stopped	Number Searched	Number of "Hits"
African American	61,688	12,016	5,134
White	106,892	5,312	3,006

Source: Khadjavi, L. S. "Driving while black in the City of Angels," *Chance,* Vol. 19, No. 2, Spring 2006, p. 45 (Tables 1 and 2). Reprinted with permission from *Chance.* © 2006 by the American Statistical Association. All rights reserved.

a. Data on stops and searches of both African Americans and white drivers are summarized in the accompanying table. Conduct a test (at $\alpha = .05$) to determine if there is a disparity in the proportions of African American and white drivers who are searched by the LA police after being stopped.

b. The LAPD defines a *hit rate* as the proportion of searches that result in a discovery of criminal activity. Use the data in the table to estimate the disparity in the hit rates for African American and white drivers using a 95% confidence interval. Interpret the results.

7.54 Angioplasty's benefits challenged. More than one million heart patients each year undergo an angioplasty. The benefits of an angioplasty were challenged in a recent study of 2,287 patients (2007 Annual Conference of the American College of Cardiology, New Orleans). All the patients had substantial blockage of the arteries but were medically stable. All were treated with medication such as aspirin and beta-blockers. However, half the patients were randomly assigned to get an angioplasty and half were not. After five years, the researchers found that 211 of the 1,145 patients in the angioplasty group had subsequent heart attacks compared to 202 of 1,142 patients in the medication only group. Do you agree with the study's conclusion, "There was no significant difference in the rate of heart attacks for the two groups"? Support your answer with a 95% confidence interval.

7.55 Entrepreneurial careers of MBA alumni. Are African American MBA students more likely to begin their careers as an entrepreneur than white MBA students? This was a question of interest to the Graduate Management Admission Council (GMAC). *GMAC Research Reports* (Oct. 3, 2005) published the results of a survey of MBA alumni. Of the 1,304 African Americans who responded to the survey, 209 reported their employment status after graduation as self-employed or a small business owner. Of the 7,120 whites who responded to the survey, 356 reported their employment status after graduation as self-employed or a small business owner. Use this information to answer the research question.

7.56 Predicting software defects. Refer to the PROMISE Software Engineering Repository data on 498 modules of software code written in "C" language for a NASA spacecraft instrument, saved in the **SWDEFECTS** file. (See Exercise 3.74, pp. 154–155) Recall that the software code in each module was evaluated for defects; 49 were classified as "true" (i.e., module has defective code), and 449 were classified as "false" (i.e., module has correct code). Consider these to be independent random samples of software code modules. Researchers predicted the defect status of each module using the simple algorithm, "If number of lines of code in the module exceeds 50, predict the module to have a defect." The SPSS printout below shows the number of modules in each of the two samples that were predicted to have defects (PRED_LOC = "yes") and predicted to have no defects (PRED_LOC = "no"). Now, define the *accuracy rate* of the algorithm as the proportion of modules that were correctly predicted. Compare the accuracy rate of the algorithm when applied to modules with defective code to the accuracy rate of the algorithm when applied to modules with correct code. Use a 99% confidence interval.

DEFECT * PRED_LOC Crosstabulation

Count

		PRED_LOC		
		no	yes	Total
DEFECT	false	400	49	449
	true	29	20	49
Total		429	69	498

Applying the Concepts—Advanced

7.57 Food craving study. Do you have an insatiable craving for chocolate or some other food? Because many North Americans apparently do, psychologists are designing scientific studies to examine the phenomenon. According to the *New York Times* (Feb. 22, 1995), one of the largest studies of food cravings involved a survey of 1,000 McMaster University (Canada) students. The survey revealed that 97% of the women in the study acknowledged specific food cravings while only 67% of the men did.

a. How large do n_1 and n_2 need to be to conclude that the true proportion of women who acknowledge having food cravings exceeds the corresponding proportion of men? Assume $\alpha = .01$.

b. Why is it dangerous to conclude from the study that women have a higher incidence of food cravings than men?

7.58 Gambling in public high schools. With the rapid growth in legalized gambling in the United States, there is concern that the involvement of youth in gambling activities is also increasing. University of Minnesota Professor Randy Stinchfield compared the rates of gambling among Minnesota public school students between 1992 and 1998 (*Journal of Gambling Studies,* Winter 2001). Based on survey data, the table shows the percentages of ninth-grade boys who gambled weekly or daily on any game (e.g., cards, sports betting, lotteries) for the 2 years.

a. Are the percentages of ninth-grade boys who gambled weekly or daily on any game in 1992 and 1998 significantly different? (Use $\alpha = .01$.)

b. Professor Stinchfield states that "because of the large sample sizes, even small differences may achieve statistical significance, so interpretations of the differences should include a judgment regarding the magnitude of the difference and its public health significance." Do you agree with this statement? If not, why not? If so, obtain a measure of the magnitude of the difference between 1992 and 1998 and attach a measure of reliability to the difference.

	1992	1998
Number of Ninth-Grade Boys in Survey	21,484	23,199
Number who Gambled Weekly/Daily	4,684	5,313

7.5 Determining the Sample Size

You can find the appropriate sample size to estimate the difference between a pair of parameters with a specified margin of error (ME) and degree of reliability by using the method described in Section 5.5 — that is, to estimate the difference between a pair of parameters correct to within ME units with confidence level $(1 - \alpha)$, let $z_{\alpha/2}$ standard deviations of the sampling distribution of the estimator equal ME. Then solve for the sample size. To do this, you have to solve the problem for a specific ratio between n_1 and n_2. Most often, you will want to have equal sample sizes—that is, $n_1 = n_2 = n$. We will illustrate the procedure with two examples.

Example 7.8

Finding the Sample Sizes for Estimating $\mu_1 - \mu_2$: Comparing Mean Crop Yields

Problem New fertilizer compounds are often advertised with the promise of increased crop yields. Suppose we want to compare the mean yield μ_1 of wheat when a new fertilizer is used to the mean yield μ_2 with a fertilizer in common use. The estimate of the difference in mean yield per acre is to be correct to within .25 bushel with a confidence coefficient of .95. If the sample sizes are to be equal, find $n_1 = n_2 = n$, the number of one-acre plots of wheat assigned to each fertilizer.

Solution To solve the problem, you need to know something about the variation in the bushels of yield per acre. Suppose from past records you know the yields of wheat possess a range of approximately 10 bushels per acre. You could then approximate $\sigma_1 = \sigma_2 = \sigma$ by letting the range equal 4σ. Thus,

$$4\sigma \approx 10 \text{ bushels}$$
$$\sigma \approx 2.5 \text{ bushels}$$

The next step is to solve the equation

$$z_{\alpha/2}\sigma_{(\bar{x}_1 - \bar{x}_2)} = \text{ME} \qquad \text{or} \qquad z_{\alpha/2}\sqrt{\frac{\sigma_1^2}{n_1} + \frac{\sigma_2^2}{n_2}} = \text{ME}$$

for n, where $n = n_1 = n_2$. Because we want the estimate to lie within ME = .25 of $(\mu_1 - \mu_2)$ with confidence coefficient equal to .95, we have $z_{\alpha/2} = z_{.025} = 1.96$. Then, letting $\sigma_1 = \sigma_2 = 2.5$ and solving for n, we have

$$1.96\sqrt{\frac{(2.5)^2}{n} + \frac{(2.5)^2}{n}} = .25$$

$$1.96\sqrt{\frac{2(2.5)^2}{n}} = .25$$

$$n = 768.32 \approx 769 \text{ (rounding up)}$$

Consequently, you will have to sample 769 acres of wheat for each fertilizer to estimate the difference in mean yield per acre to within .25 bushel.

Look Back Because $n = 769$ would necessitate extensive and costly experimentation, you might decide to allow a larger margin of error (say, ME = .50 or ME = 1) to reduce the sample size, or you might decrease the confidence coefficient. The point is that we can obtain an idea of the experimental effort necessary to achieve a specified precision in our final estimate by determining the approximate sample size *before* the experiment is started.

Now Work Exercise 7.60a

Example 7.9

Finding the Sample Sizes for Estimating $p_1 - p_2$: Comparing Defect Rates of Two Machines

Problem A production supervisor suspects a difference exists between the proportions p_1 and p_2 of defective items produced by two different machines. Experience has shown that the proportion defective for each of the two machines is in the neighborhood of .03. If the supervisor wants to estimate the difference in the proportions to within .005 using a 95% confidence interval, how many items must be randomly sampled from the production of each machine? (Assume that the supervisor wants $n_1 = n_2 = n$.)

Solution In this sampling problem, the margin of error is ME = .005, and for the specified level of reliability, $(1 - \alpha) = .95$, $z_{\alpha/2} = z_{.025} = 1.96$. Then, letting $p_1 = p_2 = .03$ and $n_1 = n_2 = n$, we find the required sample size per machine by solving the following equation for n:

$$z_{\alpha/2}\sigma_{(\hat{p}_1 - \hat{p}_2)} = \text{ME}$$

or

$$z_{\alpha/2}\sqrt{\frac{p_1q_1}{n_1} + \frac{p_2q_2}{n_2}} = \text{ME}$$

$$1.96\sqrt{\frac{(.03)(.97)}{n} + \frac{(.03)(.97)}{n}} = .005$$

$$1.96\sqrt{\frac{2(.03)(.97)}{n}} = .005$$

$$n = 8{,}943.2$$

Look Back This large n will likely result in a tedious sampling procedure. If the supervisor insists on estimating $(p_1 - p_2)$ correct to within .005 with 95% confidence, approximately 9,000 items will have to be inspected for each machine.

Now Work Exercise 7.59a

You can see from the calculations in Example 7.9 that $\sigma_{(\hat{p}_1 - \hat{p}_2)}$ (and hence the solution, $n_1 = n_2 = n$) depends on the actual (but unknown) values of p_1 and p_2. In fact, the required sample size $n_1 = n_2 = n$ is largest when $p_1 = p_2 = .5$. Therefore, if you have no prior information on the approximate values of p_1 and p_2, use $p_1 = p_2 = .5$ in the formula for $\sigma_{(\hat{p}_1 - \hat{p}_2)}$. If p_1 and p_2 are in fact close to .5, then the values of n_1 and n_2 that you have calculated will be correct. If p_1 and p_2 differ substantially from .5, then your solutions for n_1 and n_2 will be larger than needed. Consequently, using $p_1 = p_2 = .5$ when solving for n_1 and n_2 is a conservative procedure because the sample sizes n_1 and n_2 will be at least as large as (and probably larger than) needed.

The procedures for determining the sample sizes necessary for estimating $(\mu_1 - \mu_2)$ or $(p_1 - p_2)$ for the case $n_1 = n_2$ are given in the boxes on the next page.

Determination of Sample Size for Estimating $\mu_1 - \mu_2$

To estimate $(\mu_1 - \mu_2)$ with a given margin of error ME and with confidence level $(1 - \alpha)$, use the following formula to solve for equal sample sizes that will achieve the desired reliability:

$$n_1 = n_2 = \frac{(z_{\alpha/2})^2(\sigma_1^2 + \sigma_2^2)}{(\text{ME})^2}$$

You will need to substitute estimates for the values of σ_1^2 and σ_2^2 before solving for the sample size. These estimates might be sample variances s_1^2 and s_2^2 from prior sampling (e.g., a pilot sample) or from an educated (and conservatively large) guess based on the range—that is, $s \approx R/4$.

Determination of Sample Size for Estimating $(p_1 - p_2)$

To estimate $(p_1 - p_2)$ with a given margin of error ME and with confidence level $(1 - \alpha)$, use the following formula to solve for equal sample sizes that will achieve the desired reliability:

$$n_1 = n_2 = \frac{(z_{\alpha/2})^2(p_1q_1 + p_2q_2)}{(\text{ME})^2}$$

You will need to substitute estimates for the values of p_1 and p_2 before solving for the sample size. These estimates might be based on prior samples, obtained from educated guesses, or, most conservatively, specified as $p_1 = p_2 = .5$.

Exercises 7.59–7.70

Learning the Mechanics

7.59 Assuming that $n_1 = n_2$, find the sample sizes needed to estimate $(p_1 - p_2)$ for each of the following situations:

NW **a.** Margin of error = .01 with 99% confidence. Assume that $p_1 \approx .4$ and $p_2 \approx .7$.

b. A 90% confidence interval of width .05. Assume that there is no prior information available to obtain approximate values of p_1 and p_2.

c. Margin of error = .03 with 90% confidence. Assume that $p_1 \approx .2$ and $p_2 \approx .3$.

7.60 Find the appropriate values of n_1 and n_2 (assume $n_1 = n_2$) needed to estimate $(\mu_1 - \mu_2)$ for each of the following situations:

NW **a.** A margin of error equal to 3.2 with 95% confidence. From prior experience it is known that $\sigma_1 \approx 15$ and $\sigma_2 \approx 17$.

b. A margin of error equal to 8 with 99% confidence. The range of each population is 60.

c. A 90% confidence interval of width 1.0. Assume that $\sigma_1^2 \approx 5.8$ and $\sigma_2^2 \approx 7.5$.

7.61 Suppose you want to estimate the difference between two population means correct to within 1.8 with a 95% confidence interval. If prior information suggests that the population variances are approximately equal to $\sigma_1^2 = \sigma_2^2 = 14$ and you want to select independent random samples of equal size from the populations, how large should the sample sizes, n_1 and n_2, be?

7.62 Enough money has been budgeted to collect independent random samples of size $n_1 = n_2 = 100$ from populations 1 and 2 to estimate $(\mu_1 - \mu_2)$. Prior information indicates that $\sigma_1 = \sigma_2 = 10$. Have sufficient funds been allocated to construct a 90% confidence interval for $(\mu_1 - \mu_2)$ of width 5 or less? Justify your answer.

Applying the Concepts—Basic

7.63 **Homework assistance for accounting students.** Refer to the *Journal of Accounting Education* (Vol. 25, 2007) study of providing homework assistance to accounting students, Exercise 7.16 (p. 395). Recall that one group of students was given a completed homework solution and another group was given only check figures at various steps of the solution. The researchers wanted to compare the average test score improvement of the two groups. How many students should be sampled in each group to estimate the difference in the averages to within .5 point with 99% confidence? Assume that the standard deviations of the test score improvements for the two groups are approximately equal to 1.

7.64 **Bacteria counts in a plant's liquid waste discharge.** Refer to the EPA study of average bacteria counts in water specimens at two river locations, Exercise 7.22 (p. 397). How many water specimens need to be sampled at each location for a 95% confidence interval for the true mean difference in bacteria counts to yield an estimate that lies within 1.5 bacteria of the true difference? Assume equal sample sizes will be collected at each location.

7.65 **Electronic versus printed surveys.** Refer to the *Decision Line* (July 2001) study designed to compare the response rates of electronic surveys and traditional print surveys, Exercise 7.52 (pp. 414–415). Recall that the two surveys were developed for customers who had purchased products over the Internet from a leading retailer of office supplies. How many customers should be sampled to estimate the difference between the response rates of the two survey types to within .01 using a 90% confidence interval? Assume the same number of customers should be sampled for each survey.

7.66 Conducting a political poll. A pollster wants to estimate the difference between the proportions of men and women who favor a particular national candidate using a 90% confidence interval of width .04. Suppose the pollster has no prior information about the proportions. If equal numbers of men and women are to be polled, how large should the sample sizes be?

Applying the Concepts—Intermediate

7.67 Life expectancies of working women and housewives. Is housework hazardous to your health? A study in *Public Health Reports* compared the life expectancies of 25-year-old white women in the labor force to those who are housewives. How large a sample would have to be taken from each group in order to be 95% confident that the estimate of difference in average life expectancies for the two groups is within 1 year of the true difference in average life expectancies? Assume that equal sample sizes will be selected from the two groups and that the standard deviation for both groups is approximately 15 years.

7.68 Users of home shopping services. All cable companies carry at least one home shopping channel. Who uses these home shopping services? Are the shoppers primarily men or women? Suppose you want to estimate the difference in the proportions of men and women who say they have used or expect to use televised home shopping using an 80% confidence interval of width .06 or less.

a. Approximately how many people should be included in your samples?

b. Suppose you want to obtain individual estimates for the two proportions of interest. Will the sample size found in part **a** be large enough to provide estimates of each proportion correct to within .02 with probability equal to .90? Justify your response.

7.69 Angioplasty's benefits challenged. Refer to the study of patients with substantial blockage of the arteries presented at the 2007 Annual Conference of the American College of Cardiology, Exercise 7.54 (p. 415). Recall that half the patients were randomly assigned to get an angioplasty and half were not. The researchers compared the proportion of patients with subsequent heart attacks for the two groups and reported no significant difference between the two proportions. Although the study involved over 2,000 patients, the sample size may have been too small to detect a difference in heart attack rates.

a. How many patients must be sampled in each group to estimate the difference in heart attack rates to within .015 with 95% confidence? (Use summary data from Exercise 7.54 in your calculation.)

b. Comment on the practicality of carrying out the study with the sample sizes determined in part **a.**

c. Comment on the practical significance of the difference detected in the confidence interval for the study, part **a.**

7.70 Average housing space per person. Even though Japan is an economic superpower, Japanese workers are in many ways worse off than their U.S. and European counterparts. For example, in 2004 the estimated average housing space per person (in square feet) was 645 in the United States but only 344 in Japan (Diawa House Industry, Co., Japan). Suppose a team of economists and sociologists from the United Nations plans to reestimate the difference in the mean housing space per person for U.S. and Japanese workers. Assume that equal sample sizes will be used for each country and that the standard deviation is 35 square feet for Japan and 80 for the United States. How many people should be sampled in each country to estimate the difference to within 10 square feet with 95% confidence?

7.6 Comparing Two Population Variances: Independent Sampling

Many times, it is of practical interest to use the techniques developed in this chapter to compare the means or proportions of two populations. However, there are also important instances when we wish to compare two population variances. For example, when two devices are available for producing precision measurements (scales, calipers, thermometers, etc.), we might want to compare the variability of the measurements of the devices before deciding which one to purchase. Or when two standardized tests can be used to rate job applicants, the variability of the scores for both tests should be taken into consideration before deciding which test to use.

For problems like these, we need to develop a statistical procedure to compare population variances. The common statistical procedure for comparing population variances, σ_1^2 and σ_2^2, makes an inference about the ratio σ_1^2/σ_2^2. In this section, we will show how to test the null hypothesis that the ratio σ_1^2/σ_2^2 equals 1 (the variances are equal) against the alternative hypothesis that the ratio differs from 1 (the variances differ):

$$H_0: \frac{\sigma_1^2}{\sigma_2^2} = 1 \qquad (\sigma_1^2 = \sigma_2^2)$$

$$H_a: \frac{\sigma_1^2}{\sigma_2^2} \neq 1 \qquad (\sigma_1^2 \neq \sigma_2^2)$$

To make an inference about the ratio σ_1^2/σ_2^2, it seems reasonable to collect sample data and use the ratio of the sample variances, s_1^2/s_2^2. We will use the test statistic

$$F = \frac{s_1^2}{s_2^2}$$

To establish a rejection region for the test statistic, we need to know the sampling distribution of s_1^2/s_2^2. As you will subsequently see, the sampling distribution of s_1^2/s_2^2 is based on two of the assumptions already required for the t-test:

1. The two sampled populations are normally distributed.
2. The samples are randomly and independently selected from their respective populations.

When these assumptions are satisfied and when the null hypothesis is true (that is, $\sigma_1^2 = \sigma_2^2$), the sampling distribution of $F = s_1^2/s_2^2$ is the ***F*-distribution** with $(n_1 - 1)$ numerator degrees of freedom and $(n_2 - 1)$ denominator degrees of freedom, respectively. The shape of the F-distribution depends on the degrees of freedom associated with s_1^2 and s_2^2—that is, on $(n_1 - 1)$ and $(n_2 - 1)$. An F-distribution with 7 and 9 df is shown in Figure 7.17. As you can see, the distribution is skewed to the right because s_1^2/s_2^2 cannot be less than 0 but can increase without bound.

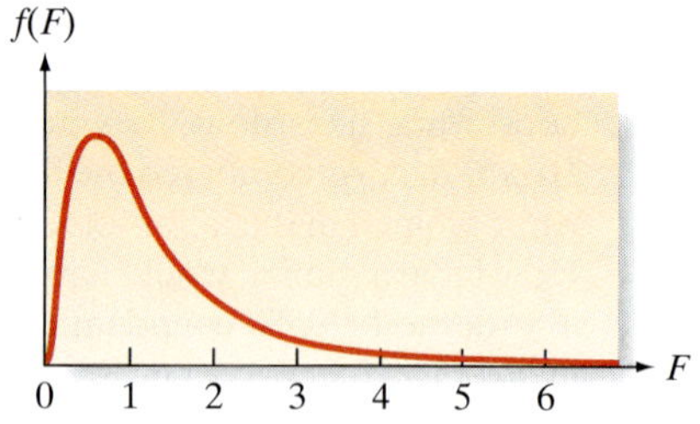

Figure 7.17
An F-distribution with 7 numerator and 9 denominator degrees of freedom

We need to be able to find F values corresponding to the tail areas of this distribution to establish the rejection region for our test of hypothesis because we expect the ratio F of the sample variances to be either very large or very small when the population variances are unequal. The upper-tail F values for $\alpha = .10, .05, .025$, and $.01$ can be found in Tables VII, VIII, IX, and X in Appendix B. Table VIII is partially reproduced in Table 7.7. It gives F values that correspond to $\alpha = .05$ upper-tail areas for different degrees of freedom ν_1 for the numerator sample variance, s_1^2, whereas the rows correspond to the degrees of freedom ν_2 for the denominator sample variance, s_2^2. Thus, if the numerator degrees of freedom is $\nu_1 = 7$ and the denominator degrees of freedom is $\nu_2 = 9$, we look in the seventh column and ninth row to find $F_{.05} = 3.29$. As shown in Figure 7.18, $\alpha = .05$ is the tail area to the right of 3.29 in the F-distribution with 7 and 9 df—that is, if $\sigma_1^2 = \sigma_2^2$, then the probability that the F-statistic will exceed 3.29 is $\alpha = .05$.

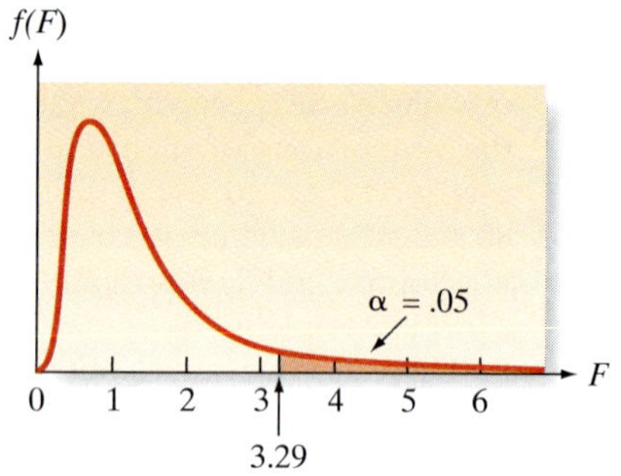

Figure 7.18
An F-distribution for $\nu_1 = 7$ df and $\nu_2 = 9$ df; $\alpha = .05$

BIOGRAPHY **GEORGE W. SNEDECOR (1882–1974)**

Snedecor's F-test

George W. Snedecor's education began at the University of Alabama, where he obtained his bachelor's degree in mathematics and physics. He went on to the University of Michigan for his master's degree in physics and finally earned his PhD in mathematics at the University of Kentucky. Snedecor learned of an opening for an assistant professor of mathematics at the University of Iowa, packed his belongings in his car, and began driving to apply for the position. By mistake, he ended up in Ames, Iowa, home of Iowa State University—then an agricultural school that had no need for a mathematics teacher. Nevertheless, Snedecor stayed and founded a statistics laboratory, eventually teaching the first course in statistics at Iowa State in 1915. In 1933, Snedecor turned the statistics laboratory into the first-ever Department of Statistics in the United States. During his tenure as chair of the department, Snedecor published his landmark textbook, *Statistical Methods* (1937). The text contained the first published reference for a test of hypothesis to compare two variances. Although Snedecor named it the F-test in honor of statistician R. A. Fisher (who had developed the F-distribution a few years earlier), many researchers still refer to it as Snedecor's F-test. Now in its ninth edition, *Statistical Methods* (with William Cochran as a coauthor) continues to be one of the most frequently cited texts in the statistics field. ■

Table 7.7 Reproduction of Part of Table VIII in Appendix B: Percentage Points of the *F*-Distribution, $\alpha = .05$

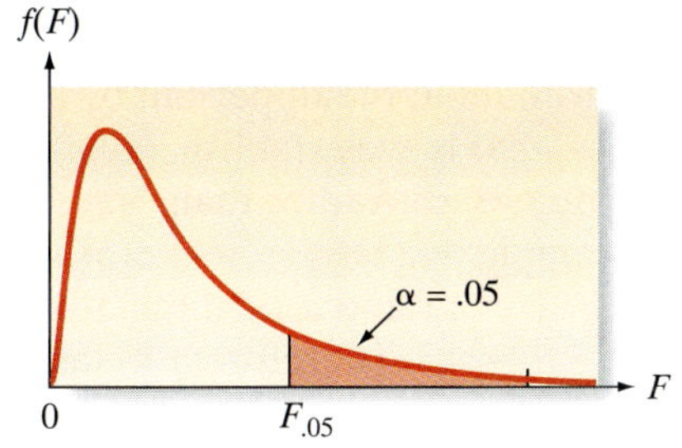

ν_2 \ ν_1 (Numerator Degrees of Freedom)	1	2	3	4	5	6	7	8	9
Denominator Degrees of Freedom									
1	161.4	199.5	215.7	224.6	230.2	234.0	236.8	238.9	240.5
2	18.51	19.00	19.16	19.25	19.30	19.33	19.35	19.37	19.38
3	10.13	9.55	9.28	9.12	9.01	8.94	8.89	8.85	8.81
4	7.71	6.94	6.59	6.39	6.26	6.16	6.09	6.04	6.00
5	6.61	5.79	5.41	5.19	5.05	4.95	4.88	4.82	4.77
6	5.99	5.14	4.76	4.53	4.39	4.28	4.21	4.15	4.10
7	5.59	4.74	4.35	4.12	3.97	3.87	3.79	3.73	3.68
8	5.32	4.46	4.07	3.84	3.69	3.58	3.50	3.44	3.39
9	5.12	4.26	3.86	3.63	3.48	3.37	3.29	3.23	3.18
10	4.96	4.10	3.71	3.48	3.33	3.22	3.14	3.07	3.02
11	4.84	3.98	3.59	3.36	3.20	3.09	3.01	2.95	2.90
12	4.75	3.89	3.49	3.25	3.11	3.00	2.91	2.85	2.80
13	4.67	3.81	3.41	3.18	3.03	2.92	2.83	2.77	2.71
14	4.60	3.74	3.34	3.11	2.96	2.85	2.76	2.70	2.65

Example 7.10
Comparing Paper Mill Production Variation Using an *F*-Test

Problem A manufacturer of paper products wants to compare the variation in daily production levels at two paper mills. Independent random samples of days are selected from each mill, and the production levels (in units) are recorded. The data are shown in Table 7.8. Do these data provide sufficient evidence to indicate a difference in the variability of production levels at the two paper mills? (Use $\alpha = .10$.)

Solution Let

$$\sigma_1^2 = \text{Population variance of production levels at mill 1}$$
$$\sigma_2^2 = \text{Population variance of production levels at mill 2}$$

The hypotheses of interest, then, are

$$H_0: \frac{\sigma_1^2}{\sigma_2^2} = 1 \quad (\sigma_1^2 = \sigma_2^2)$$

$$H_a: \frac{\sigma_1^2}{\sigma_2^2} \neq 1 \quad (\sigma_1^2 \neq \sigma_2^2)$$

Table 7.8 Production Levels at Two Paper Mills

Mill 1:	34	18	28	21	40	23	29		
	25	10	38	32	22	22			
Mill 2:	31	13	27	19	22	18	23	22	21
	18	15	24	13	19	18	19	23	13

Data Set: PAPERMILLS

The nature of the F-tables given in Appendix B affects the form of the test statistic. To form the rejection region for a two-tailed F-test, we want to make certain that the upper tail is used because only the upper-tail values of F are shown in Tables VII, VIII, IX, and X. To accomplish this, *we will always place the larger sample variance in the numerator of the* F*-test statistic.* This doubles the tabulated value for α because we double the probability that the F-ratio will fall in the upper tail by always placing the larger sample variance in the numerator—that is, we establish a one-tailed rejection region by putting the larger variance in the numerator rather than establishing rejection regions in both tails.

Thus, for our example, we have a numerator s_1^2 with df $= \nu_1 = n_1 - 1 = 12$ and a denominator s_2^2 with df $= \nu_2 = n_2 - 1 = 17$. Therefore, the test statistic will be

$$F = \frac{\text{Larger sample variance}}{\text{Smaller sample variance}} = \frac{s_1^2}{s_2^2}$$

and we will reject H_0: $\sigma_1^2 = \sigma_2^2$ for $\alpha = .10$ when the calculated value of F exceeds the tabulated value:

$$F_{\alpha/2} = F_{.05} = 2.38 \qquad \text{(see Figure 7.19)}$$

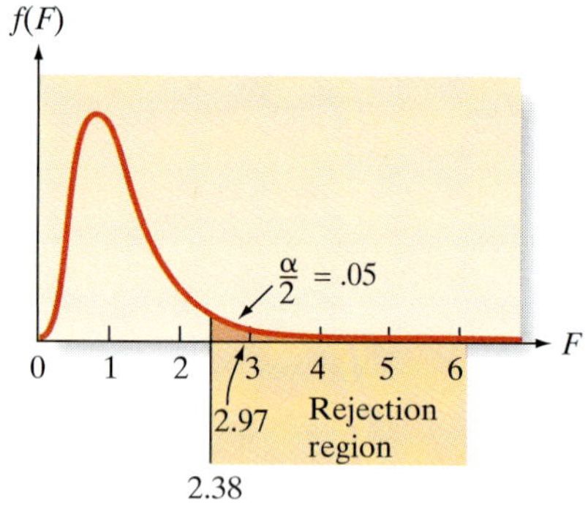

Figure 7.19
Rejection region for Example 7.10

To calculate the value of the test statistic, we require the sample variances. Summary statistics for the data in Table 7.9 are shown on the Minitab printout, Figure 7.20. The sample standard deviations (shaded) are $s_1 = 8.36$ and $s_2 = 4.85$. Therefore,

$$F = \frac{s_1^2}{s_2^2} = \frac{(8.36)^2}{(4.85)^2} = 2.97$$

When we compare this result to the rejection region shown in Figure 7.19, we see that $F = 2.97$ falls in the rejection region. Therefore, the data provide sufficient evidence to indicate that the population variances differ. It appears that the variation in production levels at mill 1 tends to be higher than the variation at mill 2.

Look Back What would you have concluded if the value of F calculated from the samples had not fallen in the rejection region? Would you conclude that the null hypothesis of equal variances is true? No, because then you risk the possibility of a Type II error (accepting H_0 if H_a is true) without knowing the value of β, the probability of accepting H_0: $\sigma_1^2 = \sigma_2^2$ if, in fact, it is false. Because we will not consider the calculation of β for specific alternatives in this text, when the F-statistic does not fall in the rejection region, we simply conclude that insufficient sample evidence exists to refute the null hypothesis that $\sigma_1^2 = \sigma_2^2$.

Descriptive Statistics: LEVEL

Variable	MILL	N	Mean	StDev	Minimum	Median	Maximum
LEVEL	1	13	26.31	8.36	10.00	25.00	40.00
	2	18	19.89	4.85	13.00	19.00	31.00

Figure 7.20
Minitab summary statistics for data in Table 7.8

Now Work Exercise 7.76a

The F-test for equal population variances is summarized in the following boxes.*

F-Test for Equal Population Variances

One-Tailed Test	Two-Tailed Test
$H_0: \sigma_1^2 = \sigma_2^2$	$H_0: \sigma_1^2 = \sigma_2^2$
$H_a: \sigma_1^2 < \sigma_2^2$ (or $H_a: \sigma_1^2 > \sigma_2^2$)	$H_a: \sigma_1^2 \neq \sigma_2^2$
Test statistic:	*Test statistic:*
$F = \frac{s_2^2}{s_1^2}$	$F = \frac{\text{Larger sample variance}}{\text{Smaller sample variance}}$
$\left(\text{or } F = \frac{s_1^2}{s_2^2} \text{ when } H_a: \sigma_1^2 > \sigma_2^2\right)$	$= \frac{s_1^2}{s_2^2} \text{ when } s_1^2 > s_2^2$ $\left(\text{or } \frac{s_2^2}{s_1^2} \text{ when } s_2^2 > s_1^2\right)$
Rejection region:	*Rejection region:*
$F > F_\alpha$	$F > F_{\alpha/2}$

where F_α and $F_{\alpha/2}$ are based on ν_1 = numerator degrees of freedom and ν_2 = denominator degrees of freedom; ν_1 and ν_2 are the degrees of freedom for the numerator and denominator sample variances, respectively.

Conditions Required for a Valid F-Test for Equal Variances

1. Both sampled populations are normally distributed.
2. The samples are random and independent.

Example 7.11 The Observed Significance Level of an F-Test

Problem Find the p-value for the test in Example 7.10 using the F-tables in Appendix B. Compare this to the exact p-value obtained from a computer printout.

Solution Because the observed value of the F-statistic in Example 7.10 was 2.97, the observed significance level of the test would equal the probability of observing a value of F at least as contradictory to $H_0: \sigma_1^2 = \sigma_2^2$ as $F = 2.97$, if, in fact, H_0 is true. Because we give the F-tables in Appendix B only for values of α equal to .10, .05, .025, and .01, we can only approximate the observed significance level. Checking Tables IX and X, we find $F_{.025} = 2.82$ and $F_{.01} = 3.46$. Because the observed value of F exceeds $F_{.025}$ but is less than $F_{.01}$, the observed significance level for the test is less than $2(.025) = .05$ but greater than $2(.01) = .02$—that is,

$$.02 < p\text{-value} < .05$$

The exact p-value of the test is shown on the Minitab printout, Figure 7.21. This value (highlighted) is .04.

Look Back We double the α value shown in Tables IX and X because this is a two-tailed test.

*Although a test of a hypothesis of equality of variances is the most common application of the F-test, it can also be used to test a hypothesis that the ratio between the population variances is equal to some specified value: $H_0: \sigma_1^2/\sigma_2^2 = k$. The test is conducted in exactly the same way as specified in the box, except that we use the test statistic

$$F = \frac{s_1^2}{s_2^2}\left(\frac{1}{k}\right)$$

Test for Equal Variances: LEVEL versus MILL

```
95% Bonferroni confidence intervals for standard deviations

MILL   N     Lower     StDev     Upper
   1  13   5.73182   8.36047   14.9507
   2  18   3.49950   4.84936    7.7455


F-Test (normal distribution)
Test statistic = 2.97, p-value = 0.040


Levene's Test (any continuous distribution)
Test statistic = 3.78, p-value = 0.062
```

Figure 7.21
Minitab F-test for equal variances

Now Work Exercise 7.76b

As a final example of an application, consider the comparison of population variances as a check of the assumption $\sigma_1^2 = \sigma_2^2$ needed for the two-sample t-test. Rejection of the null hypothesis $\sigma_1^2 = \sigma_2^2$ would indicate that the assumption is invalid. [*Note:* Nonrejection of the null hypothesis does *not* imply that the assumption is valid.] We illustrate with an example.

Example 7.12
Checking the Assumption of Equal Variances

Problem In Example 7.4 (Section 7.2) we used the two-sample t-statistic to compare the success indexes of two groups of managers. The data are repeated in Table 7.9 for convenience. The use of the t-statistic was based on the assumption that the population variances of the managerial success indexes were equal for the two groups. Conduct a test of hypothesis to check this assumption at $\alpha = .10$.

Table 7.9 Managerial Success Indexes for Two Groups of Managers

Group 1						Group 2					
Interaction with Outsiders						Few Interactions					
65	58	78	60	68	69	62	53	36	34	56	50
66	70	53	71	63	63	42	57	46	68	48	42
						52	53	43			

Data Set: MANSUCCESS

Solution We want to test

$$H_0: \sigma_1^2 = \sigma_2^2$$
$$H_a: \sigma_1^2 \neq \sigma_2^2$$

This F-test is shown on the Excel/DDXL printout, Figure 7.22. Both the test statistic, $F = .5$, and two-tailed p-value, p-value $= .2554$, are highlighted on the printout. Because $\alpha = .10$ is less than the p-value, we do not reject the null hypothesis that the population variances of the success indexes are equal. It is here that the temptation to misuse the F-test is strongest. *We cannot conclude that the data justify the use of the* t*-statistic.* This is equivalent to accepting H_0, and we have repeatedly warned against this conclusion because the probability of a Type II error, β, is unknown. The α level of .10 protects us only against rejecting H_0 if it is true. This use of the F-test may prevent us from abusing the t procedure when we obtain a value of F that leads to a rejection of the assumption that $\sigma_1^2 = \sigma_2^2$. But when the F-statistic does not fall in the rejection region, we know little more about the validity of the assumption than before we conducted the test.

Summary: SUCCESS by GROUP

Group	Count	Mean	Median	Std Dev
1	12	65.333	65.5	6.61
2	15	49.467	50	9.334

Test Results for Test of GROUP1 vs. GROUP2

Test Summary

Ho:	SD1 - SD2 =	0
Ha:	2-tailed: SD1 - SD2 ≠	0
df1:		11
df2:		14
F Statistic:		0.5
p-value:		0.2554

Test Results

Conclusion

Fail to reject Ho at alpha = 0.10

F Confidence Interval for the Ratio of Standard Deviations

Confidence Level	Lower Conf. Limit	Stan. Dev. Ratio	Upper Conf. Limit
0.90	0.442	0.708	1.172

Figure 7.22
Excel/DDXL analysis for testing assumption of equal variances

Look Back A 90% confidence interval for the ratio σ_1/σ_2 is shown at the bottom of the Excel/DDXL printont of Figure 7.22. Note that the interval (.442, 1.172) includes 1; hence, we cannot conclude that the ratio differs from 1. Thus, the confidence interval leads to the same conclusion as the two-tailed test does: There is insufficient evidence of a difference between the population variances.

Now Work Exercise 7.82

What Do You Do If the Assumption of Normal Population Distributions Is Not Satisfied?

Answer: The F-test is much less robust (i.e., much more sensitive) to departures from normality than the t-test for comparing the population means (Section 7.2). If you have doubts about the normality of the population frequency distributions, use a **nonparametric method** (e.g., *Levene's test*) for comparing the two population variances. This method can be found in the nonparametric statistics texts listed in the references for Chapter 14.

Exercises 7.71–7.85

Learning the Mechanics

7.71 NW Use Tables VII, VIII, IX, and X in Appendix B to find each of the following F-values:

a. $F_{.05}$ where $\nu_1 = 9$ and $\nu_2 = 6$
b. $F_{.01}$ where $\nu_1 = 18$ and $\nu_2 = 14$
c. $F_{.025}$ where $\nu_1 = 11$ and $\nu_2 = 4$
d. $F_{.10}$ where $\nu_1 = 20$ and $\nu_2 = 5$

7.72 Given ν_1 and ν_2, find the following probabilities:

a. $\nu_1 = 2, \nu_2 = 30, P(F \geq 5.39)$
b. $\nu_1 = 24, \nu_2 = 10, P(F < 2.74)$
c. $\nu_1 = 7, \nu_2 = 1, P(F \leq 236.8)$
d. $\nu_1 = 40, \nu_2 = 40, P(F > 2.11)$

7.73 For each of the following cases, identify the rejection region that should be used to test H_0: $\sigma_1^2 = \sigma_2^2$ against H_a: $\sigma_1^2 > \sigma_2^2$. Assume $\nu_1 = 30$ and $\nu_2 = 20$.

a. $\alpha = .10$ **b.** $\alpha = .05$
c. $\alpha = .025$ **d.** $\alpha = .01$

7.74 For each of the following cases, identify the rejection region that should be used to test H_0: $\sigma_1^2 = \sigma_2^2$ against H_a: $\sigma_1^2 \neq \sigma_2^2$. Assume $\nu_1 = 10$ and $\nu_2 = 12$.

a. $\alpha = .20$ **b.** $\alpha = .10$
c. $\alpha = .05$ **d.** $\alpha = .02$

7.75 Specify the appropriate rejection region for testing H_0: $\sigma_1^2 = \sigma_2^2$ in each of the following situations:

a. H_a: $\sigma_1^2 > \sigma_2^2$; $\alpha = .05$, $n_1 = 25$, $n_2 = 20$
b. H_a: $\sigma_1^2 < \sigma_2^2$; $\alpha = .05$, $n_1 = 10$, $n_2 = 15$
c. H_a: $\sigma_1^2 \neq \sigma_2^2$; $\alpha = .10$, $n_1 = 21$, $n_2 = 31$
d. H_a: $\sigma_1^2 < \sigma_2^2$; $\alpha = .01$, $n_1 = 31$, $n_2 = 41$
e. H_a: $\sigma_1^2 \neq \sigma_2^2$; $\alpha = .05$, $n_1 = 7$, $n_2 = 16$

7.76 NW Independent random samples were selected from each of two normally distributed populations, $n_1 = 12$ from population 1 and $n_2 = 27$ from population 2. The means and variances for the two samples are shown in the table.

Sample 1	Sample 2
$n_1 = 12$	$n_2 = 27$
$\bar{x}_1 = 31.7$	$\bar{x}_2 = 37.4$
$s_1^2 = 3.87$	$s_2^2 = 8.75$

a. Test the null hypothesis H_0: $\sigma_1^2 = \sigma_2^2$ against the alternative hypothesis H_a: $\sigma_1^2 \neq \sigma_2^2$. Use $\alpha = .10$.
b. Find and interpret the approximate p-value of the test.

7.77 Independent random samples were selected from each of two normally distributed populations, $n_1 = 6$ from population 1 and $n_2 = 5$ from population 2. The data are shown in the next table and saved in the **LM7_77** file.

Sample 1	Sample 2
3.1	2.3
4.4	1.4
1.2	3.7
1.7	8.9
.7	5.5
3.4	

a. Test H_0: $\sigma_1^2 = \sigma_2^2$ against H_a: $\sigma_1^2 < \sigma_2^2$. Use $\alpha = .01$.
b. Find and interpret the approximate p-value of the test.

Applying the Concepts—Basic

7.78 Children's recall of TV ads. Refer to the *Journal of Advertising* (Spring 2006) study of children's recall of television commercials, Exercise 7.13 (p. 394). You used a small-sample t-test to test the null hypothesis H_0: $(\mu_1 - \mu_2) = 0$, where μ_1 = mean number of ads recalled by children in the video only group and μ_1 = mean number of ads recalled by children in the A/V group. Summary statistics for the study are reproduced in the table in the next column. The validity of the inference derived from the test is based on the assumption of equal group variances, i.e., $\sigma_1^2 = \sigma_2^2$.

a. Set up the null and alternative hypotheses for testing this assumption.
b. Compute the test statistic.
c. Find the rejection region for the test using $\alpha = .10$.
d. Make the appropriate conclusion in the words of the problem.
e. Comment on the validity of the inference derived about the difference in population means in Exercise 7.13.

Video Only Group	A/V Group
$n_1 = 20$	$n_2 = 20$
$\bar{x}_1 = 3.70$	$\bar{x}_2 = 3.30$
$s_1 = 1.98$	$s_2 = 2.13$

7.79 Rating service at five-star hotels. Refer to The *Journal of American Academy of Business, Cambridge* (March 2002) study of how guests perceive the service quality at five-star hotels in Jamaica, Exercise 7.14 (p. 395). A summary of the guest perception scores, by gender, are reproduced in the accompanying table. Let σ_M^2 and σ_F^2 represent the true variances in scores for male and female guests, respectively.

Gender	Sample Size	Mean Score	Standard Deviation
Males	127	39.08	6.73
Females	114	38.79	6.94

a. Set up H_0 and H_a for determining whether σ_M^2 is less than σ_F^2.
b. Find the test statistic for the test.
c. Give the rejection region for the test if $\alpha = .10$.
d. Find the approximate p-value of the test.
e. Make the appropriate conclusion in the words of the problem.
f. What conditions are required for the test results to be valid?

7.80 Mental health of workers and the unemployed. A study in the *Journal of Occupational and Organizational Psychology* investigated the relationship of employment status and mental health. A sample of working and unemployed people was selected, and each person was given a mental health examination using the General Health Questionnaire (GHQ), a widely recognized measure of mental health. Although the article focused on comparing the mean GHQ levels, a comparison of the variability of GHQ scores for employed and unemployed men and women is of interest as well.

a. In general terms, what does the amount of variability in GHQ scores tell us about the group?
b. What are the appropriate null and alternative hypotheses to compare the variability of the mental health scores of the employed and unemployed groups? Define any symbols you use.
c. The standard deviation for a sample of 142 employed men was 3.26, while the standard deviation for 49 unemployed men was 5.10. Conduct the test you set up in part **b** using $\alpha = .05$. Interpret the results.
d. What assumptions are necessary to ensure the validity of the test?

Applying the Concepts—Intermediate

7.81 Analyzing human inspection errors. Tests of product quality using human inspectors can lead to serious inspection error problems (*Journal of Quality Technology*). To evaluate the performance of inspectors in a new company, a quality manager had a sample of 12 novice inspectors evaluate

200 finished products. The same 200 items were evaluated by 12 experienced inspectors. The quality of each item—whether defective or nondefective—was known to the manager. The table lists the number of inspection errors (classifying a defective item as nondefective or vice versa) made by each inspector. These data are saved in the **INSPECT** file.

Novice Inspectors				Experienced Inspectors			
30	35	26	40	31	15	25	19
36	20	45	31	28	17	19	18
33	29	21	48	24	10	20	21

a. Prior to conducting this experiment, the manager believed the variance in inspection errors was lower for experienced inspectors than for novice inspectors. Do the sample data support her belief? Test using $\alpha = .05$.

b. What is the appropriate p-value of the test you conducted in part **a**?

7.82 Patent infringement case. Refer to the *Chance* (Fall 2002) description of a patent infringement case against Intel Corp., Exercise 7.17 (pp. 395–396). The zinc measurements for three locations on the original inventor's notebook—on a text line, on a witness line, and on the intersection of the witness and text line—are reproduced in the table and saved in the **PATENT** file.

Text Line	.335	.374	.440			
Witness Line	.210	.262	.188	.329	.439	.397
Intersection	.393	.353	.285	.295	.319	

a. Use a test (at $\alpha = .05$) to compare the variation in zinc measurements for the text line with the corresponding variation for the intersection.

b. Use a test (at $\alpha = .05$) to compare the variation in zinc measurements for the witness line with the corresponding variation for the intersection.

c. From the results, parts **a** and **b**, what can you infer about the variation in zinc measurements at the three notebook locations?

d. What assumptions are required for the inferences to be valid? Are they reasonably satisfied? (You checked these assumptions when answering Exercise 7.17d.)

7.83 "Just-in-time" deliveries from factories to foxholes. Following the initial Persian Gulf War, the Pentagon changed its logistics processes to be more corporate-like. The extravagant "just-in-case" mentality was replaced with "just-in-time" systems. Emulating FedEx and United Parcel Service, deliveries from factories to foxholes are now expedited using bar codes, laser cards, radio tags, and databases to track supplies. The table in the next column contains order-to-delivery times (in days) for a sample of shipments from the United States to the Persian Gulf and a sample of shipments to Bosnia. These data are saved in the **ORDTIMES** file.

Persian Gulf	Bosnia
28.0	15.1
20.0	6.4
26.5	5.0
10.6	11.4
9.1	6.5
35.2	6.5
29.1	3.0
41.2	7.0
27.5	5.5

Source: Adapted from Crock, S. "The Pentagon goes to B-school," *Business Week,* December 11, 1995, p. 98.

a. Is there sufficient evidence to indicate that the variances in order-to-delivery times for Persian Gulf and Bosnia shipments differ? Use $\alpha = .05$.

b. Given your answer to part **a**, is it appropriate to construct a paired-difference confidence interval for the difference between the mean order-to-delivery times? Explain.

7.84 NW **Cooling method for gas turbines.** Refer to the *Journal of Engineering for Gas Turbines and Power* (Jan. 2005) study of gas turbines augmented with high-pressure inlet fogging, Exercise 7.23 (p. 397). Heat rate data (kilojoules per kilowatt per hour) for each of three types of gas turbines (advanced, aeroderivative, traditional) are saved in the **GASTURBINE** file. In order to compare the mean heat rates of two types of gas turbines, you assumed that the heat rate variances were equal.

a. Conduct a test (at $\alpha = .05$) for equality of heat rate variances for traditional and aeroderivative augmented gas turbines. Use the result to make a statement about the validity of the inference derived in Exercise 7.23a.

b. Conduct a test (at $\alpha = .05$) for equality of heat rate variances for advanced and aeroderivative augmented gas turbines. Use the result to make a statement about the validity of the inference derived in Exercise 7.23b.

7.85 Is honey a cough remedy? Refer to the *Archives of Pediatrics and Adolescent Medicine* (Dec. 2007) study of honey as a children's cough remedy, Exercise 7.21 (p. 397). The data (cough improvement scores) for the 35 children in the DM dosage group and the 35 children in the honey dosage group (saved in the **HONEYCOUGH** file) are reproduced in the table below. In Exercise 7.21, you used a comparison of two means to determine whether "honey may be a preferable treatment for the cough and sleep difficulty associated with childhood upper respiratory tract infection." The researchers also want to know if the variability in coughing improvement scores differs for the two groups. Conduct the appropriate analysis, using $\alpha = .10$.

Honey Dosage:	12	11	15	11	10	13	10	4	15	16	9	14	10	6	10	11	12
	12	9	11	15	10	9	13	8	12	10	8	9	5	12	8	12	8
DM Dosage:	4	6	9	4	7	7	7	9	12	10	11	3	9	7	8	12	12
	13	7	10	13	9	4	4	10	15	9	6	4	12	6	12	4	

Source: Paul, I. M., et al. "Effect of honey, dextromethorphan, and no treatment on nocturnal cough and sleep quality for coughing children and their parents," *Archives of Pediatrics and Adolescent Medicine,* Vol. 161, No. 12, Dec. 2007 (data simulated).

CHAPTER NOTES

Key Terms

Blocking 400
F-distribution 420
Nonparametric method 425
Paired difference experiment 400
Pooled sample estimator of σ^2 387
Randomized block experiment 400
Standard error 384

Key Symbols

$\mu_1 - \mu_2$	Difference between population means
μ_d	Paired difference in population means
$p_1 - p_2$	Difference between population proportions
σ_1^2/σ_2^2	Ratio of population variances
D_0	Hypothesized value of difference
$\bar{x}_1 - \bar{x}_2$	Difference between sample means
$\bar{d}$	Mean of sample differences
$\hat{p}_1 - \hat{p}_2$	Difference between sample proportions
s_1^2/s_2^2	Ratio of sample variances
$\sigma_{(\bar{x}_1 - \bar{x}_2)}$	Standard error for $\bar{x}_1 - \bar{x}_2$
$\sigma_{\bar{d}}$	Standard error for $\bar{d}$
$\sigma_{(\hat{p}_1 - \hat{p}_2)}$	Standard error tor $\hat{p}_1 - \hat{p}_2$
F_α	Critical value for F-distribution
ν_1	Numerator degrees of freedom for F-distribution
ν_2	Denominator degrees of freedom for F-distribution
ME	Margin of error in estimation

Key Ideas

Key Words for Identifying the Target Parameter

$\mu_1 - \mu_2$	Difference in means or averages
μ_d	Paired difference in means or averages
$p_1 - p_2$	Difference in proportions, fractions, percentages, rates
σ_1^2/σ_2^2	Ratio (or difference) in variances, spreads

Determining the Sample Size

Estimating $\mu_1 - \mu_2$: $n_1 = n_2 = (z_{\alpha/2})^2(\sigma_1^2 + \sigma_2^2)/(\text{ME})^2$
Estimating $p_1 - p_2$: $n_1 = n_2 = (z_{\alpha/2})^2(p_1q_1 + p_2q_2)/(\text{ME})^2$

Conditions Required for Inferences about $\mu_1 - \mu_2$

Large samples:
1. Independent random samples
2. $n_1 \geq 30$, $n_2 \geq 30$

Small samples:
1. Independent random samples
2. Both populations normal
3. $\sigma_1^2 = \sigma_2^2$

Conditions Required for Inferences about σ_1^2/σ_2^2

Large or small samples:
1. Independent random samples
2. Both populations normal

Conditions Required for Inferences about μ_d

Large samples:
1. Random sample of paired differences
2. $n_d \geq 30$

Small samples:
1. Random sample of paired differences
2. Population of differences is normal

Conditions Required for Inferences about $p_1 - p_2$

Large samples:
1. Independent random samples
2. $n_1p_1 \geq 15$, $n_1q_1 \geq 15$
3. $n_2p_2 \geq 15$, $n_2q_2 \geq 15$

Using a Confidence Interval for $(\mu_1 - \mu_2)$ or $(p_1 - p_2)$ to Determine whether a Difference Exists

1. If the confidence interval includes all *positive* numbers $(+, +)$: $\rightarrow$ Infer $\mu_1 > \mu_2$ or $p_1 > p_2$
2. If the confidence interval includes all *negative* numbers $(-, -)$: $\rightarrow$ Infer $\mu_1 < \mu_2$ or $p_1 < p_2$
3. If the confidence interval includes *0* $(-, +)$: $\rightarrow$ Infer no evidence of a difference

Guide to Selecting a Two-Sample Hypothesis Test and Confidence Interval

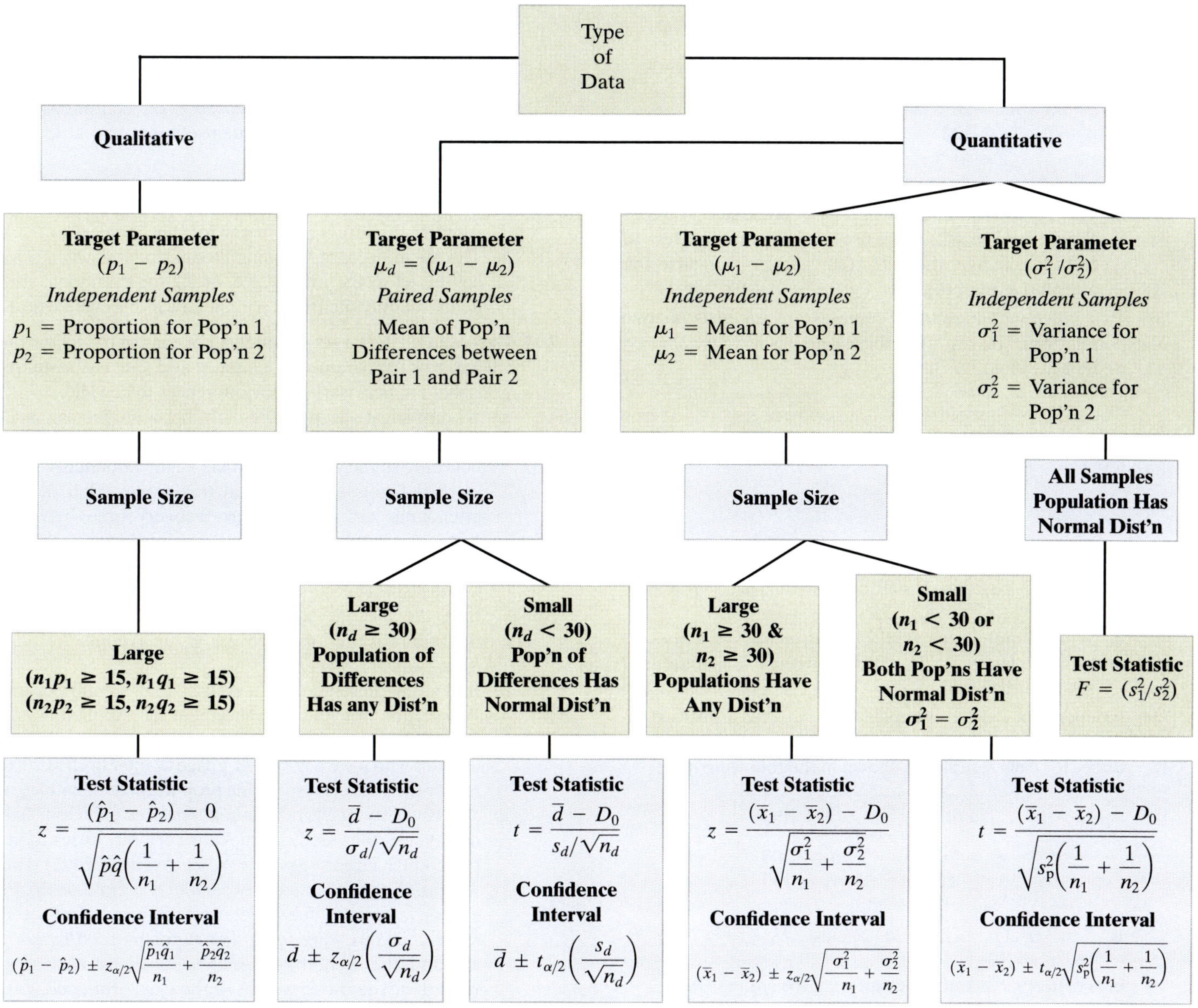

Supplementary Exercises 7.86–7.117

Learning the Mechanics

7.86 List the assumptions necessary for each of the following inferential techniques:

a. Large-sample inferences about the difference $(\mu_1 - \mu_2)$ between population means using a two-sample z-statistic

b. Small-sample inferences about $(\mu_1 - \mu_2)$ using an independent samples design and a two-sample t-statistic

c. Small-sample inferences about $(\mu_1 - \mu_2)$ using a paired difference design and a single-sample t-statistic to analyze the differences

d. Large-sample inferences about the differences $(p_1 - p_2)$ between binomial proportions using a two-sample z-statistic

e. Inferences about the ratio σ_1^2/σ_2^2 of two population variances using an F-test.

7.87 Independent random samples were selected from two normally distributed populations with means μ_1 and μ_2, respectively. The sample sizes, means, and variances are shown in the following table.

Sample 1	Sample 2
$n_1 = 12$	$n_2 = 14$
$\bar{x}_1 = 17.8$	$\bar{x}_2 = 15.3$
$s_1^2 = 74.2$	$s_2^2 = 60.5$

a. Test $H_0: (\mu_1 - \mu_2) = 0$ against $H_a: (\mu_1 - \mu_2) > 0$. Use $\alpha = .05$.

b. Form a 99% confidence interval for $(\mu_1 - \mu_2)$.

c. How large must n_1 and n_2 be if you wish to estimate $(\mu_1 - \mu_2)$ to within 2 units with 99% confidence? Assume that $n_1 = n_2$.

7.88 Two independent random samples were selected from normally distributed populations with means and variances (μ_1, σ_1^2) and (μ_2, σ_2^2), respectively. The sample sizes, means, and variances are shown in the following table.

Sample 1	Sample 2
$n_1 = 20$	$n_2 = 15$
$\bar{x}_1 = 123$	$\bar{x}_2 = 116$
$s_1^2 = 31.3$	$s_2^2 = 120.1$

a. Test $H_0: \sigma_1^2 = \sigma_2^2$ against $H_a: \sigma_1^2 \neq \sigma_2^2$. Use $\alpha = .05$.

b. Would you be willing to use a small sample t-test to test the null hypothesis $H_0: (\mu_1 - \mu_2) = 0$ against the alternative hypothesis $H_a: (\mu_1 - \mu_2) \neq 0$? Why?

7.89 Two independent random samples are taken from two populations. The results of these samples are summarized in the next table.

Sample 1	Sample 2
$n_1 = 135$	$n_2 = 148$
$\bar{x}_1 = 12.2$	$\bar{x}_2 = 8.3$
$s_1^2 = 2.1$	$s_2^2 = 3.0$

a. Form a 90% confidence interval for $(\mu_1 - \mu_2)$.

b. Test $H_0: (\mu_1 - \mu_2) = 0$ against $H_a: (\mu_1 - \mu_2) \neq 0$. Use $\alpha = .01$.

c. What sample sizes would be required if you wish to estimate $(\mu_1 - \mu_2)$ to within .2 with 90% confidence? Assume that $n_1 = n_2$.

7.90 Independent random samples were selected from two binomial populations. The sizes and number of observed successes for each sample are shown in the following table.

Sample 1	Sample 2
$n_1 = 200$	$n_2 = 200$
$x_1 = 110$	$x_2 = 130$

a. Test $H_0: (p_1 - p_2) = 0$ against $H_a: (p_1 - p_2) < 0$. Use $\alpha = .10$.

b. Form a 95% confidence interval for $(p_1 - p_2)$.

c. What sample sizes would be required if we wish to use a 95% confidence interval of width .01 to estimate $(p_1 - p_2)$?

7.91 A random sample of five pairs of observations were selected, one of each pair from a population with mean μ_1, the other from a population with mean μ_2. The data (saved in the **LM7_91** file) are shown in the accompanying table.

Pair	Value from Population 1	Value from Population 2
1	28	22
2	31	27
3	24	20
4	30	27
5	22	20

a. Test the null hypothesis $H_0: \mu_d = 0$ against $H_a: \mu_d \neq 0$, where $\mu_d = \mu_1 - \mu_2$. Use $\alpha = .05$.

b. Form a 95% confidence interval for μ_d.

c. When are the procedures you used in parts **a** and **b** valid?

Applying the Concepts—Basic

7.92 Financial incentives for college students. A study reported in *Inside Higher Education News* (May 22, 2006) found that financial incentives can improve low-income college students' grades and retention. As part of their "Opening Doors" program, a Louisiana community college offered to pay students $1,000 per semester on condition that they maintain at least half-time enrollment and at least a 2.0 grade-point average (GPA).

a. About 61% of Opening Doors students enrolled full time as opposed to about 52% of traditional students. Identify the target parameter for this comparison.

b. The mean GPA of Opening Doors students was 2.3 as compared to the mean GPA of 2.1 for traditional students. Identify the target parameter for this comparison.

7.93 Determining the target parameter. For each of the following studies, give the parameter of interest and state any assumptions that are necessary for the inferences to be valid.

a. To investigate a possible link between jet lag and memory impairment, a University of Bristol (England) neurologist recruited 20 female flight attendants who worked flights across several time zones. Half of the attendants had only a short recovery time between flights and half had a long recovery time between flights. The average size of the right temporal lobe of the brain for the short-recovery group was significantly smaller than the average size of the right temporal lobe of the brain for the long-recovery group (*Tampa Tribune*, May 23, 2001).

b. In a study presented at the March 2001 meeting of the Association for the Advancement of Applied Sport Psychology, researchers revealed that the proportion of athletes who have a good self-image of their body is 20% higher than the corresponding proportion of nonathletes.

c. A University of Florida animal sciences professor has discovered that feeding chickens corn oil causes them to produce larger eggs (*UF News*, April 11, 2001). The weight of eggs produced by each of a sample of chickens on a regular feed diet was recorded. Then, the same chickens were fed a diet supplemented by corn oil, and the weight of eggs produced by each was recorded. The mean weight of the eggs produced with corn oil was 3 grams heavier than the mean weight produced with the regular diet.

7.94 Hull failures of oil tankers. Refer to the *Marine Technology* (Jan. 1995) study of major oil spills from tankers and carriers, Exercise 2.147 (p. 106). The data for the 50 recent spills are saved in the **OILSPILL** file.

a. Construct a 90% confidence interval for the difference between the mean spillage amount of accidents caused by collision and the mean spillage amount of accidents caused by fire/explosion. Interpret the result.

b. Conduct a test of hypothesis to compare the mean spillage amount of accidents caused by grounding to the corresponding mean of accidents caused by hull failure. Use $\alpha = .05$.

c. Refer to parts **a** and **b.** State any assumptions required for the inferences derived from the analyses to be valid. Are these assumptions reasonably satisfied?

d. Conduct a test of hypothesis to compare the variation in spillage amounts for accidents caused by collision and accidents caused by grounding. Use $\alpha = .02$.

7.95 Financial incentives for college students. Do students who play virtual-reality educational software games perform better in school? If so, software designers will use this information to create more attractive and motivating educational software. In a study published in *Educational Technology & Society* (April 2005), a class of 90 elementary schoolchildren were randomly divided into two groups of 45. One group used a virtual-reality game called VR-ENGAGE to learn about geography. The other group learned geography on the computer with a simple user interface. All students were given an exam at the beginning and the end of the learning period. The difference in test scores was used to find the percentage improvement in scores for each student. The following table summarizes the results for the two groups.

	VR-ENGAGE	Simple User Interface
n:	45	45
$\bar{x}$:	43.15	32.48
s:	12.57	9.26

Source: Adapted from Virvou, M., Katsionis, G., and Manos, K. "Combining software games with education: Evaluation of its educational effectiveness," *Educational Technology & Society,* Vol. 8, No. 2, April 2005 (Table 2).

a. Visually inspect the summary results. Does it appear that the mean improvement in scores for the virtual-reality group is greater than the mean for the simple user interface group?

b. Conduct an analysis (either a confidence interval or a test of hypothesis) at $\alpha = .05$ to support your observation in part **a.**

7.96 Durability of shock absorbers. A manufacturer of automobile shock absorbers was interested in comparing the durability of its shocks with that of the shocks produced by its biggest competitor. To make the comparison, one of the manufacturer's and one of the competitor's shocks were randomly selected and installed on the rear wheels of each of six cars. After the cars had been driven 20,000 miles, the strength of each test shock was measured, coded, and recorded. Results of the examination, shown in the table, are saved in the **SHOCKABS** file.

Car Number	Manufacturer's Shock	Competitor's Shock
1	8.8	8.4
2	10.5	10.1
3	12.5	12.0
4	9.7	9.3
5	9.6	9.0
6	13.2	13.0

a. Explain why the data are collected as matched pairs.

b. Do the data present sufficient evidence to conclude that there is a difference in the mean strength of the two types of shocks after 20,000 miles of use? Use $\alpha = .05$.

c. Find the approximate observed significance level for the test and interpret its value.

d. What assumptions are necessary to apply a paired difference analysis to the data?

e. Construct a 95% confidence interval for μ_d. Interpret the confidence interval.

f. Suppose the data are based on independent random samples. Construct a 95% confidence interval for $(\mu_1 - \mu_2)$. Interpret your result.

g. Compare the confidence intervals you obtained in parts **e** and **f.** Which is wider? To what do you attribute the difference in width? Assuming in each case that the appropriate assumptions are satisfied, which interval provides you with more information about $(\mu_1 - \mu_2)$? Explain.

h. Are the results of an unpaired analysis valid if the data come from a paired experiment?

7.97 Visual search and memory study. When searching for an item (e.g., a roadside traffic sign, a misplaced file, or a tumor in a mammogram), common sense dictates that you will not reexamine items previously rejected. However, researchers at Harvard Medical School found that a visual search has no memory (*Nature,* Aug. 6, 1998). In their experiment, nine subjects searched for the letter "T" mixed among several letters "L." Each subject conducted the search under two conditions: random and static. In the random condition, the location of the letters were changed every 111 milliseconds; in the static condition, the location of the letters remained unchanged. In each trial, the reaction time (i.e., the amount of time it took the subject to locate the target letter) was recorded in milliseconds.

a. One goal of the research is to compare the mean reaction times of subjects in the two experimental conditions. Explain why the data should be analyzed as a paired-difference experiment.

b. If a visual search had no memory, then the mean reaction times in the two conditions will not differ. Specify H_0 and H_a for testing the "no memory" theory.

c. The test statistic was calculated as $t = 1.52$ with p-value $= .15$. Make the appropriate conclusion.

7.98 Diamonds sold at retail. Refer to the data for 308 diamonds saved in the **DIAMONDS** file. Two quantitative variables in the data set are number of carats and selling price. One of the qualitative variables is the independent certification body that assessed each of the stones. Three certification bodies were used: GIA, IGI, and HRD. The Minitab printout shown below gives the means and standard deviations of the quantitative variables for each certification body.

Descriptive Statistics: CARAT, PRICE

Variable	CERT	N	Mean	StDev
CARAT	GIA	151	0.6723	0.2456
	HRD	79	0.8129	0.1831
	IGI	78	0.3665	0.2163
PRICE	GIA	151	5310	3247
	HRD	79	7181	2898
	IGI	78	2267	2121

a. Construct a 95% confidence interval for the difference between the mean carat size of diamonds certified by GIA and the mean carat size of diamonds certified by HRD.

b. Interpret the result, part **a.** Specifically, which (if either) of the two population means compared is larger and by how much?

c. Construct a 95% confidence interval for the difference between the mean carat size of diamonds certified by GIA and the mean carat size of diamonds certified by IGI.

d. Interpret the result, part **c.** Specifically, which (if either) of the two population means is larger and by how much?

e. Construct a 95% confidence interval for the difference between the mean selling price of diamonds certified by HRD and the mean selling price of diamonds certified by IGI.
f. Interpret the result, part **e.** Specifically, which (if either) of the two population means is larger and by how much?
g. Conduct a test to determine whether the variation in carat size differs for diamonds certified by GIA and diamonds certified by HRD. Use $\alpha = .05$.
h. Conduct a test to determine whether the variation in carat size differs for diamonds certified by GIA and diamonds certified by IGI. Use $\alpha = .05$.
i. Conduct a test to determine whether the variation in selling price differs for diamonds certified by HRD and diamonds certified by IGI. Use $\alpha = .05$.
j. Use a statistical software package (and the data in the **DIAMONDS** file) to determine whether the assumption of normally distributed data for each certification group is reasonably satisfied.

7.99 Ages of cable TV shoppers. Refer to the International Association for Time Use Research study on cable TV viewers who purchase items from one of the home shopping channels, Exercise 6.48 (p. 342). The 1,600 sampled viewers described their motivation for watching cable TV shopping networks by giving their level of agreement (on a 5-point scale, where 1 = strongly disagree and 5 = strongly agree) with the statement, "I have nothing else to do." The researcher wanted to compare the mean responses of viewers who watch the shopping network at noon with those viewers who do not watch at noon.
a. Give the null and alternative hypotheses for determining whether the mean response of noontime watchers differs from the mean response of non-noontime watchers.
b. The researcher found the p-value for the test, part **a,** to be .02. Interpret this result, assuming $\alpha = .05$.
c. Interpret the result, part **b,** assuming $\alpha = .01$.
d. The sample means for noontime watchers and non-watchers were found to be 3.3 and 3.4, respectively. Comment on the practical significance of this result.

7.100 Career success expectations. In evaluating the usefulness and validity of a questionnaire, researchers often pretest the questionnaire on different independently selected samples of respondents. Knowledge of the differences and similarities of the samples and their respective populations is important for interpreting the questionnaire's validity. *Educational and Psychological Measurement* (Feb. 1998) reported on a newly developed questionnaire for measuring the career success expectations of employees. The instrument was tested on the two independent samples described in the following table.

	Managers and Professionals	Part-Time MBA Students
Sample Size	162	109
Gender (% males)	95.0	68.9
Marital Status (% married)	91.2	53.4

Source: Stephens, G. K., Szajna, B., and Broome, K. M., "The career success expectation scale: An exploratory and confirmatory factor analysis," *Educational and Psychological Measurement,* Vol. 58, No. 1, Feb. 1998, pp. 129–141. © 1998. Reprinted by permission of SAGE Publishing.

a. Does the population of managers and professionals from which the sample was drawn consist of more males than the part-time MBA population does? Conduct the appropriate test using $\alpha = .05$.
b. Describe any assumptions you made in conducting the test of part **a** and why you made them.
c. Does the population of managers and professionals consist of more married individuals than the part-time MBA population does? Conduct the appropriate hypothesis test using $\alpha = .01$.
d. What assumptions must hold for the test of part **c** to be valid?

7.101 Ages of commercial product managers. *Industrial Marketing Management* (Vol. 25, 1996) published a study that examined the demographics, decision-making roles, and time demands of product managers. Independent samples of $n_1 = 93$ consumer/commercial product managers and $n_2 = 212$ industrial product managers took part in the study. In the consumer/commercial group, 40% of the product managers are 40 years of age or older; in the industrial group, 54% are 40 or more years old. Make an inference about the difference between the true proportions of consumer/commercial and industrial product managers who are at least 40 years old. Justify your choice of method (confidence interval or hypothesis test) and α level. Do industrial product managers tend to be older than consumer/commercial product managers?

7.102 Turnover rates in the United States and Japan. High job turnover rates are often associated with high product defect rates because high turnover rates mean more inexperienced workers who are unfamiliar with the company's product lines (Stevenson, *Production/Operations Management,* 2000). In a recent study, five Japanese and five U.S. plants that manufacture air conditioners were randomly sampled; their turnover rates are listed in the table and saved in the **TURNOVER** file.

U.S. Plants	Japanese Plants
7.11%	3.52%
6.06	2.02
8.00	4.91
6.87	3.22
4.77	1.92

a. Do the data provide sufficient evidence to indicate that the mean annual percentage turnover for U.S. plants exceeds the corresponding mean percentage for Japanese plants? Test using $\alpha = .05$.
b. Find and interpret the observed significance level of the test you conducted in part **a.**
c. List any assumptions you made in conducting the hypothesis test of part **a.** Comment on their validity for this application.

7.103 Comparing unemployment rates. An economist wants to investigate the difference in unemployment rates between an urban industrial community and a university community in the same state. She interviews 525 potential members of the workforce in the industrial community and 375 in the university community. Of these, 47 and 22, respectively, are unemployed. Use a 95% confidence interval to estimate the difference in unemployment rates in the two communities.

Applying the Concepts—Intermediate

7.104 Sampling plan for a movie promotion. Advertising companies often try to characterize the average user of a client's product so ads can be targeted at particular segments of the buying community. A new movie is about to be released, and the advertising company wants to determine whether to aim the ad campaign at people under or over 25 years of age. It plans to arrange an advance showing of the movie to an audience from each group and then obtain an opinion about the movie from each individual. How many individuals should be included in each sample if the advertising company wants to estimate the difference in the proportions of viewers in each age group who will like the movie to within .05 with 90% confidence? Assume the sample size for each group will be the same and about half of each group will like the movie.

7.105 Ingratiatory behavior toward supervisors. *Ingratiation* is defined as a class of strategic behaviors designed to make others believe in the attractiveness of one's personal qualities. In organizational settings, individuals use such behaviors to influence superiors in order to attain personal goals. An index that measures ingratiatory behavior, called the Measure of Ingratiatory Behaviors in Organizational Settings (MIBOS) Index, was applied independently to a sample of managers employed by four manufacturing companies in the southeastern United States and to clerical personnel from a large university in the northwestern United States (*Journal of Applied Psychology,* Dec. 1998). Scores are reported on a five-point scale, with higher scores indicating more extensive ingratiatory behavior. Summary statistics are shown in the table.

Managers	Clerical Personnel
$n_1 = 288$	$n_2 = 110$
$\bar{x}_1 = 2.41$	$\bar{x}_2 = 1.90$
$s_1 = .74$	$s_2 = .59$

Source: Harrison, A. W., Hochwarter, W. A., Perrewe, P. L, and Ralston, D. A. "The ingratiation construct: An assessment of the validity of the measure of ingratiatory behaviors in organization settings (MIBOS)," *Journal of Applied Psychology,* Vol. 83, No. 6, Dec. 1998, pp. 932–943. Copyright © 1998 by the American Psychological Association. Reproduced with permission. The use of APA information does not imply endorsement by APA.

a. Specify the null and alternative hypotheses you would use to test for a difference in ingratiatory behavior between managers and clerical personnel.

b. Conduct the test of part **a** using $\alpha = .05$. Interpret the results of the test in the context of the problem.

c. Construct a 95% confidence interval for $(\mu_1 - \mu_2)$ and interpret the result. Your conclusion should agree with your answer in part **b.**

7.106 Killing moths with carbon dioxide. A University of South Florida biologist conducted an experiment to determine whether increased levels of carbon dioxide kill leaf-eating moths (*USF Magazine,* Winter 1999). Moth larvae were placed in open containers filled with oak leaves. Half the containers had normal carbon dioxide levels, while the other half had double the normal level of carbon dioxide. Ten percent of the larvae in the containers with high carbon dioxide levels died, compared to 5% in the containers with normal levels. Assume that 80 moth larvae were placed, at random, in each of the two types of containers. Do the experimental results demonstrate that an increased level of carbon dioxide is effective in killing a higher percentage of leaf-eating moth larvae? Test using $\alpha = .01$.

7.107 Comparing purchasers and nonpurchasers of toothpaste. Marketing strategists would like to predict consumer response to new products and their accompanying promotional schemes. Consequently, studies that examine the differences between buyers and nonbuyers of a product are of interest. One classic study conducted by Shuchman and Riesz (*Journal of Marketing Research*) was aimed at characterizing the purchasers and nonpurchasers of Crest toothpaste. The researchers demonstrated that both the mean household size (number of persons) and mean household income were significantly larger for purchasers than for nonpurchasers. A similar study used independent random samples of size 20 and yielded the data shown in the next table (saved in the **CREST** file) on the age of the householder primarily responsible for buying toothpaste.

Purchasers						Nonpurchasers					
34	35	23	44	52	46	28	22	44	33	55	63
28	48	28	34	33	52	45	31	60	54	53	58
41	32	34	49	50	45	52	52	66	35	25	48
29	59					59	61				

a. Do the data present sufficient evidence to conclude there is a difference in the mean age of purchasers and nonpurchasers? Use $\alpha = .10$.

b. What assumptions are necessary in order to answer part **a**?

c. Find the observed significance level for the test and interpret its value.

d. Calculate and interpret a 90% confidence interval for the difference between the mean ages of purchasers and nonpurchasers.

7.108 Switching college majors. When female undergraduates switch from science, mathematics, and engineering (SME) majors into disciplines that are not science based, such as journalism, marketing, and sociology, are their reasons different from those of their male counterparts? This question was investigated in *Science Education* (July 1995). A sample of 335 junior/senior undergraduates—172 females and 163 males—at two large research universities were identified as "switchers"—that is, they left a declared SME major for a non-SME major. Each student listed one or more factors that contributed to his/her switching decision.

a. Of the 172 females in the sample, 74 listed lack or loss of interest in SME (i.e., "turned off" by science) as a major factor, compared to 72 of the 163 males. Conduct a test (at $\alpha = .10$) to determine whether the proportion of female switchers who give "lack of interest in SME" as a major reason for switching differs from the corresponding proportion of males.

b. Thirty-three of the 172 females in the sample admitted they were discouraged or lost confidence due to low grades in SME during their early years, compared to 44 of 163 males. Construct a 90% confidence interval for the difference between the proportions of female and male switchers who lost confidence due to low grades in SME. Interpret the result.

7.109 State SAT scores. Refer to Exercise 2.29 (p. 52) and the data on average SAT scores for each of the 50 states and District of Columbia for the years 2000 and 2007. The data are saved in the **SATSCORES** file. (The first five observations and last two observations in the data set are shown in the table below.)

State	2000	2007
Alabama	1114	1119
Alaska	1034	1036
Arizona	1044	1044
Arkansas	1117	1144
California	1015	1015
⋮	⋮	⋮
Wisconsin	1181	1185
Wyoming	1090	1136

Source: "College-Bound Seniors 2008." Copyright © 2008 the College Board; www.collegeboard.com. Reproduced with permission.

a. In Exercise 2.29b, you computed the *paired differences* of SAT scores by subtracting the 2000 score from the 2007 score for each state. Find the mean of these 50 paired differences. This value is μ_d, the mean difference in SAT scores for the population of 50 states and the District of Columbia.

b. Explain why there is no need to employ the confidence interval or test procedures of this section to make an inference about μ_d.

c. Now, suppose the 50 paired differences of part **a** represent a sample of SAT score differences for 50 randomly selected high school students. Use the data in the **SATSCORES** file to make an inference about whether the true mean SAT score of high school students in 2007 differs from the true mean in 2000. Use a confidence level of .90.

7.110 Environmental impact study. Some power plants are located near rivers or oceans so that the available water can be used for cooling the condensers. Suppose that, as part of an environmental impact study, a power company wants to estimate the difference in mean water temperature between the discharge of its plant and the offshore waters. How many sample measurements must be taken at each site to estimate the true difference between means to within .2°C with 95% confidence? Assume that the range in readings will be about 4°C at each site and the same number of readings will be taken at each site.

7.111 Rat damage to sugarcane fields. Poisons are used to prevent rat damage in sugarcane fields. The U.S. Department of Agriculture is investigating whether the rat poison should be located in the middle of the field or on the outer perimeter. One way to answer this question is to determine where the greater amount of damage occurs. If damage is measured by the proportion of cane stalks that have been damaged by rats, how many stalks from each section of the field should be sampled to estimate the true difference between proportions of stalks damaged in the two sections to within .02 with 95% confidence?

7.112 Instrument precision. When new instruments are developed to perform chemical analyses of products (food, medicine, etc.), they are usually evaluated with respect to two criteria: accuracy and precision. *Accuracy* refers to the ability of the instrument to identify correctly the nature and amounts of a product's components. *Precision* refers to the consistency with which the instrument will identify the components of the same material. Thus, a large variability in the identification of a single batch of a product indicates a lack of precision. Suppose a pharmaceutical firm is considering two brands of an instrument designed to identify the components of certain drugs. As part of a comparison of precision, 10 test-tube samples of a well-mixed batch of a drug are selected and then 5 are analyzed by instrument A and 5 by instrument B. The data shown below (saved in the **INSTRAB** file) are the percentages of the primary component of the drug given by the instruments. Do these data provide evidence of a difference in the precision of the two machines? Use $\alpha = .10$.

Instrument A	Instrument B
43	46
48	49
37	43
52	41
45	48

7.113 Evaluating customer interest with a pupillometer. A *pupillometer* is a device used to observe changes in pupil dilations as the eye is exposed to different visual stimuli. Because there is a direct correlation between the amount an individual's pupil dilates and his or her interest in the stimuli, marketing organizations sometimes use pupillometers to help them evaluate potential consumer interest in new products, alternative package designs, and other factors (*Optical Engineering,* Mar. 1995). The Design and Market Research Laboratories of the Container Corporation of America used a pupillometer to evaluate consumer reaction to different silverware patterns for a client. Suppose 15 consumers were chosen at random, and each was shown two silverware patterns. Their pupillometer readings (in millimeters) are shown in the table below and saved in the **PUPILL** file.

Consumer	Pattern 1	Pattern 2
1	1.00	.80
2	.97	.66
3	1.45	1.22
4	1.21	1.00
5	.77	.81
6	1.32	1.11
7	1.81	1.30
8	.91	.32
9	.98	.91
10	1.46	1.10
11	1.85	1.60
12	.33	.21
13	1.77	1.50
14	.85	.65
15	.15	.05

a. What are the appropriate null and alternative hypotheses to test whether the mean amount of pupil dilation differs for the two patterns? Define any symbols you use.

b. Conduct the test, part **a.** Interpret the results.

c. Is the paired difference design used for this study preferable to an independent samples design? For independent samples, we could select 30 consumers, divide them into two groups of 15, and show each group a different pattern. Explain your preference.

7.114 Positive spillover effects from self-managed work teams. To improve quality, productivity, and timeliness, many American industries employ self-managed work teams

SPSS Output for Exercise 7.114

Independent Samples Test

		Levene's Test for Equality of Variances		t-test for Equality of Means					95% Confidence Interval of the Difference	
		F	Sig.	t	df	Sig. (2-tailed)	Mean Difference	Std. Error Difference	Lower	Upper
CREATIVE	Equal variances assumed	16.479	.000	8.565	112	.000	.808	.094	.621	.994
	Equal variances not assumed			8.847	108.727	.000	.808	.091	.627	.988

(SMWTs). A team typically consists of 5 to 15 workers who are collectively responsible for making decisions and performing all tasks related to a particular project. Because SMWTs require that employees be trained in interpersonal skills, they can have potential positive spillover effects on a worker's family life. The link between SMWT work characteristics and workers' perceptions of positive spillover into family life was investigated in the *Quality Management Journal* (Summer 1995). Survey data were collected from 114 AT&T employees who work in 1 of 15 SMWTs at an AT&T technical division. The workers were divided into two groups: (1) those who reported positive spillover of work skills to family life and (2) those who did not report positive work spillover. The two groups were compared on a variety of job and demographic characteristics, one of which was the use of creative ideas (measured on a 7-point scale, where the larger the number, the more of the characteristic indicated). The data (simulated from summary information provided in the *Quality Management Journal* article) are saved in the **SPILLOVER** file.

a. One comparison of interest to the researchers is whether the mean creative use of ideas scale score for employees who report positive spillover of work skills to family life differs from the mean scale score for employees who did not report positive work spillover. Give the null and alternative hypotheses that will allow the researchers to make the comparison.

b. Discuss whether it is appropriate to apply the large-sample z-test to test the hypotheses, part **a.**

c. The results of the test are shown in the above SPSS printout. Interpret these results. Make the appropriate conclusion using $\alpha = .05$.

d. A 95% confidence interval for the difference between the mean use of creative ideas scale scores is shown in the last column of the SPSS printout. Interpret this interval. Does the inference derived from the confidence interval agree with that from the hypothesis test?

e. The data also include the qualitative variable, Gender, for each worker. The researchers want to know whether the proportion of male workers in the two groups are significantly different. A Minitab printout of the analysis is shown below. Fully interpret the results in the words of the problem.

Applying the Concepts—Advanced

7.115 Impact of gender on advertising. How does gender affect the type of advertising that proves to be most effective? An article in the *Journal of Advertising Research* (May/June 1990) makes reference to numerous studies that conclude males tend to be more competitive with others than with themselves. To apply this conclusion to advertising, the author created two ads promoting a new brand of soft drink:

Ad 1: Four men are shown competing in racquetball
Ad 2: One man is shown competing against himself in racquetball

The author hypothesized that the first ad will be more effective when shown to males. To test this hypothesis, 43 males were shown both ads and asked to measure their attitude toward the advertisement (Aad), their attitude toward the brand of soft drink (Ab), and their intention to purchase the soft drink (Intention). Each variable was measured using a 7-point scale, with higher scores indicating a more favorable attitude. The results are shown in the table on the next page. Do you agree with the author's hypothesis?

Minitab Output for Exercise 7.114

Test and CI for Two Proportions: GENDER, GROUP

```
Event = MALE


GROUP      X   N  Sample p
NOSPILL   59  67  0.880597
SPILLOV   39  47  0.829787


Difference = p (NOSPILL) - p (SPILLOV)
Estimate for difference:  0.0508098
95% CI for difference:  (-0.0817519, 0.183371)
Test for difference = 0 (vs not = 0):  Z = 0.75  P-Value = 0.453
```

	Sample Means		
	Aad	Ab	Intention
Ad 1	4.465	3.311	4.366
Ad 2	4.150	2.902	3.813
Level of Significance	$p = .091$	$p = .032$	$p = .050$

7.116 Salaries of postgraduates. Refer to the *Economics of Education Review* (Vol. 21, 2002) study of the relationship between education level and earnings, Exercise 6.31 (p. 336). A National Adult Literacy Survey revealed that males with a postgraduate degree had a sample mean salary of \$61,340 (with standard error $s_{\bar{x}_M} = \$2,185$), while females with a postgraduate degree had a sample mean salary of \$32,227 (with standard error $s_{\bar{x}_F} = \$932$). Let μ_M represent the population mean salary of all males with postgraduate degrees and μ_F represent the population mean salary of all females with postgraduate degrees.

a. Set up the null and alternative hypotheses for determining whether μ_M exceeds μ_F.

b. Calculate the test statistic for the test, part **a.** [*Note:* $s_{\bar{x}_M - \bar{x}_F} = \sqrt{(s^2_{\bar{x}_M} + s^2_{\bar{x}_F})}$.]

c. Find the rejection region for the test using $\alpha = .01$.

d. Use the results, parts **b** and **c,** to make the appropriate conclusion.

Critical Thinking Challenge

7.117 Facility layout study. Facility layout and material flowpath design are major factors in the productivity analysis of automated manufacturing systems. Facility layout is concerned with the location arrangement of machines and buffers for work-in-process. Flowpath design is concerned with the direction of manufacturing material flows (e.g., unidirectional or bidirectional; Lee, Lei, and Pinedo, *Annals of Operations Research,* 1997). A manufacturer of printed circuit boards is interested in evaluating two alternative existing layout and flowpath designs. The output of each design was monitored for 8 consecutive working days. The data (shown below) are saved in the **FLOWPATH** file. Design 2 appears to be superior to Design 1. Do you agree? Explain fully.

Working Days	Design 1 (units)	Design 2 (units)
8/16	1,220	1,273
8/17	1,092	1,363
8/18	1,136	1,342
8/19	1,205	1,471
8/20	1,086	1,299
8/23	1,274	1,457
8/24	1,145	1,263
8/25	1,281	1,368

References

Freedman, D., Pisani, R., and Purves, R. *Statistics*. New York: W. W. Norton and Co., 1978.

Gibbons, J. D. *Nonparametric Statistical Inference,* 2nd ed. New York: McGraw-Hill, 1985.

Hollander, M., and Wolfe, D. A. *Nonparametric Statistical Methods.* New York: Wiley, 1973.

Mendenhall, W., Beaver, R. J., and Beaver, B. M. *Introduction to Probability and Statistics,* 13th ed. Belmont, CA: Brooks/Cole, 2009.

Satterthwaite, F. W. "An approximate distribution of estimates of variance components," *Biometrics Bulletin,* Vol. 2, 1946, pp. 110–114.

Snedecor, G. W., and Cochran, W. *Statistical Methods,* 7th ed. Ames: Iowa State University Press, 1980.

Steel, R. G. D., and Torrie, J. H. *Principles and Procedures of Statistics,* 2nd ed. New York: McGraw-Hill, 1980.

USING TECHNOLOGY

SPSS: Two-Sample Inferences

SPSS can be used to make two-sample inferences about $\mu_1 - \mu_2$ for independent samples and μ_d for paired samples but cannot currently conduct analyses about $p_1 - p_2$ or an F-test to compare population variances.

Comparing Means with Independent Samples

Step 1 Access the SPSS spreadsheet file that contains the sample data. The data file should contain one quantitative variable (which the means will be calculated on) and one qualitative variable with either two numerical coded values (e.g., 1 and 2) or two short categorical levels (e.g., "yes" and "no"). These two values represent the two groups or populations to be compared.

Step 2 Click on the "Analyze" button on the SPSS menu bar and then click on "Compare Means" and "Independent-Samples T Test," as shown in Figure 7.S.1.

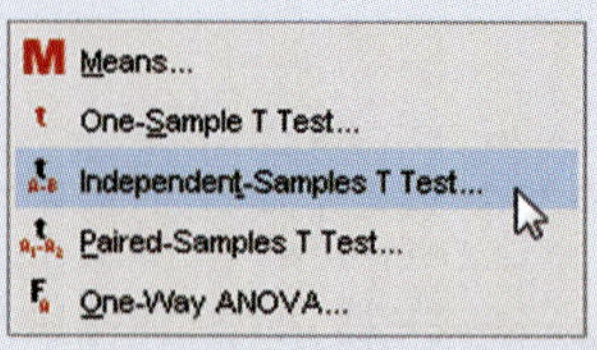

Figure 7.S.1 SPSS menu options for comparing two means

Step 3 On the resulting dialog box (shown in Figure 7.S.2), specify the quantitative variable of interest in the "Test Variable(s)" box and the qualitative variable in the "Grouping Variable" box.

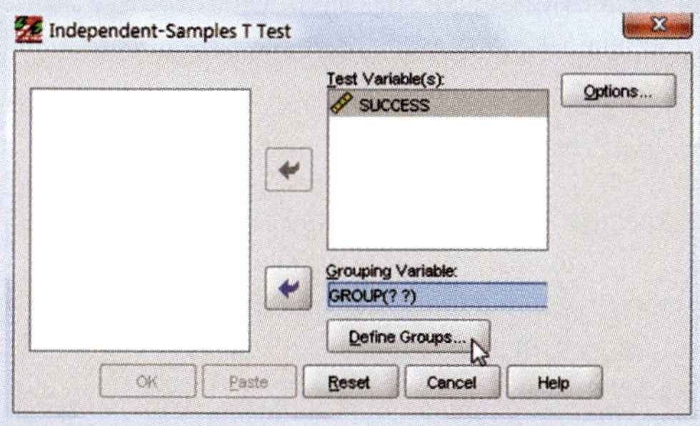

Figure 7.S.2 SPSS independent-samples T test dialog box

Step 4 Click the "Define Groups" button and specify the values of the two groups in the resulting dialog box (see Figure 7.S.3).

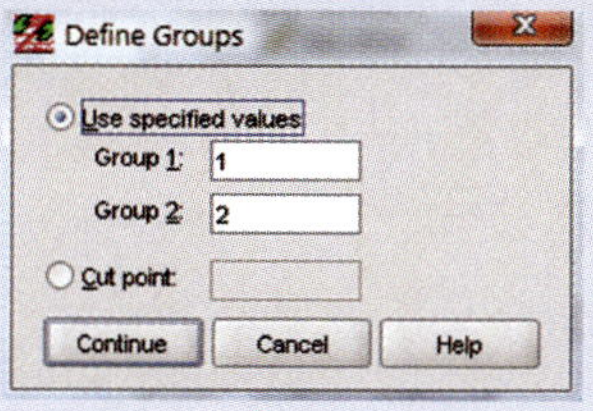

Figure 7.S.3 SPSS define groups dialog box

Step 5 Click "Continue" to return to the "Independent-Samples T Test" dialog screen. Without any further menu selections, SPSS will automatically conduct a two-tailed test of the null hypothesis, H_0: $\mu_1 - \mu_2 = 0$.

Step 6 If you want to generate a confidence interval for $\mu_1 - \mu_2$, click the "Options" button and specify the confidence level on the resulting menu screen, as shown in Figure 7.S.4. Click "Continue" to return to the "T Test" dialog box.

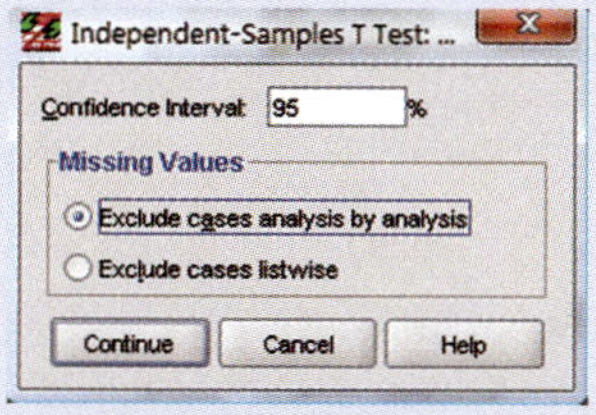

Figure 7.S.4 SPSS options dialog box

Step 7 Click "OK" to generate the SPSS printout.

Important Note: The SPSS two-sample *t*-procedure uses the *t*-statistic to conduct the test of hypothesis. When the sample sizes are small, this is the appropriate method. When the sample sizes are large, the *t*-value will be approximately equal to the large-sample *z*-value, and the resulting test will still be valid.

Comparing Means with Paired Samples

Step 1 Access the SPSS spreadsheet file that contains the sample data. The data file should contain two quantitative variables—one with the data values for the first group (or population) and one with the data values for the second group. (*Note:* The sample size should be the same for each group.)

Step 2 Click on the "Analyze" button on the SPSS menu bar and then click on "Compare Means" and "Paired-Samples T Test" (see Figure 7.S.1).

Step 3 On the resulting dialog box (shown in Figure 7.S.5), specify the two quantitative variables of interest in the "Paired Variables" box. Without any further menu selections, SPSS will automatically conduct a two-tailed test of the null hypothesis, H_0: $\mu_d = 0$.

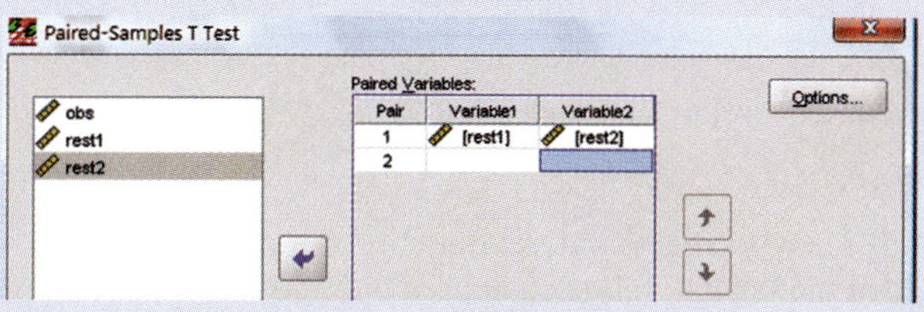

Figure 7.S.5 SPSS paired-samples dialog box

Step 4 If you want to generate a confidence interval for μ_d, click the "Options" button and specify the confidence level on the resulting menu screen (as shown in Figure 7.S.4). Click "Continue" to return to the "Paired-Samples" dialog box.

Step 5 Click "OK" to generate the SPSS printout.

Note: Although SPSS cannot perform an *F*-test to compare two variances, it will automatically provide a nonparametric test (e.g., Levene's test) for equal variances when you choose to do an independent samples *t*-test for $\mu_1 - \mu_2$.

Minitab: Two-Sample Inferences

Minitab can be used to make two-sample inferences about $\mu_1 - \mu_2$ or independent samples, μ_d for paired samples, $p_1 - p_2$ and σ_1^2/σ_2^2.

Comparing Means with Independent Samples

Step 1 Access the Minitab worksheet that contains the sample data.

Step 2 Click on the "Stat" button on the Minitab menu bar and then click on "Basic Statistics" and "2-Sample t," as shown in Figure 7.M.1. The resulting dialog box appears as shown in Figure 7.M.2.

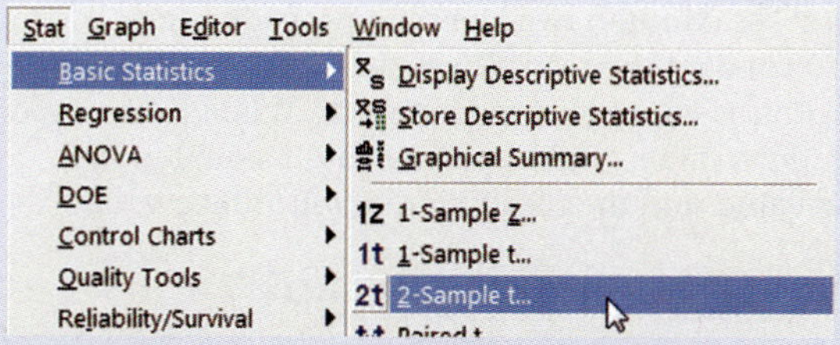

Figure 7.M.1 Minitab menu options for comparing two means

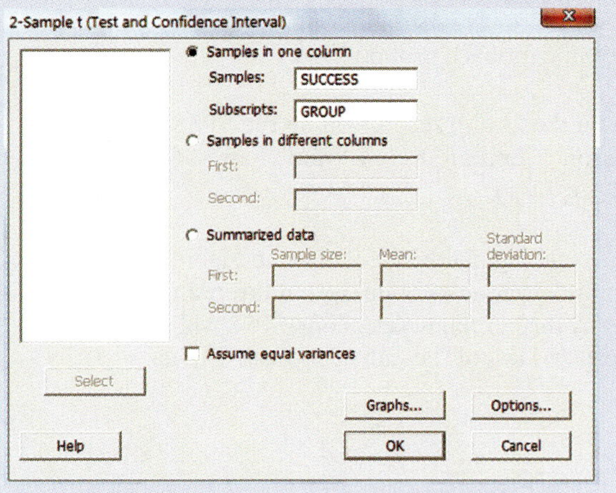

Figure 7.M.2 Minitab 2-sample t dialog box

Step 3a If the worksheet contains data for one quantitative variable (which the means will be computed on) and one qualitative variable (which represents the two groups or populations), select "Samples in one column" and then specify the quantitative variable in the "Samples" area and the qualitative variable in the "Subscripts" area. (See Figure 7.M.2.)

Step 3b If the worksheet contains the data for the first sample in one column and the data for the second sample in another column, select "Samples in different columns" and then specify the "First" and "Second" variables. Alternatively, if you have only summarized data (i.e., sample sizes, sample means, and sample standard deviations), select "Summarized data" and enter these summarized values in the appropriate boxes.

Step 4 Click the "Options" button on the Minitab "2-Sample T" dialog box. Specify the confidence level for a confidence interval, the null hypothesized value of the difference, $\mu_1 - \mu_2$, and the form of the alternative hypothesis (lower tailed, two tailed, or upper tailed) in the resulting dialog box, as shown in Figure 7.M.3.

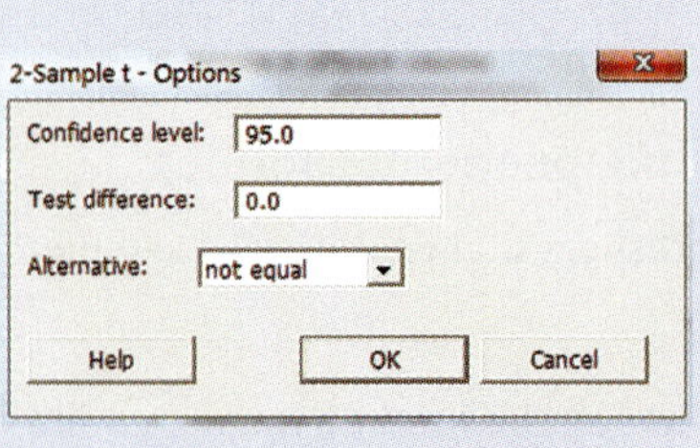

Figure 7.M.3 Minitab options dialog box

Step 5 Click "OK" to return to the "2-Sample T" dialog box and then click "OK" again to generate the Minitab printout.

Important Note: The Minitab two-sample t-procedure uses the t-statistic to conduct the test of hypothesis. When the sample sizes are small, this is the appropriate method. When the sample sizes are large, the t-value will be approximately equal to the large-sample z-value, and the resulting test will still be valid.

Comparing Means with Paired Samples

Step 1 Access the Minitab worksheet that contains the sample data. The data file should contain two quantitative variables—one with the data values for the first group (or population) and one with the data values for the second group. (*Note:* The sample size should be the same for each group.)

Step 2 Click on the "Stat" button on the Minitab menu bar and then click on "Basic Statistics" and "Paired t" (see Figure 7.M.1).

Step 3 On the resulting dialog box, select the "Samples in columns" option and specify the two quantitative variables of interest in the "First sample" and "Second sample" boxes, as shown in Figure 7.M.4. [Alternatively, if you have only summarized data of the paired differences, select the "Summarized data (differences)" option and enter the sample size, sample mean, and sample standard deviation in the appropriate boxes.]

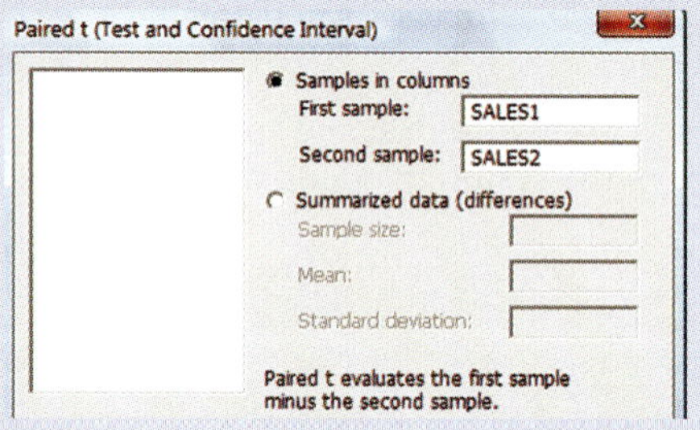

Figure 7.M.4 Minitab paired-samples t dialog box

Step 4 Click the "Options" button and specify the confidence level for a confidence interval, the null hypothesized value of the difference, μ_d, and the form of the alternative hypothesis (lower tailed, two tailed, or upper tailed) in the resulting dialog box. (See Figure 7.M.3.)

Step 5 Click "OK" to return to the "Paired t" dialog box and then click "OK" again to generate the Minitab printout.

Comparing Proportions with Large Independent Samples

Step 1 Access the Minitab worksheet that contains the sample data.

Step 2 Click on the "Stat" button on the Minitab menu bar and then click on "Basic Statistics" and "2 Proportions," as shown in Figure 7.M.1.

Step 3 On the resulting dialog box (shown in Figure 7.M.5), select the data option ("Samples in one column" or "Samples in

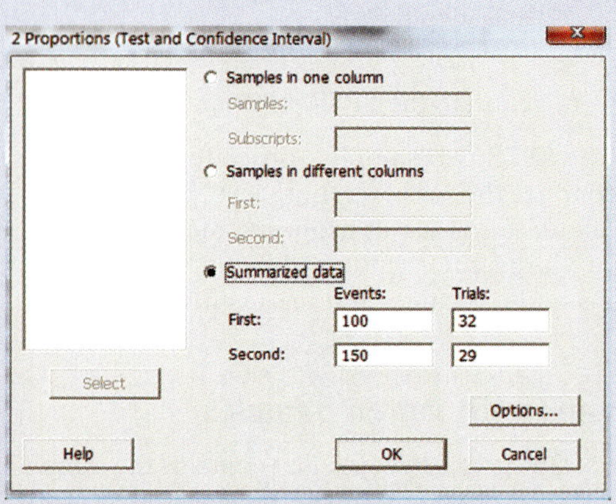

Figure 7.M.5 Minitab 2 proportions dialog box

different columns" or "Summarized data") and make the appropriate menu choices. (Figure 7.M.5 shows the menu options when you select "Summarized data.")

Step 4 Click the "Options" button and specify the confidence level for a confidence interval, the null hypothesized value of the difference, and the form of the alternative hypothesis (lower tailed, two tailed, or upper tailed) in the resulting dialog box, as shown in Figure 7.M.6. (If you desire a pooled estimate of p for the test, be sure to check the appropriate box.)

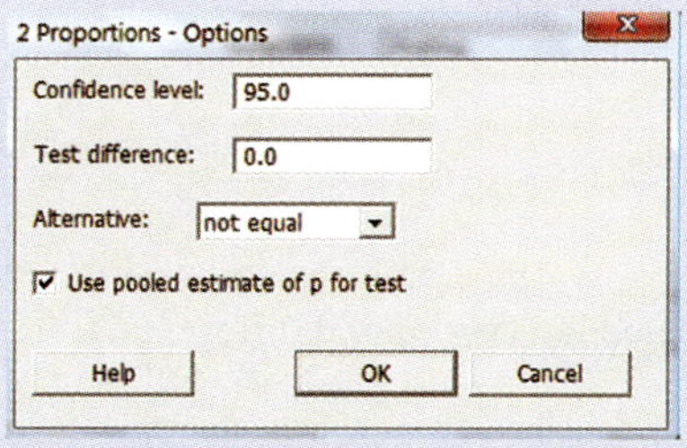

Figure 7.M.6 Minitab 2 proportions options

Step 5 Click "OK" to return to the "2 Proportions" dialog box and then click "OK" again to generate the Minitab printout.

Comparing Variances with Independent Samples

Step 1 Access the Minitab worksheet that contains the sample data.

Step 2 Click on the "Stat" button on the Minitab menu bar and then click on "Basic Statistics" and "2 Variances" (Figure 7.M.1.)

Step 3 On the resulting dialog box (shown in Figure 7.M.7), the menu selections and options are similar to those for the two-sample t-test.

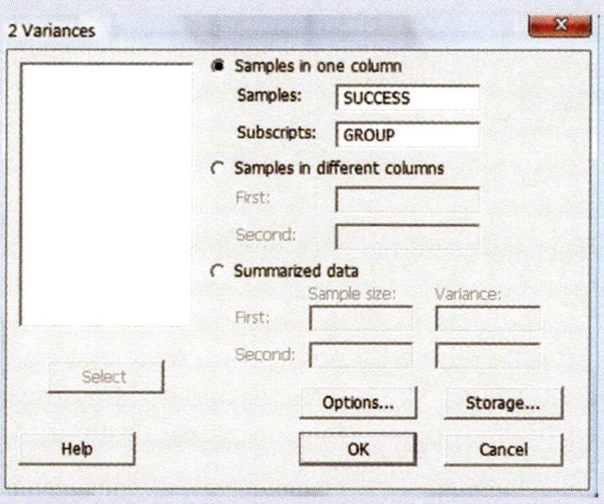

Figure 7.M.7 Minitab 2 variances dialog box

Step 4 Click "OK" to produce the Minitab F-test printout.

Excel/DDXL: Two-Sample Inferences

Comparing Means with Independent Samples

To obtain an independent-samples hypothesis test for $(\mu_1 - \mu_2)$, perform the following:

Step 1 Highlight (select) the data columns you want to analyze on the Excel spreadsheet. (*Note:* The quantitative data should be in two columns, one for the first sample and one for the second sample.)

Step 2 Click on "Add-Ins" in the main Excel menu bar and select "DDXL." On the resulting menu, select "Hypothesis Tests," as shown in Figure 7.E.1.

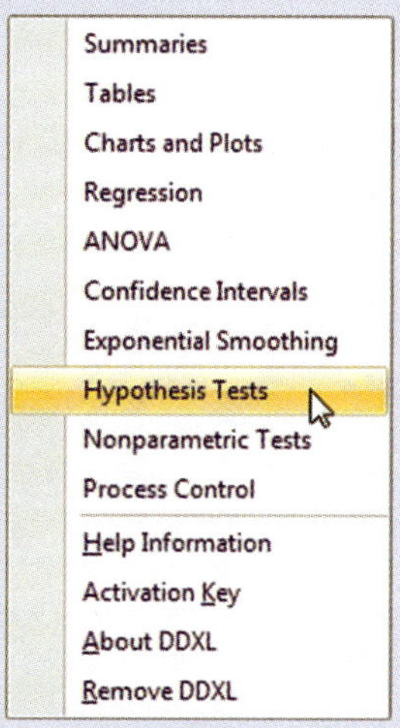

Figure 7.E.1 Excel/DDXL menu options for hypothesis tests

Step 3 On the resulting menu, select "2 Var t Test" in the Function Type box, as shown in Figure 7.E.2.

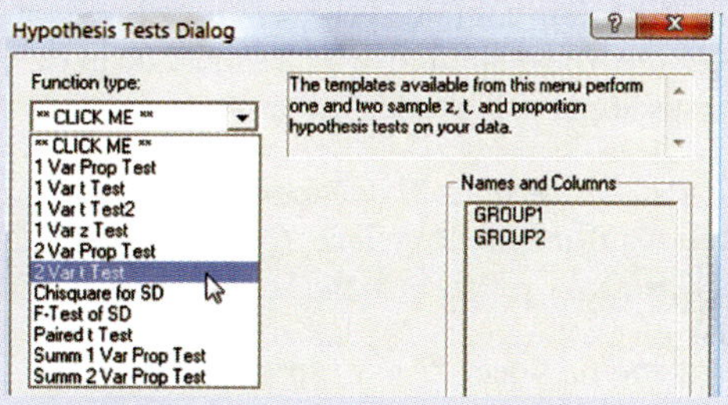

Figure 7.E.2 DDXL hypothesis tests dialog box

Step 4 Move the column for the first sample into the "1st Quantitative Variable" box and the column for the second sample into the "2nd Quantitative Variable" box, as shown in Figure 7.E.3. Then click "OK."

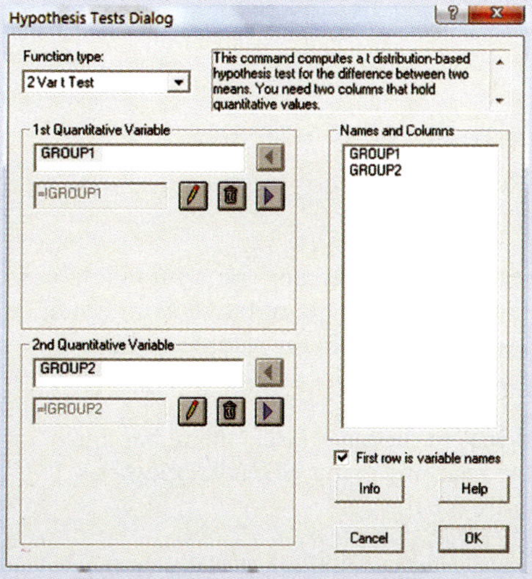

Figure 7.E.3 DDXL menu selections for test to compare means

Step 5 On the resulting menu specify the type of test (select "2-sample" for large samples and "pooled" for small samples), the hypothesized difference, α, and the form of the alternative hypothesis, as shown in Figure 7.E.4. Click "Compute" to produce the test results.

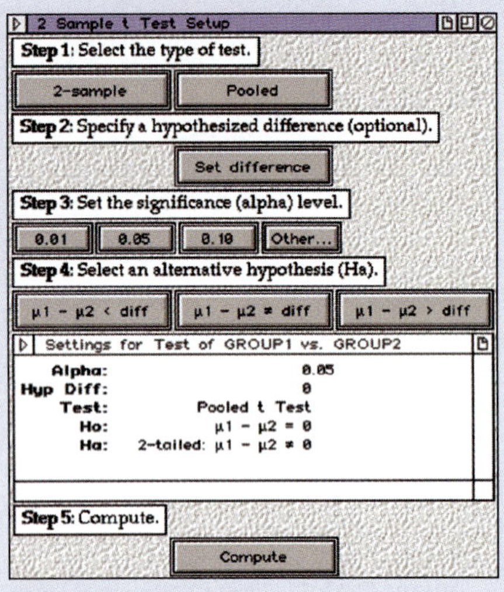

Figure 7.E.4 DDXL *t* test to compare means setup options

To obtain an independent-samples confidence interval for $(\mu_1 - \mu_2)$, perform the following:

Step 1 Highlight (select) the data columns you want to analyze on the Excel spreadsheet. (*Note:* The quantitative data should be in two columns, one for the first sample and one for the second sample.)

Step 2 Click on "Add-Ins" in the main Excel menu bar and select "DDXL." On the resulting menu, select "Confidence Intervals" (see Figure 7.E.1).

Step 3 On the resulting menu, select "2 Var *t* Interval" in the Function Type box.

Step 4 Move the column for the first sample into the "1st Quantitative Variable" box and the column for the second sample into the "2nd Quantitative Variable" box (similar to Figure 7.E.3). Then click "OK."

Step 5 On the resulting menu specify the type of interval (select "2-sample" for large samples and "pooled" for small samples) and the confidence level. Click "Compute Interval" to produce the results.

Comparing Means with Paired Samples

To obtain a paired-difference hypothesis test for $(\mu_1 - \mu_2)$, perform the following:

Step 1 Highlight (select) the data columns you want to analyze on the Excel spreadsheet. (*Note:* The quantitative data should be in two columns, one for the first pair in the sample and one for the second pair.)

Step 2 Click on "Add-Ins" in the main Excel menu bar and select "DDXL." On the resulting menu, select "Hypothesis Tests," as shown in Figure 7.E.1.

Step 3 On the resulting menu, select "Paired *t* Test" in the Function Type box (see Figure 7.E.2).

Step 4 Move the column for the first paired observation into the "1st Quantitative Variable" box and the column for the second paired observation into the "2nd Quantitative Variable" box, as shown in Figure 7.E.5. Then click "OK."

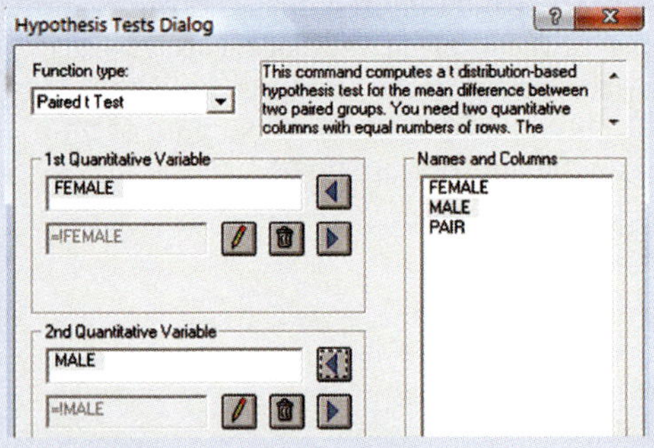

Figure 7.E.5 DDXL menu selections for paired difference test

Step 5 On the resulting menu, specify the hypothesized difference, α, and the form of the alternative hypothesis, as shown in Figure 7.E.6. Click "Compute" to produce the test results.

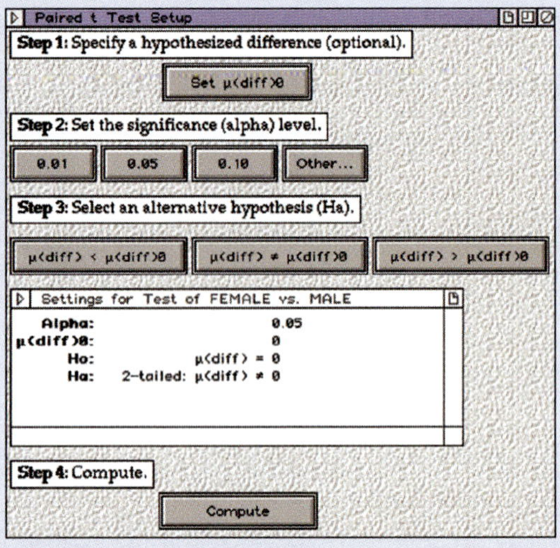

Figure 7.E.6 DDXL paired difference test setup options

To obtain a paired-difference confidence interval for $(\mu_1 - \mu_2)$, perform the following:

Step 1 Highlight (select) the data columns you want to analyze on the Excel spreadsheet. (*Note:* The quantitative data should be in two columns, one for the first pair in the sample and one for the second pair in the sample.)

Step 2 Click on "Add-Ins" in the main Excel menu bar and select "DDXL." On the resulting menu, select "Confidence Intervals" (see Figure 7.E.1).

Step 3 On the resulting menu, select "Paired *t* Interval" in the Function Type box.

Step 4 Move the column for the first paired observation into the "1st Quantitative Variable" box and the column for the second paired observation into the "2nd Quantitative Variable" box (similar to Figure 7.E.5). Then click "OK."

Step 5 On the resulting menu, specify the confidence level and then click "Compute Interval" to produce the results.

Comparing Proportions with Large Independent Samples

To obtain an independent-samples hypothesis test for $(p_1 - p_2)$, perform the following:

Step 1 Highlight (select) the data columns you want to analyze on the Excel spreadsheet. (*Note:* The summarized data should be in four columns, one for the first sample size, one for the number of successes in the first sample, one for the second sample size, and one for the number of successes in the second sample. There should only be one row for these data.)

Step 2 Click on "Add-Ins" in the main Excel menu bar and select "DDXL." On the resulting menu, select "Hypothesis Tests," as shown in Figure 7.E.1.

Step 3 On the resulting menu, select "Summ 2 Var Prop Test" in the Function Type box (see Figure 7.E.2).

Step 4 Move the appropriate columns into the "Num Successes 1" box, "Num Trials 1" box, "Num Successes 2" box, and the "Num Trials 2" box, respectively, as shown in Figure 7.E.7. Then click "OK."

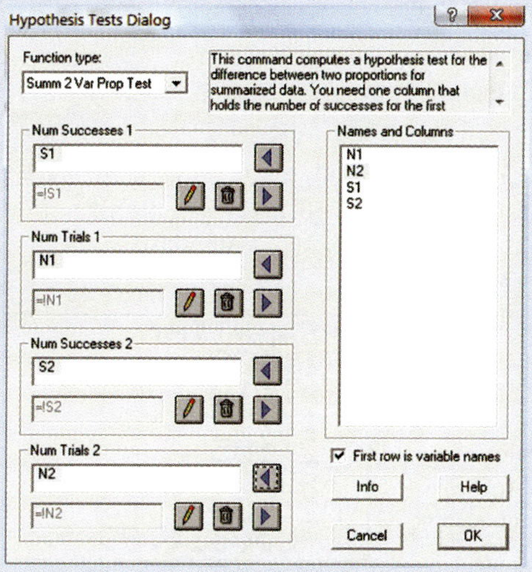

Figure 7.E.7 DDXL menu selections for test to compare proportions

Step 5 On the resulting menu specify the hypothesized difference, α, and the form of the alternative hypothesis, as shown in Figure 7.E.8. Click "Compute" to produce the test results.

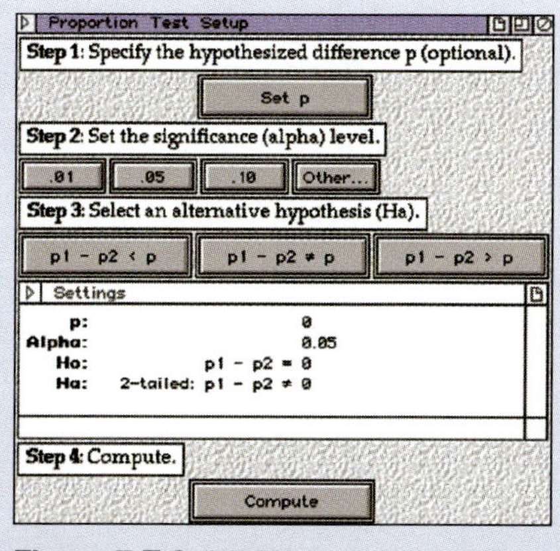

Figure 7.E.8 DDXL test to compare proportions setup options

To obtain an independent-samples confidence interval for $(p_1 - p_2)$, perform the following:

Step 1 Highlight (select) the data columns you want to analyze on the Excel spreadsheet. (*Note:* The summarized data should be in four columns, one for the first sample size, one for the number of successes in the first sample, one for the second sample size, and one for the number of successes in the second sample. There should only be one row for these data.)

Step 2 Click on "Add-Ins" in the main Excel menu bar and select "DDXL." On the resulting menu, select "Confidence Intervals" (see Figure 7.E.1).

Step 3 On the resulting menu, select "Summ 2 Var Prop Interval" in the Function Type box.

Step 4 Move the appropriate columns into the "Num Successes 1" box, "Num Trials 1" box, "Num Successes 2" box, and the "Num Trials 2" box, respectively, as shown in Figure 7.E.7. Then click "OK."

Step 5 On the resulting menu, specify the confidence level and then click "Compute Interval" to produce the results.

Comparing Variances with Independent Samples

To obtain an independent-samples hypothesis test for (σ_1^2/σ_2^2), perform the following:

Step 1 Highlight (select) the data columns you want to analyze on the Excel spreadsheet. (*Note:* The quantitative data should be in two columns, one for the first sample and one for the second sample.)

Step 2 Click on "Add-Ins" in the main Excel menu bar and select "DDXL." On the resulting menu, select "Hypothesis Tests," as shown in Figure 7.E.1.

Step 3 On the resulting menu, select "F-Test of SD" in the Function Type box (see Figure 7.E.2).

Step 4 Move the column for the first sample into the "1st Quantitative Variable" box and the column for the second sample into the "2nd Quantitative Variable" box, as shown in Figure 7.E.9. Then click "OK."

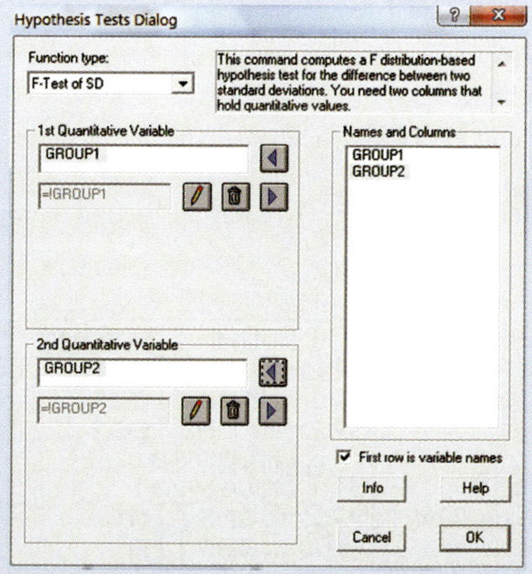

Figure 7.E.9 DDXL menu selections for comparing variances

Step 5 On the resulting menu, specify α and the form of the alternative hypothesis, as shown in Figure 7.E.10. Click "Compute" to produce the test results.

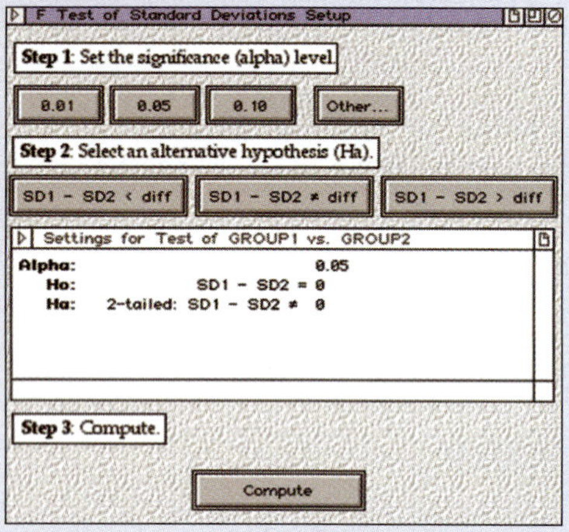

Figure 7.E.10 DDXL test to compare variances setup options

TI-83/TI-84 Graphing Calculator: Two-Sample Inferences

The TI-84 graphing calculator can be used to conduct tests and form confidence intervals for the difference between two means with independent samples, the difference between two means with matched pairs, the difference between two proportions for large independent samples, and the ratio of two variances.

Confidence Interval for $\mu_1 - \mu_2$

Step 1 *Enter the data (Skip to Step 2 if you have summary statistics, not raw data)*

- Press **STAT** and select **1:Edit**

Note: If the lists already contain data, clear the old data. Use the up **ARROW** to highlight "**L1.**"

- Press **CLEAR ENTER**
- Use the up **ARROW** to highlight "**L2**"
- Press **CLEAR ENTER**
- Use the **ARROW** and **ENTER** keys to enter the first data set into **L1**
- Use the **ARROW** and **ENTER** keys to enter the second data set into **L2**

Step 2 *Access the Statistical Tests Menu*

- Press **STAT**
- Arrow right to **TESTS**
- Arrow down to **2-SampTInt**
- Press **ENTER**

Step 3 *Choose "**Data**" or "**Stats**" ("Data" is selected when you have entered the raw data into the Lists. "Stats" is selected when you are given only the means, standard deviations, and sample sizes.)*

- Press **ENTER**
- If you selected "Data," set **List1** to **L1** and **List2** to **L2**
- Set **Freq1** to **1** and set **Freq2** to **1**
- Set **C-Level** to the confidence level
- If you are assuming that the two populations have equal variances, select **Yes** for **Pooled**
- If you are not assuming equal variances, select **No**
- Press **ENTER**
- Arrow down to "**Calculate**"
- Press **ENTER**

- If you selected "Stats," enter the means, standard deviations, and sample sizes
- Set **C-Level** to the confidence level
- If you are assuming that the two populations have equal variances, select **Yes** for **Pooled**
- If you are not assuming equal variances, select **No**
- Press **ENTER**
- Arrow down to "**Calculate**"
- Press **ENTER**

(The accompanying screen is set up for an example with a mean of 100, a standard deviation of 10, and a sample size of 15 for the first data set and a mean of 105, a standard deviation of 12, and a sample size of 18 for the second data set.)

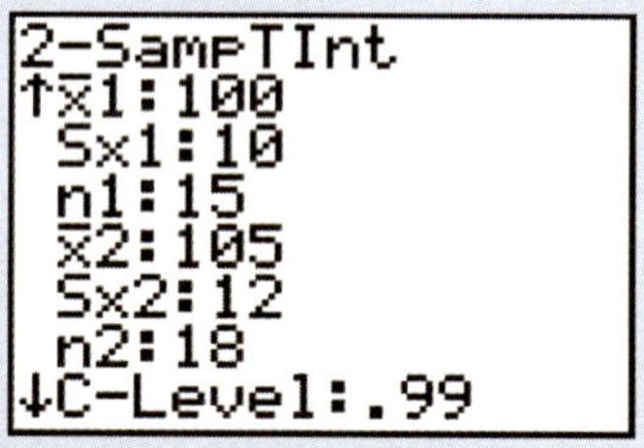

The confidence interval will be displayed with the degrees of freedom, the sample statistics, and the pooled standard deviation (when appropriate).

Hypothesis Test for $\mu_1 - \mu_2$

Step 1 *Enter the data (Skip to Step 2 if you have summary statistics, not raw data)*

- Press **STAT** and select **1:Edit**

Note: If the lists already contain data, clear the old data. Use the up **ARROW** to highlight "**L1.**"

- Press **CLEAR ENTER**
- Use the up **ARROW** to highlight "**L2**"
- Press **CLEAR ENTER**
- Use the **ARROW** and **ENTER** keys to enter the first data set into **L1**
- Use the **ARROW** and **ENTER** keys to enter the second data set into **L2**

Step 2 *Access the Statistical Tests Menu*

- Press **STAT**
- Arrow right to **TESTS**

- Arrow down to **2-SampTTest**
- Press **ENTER**

Step 3 *Choose "**Data**" or "**Stats**" ("Data" is selected when you have entered the raw data into the Lists. "Stats" is selected when you are given only the means, standard deviations, and sample sizes.)*

- Press **ENTER**
- If you selected "Data," set **List1** to **L1** and **List2** to **L2**
- Set **Freq1** to **1** and set **Freq2** to **1**
- Use the **ARROW** to highlight the appropriate alternative hypothesis
- Press **ENTER**
- If you are assuming that the two populations have equal variances, select **Yes** for **Pooled**
- If you are not assuming equal variances, select **No**
- Press **ENTER**
- Arrow down to "**Calculate**"
- Press **ENTER**
- If you selected "**Stats,**" enter the means, standard deviations, and sample sizes

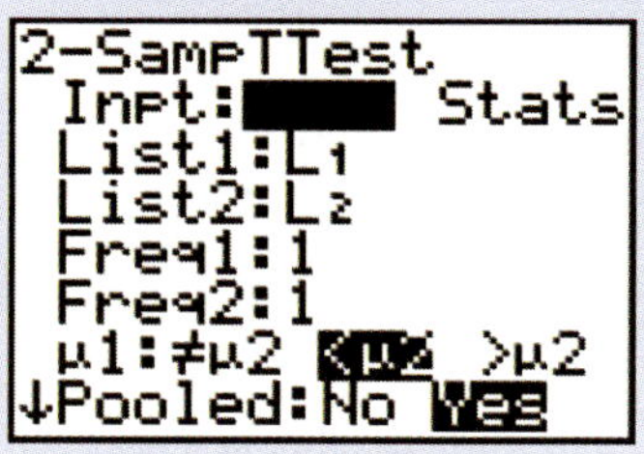

- Use the **ARROW** to highlight the appropriate alternative hypothesis
- Press **ENTER**
- If you are assuming that the two populations have equal variances, select **Yes** for **Pooled**
- If you are not assuming equal variances, select **No**
- Press **ENTER**
- Arrow down to "**Calculate**"
- Press **ENTER**

(The screen at the right is set up for an example with a mean of 100, a standard deviation of 10, and a sample size of 15 for the first data set and a mean of 120, a standard deviation of 12, and a sample size of 18 for the second data set.)

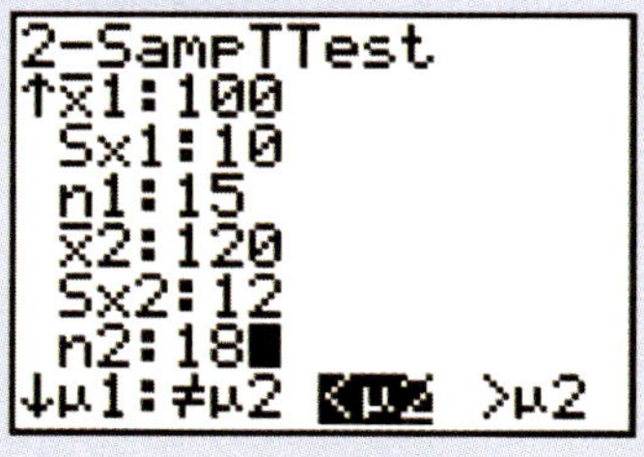

The results of the hypothesis test will be displayed with the *p*-value, degrees of freedom, the sample statistics, and the pooled standard deviation (when appropriate).

Confidence Interval for a Paired Difference Mean

Note: There is no paired difference option on the calculator. These instructions demonstrate how to calculate the differences and then use the 1-sample *t*-interval.

Step 1 *Enter the data and calculate the differences.*

- Press **STAT** and select **1:Edit**

Note: If the lists already contain data, clear the old data. Use the up **ARROW** to highlight "**L1.**"

- Press **CLEAR ENTER**
- Use the up **ARROW** to highlight "**L2**"
- Press **CLEAR ENTER**
- Use the **ARROW** and **ENTER** keys to enter the first data set into **L1**
- Use the **ARROW** and **ENTER** keys to enter the second data set into **L2**
- The differences will be calculated in **L3**
- Use the up **ARROW** to highlight "**L3**"
- Press **CLEAR**— This will clear any old data, but **L3** will remain highlighted
- To enter the equation L3 = L1 − L2, use the following keystrokes:
  - Press 2ND "**1**" (this will enter L1)
  - Press the **MINUS** button
  - Press 2ND "**2**" (this will enter L2)

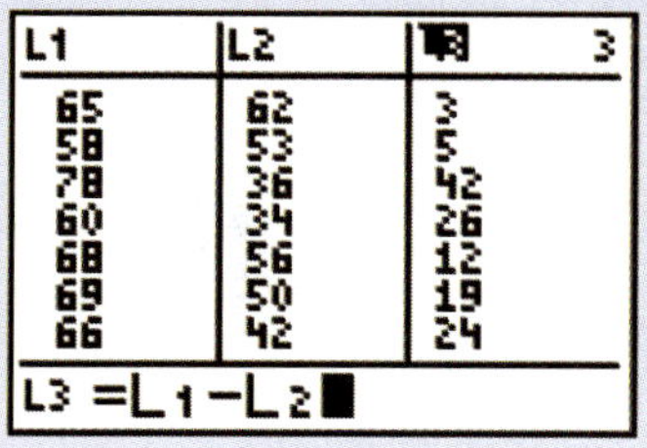

(Notice the equation at the bottom of the screen.)

- Press **ENTER** (the differences should be calculated in L3)

Step 2 *Access the Statistical Tests Menu*

- Press **STAT**
- Arrow right to **TESTS**
- Arrow down to **TInterval (even for large sample case)**
- Press **ENTER**

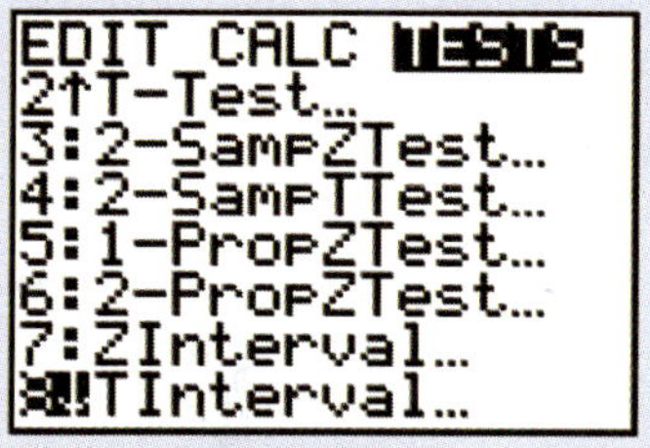

Step 3 *Choose "**Data**"*

- Press **ENTER**
- Set **List** to **L3**
- Set **Freq** to **1**
- Set **C-Level** to the confidence level
- Arrow down to "**Calculate**"
- Press **ENTER**

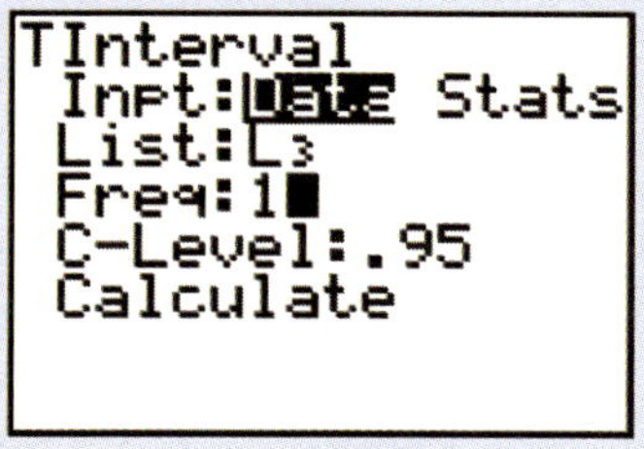

The confidence interval will be displayed with the mean, standard deviation, and sample size of the differences.

Hypothesis Test for a Paired Difference Mean

Note: There is no paired difference option on the calculator. These instructions demonstrate how to calculate the differences and then use the 1-sample *t*-test.

Step 1 *Enter the data and calculate the differences.*

- Press **STAT** and select **1:Edit**

Note: If the lists already contain data, clear the old data. Use the up **ARROW** to highlight "**L1.**"

- Press **CLEAR ENTER**
- Use the up **ARROW** to highlight "**L2**"
- Press **CLEAR ENTER**
- Use the **ARROW** and **ENTER** keys to enter the first data set into **L1**
- Use the **ARROW** and **ENTER** keys to enter the second data set into **L2**
- The differences will be calculated in **L3**
- Use the up **ARROW** to highlight "**L3**"
- Press **CLEAR**—This will clear any old data, but **L3** will remain highlighted
- To enter the equation L3 = L1 − L2, use the following keystrokes:
  - Press **2**$^{\textbf{ND}}$ "**1**" (this will enter L1)
  - Press the **MINUS** button
  - Press **2**$^{\textbf{ND}}$ "**2**" (this will enter L2)

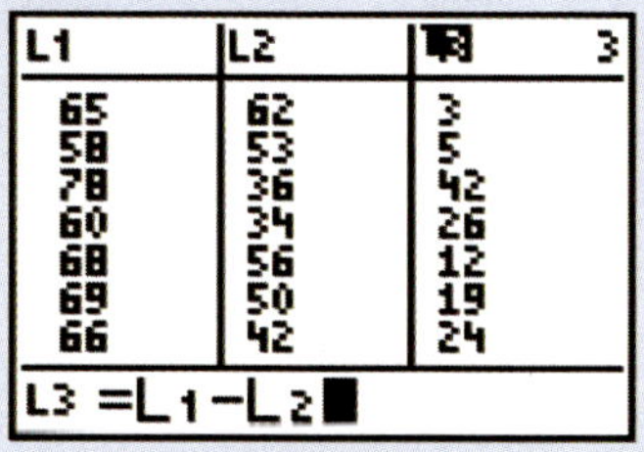

(Notice the equation at the bottom of the screen.)

- Press **ENTER** (the differences should be calculated in L3)

Step 2 *Access the Statistical Tests Menu*

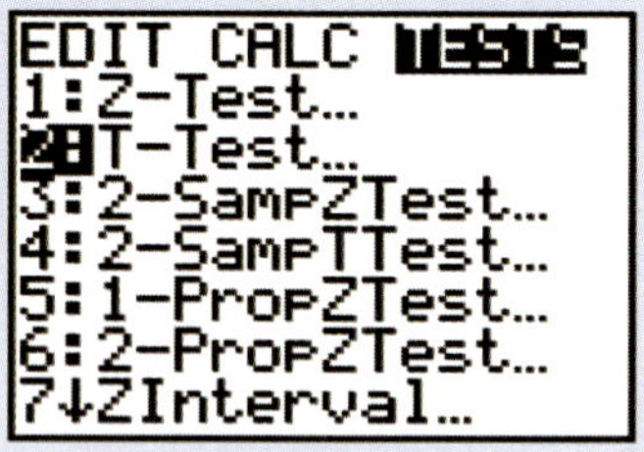

- Press **STAT**
- Arrow right to **TESTS**
- Arrow down to **T-Test (even for large sample case)**
- Press **ENTER**

Step 3 *Choose "**Data**"*

- Press **ENTER**
- Enter the values for the hypothesis test, where μ_0 = the value for μ_d in the null hypothesis
- Set **List** to **L3**
- Set **Freq** to **1**
- Use the **ARROW** to highlight the appropriate alternative hypothesis
- Press **ENTER**
- Arrow down to "**Calculate**"
- Press **ENTER**

The test statistic and the *p*-value will be displayed, as well as the sample mean, standard deviation, and sample size of the differences.

Confidence Interval for $(p_1 - p_2)$

Step 1 *Access the Statistical Tests Menu*

- Press **STAT**
- Arrow right to **TESTS**
- Arrow down to **2-PropZInt**
- Press **ENTER**

Step 2 *Enter the values from the sample information and **the confidence level***

where $\mathbf{x_1}$ = number of successes in the first sample (e.g., 53)

$\mathbf{n_1}$ = sample size for the first sample (e.g., 400)

$\mathbf{x_2}$ = number of successes in the second sample (e.g., 78)

$\mathbf{n_2}$ = sample size for the second sample (e.g., 500)

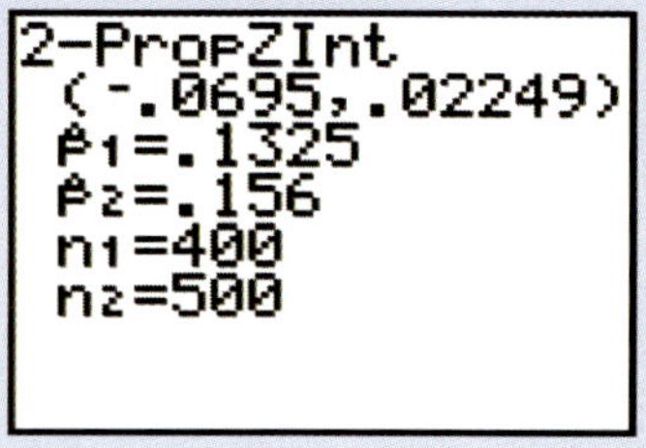

- Set **C-Level** to the confidence level
- Arrow down to "**Calculate**"
- Press **ENTER**

Hypothesis Test for $(p_1 - p_2)$

Step 1 *Access the Statistical Tests Menu*

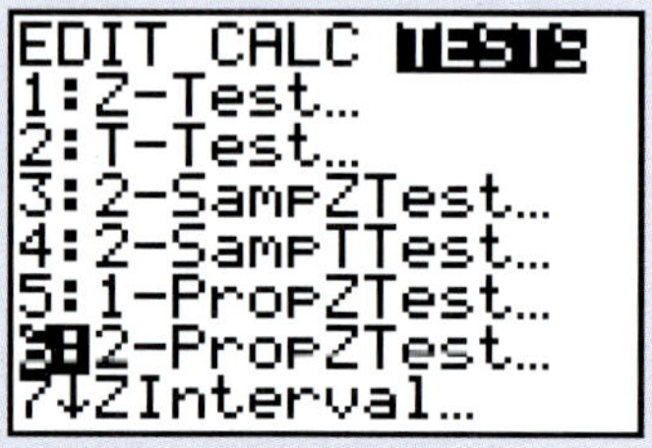

- Press **STAT**
- Arrow right to **TESTS**
- Arrow down to **2-PropZTest**
- Press **ENTER**

Step 2 *Enter the values from the sample information and **select the alternative hypothesis***

where $\mathbf{x_1}$ = number of successes in the first sample (e.g., 53)

$\mathbf{n_1}$ = sample size for the first sample (e.g., 400)

$\mathbf{x_2}$ = number of successes in the second sample (e.g., 78)

$\mathbf{n_2}$ = sample size for the second sample (e.g., 500)

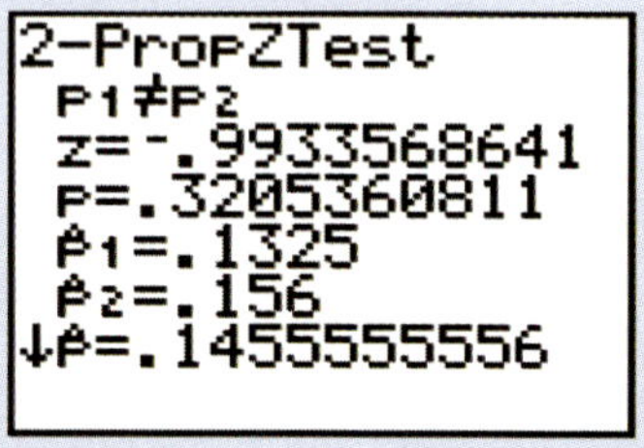

- Use the **ARROW** to highlight the appropriate alternative hypothesis
- Press **ENTER**
- Arrow down to "**Calculate**"
- Press **ENTER**

Hypothesis Test for (σ_1^2/σ_2^2)

Step 1 *Enter the data (Skip to Step 2 if you have summary statistics, not raw data)*

- Press **STAT** and select **1:Edit**

Note: If the lists already contain data, clear the old data. Use the up **ARROW** to highlight "**L1**."

- Press **CLEAR ENTER**
- Use the up **ARROW** to highlight "**L2**"
- Press **CLEAR ENTER**
- Use the **ARROW** and **ENTER** keys to enter the first data set into **L1**
- Use the **ARROW** and **ENTER** keys to enter the second data set into **L2**

Step 2 *Access the Statistical Tests Menu*

- Press **STAT**
- Arrow right to **TESTS**

- Arrow down to **2-SampFTest**
- Press **ENTER**

Step 3 *Choose "**Data**" or "**Stats**" ("Data" is selected when you have entered the raw data into the Lists. "Stats" is selected when you are given only the means, standard deviations, and sample sizes.)*

- Press **ENTER**
- If you selected "Data"
  - Set **List1** to **L1** and **List2** to **L2**
  - Set **Freq1** to **1** and set **Freq2** to **1**
  - Use the **ARROW** to highlight the appropriate alternative hypothesis
  - Press **ENTER**
  - Arrow down to "**Calculate**"
  - Press **ENTER**

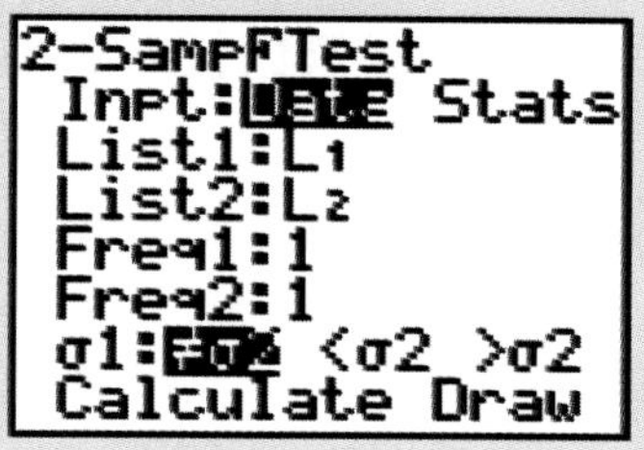

- If you selected "Stats," enter the standard deviations and sample sizes
  - Use the **ARROW** to highlight the appropriate alternative hypothesis.
  - Press **ENTER**
  - Arrow down to "**Calculate**"
  - Press **ENTER**

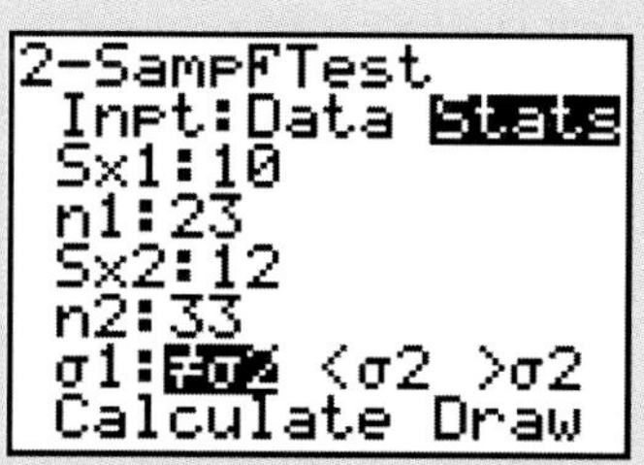

The results of the hypothesis test will be displayed with the p-value and the input data used.

The Kentucky Milk Case—Part II

Covers Chapters 5–7 – In The Kentucky Milk Case—Part I, you used graphical and numerical descriptive statistics to investigate bid collusion in the Kentucky school milk market. This case expands your previous analyses, incorporating inferential statistical methodology. The three areas of your focus are described below. (See page 114 for the file layout of the **MILK** data.) Again, you should prepare a professional document that presents the results of the analyses and any implications regarding collusionary practices in the tri-county Kentucky milk market.

1. *Incumbency rates.* Recall from Part I that market allocation (where the same dairy controls the same school districts year after year) is a common form of collusive behavior in bid-rigging conspiracies. Market allocation is typically gauged by the incumbency rate for a market in a given school year—defined as the percentage of school districts that are won by the same milk vendor who won the previous year. Past experience with milk bids in a competitive market reveals that a "normal" incumbency rate is about .7—that is, 70% of the school districts are expected to purchase their milk from the same vendor who supplied the milk the previous year. In the 13-district tri-county Kentucky market, 13 vendor transitions potentially exist each year. Over the 1985–1988 period (when bid collusion was alleged to have occurred), there are 52 potential vendor transitions. Based on the actual number of vendor transitions that occurred each year and over the 1985–1988 period, make an inference regarding bid collusion.

2. *Bid price dispersion.* Recall that in competitive sealed-bid markets, more dispersion or variability among the bids is observed than in collusive markets. (This is due to conspiring vendors sharing information about their bids.) Consequently, if collusion exists, the variation in bid prices in the tri-county market should be significantly smaller than the corresponding variation in the surrounding market. For each milk product, conduct an analysis to compare the bid price variances of the two markets each year. Make the appropriate inferences.

3. *Average winning bid price.* According to collusion theorists, the mean winning bid price in the "rigged" market will exceed the mean winning bid price in the competitive market for each year in which collusion occurs. In addition, the difference between the competitive average and the "rigged" average tends to grow over time when collusionary tactics are employed over several consecutive years. For each milk product, conduct an analysis to compare the winning bid price means of the tri-county and surrounding markets each year. Make the appropriate inferences.

CONTENTS

8 Design of Experiments and Analysis of Variance

Where We've Been

- Presented methods for estimating and testing hypotheses about a single population mean
- Presented methods for comparing two population means

Where We're Going

- Discuss the critical elements in the design of a sampling experiment
- Learn how to set up three of the more popular experimental designs for comparing more than two population means: *completely randomized, randomized block,* and *factorial designs*
- Show how to analyze data collected from a designed experiment using a technique called an *analysis of variance*

Statistics IN Action Violence, Sex, and Recall of Television Advertisements

Television advertisers seek to promote their products on TV programs that attract the most viewers. For example, the annual Super Bowl, in which the two best teams in the National Football League play for the championship, is one of the most watched programs (sports or otherwise) on television, reaching over 130 million viewers. Many of these viewers watch simply to see the creative commercials. For the 2009 Super Bowl, the National Broadcasting Company charged advertisers $3 million for a single 30-second TV commercial. Such a cost is justified, of course, only if the TV viewer actually remembers the product that is advertised.

The decision as to which TV programs to purchase commercial spots for is a critical one for marketers of a product. Research suggests that certain types of TV programs draw attention away from commercials embedded in the show. As a consequence, the viewer is less likely to remember the commercial and the product being advertised. Do TV shows with violence and sex impair memory for commercials? *(continued)*

Statistics IN Action *(continued)*

To answer this question, Iowa State University Professors B. Bushman and A. Bonacci conducted a designed experiment involving 324 adults (*Journal of Applied Psychology,* June 2002). Each participant was randomly assigned to one of three TV content groups, with 108 subjects in each group. One group watched a TV program with a violent content code (V) rating (e.g., "Tour of Duty," "World Wrestling Federation Monday Night Nitro," "La Femme Nikita"); the second group viewed a show with a sex content code (S) rating (e.g., "Strip Mall," "Howard Stern," "Son of the Beach"); and the last group watched a neutral TV program with neither a V nor an S rating (e.g., "Candid Camera," "Miracle Pets," "Encounters with the Unexplained"). Nine commercials were embedded into each TV show. After viewing the program, each participant was scored on his/her recall of the brand names in the commercial messages, with scores ranging from 0 (no brands recalled) to 9 (all brands recalled). In addition, the researchers recorded whether the subject had seen the commercial before.

The data (simulated from information provided in the article) are saved in the **TVADRECALL** file. The researchers want to (1) compare the mean recall scores of the three TV content groups and (2) determine if the differences in the mean recall scores of the three content groups depend on whether or not the commercial was seen before. In this chapter, we demonstrate how to analyze the data in the **TVADRECALL** file. We apply the statistical methodology of this chapter in three Statistics in Action Revisited examples.

Statistics IN Action Revisited

- Testing for Differences in Mean Ad Recall Scores (p. 465)
- Ranking Mean Ad Recall Scores (p. 476)
- A Two-Way Analysis of Variance (p. 502)

Data Set: TVADRECALL

Most of the data analyzed in previous chapters were collected in *observational* sampling experiments rather than *designed* sampling experiments. In *observational studies,* the analyst has little or no control over the variables under study and merely observes their values. In contrast, *designed experiments* are those in which the analyst attempts to control the levels of one or more variables to determine their effect on a variable of interest. When properly designed, such experiments allow the analyst to determine whether a change in the controlled variable *causes* a change in the response variable, that is, it allows one to infer *cause and effect.* Although many practical business situations do not present the opportunity for such control, it is instructive, even for observational experiments, to have a working knowledge of the analysis and interpretation of data that result from designed experiments and to know the basics of how to design experiments when the opportunity arises.

We first present the basic elements of an experimental design in Section 8.1. We then discuss three of the simpler, and more popular, experimental designs in Sections 8.2, 8.4, and 8.5. In Section 8.3 we show how to rank the population means, from smallest to largest.

8.1 Elements of a Designed Experiment

Certain elements are common to almost all designed experiments, regardless of the specific area of application. For example, the *response* is the variable of interest in the experiment. The response might be the SAT scores of a high school senior, the total sales of a firm last year, or the total income of a particular household this year. The response is also called the *dependent variable, y.* We use these terms interchangeably in this chapter.

The **response variable** is the variable of interest to be measured in the experiment. We also refer to the response as the **dependent variable.**

The intent of most statistical experiments is to determine the effect of one or more variables on the response. These variables, which we called the *independent variables* in regression analysis, are often referred to as the *factors* in a designed experiment. Like independent variables, factors are either *quantitative* or *qualitative,* depending on whether the variable is measured on a numerical scale or not. For example, we might want to explore the effect of the qualitative factor Gender on the response SAT score. In other words, we want to compare the SAT scores of male and female high school seniors. Or, we might wish to determine the effect of the quantitative factor Number of salespeople on the response Total sales for retail firms. Often two or more factors are of interest. For example, we might want to determine the effect of the quantitative factor Number of wage earners and the qualitative factor Location on the response Household income.

Factors are those variables whose effect on the response is of interest to the experimenter. **Quantitative factors** are measured on a numerical scale, whereas **qualitative factors** are those that are not (naturally) measured on a numerical scale. Factors are also referred to as **independent variables.**

Levels are the values of the factors that are used in the experiment. The levels of qualitative factors are usually nonnumerical. For example, the levels of Gender are Male and Female, and the levels of Location might be North, East, South, and West.* The levels of quantitative factors are numerical values. For example, the Number of salespeople may have levels 1, 3, 5, 7, and 9. The factor Years of education may have levels 8, 12, 16, and 20.

Factor levels are the values of the factor used in the experiment.

When a *single factor* is employed in an experiment, the *treatments* of the experiment are the levels of the factor. For example, if the effect of the factor Gender on the response SAT score is being investigated, the treatments of the experiment are the two levels of Gender—Female and Male. Or, if the effect of the Number of wage earners on Household income is the subject of the experiment, the numerical values assumed by the quantitative factor Number of wage earners are the treatments. If *two or more factors* are used in an experiment, the treatments are the factor-level combinations used. For example, if the effects of the factors Gender and Socioeconomic Status (SES) on the response SAT score are being investigated, the treatments are the combinations of the levels of Gender and SES used; thus (Female, high SES), (Male, high SES), and (Female, low SES) would all be treatments.

The **treatments** of an experiment are the factor-level combinations used.

*The levels of a qualitative variable may bear numerical labels. For example, the Locations could be numbered 1, 2, 3 and 4. However, in such cases, the numerical labels for a qualitative variable will usually be codes representing nonnumerical levels.

The objects on which the response variable and factors are observed are the *experimental units.* For example, SAT score, High school GPA, and Gender are all variables that can be observed on the same experimental unit—a high school senior. Or, the Total sales, the Earnings per share, and the Number of salespeople can be measured on a particular firm in a particular year, and the firm-year combination is the experimental unit. The Total income, the Number of female wage earners, and the Location can be observed for a household at a particular point in time, and the household-time combination is the experimental unit. Every experiment, whether observational or designed, has experimental units on which the variables are observed. However, the identification of the experimental units is more important in designed experiments, when the experimenter must actually sample the experimental units and measure the variables.

An **experimental unit** is the object on which the response and factors are observed or measured.*

When the specification of the treatments and the method of assigning the experimental units to each of the treatments are controlled by the analyst, the experiment is said to be *designed.* In contrast, if the analyst is just an observer of the treatments on a sample of experimental units, the experiment is *observational.* For example, if you give one randomly selected group of employees a training program and withhold it from another randomly selected group to evaluate the effect of the training on worker productivity, then you are designing an experiment. If, on the other hand, you compare the productivity of employees with college degrees with the productivity of employees without college degrees, the experiment is observational.

A **designed experiment** is one for which the analyst controls the specification of the treatments and the method of assigning the experimental units to each treatment. An **observational experiment** is one for which the analyst simply observes the treatments and the response on a sample of experimental units.

BIOGRAPHY **SIR RONALD A. FISHER (1890–1962)**

The Founder of Modern Statistics

At a young age, Ronald Fisher demonstrated special abilities in mathematics, astronomy, and biology. (Fisher's biology teacher once divided all his students for "sheer brilliance" into two groups—Fisher and the rest.) Fisher graduated from prestigious Cambridge University in London in 1912 with a BA in astronomy, and, after several years teaching mathematics, he found work at the Rothamsted Agricultural Experiment station. There, Fisher began his extraordinary career as a statistician. Many consider Fisher to be the leading founder of modern statistics. His contributions to the field include the notion of unbiased statistics, the development of *p*-values for hypothesis tests, the invention of analysis of variance for designed experiments, maximum likelihood estimation theory, and the mathematical distributions of several well-known statistics. Fisher's book *Statistical Methods for Research Workers* (written in 1925) revolutionized applied statistics, demonstrating how to analyze data and interpret the results with very readable and practical examples. In 1935, Fisher wrote *The Design of Experiments,* where he first described his famous experiment on the "lady tasting tea." (Fisher showed, through a designed experiment, that the lady really could determine whether tea poured into milk tastes better than milk poured into tea.) Before his death, Fisher was elected a Fellow of the Royal Statistical Society, was awarded numerous medals, and was knighted by the Queen of England. ■

*Recall (Chapter 1) that the set of all experimental units is the population.

The diagram in Figure 8.1 provides an overview of the experimental process and a summary of the terminology introduced in this section. Note that the experimental unit is at the core of the process. The method by which the sample of experimental units is selected from the population determines the type of experiment. The level of every factor (the treatment) and the response are all variables that are observed or measured on each experimental unit.

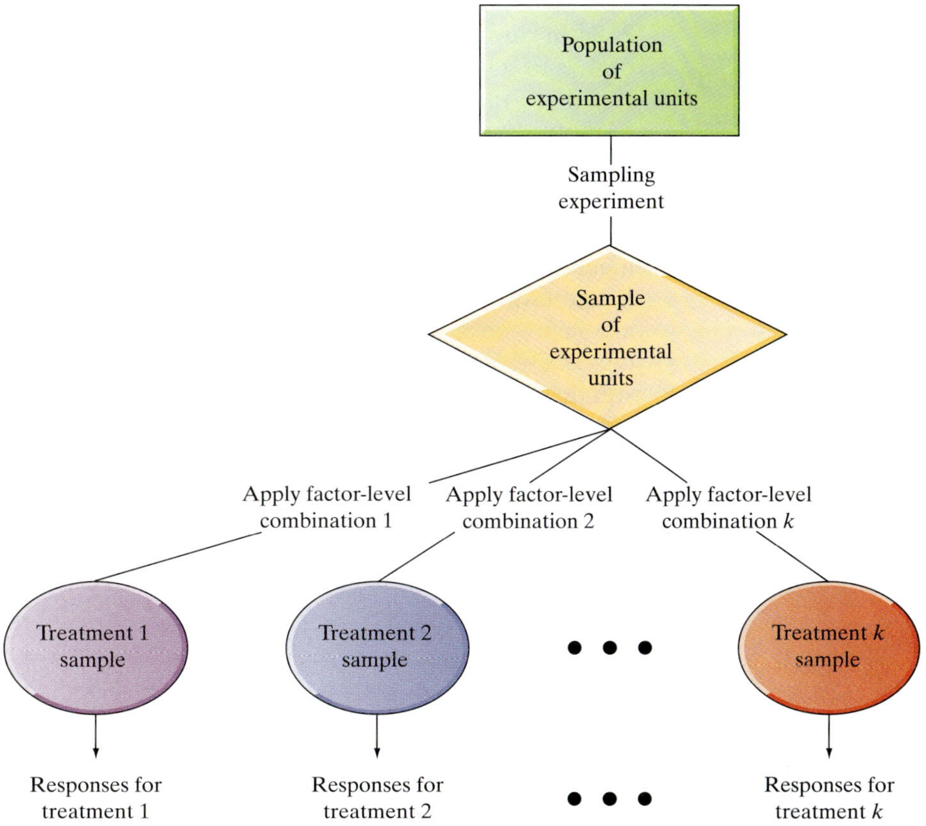

Figure 8.1
Sampling experiment: Process and terminology

Example 8.1
The Key Elements of a Designed Experiment to Test Golf Ball Brands

Problem The USGA (United States Golf Association) regularly tests golf equipment to ensure that it conforms to USGA standards. Suppose it wishes to compare the mean distance traveled by four different brands of golf balls when struck by a driver (the club used to maximize distance). The following experiment is conducted: 10 balls of each brand are randomly selected. Each is struck by "Iron Byron" (the USGA's golf robot named for the famous golfer, Byron Nelson) using a driver, and the distance traveled is recorded. Identify each of the following elements in this experiment: response, factors, factor types, levels, treatments, and experimental units.

Solution The response is the variable of interest, Distance traveled. The only factor being investigated is the Brand of golf ball, and it is nonnumerical and therefore qualitative. The four brands (say A, B, C, and D) represent the levels of this factor. Because only one factor is used, the treatments are the four levels of this factor—that is, the four brands. The experimental unit is a golf ball; more specifically, it is a golf ball at a particular position in the striking sequence, because the distance traveled can be recorded only when the ball is struck, and we would expect the distance to be different (due to random factors such as wind resistance, landing place, and so forth) if the same ball is struck a second time. Note that 10 experimental units are sampled for each treatment, generating a total of 40 observations.

Look Back This experiment, like many real applications, is a blend of designed and observational: The analyst cannot control the assignment of the brand to each golf ball (observational), but he or she can control the assignment of each ball to the position in the striking sequence (designed).

Now Work Exercise 8.5

Example 8.2

A Two-Factor Experiment to Test Golf Ball Brands

Problem Suppose the USGA is interested in comparing the mean distances the four brands of golf balls travel when struck by a five-iron and by a driver. Ten balls of each brand are randomly selected, five to be struck by the driver and five by the five-iron. Identify the elements of the experiment and construct a schematic diagram similar to Figure 8.1 to provide an overview of this experiment.

Solution The response is the same as in Example 8.1—Distance traveled. The experiment now has two factors: Brand of golf ball and Club used. There are four levels of Brand (A, B, C, and D) and two of Club (driver and five-iron, or 1 and 5). Treatments are factor-level combinations, so there are $4 \times 2 = 8$ treatments in this experiment: (A, 1), (A, 5), (B, 1), (B, 5), (C, 1), (C, 5), (D, 1), and (D, 5). The experimental units are still the golf balls. Note that five experimental units are sampled per treatment, generating 40 observations. The experiment is summarized in Figure 8.2.

Look Back Whenever there are two or more factors in an experiment, remember to combine the levels of the factors—one level from each factor—to obtain the treatments.

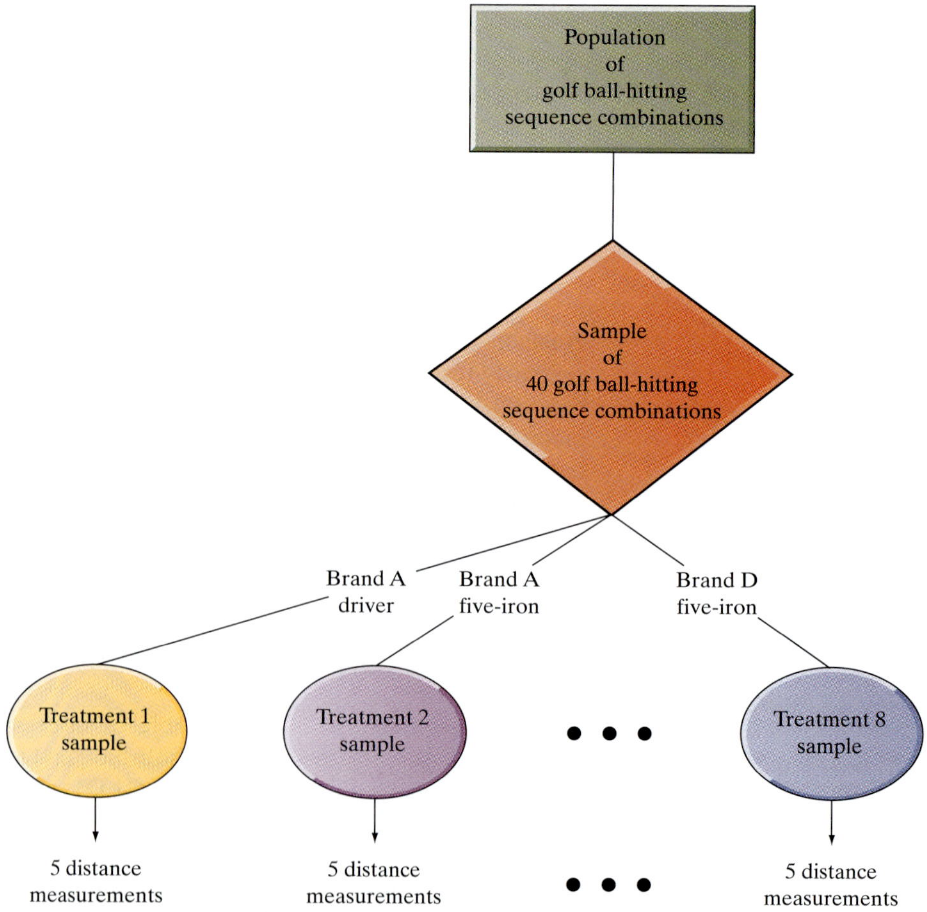

Figure 8.2
Two-factor golf experiment summary: Example 8.2

Now Work Exercise 8.10

Our objective in designing an experiment is usually to maximize the amount of information obtained about the relationship between the treatments and the response. Of course, we are almost always subject to constraints on budget, time, and even the availability of experimental units. Nevertheless, designed experiments are generally preferred to observational experiments. Not only do we have better control of the amount and quality of the information collected, but we also avoid the biases inherent in observational experiments in the selection of the experimental units representing each treatment. Inferences based on observational experiments always carry the implicit assumption that the sample has no hidden bias that was not considered in the statistical analysis. Better understanding of the potential problems with observational experiments is a by-product of our study of experimental design in the remainder of this chapter.

Activity 8.1 Designed versus Observational Experiments

In this Activity, you will revisit Activity 7.1 (p. 404) and Activity 7.2 (p. 413) and consider two new but similar experiments.

1. Explain why each of the situations in Activity 7.1 and Activity 7.2 are observational experiments. What key elements of a designed experiment are missing in each situation?
2. A movie company wishes to measure the effect of advertising on box office receipts. Thirty U.S. cities with similar demographics are chosen for an experiment. The thirty cities are randomly divided into three groups of ten cities. In each city a trailer for a new film will be run on a local cable station during prime time in the week leading up to the release of the film. In the first group of ten cities the trailer will be run 500 times, in the second group the trailer will be run 1,000 times, and in the third group it will be run 1,500 times. The company will collect the box office receipts for the opening weekend of the film in each city and compare the mean box office receipts.

   Explain why this is a designed experiment. Identify the factor and the response variable. Is the factor quantitative or qualitative? Identify the factor levels and the experimental units. What part of choosing the cities is not necessarily random? Explain why it might be difficult to randomize this part of the experiment.
3. Suppose Bank of America wants to determine whether e-mail and postcard reminders of the benefits of the *Keep the Change* program result in customers using their debit cards more often. A random sample of customers is chosen and split into four groups. The customers in one group are sent an e-mail reminder, those in the second group are sent a postcard, the customers in the third group are sent both, and those in the last group neither. The bank keeps track of how many more times each customer uses his/her debit card in the two weeks after the reminders are sent as compared to the two weeks before the reminders are sent. The means for the four groups are compared.

   Explain why this is a designed experiment. Identify the factors and the response variable. Is the factor quantitative or qualitative? Identify the factor levels, the treatments, and the experimental units. Comparing this experiment to that in Exercise 2, why is it more realistic to choose a random sample for the experiment?

Exercises 8.1–8.14

Learning the Mechanics

8.1 What are the treatments for a designed experiment that uses one qualitative factor with four levels—A, B, C, and D?

8.2 What are the treatments for a designed experiment with two factors, one qualitative with two levels (A and B) and one quantitative with five levels (50, 60, 70, 80, and 90)?

8.3 What is the difference between an observational experiment and a designed experiment?

8.4 What are the experimental units on which each of the following responses are observed?

a. College GPA
b. Household income
c. Gasoline mileage rating for an automobile model
d. Number of defective sectors on a computer disk
e. December unemployment rate for a state

Applying the Concepts—Basic

8.5 NW **Identifying the type of experiment.** Brief descriptions of a number of experiments are given next. Determine whether each is observational or designed and explain your reasoning.

a. An economist obtains the unemployment rate and gross state product for a sample of states over the past 10 years, with the objective of examining the relationship between the unemployment rate and the gross state product by census region.
b. A manager in a paper production facility installs one of three incentive programs in each of nine plants to determine the effect of each program on productivity.
c. A marketer of personal computers runs ads in each of four national publications for one quarter and keeps track of the number of sales that are attributable to each publication's ad.

d. An electric utility engages a consultant to monitor the discharge from its smokestack on a monthly basis over a 1-year period to relate the level of sulfur dioxide in the discharge to the load on the facility's generators.

e. Intrastate trucking rates are compared before and after governmental deregulation of prices changed, with the comparison also taking into account distance of haul, goods hauled, and the price of diesel fuel.

8.6 Credit card market shares. Refer to the study of credit card purchases, Exercise 1.14 (p. 22). Recall that *The Nilson Report* (2007) tracked credit card purchases made by cardholders and measured two variables: (1) the type of credit card used (VISA, MasterCard, American Express, or Discover), and (2) the amount (in dollars) of the purchase. Suppose we want to compare the mean purchase amounts of VISA, MasterCard, American Express, and Discover cardholders. Identify each of the following elements for this study:

a. Response variable **b.** Factor(s)
c. Treatments **d.** Experimental units

8.7 CT scanning for lung cancer. Refer to Exercise 1.22 (p. 23) and the University of South Florida clinical trial of 50,000 smokers to compare the effectiveness of computed tomography (CT) scans with X-rays for detecting lung cancer (*Todays' Tomorrows,* Fall 2002). Recall that each participating smoker will be randomly assigned to one of two screening methods, CT or chest X-ray, and the age (in years) at which the scanning method first detects a tumor will be determined. One goal of the study is to compare the mean ages when cancer is first detected of the two screening methods.

a. Identify the response variable of the study.
b. Identify the experimental units of the study.
c. Identify the factor(s) in the study.
d. Identify the treatments in the study.

8.8 Accounting and Machiavellianism. A study of Machiavellian traits in accountants was published in *Behavioral Research in Accounting* (January 2008). Recall (from Exercise 1.27, p. 24) that *Machiavellian* describes negative character traits such as manipulation, cunning, duplicity, deception, and bad faith. A Mach rating score was determined for each in a sample of accounting alumni of a large southwestern university. The accountants were then classified as having high, moderate, or low Mach rating scores. For one portion of the study, the researcher investigated the impact of both Mach score classification and gender on the average income of an accountant. For this experiment, identify each of the following:

a. Experimental unit
b. Response variable
c. Factors
d. Levels of each factor
e. Treatments

Applying the Concepts—Intermediate

8.9 Exam performance study. In *Teaching of Psychology* (August 1998), a study investigated whether final exam performance is affected by whether students take a practice test. Students in an introductory psychology class at Pennsylvania State University were initially divided into three groups based on their class standing: Low, Medium, or High. Within each group, students were randomly assigned to either attend a review session or take a practice test prior to the final exam. Thus, six groups were formed: (Low, Review), (Low, Practice exam), (Medium, Review), (Medium, Practice exam), (High, Review), and (High, Practice exam). One goal of the study was to compare the mean final exam scores of the six groups of students.

a. What is the experimental unit for this study?
b. Is the study a designed experiment? Why?
c. What are the factors in the study?
d. Give the levels of each factor.
e. How many treatments are in the study? Identify them.
f. What is the response variable?

8.10 NW **Value perceptions of consumers.** Refer to the *Journal of Consumer Research* study of whether between-store comparisons result in greater perceptions of value by consumers than within-store comparisons, Example 1.7 (p. 16). Recall that 50 consumers were randomly selected from all consumers in a designated market area to participate in the study. The researchers randomly assigned 25 consumers to read a within-store price promotion advertisement ("was \$100, now \$80") and 25 consumers to read a between store price promotion ("\$100 there, \$80 here"). The consumers then gave their opinion on the value of the discount offer on a 10-point scale (where 1 = lowest value and 10 = highest value). The goal is to compare the average discount values of the two groups of consumers.

a. What is the response variable for this study?
b. What are the treatments for this study?
c. What is the experimental unit for this study?

8.11 Value perceptions of consumers (cont'd). Refer to Exercise 8.10. In addition to the factor, Type of advertisement (within-store price promotion and between-store price promotion), the researchers also investigated the impact of a second factor—Location where ad is read (at home or in the store). About half of the consumers who were assigned to the within-store price promotion read the ad at home, and the other half read the ad in the store. Similarly, about half of the consumers who were assigned to the between-store price promotion read the ad at home, and the other half read the ad in the store. In this second experiment, the goal is to compare the average discount values of the groups of consumers created by combining Type of advertisement with Location.

a. How many treatments are involved in this experiment?
b. Identify the treatments.

8.12 Baker's versus brewer's yeast. The *Electronic Journal of Biotechnology* (Dec. 15, 2003) published an article on a comparison of two yeast extracts—baker's yeast and brewer's yeast. Brewer's yeast is a surplus by-product obtained from a brewery; hence it is less expensive than primary-grown baker's yeast. Samples of both yeast extracts were prepared at four different temperatures (45, 48, 51, and 54°C), and the autolysis yield (recorded as a percentage) was measured for each of the yeast-temperature combinations. The goal of the analysis was to investigate the impact of yeast extract and temperature on mean autolysis yield.

a. Identify the factors (and factor levels) in the experiment.
b. Identify the response variable.
c. How many treatments are included in the experiment?
d. What type of experimental design is employed?

Applying the Concepts—Advanced

8.13 Testing a new pain-reliever tablet. Paracetamol is the active ingredient in drugs designed to relieve mild to moderate pain and fever. To save costs, pharmaceutical companies are looking to produce paracetamol tablets from locally available materials. The properties of paracetamol tablets derived from khaya gum were studied in the *Tropical Journal of Pharmaceutical Research* (June 2003). Three factors believed to impact the properties of paracetamol tablets are (1) the nature of the binding agent, (2) the concentration of the binding agent, and (3) the relative density of the tablet. In the experiment, binding agent was set at two levels (khaya gum and PVP), binding concentration at two levels (.5% and 4.0%), and relative density at two levels (low and high). One of the dependent variables investigated in the study was tablet dissolution time (i.e., the amount of time [in minutes] for 50% of the tablet to dissolve). The goal of the study was to determine the effect of binding agent, binding concentration, and relative density on mean dissolution time.

a. Identify the dependent (response) variable in the study.
b. What are the factors investigated in the study? Give the levels of each.
c. How many treatments are possible in the study? List them.

8.14 Ethics of salespeople. Within marketing, the area of personal sales has long suffered from a poor ethical image, particularly in the eyes of college students. An article in *Journal of Business Ethics* (Vol. 15, 1996) investigated whether such opinions by college students are a function of the type of sales job (high tech versus low tech) and/or the sales task (new account development versus account maintenance). Four different samples of college students were confronted with the four different situations (new account development in a high-tech sales task, new account development in a low-tech sales task, account maintenance in a high-tech sales task, and account maintenance in a low-tech sales task) and were asked to evaluate the ethical behavior of the salesperson on a 7-point scale ranging from 1 (not a serious ethical violation) to 7 (a very serious ethical violation). Identify each of the following elements of the experiment:

a. Response **b.** Factor(s) and factor level(s)
c. Treatments **d.** Experimental units

8.2 The Completely Randomized Design: Single Factor

The simplest experimental design, a *completely randomized design,* consists of the *independent random selection* of experimental units representing each treatment. For example, we could independently select random samples of 20 female and 15 male high school seniors to compare their mean SAT scores. Or, we could independently select random samples of 30 households from each of four census districts to compare the mean income per household among the districts. In both examples, our objective is to compare treatment means by selecting random, independent samples for each treatment.

> A **completely randomized design** is a design for which independent random samples of experimental units are selected for each treatment.*

Example 8.3
Assigning Treatments in a Completely Randomized Design to Compare Bottled Water Brands

Problem Suppose we want to compare the taste preferences of consumers for three different brands of bottled water (say, Brands A, B, and C) using a random sample of 15 bottled water consumers. Set up a completely randomized design for this purpose—that is, assign the treatments to the experimental units for this design.

Solution In this study, the experimental units are the 15 consumers, and the treatments are the three brands of bottled water. One way to set up the completely randomized design is to randomly assign one of the three brands to each consumer to taste. Then we could measure (say, on a 1-to 10-point scale) the taste preference of each consumer. A good practice is to assign the same number of consumers to each brand—in this case, 5 consumers to each of the

*We use *completely randomized* design to refer to both designed and observational experiments. Thus, the only requirement is that the experimental units to which treatments are applied (designed) or on which treatments are observed (observational) are independently selected for each treatment.

CRD3brands.MTW ***

↓	C1	C2	C3	C4
	Consumer	BrandA	BrandB	BrandC
1	1	2	15	6
2	2	11	14	5
3	3	1	7	12
4	4	13	10	9
5	5	3	8	4
6	6			
7	7			

Figure 8.3
Minitab random assignments of consumers to brands

three brands. (When an equal number of experimental units are assigned to each treatment, we call the design a *balanced design.*)

A random number table (Table I in Appendix B) or computer software can be used to make the random assignments. Figure 8.3 is a Minitab worksheet showing the random assignments made with the Minitab "Random Data" function. You can see that Minitab randomly assigned consumers numbered 2, 11, 1, 13, and 3 to taste Brand A, consumers numbered 15, 14, 7, 10, and 8 to taste Brand B, and consumers numbered 6, 5, 12, 9, and 4 to taste Brand C.

Look Back In some experiments, it will not be possible to randomly assign treatments to the experimental units the units will already be associated with one of the treatments. (For example, if the treatments are Male and Female, you cannot change a person's gender.) In this case, a completely randomized design is one where you select independent random samples of experimental units from each treatment.

Now Work: Exercise 8.22d

The objective of a completely randomized design is usually to compare the treatment means. If we denote the true, or population, means of the k treatments as $\mu_1, \mu_2, \ldots, \mu_k$, then we will test the null hypothesis that the treatment means are all equal against the alternative that at least two of the treatment means differ:

$$H_0: \mu_1 = \mu_2 = \cdots = \mu_k$$

H_a: At least two of the k treatment means differ

The μ's might represent the means of *all* female and male high school seniors' SAT scores or the means of *all* households' income in each of four census regions.

To conduct a statistical test of these hypotheses, we will use the means of the independent random samples selected from the treatment populations using the completely randomized design—that is, we compare the k sample means, $\bar{x}_1, \bar{x}_2, \ldots, \bar{x}_k$.

For example, suppose you select independent random samples of five female and five male high school seniors and obtain sample mean SAT scores of 550 and 590, respectively. Can we conclude that males score 40 points higher, on average, than females? To answer this question, we must consider the amount of sampling variability among the experimental units (students). If the scores are as depicted in the dot plot shown in Figure 8.4, then the difference between the means is small relative to the sampling variability of the scores within the treatments—Female and Male. We would be inclined not to reject the null hypothesis of equal population means in this case.

In contrast, if the data are as depicted in the dot plot of Figure 8.5, then the sampling variability is small relative to the difference between the two means. We would be inclined to favor the alternative hypothesis that the population means differ in this case.

You can see that the key is to compare the difference between the treatment means to the amount of sampling variability. To conduct a formal statistical test of the hypotheses requires numerical measures of the difference between the treatment

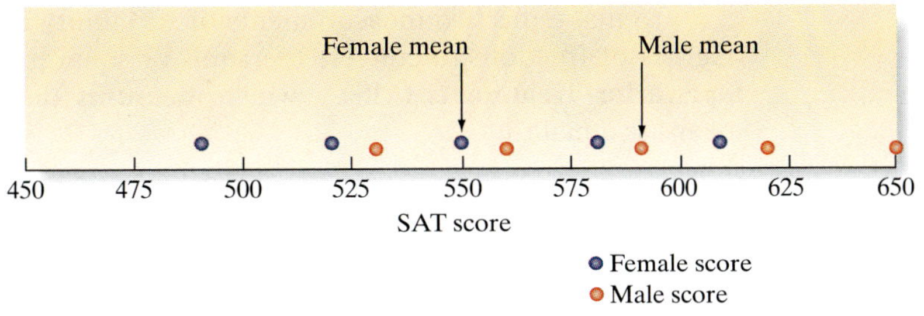

Figure 8.4
Dot plot of SAT scores: Difference between means dominated by sampling variability

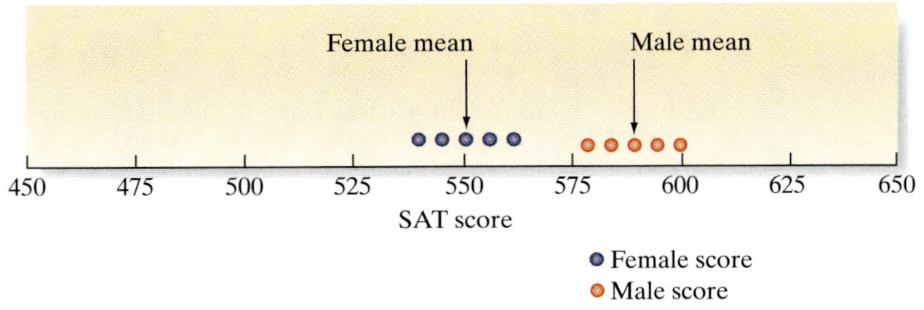

Figure 8.5
Dot plot of SAT scores: Difference between means large relative to sampling variability

means and the sampling variability within each treatment. The variation between the treatment means is measured by the **Sum of Squares for Treatments (SST),** which is calculated by squaring the distance between each treatment mean and the overall mean of *all* sample measurements, multiplying each squared distance by the number of sample measurements for the treatment and adding the results over all treatments:

$$\text{SST} = \sum_{i=1}^{k} n_i(\bar{x}_i - \bar{x})^2 = 5(550 - 570)^2 + 5(590 - 570)^2 = 4{,}000$$

where we use $\bar{x}$ to represent the overall mean response of all sample measurements—that is, the mean of the combined samples. The symbol n_i is used to denote the sample size for the ith treatment. You can see that the value of SST is 4,000 for the two samples of five female and five male SAT scores depicted in Figures 8.4 and 8.5.

Next, we must measure the sampling variability within the treatments. We call this the **Sum of Squares for Error (SSE)** because it measures the variability around the treatment means that is attributed to sampling error. Suppose the 10 measurements in the first dot plot (Figure 8.4) are 490, 520, 550, 580, and 610 for females, and 530, 560, 590, 620, and 650 for males. Then the value of SSE is computed by summing the squared distance between each response measurement and the corresponding treatment mean and then adding the squared differences over all measurements in the entire sample:

$$\text{SSE} = \sum_{j=1}^{n_1}(x_{1j} - \bar{x}_1)^2 + \sum_{j=1}^{n_2}(x_{2j} - \bar{x}_2)^2 + \cdots + \sum_{j=1}^{n_k}(x_{kj} - \bar{x}_k)^2$$

where the symbol x_{1j} is the jth measurement in sample 1, x_{2j} is the jth measurement in sample 2, and so on. This rather complex-looking formula can be simplified by recalling the formula for the sample variance, s^2, given in Chapter 2:

$$s^2 = \sum_{i=1}^{n} \frac{(x_i - \bar{x})^2}{n - 1}$$

Note that each sum in SSE is simply the numerator of s^2 for that particular treatment. Consequently, we can rewrite SSE as

$$\text{SSE} = (n_1 - 1)s_1^2 + (n_2 - 1)s_2^2 + \cdots + (n_k - 1)s_k^2$$

where $s_1^2, s_2^2, \ldots, s_k^2$ are the sample variances for the k treatments. For our samples of SAT scores, we find $s_1^2 = 2{,}250$ (for females) and $s_2^2 = 2{,}250$ (for males); then we have

$$\text{SSE} = (5 - 1)(2{,}250) + (5 - 1)(2{,}250) = 18{,}000$$

To make the two measurements of variability comparable, we divide each by the degrees of freedom to convert the sums of squares to mean squares. First, the **Mean Square for Treatments (MST),** which measures the variability among the treatment means, is equal to

$$\text{MST} = \frac{\text{SST}}{k - 1} = \frac{4{,}000}{2 - 1} = 4{,}000$$

where the number of degrees of freedom for the k treatments is $(k - 1)$. Next, the **Mean Square for Error (MSE),** which measures the sampling variability within the treatments, is

$$\text{MSE} = \frac{\text{SSE}}{n - k} = \frac{18{,}000}{10 - 2} = 2{,}250$$

Finally, we calculate the ratio of MST to MSE—an ***F*-statistic:**

$$F = \frac{\text{MST}}{\text{MSE}} = \frac{4{,}000}{2{,}250} = 1.78$$

Values of the F-statistic near 1 indicate that the two sources of variation, between treatment means and within treatments, are approximately equal. In this case, the difference between the treatment means may well be attributable to sampling error, which provides little support for the alternative hypothesis that the population treatment means differ. Values of F well in excess of 1 indicate that the variation among treatment means well exceeds that within treatments and therefore support the alternative hypothesis that the population treatment means differ.

When does F exceed 1 by enough to reject the null hypothesis that the means are equal? This depends on the degrees of freedom for treatments and for error and on the value of α selected for the test. We compare the calculated F-value to a table of F-values (Tables VII–X in Appendix B) with $v_1 = (k - 1)$ degrees of freedom in the numerator and $v_2 = (n - k)$ degrees of freedom in the denominator and corresponding to a Type I error probability of α. For the SAT score example, the F-statistic has $v_1 = (2 - 1) = 1$ numerator degree of freedom and $v_2 = (10 - 2) = 8$ denominator degrees of freedom. Thus, for $\alpha = .05$, we find (Table VIII in Appendix B)

$$F_{.05} = 5.32$$

The implication is that MST would have to be 5.32 times greater than MSE before we could conclude at the .05 level of significance that the two population treatment means differ. Because the data yielded $F = 1.78$, our initial impressions from the dot plot in Figure 8.4 are confirmed—there is insufficient information to conclude that the mean SAT scores differ for the populations of female and male high school seniors. The rejection region and the calculated F value are shown in Figure 8.6.

In contrast, consider the dot plot in Figure 8.5. Because the means are the same as in the first example, 550 and 590, respectively, the variation between the means is the same, MST = 4,000. But the variation within the two treatments appears to be considerably smaller. The observed SAT scores are 540, 545, 550, 555, and 560 for females and

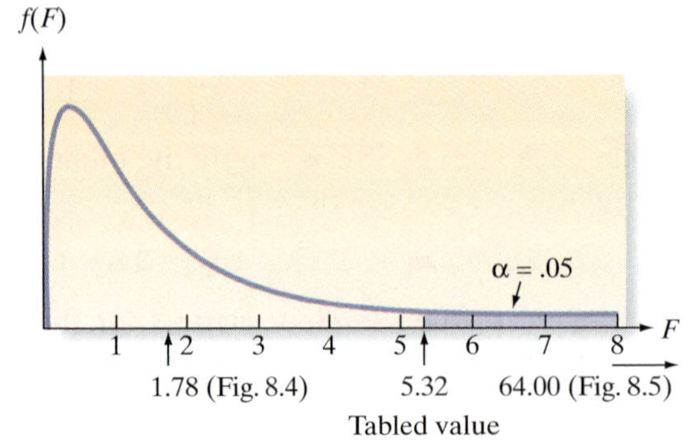

Figure 8.6
Rejection region and calculated F values for SAT score samples

580, 585, 590, 595, and 600 for males. These values yield $s_1^2 = 62.5$ and $s_2^2 = 62.5$. Thus, the variation within the treatments is measured by

$$\text{SSE} = (5 - 1)(62.5) + (5 - 1)(62.5) = 500$$

$$\text{MSE} = \frac{\text{SSE}}{n - k} = \frac{500}{8} = 62.5$$

Then the F-ratio is

$$F = \frac{\text{MST}}{\text{MSE}} = \frac{4{,}000}{62.5} = 64.0$$

Again, our visual analysis of the dot plot is confirmed statistically: $F = 64.0$ well exceeds the tabled F-value, 5.32, corresponding to the .05 level of significance. We would therefore reject the null hypothesis at that level and conclude that the SAT mean score of males differs from that of females.

Recall that we performed a hypothesis test for the difference between two means in Section 7.2, using a two-sample t-statistic for two independent samples. When two independent samples are being compared, the two-tailed t- and F-tests are equivalent. To see this, recall the formula

$$t = \frac{\bar{x}_1 - \bar{x}_2}{\sqrt{s_p^2\left(\frac{1}{n_1} + \frac{1}{n_2}\right)}} = \frac{590 - 550}{\sqrt{(62.5)\left(\frac{1}{5} + \frac{1}{5}\right)}} = \frac{40}{5} = 8$$

where we used the fact that $s_p^2 = \text{MSE}$, which you can verify by comparing the formulas. Note that the calculated F for these samples ($F = 64$) equals the square of the calculated t for the same samples ($t = 8$). Likewise, the tabled F-value (5.32) equals the square of the tabled t-value at the two-sided .05 level of significance ($t_{.025} = 2.306$ with 8 df). Because both the rejection region and the calculated values are related in the same way, the tests are equivalent. Moreover, the assumptions that must be met to ensure the validity of the t- and F-tests are the same:

1. The probability distributions of the populations of responses associated with each treatment must all be approximately normal.
2. The probability distributions of the populations of responses associated with each treatment must have equal variances.
3. The samples of experimental units selected for the treatments must be random and independent.

In fact, the only real difference between the tests is that the F-test can be used to compare *more than two* treatment means, whereas the t-test is applicable to two samples only. The F-test is summarized in the accompanying box.

ANOVA F-Test to Compare k Treatment Means: Completely Randomized Design

H_0: $\mu_1 = \mu_2 = \cdots = \mu_k$

H_a: At least two treatment means differ

Test statistic: $F = \dfrac{\text{MST}}{\text{MSE}}$

Rejection region: $F > F_\alpha$, where F_α is based on $(k - 1)$ numerator degrees of freedom (associated with MST) and $(n - k)$ denominator degrees of freedom (associated with MSE).

Conditions Required for a Valid ANOVA *F*-test: Completely Randomized Design

1. The samples are randomly selected in an independent manner from the k treatment populations. (This can be accomplished by randomly assigning the experimental units to the treatments.)
2. All k sampled populations have distributions that are approximately normal.
3. The k population variances are equal (i.e., $\sigma_1^2 = \sigma_2^2 = \sigma_3^2 = \cdots = \sigma_k^2$).

Computational formulas for MST and MSE are given in Appendix C. We will rely on some of the many statistical software packages available to compute the F statistic, concentrating on the interpretation of the results rather than their calculations.

Example 8.4
Conducting an ANOVA *F*-Test to Compare Golf Ball Brands

Problem Suppose the USGA wants to compare the mean distances associated with four different brands of golf balls when struck with a driver. A completely randomized design is employed, with Iron Byron, the USGA's robotic golfer, using a driver to hit a random sample of 10 balls of each brand in a random sequence. The distance is recorded for each hit, and the results are shown in Table 8.1, organized by brand.

a. Set up the test to compare the mean distances for the four brands. Use $\alpha = .10$.

b. Use Excel to obtain the test statistic and p-value. Interpret the results.

Table 8.1 Results of Completely Randomized Design: Iron Byron Driver

	Brand A	Brand B	Brand C	Brand D
	251.2	263.2	269.7	251.6
	245.1	262.9	263.2	248.6
	248.0	265.0	277.5	249.4
	251.1	254.5	267.4	242.0
	260.5	264.3	270.5	246.5
	250.0	257.0	265.5	251.3
	253.9	262.8	270.7	261.8
	244.6	264.4	272.9	249.0
	254.6	260.6	275.6	247.1
	248.8	255.9	266.5	245.9
Sample Means	250.8	261.1	270.0	249.3

Data Set: GOLFCRD

Solution

a. To compare the mean distances of the four brands, we first specify the hypotheses to be tested. Denoting the population mean of the ith brand by μ_i, we test

H_0: $\mu_1 = \mu_2 = \mu_3 = \mu_4$
H_a: The mean distances differ for at least two of the brands

The test statistic compares the variation among the four treatment (Brand) means to the sampling variability within each of the treatments.

Test statistic: $F = \dfrac{\text{MST}}{\text{MSE}}$

Rejection region: $F > F_\alpha = F_{.10}$

with $\nu_1 = (k - 1) = 3$ df and $\nu_2 = (n - k) = 36$ df

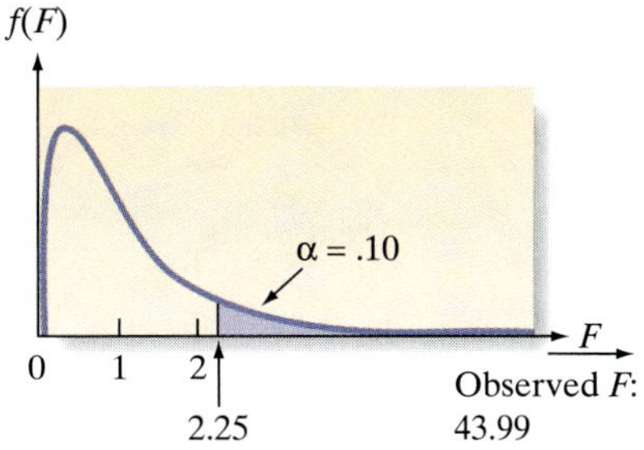

Figure 8.7
F-test for completely randomized design: Golf ball experiment

From Table VII in Appendix B, we find $F_{.10} \approx 2.25$ for 3 and 36 df. Thus, we will reject H_0 if $F > 2.25$. (See Figure 8.7.)

The assumptions necessary to ensure the validity of the test are as follows:

1. The samples of 10 golf balls for each brand are selected randomly and independently.
2. The probability distributions of the distances for each brand are normal.
3. The variances of the distance probability distributions for each brand are equal.

b. The Excel printout for the data in Table 8.1 resulting from this completely randomized design is given in Figure 8.8. At the bottom of the printout, the Total Sum of Squares is designated **Total,** and it is partitioned into the **Between Groups** (i.e., Brand) and the **Within Groups** (i.e., Error) sum of squares. The **Sum of Squares** column is headed **SS.**

The values of the mean squares, MST and MSE (highlighted on the printout), are 931.4629 and 21.17503, respectively. The *F*-ratio, 43.98875, also highlighted on the printout, exceeds the tabled value of 2.25. We therefore reject the null hypothesis at the .10 level of significance, concluding that at least two of the brands differ with respect to mean distance traveled when struck by the driver.

Look Back We can also arrive at the appropriate conclusion by noting that the observed significance level of the *F*-test (highlighted on the printout) is approximately 0. This implies that we would reject the null hypothesis that the means are equal at any reasonably-selected α level.

[*Note:* Excel uses exponential notation to display the *p*-value. The value 3.97E-12 is equal to .00000000000397.]

	A	B	C	D	E	F	G
1	Anova: Single Factor						
2							
3	SUMMARY						
4	*Groups*	*Count*	*Sum*	*Average*	*Variance*		
5	BrandA	10	2507.8	250.78	22.42178		
6	BrandB	10	2610.6	261.06	14.94711		
7	BrandC	10	2699.5	269.95	20.25833		
8	BrandD	10	2493.2	249.32	27.07289		
9							
10							
11	ANOVA						
12	*Source of Variation*	*SS*	*df*	*MS*	*F*	*P-value*	*F crit*
13	Between Groups	2794.389	3	931.4629	43.98875	3.97E-12	2.866266
14	Within Groups	762.301	36	21.17503			
15							
16	Total	3556.69	39				

Figure 8.8
Excel printout for ANOVA of golf ball distance data

Now Work Exercise 8.21

Table 8.2 General ANOVA Summary Table for a Completely Randomized Design

Source	df	SS	MS	F
Treatments	$k - 1$	SST	$\text{MST} = \frac{\text{SST}}{k-1}$	$\frac{\text{MST}}{\text{MSE}}$
Error	$n - k$	SSE	$\text{MSE} = \frac{\text{SSE}}{n-k}$	
Total	$n - 1$	SS(Total)		

The results of an **analysis of variance (ANOVA)** can be summarized in a simple tabular format similar to that obtained from the Excel printout in Example 8.4. The general form of the table is shown in Table 8.2, where the symbols df, SS, and MS stand for degrees of freedom, Sum of Squares, and Mean Square, respectively. Note that the two sources of variation, Treatments and Error, add to the Total Sum of Squares, SS(Total). The ANOVA summary table for Example 8.3 is given in Table 8.3, and the partitioning of the Total Sum of Squares into its two components is illustrated in Figure 8.9.

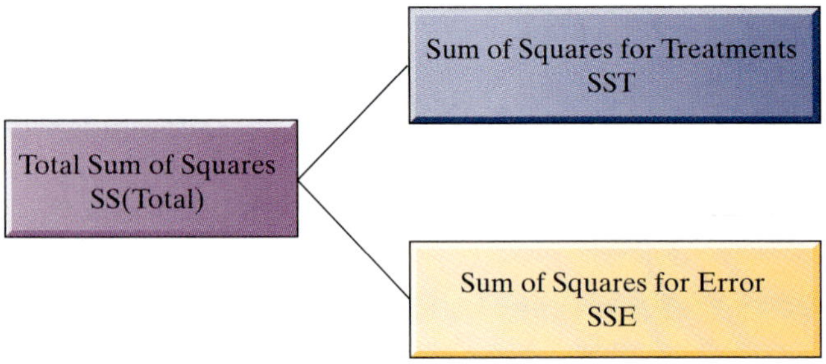

Figure 8.9
Partitioning of the Total Sum of Squares for the completely randomized design

Suppose the F-test results in a rejection of the null hypothesis that the treatment means are equal. Is the analysis complete? Usually, the conclusion that at least two of the treatment means differ leads to other questions. Which of the means differ and by how much? For example, the F-test in Example 8.4 leads to the conclusion that at least two of the brands of golf balls have different mean distances traveled when struck with a driver. Now the question is, which of the brands differ? How are the brands ranked with respect to mean distance?

One way to obtain this information is to construct a confidence interval for the difference between the means of any pair of treatments using the method of Section 7.2. For example, if a 95% confidence interval for $(\mu_A - \mu_C)$ in Example 8.4 is found to be $(-24, -13)$, we are confident that the mean distance for Brand C exceeds the mean for Brand A (because all differences in the interval are negative). Constructing these confidence intervals for all possible brand pairs will allow you to rank the brand means. A method for conducting these *multiple comparisons*—one that controls for Type I errors—is presented in Section 8.3.

Table 8.3 ANOVA Summary Table for Example 8.4

Source	df	SS	MS	F	p-value
Brands	3	2,794.39	931.46	43.99	.0000
Error	36	762.30	21.18		
Total	39	3,556.69			

Example 8.5 Checking the ANOVA Assumptions

Problem Refer to the completely randomized design ANOVA conducted in Example 8.4. Are the assumptions required for the test approximately satisfied?

Solution The assumptions for the test are repeated below.

1. The samples of golf balls for each brand are selected randomly and independently.
2. The probability distributions of the distances for each brand are normal.
3. The variances of the distance probability distributions for each brand are equal.

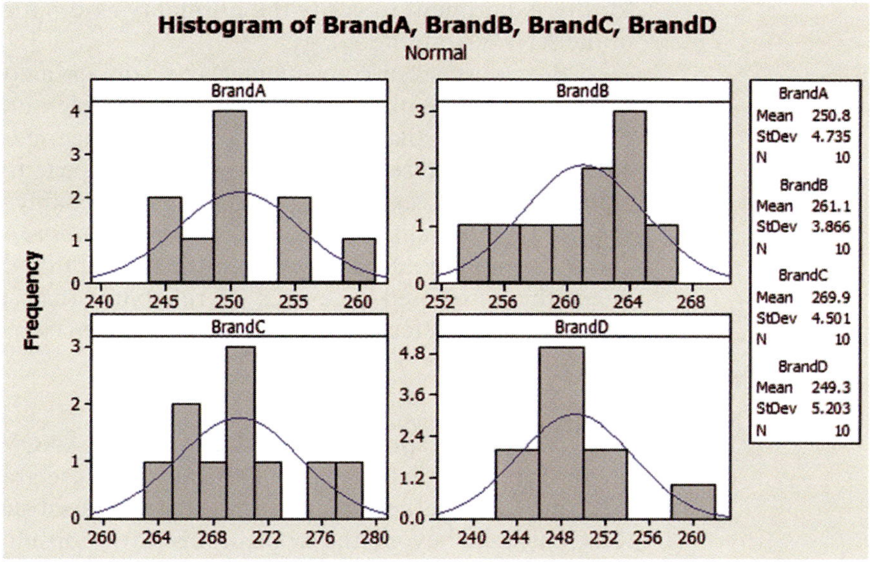

Figure 8.10
Minitab histograms for golf ball distances

Because the sample consisted of 10 randomly selected balls of each brand and the robotic golfer Iron Byron was used to drive all the balls, the first assumption of independent random samples is satisfied. To check the next two assumptions, we will employ two graphical methods presented in Chapter 2: histograms and box plots. A Minitab histogram of driving distances for each brand of golf ball is shown in Figure 8.10, followed by SPSS box plots in Figure 8.11.

The normality assumption can be checked by examining the histograms in Figure 8.10. With only 10 sample measurements for each brand, however, the displays are not very informative. More data would need to be collected for each brand before we could assess whether the distances come from normal distributions. Fortunately, analysis of variance has been shown to be a very **robust method** when the assumption of normality is not satisfied exactly—that is, *moderate departures from normality do not have much effect on the significance level of the ANOVA* F*-test or on confidence coefficients.* Rather than spend the time, energy, or money to collect additional data

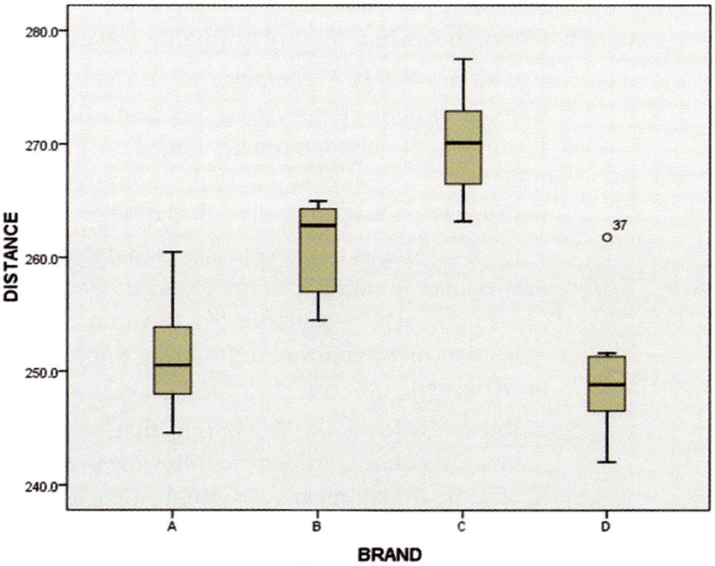

Figure 8.11
SPSS box plots for golf ball distances

for this experiment to verify the normality assumption, we will rely on the robustness of the ANOVA methodology.

Box plots are a convenient way to obtain a rough check on the assumption of equal variances. With the exception of a possible outlier for Brand D, the box plots in Figure 8.11 show that the spread of the distance measurements is about the same for each brand. Because the sample variances appear to be the same, the assumption of equal population variances for the brands is probably satisfied. Although robust with respect to the normality assumption, ANOVA is *not robust* with respect to the equal variances assumption. Departures from the assumption of equal population variances can affect the associated measures of reliability (e.g., p-values and confidence levels). Fortunately, the effect is slight when the sample sizes are equal, as in this experiment.

Now Work Exercise 8.30

Although graphs can be used to check the ANOVA assumptions, as in Example 8.5, no measures of reliability can be attached to these graphs. When you have a plot that is unclear as to whether or not an assumption is satisfied, you can use formal statistical tests, which are beyond the scope of this text. (Consult the references for information on these tests.) When the validity of the ANOVA assumptions is in doubt, nonparametric statistical methods are useful.

What Do You Do When the Assumptions Are Not Satisfied for an ANOVA for a Completely Randomized Design?

Answer: Use a nonparametric statistical method such as the Kruskal-Wallis H-test of Section 14.5.

The procedure for conducting an analysis of variance for a completely randomized design is summarized in the following box. Remember that the hallmark of this design is independent random samples of experimental units associated with each treatment. However, designs with dependent samples may be more appropriate in certain situations. Consult the references for information on the use of these designs.

Steps for Conducting an ANOVA for a Completely Randomized Design

1. Be sure the design is truly completely randomized, with independent random samples for each treatment.
2. Check the assumptions of normality and equal variances.
3. Create an ANOVA summary table that specifies the variability attributable to treatments and error, making sure that it leads to the calculation of the F-statistic for testing the null hypothesis that the treatment means are equal in the population. Use a statistical software program to obtain the numerical results. If no such package is available, use the calculation formulas in Appendix C.
4. If the F-test leads to the conclusion that the means differ,
   a. Conduct a multiple comparisons procedure for as many of the pairs of means as you wish to compare (see Section 8.3). Use the results to summarize the statistically significant differences among the treatment means.
   b. If desired, form confidence intervals for one or more individual treatment means.
5. If the F-test leads to the nonrejection of the null hypothesis that the treatment means are equal, consider the following possibilities:
   a. The treatment means are equal—that is, the null hypothesis is true.
   b. The treatment means really differ, but other important factors affecting the response are not accounted for by the completely randomized design. These factors inflate the sampling variability, as measured by MSE, resulting in

smaller values of the *F*-statistic. Either increase the sample size for each treatment or use a different experimental design (as in Section 8.4) that accounts for the other factors affecting the response.

Note: Be careful not to automatically conclude that the treatment means are equal because the possibility of a Type II error must be considered if you accept H_0.

Statistics IN Action Revisited — Testing for Differences in Mean Ad Recall Scores

Do TV shows with violence and sex impair memory for commercials? This was the question of interest in the *Journal of Applied Psychology* (June 2002) study presented in the Statistics in Action (p. 447). Recall that each of 324 adults was randomly assigned to one of three TV content groups: violent content (V), sex content (S), and neutral content (N). After viewing the program (and embedded commercials), each participant was scored on his/her recall of the brand names in the commercial messages (with scores ranging from 0 to 9). The researchers want to compare the mean recall scores of the three TV content groups.

Because the participants were randomly assigned to one of the three TV content groups, we can treat the data as coming from a completely randomized design. The response (dependent) variable is the recall score, and the single factor is TV content group. The three levels (V, S, and N) of content group represent the treatments in the experiment. Then, the appropriate null and alternative hypotheses are

H_0: $\mu_V = \mu_S = \mu_N$

H_a: At least two of the means, μ_V, μ_S, μ_N are different

A Minitab printout of the analysis of variance for the data in the **TVADRECALL** file is shown in Figure SIA8.1.

Both the *F*-value and observed significance level (*p*-value) are highlighted on the printout. Because *p*-value = .000, for any α-level we select (.01, .05, or .10), there is sufficient evidence to reject the null hypothesis. Consequently, the researchers concluded that the mean TV ad recall scores of the three content groups were significantly different.

Should the researchers now infer that TV shows with violence and sex impair memory for commercials? Possibly, but not based solely on the results of the ANOVA *F*-test. Remember, the alternative hypothesis states that at least two of the three population means are different, but it does not indicate which means in particular are different, nor does it inform us about which mean has the highest TV ad recall value or the lowest. The sample mean recall scores for the three groups (highlighted on the printout) are $\bar{x}_V = 2.08$, $\bar{x}_S = 1.71$, and $\bar{x}_N = 3.17$. In the next section, we demonstrate how to rank these means using a procedure that provides a measure of reliability for the inference.

The validity of the inferences drawn from the ANOVA *F*-test depend on whether or not the assumptions are reasonably satisfied. The assumptions are that the recall scores for each TV content (treatment) group are normally distributed with equal variances. Minitab histograms and box plots for the data are shown in Figures SIA8.2 and SIA8.3, respectively. First, note the nonsymmetrical shapes in the histograms for groups V and S, Figure SIA8.2. These

Descriptive Statistics: RECALL

Variable	CONTENT	N	Mean	StDev	Minimum	Maximum
RECALL	NEUTRAL	108	3.167	1.811	0.000	7.000
	SEX	108	1.713	1.664	0.000	7.000
	VIOLENT	108	2.083	1.730	0.000	7.000

One-way ANOVA: RECALL versus CONTENT

Source	DF	SS	MS	F	P
CONTENT	2	123.27	61.63	20.45	0.000
Error	321	967.35	3.01		
Total	323	1090.62			

Figure SIA8.1
Minitab 1-way ANOVA for TV ad recall data

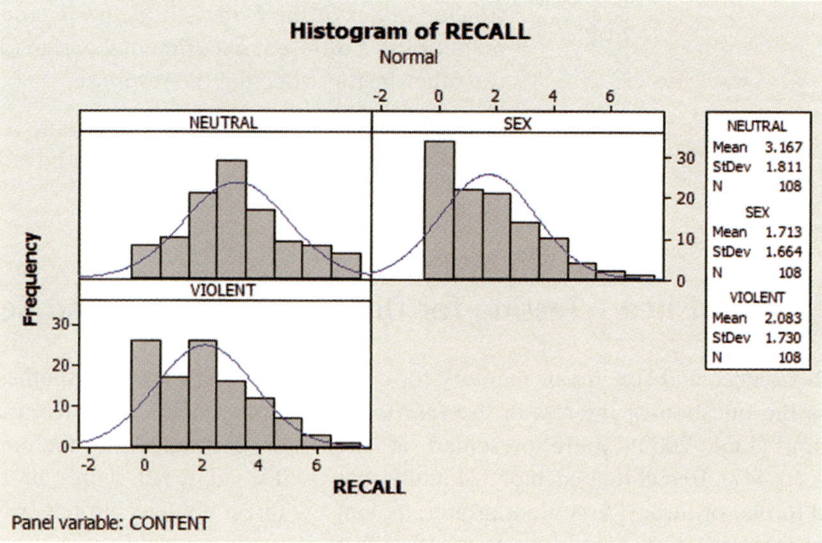

Figure SIA8.2
Minitab histograms for the TV ad recall data

graphs do not appear to support the assumption of normality for these treatments. However, this assumption does not need to be satisfied exactly for the ANOVA to yield valid results. Second, examine the variability in the data shown on the box plots, Figure SIA8.3. The spread of the recall scores appears to be about the same for the three groups. If, however, you are concerned about the possible greater variance for group S, then you should consider running a nonparametric analysis of the data—an analysis that does not rely on the assumptions of normality and equal variances.

Data Set: TVADRECALL

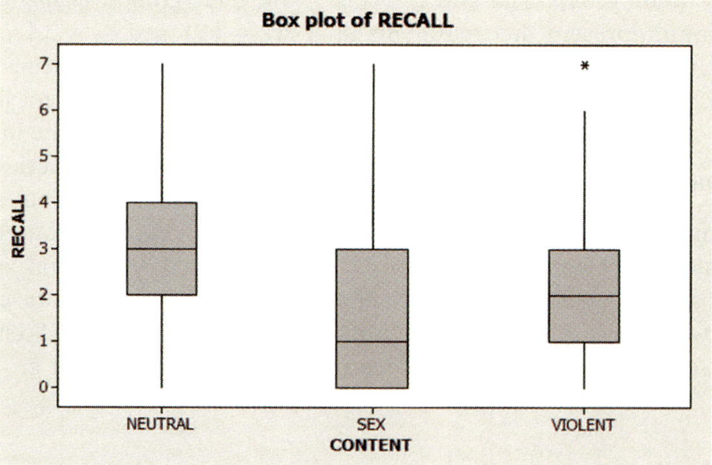

Figure SIA8.3
Minitab box plots for the TV ad recall data

Exercises 8.15–8.34

Learning the Mechanics

8.15 Use Tables VII, VIII, IX, and X in Appendix B to find each of the following F values:

a. $F_{.05}$, $v_1 = 4$, $v_2 = 4$

b. $F_{.01}$, $v_1 = 4$, $v_2 = 4$

c. $F_{.10}$, $v_1 = 30$, $v_2 = 40$

d. $F_{.025}$, $v_1 = 15$, $v_2 = 12$

8.16 Find the following probabilities:

a. $P(F \leq 3.48)$ for $v_1 = 5$, $v_2 = 9$

b. $P(F > 3.09)$ for $v_1 = 15$, $v_2 = 20$

c. $P(F > 2.40)$ for $v_1 = 15$, $v_2 = 15$

d. $P(F \leq 1.83)$ for $v_1 = 8$, $v_2 = 40$

8.17 Consider dot plots 1 and 2 shown on the next page. Assume that the two samples represent independent, random samples corresponding to two treatments in a completely randomized design.

a. In which plot is the difference between the sample means small relative to the variability within the sample observations? Justify your answer.

Dot Plots for Exercise 8.17

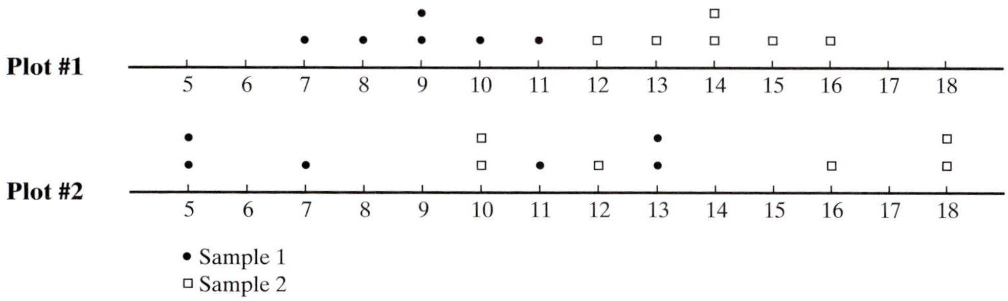

b. Calculate the treatment means (i.e., the means of samples 1 and 2) for both dot plots.

c. Use the means to calculate the Sum of Squares for Treatments (SST) for each dot plot.

d. Calculate the sample variance for each sample and use these values to obtain the Sum of Squares for Error (SSE) for each dot plot.

e. Calculate the Total Sum of Squares [SS(Total)] for the two dot plots by adding the Sums of Squares for Treatments and Error. What percentage of SS(Total) is accounted for by the treatments—that is, what percentage of the Total Sum of Squares is the Sum of Squares for Treatments—in each case?

f. Convert the Sum of Squares for Treatments and Error to mean squares by dividing each by the appropriate number of degrees of freedom. Calculate the F-ratio of the Mean Square for Treatments (MST) to the Mean Square for Error (MSE) for each dot plot.

g. Use the F-ratios to test the null hypothesis that the two samples are drawn from populations with equal means. Use $\alpha = .05$.

h. What assumptions must be made about the probability distributions corresponding to the responses for each treatment to ensure the validity of the F-tests conducted in part **g**?

8.18 Refer to Exercise 8.17. Conduct a two-sample t-test (Section 7.2) of the null hypothesis that the two treatment means are equal for each dot plot. Use $\alpha = .05$ and two-tailed tests. In the course of the test, compare each of the following with the F-tests in Exercise 8.17:

a. The pooled variances and the MSEs

b. The t- and F-test statistics

c. The tabled values of t and F that determine the rejection regions

d. The conclusions of the t- and F-tests

e. The assumptions that must be made to ensure the validity of the t- and F-tests

8.19 Refer to Exercises 8.17 and 8.18. Complete the following ANOVA table for each of the two dot plots:

Source	df	SS	MS	F
Treatments				
Error				
Total				

8.20 A partially completed ANOVA table for a completely randomized design is shown here:

Source	df	SS	MS	F
Treatments	6	17.5	—	—
Error	—	—	—	
Total	41	46.5		

a. Complete the ANOVA table.

b. How many treatments are involved in the experiment?

c. What is the total sample size, n, for the experiment?

d. Use a random number generator to randomly assign each experimental unit to one of the treatments. Assume the sample size will be the same for each treatment.

e. Do the data provide sufficient evidence to indicate a difference among the population means? Test using $\alpha = .10$.

f. Find the approximate observed significance level for the test in part **c** and interpret it.

g. Suppose that $\bar{x}_1 = 3.7$ and $\bar{x}_2 = 4.1$. Do the data provide sufficient evidence to indicate a difference between μ_1 and μ_2? Assume that there are six observations for each treatment. Test using $\alpha = .10$.

h. Refer to part **g.** Find a 90% confidence interval for $(\mu_1 - \mu_2)$. [Hint: Use $s = \sqrt{\text{MSE}}$ as an estimate of both σ_1 and σ_2.]

i. Refer to part **g.** Find a 90% confidence interval for μ_1. [Hint: Use $s = \sqrt{\text{MSE}}$ as an estimate of σ_1.]

8.21 NW The data in the next table (saved in the **LM8_21** file) resulted from an experiment that used a completely randomized design.

Treatment 1	Treatment 2	Treatment 3
3.8	5.4	1.3
1.2	2.0	0.7
4.1	4.8	2.2
5.5	3.8	
2.3		

a. Use statistical software (or the appropriate calculation formulas in Appendix C) to complete the following ANOVA table:

Source	df	SS	MS	F
Treatments				
Error				
Total				

b. Test the null hypothesis that $\mu_1 = \mu_2 = \mu_3$, where μ_i represents the true mean for treatment i, against the alternative that at least two of the means differ. Use $\alpha = .01$.

Applying the Concepts—Basic

8.22 Making high-stakes insurance decisions. The *Journal of Economic Psychology* (Sep. 2008) published the results of a high stakes experiment where subjects (university students) were asked how much they would pay for insuring a valuable painting. The painting was threatened by both fire and theft, hence, the need for insurance. Of interest was the amount the subject was willing to pay (WTP) for insurance (thousands of dollars). For one part of the experiment, a total of 252 subjects were randomly assigned to one of three groups. Group 1 subjects ($n_1 = 84$) were informed of the hazards (both fire and theft) but were not told the exact probabilities of the hazards occurring. These subjects provided a separate WTP value for fire and theft. Group 2 subjects ($n_2 = 84$) were also informed of the hazards (fire/theft) and were not told the exact probabilities of the hazards occurring. However, these subjects provided a single WTP value covering both fire and theft. Group 3 subjects ($n_3 = 84$) were told of the hazards in sequential order (fire first, then theft). After being given the exact probability of fire occurring, the subjects provided a WTP value for fire. Then they were given the exact probability of theft occurring and were asked to provide a WTP value for theft. The researchers investigated whether the mean total WTP value differed for the three groups.

a. Explain why the experimental design employed is a completely randomized design.

b. Identify the dependent (response) variable and treatments for the design.

c. Give the null and alternative hypotheses of interest to the researchers.

NW **d.** Use a random number generator to randomly assign each of the 252 subjects to one of the three groups. Be sure to assign 84 subjects to each group.

8.23 Contingent valuation of homes in contaminated areas. Contingent valuation (CV) is a method of estimating property values that uses survey responses from potential home owners. CV surveys were employed to determine the impact of contamination on property values in the *Journal of Real Estate Research* (Vol. 27, 2005). Home owners were randomly selected from each of seven states—Kentucky, Pennsylvania, Ohio, Alabama, Illinois, South Carolina, and Texas. Each home owner was asked to estimate the property value of a home located in an area contaminated by petroleum leaking from underground storage tanks (LUST). The dependent variable of interest was the LUST discount percentage (i.e., the difference between the current home value and estimated LUST value, as a percentage). The researchers were interested in comparing the mean LUST discount percentages across the seven states.

a. Give the null and alternative hypotheses of interest to the researchers.

b. An ANOVA summary table is shown in the next column. Use the information provided to conduct the hypothesis test, part **a.** Use $\alpha = .10$.

Source	df	SS	MS	*F*-value	*p*-value
States	6	.1324	.0221	1.60	0.174
Error	59	.8145	.0138		
Total	65	.9469			

8.24 College tennis recruiting with a team Web site. Most university athletic programs now have a Web site with information on individual sports and a Prospective Student Athlete Form that allows high school athletes to submit their academic and sports achievements directly to the college coach. The *Sport Journal* (Winter 2004) published a study of how important team Web sites are to the recruitment of college tennis players. A survey was conducted of NCAA tennis coaches, of which 53 were from Division I schools, 20 were from Division II schools, and 53 were from Division III schools. Coaches were asked to respond to a series of statements, including "the Prospective Student Athlete Form on the Web site contributes very little to the recruiting process." Responses were measured on a 7-point scale (where 1 = strongly disagree and 7 = strongly agree). In order to compare the mean responses of tennis coaches from the three NCAA divisions, the data were analyzed with a completely randomized design ANOVA.

a. Identify the experimental unit, dependent (response) variable, factor, and treatments for this study.

b. Give the null and alternative hypotheses for the ANOVA *F*-test.

c. The observed significance level of the test was found to be p-value $< .003$. What conclusion can you draw if you want to test at $\alpha = .05$?

8.25 Robots trained to behave like ants. Robotic researchers investigated whether robots could be trained to behave like ants in an ant colony (*Nature,* Aug. 2000). Robots were trained and randomly assigned to "colonies" (i.e., groups) consisting of 3, 6, 9, or 12 robots. The robots were assigned the task of foraging for "food" and to recruit another robot when they identified a resource-rich area. One goal of the experiment was to compare the mean energy expended (per robot) of the four different colony sizes.

a. What type of experimental design was employed?

b. Identify the treatments and the dependent variable.

c. Set up the null and alternative hypothesis of the test.

d. The following ANOVA results were reported: $F = 7.70$, numerator df $= 3$, denominator df $= 56$, p-value $< .001$. Conduct the test at a significance level of $\alpha = .05$ and interpret the result.

8.26 Study of mutual fund performance. Mutual funds are classified as large-cap funds, medium-cap funds, or small-cap funds, depending on the capitalization of the companies in the fund. Hawaii Pacific University researchers S. Shi and M. Seiler investigated whether the average performance of a mutual fund is related to capitalization size (*American Business Review,* Jan. 2002). Independent random samples of 30 mutual funds were selected from each of the three fund groups, and the 90-day rate of return was determined for each fund. The data for the 90 funds were subjected to an analysis of variance, with the results shown in the ANOVA summary table on the next page.

Source	df	SS	MS	*F*	*p*-value
Fund group	2	409.566	204.783	6.965	.002
Error	87	2,557.860	29.401		
Total	89	2,967.426			

Source: Shi, S. W. W., and Seiler, M. J. "Growth and value style comparison of U.S. stock mutual funds," *American Business Review,* January 2002 (Table 3).

a. State the null and alternative hypotheses for the ANOVA.

b. Give the rejection region for the test using $\alpha = .01$.

c. Make the appropriate conclusion using either the test statistic or the *p*-value.

8.27 Income and road rage. Is a driver's propensity to engage in road rage related to his or her income? Researchers at Mississippi State University attempted to answer this question by conducting a survey of a representative sample of over 1,000 U.S. adult drivers (*Accident Analysis and Prevention,* Vol. 34, 2002). Based on how often each driver engaged in certain road rage behaviors (e.g., making obscene gestures at, tailgating, and thinking about physically hurting another driver), a road rage score was assigned. (Higher scores indicate a greater pattern of road rage behavior.) The drivers were also grouped by annual income: under \$30,000, between \$30,000 and \$60,000, and over \$60,000. The data were subjected to an analysis of variance, with the results summarized in the table. Is there evidence to indicate that the mean road rage score differs for the three income groups? Test using $\alpha = .05$.

Income Group	Sample Size	Mean Road Rage Score
Under \$30,000	379	4.60
\$30,000 to \$60,000	392	5.08
Over \$60,000	267	5.15

ANOVA Results: *F*-value = 3.90 *p*-value < .01

Source: Wells-Parker, E., et al. "An exploratory study of the relationship between road rage and crash experience in a representative sample of US Drivers," *Accident Analysis and Prevention,* Vol. 34, 2002 (Table 2).

Applying the Concepts—Intermediate

8.28 Homework assistance for accounting students. Refer to the *Journal of Accounting Education* (Vol. 25, 2007) study of assisting accounting students with their homework, Exercise 7.16 (p. 395). A total of 75 junior-level accounting majors who were enrolled in Intermediate Financial Accounting participated in the experiment. Recall that students took a pretest on a topic not covered in class and then each was given a homework problem to solve on the same topic. A completely randomized design was employed, with students randomly assigned to receive one of three different levels of assistance on the homework: (1) the completed solution, (2) check figures at various steps of the solution, and (3) no help at all. After finishing the homework, each student was all given a posttest on the subject. The response variable of interest to the researchers was the knowledge gain (or, test score improvement), measured as the difference between the posttest and pretest scores. The data (simulated from descriptive statistics published in the article) are saved in the **ACCHW** file.

a. Give the null and alternative hypotheses tested in an analysis of variance of the data.

b. Summarize the results of the analysis in an ANOVA table.

c. Interpret the results, practically. Does your conclusion agree with the inferences drawn in Exercise 7.16?

8.29 Ethics of downsizing. A major strategic alternative for many U.S. firms is to reduce the size of its workforce, i.e., to "downsize." The ethics of downsizing decisions from the employee's perspective was investigated in the *Journal of Business Ethics* (Vol. 18, 1999). The researchers surveyed a sample of 209 employees who were enrolled in an Executive MBA Program or weekend program at one of three Colorado universities. These individuals were divided into five distinct groups, depending on their job situation at a previous or current firm. The groups were named (1) Casualties, (2) Survivors, (3) Implementors/casualties, (4) Implementors/survivors, and (5) Formulators. The sampled employees completed a questionnaire on their ethical perceptions of downsizing. One item asked employees to respond to the statement: "It is unethical for a downsizing

Minitab Output for Exercise 8.29

One-way ANOVA: CASUAL, SURVIVE, IMPCAS, IMPSUR, FORMUL

```
Source   DF      SS     MS     F      P
Factor    4   40.84  10.21  9.85  0.000
Error   204  211.35   1.04
Total   208  252.19

S = 1.018   R-Sq = 16.19%   R-Sq(adj) = 14.55%


                                 Individual 95% CIs For Mean Based on Pooled StDev
Level     N   Mean  StDev        +---------+---------+---------+---------
CASUAL   47  1.787  0.832             (----*----)
SURVIVE  71  1.845  1.023               (---*---)
IMPCAS   27  1.593  0.636        (------*-----)
IMPSUR   33  2.545  1.301                         (----*-----)
FORMUL   31  2.871  1.176                              (-----*-----)
                                 +---------+---------+---------+---------
                               1.20      1.80      2.40      3.00

Pooled StDev = 1.018
```

decision to be announced or implemented on or prior to a major holiday." Responses were measured using a 5-point Likert scale, where 1 = strongly agree, 2 = agree, 3 = neutral, 4 = disagree, and 5 = strongly disagree. Data on both the qualitative variable "Group" and the quantitative variable "Ethics response" are saved in the **DOWNSIZE** file. The researchers' goal was to determine if any differences exist among the mean ethics scores for the five groups.

a. The data were analyzed using an ANOVA for a completely randomized design. Identify the factor, treatments, response variable, and experimental units for this design.

b. Specify the null and alternative hypotheses tested.

c. A Minitab printout of the ANOVA is displayed on page 469. Can you conclude that the mean ethics scores of the five groups of employees are significantly different? Explain.

d. Access the **DOWNSIZE** file and check that the assumptions required for the ANOVA F-test are reasonably satisfied.

8.30 NW **Diamonds sold at retail.** Refer to the *Journal of Statistics Education* study of 308 diamonds for sale on the open market, Exercise 2.25 (p. 50). Recall that the **DIAMONDS** file contains information on the quantitative variables, size (number of carats) and price (in dollars), and on the qualitative variables, color (D, E, F, G, H, and I), clarity (IF, VS1, VS2, VVS1, and VVS2), and independent certification group (GIA, HRD, or IGI). Select one of the quantitative variables and one of the qualitative variables.

a. Set up the null and alternative hypotheses for determining whether the means of the quantitative variable differ for the levels of the qualitative variable.

b. Use the data in the **DIAMONDS** file to conduct the test, part **a,** at α = 10. State the conclusion in the words of the problem.

c. Check any assumptions required for the methodology used in part **b** to be valid.

8.31 **Effectiveness of sales closing techniques.** Industrial sales professionals have long debated the effectiveness of various sales closing techniques. University of Akron researchers S. Hawes, J. Strong, and B. Winick investigated the impact of five different closing techniques and a no-close condition on the level of a sales prospect's trust in the salesperson (*Industrial Marketing Management*, Sept. 1996). Two of the five closing techniques were the *assumed close* and the *impending event technique.* In the former, the salesperson simply writes up the order or behaves as if the sale has been made. In the latter, the salesperson encourages the buyer to buy now before some future event occurs that makes the terms of the sale less favorable for the buyer. Sales scenarios were presented to a sample of 237 purchasing executives. Each subject received one of the five closing techniques or a scenario in which no close was achieved. After reading the sales scenario, each executive was asked to rate his/her level of trust in the salesperson on a 7-point scale. The next table reports the six treatments employed in the study and the number of subjects receiving each treatment.

Treatments: Closing Techniques	Sample Size
1. No close	38
2. Impending event	36
3. Social validation	29
4. If-then	42
5. Assumed close	36
6. Either-or	56

a. The investigator's hypotheses were

H_0: The salesperson's level of prospect trust is not influenced by the choice of closing method.

H_a: The salesperson's level of prospect trust is influenced by the choice of closing method.

Rewrite these hypotheses in the form required for an analysis of variance.

b. The researchers reported the ANOVA F-statistic as $F = 2.21$. Is there sufficient evidence to reject H_0 at $\alpha = .05$?

c. What assumptions must be met for the test of part **a** to be valid?

d. Would you classify this experiment as observational or designed? Explain.

8.32 **Is honey a cough remedy?** Pediatric researchers at Pennsylvania State University carried out a designed study to test whether a teaspoon of honey before bed calms a child's cough and published their results in *Archives of Pediatrics and Adolescent Medicine* (Dec. 2007). (This experiment was first described in Exercise 2.28, p. 51.) A sample of 105 children who were ill with an upper respiratory tract infection and their parents participated in the study. On the first night, the parents rated their children's cough symptoms on a scale from 0 (no problems at all) to 6 (extremely severe) in five different areas. The total symptoms score (ranging from 0 to 30 points) was the variable of interest for the 105 patients. On the second night, the parents were instructed to give their sick children a dosage of liquid "medicine" prior to bedtime. Unknown to the parents, some were given a dosage of dextromethorphan (DM)—an over-the-counter cough medicine—while others were given a similar dose of honey. Also, a third group of

Honey Dosage:	12 11 15 11 10 13 10 4 15 16 9 14 10 6 10 8 11 12 12 8 12 9 11 15 10 15 9 13 8 12 10 8 9 5 12
DM Dosage:	4 6 9 4 7 7 7 9 12 10 11 6 3 4 9 12 7 6 8 12 12 4 12 13 7 10 13 9 4 4 10 15 9
No Dosage (Control):	5 8 6 1 0 8 12 8 7 7 1 6 7 7 12 7 9 7 9 5 11 9 5 6 8 8 6 7 10 9 4 8 7 3 1 4 3

Source: Paul, I. M., et al. "Effect of honey, dextromethorphan, and no treatment on nocturnal cough and sleep quality for coughing children and their parents," *Archives of Pediatrics and Adolescent Medicine*, Vol. 161, No. 12, Dec. 2007 (data simulated).

parents (the control group) gave their sick children no dosage at all. Again, the parents rated their children's cough symptoms, and the improvement in total cough symptoms score was determined for each child. The data (improvement scores) for the study are shown in the accompanying table and saved in the **HONEYCOUGH** file. The goal of the researchers was to compare the mean improvement scores for the three treatment groups.

a. Identify the type of experimental design employed. What are the treatments?

b. Conduct an analysis of variance on the data and interpret the results.

8.33 Commercial eggs produced from different housing systems. In the production of commercial eggs in Europe, four different types of housing systems for the chickens are used: cage, barn, free range, and organic. The characteristics of eggs produced from the four housing systems were investigated in *Food Chemistry* (Vol. 106, 2008). Twenty-eight commercial grade A eggs were randomly selected from supermarkets—10 of which were produced in cages, 6 in barns, 6 with free range, and 6 organic. A number of quantitative characteristics were measured for each egg, including shell thickness (millimeters), whipping capacity (percent overrun), and penetration strength (Newtons). The data (simulated from summary statistics provided in the journal article) are saved in the **EGGS** file. For each characteristic, the researchers compared the means of the four housing systems. Minitab descriptive statistics and ANOVA printouts for each characteristic are shown below. Fully interpret the results. Identify the characteristics for which housing systems differ.

Descriptive Statistics: THICKNESS, OVERRUN, STRENGTH

Variable	HOUSING	N	Mean	StDev	Minimum	Maximum
THICKNESS	BARN	6	0.50000	0.01414	0.48000	0.52000
	CAGE	10	0.4230	0.0350	0.3700	0.4700
	FREE	6	0.5017	0.0279	0.4700	0.5500
	ORGANIC	6	0.4817	0.0387	0.4300	0.5200
OVERRUN	BARN	6	513.33	8.38	501.00	526.00
	CAGE	10	480.60	12.91	462.00	502.00
	FREE	6	517.50	8.17	510.00	531.00
	ORGANIC	6	529.17	10.65	511.00	544.00
STRENGTH	BARN	6	39.333	1.120	37.600	40.300
	CAGE	10	37.320	2.127	33.000	40.200
	FREE	6	37.17	3.79	31.50	40.60
	ORGANIC	6	35.97	3.04	32.60	40.20

One-way ANOVA: THICKNESS versus HOUSING

Source	DF	SS	MS	F	P
HOUSING	3	0.034291	0.011430	11.74	0.000
Error	24	0.023377	0.000974		
Total	27	0.057668			

One-way ANOVA: OVERRUN versus HOUSING

Source	DF	SS	MS	F	P
HOUSING	3	10788	3596	31.36	0.000
Error	24	2752	115		
Total	27	13540			

One-way ANOVA: STRENGTH versus HOUSING

Source	DF	SS	MS	F	P
HOUSING	3	35.12	11.71	1.70	0.193
Error	24	164.82	6.87		
Total	27	199.94			

Applying the Concepts—Advanced

8.34 Animal-assisted therapy for heart patients. Chief executive officers of hospitals are adopting unconventional therapeutic methods to shorten the length of stay of patients. At an *American Heart Association Conference* (Nov. 2005), a study to gauge whether animal-assisted therapy can improve the physiological responses of heart failure patients was presented. In the study, 76 heart patients were randomly assigned to one of three groups. Each patient in group T was visited by a human volunteer accompanied by a trained dog; each patient in group V was visited by a volunteer only; and, the patients in group C were not visited at all. The anxiety level of each patient was measured (in points) both before and after the visits. The table below gives summary statistics for the drop in anxiety level for patients in the three groups. The mean drops in anxiety levels of the three groups of patients were compared using an analysis of variance. Although the ANOVA table was not provided in the article, sufficient information is provided to reconstruct it.

	Sample Size	Mean Drop	Std. Dev.
Group T: Volunteer + Trained Dog	26	10.5	7.6
Group V: Volunteer only	25	3.9	7.5
Group C: Control group (no visit)	25	1.4	7.5

Source: Cole, K., et al. "Animal assisted therapy decreases hemodynamics, plasma epinephrine and state anxiety in hospitalized heart failure patients," *American Heart Association Conference,* Dallas, Texas, Nov. 2005.

a. Compute SST for the ANOVA, using the formula (see p. 457)

$$SST = \sum_{i=1}^{3} n_i(\bar{x}_i - \bar{x})^2$$

where $\bar{x}$ is the overall mean drop in anxiety level of all 76 subjects. [Hint: $\bar{x} = (\sum_{i=1}^{3} n_i(\bar{x}_i)/76$.]

b. Recall that SSE for the ANOVA can be written as

$$SSE = (n_1 - 1)s_1^2 + (n_2 - 1)s_2^2 + (n_3 - 1)s_3^2$$

where s_1^2, s_2^2, and s_3^2 are the sample variances associated with the three treatments. Compute SSE for the ANOVA.

c. Use the results from parts **a** and **b** to construct the ANOVA table.

d. Is there sufficient evidence (at $\alpha = .01$) of differences among the mean drops in anxiety levels by the patients in the three groups?

e. Comment on the validity of the ANOVA assumptions. How might this affect the results of the study?

8.3 Multiple Comparisons of Means

Consider a completely randomized design with three treatments, A, B, and C. Suppose we determine that the treatment means are statistically different via the ANOVA F-test of Section 8.2. To complete the analysis, we want to rank the three treatment means. As mentioned in Section 8.2, we start by placing confidence intervals on the difference between various pairs of treatment means in the experiment. In the three-treatment experiment, for example, we would construct confidence intervals for the following differences: $\mu_A - \mu_B$, $\mu_A - \mu_C$, and $\mu_B - \mu_C$.

Determining the Number of Pairwise Comparisons of Treatment Means

In general, if there are k treatment means, there are

$$c = \frac{k(k-1)}{2}$$

pairs of means that can be compared.

If we want to have $100(1 - \alpha)\%$ confidence that each of the c confidence intervals contains the true difference it is intended to estimate, we must use a smaller value of α for each individual confidence interval than we would use for a single interval. For example, suppose we want to rank the means of the three treatments, A, B, and C, with 95% confidence that all three confidence intervals comparing the means contain the true differences between the treatment means. Then each individual confidence interval will need to be constructed using a level of significance smaller than $\alpha = .05$ in order to have 95% confidence that the three intervals collectively include the true differences.*

To make **multiple comparisons of a set of treatment means,** we can use a number of procedures that, under various assumptions, ensure that the overall confidence level associated with all the comparisons remains at or above the specified $100(1 - \alpha)\%$ level. Three widely used techniques are the Bonferroni, Scheffé, and

BIOGRAPHY **CARLO E. BONFERRONI (1892–1960)**

Bonferroni Inequalities

During his childhood years in Turin, Italy, Carlo Bonferroni developed an aptitude for mathematics while studying music. He went on to obtain a degree in mathematics at the University of Turin. Bonferroni's first appointment as a professor of mathematics was at the University of Bari in 1923. Ten years later, he became chair of financial mathematics at the University of Florence, where he remained until his death. Bonferroni was a prolific writer, authoring over 65 research papers and books. His interest in statistics included various methods of calculating a mean and a correlation coefficient. Among statisticians, however, Bonferroni is most well known for developing his Bonferroni inequalities in probability theory in 1935. Later, other statisticians proposed using these inequalities for finding simultaneous confidence intervals, which led to the development of the Bonferroni multiple comparisons method in ANOVA. Bonferroni balanced these scientific accomplishments with his music, becoming an excellent pianist and composer. ■

*The reason each interval must be formed at a higher confidence level than that specified for the collection of intervals can be demonstrated as follows:

$$\begin{aligned} &P\{\text{At least one of } c \text{ intervals fails to contain the true difference}\} \\ &= 1 - P\{\text{All } c \text{ intervals contain the true differences}\} \\ &= 1 - (1 - \alpha)^c \geq \alpha \end{aligned}$$

Thus, to make this probability of at least one failure equal to α, we must specify the individual levels of significance to be less than α.

Tukey methods. For each of these procedures, the risk of making a Type I error applies to the comparisons of the treatment means in the experiment; thus, the value of α selected is called an **experimentwise error rate** (in contrast to a **comparisonwise error rate**).

The choice of a multiple comparisons method in ANOVA will depend on the type of experimental design used and the comparisons of interest to the analyst. For example, **Tukey** (1949) developed his procedure specifically for pairwise comparisons when the sample sizes of the treatments are equal. The **Bonferroni method** (see Miller, 1981), like the Tukey procedure, can be applied when pairwise comparisons are of interest; however, Bonferroni's method does not require equal sample sizes. **Scheffé** (1953) developed a more general procedure for comparing all possible linear combinations of treatment means (called *contrasts*). Consequently, when making pairwise comparisons, the confidence intervals produced by Scheffé's method will generally be wider than the Tukey or Bonferroni confidence intervals.

The formulas for constructing confidence intervals for differences between treatment means using the Tukey, Bonferroni, or Scheffé method are provided in Appendix C. However, because these procedures (and many others) are available in the ANOVA programs of most statistical software packages, we use the computer to conduct the analysis. The programs generate a confidence interval for the difference between two treatment means for all possible pairs of treatments based on the experimentwise error rate (α) selected by the analyst.

Example 8.6

Ranking Treatment Means in the Golf Ball Experiment

Problem Refer to the completely randomized design of Example 8.4, in which we concluded that at least two of the four brands of golf balls are associated with different mean distances traveled when struck with a driver.

a. Use Tukey's multiple comparisons procedure to rank the treatment means with an overall confidence level of 95%.

b. Estimate the mean distance traveled for balls manufactured by the brand with the highest rank.

Solution

a. To rank the treatment means with an overall confidence level of .95, we require the experimentwise error rate of $\alpha = .05$. The confidence intervals generated by Tukey's method appear at the top of the SPSS printout, Figure 8.12. [*Note:* SPSS uses the

Multiple Comparisons

Dependent Variable: DISTANCE
Tukey HSD

(I) BRANDNUM	(J) BRANDNUM	Mean Difference (I-J)	Std. Error	Sig.	95% Confidence Interval Lower Bound	95% Confidence Interval Upper Bound
1	2	-10.2800*	2.0579	.000	-15.822	-4.738
	3	-19.1700*	2.0579	.000	-24.712	-13.628
	4	1.4600	2.0579	.893	-4.082	7.002
2	1	10.2800*	2.0579	.000	4.738	15.822
	3	-8.8900*	2.0579	.001	-14.432	-3.348
	4	11.7400*	2.0579	.000	6.198	17.282
3	1	19.1700*	2.0579	.000	13.628	24.712
	2	8.8900*	2.0579	.001	3.348	14.432
	4	20.6300*	2.0579	.000	15.088	26.172
4	1	-1.4600	2.0579	.893	-7.002	4.082
	2	-11.7400*	2.0579	.000	-17.282	-6.198
	3	-20.6300*	2.0579	.000	-26.172	-15.088

*. The mean difference is significant at the .05 level.

Figure 8.12 *(Continued)*

DISTANCE

Tukey HSD[a]

BRANDNUM	N	Subset for alpha = .05		
		1	2	3
4	10	249.320		
1	10	250.780		
2	10		261.060	
3	10			269.950
Sig.		.893	1.000	1.000

Means for groups in homogeneous subsets are displayed.

a. Uses Harmonic Mean Sample Size = 10.000.

Figure 8.12
SPSS printout of Tukey's multiple comparisons for the golf ball data

number 1 for Brand A, 2 for Brand B, etc.] For any pair of means, μ_i and μ_j, SPSS computes two confidence intervals—one for $(\mu_i - \mu_j)$ and one for $(\mu_j - \mu_i)$. Only one of these intervals is necessary to decide whether the means differ significantly.

In this example, we have $k = 4$ brand means to compare. Consequently, the number of relevant pairwise comparisons—that is, the number of nonredundant confidence intervals—is $c = 4(3)/2 = 6$. These six intervals, highlighted in Figure 8.12, are given in Table 8.4.

We are 95% confident that the intervals *collectively* contain all the differences between the true brand mean distances. Note that intervals that contain 0, such as the Brand A–Brand D interval from −4.08 to 7.00, do not support a conclusion that the true brand mean distances differ. If both endpoints of the interval are positive, as with the Brand B–Brand D interval from 6.20 to 17.28, the implication is that the first Brand (B) mean distance exceeds the second (D). Conversely, if both endpoints of the interval are negative, as with the Brand A–Brand C interval from −24.71 to −13.63, the implication is that the second Brand (C) mean distance exceeds the first Brand (A) mean distance.

A convenient summary of the results of the Tukey multiple comparisons is a listing of the brand means from highest to lowest, with a solid line connecting those that are *not* significantly different. This summary is shown in Figure 8.13. A similar summary is shown at the bottom of the SPSS printout, Figure 8.12. The interpretation is that Brand C's mean distance exceeds all others; Brand B's mean exceeds that of Brands A and D; and the means of Brands A and D do not differ significantly. All

Table 8.4 Pairwise Comparisons for Example 8.6

Brand Comparison	Confidence Interval
$(\mu_A - \mu_B)$	(−15.82, −4.74)
$(\mu_A - \mu_C)$	(−24.71, −13.63)
$(\mu_A - \mu_D)$	(−4.08, 7.00)
$(\mu_B - \mu_C)$	(−14.43, −3.35)
$(\mu_B - \mu_D)$	(6.20, 17.28)
$(\mu_C - \mu_D)$	(15.09, 26.17)

Figure 8.13
Summary of Tukey multiple comparisons

Mean:	249.3	250.8	261.1	270.0
Brand:	D	A	B	C

these inferences are made simultaneously with 95% confidence, the overall confidence level of the Tukey multiple comparisons.

b. Brand C is ranked highest; thus, we want a confidence interval for μ_C. Because the samples were selected independently in a completely randomized design, a confidence interval for an individual treatment mean is obtained with the one-sample t confidence interval of Section 5.3, using the mean square for error, MSE, as the measure of sampling variability for the experiment. A 95% confidence interval on the mean distance traveled by Brand C (apparently the "longest ball" of those tested), is

$$\bar{x}_C \pm t_{.025}\sqrt{\frac{\text{MSE}}{n}}$$

where $n = 10$, $t_{.025} \approx 2$ (based on 36 degrees of freedom), and MSE $= 21.175$ (obtained from the Excel printout, Figure 8.8). Substituting, we obtain

$$270.0 \pm (2)\sqrt{\frac{21.175}{10}} = 270.0 \pm 2.9 \text{ or } (267.1, 272.9)$$

Thus, we are 95% confident that the true mean distance traveled for brand C is between 267.1 and 272.9 yards, when hit with a driver by Iron Byron.

Look Back The easiest way to create a summary table like the one in Figure 8.13 is to first list the treatment means in rank order. Begin with the largest mean and compare it to (in order) the second largest mean, the third largest mean, and so on, by examining the appropriate confidence intervals shown on the computer printout. If a confidence interval contains 0, then connect the two means with a line. (These two means are not significantly different.) Continue in this manner by comparing the second largest mean with the third largest, fourth largest, and so on, until all possible $c = (k)(k + 1)/2$ comparisons are made.

Now Work Exercise 8.39

Remember that the Tukey method—designed for comparing pairs of treatment means with equal sample sizes—is just one of numerous multiple comparisons procedures available. Another technique may be more appropriate for the experimental design you employ. Consult the references for details on these other methods and when they should be applied. Guidelines for using the Tukey, Bonferroni, and Scheffé methods are given in the box.

Ethics IN Statistics

Running several different multiple comparisons methods and reporting only the one that produces a desired outcome, without regard to the experimental design, is considered *unethical statistical practice*.

Guidelines for Selecting a Multiple Comparisons Method in ANOVA

Method	Treatment Sample Sizes	Types of Comparisons
Tukey	Equal	Pairwise
Bonferroni	Equal or unequal	Pairwise or general contrasts (number of contrasts known)
Scheffé	Equal or unequal	General contrasts

Note: For equal sample sizes and pairwise comparisons, Tukey's method will yield simultaneous confidence intervals with the smallest width, and the Bonferroni intervals will have smaller widths than the Scheffé intervals.

Statistics IN Action Revisited — Ranking Mean Ad Recall Scores

In the previous Statistics in Action Revisited (p. 465), we used an ANOVA for a completely randomized design to compare the mean TV ad recall scores of the three TV content groups—violent content (V), sex content (S), and neutral content (N). The ANOVA F-test led us to conclude that the mean recall scores were significantly different. Now, we follow up the ANOVA with multiple comparisons of the group (treatment) means. Because 108 subjects were randomly assigned to each content group, the sample sizes for the groups are equal. Consequently, the method with the highest power (i.e., the one with the greatest chance of detecting a difference when differences actually exist) is the Tukey multiple comparisons methods.

The Minitab printout for this analysis, based on an experimentwise error rate of .05, is displayed in Figure SIA8.4. The simultaneous confidence intervals for all possible differences between treatment means are highlighted on the printout.

Note that there are $c = (k)(k - 1)/2 = (3)(2)/2 = 3$ comparisons of interest. A confidence interval that does not include a value of 0 indicates that the two means being compared are significantly different. From Figure SIA8.4, you can see that confidence intervals for $(\mu_S - \mu_N)$, $(\mu_V - \mu_N)$, and $(\mu_V - \mu_S)$, are $(-2.01, -.90)$, $(-1.64, -.53)$, and $(-.18, .92)$, respectively. The first interval implies that $\mu_S < \mu_N$; the second implies that $\mu_V < \mu_N$; and, because the third interval includes 0, there is no evidence of a difference between μ_S and μ_V. Consequently, we rank the TV content group means as follows:

Recall mean:	1.71	2.08	3.17
TV content:	Sex	Violence	Neutral

Because its clear that the Neutral content group had the highest mean recall score, our conclusion agrees with those of the researchers, who state in the journal article that "memory for [television] commercials is impaired after watching violent or sexual programming."

```
Tukey 95% Simultaneous Confidence Intervals
All Pairwise Comparisons among Levels of CONTENT

Individual confidence level = 98.01%


CONTENT = NEUTRAL subtracted from:

CONTENT    Lower   Center    Upper   -----+---------+---------+---------+
SEX       -2.007   -1.454   -0.901   (------*------)
VIOLENT   -1.636   -1.083   -0.530        (-----*------)
                                     -----+---------+---------+---------+
                                       -1.60     -0.80     -0.00      0.8


CONTENT = SEX subtracted from:

CONTENT    Lower   Center   Upper   -----+---------+---------+---------+-
VIOLENT   -0.183    0.370   0.923                         (------*------
                                    -----+---------+---------+---------+-
                                      -1.60     -0.80     -0.00      0.80
```

Figure SIA8.4
Minitab multiple comparisons of the three recall score means

Exercises 8.35–8.49

Learning the Mechanics

8.35 Consider a completely randomized design with k treatments. Assume all pairwise comparisons of treatment means are to be made using a multiple comparisons procedure. Determine the total number of pairwise comparisons for the following values of k.

a. $k = 3$ **b.** $k = 5$

c. $k = 4$ **d.** $k = 10$

8.36 Define an experimentwise error rate.

8.37 Define a comparisonwise error rate.

8.38 Consider a completely randomized design with five treatments, A, B, C, D, and E. The ANOVA F-test revealed significant differences among the means. A multiple comparisons procedure was used to compare all possible pairs of treatment means at $\alpha = .05$. The ranking of the five treatment means is summarized in each part on the next page. Identify which pairs of means are significantly different.

a. A C E B D **b.** A C E B D
c. A C E B D **d.** A C E B D

8.39 A multiple-comparison procedure for comparing four treatment means produced the confidence intervals shown here. Rank the means from smallest to largest. Which means are significantly different?
NW

$(\mu_1 - \mu_2)$: (2, 15)
$(\mu_1 - \mu_3)$: (4, 7)
$(\mu_1 - \mu_4)$: (−10, 3)
$(\mu_2 - \mu_3)$: (−5, 11)
$(\mu_2 - \mu_4)$: (−12, −6)
$(\mu_3 - \mu_4)$: (−8, −5)

Applying the Concepts—Basic

8.40 Guilt in decision making. The effect of guilt emotion on how a decision maker focuses on a problem was investigated in the Jan. 2007 issue of the *Journal of Behavioral Decision Making* (see Exercise 3.40, p. 140). A sample of 77 volunteer students participated in one portion of the experiment, where each was randomly assigned to one of three emotional states (guilt, anger, or neutral) through a reading/writing task. Immediately after the task, students were presented with a decision problem where the stated option has predominantly negative features (e.g., spending money on repairing a very old car). Prior to making the decision, the researchers asked each subject to list possible, more attractive alternatives. The researchers then compared the mean number of alternatives listed across the three emotional states with an analysis of variance for a completely randomized design. A partial ANOVA summary table is shown below.

Source	df	*F*-value	*p*-value
Emotional State	2	22.68	0.001
Error	74		
Total	76		

a. What conclusion can you draw from the ANOVA results?
b. A multiple comparisons of means procedure was applied to the data using an experimentwise error rate of .05. Explain what the .05 represents.
c. The multiple comparisons yielded the following results. What conclusion can you draw?

Sample mean:	1.90	2.17	4.75
Emotional state:	Angry	Neutral	Guilt

8.41 College tennis recruiting with a team Web site. Refer to the *Sport Journal* (Winter 2004) study to compare the attitudes of Division I, Division II, and Division III college tennis coaches toward team Web sites as recruiting tools, Exercise 8.24 (p. 468). The mean responses (measured on a 7-point scale) to the statement, "the Prospective Student-Athlete Form on the Web site contributes very little to the recruiting process," are listed and ranked in the next column. The results were obtained using a multiple comparisons procedure with an experimentwise error rate of .05. Interpret the results practically.

Mean:	4.51	3.60	3.21
Division:	I	II	III

8.42 Robots trained to behave like ants. Refer to the *Nature* (Aug. 2000) study of robots trained to behave like ants, Exercise 8.25 (p. 468). Multiple comparisons of mean energy expended for the four colony sizes were conducted using an experimentwise error rate of .05. The results are summarized below.

Sample mean:	.97	.95	.93	.80
Group size:	3	6	9	12

a. How many pairwise comparisons are conducted in this analysis?
b. Interpret the results shown in the table.

8.43 Study of mutual fund performance. Refer to the *American Business Review* (Jan. 2002) comparison of large-cap, medium-cap, and small-cap mutual funds, Exercise 8.26 (p. 468). Using an experimentwise error rate of .05, Tukey confidence intervals for the difference between mean rates of return for all possible pairs of fund types are given below.

Comparison	Tukey confidence interval
$\mu_{\text{Large}} - \mu_{\text{Medium}}$	(−.1847, 5.3807)
$\mu_{\text{Large}} - \mu_{\text{Small}}$	(2.4426, 8.0080)
$\mu_{\text{Medium}} - \mu_{\text{Small}}$	(−.1554, 5.4100)

a. Why is the Tukey multiple comparisons method preferred over another method?
b. Is there a significant difference between the treatment means for large-cap and medium-cap mutual funds? Explain.
c. Is there a significant difference between the treatment means for large-cap and small-cap mutual funds? Explain.
d. Is there a significant difference between the treatment means for medium-cap and small-cap mutual funds? Explain.
e. Use your answers to parts **b–d** to rank the treatment means.
f. Give a measure of reliability for the inference in part **e.**

8.44 Ethics of downsizing. Refer to the *Journal of Business Ethics* (Vol. 18, 1999) study of the ethics of downsizing, Exercise 8.29 (p. 469). You used an ANOVA to discover differences in the mean ethics scores of five groups of employees: (1) Casualties, (2) Survivors, (3) Implementors/casualties, (4) Implementors/survivors, and (5) Formulators. Multiple comparisons of the treatment (group) means were conducted using the Bonferroni method with an experimentwise error rate of .05.
a. Explain why the Bonferroni method is preferred over another multiple comparisons method (e.g., Tukey or Scheffé).
b. Determine the number of pairwise comparisons for this analysis.
c. The sample mean ethics scores for the five groups and Bonferroni rankings are summarized on the next page. Identify the groups with the significantly largest mean ethics scores.

1.59	1.79	1.84	2.45	2.87
Implementors/ casualties Group 3	Casualties Group 1	Survivors Group 2	Implementors/ survivors Group 4	Formulators Group 5

Applying the Concepts—Intermediate

8.45 Income and road rage. Refer to the *Accident Analysis and Prevention* (Vol. 34, 2002) study of road rage, Exercise 8.27 (p. 469). Recall that the mean road rage scores of drivers in the three income groups, under \$30,000, between \$30,000 and \$60,000, and over \$60,000, were 4.60, 5.08, and 5.15, respectively.

a. An experimentwise error rate of .01 was used to rank the three means. Give a practical interpretation of this error rate.

b. How many pairwise comparisons are necessary to compare the three means? List them.

c. A multiple comparisons procedure revealed that the means for the income groups \$30,000–\$60,000 and over \$60,000 were not significantly different. All other pairs of means were found to be significantly different. Summarize these results in table form.

d. Which of the comparisons of part **b** will yield a confidence interval that does not contain 0?

8.46 Diamonds sold at retail. Refer to the *Journal of Statistics Education* data on diamonds for sale on the open market saved in the **DIAMONDS** file. In Exercise 8.30 (p. 470) you ran an ANOVA to compare levels of a qualitative variable (either color, clarity, or certification group) on the mean of a quantitative variable (either carat size or price). Follow up the analysis with multiple comparisons of the treatment means. Use an experimentwise error rate of .05. Interpret the results practically.

8.47 Effectiveness of sales closing techniques. Refer to the *Industrial Marketing Management* (Sept. 1996) comparison of six sales closing techniques, Exercise 8.31 (p. 470). The "level of trust" means for prospects of salespeople using each of the six closing techniques are listed in the table. A multiple comparisons of means analysis was conducted (at $\alpha = .05$), with the results shown in the third column of the table. Fully interpret the results.

Treatments: Closing Technique	Mean Level of Trust	Differing Treatments
1. No close	4.67	1 from 5 and 6
2. Impending event	4.48	2 from 6
3. Social validation	4.40	No differences
4. If-then	4.33	No differences
5. Assumed close	4.04	5 from 1
6. Either-or	3.98	6 from 1 and 2

Source: Reprinted from Hawes, S. M., Strong, J. T., and Winick, B. S. "Do closing techniques diminish prospect trust?" *Industrial Marketing Management,* Vol. 25, No. 5, Sept. 1996, p. 355. ©1996 with permission of Elsevier.

8.48 Is honey a cough remedy? Refer to the *Archives of Pediatrics and Adolescent Medicine* (Dec. 2007) study of treatments for children's cough symptoms, Exercise 8.32 (p. 470). Recall that the data are available in the **HONEYCOUGH** file. Do you agree with the statement (extracted from the article), "Honey may be a preferable treatment for the cough and sleep difficulty associated with childhood upper respiratory tract infection"? Perform a multiple comparisons of means to answer the question.

8.49 Commercial eggs produced from different housing systems. Refer to the *Food Chemistry* (Vol. 106, 2008) study of four different types of egg housing systems, Exercise 8.33 (p. 471). Recall that you analyzed the data in the **EGGS** file and discovered that the mean shell thickness (millimeters) differed for cage, barn, free range, and organic egg housing systems. A multiple comparisons of means was conducted using the Bonferroni method with an experimentwise error rate of .05. The results are displayed in the accompanying SPSS printout.

SPSS Output for Exercise 8.49

Multiple Comparisons

THICKNESS
Bonferroni

(I) HOUSE	(J) HOUSE	Mean Difference (I-J)	Std. Error	Sig.	95% Confidence Interval Lower Bound	95% Confidence Interval Upper Bound
CAGE	BARN	-.07867*	.01612	.000	-.1250	-.0323
	FREE	-.07700*	.01612	.000	-.1233	-.0307
	ORGANIC	-.05867*	.01612	.008	-.1050	-.0123
BARN	CAGE	.07867*	.01612	.000	.0323	.1250
	FREE	.00167	.01802	1.000	-.0501	.0535
	ORGANIC	.02000	.01802	1.000	-.0318	.0718
FREE	CAGE	.07700*	.01612	.000	.0307	.1233
	BARN	-.00167	.01802	1.000	-.0535	.0501
	ORGANIC	.01833	.01802	1.000	-.0335	.0701
ORGANIC	CAGE	.05867*	.01612	.008	.0123	.1050
	BARN	-.02000	.01802	1.000	-.0718	.0318
	FREE	-.01833	.01802	1.000	-.0701	.0335

*. The mean difference is significant at the 0.05 level.

a. Locate the confidence interval for $(\mu_{\text{CAGE}} - \mu_{\text{BARN}})$ on the printout and interpret the result.
b. Locate the confidence interval for $(\mu_{\text{CAGE}} - \mu_{\text{FREE}})$ on the printout and interpret the result.
c. Locate the confidence interval for $(\mu_{\text{CAGE}} - \mu_{\text{ORGANIC}})$ on the printout and interpret the result.
d. Locate the confidence interval for $(\mu_{\text{BARN}} - \mu_{\text{FREE}})$ on the printout and interpret the result.
e. Locate the confidence interval for $(\mu_{\text{BARN}} - \mu_{\text{ORGANIC}})$ on the printout and interpret the result.
f. Locate the confidence interval for $(\mu_{\text{FREE}} - \mu_{\text{ORGANIC}})$ on the printout and interpret the result.
g. Based on the results, parts **a–f,** provide a ranking of the housing system means. Include the experimentwise error rate as a statement of reliability.

8.4 The Randomized Block Design

If the completely randomized design results in nonrejection of the null hypothesis that the treatment means differ because the sampling variability (as measured by MSE) is large, we may want to consider an experimental design that better controls the variability. In contrast to the selection of independent samples of experimental units specified by the completely randomized design, the *randomized block design* uses experimental units that are *matched sets,* assigning one from each set to each treatment. The matched sets of experimental units are called *blocks.* The theory behind the randomized block design is that the sampling variability of the experimental units in each block will be reduced, in turn reducing the measure of error, MSE.

The **randomized block design** consists of a two-step procedure:

1. Matched sets of experimental units, called **blocks,** are formed, each block consisting of k experimental units (where k is the number of treatments). The b blocks should consist of experimental units that are as similar as possible.
2. One experimental unit from each block is randomly assigned to each treatment, resulting in a total of $n = bk$ responses.

For example, if we wish to compare the SAT scores of female and male high school seniors, we could select independent random samples of five females and five males, and analyze the results of the completely randomized design as outlined in Section 8.2. Or, we could select matched pairs of females and males according to their scholastic records and analyze the SAT scores of the pairs. For instance, we could select pairs of students with approximately the same GPAs from the same high school. Five such pairs (blocks) are depicted in Table 8.5. Note that this is just a *paired difference experiment,* first discussed in Section 7.3.

As before, the variation between the treatment means is measured by squaring the distance between each treatment mean and the overall mean, multiplying each squared distance by the number of measurements for the treatment and summing over treatments:

$$\begin{aligned}\text{SST} &= \sum_{i=1}^{k} b(\bar{x}_{T_i} - \bar{x})^2 \\ &= 5(606 - 600)^2 + 5(594 - 600)^2 = 360\end{aligned}$$

where $\bar{x}_{T_i}$ represents the sample mean for the ith treatment, b (the number of blocks) is the number of measurements for each treatment, and k is the number of treatments.

Table 8.5 Randomized Block Design: SAT Score Comparison

Block	Female SAT Score	Male SAT Score	Block Mean
1 (School A, 2.75 GPA)	540	530	535
2 (School B, 3.00 GPA)	570	550	560
3 (School C, 3.25 GPA)	590	580	585
4 (School D, 3.50 GPA)	640	620	630
5 (School E, 3.75 GPA)	690	690	690
Treatment Mean	606	594	

The blocks also account for some of the variation among the different responses—that is, just as SST measures the variation between the female and male means, we can calculate a measure of variation among the five block means representing different schools and scholastic abilities. Analogous to the computation of SST, we sum the squares of the differences between each block mean and the overall mean, multiplying each squared difference by the number of measurements for each block, and sum over blocks to calculate the **Sum of Squares for Blocks (SSB):**

$$\begin{aligned} \text{SSB} &= \sum_{i=1}^{b} k(\bar{x}_{B_i} - \bar{x})^2 \\ &= 2(535 - 600)^2 + 2(560 - 600)^2 + 2(585 - 600)^2 \\ &\quad + 2(630 - 600)^2 + 2(690 - 600)^2 \\ &= 30{,}100 \end{aligned}$$

where $\bar{x}_{B_i}$ represents the sample mean for the ith block and k (the number of treatments) is the number of measurements in each block. As we expect, the variation in SAT scores attributable to schools and levels of scholastic achievement is apparently large.

Now, we want to compare the variability attributed to treatments with that which is attributed to sampling variability. In a randomized block design, the sampling variability is measured by subtracting that portion attributed to treatments and blocks from the Total Sum of Squares, SS(Total). The total variation is the sum of squared differences of each measurement from the overall mean:

$$\begin{aligned} \text{SS(Total)} &= \sum_{i=1}^{n} (x_i - \bar{x})^2 \\ &= (540 - 600)^2 + (530 - 600)^2 + (570 - 600)^2 + (550 - 600)^2 \\ &\quad + \cdots + (690 - 600)^2 \\ &= 30{,}600 \end{aligned}$$

Then the variation attributable to sampling error is found by subtraction:

$$\text{SSE} = \text{SS(Total)} - \text{SST} - \text{SSB} = 30{,}600 - 360 - 30{,}100 = 140$$

In summary, the Total Sum of Squares—30,600—is divided into three components: 360 attributed to treatments (Gender), 30,100 attributed to blocks (Scholastic ability and School), and 140 attributed to sampling error.

The mean squares associated with each source of variability are obtained by dividing the sum of squares by the appropriate number of degrees of freedom. The partitioning of the Total Sum of Squares and the total degrees of freedom for a randomized block experiment is summarized in Figure 8.14.

To determine whether we can reject the null hypothesis that the treatment means are equal in favor of the alternative that at least two of them differ, we calculate

$$\text{MST} = \frac{\text{SST}}{k - 1} = \frac{360}{2 - 1} = 360$$

$$\text{MSE} = \frac{\text{SSE}}{n - b - k + 1} = \frac{140}{10 - 5 - 2 + 1} = 35$$

The F-ratio that is used to test the hypothesis is

$$F = \frac{360}{35} = 10.29$$

Comparing this ratio to the tabled F value corresponding to $\alpha = .05$, $v_1 = (k - 1) = 1$ degree of freedom in the numerator, and $v_2 = (n - b - k + 1) = 4$ degrees of freedom in the denominator, we find that

$$F = 10.29 > F_{.05} = 7.71$$

which indicates that we should reject the null hypothesis and conclude that the mean SAT scores differ for females and males.

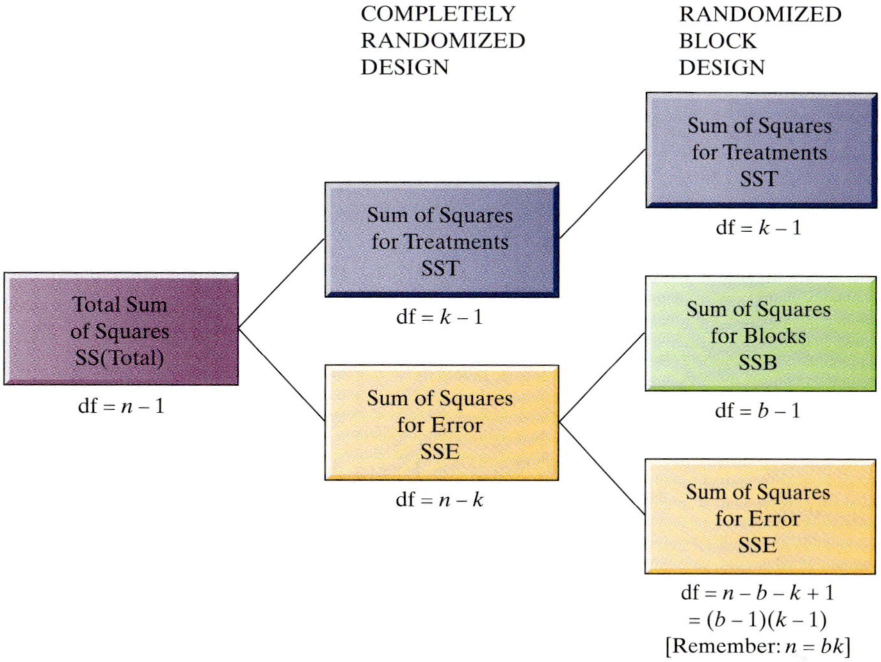

Figure 8.14
Partitioning of the total sum of squares for the randomized block design

If you review Section 7.3, you will find that the analysis of a paired difference experiment results in a one-sample t-test on the differences between the treatment responses within each block. Applying the procedure to the differences between female and male scores in Table 8.5, we find

$$t = \frac{\bar{x}_d}{s_d/\sqrt{n_d}} = \frac{12}{\sqrt{70}/\sqrt{5}} = 3.207$$

At the .05 level of significance with $(n_d - 1) = 4$ degrees of freedom,

$$t = 3.21 > t_{.025} = 2.776$$

Because $t^2 = (3.207)^2 = 10.29$ and $t^2_{.025} = (2.776)^2 = 7.71$, we find that the paired difference t-test and the ANOVA F-test are equivalent, with both the calculated test statistics and the rejection region related by the formula $F = t^2$. The difference between the tests is that the paired difference t-test can be used to compare only two treatments in a randomized block design, whereas the F-test can be applied to *two or more* treatments in a randomized block design. The F-test is summarized in the following box.

ANOVA F-Test to Compare k Treatment Means: Randomized Block Design

H_0: $\mu_1 = \mu_2 = \cdots = \mu_k$

H_a: At least two treatment means differ

Test statistic: $F = \dfrac{\text{MST}}{\text{MSE}}$

Rejection region: $F > F_\alpha$, where F_α is based on $(k - 1)$ numerator degrees of freedom and $(n - b - k + 1)$ denominator degrees of freedom.

Conditions Required for a Valid ANOVA F-test: Randomized Block Design

1. The b blocks are randomly selected, and all k treatments are applied (in random order) to each block.
2. The distributions of observations corresponding to all bk block-treatment combinations are approximately normal.
3. The bk block-treatment distributions have equal variances.

Note that the assumptions concern the probability distributions associated with each block-treatment combination. The experimental unit selected for each combination is assumed to have been randomly selected from all possible experimental units for that combination, and the response is assumed to be normally distributed with the same variance for each of the block-treatment combinations. For example, the F-test comparing female and male SAT score means requires the scores for each combination of gender and scholastic ability (e.g., females with 3.25 GPA from School C) to be normally distributed with the same variance as the other combinations employed in the experiment.

The calculation formulas for randomized block designs are given in Appendix C. We will rely on statistical software packages to analyze randomized block designs and to obtain the necessary ingredients for testing the null hypothesis that the treatment means are equal.

Example 8.7 Experimental Design Principles

Problem Refer to Examples 8.4–8.6. Suppose the USGA wants to compare the mean distances associated with the four brands of golf balls when struck by a driver but wishes to employ human golfers rather than the robot Iron Byron. Assume that 10 balls of each brand are to be used in the experiment.

a. Explain how a completely randomized design could be employed.

b. Explain how a randomized block design could be employed.

c. Which design is likely to provide more information about the differences among the brand mean distances?

Solution

a. Because the completely randomized design calls for independent samples, we can employ such a design by randomly selecting 40 golfers and then randomly assigning 10 golfers to each of the four brands. Finally, each golfer will strike the ball of the assigned brand, and the distance will be recorded. The design is illustrated in Figure 8.15a.

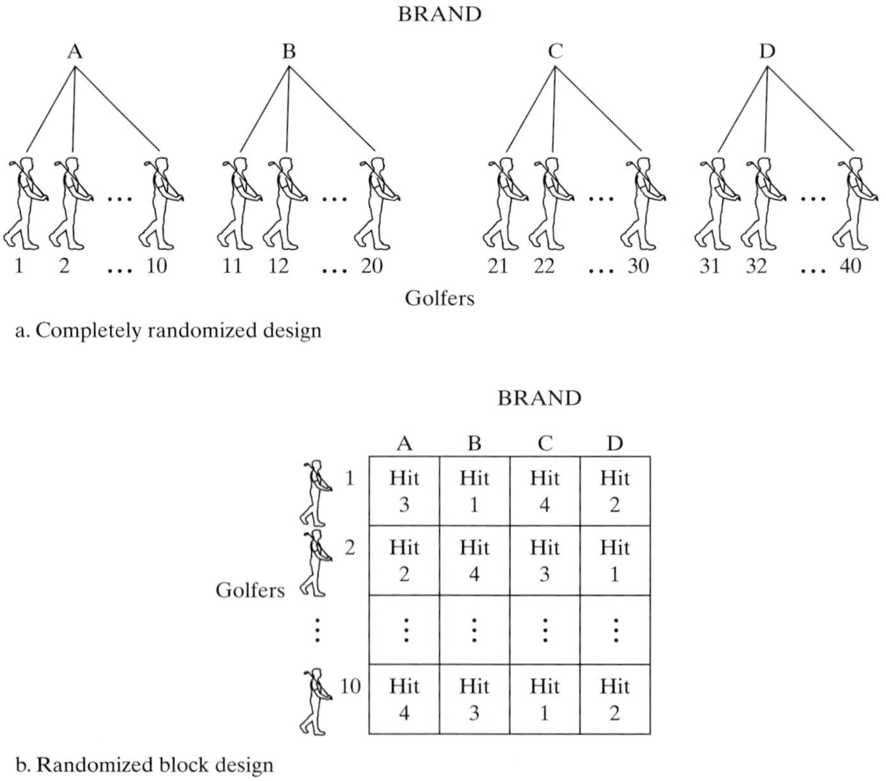

Figure 8.15
Illustration of completely randomized design and randomized block design: comparison of four golf ball brands

b. The randomized block design employs blocks of relatively homogeneous experimental units. For example, we could randomly select 10 golfers and permit each golfer to hit four balls, one of each brand, in a random sequence. Then each golfer is a block, with each treatment (brand) assigned to each block (golfer). The design is summarized in Figure 8.15b.

c. Because we expect much more variability among distances generated by "real" golfers than by Iron Byron, we would expect the randomized block design to control the variability better than the completely randomized design—that is, with 40 different golfers, we would expect the sampling variability among the measured distances within each brand to be greater than that among the four distances generated by each of 10 golfers hitting one ball of each brand.

Now Work Exercise 8.56a–b

Example 8.8
Comparing Golf Ball Brands with a Randomized Block Design

Problem Refer to Example 8.7. Suppose the randomized block design of part **b** is employed, employing a random sample of 10 golfers, with each golfer using a driver to hit four balls, one of each brand, in a random sequence.

a. Set up a test of the research hypothesis that the brand mean distances differ. Use $\alpha = .05$.

b. The data for the experiment are given in Table 8.6. Use statistical software to analyze the data and conduct the test set up in part **a.**

Solution

a. We want to test whether the data in Table 8.6 provide sufficient evidence to conclude that the brand mean distances differ. Denoting the population mean of the ith brand by μ_i, we test

H_0: $\mu_1 = \mu_2 = \mu_3 = \mu_4$

H_a: The mean distances differ for at least two of the brands.

The test statistic compares the variation among the four treatment (brand) means to the sampling variability within each of the treatments.

$$\textit{Test statistic: } F = \frac{\text{MST}}{\text{MSE}}$$

Rejection region: $F > F_\alpha = F_{.05}$, with $v_1 = (k - 1) = 3$ numerator degrees of freedom and $v_2 = (n - k - b + 1) = 27$ denominator degrees of freedom. From Table VIII in Appendix B, we find $F_{.05} = 2.96$. Thus, we will reject H_0 if $F > 2.96$. The assumptions necessary to ensure the validity of the test are as follows: (1) The probability distributions of the distances for each brand–golfer combination are

Table 8.6 Distance Data for Randomized Block Design

Golfer (Block)	Brand A	Brand B	Brand C	Brand D
1	202.4	203.2	223.7	203.6
2	242.0	248.7	259.8	240.7
3	220.4	227.3	240.0	207.4
4	230.0	243.1	247.7	226.9
5	191.6	211.4	218.7	200.1
6	247.7	253.0	268.1	244.0
7	214.8	214.8	233.9	195.8
8	245.4	243.6	257.8	227.9
9	224.0	231.5	238.2	215.7
10	252.2	255.2	265.4	245.2
Sample Means	227.0	233.2	245.3	220.7

Data Set: GOLFRBD

Two-way ANOVA: DISTANCE versus BRAND, GOLFER

```
Source  DF       SS       MS      F      P
BRAND    3   3298.7  1099.55  54.31  0.000
GOLFER   9  12073.9  1341.54  66.26  0.000
Error   27    546.6    20.25
Total   39  15919.2

S = 4.499   R-Sq = 96.57%   R-Sq(adj) = 95.04%
```

Figure 8.16
Minitab randomized block design ANOVA: golf ball brand comparison

normal. (2) The variances of the distance probability distributions for each brand–golfer combination are equal.

b. Minitab was used to analyze the data in Table 8.6, and the result is shown in Figure 8.16. The values of MST and MSE (highlighted on the printout) are 1,099.6 and 20.2, respectively. The *F*-ratio for Brand (also highlighted on the printout) is $F = 54.31$, which exceeds the tabled value of 2.96. We therefore reject the null hypothesis at the $\alpha = .05$ level of significance, concluding that at least two of the brands differ with respect to mean distance traveled when struck by the driver.

Look Back The result of part **b** can also be obtained by noting that the observed significance level of the test, highlighted on the printout, is $p \approx 0$.

Now Work Exercise 8.56c

The results of an ANOVA can be summarized in a simple tabular format, similar to that used for the completely randomized design in Section 8.2. The general form of the table is shown in Table 8.7, and that for Example 8.8 is given in Table 8.8. Note that the randomized block design is characterized by three sources of variation—Treatments, Blocks, and Error—which sum to the Total Sum of Squares. We hope that employing blocks of experimental units will reduce the error variability, thereby making the test for comparing treatment means more powerful.

When the *F*-test results in the rejection of the null hypothesis that the treatment means are equal, we will usually want to compare the various pairs of treatment means to determine which specific pairs differ. We can employ a multiple comparisons procedure as in Section 8.3. The number of pairs of means to be compared will again be $c = k(k - 1)/2$, where k is the number of treatment means. In Example 8.8, $c = 4(3)/2 = 6$; that is, there are six pairs of golf ball brand means to be compared.

Table 8.7 General ANOVA Summary Table for a Randomized Block Design

Source	df	SS	MS	*F*
Treatment	$k - 1$	SST	MST	MST/MSE
Block	$b - 1$	SSB	MSB	
Error	$n - k - b + 1$	SSE	MSE	
Total	$n - 1$	SS(Total)		

Table 8.8 ANOVA Table for Example 8.8

Source	df	SS	MS	*F*	*p*
Treatment (brand)	3	3,298.7	1,099.6	54.31	.000
Block (golfer)	9	12,073.9	1,341.5		
Error	27	546.6	20.2		
Total	39	15,919.2			

Example 8.9

Ranking Golf Ball Brand Means in a Randomized Block Design

Problem Bonferroni's procedure is used to compare the mean distances of the four golf ball brands in Example 8.8. The resulting confidence intervals, with an experimentwise error rate of $\alpha = .05$, are shown in the SPSS printout, Figure 8.17. Interpret the results.

Multiple Comparisons

Dependent Variable: DISTANCE
Bonferroni

(I) BRAND	(J) BRAND	Mean Difference (I-J)	Std. Error	Sig.	95% Confidence Interval Lower Bound	95% Confidence Interval Upper Bound
A	B	-6.1300*	2.01222	.031	-11.8586	-.4014
	C	-18.2800*	2.01222	.000	-24.0086	-12.5514
	D	6.3200*	2.01222	.024	.5914	12.0486
B	A	6.1300*	2.01222	.031	.4014	11.8586
	C	-12.1500*	2.01222	.000	-17.8786	-6.4214
	D	12.4500*	2.01222	.000	6.7214	18.1786
C	A	18.2800*	2.01222	.000	12.5514	24.0086
	B	12.1500*	2.01222	.000	6.4214	17.8786
	D	24.6000*	2.01222	.000	18.8714	30.3286
D	A	-6.3200*	2.01222	.024	-12.0486	-.5914
	B	-12.4500*	2.01222	.000	-18.1786	-6.7214
	C	-24.6000*	2.01222	.000	-30.3286	-18.8714

Based on observed means.

*. The mean difference is significant at the .05 level.

Figure 8.17
SPSS listing of Bonferroni confidence intervals: follow-up to randomized block design ANOVA

Solution Note that 12 confidence intervals are shown in Figure 8.17, rather than 6. Recall that SPSS computes intervals for both $\mu_i - \mu_j$ and $\mu_j - \mu_i$, $i \neq j$. Only half of these are necessary to conduct the analysis, and these are highlighted on the printout. The intervals (rounded) are summarized below:

$$(\text{A} - \text{B}): \ (-11.9, -.4)$$
$$(\text{A} - \text{C}): \ (-24.0, -12.6)$$
$$(\text{A} - \text{D}): \ (.6, 12.0)$$
$$(\text{B} - \text{C}): \ (-17.9, -6.4)$$
$$(\text{B} - \text{D}): \ (6.7, 18.2)$$
$$(\text{C} - \text{D}): \ (18.9, 30.3)$$

Note that we are 95% confident that all the brand means differ because none of the intervals contains 0. The listing of the brand means in Figure 8.18 has no lines connecting them because there are no nonsignificant differences at the .05 level.

Mean:	220.7	227.0	233.2	245.3
Brand:	D	A	B	C

Figure 8.18
Listing of brand means for randomized block design [*Note:* All differences are statistically significant.]

Now Work Exercise 8.56d

Unlike the completely randomized design, the randomized block design cannot, in general, be used to estimate individual treatment means. Whereas the completely randomized design employs a random sample for each treatment, the randomized block design does not necessarily employ a random sample of experimental units for each treatment. The experimental units within the blocks are assumed to be randomly selected, but the blocks themselves may not be randomly selected.

We can, however, test the hypothesis that the block means are significantly different. We simply compare the variability attributable to differences among the block means to that associated with sampling variability. The ratio of MSB to MSE is an F-ratio similar to

Table 8.9 ANOVA Table for Randomized Block Design: Test for Blocks Included

Source	df	SS	MS	F	p
Treatments (brands)	3	3,298.7	1,099.6	54.31	.000
Blocks (golfers)	9	12,073.9	1,341.5	66.26	.000
Error	27	546.6	20.2		
Total	39	15,919.2			

that formed in testing treatment means. The F-statistic is compared to a tabled value for a specific value of α, with numerator degrees of freedom $(b - 1)$ and denominator degrees of freedom $(n - k - b + 1)$. The test is usually given on the same printout as the test for treatment means. Refer to the Minitab printout in Figure 8.18 and note that the test statistic for comparing the block means is

$$F = \frac{\text{MSB}}{\text{MSE}} = \frac{\text{MS(Golfers)}}{\text{MS(Error)}} = \frac{1{,}341.5}{20.2} = 66.26$$

with a p-value of .000. Because $\alpha = .05$ exceeds this p-value, we conclude that the block means are different. The results of the test are summarized in Table 8.9.

In the golf example, the test for block means confirms our suspicion that the golfers vary significantly; therefore, use of the block design was a good decision. However, be careful not to conclude that the block design was a mistake if the F-test for blocks does not result in rejection of the null hypothesis that the block means are the same. Remember that the possibility of a Type II error exists, and we are not controlling its probability as we are the probability α of a Type I error. If the experimenter believes that the experimental units are more homogeneous within blocks than between blocks, then he or she should use the randomized block design regardless of the results of a single test comparing the block means.

The procedure for conducting an analysis of variance for a randomized block design is summarized in the next box. Remember that the hallmark of this design is using blocks of homogeneous experimental units in which each treatment is represented.

Steps for Conducting an ANOVA for a Randomized Block Design

1. Be sure the design consists of blocks (preferably, blocks of homogeneous experimental units) and that each treatment is randomly assigned to one experimental unit in each block.
2. If possible, check the assumptions of normality and equal variances for all block-treatment combinations. [*Note:* This may be difficult to do because the design will likely have only one observation for each block-treatment combination.]
3. Create an ANOVA summary table that specifies the variability attributable to Treatments, Blocks, and Error, which leads to the calculation of the F-statistic to test the null hypothesis that the treatment means are equal in the population. Use a statistical software package or the calculation formulas in Appendix C to obtain the necessary numerical ingredients.
4. If the F-test leads to the conclusion that the means differ, use the Bonferroni, Tukey, or similar procedure to conduct multiple comparisons of as many of the pairs of means as you wish. Use the results to summarize the statistically significant differences among the treatment means. Remember that, in general, the randomized block design cannot be used to form confidence intervals for individual treatment means.
5. If the F-test leads to the nonrejection of the null hypothesis that the treatment means are equal, several possibilities exist:
   a. The treatment means are equal—that is, the null hypothesis is true.
   b. The treatment means really differ, but other important factors affecting the response are not accounted for by the randomized block design. These factors

inflate the sampling variability, as measured by MSE, resulting in smaller values of the F-statistic. Either increase the sample size for each treatment or conduct an experiment that accounts for the other factors affecting the response (as in Section 8.5). Do not automatically reach the former conclusion because the possibility of a Type II error must be considered if you accept H_0.

6. If desired, conduct the F-test of the null hypothesis that the block means are equal. Rejection of this hypothesis lends statistical support to using the randomized block design.

Note: It is often difficult to check whether the assumptions for a randomized block design are satisfied. There is usually only one observation for each block-treatment combination. When you feel these assumptions are likely to be violated, a nonparametric procedure is advisable.

What Do You Do When Assumptions Are Not Satisfied for the Analysis of Variance for a Randomized Block Design?

Answer: Use a nonparametric statistical method such as the Friedman F_r test of Section 14.6.

Activity 8.2 Randomized Block Design

In this Activity, you will revisit Activity 8.1 (p. 453). For each of the designed experiments in Exercises 2 and 3, you will rework the experiment to have a randomized block design. Explain how you will choose your experimental units from each population. Then describe the criteria you would use to split the experimental units into matched sets. How do you determine which experimental unit in each matched set receives each level of treatment? What data are collected and how are the data compared? Do you believe that there is any benefit to the block design? Explain.

Exercises 8.50–8.61

Learning the Mechanics

8.50 A randomized block design yielded the following ANOVA table.

Source	df	SS	MS	F
Treatments	4	501	125.25	9.109
Blocks	2	225	112.50	8.182
Error	8	110	13.75	
Total	14	836		

a. How many blocks and treatments were used in the experiment?

b. How many observations were collected in the experiment?

c. Specify the null and alternative hypotheses you would use to compare the treatment means.

d. What test statistic should be used to conduct the hypothesis test of part **c**?

e. Specify the rejection region for the test of parts **c** and **d.** Use $\alpha = .01$.

f. Conduct the test of parts **c–e** and state the proper conclusion.

g. What assumptions are necessary to ensure the validity of the test you conducted in part **f**?

8.51 An experiment was conducted using a randomized block design. The data from the experiment are displayed in the following table and saved in the **LM8_51** file.

Treatment	Block 1	Block 2	Block 3
1	2	3	5
2	8	6	7
3	7	6	5

a. Fill in the missing entries in the ANOVA table.

Source	df	SS	MS	F
Treatments	2	21.5555		
Blocks	2			
Error	4			
Total	8	30.2222		

b. Specify the null and alternative hypotheses you would use to investigate whether a difference exists among the treatment means.

c. What test statistic should be used in conducting the test of part **b**?

d. Describe the Type I and Type II errors associated with the hypothesis test of part **b.**

e. Conduct the hypothesis test of part **b** using $\alpha = .05$.

8.52 A randomized block design was used to compare the mean responses for three treatments. Four blocks of three homogeneous experimental units were selected, and each treatment was randomly assigned to one experimental unit within each block. The data are shown in the next table (saved in the **LM8_52** file), and the SPSS ANOVA printout for this experiment is shown below.

	Block			
Treatment	1	2	3	4
A	3.4	5.5	7.9	1.3
B	4.4	5.8	9.6	2.8
C	2.2	3.4	6.9	.3

a. Use the printout to fill in the entries in the following ANOVA table.

Source	df	SS	MS	F
Treatments				
Blocks				
Error				
Total				

b. Do the data provide sufficient evidence to indicate that the treatment means differ? Use $\alpha = .05$.

c. Do the data provide sufficient evidence to indicate that blocking was effective in reducing the experimental error? Use $\alpha = .05$.

d. Use the printout to rank the treatment means at $\alpha = .05$.

e. What assumptions are necessary to ensure the validity of the inferences made in parts **b, c,** and **d**?

8.53 Suppose an experiment employing a randomized block design has four treatments and nine blocks, for a total of $4 \times 9 = 36$ observations. Assume that the Total Sum of Squares for the response is SS(Total) = 500. For each of the following partitions of SS(Total), test the null hypothesis that the treatment means are equal and test the null hypothesis that the block means are equal. Use $\alpha = .05$ for each test.

a. The Sum of Squares for Treatments (SST) is 20% of SS(Total), and the Sum of Squares for Blocks (SSB) is 30% of SS(Total).

b. SST is 50% of SS(Total), and SSB is 20% of SS(Total).

c. SST is 20% of SS(Total), and SSB is 50% of SS(Total).

d. SST is 40% of SS(Total), and SSB is 40% of SS(Total).

e. SST is 20% of SS(Total), and SSB is 20% of SS(Total).

SPSS output for Exercise 8.52

Tests of Between-Subjects Effects

Dependent Variable: RESPONSE

Source	Type III Sum of Squares	df	Mean Square	F	Sig.
Corrected Model	83.781[a]	5	16.756	141.935	.000
Intercept	238.521	1	238.521	2020.412	.000
TRTMENT	12.032	2	6.016	50.958	.000
BLOCK	71.749	3	23.916	202.586	.000
Error	.708	6	.118		
Total	323.010	12			
Corrected Total	84.489	11			

a. R Squared = .992 (Adjusted R Squared = .985)

Multiple Comparisons

Dependent Variable: RESPONSE
Tukey HSD

(I) TRTMENT	(J) TRTMENT	Mean Difference (I-J)	Std. Error	Sig.	95% Confidence Interval Lower Bound	95% Confidence Interval Upper Bound
A	B	-1.125*	.2430	.009	-1.870	-.380
	C	1.325*	.2430	.004	.580	2.070
B	A	1.125*	.2430	.009	.380	1.870
	C	2.450*	.2430	.000	1.705	3.195
C	A	-1.325*	.2430	.004	-2.070	-.580
	B	-2.450*	.2430	.000	-3.195	-1.705

Based on observed means.

*. The mean difference is significant at the .05 level.

Applying the Concepts—Basic

8.54 Making high-stakes insurance decisions. Refer to the *Journal of Economic Psychology* (Sep. 2008) study on high-stakes insurance decisions, Exercise 8.22 (p. 468). A second experiment involved only the group 2 subjects. In part A of the experiment, these 84 subjects were informed of the hazards (both fire and theft) of owning a valuable painting but were not told the exact probabilities of the hazards occurring. The subjects then provided an amount they were willing to pay (WTP) for insuring the painting. In part B of the experiment, these same subjects were informed of the exact probabilities of the hazards (fire and theft) of owning a valuable sculpture. The subjects then provided a WTP amount for insuring the sculpture. The researchers were interested in comparing the mean WTP amounts for the painting and the sculpture.

a. Explain why the experimental design employed is a randomized block design.

b. Identify the dependent (response) variable, treatments, and blocks for the design.

c. Give the null and alternative hypotheses of interest to the researchers.

8.55 Peer mentor training at a firm. Peer mentoring occurs when a more experienced employee provides one-on-one support and knowledge sharing with a less experienced employee. The *Journal of Managerial Issues* (Spring 2008) published a study of the impact of peer mentor training at a large software company. Participants were 222 employees who volunteered to attend a one-day peer mentor training session. One variable of interest was the employee's level of competence in peer mentoring (measured on a 7-point scale). The competence level of each trainee was measured at three different times in the study: one week before training, two days after training, and two months after training. One goal of the experiment was to compare the mean competence levels of the three time periods.

a. Explain why these data should be analyzed using a randomized block design. As part of your answer, identify the blocks and the treatments.

b. A partial ANOVA table for the experiment is shown below. Explain why there is enough information in the table to make conclusions.

Source	df	SS	MS	F-value	p-value
Time Period	2	___	___	___	0.001
Blocks	221	___	___	___	0.001
Error	442	___	___		
Total	665	___			

c. State the null hypothesis of interest to the researcher.

d. Make the appropriate conclusion.

e. A multiple comparisons of means for the three time periods (using an experimentwise error rate of .10) is summarized below. Fully interpret the results.

Sample mean:	3.65	4.14	4.17
Time period:	*before*	*2 months after*	*2 days after*

8.56 NW **Flower production of dwarf shrubs.** Dwarf shrubs are popular with model home landscapers. Stetson University researchers conducted an experiment to determine the effects of fire on the shrub's growth (*Florida Scientist,* Spring 1997). Twelve experimental plots of land were selected in a pasture where the shrub is abundant. Within each plot, three dwarf shrubs were randomly selected and treated as follows: one shrub was subjected to fire, another to clipping, and the third was left unmanipulated (a control). After 5 months, the number of flowers produced by each of the 36 shrubs was determined. The objective of the study was to compare the mean number of flowers produced by dwarf shrubs for the three treatments (fire, clipping, and control).

a. Identify the type of experimental design employed, including the treatments, response variable, and experimental units.

b. Illustrate the layout of the design using a graphic similar to Figure 8.15.

c. The ANOVA of the data resulted in a test statistic of $F = 5.42$ for treatments, with an associated p-value of $p = .009$. Interpret this result.

d. The three treatment means were compared using Tukey's method at $\alpha = .05$. Interpret the results shown below.

Mean number of flowers:	1.17	10.58	17.08
Treatment:	*Control*	*Clipping*	*Burning*

8.57 Rotary oil rigs. An economist wants to compare the average monthly number of rotary oil rigs running in three states—California, Utah, and Alaska. In order to account for month-to-month variation, 3 months were randomly selected over a 2-year period, and the number of oil rigs running in each state in each month was obtained from data provided from *World Oil* (Jan. 2002) magazine. The data, reproduced in the accompanying table and saved in the **OILRIGS** file, were analyzed using a randomized block design.

Month/Year	California	Utah	Alaska
Nov. 2000	27	17	11
Oct. 2001	34	20	14
Nov. 2001	36	15	14

a. Why is a randomized block design preferred over a completely randomized design for comparing the mean number of oil rigs running monthly in California, Utah, and Alaska?

b. Identify the treatments for the experiment.

c. Identify the blocks for the experiment.

d. State the null hypothesis for the ANOVA F-test.

e. Locate the test statistic and p-value on the Minitab printout shown on the next page. Interpret the results.

f. A Tukey multiple comparisons of means (at $\alpha = .05$) is summarized in the SPSS printout on page 490. Which state(s) have the significantly largest mean number of oil rigs running monthly?

Applying the Concepts—Intermediate

8.58 Reducing on-the-job stress. Plant therapists believe that plants can reduce on-the-job stress. A Kansas State University study was conducted to investigate this phenomenon. Two weeks prior to final exams, 10 undergraduate students took part in an experiment to determine what effect the presence of a live plant, a photo of a plant, or absence of a plant has on a student's ability to relax while isolated in a dimly lit room. Each student participated in

Minitab output for Exercise 8.57

Two-way ANOVA: NumRigs versus State, Month/Year

```
Source      DF       SS       MS      F      P
State        2  617.556  308.778  38.07  0.002
Month/Year   2   30.889   15.444   1.90  0.262
Error        4   32.444    8.111
Total        8  680.889

S = 2.848   R-Sq = 95.23%   R-Sq(adj) = 90.47%


                    Individual 95% CIs For Mean Based on
                    Pooled StDev
State      Mean     ---------+---------+---------+---------+
AL      13.0000     (----*-----)
CAL     32.3333                             (----*-----)
UT      17.3333          (-----*----)
                    ---------+---------+---------+---------+
                          16.0      24.0      32.0      40.0
```

SPSS output for Exercise 8.57

NUMRIGS

Tukey HSD[a,b]

STATE	N	Subset 1	Subset 2
AL	3	13.00	
UT	3	17.33	
CAL	3		32.33
Sig.		.262	1.000

Means for groups in homogeneous subsets are displayed.
Based on Type III Sum of Squares
The error term is Mean Square(Error) = 8.111.

a. Uses Harmonic Mean Sample Size = 3.000.

b. Alpha = .05.

Student	Live Plant	Plant Photo	No Plant (control)
1	91.4	93.5	96.6
2	94.9	96.6	90.5
3	97.0	95.8	95.4
4	93.7	96.2	96.7
5	96.0	96.6	93.5
6	96.7	95.5	94.8
7	95.2	94.6	95.7
8	96.0	97.2	96.2
9	95.6	94.8	96.0
10	95.6	92.6	96.6

Source: Elizabeth Schreiber. Department of Statistics, Kansas State University, Manhattan, Kansas.

three sessions—one with a live plant, one with a plant photo, and one with no plant (control).* During each session, finger temperature was measured at 1-minute intervals for 20 minutes. Because increasing finger temperature indicates an increased level of relaxation, the maximum temperature (in degrees) was used as the response variable. The data for the experiment (saved in the **PLANTS** file) are provided in the table at the left. Conduct an ANOVA and make the proper inferences at $\alpha = .10$.

8.59 Absentee rates at a jeans plant. A plant that manufactures denim jeans in the United Kingdom recently introduced a computerized automated handling system. The new system delivers garments to the assembly line operators by means of an overhead conveyor. Although the automated system minimizes operator handling time, it inhibits operators from working ahead and taking breaks from their machine. A study in *New Technology, Work, and Employment* (July 2001) investigated the impact of the new handling system on worker absentee rates at the jeans plant. One theory is that the mean absentee rate will vary by day of the week, as operators decide to indulge in 1-day absences to relieve work pressure. Nine weeks were randomly selected, and the absentee rate (percentage of workers absent) determined for each day (Monday through Friday) of the workweek. The data (saved in the **JEANS** file) are listed in the table. Conduct a complete analysis of the data to determine whether the mean absentee rate differs across the 5 days of the workweek.

Week	Mon	Tues	Wed	Thur	Fri
1	5.3	0.6	1.9	1.3	1.6
2	12.9	9.4	2.6	0.4	0.5
3	0.8	0.8	5.7	0.4	1.4
4	2.6	0.0	4.5	10.2	4.5
5	23.5	9.6	11.3	13.6	14.1
6	9.1	4.5	7.5	2.1	9.3
7	11.1	4.2	4.1	4.2	4.1
8	9.5	7.1	4.5	9.1	12.9
9	4.8	5.2	10.0	6.9	9.0

Source: Boggis, J. J. "The eradication of leisure," *New Technology, Work, and Employment,* Vol. 16, No. 2, July 2001, pp. 118–129 (Table 3). Reprinted with permission of John Wiley & Sons, Inc.

8.60 Participation in a company's walking program. A study was conducted to investigate the effect of prompting in a walking program instituted at a large corporation (*Health Psychology,* Mar. 1995). Five groups of walkers—27 in each group—agreed to participate by walking for 20 minutes at least one day per week over a 24-week period. The participants were prompted to walk each week via telephone calls,

*The experiment is simplified for this exercise. The actual experiment involved 30 students who participated in 12 sessions.

but different prompting schemes were used for each group. Walkers in the control group received no prompting phone calls; walkers in the "frequent/low" group received a call once a week with low structure (i.e., "just touching base"); walkers in the "frequent/high" group received a call once a week with high structure (i.e., goals are set); walkers in the "infrequent/low" group received a call once every 3 weeks with low structure; and walkers in the "infrequent/high" group received a call once every 3 weeks with high structure. The table at the bottom of the page lists the number of participants in each group who actually walked the minimum requirement each week for weeks 1, 4, 8, 12, 16, and 24. The data (saved in the **WALKERS** file) were subjected to an analysis of variance for a randomized block design, with the five walker groups representing the treatments and the six time periods (weeks) representing the blocks.

Source	df	SS	MS	*F*	*p*-value
Prompt	4	1185.000	—	—	0.0000
Week	—	386.400	77.28000	10.40	0.0001
Error	20	148.600	7.43000		
Total	29	1720.00			

a. What is the purpose of blocking on weeks in this study?

b. Fill in the missing entries on the ANOVA summary table shown above.

c. Is there sufficient evidence of a difference in the mean number of walkers per week among the five walker groups? Use $\alpha = .05$.

d. Tukey's technique was used to compare all pairs of treatment means with an experimentwise error rate of $\alpha = .05$. The rankings are shown at the bottom of the page. Interpret these results.

e. What assumptions must hold to ensure the validity of the inferences in parts **c** and **d**?

Applying the Concepts—Advanced

8.61 Anticorrosive behavior of steel coated with epoxy. Organic coatings that use epoxy resins are widely used for protecting steel and metal against weathering and corrosion. Researchers at National Technical University (Athens, Greece) examined the steel anticorrosive behavior of different epoxy coatings formulated with zinc pigments in an attempt to find the epoxy coating with the best corrosion inhibition (*Pigment & Resin Technology,* Vol. 32, 2003). The experimental units were flat, rectangular panels cut from steel sheets. Each panel was coated with one of four different coating systems, S1, S2, S3, and S4. Three panels were prepared for each coating system. (These panels are labeled S1-A, S1-B, S1-C, S2-A, S2-B, . . . , S4-C.) The characteristics of the four coating systems are listed on page 492.

Each coated panel was immersed in deionized and deaerated water and then tested for corrosion. Because exposure time is likely to have a strong influence on anticorrosive behavior, the researchers attempted to remove this extraneous source of variation through the experimental design. Exposure times were fixed at 24 hours, 60 days, and 120 days. For each of the coating systems, one panel was exposed to water for 24 hours, one exposed to water for 60 days, and one exposed to water for 120 days in random order. The design is illustrated in the accompanying table.

Exposure Time	Coating System/Panel Exposed
24 hours	S1-A, S2-C, S3-C, S4-B
60 days	S1-C, S2-A, S3-B, S4-A
120 days	S1-B, S2-B, S3-A, S4-C

Following exposure, the corrosion rate (nanoamperes per square centimeter) was determined for each panel. The lower the corrosion rate, the greater the anticorrosion performance of the coating system. The data are shown in the next table and saved in the **EPOXY** file. Are there differences among the epoxy treatment means? If so, which of the epoxy coating systems yields the lowest corrosion rate?

Exposure Time	System S1	System S2	System S3	System S4
24 hours	6.7	7.5	8.2	6.1
60 days	8.7	9.1	10.5	8.3
120 days	11.8	12.6	14.5	11.8

Source: Kouloumbi, N., Pantazopoulou, P., and Moundoulas, P. "Anticorrosion performance of epoxy coatings on steel surface exposed to deionized water," *Pigment & Resin Technology,* Vol. 32, No. 2, 2003, pp. 89–99 (Table II).

Table for Exercise 8.60

Week	Control	Frequent/Low	Frequent/High	Infrequent/Low	Infrequent/High
1	7	23	25	21	19
4	2	19	25	10	12
8	2	18	19	9	9
12	2	7	20	8	2
16	2	18	18	8	7
24	1	17	17	7	6

Source: Lombard, D. N., Lombard, T. N., and Winett, R. A. "Walking to meet health guidelines: The effect of prompting frequency and prompt structure," *Health Psychology,* Vol. 14, No. 2, Mar. 1995, p. 167 (Table 2).

Tukey Rankings for Exercise 8.60

Mean:	2.67	9.17	10.50	17.00	20.67
Prompt:	*Control*	*Infr./High*	*Infr./Low*	*Frequent/Low*	*Frequent/High*

Coating Systems Characteristics for Exercise 8.61

Coating System	1st Layer	2nd Layer
S1	Zinc dust	Epoxy paint, 100 micrometers thick
S2	Zinc phosphate	Epoxy paint, 100 micrometers thick
S3	Zinc phosphate with mica	Finish layer, 100 micrometers thick
S4	Zinc phosphate with mica	Finish layer, 200 micrometers thick

8.5 Factorial Experiments: Two Factors

All of the experiments discussed in Sections 8.2–8.4 were **single-factor experiments.** The treatments were levels of a single factor, with the sampling of experimental units performed using either a completely randomized or a randomized block design. However, most responses are affected by more than one factor, and we will therefore often wish to design experiments involving more than one factor.

Consider an experiment in which the effects of two factors on the response are being investigated. Assume that factor A is to be investigated at a levels and factor B at b levels. Recalling that treatments are factor-level combinations, you can see that the experiment has, potentially, ab treatments that could be included in the experiment. A *complete factorial experiment* is one in which all possible ab treatments are employed.

A **complete factorial experiment** is one in which every factor-level combination is employed—that is, the number of treatments in the experiment equals the total number of factor-level combinations.

For example, suppose the USGA wants to determine not only the relationship between distance and brand of golf ball but also between distance and the club used to hit the ball. If they decide to use four brands and two clubs (say, driver and five-iron) in the experiment, then a complete factorial would call for employing all $4 \times 2 = 8$ Brand-Club combinations. This experiment is referred to more specifically as a *complete* 4×2 *factorial.* A layout for a two-factor factorial experiment (we are henceforth referring to a *complete factorial* when we use the term *factorial*) is given in Table 8.10. The factorial experiment is also referred to as a **two-way classification** because it can be arranged in the row-column format exhibited in Table 8.10.

In order to complete the specification of the experimental design, the treatments must be assigned to the experimental units. If the assignment of the ab treatments in the factorial experiment is random and independent, the design is completely randomized. For example, if the machine Iron Byron is used to hit 80 golf balls, 10 for each of the eight Brand-Club combinations, in a random sequence, the design would be completely randomized. In the remainder of this section, we confine our attention to factorial experiments employing completely randomized designs.

If we employ a completely randomized design to conduct a factorial experiment with ab treatments, we can proceed with the analysis in exactly the same way as we did in Section 8.2—that is, we calculate (or let the computer calculate) the measure of treatment

Table 8.10 Schematic Layout of Two-Factor Factorial Experiment

		Factor B at b Levels				
	Level	1	2	3	...	b
	1	Trt. 1	Trt. 2	Trt. 3	...	Trt. b
Factor A	2	Trt. $b+1$	Trt. $b+2$	Trt. $b+3$	...	Trt. $2b$
at a Levels	3	Trt. $2b+1$	Trt. $2b+2$	Trt. $2b+3$	...	Trt. $3b$
	⋮	⋮	⋮	⋮	...	⋮
	a	Trt. $(a-1)b+1$	Trt. $(a-1)b+2$	Trt. $(a-1)b+3$	...	Trt. ab

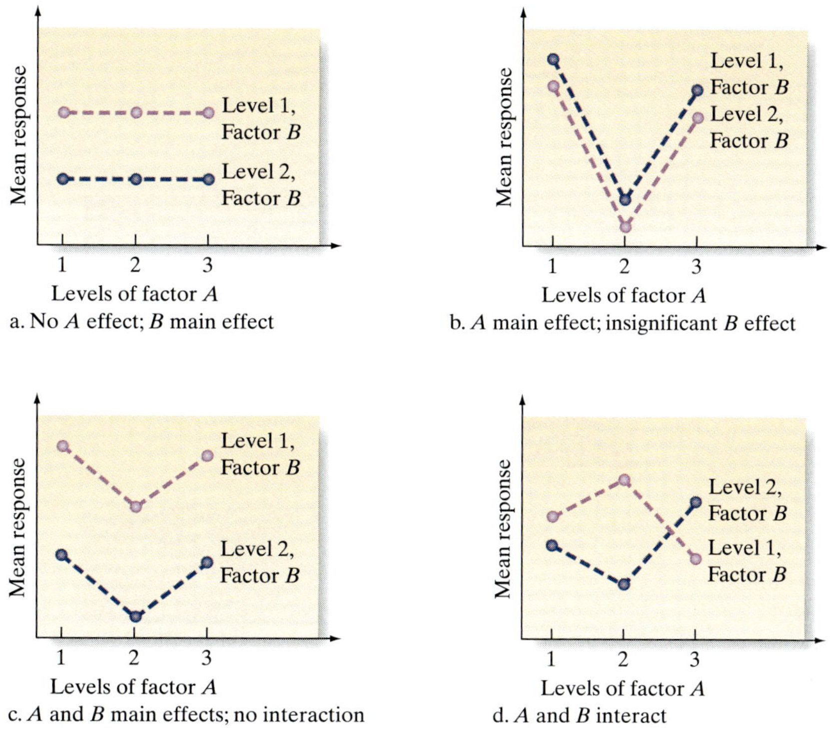

Figure 8.19
Illustration of possible treatment effects: factorial experiment

mean variability (MST) and the measure of sampling variability (MSE) and use the F-ratio of these two quantities to test the null hypothesis that the treatment means are equal. However, if this hypothesis is rejected, so that we conclude some differences exist among the treatment means, important questions remain. Are both factors affecting the response or only one? If both, do they affect the response independently, or do they interact to affect the response?

For example, suppose the distance data indicate that at least two of the eight treatment (Brand-Club combinations) means differ in the golf experiment. Does the brand of ball (factor A) or the club used (factor B) affect mean distance, or do both affect it? Several possibilities are shown in Figure 8.19. In Figure 8.19a, the brand means are equal (only three are shown for the purpose of illustration), but the distances differ for the two levels of factor B (Club). Thus, there is no effect of Brand on distance, but a Club main effect is present. In Figure 8.19b, the Brand means differ, but the Club means are equal for each Brand. Here a Brand main effect is present, but no effect of Club is present.

Figures 8.19c and 8.19d illustrate cases in which both factors affect the response. In Figure 8.19c, the mean distances between Clubs does not change for the three Brands, so that the effect of Brand on distance is independent of Club—that is, the two factors Brand and Club *do not interact.* In contrast, Figure 8.19d shows that the difference between mean distances between Clubs varies with Brand. Thus, the effect of Brand on distance depends on Club, and therefore the two factors *do interact.*

To determine the nature of the treatment effect, if any, on the response in a factorial experiment, we need to break the treatment variability into three components: Interaction between Factors A and B, Main Effect of Factor A, and Main Effect of Factor B. The **Factor Interaction** component is used to test whether the factors combine to affect the response, while the **Factor Main Effect** components are used to determine whether the factors separately affect the response.

The partitioning of the Total Sum of Squares into its various components is illustrated in Figure 8.20. Notice that at stage 1, the components are identical to those in the one-factor, completely randomized designs of Section 8.2; the Sums of Squares for Treatments and Error sum to the Total Sum of Squares. The degrees of freedom for treatments is equal to $(ab - 1)$, one less than the number of treatments. The degrees of freedom for error is equal to $(n - ab)$, the total sample size minus the number of treatments. Only at stage 2 of the partitioning does the factorial experiment differ from those previously discussed. Here we divide the Treatments Sum of Squares into its three

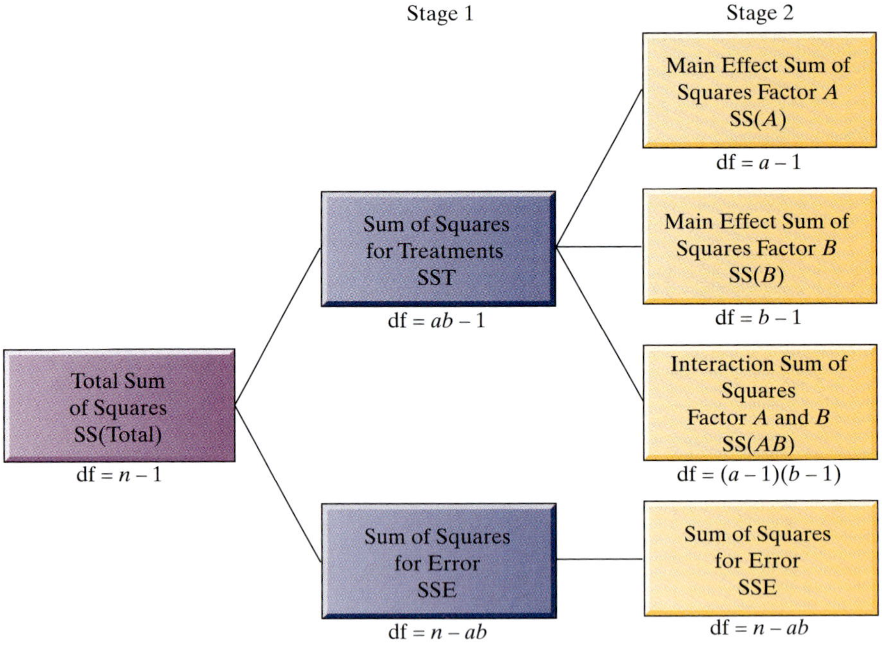

Figure 8.20
Partitioning the Total Sum of Squares for a two-factor factorial

components: Interaction and the two Main Effects. These components can then be used to test the nature of the differences, if any, among the treatment means.

There are a number of ways to proceed in the testing and estimation of factors in a factorial experiment. We present one approach in the next box.

Procedure for Analysis of Two-Factor Factorial Experiment

1. Partition the Total Sum of Squares into the Treatments and Error components (stage 1 of Figure 8.20). Use either a statistical software package or the calculation formulas in Appendix C to accomplish the partitioning.
2. Use the F-ratio of Mean Square for Treatments to Mean Square for Error to test the null hypothesis that the treatment means are equal.*
   a. If the test results in nonrejection of the null hypothesis, consider refining the experiment by increasing the number of replications or introducing other factors. Also consider the possibility that the response is unrelated to the two factors.
   b. If the test results in rejection of the null hypothesis, then proceed to step 3.
3. Partition the Treatments Sum of Squares into the Main Effect and Interaction Sum of Squares (stage 2 of Figure 8.20). Use either a statistical software package or the calculation formulas in Appendix C to accomplish the partitioning.
4. Test the null hypothesis that factors A and B do not interact to affect the response by computing the F-ratio of the Mean Square for Interaction to the Mean Square for Error.
   a. If the test results in nonrejection of the null hypothesis, proceed to step 5.
   b. If the test results in rejection of the null hypothesis, conclude that the two factors interact to affect the mean response. Then proceed to step 6a.
5. Conduct tests of two null hypotheses that the mean response is the same at each level of factor A and factor B. Compute two F-ratios by comparing the Mean Square for each Factor Main Effect to the Mean Square for Error.
   a. If one or both tests result in rejection of the null hypothesis, conclude that the factor affects the mean response. Proceed to step 6b.

*Some analysts prefer to proceed directly to test the interaction and main effect components, skipping the test of treatment means. We begin with this test to be consistent with our approach in the one-factor completely randomized design.

b. If both tests result in nonrejection, an apparent contradiction has occurred. Although the treatment means apparently differ (step 2 test), the interaction (step 4) and main effect (step 5) tests have not supported that result. Further experimentation is advised.

6. Compare the means:

a. If the test for interaction (step 4) is significant, use a multiple comparisons procedure to compare any or all pairs of the treatment means.

b. If the test for one or both main effects (step 5) is significant, use a multiple comparisons procedure to compare the pairs of means corresponding to the levels of the significant factor(s).

We assume the completely randomized design is a **balanced design,** meaning that the same number of observations are made for each treatment—that is, we assume that r experimental units are randomly and independently selected for each treatment. The numerical value of r must exceed 1 in order to have any degrees of freedom with which to measure the sampling variability. [Note that if $r = 1$, then $n = ab$, and the degrees of freedom associated with Error (Figure 8.20) is $\text{df} = n - ab = 0$.] The value of r is often referred to as the number of **replicates** of the factorial experiment because we assume that all ab treatments are repeated, or replicated, r times. Whatever approach is adopted in the analysis of a factorial experiment, several tests of hypotheses are usually conducted. The tests are summarized in the next box.

ANOVA Tests Conducted for Factorial Experiments: Completely Randomized Design, *r* Replicates per Treatment

Test for Treatment Means

H_0: No difference among the ab treatment means
H_a: At least two treatment means differ

Test statistic: $F = \dfrac{\text{MST}}{\text{MSE}}$

Rejection region: $F \geq F_\alpha$, based on $(ab - 1)$ numerator and $(n - ab)$ denominator degrees of freedom [*Note:* $n = abr$.]

Test for Factor Interaction

H_0: Factors A and B do not interact to affect the response mean.
H_a: Factors A and B do interact to affect the response mean.

Test statistic: $F = \dfrac{\text{MS}(AB)}{\text{MSE}}$

Rejection region: $F \geq F_\alpha$, based on $(a - 1)(b - 1)$ numerator and $(n - ab)$ denominator degrees of freedom

Test for Main Effect of Factor A

H_0: No difference among the a mean levels of factor A
H_a: At least two factor A mean levels differ

Test statistic: $F = \dfrac{\text{MS}(A)}{\text{MSE}}$

Rejection region: $F \geq F_\alpha$, based on $(a - 1)$ numerator and $(n - ab)$ denominator degrees of freedom

Test for Main Effect of Factor B

H_0: No difference among the b mean levels of factor B

H_a: At least two factor B mean levels differ

Test statistic: $F = \dfrac{\text{MS}(B)}{\text{MSE}}$

Rejection region: $F \geq F_\alpha$, based on $(b - 1)$ numerator and $(n - ab)$ denominator degrees of freedom

Conditions Required for Valid *F*-Tests in Factorial Experiments

1. The response distribution for each factor-level combination (treatment) is normal.
2. The response variance is constant for all treatments.
3. Random and independent samples of experimental units are associated with each treatment.

Example 8.10

Conducting a Factorial ANOVA of the Golf Ball Brand Data

Problem The USGA is interested in knowing whether the difference in distance traveled by any two golf ball brands depends on the club used. Consequently, the USGA tests four different brands (A, B, C, D) of golf balls and two different clubs (driver, five-iron) in a completely randomized design. Each of the eight Brand-Club combinations (treatments) is randomly and independently assigned to four experimental units, each experimental unit consisting of a specific position in the sequence of hits by Iron Byron. The distance response is recorded for each of the 32 hits, and the results are shown in Table 8.11.

a. Use a statistical software package to partition the Total Sum of Squares into the components necessary to analyze this 4 × 2 factorial experiment.

b. Conduct the appropriate ANOVA tests and interpret the results of your analysis. Use $\alpha = .10$ for each test you conduct.

c. If appropriate, conduct multiple comparisons of the treatment means. Use an experimentwise error rate of .10. Illustrate the comparisons with a graph.

Solution

a. The SPSS printout that partitions the Total Sum of Squares [i.e., SS(Total)] for this factorial experiment is given in Figure 8.21. The value SS(Total) = 34,482.049, shown as "Corrected Total SS" at the bottom of the printout, is partitioned into the "Corrected Model" (i.e., Treatment) and Error Sums of Squares. Note that SST = 33,659.09 (with 7 df) and SSE = 822.24 (with 24 df) add to SS(Total) (with

Table 8.11 Distance Data for 4 × 2 Factorial Golf Experiment

		Brand			
		A	B	C	D
Club	Driver	226.4	238.3	240.5	219.8
		232.6	231.7	246.9	228.7
		234.0	227.7	240.3	232.9
		220.7	237.2	244.7	237.6
	Five-Iron	163.8	184.4	179.0	157.8
		179.4	180.6	168.0	161.8
		168.6	179.5	165.2	162.1
		173.4	186.2	156.5	160.3

Data Set: GOLFFAC1

Tests of Between-Subjects Effects

Dependent Variable: DISTANCE

Source	Type III Sum of Squares	df	Mean Square	F	Sig.
Corrected Model	33659.809[a]	7	4808.544	140.354	.000
Intercept	1306778.61	1	1306778.611	38142.98	.000
BRAND	800.736	3	266.912	7.791	.001
CLUB	32093.111	1	32093.111	936.752	.000
BRAND * CLUB	765.961	3	255.320	7.452	.001
Error	822.240	24	34.260		
Total	1341260.66	32			
Corrected Total	34482.049	31			

a. R Squared = .976 (Adjusted R Squared = .969)

Figure 8.21
SPSS ANOVA for factorial experiment on golf ball data

31 df). The Treatment Sum of Squares, SST, is further divided into Main Effect (Brand and Club) and Interaction Sum of Squares. These values, highlighted on Figure 8.21, are SS(Brand) = 800.7 (with 3 df), SS(Club) = 32,093.1 (with 1 df), and SS(Brand × Club) = 766.0 (with 3 df).

b. Once partitioning is accomplished, our first test is

H_0: The eight treatment means are equal.
H_a: At least two of the eight means differ.

Test statistic: $F = \dfrac{\text{MST}}{\text{MSE}} = 140.354$ (top line of printout)

Observed significance level: $p = .000$ (top line of printout)

Because $\alpha = .10$ exceeds p, we reject this null hypothesis and conclude that at least two of the Brand-Club combinations differ in mean distance.

After accepting the hypothesis that the treatment means differ, and therefore that the factors Brand and/or Club somehow affect the mean distance, we want to determine how the factors affect the mean response. We begin with a test of interaction between Brand and Club:

H_0: The factors Brand and Club do not interact to affect the mean response.
H_a: Brand and Club interact to affect mean response.

$$\textit{Test statistic: } F = \frac{\text{MS}(AB)}{\text{MSE}} = \frac{\text{MS(Brand} \times \text{Club)}}{\text{MSE}}$$

$$= \frac{255.32}{34.26} = 7.452 \quad \text{(bottom of printout)}$$

Observed significance level: $p = .001$ (bottom of printout)

Because $\alpha = .10$ exceeds the p-value, we conclude that the factors Brand and Club interact to affect mean distance.

Because the factors interact, we do not test the main effects for Brand and Club. Instead, we compare the treatment means in an attempt to learn the nature of the interaction in part **c.**

c. Rather than compare all 8(7)/2 = 28 pairs of treatment means, we test for differences only between pairs of brands within each club. That differences exist *between* clubs can be assumed. Therefore, only 4(3)/2 = 6 pairs of means need to be compared for each club, or a total of 12 comparisons for the two clubs. The results of these comparisons using Tukey's method with an experimentwise error rate of $\alpha = .10$ for each club are displayed in the SPSS printout, Figure 8.22. For each club, the brand means are listed in descending order in Figure 8.22, and those not significantly different are listed in the same "Homogeneous Subset" column.

CLUB=5IRON

Tukey HSD[a,b]

BRAND	N	Subset 1	Subset 2
D	4	160.500	
C	4	167.175	
A	4	171.300	
B	4		182.675
Sig.		.103	1.000

Means for groups in homogeneous subsets are displayed.
Based on Type III Sum of Squares
The error term is Mean Square(Error) = 36.108.
a. Uses Harmonic Mean Sample Size = 4.000.
b. Alpha = .10.

CLUB=DRIVER

Tukey HSD[a,b]

BRAND	N	Subset 1	Subset 2
A	4	228.425	
D	4	229.750	
B	4	233.725	233.725
C	4		243.100
Sig.		.570	.146

Means for groups in homogeneous subsets are displayed.
Based on Type III Sum of Squares
The error term is Mean Square(Error) = 32.412.
a. Uses Harmonic Mean Sample Size = 4.000.
b. Alpha = .10.

Figure 8.22
SPSS ranking of brand means for each level of club

As shown in Figure 8.22, the picture is unclear with respect to Brand means. For the five-iron (top of Figure 8.22), the brand B mean significantly exceeds all other brands. However, when hit with a driver (bottom of Figure 8.22), brand B's mean is not significantly different from any of the other brands. The Club × Brand interaction can be seen in the SPSS plot of means in Figure 8.23. Note that the difference between the mean distances of the two clubs (driver and five-iron) varies depending on brand. The biggest difference appears for brand C, while the smallest difference is for brand B.

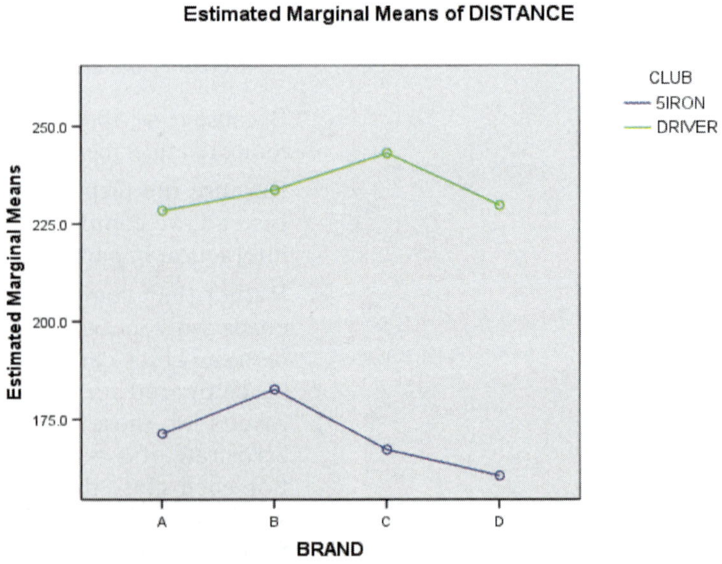

Figure 8.23
SPSS means plot for factorial golf ball experiment

Look Back Note the nontransitive nature of the multiple comparisons. For example, for the driver, the Brand C mean can be "the same" as the Brand B mean, and the Brand B mean can be "the same" as the Brand D mean, and yet the Brand C mean can significantly exceed the Brand D mean. The reason lies in the definition of "the same"—we must be careful not to conclude two means are equal simply because they are placed in the same subgroup or connected by a vertical line. The line indicates only that *the connected means are not significantly different.* You should conclude (at the overall α level of significance) only that means not connected are different, while withholding judgment on those that are connected. The picture of which means differ and by how much will become clearer as we increase the number of replicates of the factorial experiment.

Now Work Exercise 8.68

As with completely randomized and randomized block designs, the results of a factorial ANOVA are typically presented in an ANOVA summary table. Table 8.12 gives the general form of the ANOVA table, while Table 8.13 gives the ANOVA table for the golf ball data analyzed in Example 8.10. A two-factor factorial is characterized by four sources of variation—Factor A, Factor B, $A \times B$ interaction, and Error—which sum to the total Sum of Squares.

Table 8.12 General ANOVA Summary Table for a Two-Factor Factorial Experiment with *r* Replicates, where Factor *A* Has *a* levels and Factor *B* Has *b* Levels

Source	df	SS	MS	*F*
A	$a - 1$	SSA	MSA	MSA/MSE
B	$b - 1$	SSB	MSB	MSB/MSE
AB	$(a - 1)(b - 1)$	SSAB	MSAB	MSAB/MSE
Error	$ab(r - 1)$	SSE	MSE	
Total	$n - 1$	SS(Total)		

Table 8.13 ANOVA Summary Table for Example 8.10

Source	df	SS	MS	*F*
Brand	1	32,093.11	32,093.11	936.75
Club	3	800.74	266.91	7.79
Interaction	3	765.96	255.32	7.45
Error	24	822.24	34.26	
Total	31	34,482.05		

Example 8.11 More Practice on Conducting a Factorial Analysis

Problem Refer to Example 8.10. Suppose the same factorial experiment is performed on four other brands (E, F, G, and H), and the results are as shown in Table 8.14. Repeat the factorial analysis and interpret the results.

Solution The Minitab printout for the second factorial experiment is shown in Figure 8.24. Note that Minitab (unlike SPSS) does not automatically conduct the *F*-test for treatment differences. Consequently, to conduct this test, we must first calculate the Sum of Squares for Treatments. Using the Sums of Squares for Brands, Clubs, and Interaction shown on the printout, we obtain

$$\begin{aligned} \text{SS(Treatments)} &= \text{SS(Clubs)} + \text{SS(Brands)} + \text{SS(Interaction)} \\ &= 46{,}443.9 + 3{,}410.3 + 105.2 = 49{,}959.4 \end{aligned}$$

Table 8.14 Distance Data for Second Factorial Golf Experiment

		Brand			
		E	F	G	H
Club	Driver	238.6 241.9 236.6 244.9	261.4 261.3 254.0 259.9	264.7 262.9 253.5 255.6	235.4 239.8 236.2 237.5
	Five-Iron	165.2 156.9 172.2 163.2	179.2 171.0 178.0 182.7	189.0 191.2 191.3 180.5	171.4 159.3 156.6 157.4

Data Set: GOLFFAC2

Two-way ANOVA: DISTANCE versus BRAND, CLUB

```
Source       DF       SS       MS        F      P
BRAND         3   3410.3   1136.8    46.21  0.000
CLUB          1  46443.9  46443.9  1887.94  0.000
Interaction   3    105.2     35.1     1.42  0.260
Error        24    590.4     24.6
Total        31  50549.8

S = 4.960   R-Sq = 98.83%   R-Sq(adj) = 98.49%
```

Figure 8.24
Minitab analysis for second factorial golf experiment

For this 4 $\times$ 2 factorial experiment, there are 8 treatments. Then

$$\text{MS(Treatments)} = \text{SS(Treatments)}/(8 - 1) = 49{,}959.4/7 = 7{,}137.1$$

The test statistic is

$$F = \text{MS(Treatments)}/\text{MSE} = 7{,}137.1/24.6 = 290.1$$

Because this F-value exceeds the critical value of $F_{.10} = 1.98$ (obtained from Table VII in Appendix B), we reject the null hypothesis of no treatment differences and conclude that at least two of the Brand-Club combinations have significantly different mean distances.

Now, we test for interaction between Brand and Club:

$$F = \frac{\text{MS(Brand} \times \text{Club)}}{\text{MSE}} = 1.42 \text{ (highlighted on the printout)}$$

Because this F-ratio does not exceed the tabled value of $F_{.10} = 2.33$ with 3 and 24 df (obtained from Table VII in Appendix B) we cannot conclude at the .10 level of significance that the factors interact. In fact, note that the observed significance level (on the Minitab printout) for the test of interaction is .26. Thus, at any level of significance lower than $\alpha = .26$, we could not conclude that the factors interact. We therefore test the main effects for Brand and Club.

We first test the Brand main effect:

H_0: No difference exists among the true Brand mean distances.

H_a: At least two Brand mean distances differ.

$$\textit{Test statistic: } F = \frac{\text{MS(Brand)}}{\text{MSE}} = \frac{1{,}136.77}{24.60} = 46.21 \text{ (highlighted on the printout)}$$

Observed significance level: $p = .000$

Because $\alpha = .10$ exceeds the p-value, we conclude that at least two of the brand means differ. We will subsequently determine which brand means differ using Tukey's multiple comparisons procedure. But first, we want to test the Club main effect:

H_0: No differences exist between the Club mean distances.

H_a: The Club mean distances differ.

$$\textit{Test statistic: } F = \frac{\text{MS(Club)}}{\text{MSE}} = \frac{46{,}443.9}{24.60} = 1{,}887.94$$

Observed significance level: $p = .000$

Because $\alpha = .10$ exceeds the p-value, we conclude that the two clubs are associated with different mean distances. Because only two levels of Club were used in experiment, this F-test leads to the inference that the mean distance differs for two clubs. It is no surprise (to golfers) that the mean distance for balls hit with the driver is significantly greater than the mean distance for those hit with the five-iron.

To determine which of the Brands' mean distances differ, we want to compare the $k = 4$ Brand means using Tukey's method at $\alpha = .10$. The results of these multiple comparisons are displayed in the Minitab printout, Figure 8.25. Minitab computes simultaneous 90% confidence intervals for the $c = 4(3)/2 = 6$ possible comparisons of the form $\mu_i - \mu_j$. These intervals (highlighted on the printout) are summarized in Table 8.15. Any interval that does not include 0 implies a significant difference between the two treatment means. You can see that Brands G and F are associated with significantly greater mean distances than Brands E and H, but we cannot distinguish between Brands G and F or between Brands E and H.

```
Tukey 90.0% Simultaneous Confidence Intervals
Response Variable DISTANCE
All Pairwise Comparisons among Levels of BRAND
BRAND = E  subtracted from:

BRAND   Lower   Center   Upper  ---------+---------+---------+-------
F      10.003   16.000  21.997                          (---*---)
G      15.153   21.150  27.147                             (---*---)
H      -9.235   -3.237   2.760              (---*---)
                                ---------+---------+---------+-------
                                       -16         0        16


BRAND = F  subtracted from:

BRAND   Lower   Center   Upper  ---------+---------+---------+-------
G       -0.85     5.15   11.15                    (---*---)
H      -25.23   -19.24  -13.24     (---*---)
                                ---------+---------+---------+-------
                                       -16         0        16


BRAND = G  subtracted from:

BRAND   Lower   Center   Upper  ---------+---------+---------+-------
H      -30.38   -24.39  -18.39  (---*---)
                                ---------+---------+---------+-------
                                       -16         0        16
```

Figure 8.25
Minitab Tukey multiple comparisons for brand in second factorial golf experiment

Table 8.15 Summary of Tukey's Multiple Comparisons

Comparison	90% Confidence Interval	Inference
$\mu_F - \mu_E$	(10.00, 21.99)	$\mu_F > \mu_E$
$\mu_G - \mu_E$	(15.15, 27.15)	$\mu_G > \mu_E$
$\mu_H - \mu_E$	(−9.24, 2.76)	No significant difference
$\mu_G - \mu_F$	(−.85, 11.15)	No significant difference
$\mu_H - \mu_F$	(−25.23, −13.24)	$\mu_H < \mu_F$
$\mu_H - \mu_G$	(−30.38, −18.39)	$\mu_H < \mu_G$

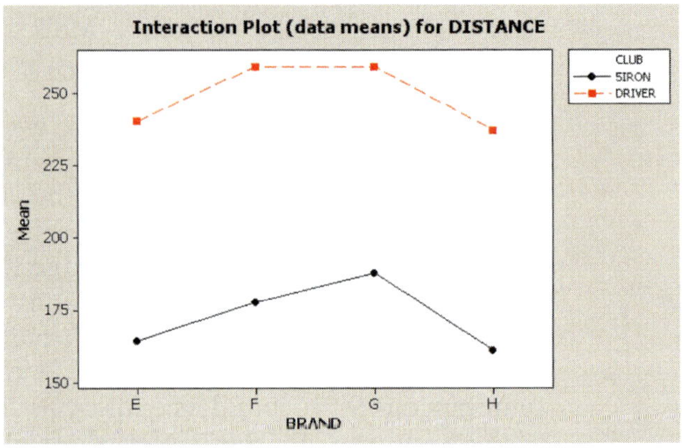

Figure 8.26
Minitab means plot for second factorial golf ball experiment

Look Back Because the interaction between Brand and Club was not significant, we conclude that this difference among brands applies to both clubs. The sample means for all Club-Brand combinations are shown in Figure 8.26 and appear to support the conclusions of the tests and comparisons. Note that the Brand means maintain their relative positions for each Club—brands F and G dominate brands E and H for both the driver and the five-iron.

Now Work Exercise 8.70

The analysis of factorial experiments can become complex if the number of factors is increased. Even the two-factor experiment becomes more difficult to analyze if some factor combinations have different numbers of observations than others. We have provided an introduction to these important experiments using two-factor factorials with equal numbers of observations for each treatment. Although similar principles apply to most factorial experiments, you should consult the references for this chapter and at the end of the book if you need to design and analyze more complex factorials.

Statistics IN Action Revisited — A Two-Way Analysis of Variance

We now return to the study of the effect of violence and sex on a television viewer's ability to recall a TV commercial (p. 477). Recall that 324 adults were randomly assigned to one of three TV content groups, with 108 subjects in each group. One group watched a TV program with a violent content code (V) rating; the second group viewed a show with a sex content code (S) rating; and the last group watched a neutral TV program. Commercials were embedded into each TV show, and, after viewing the show, the advertisement recall score was measured for each participant. In addition, the researchers recorded whether or not the subject had previously seen the commercial. The layout for the full experimental design is shown in Figure SIA8.5. Note that there are two factors in this experiment—TV content group at 3 levels and watched commercial before status at 2 levels—so the design is a 3×2 factorial.

The researchers want to know whether the two factors, TV content group and watched commercial before status, impact mean recall score. To answer this question, we conduct a two-way factorial analysis of variance on the data saved in the **TVADRECALL** file.

A Minitab printout of the ANOVA is displayed in Figure SIA8.6. First, note that the *p*-value for the test for

	TV Content Group		
	Violent (V)	*Sex (S)*	*Neutral (N)*
Watched commercial before: *Yes*	n = 48	n = 60	n = 54
No	n = 60	n = 48	n = 54

Figure SIA8.5
Layout of experimental design for TV ad recall study

```
Factor    Type    Levels  Values
CONTENT   fixed        3  NEUTRAL, SEX, VIOLENT
BEFORE    fixed        2  NO, YES


Analysis of Variance for RECALL, using Adjusted SS for Tests

Source           DF    Seq SS    Adj SS   Adj MS      F      P
CONTENT           2   123.265   120.004   60.002  20.01  0.000
BEFORE            1     6.458     6.393    6.393   2.13  0.145
CONTENT*BEFORE    2     7.472     7.472    3.736   1.25  0.289
Error           318   953.421   953.421    2.998
Total           323  1090.617
```

Figure SIA8.6
Minitab two-way ANOVA for the TV ad recall data

factor interaction (highlighted on the printout) is .289. Thus, there is insufficient evidence (at $\alpha = .05$) of interaction between the two factors. This implies that the effect of one factor (say, TV content group) on mean recall score does not depend on the level of the other factor (watched commercial status). With no evidence of interaction, it is appropriate to conduct the main-effect tests on the two factors.

The p-values for the main effects of content group and watched commercial status (both highlighted on the printout) are .000 and .145, respectively. The first p-value implies that differences in mean recall scores exist for the three TV content groups (at $\alpha = .05$)—a result consistent with the one-way ANOVA conducted in an earlier Statistics in Action Revisited (p. 465). Based on the multiple comparisons of means conducted earlier in this chapter (p. 476), the researchers also know that the neutral TV group has a significantly higher mean recall score than either of the other two TV content groups. The second p-value implies that there is not a significant difference (at $\alpha = .05$) between the mean recall scores of the those who have previously watched the commercial and those who had not.

The final conclusions of the researchers are (1) that the Neutral TV content group has the highest mean recall score, but that there is no significant difference between the mean recall scores of the Violent and Sex content groups; and (2) there is no significant difference between the mean recall scores of the those who have previously watched the commercial and those who had not.

Data Set: TVADRECALL

Exercises 8.62–8.77

Learning the Mechanics

8.62 Suppose you conduct a 4 × 3 factorial experiment.
- **a.** How many factors are used in the experiment?
- **b.** Can you determine the factor type(s)—qualitative or quantitative—from the information given? Explain.
- **c.** Can you determine the number of levels used for each factor? Explain.
- **d.** Describe a treatment for this experiment and determine the number of treatments used.
- **e.** What problem is caused by using a single replicate of this experiment? How is the problem solved?

8.63 The partially completed ANOVA for a 3 × 4 factorial experiment with two replications is shown below.

Source	df	SS	MS	F
A	–	.8	–	–
B	–	5.3	–	–
AB	–	9.6	–	–
Error	–	–	–	
Total	–	17.0		

a. Complete the ANOVA table.

b. Which sums of squares are combined to find the Sum of Squares for Treatments? Do the data provide sufficient evidence to indicate that the treatment means differ? Use $\alpha = .05$.

c. Does the result of the test in part **b** warrant further testing? Explain.

d. What is meant by *factor interaction*, and what is the practical implication if it exists?

e. Test to determine whether these factors interact to affect the response mean. Use $\alpha = .05$ and interpret the result.

f. Does the result of the interaction test warrant further testing? Explain.

8.64 The partially complete ANOVA table given here is for a two-factor factorial experiment.

Source	df	SS	MS	F
Treatments	7	4.1	–	–
A	3	–	.75	–
B	1	.95	–	–
AB	–	–	.30	–
Error	–	–	–	
Total	23	6.5		

a. Give the number of levels for each factor.

b. How many observations were collected for each factor-level combination?

c. Complete the ANOVA table.

d. Test to determine whether the treatment means differ. Use $\alpha = .10$.

e. Conduct the tests of factor interaction and main effects, each at the $\alpha = .10$ level of significance. Which of the tests are warranted as part of the factorial analysis? Explain.

8.65 The following two-way table gives data for a 2×3 factorial experiment with two observations for each factor-level combination. The data are saved in the **LM8_65** file.

		Factor *B*		
	Level	1	2	3
Factor *A*	1	3.1, 4.0	4.6, 4.2	6.4, 7.1
	2	5.9, 5.3	2.9, 2.2	3.3, 2.5

a. Identify the treatments for this experiment. Calculate and plot the treatment means, using the response variable as the *y*-axis and the levels of factor *B* as the *x*-axis. Use the levels of factor *A* as plotting symbols. Do the treatment means appear to differ? Do the factors appear to interact?

b. The Minitab ANOVA printout for this experiment is shown in the next column. Sum the appropriate sums of squares and test to determine whether the treatment means differ at the $\alpha = .05$ level of significance. Does the test support your visual interpretation from part **a**?

c. Does the result of the test in part **b** warrant a test for interaction between the two factors? If so, perform it using $\alpha = .05$.

d. Do the results of the previous tests warrant tests of the two factor main effects? If so, perform them using $\alpha = .05$.

e. Interpret the results of the tests. Do they support your visual interpretation from part **a**?

Two-way ANOVA: RESPONSE versus A, B

```
Source        DF       SS        MS       F       P
A              1   4.4408   4.44083   18.06   0.005
B              2   4.1267   2.06333    8.39   0.018
Interaction    2  18.0067   9.00333   36.62   0.000
Error          6   1.4750   0.24583
Total         11  28.0492

S = 0.4958   R-Sq = 94.74%   R-Sq(adj) = 90.36%
```

8.66 The next table gives data for a 2×2 factorial experiment with two observations per factor-level combination. These data are saved in the **LM8_66** file.

		Factor *B*	
	Level	1	2
Factor *A*	1	29.6, 35.2	47.3, 42.1
	2	12.9, 17.6	28.4, 22.7

a. Identify the treatments for this experiment. Calculate and plot the treatment means, using the response variable as the *y*-axis and the levels of factor *B* as the *x*-axis. Use the levels of factor *A* as plotting symbols. Do the treatment means appear to differ? Do the factors appear to interact?

b. Use the computational formulas in Appendix C to create an ANOVA table for this experiment.

c. Test to determine whether the treatment means differ at the $\alpha = .05$ level of significance. Does the test support your visual interpretation from part **a**?

d. Does the result of the test in part **b** warrant a test for interaction between the two factors? If so, perform it using $\alpha = .05$.

e. Do the results of the previous tests warrant tests of the two factor main effects? If so, perform them using $\alpha = .05$. Yes.

f. Interpret the results of the tests. Do they support your visual interpretation from part **a**?

g. Given the results of your tests, which pairs of means, if any, should be compared?

8.67 Suppose a 3×3 factorial experiment is conducted with three replications. Assume that SS(Total) = 1,000. For each of the following scenarios, form an ANOVA table, conduct the appropriate tests, and interpret the results.

a. The Sum of Squares of factor *A* main effect [SS(*A*)] is 20% of SS(Total), the Sum of Squares for factor *B* main effect [SS(*B*)] is 10% of SS(Total), and the Sum of Squares for interaction [SS(*AB*)] is 10% of SS(Total).

b. SS(*A*) is 10%, SS(*B*) is 10%, and SS(*AB*) is 50% of SS(Total).

c. SS(*A*) is 40%, SS(*B*) is 10%, and SS(*AB*) is 20% of SS(Total).

d. SS(*A*) is 40%, SS(*B*) is 40%, and SS(*AB*) is 10% of SS(Total).

Applying the Concepts—Basic

8.68 NW **Baker's versus brewer's yeast.** The *Electronic Journal of Biotechnology* (Dec. 15, 2003) published an article on a comparison of two yeast extracts, baker's yeast and brewer's yeast. Brewer's yeast is a surplus by-product obtained from a brewery, hence it is less expensive than primary-grown baker's yeast. Samples of both yeast extracts were prepared at four different temperatures (45, 48, 51, and 54°C); thus, a 2×4 factorial design with yeast extract at two levels and temperature at four levels was employed. The response variable was the autolysis yield (recorded as a percentage).

a. How many treatments are included in the experiment?
b. An ANOVA found sufficient evidence of factor interaction at $\alpha = .05$. Interpret this result practically.
c. Give the null and alternative hypotheses for testing the main effects of yeast extract and temperature.
d. Explain why the tests, part **c,** should not be conducted.
e. Multiple comparisons of the four temperature means were conducted for each of the two yeast extracts. Interpret the results shown below.

Baker's yeast:	Mean yield (%):	41.1	47.5	48.6	50.3
	Temperature (°C):	54	45	48	51
Brewer's yeast:	Mean yield (%):	39.4	47.3	49.2	49.6
	Temperature (°C):	54	51	48	45

8.69 Removing bacteria from water. A coagulation-microfiltration process for removing bacteria from water was investigated in *Environmental Science & Engineering* (Sept. 1, 2000). Chemical engineers at Seoul National University performed a designed experiment to estimate the effect of both the level of the coagulant and acidity (pH) level on the coagulation efficiency of the process. Six levels of coagulant (5, 10, 20, 50, 100, and 200 milligrams per liter) and six pH levels (4.0, 5.0, 6.0, 7.0, 8.0, and 9.0) were employed. Water specimens collected from the Han River in Seoul, Korea, were placed in jars, and each jar was randomly assigned to receive one of the $6 \times 6 = 36$ combinations of coagulant level and pH level.

a. What type of experimental design was applied in this study?
b. Give the factors, factor levels, and treatments for the study.

8.70 NW **Insomnia and education.** Many workers suffer from stress and chronic insomnia. Is insomnia related to education status? Researchers at the Universities of Memphis, Alabama at Birmingham, and Tennessee investigated this question in the *Journal of Abnormal Psychology* (Feb. 2005). Adults living in Tennessee were selected to participate in the study using a random-digit telephone dialing procedure. In addition to insomnia status (normal sleeper or chronic insomnia), the researchers classified each participant into one of four education categories (college graduate, some college, high school graduate, and high school dropout). One dependent variable of interest to the researchers was a quantitative measure of daytime functioning called the Fatigue Severity Scale (FSS). The data were analyzed as a 2×4 factorial experiment, with Insomnia status and Education level as the two factors.

a. Determine the number of treatments for this study. List them.
b. The researchers reported that "the Insomnia $\times$ Education interaction was not statistically significant." Practically interpret this result. (Illustrate with a graph.)
c. The researchers discovered that the sample mean FSS for people with insomnia was greater than the sample mean FSS for normal sleepers, and this difference was statistically significant. Practically interpret this result.
d. The researchers reported that the main effect of Education was statistically significant. Practically interpret this result.
e. Refer to part **d.** In a follow-up analysis, the sample mean FSS values for the four Education levels were compared using Tukey's method ($\alpha = .05$), with the results shown below. What do you conclude?

Mean:	3.3	3.6	3.7	4.2
Education:	*College graduate*	*Some college*	*HS graduate*	*HS dropout*

8.71 Impact of paper color on exam scores. A study published in *Teaching Psychology* (May 1998) examined how external clues influence student performance. Undergraduate students were randomly assigned to one of four different midterm examinations. Form 1 was printed on blue paper and contained difficult questions, while form 2 was also printed on blue paper but contained simple questions. Form 3 was printed on red paper, with difficult questions; form 4 was printed on red paper with simple questions. The researchers were interested in the impact that Color (red or blue) and Question (simple or difficult) had on mean exam score.

Form	Color	Question	Mean Score
1	Blue	Difficult	53.3
2	Blue	Simple	80.0
3	Red	Difficult	39.3
4	Red	Simple	73.6

a. What experimental design was employed in this study? Identify the factors and treatments.
b. The researchers conducted an ANOVA and found a significant interaction between Color and Question (p-value $< .03$). Interpret this result.
c. The sample mean scores (percentage correct) for the four exam forms are listed in the table below. Plot the four means on a graph to illustrate the Color $\times$ Question interaction.

Applying the Concepts—Intermediate

8.72 Factors that impact a customer's willingness to buy. Advancements in information technology have yielded services that compete against products, with each providing roughly the same benefits to the consumer (e.g., home answering machines and voice-mail services). With the advent of such services, consumers also face different types of pricing schemes. Using a 2×2 factorial design, D. Fortin and T. Greenlee of the University of Rhode Island investigated the effects of the type of message retrieval system (answering machine vs. voice-mail service) and the type of pricing (lump sum amount for 5 years of use vs. monthly cost for 5 years of use) on consumers' willingness to buy (*Journal of Business Research,* Vol. 41, 1998). The first pricing option requires the consumer to do mental arithmetic to determine the total cost of the system; the second provides the true full cost. Thirty subjects were randomly assigned to each of the four treatments. Each was exposed to a purchase situation involving the relevant product or service and payment description and was asked to indicate his or her willingness to buy the item on a 5-point scale (1 = definitely would not buy; 5 = definitely would buy). The results are presented in the incomplete ANOVA table on the next page.

a. Fill in the degrees of freedom (df) column in the ANOVA table.

b. Specify the null and alternative hypotheses that should be used in testing for interaction effects between type of message retrieval system and pricing option.
c. Conduct the test of part **b** using $\alpha = .05$. Interpret the results in the context of the problem.
d. Given the results of part **c**, is it advisable to conduct main effects tests? Why or why not? If so, perform the appropriate main effects tests using $\alpha = .05$.

Source	df	SS	MS	F
Type of message retrieval system	–	–	–	2.001
Pricing option	–	–	–	5.019
Type of system × pricing option	–	–	–	4.986
Error	–	–	–	
Total	119	–		

Source: Reprinted from Fortin, D., and Greenlee, T. "Using a product/service evaluation frame: An experiment on the economic equivalence of product versus service alternatives for message retrieval systems," *Journal of Business Research*, Vol. 41, 1998, pp. 205–214. © 1998, adapted with permission from Elsevier.

8.73 Commercial eggs produced from different housing systems. Refer to the *Food Chemistry* (Vol. 106, 2008) study of four different types of egg housing systems, Exercise 8.33 (p. 471). Recall that the four housing systems were cage, barn, free range, and organic. In addition to housing system, the researchers also determined the weight class (medium or large) for each sampled egg. The data on whipping capacity (percent overrun) for the 28 sampled eggs are shown in the accompanying table and saved in the **EGGS** file. The researchers want to investigate the effect of both housing system and weight class on the mean whipping capacity of the eggs. In particular, they want to know whether the difference between the mean whipping capacity of medium and large eggs depends on the housing system.

Housing	Wtclass	Overrun (%)
Cage	M	495, 462, 488, 471, 471
	L	502, 472, 474, 492, 479
Free	M	513, 510, 510
	L	520, 531, 521
Barn	M	515, 516, 514
	L	526, 501, 508
Organic	M	532, 511, 527
	L	530, 544, 531

a. Identify the factors and treatments for this experiment.
b. Use statistical software to conduct an ANOVA on the data. Report the results in an ANOVA table.
c. Is there evidence of interaction between housing system and weight class? Test using $\alpha = .05$. [*Hint:* Due to an unbalanced design, you will need to analyze the data using the general linear model procedure of your statistical software.] What does this imply, practically?
d. Interpret the main effect test for housing system (using $\alpha = .05$). What does this imply, practically?
e. Interpret the main effect test for weight class (using $\alpha = .05$). What does this imply, practically?

8.74 Manager's trust and job-related tension. Research published in *Accounting, Organizations and Society* (Vol. 19, 1994) investigated whether the effects of different performance evaluation styles (PES) on the level of job-related tension is affected by trust. Three performance evaluation styles were considered. Each is related to the way in which accounting information is used for the purpose of evaluation. The three styles are budget-constrained (BC), profit-conscious (PC), and the nonaccounting style (NA), which focuses on factors such as quality of output and attitude toward the job. A questionnaire was administered to 215 managers working in 18 Australian organizations. It measured the performance evaluation style of each manager's superior, the manager's job-related tension, and the manager's level of trust (low, medium, and high) in his or her superior. These data were used to produce the partial ANOVA table and table of treatment means shown next.

Source	df	SS	MS	F
PES	2	2.1774	–	–
Trust	–	7.6367	–	–
PES × trust	4	1.7380	–	–
Error	206	–		
Total	214	161.1162		

		Performance Evaluation Style		
		BC	PC	NA
	Low	3.2350 ($n = 32$)	3.111 ($n = 24$)	3.2290 ($n = 16$)
Trust	Medium	2.7601 ($n = 26$)	2.8530 ($n = 31$)	2.6373 ($n = 14$)
	High	2.3067 ($n = 30$)	2.4436 ($n = 26$)	3.1810 ($n = 16$)

Source: Ross, A. "Trust as a moderator of the effect of performance evaluation style on job-related tension: A research note," *Accounting, Organizations and Society*, Vol. 19, No. 7, 1994, pp. 629–635 (Tables 3 and 4). © 1994 with permission from Elsevier.

a. Describe the treatments of this study.
b. Complete the ANOVA table.
c. Investigate the presence of an interaction effect by conducting the appropriate hypothesis test using $\alpha = .05$.
d. Use a plot of treatment means to investigate the interaction effect. Interpret your results. Are your results of parts **c** and **d** consistent?
e. Given your answers to parts **c** and **d**, explain why the F-tests for the two main effects are irrelevant.

8.75 Testing a new pain-reliever tablet. Refer to the *Tropical Journal of Pharmaceutical Research* (June 2003) study of the impact of binding agent, binding concentration, and relative density on the mean dissolution time of pain-relief tablets, Exercise 8.13 (p. 455). Recall that the binding agent was set at two levels (khaya gum and PVP), binding concentration at two levels (.5% and 4.0%), and relative density at two levels (low and high); thus, a 2 × 2 × 2 factorial design was employed. The sample mean dissolution times for the treatments associated with the factors binding agent and relative density when the other factor (binding concentration) is held fixed at .5% are $\bar{x}_{\text{Gum/Low}} = 4.70$, $\bar{x}_{\text{Gum/High}} = 7.95$, $\bar{x}_{\text{PVP/Low}} = 3.00$, and $\bar{x}_{\text{PVP/High}} = 4.10$. Do the results suggest there is an interaction between binding agent and relative density? Explain.

Applying the Concepts—Advanced

8.76 On the trail of the cockroach. Knowledge of how cockroaches forage for food is valuable for companies that develop and manufacture roach bait and traps. Many entomologists believe, however, that the navigational behavior of cockroaches scavenging for food is random. D. Miller of Virginia Tech University challenged the "random-walk" theory by designing an experiment to test a cockroach's ability to follow a trail of their fecal material (*Explore,* Research at the University of Florida, Fall 1998).

A methanol extract from roach feces—called a *pheromone*—was used to create a chemical trail. German cockroaches were released at the beginning of the trail, one at a time, and a video surveillance camera was used to monitor the roach's movements. In addition to the trail containing the fecal extract (the treatment), a trail using methanol only (the control) was created. To determine if trail-following ability differed among cockroaches of different age, sex, and reproductive status, four roach groups were used in the experiment: adult males, adult females, gravid (pregnant) females, and nymphs (immatures). Twenty roaches of each type were randomly assigned to the treatment trail, and 10 of each type were randomly assigned to the control trail. Thus, a total of 120 roaches were used in the experiment. The movement pattern of each cockroach was measured (in "pixels") as the average trail deviation. The data for the 120 cockroaches in the study are stored in the data file named **ROACH.** (The first 5 and last 5 observations in the data set are listed here.) Conduct a complete analysis of the data. Determine whether roaches can distinguish between the fecal extract and control trail and whether trail-following ability differs according to age, sex, and reproductive status.

Trail Deviation	Roach Group	Trail
3.1	Adult Male	Extract
42.0	Adult Male	Control
6.2	Adult Male	Extract
22.7	Adult Male	Control
34.0	Adult Male	Extract
⋮	⋮	⋮
23.8	Nymph	Extract
5.1	Nymph	Extract
3.8	Nymph	Extract
3.1	Nymph	Extract
2.8	Nymph	Extract

8.77 Impact of flavor name on consumer choice. Do consumers react favorably to products with ambiguous colors or names? Marketing Professors E. G. Miller and B. E. Kahn investigated this phenomenon in the *Journal of Consumer Research* (June 2005). As a "reward" for participating in an unrelated experiment, 100 consumers were told they could have some jelly beans available in several cups on a table. Half the consumers were assigned to take jelly beans with common descriptive flavor names (e.g., watermelon green), while the other half were assigned to take jelly beans with ambiguous flavor names (e.g., monster green). Within each group, half of the consumers took the jelly beans and left (low cognitive load condition), while the other half were distracted with additional questions designed to distract them while they were taking their jelly beans (high cognitive load condition). Consequently, a 2×2 factorial experiment was employed—with Flavor Name (common or ambiguous) and Cognitive Load (low or high) as the two factors—with 25 consumers assigned to each of four treatments. The dependent variable of interest was the number of jelly beans taken by each consumer. The means and standard deviations of the four treatments are shown in the accompanying table.

	Ambiguous		Common	
	Mean	Std. Dev.	Mean	Std. Dev.
Low load	18.0	15.0	7.8	9.5
High load	6.1	9.5	6.3	10.0

Source: Miller, E. G., and Kahn, B. E. "Shades of meaning: The effect of color and flavor names on consumer choice," *Journal of Consumer Research,* Vol. 32, No. 1, June 2005 (Table 1). © 2005 the University of Chicago Press.

a. Calculate the total of the $n = 25$ measurements for each of the four categories in the 2×2 factorial experiment.
b. Calculate the correction for mean, CM. (See Appendix C for computational formulas.)
c. Use the results of parts **a** and **b** to calculate the sums of squares for Load, Name, and Load $\times$ Name interaction.
d. Calculate the sample variance for each treatment. Then calculate the sum of squares of deviations within each sample for the four treatments.
e. Calculate SSE. (*Hint:* SSE is the pooled sum of squares for the deviations calculated in part **d.**)
f. Now that you know SS(Load), SS(Name), SS(Load $\times$ Name), and SSE, find SS(Total).
g. Summarize the calculations in an ANOVA table.
h. The researchers reported the F-value for Load $\times$ Name interaction as $F = 5.34$. Do you agree?
i. Conduct a complete analysis of these data. Use $\alpha = .05$ for any inferential techniques you employ. Illustrate your conclusions graphically.
j. What assumptions are necessary to ensure the validity of the inferential techniques you used? State them in terms of this experiment.

CHAPTER NOTES

Key Terms

Key Symbols/Notation

ANOVA	Analysis of variance
SST	Sum of Squares for Treatments
MST	Mean Square for Treatments
SSB	Sum of Squares for Blocks
MSB	Mean Square for Blocks
SSE	Sum of Squares for Error
MSE	Mean Square for Error
$a \times b$ factorial	Factorial design with one factor at a levels and the other factor at b levels
SS(A)	Sum of Squares for main effect factor A
MS(A)	Mean Square for main effect factor A
SS(B)	Sum of Squares for main effect factor B
MS(B)	Mean Square for main effect factor B
SS(AB)	Sum of Squares for factor $A \times B$ interaction
MS(AB)	Mean Square for factor $A \times B$ interaction

Key Ideas

Key Elements of a Designed Experiment

1. *Response (dependent) variable*—quantitative
2. *Factors (independent variables)*—quantitative or qualitative
3. *Factor levels (values of factors)*—selected by the experimenter
4. *Treatments*—combinations of factor levels
5. *Experimental units*—assign treatments to experimental units and measure response for each

Balanced Design
Sample sizes for each treatment are equal.

Tests for main effects in a factorial design
Only appropriate if the test for factor interaction is nonsignificant.

Robust method
Slight to moderate departures from normality do not have impact on validity of the ANOVA results.

Conditions Required for Valid *F*-Test in a Completely Randomized Design

1. All k treatment populations are approximately normal.
2. $\sigma_1^2 = \sigma_2^2 = \cdots = \sigma_k^2$

Conditions Required for Valid *F*-Tests in a Randomized Block Design

1. All treatment-block populations are approximately normal.
2. All treatment-block populations have the same variance.

Guide to Selecting the Experimental Design

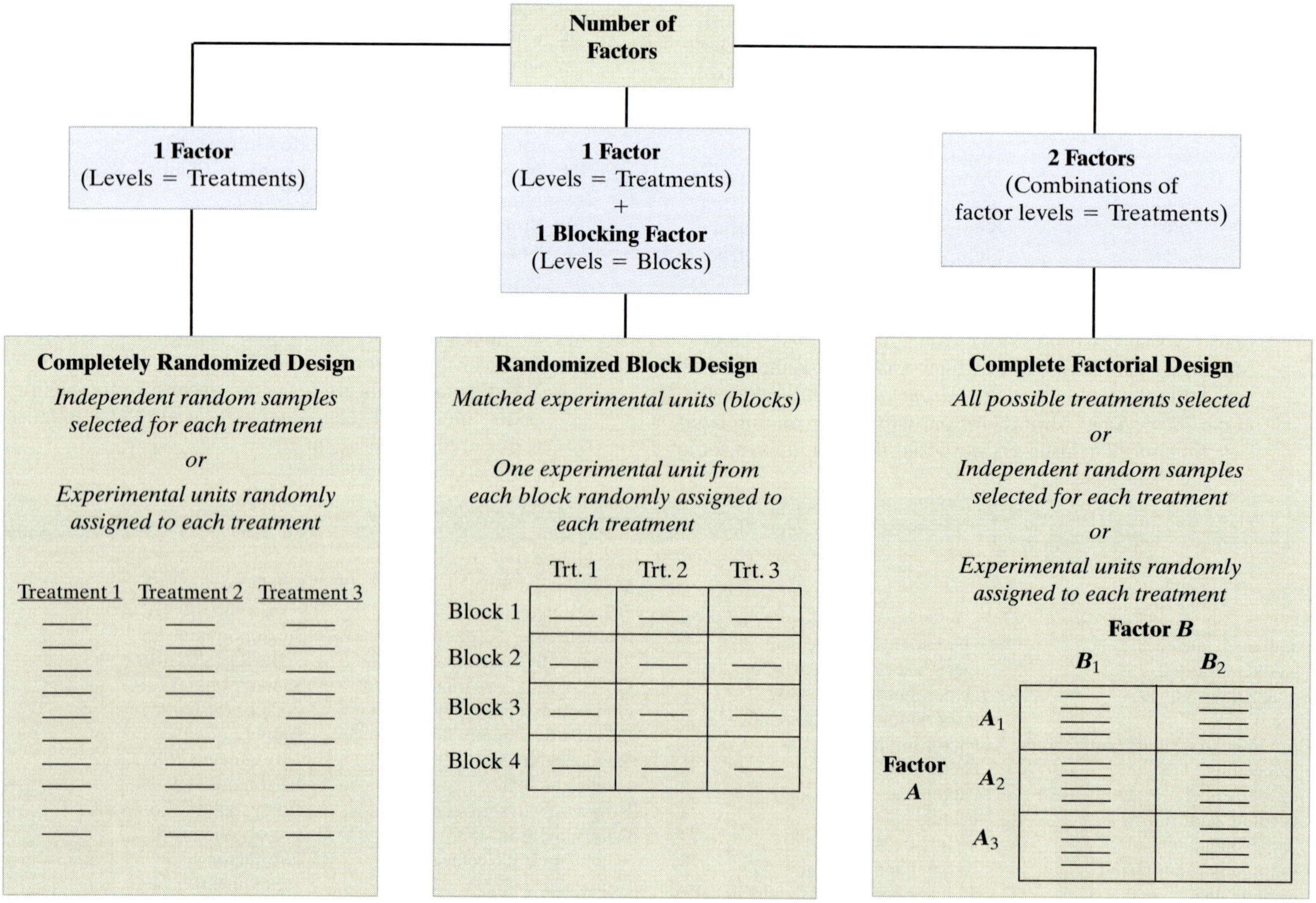

Conditions Required for Valid F-Tests in a Complete Factorial Design

1. All treatment populations are approximately normal.
2. All treatment populations have the same variance.

Experimentwise error rate
Risk of making at least one Type I error when making multiple comparisons of means in ANOVA

Mulitiple Comparisons of Means Methods

Number of pairwise comparisons with k treatment means
$c = k(k-1)/2$

Tukey method:
1. Balanced design
2. Pairwise comparisons of means

Bonferroni method:
1. Either balanced or unbalanced design
2. Pairwise comparisons of means

Scheffé method:
1. Either balanced or unbalanced design
2. General contrasts of means

Guide to Conducting ANOVA F-Tests

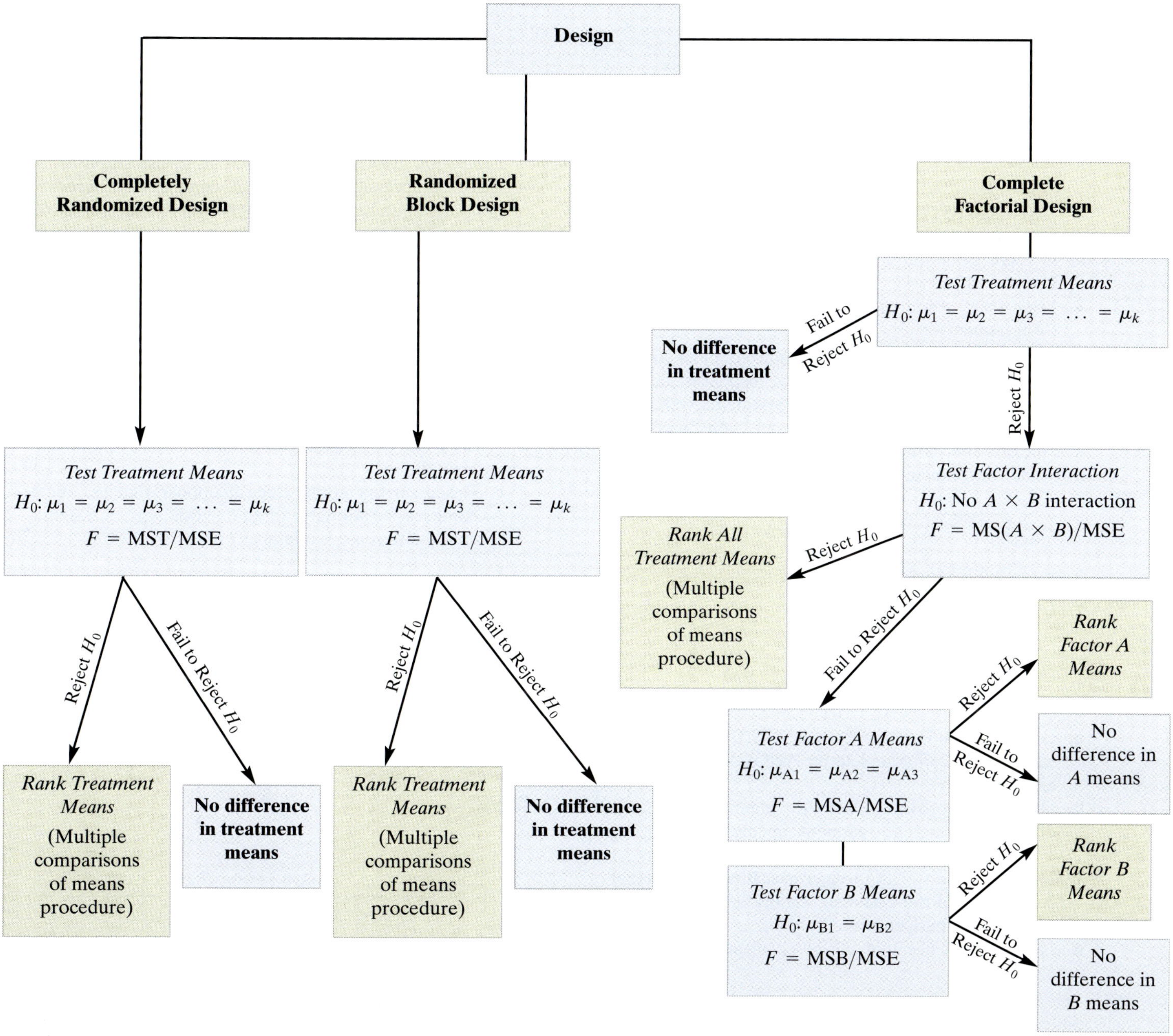

Supplementary Exercises 8.78–8.103

Learning the Mechanics

8.78 What is the difference between a one-way ANOVA and a two-way ANOVA?

8.79 Explain the difference between an experiment that employs a completely randomized design and one that employs a randomized block design.

8.80 What are the treatments in a two-factor experiment, with factor A at three levels and factor B at two levels?

8.81 Why does the experimentwise error rate of a multiple comparisons procedure differ from the significance level for each comparison (assuming the experiment has more than two treatments)?

8.82 A completely randomized design is used to compare four treatment means. The data are saved in the **LM8_82** file and shown in the table.

Treatment 1	Treatment 2	Treatment 3	Treatment 4
8	6	9	12
10	9	10	13
9	8	8	10
10	8	11	11
11	7	12	11

a. Given that SST = 36.95 and SS(Total) = 62.55, complete an ANOVA table for this experiment.
b. Is there evidence that the treatment means differ? Use $\alpha = .10$.
c. Place a 90% confidence interval on the mean response for treatment 4.

8.83 An experiment employing a randomized block design was conducted to compare the mean responses for four treatments—A, B, C, and D. The treatments were randomly assigned to the four experimental units in each of five blocks. The data are saved in the **LM8_83** file and shown in the following table.

	Block				
Treatment	1	2	3	4	5
A	8.6	7.5	8.7	9.8	7.4
B	7.3	6.3	7.3	8.4	6.3
C	9.1	8.3	9.0	9.9	8.2
D	9.3	8.2	9.2	10.0	8.4

a. Given that SS(Total) = 22.31 and SS(Block) = 10.688 and SSE = .288 complete an ANOVA table for the experiment.
b. Do the data provide sufficient evidence to indicate a difference among treatment means? Test using $\alpha = .05$.
c. Does the result of the test in part **b** warrant further comparison of the treatment means? If so, how many pairwise comparisons need to be made?
d. Is there evidence that the block means differ? Use $\alpha = .05$.

8.84 The next table shows a partially completed ANOVA table for a two-factor factorial experiment.

a. Complete the ANOVA table.
b. How many levels were used for each factor? How many treatments were used? How many replications were performed?

Source	df	SS	MS	F
A	3	2.6	–	–
B	5	9.2	–	–
$A \times B$	–	–	3.1	–
Error	–	18.7		
Total	47			

c. Find the value of the Sum of Squares for Treatments. Test to determine whether the data provide evidence that the treatment means differ. Use $\alpha = .05$.
d. Is further testing of the nature of factor effects warranted? If so, test to determine whether the factors interact. Use $\alpha = .05$. Interpret the result.

Applying the Concepts—Basic

8.85 Strength of fiberboard boxes. The *Journal of Testing and Evaluation* (July 1992) published an investigation of the mean compression strength of corrugated fiberboard shipping containers. Comparisons were made for boxes of five different sizes: A, B, C, D, and E. Twenty identical boxes of each size were tested, and the peak compression strength (pounds) was recorded for each box. The figure below shows the sample means for the five box types as well as the variation around each sample mean.
a. Explain why the data are collected as a completely randomized design.
b. Refer to box types B and D. Based on the graph, does it appear that the mean compressive strengths of these two box types are significantly different? Explain.
c. Based on the graph, does it appear that the mean compressive strengths of all five box types are significantly different? Explain.

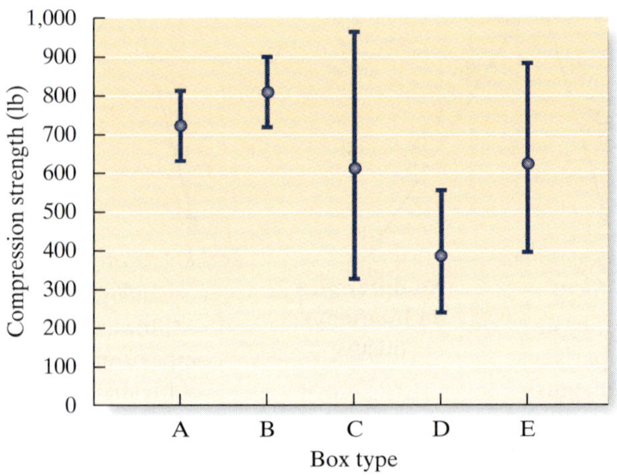

Source: Singh, S. P., et al. "Compression of single-wall corrugated shipping containers using fixed and floating test platens," *Journal of Testing and Evaluation,* Vol. 20, No. 4, July 1992, p. 319 (Figure 3). Copyright American Society for Testing and Materials. Reprinted with permission.

8.86 Study on how to ask survey questions. Marketing researchers generally agree that how you ask a question of consumers is as important as what you ask. An article in the *Journal of the Market Research Society* (July 1996) investigated the effects of positively versus negatively worded questions on consumer responses. Consumers were asked to express their level of agreement with positively worded statements on a scale of 1 (strongly disagree) to 5 (strongly agree). The scale values were reversed for negatively worded questions. A total of 154 subjects from the Midwest were randomly assigned to four treatment conditions defined by the two levels of the variables involvement in topic (high and low) and question wording (positive/negative). Each subject was asked to respond to a question (positively worded for one group and negatively worded for the other group). The 5-point scales described above were used. Those in the high-involvement condition received a question about the Gulf War; those in the low-involvement condition received a general question about shopping and products. The resulting data were analyzed using ANOVA.

a. What type of experimental design was used by the researcher?
b. Identify the factors used in the experiment. Are they quantitative or qualitative?
c. Describe the levels of each factor.
d. Describe the experiment's treatments.
e. What is the experiment's dependent variable?

8.87 Fortune's E-50 list. The Fortune E-50 is a listing of the top 50 electronic commerce and Internet-based companies, as determined by *Fortune* magazine each year. *Fortune* groups the companies into four categories: (1) e-companies, (2) Internet software and service, (3) Internet hardware, and (4) Internet communication. Consider a study to compare the mean rates of return for the stock of companies in the four *Fortune* categories. Because the age of an electronic commerce or Internet-based company may have an impact on rate of return, the study is designed to remove any age variation. Four 1-year-old companies, four 3-year-old companies, and four 5-year-old companies were selected; within each age group, one company was randomly selected from Category 1, one from Category 2, one from Category 3, and one from Category 4.

a. What type of experimental design is employed?
b. Identify the key elements of the experiment (i.e., treatments, blocks, response variable, and experimental unit).

8.88 Safety of nuclear power plants. Researchers at Pennsylvania State University and Iowa State University jointly studied the attitudes of three groups of professionals who influence U.S. policy governing new technologies: scientists, journalists, and federal government policymakers (*American Journal of Political Science,* Jan. 1998). Random samples of 100 scientists, 100 journalists, and 100 government officials were asked about the safety of nuclear power plants. Responses were made on a 7-point scale, where 1 = very unsafe and 7 = very safe. The mean safety scores for the groups are scientists, 4.1; journalists, 3.7; and government officials, 4.2.

a. Identify the response variable for this study.
b. How many treatments are included in this study? Describe them.
c. Specify the null and alternative hypotheses that should be used to investigate whether there are differences in the attitudes of scientists, journalists, and government officials regarding the safety of nuclear power plants.
d. For the sample data, MSE = 2.355 and MST = 11.28. Find the value of the test statistic.
e. If $\alpha = .05$, what is your conclusion?
f. The mean safety scores for the groups were

Government officials	4.2
Scientists	4.1
Journalists	3.7

Determine the number of pairwise comparisons of treatment means that can be made in this study.
g. Using an experimentwise error rate of $\alpha = .05$, Tukey's minimum significant difference for comparing means is .23—that is, if the difference between the sample means exceeds .23, the treatment means are statistically different. Use this information to conduct a multiple comparisons procedure of the safety score means. Fully interpret the results.

8.89 Leadership style and subordinate behavior. Leadership acts occur when one person tries to influence the behavior of others toward the attainment of some goal. The effects of leadership style on the behavior of subordinates were investigated in *Accounting, Organizations and Society* (Vol. 20, 1995). Four types of leadership style were defined based on two variables: the degree of control applied (high or low) and the level of consideration shown for subordinates (high or low). A sample of 257 senior auditors in Big-Six accounting firms yielded the following distribution of leadership styles for the auditors' leaders:

Leadership Style	*n*
A. High control, low consideration	51
B. Low control, low consideration	63
C. High control, high consideration	79
D. Low control, high consideration	64
Total	257

All subjects were asked to indicate (confidentially) how frequently their auditing fieldwork had been intentionally substandard in a particular way. They were asked to respond using a scale that ranged from 1 (never) to 5 (always). These data are summarized in the following table. An ANOVA conducted to test for differences in the four treatment means yielded $F = 30.4$.

Leadership Style	Mean	Standard Deviation	Bonferroni Analysis: Significantly Smaller Means
A	4.27	1.13	B, C, D
B	2.83	1.18	D
C	2.54	1.24	None
D	2.87	1.31	None
Overall	3.04	1.42	

Source: Otley, D. T., and Pierce, B. J. "The control problem in public accounting firms: An empirical study of the impact of leadership style," *Accounting Organizations and Society,* Vol. 20, No. 5, 1995, pp. 405–420. © 1995 with permission from Elsevier.

a. Do the data indicate that leadership style affects the behavior of subordinates? Test using $\alpha = .05$.

b. The Bonferroni multiple comparisons procedure was used to rank the four treatment means at an experimentwise error rate of $\alpha = .05$. Carefully interpret the results shown in the table.

c. What assumptions must hold to ensure the validity of the Bonferroni procedure?

8.90 Effectiveness of geese decoys. What type of decoy should you purchase for hunting waterfowl? A study in the *Journal of Wildlife Management* (July 1995) compared the effectiveness of three different decoy types—taxidermy-mounted decoys, plastic shell decoys, and full-bodied plastic decoys—in attracting Canada geese to sunken pit blinds. In order to account for an extraneous source of variation, three pit blinds were used as blocks in the experiment. Thus, a randomized block design with three treatments (decoy types) and three blocks (pit blinds) was employed. The response variable was the percentage of a goose flock to approach within 46 meters of the pit blind on a given day. The data are given in the table* and saved in the **DECOY** file. A Minitab printout of the analysis follows. Locate the p-value for treatments on the printout and interpret the result.

Two-way ANOVA: PERCENT versus DECOY, BLIND

```
Source  DF       SS       MS     F      P
DECOY    2   30.069  15.0344  0.61  0.589
BLIND    2   44.149  22.0744  0.89  0.479
Error    4   99.338  24.8344
Total    8  173.556

S = 4.983   R-Sq = 42.76%   R-Sq(adj) = 0.00%
```

Blind	Shell	Full-Bodied	Taxidermy-Mounted
1	7.3	13.6	17.8
2	12.6	10.4	17.0
3	16.4	23.4	13.6

Source: Harrey, W. F., Hindman, L. J., and Rhodes. W. E. "Vulnerability of Canada geese to taxidermy-mounted decoys," *Journal of Wildlife Management*, Vol. 59, No. 3, July 1995, p. 475 (Table 1).

8.91 Students' use of the computer. A computer lab at the University of Oklahoma is open 24 hours a day, 7 days a week. In *Production and Inventory Management Journal* (3rd Qtr., 1999), S. Barman investigated whether usage differed significantly (1) among the days of the week and (2) among the hours of the day. Using student log-on records, data on hourly student loads (number of users per hour) were collected during a 7-week period. A factorial ANOVA was used to analyze the data, with the results presented in the table in the next column.

*The actual design employed in the study was more complex than the randomized block design shown here. In the actual study, each number in the table represents the mean daily percentage of goose flocks attracted to the blind, averaged over 13–17 days.

Source	df	F	p
Model	167	25.06	.001
Error	1,004		
Total	1,171		

$R^2 = .8065$

Source	df	F	p
Day	6	68.39	.0001
Time	23	156.80	.0001
Day × time	138	1.22	.0527

Source: Barman, S. "A Statistical analysis of the attendance pattern of a computer laboratory," *Production and Inventory Management Journal,* Third Quarter, 1999, pp. 26–30.

a. Is this an observational or a designed experiment? Explain.

b. What are the two factors of the experiment and how many levels of each factor are used?

c. This is an $a \times b$ factorial experiment. What are a and b?

d. Conduct a test to determine whether any of the $a \times b$ treatment means significantly differ. Use $\alpha = .01$.

e. Specify the null and alternative hypotheses that should be used to test for an interaction effect between the two factors of the study.

f. Conduct the test of part **e** using $\alpha = .01$. Interpret your result in the context of the problem.

g. If appropriate, conduct main effects tests for both day and time. Use $\alpha = .01$. Interpret your results in the context of the problem.

Applying the Concepts—Intermediate

8.92 A managerial decision problem. A direct-mail company assembles and stores paper products (envelopes, letters, brochures, order cards, etc.) for its customers. The company estimates the total number of pieces received in a shipment by estimating the weight per piece and then weighing the entire shipment. The company is unsure whether the sample of pieces used to estimate the mean weight per piece should be drawn from a single carton, or whether it is worth the extra time required to pull a few pieces from several cartons. To aid management in making a decision, eight brochures were pulled from each of five cartons of a typical shipment and weighed. The weights (in pounds) are shown in the table and saved in the **CARTONS** file.

Carton 1	Carton 2	Carton 3	Carton 4	Carton 5
.01851	.01872	.01869	.01899	.01882
.01829	.01861	.01853	.01917	.01895
.01844	.01876	.01876	.01852	.01884
.01859	.01886	.01880	.01904	.01835
.01854	.01896	.01880	.01923	.01889
.01853	.01879	.01882	.01905	.01876
.01844	.01879	.01862	.01924	.01891
.01833	.01879	.01860	.01893	.01879

a. Identify the response, factor(s), treatments, and experimental units.

b. Do these data provide sufficient evidence to indicate differences in the mean weight per brochure among the five cartons?

c. What assumptions must be satisfied in order for the test of part **b** to be valid?

d. Use Tukey's method to compare all pairs of means, with $\alpha = .05$ as the overall level of significance.

e. Given the results, make a recommendation to management about whether to sample from one carton or from many cartons.

8.93 Activities of entrepreneurs at start-up. On average, over a million new businesses are started in the United States every year. An article in the *Journal of Business Venturing* (Vol. 11, 1996) reported on the activities of entrepreneurs during the organization creation process. Among the questions investigated were what activities and how many activities do entrepreneurs initiate in attempting to establish a new business? A total of 71 entrepreneurs were interviewed and divided into three groups: those who were successful in founding a new firm (34), those still actively trying to establish a firm (21), and those who tried to start a new firm but eventually gave up (16). The total number of activities undertaken (i.e., developed a business plan, sought funding, looked for facilities, etc.) by each group over a specified time period during organization creation was measured, and the following incomplete ANOVA table was produced:

Source	df	SS	MS	F
Groups	–	128.70	–	–
Error	–	27,124.52	–	

Source: Reprinted from Carter, N., Garner, W., and Reynolds, P. "Exploring start-up event sequences," *Journal of Business Venturing,* Vol. 11, 1996, p. 159. © 1996 with permission from Elsevier.

a. Complete the ANOVA table.

b. Do the data provide sufficient evidence to indicate that the total number of activities undertaken differed among the three groups of entrepreneurs? Test using $\alpha = .05$.

c. What is the p-value of the test you conducted in part **b**?

d. One of the conclusions of the study was that the behaviors of entrepreneurs who have successfully started a new company can be differentiated from the behaviors of entrepreneurs who failed. Do you agree? Justify your answer.

8.94 Study of anticoagulant drugs. Three anticoagulant drugs are studied to compare their effectiveness in dissolving blood clots. Each of five subjects receives the drugs at equally spaced time intervals and in random order. Time periods between drug applications permit a drug to be passed out of a subject's body before the subject receives the next drug. After each drug is in the bloodstream, the length of time (in seconds) required for a cut of specified size to stop bleeding is recorded. The results are shown in the following table and saved in the **CLOTS** file.

a. What type of experimental design was used in this study? Identify the response, factor(s), factor type(s), treatments, and experimental units.

b. Is there evidence of a difference in mean clotting time among the three drugs? Test using $\alpha = .10$.

	Drug		
Person	A	B	C
1	127.5	129.0	135.5
2	130.6	129.1	138.0
3	118.3	111.7	110.1
4	155.5	144.3	162.3
5	180.7	174.4	181.8

c. What is the observed significance level of the test you conducted in part **a**? Interpret it.

d. Was blocking effective in reducing the variation among the data? That is, do the data support the contention that the mean clotting time varies from person to person?

e. If warranted, use a multiple comparisons technique to determine whether one of the drugs is most effective. Use an overall significance level of $\alpha = .10$.

8.95 Steel ingot quality study. A quality control supervisor measures the quality of a steel ingot on a scale of 0 to 10. He designs an experiment in which three different temperatures (ranging from 1,100 to 1,200°F) and five different pressures (ranging from 500 to 600 psi) are used, with 20 ingots produced at each Temperature-Pressure combination. Identify the following elements of the experiment:

a. Response **b.** Factor(s) and factor type(s)

c. Treatments **d.** Experimental units

8.96 Accountants' attitudes toward risk. Louisiana State University Professor D. C. Kim investigated the effects of accountants' attitudes toward risk and recent budget-setting performance (relative to their colleagues) on their current budgetary decisions (*Accounting Review,* Apr. 1992). Eighty-one undergraduate students were used as subjects in the study. Using standard risk measurement methods, 40 subjects were found to be risk-seeking, and 41 were risk-averse. Subjects were asked to play the role of an entry-level accountant in a public accounting firm. Each was asked to read a hypothetical audit engagement case. Each subject had either a case in which his or her recent performance at budgeting audit costs (billable hours) was either favorable or unfavorable relative to colleagues. After reading the case, each subject was asked to establish either a tight (risky) budget or a safe (riskless) budget and to indicate the strength of his or her preference for the budget. Subjects responded on an 11-point scale that varied from −5 (strong preference for a risky choice) to +5 (strong preference for a riskless choice). From these data, Kim generated the following partial ANOVA table:

Source	df	SS	MS	F
Recent performance (A)	1	243.2	–	–
Risk assessment (B)	1	57.8	–	–
AB	1	–	–	–
Error	77	670.8	–	
Total	80	976.3		

Source: Kim, D. C. "Risk preferences in participative budgeting," *Accounting Review,* Vol. 67, Apr. 1992, pp. 303–318.

a. Complete the ANOVA table.

b. Do the data indicate that factors A and B interact? Test using $\alpha = .05$.

c. Does an individual's risk attitude affect his or her budgetary decisions? Test using $\alpha = .05$.

d. Does recent budgeting performance affect current budgeting decisions? Test using $\alpha = .01$.

8.97 Product tampering study. A study in the *Journal of Psychology and Marketing* (Jan. 1992) investigated consumer attitudes toward product tampering. One variable considered was the education level of the consumer. Consumers were divided into five educational classifications and asked to rate their concern about product tampering on a scale of 1 (little or no concern) to 9 (very concerned). The table gives the education levels and the means.

Education Level	Mean	Sample Size
Non–high school graduate	3.731	26
High school graduate	3.224	49
Some college completed	3.330	94
College graduate	3.167	60
Some postgraduate work	4.341	86

a. Identify the type of ANOVA design used in this experiment. Identify the treatments in this experiment.

b. The article compared the mean concern ratings for the five education levels. The *F*-statistic for this test was reported to be 3.298. Conduct a test of hypothesis (at $\alpha = .05$) to determine whether the mean concern ratings differ for at least two of the education levels. [*Hint:* Calculate the degrees of freedom for Treatments and Error from the information given.]

c. Using $\alpha = .05$, a Bonferroni analysis yielded the result shown at the bottom of the page. Interpret the result.

8.98 Comparing two glaucoma drugs. Two drugs, A and B, used for the treatment of glaucoma (an eye disease) were tested for effectiveness on 10 diseased dogs. Drug A was administered to one eye (chosen randomly) of each dog and drug B to the other eye. Pressure measurements were taken 1 hour later on both eyeballs of each dog. The 10 diseased dogs serve as the blocks for comparing the two treatments, drugs A and B. Pressure measurements are given in the following table and saved in the **EYEDOGS** file. (The smaller the measurement, the less serious the eye disease.)

Dog	Drug A	Drug B
1	.17	.15
2	.20	.18
3	.14	.13
4	.18	.18
5	.23	.19
6	.19	.12
7	.12	.07
8	.10	.09
9	.16	.14
10	.13	.08

a. Perform an analysis of variance for these data. Do the data provide sufficient evidence to indicate a difference in mean pressure readings for the two treatments (i.e., is one of the glaucoma drugs better than the other)? Use $\alpha = .05$.

b. What is the purpose of using the dogs as blocks in this experiment?

c. Recall that a randomized block design with $p = 2$ treatments is a paired difference experiment (Chapter 7). Analyze the data as a paired difference experiment using a *t*-test to compare the treatment means. Use $\alpha = .05$.

d. Compare the computed *F* and *t* values from parts **a** and **c** and verify that $F = t^2$. Also verify that for the rejection region values of *F* and *t*, $F_\alpha = t^2_{\alpha/2}$.

e. Find the approximate observed significance level for the test in part **a** and interpret its value.

8.99 Improving a can-filling process. *Quality Engineering* (Vol. 2, 1990) published the results of an experiment that was conducted by a dog food manufacturer to improve a filling process in which ground meat is packed into cans. The process uses a rotary filling machine with six cylinders, each of which dispenses ground meat. The company wanted to study the effects of differences in batches of meat and differences in cylinders on the weight of the final product. Five batches of meat were used in the experiment. Three filled cans were randomly selected from each cylinder while each batch was being run. The cans were weighed, and the weights were recorded. To simplify the analysis, the weights were coded by subtracting 12 ounces from each weight. The coded data appear in the table below and are saved in the **DOGFOOD** file.

		Batch				
		1	2	3	4	5
Cylinder	1	1	4	6	3	1
		1	3	3	1	3
		2	5	7	3	3
	2	−1	−2	3	2	1
		3	1	1	0	0
		−1	0	5	1	1
	3	1	2	2	1	3
		1	0	4	3	3
		1	1	3	3	3
	4	−2	−2	3	0	0
		3	0	3	0	1
		0	1	4	2	1
	5	1	2	0	1	−2
		1	1	1	0	3
		−1	5	2	−1	1
	6	0	0	3	3	3
		1	0	3	0	1
		1	3	4	2	2

Source: Griffith, B. A., Westman, A. E. R., and Lloyd, B. H. "Analysis of variance," *Quality Engineering*, Vol 2. No. 2. 1989, pp. 195–226. Reprinted with permission of the publisher (Taylor & Francis Group, www.informaworld.com).

Multiple Comparisons for Exercise 8.97

Mean	3.167	3.224	3.330	3.731	4.3441
Education Level	*College Graduate*	*High school graduate*	*Some college*	*Non–high school graduate*	*Postgraduate*

a. What type of experimental design was used by the dog food manufacturer?
b. Identify the factors used in the study and their levels.
c. How many different treatments were used in the experiment?
d. Analyze the data with an analysis of variance. Summarize the results in an ANOVA table.
e. In the context of the problem, explain what it means to say that batch and cylinder interact. Speculate on what could cause such an interaction.
f. Test for an interaction between batches and cylinders. Use $\alpha = .05$.
g. If appropriate, test for main effects using $\alpha = .05$.

8.100 Testing the effectiveness of supermarket sales strategies. Factorial designs are commonly employed in marketing research to evaluate the effectiveness of sales strategies. At one supermarket, two of the factors were Price level (regular, reduced price, cost to supermarket) and Display level (normal display space, normal display space plus end-of-aisle display, twice the normal display space). A 3×3 complete factorial design was employed, where each treatment was applied three times to a particular product at a particular supermarket. The dependent variable of interest was unit sales for the week. (To minimize treatment carryover effects, each treatment was preceded and followed by a week in which the product was priced at its regular price and was displayed in its normal manner.) The table below reports the data collected (saved in the **SUPERMKT** file).
a. How many treatments are considered in this study?
b. Do the data indicate that the mean sales differ among the treatments? Test using $\alpha = .10$.
c. Is the test of interaction between the factors Price and Display warranted as a result of the test in part **b**? If so, conduct the test using $\alpha = .10$.
d. Are the tests of the main effects for Price and Display warranted as a result of the previous tests? If so, conduct them using $\alpha = .10$.
e. Which pairs of treatment means should be compared as a result of the tests in parts **b–d**?

		Price		
		Regular	Reduced	Cost to Supermarket
	Normal	989 1,025 1,030	1,211 1,215 1,182	1,577 1,559 1,598
Display	Normal Plus	1,191 1,233 1,221	1,860 1,910 1,926	2,492 2,527 2,511
	Twice Normal	1,226 1,202 1,180	1,516 1,501 1,498	1,801 1,833 1,852

Applying the Concepts—Advanced

8.101 Testing a new insect repellent. Traditionally, people protect themselves from mosquito bites by applying insect repellent to their skin and clothing. Recent research suggests that permethrin, an insecticide with low toxicity to humans, can provide protection from mosquitoes. A study in the *Journal of the American Mosquito Control Association* (Mar. 1995) investigated whether a tent sprayed with a commercially available 1% permethrin formulation would protect people, both inside and outside the tent, against biting mosquitoes. Two canvas tents—one treated with permethrin, the other untreated—were positioned 25 meters apart on flat, dry ground in an area infested with mosquitoes. Eight people participated in the experiment, with four randomly assigned to each tent. Of the four stationed at each tent, two were randomly assigned to stay inside the tent (at opposite corners) and two to stay outside the tent (at opposite corners). During a specified 20-minute period during the night, each person kept count of the number of mosquito bites received. The goal of the study was to determine the effect of both Tent type (treated or untreated) and Location (inside or outside the tent) on the mean mosquito bite count.
a. What type of design was employed in the study?
b. Identify the factors and treatments.
c. Identify the response variable.
d. The study found statistical evidence of interaction between Tent type and Location. Give a practical interpretation of this result.

8.102 Improving the output of an industrial lathe. *Quality Engineering* (Vol. 6, 1994) reported the results of an experiment that was designed to find ways to improve the output of an industrial lathe. The lathe is controlled by a computer that automatically feeds bar stock, cuts the stock, machines the surface finish, and releases the part. As it is machined, the bar stock spins and is held in place by a collet. The lathe operator sets the feed (the rate at which bars are machined) and the speed (spin rate). The product characteristic of interest is surface finish. It is measured on a gauge that records the vertical distance a probe travels as it moves along a given horizontal distance on the bar. The rougher the surface, the higher the gauge measurement. The factors that were manipulated in the experiment were speed, feed, collet tightness, and tool wear. The table on the next page (saved in the **LATHE** file) reports the factor-level settings and the resulting surface-finish measurements (H = High; L = Low).
a. What type of experimental design was used?
b. How many different treatments were applied?
c. Perform an ANOVA for these data.
d. Do significant interaction effects exist? Test using $\alpha = .05$. Interpret your results.
e. Is it necessary to perform main effect tests? Why or why not? If so, perform the tests using $\alpha = .05$.
f. What assumptions must hold to ensure the validity of your results in parts **c, d,** and **e**?

Critical Thinking Challenge

8.103 Exam performance study. Refer to the *Teaching of Psychology* (August 1998) study of whether a practice test helps students prepare for a final exam, Exercise 8.9 (p. 454). Recall that undergraduate students were grouped according to their class standing and whether they attended a review session or took a practice test prior to the final exam. The experimental design was a 3×2 factorial design, with Class Standing at 3 levels (low, medium, or high) and Exam Preparation at 2 levels (practice exam or review session). There were 22 students in each of the

Table for Exercise 8.102

Speed	Feed	Collet Tightness	Surface Tool Wear	Finish
H	H	H	H	216
L	H	H	H	212
H	L	H	H	48
L	L	H	H	40
H	H	L	H	232
L	H	L	H	248
H	L	L	H	514
L	L	L	H	298
H	H	H	L	238
L	H	H	L	219
H	L	H	L	40
L	L	H	L	33
H	H	L	L	230
L	H	L	L	253
H	L	L	L	273
L	L	L	L	101
H	H	H	H	217
L	H	H	H	221
H	L	H	H	39
L	L	H	H	31
H	H	L	H	235
L	H	L	H	238
H	L	L	H	437
L	L	L	H	87
H	H	H	L	245
L	H	H	L	226
H	L	H	L	51
L	L	H	L	33
H	H	L	L	226
L	H	L	L	214
H	L	L	L	691
L	L	L	L	130

Source: Collins, W. H., and Colins, C. B. "Including residual analysis in designed experiments: Case studies," *Quality Engineering*, Vol. 6, No. 4, 1994, pp. 547–565. Reprinted by permission of the publisher (Taylor & Francis Group, www.informaworld.com).

$3 \times 2 = 6$ treatment groups. After completing the final exam, each student rated his/her exam preparation on an 11-point scale ranging from 0 (not helpful at all) to 10 (extremely helpful). The data for this experiment (simulated from summary statistics provided in the article) are saved in the **PRACEXAM** file. The first 5 and last 5 observations in the data set are listed below. Conduct a complete analysis of variance of the helpfulness ratings data, including (if warranted) multiple comparisons of means. Do your findings support the research conclusion that "students at all levels of academic ability benefit from a...practice exam"?

Exam Preparation	Class Standing	Helpfulness Rating
Practice	Low	6
Practice	Low	7
Practice	Low	7
Practice	Low	5
Practice	Low	3
⋮	⋮	⋮
Review	High	5
Review	High	2
Review	High	5
Review	High	4
Review	High	3

Source: Balch, W. R. "Practice versus review exams and final exam performance," *Teaching of Psychology*, Vol. 25, No. 3, August 1998 (adapted from Table 1).

References

Cochran, W. G., and Cox, G. M. *Experimental Designs*, 2nd ed. New York: Wiley, 1957.

Hsu, J. C. *Multiple Comparisons: Theory and Methods.* London: Chapman & Hall, 1996.

Kramer, C. Y. "Extension of multiple range tests to group means with unequal number of replications," *Biometrics*, Vol. 12, 1956, pp. 307–310.

Mason, R. L., Gunst, R. F., and Hess, J. L. *Statistical Design and Analysis of Experiments.* New York: Wiley, 1989.

Mendenhall, W. *Introduction to Linear Models and the Design and Analysis of Experiments.* Belmont, Calif.: Wadsworth, 1968.

Miller, R. G., Jr. *Simultaneous Statistical Inference.* New York: Springer-Verlag, 1981.

Neter, J., Kutner, M., Nachtsheim, C., and Wasserman, W. *Applied Linear Statistical Models*, 4th ed. Homewood, Ill.: Richard D. Irwin, 1996.

Scheffé, H. "A method for judging all contrasts in the analysis of variance," *Biometrica*, Vol. 40, 1953, pp. 87–104.

Scheffé, H. *The Analysis of Variance.* New York: Wiley, 1959.

Snedecor, G. W., and Cochran, W. G. *Statistical Methods*, 7th ed. Ames: Iowa State University Press, 1980.

Steele, R. G. D., and Torrie, J. H. *Principles and Procedures of Statistics: A Biometrical Approach*, 2nd ed. New York: McGraw-Hill, 1980.

Tukey, J. "Comparing individual means in the analysis of variance," *Biometrics*, Vol. 5, 1949, pp. 99–114.

Winer, B. J. *Statistical Principles in Experimental Design*, 2nd ed. New York: McGraw-Hill, 1971.

USING TECHNOLOGY

SPSS: Analysis of Variance

SPSS can conduct ANOVAs for all three types of experimental designs discussed in this chapter: completely randomized, randomized block, and factorial designs.

Completely Randomized Design

Step 1 Access the SPSS spreadsheet file that contains the sample data. The data file should contain one quantitative variable (the response, or dependent, variable) and one factor variable with at least two levels. (These values must be numbers, e.g., 1, 2, 3, . . .)

Step 2 Click on the "Analyze" button on the SPSS menu bar, then click on "Compare Means" and "One-Way ANOVA," as shown in Figure 8.S.1.

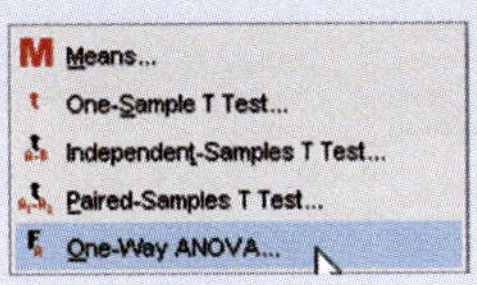

Figure 8.S.1 SPSS menu options for one-way ANOVA

Step 3 In the resulting dialog box (shown in Figure 8.S.2), specify the response variable under "Dependent List" and the factor variable under "Factor."

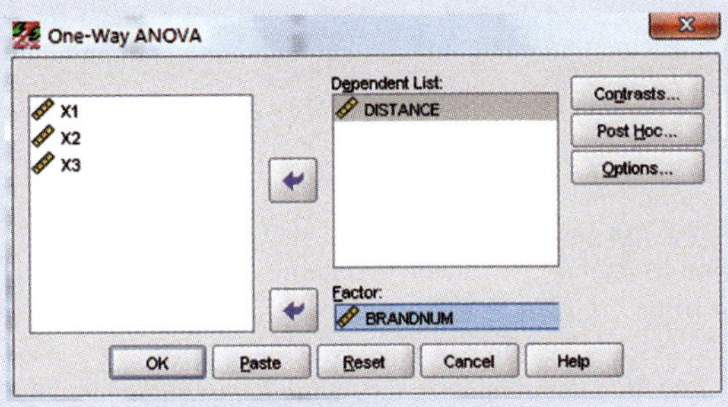

Figure 8.S.2 SPSS one-way ANOVA dialog box

Step 4 Click the "Post Hoc" button and select a multiple comparisons method and experimentwise error rate in the resulting dialog box (see Figure 8.S.3).

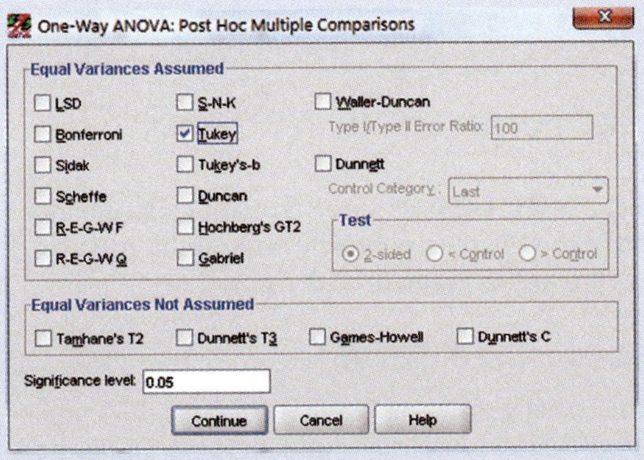

Figure 8.S.3 SPSS multiple comparisons dialog box

Step 5 Click "Continue" to return to the "One-Way ANOVA" dialog screen and then click "OK" to generate the SPSS printout.

Randomized Block and Factorial Designs

Step 1 Access the SPSS spreadsheet file that contains the sample data. The data file should contain one quantitative variable (the response, or dependent, variable) and at least two other variables that represent the factors and/or blocks.

Step 2 Click on the "Analyze" button on the SPSS menu bar and then click on "General Linear Model" and "Univariate," as shown in Figure 8.S.4.

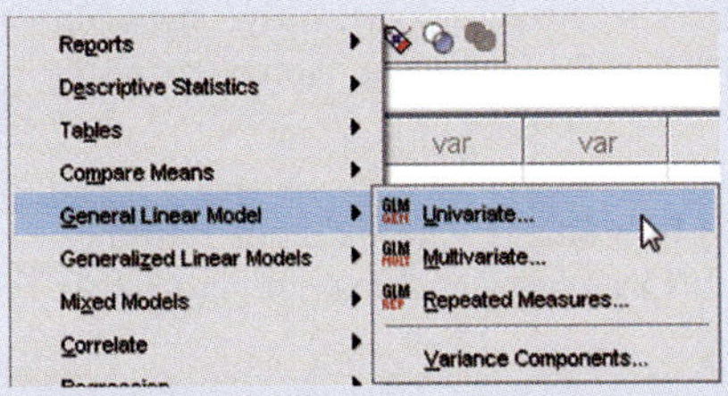

Figure 8.S.4 SPSS menu options for two-way ANOVA

Step 3 On the resulting dialog box (Figure 8.S.5.), specify the response variable under "Dependent Variable" and the factor variable(s) and block variable under "Fixed Factor(s)."

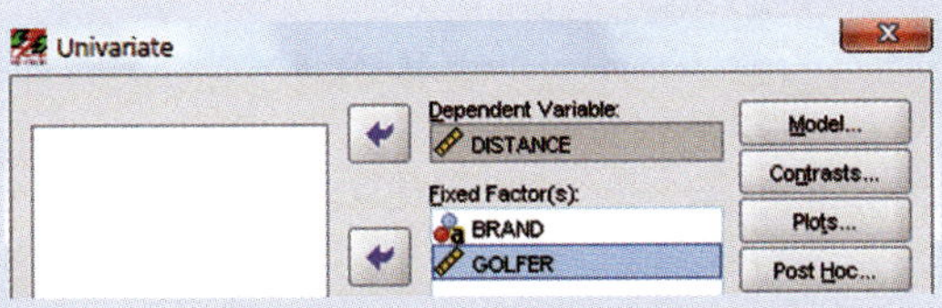

Figure 8.S.5 SPSS two-way ANOVA dialog box

Step 4 Click the "Post Hoc" button and select the factor variable of interest, a multiple comparisons method, and experimentwise error rate in the resulting dialog box (similar to Figure 8.S.3).

Step 5 Click "Continue" to return to the "Two-Way ANOVA" dialog screen and then click the "Model" button to specify the type of experimental design (randomized block or factorial) on the resulting dialog screen (as shown in Figure 8.S.6). For factorial designs, select the "Full Factorial" option; for randomized block designs, select the "Custom" option and specify the treatment and blocking factors under "Model."

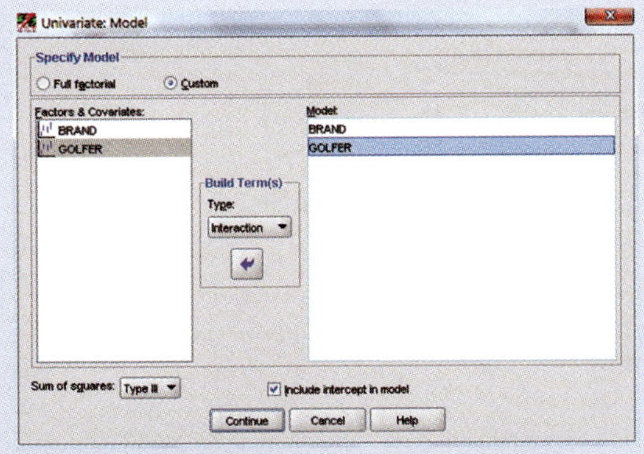

Figure 8.S.6 SPSS design (model) specification box

Step 6 Click "Continue" to return to the "Two-Way ANOVA" dialog screen and then click "OK" to generate the SPSS printout.

Minitab: Analysis of Variance

Minitab can conduct ANOVAs for all three types of experimental designs discussed in this chapter: completely randomized, randomized block, and factorial designs.

Completely Randomized Design

Step 1 Access the Minitab worksheet file that contains the sample data. The data file should contain one quantitative variable (the response, or dependent, variable) and one factor variable with at least two levels.

Step 2 Click on the "Stat" button on the Minitab menu bar and then click on "ANOVA" and "One-Way," as shown in Figure 8.M.1.

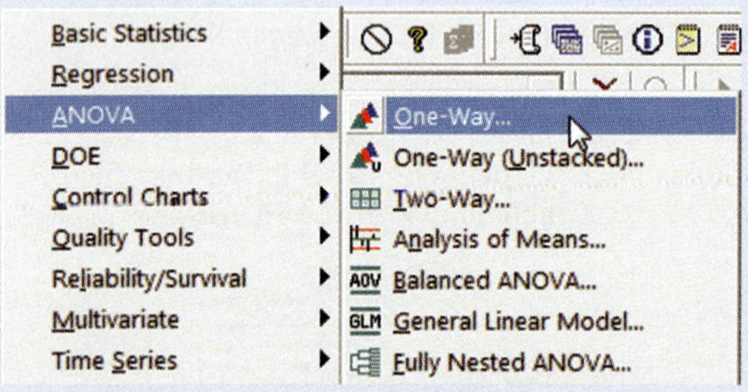

Figure 8.M.1 Minitab menu options for one-way ANOVA

Step 3 On the resulting dialog screen (Figure 8.M.2), specify the response variable in the "Response" box and the factor variable in the "Factor" box.

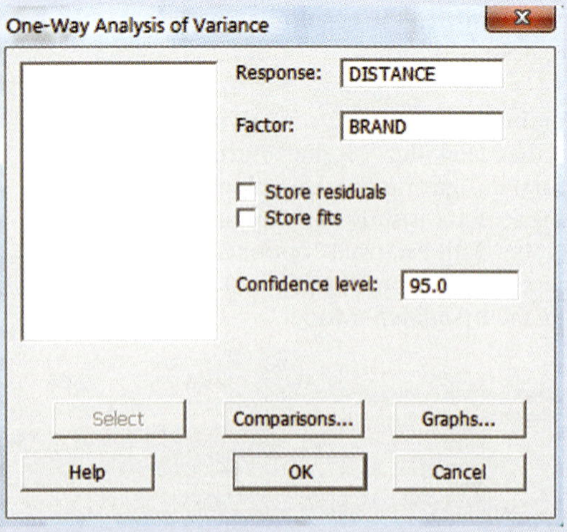

Figure 8.M.2 Minitab one-way ANOVA dialog box

Step 4 Click the "Comparisons" button and select a multiple comparisons method and experimentwise error rate in the resulting dialog box (see Figure 8.M.3).

Step 5 Click "OK" to return to the "One-Way ANOVA" dialog screen and then click "OK" to generate the Minitab printout.

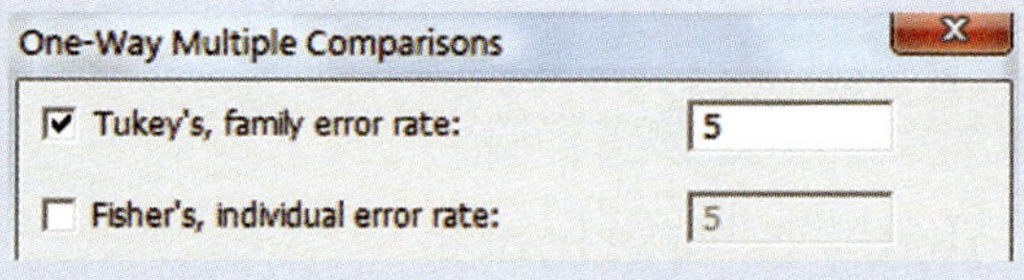

Figure 8.M.3 Minitab multiple comparisons dialog box

Randomized Block and Factorial Designs

Step 1 Access the Minitab worksheet file that contains the sample data. The data file should contain one quantitative variable (the response, or dependent, variable) and two other variables that represent the factors and/or blocks.

Step 2 Click on the "Stat" button on the Minitab menu bar and then click on "ANOVA" and "Two-Way" (see Figure 8.M.1). The resulting dialog screen appears as shown in Figure 8.M.4.

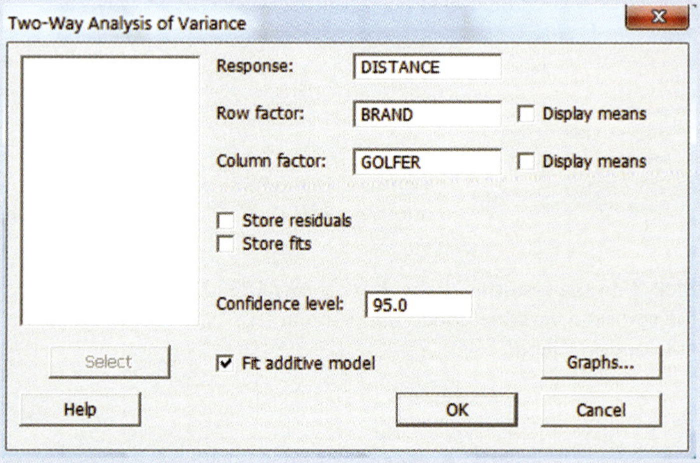

Figure 8.M.4 Minitab two-way ANOVA dialog box

Step 3 Specify the response variable in the "Response" box, the first factor variable in the "Row factor" box, and the second factor or block variable in the "Column factor" box. If the design is a randomized block, select the "Fit additive model" option, as shown in Figure 8.M.4. If the design is factorial, leave the "Fit additive model" option unselected.

Step 4 Click "OK" to generate the Minitab printout.

Note: Multiple comparisons of treatment means are obtained by selecting "Stat," then "ANOVA," and then "General Linear Model." Specify the factors in the "Model" box and then select "Comparisons" and put the factor of interest in the "Terms" box. Press "OK" twice.

Excel/DDXL: Analysis of Variance

Excel can conduct ANOVAs for all three types of experimental designs discussed in this chapter: completely randomized, randomized block, and two-factor factorial designs.

Completely Randomized Design

Step 1 Access the Excel workbook file that contains the sample data. The file should contain k columns of data for the quantitative (response) variable, one for each of the k treatments.

Step 2 Click on the "Tools" button on the Excel main menu bar and then click on "Data Analysis," as shown in Figure 8.E.1.

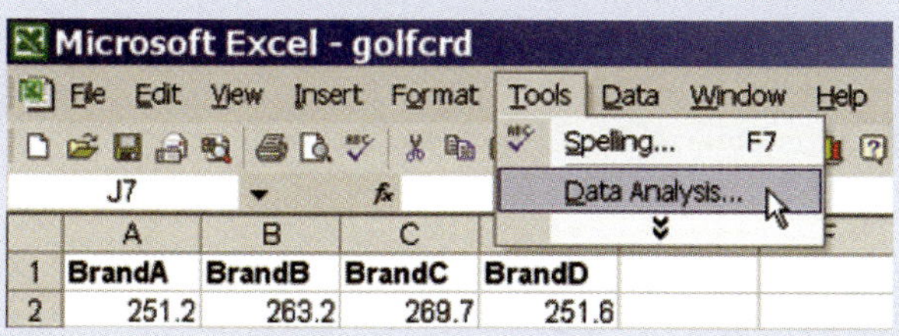

Figure 8.E.1 Excel main menu options for data analysis

Step 3 Select "Anova: Single Factor" from the Data Analysis menu, as shown in Figure 8.E.2 and click "OK." The resulting dialog box appears as shown in Figure 8.E.3.

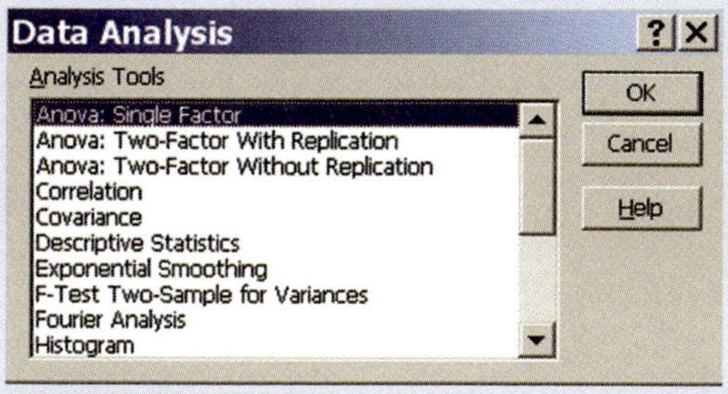

Figure 8.E.2 Excel data analysis menu option for one-way ANOVA

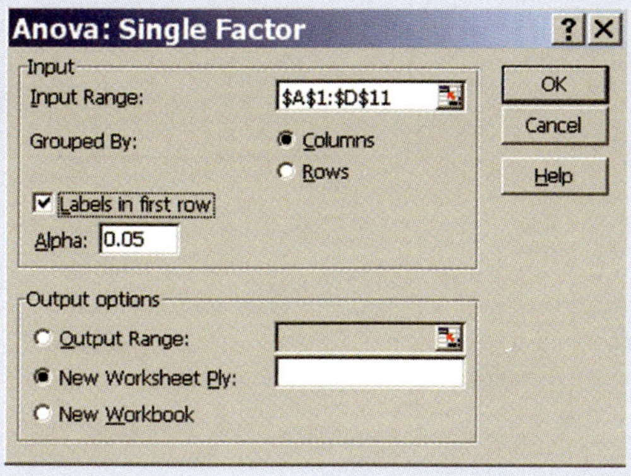

Figure 8.E.3 Excel single factor ANOVA dialog box

Step 4 Specify the "Input Range" of your data, select the "Grouped By: Columns" option, the value of α and, if the workbook contains labels, select "Labels in first row," as shown in Figure 8.E.3.

Step 5 Click "OK" to generate the Excel printout.

Multiple Comparisons of Means

You can obtain multiple comparisons of means for a one-way ANOVA using the DDXL add-in to Excel:

Step 1 Highlight (select) the data columns you want to analyze on the Excel spreadsheet. (*Note:* The response variable should be in one column and the factor variable in a second column.)

Step 2 Click on "Add-Ins" in the main Excel menu bar and select "DDXL." On the resulting menu, select "ANOVA," as shown in Figure 8.E.4.

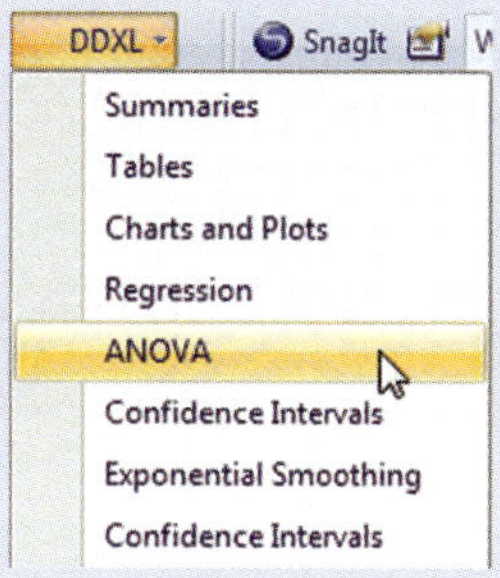

Figure 8.E.4 Excel/DDXL menu options for ANOVA

Step 3 On the resulting menu, select "1 Way ANOVA" in the Function Type box, as shown in Figure 8.E.5.

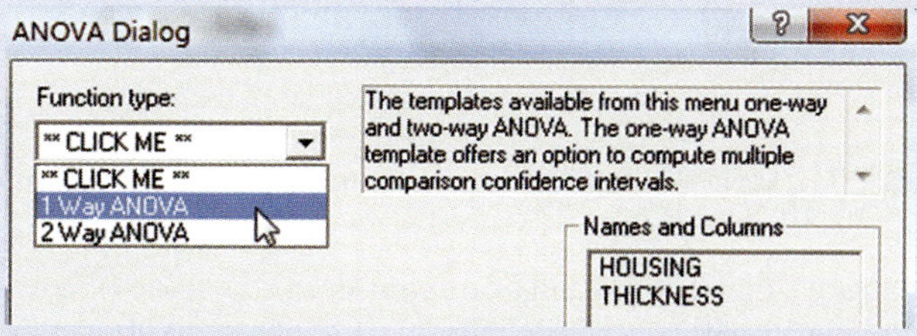

Figure 8.E.5 DDXL ANOVA dialog box

Step 4 Move the column for the response variable into the "Response Variable" box and the column for the factor into the "Factor Variable" box, as shown in Figure 8.E.6. Then click "OK" to generate the ANOVA printout.

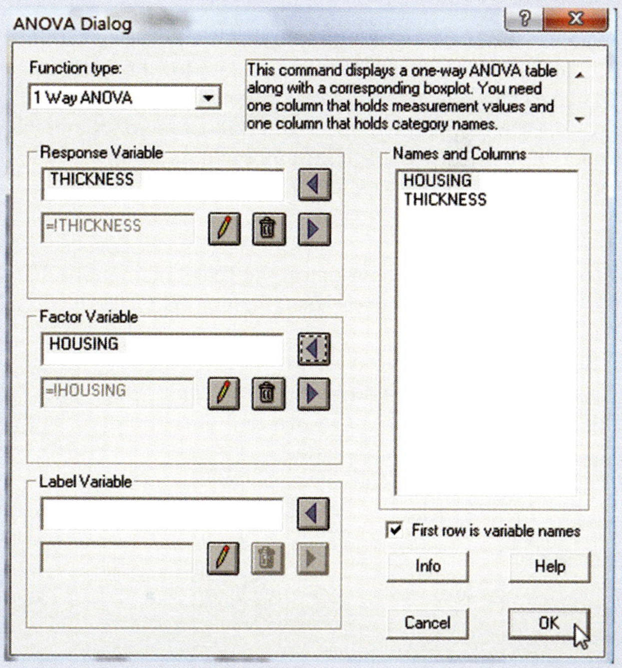

Figure 8.E.6 DDXL menu selections for 1 way ANOVA

Step 5 At the bottom of the ANOVA printout, click on the desired confidence level to obtain the Tukey multiple comparisons.

Randomized Block Design

Step 1 Access the Excel workbook file that contains the sample data. The file should contain k columns of data for the quantitative (response) variable, one for each of the k treatments, and b rows—one for each of the b blocks.

Step 2 Click on the "Tools" button on the Excel main menu bar and then click on "Data Analysis" (see Figure 8.E.1).

Step 3 Select "Anova: Two Factor Without Replication" from the Data Analysis menu (see Figure 8.E.2) and click "OK." The resulting dialog box appears as shown in Figure 8.E.7.

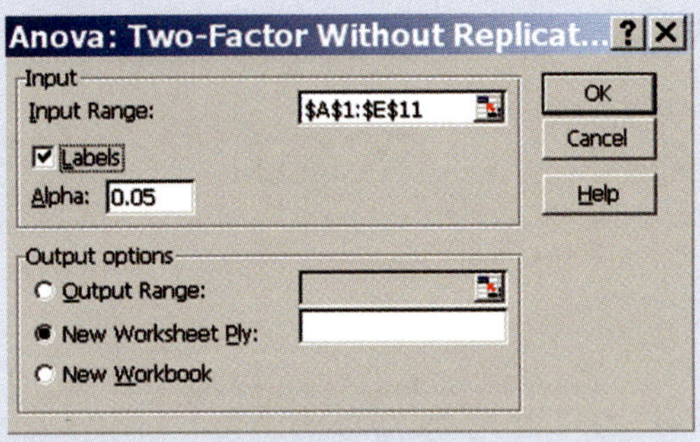

Figure 8.E.7 Excel two factor with replication ANOVA dialog box

Step 5 Specify the "Input Range" of your data, the value of α, and, if the workbook contains labels, select "Labels," as shown in Figure 8.E.7.

Step 6 Click "OK" to generate the Excel printout.

Note: Multiple comparisons of treatment means are unavailable in Excel for a randomized block design.

Two-Factor Factorial Design

Step 1 Access the Excel workbook file that contains the sample data. The file should contain a columns of data for the quantitative (response) variable, one for each of the a levels of Factor A, and br rows—where b is the number of levels of factor B and r is the number of replicates. [*Note:* Place all the data for one level of factor B in consecutive rows, followed by all the data for the next level of B, etc.]

Step 2 Click on the "Tools" button on the Excel main menu bar and then click on "Data Analysis" (see Figure 8.E.1).

Step 3 Select "Anova: Two Factor with Replication" from the Data Analysis menu (see Figure 8.E.2) and click "OK." The resulting dialog box appears as shown in Figure 8.E.8.

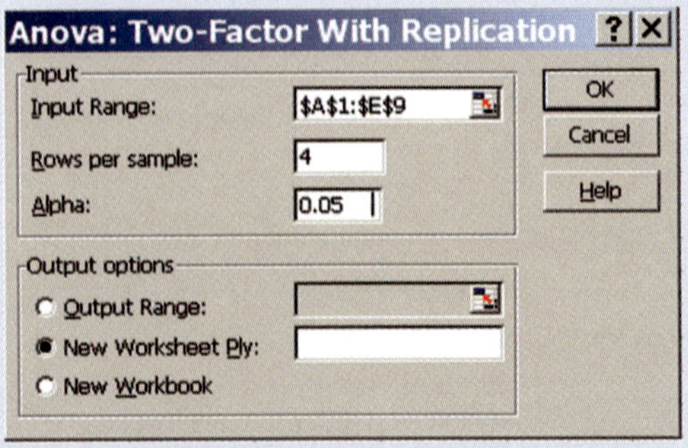

Figure 8.E.8 Excel two factor with replication ANOVA dialog box

Step 4 Specify the "Input Range" of your data, the number of "Rows per sample" (this is the number of replicates), and the value of α, as shown in Figure 8.E.8.

Step 5 Click "OK" to generate the Excel printout.

Note: Multiple comparisons of treatment means are unavailable in Excel for a factorial block design.

TI-84 Graphing Calculator: Analysis of Variance

The TI-84 graphing calculator can be used to compute a one-way ANOVA for a completely randomized design but not a two-way ANOVA for either a randomized block or factorial design.

Completely Randomized Design

Step 1 *Enter each data set into its own list (i.e., sample 1 into L1, sample 2 into L2, sample 3 into L3, etc.).*

Step 2 *Access the Statistical Test Menu*

- Press **STAT**
- Arrow right to **TESTS**
- Arrow down to **ANOVA(**
- Press **ENTER**
- Type in each List name separated by commas *(e.g., L1, L2, L3, L4)*
- Press **ENTER**

Step 3 *View Display*

The calculator will display the F-test statistic, as well as the p-value, the factor degrees of freedom, sum of squares, mean square, and by arrowing down, the Error degrees of freedom, sum of squares, mean square, and the pooled standard deviation.

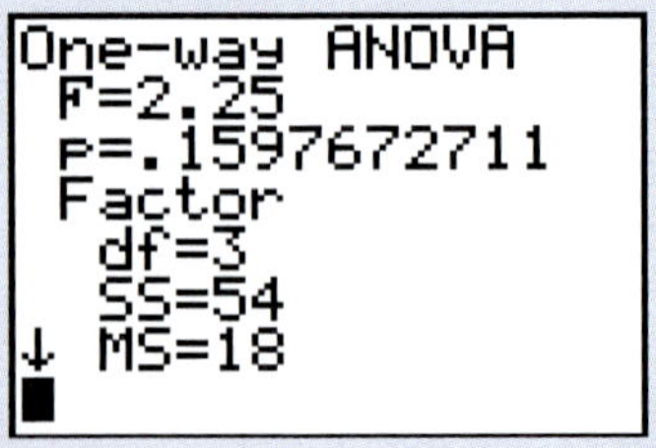

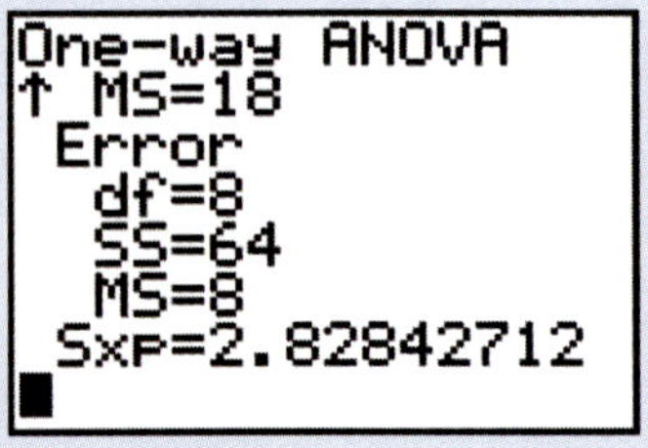

CONTENTS

9 Categorical Data Analysis

Where We've Been

- Presented methods for making inferences about the population proportion associated with a two-level qualitative variable (i.e., a binomial variable)
- Presented methods for making inferences about the difference between two binomial proportions

Where We're Going

- Discuss qualitative (i.e., categorical) data with more than two outcomes
- Present a *chi-square* hypothesis test for comparing the category proportions associated with a single qualitative variable—called a *one-way analysis*
- Present a *chi-square* hypothesis test for relating two qualitative variables—called a *two-way analysis*

Statistics IN Action A Study of Coupon Users—Mail versus the Internet

The service encounter is the critical interaction between a customer and the firm. In this encounter, the firm attempts to sell its services, reinforce its offerings, and satisfy the customer. A hot topic in marketing research is the exploration of a technology-based self-service (TBSS) encounter, where various technologies allow the customer to perform all or part of the service encounter. Examples of TBSS systems include ATMs, automated hotel checkout, banking by phone, self-scanning at retail stores, and transactions via the Internet, such as FedEx's package tracking and Charles Schwab's online brokerage services.

Marketing Professor Dan Ladik (University of Suffolk) investigated a customer's motivation to use a TBSS and compared two customer segments—one that does not require any electronic technology use for service delivery and another that relies on the Internet for service delivery. The self-service delivery system studied was one that distributes discount coupons via both the mail and the Internet. Ladik investigated whether there were differences in customer characteristics and customer *(continued)*

Statistics IN Action *(continued)*

satisfaction between the mail (nontechnology) coupon users and the Internet (TBSS) coupon users.

The data for the study were obtained from a national services firm that specializes in discount coupons. (For reasons of confidentiality, the firm is not named.) Both customers who use the nontechnology mail delivery method and customers who use the firm's Internet Web site to access the coupons were sampled. For this Statistics in Action problem, we will focus on a subset of the full data set—a sample of 440 coupon users. Using a questionnaire, several qualitative variables were measured for each user. These are listed in Table SIA9.1. The data are saved in the **COUPONS** file.

In an attempt to answer the researcher's questions, we apply the statistical methodology presented in this chapter to this data set in two Statistics in Action Revisited examples.

Statistics IN Action Revisited

- Testing Customer Category Proportions (p. 527)
- Testing Coupon Customer Characteristics (p. 538)

Table SIA9.1 Qualitative Variables Measured in the Coupon Study

Variable Name	Levels (Possible Values)
Coupon user type	Mail, Internet, or both
Gender	Male or female
Education	High school, vocational-tech/college, 4-year college degree, or graduate school
Work status	Full time, part time, not working, retired
Coupon satisfaction	Satisfied, unsatisfied, indifferent

Data Set: COUPONS

9.1 Categorical Data and the Multinomial Experiment

Recall from Section 1.5 that observations on a qualitative variable can be categorized only. For example, consider the highest level of education attained by each in a group of salespersons. Level of education is a qualitative variable, and each salesperson would fall in one and only one of the following five categories: some high school, high school diploma, some college, college degree, and graduate degree. The result of the categorization would be a count of the numbers of salespersons falling into the respective categories.

When the qualitative variable results in one of two responses (yes or no, success or failure, favor or do not favor, etc.), the data (i.e., the counts) can be analyzed using the binomial probability distribution discussed in Section 4.3. However, qualitative variables, such as level of education, that allow for more than two categories for a response are much more common, and these must be analyzed using a different method.

Qualitative data that fall in more than two categories often result from a **multinomial experiment.** The characteristics for a multinomial experiment with k outcomes are described in the box. You can see that the binomial experiment of Chapter 4 is a multinomial experiment with $k = 2$.

Properties of the Multinomial Experiment

1. The experiment consists of n identical trials.
2. There are k possible outcomes to each trial. These outcomes are called **classes, categories,** or **cells.**
3. The probabilities of the k outcomes, denoted by $p_1, p_2, \ldots, p_k$, remain the same from trial to trial, where $p_1 + p_2 + \cdots + p_k = 1$.
4. The trials are independent.
5. The random variables of interest are the **cell counts,** $n_1, n_2, \ldots, n_k$, of the number of observations that fall in each of the k classes.

Example 9.1

Identifying a Multinomial Experiment

Problem Consider the problem of determining the highest level of education attained by each of $n = 100$ salespersons at a large company. Suppose we categorize level of education into one of five categories—some high school, high school diploma, some college, college degree, and graduate degree—and count the number of the 100 salespeople that fall into each category. Is this a multinomial experiment to a reasonable degree of approximation?

Solution Checking the five properties of a multinomial experiment shown in the box, we have the following:

1. The experiment consists of $n = 100$ identical trials, where each trial is to determine the highest level of education of a salesperson.
2. There are $k = 5$ possible outcomes to each trial corresponding to the five education-level categories.
3. The probabilities of the $k = 5$ outcomes, p_1, p_2, p_3, p_4, and p_5, remain (to a reasonable degree of approximation) the same from trial to trial, where p_i represents the true probability that a salesperson attains level of education i.
4. The trials are independent (i.e., the education level attained by one salesperson does not affect the level attained by any other salesperson).
5. We are interested in the count of the number of salespeople who fall into each of the five categories. These five *cell counts* are denoted n_1, n_2, n_3, n_4, and n_5.

Thus, the properties of a multinomial experiment are satisfied.

In this chapter, we are concerned with the analysis of categorical data—specifically, the data that represent the counts for each category of a multinomial experiment. In Section 9.2, we learn how to make inferences about category probabilities for data classified according to a single qualitative (or categorical) variable. Then, in Section 9.3, we consider inferences about category probabilities for data classified according to two qualitative variables. The statistic used for these inferences is one that possesses, approximately, a chi-square distribution (see Section 6.7).

9.2 Testing Category Probabilities: One-Way Table

In this section, we consider a multinomial experiment with k outcomes that correspond to categories of a *single* qualitative variable. The results of such an experiment are summarized in a **one-way table.** The term *one-way* is used because only one variable is classified. Typically, we want to make inferences about the true proportions that occur in the k categories based on the sample information in the one-way table.

To illustrate, suppose a large supermarket chain conducts a consumer preference survey by recording the brand of bread purchased by customers in its stores. Assume the chain carries three brands of bread—two major brands (A and B) and its own store brand. The brand preferences of a random sample of 150 consumers are observed, and the number preferring each brand is tabulated; the resulting count data appear in Table 9.1.

Table 9.1 Results of Consumer Preference Survey

A	B	Store Brand
61	53	36

Note that our consumer-preference survey satisfies the properties of a multinomial experiment for the qualitative variable brand of bread. The experiment consists of randomly sampling $n = 150$ buyers from a large population of consumers containing an unknown proportion p_1 who prefer brand A, a proportion p_2 who prefer brand B, and a proportion p_3 who prefer the store brand. Each buyer represents a single trial that can result in one of three outcomes: The consumer prefers brand A, B, or the store brand with probabilities p_1, p_2, and p_3, respectively. (Assume that all consumers will have a preference.) The buyer preference of any single consumer in the sample does not affect the preference of another; consequently, the trials are independent. And, finally, you can see that the recorded data are the number of buyers in each of three consumer-preference categories. Thus, the consumer-preference survey satisfies the five properties of a multinomial experiment.

In the consumer-preference survey, and in most practical applications of the multinomial experiment, the k outcome probabilities $p_1, p_2, \ldots, p_k$ are unknown, and we typically want to use the survey data to make inferences about their values. The unknown probabilities in the consumer-preference survey are

$$p_1 = \text{Proportion of all buyers who prefer brand A}$$
$$p_2 = \text{Proportion of all buyers who prefer brand B}$$
$$p_3 = \text{Proportion of all buyers who prefer the store brand}$$

For example, to decide whether the consumers have a preference for any of the brands, we will want to test the null hypothesis that the brands of bread are equally preferred (that is, $p_1 = p_2 = p_3 = \frac{1}{3}$) against the alternative hypothesis that one brand is preferred (that is, at least one of the probabilities p_1, p_2, and p_3 exceeds $\frac{1}{3}$). Thus, we want to test

$$H_0\text{: } p_1 = p_2 = p_3 = \tfrac{1}{3} \text{ (no preference)}$$
$$H_a\text{: At least one of the proportions exceeds } \tfrac{1}{3} \text{ (a preference exists)}$$

If the null hypothesis is true and $p_1 = p_2 = p_3 = \frac{1}{3}$, the expected value (mean value) of the number of customers who prefer brand A is given by

$$E_1 = np_1 = (n)\tfrac{1}{3} = (150)\tfrac{1}{3} = 50$$

Similarly, $E_2 = E_3 = 50$ if the null hypothesis is true and no preference exists.

The following test statistic—the **chi-square test**—measures the degree of disagreement between the data and the null hypothesis:

$$\chi^2 = \frac{[n_1 - E_1]^2}{E_1} + \frac{[n_2 - E_2]^2}{E_2} + \frac{[n_3 - E_3]^2}{E_3}$$
$$= \frac{(n_1 - 50)^2}{50} + \frac{(n_2 - 50)^2}{50} + \frac{(n_3 - 50)^2}{50}$$

Note that the farther the observed numbers n_1, n_2, and n_3, are from their expected value (50), the larger χ^2 will become—that is, large values of χ^2 imply that the null hypothesis is false.

We have to know the distribution of χ^2 in repeated sampling before we can decide whether the data indicate that a preference exists. When H_0 is true, χ^2 can be shown to have (approximately) the chi-square distribution of Section 6.7. For this one-way classification, the χ^2 distribution has $(k - 1)$ degrees of freedom.* The rejection region for the consumer-preference survey for $\alpha = .05$ and $k - 1 = 3 - 1 = 2$ df is

$$\text{Rejection region: } \chi^2 > \chi^2_{.05}$$

The value of $\chi^2_{.05}$ (found in Table VI in Appendix B) is 5.99147. (See Figure 9.1.) The computed value of the test statistic is

$$\chi^2 = \frac{(n_1 - 50)^2}{50} + \frac{(n_2 - 50)^2}{50} + \frac{(n_3 - 50)^2}{50}$$
$$= \frac{(61 - 50)^2}{50} + \frac{(53 - 50)^2}{50} + \frac{(36 - 50)^2}{50} = 6.52$$

BIOGRAPHY

KARL PEARSON (1895–1980)

The Father of Statistics

While attending college, London-born Karl Pearson exhibited a wide range of interests, including mathematics, physics, religion, history, socialism, and Darwinism. After earning a law degree at Cambridge University and a PhD in political science at the University of Heidelberg (Germany), Pearson became a professor of applied mathematics at University College in London. His 1892 book, *The Grammer of Science,* illustrated his conviction that statistical data analysis lies at the foundation of all knowledge; consequently, many consider Pearson to be the "father of statistics." A few of Pearson's many contributions to the field include introducing the term *standard deviation* and its associated symbol (σ), developing the distribution of the correlation coefficient, cofounding and editing the prestigious statistics journal *Biometrika,* and (what many consider his greatest achievement) creating the first chi-square "goodness-of-fit" test. Pearson inspired his students (including his son, Egon, and William Gossett) with his wonderful lectures and enthusiasm for statistics. ■

*The derivation of the degrees of freedom for χ^2 involves the number of linear restrictions imposed on the count data. In the present case, the only constraint is that $\Sigma n_i = n$, where n (the sample size) is fixed in advance. Therefore, df $= k - 1$. For other cases, we will give the degrees of freedom for each usage of χ^2 and refer the interested reader to the references for more detail.

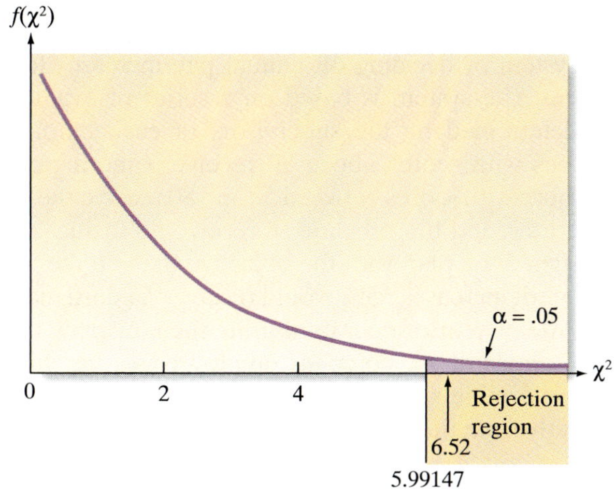

Figure 9.1
Rejection region for consumer-preference survey

Because the computed $\chi^2 = 6.52$ exceeds the critical value of 5.99147, we conclude at the $\alpha = .05$ level of significance that a consumer preference exists for one or more of the brands of bread.

Now that we have evidence to indicate that the proportions p_1, p_2, and p_3 are unequal, we can make inferences concerning their individual values using the methods of Section 5.4. [*Note:* We cannot use the methods of Section 7.4 to compare two proportions because the cell counts are dependent random variables.] The general form for a test of a hypothesis concerning multinomial probabilities is shown in the next box.

A Test of a Hypothesis about Multinomial Probabilities: One-Way Table

$$H_0\text{: } p_1 = p_{1,0}, p_2 = p_{2,0}, \ldots, p_k = p_{k,0}$$

where $p_{1,0}, p_{2,0}, \ldots, p_{k,0}$ represent the hypothesized values of the multinomial probabilities.

H_a: At least one of the multinomial probabilities does not equal its hypothesized value.

$$\textit{Test statistic: } \chi^2 = \sum \frac{[n_i - E_i]^2}{E_i}$$

where $E_i = np_{i,0}$ is the **expected cell count**—that is, the expected number of outcomes of type i assuming that H_0 is true. The total sample size is n.

$$\textit{Rejection region: } \chi^2 > \chi^2_\alpha$$

where χ^2_α has $(k - 1)$ df.

Conditions Required for a Valid χ^2 Test: One-Way Table

1. A multinomial experiment has been conducted. This is generally satisfied by taking a random sample from the population of interest.
2. The sample size n is large. This is satisfied if for every cell, the expected cell count E_i will be equal to 5 or more.*

*The assumption that all expected cell counts are at least 5 is necessary to ensure that the χ^2 approximation is appropriate. Exact methods for conducting the test of a hypothesis exist and may be used for small expected cell counts, but these methods are beyond the scope of this text. Also, some researchers relax this assumption somewhat, requiring expected cell counts to be at least 1 and no more than 20% to be less than 5.

Example 9.2
Evaluating a Firm's Merit-Increase Plan Using a One-Way χ^2 Test

Table 9.2 Distribution of Pay Increases

None	Standard	Merit
42	365	193

Data Set: PAYPLAN

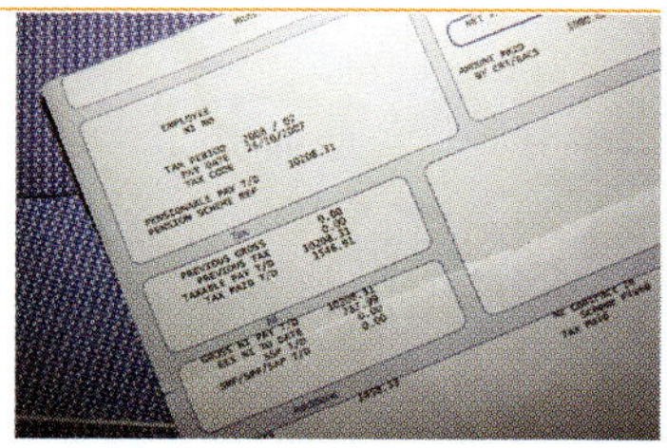

Problem A large firm has established what it hopes is an objective system of deciding on annual pay increases for its employees. The system is based on a series of evaluation scores determined by the supervisors of each employee. Employees with scores above 80 receive a merit pay increase, those with scores between 50 and 80 receive the standard increase, and those below 50 receive no increase. The firm designed the plan with the objective that, on the average, 25% of its employees would receive merit increases, 65% would receive standard increases, and 10% would receive no increase. After 1 year of operation using the new plan, the distribution of pay increases for a random sample of 600 company employees was as shown in Table 9.2. Test at the $\alpha = .01$ level to determine whether these data indicate that the distribution of pay increases differs significantly from the proportions established by the firm.

Solution Define the population proportions for the three pay increase categories to be

$$p_1 = \text{Proportion of employees who receive no pay increase}$$
$$p_2 = \text{Proportion of employees who receive a standard increase}$$
$$p_3 = \text{Proportion of employees who receive a merit increase}$$

Then the null hypothesis representing the distribution of percentages in the firm's proposed plan is

$$H_0\text{: } p_1 = .10, p_2 = .65, p_3 = .25$$

and the alternative is

H_a: At least two of the proportions differ from the firm's proposed plan.

$$\textit{Test statistic: } \chi^2 = \sum \frac{[n_i - E_i]^2}{E_i}$$

where

$$E_1 = np_{1,0} = 600(.10) = 60$$
$$E_2 = np_{2,0} = 600(.65) = 390$$
$$E_3 = np_{3,0} = 600(.25) = 150$$

Rejection region: For $\alpha = .01$ and df $= k - 1 = 2$, reject H_0 if $\chi^2 > \chi^2_{.01}$, where (from Table VI in Appendix B) $\chi^2_{.01} = 9.21034$.

We now calculate the test statistic:

$$\chi^2 = \frac{(42 - 60)^2}{60} + \frac{(365 - 390)^2}{390} + \frac{(193 - 150)^2}{150} = 19.33$$

This value exceeds the table value of χ^2 (9.21034); therefore, the data provide strong evidence ($\alpha = .01$) that the company's actual pay plan distribution differs from its proposed plan.

The χ^2 test can also be conducted using statistical software. Figure 9.2 is an Excel/DDXL printout of the analysis of the data in Table 9.2; note that the p-value of the test is reported as less than .0001. Because $\alpha = .01$ exceeds this p-value, there is sufficient evidence to reject H_0.

Look Back Note that all the expected cell counts exceed 5. Consequently, the χ^2 test is valid.

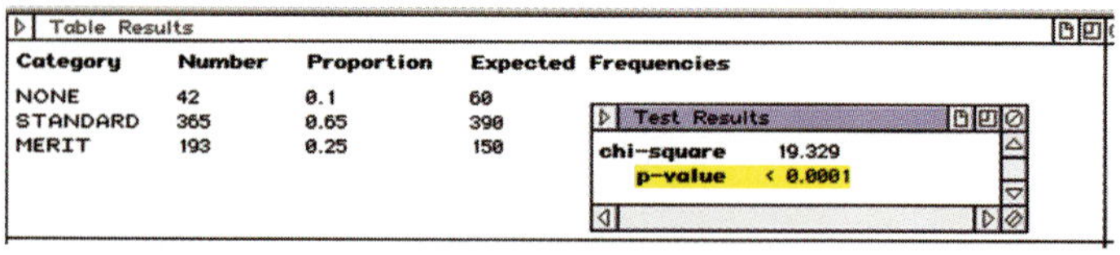

Figure 9.2
Excel/DDXL chi-square analysis of data in Table 9.2

Now Work Exercise 9.6

If we focus on one particular outcome of a multinomial experiment, we can use the methods developed in Section 5.4 for a binomial proportion to establish a confidence interval for any one of the multinomial probabilities.* For example, if we want a 95% confidence interval for the proportion of the company's employees who will receive merit increases under the new system, we calculate

$$\hat{p}_3 \pm 1.96\sigma_{\hat{p}_3} \approx \hat{p}_3 \pm 1.96\sqrt{\frac{\hat{p}_3(1-\hat{p}_3)}{n}} \quad \text{where } \hat{p}_3 = \frac{n_3}{n} = \frac{193}{600} = .32$$

$$= .32 \pm 1.96\sqrt{\frac{(.32)(1-.32)}{600}} = .32 \pm .04$$

Thus, we estimate that between 28% and 36% of the firm's employees will qualify for merit increases under the new plan. It appears that the firm will have to raise the requirements for merit increases in order to achieve the stated goal of a 25% employee qualification rate.

Statistics IN Action Revisited — Testing Category Customer Proportions

In the research on a TBSS encounter (pp. 521–522), 440 users of a firm's discount coupons were sampled and given a questionnaire to complete. One of the variables of interest to the researcher was type of coupon user. Recall (from Table SIA9.1) that the customers received the coupons in one of three ways: only through the mail (nontechnology user), only via the Internet (TBSS user), and via both the mail and Internet. What are the proportions of mail-only, Internet-only, and both users, and are these proportions statistically different?

To answer this question, we used SPSS to analyze the type of user variable in the **COUPONS** file. Figure SIA9.1 shows summary statistics and a graph to describe the three categories. From the summary table at the top of the printout, you can see that 262 (or 59.5%) of the customers are mail-only coupon users, 43 (or 9.8%) are Internet-only users, and the remainder (30.7%) use both mail and the Internet. These sample percentages are illustrated in the bar graph at the bottom of Figure SIA9.1. In this sample of customers, the majority (almost 60%) obtain their coupons strictly through the mail.

Is this sufficient evidence to indicate that the true proportions in the population of customers are different? Letting p_1, p_2, and p_3 represent the true proportions for the mail-only, Internet-only, and both categories, respectively, we tested H_0: $p_1 = p_2 = p_3 = \frac{1}{3}$ using SPSS. The printout is displayed in Figure SIA9.2. The cell frequencies and expected numbers are shown in the top table of the figure, while the chi-square test statistic (164.895) and p-value (.000) are shown in the bottom table. At any reasonably selected α-level (say, $\alpha = .01$), the small p-value indicates that there is sufficient evidence to reject the null hypothesis and conclude that the true proportions associated with the three user type categories are indeed statistically different.

Data Set: COUPONS

USER

		Frequency	Percent	Valid Percent	Cumulative Percent
Valid	Mail	262	59.5	59.5	59.5
	Net	43	9.8	9.8	69.3
	Both	135	30.7	30.7	100.0
	Total	440	100.0	100.0	

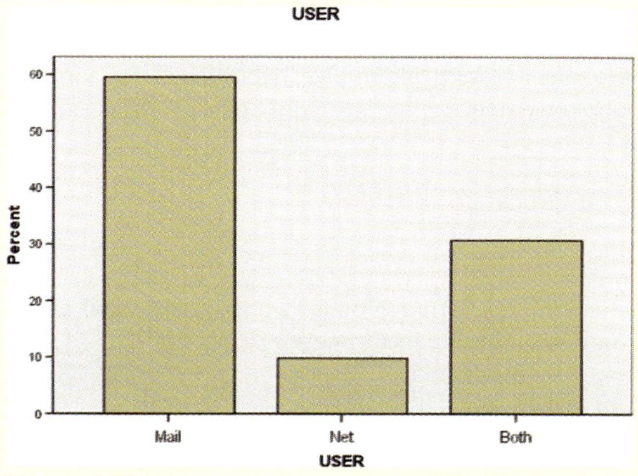

Figure SIA9.1
SPSS descriptive statistics and graph for type of coupon user

*Note that focusing on one outcome has the effect of lumping the other $(k - 1)$ outcomes into a single group. Thus, we obtain, in effect, two outcomes—or a binomial experiment.

Chi-Square Test

Frequencies

USER

	Observed N	Expected N	Residual
Mail	262	146.7	115.3
Net	43	146.7	-103.7
Both	135	146.7	-11.7
Total	440		

Test Statistics

	USER
Chi-Square[a]	164.895
df	2
Asymp. Sig.	.000

a. 0 cells (.0%) have expected frequencies less than 5. The minimum expected cell frequency is 146.7.

Figure SIA9.2
SPSS chi-square test of user type categories

Activity 9.1 Binomial versus Multinomial Experiments

In this activity, you will study the difference between binomial and multinomial experiments.

1. A television station has hired an independent research group to determine whether television viewers in the area prefer its local news program to the news programs of two other stations in the same city. Explain why a multinomial experiment would be appropriate and design a poll that satisfies the five properties of a multinomial experiment. State the null and alternative hypotheses for the corresponding χ^2-test.
2. Suppose the television station believes that a majority of local viewers prefer its news program to those of its two competitors. Explain why a binomial experiment would be appropriate to support this claim and design a poll that satisfies the five properties of a binomial experiment. State the null and alternative hypotheses for the corresponding test.
3. Generalize the situations in Exercises 1 and 2 to describe conditions under which a multinomial experiment can be rephrased as a binomial experiment. Is there any advantage in doing so? Explain.

Exercises 9.1–9.16

Learning the Mechanics

9.1 Find the rejection region for a one-dimensional χ^2 test of a null hypothesis concerning $p_1, p_2, \ldots, p_k$ if

a. $k = 3; \alpha = .05$
b. $k = 5; \alpha = .10$
c. $k = 4; \alpha = .01$

9.2 What are the characteristics of a multinomial experiment? Compare the characteristics to those of a binomial experiment.

9.3 What conditions must n satisfy to make the χ^2 test valid?

9.4 A multinomial experiment with $k = 3$ cells and $n = 320$ produced the data shown in the following one-way table. Do these data provide sufficient evidence to contradict the null hypothesis that $p_1 = .25$, $p_2 = .25$, and $p_3 = .50$? Test using $\alpha = .05$.

	Cell		
	1	2	3
n_i	78	60	182

9.5 A multinomial experiment with $k = 4$ cells and $n = 205$ produced the data shown in the one-way table below.

	Cell			
	1	2	3	4
n_i	43	56	59	47

a. Do these data provide sufficient evidence to conclude that the multinomial probabilities differ? Test using $\alpha = .05$.

b. What are the Type I and Type II errors associated with the test of part **a**?

c. Construct a 95% confidence interval for the multinomial probability associated with cell 3.

Applying the Concepts—Basic

9.6 **Location of major sports venues.** There has been a recent NW trend for professional sports franchises in Major League Baseball (MLB), the National Football League (NFL), the National Basketball Association (NBA), and the National Hockey League (NHL) to build new stadiums and ballparks in urban, downtown venues. An article in *Professional Geographer* (Feb. 2000) investigated whether there has been a significant suburban-to-urban shift in the location of major sport facilities. In 1985, 40% of all major sport facilities were located downtown, 30% in central city, and 30% in suburban areas. In contrast, of the 113 major sports franchises that existed in 1997, 58 were built downtown, 26 in a central city, and 29 in a suburban area.

a. Describe the qualitative variable of interest in the study. Give the levels (categories) associated with the variable.

b. Give the null hypothesis for a test to determine whether the proportions of major sports facilities in downtown, central city, and suburban areas in 1997 are the same as in 1985.

c. If the null hypothesis, part **b,** is true, how many of the 113 sports facilities in 1997 would you expect to be located in downtown, central city, and suburban areas, respectively?

d. Find the value of the chi-square statistic for testing the null hypothesis, part **b.**

e. Find the (approximate) p-value of the test and give the appropriate conclusion in the words of the problem. Assume $\alpha = .05$.

9.7 **"Made in the USA" survey.** Refer to the *Journal of Global Business* (Spring 2002) study of what "Made in the USA" on product labels means to the typical consumer, Exercise 2.8 (pp. 39–40). Recall that 106 shoppers participated in the survey. Their responses, given as a percentage of U.S. labor and materials in four categories, are summarized in the table. Suppose a consumer advocate group claims that half of all consumers believe that "Made in the USA" means "100%" of labor and materials are produced in the United States, one-fourth believe that "75% to 99%" are produced in the United States, one-fifth believe that "50% to 74%" are produced in the United States, and 5% believe that "less than 50%" are produced in the United States.

Response to "Made in the USA"	Number of shoppers
100%	64
75% to 99%	20
50% to 74%	18
Less than 50%	4

Source: "'Made in the USA': Consumer perceptions, deception and policy alternatives," *Journal of Global Business,* Vol. 13, No. 24, Spring 2002 (Table 3).

a. Describe the qualitative variable of interest in the study. Give the levels (categories) associated with the variable.

b. What are the values of p_1, p_2, p_3, and p_4, the probabilities associated with the four response categories hypothesized by the consumer advocate group?

c. Give the null and alternative hypotheses for testing the consumer advocate group's claim.

d. Compute the test statistic for testing the hypotheses, part **c.**

e. Find the rejection region of the test at $\alpha = .10$.

f. State the conclusion in the words of the problem.

g. Find and interpret a 90% confidence interval for the true proportion of consumers who believe "Made in the USA" means "100%" of labor and materials are produced in the United States.

9.8 **Rankings of MBA programs.** Refer to the *Business Ethics* (Fall 2005) rankings of master in business administration (MBA) programs worldwide, Exercise 2.14 (p. 41). Recall that each of 30 business schools was rated according to student exposure to social and environmental issues in the classroom. Ratings ranged from 1 star (lowest rated group) to 5 stars (highest rated group). A summary of the star ratings assigned to the 30 MBA programs is reproduced in the table.

Criteria	5 stars	4 stars	3 stars	2 stars	1 star	Total
Student Exposure	2	9	14	5	0	30

Source: Adapted from Stewardship at MBA programs. Biello, D. "MBA programs for social and environmental stewardship," *Business Ethics,* Fall 2005, p. 25. Used with permission of the Aspen Institute.

a. Identify the categorical variable (and its levels) measured in this study.

b. How many of the sampled MBA programs would you expect to observe in each star rating category if there are no differences in the category proportions in the population of all MBA programs?

c. Specify the null and alternative hypotheses for testing whether there are differences in the star rating category proportions in the population of all MBA programs.

d. Calculate the χ^2 test statistic for testing the hypotheses in part **c.**

e. Give the rejection region for the test using $\alpha = .05$.

f. Use the results, parts **d** and **e,** to make the appropriate conclusion.

g. Find and interpret a 95% confidence interval for the proportion of all MBA programs that are ranked in the three-star category.

9.9 **Survey on giving and volunteering.** The *National Tax Journal* (Dec. 2001) published a study of charitable givers based on data collected from the Independent Sector Survey on Giving and Volunteering. A total of 1,072 charitable givers reported that their charitable contributions were motivated by tax considerations. The number of these 1,072 givers in each of 10 household income categories (saved in the **GIVERS** file) is shown in the table on the next page.

Household Income Group	Number of Charitable Givers
Under $10,000	42
$10,000–$20,000	93
$20,000–$30,000	99
$30,000–$40,000	153
$40,000–$50,000	91
$50,000–$60,000	114
$60,000–$70,000	157
$70,000–$80,000	101
$80,000–$100,000	95
Over $100,000	127

Source: Tiehen, L. "Tax policy and charitable contributions of money," *National Tax Journal,* Vol. 54, No. 4, Dec. 2001, p. 717 (adapted from Table 5).

a. If the true proportions of charitable givers in each household income group are the same, how many of the 1,072 sampled givers would you expect to find in each income category?

b. Give the null hypothesis for testing whether the true proportions of charitable givers in each household income group are the same.

c. Compute the chi-square test statistic for testing the null hypothesis, part **b.**

d. Find the rejection region for the test if $\alpha = .10$.

e. Give the appropriate conclusion for the test in the words of the problem.

Applying the Concepts—Intermediate

9.10 Top Internet search engines. Nielsen/NetRatings is a global leader in Internet media and market research. In May 2006, the firm reported on the "search" shares (i.e., percentage of all Internet searches) for the most popular search engines available on the Web. Google accounted for 50% of all searches, Yahoo! for 22%, MSN for 11%, and all other search engines for 17%. Suppose that in a random sample of 1,000 recent Internet searches, 487 used Google, 245 used Yahoo!, 121 used MSN, and 147 used another search engine.

a. Do the sample data disagree with the percentages reported by Nielsen/NetRatings? Test using $\alpha = .05$.

b. Find and interpret a 95% confidence interval for the percentage of all Internet searches that use the Google search engine.

9.11 Who is to blame for rising health-care costs? Refer to Exercise 2.4 (pp. 38–39) and *The Harris Poll* (Oct. 28, 2008) on who is to blame for the rising costs of health care. Recall that a nationwide survey of 2,119 U.S. adults answered the question "When you think of the rising costs of health care, who do you think is most responsible?" The responses are summarized in the next table. One theory is that 50% of adults blame insurance companies, 10% blame pharmaceutical companies, 10% blame government, 10% blame hospitals, 10% blame physicians, 5% blame some other entity, and 5% are unsure.

a. Explain why the data come from a multinomial experiment.

b. Specify the null hypothesis for a test of the theory.

c. Use statistical software to conduct the test using $\alpha = .01$. What do you conclude?

Most Responsible for Rising Health-Care Costs	Number Responding
Insurance companies	869
Pharmaceutical companies	339
Government	338
Hospitals	127
Physicians	85
Other	128
Not sure	233
Total	2,119

9.12 Attitudes toward top corporate managers. Recent scandals involving large U.S. corporations (e.g., Enron, WorldCom, and Adelphia) apparently have had a major impact on the public's attitude toward business managers. In a Harris Poll administered immediately after the Enron scandal, a national sample of 2,023 adults were asked to agree or disagree with the following statement: "Top company managers have become rich at the expense of ordinary workers" (*The Harris Poll,* #55, Oct. 18, 2002). The response categories (and number of respondents in each) were strongly agree (1,173), somewhat agree (587), somewhat disagree (182), and strongly disagree (81). Suppose that prior to the Enron scandal the percentages of all U.S. adults falling into the four response categories were 45%, 35%, 15%, and 5%, respectively. Is there evidence to infer that the percentages of all U.S. adults falling into the four response categories changed after the Enron scandal? Test using $\alpha = .01$.

9.13 Management system failures. Refer to the *Process Safety Progress* (Dec. 2004) and U.S. Chemical Safety and Hazard Investigation Board study of industrial accidents caused by management system failures, Exercise 2.6 (p. 39). The table below (saved in the **MSFAIL** file) gives a breakdown of the root causes of a sample of 83 incidents. Are there significant differences in the percentage of incidents in the four cause categories? Test using $\alpha = .05$.

Management System Cause	Number of Incidents
Engineering & Design	27
Procedures & Practices	24
Management & Oversight	22
Training & Communication	10
Total	83

Source: Blair, A. S. "Management system failures identified in incidents investigated by the U.S. Chemical Safety and Hazard Investigation Board," *Process Safety Progress,* Vol. 23, No. 4, Dec. 2004, pp. 232–236 (Table 1). Reprinted with permission of John Wiley & Sons, Inc.

9.14 Cell phone user survey. If you subscribe to a cell phone plan, how many different cell phone numbers do you own? This was one question of interest in *Public Opinion Quarterly* (Vol. 70, No. 5, 2006). According to the Current Population Survey (CPS) Cell Phone Supplement, 51% of cell phone plans have only one cell number, 37% have two numbers, 9% have three numbers, and 3% have four or more numbers. An independent survey of 943 randomly selected cell phone users found that 473 pay for only one number, 334 pay for two numbers, 106 pay for three numbers, and 30 pay for four or more numbers. Conduct a test to determine if the data from the independent survey contradict the percentages reported by the CPS Cell Phone Supplement. Use $\alpha = .10$.

Applying the Concepts—Advanced

9.15 Overloading in the trucking industry. Although illegal, overloading is common in the trucking industry. A state highway planning agency (Minnesota Department of Transportation) monitored the movements of overweight trucks on an interstate highway using an unmanned, computerized scale that is built into the highway. Unknown to the truckers, the scale weighed their vehicles as they passed over it. Each day's proportion of one week's total truck traffic (five-axle tractor truck semitrailers) is shown in the first table below. During the same week, the number of overweight trucks per day is given in the second table. This information is saved in the **OVERLOAD** file. The planning agency would like to know whether the number of overweight trucks per week is distributed over the seven days of the week in direct proportion to the volume of truck traffic. Test using $\alpha = .05$.

Monday	.191
Tuesday	.198
Wednesday	.187
Thursday	.180
Friday	.155
Saturday	.043
Sunday	.046

Monday	90
Tuesday	82
Wednesday	72
Thursday	70
Friday	51
Saturday	18
Sunday	31

9.16 Political representation of religious groups. Do those elected to the U.S. House of Representatives really "represent" their constituents demographically? This was a question of interest in *Chance* (Summer 2002). One of several demographics studied was religious affiliation. The accompanying table (saved in the **USHOUSE** file) gives the proportion of the U.S. population for several religions, as well as the number of the 435 seats in the House of Representatives that are affiliated with that religion. Give your opinion on whether or not the House of Representatives is statistically representative of the religious affiliations of their constituents in the United States.

Religion	Proportion of U.S. Population	Number of Seats in House
Catholic	.28	117
Methodist	.04	61
Jewish	.02	30
Other	.66	227
Totals	1.00	435

9.3 Testing Category Probabilities: Two-Way (Contingency) Table

In Section 9.1, we introduced the multinomial probability distribution and considered data classified according to a single qualitative criterion. We now consider multinomial experiments in which the data are classified according to two criteria—that is, *classification with respect to two qualitative factors.*

For example, consider a study published in the *Journal of Marketing* on the impact of using celebrities in television advertisements. The researchers investigated the relationship between gender of a viewer and the viewer's brand awareness. Three hundred TV viewers were asked to identify products advertised by male celebrity spokespersons. The data are summarized in the **two-way table** shown in Table 9.3. This table is called a **contingency table;** it presents multinomial count data classified on two scales, or **dimensions, of classification**—namely, gender of viewer and brand awareness.

The symbols representing the cell counts for the multinomial experiment in Table 9.3 are shown in Table 9.4a; and the corresponding cell, row, and column

Table 9.3 Contingency Table for Marketing Example

		Gender		
		Male	Female	Totals
Brand Awareness	Could Identify Product	95	41	136
	Could Not Identify Product	55	109	164
	Totals	150	150	300

Table 9.4a Observed Counts for Contingency Table 9.3

		Gender		
		Male	Female	Totals
Brand Awareness	Could Identify Product	n_{11}	n_{12}	R_1
	Could Not Identify Product	n_{21}	n_{22}	R_2
	Totals	C_1	C_2	n

Table 9.4b Probabilities for Contingency Table 9.3

		Gender		
		Male	Female	Totals
Brand Awareness	Could Identify Product	p_{11}	p_{12}	p_{r1}
	Could Not Identify Product	p_{21}	p_{22}	p_{r2}
	Totals	p_{c1}	p_{c2}	1

probabilities are shown in Table 9.4b. Thus, n_{11} represents the number of viewers who are male and could identify the brand, and p_{11} represents the corresponding cell probability. Note the symbols for the row and column totals and also the symbols for the probability totals. The latter are called **marginal probabilities** for each row and column. The marginal probability p_{r1} is the probability that a TV viewer identifies the product; the marginal probability p_{c1} is the probability that the TV viewer is male. Thus,

$$p_{r1} = p_{11} + p_{12} \quad \text{and} \quad p_{c1} = p_{11} + p_{21}$$

Thus, we can see that this really is a multinomial experiment with a total of 300 trials, (2)(2) = 4 cells or possible outcomes, and probabilities for each cell as shown in Table 9.4b. If the 300 TV viewers are randomly chosen, the trials are considered independent, and the probabilities are viewed as remaining constant from trial to trial.

Suppose we want to know whether the two classifications, gender and brand awareness, are dependent—that is, if we know the gender of the TV viewer, does that information give us a clue about the viewer's brand awareness?

In a probabilistic sense, we know (Chapter 3) that independence of events A and B implies $P(\text{AB}) = P(\text{A})P(\text{B})$. Similarly, in the contingency table analysis, if the **two classifications are independent,** the probability that an item is classified in any particular cell of the table is the product of the corresponding marginal probabilities. Thus, under the hypothesis of independence, in Table 9.4b, we must have

$$p_{11} = p_{r1}p_{c1}$$
$$p_{12} = p_{r1}p_{c2}$$

and so forth.

To test the hypothesis of independence, we use the same reasoning employed in the one-dimensional tests of Section 9.2. First, we calculate the *expected,* or *mean, count in each cell,* assuming that the null hypothesis of independence is true. We do this by noting that the expected count in a cell of the table is just the total number of multinomial trials, n, times the cell probability. Recall that n_{ij} represents the **observed count** in the cell located in the ith row and jth column. Then the expected cell count for the upper left-hand cell (first row, first column) is

$$E_{11} = np_{11}$$

or, when the null hypothesis (the classifications are independent) is true,

$$E_{11} = np_{r1}p_{c1}$$

Because these true probabilities are not known, we estimate p_{r1} and p_{c1} by the proportions $\hat{p}_{r1} = R_1/n$ and $\hat{p}_{c1} = C_1/n$, where R_1 and C_1 represent the totals for row 1 and column 1, respectively. Thus, the estimate of the expected value E_{11} is

$$\hat{E}_{11} = n\left(\frac{R_1}{n}\right)\left(\frac{C_1}{n}\right) = \frac{R_1C_1}{n}$$

Similarly, for each i, j,

$$\hat{E}_{ij} = \frac{\text{(Row total)(Column total)}}{\text{Total sample size}}$$

Thus,

$$\hat{E}_{12} = \frac{R_1C_1}{n}$$

$$\hat{E}_{21} = \frac{R_2C_1}{n}$$

$$\hat{E}_{22} = \frac{R_2C_2}{n}$$

Finding Expected Cell Counts for a Two-Way Contingency Table

The estimate of the expected number of observations falling into the cell in row i and column j is given by

$$\hat{E}_{ij} = \frac{R_iC_j}{n}$$

where R_i = total for row i, C_j = total for column j, and n = sample size.

Using the data in Table 9.3, we find

$$\hat{E}_{11} = \frac{R_1C_1}{n} = \frac{(136)(150)}{300} = 68$$

$$\hat{E}_{12} = \frac{R_1C_2}{n} = \frac{(136)(150)}{300} = 68$$

$$\hat{E}_{21} = \frac{R_2C_1}{n} = \frac{(164)(150)}{300} = 82$$

$$\hat{E}_{22} = \frac{R_2C_2}{n} = \frac{(164)(150)}{300} = 82$$

The observed data and the estimated expected values (in parentheses) are shown in Table 9.5.

Table 9.5 Observed and Estimated Expected (in Parentheses) Counts

		Gender		
		Male	Female	Totals
Brand Awareness	Could Identify Product	95(68)	41(68)	136
	Could Not Identify Product	55(82)	109(82)	164
	Totals	150	150	300

We now use the χ^2 statistic to compare the observed and expected (estimated) counts in each cell of the contingency table:

$$\chi^2 = \frac{[n_{11} - \hat{E}_{11}]^2}{\hat{E}_{11}} + \frac{[n_{12} - \hat{E}_{12}]^2}{\hat{E}_{12}} + \frac{[n_{21} - \hat{E}_{21}]^2}{\hat{E}_{21}} + \frac{[n_{22} - \hat{E}_{22}]^2}{\hat{E}_{22}}$$

$$= \sum \frac{[n_{ij} - \hat{E}_{ij}]^2}{\hat{E}_{ij}}$$

Note: The use of Σ in the context of a contingency table analysis refers to a sum over all cells in the table.

Substituting the data of Table 9.5 into this expression, we get

$$\chi^2 = \frac{(95 - 68)^2}{68} + \frac{(41 - 68)^2}{68} + \frac{(55 - 82)^2}{82} + \frac{(109 - 82)^2}{82} = 39.22$$

Large values of χ^2 imply that the observed counts do not closely agree, and hence, the hypothesis of independence is false. To determine how large χ^2 must be before it is too large to be attributed to chance, we make use of the fact that the sampling distribution of χ^2 is approximately an χ^2 probability distribution when the classifications are independent.

When testing the null hypothesis of independence in a two-way contingency table, the appropriate degrees of freedom will be $(r - 1)(c - 1)$, where r is the number of rows and c is the number of columns in the table.

For the brand awareness example, the degrees of freedom for χ^2 are $(r - 1)(c - 1) = (2 \quad 1)(2 - 1) = 1$. Then, for $\alpha = .05$, we reject the hypothesis of independence when

$$\chi^2 > \chi^2_{.05} = 3.8146$$

Because the computed $\chi^2 = 39.22$ exceeds the value 3.84146, we conclude that viewer gender and brand awareness are dependent events.

The pattern of **dependence** can be seen more clearly by expressing the data as percentages. We first select one of the two classifications to be used as the base variable. In the preceding example, suppose we select gender of the TV viewer as the classificatory variable to be the base. Next, we represent the responses for each level of the second categorical variable (brand awareness in our example) as a percentage of the subtotal for the base variable. For example, from Table 9.5, we convert the response for males who identify the brand (95) to a percentage of the total number of male viewers (150)—that is,

$$(95/150)100\% = 63.3\%$$

The conversions of all Table 9.5 entries are similarly computed, and the values are shown in Table 9.6. The value shown at the right of each row is the row's total expressed as a percentage of the total number of responses in the entire table. Thus, the percentage of TV viewers who identify the product is $\left(\frac{136}{300}\right)100\% = 45.3\%$ (rounded to the nearest 10th of a percent).

If the gender and brand awareness variables are independent, then the percentages in the cells of the table are expected to be approximately equal to the corresponding row percentages. Thus, we would expect the percentages who identify the brand for each gender to be approximately 45% if the two variables are independent. The extent to which each gender's percentage departs from this value determines the dependence of

Table 9.6 Percentage of TV Viewers Who Identify Brand, by Gender

		Gender		
		Male	Female	Totals
Brand Awareness	Could Identify Product	63.3	27.3	45.3
	Could Not Identify Product	36.7	72.7	54.7
	Totals	100	100	100

the two classifications, with greater variability of the row percentages meaning a greater degree of dependence. A plot of the percentages helps summarize the observed pattern. In the SPSS bar graph in Figure 9.3, we show the gender of the viewer (the base variable) on the horizontal axis, and the percentages of TV viewers who identify the brand on the vertical axis. The "expected" percentage under the assumption of independence is shown as a dashed horizontal line.

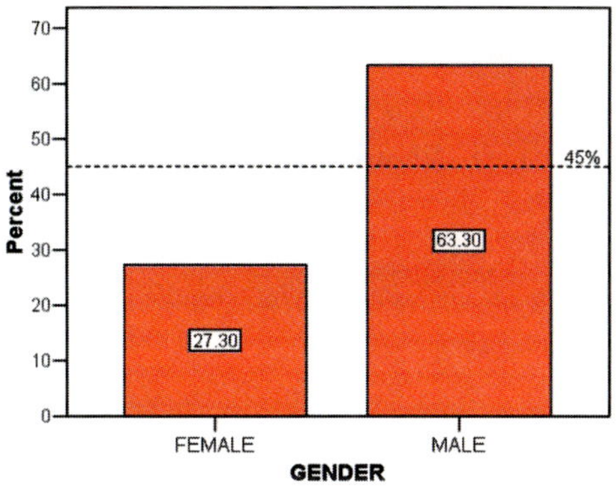

Figure 9.3
SPSS bar graph showing percentage of viewers who identified the TV product

Figure 9.3 clearly indicates the reason that the test resulted in the conclusion that the two classifications in the contingency table are dependent. The percentage of male TV viewers who identify the brand promoted by a male celebrity is more than twice as high as the percentage of female TV viewers who identify the brand.

Statistical measures of the degree of dependence and procedures for making comparisons of pairs of levels for classifications are available. They are beyond the scope of this text but can be found in the references. We will, however, use descriptive summaries such as Figure 9.3 to examine the degree of dependence exhibited by the sample data.

The general form of a two-way contingency table containing r rows and c columns (called an $r \times c$ contingency table) is shown in Table 9.7. Note that the observed count in the (ij) cell is denoted by n_{ij}, the ith row total is R_i, the jth column total is C_j, and the total sample size is n. Using this notation, we give the general form of the contingency table test for independent classifications in the next box.

Table 9.7 General $r \times c$ Contingency Table

	Column	1	2	$\cdots$	c	Row Totals
Row	1	n_{11}	n_{12}	$\cdots$	n_{1c}	R_1
	2	n_{21}	n_{22}	$\cdots$	n_{2c}	R_2
	$\vdots$	$\vdots$	$\vdots$		$\vdots$	$\vdots$
	r	n_{r1}	n_{r2}	$\cdots$	n_{rc}	R_r
	Column Totals	C_1	C_2	$\cdots$	C_c	n

General Form of a Contingency Table Analysis: χ^2-Test for Independence

H_0: The two classifications are independent.

H_a: The two classifications are dependent.

Test statistic: $\chi^2 = \sum \frac{[n_{ij} - \hat{E}_{ij}]^2}{\hat{E}_{ij}}$

where $\hat{E}_{ij} = \frac{R_i C_j}{n}$

Rejection region: $\chi^2 > \chi^2_\alpha$, where χ^2_α has $(r - 1)(c - 1)$ df.

Conditions Required for a Valid χ^2-Test: Contingency Table

1. The n observed counts are a random sample from the population of interest. We may then consider this to be a multinomial experiment with $r \times c$ possible outcomes.
2. The sample size, n, will be large enough so that, for every cell, the estimated expected count, $\hat{E}_{ij}$, will be equal to 5 or more.

Example 9.3

Conducting a Two-Way Analysis for a Brokerage Firm

Problem A large brokerage firm wants to determine whether the service it provides to affluent clients differs from the service it provides to lower-income clients. A sample of 500 clients is selected, and each client is asked to rate his or her broker. The results are shown in Table 9.8.

a. Test to determine whether there is evidence that broker rating and customer income are dependent. Use $\alpha = .10$.

b. Plot the data and describe the patterns revealed. Is the result of the test supported by the plot?

Solution

a. The first step is to obtain estimated expected cell frequencies under the assumption that the classifications are independent. The Minitab printout of the analysis of Table 9.8 is displayed in Figure 9.4. Each cell in Figure 9.4 contains the observed (top) and expected (bottom) frequency in that cell. Note that $\hat{E}_{11}$, the estimated expected

Table 9.8 Survey Results (Observed Clients), Example 9.3

		Client's Income			
		Under $30,000	$30,000–$60,000	Over $60,000	Totals
Broker Rating	Outstanding	48	64	41	153
	Average	98	120	50	268
	Poor	30	33	16	79
	Totals	176	217	107	500

Data Set: BROKERAGE

```
Rows: RATING   Columns: INCOME

            1:UND30K  2:30K-60K  3:OVR60K     All

1:OUTSTAN        48         64        41     153
              27.27      29.49     38.32   30.60
              53.86      66.40     32.74  153.00

2:AVERAGE        98        120        50     268
              55.68      55.30     46.73   53.60
              94.34     116.31     57.35  268.00

3:POOR           30         33        16      79
              17.05      15.21     14.95   15.80
              27.81      34.29     16.91   79.00

All             176        217       107     500
             100.00     100.00    100.00  100.00
             176.00     217.00    107.00  500.00

Cell Contents:      Count
                    % of Column
                    Expected count

Pearson Chi-Square = 4.278, DF = 4, P-Value = 0.370
Likelihood Ratio Chi-Square = 4.184, DF = 4, P-Value = 0.382
```

Figure 9.4
Minitab contingency table analysis for brokerage data

count for the Outstanding, Under $30,000 cell is 53.86. Similarly, the estimated expected count for the Outstanding, $30,000–$60,000 cell is $\hat{E}_{12} = 66.40$. Because all the estimated expected cell frequencies are greater than 5, the χ^2 approximation for the test statistic is appropriate. Assuming the clients chosen were randomly selected from all clients of the brokerage firm, the characteristics of the multinomial probability distribution are satisfied. The null and alternative hypotheses we want to test are

H_0: The rating a client gives his or her broker is independent of client's income.

H_a: Broker rating and client income are dependent.

The test statistic, $\chi^2 = 4.278$, which is highlighted at the bottom of the printout, is the observed significance level (p-value) of the test. Because $\alpha = .10$ is less than $p = .370$, we fail to reject H_0. This survey does not support the firm's alternative hypothesis that affluent clients receive different broker service than lower-income clients.

b. The broker rating frequencies are expressed as percentages of income category frequencies and highlighted in the Minitab printout, Figure 9.4. The expected percentages under the assumption of independence are shown in the "All" column of the printout. A Minitab side-by-side bar graph of the data is shown in Figure 9.5. Note that the response percentages deviate only slightly from those expected under the assumption of independence, supporting the result of the test in part **a**—that is, neither the descriptive plot nor the statistical test provides evidence that the rating given for broker services depends on (varies with) the customer's income.

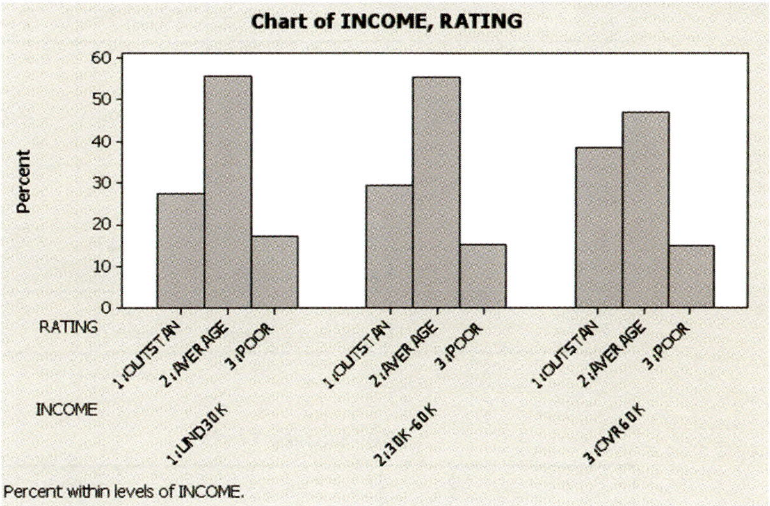

Figure 9.5
Minitab side-by-side bar graph for brokerage data

Now Work Exercise 9.22

Statistics IN Action Revisited — Testing Coupon Customer Characteristics

In his study of a firm's discount coupon users (pp. 521–522), marketing Professor Dan Ladik wanted to know whether there are differences in customer characteristics (i.e., gender, education, work status, and satisfaction) among the three types of coupon users: mail-only (nontechnology) users, Internet-only (TBSS) users, and users of both mail and Internet coupons. One approach to analyzing these data is to determine whether each of the four customer characteristic variables is related to coupon user type. Because all the variables measured on the sample of 440 coupon users are qualitative, a contingency table analysis is appropriate.

Figures SIA9.3–SIA9.6 show the SPSS contingency table analyses relating each customer characteristic to coupon user type. The p-values for the chi-square tests for the variables gender, education, satisfaction, and work status are .033, .361, .000, and .069, respectively. If we conduct each test at $\alpha = .01$

GENDER * USER Crosstabulation

			USER			
			Mail	Net	Both	Total
GENDER	Male	Count	84	7	31	122
		Expected Count	72.6	11.9	37.4	122.0
	Female	Count	178	36	104	318
		Expected Count	189.4	31.1	97.6	318.0
Total		Count	262	43	135	440
		Expected Count	262.0	43.0	135.0	440.0

Chi-Square Tests

	Value	df	Asymp. Sig. (2-sided)
Pearson Chi-Square	6.797[a]	2	.033
Likelihood Ratio	7.105	2	.029
Linear-by-Linear Association	4.371	1	.037
N of Valid Cases	440		

a. 0 cells (.0%) have expected count less than 5. The minimum expected count is 11.92.

Figure SIA9.3
SPSS contingency table analysis—gender versus user type

EDUC * USER Crosstabulation

			USER			
			Mail	Net	Both	Total
EDUC	HS	Count	34	7	19	60
		Expected Count	35.7	5.9	18.4	60.0
	VT/COLL	Count	96	20	62	178
		Expected Count	106.0	17.4	54.6	178.0
	COLL4	Count	85	9	38	132
		Expected Count	78.6	12.9	40.5	132.0
	GRAD	Count	47	7	16	70
		Expected Count	41.7	6.8	21.5	70.0
Total		Count	262	43	135	440
		Expected Count	262.0	43.0	135.0	440.0

Chi-Square Tests

	Value	df	Asymp. Sig. (2-sided)
Pearson Chi-Square	6.587[a]	6	.361
Likelihood Ratio	6.786	6	.341
Linear-by-Linear Association	3.546	1	.060
N of Valid Cases	440		

a. 0 cells (.0%) have expected count less than 5. The minimum expected count is 5.86.

Figure SIA9.4
SPSS contingency table analysis—education versus user type

(we purposely chose a small α to minimize the chance of making a Type I error), the only significant result is for customer satisfaction—that is, the data provide evidence to indicate that the level of customer satisfaction depends on the type of coupon user; however, there is not sufficient evidence to say that any of the other customer characteristics (gender, education, or work status) are related to the type of coupon user.

The column percentages highlighted in the contingency table of Figure SIA9.5 reveal the differences in satisfaction levels of the three user types. The percentages of satisfied customers for mail-only and Internet-only coupon users are 65.6% and 69.8%, respectively. However, for users of both mail and the Internet, 91.1% are satisfied. This information was used by the coupon firm to develop a marketing strategy aimed at mail-only and Internet-only coupon users.

COUPSAT * USER Crosstabulation

			USER			Total
			Mail	Net	Both	
COUPSAT	Satisfied	Count	172	30	123	325
		Expected Count	193.5	31.8	99.7	325.0
		% within USER	65.6%	69.8%	91.1%	73.9%
	Indiff	Count	62	9	9	80
		Expected Count	47.6	7.8	24.5	80.0
		% within USER	23.7%	20.9%	6.7%	18.2%
	Unsatis	Count	28	4	3	35
		Expected Count	20.8	3.4	10.7	35.0
		% within USER	10.7%	9.3%	2.2%	8.0%
Total		Count	262	43	135	440
		Expected Count	262.0	43.0	135.0	440.0
		% within USER	100.0%	100.0%	100.0%	100.0%

Chi-Square Tests

	Value	df	Asymp. Sig. (2-sided)
Pearson Chi-Square	30.418[a]	4	.000
Likelihood Ratio	34.934	4	.000
Linear-by-Linear Association	25.717	1	.000
N of Valid Cases	440		

a. 1 cells (11.1%) have expected count less than 5. The minimum expected count is 3.42.

Figure SIA9.5
SPSS contingency table analysis—education versus user type

WORK * USER Crosstabulation

			USER			Total
			Mail	Net	Both	
WORK	FULL	Count	148	29	90	267
		Expected Count	159.0	26.1	81.9	267.0
	PART	Count	31	8	13	52
		Expected Count	31.0	5.1	16.0	52.0
	NONE	Count	31	3	17	51
		Expected Count	30.4	5.0	15.6	51.0
	RETIRED	Count	52	3	15	70
		Expected Count	41.7	6.8	21.5	70.0
Total		Count	262	43	135	440
		Expected Count	262.0	43.0	135.0	440.0

Chi-Square Tests

	Value	df	Asymp. Sig. (2-sided)
Pearson Chi-Square	11.687[a]	6	.069
Likelihood Ratio	12.208	6	.057
Linear-by-Linear Association	5.619	1	.018
N of Valid Cases	440		

a. 1 cells (8.3%) have expected count less than 5. The minimum expected count is 4.98.

Figure SIA9.6
SPSS contingency table analysis—work status versus user type

Activity 9.2 Contingency Tables

In this Activity, you will revisit Activity 3.1, *Exit Polls* (pp. 150–151). For convenience, the table shown in that activity is repeated here.

2008 Presidential Election, Vote by Gender

	Obama	McCain	Other
Male (47%)	49%	48%	3%
Female (53%)	56%	43%	1%

Source: www.cnn.com

1. Determine whether the table above and the similar tables that you found for Activity 3.1 are contingency tables. If not, do you have enough information to create a contingency table for the data? If you need more information, state specifically what information you need.
2. Choose one of your examples from the previous activity if it contains a contingency table or enough information to create one or use the Internet or some other source to find a new example with a contingency table given. Determine whether the conditions for a valid χ^2-test are met. If not, choose a different example where the conditions are met.
3. Perform an χ^2-test for independence for the example chosen in Exercise 2. Are the results what you would expect in the given situation? Explain.

Exercises 9.17–9.33

Learning the Mechanics

9.17 Find the rejection region for a test of independence of two classifications where the contingency table contains r rows and c columns.

a. $r = 5, c = 5, \alpha = .05$
b. $r = 3, c = 6, \alpha = .10$
c. $r = 2, c = 3, \alpha = .01$

9.18 Consider the 2×3 (i.e., $r = 2$ and $c = 3$) contingency table shown below and saved in the **LM9_18** file.

		Column		
		1	2	3
Row	1	9	34	53
	2	16	30	25

a. Specify the null and alternative hypotheses that should be used in testing the independence of the row and column classifications.
b. Specify the test statistic and the rejection region that should be used in conducting the hypothesis test of part **a.** Use $\alpha = .01$.
c. Assuming the row classification and the column classification are independent, find estimates for the expected cell counts.
d. Conduct the hypothesis test of part **a.** Interpret your result.

9.19 Refer to Exercise 9.18.

a. Convert the frequency responses to percentages by calculating the percentage of each column total falling in each row. Also convert the row totals to percentages of the total number of responses. Display the percentages in a table.
b. Create a bar graph with row 1 percentage on the vertical axis and column number on the horizontal axis. Show the row 1 total percentage as a horizontal line on the graph.
c. What pattern do you expect to see if the rows and columns are independent? Does the plot support the result of the test of independence in Exercise 9.18?

9.20 Test the null hypothesis of independence of the two classifications, A and B, in the 3×3 contingency table shown in the next column. Test using $\alpha = .05$. The data are saved in the **LM9_20** file.

		B		
		B_1	B_2	B_3
A	A_1	40	72	42
	A_2	63	53	70
	A_3	31	38	30

9.21 Refer to Exercise 9.20.

a. Convert the responses to percentages by calculating the percentage of each B class total falling into each A classification.
b. Calculate the percentage of the total number of responses that constitute each of the A classification totals.
c. Create a bar graph with row A_1 percentage on the vertical axis and B classification on the horizontal axis. Does the graph support the result of the test of hypothesis in Exercise 9.20? Explain.
d. Repeat part **c** for the row A_2 percentages.
e. Repeat part **c** for the row A_3 percentages.

Applying the Concepts—Basic

9.22 NW **Stereotyping in deceptive and authentic news stories.** Major newspapers lose their credibility (and subscribers) when they are found to have published deceptive or misleading news stories. In *Journalism and Mass Communication Quarterly* (Summer 2007), University of Texas researchers investigated whether certain stereotypes (e.g., negative references to certain nationalities) occur more often in deceptive news stories than in authentic news stories. The researchers analyzed 183 news stories that were proven to be deceptive in nature and 128 news stories that were considered authentic. Specifically, the researchers determined whether each story was negative, neutral, or positive in tone. The accompanying table (saved in the **NEWSSTORY** file) gives the number of news stories found in each tone category.

	Authentic News Stories	Deceptive News Stories
Negative Tone	59	111
Neutral Tone	49	61
Positive Tone	20	11
Totals	128	183

Source: Lasorsa, D., and Dai, J. "When news reporters deceive: The production of stereotypes," *Journalism and Mass Communication Quarterly,* Vol. 84, No. 2, Summer 2007 (Table 2). Used by permission of *J & MC/Quarterly* School of Journalism and Mass Communication.

Excel/DDXL Output for Exercise 9.22

Table Results

TONE	STORY	NUMBER	Row Total	Column Total	Expected Frequencies
Negative	Authentic	59	170	128	69.968
Neutral	Authentic	49	110	128	45.273
Positive	Authentic	20	31	128	12.759
Negative	Deceptive	111	170	183	100.032
Neutral	Deceptive	61	110	183	64.727
Positive	Deceptive	11	31	183	18.241

Assumptions

All Exp. Freqs >= 1?	Assumption Met
At Most 20% of Exp. Freqs < 5?	Assumption Met

Test Results for Test of TONE vs. STORY

chi-square	10.427
p-value	0.0054

a. Find the sample proportion of negative tone news stories that are deceptive.

b. Find the sample proportion of neutral news stories that are deceptive.

c. Find the sample proportion of positive news stories that are deceptive.

d. Compare the sample proportions, parts **a–c.** Does it appear that the proportion of news stories that are deceptive depends on story tone?

e. Give the null hypothesis for testing whether the authenticity of a news story depends on tone.

f. Use the Excel/DDXL printout above to conduct the test, part **e.** Test at $\alpha = .05$.

9.23 Are travel professionals equitably paid? *Business Travel News* (July 17, 2006) reported the results of its annual Travel Manager Salary & Attitude survey. A total of 277 travel professionals, 103 males and 174 females, participated in the 2005 survey. One question asked for the travel professional's opinion on the fairness of his/her salary. Responses were classified as "salary too low," "equitable/fair," or "paid well." The table below gives a breakdown of the responses in each category by gender. This summary information is saved in the **TRAVELPRO** file.

	Males	Females
Salary too low	29	89
Equitable/fair	58	64
Paid well	16	21
Totals	103	174

a. Find the proportion of male travel professionals who believe their salary is too low and compare it to the proportion of female travel professionals who believe their salary is too low.

b. Repeat part **a** but compare the proportions who believe their salary is equitable/fair.

c. Repeat part **a** but compare the proportions who believe they are paid well.

d. Based on the comparisons, parts **a–c,** do you think opinion on the fairness of a travel professional's salary differs for males and females?

e. Refer to part **d.** Conduct the appropriate statistical test using $\alpha = .10$.

f. Construct and interpret a 90% confidence interval for the difference between the proportions of part **a.**

9.24 Survey on giving and volunteering. Refer to the study of charitable givers published in the *National Tax Journal* (Dec. 2001), Exercise 9.9 (pp. 529–530). In addition to the 1,072 charitable givers who reported that their charitable contributions were motivated by tax considerations, another 1,693 givers reported no tax motivation, giving a total sample of 2,765 charitable givers. Of the 1,072 who were motivated by tax considerations, 691 itemized deductions on their income tax returns. Of the 1,693 who were not motivated by tax considerations, 794 itemized deductions.

a. Consider the two categorical variables, tax motivation (yes or no) and itemize deductions (yes or no). Form a 2×2 contingency table for these variables.

b. Compute the expected cell counts for the contingency table, part **a.**

c. Compute the value of χ^2 for a test of independence.

d. At $\alpha = .05$, what inference can you make about whether the two variables, tax motivation and itemize deductions, are related for charitable givers?

e. Create a bar graph that will visually support your conclusion in part **d.**

9.25 Dust plumes from farm equipment. Fugitive dust plumes generated by farm equipment can be hazardous to human health. In the *Journal of Agricultural, Biological, and Environmental Sciences* (Mar. 2001), environmental engineers developed a model for dust particle concentrations in plumes produced by a tractor operating in a wheat field. The tractor traveled along six parallel, equi-length paths in the field. A remote sensing instrument with a laser beam, placed at the edge of the field, measured the particulate matter in the dust every .5 seconds. Unfortunately, a few of the measurements were censored (i.e., higher than the signal level of the instrument). This usually occurred when the tractor was a short distance from the instrument's laser beam. The table on the next page (saved in the **DUSTCENSOR** file) shows the number of censored measurements for each of the six tractor lines.

a. Calculate and compare the sample proportion of censored measurements for the six tractor lines.

b. Do the data provide sufficient evidence to indicate that the proportion of censored measurements differs for the six tractor lines? Test using $\alpha = .01$.

c. Comment on the practical versus statistical significance of the test.

Applying the Concepts—Intermediate

9.26 Guilt in decision making. The effect of guilt emotion on how a decision maker focuses on the problem was investigated in the Jan. 2007 issue of the *Journal of Behavioral Decision Making* (see Exercises 1.26 and 3.40, pp. 24 and 139–140). A total of 171 volunteer students participated in the experiment,

Table for Exercise 9.25

Tractor Line	Uncensored Measurements	Censored Measurements	Totals
1	6,047	175	6,222
2	4,456	236	4,692
3	6,821	319	7,140
4	5,889	231	6,120
5	9,873	480	10,353
6	4,607	187	4,794
Totals	37,693	1,628	39,321

Source: Johns, C., Holmen, B., Niemeier, A., and Shumway, R. "Nonlinear regression for modeling censored one-dimensional concentration profiles of fugitive dust plumes," *Journal of Agricultural, Biological, and Environmental Sciences,* Vol. 6, No. 1, March 2001 (from data file provided by coauthor Brit Holmen).

where each was randomly assigned to one of three emotional states (guilt, anger, or neutral) through a reading/writing task. Immediately after the task, students were presented with a decision problem where the stated option has predominantly negative features (e.g., spending money on repairing a very old car). The results (number responding in each category) are summarized in the accompanying table. Is there sufficient evidence (at $\alpha = .10$) to claim that the option choice depends on emotional state? Use the data saved in the **GUILT** file to answer the question.

Emotional State	Choose Stated Option	Do Not Choose Stated Option	Totals
Guilt	45	12	57
Anger	8	50	58
Neutral	7	49	56
Totals	60	111	171

Source: Gangemi, A., and Mancini, F. "Guilt and focusing in decision-making," *Journal of Behavioral Decision Making,* Vol. 20, Jan. 2007 (Table 2).

9.27 Software defects. The PROMISE Software Engineering Repository at the University of Ottawa provides researchers with data sets for building predictive software models. (See Exercise 2.148, p. 106.) Data on 498 modules of software code written in "C" language for a NASA spacecraft instrument are saved in the **SWDEFECTS** file. Recall that each module was analyzed for defects and classified as "true" if it contained defective code and "false" if not. One algorithm for predicting whether or not a module has defects is "essential complexity" (denoted EVG), where a module with at least 15 subflow graphs with D-structured primes is predicted to have a defect. When the method predicts a defect, the predicted EVG value is "yes"; otherwise, it is "no." A contingency table for the two variables, actual defective status and predicted EVG, is shown in the SPSS printout below. Interpret the results. Would you recommend the essential complexity algorithm as a predictor of defective software modules? Explain.

9.28 Pig farm study. An article in *Sociological Methods & Research* (May 2001) analyzed the data presented in the table. A sample of 262 Kansas pig farmers were classified according to their education level (college or not) and size of their pig farm (number of pigs). The data are saved in the **PIGFARM** file. Conduct a test to determine whether a pig farmer's education level has an impact on the size of the pig farm. Use $\alpha = .05$ and support your answer with a graph.

	Education Level		
Farm Size	No College	College	Totals
$<$ 1,000 pigs	42	53	95
1,000–2,000 pigs	27	42	69
2,000–5,000 pigs	22	20	42
$>$ 5,000 pigs	27	29	56
Totals	118	144	262

Source: Agresti, A., and Liu, I. "Strategies for modeling a categorical variable allowing multiple category choices," *Sociological Methods & Research,* Vol. 29, No. 4, May 2001 (Table 1). Copyright © 2001. Reprinted by permission of SAGE Publications.

SPSS Output for Exercise 9.27

DEFECT * PRED_EVG Crosstabulation

Count

		PRED_EVG		
		no	yes	Total
DEFECT	false	441	8	449
	true	47	2	49
Total		488	10	498

Chi-Square Tests

	Value	df	Asymp. Sig. (2-sided)	Exact Sig. (2-sided)	Exact Sig. (1-sided)
Pearson Chi-Square	1.188[b]	1	.276		
Continuity Correction[a]	.306	1	.580		
Likelihood Ratio	.948	1	.330		
Fisher's Exact Test				.257	.257
N of Valid Cases	498				

a. Computed only for a 2x2 table

b. 1 cells (25.0%) have expected count less than 5. The minimum expected count is .98.

9.29 Creating menus to influence others. Refer to the *Journal of Consumer Research* (Mar. 2003) study on influencing the choices of others by offering undesirable alternatives, Exercise 6.78 (p. 356). In another experiment conducted by the researcher, 96 subjects were asked to imagine that they had just moved to an apartment with two others and that they were shopping for a new appliance (e.g., television, microwave oven). Each subject was asked to create a menu of three brand choices for their roommates; then subjects were randomly assigned (in equal numbers) to one of three different "goal" conditions—(1) create the menu in order to influence roommates to buy a preselected brand, (2) create the menu in order to influence roommates to buy a brand of your choice, and (3) create the menu with no intent to influence roommates. The researcher theorized that the menus created to influence others will likely include undesirable alternative brands. Consequently, the number of menus in each goal condition that was consistent with the theory was determined. The data (saved in the **MENU3** file) are summarized in the table below. Analyze the data for the purpose of determining whether the proportion of subjects who select menus consistent with the theory depends on goal condition. Use $\alpha = .01$.

Goal Condition	Number Consistent with Theory	Number Not Consistent with Theory	Totals
Influence/preselected brand	15	17	32
Influence/own brand	14	18	32
No influence	3	29	32

Source: Hamilton, R. W. "Why do people suggest what they do not want? Using context effects to influence others' choices," *Journal of Consumer Research*, Vol. 29, No. 4, March 2003 (Table 1). Copyright © 2003 JCR, Inc.

9.30 History of corporate acquisitions. Refer to the *Academy of Management Journal* (Aug. 2008) investigation of the performance and timing of corporate acquisitions, Exercise 2.12 (p. 40). Data on the number of firms sampled and number that announced one or more acquisitions during the year from 1980 to 2000 are saved in the **ACQUISITIONS2** file. Suppose you want to determine if the proportion of firms with acquisitions differs annually from 1990 to 2000, that is, you want to determine if year and acquisition status are dependent from 1990 to 2000.

a. Identify the two qualitative variables (and their respective categories) to be analyzed.

b. Set up the null and alternative hypotheses for the test.

c. Use the Minitab printout below to conduct the test at $\alpha = .05$.

9.31 Classifying air threats with heuristics. The *Journal of Behavioral Decision Making* (Jan. 2007) published a study on the use of heuristics to classify the threat level of approaching aircraft. Of special interest was the use of a fast and frugal heuristic—a computationally simple procedure for making judgments with limited information—named "Take-the-Best-for-Classification" (TTBC). Subjects were 48 men and women; some were from a Canadian Forces reserve unit, and others were university students. Each subject was presented with a radar screen on which simulated approaching aircraft were identified with asterisks. By using the computer mouse to click on the asterisk, further information about the aircraft was provided. The goal was to identify the aircraft as "friend" or "foe" as fast as possible. Half the subjects were given cue-based instructions for determining the type of aircraft, while the other half were given pattern-based instructions. The researcher also classified the heuristic strategy used by the subject as TTBC, Guess, or Other. Data on the two variables, instruction type and strategy, measured for each of the 48 subjects are saved in the **AIRTHREAT** file. (Data for the first five and last five subjects are shown in the table on the next page.) Do the data provide sufficient evidence (at $\alpha = .05$) to indicate that choice of heuristic strategy depends on type of instruction provided? At $\alpha = .01$?

Minitab Output for Exercise 9.30

Tabulated statistics: ACQUISITION, YEAR

```
Using frequencies in NUMBER


Rows: ACQUISITION   Columns: YEAR

       1990  1991  1992  1993  1994  1995  1996  1997  1998  1999  2000    All

No     1847  1891  1936  2050  2149  2238  2319  2300  2047  2049  2030  22856
       1689  1738  1817  1985  2134  2222  2360  2383  2240  2152  2136  22856

Yes     350   370   427   532   626   652   751   799   866   750   748   6871
        508   523   546   597   641   668   710   716   673   647   642   6871

All    2197  2261  2363  2582  2775  2890  3070  3099  2913  2799  2778  29727
       2197  2261  2363  2582  2775  2890  3070  3099  2913  2799  2778  29727

Cell Contents:      Count
                    Expected count


Pearson Chi-Square = 297.048, DF = 10, P-Value = 0.000
Likelihood Ratio Chi-Square = 303.612, DF = 10, P-Value = 0.000
```

Data for Exercise 9.31

Instruction Strategy	
Pattern	Other
Pattern	Other
Pattern	Other
Cue	TTBC
Cue	TTBC
⋮	⋮
Pattern	TTBC
Cue	Guess
Cue	TTBC
Cue	Guess
Pattern	Guess

Source: Bryant, D. J. "Classifying simulated air threats with fast and frugal heuristics," *Journal of Behavioral Decision Making,* Vol. 20, January 2007 (Appendix C).

Applying the Concepts—Advanced

9.32 Examining the "Monty Hall Dilemma." In Exercise 3.131 (p. 169) you solved the game show problem of whether or not to switch your choice of three doors—one of which hides a prize—after the host reveals what is behind a door not chosen. (Despite the natural inclination of many to keep one's first choice, the correct answer is that you should switch your choice of doors.) This problem is sometimes called the "Monty Hall Dilemma," named for Monty Hall, the host of the popular TV game show "Let's Make a Deal." In *Thinking & Reasoning* (Oct. 2006), Wichita State University professors set up an experiment designed to influence subjects to switch their original choice of doors. Each subject participated in 23 trials. In trial #1, 3 doors (boxes) were presented on a computer screen, only one of which hid a prize. In each subsequent trial, an additional box was presented, so that in trial #23, 25 boxes were presented. After selecting a box in each trial, all the remaining boxes except for one were either (1) shown to be empty (*Empty* condition), (2) disappeared (*Vanish* condition), (3) disappeared and the chosen box enlarged (*Steroids* condition), or (4) disappeared and the remaining box not chosen enlarged (*Steroids2* condition). A total of 27 subjects were assigned to each condition. The number of subjects who ultimately switched boxes is tallied, by condition, in the accompanying table for both the first trials and the last trial. The summary information is saved in the **MONTYHALL** file.

	First Trial (#1)		Last Trial (#23)	
Condition	Switch Boxes	No Switch	Switch Boxes	No Switch
Empty	10	17	23	4
Vanish	3	24	12	15
Steroids	5	22	21	6
Steroids2	8	19	19	8

Source: Howard, J. N., Lambdin, C. G., and Datteri, D. L. "Let's make a deal: Quality and availability of second-stage information as a catalyst for change," *Thinking & Reasoning,* Vol. 13, No. 3, August 2007, pp. 248–272 (Table 2). Reprinted by permission of the publisher (Taylor & Francis Group, www.informaworld.com).

a. For a selected trial, does the likelihood of switching boxes depend on condition?

b. For a given condition, does the likelihood of switching boxes depend on trial number?

c. Based on the results, parts **a** and **b,** what factors influence a subject to switch choices?

9.33 Efficacy of an HIV vaccine. New, effective AIDS vaccines are now being developed using the process of "sieving" (i.e., sifting out infections with some strains of HIV). Harvard School of Public Health Statistician Peter Gilbert demonstrated how to test the efficacy of an HIV vaccine in *Chance* (Fall 2000). As an example, Gilbert reported the results of VaxGen's preliminary HIV vaccine trial using the 2 × 2 table on the next page. The vaccine was designed to eliminate a particular strain of the virus, called the "MN strain." The trial consisted of 7 AIDS patients vaccinated with the new drug and 31 AIDS patients who were treated with a placebo (no vaccination). The table on the next page (saved in the **VAXGEN1** file) shows the number of patients who tested positive and negative for the MN strain in the trial follow-up period.

SPSS Output for Exercise 9.33

GROUP * MNSTRAIN Crosstabulation

			MNSTRAIN		
			NEG	POS	Total
GROUP	UNVAC	Count	9	22	31
		Expected Count	11.4	19.6	31.0
	VACC	Count	5	2	7
		Expected Count	2.6	4.4	7.0
Total		Count	14	24	38
		Expected Count	14.0	24.0	38.0

Chi-Square Tests

	Value	df	Asymp. Sig. (2-sided)	Exact Sig. (2-sided)	Exact Sig. (1-sided)
Pearson Chi-Square	4.411[b]	1	.036		
Continuity Correction[a]	2.777	1	.096		
Likelihood Ratio	4.289	1	.038		
Fisher's Exact Test				.077	.050
N of Valid Cases	38				

a. Computed only for a 2x2 table

b. 2 cells (50.0%) have expected count less than 5. The minimum expected count is 2.58.

MN Strain			
Patient Group	Positive	Negative	Totals
Unvaccinated	22	9	31
Vaccinated	2	5	7
Totals	24	14	38

Source: Gilbert, P. "Developing an AIDS vaccine by sieving," *Chance*, Vol. 13, No. 4, Fall 2000, pp. 16–21 (adapted Table 1, p. 19). Reprinted with permission from *Chance*.

a. Conduct a test to determine whether the vaccine is effective in treating the MN strain of HIV. Use $\alpha = .05$.

b. Are the assumptions for the test, part **a,** satisfied? What are the consequences if the assumptions are violated?

c. In the case of a 2×2 contingency table, R. A. Fisher (1935) developed a procedure for computing the exact p-value for the test (called *Fisher's exact test*). The method uses the *hypergeometric probability distribution* (a discrete probability distribution not covered in Chapter 4). Consider the hypergeometric probability

$$\frac{\binom{7}{2}\binom{31}{22}}{\binom{38}{24}}$$

This represents the probability that 2 out of 7 vaccinated AIDS patients test positive and 22 out of 31 unvaccinated patients test positive (i.e., the probability of the table result given the null hypothesis of independence is true). Compute this probability (called the *probability of the contingency table*).

d. Refer to part **c.** Two contingency tables (with the same marginal totals as the original table) that are more contradictory to the null hypothesis of independence than the observed table follow and are saved in the **VAXGEN2** and **VAXGEN3** files, respectively. First, explain why these tables provide more evidence to reject H_0 than the original table; then compute the probability of each table using the hypergeometric formula.

MN Strain			
Patient Group	Positive	Negative	Totals
Unvaccinated	23	8	31
Vaccinated	1	6	7
Totals	24	14	38

MN Strain			
Patient Group	Positive	Negative	Totals
Unvaccinated	24	7	31
Vaccinated	0	7	7
Totals	24	14	38

e. The p-value of Fisher's exact test is the probability of observing a result at least as contradictory to the null hypothesis as the observed contingency table, given the same marginal totals. Sum the probabilities of parts **c** and **d** to obtain the p-value of Fisher's exact test. (To verify your calculations, check the p-value at the bottom of the SPSS printout on page 544.) Interpret this value in the context of the vaccine trial.

9.4 A Word of Caution about Chi-Square Tests

Because the χ^2 statistic for testing hypotheses about multinomial probabilities is one of the most widely applied statistical tools, it is also one of the most abused statistical procedures. Consequently, the user should always be certain that the experiment satisfies the assumptions given with each procedure. Furthermore, the user should be certain that the sample is drawn from the correct population—that is, from the population about which the inference is to be made.

The use of the χ^2 probability distribution as an approximation to the sampling distribution for χ^2 should be avoided when the expected counts are very small. The approximation can become very poor when these expected counts are small, and thus the true α level may be quite different from the tabled value. As a rule of thumb, an expected cell count of at least 5 means that the χ^2 probability distribution can be used to determine an approximate critical value.

If the χ^2 value does not exceed the established critical value of χ^2, *do not accept the hypothesis of independence.* You would be risking a Type II error (accepting H_0 if it is false), and the probability β of committing such an error is unknown. The usual alternative hypothesis is that the classifications are dependent. Because the number of ways in which two classifications can be dependent is virtually infinite, it is difficult to calculate one or even several values of β to represent such a broad alternative hypothesis. Therefore, we avoid concluding that two classifications are independent, even when χ^2 is small.

Ethics IN Statistics

Using the results of a chi-square analysis to make a desired influence when you are fully aware that the sample is too small or that the assumptions are violated is considered *unethical statistical practice*.

Finally, if a contingency table χ^2 value does exceed the critical value, we must be careful to avoid inferring that a *causal* relationship exists between the classifications. Our alternative hypothesis states that the two classifications are statistically dependent—and a statistical dependence does not imply causality. Therefore, *the existence of a causal relationship cannot be established by a contingency table analysis.*

CHAPTER NOTES

Key Terms

Categories 522
Cells 522
Cell counts 522
Chi-square test 524
Classes 522
Contingency table 531
Dependence 534
Dimensions of classification 531
Expected cell count 525
Independence of two classifications 532
Marginal probabilities 532
Multinomial experiment 522
Observed cell count 532
One-way table 523
Two-way table 531

Key Symbols/Notation

$p_{i,0}$	Value of multinomial probability p_i hypothesized in H_0
χ^2	Chi-square test statistic used in analysis of categorical data
n_i	Number of observed outcomes in cell i of a one-way table
E_i	Expected number of outcomes in cell i of a one-way table
p_{ij}	Probability of an outcome in row i and column j of a two-way table
n_{ij}	Number of observed outcomes in row i and column j of a two-way table
$\hat{E}_{ij}$	Estimated expected number of outcomes in row i and column j of a two-way table
R_i	Total number of outcomes in row i of a two-way table
C_j	Total number of outcomes in column j of a two-way table

Key Ideas

Multinomial Data
Qualitative data that fall into more than two categories (or classes)

Properties of a Multinomial Experiment

1. n identical trials
2. k possible outcomes to each trial
3. probabilities of the k outcomes $(p_1, p_2, \ldots, p_k)$ remain the same from trial to trial, where $p_1 + p_2 + \cdots + p_k = 1$
4. trials are independent
5. variables of interest: *cell counts* (i.e., number of observations falling into each outcome category), denoted $n_1, n_2, \ldots, n_k$

One-Way Table
Summary table for a *single* qualitative variable

Two-Way (Contingency) Table
Summary table for *two* qualitative variables

Chi-Square (χ^2) Statistic
used to test category probabilities in one-way and two-way tables

Chi-square tests for independence
should **not** be used to *infer a causal relationship between 2 QLs*

Conditions Required for Valid χ^2-Tests

1. multinomial experiment
2. sample size n is large (expected cell counts are all greater than or equal to 5)

Categorical Data Analysis Guide

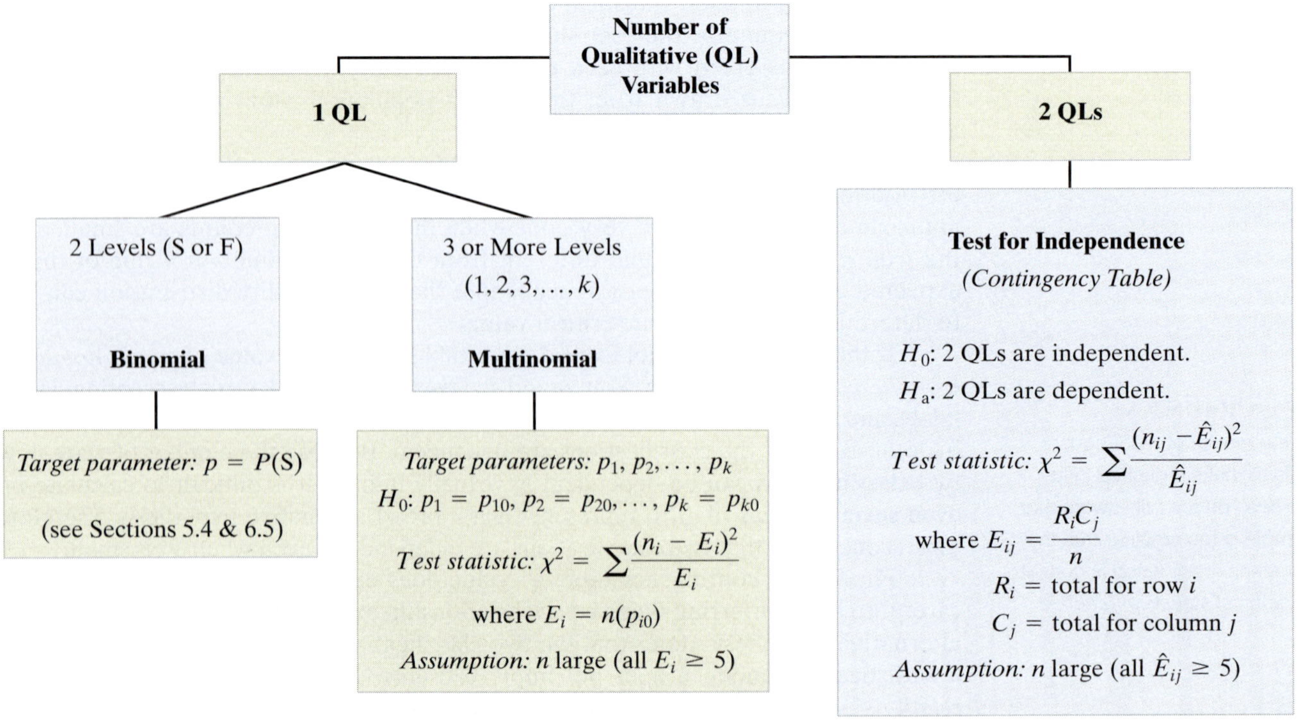

Supplementary Exercises 9.34–9.53

Learning the Mechanics

9.34 A random sample of 250 observations was classified according to the row and column categories shown in the table below. The summary information is saved in the **LM9_34** file.

		Column		
		1	2	3
Row	1	20	20	10
	2	10	20	70
	3	20	50	30

a. Do the data provide sufficient evidence to conclude that the rows and columns are dependent? Test using $\alpha = .05$.

b. Would the analysis change if the row totals were fixed before the data were collected?

c. Do the assumptions required for the analysis to be valid differ according to whether the row (or column) totals are fixed? Explain.

d. Convert the table entries to percentages by using each column total as a base and calculating each row response as a percentage of the corresponding column total. In addition, calculate the row totals and convert them to percentages of all 250 observations.

e. Create a bar graph with row 1 percentage on the vertical axis against the column number on the horizontal axis. Draw horizontal lines corresponding to the row 1 percentages. Does the graph support the result of the test conducted in part **a**?

9.35 A random sample of 150 observations was classified into the categories shown in the table below. Summary information is saved in the **LM9_35** file.

	Category				
	1	2	3	4	5
n_i	28	35	33	25	29

a. Do the data provide sufficient evidence that the categories are not equally likely? Use $\alpha = .10$.

b. Form a 90% confidence interval for p_2, the probability that an observation will fall in category 2.

Applying the Concepts—Basic

9.36 **Consumers' least favorite vegetables.** *Bon Appetit* magazine polled 200 of its readers concerning which of the four vegetables—brussel sprouts, okra, lima beans, and cauliflower—is their least favorite. The results (adapted from *Adweek,* Feb. 21, 2000) are presented in the table and saved in the **BONAPP** file. Let p_1, p_2, p_3, and p_4 represent the proportions of all *Bon Appetit* readers who indicate brussel sprouts, okra, lima beans, and cauliflower, respectively, as their least favorite vegetable.

Brussel Sprouts	Okra	Lima Beans	Cauliflower
46	76	44	34

a. If, in general, *Bon Appetit* readers do not have a preference for their least favorite vegetable, what are the values of p_1, p_2, p_3, and p_4?

b. Specify the null and alternative hypotheses that should be used to determine whether *Bon Appetit* readers have a preference for one of the vegetables as "least favorite."

c. Conduct the test you described in part **b** using $\alpha = .05$. Report your conclusion in the context of the problem.

d. What assumptions must hold to ensure the validity of the test you conducted in part **c**? Which, if any, of these assumptions may be a concern in this application?

9.37 **Opinions on national health insurance.** In order to study consumer preferences for health-care reform in the United States, researchers from the University of Michigan surveyed 500 U.S. households (*Journal of Consumer Affairs,* Winter 1999). Heads of household were asked whether they are in favor of, neutral about, or opposed to a national health insurance program in which all Americans are covered and costs are paid by tax dollars. The 434 usable responses are summarized in the table and saved in the **HEALTH** file.

Favor	Neutral	Oppose
234	119	81

Source: Hong, G., and White-Means, S. "Consumer preferences for health care reform," *Journal of Consumer Affairs,* Vol. 33, No. 2, Winter 1999, pp. 237–253.

a. Specify H_0 and H_a for a test to determine if the opinions of heads of households are not evenly divided on the issue of national health insurance.

b. Conduct the test in part **a,** using $\alpha = .01$.

c. Construct a 95% confidence interval for the proportion of heads of household in the U.S. population who favor national health insurance.

9.38 **JAMA study of heart patients.** The *Journal of the American Medical Association* (Apr. 18, 2001) published the results of a study of alcohol consumption in patients suffering from acute myocardial infarction (AMI). The patients were classified according to average number of alcoholic drinks per week and whether or not they had congestive heart failure. A summary of the results for 1,913 AMI patients is shown in the table and saved in the **AMAAMI** file.

	Alcohol Consumption		
Congestive Heart Failure	Abstainers	Less than 7 drinks/week	7 or more drinks/week
Yes	146	106	29
No	750	590	292
Totals	896	696	321

Source: Mukamal, K. J., et al. "Prior alcohol consumption and mortality following acute myocardial infarction," *Journal of the American Medical Association,* Vol. 285, No. 15, April 18, 2001 (Table 1).

a. Find the sample proportion of abstainers with congestive heart failure.

b. Find the sample proportion of moderate drinkers (patients who have less than 7 drinks per week) with congestive heart failure.

c. Find the sample proportion of heavy drinkers (patients who have 7 or more drinks per week) with congestive heart failure.

d. Compare the sample proportions, parts **a–c.** Does it appear that the proportion of AMI patients with congestive heart failure depends on alcohol consumption?

e. Give the null hypothesis for testing whether the proportion of AMI patients with congestive heart failure depends on alcohol consumption.

f. Conduct the test, part **e**, using $\alpha = .05$. What do you conclude?

9.39 Where do you get your travel information? In order to create a behavioral profile of pleasure travelers, M. Bonn (Florida State University), L. Forr (Georgia Southern University), and A. Susskind (Cornell University) interviewed 5,026 pleasure travelers in the Tampa Bay region (*Journal of Travel Research,* May 1999). Two of the characteristics they investigated were the travelers' education level and their use of the Internet to seek travel information. The table below (saved in the **NETRAVEL** file) summarizes the results of the interviews. The researchers concluded that travelers who use the Internet to search for travel information are likely to be people who are college educated. Do you agree? Test using $\alpha = .05$. What assumptions must hold to ensure the validity of your test?

	Use Internet	
Education	**Yes**	**No**
College Degree or More	1,072	1,287
Less than a College Degree	640	2,027

Source: Bonn, M., Furr, L., and Susskind, A. "Predicting a behavioral profile for pleasure travelers on the basis of internet use segmentation," *Journal of Travel Research,* Vol. 37, No. 4, May 1999, pp. 333–340. Reprinted by permission of SAGE Publications..

9.40 Colors of M&M's candies. M&M's plain chocolate candies come in six different colors: dark brown, yellow, red, orange, green, and blue. According to the manufacturer (Mars, Inc.), the color ratio in each large production batch is 30% brown, 20% yellow, 20% red, 10% orange, 10% green, and 10% blue. To test this claim, a professor at Carleton College (Minnesota) had students count the colors of M&M's found in "fun size" bags of the candy (*Teaching Statistics,* Spring 1993). The results for 370 M&M's are displayed in the table and saved in the **M&M** file.

Brown	Yellow	Red	Orange	Green	Blue	Total
84	79	75	49	36	47	370

Source: Johnson, R.W. "Testing colour proportions of M&M's," *Teaching Statistics,* Vol. 15, No. 1, Spring 1993, p. 2 (Table 1).

a. Assuming the manufacturer's stated percentages are accurate, calculate the expected numbers falling into the six categories.

b. Calculate the value of χ^2 for testing the manufacturer's claim.

c. Conduct a test to determine whether the true percentages of the colors produced differ from the manufacturer's stated percentages. Use $\alpha = .05$.

Applying the Concepts—Intermediate

9.41 Ethical behavior of accountants. University of Louisville Professor Julia Karcher conducted an experiment to investigate the ethical behavior of accountants (*Journal of Business Ethics,* Vol. 15, 1996). She focused on auditor abilities to detect ethical problems that may not be obvious. Seventy auditors from Big-Six accounting firms were given a detailed case study that contained several problems, including tax evasion by the client. In 35 of the cases, the tax evasion issue was severe; in the other 35 cases, it was moderate. The auditors were asked to identify any problems they detected in the case. The following table (saved in the **ACCETHIC** file) summarizes the results for the ethical issue.

	Severity of Ethical Issue	
	Moderate	**Severe**
Ethical Issue Identified	27	26
Ethical Issue Not Identified	8	9

Source: Karcher, J. "Auditors' ability to discern the presence of ethical problems," *Journal of Business Ethics,* Vol. 15, 1996, p. 1041 (Table V). Copyright © 1996 Springer.

a. Did the severity of the ethical issue influence whether the issue was identified or not by the auditors? Test using $\alpha = .05$.

b. Suppose the left-hand column of the table contained the counts 35 and 0 instead of 27 and 8. Should the test of part **a** still be conducted? Explain.

c. Keeping the sample size the same, change the numbers in the contingency table so that the answer you would get for the question posed in part **a** changes.

9.42 Insider ownership and firm size. Because shareholders control the firm, they can transfer wealth from the firm's bondholders to themselves through several different dividend strategies. This potential conflict of interest between shareholders and bondholders can be reduced through the use of debt covenants. Accountants E. Griner and H. Huss of Georgia State University investigated the effects of insider ownership and the size of the firm on the types of debt covenants required by a firm's bondholders (*Journal of Applied Business Research,* Vol. 11, 1995). As part of the study, they examined a sample of 31 companies whose bondholders required covenants based on tangible assets rather than on liquidity or net assets or retained earnings. Characteristics of those 31 firms are summarized below and saved in the **INSIDOWN** file. The objective of the study is to determine if there is a relationship between the extent of insider ownership and the size of the firm for firms with tangible asset covenants.

		Size	
		Small	**Large**
Inside Ownership	Low	3	17
	High	8	3

Source: Griner, E., and Huss, H. "Firm size, insider ownership, and accounting-based debt covenants," *Journal of Applied Business Research,* Vol. 11, No. 4, 1995, p. 7 (Table 4).

a. Assuming the null hypothesis of independence is true, how many firms are expected to fall in each cell of the table above?

b. The researchers were unable to use the chi-square test to analyze the data. Show why.

c. A test of the null hypothesis can be conducted using *Fisher's exact test.* (See Exercise 9.33, p. 545.) This method calculates the exact probability (p-value) of observing sample results at least as contradictory to the null hypothesis as those observed for the researchers' data. The researchers reported the p-value for this test as .0043. Interpret this result.

d. Investigate the nature of the dependence exhibited by the contingency table by plotting the appropriate contingency table percentages. Describe what you find.

9.43 Dosing errors at hospitals. Each year, approximately 1.3 million Americans suffer adverse drug effects (ADEs)—that is, unintended injuries caused by prescribed medication. A study in the *Journal of the American Medical Association* (July 5, 1995) identified the cause of 247 ADEs that occurred at two Boston hospitals. The researchers found that dosing errors (that is, wrong dosage prescribed and/or dispensed) were the most common. The next table (saved in the **ADE** file) summarizes the proximate cause of 95 ADEs that resulted from a dosing error. Conduct a test (at $\alpha = .10$) to determine whether the true percentages of ADEs in the five "cause" categories are different.

Wrong Dosage Cause	Number of ADEs
Lack of knowledge of drug	29
Rule violation	17
Faulty dose checking	13
Slips	9
Other	27

9.44 Occupations of fathers and sons. An economist was interested in knowing whether sons have a tendency to choose the same occupation as their fathers. To investigate this question, 500 males were polled, and each was questioned concerning his occupation and the occupation of his father. A summary of the numbers of father-son pairs falling into each occupational category (saved in the **FATHSON** file) is shown in the table at the bottom of the page. Do the data provide sufficient evidence at $\alpha = .05$ to indicate a dependence between a son's choice of occupation and his father's occupation?

9.45 Performance of solder joint inspectors. Westinghouse Electric Company has experimented with different means of evaluating the performance of solder joint inspectors. One approach involves comparing an individual inspector's classifications with those of the group of experts that comprise Westinghouse's Work Standards Committee. In one experiment conducted by Westinghouse, 153 solder connections were evaluated by the committee, and 111 were classified as acceptable. An inspector evaluated the same 153 connections and classified 124 as acceptable. Of the items rejected by the inspector, the committee agreed with 19.

a. Construct a contingency table that summarizes the classifications of the committee and the inspector.

b. Based on a visual examination of the table you constructed in part **a,** does it appear that there is a relationship between the inspector's classifications and the committee's? Explain. (A graph of the percentage rejected by committee and inspector will aid your examination.)

c. Conduct a chi-square test of independence for these data. Use $\alpha = .05$. Carefully interpret the results of your test in the context of the problem.

9.46 Firms that practice TQM. To better understand whether and how Total Quality Management (TQM) is practiced in U.S. companies, University of Scranton researchers N. Tamimi and R. Sebastianelli interviewed one manager in each of a sample of 86 companies in Pennsylvania, New York, and New Jersey (*Production and Inventory Management Journal,* 1996). Concerning whether or not the firms were involved with TQM, the following data were obtained and are saved in the **TQM** file.

	Service Firms	Manufacturing Firms
Number practicing TQM	34	23
Number not practicing TQM	18	11
Total	52	34

Source: Adapted from Tamimi, N., and Sebastianelli, R. "How firms define and measure quality," *Production and Inventory Management Journal,* Third Quarter, 1996, p. 35.

a. The researchers concluded that "manufacturing firms were not significantly more likely to be involved with TQM than service firms." Do you agree? Test using $\alpha = .05$.

b. Find and interpret the approximate p-value for the test you conducted in part **a.**

c. What assumptions must hold in order for your test of part **a** and your p-value of part **b** to be valid?

9.47 Peanut butter market shares. Data from supermarket scanners are used by researchers to understand the purchasing patterns and preferences of consumers. Researchers frequently study the purchases of a sample of households, called a *scanner panel.* When shopping, these households present a magnetic identification card that permits their purchase data to be identified and aggregated. Marketing researchers recently studied the

Table for Exercise 9.44

		Son			
		Professional or Business	Skilled	Unskilled	Farmer
Father	Professional or Business	55	38	7	0
	Skilled	79	71	25	0
	Unskilled	22	75	38	10
	Farmer	15	23	10	32

extent to which panel households' purchase behavior is representative of the population of households shopping at the same stores (*Marketing Research,* Nov. 1996). The table below (saved in the **SCANNER** file) reports the peanut butter purchase data collected by A. C. Nielsen Company for a panel of 2,500 households in Sioux Falls, SD, over a 102-week period. The market share percentages in the right column are derived from all peanut butter purchases at the same 15 stores at which the panel shopped during the same 102-week period.

Brand	Size	Number of Purchases By Household Panel	Market Share
Jif	18 oz.	3,165	20.10%
Jif	28	1,892	10.10
Jif	40	726	5.42
Peter Pan	10	4,079	16.01
Skippy	18	6,206	28.65
Skippy	28	1,627	12.38
Skippy	40	1,420	7.32
Total		19,115	

Source: Gupta, S., et. al. "Do household scanner data provide representative inferences from brand choices? A comparison with store data," *Journal of Marketing Research,* Vol. 33, Nov. 1996, p. 393 (Table 6). Copyright 1996 by American Marketing Association. Reproduced with permission of American Marketing Association in the format Textbook via Copyright Clearance Center.

a. Do the data provide sufficient evidence to conclude that the purchases of the household panel are not representative of the population of households? Test using $\alpha = .05$.

b. What assumptions must hold to ensure the validity of the testing procedure you used in part **a**?

c. Find the approximate p-value for the test of part **a** and interpret it in the context of the problem.

9.48 Multiple sclerosis drug. Interferons are proteins produced naturally by the human body that help fight infections and regulate the immune system. A drug developed from interferons, called Avonex, is now available for treating patients with multiple sclerosis (MS). In a clinical study, 85 MS patients received weekly injections of Avonex over a 2-year period. The number of exacerbations (i.e., flare-ups of symptoms) was recorded for each patient and is summarized in the accompanying table. (These data are saved in the **AVONEX** file.) For MS patients who take a placebo (no drug) over a similar two-week period, it is known from previous studies that 26% will experience no exacerbations, 30% one exacerbation, 11% two exacerbations, 14% three exacerbations, and 19% four or more exacerbations.

Number of Exacerbations	Number of Patients
0	32
1	26
2	15
3	6
4 or more	6

Source: Biogen, Inc., 1997.

a. Conduct a test to determine whether the exacerbation distribution of MS patients who take Avonex differs from the percentages reported for placebo patients. Test using $\alpha = .05$.

b. Find a 95% confidence interval for the true proportion of Avonex MS patients who are exacerbation free during a 2-year period.

c. Refer to part **b.** Is there evidence that Avonex patients are more likely to have no exacerbations than placebo patients? Explain.

9.49 Flight response of geese to helicopter traffic. Offshore oil drilling near an Alaskan estuary has led to increased air traffic—mostly large helicopters—in the area. The U.S. Fish and Wildlife Service commissioned a study to investigate the impact these helicopters have on the flocks of Pacific brant geese that inhabit the estuary in fall before migrating (*Statistical Case Studies: A Collaboration between Academe and Industry,* 1998). Two large helicopters were flown repeatedly over the estuary at different altitudes and lateral distances from the flock. The flight responses of the geese (recorded as "low" or "high"), altitude (hundreds of meters), and lateral distance (hundreds of meters) for each of the 464 helicopter overflights were recorded and are saved in the **PACGEESE** file. (The data for the first 10 overflights are shown in the following table.)

Overflight	Altitude	Lateral Distance	Flight Response
1	0.91	4.99	High
2	0.91	8.21	High
3	0.91	3.38	High
4	9.14	21.08	Low
5	1.52	6.60	High
6	0.91	3.38	High
7	3.05	0.16	High
8	6.10	3.38	High
9	3.05	6.60	High
10	12.19	6.60	High

Source: Erickson, W., Nick, T., and Ward, D. "Investigating flight response of pacific brant to helicopters at Izembek Lagoon, Alaska, by using logistic regression," *Statistical Case Studies: A Collaboration between Academe and Industry,* ASA-SIAM Series on Statistics and Applied Probability, pp. 155–170, 1998. © 1998 Society for Industrial and Applied Mathematics. Reprinted with permission. All rights reserved.

a. The researchers categorized altitude as follows: less than 300 meters, 300–600 meters, and 600 or more meters. Summarize the data in the **PACGEESE** file by creating a contingency table for altitude category and flight response.

b. Conduct a test to determine if flight response of the geese depends on altitude of the helicopter. Test using $\alpha = .01$.

c. The researchers categorized lateral distance as follows: less than 1,000 meters, 1,000–2,000 meters, 2,000–3,000 meters, and 3,000 or more meters. Summarize the data in the **PACGEESE** file by creating a contingency table for lateral distance category and flight response.

d. Conduct a test to determine if flight response of the geese depends on lateral distance of helicopter from the flock. Test using $\alpha = .01$.

e. The current Federal Aviation Authority (FAA) minimum altitude standard for flying over the estuary is 2,000 feet (approximately 610 meters). Based on the results, parts **a–d,** what changes to the FAA regulations do you recommend to minimize the effects to Pacific brant geese?

9.50 "Fitness for use" of gasoline filters. Product or service quality is often defined as *fitness for use.* This means the product or service meets the customer's needs. Generally speaking, fitness for use is based on five quality characteristics: technological (e.g., strength, hardness), psychological (taste, beauty), time-oriented (reliability), contractual (guarantee provisions), and ethical (courtesy, honesty). The quality of a service may involve all these characteristics, while the quality of a manufactured product generally depends on technological and time-oriented characteristics (Schroeder, *Operations Management,* 2008). After a barrage of customer complaints about poor quality, a manufacturer of gasoline filters for cars had its quality inspectors sample 600 filters—200 per work shift—and check for defects. The data in the table resulted and are saved in the **FILTER** file.

Shift	Defectives Produced
First	25
Second	35
Third	80

a. Do the data indicate that the quality of the filters being produced may be related to the shift producing the filter? Test using $\alpha = .05$.

b. Estimate the proportion of defective filters produced by the first shift. Use a 95% confidence interval.

Applying the Concepts—Advanced

9.51 Goodness-of-fit test. A statistical analysis is to be done on a set of data consisting of 1,000 monthly salaries. The analysis requires the assumption that the sample was drawn from a normal distribution. A preliminary test, called the χ^2 *goodness-of-fit test,* can be used to help determine whether it is reasonable to assume that the sample is from a normal distribution. Suppose the mean and standard deviation of the 1,000 salaries are hypothesized to be \$1,200 and \$200, respectively. Using the standard normal table, we can approximate the probability of a salary being in the intervals listed in the below table. The third column represents the expected number of the 1,000 salaries to be found in each interval if the sample was drawn from a normal distribution with $\mu = \$1,200$ and $\Sigma = \$200$. Suppose the last column contains the actual observed frequencies in the sample. Large differences between the observed and expected frequencies cast doubt on the normality assumption.

a. Compute the χ^2 statistic based on the observed and expected frequencies—just as you did in Section 9.2.

b. Find the tabulated χ^2 value when $\alpha = .05$ and there are 5 degrees of freedom. (There are $k - 1 = 5$ df associated with this χ^2 statistic.)

c. Based on the χ^2 statistic and the tabulated χ^2 value, is there evidence that the salary distribution is nonnormal?

d. Find an approximate observed significance level for the test in part **c.**

9.52 Analysis of a Scrabble game. In the board game Scrabble™, a player initially draws a "hand" of seven tiles at random from 100 tiles. Each tile has a letter of the alphabet and the player attempts to form a word from the letters in his/her hand. In *Chance* (Winter 2002), C. J. Robinove investigated whether a handheld electronic version of the game, called ScrabbleExpress™, produces too few vowels in the 7-letter draws. For each of the 26 letters (and "blank" for any letter), the next table gives the true relative frequency of the letter in the board game as well as the frequency of occurrence of the letter in a sample of 700 tiles (i.e., 100 "hands") randomly drawn using the electronic game. This summary information is saved in the **SCRABBLE** file.

a. Do the data support the scientist's contention that ScrabbleExpress™ "presents the player with unfair word selection opportunities" that are not

Letter	Relative Frequency in Board Game	Frequency in Electronic Game
A	.09	39
B	.02	18
C	.02	30
D	.04	30
E	.12	31
F	.02	21
G	.03	35
H	.02	21
I	.09	25
J	.01	17
K	.01	27
L	.04	18
M	.02	31
N	.06	36
O	.08	20
P	.02	27
Q	.01	13
R	.06	27
S	.04	29
T	.06	27
U	.04	21
V	.02	33
W	.02	29
X	.01	15
Y	.02	32
Z	.01	14
# (blank)	.02	34
Total		700

Source: Robinove, C. J. "Letter-frequency bias in an electronic Scrabble game," *Chance,* Vol. 15, No. 1, Winter 2002, p. 31 (Table 3). Reprinted with permission from *Chance.* © 2002 by the American Statistical Association. All rights reserved.

Table for Exercise 9.51

Interval	Probability	Expected Frequency	Observed Frequency
Less than \$800	.023	23	26
Between \$800 and \$1,000	.136	136	146
Between \$1,000 and \$1,200	.341	341	361
Between \$1,200 and \$1,400	.341	341	311
Between \$1,400 and \$1,600	.136	136	143
Above \$1,600	.023	23	13

the same as the Scrabble™ board game? Test using $\alpha = .05$.

b. Estimate the true proportion of letters drawn in the electronic game that are vowels using a 95% confidence interval. Compare the results to the true relative frequency of a vowel in the board game.

Critical Thinking Challenge

9.53 A "rigged" election? *Chance* (Spring 2004) presented data from a recent election held to determine the board of directors of a local community. There were 27 candidates for the board, and each of 5,553 voters was allowed to choose 6 candidates. The claim was that "a fixed vote with fixed percentages (was) assigned to each and every candidate making it impossible to participate in an honest election." Votes were tallied in six time periods: after 600 total votes were in, after 1,200, after 2,444, after 3,444, after 4,444, and after 5,553 votes. The data for three of the candidates (Smith, Coppin, and Montes) are shown in the following table and saved in the **RIGVOTE** file. A residential organization believes that "there was nothing random about the count and tallies each time period and specific unnatural or rigged percentages were being assigned to each and every candidate." Give your opinion. Is the probability of a candidate receiving votes independent of the time period? If so, does this imply a rigged election?

	Time Period					
	1	2	3	4	5	6
Votes for Smith	208	208	451	392	351	410
Votes for Coppin	55	51	109	98	88	104
Votes for Montes	133	117	255	211	186	227
Total Votes	600	600	1,244	1,000	1,000	1,109

Source: Gelman, A. "55,000 residents desperately need your help!" *Chance,* Vol. 17, No. 2, Spring 2004, p. 32 (Figures 1 and 5, p. 34). Reprinted with permission from *Chance.* © 2004 by the American Statistical Association. All rights reserved.

References

Agresti, A. *Categorical Data Analysis.* New York: Wiley, 1990.

Cochran, W. G. "The χ^2 test of goodness of fit," *Annuls of Mathematical Statistics,* 1952, 23.

Conover, W. J. *Practical Nonparametric Statistics,* 2nd ed. New York: Wiley, 1980.

DeGroot, M. H., Fienberg, S. E., and Kadane, J. B., eds. *Statistics and the Law.* New York: Wiley, 1986.

Fisher, R. A. "The logic of inductive inference (with discussion)," *Journal of the Royal Statistical Society,* Vol. 98, 1935, pp. 39–82.

Hollander, M., and Wolfe, D. A. *Nonparametric Statistical Methods.* New York: Wiley, 1973.

Savage, I. R. "Bibliography of nonparametric statistics and related topics," *Journal of the American Statistical Association,* 1953, 48.

USING TECHNOLOGY

SPSS: Chi-Square Analyses

SPSS can conduct chi-square tests for both one-way and two-way (contingency) tables.

One-Way Table

Step 1 Access the SPSS spreadsheet file that contains the variable with category values for each of the n observations in the data set. (*Note:* SPSS requires that these categories be specified numerically, e.g., 1, 2, 3.)

Step 2 Click on the "Analyze" button on the SPSS menu bar and then click on "Nonparametric Tests" and "Chi-Square," as shown in Figure 9.S.1. The resulting dialog box appears as shown in Figure 9.S.2.

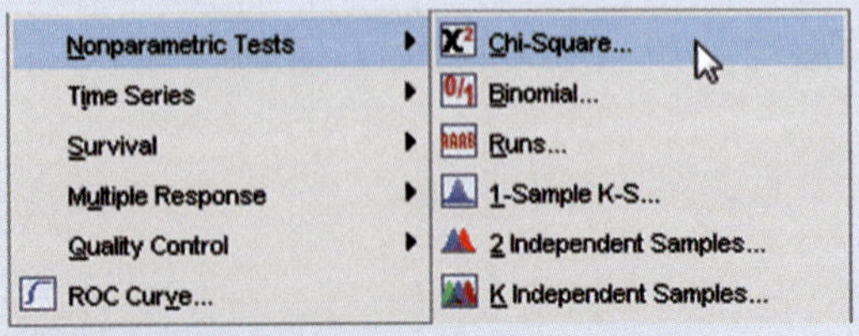

Figure 9.S.1 SPSS menu options for one-way chi-square analysis

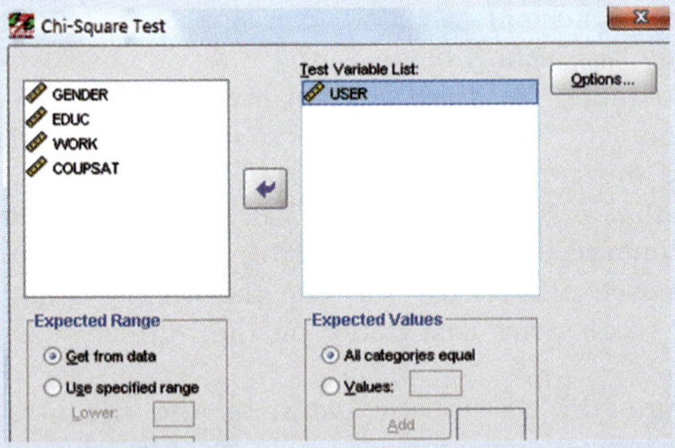

Figure 9.S.2 SPSS one-way chi-square dialog box

Step 3 Specify the qualitative variable of interest in the "Test Variable List" box.

Step 4 If you want to test for equal cell probabilities in the null hypothesis, then select the "All categories equal" option under the "Expected Values" box (as shown in Figure 9.S.2). If the null hypothesis specifies unequal cell probabilities, then select the "Values" option under the "Expected Values" box. Enter the hypothesized cell probabilities in the adjacent box, one at a time, clicking "Add" after each specification.

Step 5 Click "OK" to generate the SPSS printout.

Two-Way Table

Step 1 Access the SPSS spreadsheet file that contains the sample data. The data file should contain two qualitative variables, with category values for each of the n observations in the data set.

Step 2 Click on the "Analyze" button on the SPSS menu bar and then click on "Descriptive Statistics" and "Crosstabs," as shown in Figure 9.S.3. The resulting dialog box appears as shown in Figure 9.S.4.

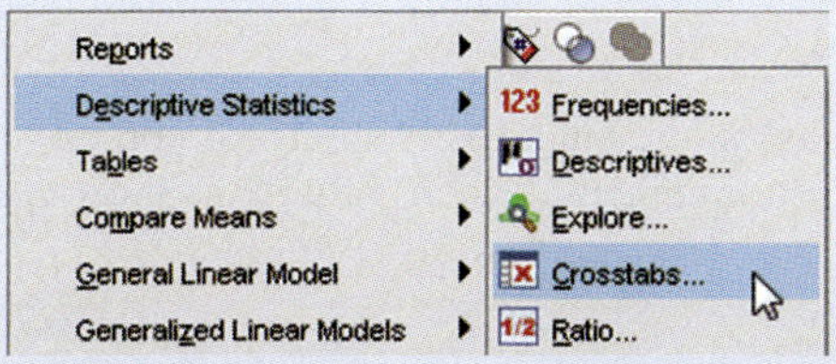

Figure 9.S.3 SPSS menu options for two-way chi-square analysis

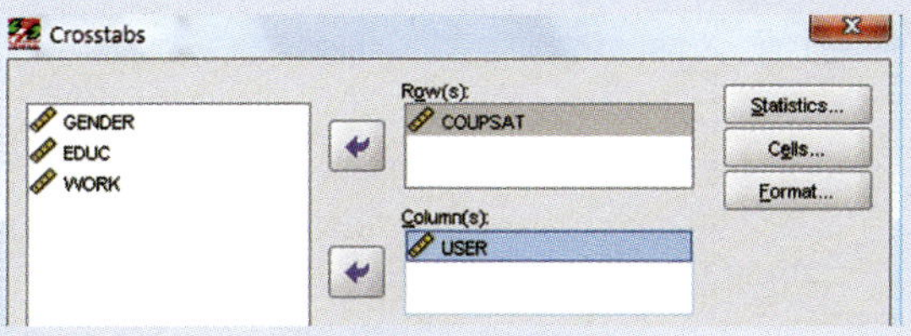

Figure 9.S.4 SPSS crosstabs dialog box

Step 3 Specify one qualitative variable in the "Row(s)" box and the other qualitative variable in the "Column(s)" box.

Step 4 Click the "Statistics" button and select the "Chi-square" option, as shown in Figure 9.S.5.

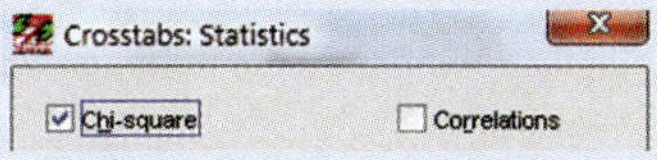

Figure 9.S.5 SPSS statistics menu selections for the two-way analysis

Step 5 Click "Continue" to return to the "Crosstabs" dialog box. If you want the contingency table to include expected values, row percentages, and/or column percentages, click the "Cells" button and make the appropriate menu selections.

Step 6 When you return to the "Crosstabs" menu screen, click "OK" to generate the SPSS printout.

Note: If your SPSS spreadsheet contains summary information (i.e., the cell counts for the contingency table) rather than the actual categorical data values for each observation, you must weight each observation in your data file by the cell count for that observation prior to running the chi-square analysis. Do this by selecting the "Data" button on the SPSS menu bar and then click on "Weight Cases" and specify the variable that contains the cell counts.

Minitab: Chi-Square Analyses

Minitab can conduct chi-square tests for both one-way and two-way (contingency) tables.

One-Way Table

Step 1 Access the Minitab worksheet file that contains the sample data for the qualitative variable of interest. [*Note:* The data file can have actual values (levels) of the variable for each observation, or, alternatively, two columns—one column listing the levels of the qualitative variable and the other column with the observed counts for each level.]

Step 2 Click on the "Stat" button on the Minitab menu bar and then click on "Tables" and "Chi-Square Goodness-of-Fit Test (One Variable)," as shown in Figure 9.M.1. The resulting dialog box appears as shown in Figure 9.M.2.

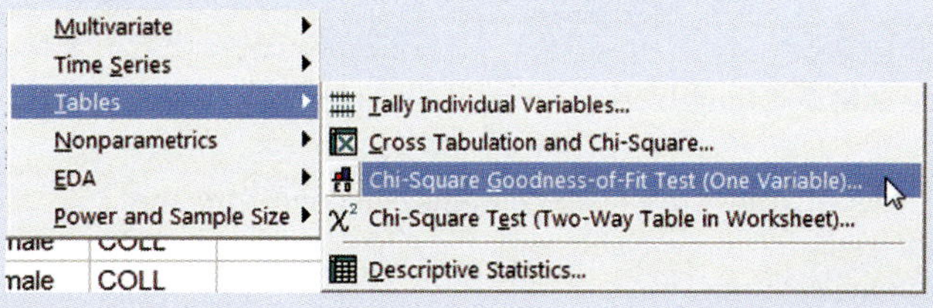

Figure 9.M.1 Minitab menu options for a one-way chi-square analysis

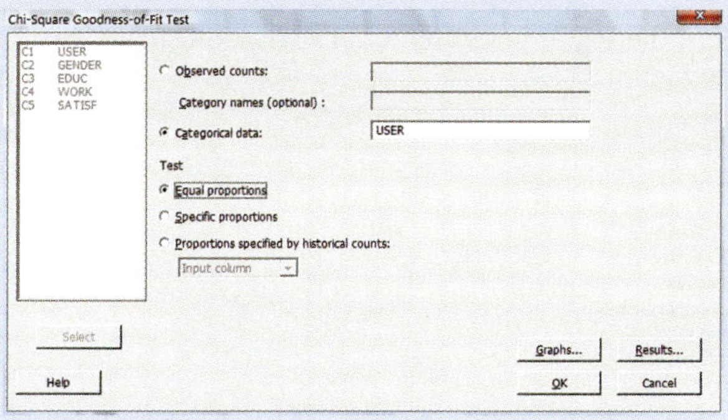

Figure 9.M.2 Minitab one-way chi-square dialog box

Step 3 If your data has one column of values for your qualitative variable, select "Categorical data" and specify the variable name (or column) in the box. If your data has summary information in two columns (see above), select "Observed counts" and specify the column with the counts and the column with the variable names in the respective boxes.

Step 4 Select "Equal proportions" for a test of equal proportions or select "Specific proportions" and enter the hypothesized proportion next to each level in the resulting box.

Step 5 Click "OK" to generate the Minitab printout.

Two-Way Table

Step 1 Access the Minitab worksheet file that contains the sample data. The data file should contain two qualitative variables, with category values for each of the n observations in the data set. Alternatively, the worksheet can contain the cell counts for each of the categories of the two qualitative variables.

Step 2 Click on the "Stat" button on the Minitab menu bar and then click on "Tables" and "Cross Tabulation and Chi-Square," (see Figure 9.M.1). The resulting dialog box appears as shown in Figure 9.M.3.

Step 3 Specify one qualitative variable in the "For rows" box and the other qualitative variable in the "For columns"

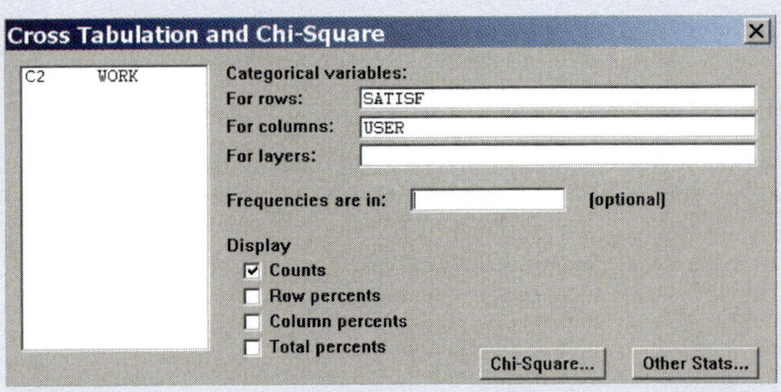

Figure 9.M.3 Minitab cross tabulation dialog box

box. [*Note:* If your worksheet contains cell counts for the categories, enter the variable with the cell counts in the "Frequencies are in" box.]

Step 4 Select the summary statistics (e.g., counts, percentages) you want to display in the contingency table.

Step 5 Click the "Chi-square" button. The resulting dialog box is shown in Figure 9.M.4.

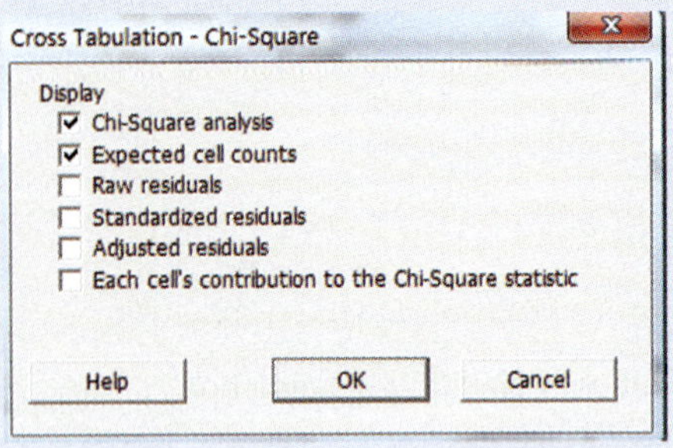

Figure 9.M.4 Minitab chi-square dialog box

Step 6 Select "Chi-Square analysis" and "Expected cell counts" and click "OK."

Step 7 When you return to the "Cross Tabulation" menu screen, click "OK" to generate the Minitab printout.

Note: If your Minitab worksheet contains only the cell counts for the contingency table in columns, click the "Chi-Square Test (Table in Worksheet)" menu option (see Figure 9.M.1) and specify the columns in the "Columns containing the table" box. Click "OK" to produce the Minitab printout.

Excel/DDXL: Chi-Square Analyses

Excel with the DDXL add-in can conduct chi-square tests for both one-way and two-way (contingency) tables; however, you must have previously computed the cell counts for the table.

One-Way Table

Step 1 Create a workbook with columns representing the levels (categories) of the qualitative variable, the cell (category) counts, and the hypothesized proportions for the levels, as shown in Figure 9.E.1.

	A	B	C	D
1	Category	Number	Proportion	
2	NONE	42	0.1	
3	STANDARD	365	0.65	
4	MERIT	193	0.25	
5				

Figure 9.E.1 Excel workbook format for one-way chi-square analysis

Step 2 Highlight (select) these data columns on the Excel spreadsheet.

Step 3 Click on "Add-Ins" in the main Excel menu bar and select "DDXL." On the resulting menu, select "Tables," as shown in Figure 9.E.2.

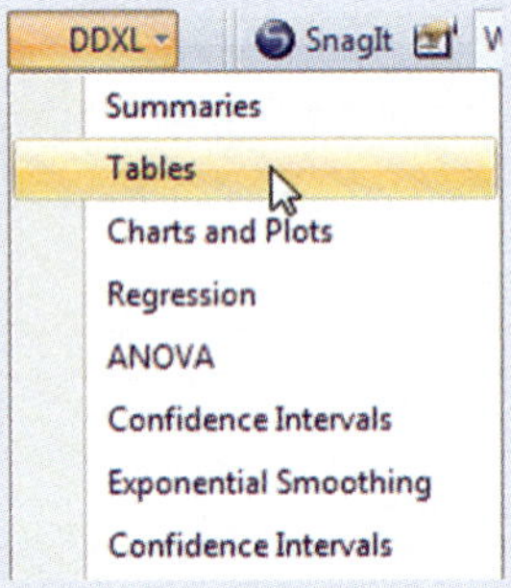

Figure 9.E.2 Excel/DDXL menu options for chi-square analyses

Step 4 On the resulting menu, select "Goodness of Fit" in the Function Type box, as shown in Figure 9.E.3.

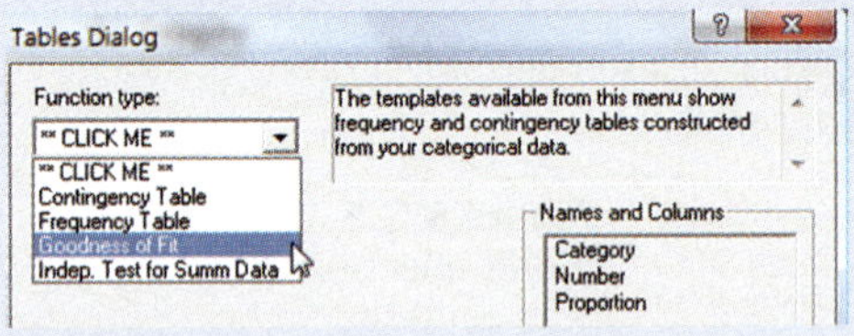

Figure 9.E.3 DDXL tables dialog box

Step 5 Move the column with the categories of the qualitative variable into the "Category Names" box, the column with the cell counts into the "Observed Counts" box, and the column with the hypothesized proportions into the "Test Distribution" box, as shown in Figure 9.E.4.

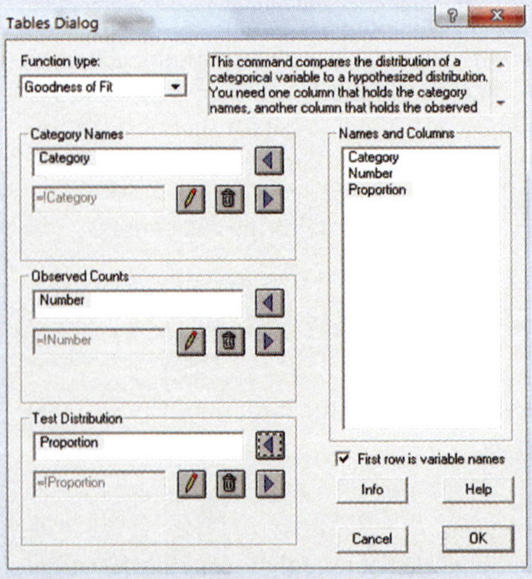

Figure 9.E.4 DDXL tables menu selections for one-way chi-square analysis

Step 6 Click "OK" to generate the 1-way chi-square printout.

Note: If your data are not summarized (i.e., you have categorical outcomes for each observation in the data set), you can use DDXL to generate a one-way summary table, as follows.

Step 7 Highlight (select) the column that contains the categorical outcomes on the Excel spreadsheet.

Step 8 Click on "Add-Ins" in the main Excel menu bar and select "DDXL." On the resulting menu, select "Tables" (see Figure 9.E.2).

Step 9 On the resulting menu, select "Frequency Table" in the Function Type box (see Figure 9.E.3).

Step 10 Move the column with the categorical outcomes into the "Categorical Variable" box and then click "OK" to generate the summary table.

Step 11 Use the information in the summary table to create a new Excel workbook and follow the instructions above.

Two-Way Table

Step 1 Create a workbook with columns representing the levels (categories) of the first qualitative variable, columns representing the levels (categories) of the second qualitative variable, and the cell counts, as shown in Figure 9.E.5.

	A	B	C	D
1	TONE	STORY	NUMBER	
2	Negative	Authentic	59	
3	Neutral	Authentic	49	
4	Positive	Authentic	20	
5	Negative	Deceptive	111	
6	Neutral	Deceptive	61	
7	Positive	Deceptive	11	

Figure 9.E.5 Excel workbook format for two-way chi-square analysis

Step 2 Highlight (select) these data columns on the Excel spreadsheet.

Step 3 Click on "Add-Ins" in the main Excel menu bar and select "DDXL." On the resulting menu, select "Tables," as shown in Figure 9.E.2.

Step 4 On the resulting menu, select "Independent Test for Summ Data" in the Function Type box (see Figure 9.E.3).

Step 5 Move the column with the categories of the first qualitative variable into the "Variable One Names" box, the column with the categories of the second qualitative variable into the "Variable Two Names" box, and the column with the cell counts into the "Counts" box, as shown in Figure 9.E.6.

Step 6 Click "OK" to generate the two-way chi-square printout.

Note: If your data are not summarized (i.e., you have categorical outcomes for each qualitative variable), you can use DDXL to generate a two-way summary (contingency) table, as follows:

Step 1 Highlight (select) the columns that contain the categorical outcomes for the two qualitative variables on the Excel spreadsheet.

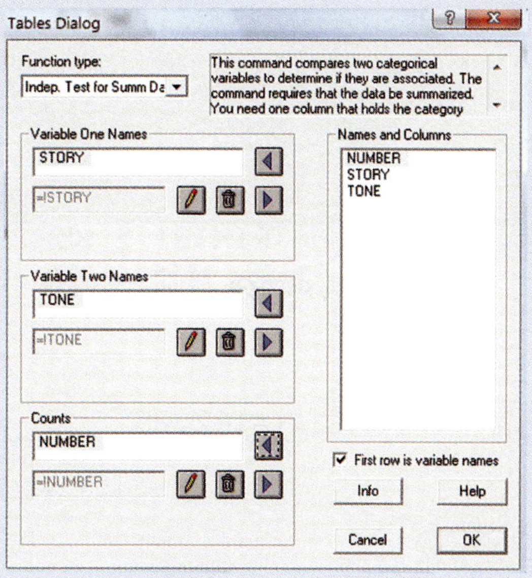

Figure 9.E.6 DDXL tables menu selections for two-way chi-square analysis

Step 2 Click on "Add-Ins" in the main Excel menu bar and select "DDXL." On the resulting menu, select "Tables" (see Figure 9.E.2).

Step 3 On the resulting menu, select "Contingency Table" in the Function Type box (see Figure 9.E.3).

Step 4 Move the column with the outcomes for the first qualitative variable into the "1st Categorical Variable" box and the column with the outcomes for the second qualitative variable into the "2nd Categorical Variable" box. Then click "OK" to generate the contingency table.

Step 5 Use the information in the contingency table to create a new Excel workbook and follow the instructions above.

TI-83/TI-84 Graphing Calculator: Chi-Square Analyses

The TI-83/TI-84 graphing calculator can be used to conduct a chi-square test for a two-way (contingency) table but cannot conduct a chi-square test for a one-way table.

Two-Way (Contingency) Table

Step 1 *Access the Matrix menu to enter the observed values*

- Press **2nd** $\mathbf{x^{-1}}$ for **MATRX**
- Arrow right to **EDIT**
- Press **ENTER**
- Use the **ARROW** key to enter the row and column dimensions of your observed Matrix
- Use the **ARROW** key to enter your observed values into Matrix [A]

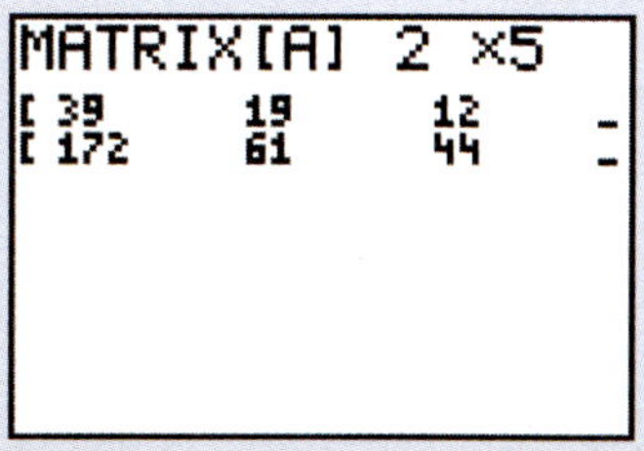

Step 2 *Access the Matrix menu to enter the expected values*

- Press **2nd x^{-1}** for **MATRX**
- Arrow right to **EDIT**
- Arrow down to **2:[B]**
- Press **ENTER**
- Use the **ARROW** key to enter the row and column dimensions of your expected matrix (The dimensions will be the same as in Matrix A)
- Use the **ARROW** key to enter your expected values into Matrix [B]

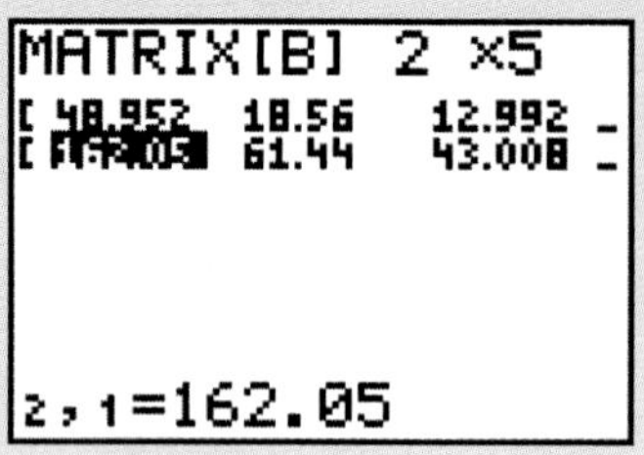

Step 3 *Access the Statistical Tests menu and perform the Chi-square test.*

- Press **STAT**
- Arrow right to **TESTS**
- Arrow down to χ^2 **Test**
- Press **ENTER**
- Arrow down to **Calculate**
- Press **ENTER**

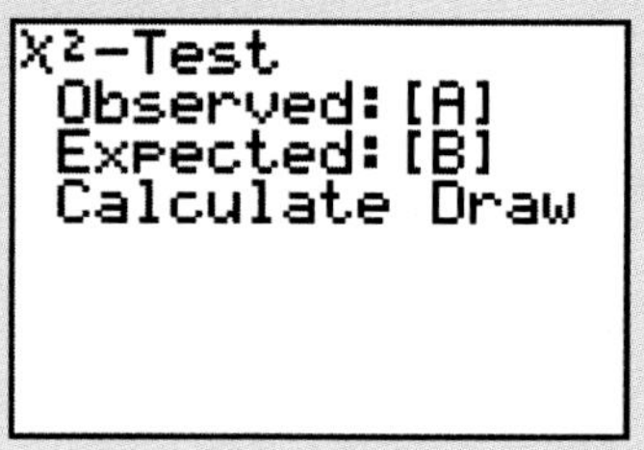

MAKING BUSINESS DECISIONS

Discrimination in the Workplace

Covers Chapters 8 and 9 — Title VII of the Civil Rights Act of 1964 prohibits discrimination in the workplace on the basis of race, color, religion, gender, or national origin. The Age Discrimination in Employment Act of 1967 (ADEA) protects workers age 40 to 70 against discrimination based on age. The potential for discrimination exists in such processes as hiring, promotion, compensation, and termination.

In 1971 the U.S. Supreme Court established that employment discrimination cases fall into two categories: **disparate treatment** and **disparate impact.** In the former, the issue is whether the employer intentionally discriminated against a worker. For example, if the employer considered an individual's race in deciding whether to terminate him, the case is one of disparate treatment. In a disparate impact case, the issue is whether employment practices have an adverse impact on a protected group or class of people, even when the employer does not intend to discriminate.

Part I: Downsizing at a Computer Firm

Disparate impact cases almost always involve the use of statistical evidence and expert testimony by professional statisticians. Attorneys for the plaintiffs frequently use hypothesis test results in the form of p-values in arguing the case for their clients.

Table C4.1 was recently introduced as evidence in a race case that resulted from a round of layoffs during the downsizing of a division of a computer manufacturer. The company had selected 51 of the division's 1,215 employees to lay off. The plaintiffs—in this case 15 of the 20 African Americans who were laid off—were suing the company for $20 million in damages.

The company's lawyers argued that the selections followed from a performance-based ranking of all employees. The plaintiffs' legal team and their expert witnesses, citing the results of a statistical test of hypothesis, argued that layoffs were a function of race.

The validity of the plaintiffs' interpretation of the data is dependent on whether the assumptions of the test are met in this situation. In particular, like all hypothesis tests presented in this text, the assumption of random sampling must hold. If it does not, the results of the test may be due to the violation of this assumption rather than to discrimination. In general, the appropriateness of the testing procedure is dependent on the test's ability to capture the relevant aspects of the employment process in question (DeGroot, Fienberg, and Kadane, *Statistics and the Law,* 1986).

Prepare a document to be submitted as evidence in the case (i.e., an exhibit), in which you evaluate the validity of the plaintiffs' interpretation of the data. Your evaluation should be based in part on your knowledge of the processes companies use to lay off employees and how well those processes are reflected in the hypothesis-testing procedure employed by the plaintiffs.

Table C4.1 Summary of Downsizing Data for Race Case

		Decision	
		Retained	Laid off
Race	White	1,051	31
	Black	113	20

Source: Confidential personal communication with P. George Benson.

Data Set: LAYOFFS

Part II: Age Discrimination—You Be the Judge

In 1996, as part of a significant restructuring of product lines, AJAX Pharmaceuticals (a fictitious name for a real company) laid off 24 of 55 assembly-line workers in its Pittsburgh manufacturing plant. Citing the ADEA, 11 of the laid-off workers claimed they were discriminated against on the basis of age and sued AJAX for $5,000,000. Management disputed the claim, saying that because the workers were essentially interchangeable, they had used random sampling to choose the 24 workers to be terminated.

Table C4.2 Data for Age Discrimination Case

Employee	Yearly Wages	Age	Employment Status
*Adler, C. J.	$41,200	45	Terminated
Alario, B. N.	39,565	43	Active
Anders, J. M.	30,980	41	Active
Bajwa, K. K.	23,225	27	Active
Barny, M. L.	21,250	26	Active
*Berger, R. W.	41,875	45	Terminated
Brenn, L. O.	31,225	41	Active
Cain, E. J.	30,135	36	Terminated
Carle, W. J.	29,850	32	Active
Castle, A. L.	21,850	22	Active
Chan, S. D.	43,005	48	Terminated
Cho, J. Y.	34,785	41	Active
Cohen, S. D.	25,350	27	Active
Darel, F. E.	36,300	42	Active
*Davis, D. E.	40,425	46	Terminated
*Dawson, P. K.	39,150	42	Terminated
Denker, U. H.	19,435	19	Active
Dorando, T. R.	24,125	28	Active
Dubois, A. G.	30,450	40	Active
England, N.	24,750	25	Active
Estis, K. B.	22,755	23	Active
Fenton, C. K.	23,000	24	Active
Finer, H. R.	42,000	46	Terminated
*Frees, O. C.	44,100	52	Terminated
Gary, J. G.	44,975	55	Terminated
Gillen, D. J.	25,900	27	Active
Harvey, D. A.	40,875	46	Terminated
Higgins, N. M.	38,595	41	Active
*Huang, T. J.	42,995	48	Terminated
Jatho, J. A.	31,755	40	Active
Johnson, C. H.	29,540	32	Active
Jurasik, T. B.	34,300	41	Active
Klein, K. L.	43,700	51	Terminated
Lang, T. F.	19,435	22	Active
Liao, P. C.	28,750	32	Active
*Lostan, W. J.	44,675	52	Terminated
Mak, G. L.	35,505	38	Terminated
Maloff, V. R.	33,425	38	Terminated
McCall, R. M.	31,300	36	Terminated
*Nadeau, S. R.	42,300	46	Terminated
*Nguyen, O. L.	43,625	50	Terminated

(continued)

Oas, R. C.	37,650	42	Active
*Patel, M. J.	38,400	43	Terminated
Porter, K. D.	32,195	35	Terminated
Rosa, L. M.	19,435	21	Active
Roth, J. H.	32,785	39	Terminated
Sayino, G. L.	37,900	42	Active
Scott, I. W.	29,150	30	Terminated
Smith, E. E.	35,125	41	Active
Teel, Q. V.	27,655	33	Active
*Walker, F. O.	42,545	47	Terminated
Wang, T. G.	22,200	32	Active
Yen, D. O.	40,350	44	Terminated
Young, N. L.	28,305	34	Active
Zeitels, P. W.	36,500	42	Active

**Denotes plaintiffs*

Table C4.2 lists the 55 assembly-line workers and identifies which were terminated and which remained active. Plaintiffs are denoted by an asterisk. These data were used by both the plaintiffs and the defendants to determine whether the layoffs had an adverse impact on workers age 40 and over and to establish the credibility of management's random sampling claim.

Using whatever statistical methods you think are appropriate, build a case that supports the plaintiffs' position. (Call documents related to this issue Exhibit A.) Similarly, build a case that supports the defendants' position. (Call these documents Exhibit B.) Then discuss which of the two cases is more convincing and why. [*Note:* The data for this case are available in the **DISCRIM** file, described in the table.]

Variable	Type
LASTNAME	QL
WAGES	QN
AGE	QN
STATUS	QL

(Number of Observations: 55) *Data Set:* DISCRIM

CONTENTS

10 Simple Linear Regression

Where We've Been

- Presented methods for estimating and testing population parameters (e.g., the mean, proportion, and variance) for a single sample
- Extended these methods to allow for a comparison of population parameters for multiple samples

Where We're Going

- Introduce the straight-line (*simple linear regression*) model as a means of relating one quantitative variable to another quantitative variable
- Introduce the *correlation coefficient* as a means of relating one quantitative variable to another quantitative variable
- Assess how well the simple linear regression model fits the sample data
- Employ the simple linear regression model for predicting the value of one variable from a specified value of another variable

Statistics IN Action Legal Advertising—Does It Pay?

According to the American Bar Association, there are over 1 million lawyers competing for your business. To gain a competitive edge, these lawyers aggressively advertise their services. The advertising of legal services has long been a controversial subject, with many believing that it constitutes an unethical (and in some cases even illegal) practice. Nonetheless, legal advertisements appear in nearly all media, ranging from the covers of telephone directories to infomercials on television, as well as a significant presence on the Internet. In fact, Erickson Marketing, Inc., reports that "attorneys are the #1 category of advertising in the Yellow Pages."

For this *Statistics in Action,* we present an actual recent case involving two former law partners. One partner (A) sued the other (B) over who should pay what share of the expenses of their former partnership. Partner A handled personal injury (PI) cases, while partner B handled only worker's compensation (WC) cases. The firm's advertising was focused on personal injury only, but partner A claimed that the ads resulted in the firm getting more WC cases for partner B, and therefore partner B should share the advertising expenses. *(continued)*

Statistics IN Action *(continued)*

Table SIA10.1 shows the firm's new PI and WC cases each month over a 42-month period for this partnership. Also shown is the total expenditure on advertising each month and over the previous six months. Do these data provide support for the hypothesis that increased advertising expenditures are associated with more PI cases? With more WC cases?

If advertising expenditures have a statistically significant association with the number of cases, does this necessarily mean that there is a causal relationship, that is, that spending more on advertising causes an increase in the number of cases? Based

Table SIA 10.1 Legal Advertising Data

Month	Advertising Expenditure ($)	New PI Cases	New WC Cases	6 Months Cumulative Adv. Exp. ($)
1	9,221.55	7	26	n/a
2	6,684.00	9	33	n/a
3	200.00	12	18	n/a
4	14,546.75	27	15	n/a
5	5,170.14	9	19	n/a
6	5,810.30	13	26	n/a
7	5,816.20	11	24	41,632.74
8	8,236.38	7	22	38,227.39
9	−2,089.55	13	12	39,779.77
10	29,282.24	7	15	37,490.22
11	9,193.58	9	21	52,225.71
12	9,499.18	8	24	56,249.15
13	11,128.76	18	25	59,938.03
14	9,057.64	9	19	65,250.59
15	13,604.54	25	12	66,071.85
16	14,411.76	26	33	81,765.94
17	13,724.28	27	32	66,895.46
18	13,419.42	12	21	71,426.16
19	17,372.33	14	18	75,346.40
20	6,296.35	5	25	81,589.97
21	13,191.59	22	12	78,828.68
22	26,798.80	15	7	78,415.73
23	18,610.95	12	22	90,802.77
24	829.53	18	27	95,689.44
25	16,976.53	20	25	83,099.55
26	14,076.98	38	26	82,703.75
27	24,791.75	13	28	90,484.38
28	9,691.25	18	31	102,084.54
29	28,948.25	21	40	84,976.99
30	21,373.52	7	39	95,314.29
31	9,675.25	16	41	115,858.28
32	33,213.55	12	48	108,557.00
33	19,859.85	15	28	127,693.57
34	10,475.25	18	29	122,761.67
35	24,790.84	30	20	123,545.67
36	36,660.94	12	27	119,388.26
37	8,812.50	30	26	134,675.68
38	41,817.75	20	45	133,812.93
39	27,399.33	19	30	142,417.13
40	25,723.10	29	33	149,956.61
41	16,312.10	58	24	165,204.46
42	26,332.78	42	40	156,725.72
43	60,207.58	24	36	146,397.56
44	42,485.39	47	29	197,792.64
45	35,601.92	24	17	198,460.28
46	72,071.50	14	13	206,662.87
47	12,797.11	31	15	253,011.27
48	12,310.50	26	16	249,496.28

Source: Info Tech, Inc., Gainesville, Florida.

Data Set: LEGALADV

on these data, should partner A or partner B bear the brunt of the advertising expenditures?

The data for the case are saved in the **LEGALADV** file. In the *Statistics in Action Revisited* sections of this chapter, we examine the relationship between monthly advertising expenditure and the number of new legal cases in an attempt to answer these questions.

Data Set: LEGALADV

Statistics IN Action Revisited

- Estimating a Straight-Line Regression Model (p. 570)
- Assessing How Well a Straight-Line Regression Model Fits Data (p. 585)
- Using the Coefficient of Correlation and the Coefficient of Determination (p. 595)
- Prediction Using the Straight-Line Model (p. 604)

In Chapters 5–8, we described methods for making inferences about population means. The mean of a population was treated as a *constant,* and we showed how to use sample data to estimate or to test hypotheses about this constant mean. In many applications, the mean of a population is not viewed as a constant but rather as a variable. For example, the mean sale price of residences sold this year in a large city might be treated as a variable that depends on the square feet of living space in the residence. For example, the relationship might be

$$\text{Mean sale price} = \$30{,}000 + \$60\ (\text{square feet})$$

This formula implies that the mean sale price of 1,000-square-foot homes is \$90,000, the mean sale price of 2,000-square-foot homes is \$150,000, and the mean sale price of 3,000-square-foot homes is \$210,000.

In this chapter, we discuss situations in which the mean of the population is treated as a variable, dependent on the value of another variable. The dependence of residential sale price on the square feet of living space is one illustration. Other examples include the dependence of mean sales revenue of a firm on advertising expenditure, the dependence of mean starting salary of a college graduate on the student's GPA, and the dependence of mean monthly production of automobiles on the total number of sales in the previous month.

We begin our discussion with the simplest of all models relating a population mean to another variable—*the straight-line model.* We show how to use the sample data to estimate the straight-line relationship between the mean value of one variable, y, as it relates to a second variable, x. The methodology of estimating and using a straight-line relationship is referred to as *simple linear regression analysis.*

10.1 Probabilistic Models

An important consideration in merchandising a product is the amount of money spent on advertising. Suppose you want to model the monthly sales revenue of an appliance store as a function of the monthly advertising expenditure. The first question to be answered is this: "Do you think an exact relationship exists between these two variables?" That is, do you think it is possible to state the exact monthly sales revenue if the amount spent on advertising is known? We think you will agree with us that this is *not* possible for several reasons. Sales depend on many variables other than advertising expenditure—for example, time of year, the state of the general economy, inventory, and price structure. Even if many variables are included in a model (the topic of Chapter 11), it is still unlikely that we would be able to predict the monthly sales *exactly.* There will almost certainly be some variation in monthly sales due strictly to *random phenomena* that cannot be modeled or explained.

If we were to construct a model that hypothesized an exact relationship between variables, it would be called a **deterministic model.** For example, if we believe that y, the monthly sales revenue, will be exactly 15 times x, the monthly advertising expenditure, we write

$$y = 15x$$

This represents a *deterministic relationship* between the variables y and x. It implies that y can always be determined exactly when the value of x is known. *There is no allowance for error in this prediction.*

If, on the other hand, we believe there will be unexplained variation in monthly sales—perhaps caused by important but unincluded variables or by random phenomena—we discard the deterministic model and use a model that accounts for this **random error.** This **probabilistic model** includes both a deterministic component and a random error component. For example, if we hypothesize that the sales y are related to advertising expenditure x by

$$y = 15x + \text{Random error}$$

we are hypothesizing a *probabilistic relationship* between y and x. Note that the deterministic component of this probabilistic model is $15x$.

Figure 10.1a shows the possible values of y and x for five different months, when the model is deterministic. All the pairs of (x, y) data points must fall exactly on the line because a deterministic model leaves no room for error.

Figure 10.1b shows a possible set of points for the same values of x when we are using a probabilistic model. Note that the deterministic part of the model (the straight line itself) is the same. Now, however, the inclusion of a random error component allows the monthly sales to vary from this line. Because we know that the sales revenue does vary randomly for a given value of x, the probabilistic model provides a more realistic model for y than does the deterministic model.

General Form of Probabilistic Models

$$y = \text{Deterministic component} + \text{Random error}$$

where y is the variable of interest. We always assume that the mean value of the random error equals 0. This is equivalent to assuming that the mean value of y, $E(y)$, equals the deterministic component of the model; that is,

$$E(y) = \text{Deterministic component}$$

In this chapter, we present the simplest of probabilistic models—the **straight-line model**—which derives its name from the fact that the deterministic portion of the model graphs as a straight line. Fitting this model to a set of data is an example of **regression analysis,** or **regression modeling.** The elements of the straight-line model are summarized in the next box.

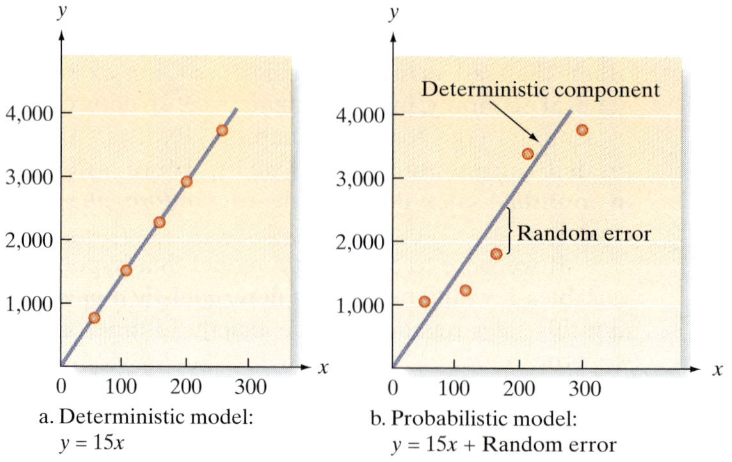

Figure 10.1
Possible sales revenues, y, for five different months, x

A First-Order (Straight-Line) Probabilistic Model

$$y = \beta_0 + \beta_1 x + \varepsilon$$

where

y = **Dependent** *or* **response variable** (variable to be modeled)

x = **Independent** *or* **predictor variable** (variable used as a predictor of y)*

$E(y) = \beta_0 + \beta_1 x$ = Deterministic component

ε (epsilon) = Random error component

β_0 (beta zero) = **y-intercept of the line,** that is, the point at which the line *intercepts or cuts through the* y-axis (see Figure 10.2)

β_1 (beta one) = **Slope of the line,** that is, the change (amount of increase or decrease) in the deterministic component of y for every 1-unit increase in x

[*Note:* A *positive* slope implies that $E(y)$ *increases* by the amount β_1 for each unit increase in x (see Figure 10.2). A *negative* slope implies that $E(y)$ *decreases* by the amount β_1.]

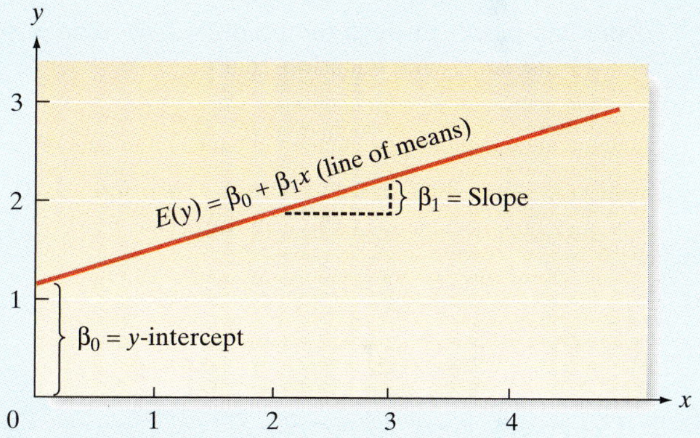

Figure 10.2
The straight-line model

BIOGRAPHY

FRANCIS GALTON (1822–1911)

The Law of Universal Regression

Francis Galton was the youngest of seven children born to a middle-class English family of Quaker faith. A cousin of Charles Darwin, Galton attended Trinity College (Cambridge, England) to study medicine. Due to the death of his father, Galton was unable to obtain his degree. His competence in both medicine and mathematics, however, led Galton to pursue a career as a scientist. Galton made major contributions to the fields of genetics, psychology, meteorology, and anthropology. Some consider Galton to be the first social scientist for his applications of the novel statistical concepts of the time—in particular, regression and correlation. While studying natural inheritance in 1886, Galton collected data on the heights of parents and adult children. He noticed the tendency for tall (or short) parents to have tall (or short) children, but the children were not as tall (or short), on average, as their parents. Galton called this phenomenon the "law of universal regression," for the average heights of adult children tended to "regress" to the mean of the population. Galton, with the help of his friend and disciple, Karl Pearson, applied the straight-line model to the height data, and the term *regression* model was coined. ■

In the probabilistic model, the deterministic component is referred to as the **line of means** because the mean of y, $E(y)$, is equal to the straight-line component of the model—that is,

$$E(y) = \beta_0 + \beta_1 x$$

Note that the Greek symbols β_0 and β_1, respectively, represent the y-intercept and slope of the model. They are population parameters that will be known only if we have access to the entire population of (x, y) measurements. Together with a specific value of the independent variable x, they determine the mean value of y, which is a specific point on the line of means (Figure 10.2).

The values of β_0 and β_1 will be unknown in almost all practical applications of regression analysis. The process of developing a model, estimating the unknown parameters, and using the model can be viewed as the five-step procedure shown in the next box.

Step 1: Hypothesize the deterministic component of the model that relates the mean, $E(y)$, to the independent variable x (Section 10.1).

Step 2: Use the sample data to estimate unknown parameters in the model (Section 10.2).

Step 3: Specify the probability distribution of the random error term and estimate the standard deviation of this distribution (Section 10.3).

*The word *independent* should not be interpreted in a probabilistic sense, as defined in Chapter 3. The phrase *independent variable* is used in regression analysis to refer to a predictor variable for the response y.

Step 4: Statistically evaluate the usefulness of the model (Sections 10.4 and 10.5).
Step 5: When satisfied that the model is useful, use it for prediction, estimation, and other purposes (Section 10.6).

Exercises 10.1–10.9

Learning the Mechanics

10.1 In each case, graph the line that passes through the given points.
a. (1, 1) and (5, 5) **b.** (0, 3) and (3, 0)
c. (−1, 1) and (4, 2) **d.** (−6, −3) and (2, 6)

10.2 Give the slope and y-intercept for each of the lines graphed in Exercise 10.1.

10.3 The equation for a straight line (deterministic model) is

$$y = \beta_0 + \beta_1 x$$

If the line passes through the point (−2, 4), then $x = -2$, $y = 4$ must satisfy the equation; that is,

$$4 = \beta_0 + \beta_1(-2)$$

Similarly, if the line passes through the point (4, 6), then $x = 4$, $y = 6$ must satisfy the equation; that is,

$$6 = \beta_0 + \beta_1(4)$$

Use these two equations to solve for β_0 and β_1; then find the equation of the line that passes through the points (−2, 4) and (4, 6).

10.4 Refer to Exercise 10.3. Find the equations of the lines that pass through the points listed in Exercise 10.1.

10.5 Plot the following lines:
a. $y = 4 + x$ **b.** $y = 5 - 2x$
c. $y = -4 + 3x$ **d.** $y = -2x$
e. $y = x$ **f.** $y = .50 + 1.5x$

10.6 Give the slope and y-intercept for each of the lines graphed in Exercise 10.5.

10.7 Why do we generally prefer a probabilistic model to a deterministic model? Give examples for when the two types of models might be appropriate.

10.8 What is the line of means?

10.9 If a straight-line probabilistic relationship relates the mean $E(y)$ to an independent variable x, does it imply that every value of the variable y will always fall exactly on the line of means? Why or why not?

10.2 Fitting the Model: The Least Squares Approach

After the straight-line model has been hypothesized to relate the mean $E(y)$ to the independent variable x, the next step is to collect data and to estimate the (unknown) population parameters, the y-intercept β_0 and the slope β_1.

To begin with a simple example, suppose an appliance store conducts a 5-month experiment to determine the effect of advertising on sales revenue. The results are shown in Table 10.1. (The number of measurements and the measurements themselves are unrealistically simple to avoid arithmetic confusion in this introductory example.) This set of data will be used to demonstrate the five-step procedure of regression modeling given in Section 10.1. In this section, we hypothesize the deterministic component of the model and estimate its unknown parameters (steps 1 and 2). The model assumptions and the random error component (step 3) are the subjects of Section 10.3, whereas Sections 10.4 and 10.5 assess the utility of the model (step 4). Finally, we use the model for prediction and estimation (step 5) in Section 10.6.

Table 10.1 Advertising-Sales Data

Month	Advertising Expenditure, x ($100s)	Sales Revenue, y ($1,000s)
1	1	1
2	2	1
3	3	2
4	4	2
5	5	4

Data Set: ADSALES

Step 1: *Hypothesize the deterministic component of the probabilistic model.* As stated before, we will consider only straight-line models in this chapter. Thus, the complete model to relate mean sales revenue $E(y)$ to advertising expenditure x is given by

$$E(y) = \beta_0 + \beta_1 x$$

Step 2: *Use sample data to estimate unknown parameters in the model.* This step is the subject of this section—namely, how can we best use the information in the sample of five observations in Table 10.1 to estimate the unknown y-intercept β_0 and slope β_1?

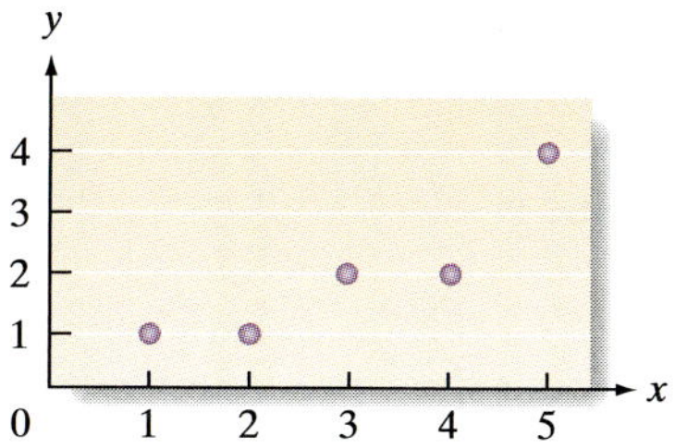

Figure 10.3
Scattergram for data in Table 10.1

To determine whether a linear relationship between y and x is plausible, it is helpful to plot the sample data in a **scattergram** (or **scatterplot**). Recall (Section 2.9) that a scattergram locates each of the five data points on a graph, as shown in Figure 10.3. Note that the scattergram suggests a general tendency for y to increase as x increases. If you place a ruler on the scattergram, you will see that a line may be drawn through three of the five points, as shown in Figure 10.4. To obtain the equation of this visually fitted line, note that the line intersects the y-axis at $y = -1$, so the y-intercept is -1. Also, y increases exactly 1 unit for every 1-unit increase in x, indicating that the slope is $+1$. Therefore, the equation is

$$\tilde{y} = -1 + 1(x) = -1 + x$$

where $\tilde{y}$ is used to denote the predicted y from the visual model.

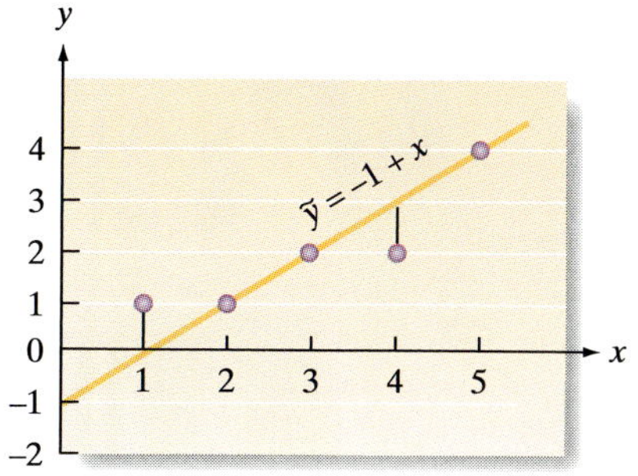

Figure 10.4
Visual straight line fitted to the data in Figure 10.3

One way to decide quantitatively how well a straight line fits a set of data is to note the extent to which the data points deviate from the line. For example, to evaluate the model in Figure 10.4, we calculate the magnitude of the *deviations* (i.e., the differences between the observed and the predicted values of y). These deviations, or **errors of prediction,** are the vertical distances between observed and predicted values (see Figure 10.4). The observed and predicted values of y, their differences, and their squared differences are shown in Table 10.2. Note that the *sum of errors* equals 0 and the *sum of squares of the errors* (SSE), which gives greater emphasis to large deviations of the points from the line, is equal to 2.

You can see by shifting the ruler around the graph that it is possible to find many lines for which the sum of errors is equal to 0, but it can be shown that there is one (and only one) line for which the SSE is a *minimum.* This line is called the **least squares line,** the **regression line,** or the **least squares prediction equation.** The methodology used to obtain this line is called the **method of least squares.**

Table 10.2 **Comparing Observed and Predicted Values for the Visual Model**

x	y	$\tilde{y} = -1 + x$	$(y - \tilde{y})$	$(y - \tilde{y})^2$
1	1	0	$(1 - 0) = 1$	1
2	1	1	$(1 - 1) = 0$	0
3	2	2	$(2 - 2) = 0$	0
4	2	3	$(2 - 3) = -1$	1
5	4	4	$(4 - 4) = 0$	0
			Sum of errors = 0	Sum of squared errors (SSE) = 2

To find the least squares prediction equation for a set of data, assume that we have a sample of n data points consisting of pairs of values of x and y, say (x_1, y_1), $(x_2, y_2), \ldots, (x_n, y_n)$. For example, the $n = 5$ data points shown in Table 10.2 are (1, 1), (2, 1), (3, 2), (4, 2), and (5, 4). The fitted line, which we will calculate based on the five data points, is written as

$$\hat{y} = \hat{\beta}_0 + \hat{\beta}_1 x$$

The "hats" indicate that the symbols below them are estimates: $\hat{y}$ (y-hat) is an estimator of the mean value of y, $E(y)$, and a predictor of some future value of y; and $\hat{\beta}_0$ and $\hat{\beta}_1$ are estimators of β_0 and β_1, respectively.

For a given data point, say the point (x_i, y_i), the observed value of y is y_i, and the predicted value of y would be obtained by substituting x_i into the prediction equation:

$$\hat{y}_i = \hat{\beta}_0 + \hat{\beta}_1 x_i$$

And the deviation of the ith value of y from its predicted value is

$$(y_i - \hat{y}_i) = [y_i - (\hat{\beta}_0 + \hat{\beta}_1 x_i)]$$

Then the sum of squares of the deviations of the y-values about their predicted values for all the n points is

$$\text{SSE} = \sum [y_i - (\hat{\beta}_0 + \hat{\beta}_1 x_i)]^2$$

The quantities $\hat{\beta}_0$ and $\hat{\beta}_1$ that make the SSE a minimum are called the **least squares estimates** of the population parameters β_0 and β_1, and the prediction equation $\hat{y} = \hat{\beta}_0 + \hat{\beta}_1 x$ is called the *least squares line.*

The **least squares line** $\hat{y} = \hat{\beta}_0 + \hat{\beta}_1 x$ is one that has the following two properties:

1. The sum of the errors equals 0, i.e., mean error = 0.
2. The sum of squared errors (SSE) is smaller than for any other straight-line model, i.e., the error variance is minimum.

The values of $\hat{\beta}_0$ and $\hat{\beta}_1$ that minimize the SSE are (proof omitted) given by the formulas in the box.*

Formulas for the Least Squares Estimates

Slope: $\hat{\beta}_1 = \dfrac{\text{SS}_{xy}}{\text{SS}_{xx}}$

y-intercept: $\hat{\beta}_0 = \bar{y} - \hat{\beta}_1 \bar{x}$

where** $\text{SS}_{xy} = \Sigma(x_i - \bar{x})(y_i - \bar{y})$

$\text{SS}_{xx} = \Sigma(x_i - \bar{x})^2$

n = Sample size

*Students who are familiar with calculus should note that the values of β_0 and β_1 that minimize $\text{SSE} = \Sigma(y_i - \hat{y}_i)^2$ are obtained by setting the two partial derivatives $\partial\text{SSE}/\partial\beta_0$ and $\partial\text{SSE}/\partial\beta_1$ equal to 0. The solutions to these two equations yield the formulas shown in the box. Furthermore, we denote the *sample* solutions to the equations by $\hat{\beta}_0$ and $\hat{\beta}_1$, where the "hat" denotes that these are sample estimates of the true population intercept β_0 and true population slope β_1.

**Alternatively, you can use the following "shortcut" formulas:

$$\text{SS}_{xy} = \Sigma x_i y_i - \frac{(\Sigma x_i)(\Sigma y_i)}{n}$$

$$\text{SS}_{xx} = \Sigma x_i^2 - \frac{(\Sigma x_i)^2}{n}$$

Example 10.1
Applying the Method of Least Squares to the Advertising–Sales Data

Problem Refer to the advertising monthly-sales data presented in Table 10.1. Consider the straight-line model, $E(y) = \beta_0 + \beta_1 x$, where y = sales revenue (thousands of dollars) and x = advertising expenditure (hundreds of dollars).

a. Use the method of least squares to estimate the values of β_0 and β_1.

b. Predict the sales revenue when advertising expenditure is \$200 (i.e., when $x = 2$).

c. Find SSE for the analysis.

d. Give practical interpretations to β_0 and β_1.

Solution

a. We used Excel to make the preliminary computations for finding the least squares line. The Excel spreadsheet is shown in Figure 10.5. Using the values on the spreadsheet, we find

$$\bar{x} = \frac{\sum x}{5} = \frac{15}{5} = 3$$

$$\bar{y} = \frac{\sum y}{5} = \frac{10}{5} = 2$$

$$SS_{xy} = \sum(x - \bar{x})(y - \bar{y}) = \sum(x - 3)(y - 2) = 7$$

$$SS_{xx} = \sum(x - \bar{x})^2 = \sum(x - 3)^2 = 10$$

Then the slope of the least squares line is

$$\hat{\beta}_1 = \frac{SS_{xy}}{SS_{xx}} = \frac{7}{10} = .7$$

and the y-intercept is

$$\hat{\beta}_0 = \bar{y} - \hat{\beta}_1\bar{x} = 2 - (.7)(3) = 2 - 2.1 = -.1$$

The least squares line is thus

$$\hat{y} = \hat{\beta}_0 + \hat{\beta}_1 x = -.1 + .7x$$

The graph of this line is shown in Figure 10.6.

b. The predicted value of y for a given value of x can be obtained by substituting into the formula for the least squares line. Substituting $x = 2$ into the least squares equation yields

$$\hat{y} = -.1 + .7x = -.1 + .7(2) = 1.3$$

Thus, when advertising expenditure is \$200, we predict monthly sales revenue to be \$1,300. We show how to find a prediction interval for y in Section 10.6.

	A	B	C	D	E	F	G	H
1		AdvExp (X)	SalesRev (Y)	(X - 3)	(Y - 2)	(X-3)(Y-2)	(X-3)(X-3)	(Y-2)(Y-2)
2		1	1	-2	-1	2	4	1
3		2	1	-1	-1	1	1	1
4		3	2	0	0	0	0	0
5		4	2	1	0	0	1	0
6		5	4	2	2	4	4	4
7								
8	Totals	15	10	0	0	7	10	6
9	Mean	3	2					

Figure 10.5
Excel spreadsheet showing calculations for advertising-sales example

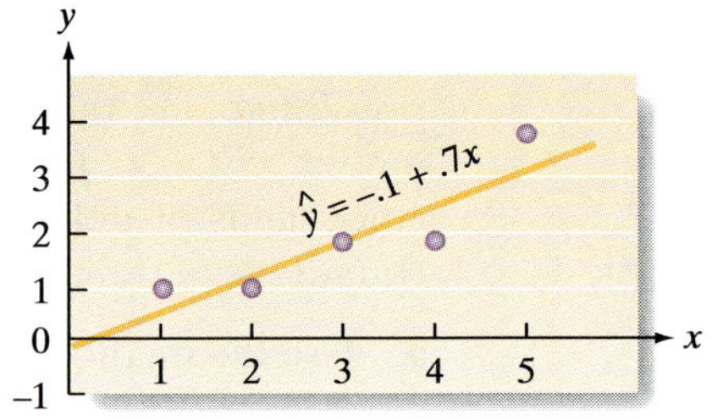

Figure 10.6
The line $\hat{y} = -.1 + .7x$ fitted to the data

c. The observed and predicted values of y, the deviations of the y values about their predicted values, and the squares of these deviations are shown in the Excel spreadsheet, Figure 10.7. Note that SSE = 1.10, and (as we would expect) this is less than the SSE = 2.0 obtained in Table 10.2 for the visually fitted line.

	A	B	C	D	E	F
1	AdvExp (X)	SalesRev (Y)	Predicted Y=-.1+.7X	(Y-Yhat)	(Y-Yhat)(Y-Yhat)	
2	1	1	0.6	0.4	0.16	
3	2	1	1.3	-0.3	0.09	
4	3	2	2	0	0	
5	4	2	2.7	-0.7	0.49	
6	5	4	3.4	0.6	0.36	
7						
8			Sum	0	1.1	
9						
10						

Figure 10.7
Excel spreadsheet comparing observed and predicted values for advertising-sales example

Model Summary

Model	R	R Square	Adjusted R Square	Std. Error of the Estimate
1	.904[a]	.817	.756	.606

a. Predictors: (Constant), ADVEXP_X

ANOVA[b]

Model		Sum of Squares	df	Mean Square	F	Sig.
1	Regression	4.900	1	4.900	13.364	.035[a]
	Residual	1.100	3	.367		
	Total	6.000	4			

a. Predictors: (Constant), ADVEXP_X
b. Dependent Variable: SALES_Y

Coefficients[a]

Model		Unstandardized Coefficients		Standardized Coefficients	t	Sig.
		B	Std. Error	Beta		
1	(Constant)	-.100	.635		-.157	.885
	ADVEXP_X	.700	.191	.904	3.656	.035

a. Dependent Variable: SALES_Y

Figure 10.8a
SPSS printout for the advertising-sales regression

d. The estimated y-intercept, $\hat{\beta}_0 = -.1$, appears to imply that the estimated mean sales revenue is equal to $-.1$, or $-\$100$, when the advertising expenditure, x, is equal to \$0. Because negative sales revenues are not possible, this seems to make the model nonsensical. However, *the model parameters should be interpreted only within the sampled range of the independent variable*—in this case, for advertising expenditures between \$100 and \$500. Thus, the y-intercept—which is, by definition, at $x = 0$ (\$0 advertising expenditure)—is not within the range of the sampled values of x and is not subject to meaningful interpretation.

The slope of the least squares line, $\hat{\beta}_1 = .7$, implies that for every unit increase of x, the mean value of y is estimated to increase by .7 unit. In terms of this example, for every \$100 increase in advertising, the mean sales revenue is estimated to increase by \$700 *over the sampled range of advertising expenditures from \$100 to \$500.* Thus, the model does not imply that increasing the advertising expenditures from \$500 to \$1,000 will result in an increase in mean sales of \$3,500, because the range of x in the sample does not extend to \$1,000 ($x = 10$). Be careful to interpret the estimated parameters only within the sampled range of x.

Look Back The calculations required to obtain $\hat{\beta}_0$, $\hat{\beta}_1$, and SSE in simple linear regression, although straightforward, can become rather tedious. Even with the use of an Excel spreadsheet, the process is laborious, especially when the sample size is large. Fortunately, a statistical software package can significantly reduce the labor involved in regression calculations. The SPSS, Minitab, and Excel/DDXL outputs for the simple linear regression of the data in Table 10.1 are displayed in Figures 10.8a–c. The values of $\hat{\beta}_0$

Regression Analysis: SALES_Y versus ADVEXP_X

```
The regression equation is
SALES_Y = - 0.100 + 0.700 ADVEXP_X

Predictor      Coef  SE Coef      T      P
Constant    -0.1000   0.6351  -0.16  0.885
ADVEXP_X     0.7000   0.1915   3.66  0.035

S = 0.605530   R-Sq = 81.7%   R-Sq(adj) = 75.6%

Analysis of Variance

Source          DF      SS      MS      F      P
Regression       1  4.9000  4.9000  13.36  0.035
Residual Error   3  1.1000  0.3667
Total            4  6.0000
```

Figure 10.8b
Minitab printout for the advertising-sales regression

```
Regression: AdvExp (X) by SalesRev (Y)
Dependent variable is:   SalesRev (Y)
No Selector
R squared = 81.7%     R squared (adjusted) = 75.6%
s = 0.6055  with  5 - 2 = 3  degrees of freedom

Source      Sum of Squares   df   Mean Square   F-ratio
Regression  4.9               1   4.9           13.4
Residual    1.1               3   0.366667

Variable    Coefficient   s.e. of Coeff   t-ratio   prob
Constant    -0.1          0.6351          -0.157    0.8849
AdvExp (X)  0.7           0.1915          3.66      0.0354
```

Figure 10.8c
Excel/DDXL printout for the advertising-sales regression

and $\hat{\beta}_1$ are highlighted on the printouts. These values, $\hat{\beta}_0 = -.1$ and $\hat{\beta}_1 = .7$, agree exactly with our calculated values. The value of SSE = 1.10 is also highlighted on the printouts.

Now Work Exercise 10.16

Interpreting the Estimates of β_0 and β_1 in Simple Linear Regression

y-intercept: $\hat{\beta}_0$ represents the predicted value of y when $x = 0$ (*Caution:* This value will not be meaningful if the value $x = 0$ is nonsensical or outside the range of the sample data.)

slope: $\hat{\beta}_1$ represents the increase (or decrease) in y for every 1-unit increase in x (*Caution:* This interpretation is valid only for x-values within the range of the sample data.)

Even when the interpretations of the estimated parameters in a simple linear regression are meaningful, we need to remember that they are only estimates based on the sample. As such, their values will typically change in repeated sampling. How much confidence do we have that the estimated slope, $\hat{\beta}_1$ accurately approximates the true slope, β_1? This requires statistical inference, in the form of confidence intervals and tests of hypotheses, which we address in Section 10.4.

To summarize, we defined the best-fitting straight line to be the one that minimizes the sum of squared errors around the line, and we called it the *least squares line*. We should interpret the least squares line only within the sampled range of the independent variable. In subsequent sections, we show how to make statistical inferences about the model.

Statistics IN Action Revisited — Estimating a Straight-Line Regression Model

We return to the legal advertising case involving two former law partners (p. 559). Recall that partner A (who handled PI cases) sued partner B (who handled WC cases) over who should pay what share of the advertising expenses of their former partnership. Monthly data were collected on the number of new PI cases, number of new WC cases, and the total amount spent (in thousands of dollars) on advertising over the previous six months. These data (shown in Table SIA10.1, p. 560) are saved in the **LEGALADV** file. Do these data provide support for the hypothesis that increased advertising expenditures are associated with more PI cases? Define y as the number of new PI cases per month and x as the cumulative 6-month advertising expenditures (in thousands of dollars). One way to investigate the link between these two variables is to fit the straight-line model, $E(y) = \beta_0 + \beta_1 x$, to the data in Table SIA10.1.

A Minitab scatterplot of the data and a simple linear regression printout are shown in Figure SIA10.1. Note that the least squares line is also displayed on the scatterplot. You can see that the line has a positive slope and, although there is some variation of the data points around the line, it appears that advertising expenditure (x) is fairly strongly related to the number of new PI cases (y). The estimated slope of the line (highlighted on Figure SIA10.1) is $\hat{\beta}_1 = .113$. Thus, we estimate a .113 increase in the number of new PI cases for every $1,000 increase in cumulative advertising expenditures.

Does such a relationship also exist between number of new WC cases and advertising expenditure? Now let y = number of new WC cases per month and x = cumulative 6-month advertising expenditures. A Minitab scatterplot and simple linear regression analysis for these two variables are shown in Figure SIA10.2. Compared to the previous scatterplot, the slope of the least squares line shown in Figure SIA10.2 is much flatter, and the variation of the data points around the line is much larger. Consequently, it does not appear that number of new WC cases is very strongly related to advertising expenditure. In fact, the estimated slope (highlighted in Figure SIA10.2) implies that a $1,000 increase in cumulative advertising expenditures will lead to only a $\hat{\beta}_1 = .0098$ increase in the number of new WC cases per month.

Based on these descriptive statistics (scatterplots and least squares lines), it appears that partner A's argument that partner B should share the advertising expenses is weak, at best. In the *Statistics in Action Revisited* sections that follow, we will provide a measure of reliability to this inference and investigate the legal advertising data further.

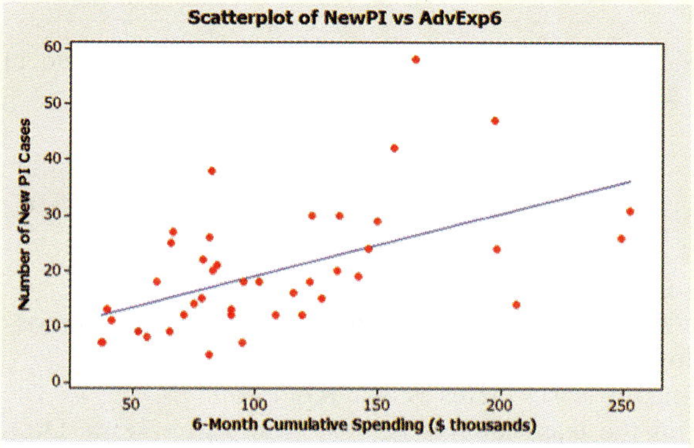

Regression Analysis: NewPI versus AdvExp6

```
The regression equation is
NewPI = 7.77 + 0.113 AdvExp6

Predictor      Coef  SE Coef     T      P
Constant      7.767    3.385  2.29  0.027
AdvExp6     0.11289  0.02793  4.04  0.000

S = 9.67521   R-Sq = 29.0%   R-Sq(adj) = 27.2%

Analysis of Variance

Source          DF      SS      MS      F      P
Regression       1  1529.5  1529.5  16.34  0.000
Residual Error  40  3744.4    93.6
Total           41  5273.9
```

Figure SIA10.1
Minitab analysis of new PI cases vs. 6-month cumulative advertising expenditure

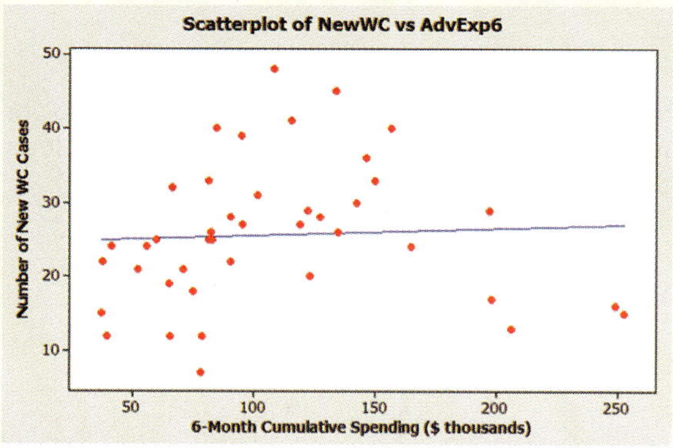

Regression Analysis: NewWC versus AdvExp6

```
The regression equation is
NewWC = 24.6 + 0.0098 AdvExp6

Predictor      Coef  SE Coef     T      P
Constant     24.574    3.367  7.30  0.000
AdvExp6     0.00982  0.02778  0.35  0.725

S = 9.62296   R-Sq = 0.3%   R-Sq(adj) = 0.0%

Analysis of Variance

Source          DF       SS     MS     F      P
Regression       1    11.58  11.58  0.13  0.725
Residual Error  40  3704.06  92.60
Total           41  3715.64
```

Figure SIA10.2
Minitab analysis of new WC cases vs. 6-month cumulative advertising expenditure

Activity 10.1 *Keep the Change:* Least Squares Models

In this activity, you will once again use data collected in Activity 1.1 *Keep the Change: Collecting Data* (p. 16). For each student in your class, collect the sum of the data set *Purchase Totals,* the sum of the data set *Amounts Transferred,* and the number of purchases (the number of data items in the set *Purchase Totals*).

1. For each student in your class, form an ordered pair

   (Sum of Purchase Totals, Sum of Amounts Transferred)

   Use these ordered pairs to create a scattergram. Then use the ordered pairs to find the values of $\hat{\beta}_0$ and $\hat{\beta}_1$ in the least squares line for the data.

2. Suppose that three customers have made only one debit-card purchase each, as shown in the following table.

Customer	Purchases	Actual Amount Transferred	Estimated Amount Transferred	Difference
A	$3.49			
B	$30.49			
C	$300.49			

   Complete the table by first finding the actual amount that will be transferred for each customer. Then use the least squares line $y = \hat{\beta}_0 + \hat{\beta}_1 x$ from Exercise 1 to estimate the amount transferred. Finally, calculate the difference between the actual and estimated amounts for each customer.

3. Based on your results in Exercise 2, comment on the usefulness of using your model from Exercise 1 to estimate the transferred amounts. Do you believe that the least squares model is an appropriate model in this situation? Explain.

4. For each student in your class, form an ordered pair

   (Number of Purchases, Sum of Amounts Transferred)

   Use these ordered pairs to create a scattergram. Then use the ordered pairs to find the values of $\hat{\beta}_0$ and $\hat{\beta}_1$ in the least squares line for the data. Create your own hypothetical data as in Exercise 2 to test your model. Do you believe that this model is more or less useful than the model in Exercise 1? Explain.

Exercises 10.10–10.24

Learning the Mechanics

10.10 The following table is similar to Table 10.3. It is used for making the preliminary computations for finding the least squares line for the given pairs of x and y values.

	x_i	y_i	x_i^2	$x_i y_i$
	7	2		
	4	4		
	6	2		
	2	5		
	1	7		
	1	6		
	3	5		
Totals	$\sum x_i =$	$\sum y_i =$	$\sum x_i^2 =$	$\sum x_i y_i =$

a. Complete the table.
b. Find SS_{xy}.
c. Find SS_{xx}.
d. Find $\hat{\beta}_1$.
e. Find $\bar{x}$ and $\bar{y}$.
f. Find $\hat{\beta}_0$.
g. Find the least squares line.

10.11 Refer to Exercise 10.10. After the least squares line has been obtained, the table below (which is similar to Table 10.4) can be used for (1) comparing the observed and the predicted values of y and (2) computing SSE.

x	y	$\hat{y}$	$(y - \hat{y})$	$(y - \hat{y})^2$
7	2			
4	4			
6	2			
2	5			
1	7			
1	6			
3	5			
			$\sum(y - \hat{y}) =$	$SSE = \sum(y - \hat{y})^2 =$

a. Complete the table.
b. Plot the least squares line on a scattergram of the data. Plot the following line on the same graph: $\hat{y} = 14 - 2.5x$.
c. Show that SSE is larger for the line in part **b** than it is for the least squares line.

10.12 Construct a scattergram for the data in the following table.

x	.5	1	1.5
y	2	1	3

a. Plot the following two lines on your scattergram:

$$y = 3 - x \quad \text{and} \quad y = 1 + x$$

b. Which of these lines would you choose to characterize the relationship between x and y? Explain.
c. Show that the SSE for both of these lines equals 0.
d. Which of these lines has the smaller SSE?
e. Find the least squares line for the data and compare it to the two lines described in part **a.**

10.13 Consider the following pairs of measurements (saved in the **LM10_13** file):

x	8	5	4	6	2	5	3
y	1	3	6	3	7	2	5

a. Construct a scattergram for these data.
b. What does the scattergram suggest about the relationship between x and y?
c. Find the least squares estimates of β_0 and β_1.
d. Plot the least squares line on your scattergram. Does the line appear to fit the data well? Explain.

Applet Exercise 10.1

Use the applet *Regression by Eye* to explore the relationship between the pattern of data in a scattergram and the corresponding least squares model.

a. Run the applet several times. For each time, attempt to move the green line into a position that appears to minimize the vertical distances of the points from the line. Then click *Show regression line* to see the actual regression line. How close is your line to the actual line? Click *New data* to reset the applet.
b. Click the trash can to clear the graph. Use the mouse to place five points on the scattergram that are approximately in a straight line. Then move the green line to approximate the regression line. Click *Show regression line* to see the actual regression line. How close were you this time?
c. Continue to clear the graph and plot sets of five points with different patterns among the points. Use the green line to approximate the regression line. How close do you come to the actual regression line each time?
d. Based on your experiences with the applet, explain why we need to use more reliable methods of finding the regression line than just "eyeing" it.

Applying the Concepts—Basic

10.14 **In business, do nice guys finish first or last?** In baseball, there is an old saying that "nice guys finish last." Is this true in the business world? Researchers at Harvard University attempted to answer this question and reported their results in *Nature* (March 20, 2008). In the study, Boston-area college students repeatedly played a version of the game "prisoner's dilemma," where competitors choose cooperation, defection, or costly punishment. (Cooperation meant paying 1 unit for the opponent to receive 2 units, defection meant gaining 1 unit at a cost of 1 unit for the opponent, and punishment meant paying 1 unit for the opponent to lose 4 units.) At the conclusion of the games, the researchers recorded the average payoff and the number of times cooperation, defection, and punishment were used for each player. The scattergrams on the next page plot average payoff (y) against level of cooperation use, defection use, and punishment use, respectively.

a. Consider cooperation use (x) as a predictor of average payoff (y). Based on the scattergram, is there evidence of a linear trend?
b. Consider defection use (x) as a predictor of average payoff (y). Based on the scattergram, is there evidence of a linear trend?

c. Consider punishment use (x) as a predictor of average payoff (y). Based on the scattergram, is there evidence of a linear trend?

d. Refer to part **c.** Is the slope of the line relating punishment use (x) to average payoff (y) positive or negative?

e. The researchers concluded that "winners don't punish." Do you agree? Explain.

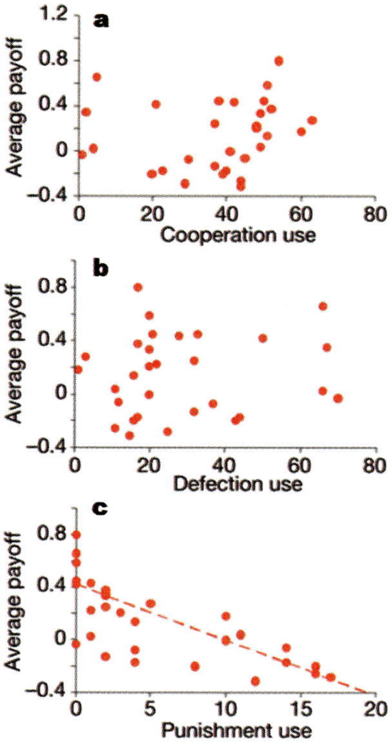

10.15 State SAT scores. Refer to the data on average state SAT scores for 2000 and 2007, Exercise 2.29 (p. 52). The first five observations and last two observations in the **SATSCORES** file are reproduced in the next table. In Exercise 2.120 (p. 93), you examined the relationship between the 2000 SAT scores and the 2007 SAT scores with a scatterplot.

State	2000	2007
Alabama	1114	1119
Alaska	1034	1036
Arizona	1044	1044
Arkansas	1117	1144
California	1015	1015
⋮	⋮	⋮
Wisconsin	1181	1185
Wyoming	1090	1136

Source: "College-Bound Seniors 2008." Copyright © 2008 the College Board. www.collegeboard.com. Reproduced with permission.

a. Write the equation of a straight-line model relating 2007 SAT scores (y) to 2000 SAT scores (x).

b. An SPSS simple linear regression printout for the data is shown at the bottom of the page. Find the least squares prediction equation.

c. Give a practical interpretation of the y-intercept of the least squares line. If a practical interpretation is not possible, explain why.

d. Give a practical interpretation of the slope of the least squares line. Over what range of x is the interpretation meaningful?

10.16 Characteristics of diamonds sold at retail. Refer to the *Journal of Statistics Education* data on 308 diamonds for sale on the open market, saved in the **DIAMONDS** file. In Exercise 2.121 (p. 93), you related the size of the diamond (number of carats) to the asking price (dollars) using a scatterplot.

NW

a. Write the equation of a straight-line model relating asking price (y) to number of carats (x).

b. A Minitab simple linear regression printout for the data is shown on page 574. Find the equation of the least squares line.

c. Give a practical interpretation of the y-intercept of the least squares line. If a practical interpretation is not possible, explain why.

d. Give a practical interpretation of the slope of the least squares line. Over what range of x is the interpretation meaningful?

e. Use the least squares line to predict the asking price of a .52-carat diamond.

10.17 Drug controlled-release rate study. Researchers at Dow Chemical Co. investigated the effect of tablet surface area and volume on the rate at which a drug is released in a controlled-release dosage (*Drug Development and Industrial Pharmacy,* Vol. 28, 2002). Six similarly shaped tablets were prepared with different weights and thicknesses, and the ratio of surface area to volume was measured for each. Using a dissolution apparatus, each tablet was placed in 900 milliliters of deionized water, and the diffusional drug release

Drug Release Rate (% released/$\sqrt{\text{time}}$)	Surface Area to Volume (mm^2/mm^3)
60	1.50
48	1.05
39	.90
33	.75
30	.60
29	.65

Source: Reynolds, T., Mitchell, S., and Balwinski, K. M. "Investigation of the effect of tablet surface area/volume on drug release from hydroxypropylmethylcellulose controlled-release matrix tablets," *Drug Development and Industrial Pharmacy,* Vol. 28, No. 4, 2002, pp. 457–466 (Figure 3). Reprinted by permission of the publisher (Taylor & Francis Group, www.informaworld.com).

SPSS Output for Exercise 10.15

Coefficients[a]

Model		Unstandardized Coefficients B	Unstandardized Coefficients Std. Error	Standardized Coefficients Beta	t	Sig.
1	(Constant)	-146.856	46.981		-3.126	.003
	SAT2000	1.144	.044	.966	26.007	.000

a. Dependent Variable: SAT2007

Minitab Output for Exercise 10.16

Regression Analysis: PRICE versus CARAT

```
The regression equation is
PRICE = - 2298 + 11599 CARAT


Predictor      Coef   SE Coef       T      P
Constant    -2298.4     158.5  -14.50  0.000
CARAT       11598.9     230.1   50.41  0.000


S = 1117.56   R-Sq = 89.3%   R-Sq(adj) = 89.2%


Analysis of Variance

Source           DF          SS          MS        F      P
Regression        1  3173248722  3173248722  2540.73  0.000
Residual Error  306   382178624     1248950
Total           307  3555427347
```

rate (percentage of drug released divided by the square root of time) was determined. The experimental data are listed in the table (p. 573) and saved in the **DOWDRUG** file.

a. Fit the simple linear model, $E(y) = \beta_0 + \beta_1 x$, where y = release rate and x = surface-area-to-volume ratio.

b. Interpret the estimates of β_0 and β_1.

c. Predict the drug release rate for a tablet that has a surface area/volume ratio of .50.

d. Comment on the reliability of the prediction in part **c.**

10.18 Extending the life of an aluminum smelter pot. An investigation of the properties of bricks used to line aluminum smelter pots was published in *The American Ceramic Society Bulletin* (Feb. 2005). Six different commercial bricks were evaluated. The life length of a smelter pot depends on the porosity of the brick lining (the less porosity, the longer the life); consequently, the researchers measured the apparent porosity of each brick specimen, as well as the mean pore diameter of each brick. The data are given in the accompanying table and saved in the **SMELTPOT** file.

Brick	Apparent Porosity (%)	Mean Pore Diameter (micrometers)
A	18.8	12.0
B	18.3	9.7
C	16.3	7.3
D	6.9	5.3
E	17.1	10.9
F	20.4	16.8

Source: Bonadia, P., et al. "Aluminosilicate refractories for aluminum cell linings," *The American Ceramic Society Bulletin*, Vol. 84, No. 2, Feb. 2005, pp. 26–31 (Table II). Reproduced with permission of American Ceramic Society, Inc. in the format Textbook and electronic usage via Copyright Clearance Center.

a. Find the least squares line relating porosity (y) to mean pore diameter (x).

b. Interpret the y-intercept of the line.

c. Interpret the slope of the line.

d. Predict the apparent porosity percentage for a brick with a mean pore diameter of 10 micrometers.

Applying the Concepts—Intermediate

10.19 Ranking driving performance of professional golfers. Refer to *The Sport Journal* (Winter 2007) study of a new method for ranking the total driving performance of golfers on the Professional Golf Association (PGA) tour, Exercise 2.50 (p. 63). Recall that the method computes a driving performance index based on a golfer's average driving distance (yards) and driving accuracy (percent of drives that land in the fairway). Data for the top 40 PGA golfers (as ranked by the new method) are saved in the **PGADRIVER** file. (The first five and last five observations are listed in the table.)

Rank	Player	Driving Distance (yards)	Driving Accuracy (%)	Driving Performance Index
1	Woods	316.1	54.6	3.58
2	Perry	304.7	63.4	3.48
3	Gutschewski	310.5	57.9	3.27
4	Wetterich	311.7	56.6	3.18
5	Hearn	295.2	68.5	2.82
⋮	⋮	⋮	⋮	⋮
36	Senden	291	66	1.31
37	Mickelson	300	58.7	1.30
38	Watney	298.9	59.4	1.26
39	Trahan	295.8	61.8	1.23
40	Pappas	309.4	50.6	1.17

Source: Wiseman, F., et al. "A new method for ranking total driving performance on the PGA Tour," *The Sport Journal*, Vol. 10, No. 1, Winter 2007 (Table 2).

a. Write the equation of a straight-line model relating driving accuracy (y) to driving distance (x).

b. Fit the model, part **a,** to the data using simple linear regression. Give the least squares prediction equation.

c. Interpret the estimated y-intercept of the line.

d. Interpret the estimated slope of the line.

e. In Exercise 2.126 (p. 95), you were informed that a professional golfer, practicing a new swing to increase his average driving distance, is concerned that his driving accuracy will be lower. Which of the two estimates, y-intercept or slope, will help you determine if the golfer's concern is a valid one? Explain.

10.20 Sweetness of orange juice. The quality of the orange juice produced by a manufacturer (e.g., Minute Maid, Tropicana) is constantly monitored. There are numerous sensory and chemical components that combine to make the best-tasting orange juice. For example, one manufacturer has developed a quantitative index of the "sweetness" of orange juice. (The higher the index, the sweeter the juice.) Is there a relationship between the sweetness index and a chemical measure such as the amount of water-soluble pectin (parts per million) in the orange juice? Data collected on these two variables for 24 production runs at a juice manufacturing plant are shown in the table and saved in the **OJUICE** file. Suppose a manufacturer wants to use simple linear regression to predict the sweetness (y) from the amount of pectin (x).

Run	Sweetness Index	Pectin (ppm)
1	5.2	220
2	5.5	227
3	6.0	259
4	5.9	210
5	5.8	224
6	6.0	215
7	5.8	231
8	5.6	268
9	5.6	239
10	5.9	212
11	5.4	410
12	5.6	256
13	5.8	306
14	5.5	259
15	5.3	284
16	5.3	383
17	5.7	271
18	5.5	264
19	5.7	227
20	5.3	263
21	5.9	232
22	5.8	220
23	5.8	246
24	5.9	241

Note: The data in the table are authentic. For confidentiality reasons, the manufacturer cannot be disclosed.

a. Find the least squares line for the data.
b. Interpret $\hat{\beta}_0$ and $\hat{\beta}_1$ in the words of the problem.
c. Predict the sweetness index if amount of pectin in the orange juice is 300 ppm. [*Note:* A measure of reliability of such a prediction is discussed in Section 10.6.]

10.21 Most generous philanthropists. Intel Corp. founder Gordon Moore donated over \$7 million to environmental causes while Microsoft founder Bill Gates donated over \$5 million to health, education, and libraries. These two entrepreneurs were the top two givers in *Business Week's* (Nov. 28, 2005) annual list of the 50 Most Generous Philanthropists. Data on total amount pledged and remaining net worth (in millions of dollars) for the 50 top givers are saved in the **TOPGIVERS** file. (The data for the first 10 names on the list are given in the table at the bottom of the page.)

a. Propose a straight-line model relating amount pledged (y) to net worth (x).
b. Fit the model to the data in the **TOPGIVERS** file using the method of least squares.
c. Graph the least squares line on a scattergram of the data. Is there visual evidence of a linear relationship between the two variables? Is the relationship positive or negative?
d. Interpret the estimates of the y-intercept and slope in the words of the problem.

10.22 FCAT scores and poverty. In the state of Florida, elementary school performance is based on the average score obtained by students on a standardized exam, called the Florida Comprehensive Assessment Test (FCAT). An analysis of the link between FCAT scores and sociodemographic factors was published in the *Journal of Educational and Behavioral Statistics* (Spring 2004). Data on average math and reading FCAT scores of third-graders, as well as the percentage of students below the poverty level, for a sample of 22 Florida elementary schools are listed in the table on the next page.

a. Propose a straight-line model relating math score (y) to percentage (x) of students below the poverty level.
b. Fit the model to the data in the **FCAT** file using the method of least squares.
c. Graph the least squares line on a scattergram of the data. Is there visual evidence of a relationship between the two variables? Is the relationship positive or negative?
d. Interpret the estimates of the y-intercept and slope in the words of the problem.
e. Now consider a model relating reading score (y) to percentage (x) of students below the poverty level. Repeat parts **a–d** for this model.

Table for Exercise 10.21 (Selected data)

Donor	Company/Background	Pledged ($ millions)	Net Worth ($ millions)
Gordon Moore	Intel	7,046	4,600
Bill Gates	Microsoft	5,458	51,000
Warren Buffett	Berkshire Hathaway	2,622	40,000
George Soros	Investor	2,367	7,200
Eli Broad	SunAmerica	1,475	5,500
James Stowers	American Century	1,205	716
Walton Family	Wal-Mart	1,100	82,700
Alfred Mann	Medical devices	993	2,100
Michael Dell	Dell	933	18,000
George Kaiser	Oil and gas	617	4,500

Source: Business Week, Nov. 28, 2005, p. 61.

Elementary School	FCAT-Math	FCAT-Reading	% Below Poverty
1	166.4	165.0	91.7
2	159.6	157.2	90.2
3	159.1	164.4	86.0
4	155.5	162.4	83.9
5	164.3	162.5	80.4
6	169.8	164.9	76.5
7	155.7	162.0	76.0
8	165.2	165.0	75.8
9	175.4	173.7	75.6
10	178.1	171.0	75.0
11	167.1	169.4	74.7
12	177.0	172.9	63.2
13	174.2	172.7	52.9
14	175.6	174.9	48.5
15	170.8	174.8	39.1
16	175.1	170.1	38.4
17	182.8	181.4	34.3
18	180.3	180.6	30.3
19	178.8	178.0	30.3
20	181.4	175.9	29.6
21	182.8	181.6	26.5
22	186.1	183.8	13.8

Source: Tekwe, C.D., et al. "An empirical comparison of statistical models for value-added assessment of school performance," *Journal of Educational and Behavioral Statistics,* Vol. 29, No. 1, Spring 2004, pp. 11–36 (Table 2). © 2004. Reprinted by permission of SAGE Publications.

a. Select one of the variables as the dependent variable, y, and another as the independent variable, x. Use your knowledge of the subject area and common sense to help you select the variables.

b. Fit the simple linear model, $E(y) = \beta_0 + \beta_1 x$, to the data in the **BSCHOOL** file. Interpret the estimates of the slope and y-intercept.

Applying the Concepts—Advanced

10.24 Spreading rate of spilled liquid. Refer to the *Chemical Engineering Progress* (Jan. 2005) study of the rate at which a spilled volatile liquid will spread across a surface, Exercise 2.124 (p. 94). Recall that a DuPont Corp. engineer calculated the mass (in pounds) of a 50-gallon methanol spill after a period of time ranging from 0 to 60 minutes. Do the data in the table below (saved in the **LIQUIDSPILL** file) indicate that the mass of the spill tends to diminish as time increases? If so, how much will the mass diminish each minute?

Time (minutes)	Mass (pounds)	Time (minutes)	Mass (pounds)
0	6.64	22	1.86
1	6.34	24	1.60
2	6.04	26	1.37
4	5.47	28	1.17
6	4.94	30	0.98
8	4.44	35	0.60
10	3.98	40	0.34
12	3.55	45	0.17
14	3.15	50	0.06
16	2.79	55	0.02
18	2.45	60	0.00
20	2.14		

Source: Barry, J. "Estimating rates of spreading and evaporation of volatile liquids," *Chemical Engineering Progress,* Vol. 101, No. 1, Jan. 2005. Reproduced with permission. Copyright © 2005 AICHE.

10.23 Survey of the top business schools. Each year, the *Wall Street Journal* and *Harris Interactive* track the opinions and experiences of college recruiters for large corporations and summarize the results in the Business School Survey. In 2005, the survey included rankings of 76 business schools. Survey data for the top 10 business schools are given in the table below. All the data are saved in the **BSCHOOL** file.

School	Enrollment (# full-time students)	Annual Tuition ($)	Mean GMAT	% with Job Offer	Avg. Salary ($)
Dartmouth	503	38,400	704	—	119,800
Michigan	1,873	33,076	690	91	105,986
Carnegie Mellon	661	38,800	691	93	95,531
Northwestern	2,650	38,844	700	94	117,060
Yale	468	36,800	696	86	104,018
Pennsylvania	1,840	40,458	716	92	117,471
Cal., Berkeley	1,281	21,512	701	92	112,699
Columbia	1,796	38,290	709	95	126,319
North Carolina	855	16,375	652	86	92,565
Southern Cal.	1,588	37,558	685	82	88,839

Source: "Wall Street Journal's annual rankings of business schools," *The Wall Street Journal,* Sep. 21, 2005. Copyright 2005 by Dow Jones & Company, Inc. in the format Textbook via Copyright Clearance Center.

10.3 Model Assumptions

In Section 10.2, we assumed that the probabilistic model relating the firm's sales revenue y to the advertising dollars is

$$y = \beta_0 + \beta_1 x + \varepsilon$$

We also recall that the least squares estimate of the deterministic component of the model, $\beta_0 + \beta_1 x$, is

$$\hat{y} = \hat{\beta}_0 + \hat{\beta}_1 x = -.1 + .7x$$

Now we turn our attention to the random component ε of the probabilistic model and its relation to the errors in estimating β_0 and β_1. We will use a probability distribution to characterize the behavior of ε. We will see how the probability distribution of ε determines how well the model describes the relationship between the dependent variable y and the independent variable x.

Step 3 in a regression analysis requires us to specify the probability distribution of the random error ε. We will make four basic assumptions about the general form of this probability distribution:

Assumption 1: The mean of the probability distribution of ε is 0—that is, the average of the values of ε over an infinitely long series of experiments is 0 for each setting of the independent variable x. This assumption implies that the mean value of y, $E(y)$, for a given value of x is $E(y) = \beta_0 + \beta_1 x$.

Assumption 2: The variance of the probability distribution of ε is constant for all settings of the independent variable x. For our straight-line model, this assumption means that the variance of ε is equal to a constant, say σ^2, for all values of x.

Assumption 3: The probability distribution of ε is normal.

Assumption 4: The values of ε associated with any two observed values of y are independent—that is, the value of ε associated with one value of y has no effect on the values of ε associated with other y values.

The implications of the first three assumptions can be seen in Figure 10.9, which shows distributions of errors for three values of x, namely, x_1, x_2, and x_3. Note that the relative frequency distributions of the errors are normal with a mean of 0 and a constant variance σ^2. (All the distributions shown have the same amount of spread or variability.) The straight line shown in Figure 10.9 is the line of means. It indicates the mean value of y for a given value of x. We denote this mean value as $E(y)$. Then, the line of means is given by the equation

$$E(y) = \beta_0 + \beta_1 x$$

These assumptions make it possible for us to develop measures of reliability for the least squares estimators and to develop hypothesis tests for examining the usefulness of the least squares line. We have various techniques for checking the validity of these assumptions, and we have remedies to apply when they appear to be invalid. Several of these remedies are discussed in Chapter 11. Fortunately, the assumptions need not hold exactly in order for least squares estimators to be useful. The assumptions will be satisfied adequately for many applications encountered in practice.

It seems reasonable to assume that the greater the variability of the random error ε (which is measured by its variance σ^2), the greater will be the errors in the estimation of the model parameters β_0 and β_1 and in the error of prediction when $\hat{y}$ is used to predict y for some value of x. Consequently, you should not be surprised, as we proceed through this chapter, to find that σ^2 appears in the formulas for all confidence intervals and test statistics that we will be using.

In most practical situations, σ^2 is unknown, and we must use our data to estimate its value. The best estimate of σ^2, denoted by s^2, is obtained by dividing the sum of squares of the deviations of the y values from the prediction line,

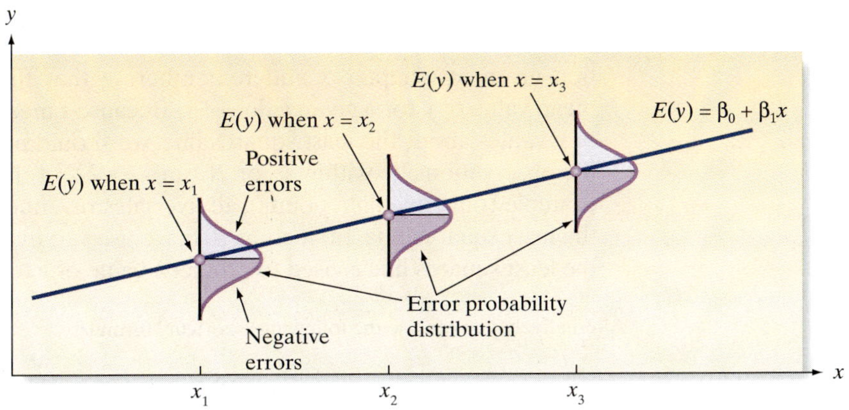

Figure 10.9
The probability distribution of ε

$$SSE = \sum (y_i - \hat{y}_i)^2$$

by the number of degrees of freedom associated with this quantity. We use 2 df to estimate the two parameters β_0 and β_1 in the straight-line model, leaving $(n - 2)$ df for the error variance estimation.

Estimation of σ^2 for a (First-Order) Straight-Line Model

$$s^2 = \frac{\text{SSE}}{\text{Degrees of freedom for error}} = \frac{\text{SSE}}{n - 2}$$

where $\text{SSE} = \Sigma(y_i - \hat{y}_i)^2 = \text{SS}_{yy} - \hat{\beta}_1 \text{SS}_{xy}$

$$\text{SS*}_{yy} = \Sigma(y_i - \bar{y})^2$$

To estimate the standard deviation σ of ε, we calculate

$$s = \sqrt{s^2} = \sqrt{\frac{\text{SSE}}{n - 2}}$$

We will refer to s as the **estimated standard error of the regression model.**

CAUTION When performing these calculations, you may be tempted to round the calculated values of SS_{yy}, $\hat{\beta}_1$, and SS_{xy}. Be certain to carry at least six significant figures for each of these quantities to avoid substantial errors in calculation of the SSE. ▲

Example 10.2
Estimating σ for the Advertising-Sales Regression

Problem Refer to Example 10.1 and the simple linear regression of the advertising-sales data in Table 10.1.

a. Compute an estimate of σ.

b. Give a practical interpretation of the estimate.

Solution

a. We previously calculated SSE = 1.10 for the least squares line $\hat{y} = -.1 + .7x$. Recalling that there were $n = 5$ data points, we have $n - 2 = 5 - 2 = 3$ df for estimating σ^2. Thus,

$$s^2 = \frac{\text{SSE}}{n - 2} = \frac{1.10}{3} = .367$$

is the estimated variance, and

$$s = \sqrt{.367} = .61$$

is the standard error of the regression model.

b. You may be able to grasp s intuitively by recalling the interpretation of a standard deviation given in Chapter 2 and remembering that the least squares line estimates the mean value of y for a given value of x. Because s measures the spread of the distribution of y values about the least squares line, we should not be surprised to find that most of the observations lie within $2s$, or $2(.61) = 1.22$, of the least squares line. For this simple example (only five data points), all five sales revenue values fall within $2s$ (or \$1,220) of the least squares line. In Section 10.6, we use s to evaluate the error of prediction when the least squares line is used to predict a value of y to be observed for a given value of x.

*Alternatively, you can use the following "shortcut" formula:

$$\text{SS}_{yy} = \Sigma y^2 - \frac{(\Sigma y)^2}{n}$$

Regression Analysis: SALES_Y versus ADVEXP_X

```
The regression equation is
SALES_Y = - 0.100 + 0.700 ADVEXP_X


Predictor      Coef   SE Coef      T      P
Constant    -0.1000    0.6351  -0.16  0.885
ADVEXP_X     0.7000    0.1915   3.66  0.035


S = 0.605530    R-Sq = 81.7%    R-Sq(adj) = 75.6%


Analysis of Variance

Source          DF      SS      MS      F      P
Regression       1  4.9000  4.9000  13.36  0.035
Residual Error   3  1.1000  0.3667
Total            4  6.0000
```

Figure 10.10
Minitab printout for the advertising-sales regression

Look Back The values of s^2 and s can also be obtained from a simple linear regression printout. The Minitab printout for the advertising-sales example is reproduced in Figure 10.10. The value of s^2 is highlighted at the bottom of the printout in the **MS** (Mean Square) column in the row labeled **Residual Error.** (In regression, the estimate of σ^2 is called Mean Square for Error, or MSE.) The value, $s^2 = .3667$, agrees with the one calculated by hand. The value of s is also highlighted in Figure 10.10. This value, $s = .60553$, agrees (except for rounding) with our hand-calculated value.

Now Work Exercise 10.30

Interpretation of s, the Estimated Standard Deviation of ε

We expect most ($\approx$95%) of the observed y values to lie within $2s$ of their respective least squares predicted values, $\hat{y}$.

Exercises 10.25–10.36

Learning the Mechanics

10.25 Visually compare the scattergrams shown below. If a least squares line were determined for each data set, which do you think would have the smallest variance, s^2? Explain.

10.26 Calculate SSE and s^2 for each of the following cases:

a. $n = 20$, $SS_{yy} = 95$, $SS_{xy} = 50$, $\hat{\beta}_1 = .75$

b. $n = 40$, $\sum y^2 = 860$, $\sum y = 50$, $SS_{xy} = 2{,}700$, $\hat{\beta}_1 = .2$

c. $n = 10$, $\sum(y_i - \bar{y})^2 = 58$, $SS_{xy} = 91$, $SS_{xx} = 170$

10.27 Suppose you fit a least squares line to 26 data points and the calculated value of SSE is 8.34.

a. Find s^2, the estimator of σ^2 (the variance of the random error term ε).

Figure for Exercise 10.25

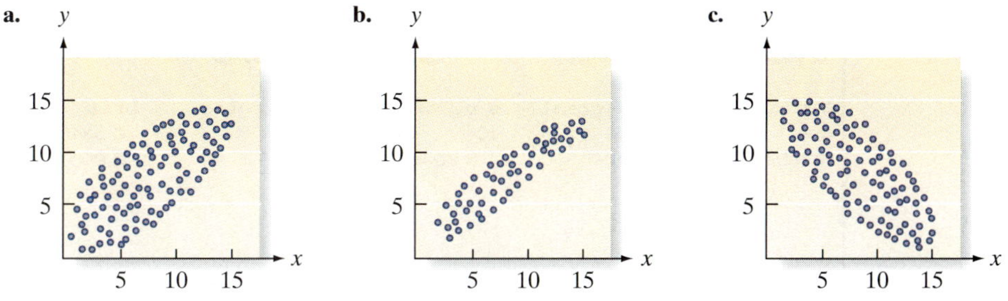

SPSS Output for Exercise 10.29

Model Summary

Model	R	R Square	Adjusted R Square	Std. Error of the Estimate
1	.966[a]	.932	.931	20.508

a. Predictors: (Constant), SAT2000

ANOVA[b]

Model		Sum of Squares	df	Mean Square	F	Sig.
1	Regression	284470.200	1	284470.200	676.377	.000[a]
	Residual	20608.388	49	420.579		
	Total	305078.588	50			

a. Predictors: (Constant), SAT2000

b. Dependent Variable: SAT2007

b. What is the largest deviation that you might expect between any one of the 26 points and the least squares line?

10.28 Refer to Exercise 10.10 (p. 572). Calculate SSE, s^2, and s for the least squares line. Use the value of s to determine where most of the errors of prediction lie.

Applying the Concepts—Basic

10.29 State SAT scores. Refer to the simple linear regression relating y = 2007 SAT scores to x = 2000 SAT scores, Exercise 10.15 (p. 573). The data are saved in the **SATSCORES** file. A portion of the SPSS printout of the analysis is shown above.

a. Locate the values of SSE, s^2, and s on the SPSS printout.

b. Give a practical interpretation of the value of s.

10.30 Characteristics of diamonds sold at retail. NW Refer to the simple linear regression relating y = asking price (dollars) to x = number of carats for diamonds sold on the open market, Exercise 10.16 (p. 573). The data are saved in the **DIAMONDS** file.

a. Locate the values of SSE, s^2, and s on the Minitab printout shown on p. 574.

b. Give a practical interpretation of the value of s.

10.31 Drug controlled-release rate study. Refer to simple linear regression relating the drug release rate (y) to surface area-to-volume ratio (x), Exercise 10.17 (p. 573): The data are saved in the **DOWDRUG** file.

a. Find the value of s for the straight-line model.

b. Give a practical interpretation of the value of s.

10.32 Structurally deficient highway bridges. Refer to Exercise 1.24 (p. 24) and the data on structurally deficient highway bridges compiled by the Federal Highway Administration (FHWA) into the National Bridge Inventory (NBI). For each state, the NBI lists the number of structurally deficient bridges and the total area (thousands of square feet) of the deficient bridges. The data for the 50 states (plus the District of Columbia and Puerto Rico) are saved in the **FHWABRIDGE** file. (The first five and last five observations are listed in the table on the next page.) For future planning and budgeting, the FHWA wants to estimate the total area of structurally deficient bridges in a state based on the number of deficient bridges.

a. Write the equation of a straight-line model relating total area (y) to number of structurally deficient bridges (x).

b. The model, part **a,** was fit to the data using Excel/DDXL, as shown below. Find the least squares prediction equation on the printout.

c. List the assumptions required for the regression analysis.

d. Locate the estimated standard error of the regression model, s, on the printout.

e. Use the value of s to find a range where most (about 95%) of the errors of prediction will fall.

Excel/DDXL Output for Exercise 10.32

Regression: NumberSD by SDArea

Dependent variable is: **SDArea**

No Selector

R squared = 38.7%   R squared (adjusted) = 37.4%

s = 635.2 with 52 - 2 = 50 degrees of freedom

Source	Sum of Squares	df	Mean Square	F-ratio
Regression	12.7101e6	1	12.7101e6	31.5
Residual	20.1731e6	50	403462	

Variable	Coefficient	s.e. of Coeff	t-ratio	prob
Constant	119.863	123	0.974	0.3346
NumberSD	0.345604	0.06158	5.61	≤ 0.0001

Data for Exercise 10.32 (selected observations)

State	Number	Area (thousands of sq. ft.)
Alabama	1,899	432.7
Alaska	155	60.9
Arizona	181	110.5
Arkansas	997	347.3
California	3,140	5,177.9
⋮	⋮	⋮
Washington	400	502.0
West Virginia	1,058	331.5
Wisconsin	1,302	399.8
Wyoming	389	143.4
Puerto Rico	241	195.4

Applying the Concepts—Intermediate

10.33 Sweetness of orange juice. Refer to the study of the quality of orange juice produced at a juice manufacturing plant, Exercise 10.20 (p. 575). The data are saved in the **OJUICE** file. Recall that simple linear regression was used to predict the sweetness index (y) from the amount of pectin (x) in the orange juice.

a. Find the values of SSE, s^2, and s for this regression.

b. Explain why it is difficult to give a practical interpretation to s^2.

c. Give a practical interpretation of the value of s.

10.34 Are geography journals worth their cost? Refer to the *Geoforum* (Vol. 37, 2006) study of whether the price of a geography journal is correlated with quality, Exercise 2.122 (p. 94). Several quantitative variables were recorded for each in a sample of 28 geography journals: cost of a 1-year subscription (dollars); Journal Impact Factor (JIF)—the average number of times articles from the journal have been cited; number of citations for the journal over the past five years; and, Relative Price Index (RPI). The data for the 28 journals are saved in the **GEOJRNL** file. (Selected observations are listed below.)

Journal	Cost ($)	JIF	Cites	RPI
J. Econ. Geogr.	468	3.139	207	1.16
Prog. Hum. Geog.	624	2.943	544	0.77
T. I. Brit. Geogr.	499	2.388	249	1.11
Econ. Geogr.	90	2.325	173	0.30
A. A. A. Geogr.	698	2.115	377	0.93
Geogr. Anal.	213	0.902	106	0.88
Geogr. J.	223	0.857	81	0.94
Appl. Geogr.	646	0.853	74	3.38

Source: Blomley, N. "Is this journal worth US$1118?" *Geoforum*, Vol. 37, 2006.

a. Fit a straight-line model relating cost (y) to JIF (x). Find an estimate of σ, the standard deviation of the error term, and interpret its value.

b. Fit a straight-line model relating cost (y) to number of citations (x). Find an estimate of σ, the standard deviation of the error term, and interpret its value.

c. Fit a straight-line model relating cost (y) to RPI (x). Find an estimate of σ, the standard deviation of the error term, and interpret its value.

d. Which predictor of cost—JIF, number of citations, or RPI—leads to a regression line with the smallest errors of prediction?

10.35 FCAT scores and poverty. Refer to the *Journal of Educational and Behavioral Statistics* (Spring 2004) study of scores on the Florida Comprehensive Assessment Test (FCAT), Exercise 10.22 (p. 575).

a. Consider the simple linear regression relating math score (y) to percentage (x) of students below the poverty level. Find and interpret the value of s for this regression.

b. Consider the simple linear regression relating reading score (y) to percentage (x) of students below the poverty level. Find and interpret the value of s for this regression.

c. Which dependent variable, math score or reading score, can be more accurately predicted by percentage (x) of students below the poverty level? Explain.

Applying the Concepts—Advanced

10.36 Life tests of cutting tools. To improve the quality of the output of any production process, it is necessary first to understand the capabilities of the process (Gitlow, et al., *Quality Management: Tools and Methods for Improvement,* 1995). In a particular manufacturing process, the useful life of a cutting tool is related to the speed at which the tool is operated. The data in the table below (saved in the **CUTTOOLS** file) were derived from life tests for the two different brands of cutting tools currently used in the production process. For which brand would you feel more confident in using the least squares line to predict useful life for a given cutting speed? Explain.

Cutting Speed (meters per minute)	Useful Life (Hours)	
	Brand A	Brand B
30	4.5	6.0
30	3.5	6.5
30	5.2	5.0
40	5.2	6.0
40	4.0	4.5
40	2.5	5.0
50	4.4	4.5
50	2.8	4.0
50	1.0	3.7
60	4.0	3.8
60	2.0	3.0
60	1.1	2.4
70	1.1	1.5
70	.5	2.0
70	3.0	1.0

10.4 Assessing the Utility of the Model: Making Inferences about the Slope β_1

Now that we have specified the probability distribution of ε and found an estimate of the variance σ^2, we are ready to make statistical inferences about the model's usefulness for predicting the response y. This is step 4 in our regression modeling procedure.

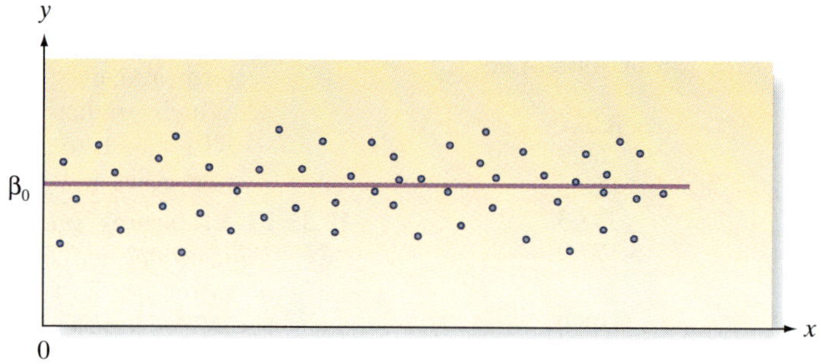

Figure 10.11
Graph of the straight-line model when slope is zero, i.e., $y = \beta_0 + \varepsilon$

Refer again to the data of Table 10.1 and suppose the appliance store's sales revenue is *completely unrelated* to the advertising expenditure. What could be said about the values of β_0 and β_1 in the hypothesized probabilistic model

$$y = \beta_0 + \beta_1 x + \varepsilon$$

if x contributes no information for the prediction of y? The implication is that the mean of y—that is, the deterministic part of the model $E(y) = \beta_0 + \beta_1 x$—does not change as x changes. In the straight-line model, this means that the true slope, β_1, is equal to 0 (see Figure 10.11). Therefore, to test the null hypothesis that the linear model contributes no information for the prediction of y against the alternative hypothesis that the linear model is useful in predicting y, we test

$$H_0\text{: } \beta_1 = 0 \text{ against } II_a\text{: } \beta_1 \neq 0$$

If the data support the alternative hypothesis, we will conclude that x does contribute information for the prediction of y using the straight-line model (although the true relationship between $E[y]$ and x could be more complex than a straight line). Thus, in effect, this is a test of the usefulness of the hypothesized model.

The appropriate test statistic is found by considering the sampling distribution of $\hat{\beta}_1$, the least squares estimator of the slope β_1, as shown in the following box.

Sampling Distribution of $\hat{\beta}_1$

If we make the four assumptions about ε (see Section 10.3), the sampling distribution of the least squares estimator $\hat{\beta}_1$ of the slope will be normal with mean β_1 (the true slope) and standard deviation

$$\sigma_{\hat{\beta}_1} = \frac{\sigma}{\sqrt{SS_{xx}}} \quad \text{(see Figure 10.12)}$$

We estimate $\sigma_{\hat{\beta}_1}$ by $s_{\hat{\beta}_1} = \dfrac{s}{\sqrt{SS_{xx}}}$ and refer to this quantity as the **estimated standard error of the least squares slope $\hat{\beta}_1$.**

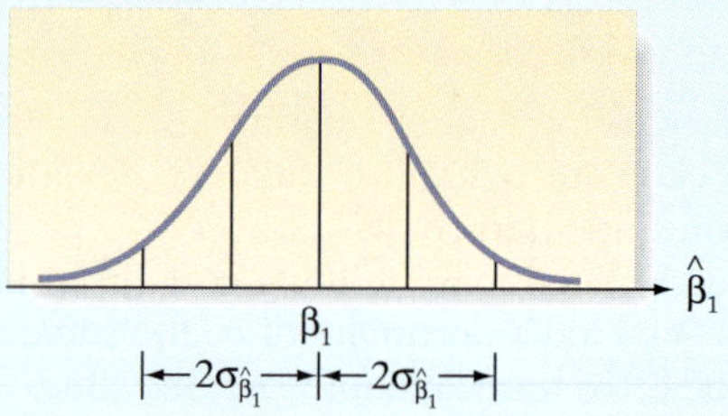

Figure 10.12
Sampling distribution of $\hat{\beta}_1$

Because σ is usually unknown, the appropriate test statistic is a t-statistic, formed as follows:

$$t = \frac{\hat{\beta}_1 - \text{Hypothesized value of } \beta_1}{s_{\hat{\beta}_1}} \qquad \text{where } s_{\hat{\beta}_1} = \frac{s}{\sqrt{SS_{xx}}}$$

Thus,

$$t = \frac{\hat{\beta}_1 - 0}{s/\sqrt{SS_{xx}}}$$

Note that we have substituted the estimator s for σ and then formed the estimated standard error $s_{\hat{\beta}_1}$ by dividing s by $\sqrt{SS_{xx}}$. The number of degrees of freedom associated with this t-statistic is the same as the number of degrees of freedom associated with s. Recall that this number is $(n - 2)$ df when the hypothesized model is a straight line (see Section 10.3). The setup of our test of the usefulness of the straight-line model is summarized in the following boxes.

A Test of Model Utility: Simple Linear Regression

One-Tailed Test	Two-Tailed Test
$H_0: \beta_1 = 0$	$H_0: \beta_1 = 0$
$H_a: \beta_1 < 0$ (or $H_a: \beta_1 > 0$)	$H_a: \beta_1 \neq 0$
	Test statistic: $t = \frac{\hat{\beta}_1}{s_{\hat{\beta}_1}} = \frac{\hat{\beta}_1}{s/\sqrt{SS_{xx}}}$
Rejection region: $t < -t_\alpha$ (or $t > t_\alpha$ when $H_a: \beta_1 > 0$)	*Rejection region:* $\lvert t \rvert > t_{\alpha/2}$

where t_α and $t_{\alpha/2}$ are based on $(n - 2)$ degrees of freedom

Conditions Required for a Valid Test: Simple Linear Regression

Refer to the four assumptions about ε listed in Section 10.3.

Example 10.3
Testing the Regression Slope, β_1—Sales Revenue Model

Problem Refer to the simple linear regression analysis of the advertising-sales data, Examples 10.1 and 10.2. Conduct a test (at $\alpha = .05$) to determine if sales revenue (y) is linearly related to advertising expenditure (x).

Solution As stated previously, we want to test $H_0: \beta_1 = 0$ against $H_a: \beta_1 \neq 0$. For this example, $n = 5$. Thus t will be based on $n - 2 = 3$ df, and the rejection region (at $\alpha = .05$) will be

$$\lvert t \rvert > t_{.025} = 3.182$$

We previously calculated $\hat{\beta}_1 = .7$, $s = .61$, and $SS_{xx} = 10$. Thus, the test statistic is

$$t = \frac{\hat{\beta}_1}{s/\sqrt{SS_{xx}}} = \frac{.7}{.61/\sqrt{10}} = \frac{.7}{.19} = 3.7$$

Because this calculated t-value falls into the upper-tail rejection region (see Figure 10.13), we reject the null hypothesis and conclude that the slope β_1 is not 0. The sample

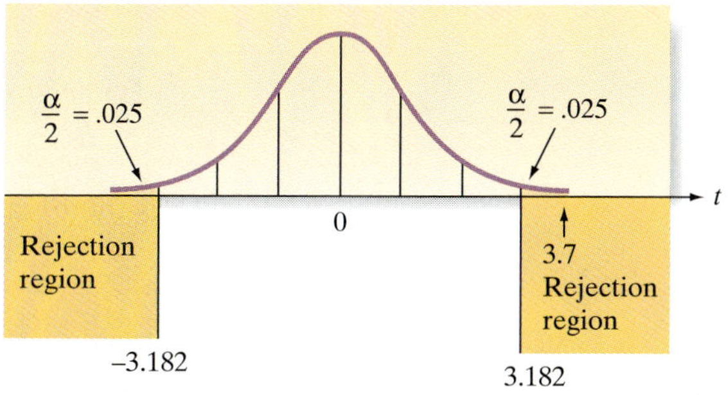

Figure 10.13
Rejection region and calculated t-value for testing $H_0: \beta_1 = 0$ versus $H_a: \beta_1 \neq 0$

Regression Analysis: SALES_Y versus ADVEXP_X

```
The regression equation is
SALES_Y = - 0.100 + 0.700 ADVEXP_X


Predictor      Coef   SE Coef      T      P
Constant    -0.1000    0.6351  -0.16  0.885
ADVEXP_X     0.7000    0.1915   3.66  0.035


S = 0.605530   R-Sq = 81.7%    R-Sq(adj) = 75.6%


Analysis of Variance

Source          DF      SS      MS      F      P
Regression       1  4.9000  4.9000  13.36  0.035
Residual Error   3  1.1000  0.3667
Total            4  6.0000
```

Figure 10.14
Minitab printout for the advertising-sales regression

evidence indicates that advertising expenditure x contributes information for the production of sales revenue y when a linear model is used.

[*Note:* We can reach the same conclusion by using the observed significance level (p-value) of the test from a computer printout. The Minitab printout for the advertising-sales example is reproduced in Figure 10.14. The test statistic and *two-tailed* p-value are highlighted on the printout. Because the p-value is smaller than $\alpha = .05$, we will reject H_0.]

Look Back What conclusion can be drawn if the calculated t-value does not fall in the rejection region or if the observed significance level of the test exceeds α? We know from previous discussions of the philosophy of hypothesis testing that such a t-value does *not* lead us to accept the null hypothesis—that is, we do not conclude that $\beta_1 = 0$. Additional data might indicate that β_1 differs from 0, or a more complex relationship may exist between x and y, requiring the fitting of a model other than the straight-line model. We discuss several such models in Chapter 11.

Now Work Exercise 10.42

Interpreting p-Values for β Coefficients in Regression

Almost all statistical computer software packages report a *two-tailed* p-value for each of the β parameters in the regression model. For example, in simple linear regression, the p-value for the two-tailed test H_0: $\beta_1 = 0$ versus H_a: $\beta_1 \neq 0$ is given on the printout. If you want to conduct a *one-tailed* test of hypothesis, you will need to adjust the p-value reported on the printout as follows:

$$\textit{Upper-tailed test } (H_a\text{: } \beta_1 > 0)\text{:} \quad p\text{-value} = \begin{cases} p/2 & \text{if } t > 0 \\ 1 - p/2 & \text{if } t < 0 \end{cases}$$

$$\textit{Lower-tailed test } (H_a\text{: } \beta_1 < 0)\text{:} \quad p\text{-value} = \begin{cases} p/2 & \text{if } t < 0 \\ 1 - p/2 & \text{if } t > 0 \end{cases}$$

where p is the p-value reported on the printout and t is the value of the test statistic.

Another way to make inferences about the slope β_1 is to estimate it using a confidence interval. This interval is formed as shown in the next box.

A 100(1 − α)% Confidence Interval for the Simple Linear Regression Slope β_1

$$\hat{\beta}_1 \pm t_{\alpha/2} s_{\hat{\beta}_1}$$

where the estimated standard error $\hat{\beta}_1$ is calculated by

$$s_{\hat{\beta}_1} = \frac{s}{\sqrt{SS_{xx}}}$$

and $t_{\alpha/2}$ is based on $(n - 2)$ degrees of freedom.

Conditions Required for a Valid Confidence Interval: Simple Linear Regression

Refer to the four assumptions about ε listed in Section 10.3.

For the simple linear regression of sales revenue (Examples 10.1–10.3), $t_{\alpha/2}$ is based on $(n - 2) = 3$ degrees of freedom, and, for $\alpha = .05$, $t_{.025} = 3.182$. Therefore, a 95% confidence interval for the slope β_1, the expected change in sales revenue for a \$100 increase in advertising expenditure, is

$$\hat{\beta}_1 \pm (t_{.025}) s_{\hat{\beta}_1} = .7 \pm 3.182\left(\frac{s}{\sqrt{SS_{xx}}}\right) = .7 \pm 3.182\left(\frac{.61}{\sqrt{10}}\right) = .7 \pm .61$$

Thus, the interval estimate of the slope parameter β_1 is .09 to 1.31. [*Note:* This interval can also be obtained using statistical software and is highlighted on the SPSS printout, Figure 10.15.] In terms of this example, the implication is that we can be 95% confident that the *true* mean increase in monthly sales revenue per additional \$100 of advertising expenditure is between \$90 and \$1,310. This inference is meaningful only over the sampled range of x—that is, from \$100 to \$500 of advertising expenditures.

Because all the values in this interval are positive, it appears that β_1 is positive and that the mean of y, $E(y)$, increases as x increases. However, the rather large width of the confidence interval reflects the small number of data points (and, consequently, a lack of information) in the experiment. Particularly bothersome is the fact that the lower end of the confidence interval implies that we are not even recovering our additional expenditure because a \$100 increase in advertising may produce as little as a \$90 increase in mean sales. If we wish to tighten this interval, we need to increase the sample size.

Coefficients[a]

Model		Unstandardized Coefficients		Standardized Coefficients	t	Sig.	95% Confidence Interval for B	
		B	Std. Error	Beta			Lower Bound	Upper Bound
1	(Constant)	-.100	.635		-.157	.885	-2.121	1.921
	ADVEXP_X	.700	.191	.904	3.656	.035	.091	1.309

a. Dependent Variable: SALES_Y

Figure 10.15
SPSS printout with 95% confidence intervals for the advertising-sales regression βs

Statistics in Action Revisited — Assessing How Well a Straight-Line Regression Model Fits Data

In the previous *Statistics in Action Revisited,* we fit the straight-line model, $E(y) = \beta_0 + \beta_1 x$, where x = cumulative 6-month advertising expenditures and y represents either the number of new PI cases or the number of new WC cases per month. The SPSS regression printouts for the two analyses are shown in Figure SIA10.3. (The regression for y = number of new PI cases is shown at the top, and the regression for y = number of new WC cases is shown at the bottom of the printout.) The objective is to determine whether one or both of the dependent variables are statistically linearly related to cumulative 6-month advertising expenditures.

The two-tailed p-values for testing the null hypothesis, H_0: $\beta_1 = 0$ (highlighted on the printouts), are p-value ≈ 0 for number of new PI cases and p-value = .725 for number of new WC cases. For y = number of new PI cases, there is sufficient evidence to reject H_0 (at $\alpha = .01$) and conclude that number of new PI cases is linearly related to cumulative 6-month advertising expenditures. In contrast, for y = number of WC cases, there is insufficient evidence to

Coefficients[a]

Model		Unstandardized Coefficients		Standardized Coefficients	t	Sig.	95% Confidence Interval for B	
		B	Std. Error	Beta			Lower Bound	Upper Bound
1	(Constant)	7.767	3.385		2.295	.027	.926	14.609
	CUM. ADV (thous)	.113	.028	.539	4.042	.000	.056	.169

a. Dependent Variable: New PI Cases

Coefficients[a]

Model		Unstandardized Coefficients		Standardized Coefficients	t	Sig.	95% Confidence Interval for B	
		B	Std. Error	Beta			Lower Bound	Upper Bound
1	(Constant)	24.574	3.367		7.299	.000	17.770	31.379
	CUM. ADV (thous)	.010	.028	.056	.354	.725	-.046	.066

a. Dependent Variable: New WC Cases

Figure SIA10.3
SPSS simple linear regressions for legal advertising data

reject H_0 (at $\alpha = .01$); thus, there is no evidence of a linear relationship between the number of new WC cases and cumulative 6-month advertising expenditures.

We can gain further insight into this phenomenon by examining a 95% confidence interval for the slope, β_1. For y = number of new PI cases, the interval (highlighted on the SPSS printout) is (.056, .169). With 95% confidence, we can state that for every $1,000 increase in monthly advertising expenditures, the number of new PI cases each month will increase between .056 and .169. Now, a more realistic increase in cumulative 6-month advertising expenditures is, say, $20,000. Multiplying the endpoints of the interval by 20, we see that this increase in advertising spending leads to an increase of anywhere between 1 and 3 new PI cases.

Now, for y = number of new WC cases, the 95% confidence interval for the slope (also highlighted on the SPSS printout) is (−.046, .066). Because the interval spans the value 0, we draw the same conclusion as we did with the hypothesis test—there is no statistical evidence of a linear relationship between the number of new WC cases and cumulative 6-month advertising expenditures.

Recall that partner A (who handled the PI cases) sued partner B (who handled the WC cases) for not paying a fair share of the advertising expenses. These results do not support partner A's argument because there is no evidence that partner B benefitted from advertising.

Exercises 10.37–10.52

Learning the Mechanics

10.37 Construct both a 95% and a 90% confidence interval for β_1 for each of the following cases:

a. $\hat{\beta}_1 = 31, s = 3, SS_{xx} = 35, n = 10$

b. $\hat{\beta}_1 = 64, SSE = 1{,}960, SS_{xx} = 30, n = 14$

c. $\hat{\beta}_1 = -8.4, SSE = 146, SS_{xx} = 64, n = 20$

10.38 Consider the following pairs of observations:

x	1	4	3	2	5	6	0
y	1	3	3	1	4	7	2

a. Construct a scattergram for the data.

b. Use the method of least squares to fit a straight line to the seven data points in the table.

c. Plot the least squares line on your scattergram of part **a.**

d. Specify the null and alternative hypotheses you would use to test whether the data provide sufficient evidence to indicate that x contributes information for the (linear) prediction of y.

e. What is the test statistic that should be used in conducting the hypothesis test of part **d**? Specify the degrees of freedom associated with the test statistic.

f. Conduct the hypothesis test of part **d** using $\alpha = .05$.

10.39 Refer to Exercise 10.38. Construct an 80% and a 98% confidence interval for β_1.

10.40 Do the accompanying data provide sufficient evidence to conclude that a straight line is useful for characterizing the relationship between x and y?

x	4	2	4	3	2	4
y	1	6	5	3	2	4

Applying the Concepts—Basic

10.41 State SAT Scores. Refer to the SPSS simple linear regression relating y = average state SAT score in 2007 with x = average state SAT score in 2000, Exercise 10.15 (p. 573) and the data saved in the **SATSCORES** file.

a. Give the null and alternative hypotheses for determining whether a positive linear relationship exists between y and x.

b. Locate the p-value of the test on the SPSS printout. Interpret the result if $\alpha = .05$.

c. Find a 95% confidence interval for the slope, β_1. Interpret the result.

10.42 Characteristics of diamonds sold at retail. Refer to the Minitab simple linear regression analysis relating y = asking price (dollars) to x = number of carats for diamonds sold on the open market, Exercise 10.16 (p. 573). Recall that the data are saved in the **DIAMONDS** file.

NW

a. Give the null and alternative hypotheses for determining whether a positive linear relationship exists between y and x.

b. Locate the p-value of the test on the Minitab printout. Interpret the result if $\alpha = .01$.

c. Find a 99% confidence interval for the slope, β_1. Interpret the result.

10.43 Drug controlled-release rate study. Refer to the simple linear regression analysis relating y = drug release rate to x = surface area-to-volume ratio, Exercise 10.17 (p. 573) and the data saved in the **DOWDRUG** file. Use the results of the regression to form a 90% confidence interval the slope, β_1. Interpret the result.

10.44 Sweetness of orange juice. Refer to the simple linear regression relating y = sweetness index of an orange juice sample with x = amount of water soluble pectin, Exercise 10.20 (p. 575) and the data saved in the **OJUICE** file. Use the results of the regression to form a 95% confidence interval the slope, β_1. Interpret the result.

Applying the Concepts—Intermediate

10.45 Ranking driving performance of professional golfers. Refer to *The Sport Journal* (Winter 2007) study of a new method for ranking the total driving performance of golfers on the PGA tour, Exercise 10.19 (p. 574). You fit a straight-line model relating driving accuracy (y) to driving distance (x) to the data saved in the **PGADRIVER** file.

a. Give the null and alternative hypotheses for testing whether driving accuracy (y) decreases linearly as driving distance (x) increases.

b. Find the test statistic and p-value of the test, part **a.**

c. Make the appropriate conclusion at $\alpha = .01$.

10.46 Are geography journals worth their cost? Refer to the *Geoforum* (Vol. 37, 2006) study of whether the price of a geography journal is correlated with quality, Exercise 10.34 (p. 581). The data are saved in the **GEOJRNL** file.

a. In Exercise 10.34a, you fit a straight-line model relating cost (y) to JIF (x). Find and interpret a 95% confidence interval for the slope of the line.

b. In Exercise 10.34b, you fit a straight-line model relating cost (y) to number of citations (x). Find and interpret a 95% confidence interval for the slope of the line.

c. In Exercise 10.34c, you fit a straight-line model relating cost (y) to RPI (x). Find and interpret a 95% confidence interval for the slope of the line.

10.47 Survey of the top business schools. Refer to the *Wall Street Journal* (Sep. 25, 2005) Business School Survey, Exercise 10.23 (p. 576). If you pay more in tuition to go to a top business school, will it necessarily result in a higher probability of a job offer at graduation? Let y = percentage of graduates with job offers and x = tuition cost; then fit the simple linear model, $E(y) = \beta_0 + \beta_1 x$, to the data in the **BSCHOOL** file. Is there sufficient evidence (at $\alpha = .10$) of a positive linear relationship between y and x?

10.48 Spreading rate of spilled liquid. Refer to the *Chemical Engineering Progress* (Jan. 2005) study of the rate at which a spilled volatile liquid will spread across a surface, Exercise 10.24 (p. 576). For the data in the **LIQUIDSPILL** file, let y = mass of the spill and x = elapsed time of the spill.

a. Is there sufficient evidence (at $\alpha = .05$) to indicate that the mass of the spill tends to diminish linearly as time increases?

b. Give an interval estimate (with 95% confidence) of the decrease in spill mass for each minute of elapsed time.

10.49 Do New Jersey banks serve minority communities? Financial institutions have a legal and social responsibility to serve all communities. Do banks adequately serve both inner-city and suburban neighborhoods, both poor and wealthy communities? In New Jersey, banks have been charged with withdrawing from urban areas with a high percentage of minorities. To examine this charge, a regional New Jersey newspaper, the *Asbury Park Press,* compiled county by county data on the number (y) of people in each county per branch bank in the county and the percentage (x) of the population in each county that is minority. These data for each of New Jersey's 21 counties are provided in the table below and saved in the **NJBANKS** file.

County	Number of People per Bank Branch	Percentage of Minority Population
Atlantic	3,073	23.3
Bergen	2,095	13.0
Burlington	2,905	17.8
Camden	3,330	23.4
Cape May	1,321	7.3
Cumberland	2,557	26.5
Essex	3,474	48.8
Gloucester	3,068	10.7
Hudson	3,683	33.2
Hunterdon	1,998	3.7
Mercer	2,607	24.9
Middlesex	3,154	18.1
Monmouth	2,609	12.6
Morris	2,253	8.2
Ocean	2,317	4.7
Passaic	3,307	28.1
Salem	2,511	16.7
Somerset	2,333	12.0
Sussex	2,568	2.4
Union	3,048	25.6
Warren	2,349	2.8

Source: D'Ambrosio, P., and Chambers, S. "No checks and balances," *Asbury Park Press,* September 10, 1995.

a. Plot the data in a scattergram. What pattern, if any, does the plot reveal?

b. Consider the linear model $E(y) = \beta_0 + \beta_1 x$. If, in fact, the charge against the New Jersey banks is true, then an increase in the percentage of minorities (x) will lead to a decrease in the number of bank branches in a county and therefore will result in an increase in the number of people (y) per branch. Will the value of β_1 be positive or negative in this situation?

c. Do these data support or refute the charge made against the New Jersey banking community? Test using $\alpha = .01$.

10.50 Evaluating managerial success. H. Mintzberg's classic book *The Nature of Managerial Work* (1973) identified the roles found in all managerial jobs. An observational study of 19 managers from a medium-sized manufacturing plant extended Mintzberg's work by investigating which activities *successful* managers actually perform (*Journal of Applied Behavioral Science,* Aug. 1985). To measure success, the researchers devised an index based on the manager's length of time in the organization and his or her level within the firm; the higher the index, the more successful the manager. The next table presents data (saved in the **MANAGERS** file) that can be used to determine whether managerial success is related to the extensiveness of a manager's network-building interactions with people outside the manager's work unit. Such interactions include phone and face-to-face meetings with customers and suppliers, attending outside meetings, and doing public relations work.

Manager	Manager Success Index, y	Number of Interactions with Outsiders, x
1	40	12
2	73	71
3	95	70
4	60	81
5	81	43
6	27	50
7	53	42
8	66	18
9	25	35
10	63	82
11	70	20
12	47	81
13	80	40
14	51	33
15	32	45
16	50	10
17	52	65
18	30	20
19	42	21

a. Construct a scattergram for the data.

b. Find the prediction equation for managerial success.

c. Find s for your prediction equation. Interpret the standard deviation s in the context of this problem.

d. Plot the least squares line on your scattergram of part **a.** Does it appear that the number of interactions with outsiders contributes information for the prediction of managerial success? Explain.

e. Conduct a formal statistical hypothesis test to answer the question posed in part **d.** Use $\alpha = .05$.

f. Construct a 95% confidence interval for β_1. Interpret the interval in the context of the problem.

Applying the Concepts—Advanced

10.51 Does elevation impact hitting performance in baseball? Refer to the *Chance* (Winter 2006) investigation of the effects of elevation on slugging percentage in Major League Baseball, Exercise 2.119 (p. 93). Recall that data were compiled on players' composite slugging percentage

City	Slug Pct.	Elevation
Anaheim	.480	160
Arlington	.605	616
Atlanta	.530	1,050
Baltimore	.505	130
Boston	.505	20
⋮	⋮	⋮
Denver	.625	5,277
⋮	⋮	⋮
Seattle	.550	350
San Francisco	.510	63
St. Louis	.570	465
Tampa	.500	10
Toronto	.535	566

Source: Schaffer, J., and Heiny, E. L. "The effects of elevation on slugging percentage in Major League Baseball," *Chance,* Vol. 19, No. 1, Winter 2006 (adapted from Figure 2, p. 30). Reprinted with permission from *Chance.* © 2006 by the American Statistical Association. All rights reserved.

Minitab Output for Exercise 10.51

Regression Analysis: SLUGPCT versus ELEVATION

```
The regression equation is
SLUGPCT = 0.515 + 0.000021 ELEVATION

Predictor         Coef       SE Coef      T      P
Constant      0.515140      0.007954  64.76  0.000
ELEVATION   0.00002074    0.00000719   2.89  0.008

S = 0.0369803    R-Sq = 23.6%    R-Sq(adj) = 20.7%

Analysis of Variance

Source           DF        SS        MS     F      P
Regression        1  0.011390  0.011390  8.33  0.008
Residual Error   27  0.036924  0.001368
Total            28  0.048314
```

at each of 29 cities for a recent season, as well as each city's elevation (feet above sea level). The data are saved in the **MLBPARKS** file. (Selected observations are shown in the table on page 588.) Consider a straight-line model relating slugging percentage (y) to elevation (x).

a. The model was fit to the data using Minitab, with the results shown in the printout on page 588. Locate the estimates of the model parameters on the printout.

b. Is there sufficient evidence (at $\alpha = .01$) of a positive linear relationship between elevation (x) and slugging percentage (y)? Use the p-value shown on the printout to make the inference.

c. Construct a scatterplot for the data and draw the least squares line on the graph. Locate the data point for Denver on the graph. What do you observe?

d. You learned in Exercise 2.119 that the Colorado Rockies, who play their home games in Coors Field, Denver, typically lead the league in team slugging percentage. Many baseball experts attribute this to the "thin air" of Denver—called the "mile-high" city due to its elevation. Remove the data point for Denver from the data set and refit the straight-line model to the remaining data. Repeat parts **a** and **b.** What conclusions can you draw about the "thin air" theory from this analysis?

10.52 Foreign investment risk. One of the most difficult tasks of developing and managing a global portfolio is assessing the risks of potential foreign investments. Duke University researcher C. R. Henry collaborated with two First Chicago Investment Management Company directors to examine the use of country credit ratings as a means of evaluating foreign investments (*Journal of Portfolio Management,* Winter 1995). To be effective, such a measure should help explain and predict the volatility of the foreign market in question. Data on annualized risk (y) and average credit rating (x) for 40 fictitious countries (based on the study results) are saved in the **GLOBRISKF** file. (The first and last five countries are shown in the table.)

Country	Annualized Risk (%)	Average Credit Rating
1	85.9	30.7
2	25.8	77.1
3	25.2	82.7
4	20.9	77.3
5	63.7	35.1
⋮	⋮	⋮
36	73.0	31.5
37	20.7	86.5
38	14.3	95.3
39	44.9	43.9
40	34.5	23.4

a. Do the data provide sufficient evidence to conclude that country credit risk (x) contributes information for the prediction of market volatility (y)?

b. Use a graph to visually locate any unusual data points (outliers).

c. Eliminate the outlier(s), part **b,** from the data set and rerun the simple linear regression analysis. Note any dramatic changes in the results.

10.5 The Coefficients of Correlation and Determination

In this section, we present two statistics that describe the adequacy of a model: the *coefficient of correlation* and the *coefficient of determination.*

Coefficient of Correlation

Recall (from Section 2.10) that a **bivariate relationship** describes a relationship between two variables, x and y. Scattergrams are used to graphically describe a bivariate relationship. In this section, we will discuss the concept of **correlation** and show how it can be used to measure the linear relationship between two variables x and y. A numerical descriptive measure of the linear association between x and y is provided by the *coefficient of correlation, r.*

The **coefficient of correlation,*** r, is a measure of the strength of the *linear* relationship between two variables x and y. It is computed (for a sample of n measurements on x and y) as follows:

$$r = \frac{SS_{xy}}{\sqrt{SS_{xx}SS_{yy}}}$$

where

$$SS_{xy} = \Sigma(x - \bar{x})(y - \bar{y})$$
$$SS_{xx} = \Sigma(x - \bar{x})^2$$
$$SS_{yy} = \Sigma(y - \bar{y})^2$$

*The value of r is often called the *Pearson correlation coefficient* to honor its developer, Karl Pearson. (See Biography p. 524.)

Note that the computational formula for the correlation coefficient r given in the definition involves the same quantities that were used in computing the least squares prediction equation. In fact, because the numerators of the expressions for $\hat{\beta}_1$ and r are identical, you can see that $r = 0$ when $\hat{\beta}_1 = 0$ (the case where x contributes no information for the prediction of y), and r is positive when the slope is positive and negative when the slope is negative. Unlike $\hat{\beta}_1$, the correlation coefficient r is *scaleless* and assumes a value between -1 and $+1$, regardless of the units of x and y.

A value of r near or equal to 0 implies little or no linear relationship between y and x. In contrast, the closer r comes to 1 or -1, the stronger the linear relationship between y and x. And if $r = 1$ or $r = -1$, all the sample points fall exactly on the least squares line. Positive values of r imply a positive linear relationship between y and x; that is, y increases as x increases. Negative values of r imply a negative linear relationship between y and x; that is, y decreases as x increases. Each of these situations is portrayed in Figure 10.16.

We demonstrate how to calculate the coefficient of correlation r using the data in Table 10.1 for the advertising-sales example. The quantities needed to calculate r are SS_{xy}, SS_{xx}, and SS_{yy}. The first two quantities have been calculated previously as $SS_{xy} = 7$ and $SS_{xx} = 10$. The calculation for $SS_{yy} = \Sigma(y - \bar{y})^2$ is shown on the last column of the Excel spreadsheet, Figure 10.5 (p. 567). The result is $SS_{yy} = 6$.

We now find the coefficient of correlation:

$$r = \frac{SS_{xy}}{\sqrt{SS_{xx}SS_{yy}}} = \frac{7}{\sqrt{(10)(6)}} = \frac{7}{\sqrt{60}} = .904$$

The fact that r is positive and near 1 in value indicates that the sales revenue y tends to increase as advertising expenditure x increases—*for this sample of five months.* This is the same conclusion we reached when we found the calculated value of the least squares slope to be positive.

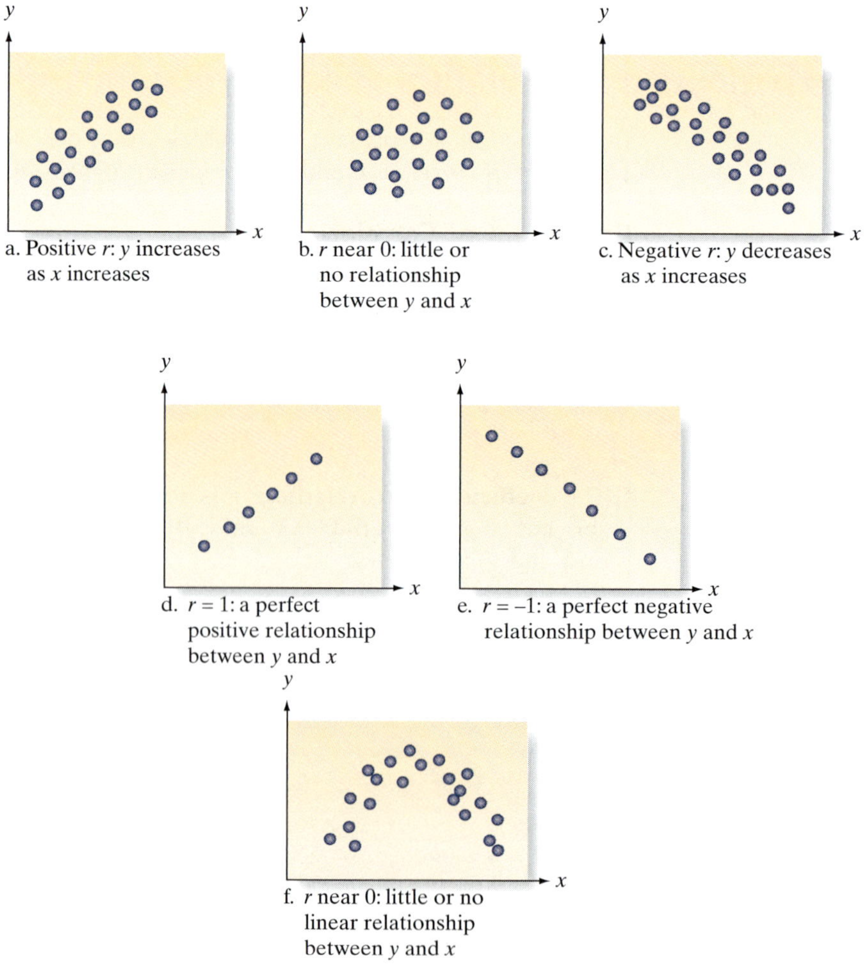

Figure 10.16
Values of r and their implications

Example 10.4

Relating Crime Rate and Casino Employment Using the Correlation Coefficient

Problem Legalized gambling is available on several riverboat casinos operated by a city in Mississippi. The mayor of the city wants to know the correlation between the number of casino employees and the yearly crime rate. The records for the past 10 years are examined, and the results listed in Table 10.3 are obtained. Calculate the coefficient of correlation r for the data. Interpret the result.

Table 10.3 Data on Casino Employees and Crime Rate, Example 10.4

Year	Number of Casino Employees, x (thousands)	Crime Rate, y (number of crimes per 1,000 population)
2000	15	1.35
2001	18	1.63
2002	24	2.33
2003	22	2.41
2004	25	2.63
2005	29	2.93
2006	30	3.41
2007	32	3.26
2008	35	3.63
2009	38	4.15

Data Set: CASINO

Solution Rather than use the computing formula given in the definition, we resort to using a statistical software package. The data of Table 10.3 were entered into a computer, and Minitab was used to compute r. The Minitab printout is shown in Figure 10.17.

The coefficient of correlation, highlighted at the top on the printout, is $r = .987$. Thus, the size of the casino workforce and crime rate in this city are very highly correlated—at least over the past 10 years. The implication is that a strong positive linear relationship exists between these variables (see the scatterplot at the bottom of Figure 10.17). We must be careful, however, not to jump to any unwarranted conclusions. For instance, the mayor may be tempted to conclude that hiring more casino workers next year will increase the crime rate—that is, that there is a *causal relationship* between the two variables. However, high correlation does not imply causality. The fact is, many things have probably contributed both to the increase in the casino workforce and to the increase in crime rate. The city's tourist trade has undoubtedly grown since riverboat casinos were legalized, and it is likely that the casinos have expanded both in services offered and in number. *We cannot infer a causal relationship on the basis of high*

Correlations: EMPLOYEES, CRIMERAT

```
Pearson correlation of EMPLOYEES and CRIMERAT = 0.987
P-Value = 0.000
```

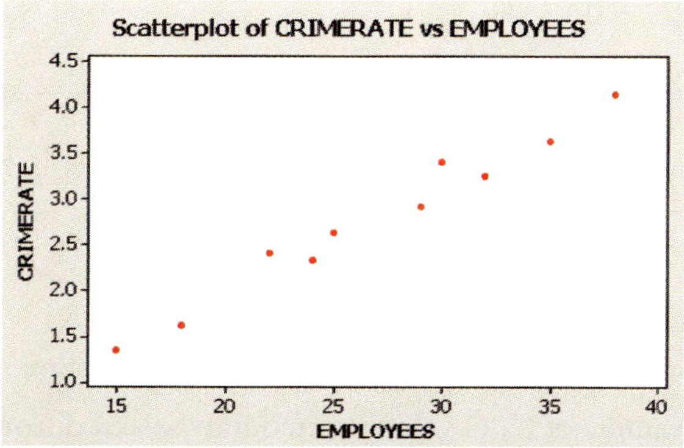

Figure 10.17
Minitab correlation printout for Example 10.4

Ethics IN Statistics

Intentionally using the correlation coefficient only to make an inference about the relationship between two variables in situations where a nonlinear relationship may exist is considered *unethical statistical practice*.

sample correlation. When a high correlation is observed in the sample data, the only safe conclusion is that a linear trend may exist between x *and* y.

Look Back Another variable, such as the increase in tourism, may be the underlying cause of the high correlation between x and y.

Now Work Exercise 10.66a,b

CAUTION When using the sample correlation coefficient, r, to infer the nature of the relationship between x and y, two caveats exist: (1) A *high correlation* does not necessarily imply that a causal relationship exists between x and y—only that a linear trend may exist; (2) a *low correlation* does not necessarily imply that x and y are unrelated—only that x and y are not strongly *linearly* related. ▲

Keep in mind that the correlation coefficient r measures the linear association between x values and y values in the sample, and a similar linear coefficient of correlation exists for the population from which the data points were selected. The **population correlation coefficient** is denoted by the symbol ρ (rho). As you might expect, ρ is estimated by the corresponding sample statistic, r. Or, instead of estimating ρ, we might want to test the null hypothesis H_0: $\rho = 0$ against H_a: $\rho \neq 0$—that is, we can test the hypothesis that x contributes no information for the prediction of y by using the straight-line model against the alternative that the two variables are at least linearly related.

However, we already performed this *identical* test in Section 10.4 when we tested H_0: $\beta_1 = 0$ against H_a: $\beta_1 \neq 0$—that is, the null hypothesis H_0: $\rho = 0$ is equivalent to the hypothesis H_0: $\beta_1 = 0$.* When we tested the null hypothesis H_0: $\beta_1 = 0$ in connection with the advertising-sales example, the data led to a rejection of the null hypothesis at the $\alpha = .05$ level. This rejection implies that the null hypothesis of a 0 correlation between the two variables (sales revenue and advertising expenditure) can also be rejected at the $\alpha = .05$ level. The only real difference between the least squares slope $\hat{\beta}_1$ and the coefficient of correlation r is the measurement scale. Therefore, the information they provide about the usefulness of the least squares model is to some extent redundant. Consequently, we will use the slope to make inferences about the existence of a positive or negative linear relationship between two variables.

For the sake of completeness, a summary of the test for linear correlation is provided in the following boxes.

A Test for Linear Correlation

One-Tailed Test	Two-Tailed Test
H_0: $\rho = 0$	H_0: $\rho = 0$
H_a: $\rho > 0$ (or H_a: $\rho < 0$)	H_a: $\rho \neq 0$

$$\textit{Test statistic: } t = \frac{r\sqrt{n-2}}{\sqrt{1-r^2}} = \frac{\hat{\beta}_1}{s_{\hat{\beta}_1}}$$

Rejection region: $t > t_\alpha$ (or $t < -t_\alpha$)	*Rejection region:* $\lvert t\rvert > t_{\alpha/2}$

where the distribution of t depends an $(n - 2)$ df.

Condition Required for a Valid Test of Correlation

The sample of (x, y) values is randomly selected from a normal population.

*The two tests are equivalent in simple linear regression only.

Coefficient of Determination

Another way to measure the usefulness of the model is to measure the contribution of x in predicting y. To accomplish this, we calculate how much the errors of prediction of y were reduced by using the information provided by x. To illustrate, consider the sample shown in the scattergram of Figure 10.18a. If we assume that x contributes no information for the prediction of y, the best prediction for a value of y is the sample mean $\bar{y}$, which is shown as the horizontal line in Figure 10.18b. The vertical line segments in Figure 10.18b are the deviations of the points about the mean $\bar{y}$. Note that the sum of squares of deviations for the prediction equation $\hat{y} = \bar{y}$ is

$$SS_{yy} = \sum (y_i - \bar{y})^2$$

Now suppose you fit a least squares line to the same set of data and locate the deviations of the points about the line as shown in Figure 10.18c. Compare the deviations about the prediction lines in Figures 10.18b and 10.18c. You can see that

1. If x contributes little or no information for the prediction of y, the sums of squares of deviations for the two lines,

$$SS_{yy} = \sum (y_i - \bar{y})^2 \quad \text{and} \quad SSE = \sum (y_i - \hat{y}_i)^2$$

   will be nearly equal.

2. If x does contribute information for the prediction of y, the SSE will be smaller than SS_{yy}. In fact, if all the points fall on the least squares line, then SSE = 0.

Then the reduction in the sum of squares of deviations that can be attributed to x, expressed as a proportion of SS_{yy}, is

$$\frac{SS_{yy} - SSE}{SS_{yy}}$$

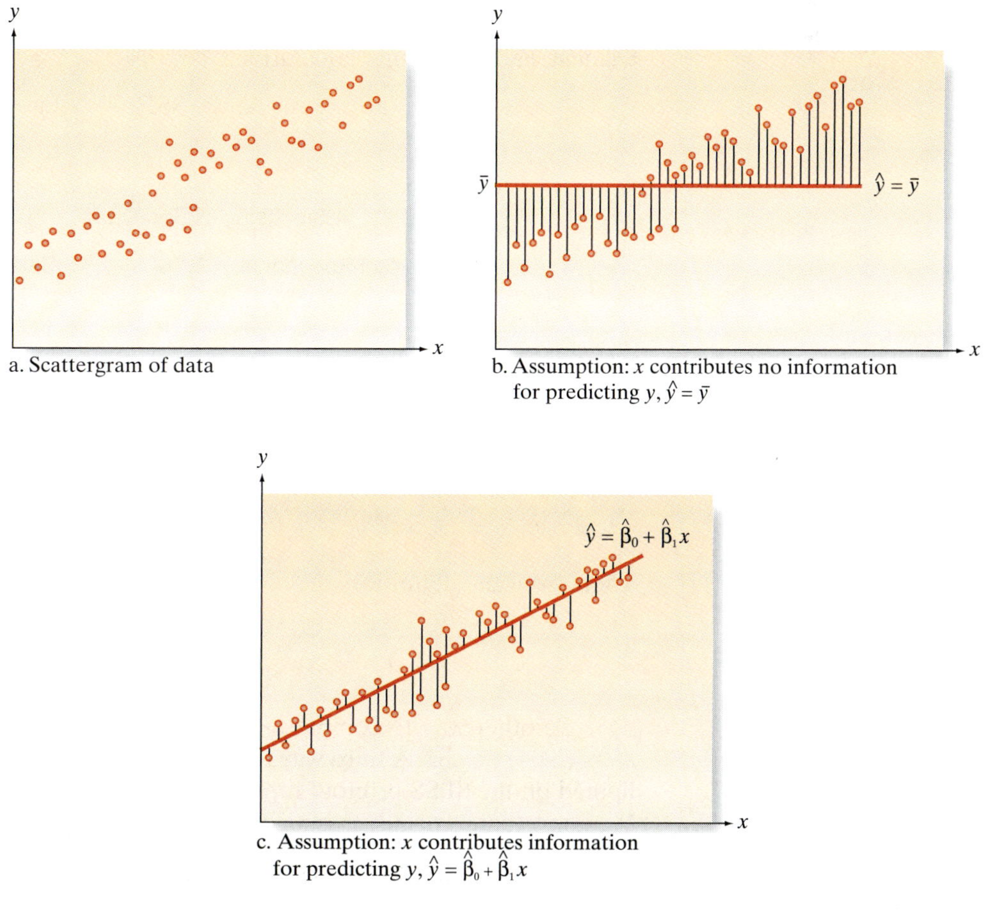

Figure 10.18
A comparison of the sum of squares of deviations for two models

Note that SS_{yy} is the "total sample variation" of the observations around the mean $\bar{y}$ and that SSE is the remaining "unexplained sample variability" after fitting the line $\hat{y}$. Thus, the difference ($SS_{yy} - SSE$) is the "explained sample variability" attributable to the linear relationship with x. Then a verbal description of the proportion is

$$\frac{SS_{yy} - SSE}{SS_{yy}} = \frac{\text{Explained sample variability}}{\text{Total sample variablity}}$$
$$= \text{Proportion of total sample variability explained by the linear relationship}$$

In simple linear regression, it can be shown that this proportion—called the *coefficient of determination*—is equal to the square of the simple linear coefficient of correlation r.

Coefficient of Determination

$$r^2 = \frac{SS_{yy} - SSE}{SS_{yy}} = 1 - \frac{SSE}{SS_{yy}}$$

It represents the proportion of the total sample variability around $\bar{y}$ that is explained by the linear relationship between y and x. (In simple linear regression, it may also be computed as the square of the coefficient of correlation r.)

Note that r^2 is always between 0 and 1 because r is between -1 and $+1$. Thus, an r^2 of .60 means that the sum of squares of deviations of the y values about their predicted values has been reduced 60% by the use of the least squares equation $\hat{y}$, instead of $\bar{y}$, to predict y.

Example 10.5

Obtaining the Value of r^2 for the Sales Revenue Model

Problem Calculate the coefficient of determination for the advertising-sales example. The data are repeated in Table 10.4 for convenience. Interpret the result.

Solution From previous calculations,

$$SS_{yy} = 6 \quad \text{and} \quad SSE = \sum(y - \hat{y})^2 = 1.10$$

Table 10.4 Advertising-Sales Data

Advertising Expenditure, x (\$100s)	Sales Revenue, y (\$1,000s)
1	1
2	1
3	2
4	2
5	4

Data Set: ADSALES

Then, from the definition, the coefficient of determination is given by

$$r^2 = \frac{SS_{yy} - SSE}{SS_{yy}} = \frac{6.0 - 1.1}{6.0} = \frac{4.9}{6.0} = .82$$

Another way to compute r^2 is to recall (Section 10.6) that $r = .904$. Then we have $r^2 = (.904)^2 = .82$. A third way to obtain r^2 is from a computer printout. This value is highlighted on the SPSS printout reproduced in Figure 10.19. Our interpretation is as follows: We know that using advertising expenditure, x, to predict y with the least squares line

$$\hat{y} = -.1 + .7x$$

Model Summary

Model	R	R Square	Adjusted R Square	Std. Error of the Estimate
1	.904a	.817	.756	.606

a. Predictors: (Constant), ADVEXP_X

Figure 10.19
Portion of SPSS printout for advertising-sales regression

accounts for 82% of the total sum of squares of deviations of the five sample y values about their mean. Or, stated another way, 82% of the sample variation in sales revenue (y) can be "explained" by using advertising expenditure (x) in a straight-line model.

Now Work Exercise 10.62a

Practical Interpretation of the Coefficient of Determination, r^2

About $100(r^2)$% of the sample variation in y (measured by the total sum of squares of deviations of the sample y values about their mean $\bar{y}$) can be explained by (or attributed to) using x to predict y in the straight-line model.

Statistics IN Action Revisited — Using the Coefficient of Correlation and the Coefficient of Determination

In the previous *Statistics in Action Revisited* (p. 585), we discovered that cumulative 6-month advertising expenditures was a statistically useful linear predictor of number of new PI cases but not a useful linear predictor of number of new WC cases. Both the coefficients of correlation and determination (highlighted on the Minitab printouts in Figures SIA10.4 and Figures SIA10.5) also support this conclusion.

For y = number of new PI cases, the correlation coefficient value of $r = .539$ is statistically significantly different from 0 and indicates a moderate positive linear relationship between the variables. The coefficient of determination, $r^2 = .29$, implies that almost 30% of the sample variation in number of new PI cases can be explained by using advertising expenditure (x) in the straight-line model. In contrast, for y = number of new WC cases, $r = .056$ is not statistically different from 0 and $r^2 = .003$ implies that only 0.3% of the sample variation in number of new WC cases can be explained by using advertising expenditure (x) in the straight-line model.

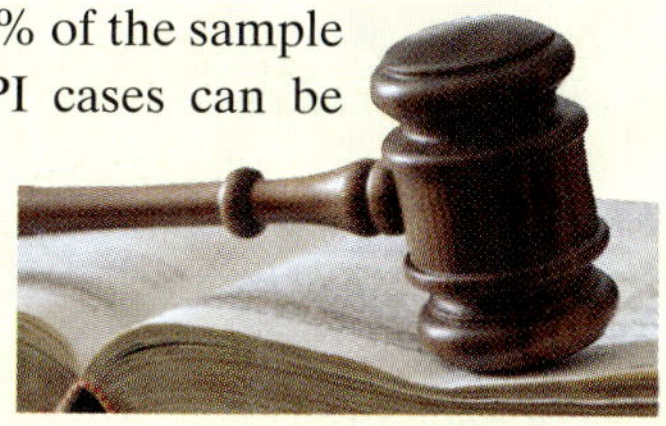

Regression Analysis: NewPI versus AdvExp6

```
The regression equation is
NewPI = 7.77 + 0.113 AdvExp6

Predictor      Coef   SE Coef     T      P
Constant      7.767     3.385  2.29  0.027
AdvExp6     0.11289   0.02793  4.04  0.000

S = 9.67521   R-Sq = 29.0%   R-Sq(adj) = 27.2%

Analysis of Variance

Source          DF      SS      MS      F      P
Regression       1  1529.5  1529.5  16.34  0.000
Residual Error  40  3744.4    93.6
Total           41  5273.9
```

Correlations: NewPI, AdvExp6

```
Pearson correlation of NewPI and AdvExp6 = 0.539
P-Value = 0.000
```

Figure SIA10.4
Minitab printout with coefficients of correlation and determination for y = number of new PI cases

Regression Analysis: NewWC versus AdvExp6

```
The regression equation is
NewWC = 24.6 + 0.0098 AdvExp6

Predictor      Coef   SE Coef     T      P
Constant     24.574     3.367  7.30  0.000
AdvExp6     0.00982   0.02778  0.35  0.725

S = 9.62296   R-Sq = 0.3%    R-Sq(adj) = 0.0%

Analysis of Variance

Source          DF       SS     MS     F      P
Regression       1    11.58  11.58  0.13  0.725
Residual Error  40  3704.06  92.60
Total           41  3715.64
```

Correlations: NewWC, AdvExp6

```
Pearson correlation of NewWC and AdvExp6 = 0.056
P-Value = 0.725
```

Figure SIA10.5
Minitab printout with coefficients of correlation and determination for y = number of new WC cases

Activity 10.2 *Keep the Change:* Correlation Coefficients

In this activity, you will use data collected in Activity 1.1, *Keep the Change: Collecting Data* (p. 16) and the results from Activity 10.1, *Keep the Change: Least Squares Models* on (p. 571) to study the strength of a linear relationship.

1. For each of the least squares models in Exercises 1 and 4 in Activity 10.1, calculate the corresponding correlation coefficient. Do these values support your conclusion about which model is more useful? Explain.
2. For each purchase in your original data set *Purchase Totals,* form the ordered pair

   (Purchase Total, Amount Transferred)

   Calculate the correlation coefficient and discuss the strength of the linear relationship between the two variables. Explain why you might expect the slope of the corresponding least squares line to be close to zero. Find the least squares model and comment on its usefulness. Does the model produce estimated values that are meaningless for large purchase totals? Explain.
3. For each student in your class, form an ordered pair

   (Sum of Amounts Transferred, Bank Matching)

   Calculate the correlation coefficient. How strong is the linear relationship between the two variables? Find the least squares model. Is the slope $\hat{\beta}_1$ approximately 13 and the y-intercept $\hat{\beta}_0$ approximately 0? Explain why the line should have this slope and y-intercept. How good are the estimates given by the model in this situation?

Exercises 10.53–10.69

Learning the Mechanics

10.53 Describe the slope of the least squares line if
a. $r = .7$ **b.** $r = -.7$
c. $r = 0$ **d.** $r^2 = .64$

10.54 Explain what each of the following sample correlation coefficients tells you about the relationship between the x and y values in the sample:
a. $r = 1$ **b.** $r = -1$ **c.** $r = 0$
d. $r = -.90$ **e.** $r = .10$ **f.** $r = -.88$

10.55 Calculate r^2 for the least squares line in each of the following exercises. Interpret their values.
a. Exercise 10.10 **b.** Exercise 10.13

10.56 Construct a scattergram for each data set. Then calculate r and r^2 for each data set. Interpret their values.

a.

x	−2	−1	0	1	2
y	−2	1	2	5	6

b.

x	−2	−1	0	1	2
y	6	5	3	2	0

c.

x	1	2	2	3	3	3	4
y	2	1	3	1	2	3	2

d.

x	0	1	3	5	6
y	0	1	2	1	0

Applet Exercise 10.2

Use the applet *Correlation by Eye* to explore the relationship between the pattern of data in a scattergram and the corresponding correlation coefficient.

a. Run the applet several times. Each time, guess the value of the correlation coefficient. Then click *Show r* to see the actual correlation coefficient. How close is your value to the actual value of r? Click *New data* to reset the applet.

b. Click the trash can to clear the graph. Use the mouse to place five points on the scattergram that are approximately in a straight line. Then guess the value of the correlation coefficient. Click *Show r* to see the actual correlation coefficient. How close were you this time?

c. Continue to clear the graph and plot sets of five points with different patterns among the points. Guess the value of r. How close do you come to the actual value of r each time?

d. Based on your experiences with the applet, explain why we need to use more reliable methods of finding the correlation coefficient than just "eyeing" it.

Applying the Concepts—Basic

10. 57 RateMyProfessors.com. A popular Web site among college students is RateMyProfessors.com (RMP). Established over 10 years ago, RMP allows students to post quantitative ratings of their instructors. In *Practical Assessment, Research & Evaluation* (May 2007), University of Maine researchers investigated whether instructor ratings posted on RMP are correlated with the formal in-class student evaluations of teaching (SET) that all universities are required to administer at the end of the semester. Data collected for $n = 426$ University of Maine instructors yielded a correlation between RMP and SET ratings of .68.

a. Give the equation of a linear model relating SET rating (y) to RMP rating (x).

b. Give a practical interpretation of the value $r = .68$.

c. Is the estimated slope of the line, part **a,** positive or negative? Explain.

d. A test of the null hypothesis H_0: $\rho = 0$ yielded a p-value of .001. Interpret this result.

e. Compute the coefficient of determination, r^2, for the regression analysis. Interpret the result.

10.58 In business, do nice guys finish first or last? Refer to the *Nature* (March 20, 2008) study of the use of punishment in cooperation games, Exercise 10.14 (p. 572). Recall that college students repeatedly played a version of the game "prisoner's dilemma," and the researchers recorded the average payoff and the number of times cooperation, defection, and punishment were used for each player.

a. A test of no correlation between cooperation use (x) and average payoff (y) yielded a p-value of .33. Interpret this result.

b. A test of no correlation between defection use (x) and average payoff (y) yielded a p-value of .66. Interpret this result.

c. A test of no correlation between punishment use (x) and average payoff (y) yielded a p-value of .001. Interpret this result.

10.59 "Metaskills" and career management. In today's business environment, effective management of one's own career requires a skill set that includes adaptability, tolerance for ambiguity, self-awareness, and ability to identify change. Management professors at Pace University (New York) used correlation coefficients to investigate the relationship between these "metaskills" and effective career management (*International Journal of Manpower,* Aug. 2000). Data were collected for 446 business graduates who had all completed a management metaskills course. Two of the many variables measured were self-knowledge skill level (x) and goal-setting ability (y). The correlation coefficient for these two variables was $r = .70$.

a. Give a practical interpretation of the value of r.

b. The p-value for a test of no correlation between the two variables was reported as p-value $= .001$. Interpret this result.

c. Find the coefficient of determination, r^2, and interpret the result.

10.60 Sports news on local TV broadcasts. *The Sports Journal* (Winter 2004) published the results of a study conducted to assess the factors that impact the time allotted to sports news on local television news broadcasts. Information on total time (in minutes) allotted to sports and audience ratings of the TV news broadcast (measured on a 100-point scale) was obtained from a national sample of 163 news directors. A correlation analysis on the data yielded $r = .43$.

a. Interpret the value of the correlation coefficient, r.

b. Find and interpret the value of the coefficient of determination, r^2.

10.61 Women in top management. An empirical analysis of women in upper management positions at U.S. firms was published in the *Journal of Organizational Culture, Communications and Conflict* (July 2007). Monthly data ($n = 252$ months) were collected for several variables, including the number of females in managerial positions, the number of females with a college degree, and the number of female high school graduates with no college degree. Similar data were collected for males.

a. The correlation coefficient relating number of females in managerial positions and number of females with a college degree was reported as $r = .983$. Interpret this result.

b. The correlation coefficient relating number of females in managerial positions and number of female high school graduates with no college degree was reported as $r = .074$. Interpret this result.

c. The correlation coefficient relating number of males in managerial positions and number of males with a college degree was reported as $r = .722$. Interpret this result.

d. The correlation coefficient relating number of males in managerial positions and number of male high school graduates with no college degree was reported as $r = .528$. Interpret this result.

Applying the Concepts—Intermediate

10.62 Characteristics of diamonds sold at retail. Refer to the Minitab simple linear regression analysis relating y = asking price (dollars) to x = number of carats for diamonds sold on the open market, Exercise 10.16 (p. 573) and the data saved in the **DIAMONDS** file.

NW **a.** Locate the coefficient of determination, r^2, on the Minitab printout and interpret the result.

b. Find the value of the coefficient of correlation, r, from the value of r^2 and the sign of the estimated slope. Interpret the result.

10.63 Sweetness of orange juice. Refer to the simple linear regression relating y = sweetness index of an orange juice sample with x = amount of water soluble pectin, Exercise 10.20 (p. 575) and the data saved in the **OJUICE** file. Find and interpret the coefficient of determination, r^2, and the coefficient of correlation, r.

10.64 Performance ratings of government agencies. The U.S. Office of Management and Budget (OMB) requires government agencies to produce annual performance and accounting reports (PARS) each year. Refer to *The Public Manager* (Summer 2008) listing of PARS evaluation scores for 24 government agencies, Exercise 2.123. (p. 94). Recall that evaluation scores ranged from 12 (lowest) to 60 (highest). The PARS evaluation scores for 2007 and 2008 are reproduced in the following table and saved in the **PARS** file.

a. Calculate and interpret the coefficient of correlation between the PARS scores in 2007 and 2008.

b. Is there sufficient evidence of a positive linear relationship between 2008 PARS score (y) and 2007 PARS score (x)? Test using $\alpha = .05$.

Agency	Score07	Score08
Transportation	55	53
Labor	53	51
Veterans	51	51
NRC	39	34
Commerce	37	36
HHS	37	35
DHS	37	30
Justice	35	37
Treasury	35	35
GSA	34	40
Agriculture	33	35
EPA	33	36
Social Security	33	33
USAID	32	42
Education	32	36
Interior	32	31
NASA	32	32
Energy	31	34
HUD	31	30
NSF	31	31
State	31	50
OPM	27	28
SBA	22	31
Defense	17	32

Source: Ellig, J., and Wray, H. "Measuring performance reporting quality," *The Public Manager,* Vol. 37, No. 2, Summer 2008 (p. 66). Copyright 2008 by *The Public Manager* in the format electronic usage via Copyright Clearance Center.

10.65 Salary linked to height. Are short people short-changed when it comes to salary? According to Business Professors T. A. Judge (University of Florida) and D. M. Cable (University of North Carolina), tall people tend to earn more money over their careers than short people (*Journal of Applied Psychology,* June 2004). Using data collected from participants in the National Longitudinal Surveys begun in 1979, the researchers computed the correlation between average earnings from 1985 to 2000 (in dollars) and height (in inches) for several occupations. The results are given in the table in the next column.

a. Interpret the value of r for people in sales occupations.

b. Compute r^2 for people in sales occupations. Interpret the result.

Occupation	Correlation, r	Sample Size, n
Sales	.41	117
Managers	.35	455
Blue Collar	.32	349
Service Workers	.31	265
Professional/Technical	.30	453
Clerical	.25	358
Crafts/Forepersons	.24	250

Source: Judge, T. A., and Cable, D. M. "The effect of physical height on workplace success and income: Preliminary test of a theoretical model," *Journal of Applied Psychology,* Vol. 89, No. 3, June 2004 (Table 5). Copyright © 2004 by the American Psychological Association. Reproduced with permission. The use of APA information does not imply endorsement by APA.

c. Give H_0 and H_a for testing whether average earnings and height are positively correlated.

d. The test statistic for testing H_0 and H_a in part **c** is

$$t = \frac{r\sqrt{n-2}}{\sqrt{1-r^2}}$$

Compute this value for people in sales occupations.

e. Use the result, part **d,** to conduct the test at $\alpha = .01$. Give the appropriate conclusion.

f. Select another occupation and repeat parts **a–e.**

10.66 The Forbes 400. The Forbes 400 is an annual ranking of the 400 wealthiest people in the United States. The top 15 billionaires on this list for 2006 are described in the table at the top of the page 599 and saved in the **FORBES400** file.

NW **a.** Construct a scattergram for these data. What does the plot suggest about the relationship between age and net worth of billionaires?

NW **b.** Find the coefficient of correlation and explain what it tells you about the relationship between age and net worth.

c. If the correlation coefficient of part **b** had the opposite sign, how would that change your interpretation of the relationship between age and net worth?

d. Find the coefficient of determination for a straight-line model relating net worth (y) to age (x). Interpret the result in the words of the problem.

10.67 Survey of the top business schools. Refer to the *Wall Street Journal* (Sep. 25, 2005) Business School Survey, Exercise 10.23 (p. 576) and the data saved in the **BSCHOOL** file. Find and interpret r and r^2 for the simple linear regression relating y = percentage of graduates with job offers and x = tuition cost and then fit the simple linear model.

10.68 Spreading rate of spilled liquid. Refer to the *Chemical Engineering Progress* (Jan. 2005) study of the rate at which a spilled volatile liquid will spread across a surface, Exercise 10.24 (p. 576) and the data saved in the **LIQUIDSPILL** file. Find and interpret r and r^2 for the simple linear regression y = mass of the spill and x = elapsed time of the spill.

Applying the Concepts—Advanced

10.69 Attitudes toward quality management. Studies of Asian (particularly Japanese) and U.S. managers in the 1970s and 1980s found sharp differences of opinion and attitude toward quality management. Do these differences continue to exist? To find out, two California State University researchers (B. F. Yavas and T. M. Burrows) surveyed 100 U.S. and 96 Asian managers in the electronics manufacturing industry

Table for Exercise 10.66

Rank	Name	Net Worth ($ billions)	Age	Marital Status	Residence	Source of Wealth
1	Gates, Bill	53.0	50	married	Medina, WA	Microsoft
2	Buffett, Warren	46.0	76	widowed	Omaha, NE	Berkshire
3	Adelson, Sheldon	20.5	73	married	Las Vegas, NV	gambling
4	Ellison, Lawrence	19.5	62	married	Redwood, CA	Oracle
5	Allen, Paul	16.0	53	single	Seattle, WA	Microsoft
6	Walton, Jim	15.7	58	married	Bentonville, AR	Wal-Mart
7	Walton, Christy	15.6	51	widowed	Jackson, WY	Wal-Mart
8	Walton, S. Robson	15.6	62	married	Bentonville, AR	Wal-Mart
9	Dell, Michael	15.5	41	married	Austin, TX	Dell Computer
10	Walton, Alice	15.5	57	divorced	Fort Worth, TX	Wal-Mart
11	Walton, Helen	15.3	86	widowed	Bentonville, AR	Wal-Mart
12	Brin, Sheldon	14.1	23	single	Palo Alto, CA	Google
13	Page, Larry	14.0	33	single	San Francisco, CA	Google
14	Taylor, Jack	13.9	84	married	St. Louis, MO	Enterprise Rent-A-Car
15	Ballmer, Steve	13.6	50	married	Bellevue, WA	Microsoft

Source: "400 wealthiest people in the United States. The top 15 billionaires," *Forbes* 400, Sept. 21, 2006. Reprinted by permission of Forbes.com. © 2009 Forbes LLC.

(*Quality Management Journal,* Fall 1994). The table (saved in the **QLAGREE** file) gives the percentages of U.S. and Asian managers who agree with each of 13 randomly selected statements regarding quality. (For example, one statement is "Quality is a problem in my company." Another is "Improving quality is expensive.")

Statement	Percentage of Managers Who Agree	
	U.S.	Asian
1	36	38
2	31	42
3	28	43
4	27	48
5	78	58
6	74	49
7	43	46
8	50	56
9	31	65
10	66	58
11	18	21
12	61	69
13	53	45

Source: Yavas, B. F., and Burrows, T. M. "A comparative study of attitudes of U.S. and Asian managers toward product quality," *Quality Management Journal,* Vol. 2, No.1, September 1994, p. 49 (Table 5). Reprinted with permission from *Quality Management Journal.* © 1994 American Society for Quality. No further distribution allowed without permission.

a. Find the coefficient of correlation r for these data.

b. Interpret r in the context of the problem.

c. Refer to part **b.** Using the coefficient of correlation r to make inferences about the difference in attitudes between U.S. and Asian managers regarding quality can be misleading. The value of r measures the strength of the linear relationship between two variables; it does not account for a difference between the means of the variables. To illustrate this, examine the hypothetical data in the following table (saved in the **QLAGREE2** file). Show that $r \approx 1$, but the Asian percentage is approximately 30 points higher for each quality statement. Would you conclude that the attitudes of U.S. and Asian managers are similar?

Quality Statement	Hypothetical Percentage of Managers Who Agree	
	U.S.	Asian
1	20	50
2	30	65
3	40	70
4	50	80
5	55	90

10.6 Using the Model for Estimation and Prediction

If we are satisfied that a useful model has been found to describe the relationship between x and y, we are ready for step 5 in our regression modeling procedure: using the model for estimation and prediction.

The most common uses of a probabilistic model for making inferences can be divided into two categories. The first is using the model to estimate the mean value of y, E(y), *for a specific value of* x. For our advertising-sales example, we may want to estimate the mean sales revenue for *all* months during which \$400 ($x = 4$) is expended on advertising.

The second use of the model entails predicting a new individual y *value for a given* x. That is, if we decide to expend \$400 in advertising next month, we may want to predict the firm's sales revenue for that month.

In the first case, we are attempting to estimate the mean value of y for a very large number of experiments at the given x value. In the second case, we are trying to predict the outcome of a single experiment at the given x value. Which of these model uses—estimating the mean value of y or predicting an individual new value of y (for the same value of x)—can be accomplished with the greater accuracy?

Before answering this question, we first consider the problem of choosing an estimator (or predictor) of the mean (or a new individual) y value. We will use the least squares prediction equation

$$\hat{y} = \hat{\beta}_0 + \hat{\beta}_1 x$$

both to estimate the mean value of y and to predict a specific new value of y for a given value of x. For our example, we found

$$\hat{y} = -.1 + .7x$$

so that the estimated mean sales revenue for all months when $x = 4$ (advertising is \$400) is

$$\hat{y} = -.1 + .7(4) = 2.7$$

or \$2,700. (Recall that the units of y are thousands of dollars.) The same value is used to predict a new y value when $x = 4$—that is, both the estimated mean and the predicted value of y are $\hat{y} = 2.7$ when $x = 4$, as shown in Figure 10.20.

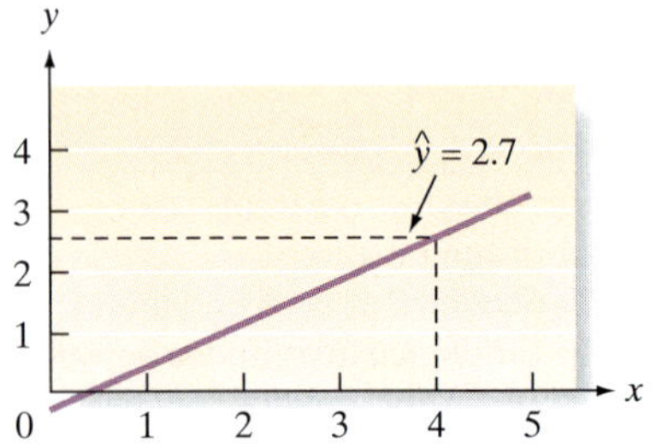

Figure 10.20
Estimated mean value and predicted individual value of sales revenue y for $x = 4$

The difference between these two model uses lies in the relative accuracy of the estimate and the prediction. These accuracies are best measured by using the sampling errors of the least squares line when it is used as an estimator and as a predictor, respectively. These errors are reflected in the standard deviations given in the next box.

Sampling Errors for the Estimator of the Mean of y and the Predictor of an Individual New Value of y

1. The *standard deviation* of the sampling distribution of the estimator $\hat{y}$ of the *mean value of y* at a specific value of x, say x_p, is

$$\sigma_{\hat{y}} = \sigma\sqrt{\frac{1}{n} + \frac{(x_p - \bar{x})^2}{SS_{xx}}}$$

where σ is the standard deviation of the random error ε. We refer to $\sigma_{\hat{y}}$ as the **standard error of $\hat{y}$.**

2. The *standard deviation* of the prediction error for the predictor $\hat{y}$ of an individual *new y value* at a specific value of x is

$$\sigma_{(y-\hat{y})} = \sigma\sqrt{1 + \frac{1}{n} + \frac{(x_p - \bar{x})^2}{SS_{xx}}}$$

where σ is the standard deviation of the random error ε. We refer to $\sigma_{(y-\hat{y})}$ as the **standard error of the prediction.**

The true value of σ is rarely known, so we estimate σ by s and calculate the estimation and prediction intervals as shown in the next two boxes.

A 100(1 − α)% Confidence Interval for the Mean Value of y at $x = x_p$

$$\hat{y} \pm t_{\alpha/2}(\text{Estimated standard error of } \hat{y})$$

or

$$\hat{y} \pm t_{\alpha/2}s\sqrt{\frac{1}{n} + \frac{(x_p - \bar{x})^2}{SS_{xx}}}$$

where $t_{\alpha/2}$ is based on $(n - 2)$ degrees of freedom.

A 100(1 − α)% Prediction Interval* for an Individual New Value of y at $x = x_p$

$$\hat{y} \pm t_{\alpha/2}(\text{Estimated standard error of prediction})$$

or

$$\hat{y} \pm t_{\alpha/2}s\sqrt{1 + \frac{1}{n} + \frac{(x_p - \bar{x})^2}{SS_{xx}}}$$

where $t_{\alpha/2}$ is based on $(n - 2)$ degrees of freedom.

Example 10.6
Estimating the Mean of y—Sales Revenue Model

Problem Refer to the sales-appraisal simple linear regression in previous examples. Find a 95% confidence interval for the mean monthly sales when the appliance store spends $400 on advertising.

Solution For a $400 advertising expenditure, $x = 4$, and the confidence interval for the mean value of y is

$$\hat{y} \pm t_{\alpha/2}s\sqrt{\frac{1}{n} + \frac{(x_p - \bar{x})^2}{SS_{xx}}} = \hat{y} \pm t_{.025}s\sqrt{\frac{1}{5} + \frac{(4 - \bar{x})^2}{SS_{xx}}}$$

where $t_{.025}$ is based on $n - 2 = 5 - 2 = 3$ degrees of freedom. Recall that $\hat{y} = 2.7$, $s = .61$, $\bar{x} = 3$, and $SS_{xx} = 10$. From Table V in Appendix B, $t_{.025} = 3.182$. Thus, we have

$$2.7 \pm (3.182)(.61)\sqrt{\frac{1}{5} + \frac{(4 - 3)^2}{10}} = 2.7 \pm (3.182)(.61)(.55)$$
$$= 2.7 \pm (3.182)(.34)$$
$$= 2.7 \pm 1.1 = (1.6, 3.8)$$

Therefore, when the store spends $400 a month on advertising, we are 95% confident that the mean sales revenue is between $1,600 and $3,800.

Look Back Note that we used a small amount of data (small in size) for purposes of illustration in fitting the least squares line. The interval would probably be narrower if more information had been obtained from a larger sample.

Now Work Exercise 10.70a–d

Example 10.7
Predicting an Individual value of y—Sales Revenue Model

Problem Refer, again, to the sales-appraisal regression. Predict the monthly sales for next month if $400 is spent on advertising. Use a 95% prediction interval.

*The term *prediction interval* is used when the interval formed is intended to enclose the value of a random variable. The term *confidence interval* is reserved for the estimation of population parameters (such as the mean).

Solution To predict the sales for a particular month for which $x_p = 4$, we calculate the 95% prediction interval as

$$\hat{y} \pm t_{\alpha/2} s \sqrt{1 + \frac{1}{n} + \frac{(x_p - \bar{x})^2}{SS_{xx}}} = 2.7 \pm (3.182)(.61)\sqrt{1 + \frac{1}{5} + \frac{(4-3)^2}{10}}$$
$$= 2.7 \pm (3.182)(.61)(1.14)$$
$$= 2.7 \pm (3.182)(.70)$$
$$= 2.7 \pm 2.2 = (.5, 4.9)$$

Therefore, we predict with 95% confidence that the sales revenue next month (a month in which we spend \$400 in advertising) will fall in the interval from \$500 to \$4,900.

Look Back Like the confidence interval for the mean value of y, the prediction interval for y is quite large. This is because we have chosen a simple example (only five data points) to fit the least squares line. The width of the prediction interval could be reduced by using a larger number of data points.

Now Work Exercise 10.70e

Both the confidence interval for $E(y)$ and prediction interval for y can be obtained using a statistical software package. Figure 10.21 is a Minitab printout showing the confidence interval and prediction interval for the data in the advertising-sales example. The 95% confidence interval for $E(y)$ when $x = 4$, highlighted under "95% CI" in Figure 10.21, is (1.645, 3.755). The 95% prediction interval for y when $x = 4$, highlighted in Figure 10.21 under "95% PI," is (.503, 4.897). Both intervals agree with the ones computed in Examples 10.6–10.7.

Note that the prediction interval for an individual new value of y is *always* wider than the corresponding confidence interval for the mean value of y. To see this, consider the following. The error in estimating the mean value of y, $E(y)$, for a given value of x, say x_p, is the distance between the least squares line and the true line of means, $E(y) = \beta_0 + \beta_1 x$. This error, $[\hat{y} - E(y)]$, is shown in Figure 10.22. In contrast,

```
Predicted Values for New Observations

New
Obs    Fit   SE Fit          95% CI              95% PI
  1  2.700    0.332  (1.645, 3.755)  (0.503, 4.897)


Values of Predictors for New Observations

New
Obs  ADVEXP_X
  1      4.00
```

Figure 10.21
Minitab printout giving 95% confidence interval for $E(y)$ and 95% prediction interval for y

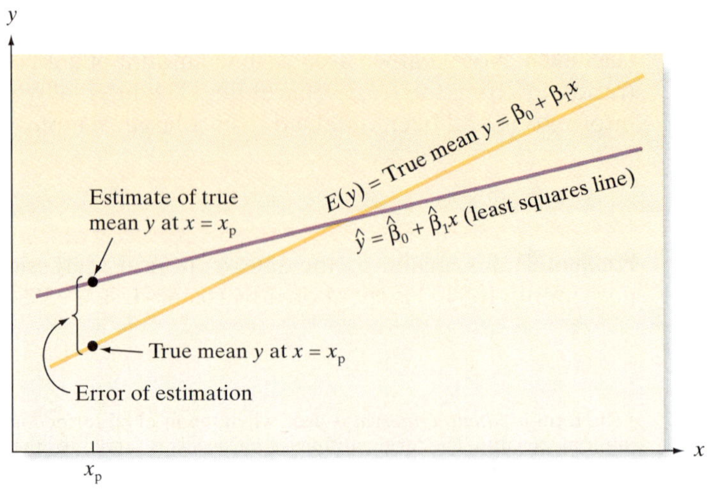

Figure 10.22
Error of estimating the mean value of y for a given value of x

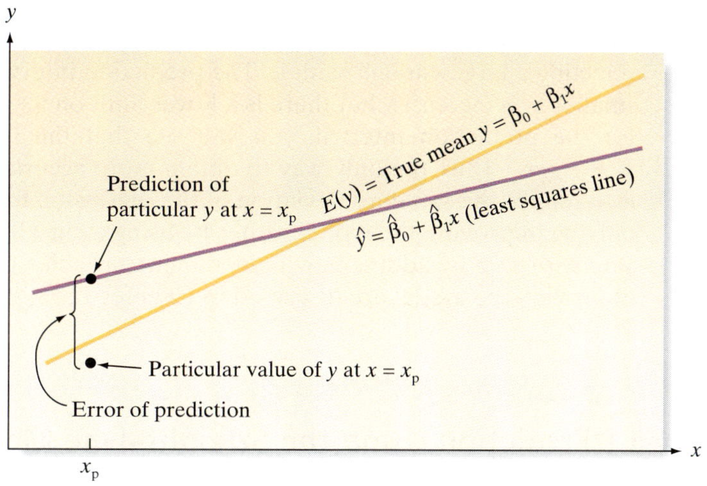

Figure 10.23
Error of predicting a future value of y for a given value of x

the error $(y_p - \hat{y})$ *in predicting some future value of* y *is the sum of two errors*—the error of estimating the mean of y, $E(y)$, plus the random error that is a component of the value of y to be predicted (see Figure 10.23). Consequently, the error of predicting a particular value of y will be larger than the error of estimating the mean value of y for a particular value of x. Note from their formulas that both the error of estimation and the error of prediction take their smallest values when $x_p = \bar{x}$. The farther x_p lies from $\bar{x}$, the larger will be the errors of estimation and prediction. You can see why this is true by noting the deviations for different values of x_p between the line of means $E(y) = \beta_0 + \beta_1 x$ and the predicted line of means $\hat{y} = \hat{\beta}_0 + \hat{\beta}_1 x$ shown in Figure 10.23. The deviation is larger at the extremes of the interval where the largest and smallest values of x in the data set occur.

Both the confidence intervals for mean values and the prediction intervals for new values are depicted over the entire range of the regression line in Figure 10.24. You can see that the confidence interval is always narrower than the prediction interval, and they are both narrowest at the mean $\bar{x}$, increasing steadily as the distance $|x - \bar{x}|$ increases. In fact, when x is selected far enough away from $\bar{x}$ so that it falls outside the range of the sample data, it is dangerous to make any inferences about $E(y)$ or y.

CAUTION Using the least squares prediction equation to estimate the mean value of y or to predict a particular value of y for values of x that fall *outside the range* of the values of x contained in your sample data may lead to errors of estimation or prediction that are much larger than expected. Although the least squares model may provide a very good fit to the data over the range of x values contained in the sample, it could give a poor representation of the true model for values of x outside this region. ▲

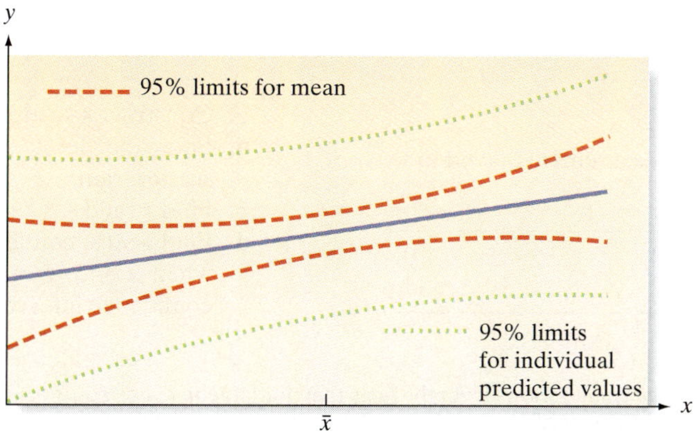

Figure 10.24
Confidence intervals for mean values and prediction intervals for new values

The confidence interval width grows smaller as n is increased; thus, in theory, you can obtain as precise an estimate of the mean value of y as desired (at any given x) by selecting a large enough sample. The prediction interval for a new value of y also grows smaller as n increases, but there is a lower limit on its width. If you examine the formula for the prediction interval, you will see that the interval can get no smaller than $\hat{y} \pm z_{\alpha/2}\sigma$.* Thus, the only way to obtain more accurate predictions for new values of y is to reduce the standard deviation of the regression model, σ. This can be accomplished only by improving the model, either by using a curvilinear (rather than linear) relationship with x or by adding new independent variables to the model, or both. Methods of improving the model are discussed in Chapter 11.

Statistics IN Action Revisited — Prediction Using the Straight-Line Model

In the two previous *Statistics in Action Revisited* sections, we demonstrated that x = cumulative 6-month advertising expenditures is a useful linear predictor of y = the number of new PI cases. The Minitab printout shown in Figure SIA10.6 gives a 95% prediction interval for the number of new PI cases in a month where cumulative spending on advertising over the past 6 months is x = \$150,000. The highlighted interval is (4.78, 44.62). Thus, we can be 95% confident that for a given month with a 6-month cumulative advertising expenditure of \$150,000, the law firm will add between 5 and 44 new PI cases that month.

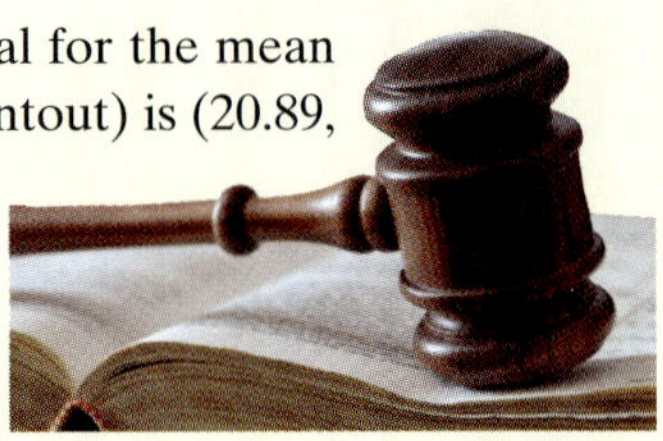

The 95% confidence interval for the mean of y (also highlighted on the printout) is (20.89, 28.51). This indicates that for all months with cumulative spending on advertising of x = \$150,000, the average number of new PI cases each month will range between 21 and 28 cases. Of course, if these projections are to be useful for future planning, other economic factors that may affect the firm's business must be similar to those during the period over which the sample data were generated.

```
Predicted Values for New Observations

New
Obs     Fit  SE Fit         95% CI              95% PI
  1   24.70    1.89  (20.89, 28.51)     (4.78, 44.62)

Values of Predictors for New Observations

New
Obs  AdvExp6
  1      150
```

Figure SIA10.6
Minitab prediction interval for legal advertising straight-line model

Exercises 10.70–10.82

Learning the Mechanics

10.70 Consider the following pairs of measurements saved in the NW **LM10_70** file:

x	−2	0	2	4	6	8	10
y	0	3	2	3	8	10	11

a. Construct a scattergram for these data.
b. Find the least squares line and plot it on your scattergram.
c. Find s^2 and s.
d. Find a 90% confidence interval for the mean value of y when $x = 3$. Plot the upper and lower bounds of the confidence interval on your scattergram.

*The result follows from the facts that, for large n, $t_{\alpha/2} \approx z_{\alpha/2}$, $s \approx \sigma$, and the last two terms under the radical in the standard error of the predictor are approximately 0.

e. Find a 90% prediction interval for a new value of y when $x = 3$. Plot the upper and lower bounds of the prediction interval on your scattergram.

f. Compare the widths of the intervals you constructed in parts **d** and **e.** Which is wider and why?

10.71 Consider the pairs of measurements shown in the table and saved in the **LM10_71** file. For these data, $SS_{xx} = 38.9000$, $SS_{yy} = 33.600$, $SS_{xy} = 32.8$, and $\hat{y} = -.414 + .843x$.

x	4	6	0	5	2	3	2	6	2	1
y	3	5	−1	4	3	2	0	4	1	1

a. Construct a scattergram for these data.

b. Plot the least squares line on your scattergram.

c. Use a 95% confidence interval to estimate the mean value of y when $x_p = 6$. Plot the upper and lower bounds of the interval on your scattergram.

d. Repeat part **c** for $x_p = 3.2$ and $x_p = 0$.

e. Compare the widths of the three confidence intervals you constructed in parts **c** and **d** and explain why they differ.

10.72 Refer to Exercise 10.71.

a. Using no information about x, estimate and calculate a 95% confidence interval for the mean value of y. [*Hint:* Use the one-sample t methodology of Section 5.4.]

b. Plot the estimated mean value and the confidence interval as horizontal lines on your scattergram.

c. Compare the confidence intervals you calculated in parts **c** and **d** of Exercise 10.71 with the one you calculated in part **a** of this exercise. Does x appear to contribute information about the mean value of y?

d. Check the answer you gave in part **c** with a statistical test of the null hypothesis H_0: $\beta_1 = 0$ against H_a: $\beta_1 \neq 0$. Use $\alpha = .05$.

10.73 In fitting a least squares line to $n = 10$ data points, the following quantities were computed:

$$SS_{xx} = 32 \quad \bar{x} = 3 \quad SS_{yy} = 26 \quad \bar{y} = 4 \quad SS_{xy} = 28$$

a. Find the least squares line.

b. Graph the least squares line.

c. Calculate SSE.

d. Calculate s^2.

e. Find a 95% confidence interval for the mean value of y when $x_p = 2.5$.

f. Find a 95% prediction interval for y when $x_p = 4$.

Applying the Concepts—Basic

10.74 In business, do nice guys finish first or last? Refer to the *Nature* (March 20, 2008) study of the use of punishment in cooperation games, Exercise 10.14 (p. 572). Recall that simple linear regression was used to model a player's average payoff (y) as a straight-line function of the number of times punishment was used (x) by the player.

a. If the researchers want to predict average payoff for a single player who used punishment 10 times, how should they proceed?

b. If the researchers want to predict the mean of the average payoffs for all players who used punishment 10 times, how should they proceed?

10.75 Characteristics of diamonds sold at retail. Refer to the Minitab simple linear regression analysis relating y = asking price (dollars) to x = number of carats for diamonds sold on the open market, Exercise 10.16 (p. 573) and the data saved in the **DIAMONDS** file. The portion of the Minitab printout giving a 95% confidence interval for $E(y)$ and a 95% prediction interval for y when $x = .52$ is shown below.

a. Locate and interpret the 95% confidence interval for $E(y)$.

b. Locate and interpret the 95% prediction interval for y.

10.76 Drug controlled-release rate study. Refer to the *Drug Development and Industrial Pharmacy* (Vol. 28, 2002) drug controlled-release rate study, Exercise 10.17 (p. 573) and the data saved in the **DOWDRUG** file. Recall that data for six drug tablets were used to fit the simple linear model, $E(y) = \beta_0 + \beta_1 x$, where y = drug release rate

Minitab output for Exercise 10.75

```
Predicted Values for New Observations

New
Obs     Fit  SE Fit        95% CI              95% PI
  1  3733.1    68.6  (3598.1, 3868.1)  (1529.8, 5936.3)

Values of Predictors for New Observations

New
Obs  CARAT
  1  0.520
```

Excel/DDXL Output for Exercise 10.76

90% Conditional Mean Confidence Intervals and Prediction Intervals

RelRate	SA-Vol	Predicted Value	Lower Cond. Mean Limit	Upper Cond. Mean Limit	Lower Prediction Limit	Upper Prediction Limit
60	1.5	61.08377	57.25719	64.910349	55.33401	66.83353
48	1.05	44.921466	42.98937	46.853562	40.215085	49.627847
39	0.9	39.534031	37.781376	41.286687	34.898427	44.169636
33	0.75	34.146597	32.172187	36.121007	29.422687	38.870507
30	0.6	28.759162	26.266678	31.251646	23.79635	33.721975
29	0.65	30.554974	28.258066	32.851881	25.687447	35.4225

SPSS Output for Exercise 10.77

	run	sweet	pectin	lower90m	upper90m
1	1	5.2	220	5.64898	5.83848
2	2	5.5	227	5.63898	5.81613
3	3	6.0	259	5.57819	5.72904
4	4	5.9	210	5.66194	5.87173
5	5	5.8	224	5.64337	5.82560
6	6	6.0	215	5.65564	5.85493
7	7	5.8	231	5.63284	5.80379
8	8	5.6	268	5.55553	5.71011
9	9	5.6	239	5.61947	5.78019
10	10	5.9	212	5.65946	5.86497
11	11	5.4	410	5.05526	5.55416
12	12	5.6	256	5.58517	5.73592

and x = surface-area-to-volume ratio. An Excel/DDXL printout giving a 90% prediction interval for each of the $n = 6$ observations in the sample is shown at the bottom of page 605. Select an observation and give a practical interpretation for this prediction interval.

10.77 Sweetness of orange juice. Refer to the simple linear regression of sweetness index y and amount of pectin x for $n = 24$ orange juice samples, Exercise 10.20 (p. 575) and the data saved in the **OJUICE** file. A 90% confidence interval for the mean sweetness index, $E(y)$, for each of the first 12 runs is shown on the SPSS spreadsheet above. Select an observation and interpret this interval.

Applying the Concepts—Intermediate

10.78 Ranking driving performance of professional golfers. Refer to *The Sport Journal* (Winter 2007) study of a new method for ranking the total driving performance of golfers on the PGA tour, Exercise 10.19 (p. 574). You fit a straight-line model relating driving accuracy (y) to driving distance (x) to the data saved in the **PGADRIVER** file. Of interest is predicting y and estimating $E(y)$ when $x = 300$ yards.

a. Find and interpret a 95% prediction interval for y.

b. Find and interpret a 95% confidence interval for $E(y)$.

c. If you are interested in knowing the average driving accuracy of all PGA golfers who have a driving distance of 300 yards, which of the intervals is relevant? Explain.

10.79 Spreading rate of spilled liquid. Refer to the *Chemical Engineering Progress* (Jan. 2005) study of the rate at which a spilled volatile liquid will spread across a surface, Exercise 10.24 (p. 576) and the data saved in the **LIQUIDSPILL** file. Recall that simple linear regression was used to model y = mass of the spill as a function of x = elapsed time of the spill.

a. Find a 99% confidence interval for the mean mass of all spills with an elapsed time of 15 minutes. Interpret the result.

b. Find a 99% prediction interval for the mass of a single spill with an elapsed time of 15 minutes. Interpret the result.

c. Compare the intervals, parts **a** and **b.** Which interval is wider? Will this always be the case? Explain.

10.80 Forecasting managerial needs. Managers are an important part of any organization's resource base. Accordingly, the organization should be just as concerned about forecasting its future managerial needs as it is with forecasting its needs for, say, the natural resources used in its production process (Northcraft and Neale, *Organizational Behavior: A Management Challenge,* 2001). A common forecasting procedure is to model the relationship between sales and the number of managers needed because the demand for managers is the result of the increases and decreases in the demand for products and services that a firm offers its customers. To develop this relationship, the data shown in the table below (saved in the **MANAGERS2** file) are collected from a firm's records.

Units Sold, x	Managers, y	Units Sold, x	Managers, y
5	10	30	22
4	11	31	25
8	10	36	30
7	10	38	30
9	9	40	31
15	10	41	31
20	11	51	32
21	17	40	30
25	19	48	32
24	21	47	32

a. Test the usefulness of the model. Use $\alpha = .05$. State your conclusion in the context of the problem.

b. The company projects that it will sell 39 units next month. Use the least squares model to construct a 90% prediction interval for the number of managers needed next month.

c. Interpret the interval in part **b.** Use the interval to determine the reliability of the firm's projection.

10.81 Predicting quit rates in manufacturing. The reasons given by workers for quitting their jobs generally fall into one of two categories: (1) worker quits to seek or take a different job, or (2) worker quits to withdraw from the labor force. Economic theory suggests that wages and quit rates are related. The table on the next page lists quit rates (quits per 100 employees) and the average hourly wage in a sample of 15 manufacturing industries. The data are saved in the **QUITTERS** file. Consider the simple linear regression of quit rate y on average wage x.

Industry	Quit Rate, y	Average Wage, x
1	1.4	$ 8.20
2	.7	10.35
3	2.6	6.18
4	3.4	5.37
5	1.7	9.94
6	1.7	9.11
7	1.0	10.59
8	.5	13.29
9	2.0	7.99
10	3.8	5.54
11	2.3	7.50
12	1.9	6.43
13	1.4	8.83
14	1.8	10.93
15	2.0	8.80

a. Do the data present sufficient evidence to conclude that average hourly wage rate contributes useful information for the prediction of quit rates? What does your model suggest about the relationship between quit rates and wages?

b. Find a 95% prediction interval for the quit rate in an industry with an average hourly wage of $9.00. Interpret the result.

c. Find a 95% confidence interval for the mean quit rate for industries with an average hourly wage of $9.00. Interpret this result.

Applying the Concepts—Advanced

10.82 Life tests of cutting tools. Refer to the data saved in the **CUTTOOLS** file, Exercise 10.36 (p. 581).

a. Use a 90% confidence interval to estimate the mean useful life of a brand A cutting tool when the cutting speed is 45 meters per minute. Repeat for brand B. Compare the widths of the two intervals and comment on the reasons for any difference.

b. Use a 90% prediction interval to predict the useful life of a brand A cutting tool when the cutting speed is 45 meters per minute. Repeat for brand B. Compare the widths of the two intervals to each other and to the two intervals you calculated in part **a.** Comment on the reasons for any differences.

c. Note that the estimation and prediction you performed in parts **a** and **b** were for a value of x that was not included in the original sample—that is, the value $x = 45$ was not part of the sample. However, the value is within the range of x values in the sample, so the regression model spans the x value for which the estimation and prediction were made. In such situations, estimation and prediction represent *interpolations.* Suppose you were asked to predict the useful life of a brand A cutting tool for a cutting speed of $x = 100$ meters per minute. Because the given value of x is outside the range of the sample x values, the prediction is an example of *extrapolation.* Predict the useful life of a brand A cutting tool that is operated at 100 meters per minute and construct a 95% prediction interval for the actual useful life of the tool. What additional assumption do you have to make in order to ensure the validity of an extrapolation?

10.7 A Complete Example

In the preceding sections, we have presented the basic elements necessary to fit and use a straight-line regression model. In this section, we will assemble these elements by applying them in an example with the aid of a computer.

Suppose a fire insurance company wants to relate the amount of fire damage in major residential fires to the distance between the burning house and the nearest fire station. The study is to be conducted in a large suburb of a major city; a sample of 15 recent fires in this suburb is selected. The amount of damage, y, and the distance between the fire and the nearest fire station, x, are recorded for each fire. The results are shown in Table 10.5 and saved in the **FIREDAM** file.

Step 1: First, we hypothesize a model to relate fire damage, y, to the distance from the nearest fire station, x. We hypothesize a straight-line probabilistic model:

$$y = \beta_0 + \beta_1 x + \varepsilon$$

Step 2: Next, we open the **FIREDAM** file and use a statistical software package to estimate the unknown parameters in the deterministic component of the hypothesized model. The Excel printout for the simple linear regression analysis is shown in Figure 10.25. The least squares estimates of the slope β_1 and intercept β_0, highlighted on the printout, are

$$\hat{\beta}_1 = 4.919331$$
$$\hat{\beta}_0 = 10.277929$$

Table 10.5 Fire Damage Data

Distance from Fire Station, x (miles)	Fire Damage, y (thousands of dollars)
3.4	26.2
1.8	17.8
4.6	31.3
2.3	23.1
3.1	27.5
5.5	36.0
.7	14.1
3.0	22.3
2.6	19.6
4.3	31.3
2.1	24.0
1.1	17.3
6.1	43.2
4.8	36.4
3.8	26.1

Data Set: FIREDAM

Regression Analysis

Regression Statistics	
Multiple R	0.960977715
R Square	0.923478169
Adjusted R Square	0.917591874
Standard Error	2.316346184
Observations	15

ANOVA

	df	SS	MS	F	Significance F
Regression	1	841.766358	841.766358	156.8861596	1.2478E-08
Residual	13	69.75097535	5.365459643		
Total	14	911.5173333			

	Coefficients	Standard Error	t Stat	P-value	Lower 95%	Upper 95%
Intercept	10.27792855	1.420277811	7.236562082	6.58556E-06	7.209605476	13.34625162
DISTANCE	4.919330727	0.392747749	12.52542054	1.2478E-08	4.070850963	5.767810491

Figure 10.25
Excel printout for fire damage regression analysis

and the least squares equation is (rounded)

$$\hat{y} = 10.278 + 4.919x$$

This prediction equation is graphed in the Minitab scatterplot, Figure 10.26.

The least squares estimate of the slope, $\hat{\beta}_1 = 4.919$, implies that the estimated mean damage increases by \$4,919 for each additional mile from the fire station. This interpretation is valid over the range of x, or from .7 to 6.1 miles from the station. The estimated y-intercept, $\hat{\beta}_0 = 10.278$, has the interpretation that a fire

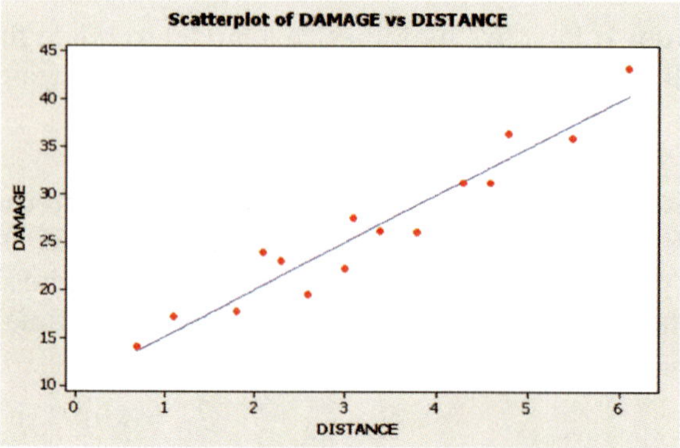

Figure 10.26
Minitab scatterplot with least squares line for fire damage regression analysis

0 miles from the fire station has an estimated mean damage of $10,278. Although this would seem to apply to the fire station itself, remember that the y-intercept is meaningfully interpretable only if $x = 0$ is within the sampled range of the independent variable. Because $x = 0$ is outside the range in this case, $\hat{\beta}_0$ has no practical interpretation.

Step 3: Now we specify the probability distribution of the random error component ε. The assumptions about the distribution are identical to those listed in Section 10.3. Although we know that these assumptions are not completely satisfied (they rarely are for practical problems), we are willing to assume they are approximately satisfied for this example. The estimate of the standard deviation σ of ε, highlighted on the Excel printout (Figure 10.25) is

$$s = 2.31635$$

This implies that most of the observed fire damage (y) values will fall within approximately $2s = 4.64$ thousand dollars of their respective predicted values when using the least squares line. [*Note:* A more precise prediction interval for y is given in Step 5.]

Step 4: We can now check the usefulness of the hypothesized model—that is, whether x really contributes information for the prediction of y using the straight-line model. First, test the null hypothesis that the slope β_1 is 0—that is, that there is no linear relationship between fire damage and the distance from the nearest fire station, against the alternative hypothesis that fire damage increases as the distance increases. We test

$$H_0\colon \beta_1 = 0$$
$$H_a\colon \beta_1 > 0$$

The two-tailed observed significance level for testing $H_a\colon \beta_1 \neq 0$, highlighted on the printout, is approximately 0. When we divide this value in half, the p-value for our one-tailed test is also approximately 0. This small p-value leaves little doubt that mean fire damage and distance between the fire and station are at least linearly related, with mean fire damage increasing as the distance increases.

We gain additional information about the relationship by forming a 95% confidence interval for the slope β_1. The lower and upper endpoints of this interval are highlighted on the Excel printout shown in Figure 10.25.

This yields the interval (4.070, 5.768). We estimate (with 95% confidence) that the interval from $4,070 to $5,768 encloses the mean increase (β_1) in fire damage per additional mile distance from the fire station.

Another measure of the utility of the model is the coefficient of determination, r^2. The value (highlighted on Figure 10.25) is $r^2 = .9235$, which implies that about 92% of the sample variation in fire damage (y) is explained by the distance (x) between the fire and the fire station.

The coefficient of correlation, r, that measures the strength of the linear relationship between y and x is not shown on the Excel printout and must be calculated. Using the facts that $r = \sqrt{r^2}$ in simple linear regression and that r and $\hat{\beta}_1$ have the same sign, we find

$$r = +\sqrt{r^2} = \sqrt{.9235} = .96$$

The high correlation confirms our conclusion that β_1 is greater than 0; it appears that fire damage and distance from the fire station are positively correlated. All signs point to a strong linear relationship between y and x.

Step 5: We are now prepared to use the least squares model. Suppose the insurance company wants to predict the fire damage if a major residential fire were to occur 3.5 miles from the nearest fire station. A 95% confidence interval for $E(y)$ and

```
Predicted Values for New Observations

New
Obs     Fit  SE Fit        95% CI                95% PI
  1  27.496   0.604  (26.190, 28.801)  (22.324, 32.667)


Values of Predictors for New Observations

New
Obs  DISTANCE
  1      3.50
```

Figure 10.27
Minitab confidence and prediction interval for fire damage regression

prediction interval for y when $x = 3.5$ are shown on the Minitab printout, Figure 10.27. The predicted value (highlighted on the printout) is $\hat{y} = 27.496$, while the 95% prediction interval (also highlighted) is (22.3239, 32.6672). Therefore, with 95% confidence we predict fire damage in a major residential fire 3.5 miles from the nearest station to be between \$22,324 and \$32,667.

CAUTION We would not use this prediction model to make predictions for homes less than .7 mile or more than 6.1 miles from the nearest fire station. A look at the data in Table 10.7 reveals that all the x values fall between .7 and 6.1. It is dangerous to use the model to make predictions outside the region in which the sample data fall. A straight line might not provide a good model for the relationship between the mean value of y and the value of x when stretched over a wider range of x values. ▲

Exercise 10.83

Applying the Concepts—Intermediate

10.83 An MBA's work-life balance. The importance of having employees with a healthy work-life balance has been recognized by U.S. companies for decades. Many business schools offer courses that assist MBA students with developing good work-life balance habits, and most large companies have developed work-life balance programs for their employees. In April 2005, the Graduate Management Admission Council (GMAC) conducted a survey of over 2,000 MBA alumni to explore the work-life balance issue. (For example, one question asked alumni to state their level of agreement with the statement "My personal and work demands are overwhelming.") Based on these responses, the GMAC determined a work-life balance scale score for each MBA alumni. Scores ranged from 0 to 100, with lower scores indicating a higher imbalance between work and life. Many other variables, including average number of hours worked per week, were also measured. The data for the work-life balance study are saved in the **GMAC** file. (The first 15 observations are listed in the accompanying table.) Let x = average number of hours worked per week and y = work-life balance scale score for each MBA alumnus. Investigate the link between these two variables by conducting a complete simple linear regression analysis of the data. Summarize your findings in a professional report.

Data for Exercise 10.83 (selected observations)

WLB Score	Hours
75.22	50
64.98	45
49.62	50
44.51	55
70.10	50
54.74	60
55.98	55
21.24	60
59.86	50
70.10	50
29.00	70
64.98	45
36.75	40
35.45	40
45.75	50

Source: "Work-life balance: An MBA alumni report," *Graduate Management Admission Council (GMAC) Research Report* (Oct. 13, 2005). Reproduced with the permission of the Graduate Management Admission Council®.

CHAPTER NOTES

Key Terms

Key Symbols/Notation

y	Dependent variable (variable to be predicted)
x	Independent variable (variable used to predict y)
$E(y)$	Expected value (mean) of y
β_0	y-intercept of true line
β_1	slope of true line
$\hat{\beta}_0$	Least squares estimate of y-intercept
$\hat{\beta}_1$	Least squares estimate of slope
ε	Random error
$\hat{y}$	Predicted value of y for a given x-value
$(y - \hat{y})$	Estimated error of prediction
SSE	Sum of squared errors of prediction
r	Coefficient of correlation
r^2	Coefficient of determination
x_p	Value of x used to predict y
$r^2 = \dfrac{SS_{yy} - SSE}{SS_{yy}}$	Coefficient of determination
$\hat{y} \pm t_{\alpha/2}s\sqrt{\dfrac{1}{n} + \dfrac{(x_p - \bar{x})^2}{SS_{xx}}}$	$100\%(1 - \alpha)$ confidence interval for $E(y)$ when $x = x_p$
$\hat{y} \pm t_{\alpha/2}s\sqrt{1 + \dfrac{1}{n} + \dfrac{(x_p - \bar{x})^2}{SS_{xx}}}$	$100\%(1 - \alpha)$ prediction interval for y when $x = x_p$

Key Ideas

Simple Linear Regression Variables

y = **Dependent** variable (quantitative)
x = **Independent** variable (quantitative)

Method of Least Squares Properties

1. average error of prediction = 0
2. sum of squared errors is minimum

Practical Interpretation of y-Intercept

Predicted y-value when $x = 0$

(no practical interpretation if $x = 0$ is either nonsensical or outside range of sample data)

Practical Interpretation of Slope

Increase (or decrease) in y for every 1-unit increase in x

First-Order (Straight-Line) Model

$$E(y) = \beta_0 + \beta_1 x$$

where $E(y)$ = mean of y

β_0 = ***y*-intercept** of line (point where line intercepts y-axis)
β_1 = **slope** of line (change in y for every 1-unit change in x)

Coefficient of Correlation, r

1. ranges between −1 and +1
2. measures strength of *linear relationship* between y and x

Coefficient of Determination, r^2

1. ranges between 0 and 1
2. measures proportion of sample variation in y "explained" by the model

Practical Interpretation of Model Standard Deviation, s

Ninety-five percent of y-values fall within 2s of their respected predicted values

Width of *confidence interval for* $E(y)$ will always be **narrower** than width of prediction interval for y

Guide to Simple Linear Regression

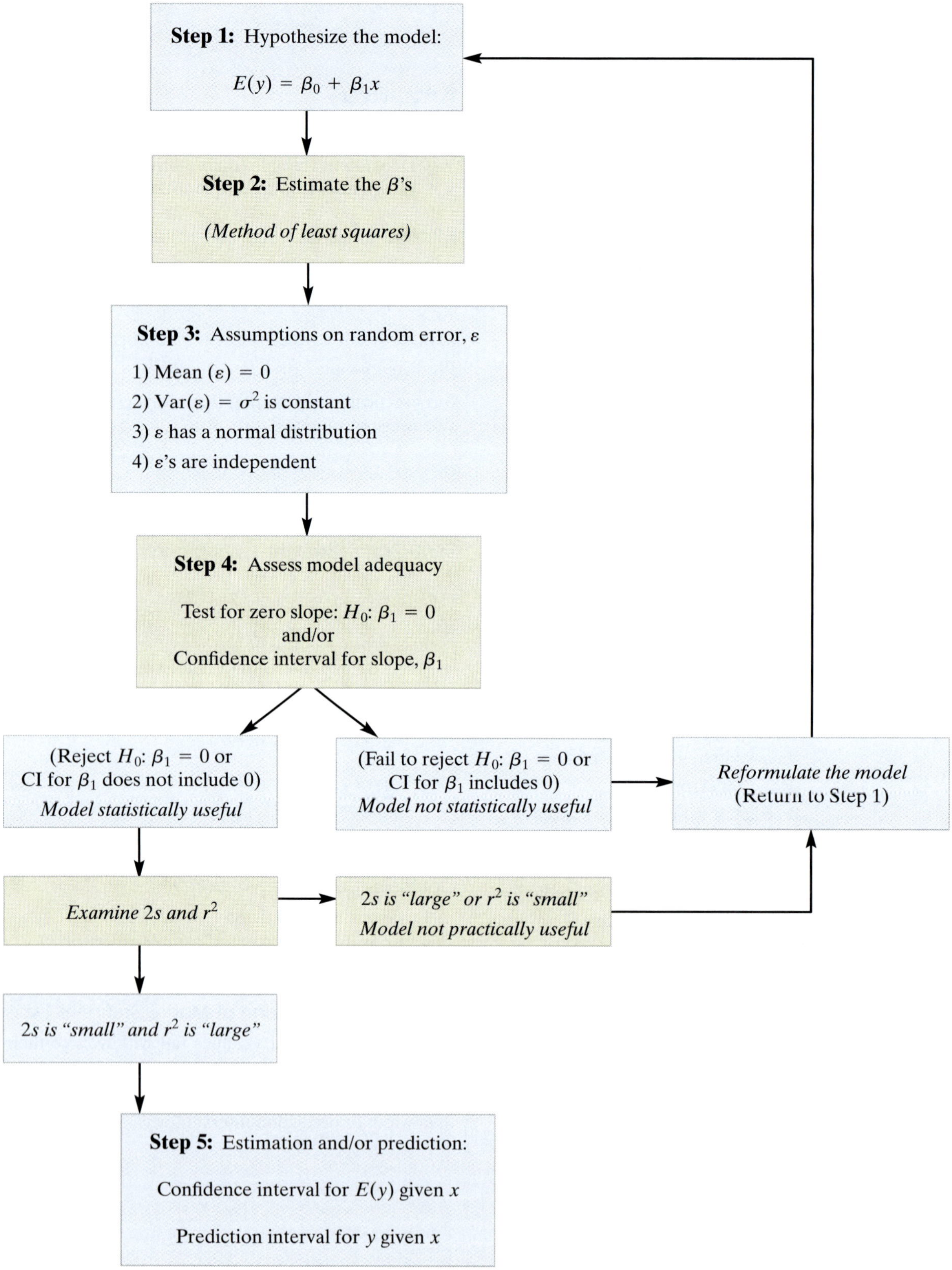

Supplementary Exercises 10.84–10.102

Learning the Mechanics

10.84 Consider the following sample data:

y	5	1	3
x	5	1	3

a. Construct a scattergram for the data.
b. It is possible to find many lines for which $\Sigma(y - \hat{y}) = 0$. For this reason, the criterion $\Sigma(y - \hat{y}) = 0$ is not used for identifying the "best-fitting" straight line. Find two lines that have $\Sigma(y - \hat{y}) = 0$.
c. Find the least squares line.
d. Compare the value of SSE for the least squares line to that of the two lines you found in part **b.** What principle of least squares is demonstrated by this comparison?

10.85 In fitting a least squares line to $n = 15$ data points, the following quantities were computed: $SS_{xx} = 55$, $SS_{yy} = 198$, $SS_{xy} = -88$, $\bar{x} = 1.3$, and $\bar{y} = 35$.
a. Find the least squares line.
b. Graph the least squares line.
c. Calculate SSE.
d. Calculate s^2.
e. Find a 90% confidence interval for β_1. Interpret this estimate.
f. Find a 90% confidence interval for the mean value of y when $x = 15$.
g. Find a 90% prediction interval for y when $x = 15$.

10.86 Consider the following 10 data points, saved in the **LM10_86** file:

x	3	5	6	4	3	7	6	5	4	7
y	4	3	2	1	2	3	3	5	4	2

a. Plot the data on a scattergram.
b. Calculate the values of r and r^2.
c. Is there sufficient evidence to indicate that x and y are linearly correlated? Test at the $\alpha = .10$ level of significance.

Applying the Concepts—Basic

10.87 Predicting water erosion of soils. The U.S. Department of Agriculture has developed and adopted the Universal Soil Loss Equation (USLE) for predicting water erosion of soils. In geographical areas where runoff from melting snow is common, the USLE requires an accurate estimate of snowmelt runoff erosion. An article in the *Journal of Soil and Water Conservation* (Mar.–Apr. 1995) used simple linear regression to develop a snowmelt erosion index. Data for 54 climatological stations in Canada were used to model the McCool winter-adjusted rainfall erosivity index, y, as a straight-line function of the once-in-5-year snowmelt runoff amount, x (measured in millimeters).
a. The data points are plotted in the scattergram shown below. Is there visual evidence of a linear trend?
b. The data for seven stations were removed from the analysis due to lack of snowfall during the study period. The simple linear regression on the remaining $n = 47$ data points yielded the following results:

$$\hat{y} = -6.72 + 1.39x; \; s_{\hat{\beta}_1} = .06$$

Use this information to construct a 90% confidence interval for β_1.
c. Interpret the interval, part **b.**

10.88 Burnout of human services professionals. Emotional exhaustion, or *burnout,* is a significant problem for people with careers in the field of human services. Regression analysis was used to investigate the relationship between burnout and aspects of the human services professional's job and job-related behavior (*Journal of Applied Behavioral Science,* Vol. 22, 1986). Emotional exhaustion was measured with the Maslach Burnout Inventory, a questionnaire. One of the independent variables considered, called *concentration,* was the proportion of social contacts with individuals who belong to a person's work group. The next table lists the values of the emotional exhaustion index (higher values indicate greater exhaustion) and concentration for a sample of 25 human services professionals who work in a large public hospital.

Scattergram for Exercise 10.87

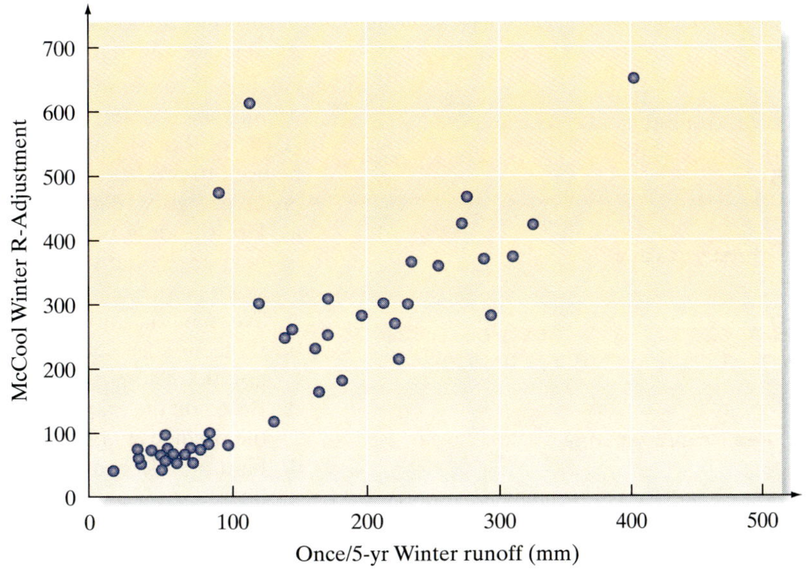

These data are saved in the **BURNOUT** file. A Minitab printout of the simple linear regression is provided below.

a. Construct a scattergram for the data. Do the variables x and y appear to be related?

b. Find the correlation coefficient for the data and interpret its value. Does your conclusion mean that concentration causes emotional exhaustion? Explain.

Exhaustion Index, y	Concentration, x	Exhaustion Index, y	Concentration, x
100	20%	493	86%
525	60	892	83
300	38	527	79
980	88	600	75
310	79	855	81
900	87	709	75
410	68	791	77
296	12	718	77
120	35	684	77
501	70	141	17
920	80	400	85
810	92	970	96
506	77		

c. Test the usefulness of the straight-line relationship with concentration for predicting burnout. Use $\alpha = .05$.

d. Find the coefficient of determination for the model and interpret it.

e. Find a 95% confidence interval for the slope β_1. Interpret the result.

Minitab output for Exercise 10.88

```
The regression equation is
EXHAUST = - 29 + 8.87 CONCEN


Predictor    Coef  SE Coef      T      P
Constant    -29.5    106.7  -0.28  0.785
CONCEN      8.865    1.471   6.03  0.000


S = 174.207   R-Sq = 61.2%   R-Sq(adj) = 59.5%


Analysis of Variance

Source           DF       SS       MS      F      P
Regression        1  1102408  1102408  36.33  0.000
Residual Error   23   698009    30348
Total            24  1800417


Predicted Values for New Observations

New
Obs    Fit  SE Fit        95% CI            95% PI
  1  679.7    38.7  (599.7, 759.8)  (310.6, 1048.9)


Values of Predictors for New Observations

New
Obs  CONCEN
  1    80.0
```

f. Use a 95% confidence interval to estimate the mean exhaustion level for all professionals who have 80% of their social contacts within their work groups. Interpret the interval.

10.89 Retaliation against company "whistle-blowers." Individuals who report perceived wrongdoing of a corporation or public agency are known as *whistle-blowers.* Two researchers developed an index to measure the extent of retaliation against a whistle-blower (*Journal of Applied Psychology,* 1986). The index was based on the number of forms of reprisal actually experienced, the number of forms of reprisal threatened, and the number of people within the organization (e.g., coworkers or immediate supervisor) who retaliated against them. The table below lists the retaliation index (higher numbers indicate more extensive retaliation) and salary for a sample of 15 whistle-blowers from federal agencies. The data are saved in the **RETAL** file.

Retaliation Index	Salary	Retaliation Index	Salary
301	$62,000	535	$19,800
550	36,500	455	44,000
755	21,600	615	46,600
327	24,000	700	15,100
500	30,100	650	70,000
377	35,000	630	21,000
290	47,500	360	16,900
452	54,000		

Source: Data adapted from Near, J. P., and Miceli, M. P. "Retaliation against whistle blowers: Predictors and effects," *Journal of Applied Psychology,* Vol. 71, No. 1, 1986, pp. 137–145.

a. Construct a scattergram for the data. Does it appear that the extent of retaliation increases, decreases, or stays the same with an increase in salary? Explain.

b. Use the method of least squares to fit a straight line to the data.

c. Graph the least squares line on your scattergram. Does the least squares line support your answer to the question in part **a**? Explain.

d. Interpret the y-intercept, $\hat{\beta}_0$, of the least squares line in terms of this application. Is the interpretation meaningful?

e. Interpret the slope, $\hat{\beta}_1$, of the least squares line in terms of this application. Over what range of x is this interpretation meaningful?

f. Test the adequacy of the model using $\alpha = .05$.

10.90 Buying income of households. *Sales and Marketing Management* determined the "effective buying income" (EBI) of the average household in a state. Can the EBI be used to predict retail sales per household in the store-group category "eating and drinking places"?

a. Use the data for 13 states given in the table on the next page (and saved in the **EBI** file) to find the least squares line relating retail sales per household (y) to average household EBI (x).

b. Plot the least squares line, as well as the actual data points, on a scattergram.

c. Based on the graph, part **b,** give your opinion regarding the predictive ability of the least squares line.

d. Find a 95% confidence interval for the slope of the line.

e. Use the results, part **d,** to assess the adequacy of the straight-line model.

State	Average Household Buying Income ($)	Retail Sales: Eating and Drinking Places ($ per household)
Connecticut	60,998	2,553.8
New Jersey	63,853	2,154.8
Michigan	46,915	2,523.3
Minnesota	44,717	2,278.6
Florida	42,442	2,475.8
South Carolina	37,848	2,358.4
Mississippi	34,490	1,538.4
Oklahoma	34,830	2,063.1
Texas	44,729	2,363.5
Colorado	44,571	3,214.9
Utah	43,421	2,653.8
California	50,713	2,215.0
Oregon	40,597	2,144.0

Source: Table adapted from "The survey of buying power," *Sales and Marketing Management,* 1995.

10.91 Foreign vs. U.S. stock rates of return. If the economies of the world were tightly interconnected, the stock markets of different countries would move together. If they did, there would be no reason for investors to diversify their stock portfolios with stocks from a variety of countries (Sharpe, Alexander, and Bailey, *Investments,* 1999). The table below lists the correlations of returns on stocks in each of six countries with the returns of U.S. stocks.

Country	Correlation between Foreign and U.S. Stocks
Australia	.48
Canada	.74
France	.50
Germany	.43
Japan	.41
United Kingdom	.58

Source: Sharpe, W. F., Alexander, G. J., and Bailey, Jeffery V. *Investments.* Upper Saddle River, N.J.: Prentice Hall, 1999, p. 887.

a. Interpret the Australia/U.S. correlation. What does it suggest about the linear relationship between the stocks of the two countries?

b. Sketch a scattergram that is roughly consistent with the magnitude of the France/U.S. correlation.

c. Why must we be careful not to conclude from the information in the table that the country that is most tightly integrated with the United States is Canada?

10.92 Snow geese feeding trial. Researchers at the University of Toronto conducted a series of experiments to investigate whether a commercially sold pet food could serve as a substitute diet for baby snow geese (*Journal of Applied Ecology,* Vol. 32, 1995). Goslings were deprived of food until their guts were empty and then were allowed to feed for 6 hours on a diet of plants or Purina Duck Chow. For each feeding trial, the change in the weight of the gosling after 2.5 hours was recorded as a percentage of initial weight. Two other variables recorded were digestion efficiency (measured as a percentage) and amount of acid-detergent fiber in the digestive tract (also measured as a percentage). The data for 42 feeding trials are saved in the **SNOWGEESE** file. (The first and last five observations are shown in the table at the bottom of the page.)

a. The researchers were interested in the correlation between weight change (y) and digestion efficiency (x). Plot the data for these two variables in a scattergram. Do you observe a trend?

b. Find the coefficient of correlation relating weight change y to digestion efficiency x. Interpret this value.

c. Conduct a test (at $\alpha = .01$) to determine whether weight change y is correlated with digestion efficiency x.

d. Repeat parts **b** and **c** but exclude the data for trials that used duck chow from the analysis. What do you conclude?

e. The researchers were also interested in the correlation between digestion efficiency (y) and acid-detergent fiber (x). Repeat parts **a–d** for these two variables.

10.93 Evaluating managerial success. Refer to Exercise 10.50 (p. 588), in which managerial success, y, was modeled as a function of the number of contacts a manager makes with people outside his or her work unit, x, during a specific period of time. The data are saved in the **MANAGERS** file.

a. A particular manager was observed for 2 weeks, as in the *Journal of Applied Behavioral Science* study. She made 55 contacts with people outside her work unit. Predict the value of the manager's success index. Use a 90% prediction interval.

Table for Exercise 10.92

Feeding Trial	Diet	Weight Change (%)	Digestion Efficiency (%)	Acid-Detergent Fiber (%)
1	Plants	−6	0	28.5
2	Plants	−5	2.5	27.5
3	Plants	−4.5	5	27.5
4	Plants	0	0	32.5
5	Plants	2	0	32
⋮	⋮	⋮	⋮	⋮
38	Duck Chow	9	59	8.5
39	Duck Chow	12	52.5	8
40	Duck Chow	8.5	75	6
41	Duck Chow	10.5	72.5	6.5
42	Duck Chow	14	69	7

Source: Gadallah, F. L., and Jefferies, R. L. "Forage quality in brood rearing areas of the lesser snow goose and the growth of captive goslings," *Journal of Applied Biology,* Vol. 32, No. 2, 1995, pp. 281–282 (adapted from Figures 2 and 3). Reprinted with permission of John Wiley & Sons, Inc.

b. A second manager was observed for 2 weeks. This manager made 110 contacts with people outside his work unit. Why should caution be exercised in using the least squares model developed from the given data set to construct a prediction interval for this manager's success index?

c. In the context of this problem, determine the value of x for which the associated prediction interval for y is the narrowest.

Applying the Concepts—Intermediate

10.94 Hospital charges vs. length of stay. *Statistical Bulletin* (Oct.–Dec. 1999) reported the average hospital charge and the average length of hospital stay for patients undergoing radical prostatectomies in a sample of 12 states. The data are listed in the accompanying table and saved in the **HOSPITAL** file.

State	Average Hospital Charge ($)	Average Length of Stay (days)
Massachusetts	11,680	3.64
New Jersey	11,630	4.20
Pennsylvania	9,850	3.84
Minnesota	9,950	3.11
Indiana	8,490	3.86
Michigan	9,020	3.54
Florida	13,820	4.08
Georgia	8,440	3.57
Tennessee	8,790	3.80
Texas	10,400	3.52
Arizona	12,860	3.77
California	16,740	3.78

Source: Statistical Bulletin, Vol. 80, No. 4, Oct.–Dec. 1999, p. 13.

a. Plot the data on a scattergram.

b. Use the method of least squares to model the relationship between average hospital charge (y) and length of hospital stay (x).

c. Conduct a test of model adequacy using $\alpha = .10$.

d. Find the estimated standard error of the regression model.

e. What fraction of the states in the sample have average hospital charges within $\pm 2s$ of the least squares line?

f. Find and interpret the values of r and r^2 for the data.

10.95 Collecting Beanie Babies. Refer to Exercise 2.144 (p. 105) and the data on 50 Beanie Babies collector's items published in *Beanie World Magazine*. Can age (in months as of Sept. 1998) of a Beanie Baby be used to accurately predict its market value? Answer this question by conducting a complete simple linear regression analysis on the data saved in the **BEANIE** file. (The first and last five observations are shown in the table.)

Name	Age (months) as of Sept. 1998	Value ($)
1. Ally the Alligator	52	55.00
2. Batty the Bat	12	12.00
3. Bongo the Brown Monkey	28	40.00
4. Blackie the Bear	52	10.00
5. Bucky the Beaver	40	45.00
⋮	⋮	⋮
46. Stripes the Tiger (Gold/Black)	40	400.00
47. Teddy the 1997 Holiday Bear	12	50.00
48. Tuffy the Terrier	17	10.00
49. Tracker the Basset Hound	5	15.00
50. Zip the Black Cat	28	40.00

Source: Beanie World Magazine, Sept. 1998.

10.96 Monetary values of NFL teams. Refer to the *Forbes* magazine (Sep. 11, 2008) report on the financial standings of each team in the National Football League (NFL), Exercise 2.24 (p. 49). The table listing the current value (without deduction for debt, except stadium debt) and operating income for each team is reproduced below. These data are saved in the **NFLTEAMVALUES** file.

Rank	Team	Current Value ($ mil)	Operating Income ($ mil)
1	Dallas Cowboys	1,612	30.6
2	Washington Redskins	1,538	58.1
3	New England Patriots	1,324	39.2
4	New York Giants	1,178	41.2
5	New York Jets	1,170	25.9
6	Houston Texans	1,125	43.9
7	Philadelphia Eagles	1,116	33.5
8	Indianapolis Colts	1,076	16.1
9	Chicago Bears	1,064	33.7
10	Baltimore Ravens	1,062	23
11	Denver Broncos	1,061	18.8
12	Tampa Bay Buccaneers	1,053	39.3
13	Miami Dolphins	1,044	36.1
14	Carolina Panthers	1,040	22.3
15	Cleveland Browns	1,035	19.3
16	Green Bay Packers	1,023	21.9
17	Kansas City Chiefs	1,016	11.9
18	Pittsburgh Steelers	1,015	14.4
19	Seattle Seahawks	1,010	8.9
20	Tennessee Titans	994	24.5
21	Cincinnati Bengals	941	22
22	New Orleans Saints	937	21.5
23	St. Louis Rams	929	26.4
24	Detroit Lions	917	−3.1
25	Arizona Cardinals	914	19.7
26	San Diego Chargers	888	19
27	Buffalo Bills	885	12.4
28	Jacksonville Jaguars	876	27.6
29	Atlanta Falcons	872	30.9
30	San Francisco 49ers	865	4.1
31	Oakland Raiders	861	27
32	Minnesota Vikings	839	18.9

Source: "NFL value," *Forbes,* September 11, 2008. First 32 observations—financial standings of the NFL. Reprinted by permission of Forbes.com. © 2009 Forbes LLC.

a. Propose a straight-line model relating an NFL team's current value (y) to its operating income (x).

b. Fit the model to the data using the method of least squares.

c. Interpret the least squares estimates of the slope and y-intercept in the words of the problem.

d. Statistically assess the adequacy of the model. Do you recommend using it to predict an NFL team's value?

10.97 Evaluating a truck weigh-in-motion program. The Minnesota Department of Transportation installed a state-of-the-art weigh-in-motion scale in the concrete surface of the eastbound lanes of Interstate 494 in Bloomington,

Minnesota. After installation, a study was undertaken to determine whether the scale's readings correspond with the static weights of the vehicles being monitored. (Studies of this type are known as *calibration studies.*) After some preliminary comparisons using a two-axle, six-tire truck carrying different loads (see the table below saved in the **TRUCKWTS** file), calibration adjustments were made in the software of the weigh-in-motion system, and the scales were reevaluated.

Trial Number	Static Weight of Truck, x (thousand pounds)	Weigh-in-Motion Reading Prior to Calibration Adjustment, y_1 (thousand pounds)	Weigh-in-Motion Reading After Calibration Adjustment, y_2 (thousand pounds)
1	27.9	26.0	27.8
2	29.1	29.9	29.1
3	38.0	39.5	37.8
4	27.0	25.1	27.1
5	30.3	31.6	30.6
6	34.5	36.2	34.3
7	27.8	25.1	26.9
8	29.6	31.0	29.6
9	33.1	35.6	33.0
10	35.5	40.2	35.0

Source: Adapted from data in Wright, J. L., Owen, F., and Pena, D. "Status of MN/DOT's weigh-in-motion program," St. Paul: Minnesota Department of Transportation, January 1983.

a. Construct two scattergrams, one of y_1 versus x and the other of y_2 versus x.

b. Use the scattergrams of part **a** to evaluate the performance of the weigh-in-motion scale both before and after the calibration adjustment.

c. Calculate the correlation coefficient for both sets of data and interpret their values. Explain how these correlation coefficients can be used to evaluate the weigh-in-motion scale.

d. Suppose the sample correlation coefficient for y_2 and x was 1. Could this happen if the static weights and the weigh-in-motion readings disagreed? Explain.

10.98 Energy efficiency of buildings. Firms planning to build new plants or make additions to existing facilities have become very conscious of the energy efficiency of proposed new structures and are interested in the relation between yearly energy consumption and the number of square feet of building shell. The accompanying table (saved in the **BTU** file) lists the energy consumption in British thermal units (a BTU is the amount of heat required to raise 1 pound of water 1°F) for 22 buildings that were all subjected to the same climatic conditions. Consider a straight-line model relating BTU consumption, y, to building shell area, x.

a. Find the least squares estimates of the intercept β_0 and the slope β_1.

b. Investigate the usefulness of the model you developed in part **a.** Is yearly energy consumption positively linearly related to the shell area of the building? Test using $\alpha = .10$.

c. Find the observed significance level of the test of part **b.** Interpret its value.

d. Find the coefficient of determination r^2 and interpret its value.

e. A company wishes to build a new warehouse that will contain 8,000 square feet of shell area. Find the predicted value of energy consumption and associated 95% prediction interval. Comment on the usefulness of this interval.

f. The application of the model you developed in part **a** to the warehouse problem of part **e** is appropriate only if certain assumptions can be made about the new warehouse. What are these assumptions?

BTU/Year (thousands)	Shell Area (square feet)
3,870,000	30,001
1,371,000	13,530
2,422,000	26,060
672,200	6,355
233,100	4,576
218,900	24,680
354,000	2,621
3,135,000	23,350
1,470,000	18,770
1,408,000	12,220
2,201,000	25,490
2,680,000	23,680
337,500	5,650
567,500	8,001
555,300	6,147
239,400	2,660
2,629,000	19,240
1,102,000	10,700
423,500	9,125
423,500	6,510
1,691,000	13,530
1,870,000	18,860

10.99 Top Florida law firms. Refer to *Florida Trend Magazine's* (April 2002) data on law firms with headquarters in the state of Florida, Exercise 2.145 (p. 105). Data on the number of lawyers and number of law offices, saved in the **FLALAW** file, are reproduced in the table on the next page. Suppose you want to predict the number of law offices (y) based on the number of lawyers (x) at the firm.

a. Use the method of least squares (and computer software) to find the best-fitting line to the data.

b. Evaluate the line by conducting a complete simple linear regression analysis.

c. A firm with 300 lawyers is planning on building its headquarters in Florida. How many law offices should the firm expect to build?

Applying the Concepts—Advanced

10.100 Regression through the origin. Sometimes it is known from theoretical considerations that the straight-line relationship between two variables, x and y, passes through the origin of the xy-plane. Consider the relationship between the total weight of a shipment of 50-pound bags of flour, y, and the number of bags in the shipment, x. Because a shipment containing $x = 0$ bags (i.e., no shipment at all) has a total weight of $y = 0$, a straight-line model of the relationship between x and y should pass through the point $x = 0, y = 0$. In such a case, you could assume

Table for Exercise 10.99

Rank	Firm	Headquarters	Number of Lawyers	Number of Offices
1	Holland & Knight	Tallahassee	529	11
2	Akerman Senterfit	Orlando	355	9
3	Greenberg Traurig	Miami	301	6
4	Carlton Fields	Tampa	207	6
5	Ruden McClosky Smit	Ft. Lauderdale	175	9
6	Fowler White Boggs	Tampa	175	7
7	Foley & Lardner	Orlando	159	5
8	GrayHarris	Orlando	158	6
9	Broad and Cassel	Orlando	150	7
10	Shutts & Bowen	Miami	144	5
11	Steel Hector & Davis	Miami	141	5
12	Gunster Yoakley	W. Palm Beach	140	6
13	Adorno & Zeder	Miami	105	4
14	Becker & Poliakoff	Ft. Lauderdale	100	12
15	Lowndes Drosdick	Orlando	100	1
16	Conroy Simberg Ganon	Hollywood	91	6
17	Stearns Weaver	Miami	85	3
18	Wicker Smith O'Hara	Miami	85	6
19	Rogers Towers Bailey	Jacksonville	80	2
20	Butler Burnette	Tampa	77	3
21	Bilzin Sumberg Dunn	Miami	70	1
22	Morgan Colling	Orlando	70	4
23	White & Case	Miami	70	1
24	Fowler White Burnett	Miami	64	4
25	Rissman Weisberg	Orlando	63	3
26	Rumberger Kirk	Orlando	63	4

Source: Florida Trend Magazine, April 2002, p. 105.

$\beta_0 = 0$ and characterize the relationship between x and y with the following model:

$$y = \beta_1 x + \varepsilon$$

The least squares estimate of β_1 for this model is

$$\hat{\beta}_1 = \frac{\sum x_i y_i}{\sum x_i^2}$$

From the records of past flour shipments, 15 shipments were randomly chosen, and the data shown in the table below were recorded. These data are saved in the **FLOUR2** file.

Weight of Shipment	Number of 50-Pound Bags in Shipment
5,050	100
10,249	205
20,000	450
7,420	150
24,685	500
10,206	200
7,325	150
4,958	100
7,162	150
24,000	500
4,900	100
14,501	300
28,000	600
17,002	400
16,100	400

a. Find the least squares line for the given data under the assumption that $\beta_0 = 0$. Plot the least squares line on a scattergram of the data.

b. Find the least squares line for the given data using the model

$$y = \beta_0 + \beta_1 x + \varepsilon$$

(i.e., do not restrict β_0 to equal 0). Plot this line on the same scatterplot you constructed in part **a.**

c. Refer to part **b.** Why might $\hat{\beta}_0$ be different from 0 even though the true value of β_0 is known to be 0?

d. The estimated standard error of $\hat{\beta}_0$ is equal to

$$s\sqrt{\frac{1}{n} + \frac{\bar{x}^2}{SS_{xx}}}$$

Use the t-statistic

$$t = \frac{\hat{\beta}_0 - 0}{s\sqrt{(1/n) + (\bar{x}^2/SS_{xx})}}$$

to test the null hypothesis H_0: $\beta_0 = 0$ against the alternative H_a: $\beta_0 \neq 0$. Use $\alpha = .10$. Should you include β_0 in your model?

Critical Thinking Challenges

10.101 Comparing cost functions. Managers are interested in modeling past cost behavior in order to make more accurate predictions of future costs. Models of past cost behavior are called *cost functions.* Factors that influence costs are called *cost drivers* (Horngren, Foster, and Datar, *Cost Accounting,* 1994). The cost data shown on the next page (and saved in the **RUG** file) are from a rug manufacturer.

Indirect manufacturing labor costs consist of machine maintenance costs and setup labor costs. Machine-hours and direct manufacturing labor-hours are cost drivers. Your task is to estimate and compare two alternative cost functions for indirect manufacturing labor costs. In the first, machine-hours is the independent variable; in the second, direct manufacturing labor-hours is the independent variable. Prepare a report that compares the two cost functions and recommends which should be used to explain and predict indirect manufacturing labor costs. Be sure to justify your choice.

Week	Indirect Manufacturing Labor Costs	Machine-Hours	Direct Manufacturing Labor-Hours
1	$1,190	68	30
2	1,211	88	35
3	1,004	62	36
4	917	72	20
5	770	60	47
6	1,456	96	45
7	1,180	78	44
8	710	46	38
9	1,316	82	70
10	1,032	94	30
11	752	68	29
12	963	48	38

Source: Data and exercise adapted from Horngren, C. T., Foster, G., and Datar, S. M. *Cost Accounting.* Englewood Cliffs, N.J.: Prentice Hall, 1994.

10.102 Spall damage in bricks. A recent civil suit revolved around a 5-building brick apartment complex located in the Bronx, New York, which began to suffer *spalling* damage (i.e., a separation of some portion of the face of a brick from its body). The owner of the complex alleged that the bricks were defectively manufactured. The brick manufacturer countered that poor design and shoddy management led to the damage. To settle the suit, an estimate of the rate of damage per 1,000 bricks, called the *spall rate,* was required (*Chance,* Summer 1994). The owner estimated the spall rate using several *scaffold-drop* surveys. (With this method, an engineer lowers a scaffold down at selected places on building walls and counts the number of visible spalls for every 1,000 bricks in the observation area.) The brick manufacturer conducted its own survey by dividing the walls of the complex into 83 wall segments and taking a photograph of each wall segment. (The number of spalled bricks that could be made out from each photo was recorded, and the sum over all 83 wall segments was used as an estimate of total spall damage.) In this court case, the jury was faced with the following dilemma: The scaffold-drop survey provided the most accurate estimate of spall rates in a given wall segment. Unfortunately, the drop areas were not selected at random from the entire complex; rather, drops were made at areas with high spall concentrations, leading to an overestimate of the total damage. On the other hand, the photo survey was complete in that all 83 wall segments in the complex were checked for spall damage. But the spall rate estimated by the photos, at least in areas of high spall concentration, was biased low (spalling damage cannot always be seen from a photo), leading to an underestimate of the total damage.

Drop Location	Drop Spall Rate (per 1,000 bricks)	Photo Spall Rate (per 1,000 bricks)
1	0	0
2	5.1	0
3	6.6	0
4	1.1	.8
5	1.8	1.0
6	3.9	1.0
7	11.5	1.9
8	22.1	7.7
9	39.3	14.9
10	39.9	13.9
11	43.0	11.8

Source: Fairley, W. B., et al. "Bricks, buildings, and the Bronx: Estimating masonry deterioration," *Chance,* Vol. 7. No. 3, Summer 1994, p. 36 (Figure 3). Reprinted with permission from *Chance.* © 1994 by the American Statistical Association. All rights reserved. [*Note:* The data points are estimated from the points shown on a scatterplot.]

The data in the table are the spall rates obtained using the two methods at 11 drop locations. Use the data (saved in the **BRICKS** file), as did expert statisticians who testified in the case, to help the jury estimate the true spall rate at a given wall segment. Then explain how this information, coupled with the data (not given here) on all 83 wall segments, can provide a reasonable estimate of the total spall damage (i.e., total number of damaged bricks).

References

Chatterjee, S., and Price, B. *Regression Analysis by Example,* 2nd ed. New York: Wiley, 1991.

Draper, N., and Smith, H. *Applied Regression Analysis,* 3rd ed. New York: Wiley, 1987.

Gitlow, H., Oppenheim, A., and Oppenheim, R. *Quality Management: Tools and Methods for Improvement,* 2nd ed. Burr Ridge, Ill.: Irwin, 1995.

Graybill, F. *Theory and Application of the Linear Model.* North Scituate, Mass.: Duxbury, 1976.

Kleinbaum, D., and Kupper, L. *Applied Regression Analysis and Other Multivariable Methods,* 2nd ed. North Scituate, Mass.: Duxbury, 1997.

Mendenhall, W. *Introduction to Linear Models and the Design and Analysis of Experiments.* Belmont, CA.: Wadsworth, 1968.

Mendenhall, W., and Sincich, T. A. *Second Course in Statistics: Regression Analysis,* 6th ed. Upper Saddle River, N.J.: Prentice Hall, 2003.

Montgomery, D., Peck, E., and Vining, G. *Introduction to Linear Regression Analysis,* 3rd ed. New York: Wiley, 2001.

Mosteller, F., and Tukey, J. W. *Data Analysis and Regression: A Second Course in Statistics.* Reading, Mass.: Addison-Wesley, 1977.

Neter, J., Kutner, M., Nachtsheim, C., and Wasserman, W. *Applied Linear Statistical Models,* 4th ed. Hornewood, Ill.: Richard Irwin, 1996.

Rousseeuw, P. J., and Leroy, A. M. *Robust Regression and Outlier Detection.* New York: Wiley, 1987.

Weisburg, S. *Applied Linear Regression,* 2nd ed. New York: Wiley, 1985.

USING TECHNOLOGY

SPSS: Simple Linear Regression

Regression Analysis

Step 1 Access the SPSS spreadsheet file that contains the two quantitative variables (dependent and independent variables).

Step 2 Click on the "Analyze" button on the SPSS menu bar and then click on "Regression" and "Linear," as shown in Figure 10.S.1.

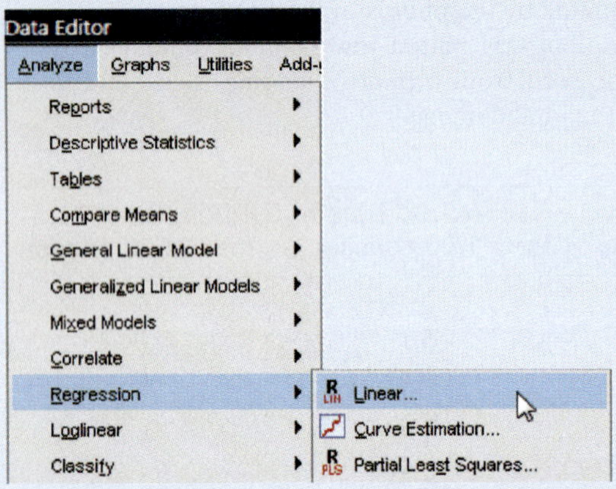

Figure 10.S.1 SPSS menu options for regression

Step 3 On the resulting dialog box (see Figure 10.S.2), specify the dependent variable in the "Dependent" box and the independent variable in the "Independent(s)" box. Be sure to select "Enter" in the "Method" box.

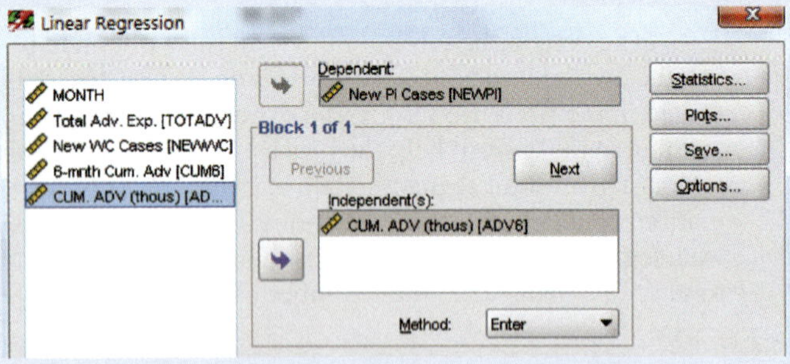

Figure 10.S.2 SPSS linear regression dialog box

Step 4 To produce confidence intervals for the model parameters, click the "Statistics" button and check the appropriate menu items in the resulting menu list.

Step 5 To obtain prediction intervals for y and confidence intervals for $E(y)$, click the "Save" button and check the appropriate items in the resulting menu list, as shown in Figure 10.S.3. (The prediction intervals will be added as new columns to the SPSS data spreadsheet.)

Step 6 To return to the main Regression dialog box from any of these optional screens, click "Continue." Click "OK" on the Regression dialog box to view the linear regression results.

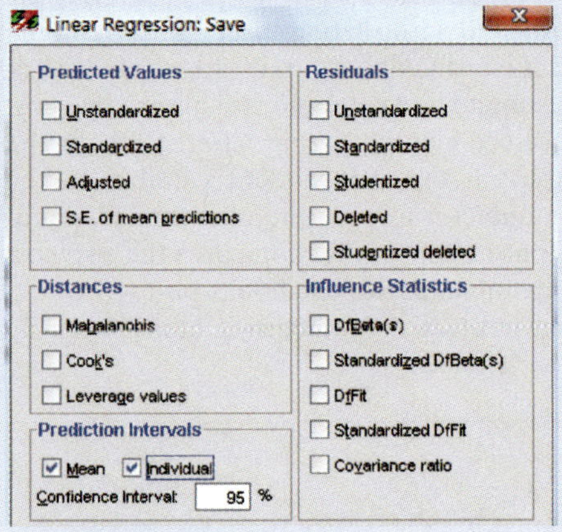

Figure 10.S.3 SPSS linear regression save dialog box

Correlation Analysis

Step 1 Click on the "Analyze" button on the main menu bar and then click on "Correlate" (see Figure 10.S.1).

Step 2 Click on "Bivariate." The resulting dialog box appears in Figure 10.S.4.

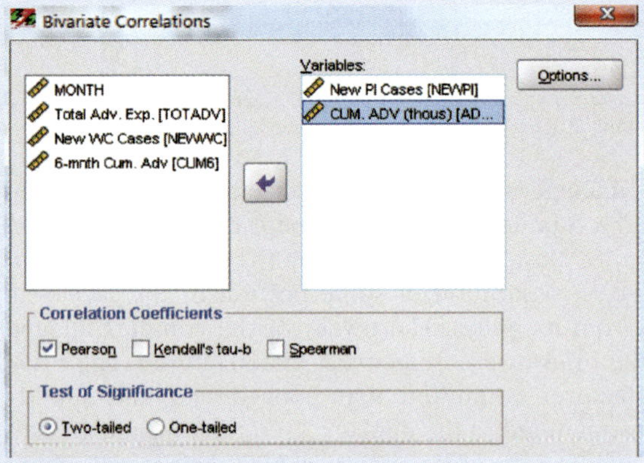

Figure 10.S.4 SPSS correlation dialog box

Step 3 Enter the variables of interest in the "Variables" box and check the "Pearson" option.

Step 4 Click "OK" to obtain a printout of the correlation.

Minitab: Simple Linear Regression

Regression Analysis

Step 1 Access the Minitab worksheet file that contains the two quantitative variables (dependent and independent variables).

Step 2 Click on the "Stat" button on the Minitab menu bar and then click on "Regression" and "Regression" again, as shown in Figure 10.M.1.

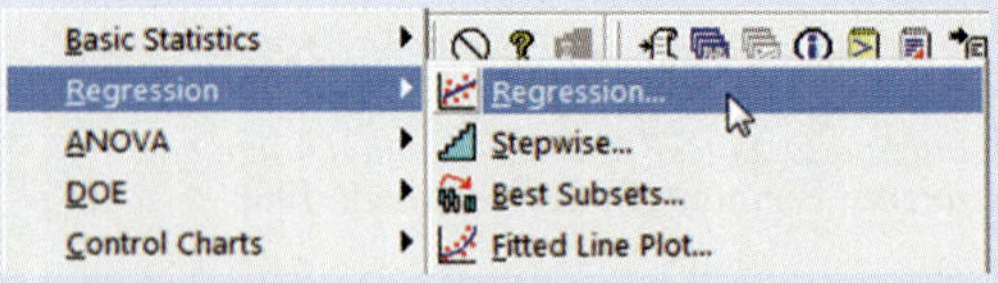

Figure 10.M.1 Minitab menu options for regression

Step 3 On the resulting dialog box (see Figure 10.M.2), specify the dependent variable in the "Response" box and the independent variable in the "Predictors" box.

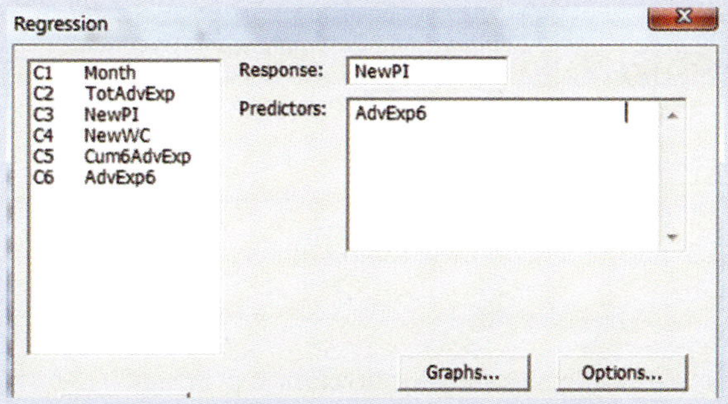

Figure 10.M.2 Minitab regression dialog box

Step 4 To produce prediction intervals for y and confidence intervals for $E(y)$, click the "Options" button. The resulting dialog box is shown in Figure 10.M.3.

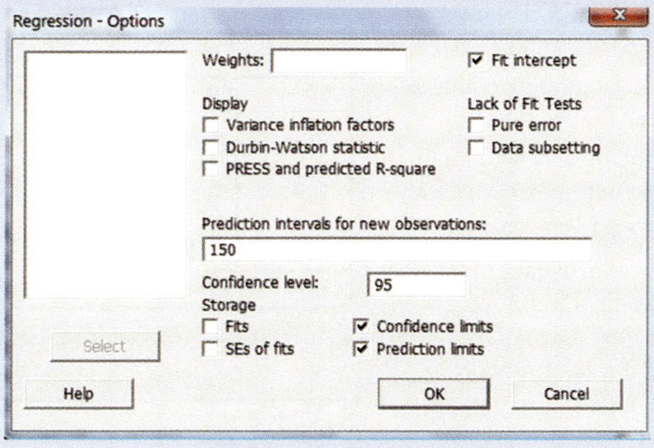

Figure 10.M.3 Minitab regression options

Step 5 Check "Confidence limits" and/or "Prediction limits," specify the "Confidence level," and enter the value of x in the "Prediction intervals for new observations" box.

Step 6 Click "OK" to return to the main Regression dialog box and then click "OK" again to produce the Minitab simple linear regression printout.

Correlation Analysis

Step 1 Click on the "Stat" button on the Minitab main menu bar, then click on "Basic Statistics," and then click on "Correlation," as shown in Figure 10.M.4.

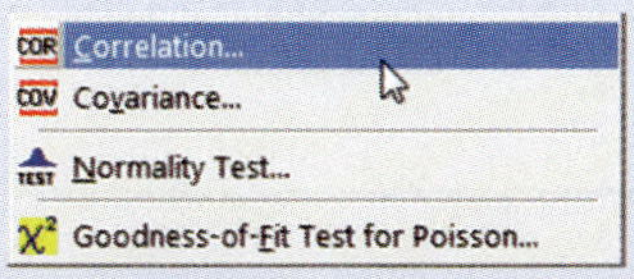

Figure 10.M.4 Minitab menu options for correlation

Step 2 On the resulting dialog box (see Figure 10.M.5), enter the two variables of interest in the "Variables" box.

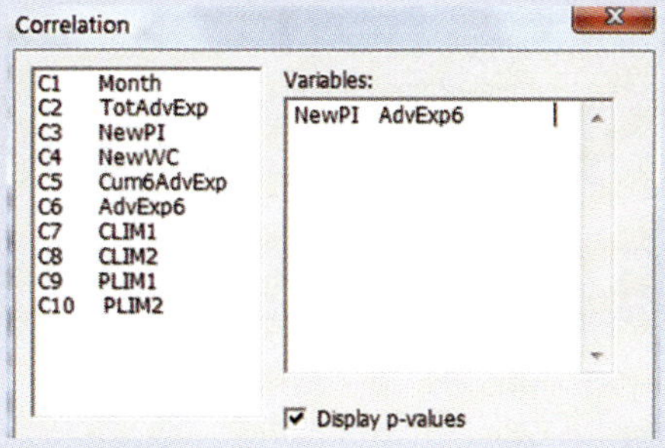

Figure 10.M.5 Minitab correlation dialog box

Step 3 Click "OK" to obtain a printout of the correlation.

Excel/DDXL: Simple Linear Regression

Regression Analysis

Step 1 Create a workbook with two columns, one representing the dependent variable and one the independent variable.

Step 2 Highlight (select) these data columns on the Excel spreadsheet.

Step 3 Click on "Add-Ins" in the main Excel menu bar and select "DDXL." On the resulting menu, select "Regression," as shown in Figure 10.E.1.

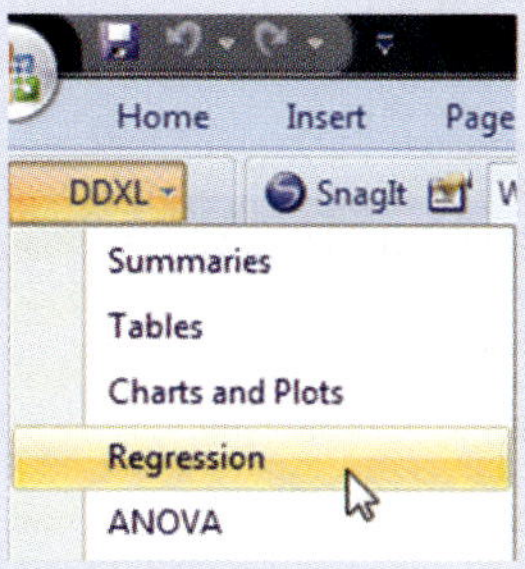

Figure 10.E.1 Excel/DDXL menu options for regression

Step 4 On the resulting menu, select "Simple Regression" in the Function Type box, as shown in Figure 10.E.2.

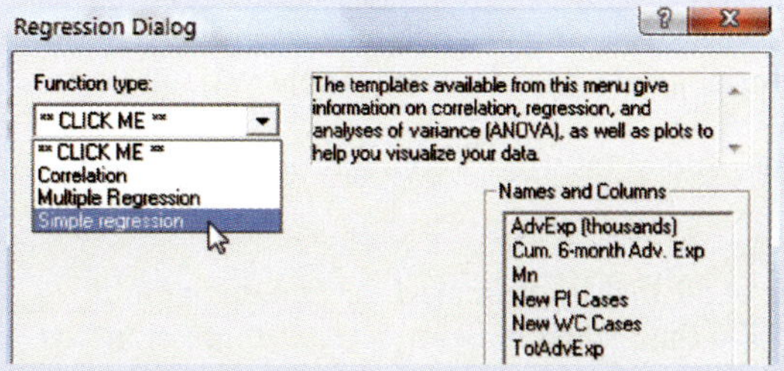

Figure 10.E.2 DDXL regression options

Step 5 Move the column with the values of the dependent variable into the "Response Variable" box and the column with the values of the independent variable into the "Explanatory Variable" box, as shown in Figure 10.E.3.

Step 6 Click "OK" to generate the simple linear regression printout.

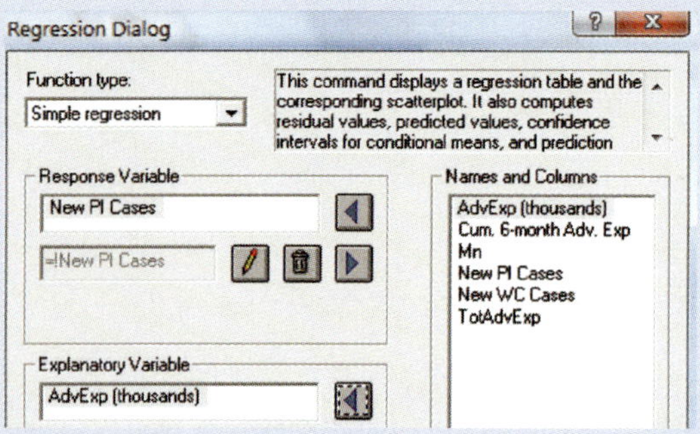

Figure 10.E.3 DDXL regression dialog box

Step 7 Under the regression results are more options. One option is "95% Confidence and Prediction Intervals." (See Figure 10.E.4.) Selecting this option will generate a confidence interval for $E(y)$ and a prediction interval for y for each observation in the workbook.

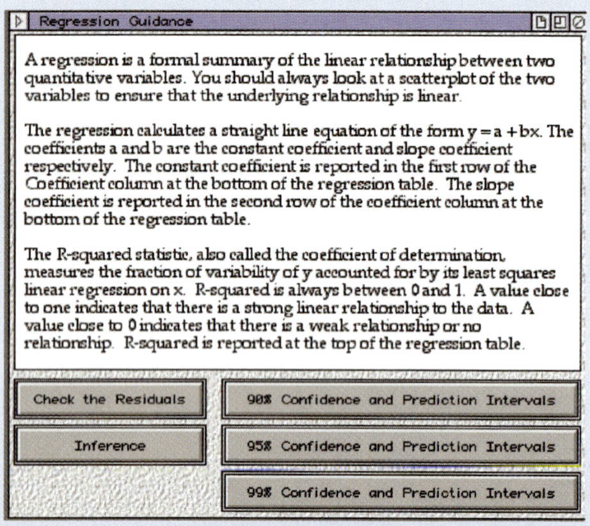

Figure 10.E.4 DDXL menu options for prediction intervals

Correlation Analysis

Step 1 Create a workbook with two columns, one representing the dependent variable and one the independent variable.

Step 2 Highlight (select) these data columns on the Excel spreadsheet.

Step 3 Click on "Add-Ins" in the main Excel menu bar and select "DDXL." On the resulting menu, select "Regression," as shown in Figure 10.E.1.

Step 4 On the resulting menu, select "Correlation" in the Function Type box (see Figure 10.E.2).

Step 5 Move the column with the values of the independent variable into the "x-Axis Quantitative Variable" box and the column with the values of the dependent variable into the "y-Axis Quantitative Variable" box, as shown in Figure 10.E.5.

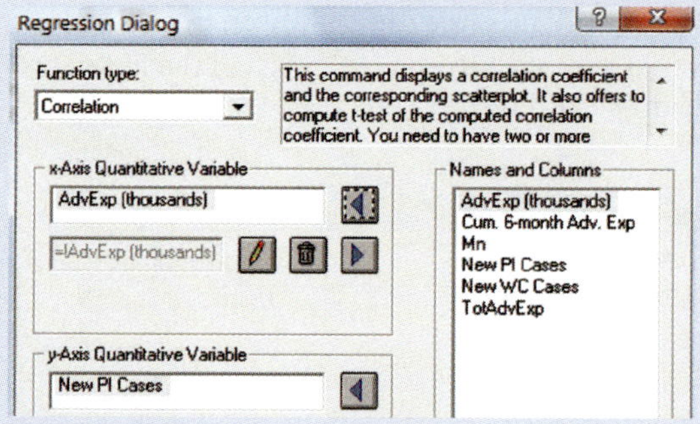

Figure 10.E.5 DDXL correlation dialog box

Step 6 Click "OK" to generate the correlation printout.

TI-84 Graphing Calculator: Simple Linear Regression

Finding the Least Squares Regression Equation

Step 1 *Enter the data*

- Press **STAT** and select **1:Edit**

Note: If a list already contains data, clear the old data.

- Use the up arrow to highlight the list name, "**L1**" or "**L2**"
- Press **CLEAR ENTER**
- Enter your x-data in **L1** and your y-data in **L2**

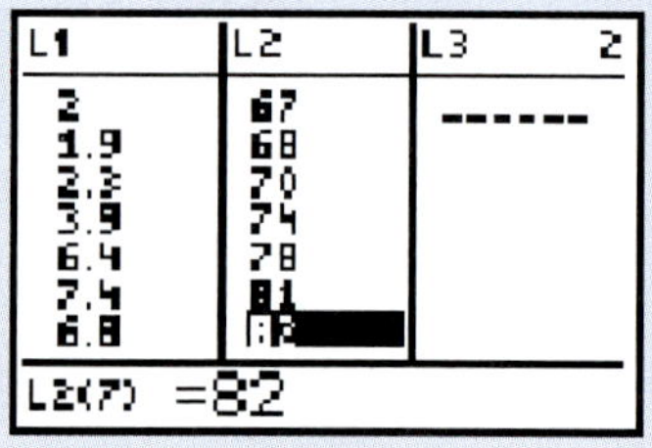

Step 2 *Find the equation*

- Press **STAT** and highlight **CALC**
- Press **4** for **LinReg(ax + b)**
- Press **ENTER**
- The screen will show the values for a and b in the equation $y = ax + b$.

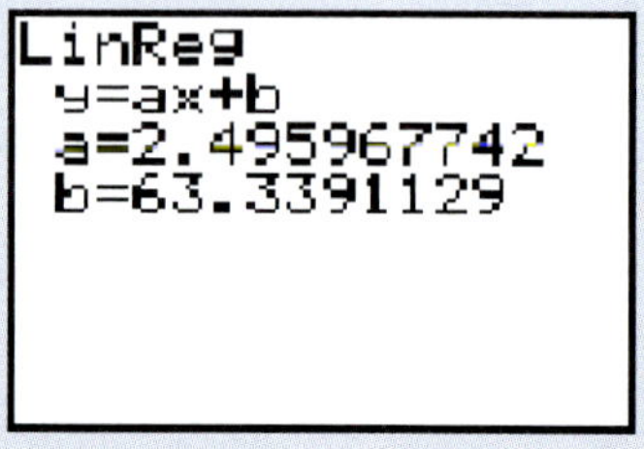

Finding r and r^2

Use this procedure if r and r^2 do not already appear on the LinReg screen from part I:

Step 1 *Turn the diagnostics feature on*

- Press **2nd 0** for **CATALOG**
- Press the **ALPHA** key and $\mathbf{x^{-1}}$ for **D**
- Press the down **ARROW** until **DiagnosticsOn** is highlighted
- Press **ENTER** twice

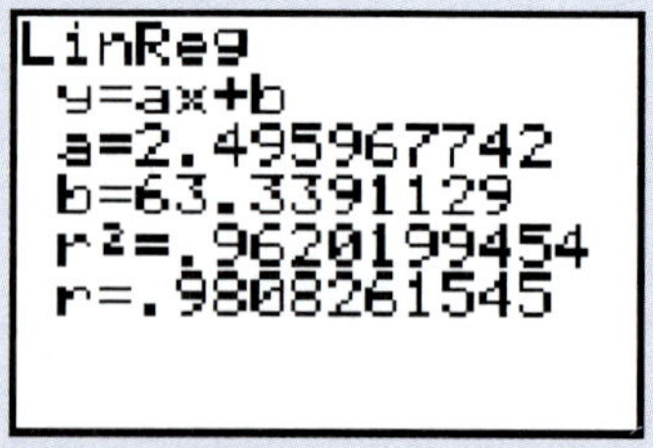

Step 2 *Find the regression equation as shown in part I above*

The values for r and r^2 will appear on the screen as well.

Graphing the Least Squares Line with the Scatterplot

Step 1 *Enter the data as shown in part I above*

Step 2 *Set up the data plot*

- Press **Y=** and **CLEAR** all functions from the Y registers
- Press **2ndY=** for **STAT PLOT**
- Press **1** for **Plot1**
- Set the cursor so that **ON** is flashing and press **ENTER**
- For **Type,** use the **ARROW** and **ENTER** keys to highlight and select the scatterplot (first icon in the first row)
- For **Xlist,** choose the column containing the *x*-data
- For **Ylist,** choose the column containing the *y*-data

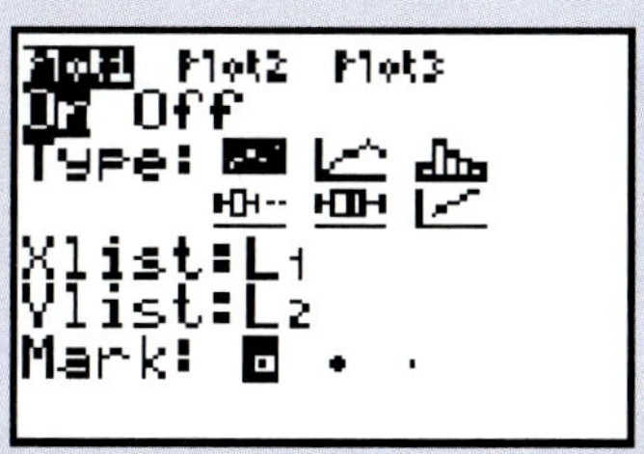

Step 3 *Find the regression equation and store the equation in Y1*

- Press **STAT** and highlight **CALC**
- Press **4** for **LinReg(ax + b)** (*Note:* Don't press **ENTER** here because you want to store the regression equation in Y1.)
- Press **VARS**
- Use the right arrow to highlight **Y-VARS**
- Press **ENTER** to select **1:Function**
- Press **ENTER** to select **1:Y1**
- Press **ENTER**

Step 4 *View the scatterplot and regression line*

- Press **ZOOM** and then press **9** to select **9:ZoomStat**

You should see the data graphed along with the regression line.

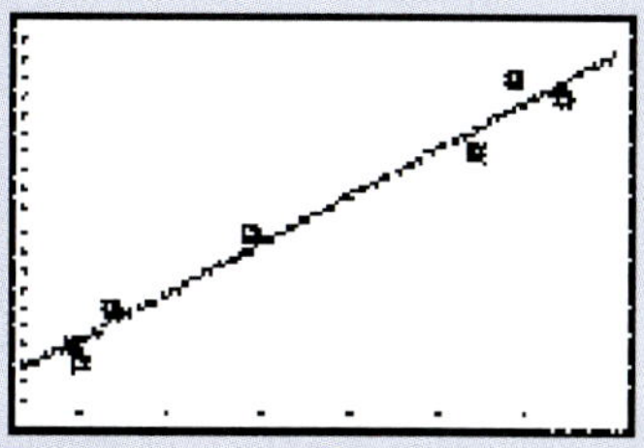

CONTENTS

11 Multiple Regression and Model Building

Where We've Been

- Introduced the straight-line model relating a dependent variable y to a single independent variable x
- Demonstrated how to estimate the parameters of the straight-line model using the method of least squares
- Showed how to statistically assess the adequacy of the model
- Showed how to use the model to Estimate $E(y)$ and predict y for a given value of x

Statistics IN Action Bid-Rigging in the Highway Construction Industry

In the United States, commercial contractors bid for the right to construct state highways and roads. A state government agency, usually the Department of Transportation (DOT), notifies various contractors of the state's intent to build a highway. Sealed bids are submitted by the contractors, and the contractor with the lowest bid (building cost) is awarded the road construction contract. The bidding process works extremely well in competitive markets but has the potential to increase construction costs if the markets are noncompetitive or if collusive practices are present. The latter occurred in the 1980s in Florida. Numerous road contractors either admitted or were found guilty of price fixing (i.e., setting the cost of construction above the fair, or competitive, cost through bid-rigging or other means).

This Statistics in Action involves data collected by the Florida attorney general shortly following the price-fixing crisis. The attorney general's objective is to build a model for the cost (y) of a road construction contract awarded using the sealed-bid system. The **FLAG** file contains data for a sample of 235 road contracts. *(continued)*

Where We're Going

- Introduce a *multiple regression* model as a means of relating a dependent variable *y* to two or more independent variables
- Present several different multiple regression models involving both quantitative and qualitative independent variables
- Assess how well the multiple regression model fits the sample data
- Show how an analysis of the model's *residuals* can aid in detecting violations of model assumptions and in identifying model modifications

The variables measured for each contract are listed in Table SIA11.1. Ultimately, the attorney general wants to use the model to predict the costs of future road contracts in the state.

In several Statistics in Action Revisited sections (see below), we show how to analyze the data using a multiple regression analysis.

Statistics IN Action Revisited

- Evaluating a First-Order Model (p. 646)
- Variable Screening and Model Building (p. 693)
- A Residual Analysis (p. 710)

Table SIA11.1 Variables in the FLAG Data File

Variable Name	Type	Description
CONTRACT	Quantitative	Road contract number
COST	Quantitative	Low bid contract cost (thousands of dollars)
DOTEST	Quantitative	DOT engineer's cost estimate (thousands of dollars)
STATUS	Qualitative	Bid status (1 = fixed, 0 = competitive)
B2B1RAT	Quantitative	Ratio of second lowest bid to low bid
B3B1RAT	Quantitative	Ratio of third lowest bid to low bid
BHB1RAT	Quantitative	Ratio of highest bid to low bid
DISTRICT	Qualitative	Location of road (1 = south Florida, 0 = north Florida)
BTPRATIO	Quantitative	Ratio of number of bidders to number of plan holders
DAYSEST	Quantitative	DOT engineer's estimate of number of workdays required

Data Set: FLAG

11.1 Multiple Regression Models

Most practical applications of regression analysis employ models that are more complex than the simple straight-line model. For example, a realistic probabilistic model for reaction time stimulus would include more than just the amount of a particular drug in the bloodstream. Factors such as age, a measure of visual perception, and sex of the subject are some of the many variables that might be related to reaction time. Thus, we would want to incorporate these and other potentially important independent variables into the model in order to make accurate predictions.

Probabilistic models that include more than one independent variable are called **multiple regression models.** The general form of these models is

$$y = \beta_0 + \beta_1 x_1 + \beta_2 x_2 + \cdots + \beta_k x_k + \varepsilon$$

The dependent variable y is now written as a function of k independent variables, $x_1, x_2, \ldots, x_k$. The random error term is added to make the model probabilistic rather than deterministic. The value of the coefficient β_i determines the contribution of the independent variable x_i, and β_0 is the y-intercept. The coefficients $\beta_0, \beta_1, \ldots, \beta_k$ are usually unknown because they represent population parameters.

At first glance it might appear that the regression model shown above would not allow for anything other than straight-line relationships between y and the independent variables, but this is not true. Actually, $x_1, x_2, \ldots, x_k$ can be functions of variables as long as the functions do not contain unknown parameters. For example, the reaction time, y, of a subject to a visual stimulus could be a function of the independent variables

$$x_1 = \text{Age of the subject}$$
$$x_2 = (\text{Age})^2 = x_1^2$$
$$x_3 = 1 \text{ if male subject, } 0 \text{ if female subject}$$

The x_2 term is called a **higher-order term** because it is the value of a quantitative variable (x_1) squared (i.e., raised to the second power). The x_3 term is a **coded variable** representing a qualitative variable (gender). The multiple regression model is quite versatile and can be made to model many different types of response variables.

The General Multiple Regression Model

$$y = \beta_0 + \beta_1 x_1 + \beta_2 x_2 + \cdots + \beta_k x_k + \varepsilon$$

where

y is the dependent variable.

$x_1, x_2, \ldots, x_k$ are the independent variables.

$E(y) = \beta_0 + \beta_1 x_1 + \beta_2 x_2 + \cdots + \beta_k x_k$ is the deterministic portion of the model.

β_i determines the contribution of the independent variable x_i.

Note: The symbols $x_1, x_2, \ldots, x_k$ may represent higher-order terms for quantitative predictors or terms that represent qualitative predictors.

As shown in the box, the steps used to develop the multiple regression model are similar to those used for the simple linear regression model.

Analyzing a Multiple Regression Model

Step 1 Hypothesize the deterministic component of the model. This component relates the mean, $E(y)$, to the independent variables $x_1, x_2, \ldots, x_k$. This involves the choice of the independent variables to be included in the model (Sections 11.2, 11.5–11.10).

Step 2 Use the sample data to estimate the unknown model parameters $\beta_0, \beta_1, \beta_2, \ldots, \beta_k$ in the model (Section 11.2).

Step 3 Specify the probability distribution of the random error term, ε, and estimate the standard deviation of this distribution, σ (Section 11.3).

Step 4 Check that the assumptions on ε are satisfied and make model modifications if necessary (Section 11.11).

Step 5 Statistically evaluate the usefulness of the model (Sections 11.3).

Step 6 When satisfied that the model is useful, use it for prediction, estimation, and other purposes (Section 11.4).

The assumptions we make about the random error ε of the multiple regression model are also similar to those in a simple linear regression (see Section 10.3). These are summarized below.

Assumptions for Random Error ε

For any given set of values of $x_1, x_2, \ldots, x_k$, the random error ε has a probability distribution with the following properties:

1. Mean equal to 0
2. Variance equal to σ^2
3. Normal distribution
4. Random errors are independent (in a probabilistic sense).

Throughout this chapter, we introduce several different types of models that form the foundation of **model building** (or useful model construction). In the next several sections, we consider the most basic multiple regression model, called the *first-order model.*

PART I: FIRST ORDER MODELS WITH QUANTITATIVE INDEPENDENT VARIABLES

11.2 Estimating and Making Inferences about the β Parameters

BIOGRAPHY

GEORGE U. YULE (1871–1951)

Yule Processes

Born on a small farm in Scotland, George Yule received an extensive childhood education. After graduating from University College (London), where he studied civil engineering, Yule spent a year employed in engineering workshops. However, he made a career change in 1893, accepting a teaching position back at University College under the guidance of statistician Karl Pearson (see p. 524). Inspired by Pearson's work, Yule produced a series of important articles on the statistics of regression and correlation. Yule is considered the first to apply the method of least squares in regression analysis, and he developed the theory of multiple regression. He eventually was appointed a lecturer in statistics at Cambridge University and later became the president of the prestigious Royal Statistical Society. Yule made many other contributions to the field, including the invention of time series analysis and the development of Yule processes and the Yule distribution. ■

A model that includes terms only for *quantitative* independent variables, called a **first-order model,** is described in the box. Note that the first-order model does not include any higher-order terms (such as x_1^2).

A First-Order Model in Five Quantitative Independent (Predictor) Variables*

$$E(y) = \beta_0 + \beta_1 x_1 + \beta_2 x_2 + \beta_3 x_3 + \beta_4 x_4 + \beta_5 x_5$$

where $x_1, x_2, \ldots, x_5$ are all quantitative variables that *are not* functions of other independent variables.

Note: β_i represents the slope of the line relating y to x_i when all the other x's are held fixed.

The method of fitting first-order models—and multiple regression models in general—is identical to that of fitting the simple straight-line model: the **method of least squares**—that is, we choose the estimated model

$$\hat{y} = \hat{\beta}_0 + \hat{\beta}_1 x_1 + \cdots + \hat{\beta}_k x_k$$

that (1) has an average error of prediction of 0, i.e., $\Sigma(y - \hat{y}) = 0$
and (2) minimizes $SSE = \Sigma(y - \hat{y})^2$

As in the case of the simple linear model, the sample estimates $\hat{\beta}_0, \hat{\beta}_1, \ldots, \hat{\beta}_k$ are obtained as a solution to a set of simultaneous linear equations.†

The primary difference between fitting the simple and multiple regression models is computational difficulty. The $(k + 1)$ simultaneous linear equations that must be solved to find the $(k + 1)$ estimated coefficients $\hat{\beta}_0, \hat{\beta}_1, \ldots, \hat{\beta}_k$ cannot be written in simple equation form; rather, the estimates are obtained using matrices and matrix algebra. Instead of presenting the complex matrix algebra required to fit the models, we resort to statistical software and present output from SPSS, Minitab, and Excel.

Example 11.1 Fitting a First-Order Model: Price of an Antique Clock

Problem A collector of antique grandfather clocks sold at auction believes that the price received for the clocks depends on both the age of the clocks and the number of bidders at the auction. Thus, he hypothesizes the first-order model

$$y = \beta_0 + \beta_1 x_1 + \beta_2 x_2 + \varepsilon$$

where

y = Auction price (dollars)
x_1 = Age of clock (years)
x_2 = Number of bidders

A sample of 32 auction prices of grandfather clocks, along with their age and the number of bidders, is given in Table 11.1.

a. Use scattergrams to plot the sample data. Interpret the plots.

b. Use the method of least squares to estimate the unknown parameters β_0, β_1, and β_2 of the model.

*The terminology *first order* is derived from the fact that each x in the model is raised to the first power.

†Students who are familiar with calculus should note that $\hat{\beta}_0, \hat{\beta}_1, \ldots, \hat{\beta}_k$ are the solutions to the set of equations $\partial SSE/\partial\hat{\beta}_0 = 0$, $\partial SSE/\partial\hat{\beta}_1 = 0, \ldots, \partial SSE/\partial\hat{\beta}_k = 0$. The solution is usually given in matrix form, but we do not present the details here. See the references for details.

Table 11.1 Auction Price Data

Age, x_1	Number of Bidders, x_2	Auction Price, y	Age, x_1	Number of Bidders, x_2	Auction Price, y
127	13	\$1,235	170	14	\$2,131
115	12	1,080	182	8	1,550
127	7	845	162	11	1,884
150	9	1,522	184	10	2,041
156	6	1,047	143	6	845
182	11	1,979	159	9	1,483
156	12	1,822	108	14	1,055
132	10	1,253	175	8	1,545
137	9	1,297	108	6	729
113	9	946	179	9	1,792
137	15	1,713	111	15	1,175
117	11	1,024	187	8	1,593
137	8	1,147	111	7	785
153	6	1,092	115	7	744
117	13	1,152	194	5	1,356
126	10	1,336	168	7	1,262

Data Set: GFCLOCKS

c. Find the value of SSE that is minimized by the least squares method.

d. Estimate σ, the standard deviation of the model, and interpret the result.

Solution

a. Minitab side-by-side scatterplots for examining the bivariate relationships between y and x_1, and between y and x_2, are shown in Figure 11.1. Of the two variables, age (x_1) appears to have the stronger linear relationship with auction price (y).

b. The model hypothesized is fit to the data of Table 11.1 with Minitab. A portion of the printout is reproduced in Figure 11.2. The least squares estimates of the β parameters (highlighted) are $\hat{\beta}_0 = -1{,}339$, $\hat{\beta}_1 = 12.74$, and $\hat{\beta}_2 = 85.95$. Therefore, the equation that minimizes SSE for this data set (i.e., the **least squares prediction equation**) is

$$\hat{y} = -1{,}339 + 12.74x_1 + 85.95x_2$$

c. The minimum value of the sum of the squared errors, also highlighted in Figure 11.2, is SSE = 516,727.

d. Recall that the estimator of σ^2 for the straight-line model is $s^2 = \text{SSE}/(n - 2)$, and note that the denominator is (n – number of estimated β parameters), which is $(n - 2)$ in the straight-line model. Because we must estimate the three parameters β_0, β_1 and β_2 for the first-order model, the estimator of σ^2 is

$$s^2 = \frac{\text{SSE}}{n - 3} = \frac{\text{SSE}}{32 - 3} = \frac{516{,}727}{29} = 17{,}818$$

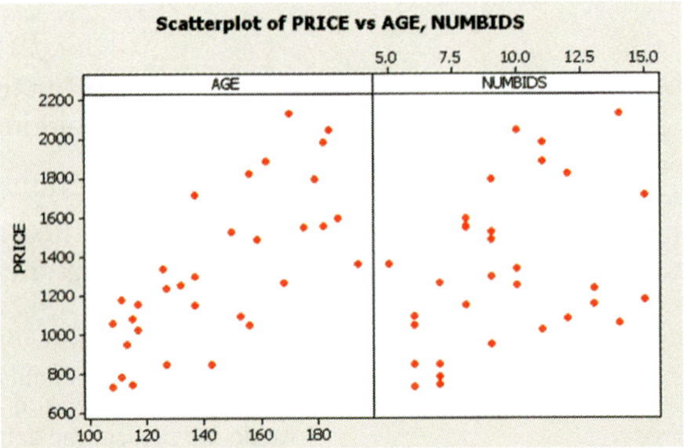

Figure 11.1
Minitab side-by-side scatterplots for the data of Table 11.1

```
Regression Analysis: PRICE versus AGE, NUMBIDS

The regression equation is
PRICE = - 1339 + 12.7 AGE + 86.0 NUMBIDS


Predictor      Coef   SE Coef      T      P
Constant    -1339.0     173.8  -7.70  0.000
AGE         12.7406    0.9047  14.08  0.000
NUMBIDS      85.953     8.729   9.85  0.000


S = 133.485    R-Sq = 89.2%    R-Sq(adj) = 88.5%


Analysis of Variance

Source          DF       SS       MS       F      P
Regression       2  4283063  2141531  120.19  0.000
Residual Error  29   516727    17818
Total           31  4799790
```

Figure 11.2
Minitab analysis of the auction price model

This value, often called the **mean square for error (MSE)** is also highlighted at the bottom of the Minitab printout in Figure 11.2. The estimate of σ, then, is

$$s = \sqrt{17{,}818} = 133.5$$

which is highlighted in the middle of the Minitab printout in Figure 11.2. One useful interpretation of the estimated standard deviation s is that the interval $\pm 2s$ will provide a rough approximation to the accuracy with which the model will predict future values of y for given values of x. Thus, we expect the model to provide predictions of auction price to within about $\pm 2s = \pm 2(133.5) = \pm 267$ dollars.*

Look Back As with simple linear regression, we will use the estimator of σ^2 both to check the utility of the model (Section 11.3) and to provide a measure of reliability of predictions and estimates when the model is used for those purposes (Section 11.4). Thus, you can see that the estimation of σ^2 plays an important part in the development of a regression model.

Now Work Exercise 11.2a–c

Estimator of σ^2 for a Multiple Regression Model with k Independent Variables

$$s^2 = \frac{\text{SSE}}{n - \text{number of estimated } \beta \text{ parameters}} = \frac{\text{SSE}}{n - (k + 1)}$$

After obtaining the least squares prediction equation, the analyst will usually want to make meaningful interpretations of the β estimates. Recall that in the straight-line model (Chapter 10)

$$y = \beta_0 + \beta_1 x + \varepsilon$$

β_0 represents the y-intercept of the line, and β_1 represents the slope of the line. From our discussion in Chapter 10, β_1 has a practical interpretation—it represents the mean change in y for every 1-unit increase in x. When the independent variables are quantitative, the β parameters in the first-order model specified in Example 11.1 have similar interpretations. The

* The $\pm 2s$ approximation will improve as the sample size is increased. We will provide more precise methodology for the construction of prediction intervals in Section 11.4.

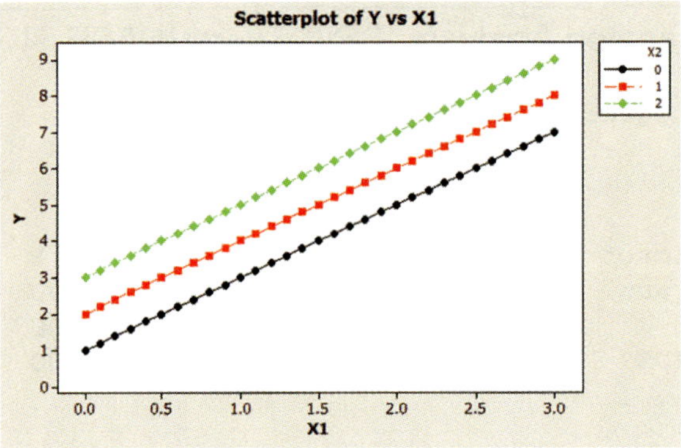

Figure 11.3
Minitab graph of $E(y) = 1 + 2x_1 + x_2$ for $x_2 = 0, 1, 2$

difference is that when we interpret the β that multiplies one of the variables (e.g., x_1), we must be certain to hold the values of the remaining independent variables (e.g., x_2, x_3) fixed.

To see this, suppose that the mean $E(y)$ of a response y is related to two quantitative independent variables, x_1 and x_2, by the first-order model

$$E(y) = 1 + 2x_1 + x_2$$

In other words, $\beta_0 = 1$, $\beta_1 = 2$, and $\beta_2 = 1$.

Now, when $x_2 = 0$, the relationship between $E(y)$ and x_1 is given by

$$E(y) = 1 + 2x_1 + (0) = 1 + 2x_1$$

A Minitab graph of this relationship (a straight line) is shown in Figure 11.3. Similar graphs of the relationship between $E(y)$ and x_1 for $x_2 = 1$,

$$E(y) = 1 + 2x_1 + (1) = 2 + 2x_1$$

and for $x_2 = 2$,

$$E(y) = 1 + 2x_1 + (2) = 3 + 2x_1$$

also are shown in Figure 11.3. Note that the slopes of the three lines are all equal to $\beta_1 = 2$, the coefficient that multiplies x_1.

Figure 11.3 exhibits a characteristic of all first-order models: If you graph $E(y)$ versus any one variable—say, x_1—for fixed values of the other variables, the result will always be a *straight line* with slope equal to β_1. If you repeat the process for other values of the fixed independent variables, you will obtain a set of *parallel* straight lines. This indicates that the effect of the independent variable x_i on $E(y)$ is independent of all the other independent variables in the model, and this effect is measured by the slope β_i (see the box on p. 627).

A Minitab three-dimensional graph of the model $E(y) = 1 + 2x_1 + x_2$ is shown in Figure 11.4. Note that the model graphs as a plane. If you slice the plane at a particular value of x_2 (say, $x_2 = 0$), you obtain a straight line relating $E(y)$ to x_1 (e.g., $E[y] = 1 + 2x_1$). Similarly, if you slice the plane at a particular value of x_1, you obtain a straight line relating $E(y)$ to x_2. Because it is more difficult to visualize three-dimensional and, in general, k-dimensional surfaces, we will graph all the models presented in this

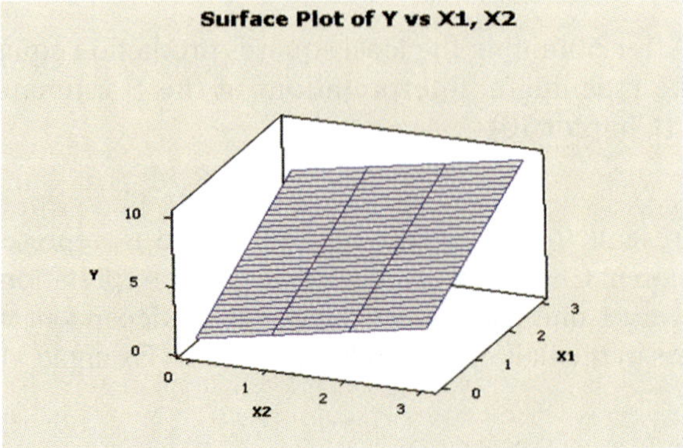

Figure 11.4
Minitab 3-dimensional graph of $E(y) = 1 + 2x_1 + x_2$

chapter in two dimensions. *The key to obtaining these graphs is to hold fixed all but one of the independent variables in the model.*

Example 11.2

Interpreting the β Estimates: Clock Auction Price Model

Problem Refer to the first-order model for auction price (y) considered in Example 11.1. Interpret the estimates of the β parameters in the model.

Solution The least squares prediction equation, as given in Example 11.1, is $\hat{y} = -1{,}339 + 12.74x_1 + 85.95x_2$. We know that with first-order models, β_1 represents the slope of the line relating y to x_1 for fixed x_2. That is, β_1 measures the change in $E(y)$ for every one-unit increase in x_1 when the other independent variable in the model is held fixed. A similar statement can be made about β_2: β_2 measures the change in $E(y)$ for every one-unit increase in x_2 when the other x in the model is held fixed. Consequently, we obtain the following interpretations:

$\hat{\beta}_1 = 12.74$: We estimate the mean auction price $E(y)$ of an antique clock to increase \$12.74 for every 1-year increase in age (x_1) when the number of bidders (x_2) is held fixed.

$\hat{\beta}_2 = 85.95$: We estimate the mean auction price $E(y)$ of an antique clock to increase \$85.95 for every 1-bidder increase in the number of bidders (x_2) when age (x_1) is held fixed.

The value $\hat{\beta}_0 = -1{,}339$ does not have a meaningful interpretation in this example. To see this, note that $\hat{y} = \hat{\beta}_0$ when $x_1 = x_2 = 0$. Thus, $\hat{\beta}_0 = -1{,}339$ represents the estimated mean auction price when the values of all the independent variables are set equal to 0. Because an antique clock with these characteristics—an age of 0 years and 0 bidders on the clock—is not practical, the value of $\hat{\beta}_0$ has no meaningful interpretation.

Look Back In general, $\hat{\beta}_0$ will not have a practical interpretation unless it makes sense to set the values of the x's simultaneously equal to 0.

Now Work Exercise 11.14a,b

CAUTION The interpretation of the β parameters in a multiple regression model will depend on the terms specified in the model. The interpretations above are for a first-order linear model only. In practice, you should be sure that a first-order model is the correct model for $E(y)$ before making these β interpretations. (We discuss alternative models for $E(y)$ in Sections 11.5–11.8.) ▲

Inferences about the individual β parameters in a model are obtained using either a confidence interval or a test of hypothesis, as outlined in the following two boxes.*

A 100(1 − α)% Confidence Interval for a β Parameter

$$\hat{\beta}_i \pm t_{\alpha/2} s_{\hat{\beta}_i}$$

where $t_{\alpha/2}$ is based on $n - (k + 1)$ degrees of freedom and

n = Number of observations

$k + 1$ = Number of β parameters in the model

*The formulas for computing $\hat{\beta}_i$ and its standard error are so complex that the only reasonable way to present them is by using matrix algebra. We do not assume a prerequisite of matrix algebra for this text, and, in any case, we think the formulas can be omitted in an introductory course without serious loss. They are programmed into almost all statistical software packages with multiple regression routines and are presented in some of the texts listed in the references.

Test of an Individual Parameter Coefficient in the Multiple Regression Model

One-Tailed Test	Two-Tailed Test
H_0: $\beta_i = 0$	H_0: $\beta_i = 0$
H_a: $\beta_i < 0$ [or H_a: $\beta_i > 0$]	H_a: $\beta_i \neq 0$

$$\textit{Test statistic: } t = \frac{\hat{\beta}_i}{s_{\hat{\beta}_i}}$$

Rejection region: $t < -t_\alpha$ [or $t > t_\alpha$ when H_a: $\beta_i > 0$]	*Rejection region:* $|t| > t_{\alpha/2}$

where t_α and $t_{\alpha/2}$ are based on $n - (k + 1)$ degrees of freedom and

n = Number of observations

$k + 1$ = Number of β parameters in the model

Conditions Required for Valid Inferences about the β Parameters

Refer to the four assumptions about the probability distribution for the random error ε. (p. 626).

We illustrate these methods with another example.

Example 11.3

Inferences about the β Parameters—Auction Price Model

Problem Refer to Examples 11.1 and 11.2. The collector of antique grandfather clocks knows that the price (y) received for the clocks increases linearly with the age (x_1) of the clocks. Moreover, the collector hypothesizes that the auction price (y) will increase linearly as the number of bidders (x_2) increases. Use the information on the Minitab printout shown in Figure 11.2 (p. 629) to

a. Test the hypothesis that the mean auction price of a clock increases as the number of bidders increases when age is held constant, that is, test $\beta_2 > 0$. Use $\alpha = .05$.

b. Form a 90% confidence interval for β_1 and interpret the result.

Solution

a. The hypotheses of interest concern the parameter β_2. Specifically,

$$H_0\text{: } \beta_2 = 0$$
$$H_a\text{: } \beta_2 > 0$$

The test statistic is a t-statistic formed by dividing the sample estimate $\hat{\beta}_2$ of the parameter β_2 by estimated standard error of $\hat{\beta}_2$ (denoted $s_{\hat{\beta}_2}$). These estimates, $\hat{\beta}_2 = 85.953$ and $s_{\hat{\beta}_2} = 8.729$, as well as the calculated t-value,

$$\textit{Test statistic: } t = \frac{\hat{\beta}_2}{s_{\hat{\beta}_2}} = \frac{85.953}{8.729} = 9.85$$

are highlighted on the Minitab printout in Figure 11.2.

The rejection region for the test is found in exactly the same way as the rejection regions for the t-tests in previous chapters—that is, we consult Table V in Appendix B to obtain an upper-tail value of t. This is a value t_α such that $P(t > t_\alpha) = \alpha$. We can then use this value to construct rejection regions for either one-tailed or two-tailed tests.

For $\alpha = .05$ and $n - (k + 1) = 32 - (2 + 1) = 29$ df, the critical t-value obtained from Table V is $t_{.05} = 1.699$. Therefore,

Rejection region: $t > 1.699$ (see Figure 11.5)

Because the test statistic value, $t = 9.85$, falls in the rejection region, we have sufficient evidence to reject H_0. Thus, the collector can conclude that the mean auction price of a clock increases as the number of bidders increases, when age is held constant. Note that the two-tailed observed significance level of the test is also highlighted on the printout. Because the one-tailed p-value (half this value) is p-value ≈ 0, any nonzero α (e.g., $\alpha = .01$) will lead us to reject H_0.

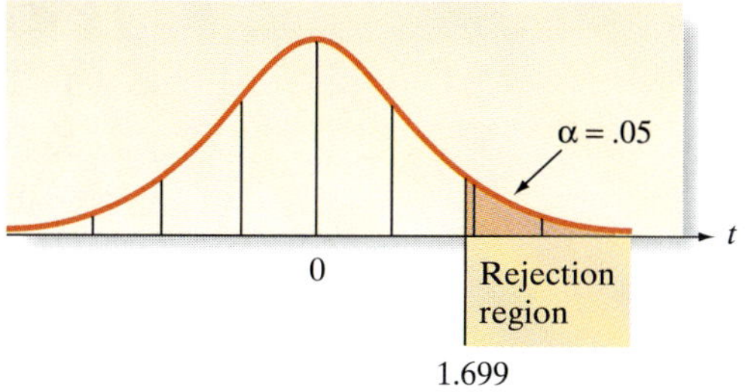

Figure 11.5
Rejection region for H_0: $\beta_2 = 0$ vs. H_a: $\beta_2 > 0$

b. A 90% confidence interval for β_1 is:

$$\hat{\beta}_1 \pm t_{\alpha/2} s_{\hat{\beta}_1} = \hat{\beta}_1 \pm t_{.05} s_{\hat{\beta}_1}$$

Substituting $\hat{\beta}_1 = 12.74$, $s_{\hat{\beta}_1} = .905$ (both obtained from the Minitab printout, Figure 11.2) and $t_{.05} = 1.699$ (from part **a**) into the equation, we obtain

$$12.74 \pm 1.699(.905) = 12.74 \pm 1.54$$

or (11.20, 14.28). Thus, we are 90% confident that β_1 falls between 11.20 and 14.28. Because β_1 is the slope of the line relating auction price (y) to age of the clock (x_1), we conclude that the price increases between \$11.20 and \$14.28 for every 1-year increase in age, holding number of bidders (x_2) constant.

Look Back When interpreting the β multiplied by one x, be sure to hold fixed the values of the other x's in the model.

Now Work Exercise 11.14c,d

Activity 11.1 *Insurance Premiums:* Collecting Data for Several Variables

Premiums for life insurance, health insurance, homeowners insurance, and car insurance are based on more than one factor. In this activity, you will consider factors that influence the price that an individual pays for insurance and how you might collect data that would be useful in studying insurance premiums. You may wish to look over statements from your own insurance policies to determine how much coverage you have and how much it costs you.

1. Suppose that you host an independent Web site that provides information about insurance coverage and costs. You would like to add a feature to your site where a person answers a few simple questions in order to receive an estimate of what that person should expect to pay for insurance with a major carrier. Pick one type of insurance and determine what five questions you would most like to ask the person in order to prepare an estimate.
2. Use your five questions from Exercise 1 to define five independent variables. What is the dependent variable in this situation? What would a first-order model for this situation look like?
3. The estimates provided on your Web site will be based on a random sample of policies recently purchased through major carriers. Because insurance rates change frequently, your Web site will need to be frequently updated through new samples. Design a sampling method that can easily be repeated as needed. Be specific as to how you will identify people who have recently purchased policies, how you will choose your sample, what data you will gather, and what method you will use to gather the data.
4. Once the data are collected, describe how you will organize it and use it to complete the model in Exercise 3.

11.3 Evaluating Overall Model Utility

In Section 11.2, we demonstrated the use of t-tests in making inferences about β parameters in the multiple regression model. There are caveats, however, with conducting these t-tests for the purposes of determining which x's are useful for predicting y. Several of these are listed next.

Use Caution When Conducting *t*-tests on the β Parameters

It is dangerous to conduct *t*-tests on the individual β parameters in a *first-order linear model* for the purpose of determining which independent variables are useful for predicting y and which are not. If you fail to reject H_0: $\beta_i = 0$, several conclusions are possible:

1. There is no relationship between y and x_i.
2. A straight-line relationship between y and x exists (holding the other x's in the model fixed), but a Type II error occurred.
3. A relationship between y and x_i (holding the other x's in the model fixed) exists but is more complex than a straight-line relationship (e.g., a curvilinear relationship may be appropriate). The most you can say about a β parameter test is that there is either sufficient (if you reject H_0: $\beta_i = 0$) or insufficient (if you do not reject H_0: $\beta_i = 0$) evidence of a *linear* (*straight-line*) relationship between y and x_i.

In addition, conducting *t*-tests on each β parameter in a model is *not* the best way to determine whether the overall model is contributing information for the prediction of y. If we were to conduct a series of *t*-tests to determine whether the independent variables are contributing to the predictive relationship, we would be very likely to make one or more errors in deciding which terms to retain in the model and which to exclude.

For example, suppose you fit a first-order model in 10 quantitative x variables and decide to conduct *t*-tests on all 10 of the individual β's in the model, each at $\alpha = .05$. Even if all the β parameters (except β_0) are equal to 0, approximately 40% of the time you will incorrectly reject the null hypothesis at least once and conclude that some β parameter differs from 0.* Thus, in multiple regression models for which a large number of independent variables are being considered, conducting a series of *t*-tests may include a large number of insignificant variables and exclude some useful ones. To test the utility of a multiple regression model, we need a *global test* (one that encompasses all the β parameters). We would also like to find some statistical quantity that measures how well the model fits the data.

We commence with the easier problem—finding a measure of how well a linear model fits a set of data. For this we use the multiple regression equivalent of r^2, the coefficient of determination for the straight-line model (Chapter 10), as shown in the box.

The **multiple coefficient of determination, R^2,** is defined as

$$R^2 = 1 - \frac{\text{SSE}}{\text{SS}_{yy}} = \frac{\text{SS}_{yy} - \text{SSE}}{\text{SS}_{yy}} = \frac{\text{Explained variablilty}}{\text{Total variability}}$$

Just like r^2 in the simple linear model, R^2 represents the fraction of the sample variation of the y values (measured by SS_{yy}) that is explained by the least squares prediction equation. Thus, $R^2 = 0$ implies a complete lack of fit of the model to the data, and $R^2 = 1$ implies a perfect fit with the model passing through every data point. In general, the larger the value of R^2, the better the model fits the data.

To illustrate, the value $R^2 = .892$ for the auction price model of Examples 11.1–11.3 is highlighted on the SPSS printout for the analysis shown in Figure 11.6. This high value of R^2 implies that using the independent variables age and number of bidders in a first-order

*The proof of this result (assuming independence of tests) proceeds as follows:

$P(\text{Reject } H_0 \text{ at least once} | \beta_1 = \beta_2 = \cdots = \beta_{10} = 0)$
$= 1 - P(\text{Reject } H_0 \text{ no times} | \beta_1 = \beta_2 = \cdots = \beta_{10} = 0)$
$\leq 1 - [P(\text{Accept } H_0\text{: } \beta_1 = 0 | \beta_1 = 0) \cdot P(\text{Accept } H_0\text{: } \beta_2 = 0 | \beta_2 = 0) \cdot \cdots \cdot P(\text{Accept } H_0\text{: } \beta_{10} = 0 | \beta_{10} = 0)]$
$= 1 - [(1 - \alpha)^{10}] = 1 - (.05)^{10} = .401$

For dependent tests, the Bonferroni inequality states that

$P(\text{Reject } H_0 \text{ at least once} | \beta_1 = \beta_2 = \cdots = \beta_{10} = 0) \leq 10(\alpha) = 10(.05) = .50.$

Model Summary

Model	R	R Square	Adjusted R Square	Std. Error of the Estimate
1	.945[a]	.892	.885	133.485

a. Predictors: (Constant), NUMBIDS, AGE

ANOVA[b]

Model		Sum of Squares	df	Mean Square	F	Sig.
1	Regression	4283062.960	2	2141531.480	120.188	.000[a]
	Residual	516726.540	29	17818.157		
	Total	4799789.500	31			

a. Predictors: (Constant), NUMBIDS, AGE

b. Dependent Variable: PRICE

Coefficients[a]

Model		Unstandardized Coefficients		Standardized Coefficients	t	Sig.	95% Confidence Interval for B	
		B	Std. Error	Beta			Lower Bound	Upper Bound
1	(Constant)	-1338.951	173.809		-7.704	.000	-1694.432	-983.471
	AGE	12.741	.905	.887	14.082	.000	10.890	14.591
	NUMBIDS	85.953	8.729	.620	9.847	.000	68.101	103.805

a. Dependent Variable: PRICE

Figure 11.6
SPSS analysis of the auction price model

model explains 89.2% of the total *sample variation* (measured by SS_{yy}) in auction price (y). Thus, R^2 is a sample statistic that tells how well the model fits the data and thereby represents a measure of the usefulness of the entire model.

A large value of R^2 computed from the *sample* data does not necessarily mean that the model provides a good fit to all of the data points in the *population*. For example, a first-order linear model that contains three parameters will provide a perfect fit to a sample of three data points, and R^2 will equal 1. Likewise, you will always obtain a perfect fit ($R^2 = 1$) to a set of n data points if the model contains exactly n parameters. Consequently, if you want to use the value of R^2 as a measure of how useful the model will be for predicting y, it should be based on a sample that contains substantially more data points than the number of parameters in the model.

CAUTION In a multiple regression analysis, use the value of R^2 as a measure of how useful a linear model will be for predicting y only if the sample contains substantially more data points than the number of β parameters in the model. ▲

As an alternative to using R^2 as a measure of model adequacy, the *adjusted multiple coefficient of determination,* denoted R_a^2, is often reported. The formula for R_a^2 is shown in the box.

The **adjusted multiple coefficient of determination** is given by

$$R_a^2 = 1 - \left[\frac{(n-1)}{n-(k+1)}\right]\left(\frac{\text{SSE}}{\text{SS}_{yy}}\right)$$

$$= 1 - \left[\frac{(n-1)}{n-(k+1)}\right](1 - R^2)$$

Note: $R_a^2 \leq R^2$

R^2 and R_a^2 have similar interpretations. However, unlike R^2, R_a^2 takes into account ("adjusts" for) both the sample size n and the number of β parameters in the model. R_a^2 will always be smaller than R^2 and, more importantly, cannot be "forced" to 1 by simply adding more and more independent variables to the model. Consequently, analysts prefer the more conservative R_a^2 when choosing a measure of model adequacy.

Despite their utility, R^2 and R_a^2 are only sample statistics. Therefore, it is dangerous to judge the global usefulness of the model based solely on these values. A better method is to conduct a test of hypothesis involving *all* the β parameters (except β_0) in a model. In particular, for the general multiple regression model $E(y) = \beta_0 + \beta_1 x_1 + \beta_2 x_2 + \cdots + \beta_k x_k$, we would test

$$H_0: \beta_1 = \beta_2 = \cdots = \beta_k = 0$$

H_a: At least one of the coefficients is nonzero.

The test statistic used to test this hypothesis is an F-statistic, and several equivalent versions of the formula can be used (although we will usually rely on the computer to calculate the F-statistic):

$$\textit{Test statistic: } F = \frac{\text{SS}_{yy} - \text{SSE}/k}{\text{SSE}/[n - (k + 1)]} = \frac{\text{Mean Square (Model)}}{\text{Mean Square (Error)}} = \frac{R^2/k}{(1 - R^2)/[n - (k + 1)]}$$

These formulas indicate that the F-statistic is the ratio of the *explained* variability divided by the model degrees of freedom to the *unexplained* variability divided by the error degrees of freedom. Thus, the larger the proportion of the total variability accounted for by the model, the larger the F-statistic.

To determine when the ratio becomes large enough that we can confidently reject the null hypothesis and conclude that the model is more useful than no model at all for predicting y, we compare the calculated F-statistic to a tabulated F value with k df in the numerator and $[n - (k + 1)]$ df in the denominator. Recall that tabulations of the F-distribution for various values of α are given in Tables VII, VIII, IX, and X in Appendix B.

Rejection region: $F > F_\alpha$, where F is based on k numerator and $n - (k + 1)$ denominator degrees of freedom.

The analysis of variance F-test for testing the usefulness of the model is summarized in the next box.

Testing Global Usefulness of the Model: The Analysis of Variance *F*-Test

$H_0: \beta_1 = \beta_2 = \cdots = \beta_k = 0$ (All model terms are unimportant for predicting y.)

H_a: At least one $\beta_i \neq 0$ (At least one model term is useful for predicting y.)

$$\textit{Test statistic: } F = \frac{(\text{SS}_{yy} - \text{SSE})/k}{\text{SSE}/[n - (k + 1)]} = \frac{R^2/k}{(1 - R^2)/[n - (k + 1)]} = \frac{\text{Mean Square (Model)}}{\text{Mean Square (Error)}}$$

where n is the sample size and k is the number of terms in the model.

Rejection region: $F > F_\alpha$, with k numerator degrees of freedom and $[n - (k + 1)]$ denominator degrees of freedom.

Conditions Required for the Global *F*-Test in Regression to Be Valid

Refer to the standard regression assumptions about the random error component (Section 11.1).

CAUTION A rejection of the null hypothesis $H_0: \beta_1 = \beta_2 = \cdots = \beta_k$ in the *global F-test* leads to the conclusion [with $100(1 - \alpha)\%$ confidence] that the model is statistically useful. However, statistically "useful" does not necessarily mean "best." Another model may prove even more useful in terms of providing more reliable estimates and predictions. This global F-test is usually regarded as a test that the model *must* pass to merit further consideration. ▲

Example 11.4
Assessing Overall Adequacy—Clock Auction Price Model

Problem Refer to Examples 11.1–11.3, in which an antique collector modeled the auction price (y) of grandfather clocks as a function of the age of the clock (x_1) and the number of bidders (x_2). Recall that the hypothesized first-order model is

$$y = \beta_0 + \beta_1 x_1 + \beta_2 x_2 + \varepsilon$$

The SPSS printout for the analysis is shown in Figure 11.6 (p. 635).

a. Find and interpret the adjusted coefficient of determination R_a^2.

b. Conduct the global F-test of model usefulness at the $\alpha = .05$ level of significance.

Solution

a. The R_a^2 value (highlighted in Figure 11.6) is .885. This implies that the least squares model has explained about 88.5% of the total sample variation in y values (auction prices), after adjusting for sample size and number of independent variables in the model.

b. The elements of the global test of the model follow:

$H_0: \beta_1 = \beta_2 = 0$ (*Note:* $k = 2$)

H_a: At least one of the two model coefficients, β_1 and β_2, is nonzero.

Test statistic: $F = \dfrac{\text{MS (Model)}}{\text{MSE}} = \dfrac{2{,}141{,}531}{17{,}818} = 120.19$ (see Figure 11.6)

p-value ≈ 0

Conclusion: Because $\alpha = .05$ exceeds the observed significance level, ($p \approx 0$), the data provide strong evidence that at least one of the model coefficients is nonzero. The overall model appears to be statistically useful for predicting auction prices.

Look Back Can we be sure that the best prediction model has been found if the global F-test indicates that a model is useful? Unfortunately, we cannot. The addition of other independent variables may improve the usefulness of the model. (See the box on page 636.) We consider more complex multiple regression models in Sections 11.5–11.8.

Now Work Exercise 11.15d–h

In this section, we discussed several different statistics for assessing the utility of a multiple regression model: t-tests on individual β parameters, R^2, R_a^2, and the global F-test. Both R^2 and R_a^2 are indicators of how well the prediction equation fits the data. Intuitive evaluations of the contribution of the model based on R^2 must be examined with care. Unlike R_a^2, the value of R^2 increases as more and more variables are added to the model. Consequently, you could force R^2 to take a value very close to 1 even though the model contributes no information for the prediction of y. In fact, R^2 equals 1 when the number of terms in the model (including β_0) equals the number of data points. Therefore, you should not rely solely on the value of R^2 (or even R_a^2) to tell you whether the model is useful for predicting y.

Conducting t-tests on all of the individual β parameters is also not the best method of testing the global utility of the model because these multiple tests result in a high probability of making at least one Type I error. Use the F-test for testing the global utility of the model.

After we have determined that the overall model is useful for predicting y using the F-test, we may elect to conduct one or more t-tests on the individual β parameters. However, the test (or tests) to be conducted should be decided a priori—that is, prior to fitting the model. Also, we should limit the number of t-tests conducted to avoid the potential problem of making too many Type I errors. Generally, the regression analyst will conduct t-tests only on the "most important" β's. We provide insight in identifying the most important β's in a linear model in Sections 11.5–11.8.

Recommendation for Checking the Utility of a Multiple Regression Model

1. First, conduct a test of overall model adequacy using the F-test—that is, test

$$H_0: \beta_1 = \beta_2 = \cdots = \beta_k = 0$$

If the model is deemed adequate (that is, if you reject H_0), then proceed to step 2. Otherwise, you should hypothesize and fit another model. The new model may include more independent variables or higher-order terms.

2. Conduct t-tests on those β parameters in which you are particularly interested (that is, the "most important" β's). As we will see in Sections 11.5 and 11.6, these usually involve only the β's associated with higher-order terms (x^2, x_1x_2, etc.). However, it is a safe practice to limit the number of β's that are tested. Conducting a series of t-tests leads to a high overall Type I error rate α.

Exercises 11.1–11.24

Learning the Mechanics

11.1 Write a first-order model relating $E(y)$ to
- **a.** two quantitative independent variables.
- **b.** four quantitative independent variables.
- **c.** five quantitative independent variables.

11.2 Minitab was used to fit the model $E(y) = \beta_0 + \beta_1x_1 + \beta_2x_2$ to $n = 20$ data points, and the printout shown below was obtained.
- **a.** What are the sample estimates of β_0, β_1, and β_2?
- **b.** What is the least squares prediction equation?
- **c.** Find SSE, MSE, and s. Interpret the standard deviation in the context of the problem.
- **d.** Test H_0: $\beta_1 = 0$ against H_0: $\beta_1 \neq 0$. Use $\alpha = .05$.
- **e.** Use a 95% confidence interval to estimate β_2.
- **f.** Find R^2 and R_a^2 and interpret these values.
- **g.** Find the test statistic for testing H_0: $\beta_1 = \beta_2 = 0$.
- **h.** Find the observed significance level of the test, part **g.** Interpret the result.

```
The regression equation is
Y = 506.35 - 941.9 X1 - 429.1 X2

Predictor      Coef   SE Coef      T      P
Constant    506.346     45.17  11.21  0.000
X1         -941.900    275.08  -3.42  0.003
X2         -429.060    379.83  -1.13  0.274

S = 94.251    R-Sq = 45.9%    R-Sq(adj) = 39.6%

Analysis of Variance

Source           DF      SS      MS      F      P
Regression        2  128329   64165   7.22  0.005
Residual Error   17  151016    8883
Total            19  279345
```

11.3 Suppose you fit the multiple regression model

$$y = \beta_0 + \beta_1x_1 + \beta_2x_2 + \beta_3x_3 + \varepsilon$$

to $n = 30$ data points and obtain the following result:

$$\hat{y} = 3.4 - 4.6x_1 + 2.7x_2 + .93x_3$$

The estimated standard errors of $\hat{\beta}_2$ and $\hat{\beta}_3$ are 1.86 and .29, respectively.
- **a.** Test the null hypothesis H_0: $\beta_2 = 0$ against the alternative hypothesis H_a: $\beta_2 \neq 0$. Use $\alpha = .05$.
- **b.** Test the null hypothesis H_0: $\beta_3 = 0$ against the alternative hypothesis H_a: $\beta_3 \neq 0$. Use $\alpha = .05$.
- **c.** The null hypothesis H_0: $\beta_2 = 0$ is not rejected. In contrast, the null hypothesis H_0: $\beta_3 = 0$ is rejected. Explain how this can happen even though $\hat{\beta}_2 > \hat{\beta}_3$.

11.4 Suppose you fit the first-order multiple regression model

$$y = \beta_0 + \beta_1x_1 + \beta_2x_2 + \varepsilon$$

to $n = 25$ data points and obtain the prediction equation

$$\hat{y} = 6.4 + 3.1x_1 + .92x_2$$

The estimated standard deviations of the sampling distributions of $\hat{\beta}_1$ and $\hat{\beta}_2$ are 2.3 and .27, respectively.
- **a.** Test H_0: $\beta_1 = 0$ against H_a: $\beta_1 > 0$. Use $\alpha = .05$.
- **b.** Test H_0: $\beta_2 = 0$ against H_a: $\beta_2 \neq 0$. Use $\alpha = .05$.
- **c.** Find a 90% confidence interval for β_1. Interpret the interval.
- **d.** Find a 99% confidence interval for β_2. Interpret the interval.

11.5 How is the number of degrees of freedom available for estimating σ^2 (the variance of ε) related to the number of independent variables in a regression model?

11.6 Consider the first-order model equation in three quantitative independent variables

$$E(y) = 1 + 2x_1 + x_2 - 3x_3$$

- **a.** Graph the relationship between y and x_1 for $x_2 = 1$ and $x_3 = 3$.
- **b.** Repeat part **a** for $x_2 = -1$ and $x_3 = 1$.
- **c.** How do the graphed lines in parts **a** and **b** relate to each other? What is the slope of each line?

d. If a linear model is first-order in three independent variables, what type of geometric relationship will you obtain when $E(y)$ is graphed as a function of one of the independent variables for various combinations of values of the other independent variables?

11.7 Suppose you fit the first-order model

$$y = \beta_0 + \beta_1 x_1 + \beta_2 x_2 + \beta_3 x_3 + \beta_4 x_4 + \beta_5 x_5 + \varepsilon$$

to $n = 30$ data points and obtain

$$SSE = .33 \quad R^2 = .92$$

a. Do the values of SSE and R^2 suggest that the model provides a good fit to the data? Explain.

b. Is the model of any use in predicting y? Test the null hypothesis H_0: $\beta_1 = \beta_2 = \beta_3 = \beta_4 = \beta_5 = 0$ against the alternative hypothesis H_a: At least one of the parameters $\beta_1, \beta_2, \ldots, \beta_5$ is nonzero. Use $\alpha = .05$.

11.8 If the analysis of variance F-test leads to the conclusion that at least one of the model parameters is nonzero, can you conclude that the model is the best predictor for the dependent variable y? Can you conclude that all of the terms in the model are important for predicting y? What is the appropriate conclusion?

Applying the Concepts—Basic

11.9 Accounting and Machiavellianism. Refer to the *Behavioral Research in Accounting* (Jan. 2008) study of Machiavellian traits (e.g., manipulation, cunning, duplicity, deception, and bad faith) in accountants, Exercise 8.8 (p. 454). Recall that a Mach rating score was determined for each in a sample of accounting alumni of a large southwestern university. For one portion of the study, the researcher modeled an accountant's Mach score (y) as a function of age, gender, education, and income. Data on $n = 198$ accountants yielded the results shown in the table.

Independent Variable	t value for H_0: $\beta_i = 0$	p-value
Age (x_1)	0.10	$> .10$
Gender (x_2)	−0.55	$> .10$
Education (x_3)	1.95	$< .01$
Income (x_4)	0.52	$> .10$

Overall model: $R^2 = 13$, $F = 4.74$ (p-value $< .01$)

a. Conduct a test of overall model utility. Use $\alpha = .05$.

b. Interpret the coefficient of determination, R^2.

c. Is there sufficient evidence (at $\alpha = .05$) to say that income is a statistically useful predictor of Mach score?

11.10 Characteristics of lead users. During new product development, companies often involve "lead users," i.e., creative individuals who are on the leading edge of an important market trend. *Creativity and Innovation Management* (Feb. 2008) published an article on identifying the social network characteristics of lead users of children's computer games. Data were collected for $n = 326$ children, and the following variables were measured: lead-user rating (y, measured on a 5-point scale), gender ($x_1 = 1$ if female, 0 if male), age (x_2, years), degree of centrality (x_3, measured as the number of direct ties to other peers in the network), and betweenness centrality (x_4, measured as the number of shortest paths between peers). A first-order model for y was fit to the data, yielding the following least squares prediction equation:

$$\hat{y} = 3.58 + .01x_1 - .06x_2 - .01x_3 + .42x_4$$

a. Give two properties of the errors of prediction that result from using the method of least squares to obtain the parameter estimates.

b. Give a practical interpretation of the estimate of β_4 in the model.

c. A test of H_0: $\beta_4 = 0$ resulted in a p-value of .002. Make the appropriate conclusion at $\alpha = .05$.

11.11 Highway crash data analysis. Researchers at Montana State University have written a tutorial on an empirical method for analyzing before and after highway crash data (Montana Department of Transportation, Research Report, May 2004). The initial step in the methodology is to develop a Safety Performance Function (SPF)—a mathematical model that estimates crash occurrence for a given roadway segment. Using data collected for over 100 roadway segments, the researchers fit the model, $E(y) = \beta_0 + \beta_1 x_1 + \beta_2 x_2$, where y = number of crashes per 3 years, x_1 = roadway length (miles), and x_2 = AADT = average annual daily traffic (number of vehicles). The results are shown in the following tables.

Interstate Highways

Variable	Parameter Estimate	Standard Error	t-value
Intercept	1.81231	.50568	3.58
Length (x_1)	.10875	.03166	3.44
AADT (x_2)	.00017	.00003	5.19

Noninterstate Highways

Variable	Parameter Estimate	Standard Error	t-value
Intercept	1.20785	.28075	4.30
Length (x_1)	.06343	.01809	3.51
AADT (x_2)	.00056	.00012	4.86

a. Give the least squares prediction equation for the interstate highway model.

b. Give practical interpretations of the β estimates, part **a.**

c. Refer to part **a.** Find a 99% confidence interval for β_1 and interpret the result.

d. Refer to part **a.** Find a 99% confidence interval for β_2 and interpret the result.

e. Repeat parts **a–d** for the noninterstate highway model.

11.12 Trust in e-retailers. Electronic commerce (or "e-commerce") describes the use of electronic networks to simplify a business operation. With e-commerce, retailers now can advertise and sell their products easily over the Web. In *Internet Research: Electronic Networking Applications and Policy* (Vol. 11, 2001), Canadian researchers investigated the factors that impact the level of trust in Web e-retailers. Five quantitative independent variables were used to model level of trust (y):

x_1 = ease of navigation on the Web site

x_2 = consistency of the Web site

x_3 = ease of learning the Web interface

x_4 = perception of the interface design
x_5 = level of support available to the user

a. Write a first-order model for level of trust as a function of the five independent variables.
b. The model, part **a,** was fit to data collected for $n = 66$ visitors to e-retailers' Web sites and yielded a coefficient of determination of $R^2 = .58$. Interpret this result.
c. Compute the F-statistic used to test the global utility of the model.
d. Using $\alpha = .10$, give the appropriate conclusion for the test, part **c.**

11.13 Urban population estimation using satellite images. Can the population of an urban area be estimated without taking a census? In *Geographical Analysis* (Berry, K. A., et al. "Interpreting what is rural and urban for western U.S. counties," *Professional Geographer*, Vol. 52, No. 1, Feb. 2000, pp. 56–59 (Table 2). Reprinted with permission from *Chance*. © 2000 by the American Statistical Association. All rights reserved.) geography professors at the University of Wisconsin–Milwaukee and Ohio State University demonstrated the use of satellite image maps for estimating urban population. A portion of Columbus, Ohio, was partitioned into $n = 125$ census block groups, and satellite imagery was obtained. For each census block, the following variables were measured: population density (y), proportion of block with low-density residential areas (x_1), and proportion of block with high-density residential areas (x_2). A first-order model for y was fit to the data with the following results:

$$\hat{y} = -.0304 + 2.006x_1 + 5.006x_2,\ R^2 = .686$$

a. Give a practical interpretation of each β estimate in the model.
b. Give a practical interpretation of the coefficient of determination, R^2.
c. State H_0 and H_a for a test of overall model adequacy.
d. Refer to part **c.** Compute the value of the test statistic.
e. Refer to parts **c** and **d.** Make the appropriate conclusion at $\alpha = .01$.

11.14 Predicting runs scored in baseball. In *Chance* (Fall 2000), statistician Scott Berry built a multiple regression model for predicting total number of runs scored by a Major League Baseball team during a season. Using data on all teams over a 9-year period (a sample of $n = 234$), the results in the next table were obtained.
NW

Independent Variable	β Estimate	Standard Error
Intercept	3.70	15.00
Walks (x_1)	.34	.02
Singles (x_2)	.49	.03
Doubles (x_3)	.72	.05
Triples (x_4)	1.14	.19
Home runs (x_5)	1.51	.05
Stolen bases (x_6)	.26	.05
Caught stealing (x_7)	−.14	.14
Strikeouts (x_8)	−.10	.01
Outs (x_9)	−.10	.01

Source: Berry, S. M. "A statistician reads the sports pages: Modeling offensive ability in baseball," *Chance*, Vol. 13, No. 4, Fall 2000, pp. 56–59 (Table 2). Reprinted with permission from *Chance*. © 2000 by the American Statistical Association. All rights reserved.

a. Write the least squares prediction equation for y = total number of runs scored by a team in a season.
b. Interpret, practically, the β estimates in the model.
c. Conduct a test of H_0: $\beta_7 = 0$ against H_a: $\beta_7 < 0$ at $\alpha = .05$. Interpret the results.
d. Form a 95% confidence interval for β_5. Interpret the results.

Applying the Concepts—Intermediate

11.15 Novelty of a vacation destination. Many tourists choose a vacation destination based on the newness or uniqueness (i.e., the novelty) of the itinerary. Texas A&M University Professor J. Petrick investigated the relationship between novelty and vacationing golfers' demographics (*Annals of Tourism Research*, Vol. 29, 2002). Data were obtained from a mail survey of 393 golf vacationers to a large coastal resort in the southeastern United States. Several measures of novelty level (on a numerical scale) were obtained for each vacationer, including "change from routine," "thrill," "boredom-alleviation," and "surprise." The researcher employed four independent variables in a regression model to predict each of the novelty measures. The independent variables were x_1 = number of rounds of golf per year, x_2 = total number of golf vacations taken, x_3 = number of years played golf, and x_4 = average golf score.
NW

a. Give the hypothesized equation of a first-order model for y = change from routine.
b. A test of H_0: $\beta_3 = 0$ versus H_a: $\beta_3 < 0$ yielded a p-value of .005. Interpret this result if $\alpha = .01$.
c. The estimate of β_3 was found to be negative. Based on this result (and the result of part **b**), the researcher concluded that "those who have played golf for more years are less apt to seek change from their normal routine in their golf vacations." Do you agree with this statement? Explain.
d. The regression results for three dependent novelty measures, based on data collected for $n = 393$ golf vacationers, are summarized in the table below. Give the null hypothesis for testing the overall adequacy of the first-order regression model.
e. Give the rejection region for the test, part **d,** for $\alpha = .01$.
f. Use the test statistics reported in the table and the rejection region from part **e** to conduct the test for each of the dependent measures of novelty.
g. Verify that the p-values reported in the table support your conclusions in part **f.**
h. Interpret the values of R^2 reported in the table.

Dependent Variable	F-value	p-value	R^2
Thrill	5.56	<.001	.055
Change from routine	3.02	.018	.030
Surprise	3.33	.011	.023

Source: Petrick, J. F. "An examination of golf vacationers' novelty," *Annals of Tourism Research*, Vol. 29, No. 2, 2002, pp. 384–400. © 2002 with permission from Elsevier.

11.16 Arsenic in groundwater. *Environmental Science & Technology* (Jan. 2005) reported on a study of the reliability of a commercial kit to test for arsenic in groundwater. The field kit was used to test a sample of 328 groundwater wells in Bangladesh. In addition to the arsenic level (micrograms per liter), the latitude (degrees), longitude (degrees), and depth (feet) of each well was measured. The data are saved in the **ASWELLS** file. (The first and last 5 observations are listed in the table on the next page.)

Table for Exercise 11.16 (selected observations)

Well ID	Latitude	Longitude	Depth	Arsenic
10	23.7887	90.6522	60	331
14	23.7886	90.6523	45	302
30	23.7880	90.6517	45	193
59	23.7893	90.6525	125	232
85	23.7920	90.6140	150	19
⋮	⋮	⋮	⋮	⋮
7353	23.7949	90.6515	40	48
7357	23.7955	90.6515	30	172
7890	23.7658	90.6312	60	175
7893	23.7656	90.6315	45	624
7970	23.7644	90.6303	30	254

a. Write a first-order model for arsenic level (y) as a function of latitude, longitude, and depth.
b. Fit the model to the data using the method of least squares.
c. Give practical interpretations of the β estimates.
d. Find the model standard deviation, s, and interpret its value.
e. Find and interpret the values of R^2 and R^2_a
f. Conduct a test of overall model utility at $\alpha = .05$.
g. Based on the results, parts **d–f,** would you recommend using the model to predict arsenic level (y)? Explain.

11.17 Incomes of Mexican street vendors. Detailed interviews were conducted with over 1,000 street vendors in the city of Puebla, Mexico, in order to study the factors influencing vendors' incomes (*World Development,* Feb. 1998). Vendors were defined as individuals working in the street and included vendors with carts and stands on wheels and excluded beggars, drug dealers, and prostitutes. The researchers collected data on gender, age, hours worked per day, annual earnings, and education level. A subset of these data appear in the table and are saved in the **STREETVN** file.

Vendor Number	Annual Earnings, y	Age, x_1	Hours Worked per Day, x_2
21	$2,841	29	12
53	1,876	21	8
60	2,934	62	10
184	1,552	18	10
263	3,065	40	11
281	3,670	50	11
354	2,005	65	5
401	3,215	44	8
515	1,930	17	8
633	2,010	70	6
677	3,111	20	9
710	2,882	29	9
800	1,683	15	5
914	1,817	14	7
997	4,066	33	12

Source: Reprinted from Smith, P. A., and Metzger, M. R. "The return to education: Street vendors in Mexico," *World Development,* Vol. 26, No. 2, Feb. 1998, pp. 289–296. © 1998 with permission from Elsevier.

a. Write a first-order model for mean annual earnings, $E(y)$, as a function of age (x_1) and hours worked (x_2).
b. Fit the model to the data and give the least squares prediction equation.
c. Interpret the estimated β coefficients in your model.
d. Is age (x_1) a statistically useful predictor of annual earnings? Test using $\alpha = .01$.
e. Construct a 95% confidence interval for β_2. Interpret the interval in the words of the problem.
f. Find and interpret the value of R^2.
g. Find and interpret the value of R^2_a. Explain the relationship between R^2 and R^2_a.
h. Conduct a test of the global utility of the model at $\alpha = .01$.

11.18 Contamination from a plant's discharge. Refer to the U.S. Army Corps of Engineers data (Ex. 1.5, p. 13) on fish contaminated from the toxic discharges of a chemical plant located on the banks of the Tennessee River in Alabama. Recall that the engineers measured the length (in centimeter), weight (in grams), and DDT level (in parts per million) for 144 captured fish. In addition, the number of miles upstream from the river was recorded. The data are saved in the **DDT** file. (The first and last five observations are shown in the accompanying table.)

River	Mile	Species	Length	Weight	DDT
FC	5	CHANNELCATFISH	42.5	732	10.00
FC	5	CHANNELCATFISH	44.0	795	16.00
FC	5	CHANNELCATFISH	41.5	547	23.00
FC	5	CHANNELCATFISH	39.0	465	21.00
FC	5	CHANNELCATFISH	50.5	1,252	50.00
⋮	⋮	⋮	⋮	⋮	⋮
TR	345	LARGEMOUTHBASS	23.5	358	2.00
TR	345	LARGEMOUTHBASS	30.0	856	2.20
TR	345	LARGEMOUTHBASS	29.0	793	7.40
TR	345	LARGEMOUTHBASS	17.5	173	0.35
TR	345	LARGEMOUTHBASS	36.0	1,433	1.90

a. Fit the first-order model, $E(y) = \beta_0 + \beta_1 x_1 + \beta_2 x_2 + \beta_3 x_3$, to the data, where y = DDT level, x_1 = mile, x_2 = length, and x_3 = weight. Report the least squares prediction equation.
b. Find the estimate of the standard deviation of ε for the model and give a practical interpretation of its value.
c. Conduct a test of the global utility of the model. Use $\alpha = .05$.
d. Do the data provide sufficient evidence to conclude that DDT level increases as length increases? Report the observed significance level of the test and reach a conclusion using $\alpha = .05$.
e. Find and interpret a 95% confidence interval for β_3.

11.19 Cooling method for gas turbines. Refer to the *Journal of Engineering for Gas Turbines and Power* (Jan. 2005) study of a high-pressure inlet fogging method for a gas turbine engine, Exercise 6.28 (p. 335). Recall that the heat rate (kilojoules per kilowatt per hour) was measured for each in a sample of 67 gas turbines augmented with high-pressure inlet fogging. In addition, several other variables were measured, including cycle speed (revolutions per minute), inlet temperature (°C), exhaust gas temperature (°C), cycle pressure ratio, and air mass flow rate (kilograms per second). The data are saved in the **GASTURBINE** file. (The first and last 5 observations are listed in the table on the next page.)

Table for Exercise 11.19 (selected observations)

RPM	CP Ratio	Inlet Temp.	Exhaust Temp.	Airflow	Heat Rate
27,245	9.2	1,134	602	7	14,622
14,000	12.2	950	446	15	13,196
17,384	14.8	1,149	537	20	11,948
11,085	11.8	1,024	478	27	11,289
14,045	13.2	1,149	553	29	11,964
⋮	⋮	⋮	⋮	⋮	⋮
18,910	14.0	1,066	532	8	12,766
3,600	35.0	1,288	448	152	8,714
3,600	20.0	1,160	456	84	9,469
16,000	10.6	1,232	560	14	11,948
14,600	13.4	1,077	536	20	12,414

Source: Bhargava, R., and Meher-Homji, C. B. "Parametric analysis of existing gas turbines with inlet evaporative and overspray fogging," *Journal of Engineering for Gas Turbines and Power,* Vol. 127, No. 1, Jan. 2005, pp. 145–158. Reprinted with permission from ASME.

a. Write a first-order model for heat rate (y) as a function of speed, inlet temperature, exhaust temperature, cycle pressure ratio, and airflow rate.
b. Fit the model to the data using the method of least squares.
c. Give practical interpretations of the β estimates.
d. Find the model standard deviation, s, and interpret its value.
e. Conduct a test for overall model utility using $\alpha = .01$.
f. Find and interpret R_a^2.
g. Is there sufficient evidence (at $\alpha = .01$) to indicate that heat rate (y) is linearly related to inlet temperature?

11.20 Extracting water from oil. In the oil industry, water that mixes with crude oil during production and transportation must be removed. Chemists have found that the oil can be extracted from the water/oil mix electrically. Researchers at the University of Bergen (Norway) conducted a series of experiments to study the factors that influence the voltage (y) required to separate the water from the oil (*Journal of Colloid and Interface Science,* Aug. 1995). The seven independent variables investigated in the study are listed in the table below. (Each variable was measured at two levels—a "low" level and a "high" level.) Sixteen water/oil mixtures were prepared using different combinations of the independent variables; then each emulsion was exposed to a high electric field. In addition, three mixtures were tested when all independent variables were set to 0. The data for all 19 experiments are saved in the **WATEROIL** file (selected observations are shown in the table below).

a. Propose a first-order model for y as a function of all seven independent variables.
b. Use a statistical software package to fit the model to the data in the table.
c. Fully interpret the β estimates.
d. Evaluate the overall utility of the model at $\alpha = .10$.

11.21 Occupational safety study. An important goal in occupational safety is "active caring." Employees demonstrate active caring (AC) about the safety of their coworkers when they identify environmental hazards and unsafe work practices and then implement appropriate corrective actions for these unsafe conditions or behaviors. Three factors hypothesized to increase the propensity for an employee to actively care for safety are (1) high self-esteem, (2) optimism, and (3) group cohesiveness. *Applied & Preventive Psychology* (Winter 1995) attempted to establish empirical support for the AC hypothesis by fitting the model $E(y) = \beta_0 + \beta_1 x_1 + \beta_2 x_2 + \beta_3 x_3$, where

y = AC score (active caring score, 15-point scale)
x_1 = Self-esteem score
x_2 = Optimism score
x_3 = Group cohesion score

The regression analysis, based on data collected for $n = 31$ hourly workers at a large fiber-manufacturing plant, yielded a multiple coefficient of determination of $R^2 = .362$.

Table for Exercise 11.20 (selected observations)

Experiment Number	Voltage, y (kw/cm)	Disperse Phase Volume, x_1 (%)	Salinity, x_2 (%)	Temperature, x_3 (°C)	Time Delay, x_4 (hours)	Surfactant Concentration, x_5 (%)	Span: Triton, x_6	Solid Particles, x_7 (%)
1	.64	40	1	4	.25	2	.25	.5
2	.80	80	1	4	.25	4	.25	2
3	3.20	40	4	4	.25	4	.75	.5
4	.48	80	4	4	.25	2	.75	2
5	1.72	40	1	23	.25	4	.75	2
⋮	⋮	⋮	⋮	⋮	⋮	⋮	⋮	⋮
16	.72	80	4	23	24	4	.75	2
17	1.08	0	0	0	0	0	0	0
18	1.08	0	0	0	0	0	0	0
19	1.04	0	0	0	0	0	0	0

Source: Førdedal, H., et al. "A multivariate analysis of W/O emulsions in high external electric fields as studied by means of dielectric time domain spectroscopy," *Journal of Colloid and Interface Science,* Vol. 173, No. 2, Aug. 1995, p. 398 (Table 2). © 1995 with permission from Elsevier.

a. Interpret the value of R^2.
b. Use the R^2 value to test the global utility of the model. Use $\alpha = .05$.

11.22 R^2 and model fit. Because the coefficient of determination R^2 always increases when a new independent variable is added to the model, it is tempting to include many variables in a model to force R^2 to be near 1. However, doing so reduces the degrees of freedom available for estimating σ^2, which adversely affects our ability to make reliable inferences. Suppose you want to use 18 economic indicators to predict next year's gross domestic product (GDP). You fit the model

$$y = \beta_0 + \beta_1 x_1 + \beta_2 x_2 + \cdots + \beta_{17} x_{17} + \beta_{18} x_{18} + \varepsilon$$

where y = GDP and $x_1, x_2, \ldots, x_{18}$ are the economic indicators. Only 20 years of data ($n = 20$) are used to fit the model, and you obtain $R^2 = .95$. Test to see whether this impressive-looking R^2 is large enough for you to infer that the model is useful—that is, that at least one term in the model is important for predicting GDP. Use $\alpha = .05$.

Applying the Concepts—Advanced

11.23 Bordeaux wine sold at auction. The vineyards in the Bordeaux region of France are known for producing excellent red wines. However, the uncertainty of the weather during the growing season, the phenomenon that wine tastes better with age, and the fact that some Bordeaux vineyards produce better wines than others, encourages speculation concerning the value of a case of wine produced by a certain vineyard during a certain year (or vintage). As a result, many wine experts attempt to predict the auction price of a case of Bordeaux wine. The publishers of a newsletter titled *Liquid Assets: The International Guide to Fine Wine* discussed a multiple regression approach to predicting the London auction price of red Bordeaux wine in *Chance* (Fall 1995). The natural logarithm of the price y (in dollars) of a case containing a dozen bottles of red wine was modeled as a function of weather during growing season and age of vintage using data collected for the vintages of 1952–1980. Three models were fit to the data. The results of the regressions are summarized in the table below.

a. For each model, conduct a t-test (at $\alpha = .05$) for each of the β parameters in the model. Interpret the results.
b. When the natural log of y is used as a dependent variable, the antilogarithm of a β coefficient minus 1—that is $e^{\beta_i} - 1$—represents the percentage change in y for every 1-unit increase in the associated x value. Use this information to interpret the β estimates of each model.
c. Based on the values of R^2 and s, which of the three models would you recommend for predicting Bordeaux wine prices? Explain.

11.24 Cost analysis for a shipping department. Multiple regression is used by accountants in cost analysis to shed light on the factors that cause costs to be incurred and the magnitudes of their effects. The independent variables of such a regression model are the factors believed to be related to cost, the dependent variable. In some instances, however, it is desirable to use physical units instead of cost as the dependent variable in a cost analysis. This would be the case if most of the cost associated with the activity of interest is a function of some physical unit, such as hours of labor. The advantage of this approach is that the regression model will provide estimates of the number of labor hours required under different circumstances, and these hours can then be costed at the current labor rate (Horngren, Foster, and Datar, *Cost Accounting,* 2006). The sample data shown in the table on the next page have been collected from a firm's accounting and production records to provide cost information about the firm's shipping department. These data are saved in the **SHIPDEPT** file. Consider the model

$$y = \beta_0 + \beta_1 x_1 + \beta_2 x_2 + \beta_3 x_3 + \varepsilon$$

a. Find the least squares prediction equation.
b. Use an F-test to investigate the usefulness of the model specified in part **a.** Use $\alpha = .01$ and state your conclusion in the context of the problem.
c. Test H_0: $\beta_2 = 0$ versus H_a: $\beta_2 \neq 0$ using $\alpha = .05$. What do the results of your test suggest about the magnitude of the effects of x_2 on labor costs?
d. Find R^2 and interpret its value in the context of the problem.
e. If shipping department employees are paid \$7.50 per hour, how much less, on average, will it cost the company per week if the average number of pounds per shipment increases from a level of 20 to 21? Assume that x_1 and x_2 remain unchanged. Your answer is an estimate of what is known in economics as the *expected marginal cost* associated with a 1-pound increase in x_3.
f. With what approximate precision can this model be used to predict the hours of labor? [*Note:* The precision of multiple regression predictions is discussed in Section 11.4.]
g. Can regression analysis alone indicate what factors *cause* costs to increase? Explain.

Table for Exercise 11.23

	Beta Estimates (Standard Errors)		
Independent Variables	Model 1	Model 2	Model 3
x_1 = Vintage year	.0354(.0137)	.0238 (.00717)	.0240(.00747)
x_2 = Average growing season temperature(°C)	(not included)	.616 (.0952)	.608 (.116)
x_3 = Sept./Aug. rainfall (cm)	(not included)	−.00386 (.00081)	−.00380 (.00095)
x_4 = Rainfall in months preceding vintage (cm)	(not included)	.0001173 (.000482)	.00115 (.000505)
x_5 = Average Sept. temperature (°C)	(not included)	(not included)	.00765 (.565)
	$R^2 = .212$	$R^2 = .828$	$R^2 = .828$
	$s = .575$	$s = .287$	$s = .293$

Source: Ashenfelter, O., Ashmore, D., and LaLonde, R. "Bordeaux wine vintage quality and weather," *Chance*, Vol. 8, No. 4, Fall 1995, p. 116 (Table 2). Reprinted with permission from *Chance*.

Table for Exercise 11.24

Week	Labor, y (hr)	Pounds Shipped, x_1 (1,000s)	Percentage of Units Shipped by Truck, x_2	Average Shipment Weight, x_3 (lb)
1	100	5.1	90	20
2	85	3.8	99	22
3	108	5.3	58	19
4	116	7.5	16	15
5	92	4.5	54	20
6	63	3.3	42	26
7	79	5.3	12	25
8	101	5.9	32	21
9	88	4.0	56	24
10	71	4.2	64	29
11	122	6.8	78	10
12	85	3.9	90	30
13	50	3.8	74	28
14	114	7.5	89	14
15	104	4.5	90	21
16	111	6.0	40	20
17	110	8.1	55	16
18	100	2.9	64	19
19	82	4.0	35	23
20	85	4.8	58	25

11.4 Using the Model for Estimation and Prediction

In Section 10.6, we discussed the use of the least squares line for estimating the mean value of y, $E(y)$, for some particular value of x, say $x = x_p$. We also showed how to use the same fitted model to predict, when $x = x_p$, some new value of y to be observed in the future. Recall that the least squares line yielded the same value for both the estimate of $E(y)$ and the prediction of some future value of y—that is, both are the result of substituting x_p into the prediction equation $\hat{y} = \hat{\beta}_0 + \hat{\beta}_1 x$ and calculating $\hat{y}_p$. There the equivalence ends. The confidence interval for the mean $E(y)$ is narrower than the prediction interval for y because of the additional uncertainty attributable to the random error ε when predicting some future value of y.

These same concepts carry over to the multiple regression model. Consider a first-order model relating sale price (y) of a residential property to land value (x_1), appraised improvements value (x_2), and home size (x_3). Suppose we want to estimate the mean sale price for a given property with $x_1 = \$15{,}000$, $x_2 = \$50{,}000$, and $x_3 = 1{,}800$ square feet. Assuming that the first-order model represents the true relationship between sale price and the three independent variables, we want to estimate

$$E(y) = \beta_0 + \beta_1 x_1 + \beta_2 x_2 + \beta_3 x_3 = \beta_0 + \beta_1(15{,}000) + \beta_2(50{,}000) + \beta_3(1{,}800)$$

After obtaining the least squares estimates $\hat{\beta}_0$, $\hat{\beta}_1$, $\hat{\beta}_2$ and $\hat{\beta}_3$, the estimate of $E(y)$ will be

$$\hat{y} = \hat{\beta}_0 + \hat{\beta}_1(15{,}000) + \hat{\beta}_2(50{,}000) + \hat{\beta}_3(1{,}800)$$

To form a confidence interval for the mean, we need to know the standard deviation of the sampling distribution for the estimator $\hat{y}$. For multiple regression models, the form of this standard deviation is rather complex. However, the regression routines of statistical computer software packages allow us to obtain the confidence intervals for mean values of y for any given combination of values of the independent variables. We illustrate with an example.

Example 11.5
Estimating $E(y)$ and Predicting y—Auction Price Model

Problem Refer to Examples 11.1–11.4 and the first-order model, $E(y) = \beta_0 + \beta_1 x_1 + \beta_2 x_2$, where y = auction price of a grandfather clock, x_1 = age of the clock, and x_2 = number of bidders.

a. Estimate the average auction price for all 150-year-old clocks sold at auctions with 10 bidders using a 95% confidence interval. Interpret the result.

b. Predict the auction price for a single 150-year old clock sold at an auction with 10 bidders using a 95% prediction interval. Interpret the result.

c. Suppose you want to predict the auction price for one clock that is 50 years old and has 2 bidders. How should you proceed?

Solution

a. Here, the key words *average* and *for all* imply we want to estimate the mean of y, $E(y)$. We want a 95% confidence interval for $E(y)$ when $x_1 = 150$ years and $x_2 = 10$ bidders. A Minitab printout for this analysis is shown in Figure 11.7. The confidence interval (highlighted under **"95% CI"**) is (1,381.4, 1,481.9). Thus, we are 95% confident that the mean auction price for all 150-year-old clocks sold at an auction with 10 bidders lies between $1,381.40 and $1,481.90.

b. The key words *predict* and *for a single* imply that we want a 95% prediction interval for y when $x_1 = 150$ years and $x_2 = 10$ bidders. This interval (highlighted under "95% PI" on the Minitab printout, Figure 11.7) is (1,154.1, 1,709.3). We say, with 95% confidence, that the auction price for a single 150-year-old clock sold at an auction with 10 bidders falls between $1,154.10 and $1,709.30.

c. Now, we want to predict the auction price, y, for a single (*one*) grandfather clock when $x_1 = 50$ years and $x_2 = 2$ bidders. Consequently, we desire a 95% prediction interval for y. However, before we form this prediction interval, we should check to make sure that the selected values of the independent variables, $x_1 = 50$ and $x_2 = 2$, are both reasonable and within their respective sample ranges. If you examine the sample data shown in Table 11.1 (p. 628), you will see that the range for age is $108 \leq x_1 \leq 194$, and the range for number of bidders is $5 \leq x_2 \leq 15$. Thus, both selected values fall well *outside* their respective ranges. Recall the *Caution* box in Section 10.6 (p. 603) warning about the dangers of using the model to predict y for a value of an independent variable that is not within the range of the sample data. Doing so may lead to an unreliable prediction.

Regression Analysis: PRICE versus AGE, NUMBIDS

```
The regression equation is
PRICE = - 1339 + 12.7 AGE + 86.0 NUMBIDS

Predictor      Coef  SE Coef      T      P
Constant    -1339.0    173.8  -7.70  0.000
AGE         12.7406   0.9047  14.08  0.000
NUMBIDS      85.953    8.729   9.85  0.000

S = 133.485    R-Sq = 89.2%    R-Sq(adj) = 88.5%

Analysis of Variance

Source          DF       SS       MS       F      P
Regression       2  4283063  2141531  120.19  0.000
Residual Error  29   516727    17818
Total           31  4799790

Predicted Values for New Observations

New
Obs     Fit  SE Fit         95% CI              95% PI
  1  1431.7    24.6  (1381.4, 1481.9)  (1154.1, 1709.3)

Values of Predictors for New Observations

New
Obs  AGE  NUMBIDS
  1  150     10.0
```

Figure 11.7
Minitab printout with 95% confidence intervals for grandfather clock model

Look Back If we want to make the prediction requested in part **c,** we would need to collect additional data on clocks with the requested characteristics (i.e., $x_1 = 50$ years and $x_2 = 2$ bidders) and then refit the model.

Now Work Exercise 11.27

Statistics IN Action Revisited Evaluating a First-Order Model

The Florida attorney general wants to develop a model for the cost (y) of a road construction contract awarded using the sealed-bid system and to use the model to predict the costs of future road contracts in the state. In addition to contract cost, the **FLAG** file contains data on eight potential predictor variables for a sample of 235 road contracts. (See Table SIA11.1 on p. 625.) Minitab scatterplots (with the dependent variable, COST, plotted against each of the potential predictors) for the data are shown in Figure SIA11.1. From the scatterplots, it appears that the DOT engineer's cost estimate (DOTEST) and estimate of work days (DAYSEST) would be good predictors of contract cost. [In a future Statistics in Action Revisited section (p. 693), we will learn that the two best predictors of contract cost are actually DOTEST and the fixed or competitive status (STATUS) of the contract.] However, in this section, we will fit the first-order regression model using all eight independent variables.

The Minitab printout for the regression analysis is shown in Figure SIA11.2. The global F-statistic ($F = 1{,}166.68$) and associated p-value (.000) shown on the printout indicate that the overall model is statistically useful for predicting construction cost. The value of R^2 indicates that the model can explain 97.6% of the sample variation in contract cost. Both of these results provide strong statistical support for using the model for estimation and prediction.

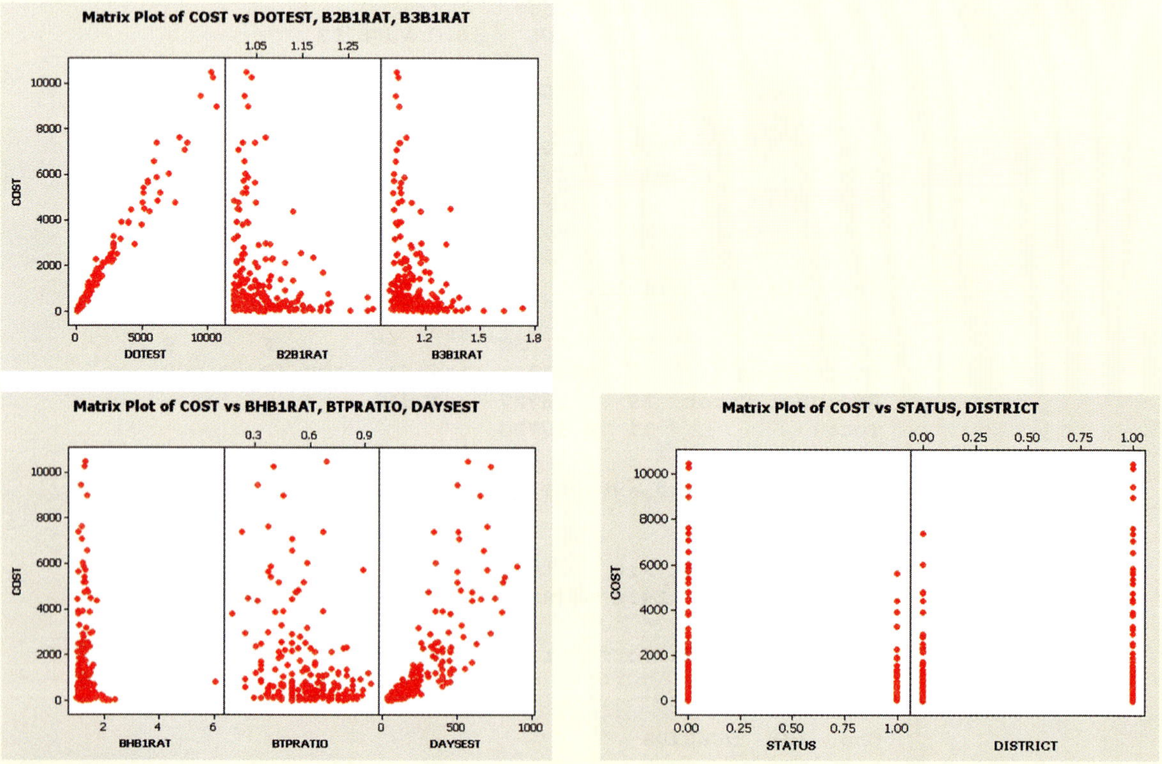

Figure SIA11.1
Minitab scatterplots for **FLAG** data

```
The regression equation is
COST = 124 + 0.906 DOTEST - 147 B2B1RAT - 84 B3B1RAT - 59.1 BHB1RAT + 148 STA>>
       + 37.8 DISTRICT + 218 BTPRATIO + 0.344 DAYSEST


Predictor       Coef   SE Coef       T      P
Constant       123.9     426.8    0.29  0.772
DOTEST       0.90647   0.01659   54.64  0.000
B2B1RAT       -147.1     419.0   -0.35  0.726
B3B1RAT        -83.8     245.4   -0.34  0.733
BHB1RAT       -59.09     54.90   -1.08  0.283
STATUS        148.16     51.48    2.88  0.004
DISTRICT       37.75     43.07    0.88  0.382
BTPRATIO       217.6     139.6    1.56  0.120
DAYSEST       0.3439    0.1803    1.91  0.058


S = 304.589   R-Sq = 97.6%   R-Sq(adj) = 97.6%


Analysis of Variance

Source           DF         SS          MS        F      P
Regression        8  865907936   108238492  1166.68  0.000
Residual Error  226   20967037       92774
Total           234  886874973
```

Figure SIA11.2
Minitab regression output for first-order model of road construction cost

[*Note:* Not all of the independent variables have statistically significant t-values. However, we caution against dropping the insignificant variables from the model at this stage. One reason (discussed in section 11.3) is that performing a large number of t-tests will yield an inflated probability of at least one Type I error. In later sections of this chapter, we develop other reasons for why the multiple t-test approach is not a good strategy for determining which independent variables to keep in the model.]

The Minitab printout shown in Figure SIA11.3 gives a 95% prediction interval for cost and a 95% confidence interval for the mean cost for the x-values associated with the last observation (contract) in the **FLAG** file. These x-values are engineer's cost estimate (DOTEST) = 497 thousand dollars, ratio of second lowest bid to lowest bid (B2B1RAT) = 1.07, ratio of third lowest bid to lowest bid (B3B1RAT) = 1.08, ratio of highest bid to lowest bid (BHB1RAT) = 1.19, competitive bid (STATUS = 0), south Florida contract (DISTRICT = 1), bidders to planholders ratio (BTPRATIO) = 0.5, and estimated work days (DAYSEST) = 90. The 95% confidence interval of (339.2, 528.2) implies that for all road contracts with these x-values, the mean contract cost falls between 339.2 and 528.2 thousand dollars, with 95% confidence.

```
Predicted Values for New Observations

New
Obs    Fit  SE Fit        95% CI               95% PI
  1  433.7    48.0  (339.2, 528.2)  (-173.9, 1041.2)


Values of Predictors for New Observations

New
Obs  DOTEST  B2B1RAT  B3B1RAT  BHB1RAT    STATUS  DISTRICT  BTPRATIO  DAYSEST
  1     497     1.07     1.08     1.19  0.000000      1.00     0.500     90.0
```

Figure SIA11.3
Minitab printout with 95% confidence and prediction intervals

The 95% prediction interval of (−173.9, 1,041.2) implies that for an individual road contract with these x-values, the contract cost falls between 0 (because cost cannot be negative) and 1,041.2 thousand dollars, with 95% confidence. Note the wide range of the prediction interval. This is due to the large magnitude of the model standard deviation, $s = 305$ thousand dollars. Although the model is deemed statistically useful for predicting contract cost, it may not be "practically" useful. To reduce the magnitude of s, we will need to improve the model's predictive ability. (We consider such a model in the next Statistics in Action Revisited section.)

Data Set: FLAG

Exercises 11.25–11.33

Applying the Concepts—Basic

11.25 Characteristics of lead users. Refer to the *Creativity and Innovation Management* (Feb. 2008) study of lead users of children's computer games, Exercise 11.10 (p. 639). Recall that the researchers modeled lead-user rating (y, measured on a 5-point scale) as a function of gender ($x_1 = 1$ if female, 0 if male), age (x_2, years), degree of centrality (x_3, measured as the number of direct ties to other peers in the network), and betweenness centrality (x_4, measured as the number of shortest paths between peers). The least squares prediction equation was $\hat{y} = 3.58 + .01x_1 - .06x_2 - .01x_3 + .42x_4$.

a. Compute the predicted lead-user rating of a 10-year old female child with 5 direct ties to other peers in her social network and with 2 shortest paths between peers.

b. Compute an estimate for the mean lead-user rating of all 8-year old male children with 10 direct ties to other peers and with 4 shortest paths between peers.

11.26 Predicting runs scored in baseball. Refer to the *Chance* (Fall 2000) study of runs scored in Major League Baseball games, Exercise 11.14 (p. 640). Multiple regression was used to model total number of runs scored (y) of a team during the season as a function of number of walks (x_1), number of singles (x_2), number of doubles (x_3), number of triples (x_4), number of home runs (x_5), number of stolen bases (x_6), number of times caught stealing (x_7), number of strikeouts (x_8), and total number of outs (x_9). Using the β estimates given in Exercise 11.14, predict the number of runs scored by your favorite Major League Baseball team last year. How close is the predicted value to the actual number of runs scored by your team? [*Note:* You can find data on your favorite team on the Internet at www.mlb.com.]

11.27 Incomes of Mexican street vendors. NW Refer to the *World Development* (Feb. 1998) study of street vendors' earnings, y, Exercise 11.17 (p. 641). (Recall that the data are saved in the **STREETVN** file.) The SPSS spreadsheet below shows both a 95% prediction interval for y (right side) and a 95% confidence interval for $E(y)$ (left side) for a 45-year-old vendor who works 10 hours a day (i.e., for $x_1 = 45$ and $x_2 = 10$).

a. Interpret the 95% prediction interval for y in the words of the problem.

b. Interpret the 95% confidence interval for $E(y)$ in the words of the problem.

c. Note that the interval of part **a** is wider than the interval of part **b.** Will this always be true? Explain.

11.28 Chemical plant contamination. Refer to Exercise 11.18 (p. 641) and the U.S. Army Corps of Engineers data saved in the **DDT** file. You fit the first-order model, $E(y) = \beta_0 + \beta_1x_1 + \beta_2x_2 + \beta_3x_3$, to the data, where y = DDT level (parts per million), x_1 = number of miles upstream, x_2 = length (centimeters), and x_3 = weight (grams). Use the SPSS printout below to predict, with 90% confidence, the DDT level of a fish caught 300 miles upstream with a length of 40 centimeters and a weight of 1,000 grams. Interpret the result.

11.29 Cooling method for gas turbines. Refer to the *Journal of Engineering for Gas Turbines and Power* (Jan. 2005) study of a high-pressure inlet fogging method for a gas turbine engine, Exercise 11.19 (p. 641). Recall that you fit a first-order model for heat rate (y) as a function of speed (x_1), inlet temperature (x_2), exhaust temperature (x_3), cycle pressure ratio (x_4), and airflow rate (x_5) to data saved in the **GASTURBINE** file. A Minitab printout with both a 95% confidence interval for $E(y)$ and prediction interval for y for selected values of the x's is shown on the next page.

SPSS Output for Exercise 11.27

age	hours	ci95low	ci95upp	pi95low	pi95upp
45	10	2620.25197	3414.87349	1759.74674	4275.37871

SPSS Output for Exercise 11.28

MILE	LENGTH	WEIGHT	DDT	PI90LOW	PI90UPP
300	40.0	1000	.	-143.21783	180.97841

Minitab Output for Exercise 11.29

```
Predicted Values for New Observations

New
Obs      Fit  SE Fit          95% CI                95% PI
  1  12632.5   237.3  (12157.9, 13107.1)  (11599.6, 13665.5)


Values of Predictors for New Observations

New
Obs   RPM  INLET-TEMP  EXH-TEMP  CPRATIO  AIRFLOW
  1  7500        1000       525     13.5     10.0
```

a. Interpret the 95% prediction interval for y in the words of the problem.

b. Interpret the 95% confidence interval for $E(y)$ in the words of the problem.

c. Will the confidence interval for $E(y)$ always be narrower than the prediction interval for y? Explain.

Applying the Concepts—Intermediate

11.30 Arsenic in groundwater. Refer to the *Environmental Science & Technology* (Jan. 2005) study of the reliability of a commercial kit to test for arsenic in groundwater, Exercise 11.16 (p. 640). Using the data in the **ASWELLS** file, you fit a first-order model for arsenic level (y) as a function of latitude, longitude, and depth. Based on the model statistics, the researchers concluded that the arsenic level is highest at a low latitude, high longitude, and low depth. Do you agree? If so, find a 95% prediction interval for arsenic level for the lowest latitude, highest longitude, and lowest depth that are within the range of the sample data. Interpret the result.

11.31 Pay-for-performance efficiency of CEOs. In Chapter 1 (p. 4), we presented *Forbes* magazine's "Executive Compensation Scoreboard." *Forbes* (April 30, 2008) determined which CEOs are worth their pay based on a comparison of the firm's stock performance and the CEO's annual salary. Data for 175 of the top performers are saved in the **CEOPAYTOP175** file. Some of the variables measured for each CEO include pay-for-performance efficiency rating (y), average stock return during the CEO's tenure (x_1), average stock return during the previous 6 years (x_2), and the CEO's average salary over the past 6 years (x_3). The data for the first 5 CEOs in the file are shown below. (Recall that the lower the efficiency rating, the more the CEO is worth his/her pay.) Consider the first-order model $E(y) = \beta_0 + \beta_1x_1 + \beta_2x_2 + \beta_3x_3$.

a. Fit the model to the data and give the least squares prediction equation.

b. Conduct a test of overall model adequacy using $\alpha = .05$.

c. Predict, with 95% confidence, the efficiency rating of a CEO with $x_1 = 40\%$, $x_2 = 32\%$, and $x_3 = \$1$ million. (*Note:* These values represent the data for J. Bezos, CEO of Amazon.com.) Interpret the result.

11.32 Extracting water from oil. Refer to Exercise 11.20 (p. 642) and the data saved in the **WATEROIL** file. The researchers concluded that "in order to break a water-oil mixture with the lowest possible voltage, the volume fraction of the disperse phase (x_1) should be high, while the salinity (x_2) and the amount of surfactant (x_5) should be low." Use this information and the first-order model of Exercise 11.20 to find a 95% prediction interval for this "low" voltage (y). Interpret the interval.

11.33 Boiler drum production. In a production facility, an accurate estimate of man-hours needed to complete a task is crucial to management in making such decisions as the proper number of workers to hire, an accurate deadline to quote a client, or cost-analysis decisions regarding budgets. A manufacturer of boiler drums wants to use regression to predict the number of man-hours needed to erect the drums in future projects. To accomplish this, data for 35 boilers were collected. In addition to man-hours (y), the variables measured were boiler capacity (x_1 = lb/hr), boiler design pressure (x_2 = pounds per square inch or psi), boiler type (x_3 = 1 if industry field erected, 0 if utility field erected), and drum type (x_4 = 1

Table for Exercise 11.31 (first 5 observations)

Rank	CEO	Company	Return during Tenure, x_1 (%)	Return last 6 years, x_2 (%)	Average Salary, x_1 ($ millions)	Efficiency (y)
1	J. Bezos	Amazon.com	40.0	32.0	1.0	32.00
2	R. Fontaine	GameStop	32.0	33.0	3.6	9.17
3	J. Kelly	Crown Castle Intl	26.0	32.0	4.8	6.67
4	J. Bucksbaum	General Growth	19.0	22.0	0.7	31.43
5	B. Nordstrom	Nordstrom	22.0	20.0	3.8	5.26

Source: "First 5 observations of Executive Compensation Scoreboard", *Forbes,* April 30, 2008. Reprinted by permission of Forbes.com. © 2009 Forbes LLC.

if steam, 0 if mud). The data are saved in the **BOILERS** file. (The first five and last five observations are listed in the accompanying table.)

a. Fit the model $E(y) = \beta_0 + \beta_1 x_1 + \beta_2 x_2 + \beta_3 x_3 + \beta_4 x_4$ to the data. Give the estimates of the β's.

b. Conduct a test for the global utility of the model. Use $\alpha = .01$.

c. Find a 95% confidence interval for $E(y)$ when $x_1 = 150{,}000$, $x_2 = 500$, $x_3 = 1$, and $x_4 = 0$. Interpret the result.

Table for Exercise 11.33 (selected observations)

Man-Hours, y	Boiler Capacity, x_1	Design Pressure, x_2	Boiler Type, x_3	Drum Type, x_4
3,137	120,000	375	1	1
3,590	65,000	750	1	1
4,526	150,000	500	1	1
10,825	1,073,877	2,170	0	1
4,023	150,000	325	1	1
⋮	⋮	⋮	⋮	⋮
4,206	441,000	410	1	0
4,006	441,000	410	1	0
3,728	627,000	1,525	0	0
3,211	610,000	1,500	0	0
1,200	30,000	325	1	0

Source: Dr. Kelly Uscategui, University of Connecticut.

PART II: MODEL BUILDING IN MULTIPLE REGRESSION

11.5 Interaction Models

In Section 11.2, we demonstrated the relationship between $E(y)$ and the independent variables in a first-order model. When $E(y)$ is graphed against any one variable (say, x_1) for fixed values of the other variables, the result is a set of *parallel* straight lines (see Figure 11.3, p. 630). When this situation occurs (as it always does for a first-order model), we say that the relationship between $E(y)$ and any one independent variable *does not depend* on the values of the other independent variables in the model.

However, if the relationship between $E(y)$ and x_1 does, in fact, depend on the values of the remaining x's held fixed, then the first-order model is not appropriate for predicting y. In this case, we need another model that will take into account this dependence. Such a model includes the *cross products* of two or more x's.

For example, suppose that the mean value $E(y)$ of a response y is related to two quantitative independent variables, x_1 and x_2, by the model

$$E(y) = 1 + 2x_1 - x_2 + 3x_1x_2$$

A graph of the relationship between $E(y)$ and x_1 for $x_2 = 0, 1,$ and 2 is displayed in the Minitab graph, Figure 11.8.

Note that the graph shows three nonparallel straight lines. You can verify that the slopes of the lines differ by substituting each of the values $x_2 = 0, 1,$ and 2 into the equation.

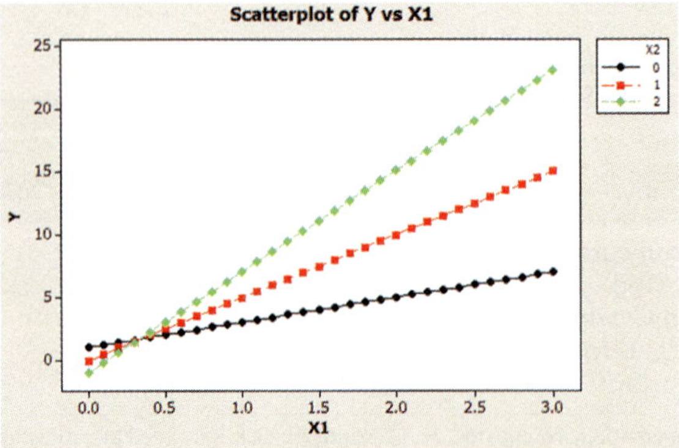

Figure 11.8
Minitab graph of $1 + 2x_1 - x_2 + 3x_1x_2$ for $x_2 = 0, 1, 2$

For $x_2 = 0$:

$$E(y) = 1 + 2x_1 - (0) + 3x_1(0) = 1 + 2x_1 \qquad \text{(slope} = 2)$$

For $x_2 = 1$,

$$E(y) = 1 + 2x_1 - (1) + 3x_1(1) = 0 + 5x_1 \qquad \text{(slope} = 5)$$

For $x_2 = 2$,

$$E(y) = 1 + 2x_1 - (2) + 3x_1(2) = -1 + 8x_1 \qquad \text{(slope} = 8)$$

Note that the slope of each line is represented by $\beta_1 + \beta_3 x_2 = 2 + 3x_2$. Thus, the effect on $E(y)$ of a change in x_1 (i.e., the slope) now *depends* on the value of x_2. When this situation occurs, we say that x_1 and x_2 **interact.** The cross-product term, x_1x_2, is called an **interaction term,** and the model $E(y) = \beta_0 + \beta_1x_1 + \beta_2x_2 + \beta_3x_1x_2$ is called an **interaction model** with two quantitative variables.

An Interaction Model Relating $E(y)$ to Two Quantitative Independent Variables

$$E(y) = \beta_0 + \beta_1x_1 + \beta_2x_2 + \beta_3x_1x_2$$

where

$(\beta_1 + \beta_3x_2)$ represents the change in $E(y)$ for every 1-unit increase in x_1, holding x_2 fixed

$(\beta_2 + \beta_3x_1)$ represents the change in $E(y)$ for every 1-unit increase in x_2, holding x_1 fixed

A three-dimensional graph (generated in Minitab) of the interaction model in two quantitative x's is shown in Figure 11.9. Unlike the flat planar surface displayed in Figure 11.4, the interaction model traces a ruled surface (twisted plane) in three-dimensional space. If we slice the twisted plane at a fixed value of x_2, we obtain a straight line relating $E(y)$ to x_1; however, the slope of the line will change as we change the value of x_2. Consequently, an interaction model is appropriate when the linear relationship between y and one independent variable depends on the value of another independent variable. The next example illustrates this idea.

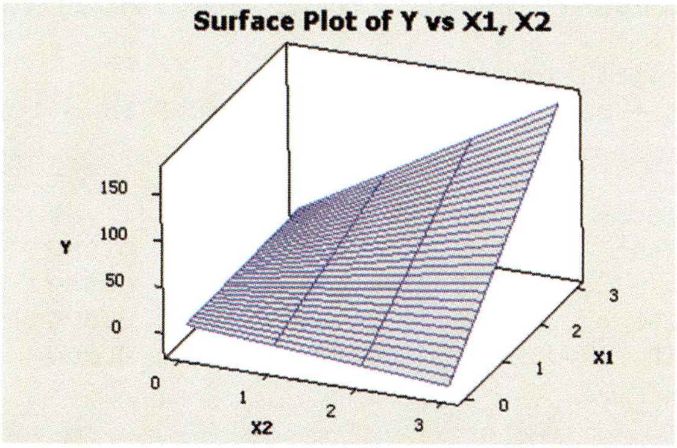

Figure 11.9
Minitab 3-dimensional graph of $1 + 2x_1 - x_2 + 3x_1x_2$

Example 11.6
Fitting an Interaction Model

Problem Refer to Examples 11.1–11.4. Suppose the collector of grandfather clocks, having observed many auctions, believes that the *rate of increase* of the auction price with age will be driven upward by a large number of bidders. Thus, instead of a relationship like that shown in Figure 11.10a, in which the rate of increase in price with age is the same for any number of bidders, the collector believes the relationship is like that shown in Figure 11.10b. Note that as the number of bidders increases from 5 to 15, the slope of the price versus age line increases.

Consequently, the interaction model is proposed:

$$y = \beta_0 = \beta_1x_1 + \beta_2x_2 + \beta_3x_1x_2 + \varepsilon$$

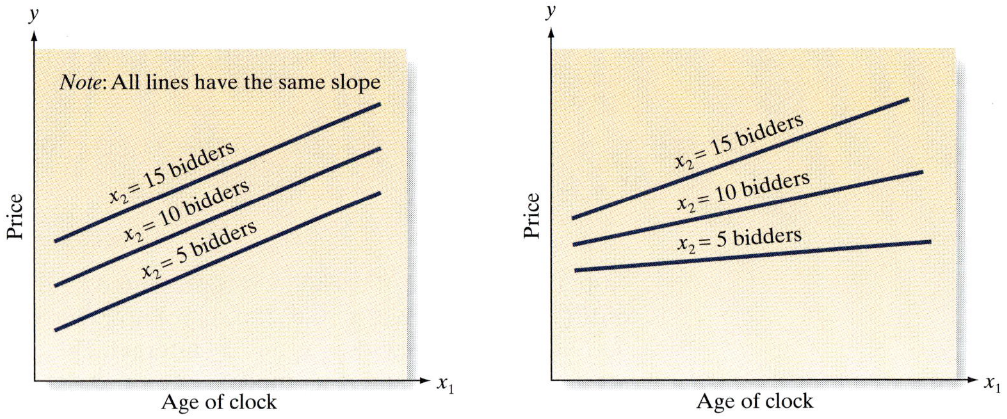

a. No interaction between x_1 and x_2 b. Interaction between x_1 and x_2

Figure 11.10
Examples of no-interaction and interaction models

The 32 data points listed in Table 11.1 (p. 628) were used to fit the model with interaction. A portion of the Minitab printout is shown in Figure 11.11.

a. Test the overall utility of the model using the global F-test at $\alpha = .05$.

b. Test the hypothesis (at $\alpha = .05$) that the price-age slope increases as the number of bidders increases—that is, that age and number of bidders, x_2, interact positively.

c. Estimate the change in auction price of a 150-year-old grandfather clock, y, for each additional bidder.

Solution

a. The global F-test is used to test the null hypothesis

$$H_0: \beta_1 = \beta_2 = \beta_3 = 0$$

Regression Analysis: PRICE versus AGE, NUMBIDS, AGEBID

```
The regression equation is
PRICE = 320 + 0.88 AGE - 93.3 NUMBIDS + 1.30 AGEBID


Predictor      Coef  SE Coef      T      P
Constant      320.5    295.1   1.09  0.287
AGE           0.878    2.032   0.43  0.669
NUMBIDS      -93.26    29.89  -3.12  0.004
AGEBID       1.2978   0.2123   6.11  0.000


S = 88.9145   R-Sq = 95.4%   R-Sq(adj) = 94.9%


Analysis of Variance

Source          DF       SS       MS       F      P
Regression       3  4578427  1526142  193.04  0.000
Residual Error  28   221362     7906
Total           31  4799790
```

Figure 11.11
Minitab printout for interaction model of auction price

The test statistic and p-value of the test (highlighted on the Minitab printout) are $F = 193.04$ and p-value $= 0$, respectively. Because $\alpha = .05$ exceeds the p-value, there is sufficient evidence to conclude that the model fit is a statistically useful predictor of auction price, y.

b. The hypotheses of interest to the collector concern the interaction parameter β_3. Specifically,

$$H_0: \beta_3 = 0$$
$$H_a: \beta_3 > 0$$

Because we are testing an individual β parameter, a t-test is required. The test statistic and two-tailed p-value (highlighted on the printout) are $t = 6.11$ and p-value $= 0$, respectively. The upper-tailed p-value, obtained by dividing the two-tailed p-value in half, is $0/2 = 0$. Because $\alpha = .05$ exceeds the p-value, the collector can reject H_0 and conclude that the rate of change of the mean price of the clocks with age increases as the number of bidders increases; that is, x_1 and x_2 interact positively. Thus, it appears that the interaction term should be included in the model.

c. To estimate the change in auction price, y, for every 1-unit increase in number of bidders, x_2, we need to estimate the slope of the line relating y to x_2 when the age of the clock, x_1, is 150 years old. An analyst who is not careful may estimate this slope as $\hat{\beta}_2 = -93.26$. Although the coefficient of x_2 is negative, this does *not* imply that auction price decreases as the number of bidders increases. Because interaction is present, the rate of change (slope) of mean auction price with the number of bidders *depends* on x_1, the age of the clock. Thus, the estimated rate of change of y for a 1-unit increase in x_2 (one new bidder) for a 150-year-old clock is

$$\text{Estimated slope of the } y \text{ versus } x_2 \text{ line} = \hat{\beta}_2 + \hat{\beta}_3 x_1 = -93.26 + 1.30(150) = 101.74$$

In other words, we estimate that the auction price of a 150-year-old clock will *increase* by about $101.74 for every additional bidder.

Look Back Although the rate of increase will vary as x_1 is changed, it will remain positive for the range of values of x_1 included in the sample. Extreme care is needed in interpreting the signs and sizes of coefficients in a multiple regression model.

Now Work Exercise 11.38

Example 11.6 illustrates an important point about conducting t-tests on the β parameters in the interaction model. The key β parameter in this model is the interaction β, β_3. [Note that this β is also the one associated with the highest-order term in the model, x_1x_2.]* Consequently, we will want to test H_0: $\beta_3 = 0$ after we have determined that the overall model is useful for predicting y. Once interaction is detected (as in Example 11.6), however, tests on the first-order terms x_1 and x_2 should *not* be conducted because they are meaningless tests; the presence of interaction implies that both x's are important.

CAUTION Once interaction has been deemed important in the model $E(y) = \beta_0 + \beta_1x_1 + \beta_2x_2 + \beta_3x_1x_2$, do not conduct t-tests on the β coefficients of the first-order terms x_1 and x_2. These terms should be kept in the model regardless of the magnitude of their associated p-values shown on the printout. ▲

*The order of a term is equal to the sum of the exponents of the quantitative variables included in the term. Thus, when x_1 and x_2 are both quantitative variables, the cross-product, x_1x_2, is a second-order term.

Exercises 11.34–11.46

Learning the Mechanics

11.34 Write an interaction model relating the mean value of y, $E(y)$ to

a. two quantitative independent variables

b. three quantitative independent variables [*Hint:* Include all possible two-way cross-product terms.]

11.35 Suppose the true relationship between $E(y)$ and the quantitative independent variables x_1 and x_2 is

$$E(y) = 3 + x_1 + 2x_2 - x_1x_2$$

a. Describe the corresponding three-dimensional response surface.

b. Plot the linear relationship between y and x_2 for $x_2 = 0, 1, 2$, where $0 \leq x_2 \leq 5$.

c. Explain why the lines you plotted in part **b** are not parallel.

d. Use the lines you plotted in part **b** to explain how changes in the settings of x_1 and x_2 affect $E(y)$.

e. Use your graph from part **b** to determine how much $E(y)$ changes when x_1 is changed from 2 to 0 and x_2 is simultaneously changed from 4 to 5.

11.36 Suppose you fit the interaction model

$$y = \beta_0 + \beta_1x_1 + \beta_2x_2 + \beta_3x_1x_2 + \varepsilon$$

to $n = 32$ data points and obtain the following results:

$$SS_{yy} = 479 \qquad SSE = 21 \qquad \hat{\beta}_3 = 10 \qquad s_{\hat{\beta}_3} = 4$$

a. Find R^2 and interpret its value.

b. Is the model adequate for predicting y? Test at $\alpha = .05$.

c. Use a graph to explain the contribution of the x_1x_2 term to the model.

d. Is there evidence that x_1 and x_2 interact? Test at $\alpha = .05$.

11.37 The Minitab printout below was obtained from fitting the model

$$y = \beta_0 + \beta_1x_1 + \beta_2x_2 + \beta_3x_1x_2 + \varepsilon$$

to $n = 15$ data points.

a. What is the prediction equation for the response surface?

b. Describe the geometric form of the response surface of part **a.**

c. Plot the prediction equation for the case when $x_2 = 1$. Do this twice more on the same graph for the cases when $x_2 = 3$ and $x_2 = 5$.

d. Explain what it means to say that x_1 and x_2 interact. Explain why your graph of part **c** suggests that x_1 and x_2 interact.

e. Specify the null and alternative hypotheses you would use to test whether x_1 and x_2 interact.

f. Conduct the hypothesis test of part **e** using $\alpha = .01$.

Minitab Output for Exercise 11.37

```
The regression equation is
Y = -2.55 + 3.82 X1 + 2.63 X2 -1.29 X1X2


Predictor     Coef   SE Coef       T      P
Constant    -2.550     1.142   -2.23  0.043
X1           3.815     0.529    7.22  0.000
X2           2.630     0.344    7.64  0.000
X1X2        -1.285     0.159   -8.06  0.000


S = 0.713      R-Sq = 85.6%    R-Sq(adj) = 81.6%


Analysis of Variance

Source          DF        SS        MS       F      P
Regression       3    33.149    11.050   21.75  0.000
Residual Error  11     5.587     0.508
Total           14    38.736
```

Applying the Concepts—Basic

11.38 Role of retailer interest on shopping behavior. Retail interest is defined by marketers as the level of interest a consumer has in a given retail store. Marketing professors at the University of Tennessee at Chattanooga and the University of Alabama investigated the role of retailer interest in consumers' shopping behavior (*Journal of Retailing*, Summer 2006). Using survey data collected for $n = 375$ consumers, the professors developed an interaction model for y = willingness of the consumer to shop at a retailer's store in the future (called *repatronage intentions*) as a function of x_1 = consumer satisfaction and x_2 = retailer interest. The regression results are shown below.

NW

Variable	$\hat{\beta}$	*t*-value	*p*-value
Satisfaction (x_1)	.426	7.33	$< .01$
Retailer interest (x_2)	.044	0.85	$> .10$
Interaction (x_1x_2)	−.157	−3.09	$< .01$

$R^2 = .65$, $F = 226.35$, p-value $< .001$

a. Is the overall model statistically useful for predicting y? Test using $\alpha = .05$.

b. Conduct a test for interaction at $\alpha = .05$.

c. Use the β-estimates to sketch the estimated relationship between repatronage intentions (y) and satisfaction (x_1) when retailer interest is $x_2 = 1$ (a low value).

d. Repeat part **c** when retailer interest is $x_2 = 7$ (a high value).

e. Sketch the two lines, parts **c** and **d,** on the same graph to illustrate the nature of the interaction.

11.39 Defects in nuclear missile housing parts. The technique of multivariable testing (MVT) was discussed in the *Journal of the Reliability Analysis Center* (First Quarter, 2004). MVT was shown to improve the quality of carbon-foam rings used in nuclear missile housings. The rings are produced via a casting process that involves mixing ingredients, oven curing, and carving the finished part. One type of defect analyzed was the number y of black streaks in the manufactured ring. Two variables found to impact the number of defects were turntable speed (revolutions per minute), x_1, and cutting-blade position (inches from center), x_2.

a. The researchers discovered "an interaction between blade position and turntable speed." Hypothesize a regression model for $E(y)$ that incorporates this interaction.

b. The researchers reported a positive linear relationship between number of defects (y) and turntable speed (x_1) but found that the slope of the relationship was much steeper for lower values of cutting-blade position (x_2). What does this imply about the interaction term in the model, part **a**? Explain.

11.40 Consumer behavior while waiting in line. While waiting in a long line for service (e.g., to use an ATM or at the post office), at some point you may decide to leave the queue. The *Journal of Consumer Research* (Nov. 2003) published a study of consumer behavior while waiting in a queue. A sample of $n = 148$ college students was asked to imagine that they were waiting in line at a post office to mail a package and that the estimated waiting time is 10 minutes or less. After a 10-minute wait, students were asked about their level of negative feelings (annoyed, anxious) on a scale of 1 (strongly disagree) to 9 (strongly agree). Before answering, however, the students were informed about how many people were ahead of them and behind them in the line. The researchers used regression to relate negative feelings score (y) to number ahead in line (x_1) and number behind in line (x_2).

a. The researchers fit an interaction model to the data. Write the hypothesized equation of this model.

b. In the words of the problem, explain what it means to say that "x_1 and x_2 interact to effect y."

c. A t-test for the interaction β in the model resulted in a p-value greater than .25. Interpret this result.

d. From their analysis, the researchers concluded that "the greater the number of people ahead, the higher the negative feeling score" and "the greater the number of people behind, the lower the negative feeling score." Use this information to determine the signs of β_1 and β_2 in the model.

Applying the Concepts—Intermediate

11.41 Incomes of Mexican street vendors. Refer to the *World Development* (Feb. 1998) study of street vendors in the city of Puebla, Mexico, Exercise 11.17 (p. 641). Recall that the vendors' mean annual earnings, $E(y)$, was modeled as a first-order function of age (x_1) and hours worked (x_2). Now, fit the interaction model $E(y) = \beta_0 + \beta_1 x_1 + \beta_2 x_2 + \beta_3 x_1 x_2$ to the data in the **STREETVN** file.

a. Find the least squares prediction equation.

b. What is the estimated slope relating annual earnings (y) to age (x_1) when number of hours worked (x_2) is 10? Interpret the result.

c. What is the estimated slope relating annual earnings (y) to hours worked (x_2) when age (x_1) is 40? Interpret the result.

d. Give the null hypothesis for testing whether age (x_1) and hours worked (x_2) interact.

e. Find the p-value of the test, part **d.**

f. Refer to part **e.** Give the appropriate conclusion in the words of the problem.

11.42 Factors that impact an auditor's judgment. A study was conducted to determine the effects of linguistic delivery style and client credibility on auditors' judgments (*Advances in Accounting and Behavioral Research,* 2004). Two hundred auditors from Big 5 accounting firms were asked to assume that he or she was an audit team supervisor of a new manufacturing client and was performing an analytical review of the client's financial statement. The researchers gave the auditors different information on the client's credibility and linguistic delivery style of the client's explanation. Each auditor then provided an assessment of the likelihood that the client-provided explanation accounts for the fluctuation in the financial statement. The three variables of interest—credibility (x_1), linguistic delivery style (x_2), and likelihood (y)—were all measured on a numerical scale. Regression analysis was used to fit the interaction model, $y = \beta_0 + \beta_1 x_1 + \beta_2 x_2 + \beta_3 x_1 x_2 + \varepsilon$. The results are summarized in the table below.

a. Interpret the phrase *client credibility and linguistic delivery style interact* in the words of the problem.

Table for Exercise 11.42

	Beta Estimate	Std Error	t-statistic	p-value
Constant	15.865	10.980	1.445	0.150
Client credibility (x_1)	0.037	0.339	0.110	0.913
Linguistic delivery style (x_2)	−0.678	0.328	−2.064	0.040
Interaction ($x_1 x_2$)	0.036	0.009	4.008	< 0.005

F-statistic $= 55.35$ ($p < 0.0005$): $R_a^2 = .450$

b. Give the null and alternative hypotheses for testing the overall adequacy of the model.

c. Conduct the test, part **b,** using the information in the table.

d. Give the null and alternative hypotheses for testing whether client credibility and linguistic delivery style interact.

e. Conduct the test, part **d,** using the information in the table.

f. The researchers estimated the slope of the likelihood–linguistic delivery style line at a low level of client credibility ($x_1 = 22$). Obtain this estimate and interpret it in the words of the problem.

g. The researchers also estimated the slope of the likelihood–linguistic delivery style line at a high level of client credibility ($x_1 = 46$). Obtain this estimate and interpret it in the words of the problem.

11.43 **Arsenic in groundwater.** Refer to the *Environmental Science & Technology* (Jan. 2005) study of the reliability of a commercial kit to test for arsenic in groundwater, Exercise 11.16 (p. 640). Recall that you fit a first-order model for arsenic level (y) as a function of latitude (x_1), longitude (x_2), and depth (x_3) to data saved in the **ASWELLS** file.

a. Write a model for arsenic level (y) that includes first-order terms for latitude, longitude, and depth, as well as terms for interaction between latitude and depth and interaction between longitude and depth.

b. Use statistical software to fit the interaction model, part **a,** to the data in the **ASWELLS** file. Give the least squares prediction equation.

c. Conduct a test (at $\alpha = .05$) to determine whether latitude and depth interact to affect arsenic level.

d. Conduct a test (at $\alpha = .05$) to determine whether longitude and depth interact to affect arsenic level.

e. Practically interpret the results of the tests, parts **c** and **d.**

11.44 **Cooling method for gas turbines.** Refer to the *Journal of Engineering for Gas Turbines and Power* (Jan. 2005) study of a high-pressure inlet fogging method for a gas turbine engine, Exercise 11.19 (p. 641). Recall that you fit a first-order model for heat rate (y) as a function of speed (x_1), inlet temperature (x_2), exhaust temperature (x_3), cycle pressure ratio (x_4), and air flow rate (x_5) to data saved in the **GASTURBINE** file.

a. Researchers hypothesize that the linear relationship between heat rate (y) and temperature (both inlet and exhaust) depends on airflow rate. Write a model for heat rate that incorporates the researchers' theories.

b. Use statistical software to fit the interaction model, part **a,** to the data in the **GASTURBINE** file. Give the least squares prediction equation.

c. Conduct a test (at $\alpha = .05$) to determine whether inlet temperature and air flow rate interact to effect heat rate.

d. Conduct a test (at $\alpha = .05$) to determine whether exhaust temperature and air flow rate interact to effect heat rate.

e. Practically interpret the results of the tests, parts **c** and **d.**

11.45 **Extracting water from oil.** Refer to the *Journal of Colloid and Interface Science* study of water/oil mixtures, Exercise 11.20 (p. 642). Recall that three of the seven variables used to predict voltage (y) were volume (x_1), salinity (x_2), and surfactant concentration (x_5). The model the researchers fit is

$$E(y) = \beta_0 + \beta_1 x_1 + \beta_2 x_2 + \beta_3 x_5 + \beta_4 x_1 x_2 + \beta_5 x_1 x_5$$

a. Note that the model includes interaction between disperse phase volume (x_1) and salinity (x_2) as well as interaction between disperse phase volume (x_1) and surfactant concentration (x_5). Discuss how these interaction terms affect the hypothetical relationship between y and x_1. Draw a sketch to support your answer.

b. Fit the interaction model to the data in the **WATEROIL** file. Does this model appear to fit the data better than the first-order model in Exercise 11.20? Explain.

c. Interpret the β estimates of the interaction model.

11.46 **Therapists' reactions to child-abuse reports.** Licensed therapists are mandated by law to report child abuse by their clients. This requires the therapist to breach confidentiality and possibly lose the client's trust. A national survey of licensed psychotherapists was conducted to investigate clients' reactions to legally mandated child-abuse reports (*American Journal of Orthopsychiatry,* Jan. 1997). The sample consisted of 303 therapists who had filed a child-abuse report against one of their clients. The researchers were interested in finding the best predictors of a client's reaction (y) to the report, where y is measured on a 30-point scale. (The higher the value, the more favorable the client's response to the report.) The independent variables found to have the most predictive power are listed here.

x_1 = Therapist's age (years)
x_2 = Therapist's gender (1 if male, 0 if female)
x_3 = Degree of therapist's role strain (25-point scale)
x_4 = Strength of client-therapist relationship (40-point scale)
x_5 = Type of case (1 if family, 0 if not)
$x_1 x_2$ = Age × gender interaction

a. Hypothesize a first-order model relating y to each of the five independent variables.

b. Give the null hypothesis for testing the contribution of x_4, strength of client-therapist relationship, to the model.

c. The test statistic for the test, part **b,** was $t = 4.408$ with an associated p-value of .001. Interpret this result.

d. The estimated β coefficient for the $x_1 x_2$ interaction term was positive and highly significant ($p < .001$). According to the researchers, "this interaction suggests that ... as the age of the therapist increased, ... male therapists were less likely to get negative client reactions than were female therapists." Do you agree?

e. For this model, $R^2 = .2946$. Interpret this value.

11.6 Quadratic and Other Higher-Order Models

All of the models discussed in the previous sections proposed straight-line relationships between $E(y)$ and each of the independent variables in the model. In this section, we consider models that allow for curvature in the 2-dimensional relationship between y

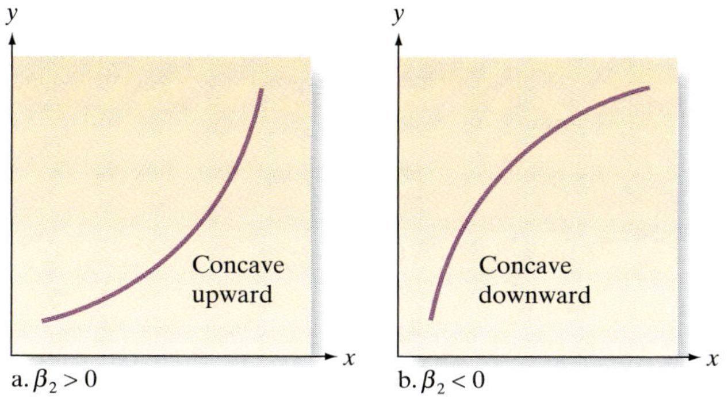

Figure 11.12
Graphs for two quadratic models

and an independent variable. Each of these models is a **second-order model** because it will include an x^2 term.

First, we consider a model that includes only one independent variable x. The form of this model, called the **quadratic model,** is

$$y = \beta_0 + \beta_1 x + \beta_2 x^2 + \varepsilon$$

The term involving x^2, called a **quadratic term** (or **second-order term**), enables us to hypothesize curvature in the graph of the response model relating y to x. Graphs of the quadratic model for two different values of β_2 are shown in Figure 11.12. When the curve opens upward, the sign of β_2 is positive (see Figure 11.12a); when the curve opens downward, the sign of β_2 is negative (see Figure 11.12b).

A Quadratic (Second-Order) Model in a Single Quantitative Independent Variable

$$E(y) = \beta_0 + \beta_1 x + \beta_2 x^2$$

where

β_0 is the y-intercept of the curve.

β_1 is a shift parameter.

β_2 is the rate of curvature.

Example 11.7
Analyzing a Quadratic Model for Electrical Usage

Problem In all-electric homes, the amount of electricity expended is of interest to consumers, builders, and groups involved with energy conservation. Suppose we wish to investigate the monthly electrical usage, y, in all-electric homes and its relationship to the size, x, of the home. Moreover, suppose we think that monthly electrical usage in all-electric homes is related to the size of the home by the quadratic model

$$y = \beta_0 + \beta_1 x + \beta_2 x^2 + \varepsilon$$

To fit the model, the values of y and x are collected for 15 homes during a particular month. The data are shown in Table 11.2.

a. Construct a scatterplot for the data. Is there evidence to support the use of a quadratic model?

b. Use the method of least squares to estimate the unknown parameters β_0, β_1, and β_2 in the quadratic model.

c. Graph the prediction equation and assess how well the model fits the data, both visually and numerically.

d. Interpret the β estimates.

Table 11.2 Home Size–Electrical Usage Data

Size of Home, x (sq. ft.)	Monthly Usage, y (kilowatt-hours)
1,290	1,182
1,350	1,172
1,470	1,264
1,600	1,493
1,710	1,571
1,840	1,711
1,980	1,804
2,230	1,840
2,400	1,986
2,710	2,007
2,930	1,984
3,000	1,960
3,210	2,001
3,240	1,928
3,520	1,945

Data Set: ELECTRIC

e. Is the overall model useful (at $\alpha = .01$) for predicting electrical usage y?

f. Is there sufficient evidence of downward curvature in the home size–electrical usage relationship? Test using $\alpha = .01$.

Solution

a. A Minitab scattergram for the data in Table 11.2 is shown in Figure 11.13. The figure illustrates that electrical usage appears to increase in a curvilinear manner with the size of the home. This provides some support for the inclusion of the quadratic term x^2 in the model.

b. We used Excel to fit the model to the data in Table 11.2. Part of the Excel regression output is displayed in Figure 11.14. The least squares estimates of the β parameters (highlighted) are $\hat{\beta}_0 = -806.7$, $\hat{\beta}_1 = 1.9616$, and $\hat{\beta}_2 = -.00034$. Therefore, the equation that minimizes the SSE for the data is

$$\hat{y} = -806.7 + 1.9616x - .00034x^2$$

c. Figure 11.15 is a Minitab graph of the least squares prediction equation. Note that the graph provides a good fit to the data of Table 11.2. A numerical measure of fit is obtained with the adjusted coefficient of determination, R_a^2. This value (highlighted on Figure 11.14) is $R_a^2 = .973$. This implies that about 97% of the sample variation in electrical usage (y) can be explained by the quadratic model (after adjusting for sample size and degrees of freedom).

d. The interpretation of the estimated coefficients in a quadratic model must be undertaken cautiously. First, the estimated y-intercept, $\hat{\beta}_0$, can be meaningfully interpreted only if the range of the independent variable includes zero—that is, if $x = 0$ is

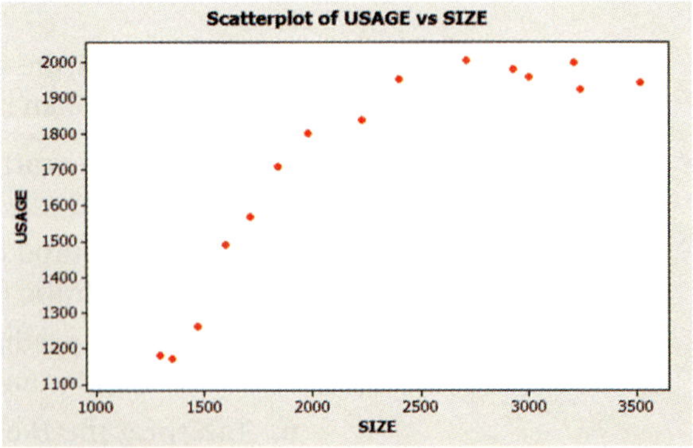

Figure 11.13
Minitab scatterplot for electrical usage data

SUMMARY OUTPUT

Regression Statistics	
Multiple R	0.988576
R Square	0.977283
Adjusted R Square	0.973496
Standard Error	50.19975
Observations	15

ANOVA

	df	*SS*	*MS*	*F*	*Significance F*
Regression	2	1300900.218	650450.1	258.1135719	0.0000000001
Residual	12	30240.18169	2520.015		
Total	14	1331140.4			

	Coefficients	*Standard Error*	*t Stat*	*P-value*	*Lower 95%*	*Upper 95%*
Intercept	-806.717	166.8720911	-4.83434	0.0004089596	-1170.299712	-443.134
SIZE	1.961617	0.152524384	12.861	0.0000000223	1.629294434	2.293939
SIZESQ	-0.00034	0.00003212	-10.599	0.0000001903	-0.000410426	-0.00027

Figure 11.14
Excel regression output for electrical usage model

included in the sampled range of x. Although $\hat{\beta}_0 = -806.7$ seems to imply that the estimated electrical usage is negative when $x = 0$, this zero point is not in the range of the sample (the lowest value of x is 1,290 square feet), and the value is nonsensical (a home with 0 square feet); thus the interpretation of $\hat{\beta}_0$ is not meaningful.

The estimated coefficient of x is $\hat{\beta}_1 = 1.9616$, but in the presence of the quadratic term x^2, it no longer represents a slope.* The estimated coefficient of the first-order term x will not, in general, have a meaningful interpretation in the quadratic model.

The sign of the coefficient, $\hat{\beta}_2 = -.00034$, of the quadratic term, x^2, is the indicator of whether the curve is concave downward (mound shaped) or concave upward (bowl shaped). A negative $\hat{\beta}_2$ implies downward concavity, as in this example (Figure 11.15), and a positive $\hat{\beta}_2$ implies upward concavity. Rather than interpreting the numerical value of $\hat{\beta}_2$ itself, we employ a graphical representation of the model, as in Figure 11.15, to describe the model.

Note that Figure 11.15 implies that the estimated electrical usage is leveling off as the home sizes increase beyond 2,500 square feet. In fact, the convexity of the model

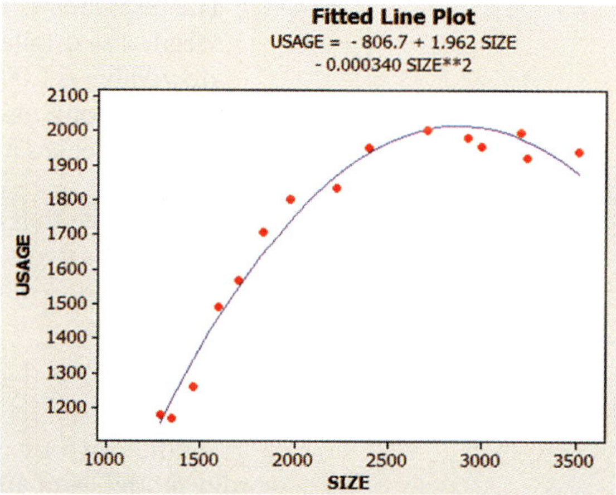

Figure 11.15
Minitab plot of least squares model for electrical usage

*For students with knowledge of calculus, note that the slope of the quadratic model is the first derivative $\partial y/\partial x = \beta_1 + 2\beta_2 x$. Thus, the slope varies as a function of x, rather than the constant slope associated with the straight-line model.

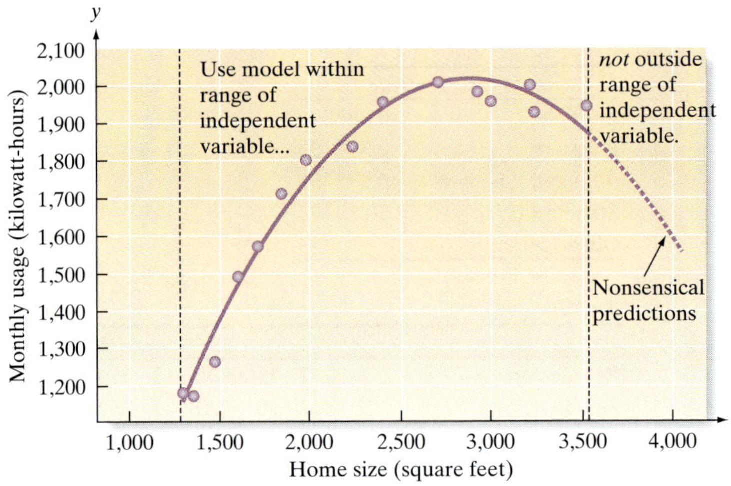

Figure 11.16
Potential misuse of quadratic model

would lead to decreasing usage estimates if we were to display the model out to 4,000 square feet and beyond (see Figure 11.16). However, model interpretations are not meaningful outside the range of the independent variable, which has a maximum value of 3,520 square feet in this example. Thus, although the model appears to support the hypothesis that the *rate of increase* per square foot *decreases* for home sizes near the high end of the sampled values, the conclusion that usage will actually begin to decrease for very large homes would be a *misuse* of the model because no homes of 3,600 square feet or more were included in the sample.

e. To test whether the quadratic model is statistically useful, we conduct the global F-test:

H_0: $\beta_1 = \beta_2 = 0$
H_a: At least one of the above coefficients is nonzero.

From the Excel printout, Figure 11.14, the test statistic (highlighted) is $F = 258.11$ with an associated p-value of approximately 0. For any reasonable α, we reject H_0 and conclude that the overall model is a useful predictor of electrical usage, y.

f. Figure 11.15 shows concave downward curvature in the relationship between size of a home and electrical usage in the sample of 15 data points. To determine if this type of curvature exists in the population, we want to test

H_0: $\beta_2 = 0$ (no curvature in the response curve)

H_a: $\beta_2 < 0$ (downward concavity exists in the response curve)

The test statistic for testing β_2 highlighted on the printout, is $t = -10.599$, and the associated two-tailed p-value is .0000002. Because this is a one-tailed test, the appropriate p-value is $(.0000002)/2 = .0000001$. Now $\alpha = .01$ exceeds this p-value. Thus, there is very strong evidence of downward curvature in the population—that is, electrical usage increases more slowly per square foot for large homes than for small homes.

Look Back Note that the Excel printout in Figure 11.14 also provides the t-test statistic and corresponding two-tailed p-values for the tests of H_0: $\beta_0 = 0$ and H_0: $\beta_1 = 0$. Because the interpretation of these parameters is not meaningful for this model, the tests are not of interest.

Now Work Exercise 11.52

When two or more quantitative independent variables are included in a second-order model, we can incorporate squared terms for each x in the model, as well as the interaction between the two independent variables. A model that includes all possible second-order terms in two independent variables—called a **complete second-order model**—is given in the box on the next page.

Complete Second-Order Model with Two Quantitative Independent Variables

$$E(y) = \beta_0 + \beta_1 x_1 + \beta_2 x_2 + \beta_3 x_2 x_2 + \beta_4 x_1^2 + \beta_5 x_2^2$$

Comments on the Parameters

β_0: y-intercept, the value of $E(y)$ when $x_1 = x_2 = 0$

β_1, β_2: Changing β_1 and β_2 causes the surface to shift along the x_1- and x_2-axes

β_3: Controls the rotation of the surface

β_4, β_5: Signs and values of these parameters control the type of surface and the rates of curvature.

Three types of 3-dimensional graphs (called **response surfaces**) are produced by a second-order model:* a **paraboloid** that opens upward (Figure 11.17a), a paraboloid that opens downward (Figure 11.17b), and a **saddle-shaped surface** (Figure 11.17c).

A complete second-order model is the 3-dimensional equivalent of a quadratic model in a single quantitative variable. Instead of tracing parabolas, it traces paraboloids and saddle surfaces. Because only a portion of the complete surface is used to fit the data, this model provides a very large variety of gently curving surfaces that can be used to fit data. It is a good choice for a model if you expect curvature in the response surface relating $E(y)$ to x_1 and x_2.

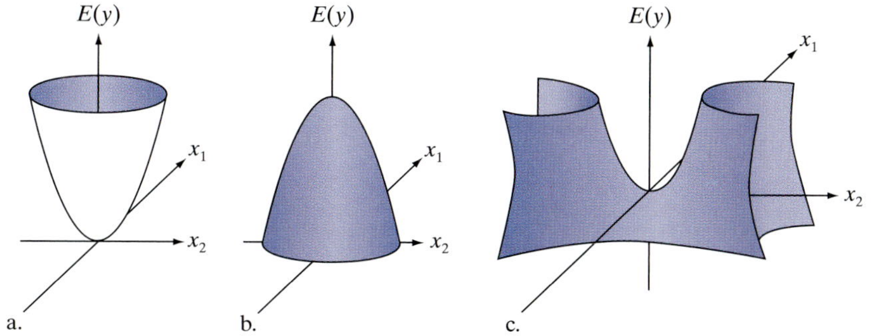

Figure 11.17
Graphs for three second-order surfaces

Example 11.8
A More Complex Second-Order Model: Hours Worked per Week

Problem A social scientist would like to relate the number of hours worked per week (outside the home) by a married woman to the number of years of formal education she has completed and the number of children in her family.

a. Identify the dependent variable and the independent variables.

b. Write the first-order model for this example.

c. Modify the model in part **b** to include an interaction term.

d. Write a complete second-order model for $E(y)$.

Solution

a. The dependent variable is

$$y = \text{Number of hours worked per week by a married woman}$$

The two independent variables, both quantitative in nature, are

x_1 = Number of years of formal education completed by the woman
x_2 = Number of children in the family

b. The first-order model is

$$E(y) = \beta_0 + \beta_1 x_1 + \beta_2 x_2$$

*The saddle-shaped surface (Figure 11.17c) is produced when $\beta_3^2 > 4\beta_4\beta_5$; the paraboloid opens upward (Figure 11.17a) when $\beta_4 + \beta_5 > 0$ and opens downward (Figure 11.17b) when $\beta_4 + \beta_5 < 0$.

This model would probably not be appropriate in this situation because x_1 and x_2 may interact and/or curvature terms corresponding to x_1^2 and x_2^2 may be needed to obtain a good model for $E(y)$.

c. Adding the interaction term, we obtain

$$E(y) = \beta_0 + \beta_1 x_1 + \beta_2 x_2 + \beta_3 x_1 x_2$$

This model should be better than the model in part **b** because we have now allowed for interaction between x_1 and x_2.

d. The complete second-order model is

$$E(y) = \beta_0 + \beta_1 x_1 + \beta_2 x_2 + \beta_3 x_1 x_2 + \beta_4 x_1^2 + \beta_5 x_2^2$$

Because it would not be surprising to find curvature in the response surface, the complete second-order model would be preferred to the models in parts **b** and **c.**

Look Back How can we tell whether the complete second-order model really does provide better predictions of hours worked than the models in parts **b** and **c**? The answers to these and similar questions are examined in Section 11.9.

Most relationships between $E(y)$ and two or more quantitative independent variables are second order and require the use of either the interactive or the complete second-order model to obtain a good fit to a data set. As in the case of a single quantitative independent variable, however, the curvature in the response surface may be very slight over the range of values of the variables in the data set. When this happens, a first-order model may provide a good fit to the data.

Exercises 11.47–11.62

Learning the Mechanics

11.47 Write a second-order model relating the mean of y, $E(y)$, to
- **a.** one quantitative independent variable
- **b.** two quantitative independent variables
- **c.** three quantitative independent variables [*Hint:* Include all possible two-way cross-product terms and squared terms.]

11.48 Suppose you fit the second-order model

$$y = \beta_0 + \beta_1 x + \beta_2 x^2 + \varepsilon$$

to $n = 25$ data points. Your estimate of β_2 is $\hat{\beta}_2 = .47$, and the estimated standard error of the estimate is .15.
- **a.** Test H_0: $\beta_2 = 0$ against H_a: $\beta_2 \neq 0$. Use $\alpha = .05$.
- **b.** Suppose you want to determine only whether the quadratic curve opens upward; that is, as x increases, the slope of the curve increases. Give the test statistic and the rejection region for the test for $\alpha = .05$. Do the data support the theory that the slope of the curve increases as x increases? Explain.

11.49 Suppose you fit the quadratic model

$$E(y) = \beta_0 + \beta_1 x + \beta_2 x^2$$

to a set of $n = 20$ data points and found $R^2 = .91$, $SS_{yy} = 29.94$, and $SSE = 2.63$.
- **a.** Is there sufficient evidence to indicate that the model contributes information for predicting y? Test using $\alpha = .05$.
- **b.** What null and alternative hypotheses would you test to determine whether upward curvature exists?
- **c.** What null and alternative hypotheses would you test to determine whether downward curvature exists?

11.50 Consider the following quadratic models:

(1) $y = 1 - 2x + x^2$
(2) $y = 1 + 2x + x^2$
(3) $y = 1 + x^2$
(4) $y = 1 - x^2$
(5) $y = 1 + 3x^2$

- **a.** Graph each of the quadratic models, side by side, on the same sheet of graph paper.
- **b.** What effect does the first-order term ($2x$) have on the graph of the curve?
- **c.** What effect does the second-order term (x^2) have on the graph of the curve?

11.51 Minitab was used to fit the complete second-order model

$$E(y) = \beta_0 + \beta_1 x_1 + \beta_2 x_2 + \beta_3 x_1 x_2 + \beta_4 x_1^2 + \beta_5 x_2^2$$

to $n = 39$ data points. The printout is shown on the next page.
- **a.** Is there sufficient evidence to indicate that at least one of the parameters—β_1, β_2, β_3, β_4, and β_5—is nonzero? Test using $\alpha = .05$.
- **b.** Test H_0: $\beta_4 = 0$ against H_a: $\beta_4 \neq 0$. Use $\alpha = .01$.
- **c.** Test H_0: $\beta_5 = 0$ against H_a: $\beta_5 \neq 0$. Use $\alpha = .01$.
- **d.** Use graphs to explain the consequences of the tests in parts **b** and **c.**

Applying the Concepts—Basic

11.52 **Catalytic converters in cars.** A quadratic model was NW applied to motor vehicle toxic emissions data collected in Mexico City (*Environmental Science & Engineering*, Sept. 1, 2000). The following equation was used to predict the percentage (y) of motor vehicles without

Minitab Output for Exercise 11.51

```
The regression equation is
Y = -24.56 + 1.12 X1 + 27.99 X2 - 0.54 X1X2 - 0.004 X1SQ + 0.002 X2SQ


Predictor        Coef      SE Coef        T       P
Constant      -24.563        6.531    -3.76   0.001
X1            1.19848       0.1103    10.86   0.000
X2             27.988       79.489     0.35   0.727
X1X2          -0.5397       1.0338    -0.52   0.605
X1SQ          -0.0043       0.0004   -10.74   0.000
X2SQ           0.0020       0.0033     0.60   0.550


S = 2.762       R-Sq = 79.7%     R-Sq(adj) = 76.6%

Analysis of Variance

Source            DF       SS       MS       F       P
Regression         5   989.30   197.86   25.93   0.000
Residual Error    33   251.81     7.63
Total             38  1241.11
```

catalytic converters in the Mexico City fleet for a given year (x):

$\hat{y} = 325{,}790 - 321.67x + .0794x^2$.

a. Explain why the value $\hat{\beta}_0 = 325{,}790$ has no practical interpretation.

b. Explain why the value $\hat{\beta}_1 = -321.67$ should not be interpreted as a slope.

c. Examine the value of $\hat{\beta}_2$ to determine the nature of the curvature (upward or downward) in the sample data.

d. The researchers used the model to estimate "that just after the year 2021 the fleet of cars with catalytic converters will completely disappear." Comment on the danger of using the model to predict y in the year 2021. (*Note:* The model was fit to data collected between 1984 and 1999.)

11.53 Testing tires for wear. Underinflated or overinflated tires can increase tire wear. A new tire was tested for wear at different pressures, with the results shown in the following table. (These data are saved in the **TIRES** file.)

Pressure, x (pounds per square inch)	Mileage, y (thousands)
30	29
31	32
32	36
33	38
34	37
35	33
36	26

a. Plot the data on a scattergram.

b. If you were given only the information for $x = 30, 31, 32, 33$, what kind of model would you suggest? For $x = 33, 34, 35, 36$? For all the data?

11.54 Assertiveness and leadership. Management professors at Columbia University examined the relationship between assertiveness and leadership (*Journal of Personality and Social Psychology,* Feb. 2007). The sample represented 388 people enrolled in a full-time MBA program. Based on answers to a questionnaire, the researchers measured two variables for each subject: assertiveness score (x) and leadership ability score (y). A quadratic regression model was fit to the data with the following results:

Independent Variable	β Estimate	t-value	p-value
x	.57	2.55	.01
x^2	–.088	–3.97	< .01
Model $R^2 = .12$			

a. Conduct a test of overall model utility. Use $\alpha = .05$.

b. The researchers hypothesized that leadership ability will increase at a decreasing rate with assertiveness. Set up the null and alternative hypotheses to test this theory.

c. Use the reported results to conduct the test, part **b.** Give your conclusion (at $\alpha = .05$) in the words of the problem.

11.55 Goal congruence in top management teams. Do chief executive officers (CEOs) and their top managers always agree on the goals of the company? Goal importance congruence between CEOs and vice presidents (VPs) was studied in the *Academy of Management Journal* (Feb. 2008). The researchers used regression to model a VP's attitude toward the goal of improving efficiency (y) as a function of the two quantitative independent variables, level of CEO leadership (x_1) and level of congruence between the CEO and the VP (x_2). A complete second-order model in x_1 and x_2 was fit to data collected for $n = 517$ top management team members at U.S. credit unions.

a. Write the complete second-order model for $E(y)$.

b. The coefficient of determination for the model, part **a,** was reported as $R^2 = .14$. Interpret this value.

c. The estimate of the β-value for the $(x_2)^2$ term in the model was found to be negative. Interpret this result, practically.

d. A t-test on the β-value for the interaction term in the model, x_1x_2, resulted in a p-value of .02. Practically interpret this result, using $\alpha = .05$.

11.56 Oil evaporation study. The *Journal of Hazardous Materials* (July 1995) presented a literature review of models designed to predict oil spill evaporation. One model discussed in the article used boiling (x_1) and API specific gravity (x_2) to predict the molecular weight (y) of the oil that is spilled. A complete second-order model for y was proposed.

a. Write the equation of the model.

b. Identify the terms in the model that allow for curvilinear relationships.

Applying the Concepts—Intermediate

11.57 Revenues of popular movies. The *Internet Movie Database* (www.imdb.com) monitors the gross revenues for all major motion pictures. The accompanying table gives both the domestic (United States and Canada) and international gross revenues for a sample of 16 popular movies. The data are saved in the **IMDB** file.

Movie Title (year)	Domestic Gross ($ millions)	International Gross ($ millions)
Titanic (1997)	600.7	1,234.6
The Dark Knight (2008)	533.2	464.0
E.T. (1982)	439.9	321.8
Pirates of the Caribbean: Dead Man's Chest (2006)	420.3	631.0
Jurassic Park (1993)	356.8	563.0
Lion King (1994)	328.4	455.0
Harry Potter and the Sorcerer's Stone (2001)	317.6	651.1
Sixth Sense (1999)	293.5	368.0
Jaws (1975)	260.0	210.6
Ghost (1990)	217.6	300.0
Saving Private Ryan (1998)	216.1	263.2
Gladiator (2000)	187.7	268.6
Dances with Wolves (1990)	184.2	240.0
The Exorcist (1973)	204.6	153.0
My Big Fat Greek Wedding (2002)	241.4	115.1
Rocky IV (1985)	127.9	172.6

Source: Information courtesy of *The Internet Movie Database* (http://www.imdb.com). Used with permission.

a. Write a first-order model for foreign gross revenues (y) as a function of domestic gross revenues (x).

b. Write a second-order model for international gross revenues y as a function of domestic gross revenues x.

c. Construct a scattergram for these data. Which of the models from parts **a** and **b** appears to be the better choice for explaining the variation in foreign gross revenues?

d. Fit the model of part **b** to the data and investigate its usefulness. Is there evidence of a curvilinear relationship between international and domestic gross revenues? Try using $\alpha = .05$.

e. Based on your analysis in part **d,** which of the models from parts **a** and **b** better explains the variation in international gross revenues? Compare your answer to your preliminary conclusion from part **c.**

11.58 Estimating change-point dosage. A standard method for studying toxic substances and their effects on humans is to observe the responses of rodents exposed to various doses of the substance over time. In the *Journal of Agricultural, Biological, and Environmental Statistics* (June 2005), researchers used least squares regression to estimate the *change-point* dosage—defined as the largest dose level that has no adverse effects. Data were obtained from a dose-response study of rats exposed to the toxic substance aconiazide. A sample of 50 rats was evenly divided into five dosage groups: 0, 100, 200, 500, and 750 milligrams per kilogram of body weight. The dependent variable y measured was the weight change (in grams) after a 2-week exposure. The researchers fit the quadratic model $E(y) = \beta_0 + \beta_1 x + \beta_2 x^2$, where x = dosage level, with the following results: $\hat{y} = 10.25 + .0053x - .0000266x^2$.

a. Construct a rough sketch of the least squares prediction equation. Describe the nature of the curvature in the estimated model.

b. Estimate the weight change (y) for a rat given a dosage of 500 mg/kg of aconiazide.

c. Estimate the weight change (y) for a rat given a dosage of 0 mg/kg of aconiazide. (This dosage is called the *control* dosage level.)

d. Of the five dosage groups in the study, find the largest dosage level x that yields an estimated weight change that is closest to but below the estimated weight change for the control group. This value is the *change-point* dosage.

11.59 Failure times of silicon wafer microchips. Researchers at National Semiconductor experimented with tin-lead solder bumps used to manufacture silicon wafer integrated circuit chips (International Wafer Level Packaging Conference, Nov. 3–4, 2005). The failure times of the microchips (in hours) was determined at different solder temperatures (degrees Celsius). The data for one experiment are given in the table and saved in the **WAFER** file. The researchers want to predict failure time (y) based on solder temperature (x).

a. Construct a scatterplot for the data. What type of relationship, linear or curvilinear, appears to exist between failure time and solder temperature?

Temperature (°C)	Time to Failure (hours)
165	200
162	200
164	1,200
158	500
158	600
159	750
156	1,200
157	1,500
152	500
147	500
149	1,100
149	1,150
142	3,500
142	3,600
143	3,650
133	4,200
132	4,800
132	5,000
134	5,200
134	5,400
125	8,300
123	9,700

Source: Gee, S., and Nguyen, L. "Mean time to failure in wafer level-CSP packages with SnPb and SnAgCu solder bmps," International Wafer Level Packaging Conference, San Jose, CA, Nov. 3–4, 2005 (adapted from Figure 7).

b. Fit the model, $E(y) = \beta_0 + \beta_1 x + \beta_2 x^2$, to the data. Give the least squares prediction equation.

c. Conduct a test to determine if there is upward curvature in the relationship between failure time and solder temperature. (Use $\alpha = .05$.)

11.60 Public perceptions of health risks. In the *Journal of Experimental Psychology: Learning, Memory, and Cognition* (July 2005), University of Basel (Switzerland) psychologists tested the ability of people to judge risk of an infectious disease. The researchers asked German college students to estimate the number of people who are infected with a certain disease in a typical year. The median estimates as well as the actual incidence rate for each in a sample of 24 infections are provided in the table and saved in the **INFECTION** file. Consider the quadratic model $E(y) = \beta_0 + \beta_1 x + \beta_2 x^2$, where y = actual incidence rate and x = estimated rate.

Infection	Incidence Rate	Estimate
Polio	0.25	300
Diphtheria	1	1,000
Trachoma	1.75	691
Rabbit fever	2	200
Cholera	3	17.5
Leprosy	5	0.8
Tetanus	9	1,000
Hemorrhagic fever	10	150
Trichinosis	22	326.5
Undulant fever	23	146.5
Well's disease	39	370
Gas gangrene	98	400
Parrot fever	119	225
Typhoid	152	200
Q fever	179	200
Malaria	936	400
Syphilis	1,514	1,500
Dysentery	1,627	1,000
Gonorrhea	2,926	6,000
Meningitis	4,019	5,000
Tuberculosis	12,619	1,500
Hepatitis	14,889	10,000
Gastroenteritis	203,864	37,000
Botulism	15	37,500

Source: Hertwig, R., Pachur, T., and Kurzenhauser, S. "Judgments of risk frequencies: Tests of possible cognitive mechanisms," *Journal of Experimental Psychology: Learning, Memory, and Cognition,* Vol. 31, No. 4, July 2005 (Table 1). Copyright © 2005 by the American Psychological Association. Reproduced with permission. The use of APA information does not imply endorsement by APA.

a. Fit the quadratic model to the data and then conduct a test to determine if incidence rate is curvilinearly related to estimated rate. (Use $\alpha = .05$.)

b. Construct a scatterplot for the data. Locate the data point for botulism on the graph. What do you observe?

c. Repeat part **a** but omit the data point for botulism from the analysis. Has the fit of the model improved? Explain.

11.61 Strength of manufactured plastic. The amount of pressure used to produce a certain plastic is thought to be related to the strength of the plastic. Researchers hypothesize that, below a certain level, increases in pressure increase the strength of the plastic; at some point, however, additional increases in pressure will have a detrimental effect on its strength. Write a model in which the relationship of the plastic strength y to pressure x reflects this hypothesis. Sketch the model.

11.62 Orange juice demand study. A chilled orange juice warehousing operation in New York City was experiencing too many out-of-stock situations with its 96-ounce containers. To better understand current and future demand for this product, the company examined the last 40 days of sales, which are shown in the next table. (The data are saved in the **NYJUICE** file.) One of the company's objectives is to model demand, y, as a function of sale day, x (where $x = 1, 2, 3, \ldots, 40$).

a. Construct a scatterplot for these data.

b. Does it appear that a second-order model might better explain the variation in demand than a first-order model? Explain.

c. Fit a first-order model to these data.

d. Fit a second-order model to these data.

e. Compare the results in parts **c** and **d** and decide which model better explains variation in demand. Justify your choice.

Sale Day, x	Demand for 96 oz. Containers, y (in cases)
1	4,581
2	4,239
3	2,754
4	4,501
5	4,016
6	4,680
7	4,950
8	3,303
9	2,367
10	3,055
11	4,248
12	5,067
13	5,201
14	5,133
15	4,211
16	3,195
17	5,760
18	5,661
19	6,102
20	6,099
21	5,902
22	2,295
23	2,682
24	5,787
25	3,339
26	3,798
27	2,007
28	6,282
29	3,267
30	4,779
31	9,000
32	9,531
33	3,915
34	8,964
35	6,984
36	6,660
37	6,921
38	10,005
39	10,153
40	11,520

Source: Personal communication from Rick Campbell, Dave Metzler, and Tom Nelson, Rutgers University.

11.7 Qualitative (Dummy) Variable Models

Multiple regression models can also be written to include **qualitative** (or **categorical**) independent variables. Qualitative variables, unlike quantitative variables, cannot be measured on a numerical scale. Therefore, we must code the values of the qualitative variable (called **levels**) as numbers before we can fit the model. These coded qualitative variables are called **dummy** (or **indicator**) **variables** because the numbers assigned to the various levels are arbitrarily selected.

To illustrate, suppose a female executive at a certain company claims that male executives earn higher salaries, on average, than female executives with the same education, experience, and responsibilities. To support her claim, she wants to model the salary y of an executive using a qualitative independent variable representing the gender of an executive (male or female).

A convenient method of coding the values of a qualitative variable at two levels involves assigning a value of 1 to one of the levels and a value of 0 to the other. For example, the dummy variable used to describe gender could be coded as follows:

$$x = \begin{cases} 1 & \text{if male} \\ 0 & \text{if female} \end{cases}$$

The choice of which level is assigned to 1 and which is assigned to 0 is arbitrary. The model then takes the following form:

$$E(y) = \beta_0 + \beta_1 x$$

The advantage of using a 0–1 coding scheme is that the β coefficients are easily interpreted. The model above allows us to compare the mean executive salary $E(y)$ for males with the corresponding mean for females.

$$\text{Males } (x = 1)\text{: } E(y) = \beta_0 + \beta_1(1) = \beta_0 + \beta_1$$
$$\text{Females } (x = 0)\text{: } E(y) = \beta_0 + \beta_1(0) = \beta_0$$

These two means are illustrated in the bar graph in Figure 11.18.

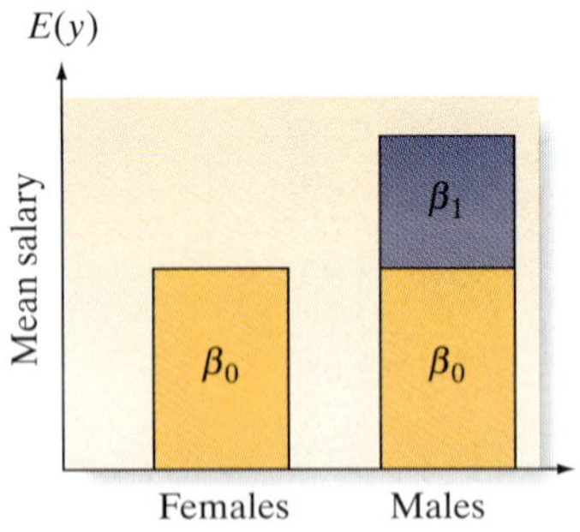

Figure 11.18
Bar graph comparing $E(y)$ for males and females

First note that β_0 represents the mean salary for females (say, μ_F). When a 0–1 coding convention is used, β_0 will always represent the mean response associated with the level of the qualitative variable assigned the value 0 (called the **base level**). The difference between the mean salary for males and the mean salary for females, $\mu_M - \mu_F$, is represented by β_1—that is,

$$\mu_M - \mu_F = (\beta_0 + \beta_1) - (\beta_0) = \beta_1$$

This difference is shown in Figure 11.18.* With a 0–1 coding convention, β_1 will always represent the difference between the mean response for the level assigned the value 1 and the mean for the base level. Thus, for the executive salary model, we have

$$\beta_0 = \mu_F$$
$$\beta_1 = \mu_M - \mu_F$$

The model relating a mean response $E(y)$ to a qualitative independent variable at two levels is summarized in the following box.

*Note that β_1 could be negative. If β_1 were negative, the height of the bar corresponding to males would be *reduced* (rather than increased) from the height of the bar for females by the amount β_1. Figure 11.18 is constructed assuming that β_1 is a positive quantity.

A Model Relating $E(y)$ to a Qualitative Independent Variable with Two Levels

$$E(y) = \beta_0 + \beta_1 x$$

where

$$x = \begin{cases} 1 & \text{if level A} \\ 0 & \text{if level B} \end{cases}$$

Interpretation of β's:

$$\beta_0 = \mu_B \text{ (mean for base level)}$$
$$\beta_1 = \mu_A - \mu_B$$

Now carefully examine the model with a single qualitative independent variable at two levels; we will use exactly the same pattern for any number of levels. Moreover, the interpretation of the parameters will always be the same. One level (say, level A) is selected as the base level. Then for the 0–1 coding* for the dummy variables,

$$\mu_A = \beta_0$$

The coding for all dummy variables is as follows: To represent the mean value of y for a particular level, let that dummy variable equal 1; otherwise, the dummy variable is set equal to 0. Using this system of coding,

$$\mu_B = \beta_0 + \beta_1$$
$$\mu_C = \beta_0 + \beta_2$$

and so on. Because $\mu_A = \beta_0$, any other model parameter will represent the difference between means for that level and the base level:

$$\beta_1 = \mu_B - \mu_A$$
$$\beta_2 = \mu_C - \mu_A$$

and so on. Consequently, each β multiplied by a dummy variable represents the difference between $E(y)$ at one level of the qualitative variable and $E(y)$ at the base level.

A Model Relating $E(y)$ to One Qualitative Independent Variable with k Levels

Always use a number of dummy variables that is one less than the number of levels of the qualitative variable. Thus, for a qualitative variable with k levels, use $k - 1$ dummy variables:

$$y = \beta_0 + \beta_1 x_1 + \beta_2 x_2 + \cdots + \beta_{k-1} x_{k-1} + \varepsilon$$

where x_i is the dummy variable for level $i + 1$ and

$$x_i = \begin{cases} 1 & \text{if } y \text{ is observed at level } i + 1 \\ 0 & \text{otherwise} \end{cases}$$

Then, for this system of coding

$$\begin{aligned} \mu_A &= \beta_0 \\ \mu_B &= \beta_0 + \beta_1 \\ \mu_C &= \beta_0 + \beta_2 \\ \mu_D &= \beta_0 + \beta_3 \end{aligned} \qquad \text{and} \qquad \begin{aligned} \beta_1 &= \mu_B - \mu_A \\ \beta_2 &= \mu_C - \mu_A \\ \beta_3 &= \mu_D - \mu_A \\ &\vdots \end{aligned}$$

*You do not have to use a 0–1 system of coding for the dummy variables. Any two-value system will work, but the interpretation given to the model parameters will depend on the code. Using the 0–1 system makes the model parameters easy to interpret.

Example 11.9

A Model with One Qualitative Independent Variable at Three Levels

Problem Suppose an economist wants to compare the mean dollar amounts owed by delinquent credit card customers in three different socioeconomic classes: (1) lower class, (2) middle class, and (3) upper class. A sample of 10 customers with delinquent accounts is selected from each group, and the amount owed by each is recorded, as shown in Table 11.3.

Table 11.3 Dollars Owed, Example 11.9

Group 1 (lower class)	Group 2 (middle class)	Group 3 (upper class)
\$148	\$513	\$335
76	264	643
393	433	216
520	94	536
236	535	128
134	327	723
55	214	258
166	135	380
415	280	594
153	304	465

Data Set: DELINQUENT

a. Hypothesize a regression model for amount owed (y) using socioeconomic class as an independent variable.

b. Interpret the β's in the model.

c. Fit the model to the data and give the least squares prediction equation.

d. Use the model to determine if the mean dollar amounts owed by customers differ significantly for the three socioeconomic groups at $\alpha = .05$.

Solution

a. Note that socioeconomic status (low, middle, upper class) is a qualitative variable (measured on an ordinal scale). For a 3-level qualitative variable, we need two dummy variables in the regression model. The model relating $E(y)$ to this single qualitative variable, socioeconomic group, is

$$E(y) = \beta_0 + \beta_1 x_1 + \beta_2 x_2$$

where (arbitrarily choosing group 1 as the base level)

$$x_1 = \begin{cases} 1 & \text{if group 2} \\ 0 & \text{if not} \end{cases} \qquad x_2 = \begin{cases} 1 & \text{if group 3} \\ 0 & \text{if not} \end{cases}$$

b. For this model

$$\beta_0 = \mu_1$$
$$\beta_1 = \mu_2 - \mu_1$$
$$\beta_2 = \mu_3 - \mu_1$$

where μ_1, μ_2, and μ_3 are the mean responses for socioeconomic groups 1, 2, and 3, respectively—that is, β_0 represents the mean amount owed for group 1 (lower class), β_1 represents the mean difference in amounts owed for group 2 (middle) and group 1 (lower), and β_2 represents the mean difference in amounts owed for group 3 (upper) and group 1 (lower).

c. The Minitab printout of the regression analysis is shown in Figure 11.19. The least squares estimates of the β's are highlighted on the printout, yielding the following least squares prediction equation:

$$\hat{y} = 229.6 + 80.3x_1 + 198.2x_2$$

Regression Analysis: Amount versus X1, X2

```
The regression equation is
Amount = 230 + 80.3 X1 + 198 X2

Predictor     Coef  SE Coef     T      P
Constant    229.60    53.43  4.30  0.000
X1           80.30    75.56  1.06  0.297
X2          198.20    75.56  2.62  0.014

S = 168.948   R-Sq = 20.5%   R-Sq(adj) = 14.6%

Analysis of Variance

Source          DF      SS     MS     F      P
Regression       2  198772  99386  3.48  0.045
Residual Error  27  770671  28543
Total           29  969443
```

Descriptive Statistics: Amount

```
Variable  Class   N   Mean  StDev  Minimum  Maximum
Amount    L      10  229.6  158.2     55.0    520.0
          M      10  309.9  147.9     94.0    535.0
          U      10  427.8  196.8    128.0    723.0
```

Figure 11.19
Minitab output for model with dummy variables

The interpretations of the β's in part **b** allow us to obtain the β estimates from the sample means associated with the different levels of the qualitative variable.*

Because $\beta_0 = \mu_1$, then the estimate of β_0 is the estimated mean delinquent amount for the lower class group. This sample mean, highlighted at the bottom of the Minitab printout, is 229.6; thus $\hat{\beta}_0 = 229.6$.

Now $\beta_1 = \mu_2 - \mu_1$; therefore, the estimate of β_1 is the difference between the sample mean delinquent amounts for the middle- and lower-class groups. Based on the sample means highlighted at the bottom of the Minitab printout, we have $\hat{\beta}_1 = 309.9 - 229.6 = 80.3$.

Similarly, the estimate of $\beta_2 = \mu_3 - \mu_1$ is the difference between the sample mean delinquent amounts for the upper- and lower-class groups. From the sample means highlighted at the bottom of the Minitab printout, we have $\hat{\beta}_2 = 427.8 - 229.6 = 198.2$.

d. Testing the null hypothesis that the means for the three groups are equal—that is, $\mu_1 = \mu_2 = \mu_3$—is equivalent to testing

$$H_0: \beta_1 = \beta_2 = 0$$

You can see this by observing that if $\beta_1 = \mu_2 - \mu_1 = 0$, then $\mu_1 = \mu_2$. Similarly, if $\beta_2 = \mu_3 - \mu_1 = 0$, then $\mu_3 = \mu_1$. Thus, if H_0 is true, then μ_1, μ_2, and μ_3 must be equal. The alternative hypothesis is

H_a: At least one of the parameters, β_1 and β_2, differs from 0, which implies that at least two of the three means (μ_1, μ_2, and μ_3) differ.

To test this hypothesis, we conduct the global F-test for the model. The value of the F-statistic for testing the adequacy of the model, $F = 3.48$, and the observed significance level of the test, p-value $= .045$, are both highlighted on Figure 11.19. Because $\alpha = .05$ exceeds the p-value, we reject H_0 and conclude that at least one of

*The least squares method and the sample means method will yield equivalent β estimates when the sample sizes associated with the different levels of the qualitative variable are equal.

the parameters, β_1 and β_2, differs from 0. Or, equivalently, we conclude that the data provide sufficient evidence to indicate that the mean indebtedness does vary from one socioeconomic group to another.

Look Back This global F-test is equivalent to the analysis of variance F-test for a completely randomized design of Chapter 8.

Now Work Exercise 11.74

CAUTION A common mistake by regression analysts is to use a single dummy variable x for a qualitative variable at k levels, where $x = 1, 2, 3, \ldots, k$. Such a regression model will have unestimable β's and β's that are difficult to interpret. Remember, when modeling $E(y)$ with a single qualitative independent variable, the number of 0–1 dummy variables to include in the model will always be one less than the number of levels of the qualitative variable. ▲

Exercises 11.63–11.77

Learning the Mechanics

11.63 Write a regression model relating the mean value of y to a qualitative independent variable that can assume two levels. Interpret all the terms in the model.

11.64 Write a regression model relating $E(y)$ to a qualitative independent variable that can assume three levels. Interpret all the terms in the model.

11.65 The following model was used to relate $E(y)$ to a single qualitative variable with four levels:

$$E(y) = \beta_0 + \beta_1 x_1 + \beta_2 x_2 + \beta_3 x_3$$

where

$$x_1 = \begin{cases} 1 \text{ if level 2} \\ 0 \text{ if not} \end{cases} \quad x_2 = \begin{cases} 1 \text{ if level 3} \\ 0 \text{ if not} \end{cases} \quad x_3 = \begin{cases} 1 \text{ if level 4} \\ 0 \text{ if not} \end{cases}$$

This model was fit to $n = 30$ data points, and the following result was obtained:

$$\hat{y} = 10.2 - 4x_1 + 12x_2 + 2x_3$$

a. Use the least squares prediction equation to find the estimate of $E(y)$ for each level of the qualitative independent variable.

b. Specify the null and alternative hypotheses you would use to test whether $E(y)$ is the same for all four levels of the independent variable.

11.66 The Minitab printout below resulted from fitting the following model to $n = 15$ data points:

$$y = \beta_0 + \beta_1 x_1 + \beta_2 x_2 + \varepsilon$$

where

$$x_1 = \begin{cases} 1 \text{ if level 2} \\ 0 \text{ if not} \end{cases} \quad x_2 = \begin{cases} 1 \text{ if level 3} \\ 0 \text{ if not} \end{cases}$$

a. Report the least squares prediction equation.

b. Interpret the values of β_1 and β_2.

c. Interpret the following hypotheses in terms of μ_1, μ_2, and μ_3:

H_0: $\beta_1 = \beta_2 = 0$

H_a: At least one of the parameters β_1 and β_2 differs from 0

d. Conduct the hypothesis test of part **c.**

Applying the Concepts—Basic

11.67 Production technologies, terroir, and quality of Bordeaux wine. In addition to state-of-the-art technologies, the

```
The regression equation is
Y = 80.0 + 16.8 X1 + 40.4 X2


Predictor      Coef   SE Coef       T      P
Constant     80.000     4.082   19.60  0.000
X1           16.800     5.774    2.91  0.013
X2           40.400     5.774    7.00  0.000


S = 9.129      R-Sq = 80.5%    R-Sq(adj) = 77.2%

Analysis of Variance

Source          DF       SS       MS       F      P
Regression       2   4118.9   2059.5   24.72  0.000
Residual Error  12   1000.0     83.3
Total           14   5118.9
```

Minitab Output for Exercise 11.66

production of quality wine is strongly influenced by the natural endowments of the grape-growing region—called the "terroir." *The Economic Journal* (May 2008) published an empirical study of the factors that yield a quality Bordeaux wine. A quantitative measure of wine quality (y) was modeled as a function of several qualitative independent variables, including grape-picking method (manual or automated), soil type (clay, gravel, or sand), and slope orientation (east, south, west, southeast, or southwest).

a. Create the appropriate dummy variables for each of the qualitative independent variables.
b. Write a model for wine quality (y) as a function of grape-picking method. Interpret the β's in the model.
c. Write a model for wine quality (y) as a function of soil type. Interpret the β's in the model.
d. Write a model for wine quality (y) as a function of slope orientation. Interpret the β's in the model.

11.68 Impact of race on football card values. University of Colorado sociologists investigated the impact of race on the value of professional football players' "rookie" cards (*Electronic Journal of Sociology,* 2007). The sample consisted of 148 rookie cards of National Football League (NFL) players who were inducted into the Football Hall of Fame. The price of the card (in dollars) was modeled as a function of several qualitative independent variables: race of player (black or white), card availability (high or low), and player position (quarterback, running back, wide receiver, tight end, defensive lineman, linebacker, defensive back, or offensive lineman).

a. Create the appropriate dummy variables for each of the qualitative independent variables.
b. Write a model for price (y) as a function of race. Interpret the β's in the model.
c. Write a model for price (y) as a function of card availability. Interpret the β's in the model.
d. Write a model for price (y) as a function of position. Interpret the β's in the model.

11.69 Accuracy of software effort estimates. Periodically, software engineers must provide estimates of their effort in developing new software. In the *Journal of Empirical Software Engineering* (Vol. 9, 2004), multiple regression was used to predict the accuracy of these effort estimates. The dependent variable, defined as the relative error in estimating effort,

$$y = (\text{actual effort} - \text{estimated effort})/(\text{actual effort})$$

was determined for each in a sample of $n = 49$ software development tasks. Several qualitative independent variables were evaluated as potential predictors of relative error. Some of these variables are described in the table.

Estimator role (developer or project leader)
Task complexity (low, medium, or high)
Contract type (fixed price or hourly rate)
Customer priority (time of delivery, cost, or quality)

a. Write a model for $E(y)$ as a function of estimator role. Interpret the β's.
b. Write a model for $E(y)$ as a function of task complexity. Interpret the β's.
c. Write a model for $E(y)$ as a function of contract type. Interpret the β's.
d. Write a model for $E(y)$ as a function of customer priority. Interpret the β's.

11.70 Buy-side vs. sell-side analysts' earnings forecasts. Refer to the *Financial Analysts Journal* (Jul./Aug. 2008) comparison of earnings forecasts of buy-side and sell-side analysts, Exercise 2.84 (p. 76). The Harvard Business School professors used regression to model the relative optimism (y) of the analysts' 3-month horizon forecasts. One of the independent variables used to model forecast optimism was the dummy variable $x = \{1$ if the analyst worked for a buy-side firm, 0 if the analyst worked for a sell-side firm$\}$.

a. Write the equation of the model for $E(y)$ as a function of type of firm.
b. Interpret the value of β_0 in the model, part **a.**
c. The professors write that the value of β_1 in the model, part **a,** "represents the mean difference in relative forecast optimism between buy-side and sell-side analysts." Do you agree?
d. The professors also argue that "if buy-side analysts make less optimistic forecasts than their sell-side counterparts, the [estimated value of β_1] will be negative." Do you agree?

11.71 Improving SAT scores. Refer to the *Chance* (Winter 2001) study of students who paid a private tutor (or coach) to help them improve their Scholastic Assessment Test (SAT) scores, Exercise 2.86 (p. 76). Multiple regression was used to estimate the effect of coaching on SAT–Mathematics scores. Data on 3,492 students (573 of whom were coached) were used to fit the model, $E(y) = \beta_0 + \beta_1 x_1 + \beta_2 x_2$, where y = SAT–Math score, x_1 = score on PSAT, and $x_2 = \{1$ if student was coached, 0 if not$\}$.

a. The fitted model had an adjusted R^2 value of .76. Interpret this result.
b. The estimate of β_2 in the model was 19, with a standard error of 3. Use this information to form a 95% confidence interval for β_2. Interpret the interval.
c. Based on the interval, part **b,** what can you say about the effect of coaching on SAT–Math scores?

Applying the Concepts—Intermediate

11.72 Deferred tax allowance study. A study was conducted to identify accounting choice variables that influence a manager's decision to change the level of the deferred tax asset allowance at the firm (*The Engineering Economist,* Jan./Feb. 2004). Data were collected for a sample of 329 firms that reported deferred tax assets in 2000. The dependent variable of interest (DTVA) is measured as the change in the deferred tax asset valuation allowance divided by the deferred tax asset. The independent variables used as predictors of DTVA are listed as follows:

LEVERAGE:	x_1 = ratio of debt book value to shareholder's equity
BONUS:	x_2 = 1 if firm maintains a management bonus plan, 0 if not
MVALUE:	x_3 = market value of common stock

BBATH: $x_4 = 1$ if operating earnings negative and lower than last year, 0 if not

EARN: $x_5 =$ change in operating earnings divided by total assets

A first-order model was fit to the data with the following results (*p*-values in parentheses):

$$R_a^2 = .280$$

$$\hat{y} = .044 + .006x_1 - .035x_2 - .001x_3 + .296x_4 + .010x_5$$
$$(.070)\ (.228)\quad (.157)\quad (.678)\quad (.001)\quad (.869)$$

a. Interpret the estimate of the β coefficient for x_4.
b. The "Big Bath" theory proposed by the researchers states that the mean DTVA for firms with negative earnings and earnings lower than last year will exceed the mean DTVA of other firms. Is there evidence to support this theory? Test using $\alpha = .05$.
c. Interpret the value of R_a^2.

11.73 REIT equity study. Robert Johnson (Association for Investment Management and Research) and Gerald Jensen (Northern Illinois University) examined the effects of the Federal Reserve's monetary policies on the rates of returns to different types of real estate assets (*Real Estate Finance,* Spring 1999). They employed the following model for the monthly rate of return (y) for the asset:

$$E(y) = \beta_0 + \beta_1 x$$

$$\text{where } x = \begin{cases} 1 \text{ if the Fed's monetary policy is restrictive} \\ 0 \text{ if the Fed's monetary policy is expansive} \end{cases}$$

In one part of the study, they fit this model using monthly rate of return data from 1972 to 1997 for an equity REIT (real estate investment trust) index and, for comparison purposes, to a Treasury Bill (T-Bill) index. They obtained the results (t statistics shown in parentheses) shown in the accompanying table.

Index	Estimate of β_0	Estimate of β_1	*F*-value	R^2
T-Bills	0.04742 (30.94)	0.001948 (8.14)	66.24	0.1819
Equity REIT	0.01863 (6.19)	−0.01582 (−3.46)	11.98	0.0387

Source: Johnson, R.R., and Jensen, G.R. "Federal Reserve monetary policy and real estate investment trust returns," *Real Estate Finance,* Vol. 16, No. 1, Spring 1999, pp. 52–59.

a. Evaluate the usefulness of each fitted model. Draw conclusions in the context of the problem.
b. Interpret the estimated value of β_1 in each model.
c. Predict the mean monthly rate of return for the equity REIT index when the Federal Reserve's monetary policy is restrictive. Repeat for an expansive policy.

11.74 Homework assistance for accounting students. Refer to the NW *Journal of Accounting Education* (Vol. 25, 2007) study of assisting accounting students with their homework, Exercise 8.28 (p. 469). Recall that 175 accounting students took a pretest on a topic not covered in class and then each was given a homework problem to solve on the same topic. The students were assigned to one of three homework assistance groups. Some students received the completed solution, some were given check figures at various steps of the solution, and some received no help at all. After finishing the homework, the students were all given a posttest on the subject. The dependent variable of interest was the knowledge gain (or test score improvement). These data are saved in the **ACCHW** file.

a. Propose a model for the knowledge gain (y) as a function of the qualitative variable, homework assistance group.
b. In terms of the β's in the model, give an expression for the difference between the mean knowledge gains of students in the "completed solution" and "no help" groups.
c. Fit the model to the data and give the least squares prediction equation.
d. Conduct the global *F*-test for model utility using $\alpha = .05$. Interpret the results, practically.
e. Show that the results, part **d,** agree with the conclusions reached in Exercise 8.28.

11.75 Comparing mosquito repellents. Which insect repellents protect best against mosquitoes? *Consumer Reports* (June 2000) tested 14 products that all claim to be an effective mosquito repellent. Each product was classified as either lotion/cream or aerosol/spray. The cost of the product (in dollars) was divided by the amount of the repellent needed to cover exposed areas of the skin (about 1/3 ounce) to obtain a cost-per-use value. Effectiveness was measured as the maximum number of hours of protection (in half-hour increments) provided when human testers exposed their arms to 200 mosquitoes. The data from the report are listed in the table and saved in the **REPELLENT** file.

Insect Repellent	Type	Cost/ Use	Maximum Protection
Amway Hour Guard 12	Lotion/Cream	$2.08	13.5 hours
Avon Skin-So-Soft	Aerosol/Spray	0.67	0.5
Avon BugGuard Plus	Lotion/Cream	1.00	2.0
Ben's Backyard Formula	Lotion/Cream	0.75	7.0
Bite Blocker	Lotion/Cream	0.46	3.0
BugOut	Aerosol/Spray	0.11	6.0
Cutter Skinsations	Aerosol/Spray	0.22	3.0
Cutter Unscented	Aerosol/Spray	0.19	5.5
Muskoll Ultra6Hours	Aerosol/Spray	0.24	6.5
Natrapel	Aerosol/Spray	0.27	1.0
Off! Deep Woods	Aerosol/Spray	1.77	14.0
Off! Skintastic	Lotion/Cream	0.67	3.0
Sawyer Deet Formula	Lotion/Cream	0.36	7.0
Repel Permanone	Aerosol/Spray	2.75	24.0

Source: "Buzz off," *Consumer Reports,* June 2000. "Insect repellants: Which keep bugs at bay?" *Consumer Reports,* June 2000. Copyright 2000 by Consumers Union of U.S., Inc. Yonkers, NY 10703-1057, a nonprofit organization. Reprinted with permission from the June 2000 issue of Consumer Reports® for educational purposes only. No commercial use or reproduction permitted. www.ConsumerReports.org.

a. Suppose you want to use repellent type to model the cost per use (y). Create the appropriate number of dummy variables for repellent type and write the model.
b. Fit the model, part **a,** to the data.
c. Give the null hypothesis for testing whether repellent type is a useful predictor of cost per use (y).
d. Conduct the test, part **c,** and give the appropriate conclusion. Use $\alpha = .10$.
e. Repeat parts **a–d** if the dependent variable is the maximum number of hours of protection (y).

Applying the Concepts—Advanced

11.76 Manipulating rates of return with stock splits. Some firms have been accused of using stock splits to manipulate their stock prices before being acquired by another firm. An article in *Financial Management* (Winter 2008) investigated the impact of stock splits on long-run stock performance for acquiring firms. A simplified version of the model fit by the researchers follows:

$$E(y) = \beta_0 + \beta_1 x_1 + \beta_2 x_2 + \beta_3 x_1 x_2,$$

where

y = Firm's 3-year buy-and-hold return rate (%)
x_1 = {1 if stock split prior to acquisition, 0 if not}
x_2 = {1 if firm's discretionary accrual is high, 0 if discretionary accrual is low}

a. In terms of the β's in the model, what is the mean buy-and-hold return rate (BAR) for a firm with no stock split and a high discretionary accrual (DA)?
b. In terms of the β's in the model, what is the mean BAR for a firm with no stock split and a low DA?
c. For firms with no stock split, find the difference between the mean BAR for firms with high and low DA. (*Hint:* Use your answers to parts **a** and **b.**)
d. Repeat part **c** for firms with a stock split.
e. Note that the differences, parts **c** and **d,** are not the same. Explain why this illustrates the notion of interaction between x_1 and x_2.
f. A test for H_0: $\beta_3 = 0$ yielded a p-value of .027. Using $\alpha = .05$, interpret this result.
g. The researchers reported that the estimated values of both β_2 and β_3 are negative. Consequently, they conclude that "high-DA acquirers perform worse compared with low-DA acquirers. Moreover, the underperformance is even greater if high-DA acquirers have a stock split before acquistion." Do you agree?

11.77 Study of recall of TV commercials. Refer to the *Journal of Applied Psychology* (June 2002) study of recall of television commercials, *Statistics in Action,* Chapter 8 (p. 447). Participants were assigned to watch one of three types of TV programs, with nine commercials embedded in each show. Group V watched a TV program with a violent content code rating (e.g., *Tour of Duty*); group S viewed a show with a sex content code rating (e.g., *Strip Mall*); and group N watched a neutral TV program with neither a V nor an S rating (e.g., *Candid Camera*). The dependent variable measured for each participant was a score (y) on his/her recall of the brand names in the commercial messages, with scores ranging from 0 (no brands recalled) to 9 (all brands recalled). The data are saved in the **TVADRECALL** file.

a. Write a model for $E(y)$ as a function of viewer group.
b. Fit the model, part **a,** to the data saved in the **TVADRECALL** file. Give the least squares prediction equation.
c. Conduct a test of overall model utility at $\alpha = .01$. Interpret the results. Show that the results agree with the analysis conducted in the Chapter 8 SIA.
d. The sample mean recall scores for the three groups were $\bar{y}_V = 2.08$, $\bar{y}_S = 1.71$, and $\bar{y}_N = 3.17$. Show how to find these sample means using only the β estimates obtained in part **b.**

11.8 Models with Both Quantitative and Qualitative Variables

Suppose you want to relate the mean monthly sales $E(y)$ of a company to monthly advertising expenditure x for three different advertising media (say newspaper, radio, and television) and you wish to use first-order (straight-line) models to model the responses for all three media. Graphs of these three relationships might appear as shown in Figure 11.20.

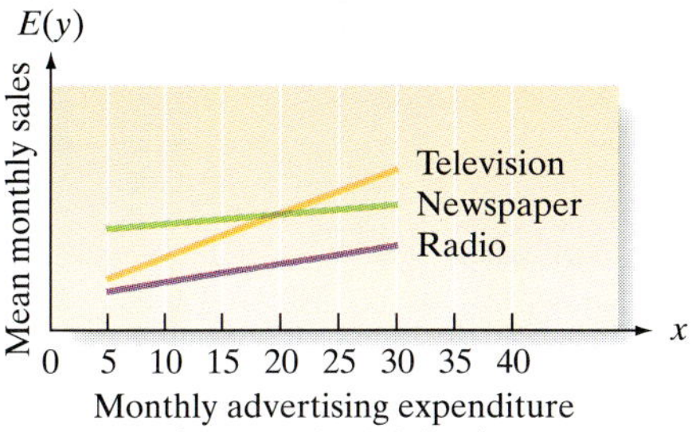

Figure 11.20
Graphs of the relationship between mean sales $E(y)$ and advertising expenditure x

Because the lines in Figure 11.20 are hypothetical, a number of practical questions arise. Is one advertising medium as effective as any other? That is, do the three mean sales lines differ for the three advertising media? Do the increases in mean sales per dollar input in advertising differ for the three advertising media? That is, do the slopes of the three lines differ? Note that the two practical questions have been rephrased into questions about the parameters that define the three lines in Figure 11.20. To answer them, we must write a single regression model that will characterize the three lines of Figure 11.20 and that, by testing hypotheses about the lines, will answer the questions.

The response described previously, monthly sales, is a function of *two* independent variables, one quantitative (advertising expenditure x_1) and one qualitative (type of medium). We will proceed, in stages, to build a model relating $E(y)$ to these variables and will show graphically the interpretation we would give to the model at each stage. This will help you to see the contributions of the various terms in the model.

1. The straight-line relationship between mean sales $E(y)$ and advertising expenditure is the same for all three media—that is, a single line will describe the relationship between $E(y)$ and advertising expenditure x_1 for all the media (see Figure 11.21).

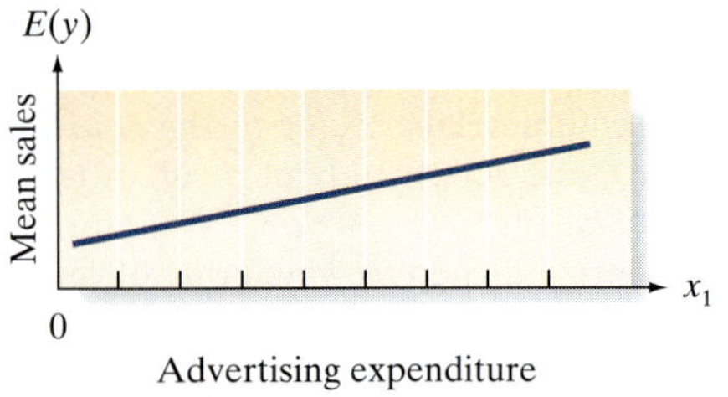

Figure 11.21
The relationship between $E(y)$ and x_1 is the same for all media

$$E(y) = \beta_0 + \beta_1 x \qquad \text{where } x_1 = \text{advertising expenditure}$$

2. The straight lines relating mean sales $E(y)$ to advertising expenditure x_1 differ from one medium to another, but the rate of increase in mean sales per increase in dollar advertising expenditure x_1 is the same for all media—that is, the lines are parallel but possess different y-intercepts (see Figure 11.22).

$$E(y) = \beta_0 + \beta_1 x_1 + \beta_2 x_2 + \beta_3 x_3$$

where

$$x_1 = \text{advertising expenditure}$$
$$x_2 = \begin{cases} 1 & \text{if radio medium} \\ 0 & \text{if not} \end{cases}$$
$$x_3 = \begin{cases} 1 & \text{if television medium} \\ 0 & \text{if not} \end{cases}$$

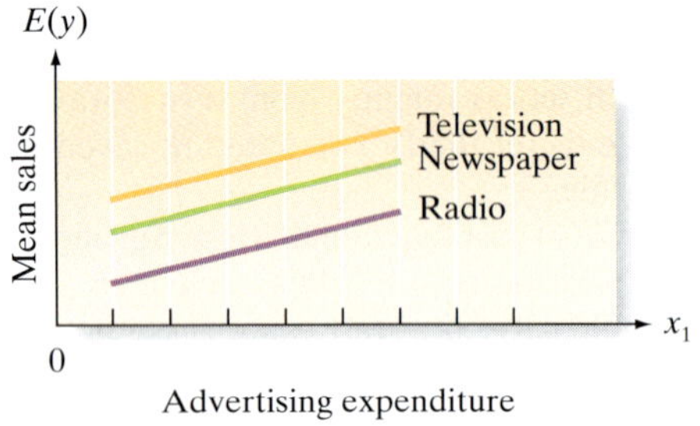

Figure 11.22
Parallel response lines for the three media

Notice that this model is essentially a combination of a first-order model with a single quantitative variable and the model with a single qualitative variable:

First-order model with a single quantitative variable: $E(y) = \beta_0 + \beta_1 x_1$

Model with single qualitative variable at three levels: $E(y) = \beta_0 + \beta_2 x_2 + \beta_3 x_3$

where x_1, x_2, and x_3 are as just defined. The model described here implies no interaction between the two independent variables, which are advertising expenditure x_1 and the qualitative variable (type of advertising medium). The change in $E(y)$ for a 1-unit increase in x_1 is identical (the slopes of the lines are equal) for all three advertising media. The terms corresponding to each of the independent variables are called **main effect terms** because they imply no interaction.

3. The straight lines relating mean sales $E(y)$ to advertising expenditure x_1 differ for the three advertising media—that is, both the line intercepts and the slopes differ (see Figure 11.23). As you will see, this interaction model is obtained by adding terms involving the cross-product terms, one each from each of the two independent variables:

$$E(y) = \beta_0 + \underbrace{\beta_1 x_1}_{\substack{\text{Main effect,}\\ \text{advertising}\\ \text{expenditure}}} + \underbrace{\beta_2 x_2 + \beta_3 x_3}_{\substack{\text{Main effect,}\\ \text{type of}\\ \text{medium}}} + \underbrace{\beta_4 x_1 x_2 + \beta_5 x_1 x_3}_{\text{Interaction}}$$

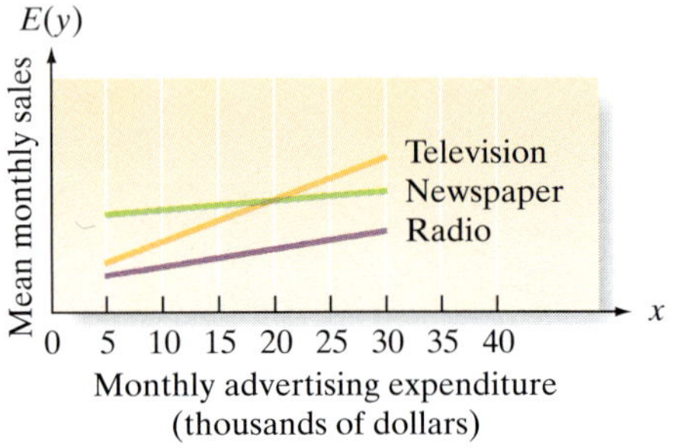

Figure 11.23
Different response lines for the three media

Note that each of the preceding models is obtained by adding terms to model 1, the single first-order model used to model the responses for all three media. Model 2 is obtained by adding the main effect terms for type of medium, the qualitative variable. Model 3 is obtained by adding the interaction terms to model 2.

Example 11.10
Interpreting the β's in a Model with Mixed Independent Variables

Problem Substitute the appropriate values of the dummy variables in model 3 (above) to obtain the equations of the three response lines in Figure 11.23.

Solution The complete model that characterizes the three lines in Figure 11.23 is

$$E(y) = \beta_0 + \beta_1 x_1 + \beta_2 x_2 + \beta_3 x_3 + \beta_4 x_1 x_2 + \beta_5 x_1 x_3$$

where

$$x_1 = \text{advertising expenditure}$$
$$x_2 = \begin{cases} 1 & \text{if radio medium} \\ 0 & \text{if not} \end{cases}$$
$$x_3 = \begin{cases} 1 & \text{if television medium} \\ 0 & \text{if not} \end{cases}$$

Examining the coding, you can see that $x_2 = x_3 = 0$ when the advertising medium is newspaper. Substituting these values into the expression for $E(y)$, we obtain the newspaper medium line:

$$E(y) = \beta_0 + \beta_1 x_1 + \beta_2(0) + \beta_3(0) + \beta_4 x_1(0) + \beta_5 x_1(0) = \beta_0 + \beta_1 x_1$$

Similarly, we substitute the appropriate values of x_2 and x_3 into the expression for $E(y)$ to obtain the radio medium line ($x_2 = 1, x_3 = 0$):

$$E(y) = \beta_0 + \beta_1 x_1 + \beta_2(1) + \beta_3(0) + \beta_4 x_1(1) + \beta_5 x_1(0)$$

$$= \overbrace{(\beta_0 + \beta_2)}^{y\text{-intercept}} + \overbrace{(\beta_1 + \beta_4)}^{\text{Slope}} x_1$$

and the television medium line: ($x_2 = 0, x_3 = 1$):

$$E(y) = \beta_0 + \beta_1 x_1 + \beta_2(0) + \beta_3(1) + \beta_4 x_1(0) + \beta_5 x_1(1)$$

$$= \overbrace{(\beta_0 + \beta_3)}^{y\text{-intercept}} + \overbrace{(\beta_1 + \beta_5)}^{\text{Slope}} x_1$$

Look Back If you were to fit model 3, obtain estimates of $\beta_0, \beta_1, \beta_2, \ldots, \beta_5$ and substitute them into the equations for the three media lines, you would obtain exactly the same prediction equations as you would obtain if you were to fit three separate straight lines, one to each of the three sets of media data. You may ask why we would not fit the three lines separately. Why bother fitting a model that combines all three lines (model 3) into the same equation? The answer is that you need to use this procedure if you wish to use statistical tests to compare the three media lines. We need to be able to express a practical question about the lines in terms of a hypothesis that a set of parameters in the model equals 0. (We demonstrate this procedure in the next section.) You could not do this if you were to perform three separate regression analyses and fit a line to each set of media data.

Now Work Exercise 11.78

Example 11.11
Testing for Two Different Slopes for Productivity

Problem An industrial psychologist conducted an experiment to investigate the relationship between worker productivity and a measure of salary incentive for two manufacturing plants: one plant operates under "disciplined management practices," and the other plant uses a traditional management style. The productivity y per worker was measured by recording the number of machined castings that a worker could produce in a 4-week period of 40 hours per week. The incentive was the amount x_1 of bonus (in cents per casting) paid for all castings produced in excess of 1,000 per worker for the 4-week period. Nine workers were selected from each plant, and three from each group of nine were assigned to receive a 20¢ bonus per casting, three a 30¢ bonus, and three a 40¢ bonus. The productivity data for the 18 workers, three for each plant type and incentive combination, are shown in Table 11.4.

Table 11.4 Productivity Data for Example 11.11

	Incentive		
Management Style	20¢/casting	30¢/casting	40¢/casting
Traditional	1,435 1,512 1,491	1,583 1,529 1,610	1,601 1,574 1,636
Disciplined	1,575 1,512 1,488	1,635 1,589 1,661	1,645 1,616 1,689

Data Set: CASTING

a. Write a model for mean productivity, $E(y)$, assuming that the relationship between $E(y)$ and incentive x_1 is first order.

b. Fit the model and graph the prediction equations for the disciplined and traditional management styles.

c. Do the data provide sufficient evidence to indicate that the rate of increase of worker productivity is different for disciplined and traditional plants? Test at $\alpha = .10$.

Solution

a. If we assume that a first-order model* is adequate to detect a change in mean productivity as a function of incentive x_1, then the model that produces two straight lines, one for each plant, is

$$E(y) = \beta_0 + \beta_1 x_1 + \beta_2 x_2 + \beta_3 x_1 x_2$$

where

$$x_1 = \text{incentive} \qquad x_2 = \begin{cases} 1 & \text{if disciplined management style} \\ 0 & \text{if traditional management style} \end{cases}$$

b. The SPSS printout for the regression analysis is shown in Figure 11.24. Reading the parameter estimates highlighted at the bottom of the printout, you can see that

$$\hat{y} = 1{,}365.833 + 6.217x_1 + 47.778x_2 + .033x_1x_2$$

Model Summary

Model	R	R Square	Adjusted R Square	Std. Error of the Estimate
1	.843[a]	.711	.649	40.839

a. Predictors: (Constant), INC_PDUM, INCENTIV, PDUMMY

ANOVA[b]

Model		Sum of Squares	df	Mean Square	F	Sig.
1	Regression	57332.39	3	19110.796	11.459	.000[a]
	Residual	23349.22	14	1667.802		
	Total	80681.61	17			

a. Predictors: (Constant), INC_PDUM, INCENTIV, PDUMMY

b. Dependent Variable: CASTINGS

Coefficients[a]

Model		Unstandardized Coefficients		Standardized Coefficients		
		B	Std. Error	Beta	t	Sig.
1	(Constant)	1365.833	51.836		26.349	.000
	INCENTIV	6.217	1.667	.758	3.729	.002
	PDUMMY	47.778	73.308	.357	.652	.525
	INC_PDUM	.033	2.358	.008	.014	.989

a. Dependent Variable: CASTINGS

Figure 11.24
SPSS printout of the complete model for the casting data

*Although the model contains a term involving x_1x_2, it is first order (graphs as a straight line) in the quantitative variable x_1. The variable x_2 is a dummy variable that introduces or deletes terms in the model. The order of a model is determined only by the quantitative variables that appear in the model.

The prediction equation for the plant using a traditional management style can be obtained (see the coding) by substituting $x_2 = 0$ into the general prediction equation. Then

$$\begin{aligned}\hat{y} &= \hat{\beta}_0 + \hat{\beta}_1 x_1 + \hat{\beta}_2(0) + \hat{\beta}_3 x_1(0) = \hat{\beta}_0 + \hat{\beta}_1 x_1 \\ &= 1{,}365.833 + 6.217x_1\end{aligned}$$

Similarly, the prediction equation for the plant with a disciplined management style can be obtained by substituting $x_2 = 1$ into the general prediction equation. Then

$$\begin{aligned}\hat{y} &= \hat{\beta}_0 + \hat{\beta}_1 x_1 + \hat{\beta}_2 x_2 + \hat{\beta}_3 x_1 x_2 \\ &= \hat{\beta}_0 + \hat{\beta}_1 x_1 + \hat{\beta}_2(1) + \hat{\beta}_3 x_1(1) \\ &= \overbrace{(\beta_0 + \beta_2)}^{y\text{-intercept}} + \overbrace{(\beta_1 + \beta_3)}^{\text{Slope}} x_1 \\ &= (1{,}365.833 + 47.778) + (6.217 + .033)x_1 \\ &= 1{,}413.611 + 6.250x_1\end{aligned}$$

A Minitab graph of these prediction equations is shown in Figure 11.25. Note that the slopes of the two lines are nearly identical (6.217 for traditional and 6.250 for disciplined).

c. If the rate of increase of productivity with incentive (i.e., the slope) for plants with the disciplined management style is different than the corresponding slope for traditional plants, then the interaction β (i.e., β_3) will differ from 0. Consequently, we want to test

$$H_0: \beta_3 = 0$$
$$H_a: \beta_3 \neq 0$$

This test is conducted using the t-test of Section 11.3. From the SPSS printout, the test statistic and corresponding p-value are

$$t = .014 \qquad p = .989$$

Because $\alpha = .10$ is less than the p-value, we fail to reject H_0. There is insufficient evidence to conclude that the traditional and disciplined shapes differ. Thus, the test supports our observation of two nearly identical slopes in part **b.**

Look Back Because interaction is not significant, we will drop the $x_1 x_2$ term from the model and use the simpler model, $E(y) = \beta_0 + \beta_1 x_1 + \beta_2 x_2$, to predict productivity.

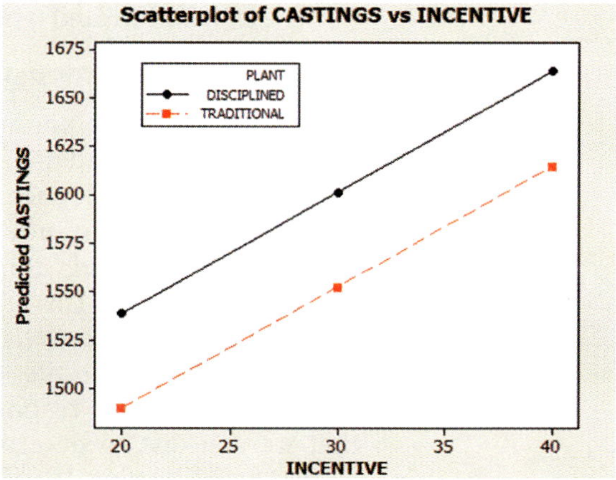

Figure 11.25
Minitab plot of prediction equations for two plants

Now Work Exercise 11.85

Models with both quantitative and qualitative x's may also include higher-order (e.g., second-order) terms. In the problem of relating mean monthly sales $E(y)$ of a company to monthly advertising expenditure x_1 and type of medium, suppose we think that the relationship between $E(y)$ and x_1 is curvilinear. We will construct the model, stage

Figure 11.26
The relationship between $E(y)$ and x_1 is the same for all media

Figure 11.27
The response curves have the same shapes but different y-intercepts

by stage, to enable you to compare the procedure with the stage-by-stage construction of the first-order model in the beginning of this section. The graphical interpretations will help you understand the contributions of the model terms.

1. The mean sales curves are identical for all three advertising media—that is, a single second-order curve will suffice to describe the relationship between $E(y)$ and x_1 for all the media (see Figure 11.26):

$$E(y) = \beta_0 + \beta_1 x_1 + \beta_2 x_1^2$$

where x_1 = advertising expenditure

2. The response curves possess the same shapes but different y-intercepts (see Figure 11.27):

$$E(y) = \beta_0 + \beta_1 x_1 + \beta_2 x_1^2 + \beta_3 x_2 + \beta_4 x_3$$

where

$$x_1 = \text{advertising expenditure}$$
$$x_2 = \begin{cases} 1 & \text{if radio medium} \\ 0 & \text{if not} \end{cases}$$
$$x_3 = \begin{cases} 1 & \text{if television medium} \\ 0 & \text{if not} \end{cases}$$

3. The response curves for the three advertising media are different (i.e., advertising expenditure and type of medium interact), as shown in Figure 11.28:

$$E(y) = \beta_0 + \beta_1 x_1 + \beta_2 x_1^2 + \beta_3 x_2 + \beta_4 x_3 + \beta_5 x_1 x_2 + \beta_6 x_1 x_3 + \beta_7 x_1^2 x_2 + \beta_8 x_1^2 x_3$$

Now that you know how to write a model with two independent variables—one qualitative and one quantitative—we ask, Why do it? Why not write a separate second-order model for each type of medium where $E(y)$ is a function of only advertising expenditure? As stated earlier, one reason we wrote the single model representing all three response curves is so that we can test to determine whether the curves are different. We illustrate this procedure in Section 11.9. A second reason for writing a single model is that we obtain a pooled estimate of σ^2, the variance of the random error component ε. If the variance of ε is truly the same for each type of medium, the pooled estimate is superior to three separate estimates calculated by fitting a separate model for each type of medium.

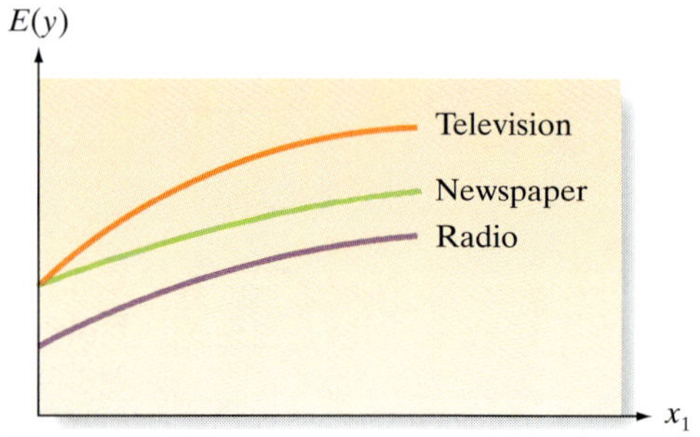

Figure 11.28
The response curves for the three media differ

Exercises 11.78–11.90

Learning the Mechanics

11.78 Consider a multiple regression model for a response y, with one quantitative independent variable x_1 and one qualitative variable at three levels.
NW

a. Write a first-order model that relates the mean response $E(y)$ to the quantitative independent variable.

b. Add the main effect terms for the qualitative independent variable to the model of part **a.** Specify the coding scheme you use.

c. Add terms to the model of part **b** to allow for interaction between the quantitative and qualitative independent variables.

d. Under what circumstances will the response lines of the model in part **c** be parallel?

e. Under what circumstances will the model in part **c** have only one response line?

11.79 Refer to Exercise 11.78.

a. Write a complete second-order model that relates $E(y)$ to the quantitative variable.

b. Add the main effect terms for the qualitative variable (at three levels) to the model of part **a.**

c. Add terms to the model of part **b** to allow for interaction between the quantitative and qualitative independent variables.

d. Under what circumstances will the response curves of the model have the same shape but different y-intercepts?

e. Under what circumstances will the response curves of the model be parallel lines?

f. Under what circumstances will the response curves of the model be identical?

11.80 Consider the model:

$$y = \beta_0 + \beta_1 x_1 + \beta_2 x_2 + \beta_3 x_3 + \varepsilon$$

where x_1 is a quantitative variable and x_2 and x_3 are dummy variables describing a qualitative variable at three levels using the coding scheme

$$x_2 = \begin{cases} 1 & \text{if level 2} \\ 0 & \text{otherwise} \end{cases} \qquad x_3 = \begin{cases} 1 & \text{if level 3} \\ 0 & \text{otherwise} \end{cases}$$

The resulting least squares prediction equation is

$$\hat{y} = 44.8 + 2.2x_1 + 9.4x_2 + 15.6x_3$$

a. What is the response line (equation) for $E(y)$ when $x_2 = x_3 = 0$? When $x_2 = 1$ and $x_3 = 0$? When $x_2 = 0$ and $x_3 = 1$?

b. What is the least squares prediction equation associated with level 1? Level 2? Level 3? Plot these on the same graph.

11.81 Consider the model:

$$y = \beta_0 + \beta_1 x_1 + \beta_2 x_1^2 + \beta_3 x_2 + \beta_4 x_3 + \beta_5 x_1 x_2 + \beta_6 x_1 x_3 + \beta_7 x_1^2 x_2 + \beta_8 x_1^2 x_3 + \varepsilon$$

where x_1 is a quantitative variable and

$$x_2 = \begin{cases} 1 & \text{if level 2} \\ 0 & \text{otherwise} \end{cases} \qquad x_3 = \begin{cases} 1 & \text{if level 3} \\ 0 & \text{otherwise} \end{cases}$$

The resulting least squares prediction equation is

$$\hat{y} = 48.8 - 3.4x_1 + .07x_1^2 - 2.4x_2 - 7.5x_3 + 3.7x_1x_2 + 2.7x_1x_3 - .02x_1^2x_2 - .04x_1^2x_3$$

a. What is the equation of the response curve for $E(y)$ when $x_2 = 0$ and $x_3 = 0$? When $x_2 = 1$ and $x_3 = 0$? When $x_2 = 0$ and $x_3 = 1$?

b. On the same graph, plot the least squares prediction equation associated with level 1, with level 2, and with level 3.

11.82 Write a model that relates $E(y)$ to two independent variables—one quantitative and one qualitative at four levels. Construct a model that allows the associated response curves to be second order but does not allow for interaction between the two independent variables.

Applying the Concepts—Basic

11.83 Impact of race on football card values. Refer to the *Electronic Journal of Sociology* (2007) study of the impact of race on the value of professional football players' "rookie" cards, Exercise 11.68 (p. 671). Recall that the sample consisted of 148 rookie cards of NFL players who were inducted into the Football Hall of Fame (HOF). The researchers modeled the natural logarithm of card price (y) as a function of the following independent variables:

Race:	x_1 = 1 if black, 0 if white
Card availability:	x_2 = 1 if high, 0 if low
Card vintage:	x_3 = year card printed
Finalist:	x_4 = natural logarithm of number of times player on final HOF ballot
Position-QB:	x_5 = 1 if quarterback, 0 if not
Position-RB:	x_7 = 1 if running back, 0 if not
Position-WR:	x_8 = 1 if wide receiver, 0 if not
Position-TE:	x_9 = 1 if tight end, 0 if not
Position-DL:	x_{10} = 1 if defensive lineman, 0 if not
Position-LB:	x_{11} = 1 if linebacker, 0 if not
Position-DB:	x_{12} = 1 if defensive back, 0 if not

[*Note:* For position, offensive lineman is the base level.]

a. The model $E(y) = \beta_0 + \beta_1 x_1 + \beta_2 x_2 + \beta_3 x_3 + \beta_4 x_4 + \beta_5 x_5 + \beta_6 x_6 + \beta_7 x_7 + \beta_8 x_8 + \beta_9 x_9 + \beta_{10} x_{10} + \beta_{11} x_{11} + \beta_{12} x_{12}$ was fit to the data with the following results: $R^2 = .705$, $R_a^2 = .681$, $F = 26.9$. Interpret the results, practically. Make an inference about the overall adequacy of the model.

b. Refer to part **a.** Statistics for the race variable were reported as follows: $\hat{\beta}_1 = -.147$, $s_{\hat{\beta}_1} = .145$, $t = -1.014$, p-value $= .312$. Use this information to make an inference about the impact of race on the value of professional football players' rookie cards.

c. Refer to part **a.** Statistics for the card vintage variable were reported as follows: $\hat{\beta}_3 = -.074$, $s_{\hat{\beta}_3} = .007$, $t = -10.92$, p-value $= .000$. Use this information to make an inference about the impact of card vintage on the value of professional football players' rookie cards.

d. Write a first-order model for $E(y)$ as a function of card vintage (x_3) and position (x_5–x_{12}) that allows for the relationship between price and vintage to vary depending on position.

11.84 Buy-side vs. sell-side analysts' earnings forecasts. Refer to the *Financial Analysts Journal* (Jul./Aug. 2008) comparison of earnings forecasts of buy-side and sell-side analysts, Exercise 11.70 (p. 671). Recall that the Harvard Business School professors used regression to model the relative optimism (y) of the analysts' 3-month horizon forecasts. The following model was fit to data collected on 11,121 forecasts: $E(y) = \beta_0 + \beta_1 x_1 + \beta_2 x_2 + \beta_3 x_3$, where

x_1 = {1 if the analyst worked for a buy-side firm, 0 if the analyst worked for a sell-side firm}

x_2 = number of days between forecast and fiscal year-end (i.e., forecast horizon)

x_3 = natural logarithm of the number of quarters the analyst had worked with the firm

a. The coefficient of determination for the model was reported as $R^2 = .069$. Interpret this value.

b. Use the value of R^2 in part **a** to conduct a test of the global utility of the model. Use $\alpha = .01$.

c. The value of β_1 was estimated as 1.07, with an associated t-value of 4.3. Use this information to test ($\alpha = .01$.) whether x_1 contributes significantly to the prediction of y.

d. The professors concluded that "after controlling for forecast horizon and analyst experience, earnings forecasts by the analysts at buy-side firms are more optimistic than forecasts made by analysts at sell-side firms." Do you agree?

11.85 Smoking and resting energy. NW The influence of cigarette smoking on resting energy expenditure (REE) in normal-weight and obese smokers was investigated (*Health Psychology,* Mar. 1995). The researchers hypothesized that the relationship between a smoker's REE and length of time since smoking differs for normal-weight and obese smokers. Consequently, the interaction model was examined:

$$E(y) = \beta_0 + \beta_1 x_1 + \beta_2 x_2 + \beta_3 x_1 x_2$$

where

y = REE, measured in kilocalories per day

x_1 = Time, in minutes after smoking, of metabolic energy reading (levels = 10, 20, and 30 minutes)

$$x_2 = \begin{cases} 1 & \text{if normal weight} \\ 0 & \text{if obese} \end{cases}$$

a. Give the equation of the hypothesized line relating mean REE to time after smoking for obese smokers. What is the slope of the line?

b. Repeat part **a** for normal-weight smokers.

c. A test for interaction resulted in an observed significance level of .044. Interpret this value.

11.86 Workplace bullying and intention to leave. Workplace bullying (e.g., work-related harassment, persistent criticism, withholding key information, spreading rumors, intimidation) has been shown to have a negative psychological effect on victims, often leading the victim to quit or resign. In *Human Resource Management Journal* (Oct. 2008), researchers employed multiple regression to examine whether perceived organizational support (POS) would moderate the relationship between workplace bullying and victims' intention to leave the firm. The dependent variable in the analysis, intention to leave (y), was measured on a quantitative scale. The two key independent variables in the study were bullying (x_1, measured on a quantitative scale) and perceived organizational support (measured qualitatively as "low," "neutral," or "high").

a. Set up the dummy variables required to represent POS in the regression model.

b. Write a model for $E(y)$ as a function of bullying and POS that hypothesizes three parallel straight lines, one for each level of POS.

c. Write a model for $E(y)$ as a function of bullying and POS that hypothesizes three nonparallel straight lines, one for each level of POS.

d. The researchers discovered that the effect of bullying on intention to leave was greater at the low level of POS than at the high level of POS. Which of the two models, parts **b** and **c,** support these findings?

Applying the Concepts—Intermediate

11.87 Chemical plant contamination. Refer to Exercise 11.18 (p. 641) and the model relating the mean DDT level $E(y)$ of contaminated fish to x_1 = miles captured upstream, x_2 = length, and x_3 = weight. Now consider a model for $E(y)$ as a function of both weight and species (channel catfish, largemouth bass, and smallmouth buffalo).

a. Set up the appropriate dummy variables for species.

b. Write the equation of a model that proposes parallel straight-line relationships between mean DDT level $E(y)$ and weight, one line for each species.

c. Write the equation of a model that proposes nonparallel straight-line relationships between mean DDT level $E(y)$ and weight, one line for each species.

d. Fit the model, part **b,** to the data saved in the **DDT** file. Give the least squares prediction equation.

e. Refer to part **d.** Interpret the value of the least squares estimate of the beta coefficient multiplied by weight.

f. Fit the model, part **c,** to the data saved in the **DDT** file. Give the least squares prediction equation.

g. Refer to part **f.** Find the estimated slope of the line relating DDT level (y) to weight for the channel catfish species.

11.88 Entrepreneurial networks and company growth. A researcher at California State University at Long Beach, E. L. Hansen, used multiple regression analysis to investigate the link between entrepreneurial networks and new organization growth (*Entrepreneurship Theory and Practice,* Summer 1995). The dependent variable, initial new organization growth (y), is defined as the monthly payroll of the firm at the end of its first year in business. One independent variable considered in the model (and found to be statistically significant) is entrepreneurial action set size (x_1), defined as the number of people in the entrepreneur's social network who are directly or indirectly involved with the founding of the new organization. A second independent variable considered in the study was a dummy variable that represented the technology level of the new firm, where

$$x_2 = \begin{cases} 1 & \text{if high-tech firm} \\ 0 & \text{if low-tech firm} \end{cases}$$

a. Write a first-order model for new firm growth (y) as a function of entrepreneurial action set size (x_1) and technology level (x_2). Assume that the effect of size on new firm growth is independent of technology level.

b. In terms of the β's of the model, part **a,** give the slope of the line relating new firm growth y to size x_1 for high-tech firms.

c. Repeat part **a** but assume that the rate of change of new firm growth y with size x_1 depends on technology level x_2.

d. In terms of the β's of the model, part **c,** give the slope of the line relating new firm growth y to size x_1 for high-tech firms.

11.89 Recently sold, single-family homes. The National Association of Realtors maintains a database consisting of sales information on homes sold in the United States. The next table (saved in the **NAR** file) lists the sale prices for a sample of 28 recently sold, single-family homes. The table also identifies the region of the country in which the home is located and the total number of homes sold in the region during the month the home sold.

Home Price	Region	Sales Volume
$218,200	NE	55,156
235,900	NE	61,025
192,888	NE	48,991
200,990	NE	55,156
345,300	NE	60,324
178,999	NE	51,446
240,855	NW	61,025
183,200	NW	94,166
165,225	NW	92,063
160,633	NW	89,485
173,900	NW	91,772
241,000	NW	99,025
188,950	NW	94,166
192,880	NW	95,688
193,000	S	155,666
211,980	S	160,000
179,500	S	153,540
185,650	S	148,668
250,900	S	163,210
190,990	S	141,822
242,790	S	163,611
258,900	W	109,083
202,420	W	101,111
365,900	W	116,983
219,900	W	108,773
228,250	W	105,106
235,300	W	107,839
269,800	W	109,026

Source: Adapted from National Association of Realtors, www.realtor.org. Copyright National Association of Realtors®. Reprinted with permission.

a. Propose a complete second-order model for the sale price of a single-family home as a function of region and sales volume.

b. Give the equation of the curve relating sale price to sales volume for homes sold in the West.

c. Repeat part **b** for homes sold in the Northwest.

d. Which β's in the model, part **a,** allow for differences among the mean sale prices for homes in the four regions?

e. Fit the model, part **a,** to the data using an available statistical software package. Is the model statistically useful for predicting sale price? Test using $\alpha = .01$.

11.90 Volatility of foreign stocks. The relationship between country credit ratings and the volatility of the countries' stock markets was examined in the *Journal of Portfolio Management* (Spring 1996). The researchers point out that this volatility can be explained by two factors: the countries' credit ratings and whether the countries in question have developed or emerging markets. Data on the volatility (measured as the standard deviation of stock returns), credit rating (measured as a percentage), and market type (developed or emerging) for a sample of 30 fictitious countries are saved in the **VOLATILE** file. (Selected observations are shown in the table below.)

a. Write a model that describes the relationship between volatility (y) and credit rating (x_1) as two nonparallel lines, one for each type of market. Specify the dummy variable coding scheme you use.

b. Plot volatility y against credit rating x_1 for all the developed markets in the sample. On the same graph, plot y against x_1 for all emerging markets in the sample. Does it appear that the model specified in part **a** is appropriate? Explain.

c. Fit the model, part **a,** to the data using a statistical software package. Report the least squares prediction equation for each of the two types of markets.

d. Plot the two prediction equations of part **c** on a scattergram of the data.

e. Is there evidence to conclude that the slope of the linear relationship between volatility y and credit rating x_1 depends on market type? Test using $\alpha = .01$.

Country	Volatility (standard deviation of return), y	Credit Rating, x_1	Developed (D) or Emerging (E), x_2
1	56.9	9.5	E
2	25.1	72.4	D
3	28.4	58.2	E
4	56.2	9.9	E
5	21.5	92.1	D
⋮	⋮	⋮	⋮
26	47.5	16.9	E
27	22.0	89.0	D
28	21.5	91.9	D
29	38.1	30.7	E
30	37.4	32.2	E

11.9 Comparing Nested Models

To be successful model builders, we require a statistical method that will allow us to determine (with a high degree of confidence) which one among a set of candidate models best fits the data. In this section, we present such a technique for *nested models.*

Two models are **nested** if one model contains all the terms of the second model and at least one additional term. The more complex of the two models is called the **complete** (or **full**) model, and the simpler of the two is called the **reduced** model.

To illustrate the concept of nested models, consider the straight-line interaction model for the mean auction price $E(y)$ of a grandfather clock as a function of two quantitative variables: age of the clock (x_1) and number of bidders (x_2). The interaction model, fit in Example 11.6, is

$$E(y) = \beta_0 + \beta_1 x_1 + \beta_2 x_2 + \beta_3 x_1 x_2$$

If we assume that the relationship between auction price (y), age (x_1), and bidders (x_2) is curvilinear, then the complete second-order model is more appropriate:

$$E(y) = \overbrace{\beta_0 + \beta_1 x_1 + \beta_2 x_2 + \beta_3 x_1 x_2}^{\text{Terms in interaction model}} + \overbrace{\beta_4 x_1^2 + \beta_5 x_2^2}^{\text{Quadratic terms}}$$

Note that the curvilinear model contains quadratic terms for x_1 and x_2, as well as the terms in the interaction model. Therefore, the models are nested models. In this case, the interaction model is nested within the more complex curvilinear model. Thus, the curvilinear model is the *complete* model, and the interaction model is the *reduced* model.

Suppose we want to know whether the more complex curvilinear model contributes more information for the prediction of y than the straight-line interaction model. This is equivalent to determining whether the quadratic terms β_4 and β_5 should be retained in the model. To test whether these terms should be retained, we test the null hypothesis

H_0: $\beta_4 = \beta_5 = 0$ (i.e., quadratic terms are not important for predicting y)

against the alternative hypothesis

H_a: At least one of the parameters β_4 and β_5 is nonzero (i.e., at least one of the quadratic terms is useful for predicting y).

Note that the terms being tested are those additional terms in the complete (curvilinear) model that are not in the reduced (straight-line interaction) model.

In Section 11.3, we presented the t-test for a single β coefficient and the global F-test for *all* the β parameters (except β_0) in the model. We now need a test for a *subset* of the β parameters in the complete model. The test procedure is intuitive. First, we use the method of least squares to fit the reduced model and calculate the corresponding sum of squares for error, SSE_R (the sum of squares of the deviations between observed and predicted y-values). Next, we fit the complete model and calculate its sum of squares for error, SSE_C. Then we compare SSE_R to SSE_C by calculating the difference, $SSE_R - SSE_C$. If the additional terms in the complete model are significant, then SSE_C should be much smaller than SSE_R, and the difference $SSE_R - SSE_C$ will be large.

Because SSE will always decrease when new terms are added to the model, the question is whether the difference $SSE_R - SSE_C$ is large enough to conclude that it is due to more than just an increase in the number of model terms and to chance. The formal statistical test uses an F-statistic, as shown in the box.

When the assumptions listed in Section 11.1 about the random error term are satisfied, this F-statistic has an F-distribution with ν_1 and ν_2 df. Note that ν_1 is the number of β parameters being tested, and ν_2 is the number of degrees of freedom associated with s^2 in the complete model.

F-Test for Comparing Nested Models

Reduced model: $E(y) = \beta_0 + \beta_1 x_1 + \cdots + \beta_g x_g$

Complete model: $E(y) = \beta_0 + \beta_1 x_1 + \cdots + \beta_g x_g + \beta_{g+1} x_{g+1} + \cdots + \beta_k x_k$

H_0: $\beta_{g+1} = \beta_{g+2} = \cdots = \beta_k = 0$

H_a: At least one of the β parameters under test is nonzero.

$$\textit{Test statistic: } F = \frac{(\text{SSE}_\text{R} - \text{SSE}_\text{C})/(k - g)}{\text{SSE}_\text{C}/[n - (k + 1)]} = \frac{(\text{SSE}_\text{R} - \text{SSE}_\text{C})/\#\beta\text{'s tested in } H_0}{\text{MSE}_\text{C}}$$

where

SSE_R = Sum of squared errors for the reduced model

SSE_C = Sum of squared errors for the complete model

MSE_C = Mean square error (s^2) for the complete model

$k - g$ = Number of β parameters specified in H_0 (i.e., number of β parameters tested)

$k + 1$ = Number of β parameters in the complete model (including β_0)

n = Total sample size

Rejection region: $F > F_\alpha$

where F is based on $v_1 = k - g$ numerator degrees of freedom and $v_2 = n - (k + 1)$ denominator degrees of freedom.

Example 11.12

Analyzing a Complete Second-Order Model for Auction Price

Problem Refer to the problem of modeling the auction price (y) of an antique grandfather clock, Examples 11.1–11.6. In these examples, we discovered that price was related to both age (x_1) of the clock and number of bidders (x_2), and age and number of bidders had an interactive effect on price. Both the first-order model of Example 11.1 and the interaction model of Example 11.6, however, propose only straight-line (linear) relationships. We did not consider the possibility that the relationship between price (y) and age (x_1) is curvilinear, or that the relationship between price (y) and number of bidders (x_2) is curvilinear.

a. Propose a complete second-order model for price (y) as a function of age (x_1) and number of bidders (x_2).

b. Fit the model to the data for the 32 clocks in Table 11.1 and give the least squares prediction equation.

c. Do the data provide sufficient evidence to indicate that the quadratic terms in the model contribute information for the prediction of price (y)? That is, is there evidence of curvature in the price-age and price-bidders relationships?

Solution

a. Because both age (x_1) and number of bidders (x_2) are quantitative variables, we write the complete second-order model as follows:

$$E(y) = \beta_0 + \beta_1 x_1 + \beta_2 x_2 + \beta_3 x_1 x_2 + \beta_4 (x_1)^2 + \beta_5 (x_2)^2$$

Note that we have added two quadratic terms, $\beta_4(x_1)^2$ and $\beta_5(x_2)^2$, to the interaction model of Example 11.6.

b. We used Minitab to fit the model to the data in Table 11.2. The Minitab output is shown in Figure 11.29. The least squares prediction equation, highlighted on the printout, is

$$\hat{y} = -332 + 3.21x_1 + 14.8x_2 + 1.12x_1x_2 - .003(x_1)^2 - 4.18(x_2)^2$$

Note that the estimates of the quadratic terms, β_4 and β_5, are both negative, implying downward curvature in both the price-age and price-bidders relationships.

Regression Analysis: PRICE versus AGE, NUMBIDS, AGE-BID, AGESQ, NUMBIDSQ

```
The regression equation is
PRICE = - 332 + 3.21 AGE + 14.8 NUMBIDS + 1.12 AGE-BID - 0.0030 AGESQ
        - 4.18 NUMBIDSQ

Predictor        Coef   SE Coef      T      P
Constant       -331.9     764.9  -0.43  0.668
AGE             3.208     8.947   0.36  0.723
NUMBIDS         14.81     62.21   0.24  0.814
AGE-BID        1.1232    0.2316   4.85  0.000
AGESQ        -0.00300   0.02748  -0.11  0.914
NUMBIDSQ       -4.179     2.145  -1.95  0.062

S = 86.1019   R-Sq = 96.0%   R-Sq(adj) = 95.2%

Analysis of Variance

Source          DF       SS      MS       F      P
Regression       5  4607038  921408  124.29  0.000
Residual Error  26   192752    7414
Total           31  4799790
```

Figure 11.29
Minitab analysis of complete second-order model for auction price

c. To determine whether the quadratic terms are useful for predicting price (y), we want to test

H_0: $\beta_4 = \beta_5 = 0$
H_a: At least one of the parameters, β_4 and β_5, differs from 0.

Because the null hypothesis involves a subset of β's, we will need to perform a nested model F-test. The complete model is the complete second-order model of part **a.** The reduced model is obtained by dropping the quadratic terms out of the complete model. Doing so yields:

$$\textit{Reduced model: } E(y) = \beta_0 + \beta_1 x_1 + \beta_2 x_2 + \beta_3 x_1 x_2$$

We fit this model in Example 11.6. The Minitab printout for this reduced model is reproduced in Figure 11.30.

To conduct this test, we first obtain the SSE values for the complete and reduced models. These values (highlighted, respectively, on Figures 11.29 and 11.30) are

$$\text{SSE}_\text{C} = 192{,}752 \text{ and } \text{SSE}_\text{R} = 221{,}362$$

Also, s^2 for the complete model (highlighted on Figure 11.29) is

$$s^2 = \text{MSE}_\text{C} = 7{,}414$$

Recall that $n = 32$ and that there are $k = 5$ terms in the complete model and $g = 3$ terms in the reduced model. Therefore, the calculated value of the test statistic is

$$\textit{Test statistic: } F = \frac{(\text{SSE}_\text{R} - \text{SSE}_\text{C})/(k - g)}{\text{MSE}_\text{C}} = \frac{(221{,}362 - 192{,}752)/2}{7{,}414} = 1.93$$

The critical F-value for the rejection region is based on $v_1 = (k - g) = 2$ numerator df and $v_2 = [n - (k + 1)] = 26$ denominator df. If we choose $\alpha = .05$, this critical value is $F_{.05} = 3.37$ and the rejection region is

$$\textit{Rejection region: } F > 3.37$$

Because the test statistic, $F = 1.93$, falls outside the rejection region, we fail to reject H_0. That is, there is insufficient evidence (at $\alpha = .05$) to conclude that at least one of the quadratic terms contributes information for the prediction of auction price (y). Consequently, there is no evidence of curvature in either the price-age or price-bidders relationship. It appears that the first-order, interaction (reduced) model is adequate for predicting auction price.

Regression Analysis: PRICE versus AGE, NUMBIDS, AGEBID

```
The regression equation is
PRICE = 320 + 0.88 AGE - 93.3 NUMBIDS + 1.30 AGEBID


Predictor      Coef   SE Coef       T      P
Constant      320.5     295.1    1.09  0.287
AGE           0.878     2.032    0.43  0.669
NUMBIDS      -93.26     29.89   -3.12  0.004
AGEBID       1.2978    0.2123    6.11  0.000


S = 88.9145    R-Sq = 95.4%    R-Sq(adj) = 94.9%


Analysis of Variance

Source           DF       SS       MS       F      P
Regression        3  4578427  1526142  193.04  0.000
Residual Error   28   221362     7906
Total            31  4799790
```

Figure 11.30
Minitab analysis of reduced model for auction price

Look Back Some statistical software packages will perform the desired nested model F-test if requested. The test statistic and p-value for the test above are highlighted on the SPSS printout, Figure 11.31. Note that p-value $= .165$ exceeds $\alpha = .05$; thus, there is insufficient evidence to reject H_0.

Now Work Exercise 11.94

Model Summary

Model	R	R Square	Adjusted R Square	Std. Error of the Estimate	Change Statistics				
					R Square Change	F Change	df1	df2	Sig. F Change
1	.977[a]	.954	.949	88.915	.954	193.041	3	28	.000
2	.980[b]	.960	.952	86.102	.006	1.930	2	26	.165

a. Predictors: (Constant), AGE_BID, AGE, NUMBIDS

b. Predictors: (Constant), AGE_BID, AGE, NUMBIDS, NUMBIDSQ, AGESQ

Figure 11.31
SPSS printout of nested model F-test for auction price

The nested model F-test can be used to determine whether *any* subset of terms should be included in a complete model by testing the null hypothesis that a particular set of β parameters simultaneously equals 0. For example, we may want to test to determine whether a set of interaction terms for quantitative variables or a set of main effect terms for a qualitative variable should be included in a model. If we reject H_0, the complete model is the better of the two nested models. If we fail to reject H_0, as in Example 11.12, we favor the reduced model. Although we must be cautious about accepting H_0, most practitioners of regression analysis adopt the principle of *parsimony*—that is, in situations where two competing models are found to have essentially the same predictive power (as in Example 11.12), the model with the fewer number of β*'s* (i.e., the more parsimonious model) is selected. Based on this principle, we would drop the two quadratic terms and select the first-order, interaction (reduced) model over the second-order (complete) model.

A **parsimonious model** is a general linear model with a small number of β parameters. In situations where two competing models have essentially the same predictive power (as determined by an F-test), choose the more parsimonious of the two.

Guidelines for Selecting Preferred Model in a Nested Model *F*-Test

Conclusion		*Preferred Model*
Reject H_0	→	Complete Model
Fail to reject H_0	→	Reduced Model

When the candidate models in model building are nested models, the F-test developed in this section is the appropriate procedure to apply to compare the models. However, if the models are not nested, this F-test is not applicable. In this situation, the analyst must base the choice of the best model on statistics such as R_a^2 and s. It is important to remember that decisions based on these and other numerical descriptive measures of model adequacy cannot be supported with a measure of reliability and are often very subjective in nature.

Exercises 11.91–11.103

Learning the Mechanics

11.91 Determine which pairs of the following models are "nested" models. For each pair of nested models, identify the complete and reduced model.

a. $E(y) = \beta_0 + \beta_1x_1 + \beta_2x_2$
b. $E(y) = \beta_0 + \beta_1x_1$
c. $E(y) = \beta_0 + \beta_1x_1 + \beta_2x_1^2$
d. $E(y) = \beta_0 + \beta_1x_1 + \beta_2x_2 + \beta_3x_1x_2$
e. $E(y) = \beta_0 + \beta_1x_1 + \beta_2x_2 + \beta_3x_1x_2 + \beta_4x_1^2 + \beta_5x_2^2$

11.92 Suppose you fit the regression model

$$y = \beta_0 + \beta_1x_1 + \beta_2x_2 + \beta_3x_1x_2 + \beta_4x_1^2 + \beta_5x_2^2 + \varepsilon$$

to $n = 30$ data points and wish to test

$$H_0\colon \beta_3 = \beta_4 = \beta_5 = 0$$

a. State the alternative hypothesis H_a.
b. Give the reduced model appropriate for conducting the test.
c. What are the numerator and denominator degrees of freedom associated with the F-statistic?
d. Suppose the SSE's for the reduced and complete models are $\text{SSE}_R = 1{,}250.2$ and $\text{SSE}_C = 1{,}125.2$. Conduct the hypothesis test and interpret the results of your test. Test using $\alpha = .05$.

11.93 The complete model

$$y = \beta_0 + \beta_1x_1 + \beta_2x_2 + \beta_3x_3 + \beta_4x_4 + \varepsilon$$

was fit to $n = 20$ data points, with SSE $= 152.66$. The reduced model, $y = \beta_0 + \beta_1x_1 + \beta_2x_2 + \varepsilon$, was also fit, with SSE $= 160.44$.

a. How many β parameters are in the complete model? The reduced model?
b. Specify the null and alternative hypotheses you would use to investigate whether the complete model contributes more information for the prediction of y than the reduced model.
c. Conduct the hypothesis test of part **d.** Use $\alpha = .05$.

Applying the Concepts—Basic

11.94 **Mental health of a community.** An article in the *Community Mental Health Journal* (Aug. 2000) used multiple regression analysis to model the level of community adjustment of clients of the Department of Mental Health and Addiction Services in Connecticut. The dependent variable, community adjustment (y), was measured quantitatively based on staff ratings of the clients. (Lower scores indicate better adjustment.) The complete model was a first-order model with 21 independent variables. The independent variables were categorized as Demographic (4 variables), Diagnostic (7 variables), Treatment (4 variables), and Community (6 variables).
NW

a. Write the equation of $E(y)$ for the complete model.
b. Give the null hypothesis for testing whether the 7 Diagnostic variables contribute information for the prediction of y.
c. Give the equation of the reduced model appropriate for the test, part **b.**
d. The test, part **b,** resulted in a test statistic of $F = 59.3$ and p-value $< .0001$. Interpret this result in the words of the problem.

11.95 **Buy-side vs. sell-side analysts' earnings forecasts.** Refer to the *Financial Analysts Journal* (Jul./Aug. 2008) comparison of earnings forecasts of buy-side and sell-side analysts, Exercise 11.84 (p. 680). Recall that the Harvard Business School professors used regression to model the relative optimism (y) of the analysts' 3-month horizon forecasts as a function of $x_1 = \{1$ if the analyst worked for a buy-side firm, 0 if the analyst worked for a sell-side firm$\}$ and x_2 = number of days between forecast and fiscal year-end (i.e., forecast horizon). Consider the complete second-order model

$$E(y) = \beta_0 + \beta_1x_1 + \beta_2x_2 + \beta_3x_1x_2 + \beta_4(x_2)^2 + \beta_5x_1(x_2)^2$$

a. What null hypothesis would you test to determine whether the quadratic terms in the model are statistically useful for predicting relative optimism (y)?
b. Give the complete and reduced models for conducting the test, part **a.**
c. What null hypothesis would you test to determine whether the interaction terms in the model are statistically useful for predicting relative optimism (y)?
d. Give the complete and reduced models for conducting the test, part **c.**
e. What null hypothesis would you test to determine whether the dummy variable terms in the model are statistically useful for predicting relative optimism (y)?
f. Give the complete and reduced models for conducting the test, part **e.**

11.96 Workplace bullying and intention to leave. Refer to the *Human Resource Management Journal* (Oct. 2008) study of workplace bullying, Exercise 11.86 (p. 680). Recall that multiple regression was used to model an employee's intention to leave (y) as a function of bullying (x_1, measured on a quantitative scale) and perceived organizational support (measured qualitatively as "low POS," "neutral POS," or "high POS"). In Exercise 11.86b, you wrote a model for $E(y)$ as a function of bullying and POS that hypothesizes three parallel straight lines, one for each level of POS. In Exercise 11.86c, you wrote a model for $E(y)$ as a function of bullying and POS that hypothesizes three nonparallel straight lines, one for each level of POS.

a. Explain why the two models are nested. Which is the complete model? Which is the reduced model?

b. Give the null hypothesis for comparing the two models.

c. If you reject H_0 in part **b,** which model do you prefer? Why?

d. If you fail to reject H_0 in part **b,** which model do you prefer? Why?

11.97 Cooling method for gas turbines. Refer to the *Journal of Engineering for Gas Turbines and Power* (Jan. 2005) study of a high-pressure inlet fogging method for a gas turbine engine, Exercise 11.19 (p. 641). Consider a model for heat rate (kilojoules per kilowatt per hour) of a gas turbine as a function of cycle speed (revolutions per minute) and cycle pressure ratio. The data are saved in the **GASTURBINE** file.

a. Write a complete second-order model for heat rate (y).

b. Give the null and alternative hypotheses for determining whether the curvature terms in the complete second-order model are statistically useful for predicting heat rate (y).

c. For the test in part **b,** identify the complete and reduced model.

d. Portions of the Minitab printouts for the two models are shown below. Find the values of SSE_R, SSE_C, and MSE_C on the printouts.

e. Compute the value of the test statistics for the test of part **b.**

f. Find the rejection region for the test of part **b** using $\alpha = .10$.

g. State the conclusion of the test in the words of the problem.

Applying the Concepts—Intermediate

11.98 Study of supervisor-targeted aggression. "Moonlighters" are workers who hold two jobs at the same time. What are the factors that impact the likelihood of a moonlighting worker becoming aggressive toward his/her supervisor? This was the research question of interest in the *Journal of Applied Psychology* (July 2005). Completed questionnaires were obtained from $n = 105$ moonlighters, and the data were used to fit several multiple regression models for

Minitab Output for Exercise 11.97

Complete Model

```
The regression equation is
HEATRATE = 15583 + 0.078 RPM - 523 CPRATIO + 0.00445 RPM_CPR - 0.000000 RPMSQ
           + 8.84 CPRSQ

S = 563.513    R-Sq = 88.5%    R-Sq(adj) = 87.5%

Analysis of Variance

Source          DF         SS        MS      F      P
Regression       5  148526859  29705372  93.55  0.000
Residual Error  61   19370350    317547
Total           66  167897208
```

Reduced Model

```
The regression equation is
HEATRATE = 12065 + 0.170 RPM - 146 CPRATIO - 0.00242 RPM_CPR

S = 633.842    R-Sq = 84.9%    R-Sq(adj) = 84.2%

Analysis of Variance

Source          DF         SS        MS       F      P
Regression       3  142586570  47528857  118.30  0.000
Residual Error  63   25310639    401756
Total           66  167897208
```

supervisor-directed aggression score (y). Two of the models (with R^2 values in parentheses) are given below:

Model 1: $E(y) = \beta_0 + \beta_1(\text{Age}) + \beta_2(\text{Gender}) +$
$\beta_3(\text{Interatctional injustice at 2nd job}) +$
$\beta_4(\text{Abusive supervisor at 2nd job})$
($R^2 = .101$)

Model 2: $E(y) = \beta_0 + \beta_1(\text{Age}) + \beta_2(\text{Gender}) +$
$\beta_3(\text{Interactional injustice at 2nd job}) +$
$\beta_4(\text{Abusive supervisor at 2nd job}) +$
$\beta_5(\text{Self-esteem}) + \beta_6(\text{History of aggression}) +$
$\beta_7(\text{Interactional injustice at primary job}) +$
$\beta_8(\text{Abusive supervisor at primary job})$
($R^2 = .555$)

a. Interpret the R^2 values for the models.
b. Give the null and alternative hypotheses for comparing the fits of Models 1 and 2.
c. Are the two models nested? Explain.
d. The nested F-test for comparing the two models resulted in $F = 42.13$ and p-value $< .001$. What can you conclude from these results?
e. A third model was fit, one that hypothesizes all possible pairs of interactions between self-esteem, history of aggression, interactional injustice at primary job, and abusive supervisor at primary job. Give the equation of this model (Model 3).
f. A nested F-test to compare Models 2 and 3 resulted in a p-value $> .10$. What can you conclude from this result?

11.99 Improving SAT scores. Refer to the *Chance* (Winter 2001) study of students who paid a private tutor (or coach) to help them improve their SAT scores, Exercise 11.71 (p. 671). Recall that the baseline model, $E(y) = \beta_0 + \beta_1 x_1 + \beta_2 x_2$, where y = SAT–Math score, x_1 = score on PSAT, and x_2 = {1 if student was coached, 0 if not}, had the following results: $R_a^2 = .76$, $\beta_2 = 19$, and $s_{\hat{\beta}_2} = 3$. As an alternative model, the researcher added several "control" variables, including dummy variables for student ethnicity (x_3, x_4, and x_5), a socioeconomic status index variable (x_6), two variables that measured high school performance (x_7 and x_8), the number of math courses taken in high school (x_9), and the overall GPA for the math courses (x_{10}).
a. Write the hypothesized equation for $E(y)$ for the alternative model.
b. Give the null hypothesis for a nested model F-test comparing the initial and alternative models.
c. The nested model F-test, part **b,** was statistically significant at $\alpha = .05$. Practically interpret this result.
d. The alternative model, part **a,** resulted in $R_a^2 = .79$, $\hat{\beta}_2 = 14$, and $s_{\hat{\beta}_2} = 3$. Interpret the value of R_a^2.
e. Refer to part **d.** Find and interpret a 95% confidence interval for β_2.
f. The researcher concluded that "the estimated effect of SAT coaching decreases from the baseline model when control variables are added to the model." Do you agree? Justify your answer.
g. As a modification to the model of part **a,** the researcher added all possible interactions between the coaching variable (x_2) and the other independent variables in the model. Write the equation for $E(y)$ for this modified model.
h. Give the null hypothesis for comparing the models, parts **a** and **g.** How would you perform this test?

11.100 Recently sold, single-family homes. Refer to the National Association of Realtors data on sales price (y), region (NE, NW, S, or W), and sales volume for 28 recently sold, single-family homes, Exercise 11.89 (p. 681). The data are saved in the **NAR** file. In Exercise 11.89, you fit a complete second-order model for $E(y)$ as a function of region and sales volume.
a. Conduct a nested-model F-test to determine whether the quadratic terms in the model are statistically useful for predicting sales price (y). Use $\alpha = .05$.
b. Based on the result, part **a,** which of the nested models (the complete or the reduced model) do you prefer to use in predicting sales price (y)? Explain.
c. Refer to part **b.** Treat the preferred model as the complete model and conduct a nested model F-test to determine whether region and sales volume interact to affect sales price (y). Use $\alpha = .05$.
d. Based on the result, part **c,** which of the nested models (the complete or the reduced model) do you prefer to use in predicting sales price (y)? Explain.

11.101 Glass as a waste encapsulant. Because glass is not subject to radiation damage, encapsulation of waste in glass is considered to be one of the most promising solutions to the problem of low-level nuclear waste in the environment. However, chemical reactions may weaken the glass. This concern led to a study undertaken jointly by the Department of Materials Science and Engineering at the University of Florida and the U.S. Department of Energy to assess the utility of glass as a waste encapsulant.* Corrosive chemical solutions (called *corrosion baths*) were prepared and applied directly to glass samples containing one of three types of waste (TDS-3A, FE, and AL); the chemical reactions were observed over time. A few of the key variables measured were

y = Amount of silicon (in parts per million) found in solution at end of experiment. (This is both a measure of the degree of breakdown in the glass and a proxy for the amount of radioactive species released into the environment.)
x_1 = Temperature (°C) of the corrosion bath
x_2 = 1 if waste type TDS-3A, 0 if not
x_3 = 1 if waste type FE, 0 if not

(Waste type AL is the base level.) Suppose we want to model amount y of silicon as a function of temperature (x_1) and type of waste (x_2, x_3).
a. Write a model that proposes parallel straight-line relationships between amount of silicon and temperature, one line for each of the three waste types.
b. Add terms for the interaction between temperature and waste type to the model of part **a.**
c. Refer to the model of part **b.** For each waste type, give the slope of the line relating amount of silicon to temperature.

*The background information for this exercise was provided by Dr. David Clark, Department of Materials Science and Engineering, University of Florida.

d. Explain how you could test for the presence of temperature–waste type interaction.

11.102 Emotional distress in firefighters. The *Journal of Human Stress* (Summer 1987) reported on a study of "psychological response of firefighters to chemical fire." It is thought that the following complete second-order model will be adequate to describe the relationship between emotional distress and years of experience for two groups of firefighters—those exposed to a chemical fire and those unexposed.

$$E(y) = \beta_0 + \beta_1 x_1 + \beta_2 x_1^2 + \beta_3 x_2 + \beta_4 x_1 x_2 + \beta_5 x_1^2 x_2$$

where

y = Emotional distress
x_1 = Experience (years)
x_2 = 1 if exposed to chemical fire, 0 if not

a. What hypothesis would you test to determine whether the *rate* of increase of emotional distress with experience is different for the two groups of firefighters?

b. What hypothesis would you test to determine whether there are differences in mean emotional distress levels that are attributable to exposure group?

c. The second-order model, fit to a sample of 200 firefighters, resulted in SSE = 783.90. The reduced model, $E(y) = \beta_0 + \beta_1 x_1 + \beta_2 x_1^2$, fit to the same data, resulted in SSE = 795.23. Is there sufficient evidence to support the claim that the mean emotional distress levels differ for the two groups of firefighters? Use $\alpha = .05$.

Applying the Concepts—Advanced

11.103 Modeling monthly collision claims. A medium-sized automobile insurance company is interested in developing a regression model to help predict the monthly collision claims of its policyholders. A company analyst has proposed modeling monthly collision claims (y) in the middle Atlantic states as a function of the percentage of claims by drivers under age 30 (x_1) and the average daily temperature during the month (x_2). She believes that as the percentage of claims by drivers under age 30 increases, claims will rise because younger drivers are usually involved in more serious accidents than older drivers. She also believes that claims will rise as the average daily temperature decreases because lower temperatures are associated with icy, hazardous driving conditions. In order to develop a preliminary model, data were collected for the state of New Jersey over a 3-year period. The data are saved in the **NJCLAIMS** file. (The first and last five observations are listed in the table below.)

a. Use a statistical software package to fit the complete second-order model

$$E(y) = \beta_0 + \beta_1 x_1 + \beta_2 x_2 + \beta_3 x_1 x_2 + \beta_4 x_1^2 + \beta_5 x_2^2$$

b. Test the hypothesis H_0: $\beta_4 = \beta_5 = 0$ using $\alpha = .05$. Interpret the results in practical terms.

c. Do the results support the analysts' beliefs? Explain. (You may need to conduct further tests of hypotheses to answer this question.)

Table for Exercise 11.103 (first and last 5 months)

Month	Monthly Collision Claims, y (\$)	Percentage of Monthly Claimants under the Age of 30, x_1	Newark, N.J., Average Daily Temperature during the Month, x_2 (°F)
1	116,250	50.0	31.5
2	217,180	60.8	33.0
3	43,436	45.1	45.0
4	159,265	56.4	53.9
5	130,308	53.3	63.9
⋮	⋮	⋮	⋮
44	136,528	53.1	76.6
45	193,608	59.8	68.6
46	38,722	45.6	62.4
47	212,309	63.9	50.0
48	118,796	52.3	42.3

Sources: Anonymous insurance company; New Jersey Department of Insurance; *Weather of U.S. Cities,* 4th ed., Gale Research Inc., Detroit, 1992.

11.10 Stepwise Regression

Consider the problem of predicting the salary y of an executive. Perhaps the biggest problem in building a model to describe executive salaries is choosing the important independent variables to be included in the model. The list of potentially important independent variables is extremely long (e.g., age, experience, tenure, education level, etc.), and we need some objective method of screening out those that are not important.

The problem of deciding which of a large set of independent variables to include in a model is a common one. Trying to determine which variables influence the profit of a firm, affect blood pressure of humans, or are related to a student's performance in college are only a few examples.

A systematic approach to building a model with a large number of independent variables is difficult because the interpretation of multivariable interactions and higher-order terms is tedious. We therefore turn to a screening procedure, available in most statistical software packages, known as **stepwise regression.**

The most commonly used stepwise regression procedure works as follows. The user first identifies the response, y, and the set of potentially important independent variables, $x_1, x_2, \ldots, x_k$, where k is generally large. [*Note:* This set of variables could include both first-order and higher-order terms. However, we may often include only the main effects of both quantitative variables (first-order terms) and qualitative variables (dummy variables) because the inclusion of second-order terms greatly increases the number of independent variables.] The response and independent variables are then entered into the computer software, and the stepwise procedure begins.

Step 1: The software program fits all possible one-variable models of the form

$$E(y) = \beta_0 + \beta_1 x_i$$

to the data, where x_i is the ith independent variable, $i = 1, 2, \ldots, k$. For each model, the test of the null hypothesis

$$H_0: \beta_1 = 0$$

against the alternative hypothesis

$$H_a: \beta_1 \neq 0$$

is conducted using the t-test (or the equivalent F-test) for a single β parameter. The independent variable that produces the largest (absolute) t-value is declared the best one-variable predictor of y.* Call this independent variable x_1.

Step 2: The stepwise program now begins to search through the remaining $(k - 1)$ independent variables for the best two-variable model of the form

$$E(y) = \beta_0 + \beta_1 x_1 + \beta_2 x_i$$

This is done by fitting all two-variable models containing x_1 and each of the other $(k - 1)$ options for the second variable x_i. The t-values for the test $H_0: \beta_2 = 0$ are computed for each of the $(k - 1)$ models (corresponding to the remaining independent variables, x_i, $i = 2, 3, \ldots, k$), and the variable having the largest t is retained. Call this variable x_2.

At this point, some software packages diverge in methodology. The better packages now go back and check the t-value of $\hat{\beta}_1$ after $\hat{\beta}_2 x_2$ has been added to the model. If the t-value has become nonsignificant at some specified α level (say $\alpha = .10$), the variable x_1 is removed and a search is made for the independent variable with a β parameter that will yield the most significant t-value in the presence of $\hat{\beta}_2 x_2$. Other packages do not recheck the significance of $\hat{\beta}_1$ but proceed directly to step 3.

The reason the t-value for x_1 may change from step 1 to step 2 is that the meaning of the coefficient $\hat{\beta}_1$ changes. In step 2, we are approximating a complex response surface in two variables with a plane. The best-fitting plane may yield a different value for $\hat{\beta}_1$ than that obtained in step 1. Thus, both the value of $\hat{\beta}_1$ and its significance usually changes from step 1 to step 2. For this reason, the software packages that recheck the t-values at each step are preferred.

Step 3: The stepwise procedure now checks for a third independent variable to include in the model with x_1 and x_2—that is, we seek the best model of the form

$$E(y) = \beta_0 + \beta_1 x_1 + \beta_2 x_2 + \beta_3 x_i$$

To do this, we fit all the $(k - 2)$ models using x_1, x_2 and each of the $(k - 2)$ remaining variables, x_i, as a possible x_3. The criterion is again to include the independent variable with the largest t-value. Call this best third variable x_3.

*In Step 1, note that the variable with the largest t-value is also the one with the largest (absolute) Pearson product moment correlation, r (Section 10.5), with y.

The better programs now recheck the t-values corresponding to the x_1 and x_2 coefficients, removing the variables with t-values that have become nonsignificant. This procedure is continued until no further independent variables can be found that yield significant t-values (at the specified α level) in the presence of the variables already in the model.

The result of the stepwise procedure is a model containing only those terms with t-values that are significant at the specified α level. Thus, in most practical situations, only several of the large number of independent variables remain. However, it is very important *not* to jump to the conclusion that all the independent variables important for predicting y have been identified or that the unimportant independent variables have been eliminated. Remember, the stepwise procedure is using only *sample estimates* of the true model coefficients (β's) to select the important variables. An extremely large number of single β parameter t-tests have been conducted, and the probability is very high that one or more errors have been made in including or excluding variables—that is, we have very probably included some unimportant independent variables in the model (Type I errors) and eliminated some important ones (Type II errors).

There is a second reason why we might not have arrived at a good model. When we choose the variables to be included in the stepwise regression, we may often omit higher-order terms (to keep the number of variables manageable). Consequently, we may have initially omitted several important terms from the model. Thus, we should recognize stepwise regression for what it is: an **objective variable screening procedure.**

Successful model builders will now consider second-order terms (for quantitative variables) and other interactions among variables screened by the stepwise procedure. It would be best to develop this response surface model with a second set of data independent of that used for the screening, so the results of the stepwise procedure can be partially verified with new data. This is not always possible, however, because in many modeling situations only a small amount of data is available.

Do not be deceived by the impressive-looking t-values that result from the stepwise procedure—it has retained only the independent variables with the largest t-values. Also, be certain to consider second-order terms in systematically developing the prediction model. Finally, if you have used a first-order model for your stepwise procedure, remember that it may be greatly improved by the addition of higher-order terms.

CAUTION Be wary of using the results of stepwise regression to make inferences about the relationship between $E(y)$ and the independent variables in the resulting first-order model. First, an extremely large number of t-tests have been conducted, leading to a high probability of making one or more Type I or Type II errors. Second, the stepwise model does not include any higher-order or interaction terms. Stepwise regression should be used only when necessary—that is, when you want to determine which of a large number of potentially important independent variables should be used in the model-building process. ▲

Example 11.13 Running a Stepwise Regression

Problem An international management consulting company develops multiple regression models for executive salaries of its client firms. The consulting company has found that models that use the natural logarithm of salary as the dependent variable have better predictive power than those using salary as the dependent variable.* A preliminary step in the construction of these models is the determination of the most important independent variables. For one firm, 10 potential independent variables (7 quantitative and 3 qualitative) were measured in a sample of 100 executives. The data, described in Table 11.5, are saved in the **EXECSAL** file. Because it would be very difficult to construct a complete second-order model with all of the 10 independent variables,

*This is probably because salaries tend to be incremented in *percentages* rather than dollar values. When a response variable undergoes percentage changes as the independent variables are varied, the logarithm of the response variable will be more suitable as a dependent variable.

Table 11.5 Independent Variables in the Executive Salary Example

Independent Variable	Description	Type
x_1	Experience (years)	Quantitative
x_2	Education (years)	Quantitative
x_3	Bonus eligibility (1 if yes, 0 if no)	Qualitative
x_4	Number of employees supervised	Quantitative
x_5	Corporate assets (millions of dollars)	Quantitative
x_6	Board member (1 if yes, 0 if no)	Qualitative
x_7	Age (years)	Quantitative
x_8	Company profits (past 12 months, millions of dollars)	Quantitative
x_9	Has international responsibility (1 if yes, 0 if no)	Qualitative
x_{10}	Company's total sales (past 12 months, millions of dollars)	Quantitative

Data Set: EXECSAL

use stepwise regression to decide which of the 10 variables should be included in the building of the final model for the natural log of executive salaries.

Solution We will use stepwise regression with the main effects of the 10 independent variables to identify the most important variables. The dependent variable y is the natural logarithm of the executive salaries. The Minitab stepwise regression printout is shown in Figure 11.32.

```
Alpha-to-Enter: 0.15  Alpha-to-Remove: 0.15

Response is Y on 10 predictors, with N = 100

Step               1        2         3         4         5
Constant      11.091   10.968    10.783    10.278     9.962

X1            0.0278   0.0273    0.0273    0.0273    0.0273
T-Value        12.62    15.13     18.80     24.68     26.50
P-Value        0.000    0.000     0.000     0.000     0.000

X3                      0.197     0.233     0.232     0.225
T-Value                  7.10     10.17     13.30     13.74
P-Value                 0.000     0.000     0.000     0.000

X4                              0.00048   0.00055   0.00052
T-Value                            7.32     10.92     11.06
P-Value                           0.000     0.000     0.000

X2                                         0.0300    0.0291
T-Value                                      8.38      8.72
P-Value                                     0.000     0.000

X5                                                  0.00196
T-Value                                                3.95
P-Value                                               0.000

S              0.161    0.131     0.106    0.0807    0.0751
R-Sq           61.90    74.92     83.91     90.75     92.06
R-Sq(adj)      61.51    74.40     83.41     90.36     91.64
Mallows C-p    343.9    195.5      93.8      16.8       3.6
```

Figure 11.32
Minitab stepwise regression printout for executive salary data

Note that the first variable included in the model is x_1, years of experience. At the second step, x_3, a dummy variable for the qualitative variable, bonus eligibility or not, is brought into the model. In steps 3, 4, and 5, the variables x_4 (number of employees supervised), x_2 (years of education), and x_5 (corporate assets), respectively, are selected for model inclusion. Minitab stops after five steps because no other independent variables met the criterion for admission into the model. As a default, Minitab uses $\alpha = .15$ in the t-tests conducted. In other words, if the p-value associated with a β coefficient exceeds $\alpha = .15$, the variable is *not* included in the model.

The results of the stepwise regression suggest that we should concentrate on these five independent variables. Models with second-order terms and interactions should be proposed and evaluated to determine the best model for predicting executive salaries.

Now Work Exercise 11.105

RECOMMENDATION Do *not* use the stepwise regression model as the *final* model for predicting y. Recall that the stepwise procedure tends to perform a large number of t-tests, inflating the overall probability of a Type I error, and does not automatically include higher-order terms (e.g., interactions and squared terms) in the final model. Use stepwise regression as a variable screening tool when there exists a large number of potential important independent variables. Then begin building models for y using the variables identified by stepwise. ▲

Statistics IN Action Revisited Variable Screening and Model Building

In the previous Statistics in Action Revisited section (p. 646), we used all eight of the independent variables in Table SIA11.1 to fit a first-order model for the cost (y) of a road construction contract awarded using the sealed-bid system. Although the model was deemed statistically useful for predicting y, the standard deviation of the model ($s = 305$ thousand dollars) was probably too large for the model to be practically useful. A more complex model—one involving higher-order terms (interactions and squared terms)—needs to be considered. A complete second-order model involving all eight of the independent variables, however, would require over 100 terms! Consequently, we'll use stepwise regression to select the best subset of independent variables and then form a complete second-order model with just these variables.

```
Alpha-to-Enter: 0.15  Alpha-to-Remove: 0.15

Response is COST on 8 predictors, with N = 235

Step              1        2        3         4
Constant      20.91   -20.54   -55.22   -212.85

DOTEST       0.9263   0.9308   0.9110    0.9132
T-Value       93.89    95.52    56.86     57.11
P-Value       0.000    0.000    0.000     0.000

STATUS                   166      167       171
T-Value                 3.38     3.40      3.50
P-Value                0.001    0.001     0.001

DAYSEST                          0.27      0.33
T-Value                          1.55      1.85
P-Value                         0.122     0.065

BTPRATIO                                    241
T-Value                                    1.81
P-Value                                   0.072

S               313      306      305       304
R-Sq          97.42    97.55    97.57     97.60
R-Sq(adj)     97.41    97.52    97.54     97.56
Mallows C-p    15.2      5.7      5.2       4.0
```

Figure SIA11.4
Minitab stepwise regression for the road cost data

Figure SIA11.4 is a Minitab printout of the stepwise regression. You can see that the DOT engineer's cost estimate (DOTEST) is the first variable selected, followed by bid status (STATUS), estimated work days (DAYSEST), and bids to plan holders ratio (BTPRATIO). Recall that Minitab uses a default α level of .15. If we reduce the significance level for entry to $\alpha = .05$, only DOTEST and STATUS are selected because the p-values (highlighted on the printout) for DAYSEST and BTPRATIO are both greater than .05.

Because bid status is a qualitative variable ($x_2 = 1$ if fixed and 0 if competitive), a complete second-order model for contract cost (y) using DOTEST (x_1) and STATUS (x_2) is given by the equation

(*Continued*)

```
The regression equation is
COST = - 3.0 + 0.916 DOTEST + 0.000001 DOTEST2 - 36.7 STATUS + 0.324 STA_DOT
       - 0.000036 STA_DOT2

Predictor            Coef       SE Coef       T       P
Constant            -2.98         30.89   -0.10   0.923
DOTEST            0.91553       0.02917   31.39   0.000
DOTEST2        0.00000072    0.00000340    0.21   0.833
STATUS             -36.72         74.77   -0.49   0.624
STA_DOT            0.3242        0.1192    2.72   0.007
STA_DOT2      -0.00003576    0.00002478   -1.44   0.150

S = 296.646   R-Sq = 97.7%   R-Sq(adj) = 97.7%

Analysis of Variance

Source           DF          SS          MS         F       P
Regression        5   866723202   173344640   1969.85   0.000
Residual Error  229    20151771       87999
Total           234   886874973
```

Figure SIA11.5
Minitab regression printout for the complete second-order model of contract cost

$$E(y) = \beta_0 + \beta_1 x_1 + \beta_2(x_1)^2 + \beta_3 x_2 + \beta_4 x_1 x_2 + \beta_5(x_1)^2 x_2$$

The Minitab printout for this model is shown in Figure SIA11.5. Note that the global F-test for the model is statistically significant (p-value $= .000$), and the model standard deviation, $s = 296.6$, is smaller than the standard deviation of the first-order model.

Are the second-order terms in the model, $\beta_2(x_1)^2$ and $\beta_5(x_1)^2 x_2$, necessary? If not, we can simplify the model by dropping these curvature terms. The hypothesis of interest is H_0: $\beta_2 = \beta_5 = 0$. To test this subset of β's, we compare the complete second-order model to a model without the curvilinear terms. The reduced model takes the form

$$E(y) = \beta_0 + \beta_1 x_1 + \beta_3 x_2 + \beta_4 x_1 x_2$$

The results of this nested model (or partial) F-test are shown at the bottom of the SPSS printout, Figure SIA11.6. The p-value of the test (highlighted on the SPSS printout) is .355. Because this p-value is greater than $\alpha = .05$, there is insufficient evidence to reject H_0—that is, there is no evidence to indicate that the two curvature terms are useful predictors of road construction cost. Consequently, the reduced model is selected as the better predictor of cost.

The Minitab printout for the reduced model is shown in Figure SIA11.7. The overall model is statistically useful (p-value $= .000$ for global F-test), explaining about 98% of the sample variation in contract costs. The model standard deviation, $s = 296.7$, implies that we can predict costs to within about 593 thousand dollars. Also, the t-test for the interaction term, $\beta_4 x_1 x_2$, is significant (p-value $= .000$), implying that the relationship between contract cost (y) and DOT cost estimate depends on bid status (fixed or competitive).

The nature of the interaction is illustrated in the Minitab graph of the least squared prediction equation for the reduced model, Figure SIA11.8. You can see that the rate of increase of contract cost (y) with the DOT engineer's estimate of cost (x_1) is steeper for fixed contracts than for competitive contracts.

Model Summary

					Change Statistics				
Model	R	R Square	Adjusted R Square	Std. Error of the Estimate	R Square Change	F Change	df1	df2	Sig. F Change
1	.988[a]	.977	.977	296.69515	.977	3281.312	3	231	.000
2	.989[b]	.977	.977	296.64238	.000	1.041	2	229	.355

a. Predictors: (Constant), STA_DOT, DOTEST, STATUS
b. Predictors: (Constant), STA_DOT, DOTEST, STATUS, DOTEST2, STA_DOT2

Figure SIA11.6
SPSS printout of test to compare the complete second-order model of contract cost to the reduced model

```
The regression equation is
COST = - 6.4 + 0.921 DOTEST + 28.7 STATUS + 0.163 STA_DOT


Predictor        Coef    SE Coef       T      P
Constant        -6.43      26.21   -0.25  0.806
DOTEST       0.921336   0.009723   94.75  0.000
STATUS          28.67      58.66    0.49  0.625
STA_DOT       0.16328    0.04043    4.04  0.000


S = 296.699   R-Sq = 97.7%    R-Sq(adj) = 97.7%


Analysis of Variance

Source           DF         SS         MS        F      P
Regression        3  866540004  288846668  3281.22  0.000
Residual Error  231   20334968      88030
Total           234  886874973
```

Figure SIA11.7
Minitab regression printout for the reduced model of contract cost

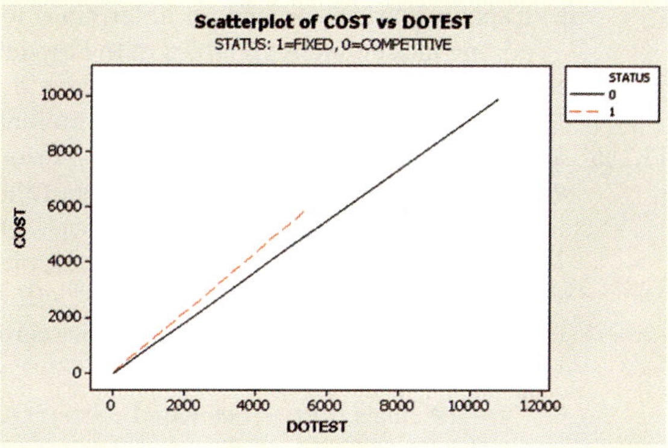

Figure SIA11.8
Minitab plot of least squares prediction equation for the reduced model of contract cost

Exercises 11.104–11.108

Learning the Mechanics

11.104 There are six independent variables, x_1, x_2, x_3, x_4, x_5, and x_6, that might be useful in predicting a response y. A total of $n = 50$ observations is available, and it is decided to employ stepwise regression to help in selecting the independent variables that appear to be useful. The software fits all possible one-variable models of the form

$$E(y) = \beta_0 + \beta_1 x_i$$

where x_i is the ith independent variable, $i = 1, 2, \ldots, 6$. The information in the table is provided from the computer printout.

a. Which independent variable is declared the best one-variable predictor of y? Explain.

b. Would this variable be included in the model at this stage? Explain.

c. Describe the next phase that a stepwise procedure would execute.

Independent Variable	$\hat{\beta}_i$	$s_{\hat{\beta}_i}$
x_1	1.6	.42
x_2	−.9	.01
x_3	3.4	1.14
x_4	2.5	2.06
x_5	−4.4	.73
x_6	.3	.35

Applying the Concepts—Basic

11.105 Entry-level job preferences. *Benefits Quarterly* (First Quarter, 1995) published a study of entry-level job preferences. A number of independent variables were used to model the job preferences (measured on a 10-point scale) of 164 business school graduates. Suppose stepwise regression is used to build a model for job preference score (y) as a function of the following independent variables:

NW

$$x_1 = \begin{cases} 1 & \text{if flextime position} \\ 0 & \text{if not} \end{cases}$$
$$x_2 = \begin{cases} 1 & \text{if day care support required} \\ 0 & \text{if not} \end{cases}$$
$$x_3 = \begin{cases} 1 & \text{if spousal transfer support required} \\ 0 & \text{if not} \end{cases}$$
$$x_4 = \text{Number of sick days allowed}$$
$$x_5 = \begin{cases} 1 & \text{if applicant married} \\ 0 & \text{if not} \end{cases}$$
$$x_6 = \text{Number of children of applicant}$$
$$x_7 = \begin{cases} 1 & \text{if male applicant} \\ 0 & \text{if female applicant} \end{cases}$$

a. How many models are fit to the data in step 1? Give the general form of these models.
b. How many models are fit to the data in step 2? Give the general form of these models.
c. How many models are fit to the data in step 3? Give the general form of these models.
d. Explain how the procedure determines when to stop adding independent variables to the model.
e. Describe two major drawbacks to using the final stepwise model as the best model for job preference score y.

11.106 Accuracy of software effort estimates. Periodically, software engineers must provide estimates of their effort in developing new software. In the *Journal of Empirical Software Engineering* (Vol. 9, 2004), multiple regression was used to predict the accuracy of these effort estimates. The dependent variable, defined as the relative error in estimating effort,

$$y = (\text{Actual effort} - \text{Estimated effort}) / (\text{Actual effort})$$

was determined for each in a sample of $n = 49$ software development tasks. Eight independent variables were evaluated as potential predictors of relative error using stepwise regression. Each of these was formulated as a dummy variable, as shown in the table.

Company role of estimator:	$x_1 = 1$ if developer, 0 if project leader
Task complexity:	$x_2 = 1$ if low, 0 if medium/high
Contract type:	$x_3 = 1$ if fixed price, 0 if hourly rate
Customer importance:	$x_4 = 1$ if high, 0 if low/medium
Customer priority:	$x_5 = 1$ if time of delivery, 0 if cost or quality
Level of knowledge:	$x_6 = 1$ if high, 0 if low/medium
Participation:	$x_7 = 1$ if estimator participates in work, 0 if not
Previous accuracy:	$x_8 = 1$ if more than 20% accurate, 0 if less than 20% accurate

a. In step 1 of the stepwise regression, how many different one-variable models are fit to the data?
b. In step 1, the variable x_1 is selected as the best one-variable predictor. How is this determined?
c. In step 2 of the stepwise regression, how many different two-variable models (where x_1 is one of the variables) are fit to the data?
d. The only two variables selected for entry into the stepwise regression model were x_1 and x_8. The stepwise regression yielded the following prediction equation:

$$\hat{y} = .12 - .28x_1 + .27x_8$$

Give a practical interpretation of the β estimates multiplied by x_1 and x_8.
e. Why should a researcher be wary of using the model, part **d**, as the final model for predicting effort (y)?

Applying the Concepts—Intermediate

11.107 Bus Rapid Transit study. Bus Rapid Transit (BRT) is a rapidly growing trend in the provision of public transportation in America. The Center for Urban Transportation Research (CUTR) at the University of South Florida conducted a survey of BRT customers in Miami (*Transportation Research Board* Annual Meeting, Jan. 2003). Data on the following variables (all measured on a 5-point scale, where 1 = very unsatisfied and 5 = very satisfied) were collected for a sample of over 500 bus riders: overall satisfaction with BRT (y), safety on bus (x_1), seat availability (x_2), dependability (x_3), travel time (x_4), cost (x_5), information/maps (x_6), convenience of routes (x_7), traffic signals (x_8), safety at bus stops (x_9), hours of service (x_{10}), and frequency of service (x_{11}). CUTR analysts used stepwise regression to model overall satisfaction (y).

a. How many models are fit at step 1 of the stepwise regression?
b. How many models are fit at step 2 of the stepwise regression?
c. How many models are fit at step 11 of the stepwise regression?
d. The stepwise regression selected the following eight variables to include in the model (in order of selection): $x_{11}, x_4, x_2, x_7, x_{10}, x_1, x_9$, and x_3. Write the equation for $E(y)$ that results from stepwise regression.
e. The model, part **d,** resulted in $R^2 = .677$. Interpret this value.
f. Explain why the CUTR analysts should be cautious in concluding that the best model for $E(y)$ has been found.

11.108 Adverse effects of hot-water runoff. A marine biologist was hired by the EPA to determine whether the hot-water runoff from a particular power plant located near a large gulf is having an adverse effect on the marine life in the area. The biologist's goal is to acquire a prediction equation for the number of marine animals located at certain designated areas, or stations, in the gulf. Based on past experience, the EPA considered the following environmental factors as predictors for the number of animals at a particular station:

x_1 = Temperature of water (TEMP)
x_2 = Salinity of water (SAL)
x_3 = Dissolved oxygen content of water (DO)
x_4 = Turbidity index, a measure of the turbidity of the water (TI)

x_5 = Depth of the water at the station (ST_DEPTH)

x_6 = Total weight of sea grasses in sampled area (TGRSWT)

As a preliminary step in the construction of this model, the biologist used a stepwise regression procedure to identify the most important of these six variables. A total of 716 samples were taken at different stations in the gulf, producing the SPSS printout shown below. (The response measured was y, the logarithm of the number of marine animals found in the sampled area.)

a. According to the SPSS printout, which of the six independent variables should be used in the model? (Use $\alpha = .10$.)

b. Are we able to assume that the marine biologist has identified all the important independent variables for the prediction of y? Why?

c. Using the variables identified in part **a,** write the first-order model with interaction that may be used to predict y.

d. How would the marine biologist determine whether the model specified in part **c** is better than the first-order model?

e. Note the small value of R^2. What action might the biologist take to improve the model?

Variables Entered/Removed[a]

Model	Variables Entered	Variables Removed	Method
1	ST_DEPTH	.	Stepwise (Criteria: Probability-of-F-to-enter <= .050, Probability-of-F-to-remove >= .100).
2	TGRSWT	.	Stepwise (Criteria: Probability-of-F-to-enter <= .050, Probability-of-F-to-remove >= .100).
3	TI	.	Stepwise (Criteria: Probability-of-F-to-enter <= .050, Probability-of-F-to-remove >= .100).

a. Dependent Variable: LOGNUM

Model Summary

Model	R	R Square	Adjusted R Square	Std. Error of the Estimate
1	.329[a]	.122	.121	.7615773
2	.427[b]	.182	.180	.7348470
3	.432[c]	.187	.184	.7348469

a. Predictors: (Constant), ST_DEPTH

b. Predictors: (Constant), ST_DEPTH, TGRSWT

c. Predictors: (Constant), ST_DEPTH, TGRSWT, TI

PART III: MULTIPLE REGRESSION DIAGNOSTICS

11.11 Residual Analysis: Checking the Regression Assumptions

When we apply regression analysis to a set of data, we never know for certain whether the assumptions of Section 11.1 are satisfied. How far can we deviate from the assumptions and still expect regression analysis to yield results that will have the reliability stated in this chapter? How can we detect departures (if they exist) from the assumptions, and what can we do about them? We provide some answers to these questions in this section.

Recall from Section 11.1 that for any given set of values of $x_1, x_2, \ldots, x_k$, we assume that the random error term ε has the following properties:

1. mean equal to 0
2. constant variance (σ^2)
3. normal probability distribution
4. probabilistically independent

It is unlikely that these assumptions are ever satisfied exactly in a practical application of regression analysis. Fortunately, experience has shown that least squares regression analysis produces reliable statistical tests, confidence intervals, and prediction intervals as long as the departures from the assumptions are not too great. In this section, we present some methods for determining whether the data indicate significant departures from the assumptions.

Because the assumptions all concern the random error component, ε, of the model, the first step is to estimate the random error. Because the actual random error associated with a particular value of y is the difference between the actual y value and its unknown mean, we estimate the error by the difference between the actual y-value and the *estimated* mean. This estimated error is called the *regression residual,* or simply the **residual,** and is denoted by $\hat{\varepsilon}$. The actual error ε and residual $\hat{\varepsilon}$ are shown in Figure 11.33.

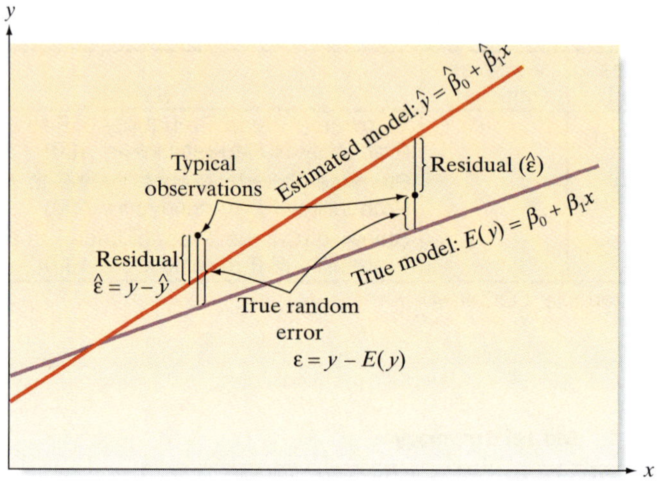

Figure 11.33
Actual random error ε and regression residual $\hat{\varepsilon}$

A **regression residual,** $\hat{\varepsilon}$, is defined as the difference between an observed y value and its corresponding predicted value:

$$\hat{\varepsilon} = (y - \hat{y}) = y - (\hat{\beta}_0 + \hat{\beta}_1 x_1 + \hat{\beta}_2 x_2 + \cdots + \hat{\beta}_k x_k)$$

Because the true mean of y (that is, the true regression model) is not known, the actual random error cannot be calculated. However, because the residual is based on the estimated mean (the least squares regression model), it can be calculated and used to estimate the random error and to check the regression assumptions. Such checks are generally referred to as **residual analyses.** Two useful properties of residuals are given in the next box.

Properties of Regression Residuals

1. The mean of the residuals is equal to 0. This property follows from the fact that the sum of the differences between the observed y values and their least squares predicted $\hat{y}$ values is equal to 0.

$$\sum(\text{Residuals}) = \sum(y - \hat{y}) = 0$$

2. The standard deviation of the residuals is equal to the standard deviation of the fitted regression model, s. This property follows from the fact that the sum of the squared residuals is equal to SSE, which when divided by the error degrees of freedom is equal to the variance of the fitted regression model, s^2. The square root of

the variance is both the standard deviation of the residuals and the standard deviation of the regression model.

$$\sum(\text{Residuals})^2 = \sum(y - \hat{y})^2 = \text{SSE}$$

$$s = \sqrt{\frac{\Sigma(\text{Residuals})^2}{n - (k + 1)}} = \sqrt{\frac{\text{SSE}}{n - (k + 1)}}$$

BIOGRAPHY **FRANCIS J. ANSCOMBE (1918–2001)**

Anscombe's Data

British citizen Frank Anscombe grew up in a small town near the English Channel. He attended Trinity College in Cambridge, England, on a merit scholarship, graduating with first class honors in mathematics in 1939. He later earned his master's degree in 1943. During World War II, Anscombe worked for the British Ministry of Supply, developing a mathematical solution for aiming antiaircraft rockets at German bombers and buzz bombs. Following the war, Anscombe worked at the Rothamsted Experimental Station, applying statistics to agriculture. There, he formed his appreciation for solving problems with social relevance. During his career as a professor of statistics, Anscombe served on the faculty of Cambridge, Princeton, and Yale universities. He was a pioneer in the application of computers to statistical analysis and was one of the original developers of residual analysis in regression. Anscombe is famous for a paper he wrote in 1973, in which he showed that one regression model could be fit by four very different data sets ("Anscombe's data"). While Anscombe published 50 research articles on statistics, he also had serious interests in classical music, poetry, and art. ■

The following examples show how a graphical analysis of regression residuals can be used to verify the assumptions associated with the model and to support improvements to the model when the assumptions do not appear to be satisfied. We rely on statistical software to generate the appropriate graphs in the examples and exercises.

Checking Assumption #1: Mean $\varepsilon = 0$

First, we demonstrate how a residual plot can detect a model in which the hypothesized relationship between $E(y)$ and an independent variable x is misspecified. The assumption of mean error of 0 is violated in these types of models.*

Example 11.14 Analyzing the Residuals from the Electrical Usage Model

Problem Refer to the problem of modeling the relationship between home size (x) and electrical usage (y) in Example 11.7 (p. 657). The data for $n = 15$ homes are repeated in Table 11.6. Minitab printouts for a straight-line model and a quadratic model fitted to the data are shown in Figures 11.34a and 11.34b, respectively. The residuals from these models are highlighted in the printouts. The residuals are then plotted on the vertical axis against the variable x, size of home, on the horizontal axis in Figures 11.35a and 11.35b, respectively.

a. Verify that each residual is equal to the difference between the observed y value and the estimated mean value, $\hat{y}$.

b. Analyze the residual plots.

Solution

a. For the straight-line model, the residual is calculated for the first y value as follows:

$$\hat{\varepsilon} = (y - \hat{y}) = 1{,}182 - 1{,}362.2 = -180.2$$

where $\hat{y}$ is the first number in the column labeled **Fit** on the Minitab printout in Figure 11.34a. Similarly, the residual for the first y value using the quadratic model (Figure 11.34b) is

$$\hat{\varepsilon} = 1{,}182 - 1{,}157.2 = 24.8$$

*For a misspecified model, the hypothesized mean of y, denoted by $E_h(y)$, will not equal the true mean of y, $E(y)$. Because $y = E_h(y) + \varepsilon$, then $\varepsilon = y - E_h(y)$ and $E(\varepsilon) = E[y - E_h(y)] = E(y) - E_h(y) \neq 0$.

Table 11.6 Home Size–Electrical Usage Data

Size of Home, x (sq. ft.)	Monthly Usage, y (kilowatt-hours)
1,290	1,182
1,350	1,172
1,470	1,264
1,600	1,493
1,710	1,571
1,840	1,711
1,980	1,804
2,230	1,840
2,400	1,956
2,710	2,007
2,930	1,984
3,000	1,960
3,210	2,001
3,240	1,928
3,520	1,945

Both residuals agree with the first values given in the column labeled **Residual** in Figures 11.34a and 11.34b, respectively. Although the residuals both correspond to the same observed y value, 1,182, they differ because the predicted mean value changes depending on whether the straight-line model or quadratic model is used. Similar calculations produce the remaining residuals.

Regression Analysis: USAGE versus SIZE

```
The regression equation is
USAGE = 903 + 0.356 SIZE

Predictor       Coef   SE Coef     T      P
Constant       903.0     132.1  6.83  0.000
SIZE         0.35594   0.05477  6.50  0.000

S = 155.251    R-Sq = 76.5%    R-Sq(adj) = 74.7%

Analysis of Variance

Source           DF       SS       MS      F      P
Regression        1  1017803  1017803  42.23  0.000
Residual Error   13   313338    24103
Total            14  1331140

Obs  SIZE   USAGE     Fit  SE Fit  Residual  St Resid
  1  1290  1182.0  1362.2    68.3    -180.2     -1.29
  2  1350  1172.0  1383.5    65.6    -211.5     -1.50
  3  1470  1264.0  1426.2    60.6    -162.2     -1.13
  4  1600  1493.0  1472.5    55.4      20.5      0.14
  5  1710  1571.0  1511.7    51.4      59.3      0.41
  6  1840  1711.0  1557.9    47.3     153.1      1.04
  7  1980  1804.0  1607.8    43.7     196.2      1.32
  8  2230  1840.0  1696.8    40.3     143.2      0.96
  9  2400  1956.0  1757.3    40.5     198.7      1.33
 10  2710  2007.0  1867.6    46.0     139.4      0.94
 11  2930  1984.0  1945.9    52.9      38.1      0.26
 12  3000  1960.0  1970.8    55.5     -10.8     -0.07
 13  3210  2001.0  2045.6    64.0     -44.6     -0.32
 14  3240  1928.0  2056.3    65.3    -128.3     -0.91
 15  3520  1945.0  2155.9    78.0    -210.9     -1.57
```

Figure 11.34a
Minitab printout for straight-line model of electrical usage

Regression Analysis: USAGE versus SIZE, SIZESQ

The regression equation is
USAGE = - 807 + 1.96 SIZE - 0.000340 SIZESQ

Predictor	Coef	SE Coef	T	P
Constant	-806.7	166.9	-4.83	0.000
SIZE	1.9616	0.1525	12.86	0.000
SIZESQ	-0.00034044	0.00003212	-10.60	0.000

S = 50.1998   R-Sq = 97.7%   R-Sq(adj) = 97.3%

Analysis of Variance

Source	DF	SS	MS	F	P
Regression	2	1300900	650450	258.11	0.000
Residual Error	12	30240	2520		
Total	14	1331140			

Obs	SIZE	USAGE	Fit	SE Fit	Residual	St Resid
1	1290	1182.0	1157.2	29.3	24.8	0.61
2	1350	1172.0	1221.0	26.2	-49.0	-1.14
3	1470	1264.0	1341.2	21.2	-77.2	-1.70
4	1600	1493.0	1460.3	18.0	32.7	0.70
5	1710	1571.0	1552.2	17.1	18.8	0.40
6	1840	1711.0	1650.1	17.6	60.9	1.30
7	1980	1804.0	1742.6	19.0	61.4	1.32
8	2230	1840.0	1874.7	21.2	-34.7	-0.76
9	2400	1956.0	1940.2	21.7	15.8	0.35
10	2710	2007.0	2009.0	20.0	-2.0	-0.04
11	2930	1984.0	2018.2	18.4	-34.2	-0.73
12	3000	1960.0	2014.1	18.4	-54.1	-1.16
13	3210	2001.0	1982.1	21.5	18.9	0.42
14	3240	1928.0	1975.1	22.5	-47.1	-1.05
15	3520	1945.0	1880.0	36.2	65.0	1.87

Figure 11.34b
Minitab printout for quadratic model of electrical usage

b. The Minitab plot of the residuals for the straight-line model (Figure 11.35a) reveals a nonrandom pattern. The residuals exhibit a curved shape, with the residuals for the small values of x below the horizontal 0 (mean of the residuals) line, the residuals corresponding to the middle values of x above the 0 line, and the residuals for the largest values of x again below the 0 line. The indication is that the mean value of the random error ε *within* each of these ranges of x (small, medium, large) may not be equal to 0. Such a pattern usually indicates that curvature needs to be added to the model.

When the second-order term is added to the model, the nonrandom pattern disappears. In Figure 11.35b, the residuals appear to be randomly distributed around the

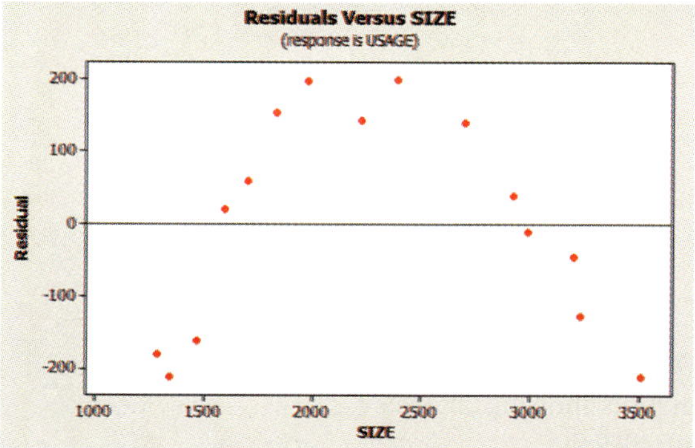

Figure 11.35a
Minitab residual plot for straight-line model of electrical usage

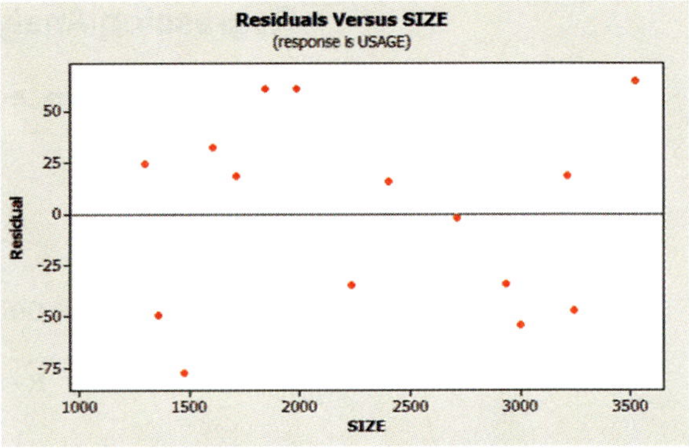

Figure 11.35b
Minitab residual plot for quadratic model of electrical usage

0 line, as expected. Note, too, that the residuals on the quadratic residual plot vary between −75 and 50, compared to between −200 and 200 on the straight-line plot. In fact, $s \approx 50$ for the quadratic model is much smaller than $s \approx 155$ for the straight-line model. The implication is that the quadratic model provides a considerably better model for predicting electrical usage.

Look Back The residual analysis verifies our conclusions from Example 11.7, where we found the t-test for the quadratic term, $\beta_2 x^2$, to be statistically significant.

Now Work Exercise 11.110a

Checking Assumption #2: Constant Error Variance

Residual plots can also be used to detect violations of the assumption of constant error variance. For example, a plot of the residuals versus the predicted value $\hat{y}$ may display one of the patterns shown in Figure 11.36. In these figures, the range in values of the residuals increases (or decreases) as $\hat{y}$ increases, thus indicating that the variance of the random error, ε, becomes larger (or smaller) as the estimate of $E(y)$ increases in value. Because $E(y)$ depends on the x values in the model, this implies that the variance of ε is not constant for all settings of the x's.

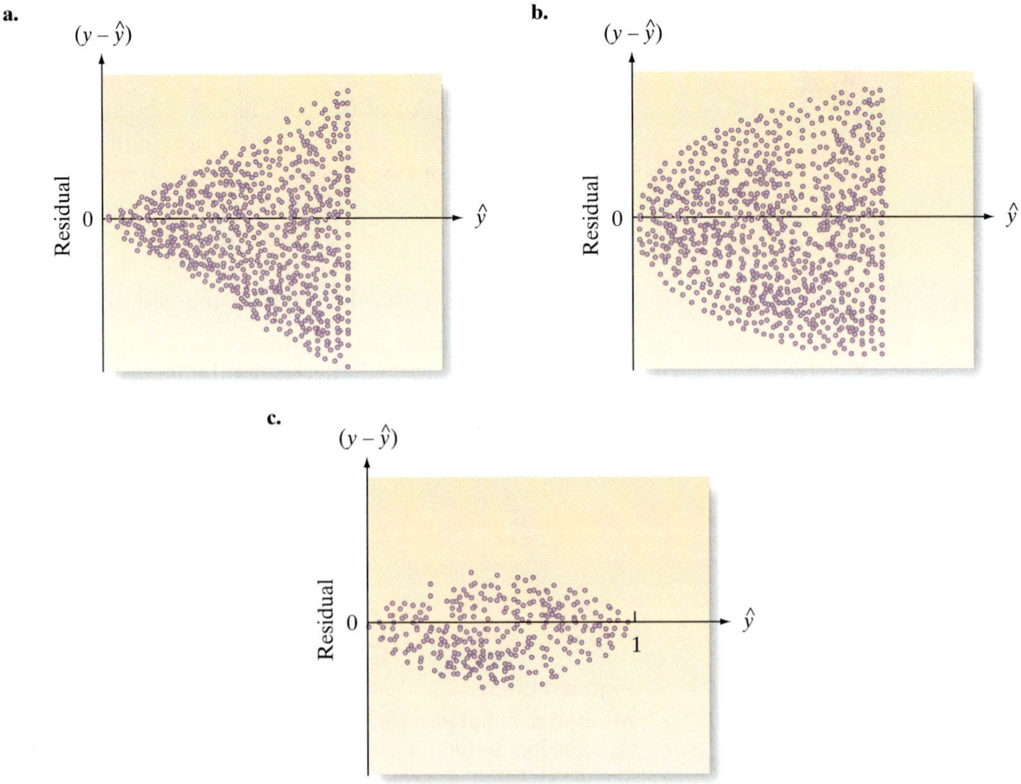

Figure 11.36
Residual plots showing changes in the variance of ε

In the next example, we demonstrate how to use this plot to detect a nonconstant variance and suggest a useful remedy.

Example 11.15 Using Residuals to Check Equal Variances

Problem The data in Table 11.7 are the salaries, y, and years of experience, x, for a sample of 50 social workers. The first-order model $E(y) = \beta_0 + \beta_1 x$ was fitted to the data using SPSS. The SPSS printout is shown in Figure 11.37, followed by a plot of the residuals versus $\hat{y}$ in Figure 11.38. Interpret the results. Make model modifications, if necessary.

Table 11.7 Salary Data for Example 11.15

Years of Experience, x	Salary, y	Years of Experience, x	Salary, y	Years of Experience, x	Salary, y
7	$26,075	21	$43,628	28	$99,139
28	79,370	4	16,105	23	52,624
23	65,726	24	65,644	17	50,594
18	41,983	20	63,022	25	53,272
19	62,308	20	47,780	26	65,343
15	41,154	15	38,853	19	46,216
24	53,610	25	66,537	16	54,288
13	33,697	25	67,447	3	20,844
2	22,444	28	64,785	12	32,586
8	32,562	26	61,581	23	71,235
20	43,076	27	70,678	20	36,530
21	56,000	20	51,301	19	52,745
18	58,667	18	39,346	27	67,282
7	22,210	1	24,833	25	80,931
2	20,521	26	65,929	12	32,303
18	49,727	20	41,721	11	38,371
11	33,233	26	82,641		

Data Set: SOCWORK

Model Summary[b]

Model	R	R Square	Adjusted R Square	Std. Error of the Estimate
1	.887[a]	.787	.782	8642.441

a. Predictors: (Constant), ESP
b. Dependent Variable: SALARY

ANOVA[b]

Model		Sum of Squares	df	Mean Square	F	Sig.
1	Regression	1.3E+10	1	1.324E+10	177.257	.000[a]
	Residual	3.6E+09	48	74691793.28		
	Total	1.7E+10	49			

a. Predictors: (Constant), ESP
b. Dependent Variable: SALARY

Coefficients[a]

Model		Unstandardized Coefficients		Standardized Coefficients	t	Sig.
		B	Std. Error	Beta		
1	(Constant)	11368.72	3160.317		3.597	.001
	ESP	2141.381	160.839	.887	13.314	.000

a. Dependent Variable: SALARY

Figure 11.37
SPSS regression printout for first-order model of salary

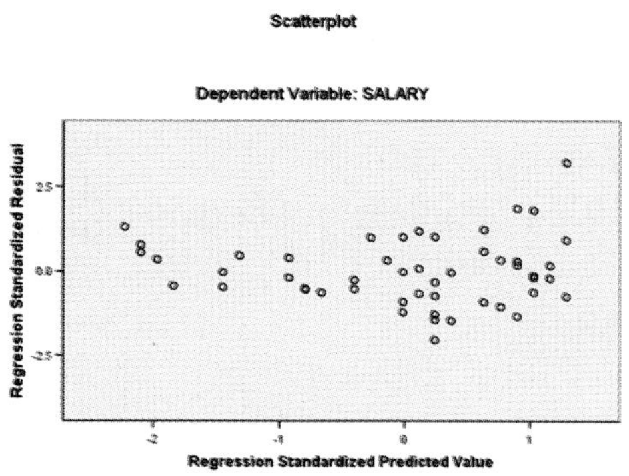

Figure 11.38
SPSS residual plot for first-order model of salary

Solution The SPSS printout, Figure 11.37, suggests that the first-order model provides an adequate fit to the data. The R^2-value indicates that the model explains 78.7% of the sample variation in salaries. The t-value for testing β_1, 13.31, is highly significant (p-value ≈ 0) and indicates that the model contributes information for the prediction of y. However, an examination of the residuals plotted against $\hat{y}$ (Figure 11.38) reveals a potential problem. Note the "cone" shape of the residual variability; the size of the residuals increases as the estimated mean salary increases, implying that the constant variance assumption is violated.

One way to stabilize the variance of ε is to refit the model using a transformation on the dependent variable y. With economic data (e.g., salaries), a useful **variance-stabilizing transformation** is the natural logarithm of y, denoted $\ln(y)$.* We fit the model

$$\ln(y) = \beta_0 + \beta_1 x + \varepsilon$$

to the data of Table 11.7. Figure 11.39 shows the SPSS regression analysis printout for the $n = 50$ measurements, while Figure 11.40 shows a plot of the residuals from the log model.

Model Summary[b]

Model	R	R Square	Adjusted R Square	Std. Error of the Estimate
1	.929[a]	.864	.861	.1541127

a. Predictors: (Constant), ESP
b. Dependent Variable: LNSALARY

ANOVA[b]

Model		Sum of Squares	df	Mean Square	F	Sig.
1	Regression	7.212	1	7.212	303.660	.000[a]
	Residual	1.140	48	.024		
	Total	8.352	49			

a. Predictors: (Constant), ESP
b. Dependent Variable: LNSALARY

Coefficients[a]

Model		Unstandardized Coefficients		Standardized Coefficients		
		B	Std. Error	Beta	t	Sig.
1	(Constant)	9.841	.056		174.631	.000
	ESP	.050	.003	.929	17.426	.000

a. Dependent Variable: LNSALARY

Figure 11.39
SPSS regression printout for model of log salary

*Other variance-stabilizing transformations that are used successfully in practice are $\sqrt{y}$ and $\sin^{-1}\sqrt{y}$. Consult the references for more details on these transformations.

You can see that the logarithmic transformation has stabilized the error variances. Note that the cone shape is gone; there is no apparent tendency of the residual variance to increase as mean salary increases. We therefore are confident that inferences using the ln(y) model are more reliable than those using the untransformed model.

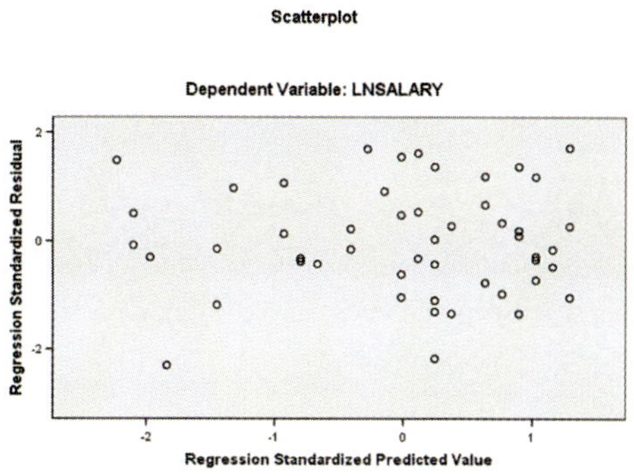

Figure 11.40
SPSS residual plot for model of ln(salary)

Look Back With transformed models, the analyst should be wary when interpreting model statistics such as $\hat{\beta}_1$ and s. These interpretations must take into account that the dependent variable is not some function of y. For example, the antilogarithm of $\hat{\beta}_1$ in the ln(y) model for salary (y) represents the percentage change in salary for every 1 year increase in experience (x).

Now Work Exercise 11.110b

Checking Assumption #3: Errors Normally Distributed

Several graphical methods are available for assessing whether the random error ε has an approximate normal distribution. Recall (Section 4.7) that stem-and-leaf displays, histograms, and normal probability plots are useful for checking whether the data are normally distributed. We illustrate these techniques in an example. But first, we discuss a related problem—using residuals to check for *outliers*.

If the assumption of normally distributed errors is satisfied, then we expect approximately 95% of the residuals to fall within 2 standard deviations of the mean of 0, and almost all of the residuals to lie within 3 standard deviations of the mean of 0. Residuals that are extremely far from 0 and disconnected from the bulk of the other residuals are called **regression outliers** and should receive special attention from the analyst.

A **regression outlier** is a residual that is larger than $3s$ (in absolute value).

Example 11.16 Identifying Outliers

Problem Refer to Example 11.6 (p. 651) in which we modeled the auction price (y) of a grandfather clock as a function of age (x_1) and number of bidders (x_2). The data for this example are repeated in Table 11.8, with one important difference: The auction price of the clock at the top of the second column has been changed from \$2,131 to \$1,131 (highlighted in Table 11.8). The interaction model

$$E(y) = \beta_0 + \beta_1 x_1 + \beta_2 x_2 + \beta_3 x_1 x_2$$

is again fit to these (modified) data, with the Minitab printout shown in Figure 11.41. The residuals are shown highlighted in the printout and then plotted against the number of bidders, x_2, in Figure 11.42. Analyze the residual plot.

Solution The residual plot dramatically reveals the one altered measurement. Note that one of the two residuals at $x_2 = 14$ bidders falls more than 3 standard deviations below 0. (This observation is highlighted on Figure 11.41.) Note that no other residual falls more than 2 standard deviations from 0.

What do we do with outliers once we identify them? First, we try to determine the cause. Were the data entered into the computer incorrectly? Was the observation recorded incorrectly when the data were collected? If so, we correct the observation and rerun the analysis. Another possibility is that the observation is not representative of the conditions we are trying to model. For example, in this case the low price may be attributable to extreme damage to the clock or to a clock of inferior quality compared to the others. In these cases, we probably would exclude the observation from the analysis. In many cases,

```
The regression equation is
PRICE = - 513 + 8.17 AGE + 19.9 NUMBIDS + 0.320 AGE_BIDS

Predictor     Coef   SE Coef      T      P
Constant    -512.8     665.9  -0.77  0.448
AGE          8.165     4.585   1.78  0.086
NUMBIDS      19.89     67.44   0.29  0.770
AGE_BIDS    0.3196    0.4790   0.67  0.510

S = 200.598   R-Sq = 72.9%   R-Sq(adj) = 70.0%

Analysis of Variance

Source          DF       SS       MS      F      P
Regression       3  3033587  1011196  25.13  0.000
Residual Error  28  1126703    40239
Total           31  4160290

Obs  AGE   PRICE     Fit  SE Fit  Residual  St Resid
  1  127  1235.0  1310.4    59.3     -75.4     -0.39
  2  115  1080.0  1105.9    62.1     -25.9     -0.14
  3  127   845.0   947.5    61.1    -102.5     -0.54
  4  150  1522.0  1322.5    37.1     199.5      1.01
  5  156  1047.0  1179.5    60.3    -132.5     -0.69
  6  182  1979.0  1831.9    82.9     147.1      0.81
  7  156  1822.0  1598.0    61.9     224.0      1.17
  8  132  1253.0  1185.8    39.7      67.2      0.34
  9  137  1297.0  1178.9    39.0     118.1      0.60
 10  113   946.0   913.9    58.6      32.1      0.17
 11  137  1713.0  1561.0    78.4     152.0      0.82
 12  117  1024.0  1072.6    53.1     -48.6     -0.25
 13  137  1147.0  1115.2    44.3      31.8      0.16
 14  153  1092.0  1149.2    59.0     -57.2     -0.30
 15  117  1152.0  1187.2    69.7     -35.2     -0.19
 16  126  1336.0  1117.6    43.4     218.4      1.12
 17  170  1131.0  1914.4   116.7    -783.4     -4.80R
 18  182  1550.0  1597.7    62.8     -47.7     -0.25
 19  162  1884.0  1598.3    57.0     285.7      1.49
 20  184  2041.0  1776.6    70.7     264.4      1.41
 21  143   845.0  1048.4    58.9    -203.4     -1.06
 22  159  1483.0  1421.8    40.6      61.2      0.31
 23  108  1055.0  1130.7    97.9     -75.7     -0.43
 24  175  1545.0  1522.7    55.4      22.3      0.12
 25  108   729.0   695.5    99.6      33.5      0.19
 26  179  1792.0  1642.7    57.6     149.3      0.78
 27  111  1175.0  1224.0   107.2     -49.0     -0.29
 28  187  1593.0  1651.3    68.6     -58.3     -0.31
 29  111   785.0   781.1    80.9       3.9      0.02
 30  115   744.0   822.7    75.5     -78.7     -0.42
 31  194  1356.0  1480.7   133.6    -124.7     -0.83 X
 32  168  1262.0  1374.0    57.7    -112.0     -0.58

R denotes an observation with a large standardized residual.
X denotes an observation whose X value gives it large influence.
```

Figure 11.41

Minitab regression printout for altered grandfather clock data

Table 11.8 Altered Auction Price Data

Age, x_1 (years)	Number of Bidders, x_2	Auction Price, y ($)	Age, x_1 (years)	Number of Bidders, x_2	Auction Price, y ($)
127	13	1,235	170	14	1,131
115	12	1,080	182	8	1,550
127	7	845	162	11	1,884
150	9	1,522	184	10	2,041
156	6	1,047	143	6	845
182	11	1,979	159	9	1,483
156	12	1,822	108	14	1,055
132	10	1,253	175	8	1,545
137	9	1,297	108	6	729
113	9	946	179	9	1,792
137	15	1,713	111	15	1,175
117	11	1,024	187	8	1,593
137	8	1,147	111	7	785
153	6	1,092	115	7	744
117	13	1,152	194	5	1,356
126	10	1,336	168	7	1,262

Data Set: GFCLOCKALT

you may not be able to determine the cause of the outlier. Even so, you may want to rerun the regression analysis excluding the outlier in order to assess the effect of that observation on the results of the analysis.

Figure 11.43 shows the printout when the outlier observation is excluded from the grandfather clock analysis, and Figure 11.44 shows the new plot of the residuals against the number of bidders. Now only one of the residuals lies beyond 2 standard deviations from 0, and none of them lies beyond 3 standard deviations. Also, the model statistics indicate a much better model without the outlier. Most notably, the s has decreased from 200.6 to 85.83, indicating a model that will provide

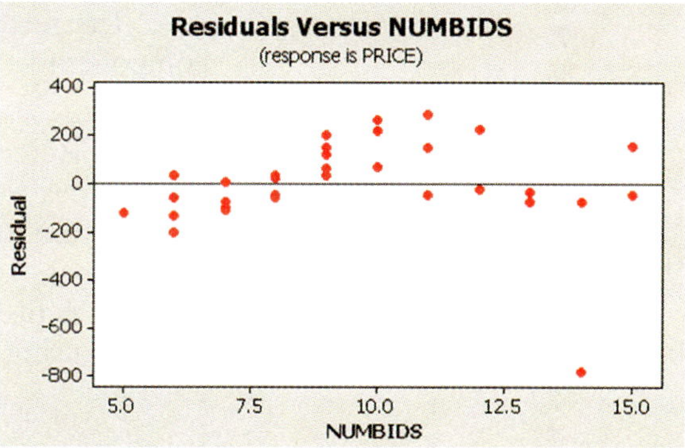

Figure 11.42
Minitab residual plot for altered grandfather clock data

```
The regression equation is
PRICE = 474 - 0.46 AGE - 114 NUMBIDS + 1.48 AGE_BIDS


Predictor       Coef   SE Coef      T      P
Constant       474.0     298.2   1.59  0.124
AGE           -0.465     2.107  -0.22  0.827
NUMBIDS      -114.12     31.23  -3.65  0.001
AGE_BIDS      1.4781    0.2295   6.44  0.000


S = 85.8286    R-Sq = 95.2%    R-Sq(adj) = 94.7%


Analysis of Variance

Source          DF       SS       MS       F      P
Regression       3  3933417  1311139  177.99  0.000
Residual Error  27   198897     7367
Total           30  4132314
```

Figure 11.43
Minitab regression printout when outlier is deleted

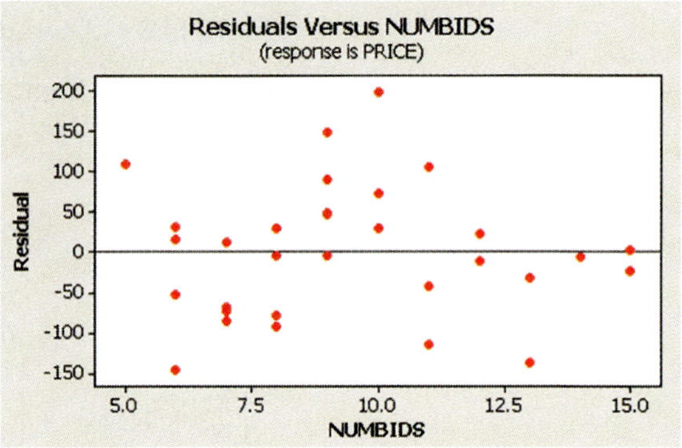

Figure 11.44
Minitab residual plot when outlier is deleted

more precise estimates and predictions (narrower confidence and prediction intervals) for clocks that are similar to those in the reduced sample.

Look Back Remember that if the outlier is removed from the analysis when in fact it belongs to the same population as the rest of the sample, the resulting model may provide misleading estimates and predictions.

Now Work Exercise 11.110c

The next example checks the assumption of the normality of the random error component.

Example 11.17
Using Residuals to Check for Normal Errors

Problem Refer to Example 11.16. Analyze the distribution of the residuals in the grandfather clock example, both before and after the outlier residual is removed. Determine whether the assumption of a normally distributed error term is reasonable.

Solution A histogram and normal probability plot for the two sets of residuals are constructed using Minitab and are shown in Figures 11.45 and 11.46. Note that the outlier

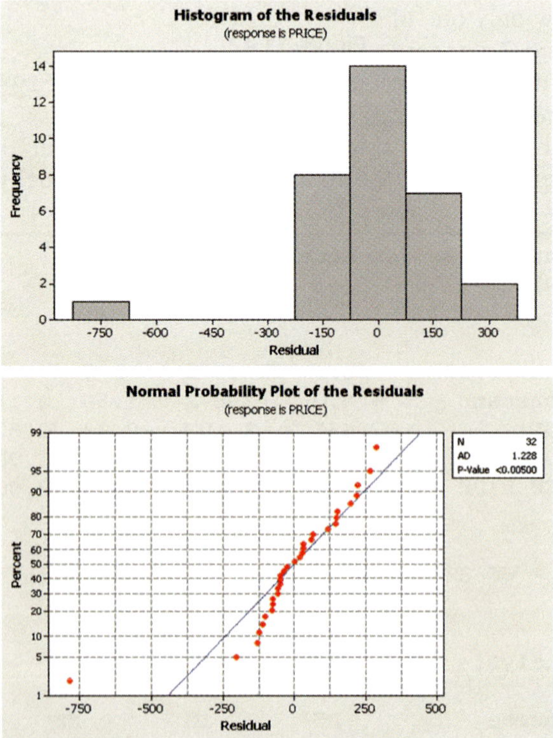

Figure 11.45
Minitab graphs of regression residuals for grandfather clock model (outlier included)

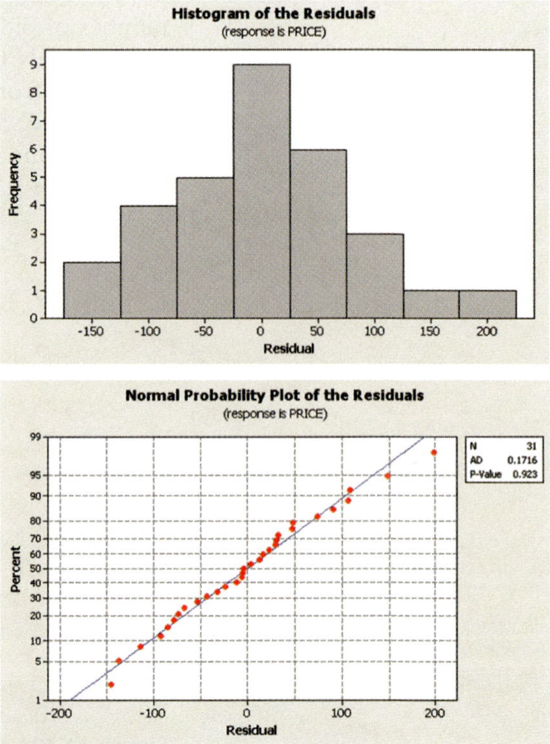

Figure 11.46
Minitab graphs of regression residuals for grandfather clock model (outlier deleted)

appears to skew the histogram in Figure 11.45, whereas the histogram in Figure 11.46 appears to be more mound shaped. Similarly, the pattern of residuals in the normal probability plot in Figure 11.46 (outlier deleted) is more nearly a straight line than the pattern in Figure 11.45 (outlier included). Thus, the normality assumption appears to be more plausible after the outlier is removed.

Look Back Although graphs do not provide formal statistical tests of normality, they do provide a descriptive display. Consult the references for methods to conduct statistical tests of normality using the residuals.

Now Work Exercise 11.110d

Of the four assumptions in Section 11.1, the assumption that the random error is normally distributed is the least restrictive when we apply regression analysis in practice—that is, moderate departures from a normal distribution have very little effect on the validity of the statistical tests, confidence intervals, and prediction intervals presented in this chapter. In this case, we say that regression analysis is **robust** with respect to nonnormal errors. However, great departures from normality cast doubt on any inferences derived from the regression analysis.

Checking Assumption #4: Errors Independent

The assumption of independent errors is violated when successive errors are correlated. This typically occurs when the data for both the dependent and independent variables are observed sequentially over a period of time—called **time series data.** Time series data have a unique characteristic; the experimental unit represents a unit of time (e.g., a year, a month, a quarter). There are both graphical and formal statistical tests available for checking the assumption of independent regression errors. For example, a simple graph is to plot the residuals against time. If the residuals tend to group alternately into positive and negative clusters (as shown in Figure 11.47), then it is likely that the errors are correlated and the assumption is violated. If correlated errors are detected, one solution is to construct a **time series model** for $E(y)$. These methods are discussed in detail in Chapter 13.

Summary

Residual analysis is a useful tool for the regression analyst, not only to check the assumptions but also to provide information about how the model can be improved. A summary of the residual analyses presented in this section to check the assumption that the random error ε is normally distributed with mean 0 and constant variance is presented in the next box.

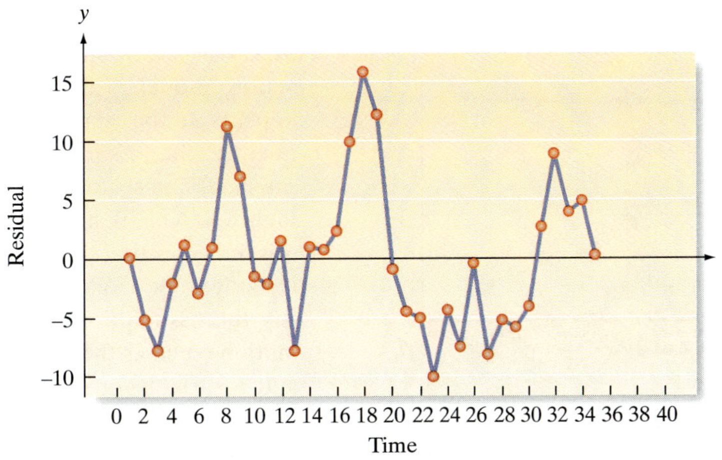

Figure 11.47
Hypothetical residual plot for time series data

Steps in a Residual Analysis

1. Check for a misspecified model by plotting the residuals against each of the quantitative independent variables. Analyze each plot, looking for a curvilinear trend. This shape signals the need for a quadratic term in the model. Try a second-order term in the variable against which the residuals are plotted.
2. Examine the residual plots for outliers. Draw lines on the residual plots at 2- and 3-standard-deviation distances below and above the 0 line. Examine residuals outside the 3-standard-deviation lines as potential outliers and check to see that no more than 5% of the residuals exceed the 2-standard-deviation lines. Determine whether each outlier can be explained as an error in data collection or transcription, corresponds to a member of a population different from that of the remainder of the sample, or simply represents an unusual observation. If the observation is determined to be an error, fix it or remove it. Even if you cannot determine the cause, you may want to rerun the regression analysis without the observation to determine its effect on the analysis.
3. Check for nonnormal errors by plotting a frequency distribution of the residuals, using a stem-and-leaf display or a histogram. Check to see if obvious departures from normality exist. Extreme skewness of the frequency distribution may be due to outliers or could indicate the need for a transformation of the dependent variable. (Normalizing transformations are beyond the scope of this book, but you can find information in the references.)
4. Check for unequal error variances by plotting the residuals against the predicted values, $\hat{y}$. If you detect a cone-shaped pattern or some other pattern that indicates that the variance of ε is not constant, refit the model using an appropriate variance-stabilizing transformation on y, such as $\ln(y)$. (Consult the references for other useful variance-stabilizing transformations.)

Statistics IN Action Revisited A Residual Analysis

In the previous Statistics in Action Revisited section (p. 693), we found the interaction model, $E(y) = \beta_0 + \beta_1x_1 + \beta_3x_2 + \beta_4x_1x_2$, to be both a statistically and practically useful model for predicting the cost (y) of a road construction contract. Recall that the two independent variables are the DOT engineer's estimate of cost (x_1) and bid status, where $x_2 = 1$ if a fixed bid and $x_2 = 0$ if a competitive bid. Before actually using the model in practice, we need to examine the residuals to be sure that the standard regression assumptions are reasonably satisfied.

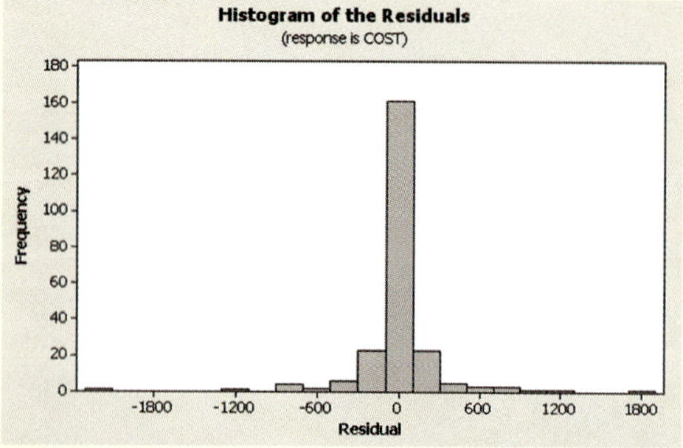

Figure SIA11.9
Minitab histogram of residuals from interaction model for road cost

Figures SIA11.9 and SIA11.10 are Minitab graphs of the residuals from the interaction model. The histogram shown in Figure SIA11.9 appears to be approximately normally distributed; consequently the assumption of normal errors is reasonably satisfied. The scatterplot of the residuals against $\hat{y}$ shown in Figure SIA11.10, however, shows a distinct "funnel" pattern; this indicates that the assumption of a constant error variance is likely to be violated. One way to modify the model to satisfy this assumption is to use a variance-stabilizing transformation (such as the natural log) on cost (y). When both the y and x variables in a regression equation are economic variables (prices, costs, salaries, etc.), it is often advantageous to transform the x variable also. Consequently, we'll modify the

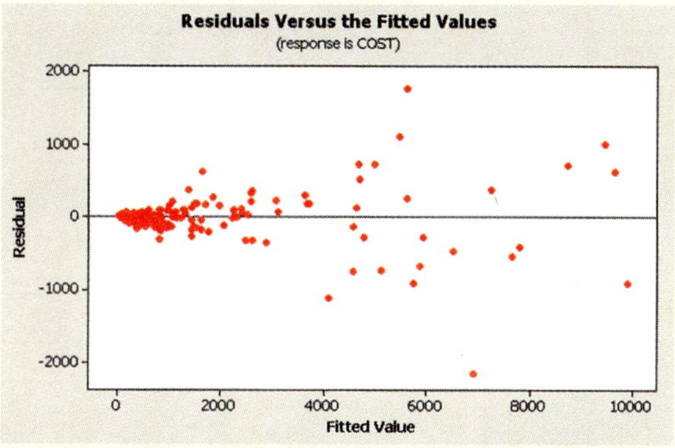

Figure SIA11.10
Minitab plot of residuals versus predicted values from interaction model for road cost

model by making a log transform on both cost (y) and DOTEST (x_1).

Our modified (log-log) interaction model takes the form

$$E(y^*) = \beta_0 + \beta_1 x_1^* + \beta_2 x_2 + \beta_3 (x_1^*) x_2$$

where $y^* = \ln(\text{COST})$ and $x_1^* = \ln(\text{DOTEST})$. The Minitab printout for this model is shown in Figure SIA11.11, followed by graphs of the residuals in Figures SIA11.12 and SIA11.13. The histogram shown in Figure SIA11.12 is approximately normal, and, more important, the scatterplot of the residuals shown in Figure SIA11.13 has no distinct trend. It appears that the log transformations successfully stabilized the error variance. Note, however, that the t-test for the interaction term in the model (highlighted in Figure SIA11.11) is no longer statistically significant (p-value = .420). Consequently, we will drop the interaction term from the model and use the simpler modified model,

$$E(y^*) = \beta_0 + \beta_1 x_1^* + \beta_2 x_2$$

to predict road construction cost.

```
The regression equation is
LNCOST = - 0.162 + 1.01 LNDOTEST + 0.324 STATUS - 0.0176 STA_LNDOT

Predictor       Coef   SE Coef       T      P
Constant    -0.16188   0.05193   -3.12  0.002
LNDOTEST     1.00780   0.00798  126.23  0.000
STATUS        0.3243    0.1356    2.39  0.018
STA_LNDOT   -0.01762   0.02181   -0.81  0.420

S = 0.154922   R-Sq = 98.8%   R-Sq(adj) = 98.7%

Analysis of Variance

Source           DF      SS      MS        F      P
Regression        3  439.64  146.55  6105.87  0.000
Residual Error  231    5.54    0.02
Total           234  445.18
```

Figure SIA11.11
Minitab regression printout for modified (log-log) model of road construction cost

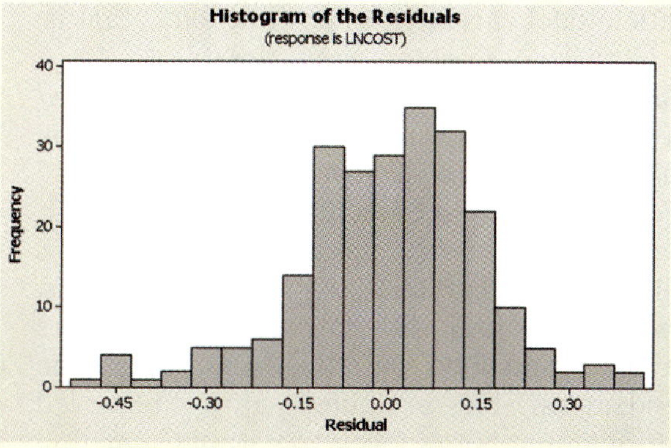

Figure SIA11.12
Minitab histogram of residuals from modified (log-log) model for road construction cost

(Continued)

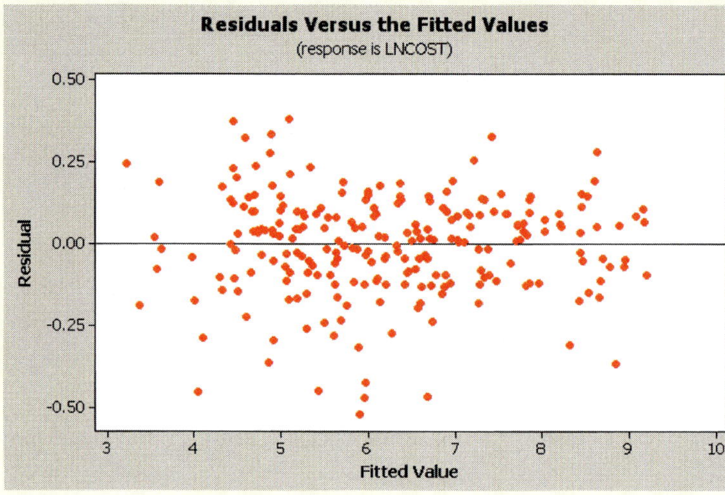

Figure SIA11.13
Minitab plot of residuals versus predicted values from modified (log-log) model for road construction cost

11.12 Some Pitfalls: Estimability, Multicollinearity, and Extrapolation

You should be aware of several potential problems when constructing a prediction model for some response y. A few of the most important are discussed in this final section.

Problem 1: Parameter Estimability

Suppose you want to fit a model relating annual crop yield y to the total expenditure for fertilizer x. We propose the first-order model

$$E(y) = \beta_0 + \beta_1 x$$

Now suppose we have three years of data and $1,000 is spent on fertilizer each year. The data are shown in Figure 11.48. You can see the problem: The parameters of the model cannot be estimated when all the data are concentrated at a single x-value. Recall that it takes two points (x-values) to fit a straight line. Thus, the parameters are not estimable when only one x is observed.

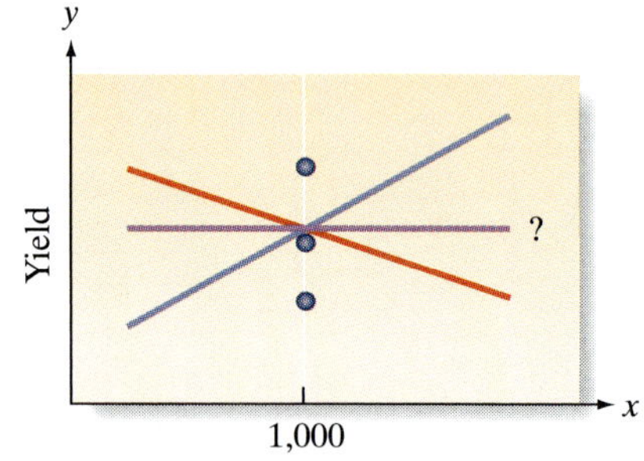

Figure 11.48
Yield and fertilizer expenditure data: Three years

A similar problem would occur if we attempted to fit the quadratic model

$$E(y) = \beta_0 + \beta_1 x + \beta_2 x^2$$

to a set of data for which only one or two different x-values were observed (see Figure 11.49). At least three different x-values must be observed before a quadratic model can be fit to a set of data (that is, before all three parameters are estimable).

In general, the number of levels of observed x-values must be one more than the order of the polynomial in x that you want to fit.

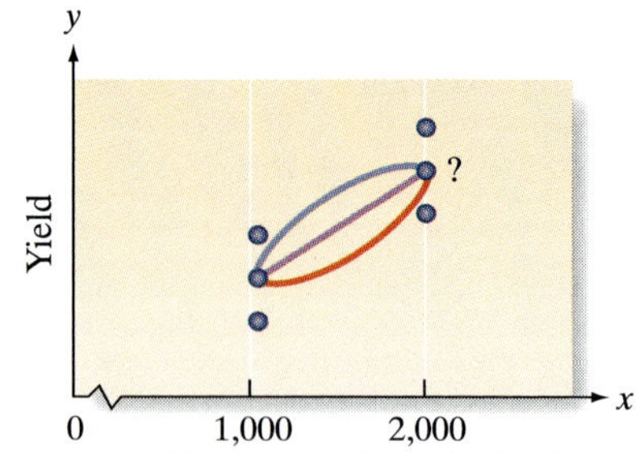

Figure 11.49
Only two x values observed: Quadratic model not estimable

For controlled experiments, the researcher can select one of the experimental designs in Chapter 8 that will permit estimation of the model parameters. Even when the values of the independent variables cannot be controlled by the researcher, the independent variables are almost always observed at a sufficient number of levels to permit estimation of the model parameters. When the statistical software you use suddenly refuses to fit a model, however, the problem is probably inestimable parameters.

Problem 2: Multicollinearity

Often, two or more of the independent variables used in a regression model contribute redundant information—that is, the independent variables are correlated with each other. For example, suppose we want to construct a model to predict the gas mileage rating of a truck as a function of its load, x_1 (in tons), and the horsepower, x_2 (in foot-pounds per second), of its engine. We would expect heavy loads to require greater horsepower and to result in lower mileage ratings. Thus, although both x_1 and x_2 contribute information for the prediction of mileage rating, y, some of the information is overlapping because x_1 and x_2 are correlated.

When the independent variables are correlated, we say *multicollinearity* exists. In practice, it is not uncommon to observe correlations among the independent variables. However, a few problems arise when serious multicollinearity is present in the regression variables.

> **Multicollinearity** exists when two or more of the independent variables used in regression are correlated.

First, high correlations among the independent variables increase the likelihood of rounding errors in the calculations of the β estimates, standard errors, and so forth. Second, and more important, the regression results may be confusing and misleading. Consider the model for gasoline mileage rating (y) of a truck:

$$E(y) = \beta_0 + \beta_1 x_1 + \beta_2 x_2$$

where x_1 = load and x_2 = horsepower. Fitting the model to a sample data set, we might find that the t-tests for testing β_1 and β_2 are both nonsignificant at the $\alpha = .05$ level, while the F-test for H_0: $\beta_1 = \beta_2 = 0$ is highly significant ($p = .001$). The tests may seem to be contradictory, but really they are not. The t-tests indicate that the contribution of one variable, say x_1 = load, is not significant after the effect of x_2 = horsepower has been accounted for (because x_2 is also in the model). The significant F-test, on the other hand, tells us that at least one of the two variables is making a contribution to the prediction of y (i.e., either β_1 or β_2 or both differ from 0). In fact, both are probably contributing, but the contribution of one overlaps with that of the other.

Multicollinearity can also have an effect on the signs of the parameter estimates. More specifically, a value of β_i may have the opposite sign from what is expected. In the truck gasoline mileage example, we expect heavy loads to result in lower mileage ratings, and we expect higher horsepowers to result in lower mileage ratings; consequently, we expect the signs of both the parameter estimates to be negative. Yet we may actually see a positive value of β_1 and be tempted to claim that heavy loads result in *higher* mileage ratings. This is the danger of interpreting a β coefficient when the independent variables are correlated. Because the variables contribute redundant information, the effect of x_1 = load on y = mileage rating is measured only partially by β_1.

How can you avoid the problems of multicollinearity in regression analysis? One way is to conduct a designed experiment (Chapter 8) so that the levels of the x variables are uncorrelated. Unfortunately, time and cost constraints may prevent you from collecting data in this manner. Consequently, most data are collected observationally. Because observational data frequently consist of correlated independent variables, you will need to recognize when multicollinearity is present and, if necessary, make modifications in the regression analysis.

Several methods are available for detecting multicollinearity in regression. A simple technique is to calculate the coefficient of correlation, r, between each pair of independent variables in the model and use the procedure outlined in Section 10.5 to test for significantly correlated variables. If one or more of the r-values is statistically different from 0, the variables in question are correlated, and a multicollinearity problem may exist.* The degree of multicollinearity will depend on the magnitude of the value of r, as shown in the box. Other indications of the presence of multicollinearity include those mentioned above—namely, nonsignificant t-tests for the individual parameter estimates when the F-test for overall model adequacy is significant and estimates with opposite signs from what is expected.†

Detecting Multicollinearity in the Regression Model

1. Significant correlations between pairs of independent variables (see next box)
2. Nonsignificant t-tests for all (or nearly all) of the individual β parameters when the F-test for overall model adequacy is significant
3. Signs opposite from what is expected in the estimated β parameters

Using the Correlation Coefficient r to Detect Multicollinearity

Extreme multicollinearity: $|r| \geq .8$

Moderate multicollinearity: $.2 \leq |r| < .8$

Low multicollinearity: $|r| < .2$

Example 11.18 Detecting Multicollinearity

Problem The Federal Trade Commission (FTC) annually ranks varieties of domestic cigarettes according to their tar, nicotine, and carbon monoxide contents. The U.S. Surgeon General considers each of these three substances hazardous to a smoker's health. Past studies have shown that increases in the tar and nicotine contents of a cigarette are accompanied by an increase in the carbon monoxide emitted from the cigarette smoke. Table 11.9 presents data on tar, nicotine, and carbon monoxide contents (in milligrams) and weight (in grams) for a sample of 25 (filter) brands tested in a recent year. Suppose we want to model carbon monoxide content, y, as a function of tar content, x_1, nicotine content, x_2, and weight, x_3, using the model

$$E(y) = \beta_0 + \beta_1 x_1 + \beta_2 x_2 + \beta_3 x_3$$

The model is fit to the 25 data points in Table 11.9, and a portion of the Minitab printout is shown in Figure 11.50. Examine the printout. Do you detect any signs of multicollinearity?

Solution First, note that the F-test for overall model utility is highly significant. The test statistic ($F = 78.98$) and observed significance level (p-value $= .000$) are highlighted on the Minitab printout, Figure 11.50. Therefore, at, say $\alpha = .01$, we can conclude that at least one of the parameters—β_1, β_2, or β_3—in the model is nonzero. The t-tests for two of three individual β's, however, are nonsignificant. (The p-values for these tests are highlighted on the printout.) Unless tar (x_1) is the only one of the three variables useful for predicting carbon monoxide content, these results are the first indication of a potential multicollinearity problem.

*Remember that r measures only the pairwise correlation between x-values. Three variables, x_1, x_2, and x_3, may be highly correlated as a group but may not exhibit large pairwise correlations. Thus, multicollinearity may be present even when all pairwise correlations are not significantly different from 0.

†More formal methods for detecting multicollinearity, such as variance-inflation factors (VIFs), are available. Independent variables with a VIF of 10 or above are usually considered to be highly correlated with one or more of the other independent variables in the model. Calculation of VIFs are beyond the scope of this introductory text. Consult the chapter references for a discussion of VIFs and other formal methods of detecting multicollinearity.

Table 11.9 FTC Cigarette Data for Example 11.18

Tar, x_1	Nicotine, x_2	Weight, x_3	Carbon Monoxide, y
14.1	.86	.9853	13.6
16.0	1.06	1.0938	16.6
29.8	2.03	1.1650	23.5
8.0	.67	.9280	10.2
4.1	.40	.9462	5.4
15.0	1.04	.8885	15.0
8.8	.76	1.0267	9.0
12.4	.95	.9225	12.3
16.6	1.12	.9372	16.3
14.9	1.02	.8858	15.4
13.7	1.01	.9643	13.0
15.1	.90	.9316	14.4
7.8	.57	.9705	10.0
11.4	.78	1.1240	10.2
9.0	.74	.8517	9.5
1.0	.13	.7851	1.5
17.0	1.26	.9186	18.5
12.8	1.08	1.0395	12.6
15.8	.96	.9573	17.5
4.5	.42	.9106	4.9
14.5	1.01	1.0070	15.9
7.3	.61	.9806	8.5
8.6	.69	.9693	10.6
15.2	1.02	.9496	13.9
12.0	.82	1.1184	14.9

Source: Federal Trade Commission

Data Set: FTC

Regression Analysis: CO versus TAR, NICOTINE, WEIGHT

```
The regression equation is
CO = 3.20 + 0.963 TAR - 2.63 NICOTINE - 0.13 WEIGHT


Predictor     Coef   SE Coef      T      P
Constant     3.202     3.462   0.93  0.365
TAR         0.9626    0.2422   3.97  0.001
NICOTINE    -2.632     3.901  -0.67  0.507
WEIGHT      -0.130     3.885  -0.03  0.974


S = 1.44573    R-Sq = 91.9%    R-Sq(adj) = 90.7%


Analysis of Variance

Source          DF      SS      MS      F      P
Regression       3  495.26  165.09  78.98  0.000
Residual Error  21   43.89    2.09
```

Correlations: TAR, NICOTINE, WEIGHT

```
             TAR  NICOTINE
NICOTINE   0.977
           0.000

WEIGHT     0.491     0.500
           0.013     0.011


Cell Contents: Pearson correlation
               P-Value
```

Figure 11.50
Minitab printout for model of carbon monoxide content, Example 11.18

The negative values for β_2 and β_3 (highlighted on the printout) are a second clue to the presence of multicollinearity. From past studies, the FTC expects carbon monoxide content (y) to increase when either nicotine content (x_2) or weight (x_3) increases—that is, the FTC expects *positive* relationships between y and x_2 and between y and x_3, not negative ones. All signs indicate that a serious multicollinearity problem exists.

Look Back To confirm our suspicions, we had Minitab produce the coefficient of correlation, r, for each of the three pairs of independent variables in the model. The resulting output is shown (highlighted) at the bottom of Figure 11.50. You can see that tar (x_1) and nicotine (x_2) are highly correlated ($r = .977$), while weight (x_3) is moderately correlated with the other two x's ($r \approx .5$). All three correlations have p-values $\approx .01$ or less; consequently, all three are significantly different from 0 at, say, $\alpha = .05$.

Now Work Exercise 11.19a

Once you have detected that multicollinearity exists, there are several alternative measures available for solving the problem. The appropriate measure to take depends on the severity of the multicollinearity and the ultimate goal of the regression analysis.

Some researchers, when confronted with highly correlated independent variables, choose to include only one of the correlated variables in the final model. If you are interested in using the model only for estimation and prediction (step 6), you may decide not to drop any of the independent variables from the model. In the presence of multicollinearity, we have seen that it is dangerous to interpret the individual β parameters. However, confidence intervals for $E(y)$ and prediction intervals for y generally remain unaffected *as long as the values of the x's used to predict y follow the same pattern of multicollinearity exhibited in the sample data*—that is, you must take strict care to ensure that the values of the x variables fall within the range of the sample data.

Solutions to Some Problems Created by Multicollinearity in Regression*

1. Drop one or more of the correlated independent variables from the model. One way to decide which variables to keep in the model is to employ stepwise regression (Section 11.10).
2. If you decide to keep all the independent variables in the model,
   a. Avoid making inferences about the individual β parameters based on the t-tests.
   b. Restrict inferences about $E(y)$ and future y values to values of the x's that fall within the range of the sample data.

Problem 3: Prediction Outside the Experimental Region

Many research economists had developed highly technical models to relate the state of the economy to various economic indices and other independent variables. Many of these models were multiple regression models, where, for example, the dependent variable y might be next year's gross domestic product (GDP) and the independent variables might include this year's rate of inflation, this year's consumer price index (CPI), and so on. In other words, the model might be constructed to predict next year's economy using this year's knowledge.

*Several other solutions are available. For example, in the case where higher-order regression models are fit, the analyst may want to code the independent variables so that higher-order terms (e.g., x^2) for a particular x variable are not highly correlated with x. One transformation that works is $z = (x - \bar{x})/s$. Other, more sophisticated procedures for addressing multicollinearity (such as *ridge regression*) are beyond the scope of the text. Consult the references at the end of this chapter.

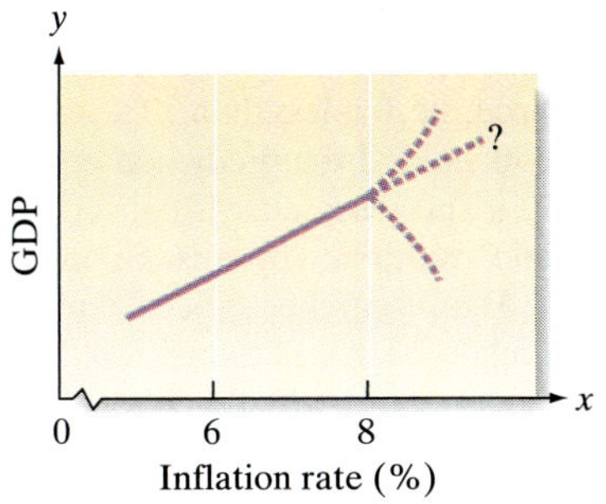

Figure 11.51
Using a regression model outside the experimental region

Unfortunately, these models were almost all unsuccessful in predicting the recession in the early 1970s and the late 1990s. What went wrong? One of the problems was that many of the regression models were used to **extrapolate** (i.e., predict y-values of the independent variables that were outside the region in which the model was developed). For example, the inflation rate in the late 1960s, when the models were developed, ranged from 6% to 8%. When the double-digit inflation of the early 1970s became a reality, some researchers attempted to use the same models to predict future growth in GDP. As you can see in Figure 11.51, the model may be very accurate for predicting y when x is in the range of experimentation, but using the model outside that range is a dangerous practice.

Exercises 11.109–11.120

Learning the Mechanics

11.109 NW Consider fitting the multiple regression model

$$E(y) = \beta_0 + \beta_1 x_1 + \beta_2 x_2 + \beta_3 x_3 + \beta_4 x_4 + \beta_5 x_5$$

A matrix of correlations for all pairs of independent variables is given below. Do you detect a multicollinearity problem? Explain.

	x_1	x_2	x_3	x_4	x_5
x_1	—	.17	.02	−.23	.19
x_2		—	.45	.93	.02
x_3			—	.22	−.01
x_4				—	.86
x_5					—

11.110 NW Identify the problem(s) in each of the residual plots shown below.

Applying the Concepts—Basic

11.111 Women in top management. Refer to the *Journal of Organizational Culture, Communications and Conflict* (July 2007) study on women in upper management positions at U.S. firms, Exercise 10.61 (p. 597). Monthly data (n = 252 months) were collected for several variables in an attempt to model the number of females in managerial positions (y). The independent variables included the number of females with a college degree (x_1), the number of female high school graduates with no college degree (x_2), the number of males in managerial positions (x_3), the number of males with a college degree (x_4), and the number of male high school graduates with no college degree (x_5). The correlations provided in Exercise 10.61 are given in each part. Determine which of the correlations results in a potential multicollinearity problem for the regression analysis.

a. The correlation relating number of females in managerial positions and number of females with a college degree: r = .983.

a.

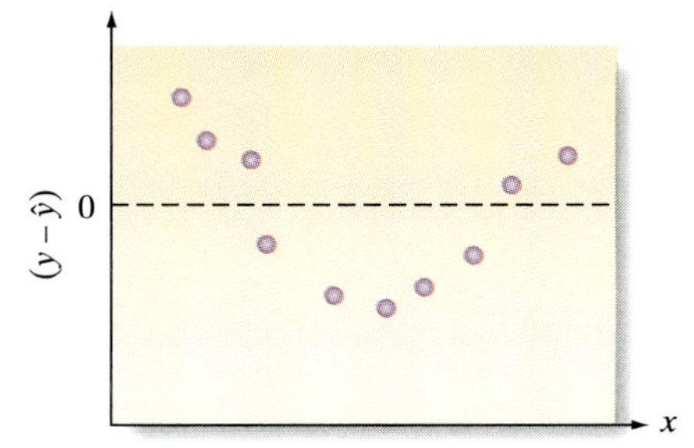

b.

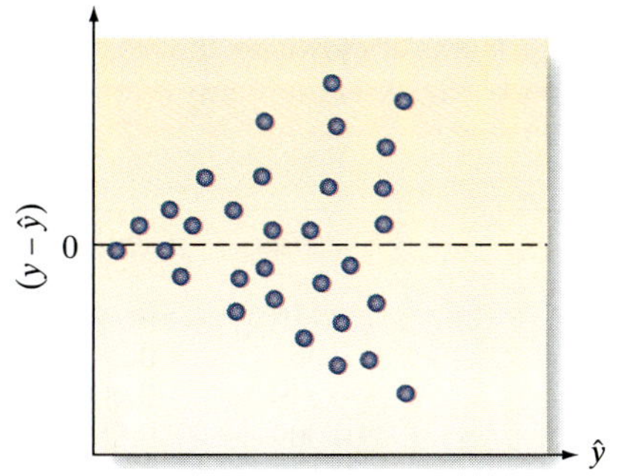

c.

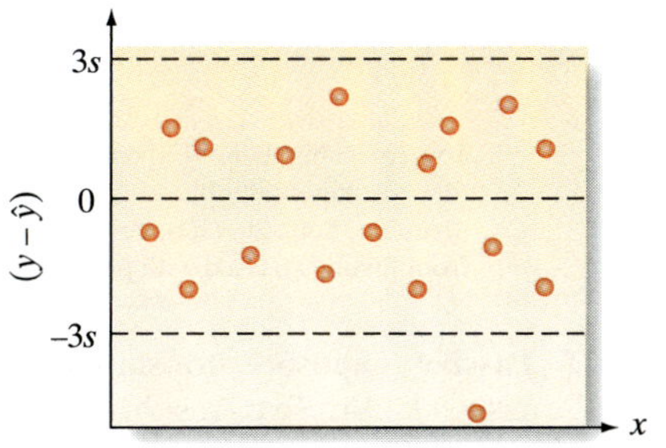

d.

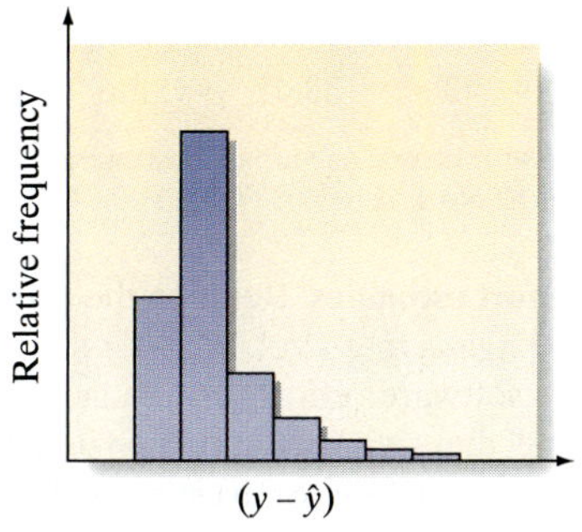

b. The correlation relating number of females in managerial positions and number of female high school graduates with no college degree: $r = 0.74$.

c. The correlation relating number of males in managerial positions and number of males with a college degree: $r = .722$.

d. The correlation relating number of males in managerial positions and number of male high school graduates with no college degree: $r = .528$.

11.112 Factors identifying urban counties. The *Professional Geographer* (Feb. 2000) published a study of urban and rural counties in the western United States. Six independent variables—total county population (x_1), population density (x_2), population concentration (x_3), population growth (x_4), proportion of county land in farms (x_5), and 5-year change in agricultural land base (x_6)—were used to model the urban/rural rating (y) of a county, where rating was recorded on a scale of 1 (most rural) to 10 (most urban). Prior to running the multiple regression analysis, the researchers were concerned about possible multicollinearity in the data. Below is a correlation matrix (i.e., a table of correlations between all pairs of the independent variables) for data collected on $n = 256$ counties.

Independent Variable	x_1	x_2	x_3	x_4	x_5
x_1: Total population					
x_2: Population density	.20				
x_3: Population concentration	.45	.43			
x_4: Population growth	−.05	−.14	−.01		
x_5: Farm land	−.16	−.15	−.07	−.20	
x_6: Agricultural change	−.12	−.12	−.22	−.06	−.06

Source: Berry, K. A., et al. "Interpreting what is rural and urban for western U.S. counties," *Professional Geographer,* Vol. 52, No. 1, Feb. 2000, pp. 93–105 (Table 2). Reprinted with permission of John Wiley & Sons, Inc.

a. Based on the correlation matrix, is there any evidence of extreme multicollinearity?

b. The first-order model with all 6 independent variables was fit, and the results are shown in the next table. Based on the reported tests, is there any evidence of extreme multicollinearity?

Independent Variable	β Estimate	p-value
x_1: Total population	0.110	0.045
x_2: Population density	0.065	0.230
x_3: Population concentration	0.540	0.000
x_4: Population growth	−0.009	0.860
x_5: Farm land	−0.150	0.003
x_6: Agricultural change	−0.027	0.580

Overall model:
$R^2 = .44$ $R^2_a = .43$ $F = 32.47$ p-value $< .001$

Source: Berry, K. A., et al. "Interpreting what is rural and urban for western U.S. counties," *Professional Geographer,* Vol. 52, No. 1, Feb. 2000 (Table 2).

11.113 Accuracy of software effort estimates. Refer to the *Journal of Empirical Software Engineering* (Vol. 9, 2004) study of the accuracy of new software effort estimates, Exercise 11.106 (p. 696). Recall that stepwise regression was used to develop a model for the relative error in estimating effort (y) as a function of company role of estimator ($x_1 = 1$ if developer, 0 if project leader) and previous accuracy ($x_8 = 1$ if more than 20% accurate, 0 if less than 20% accurate). The stepwise regression yielded the prediction equation $\hat{y} = .12 - .28x_1 + .27x_8$. The researcher is concerned that the sign of the estimated β multiplied by x_1 is the opposite from what is expected. (The researcher expects a project leader to have a smaller relative error of estimation than a developer.) Give at least one reason why this phenomenon occurred.

11.114 Passive exposure to smoke. Passive exposure to environmental tobacco smoke has been associated with growth suppression and an increased frequency of respiratory tract infections in normal children. Is this association more pronounced in children with cystic fibrosis? To answer this question, 43 children (18 girls and 25 boys) attending a 2-week summer camp for cystic fibrosis patients were studied (*The New England Journal of Medicine,* Sept. 20, 1990). Researchers investigated the correlation between a child's weight percentile (y) and the number of cigarettes smoked per day in the child's home (x). The table below (saved in the **CFSMOKE** file) lists the data for the 25 boys. Using simple linear regression, the researchers predicted the weight percentile for the last observation ($x = 44$ cigarettes) to be $\hat{y} = 29.63$. Given that the standard deviation of the model is $s = 24.68$, is this observation an outlier? Explain.

Weight Percentile, y	No. of Cigarettes Smoked per Day, x
6	0
6	15
2	40
8	23
11	20
17	7
24	3
25	0
17	25
25	20
25	15
31	23
35	10
43	0
49	0
50	0
49	22
46	30
54	0
58	0
62	0
66	0
66	23
83	0
87	44

Source: Rubin, B. K. "Exposure of children with cystic fibrosis to environmental tobacco smoke," *The New England Journal of Medicine,* Sept. 20, 1990, Vol. 323, No. 12, p. 785 (data extracted from Figure 3). Used with permission NEJM.

11.115 Passive exposure to smoke (cont'd). Refer to Exercise 11.114. Two residual plots for the simple linear regression model are shown on the next page.

a. Which graph should be used to check for normal errors? Does the assumption of normality appear to be satisfied?

b. Which graph should be used to check for unequal error variances? Does the assumption of equal variances appear to be satisfied?

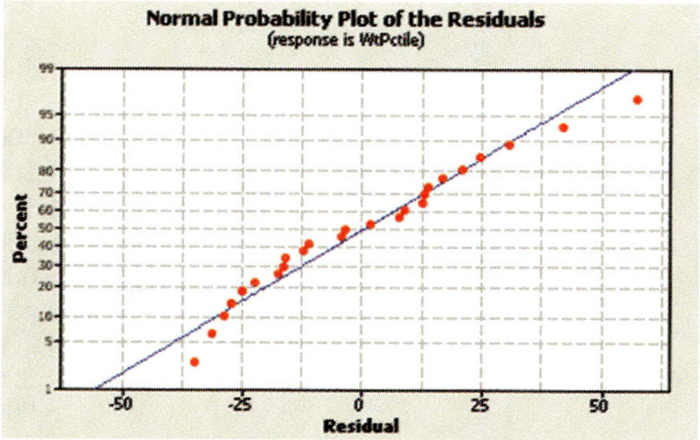

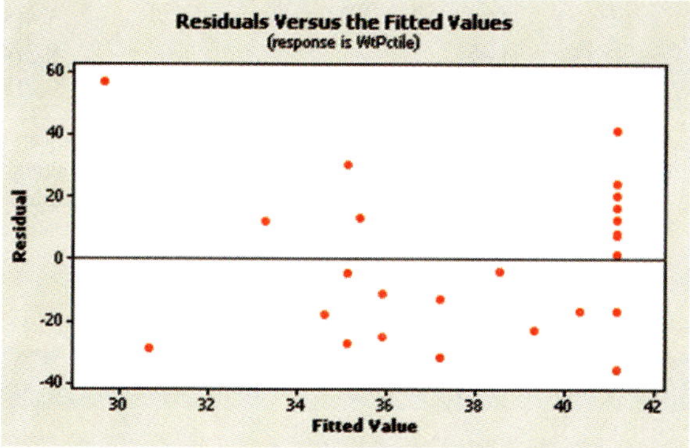

Applying the Concepts—Intermediate

11.116 Household food consumption. The data in the table below (saved in the **DCFOOD** file) were collected for a random sample of 26 households in Washington, D.C. An economist wants to relate household food consumption, y, to household income, x_1, and household size, x_2, with the first-order model

$$E(y) = \beta_0 + \beta_1 x_1 + \beta_2 x_2$$

a. Fit the model to the data. Do you detect any signs of multicollinearity in the data? Explain.

b. Is there visual evidence (from a residual plot) that a second-order model may be more appropriate for predicting household food consumption? Explain.

c. Comment on the assumption of constant error variance, using a residual plot. Does it appear to be satisfied?

d. Are there any outliers in the data? If so, identify them.

e. Based on a graph of the residuals, does the assumption of normal errors appear to be reasonably satisfied? Explain.

11.117 Arsenic in groundwater. Refer to the *Environmental Science & Technology* (Jan. 2005) study of the reliability of a commercial kit to test for arsenic in groundwater, Exercise 11.16 (p. 640). Recall that you fit a first-order model for arsenic level (y) as a function of latitude (x_1), longitude (x_2), and depth (x_3) to data saved in the **ASWELLS** file. Conduct a residual analysis of the data. Based on the results, comment on each of the following:

a. assumption of mean error = 0

b. assumption of constant error variance

c. outliers

d. assumption of normally distributed errors

e. multicollinearity

11.118 Contamination from a plant's discharge. Refer to the U.S. Army Corps of Engineers data (saved in the **DDT** file) on fish contaminated from the toxic discharges of a chemical plant located on the banks of the Tennessee River in Alabama. In Exercise 11.18 (p. 641), you fit the first-order model, $E(y) = \beta_0 + \beta_1 x_1 + \beta_2 x_2 + \beta_3 x_3$, where y = DDT level in captured fish, x_1 = miles captured upstream, x_2 = fish length, and x_3 = fish weight. Conduct a complete residual analysis for the model. Do you recommend any model modifications be made? Explain.

11.119 Failure times of silicon wafer microchips. Refer to the National Semiconductor study of manufactured silicon wafer integrated circuit chips, Exercise 11.59 (p. 664). Recall that the failure times of the microchips (in hours) was determined at different solder temperatures (degrees

Table for Exercise 11.116

Household	Food Consumption ($1,000s)	Income ($1,000s)	Household Size	Household	Food Consumption ($1,000s)	Income ($1,000s)	Household Size
1	4.2	41.1	4	14	4.1	95.2	2
2	3.4	30.5	2	15	5.5	45.6	9
3	4.8	52.3	4	16	4.5	78.5	3
4	2.9	28.9	1	17	5.0	20.5	5
5	3.5	36.5	2	18	4.5	31.6	4
6	4.0	29.8	4	19	2.8	39.9	1
7	3.6	44.3	3	20	3.9	38.6	3
8	4.2	38.1	4	21	3.6	30.2	2
9	5.1	92.0	5	22	4.6	48.7	5
10	2.7	36.0	1	23	3.8	21.2	3
11	4.0	76.9	3	24	4.5	24.3	7
12	2.7	69.9	1	25	4.0	26.9	5
13	5.5	43.1	7	26	7.5	7.3	5

Celsius). The data (saved in the **WAFER** file) are repeated in the table in the next column.

a. Fit the straight-line model $E(y) = \beta_0 + \beta_1 x$ to the data, where y = failure time and x = solder temperature.

b. Compute the residual for a microchip manufactured at a temperature of 149°C.

c. Plot the residuals against solder temperature (x). Do you detect a trend?

d. In Exercise 11.59c, you determined that failure time (y) and solder temperature (x) were curvilinearly related. Does the residual plot, part **c**, support this conclusion?

11.120 Cooling method for gas turbines. Refer to the *Journal of Engineering for Gas Turbines and Power* (Jan. 2005) study of a high-pressure inlet fogging method for a gas turbine engine, Exercise 11.97 (p. 687). Now consider the interaction model for heat rate (y) of a gas turbine as a function of cycle speed (x_1) and cycle pressure ratio (x_2), $E(y) = \beta_0 + \beta_1 x_1 + \beta_2 x_2 + \beta_3 x_1 x_2$. Use the data saved in the **GASTURBINE** file to conduct a complete residual analysis for the model. Do you recommend making model modifications?

Table for Exercise 11.119

Temperature (°C)	Time to Failure (hours)
165	200
162	200
164	1,200
158	500
158	600
159	750
156	1,200
157	1,500
152	500
147	500
149	1,100
149	1,150
142	3,500
142	3,600
143	3,650
133	4,200
132	4,800
132	5,000
134	5,200
134	5,400
125	8,300
123	9,700

Source: Gee, S., and Nguyen, L. "Mean time to failure in wafer level-CSP packages with SnPb and SnAgCu solder bmps," International Wafer Level Packaging Conference, San Jose, CA, Nov. 3-4, 2005 (adapted from Figure 7).

CHAPTER NOTES

Key Terms

Key Formulas

$s^2 = \text{MSE} = \dfrac{\text{SSE}}{n-(k+1)}$	Estimator of σ^2 for a model with k independent variables
$t = \dfrac{\hat{\beta}_i}{s_{\hat{\beta}_i}}$	Test statistic for testing H_0: β_i
$\hat{\beta}_i \pm (t_{\alpha/2}) s_{\hat{\beta}_i}$—where $t_{\alpha/2}$ depends on $n-(k+1)$ df	$100(1-\alpha)\%$ confidence interval for β_i
$R^2 = \dfrac{\text{SS}_{yy} - \text{SSE}}{\text{SS}_{yy}}$	Multiple coefficient of determination
$R_a^2 = 1 - \left[\dfrac{(n-1)}{n-(k+1)}\right](1-R^2)$	Adjusted multiple coefficient of determination
$F = \dfrac{\text{MS (Model)}}{\text{MSE}} = \dfrac{R^2/k}{(1-R^2)/[n-(k+1)]}$	Test statistic for testing H_0: $\beta_1 = \beta_2 = \cdots = \beta_k = 0$
$F = \dfrac{(\text{SSE}_\text{R} - \text{SSE}_\text{C})/\text{number of } \beta\text{'s tested}}{\text{MSE}_\text{C}}$	Test statistic for comparing reduced and complete models
$y - \hat{y}$	Regression residual

Key Symbols

x_1^2	Quadratic form for a quantitative x
x_1x_2	Interaction term
MSE	Mean square for error (estimates σ^2)
$\hat{\varepsilon}$	Estimated random error (residual)
SSE_R	Sum of squared errors, reduced model
SSE_C	Sum of squared errors, complete model
MSE_C	Mean squared error, complete model
$\ln(y)$	Natural logarithm of dependent variable

Key Ideas

Multiple Regression Variables

y = **Dependent** variable (quantitative)

$x_1, x_2, \ldots, x_k$ are **independent** variables (quantitative or qualitative)

First-Order Model in k Quantitative x's

$$E(y) = \beta_0 + \beta_1x_1 + \beta_2x_2 + \cdots + \beta_kx_k$$

Each β_i represents the change in y for every 1-unit increase in x_i, holding all other x's fixed.

Interaction Model in 2 Quantitative x's

$E(y) = \beta_0 + \beta_1x_1 + \beta_2x_2 + \beta_3x_1x_2$

$(\beta_1 + \beta_3x_2)$ represents the change in y for every 1-unit increase in x_1, for fixed value of x_2

$(\beta_2 + \beta_3x_1)$ represents the change in y for every 1-unit increase in x_2, for fixed value of x_1

Quadratic Model in 1 Quantitative x

$E(y) = \beta_0 + \beta_1x + \beta_2x^2$

β_2 represents the rate of curvature in for x

($\beta_2 > 0$ implies *upward* curvature)

($\beta_2 < 0$ implies *downward* curvature)

Complete Second-Order Model in 2 Quantitative x's

$E(y) = \beta_0 + \beta_1x_1 + \beta_2x_2 + \beta_3x_1x_2 + \beta_4x_1^2 + \beta_5x_2^2$

β_4 represents the rate of curvature in for x_1, holding x_2 fixed

β_5 represents the rate of curvature in for x_2, holding x_1 fixed

Dummy Variable Model for 1 Qualitative x

$E(y) = \beta_0 + \beta_1x_1 + \beta_2x_2 + \cdots + \beta_{k-1}x_{k-1}$

$x_1 = \{1 \text{ if level } 1, 0 \text{ if not}\}$

$x_2 = \{1 \text{ if level } 2, 0 \text{ if not}\}$

$x_{k-1} = \{1 \text{ if level } k - 1, 0 \text{ if not}\}$

$\beta_0 = E(y)$ for level k (base level) $= \mu_k$

$\beta_1 = \mu_1 - \mu_k$

$\beta_2 = \mu_2 - \mu_k$

Complete Second-Order Model in 1 Quantitative x and 1 Qualitative x (Two Levels, A and B)

$E(y) = \beta_0 + \beta_1x_1 + \beta_2x_1^2 + \beta_3x_2 + \beta_4x_1x_2 + \beta_5x_1^2x_2$

$x_2 = \{1 \text{ if level A}, 0 \text{ if level B}\}$

Adjusted Coefficient of Determination, R_a^2

Cannot be "forced" to 1 by adding independent variables to the model.

Interaction between x_1 and x_2

Implies that the relationship between y and one x depends on the other x.

Parsimonious Model

A model with a small number of β parameters.

Recommendation for Assessing Model Adequacy

1. Conduct global F-test; if significant then:
2. Conduct t-tests on only the most important β's (*interaction* or *squared terms*)
3. Interpret value of 2s
4. Interpret value of R_a^2

Recommendation for Testing Individual β's

1. If *curvature* (x^2) deemed important, do not conduct test for first-order (x) term in the model.
2. If *interaction* (x_1x_2) deemed important, do not conduct tests for first-order terms (x_1 and x_2) in the model.

Extrapolation

Occurs when you predict y for values of x's that are outside of range of sample data.

Nested Models

Are models where one model (the *complete model*) contains all the terms of another model (the *reduced model*) plus at least one additional term.

Multicollinearity

Occurs when two or more x's are correlated.

Indicators of multicollinearity:

1. Highly correlated x's
2. Significant global F-test, but all t-tests nonsignificant
3. Signs on β's opposite from expected

Problems with Using Stepwise Regression Model as the "Final" Model

1. *Extremely large number of t-tests* inflate overall probability of at least one Type I error.
2. *No higher-order terms* (interactions or squared terms) are included in the model.

Analysis of Residuals

1. Detect **misspecified model:** *plot residuals vs. quantitative x* (look for trends, e.g., curvilinear trend)
2. Detect **nonconstant error variance:** *plot residuals vs.* $\hat{y}$ (look for patterns, e.g., cone shape)
3. Detect **nonnormal errors:** *histogram, stem-leaf, or normal probability plot of residuals* (look for strong departures from normality)
4. Identify **outliers:** *residuals greater than 3s in absolute value* (investigate outliers before deleting)

Guide to Multiple Regression

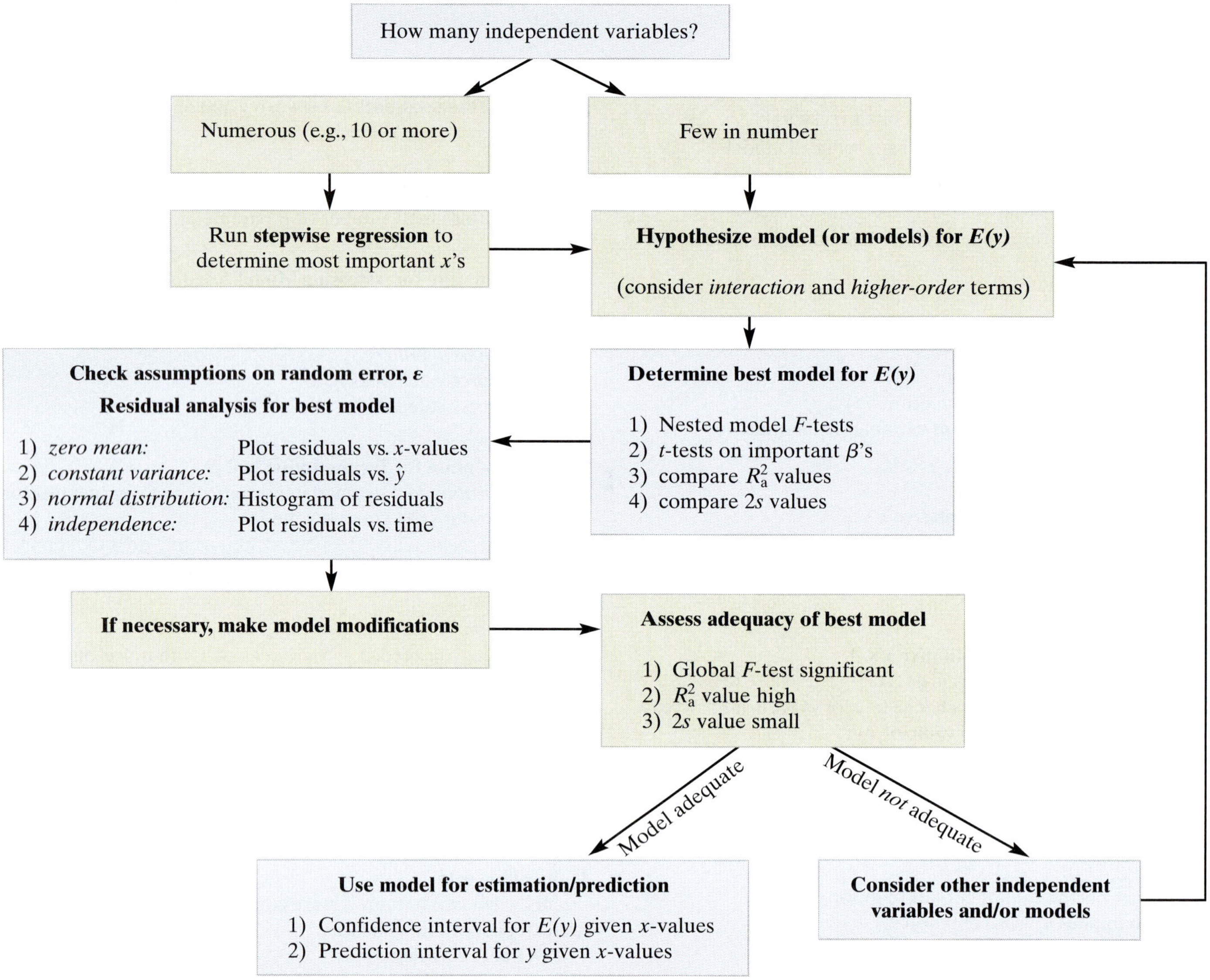

Supplementary Exercises 11.121–11.158

Learning the Mechanics

11.121 When a multiple regression model is used for estimating the mean of the dependent variable and for predicting a new value of y, which will be narrower—the confidence interval for the mean or the prediction interval for the new y value? Why?

11.122 Suppose you have developed a regression model to explain the relationship between y and x_1, x_2, and x_3. The ranges of the variables you observed were as follows: $10 \le y \le 100$, $5 \le x_1 \le 55$, $.5 \le x_2 \le 1$, and $1{,}000 \le x_3 \le 2{,}000$. Will the error of prediction be smaller when you use the least squares equation to predict y when $x_1 = 30$, $x_2 = .6$, and $x_3 = 1{,}300$, or $x_1 = 60$, $x_2 = .4$, and $x_3 = 900$? Why?

11.123 Why is the model building step the key to the success or failure of a regression analysis?

11.124 Suppose you fit the model

$$y = \beta_0 + \beta_1 x_1 + \beta_2 x_1^2 + \beta_3 x_2 + \beta_4 x_1 x_2 + \varepsilon$$

to $n = 25$ data points with the following results:

$$\hat{\beta}_0 = 1.26 \quad \hat{\beta}_1 = -2.43 \quad \hat{\beta}_2 = .05 \quad \hat{\beta}_3 = .62 \quad \hat{\beta}_4 = 1.81$$
$$s_{\hat{\beta}_1} = 1.21 \quad s_{\hat{\beta}_2} = .16 \quad s_{\hat{\beta}_3} = .26 \quad s_{\hat{\beta}_4} = 1.49$$
$$\text{SSE} = .41 \quad R^2 = .83$$

a. Is there sufficient evidence to conclude that at least one of the parameters β_1, β_2, β_3, or β_4 is nonzero? Test using $\alpha = .05$.
b. Test H_0: $\beta_1 = 0$ against H_a: $\beta_1 < 0$. Use $\alpha = .05$.
c. Test H_0: $\beta_2 = 0$ against H_a: $\beta_2 > 0$. Use $\alpha = .05$.
d. Test H_0: $\beta_3 = 0$ against H_a: $\beta_3 \ne 0$. Use $\alpha = .05$.

11.125 Suppose you used Minitab to fit the model

$$y = \beta_0 + \beta_1 x_1 + \beta_2 x_2 + \varepsilon$$

to $n = 15$ data points and obtained the printout shown on the next page.

a. What is the least squares prediction equation?
b. Find R^2 and interpret its value.

Minitab Output for Exercise 11.125

```
The regression equation is
Y  = 90.1 - 1.84 X1 + .285 X2


Predictor      Coef  SE Coef       T      P
Constant      90.10    23.10    3.90  0.002
X1           -1.836    0.367   -5.01  0.001
X2            0.285    0.231    1.24  0.465


S = 10.68      R-Sq = 91.6%    R-Sq(adj) = 90.2%

Analysis of Variance

Source          DF       SS      MS       F       P
Regression       2    14801    7400   64.91   0.001
Residual Error  12     1364     114
Total           14    16165
```

c. Is there sufficient evidence to indicate that the model is useful for predicting y? Conduct an F-test using $\alpha = .05$.

d. Test the null hypothesis H_0: $\beta_1 = 0$ against the alternative hypothesis H_a: $\beta_1 \neq 0$. Test using $\alpha = .05$. Draw the appropriate conclusions.

e. Find the standard deviation of the regression model and interpret it.

11.126 The first-order model $E(y) = \beta_0 + \beta_1 x_1$ was fit to $n = 19$ data points. A residual plot for the model is provided below. Is the need for a quadratic term in the model evident from the residual plot? Explain.

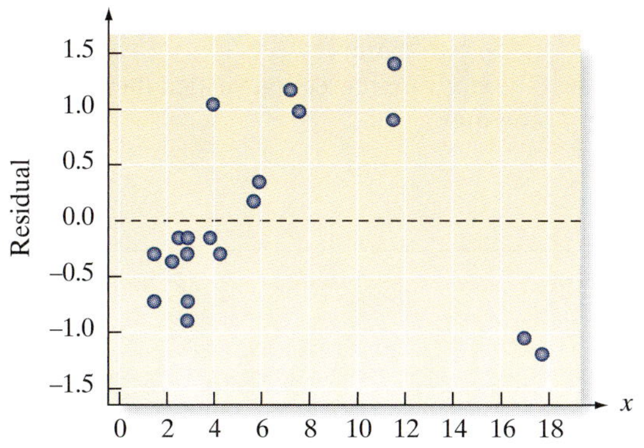

11.127 Write a model relating $E(y)$ to one qualitative independent variable that is at four levels. Define all the terms in your model.

11.128 It is desired to relate $E(y)$ to a quantitative variable x_1 and a qualitative variable at three levels.

a. Write a first-order model.

b. Write a model that will graph as three different second-order curves—one for each level of the qualitative variable.

11.129 Explain why stepwise regression is used. What is its value in the model-building process?

11.130 Consider relating $E(y)$ to two quantitative independent variables x_1 and x_2.

a. Write a first-order model for $E(y)$.

b. Write a complete second-order model for $E(y)$.

11.131 To model the relationship between y, a dependent variable, and x, an independent variable, a researcher has taken one measurement on y at each of three different x values. Drawing on his mathematical expertise, the researcher realizes that he can fit the second-order model

$$E(y) = \beta_0 + \beta_1 x + \beta_2 x^2$$

and it will pass exactly through all three points, yielding SSE = 0. The researcher, delighted with the excellent fit of the model, eagerly sets out to use it to make inferences. What problems will he encounter in attempting to make inferences?

11.132 Suppose you fit the regression model

$$E(y) = \beta_0 + \beta_1 x_1 + \beta_2 x_2 + \beta_3 x_2^2 + \beta_4 x_1 x_2 + \beta_5 x_1 x_2^2$$

to $n = 35$ data points and wish to test the null hypothesis H_0: $\beta_4 = \beta_5 = 0$.

a. State the alternative hypothesis.

b. Explain in detail how to compute the F-statistic needed to test the null hypothesis.

c. What are the numerator and denominator degrees of freedom associated with the F-statistic in part **b**?

d. Give the rejection region for the test if $\alpha = .05$.

Applying the Concepts—Basic

11.133 Comparing private and public college tuition. According to the *Chronicle of Higher Education Almanac,* 4-year private colleges charge, on average, five times as much for tuition and fees than 4-year public colleges. In order to estimate the true difference in the mean amounts charged for the 2006–2007 academic year, random samples of 40 private colleges and 40 public colleges were contacted and questioned about their tuition structures.

a. Which of the procedures described in Chapter 7 could be used to estimate the difference in mean charges between private and public colleges?

b. Propose a regression model involving the qualitative independent variable type of college that could be used to investigate the difference between the means. Be sure to specify the coding scheme for the dummy variable in the model.

c. Explain how the regression model you developed in part **b** could be used to estimate the difference between the population means.

11.134 GPAs of business students. Research scientists at the Educational Testing Service (ETS) used multiple regression analysis to model y, the final grade point average (GPA) of business and management doctoral students. A list of the potential independent variables measured for each doctoral student in the study follows:

(1) Quantitative Graduate Management Aptitude Test (GMAT) score
(2) Verbal GMAT score
(3) Undergraduate GPA
(4) First-year graduate GPA
(5) Student cohort (i.e., year in which student entered doctoral program: year 1, year 3, or year 5)

a. Identify the variables as quantitative or qualitative.

b. For each quantitative variable, give your opinion on whether the variable is positively or negatively related to final GPA.

c. For each of the qualitative variables, set up the appropriate dummy variable.

d. Write a first-order, main-effects model relating final GPA, y, to the five independent variables.

e. Interpret the β's in the model, part **d.**
f. Write a first-order model for final GPA, y, that allows for a different slope for each student cohort.
g. For each quantitative independent variable in the model, part **f,** give the slope of the line (in terms of the β's) for the year 1 cohort.

11.135 Comparing two orange juice extractors. The Florida Citrus Commission is interested in evaluating the performance of two orange juice extractors, brand A and brand B. It is believed that the size of the fruit used in the test may influence the juice yield (amount of juice per pound of oranges) obtained by the extractors. The commission wants to develop a regression model relating the mean juice yield $E(y)$ to the type of orange juice extractor (brand A or brand B) and the size of orange (diameter), x_1.
a. Identify the independent variables as qualitative or quantitative.
b. Write a model that describes the relationship between $E(y)$ and size of orange as two parallel lines, one for each brand of extractor.
c. Modify the model of part **b** to permit the slopes of the two lines to differ.
d. Sketch typical response lines for the model of part **b.** Do the same for the model of part **c.** Carefully label your graphs.
e. Specify the null and alternative hypotheses you would use to determine whether the model in part **c** provides more information for predicting yield than does the model in part **b.**
f. Explain how you would obtain the quantities necessary to compute the F-statistic that would be used in testing the hypotheses you described in part **e.**

11.136 Global warming and foreign investments. Scientists believe that a major cause of global warming is higher levels of carbon dioxide (CO_2) in the atmosphere. In the *Journal of World-Systems Research* (Summer 2003), sociologists examined the impact of foreign investment dependence on CO_2 emissions in $n = 66$ developing countries. In particular, the researchers modeled the level of CO_2 emissions in 1996 based on foreign investments made 16 years earlier and several other independent variables. The variables and the model results are listed in the first table below.
a. Interpret the value of R^2.
b. Use the value of R^2 to test the null hypothesis, $H_0: \beta_1 = \beta_2 = \cdots = \beta_7 = 0$, at $\alpha = .01$.
c. What null hypothesis would you test to determine if foreign investments in 1980 is a statistically useful predictor of CO_2 emissions in 1996?
d. Conduct the test, part **c,** at $\alpha = .05$.

11.137 Global warming and foreign investments (cont'd). Refer to Exercise 11.136. A matrix giving the correlation (r) for each pair of independent variables is shown below. Identify the independent variables that are highly correlated. What problems may result from including these highly correlated variables in the regression model?

11.138 Accuracy of software effort estimates. Refer to the *Journal of Empirical Software Engineering* (Vol. 9, 2004) study of the accuracy of software effort estimates, Exercise 11.69 (p. 671). Recall that the dependent variable (y) is measured as relative error of the effort estimate of a software development task. A total of eight independent variables were evaluated as potential predictors of relative error. Each of these was formulated as a dummy variable, as shown in the table on the next page.
a. The eight independent variables were entered into a stepwise regression, and the following two variables were selected for entry into the model: x_1 and x_8. Write the main-effects model for $E(y)$ as a function of these two variables.
b. The stepwise regression yielded the following prediction equation:

$$\hat{y} = .12 - .28x_1 + .27x_8$$

Table for Exercise 11.136

y = ln(level of CO_2 emissions in 1996)	β Estimate	t-value	p-value
x_1 = ln(foreign investments in 1980)	.79	2.52	< .05
x_2 = gross domestic investment in 1980	.01	.13	> .10
x_3 = trade exports in 1980	−.02	−1.66	> .10
x_4 = ln(GNP in 1980)	−.44	−.97	> .10
x_5 = agricultural production in 1980	−.03	−.66	> .10
x_6 = 1 if African country, 0 if not	−1.19	−1.52	> .10
x_7 = ln(level of CO_2 emissions in 1980)	.56	3.35	< .001

$R^2 = .31$

Source: Grimes, P., and Kentor, J. "Exporting the greenhouse: Foreign capital penetration and CO_2 emissions 1980–1996," *Journal of World-Systems Research,* Vol. IX, No. 2, Summer 2003 (Table 1).

Correlation matrix for Exercise 11.137

Independent Variable	x_2	x_3	x_4	x_5	x_6	x_7 = ln(level of CO_2 emissions in 1980)
x_1 = ln(foreign investments in 1980)	.13	.57	.30	−.38	.14	−.14
x_2 = gross domestic investment in 1980		.49	.36	−.47	−.14	.25
x_3 = trade exports in 1980			.43	−.47	−.06	−.07
x_4 = ln(GNP in 1980)				−.84	−.53	.42
x_5 = agricultural production in 1980					.45	−.50
x_6 = 1 if African country, 0 if not						−.47

Source: Grimes, P., and Kentor, J. "Exporting the greenhouse: Foreign capital penetration and CO_2 emissions 1980–1996," *Journal of World-Systems Research,* Vol. IX, No. 2, Summer 2003 (Appendix B).

Give a practical interpretation of the β estimates multiplied by x_1 and x_8.

c. The researcher is concerned that the sign of $\hat{\beta}_1$ in the model is the opposite from what is expected. (The researcher expects a project leader to have a smaller relative error of estimation than a developer.) Give at least one reason why this phenomenon occurred.

Company role of estimator:	x_1 = 1 if developer, 0 if project leader
Task complexity:	x_2 = 1 if low, 0 if medium/high
Contract type:	x_3 = 1 if fixed price, 0 if hourly rate
Customer importance:	x_4 = 1 if high, 0 if low/medium
Customer priority:	x_5 = 1 if time of delivery, 0 if cost or quality
Level of knowledge:	x_6 = 1 if high, 0 if low/medium
Participation:	x_7 = 1 if estimator participates in work, 0 if not
Previous accuracy:	x_8 = 1 if more than 20% accurate, 0 if less than 20% accurate

11.139 Study of fees charged by auditors. External auditors are hired to review and analyze the financial and other records of an organization and to attest to the integrity of the organization's financial statements. In recent years, the fees charged by auditors have come under increasing scrutiny. University of Melbourne (Australia) researchers investigated the effects of seven variables on the fee charged by auditors (*Journal of Business Finance and Accounting*, April 1995). The multiple regression model $E(y) = \beta_0 + \beta_1x_1 + \beta_2x_2 + \beta_3x_3 + \cdots + \beta_7x_7$ was fit to data collected for $n = 268$ companies. The results are $R^2 = .712$, $F = 111.1$.

a. Interpret the value of R^2.

b. Assess the overall fit of the model by conducting the global F-test.

c. For this model, x_1 = {1 if new auditor, 0 if incumbent auditors}. The main objective of the analysis was to determine whether new auditors charge less than incumbent auditors in a given year. If this hypothesis is true, is the true value of β_1 positive or negative? Explain.

11.140 Study of entry-level job preferences. *Benefits Quarterly* (First Quarter, 1995) published a study of entry-level job preferences. Several independent variables were used to model the job preferences (measured on a 10-point scale) of 164 business school graduates, including the following qualitative variables:

(1) Flextime of the position applied for (yes/no)
(2) Level of day care support required (none, referral, or on-premise)
(3) Spousal transfer support required (none, counseling, or active search)
(4) Marital status of applicant (married/not)
(5) Gender of applicant (male/female)

For each of the above qualitative variables, hypothesize a model for job preference (y) as a function of that variable. Interpret the β's in each model.

11.141 Prototyping new software. To meet the increasing demand for new software products, many systems development experts have adopted a prototyping methodology. The effects of prototyping on the system development life cycle (SDLC) was investigated in the *Journal of Computer Information Systems* (Spring 1993). A survey of 500 randomly selected corporate level MIS managers was conducted. Three potential independent variables were (1) *importance* of prototyping to each phase of the SDLC; (2) degree of *support* prototyping provides for the SDLC; and (3) degree to which prototyping *replaces* each phase of the SDLC. The table below gives the pairwise correlations of the three variables in the survey data for one particular phase of the SDLC. Use this information to assess the degree of multicollinearity in the survey data. Would you recommend using all three independent variables in a regression analysis? Explain.

Variable Pairs	Correlation Coefficient, r
Importance–Replace	.2682
Importance–Support	.6991
Replace–Support	−.0531

Source: Hardgrave, B. C., Doke, E. R., and Swanson, N. E. "Prototyping effects of the system development life cycle: An empirical study," *Journal of Computer Information Systems,* Vol. 33, No. 3, Spring 1993, p. 16 (Table 1).

11.142 A CEO's impact on corporate profits. Can a corporation's annual profit be predicted from information about the company's CEO? Each year *Forbes* publishes data on company profit (in \$ millions), CEO's annual income (in \$ thousands), and percentage of the company's stock owned by the CEO. Consider a model relating company profit (y) to CEO income (x_1) and stock percentage (x_2). Explain what it means to say that "CEO income x_1 and stock percentage x_2 interact to affect company profit y."

11.143 "Sun safety" study. Numerous "sun safety" products exist on the market to prevent excessive exposure to solar radiation. But many people do not practice "sun safety" or recognize the effectiveness of these products. A group of University of Arizona researchers examined the feasibility of educating preschool (4- to 5-year-old) children about sun safety (*American Journal of Public Health,* July 1995). A sample of 122 preschool children was divided into two groups, the control group and the intervention group. Children in the intervention group received a *Be Sun Safe* curriculum in preschool, while the control group did not. All children were tested for their knowledge, comprehension, and application of sun safety at two points in time: prior to the sun safety curriculum (pretest, x_1) and 7 weeks following the curriculum (posttest, y).

a. Write a first-order model for mean posttest score, $E(y)$, as a function of pretest score, x_1, and group. Assume that no interaction exists between pretest score and group.

b. For the model, part **a,** show that the slope of the line relating posttest score to pretest score is the same for both groups of children.

c. Repeat part **a** but assume that pretest score and group interact.

d. For the model, part **c,** show that the slope of the line relating posttest score to pretest score differs for the two groups of children.

Applying the Concepts—Intermediate

11.144 Promotion of supermarket vegetables. A supermarket chain is interested in exploring the relationship between the sales of its store-brand canned vegetables (y), the amount spent on promotion of the vegetables in local newspapers (x_1), and the amount of shelf space allocated to the brand (x_2). One of the chain's supermarkets was randomly selected, and over a 20-week period x_1 and x_2 were varied, as reported in the table. These data are saved in the **CANVEG** file.

Week	Sales ($)	Advertising Expenditures ($)	Shelf Space (sq. ft.)
1	2,010	201	75
2	1,850	205	50
3	2,400	355	75
4	1,575	208	30
5	3,550	590	75
6	2,015	397	50
7	3,908	820	75
8	1,870	400	30
9	4,877	997	75
10	2,190	515	30
11	5,005	996	75
12	2,500	625	50
13	3,005	860	50
14	3,480	1,012	50
15	5,500	1,135	75
16	1,995	635	30
17	2,390	837	30
18	4,390	1,200	50
19	2,785	990	30
20	2,989	1,205	30

a. Fit the following model to the data:

$$y = \beta_0 + \beta_1 x_1 + \beta_2 x_2 + \beta_3 x_1 x_2 + \varepsilon$$

b. Conduct an F-test to investigate the overall usefulness of this model. Use $\alpha = .05$.

c. Test for the presence of interaction between advertising expenditures and shelf space. Use $\alpha = .05$.

d. Explain what it means to say that advertising expenditures and shelf space interact.

e. Explain how you could be misled by using a first-order model instead of an interaction model to explain how advertising expenditures and shelf space influence sales.

f. Based on the type of data collected, comment on the assumption of independent errors.

11.145 Yield strength of steel alloy. Industrial engineers at the University of Florida used regression modeling as a tool to reduce the time and cost associated with developing new metallic alloys (*Modelling and Simulation in Materials Science and Engineering*, Vol. 13, 2005). To illustrate, the engineers built a regression model for the tensile yield strength (y) of a new steel alloy. The potential important predictors of yield strength are listed in the accompanying table.

x_1 = Carbon amount (% weight)
x_2 = Manganese amount (% weight)
x_3 = Chromium amount (% weight)
x_4 = Nickel amount (% weight)
x_5 = Molybdenum amount (% weight)
x_6 = Copper amount (% weight)
x_7 = Nitrogen amount (% weight)
x_8 = Vanadium amount (% weight)
x_9 = Plate thickness (millimeters)
x_{10} = Solution treating (milliliters)
x_{11} = Aging temperature (degrees, Celsius)

a. The engineers discovered that the variable Nickel (x_4) was highly correlated with the other potential independent variables. Consequently, Nickel was dropped from the model. Do you agree with this decision? Explain.

b. The engineers used stepwise regression on the remaining 10 potential independent variables in order to search for a parsimonious set of predictor variables. Do you agree with this decision? Explain.

c. The stepwise regression selected the following independent variables: x_1 = Carbon, x_2 = Manganese, x_3 = Chromium, x_5 = Molybdenum, x_6 = Copper, x_8 = Vanadium, x_9 = Plate thickness, x_{10} = Solution treating, and x_{11} = Aging temperature. All these variables were statistically significant in the stepwise model, with $R^2 = .94$. Consequently, the engineers used the estimated stepwise model to predict yield strength. Do you agree with this decision? Explain.

11.146 Optimizing semiconductor material processing. Fluorocarbon plasmas are used in the production of semiconductor materials. In the *Journal of Applied Physics* (Dec. 1, 2000), electrical engineers at Nagoya University (Japan) studied the kinetics of fluorocarbon plasmas in order to optimize material processing. In one portion of the study, the surface production rate of fluorocarbon radicals emitted from the production process was measured at various points in time (in milliseconds) after the radio frequency power was turned off. The data are given in the next table and saved in the **RADICALS** file. Consider a model relating surface production rate (y) to time (x).

Rate	Time	Rate	Time
1.00	0.1	0.00	1.7
0.80	0.3	−0.10	1.9
0.40	0.5	−0.15	2.1
0.20	0.7	−0.05	2.3
0.05	0.9	−0.13	2.5
0.00	1.1	−0.08	2.7
−0.05	1.3	0.00	2.9
−0.02	1.5		

Source: Takizawa, K., et al. "Characteristics of C_3 radicals in high-density C_4F_8 plasmas studied by laser-induced fluorescence spectroscopy," *Journal of Applied Physics*, Vol. 88, No. 11, Dec. 1, 2000 (Figure 7). Reprinted with permission from Journal of Applied Physics. Copyright © 2000 American Institute of Physics.

a. Graph the data in a scattergram. What trend do you observe?

b. Fit a quadratic model to the data. Give the least squares prediction equation.

c. Is there sufficient evidence of upward curvature in the relationship between surface production rate and time after turnoff? Use $\alpha = .05$.

11.147 Modeling peak-hour roadway traffic. Traffic forecasters at the Minnesota Department of Transportation (MDOT) use regression analysis to estimate weekday peak-hour traffic volumes on existing and proposed roadways. In particular, they model y, the peak-hour volume (typically, the volume between 7:00 and 8:00 A.M.), as a function of x_1, the road's total volume for the day. For one project involving the redesign of a section of Interstate 494, the forecasters collected $n = 72$ observations of peak-hour traffic volume and 24-hour weekday traffic volume using electronic sensors that count vehicles. The data are saved in the **MINNDOT** file. (The first and last five observations are listed in the table.)

Observation Number	Peak-Hour Volume	24-Hour Volume	I-35
1	1,990.94	20,070	0
2	1,989.63	21,234	0
3	1,986.96	20,633	0
4	1,986.96	20,676	0
5	1,983.78	19,818	0
⋮	⋮	⋮	
68	2,147.93	22,948	1
69	2,147.85	23,551	1
70	2,144.23	21,637	1
71	2,142.41	23,543	1
72	2,137.39	22,594	1

Source: John Sem, Director; Allan E. Pint, State Traffic Forecast Engineer; and James Page Sr., Transportation Planner, Traffic and Commodities Studies Section, Minnesota Department of Transportation, St. Paul, Minnesota.

a. Construct a scattergram for the data, plotting peak-hour volume y against 24-hour volume x_1. Note the isolated group of observations at the top of the scattergram. Investigators discovered that all of these data points were collected at the intersection of Interstate 35W and 46th Street. (These are observations 55–72 in the table.) While all other locations in the sample were three-lane highways, this location was unique in that the highway widens to four lanes just north of the electronic sensor. Consequently, the forecasters decided to include a dummy variable to account for a difference between the I-35W location and all other locations.

b. Knowing that peak-hour traffic volumes have a theoretical upper bound, the forecaster hypothesized that a second-order model should be used to explain the variation in y. Propose a complete second-order model for $E(y)$ as a function of 24-hour volume x_1 and the dummy variable for location.

c. Using an available statistical software package, fit the model of part **b** to the data. Interpret the results. Specifically, is the curvilinear relationship between peak-hour volume and 24-hour volume different at the two locations?

d. Conduct a residual analysis of the model, part **b.** Evaluate the assumptions of normality and constant error variance and determine whether any outliers exist.

11.148 Predicting percentage of problem mortgages. *Best's Review* (June 1999) compared the mortgage loan portfolios for a sample of 25 life/health insurance companies. The information in the table below, saved in the **BESTINS** file, is extracted from the article. Suppose you want to model the percentage of problem mortgages (y) of a company as a function of total mortgage loans (x_1), percentage of invested assets (x_2), percentage of commercial mortgages (x_3), and percentage of residential mortgages (x_4).

a. Write a first-order model for $E(y)$.

b. Fit the model of part **a** to the data and evaluate its overall usefulness. Use $\alpha = .05$.

c. Interpret the β estimates in the fitted model.

d. Construct scattergrams of y versus each of the four independent variables in the model. Which variables warrant inclusion in the model as second-order (i.e., squared) terms?

e. Fit the model that results from your exploratory analysis in part **d** to the data. Evaluate its overall usefulness using $\alpha = .05$.

f. Do the one or more second-order terms of your model, part **e,** contribute information for the prediction of the percentage of problem mortgages? Test using $\alpha = .05$.

11.149 Impact of advertising on market share. The audience for a product's advertising can be divided into four segments according to the degree of exposure received as a result of the advertising. These segments are groups of consumers who receive very high (VH), high (H), medium (M), or low (L) exposure to the advertising. A company is interested in exploring whether its advertising effort affects its product's market share. Accordingly, the company identifies 24 sample groups of consumers who have

Table for Exercise 11.148 (selected observations)

Company	Total Mortgage Loan, x_1	% Invested Assets, x_2	% Commercial Mortgages, x_3	% Residential Mortgages, x_4	% Problem Mortgages, y
TIAA Group	$18,803,163	20.7	100.0	0.0	11.4
Metropolitan Insurance	18,171,162	13.9	77.8	1.6	3.8
Prudential of Am Group	16,213,150	12.9	87.4	2.3	4.1
Principal Mutual IA	11,940,345	30.3	98.8	1.2	32.6
Northwestern Mutual	10,834,616	17.8	99.5	0.0	2.2
⋮		⋮	⋮	⋮	⋮
State Farm Group	2,027,648	8.6	97.6	2.4	0.1
Pacific Mutual Life	1,945,392	9.7	96.4	3.6	6.1

Source: Best's Review (Life/Health), June 1999, p. 35.

been exposed to its advertising, six groups at each exposure level. Then, the company determines its product's market share within each group.

Market Share within Group	Exposure Level	Market Share within Group	Exposure Level
10.1	L	12.2	H
10.3	L	12.1	H
10.0	L	11.8	H
10.3	L	12.6	H
10.2	L	11.9	H
10.5	L	12.9	H
10.6	M	10.7	VH
11.0	M	10.8	VH
11.2	M	11.0	VH
10.9	M	10.5	VH
10.8	M	10.8	VH
11.0	M	10.6	VH

a. Write a regression model that expresses the company's market share as a function of advertising exposure level. Define all terms in your model and list any assumptions you make about them.
b. Did you include interaction terms in your model? Why or why not?
c. The data in the table, saved in the **MKTSHR** file, were obtained by the company. Fit the model you constructed in part **a** to the data.
d. Is there evidence to suggest that the firm's expected market share differs for different levels of advertising exposure? Test using $\alpha = .05$.

11.150 Downtime of a production process. An operations manager is interested in modeling $E(y)$, the expected length of time per month (in hours) that a machine will be shut down for repairs, as a function of the type of machine (001 or 002) and the age of the machine (in years). The manager has proposed the following model:

$$E(y) = \beta_0 + \beta_1 x_1 + \beta_2 x_1^2 + \beta_3 x_2$$

where

x_1 = Age of machine
x_2 = 1 if machine type 001,0 if machine type 002

a. Use the data obtained on $n = 20$ machine breakdowns, shown in the next column (and saved in the **SHUTDOWN** file), to estimate the parameters of this model.
b. Do these data provide sufficient evidence to conclude that the second-order term (x_1^2) in the model proposed by the operations manager is necessary? Test using $\alpha = .05$.
c. Test the null hypothesis that $\beta_1 = \beta_2 = 0$ using $\alpha = .10$. Interpret the results of the test in the context of the problem.

Table for Exercise 11.150

Downtime (hours per month)	Machine Age, x_1 (years)	Machine Type	x_2
10	1.0	001	1
20	2.0	001	1
30	2.7	001	1
40	4.1	001	1
9	1.2	001	1
25	2.5	001	1
19	1.9	001	1
41	5.0	001	1
22	2.1	001	1
12	1.1	001	1
10	2.0	002	0
20	4.0	002	0
30	5.0	002	0
44	8.0	002	0
9	2.4	002	0
25	5.1	002	0
20	3.5	002	0
42	7.0	002	0
20	4.0	002	0
13	2.1	002	0

11.151 Forecasting daily admission of a water park. To determine whether extra personnel are needed for the day, the owners of a water adventure park would like to find a model that would allow them to predict the day's attendance each morning before opening based on the day of the week and weather conditions. The model is of the form

$$E(y) = \beta_0 + \beta_1 x_1 + \beta_2 x_2 + \beta_3 x_3$$

where

$$y = \text{Daily admission}$$
$$x_1 = \begin{cases} 1 & \text{if weekend} \\ 0 & \text{otherwise} \end{cases} \quad \text{(dummy variable)}$$
$$x_2 = \begin{cases} 1 & \text{if sunny} \\ 0 & \text{if overcast} \end{cases} \quad \text{(dummy variable)}$$
$$x_3 = \text{predicted daily high temperature (°F)}$$

These data were recorded for a random sample of 30 days, and a regression model was fitted to the data. The least squares analysis produced the following results:

$$\hat{y} = -105 + 25x_1 + 100x_2 + 10x_3$$

with

$$s_{\hat{\beta}_1} = 10 \quad s_{\hat{\beta}_2} = 30 \quad s_{\hat{\beta}_3} = 4 \quad R^2 = .65$$

a. Interpret the estimated model coefficients.
b. Is there sufficient evidence to conclude that this model is useful for the prediction of daily attendance? Use $\alpha = .05$.
c. Is there sufficient evidence to conclude that the mean attendance increases on weekends? Use $\alpha = .10$.
d. Use the model to predict the attendance on a sunny weekday with a predicted high temperature of 95°F.
e. Suppose the 90% prediction interval for part **d** is (645, 1,245). Interpret this interval.

11.152 Forecasting daily admission of a water park (cont'd). Refer to Exercise 11.151. The owners of the water adventure park are advised that the prediction model could probably be improved if interaction terms were added. In particular, it is thought that the *rate* at which mean attendance increases as predicted high temperature increases will be greater on weekends than on weekdays. The following model is therefore proposed:

$$E(y) = \beta_0 + \beta_1 x_1 + \beta_2 x_2 + \beta_3 x_3 + \beta_4 x_1 x_3$$

The same 30 days of data used in Exercise 11.151 are again used to obtain the least squares model

$$\hat{y} = 250 - 700x_1 + 100x_2 + 5x_3 + 15x_1x_3$$

with

$$s_{\hat{\beta}_4} = 3.0 \quad R^2 = .96$$

a. Graph the predicted day's attendance, y, against the day's predicted high temperature, x_3, for a sunny weekday and for a sunny weekend day. Plot both on the same graph for x_3 between 70°F and 100°F. Note the increase in slope for the weekend day. Interpret this.

b. Do the data indicate that the interaction term is a useful addition to the model? Use $\alpha = .05$.

c. Use this model to predict the attendance for a sunny weekday with a predicted high temperature of 95°F.

d. Suppose the 90% prediction interval for part **c** is (800, 850). Compare this result with the prediction interval for the model without interaction in Exercise 11.151, part **e.** Do the relative widths of the confidence intervals support or refute your conclusion about the utility of the interaction term (part **b**)?

e. The owners, noting that the coefficient $\hat{\beta}_1 = -700$, conclude the model is ridiculous because it seems to imply that the mean attendance will be 700 less on weekends than on weekdays. Explain why this is *not* the case.

11.153 Sale prices of apartments. A Minneapolis, Minnesota, real estate appraiser used regression analysis to explore the relationship between the sale prices of apartment buildings and various characteristics of the buildings. The **MNSALES** file contains data for a random sample of 25 apartment buildings. *Note:* Physical condition of each apartment building is coded E (excellent), G (good), or F (fair). Data for selected observations are shown in the table below.

a. Write a model that describes the relationship between sale price and number of apartment units as three parallel lines, one for each level of physical condition. Be sure to specify the dummy variable coding scheme you use.

b. Plot y against x_1 (number of apartment units) for all buildings in excellent condition. On the same graph, plot y against x_1 for all buildings in good condition. Do this again for all buildings in fair condition. Does it appear that the model you specified in part **a** is appropriate? Explain.

c. Fit the model from part **a** to the data. Report the least squares prediction equation for each of the three building condition levels.

d. Plot the three prediction equations of part **c** on a scattergram of the data.

e. Do the data provide sufficient evidence to conclude that the relationship between sale price and number of units differs depending on the physical condition of the apartments? Test using $\alpha = .05$.

f. Check the data set for multicollinearity. How does this impact your choice of independent variables to use in a model for sale price?

g. Conduct a complete residual analysis for the model to check the assumptions on ε.

11.154 Light output of a bulb. A firm that has developed a new type of lightbulb is interested in evaluating its performance in order to decide whether to market it. It is known that the light output of the bulb depends on the cleanliness of its surface area and the length of time the bulb has been in operation. Use the data in the next table (saved in the **LTBULB** file) and the procedures you learned in this chapter to build a regression model that relates drop in light output to bulb surface cleanliness and length of operation. Be sure to conduct a residual analysis also.

Drop in Light Output (% original output)	Bulb Surface (C = clean) (D = dirty)	Length of Operation (hours)
0	C	0
16	C	400
22	C	800
27	C	1,200
32	C	1,600
36	C	2,000
38	C	2,400
0	D	0
4	D	400
6	D	800
8	D	1,200
9	D	1,600
11	D	2,000
12	D	2,400

Table for Exercise 11.153 (selected observations)

Code No.	Sale Price, y ($)	No. of Apartments, x_1	Age of Structure, x_2 (years)	Lot Size, x_3 (sq. ft.)	No. of On-Site Parking Spaces, x_4	Gross Building Area, x_5 (sq. ft.)	Condition of Apartment Building
0229	90,300	4	82	4,635	0	4,266	F
0094	384,000	20	13	17,798	0	14,391	G
0043	157,500	5	66	5,913	0	6,615	G
0079	676,200	26	64	7,750	6	34,144	E
0134	165,000	5	55	5,150	0	6,120	G
⋮	⋮	⋮	⋮	⋮	⋮	⋮	⋮
0019	93,600	4	82	6,864	0	3,840	F
0074	110,000	4	50	4,510	0	3,092	G
0057	573,200	14	10	11,192	0	23,704	E
0104	79,300	4	82	7,425	0	3,876	F
0024	272,000	5	82	7,500	0	9,542	E

Source: Robinson Appraisal Co., Inc., Mankato, Minnesota.

11.155 Forecasting a job applicant's merit rating. A large research and development firm rates the performance of each member of its technical staff on a scale of 0 to 100, and this merit rating is used to determine the size of the person's pay raise for the coming year. The firm's personnel department is interested in developing a regression model to help them forecast the merit rating that an applicant for a technical position will receive after being employed three years. The firm proposes to use the following second-order model to forecast the merit ratings of applicants who have just completed their graduate studies and have no prior related job experience:

$$E(y) = \beta_0 + \beta_1x_1 + \beta_2x_2 + \beta_3x_1x_2 + \beta_4x_1^2 + \beta_5x_2^2$$

where

y = Applicant's merit rating after 3 years
x_1 = Applicant's GPA in graduate school
x_2 = Applicant's total score (verbal plus quantitative) on the Graduate Record Examination (GRE)

The model, fit to data collected for a random sample of $n = 40$ employees, resulted in SSE = 1,830.44 and SS(model) = 4,911.5. The reduced model $E(y) = \beta_0 + \beta_1x_1 + \beta_2x_2$ is also fit to the same data, resulting in SSE = 3,197.16.

a. Identify the appropriate null and alternative hypotheses to test whether the complete (second-order) model contributes information for the prediction of y.

b. Conduct the test of hypothesis given in part **a.** Test using $\alpha = .05$. Interpret the results in the context of this problem.

c. Identify the appropriate null and alternative hypotheses to test whether the complete model contributes more information than the reduced (first-order) model for the prediction of y.

d. Conduct the test of hypothesis given in part **c.** Test using $\alpha = .05$. Interpret the results in the context of this problem.

e. Which model, if either, would you use to predict y? Explain.

Applying the Concepts—Advanced

11.156 Collecting Beanie Babies. Refer to Exercise 10.95 (p. 616) and the data (saved in the **BEANIE** file) on the values of 50 Beanie Babies collector's items, published in *Beanie World Magazine*. Suppose we want to predict the market value of a Beanie Baby using age (in months as of Sept. 1998) and whether the Beanie Baby has been retired or is current (i.e., still in production).

a. Write a complete second-order model for market value as a function of age and current/retired status.

b. Specify the null hypothesis for testing whether the quadratic terms in the model, part **a,** are important for predicting market value.

c. Specify the null hypothesis for testing whether the interaction terms in the model, part **a,** are important for predicting market value.

d. Fit the three models of parts **a–c** to the data saved in the **BEANIE** file. Conduct the tests specified in parts **b** and **c.** Interpret the results.

11.157 Developing a model for college GPA. Many colleges and universities develop regression models for predicting the GPA of incoming freshmen. This predicted GPA can then be used to make admission decisions. Although most models use many independent variables to predict GPA, we will illustrate by choosing two variables:

x_1 = Verbal score on college entrance examination (percentile)
x_2 = Mathematics score on college entrance examination (percentile)

The **COLLGPA** file contains data on these variables for a random sample of 40 freshmen at one college. (Selected observations are shown in the table.) Use the data to develop a useful prediction equation for college freshman GPA (y). Be sure to conduct a residual analysis for the model.

Verbal, x_1	Mathematics, x_2	GPA, y
81	87	3.49
68	99	2.89
57	86	2.73
100	49	1.54
54	83	2.56
⋮	⋮	⋮
74	67	2.83
87	93	3.84
90	65	3.01
81	76	3.33
84	69	3.06

Critical Thinking Challenge

11.158 IQs and *The Bell Curve*. *The Bell Curve* (Free Press, 1994) written by Richard Herrnstein and Charles Murray (H&M), is a controversial book about race, genes, IQ, and economic mobility. The book heavily employs statistics and statistical methodology in an attempt to support the authors' positions on the relationships among these variables and their social consequences. The main theme of *The Bell Curve* can be summarized as follows:

(1) Measured intelligence (IQ) is largely genetically inherited.
(2) IQ is correlated positively with a variety of socioeconomic status success measures, such as prestigious job, high annual income, and high educational attainment.
(3) From 1 and 2, it follows that socioeconomic successes are largely genetically caused and therefore resistant to educational and environmental interventions (such as affirmative action).

The statistical methodology (regression) employed by the authors and the inferences derived from the statistics were critiqued in *Chance* (Summer 1995) and *The Journal of the American Statistical Association* (Dec. 1995). The following are just a few of the problems with H&M's use of regression that are identified:

Problem 1 H&M consistently use a trio of independent variables—IQ, socioeconomic status, and age—in a series of first-order models designed to predict dependent social outcome variables such as income and unemployment. (Only on a single occasion are interaction terms incorporated.) Consider, for example, the model

$$E(y) = \beta_0 + \beta_1x_1 + \beta_2x_2 + \beta_3x_3$$

where y = income, x_1 = IQ, x_2 = socioeconomic status, and x_3 = age. H&M employ t-tests on the individual

β parameters to assess the importance of the independent variables. As with most of the models considered in *The Bell Curve,* the estimate of β_1 in the income model is positive and statistically significant at $\alpha = .05$, and the associated t-value is larger (in absolute value) than the t-values associated with the other independent variables. Consequently, *H&M claim that IQ is a better predictor of income than the other two independent variables*. No attempt was made to determine whether the model was properly specified or whether the model provides an adequate fit to the data.

Problem 2 In an appendix, the authors describe multiple regression as a "mathematical procedure that yields coefficients for each of [the independent variables], indicating how much of a change in [the dependent variable] can be anticipated for a given change in any particular [independent] variable, with all the others held constant." Armed with this information and the fact that the estimate of β_1 in the model above is positive, *H&M infer that a high IQ necessarily implies (or causes) a high income, and a low IQ inevitably leads to a low income*. (Cause-and-effect inferences like this are made repeatedly throughout the book.)

Problem 3 The title of the book refers to the normal distribution and its well-known "bell-shaped" curve. There is a misconception among the general public that scores on intelligence tests (IQ) are normally distributed. In fact, most IQ scores have distributions that are decidedly skewed. Traditionally, psychologists and psychometricians have transformed these scores so that the resulting numbers have a precise normal distribution. H&M make a special point to do this. Consequently, *the measure of IQ used in all the regression models is normalized (i.e., transformed so that the resulting distribution is normal), despite the fact that regression methodology does not require predictor (independent) variables to be normally distributed.*

Problem 4 A variable that is not used as a predictor of social outcome in any of the models in *The Bell Curve* is level of education. H&M purposely omit education from the models, arguing that IQ causes education, not the other way around. Other researchers who have examined H&M's data report that *when education is included as an independent variable in the model, the effect of IQ on the dependent variable (say, income) is diminished.*

a. Comment on each of the problems identified. Why do each of these problems cast a shadow on the inferences made by the authors?

b. Using the variables specified in the model above, describe how you would conduct the multiple regression analysis. (Propose a more complex model and describe the appropriate model tests, including a residual analysis.)

References

Barnett, V., and Lewis, T. *Outliers in Statistical Data*. New York: Wiley, 1978.

Belsley, D. A., Kuh, E., and Welsch, R. E. *Regression Diagnostics: Identifying Influential Data and Sources of Collinearity*. New York: Wiley, 1980.

Chatterjee, S., and Price, B. *Regression Analysis by Example,* 2nd ed. New York: Wiley, 1991.

Draper, N., and Smith, H. *Applied Regression Analysis,* 2nd ed. New York: Wiley, 1981.

Graybill, F. *Theory and Application of the Linear Model*. North Scituate, Mass.: Duxbury, 1976.

Mendenhall, W. *Introduction to Linear Models and the Design and Analysis of Experiments*. Belmont, Calif.: Wadsworth, 1968.

Mendenhall, W., and Sincich, T. *A Second Course in Statistics: Regression Analysis*, 6th ed. Upper Saddle River, N.J.: Prentice Hall, 2003.

Mosteller, F. and Tukey, J. W. *Data Analysis and Regression: A Second Course in Statistics*. Reading, Mass.: Addison-Wesley, 1977.

Neter, J., Kutner, M., Nachtsheim, C., and Wasserman, W. *Applied Linear Statistical Models,* 4th ed. Homewood, Ill.: Richard Irwin, 1996.

Rousseeuw, P. J., and Leroy, A. M. *Robust Regression and Outlier Detection*. New York: Wiley, 1987.

Weisberg, S. *Applied Linear Regression*, 2nd ed. New York: Wiley, 1985.

USING TECHNOLOGY

SPSS: Multiple Regression

Step 1 Access the SPSS spreadsheet file that contains the dependent and independent variables.

Step 2 Click on the "Analyze" button on the SPSS menu bar and then click on "Regression" and "Linear," as shown in Figure 11.S.1. The resulting dialog box appears as shown in Figure 11.S.2.

Step 3 Specify the dependent variable in the "Dependent" box and the independent variables in the "Independent(s)" box. [*Note:* If your model includes interaction and/or squared terms, you must create and add these higher-order variables to the SPSS spreadsheet file *prior* to running a regression analysis. You can do this by clicking the "Transform" button on the SPSS main menu and selecting the "Compute" option.]

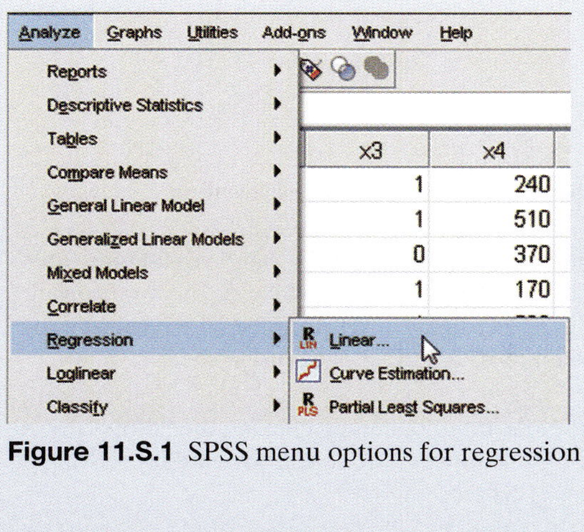

Figure 11.S.1 SPSS menu options for regression

Step 4 To perform a standard regression analysis, select "Enter" in the "Method" box. To perform a stepwise regression analysis, select "Stepwise" in the "Method" box.

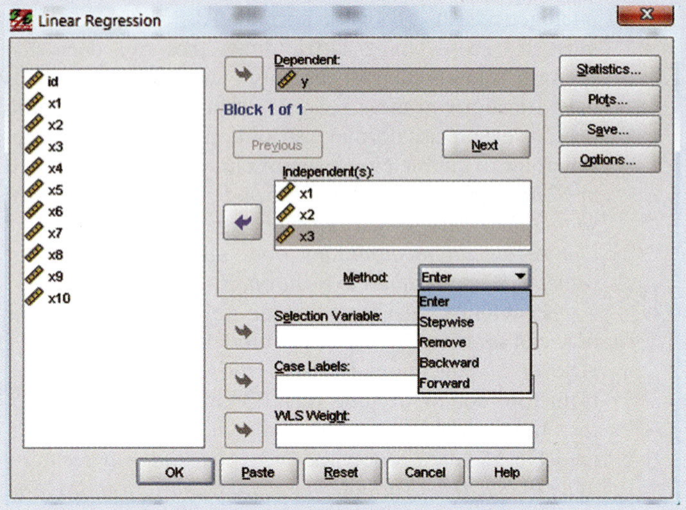

Figure 11.S.2 SPSS linear regression dialog box

Step 5 To perform a nested model F-test for additional model terms, click the "Next" button and enter the terms you want to test in the "Independent(s)" box. [*Note:* These terms, plus the terms you entered initially, form the complete model for the nested F-test.] Next, click the "Statistics" button and select "R squared change." Click "Continue" to return to the main SPSS regression dialog box.

Step 6 To produce confidence intervals for the model parameters, click the "Statistics" button and check the appropriate menu items in the resulting menu list (see Figure 11.S.3).

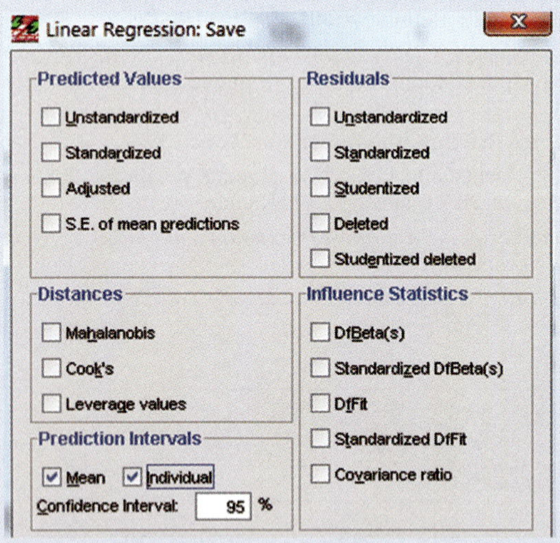

Figure 11.S.3 SPSS linear regression save options

Step 7 To obtain prediction intervals for y and confidence intervals for $E(y)$, click the "Save" button and check the appropriate items in the resulting menu list. (The prediction intervals will be added as new columns to the SPSS data spreadsheet.)

Step 8 Residual plots are obtained by clicking the "Plots" button and making the appropriate selections on the resulting menu (see Figure 11.S.4).

Step 9 To return to the main Regression dialog box from any of these optional screens, click "Continue." Click "OK" on the Regression dialog box to view the multiple regression results.

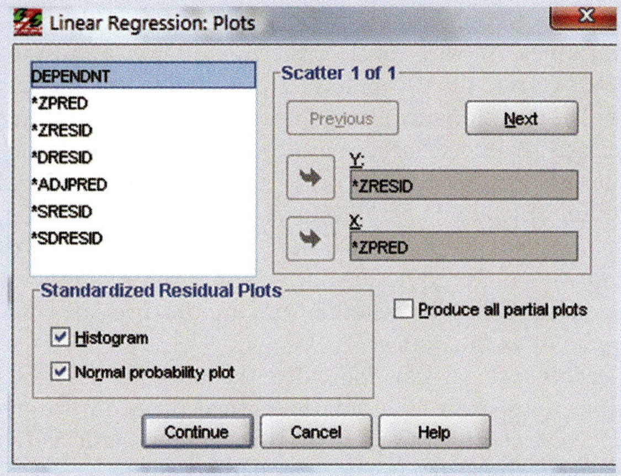

Figure 11.S.4 SPSS linear regression plots options

Minitab: Multiple Regression

Multiple Regression

Step 1 Access the Minitab worksheet file that contains the dependent and independent variables.

Step 2 Click on the "Stat" button on the Minitab menu bar and then click on "Regression" and "Regression" again, as shown in Figure 11.M.1.

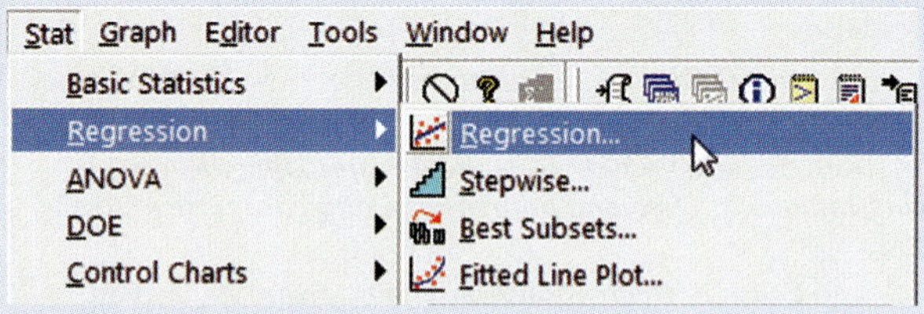

Figure 11.M.1 Minitab menu options for regression

Step 3 The resulting dialog box appears as shown in Figure 11.M.2. Specify the dependent variable in the "Response" box and the independent variables in the "Predictors" box. [*Note:* If your model includes interaction and/or squared terms, you must create and add these higher-order variables to the Minitab worksheet *prior* to running a regression analysis. You can do this by clicking the "Calc" button on the Minitab main menu and selecting the "Calculator" option.]

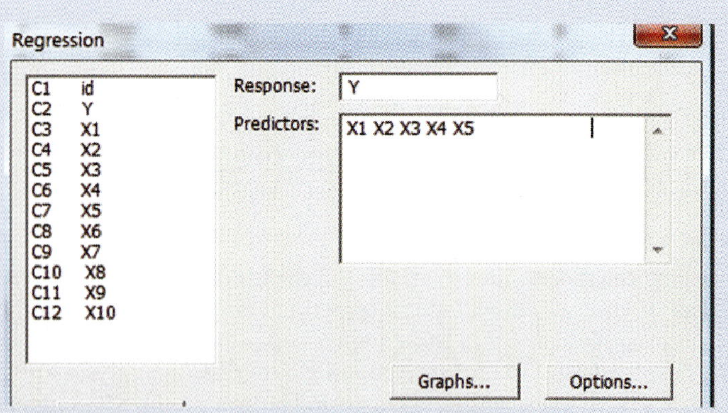

Figure 11.M.2 Minitab regression dialog box

Step 4 To produce prediction intervals for y and confidence intervals for $E(y)$, click the "Options" button and select the appropriate menu items in the resulting menu list (see Figure 11.M.3).

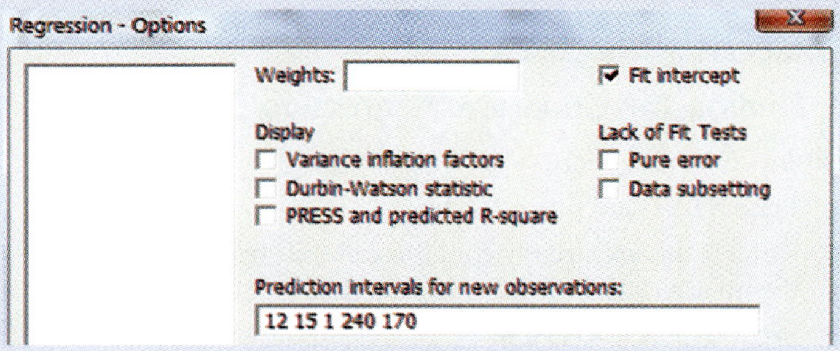

Figure 11.M.3 Minitab regression options

Step 5 Residual plots are obtained by clicking the "Graphs" button and making the appropriate selections on the resulting menu (see Figure 11.M.4).

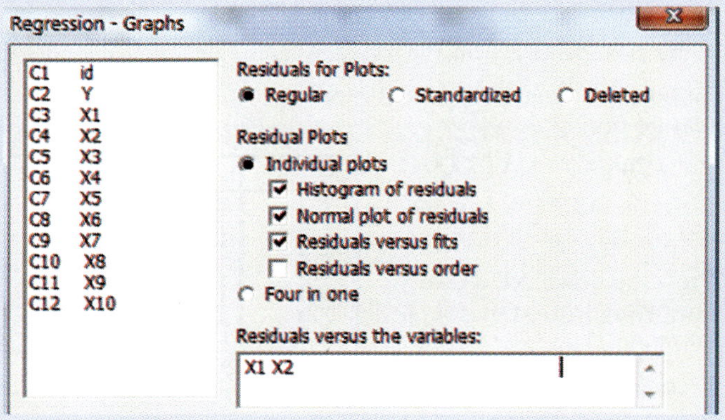

Figure 11.M.4 Minitab regression graphs options

Step 6 To return to the main Regression dialog box from any of these optional screens, click "OK."

Step 7 When you have made all your selections, click "OK" on the main Regression dialog box to produce the Minitab multiple regression printout.

Stepwise Regression

Step 1 Click on the "Stat" button on the main menu bar; then click on "Regression" and click on "Stepwise" (see Figure 11.M.1). The resulting dialog box appears like the one in Figure 11.M.2.

Step 2 Specify the dependent variable in the "Response" box and the independent variables in the stepwise model in the "Predictors" box.

Step 3 As an option, you can select the value of α to use in the analysis by clicking on the "Methods" button and specifying the value. (The default is $\alpha = .15$.)

Step 4 Click "OK" to view the stepwise regression results.

Excel: Multiple Regression

Multiple Regression in Excel

Step 1 Access the Excel worksheet file that contains the dependent and independent variables. [*Note:* If your model includes interaction and/or squared terms, you must create and add these higher-order variables to the Excel worksheet *prior* to running a regression analysis.]

Step 2 Click "Data" from the Excel menu bar and then select "Data Analysis." The resulting menu is shown in Figure 11.E.1.

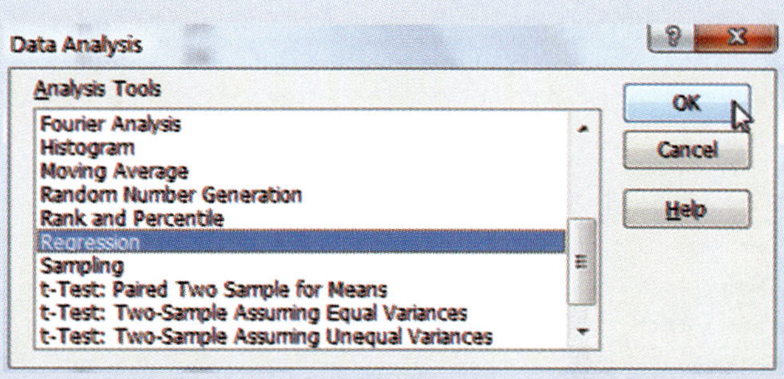

Figure 11.E.1 Excel data analysis menu options for regression

Step 3 Select "Regression" from the drop-down menu and then click "OK." The resulting dialog box is shown in Figure 11.E.2.

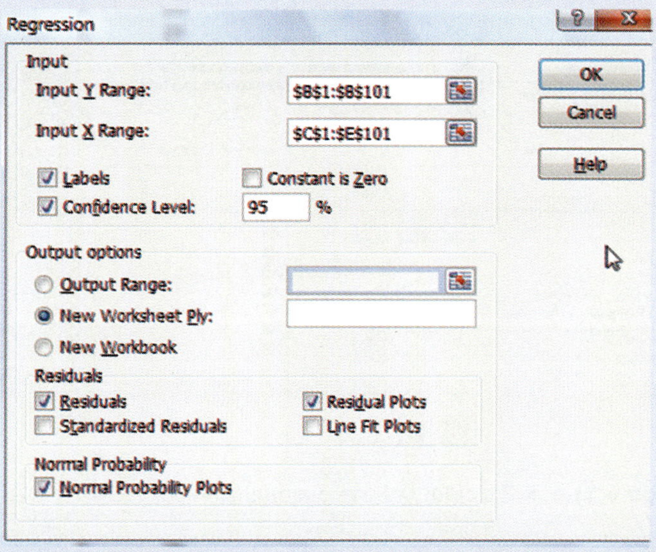

Figure 11.E.2 Excel regression dialog box

Step 4 Specify the cell ranges for the y and x variables in the "Input" area of the Multiple Regression dialog box. [*Note*: The x variables must be in adjacent columns on the Excel worksheet.]

Step 5 Select the confidence level for confidence intervals for the model parameters.

Step 6 To produce graphs of the regression residual, select the "Residual Plots" and "Normal Probability Plots" options.

Step 7 After making all your selections, click "OK" to produce the multiple regression results.

[*Note:* Stepwise regression is currently not available in either Excel or DDXL.]

Multiple Regression with Excel/DDXL

Step 1 Access the Excel spreadsheet with the data for the dependent and independent variables.

Step 2 Highlight (select) these data columns on the Excel spreadsheet.

Step 3 Click on "Add-Ins" in the main Excel menu bar and select "DDXL." On the resulting menu, select "Regression," as shown in Figure 11.E.3.

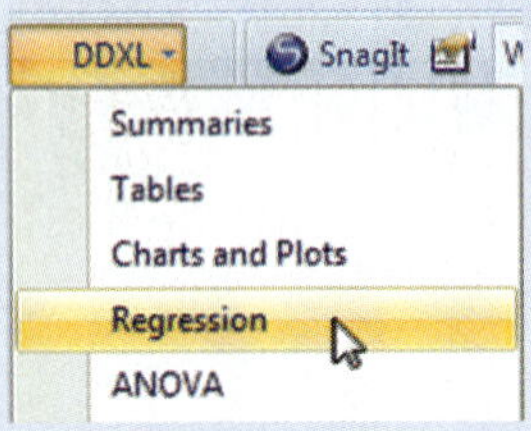

Figure 11.E.3 Excel/DDXL menu options for regression

Step 4 On the resulting menu, select "Multiple Regression" in the Function Type box, as shown in Figure 11.E.4.

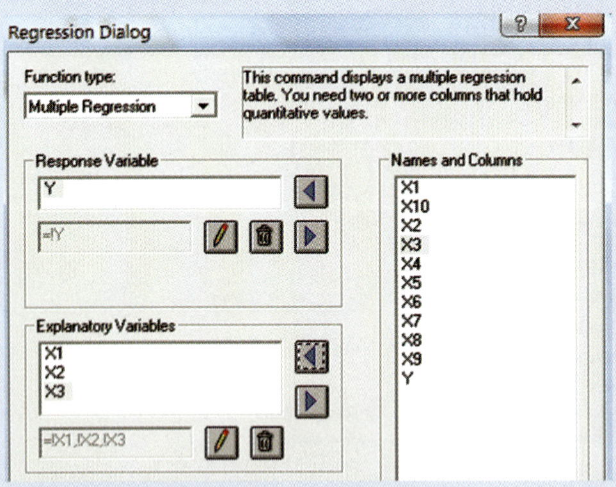

Figure 11.E.4 Excel/DDXL regression dialog box

Step 5 Move the column with the values of the dependent variable into the "Response Variable" box and the columns with the values of the independent variables into the "Explanatory Variables" box, as shown in Figure 11.E.4.

Step 6 Click "OK" to generate the multiple regression printout.

Step 7 Under the regression results are more options. One option is "Check Residuals." Selecting this option will generate a plot of the residuals against predicted and a normal probability plot.

TI-84 Graphing Calculator: Multiple Regression

Note: Only simple linear and quadratic regression models can be fit using the TI-84 graphing calculator.

Quadratic Regression

I. Finding the Quadratic Regression Equation

Step 1 *Enter the data*

- Press **STAT** and select **1:Edit**

  Note: If the list already contains data, clear the old data. Use the up arrow to highlight "**L1**" or "**L2.**"
- Press **CLEAR ENTER**
- Use the **ARROW** and **ENTER** keys to enter the data set into **L1** and **L2**

Step 2 *Find the quadratic regression equation*

- Press **STAT** and highlight **CALC**
- Press **5** for **QuadReg**
- Press **ENTER**
- The screen will show the values for a, b, and c in the equation
- If the diagnostics are on, the screen will also give the value for $\mathbf{r^2}$
- To turn the diagnostics feature on:
- Press **2nd 0** for **CATALOG**
- Press the **ALPHA** key and $\mathbf{x^{-1}}$ for **D**
- Press the down **ARROW** until **DiagnosticsOn** is highlighted
- Press **ENTER** twice

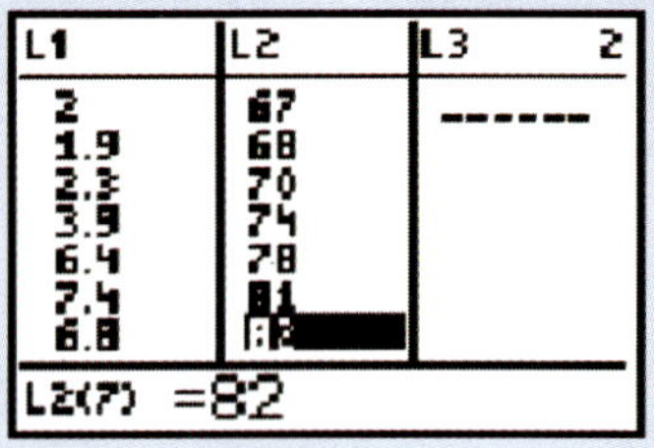

II. Graphing the Quadratic Curve with the Scatterplot

Step 1 *Enter the data as shown in part I above*

Step 2 *Set up the data plot*

- Press **Y=** and **CLEAR** all functions from the Y registers
- Press **2nd Y=** for **STAT PLOT**
- Press **1** for **Plot1**
- Set the cursor so that **ON** is flashing and press **ENTER**
- For **Type,** use the **ARROW** and **ENTER** keys to highlight and select the scatterplot (first icon in the first row)
- For **Xlist,** choose the column containing the *x*-data
- For **Freq,** choose the column containing the *y*-data

Step 3 *Find the regression equation and store the equation in Y1*

- Press **STAT** and highlight **CALC**
- Press **5** for **QuadReg** (*Note:* Don't press

ENTER here because you want to store the regression equation in Y1)

- Press **VARS**
- Use the right arrow to highlight **Y-VARS**
- Press **ENTER** to select **1:Function**
- Press **ENTER** to select **1:Y1**
- Press **ENTER**

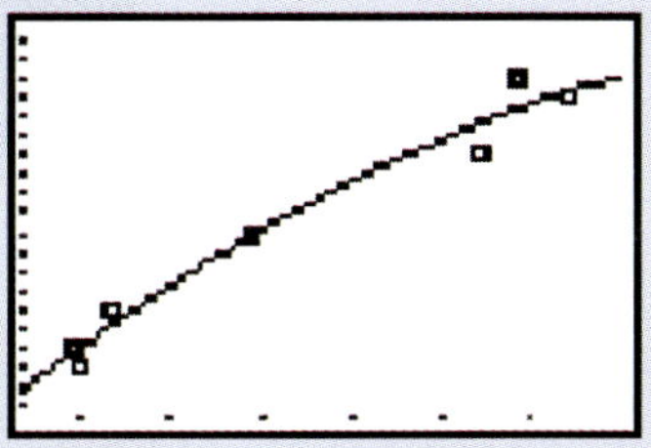

Step 4 *View the scatterplot and regression line*

- Press **ZOOM** and then press **9** to select **9:ZoomStat**

Plotting Residuals

When computing a regression equation on the TI-84, the residuals are automatically computed and saved to a list called **RESID. RESID** can be found under the **LIST menu** (**2nd STAT**).

Step 1 *Enter the data*

- Press **STAT** and select **1:Edit**

Note: If the list already contains data, clear the old data. Use the up arrow to highlight "**L1**" or "**L2.**"

- Press **CLEAR ENTER**
- Use the **ARROW** and **ENTER** keys to enter the data set into **L1** and **L2**

Step 2 *Compute the regression equation*

- Press **STAT** and highlight **CALC**
- Press **4** for **LinReg(ax + b)**
- Press **ENTER**

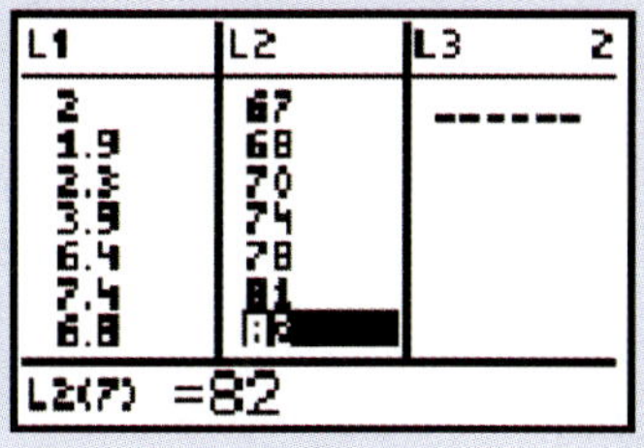

Step 3 *Set up the data plot*

- Press **Y=** and **CLEAR** all functions from the Y registers
- Press **2nd Y =** for **STATPLOT**
- Press **1** for **Plot1**
- Set the cursor so that **ON** is flashing and press **ENTER**

- For **Type,** use the **ARROW** and **ENTER** keys to highlight and select the scatterplot (first icon in the first row).
- Move the cursor to **Xlist** and choose the column containing the x-data
- Move the cursor to **Ylist** and press **2nd STAT** for **LIST**
- Use the down arrow to highlight the listname **RESID** and press **ENTER**

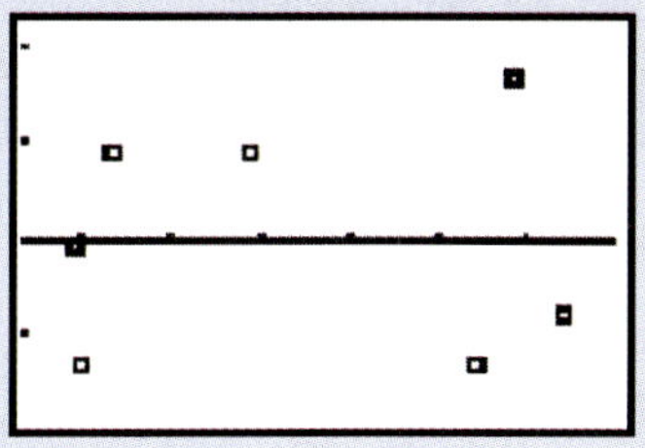

Step 4 *View the scatterplot of the residuals*

- Press **ZOOM 9** for **ZoomStat**

The Condo Sales Case

Covers Chapters 10 and 11 — This case involves an investigation of the factors that affect the sale price of oceanside condominium units. It represents an extension of an analysis of the same data by Herman Kelting (1979). Although condo sale prices have increased dramatically over the past 20 years, the relationship between these factors and sale price remain about the same. Consequently, the data provide valuable insight into today's condominium sales market.

The sales data were obtained for a new oceanside condominium complex consisting of two adjacent and connected eight-floor buildings. The complex contains 200 units of equal size (approximately 500 square feet each). The locations of the buildings relative to the ocean, the swimming pool, the parking lot, etc., are shown in the accompanying figure. There are several features of the complex that you should note:

1. The units facing south, called *ocean view,* face the beach and ocean. In addition, units in building 1 have a good view of the pool. Units to the rear of the building, called *bay-view,* face the parking lot and an area of land that ultimately borders a bay. The view from the upper floors of these units is primarily of wooded, sandy terrain. The bay is very distant and barely visible.
2. The only elevator in the complex is located at the east end of building 1, as are the office and the game room. People moving to or from the higher floor units in building 2 would likely use the elevator and move through the passages to their units. Thus, units on the higher floors and at a greater distance from the elevator would be less convenient; they would require greater effort in moving baggage, groceries, and so on and would be farther away from the game room, the office, and the swimming pool. These units also possess an advantage: there would be the least amount of traffic through the hallways in the area and hence they are the most private.
3. Lower-floor oceanside units are most suited to active people; they open onto the beach, ocean, and pool. They are within easy reach of the game room, and they are easily reached from the parking area.
4. Checking the layout of the condominium complex, you discover that some of the units in the center of the complex, units ending in numbers 11 and 14, have part of their view blocked.

5. The condominium complex was completed at the time of the 1975 recession; sales were slow, and the developer was forced to sell most of the units at auction approximately 18 months after opening. Consequently, the auction data are completely buyer specified and hence consumer oriented in contrast to most other real estate sales data that are, to a high degree, seller and broker specified.
6. Many unsold units in the complex were furnished by the developer and rented prior to the auction. Consequently, some of the units bid on and sold at auction had furniture, others did not.

This condominium complex is obviously unique. For example, the single elevator located at one end of the complex produces a remarkably high level of both inconvenience and privacy for the people occupying units on the top floors in building 2. Consequently, the developer is unsure of how the height of the unit (floor number), distance of the unit from the elevator, presence or absence of an ocean view, etc., affect the prices of the units sold at auction. To investigate these relationships, the following data (saved in the **CONDO** data file) were recorded for each of the 106 units sold at the auction:

1. *Sale price.* Measured in hundreds of dollars (adjusted for inflation).
2. *Floor height.* The floor location of the unit; the variable levels are 1, 2, . . . , 8.
3. *Distance from elevator.* This distance, measured along the length of the complex, is expressed in number of condominium

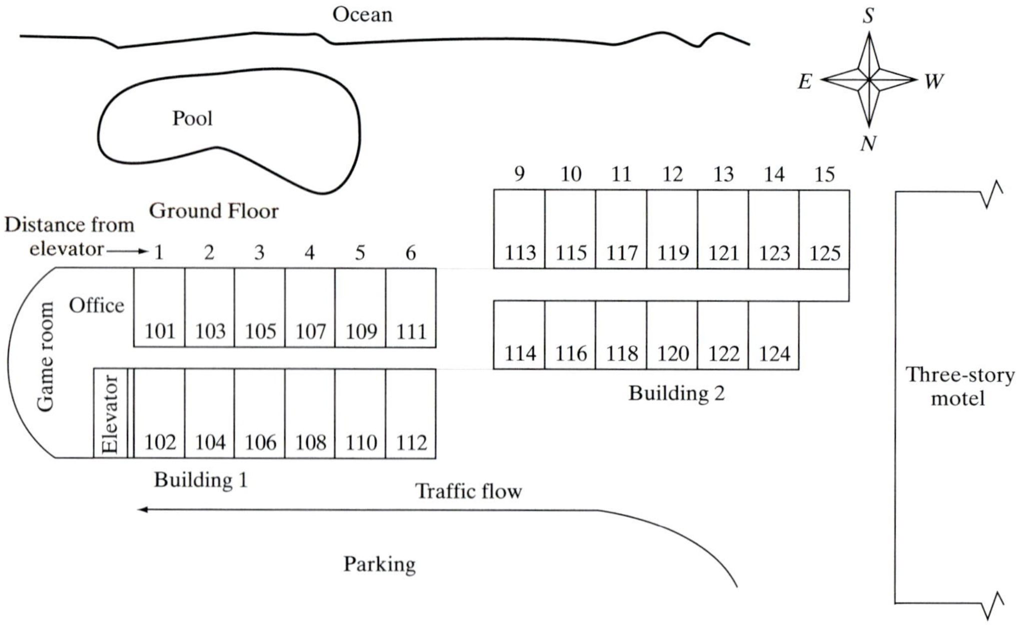

Figure C5.1
Layout of condominium complex

units. An additional two units of distance was added to the units in building 2 to account for the walking distance in the connecting area between the two buildings. Thus, the distance of unit 105 from the elevator would be 3, and the distance between unit 113 and the elevator would be 9. The variable levels are 1, 2, . . . , 15.

4. *View of ocean.* The presence or absence of an ocean view is recorded for each unit and specified with a dummy variable (1 if the unit possessed an ocean view and 0 if not). Note that units not possessing an ocean view would face the parking lot.
5. *End unit.* We expect the partial reduction of view of end units on the ocean side (numbers ending in 11) to reduce their sale price. The ocean view of these end units is partially blocked by building 2. This qualitative variable is also specified with a dummy variable (1 if the unit has a unit number ending in 11 and 0 if not).
6. *Furniture.* The presence or absence of furniture is recorded for each unit and is represented with a single dummy variable (1 if the unit was furnished and 0 if not).

Your objective for this case is to build a regression model that accurately predicts the sale price of a condominium unit sold at auction. Prepare a professional document that presents the results of your analysis. Include graphs that demonstrate how each of the independent variables in your model affects auction price. A layout of the **CONDO** data file is described below.

Variable	Type
PRICE	QN
FLOOR	QN
DISTANCE	QN
VIEW	QL
ENDUNIT	QL
FURNISH	QL

(Number of Observations: 106) *Data Set:* CONDO

Appendix A: Basic Counting Rules

Sample points associated with many experiments have identical characteristics. If you can develop a counting rule to count the number of sample points, it can be used to aid in the solution of many probability problems. For example, many experiments involve sampling n elements from a population of N. Then, as explained in Section 3.1, we can use the formula

$$\binom{N}{n} = \frac{N!}{n!(N-n)!}$$

to find the number of different samples of n elements that could be selected from the total of N elements. This gives the number of sample points for the experiment.

Here, we give you a few useful counting rules. You should learn the characteristics of the situation to which each rule applies. Then, when working a probability problem, carefully examine the experiment to see whether you can use one of the rules.

Learning how to decide whether a particular counting rule applies to an experiment takes patience and practice. If you want to develop this skill, try to use the rules to solve some of the exercises in Chapter 3. Proofs of the rules below can be found in the text by W. Feller listed in the references to Chapter 3.

Multiplicative Rule

You have k sets of different elements, n_1 in the first set, n_2 in the second set, ..., and n_k in the kth set. Suppose you want to form a sample of k elements by *taking one element from each of the* k *sets*. The number of different samples that can be formed is the product

$$n_1 \cdot n_2 \cdot n_3 \cdot \cdots \cdot n_k$$

Example A.1
Applying the Multiplicative Rule

Problem A product can be shipped by four airlines, and each airline can ship via three different routes. How many distinct ways exist to ship the product?

Solution A method of shipment corresponds to a pairing of one airline and one route. Therefore, $k = 2$, the number of airlines is $n_1 = 4$, the number of routes is $n_2 = 3$, and the number of ways to ship the product is

$$n_1 \cdot n_2 = (4)(3) = 12$$

Look Back How the multiplicative rule works can be seen by using a tree diagram, introduced in Section 3.6. The airline choice is shown by three branching lines in Figure A.1.

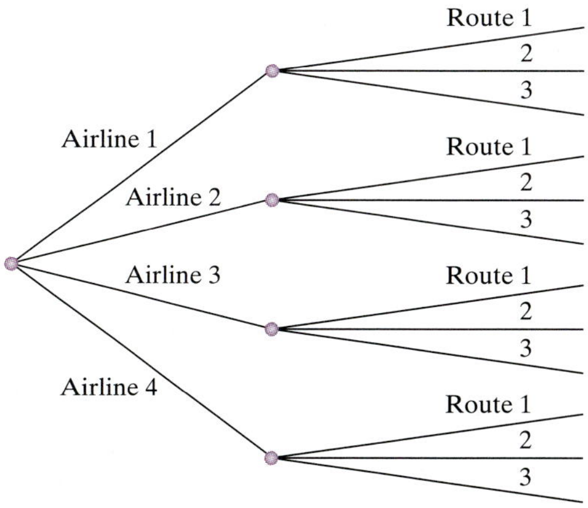

Figure A.1 Tree diagram for airline example

Example A.2 Applying the Multiplicative Rule

Problem You have 20 candidates for three different executive positions, E_1, E_2, and E_3. How many different ways could you fill the positions?

Solution For this example, there are $k = 3$ sets of elements:

Set 1: The candidates available to fill position E_1
Set 2: The candidates remaining (after filling E_1) that are available to fill E_2
Set 3: The candidates remaining (after filling E_1 and E_2) that are available to fill E_3

The numbers of elements in the sets are $n_1 = 20$, $n_2 = 19$, and $n_3 = 18$. Thus, the number of different ways to fill the three positions is

$$n_1 \cdot n_2 \cdot n_3 = (20)(19)(18) = 6{,}480$$

Partitions Rule

You have a *single* set of N distinctly different elements, and you want to partition it into k sets, the first set containing n_1 elements, the second containing n_2 elements, . . . , and the kth containing n_k elements. The number of different partitions is

$$\frac{N!}{n_1!n_2! \cdot \cdots \cdot n_k!} \text{ where } n_1 + n_2 + n_3 + \cdots + n_k = N$$

Example A.3 Applying the Partitions Rule

Problem You have 12 construction workers available for 3 job sites. Suppose you want to assign 3 workers to job 1, 4 to job 2, and 5 to job 3. How many different ways could you make this assignment?

Solution For this example, $k = 3$ (corresponding to the $k = 3$ job sites), $N = 12$, and $n_1 = 3$, $n_2 = 4$, $n_3 = 5$. Then the number of different ways to assign the workers to the job sites is

$$\frac{N!}{n_1!n_2!n_3!} = \frac{12!}{3!4!5!} = \frac{12 \cdot 11 \cdot 10 \cdot \cdots \cdot 3 \cdot 2 \cdot 1}{(3 \cdot 2 \cdot 1)(4 \cdot 3 \cdot 2 \cdot 1)(5 \cdot 4 \cdot 3 \cdot 2 \cdot 1)} = 27{,}720$$

Combinations Rule

The combinations rule given in Chapter 3 is a special case ($k = 2$) of the partitions rule—that is, sampling is equivalent to partitioning a set of N elements into $k = 2$ groups: elements that appear in the sample and those that do not. Let $n_1 = n$, the number of elements in the sample, and $n_2 = N - n$, the number of elements remaining. Then the number of different samples of n elements that can be selected from N is

$$\frac{N!}{n_1!n_2!} = \frac{N!}{n!(N-n)!} = \binom{N}{n}$$

This formula was given in Section 3.1.

Example A.4 Applying the Combinations Rule

Problem How many samples of 4 firefighters can be selected from a group of 10?

Solution We have $N = 10$ and $n = 4$; then

$$\binom{N}{n} = \binom{10}{4} = \frac{10!}{4!6!} = \frac{10 \cdot 9 \cdot 8 \cdot \cdots \cdot 3 \cdot 2 \cdot 1}{(4 \cdot 3 \cdot 2 \cdot 1)(6 \cdot 5 \cdot \cdots \cdot 2 \cdot 1)} = 210$$

Appendix B: Tables

Table I Random Numbers

Row \ Column	1	2	3	4	5	6	7	8	9	10	11	12	13	14
1	10480	15011	01536	02011	81647	91646	69179	14194	62590	36207	20969	99570	91291	90700
2	22368	46573	25595	85393	30995	89198	27982	53402	93965	34095	52666	19174	39615	99505
3	24130	48360	22527	97265	76393	64809	15179	24830	49340	32081	30680	19655	63348	58629
4	42167	93093	06243	61680	07856	16376	39440	53537	71341	57004	00849	74917	97758	16379
5	37570	39975	81837	16656	06121	91782	60468	81305	49684	60672	14110	06927	01263	54613
6	77921	06907	11008	42751	27756	53498	18602	70659	90655	15053	21916	81825	44394	42880
7	99562	72905	56420	69994	98872	31016	71194	18738	44013	48840	63213	21069	10634	12952
8	96301	91977	05463	07972	18876	20922	94595	56869	69014	60045	18425	84903	42508	32307
9	89579	14342	63661	10281	17453	18103	57740	84378	25331	12566	58678	44947	05585	56941
10	85475	36857	53342	53988	53060	59533	38867	62300	08158	17983	16439	11458	18593	64952
11	28918	69578	88231	33276	70997	79936	56865	05859	90106	31595	01547	85590	91610	78188
12	63553	40961	48235	03427	49626	69445	18663	72695	52180	20847	12234	90511	33703	90322
13	09429	93969	52636	92737	88974	33488	36320	17617	30015	08272	84115	27156	30613	74952
14	10365	61129	87529	85689	48237	52267	67689	93394	01511	26358	85104	20285	29975	89868
15	07119	97336	71048	08178	77233	13916	47564	81056	97735	85977	29372	74461	28551	90707
16	51085	12765	51821	51259	77452	16308	60756	92144	49442	53900	70960	63990	75601	40719
17	02368	21382	52404	60268	89368	19885	55322	44819	01188	65255	64835	44919	05944	55157
18	01011	54092	33362	94904	31273	04146	18594	29852	71585	85030	51132	01915	92747	64951
19	52162	53916	46369	58586	23216	14513	83149	98736	23495	64350	94738	17752	35156	35749
20	07056	97628	33787	09998	42698	06691	76988	13602	51851	46104	88916	19509	25625	58104
21	48663	91245	85828	14346	09172	30168	90229	04734	59193	22178	30421	61666	99904	32812
22	54164	58492	22421	74103	47070	25306	76468	26384	58151	06646	21524	15227	96909	44592
23	32639	32363	05597	24200	13363	38005	94342	28728	35806	06912	17012	64161	18296	22851
24	29334	27001	87637	87308	58731	00256	45834	15398	46557	41135	10367	07684	36188	18510
25	02488	33062	28834	07351	19731	92420	60952	61280	50001	67658	32586	86679	50720	94953
26	81525	72295	04839	96423	24878	82651	66566	14778	76797	14780	13300	87074	79666	95725
27	29676	20591	68086	26432	46901	20849	89768	81536	86645	12659	92259	57102	80428	25280
28	00742	57392	39064	66432	84673	40027	32832	61362	98947	96067	64760	64584	96096	98253
29	05366	04213	25669	26422	44407	44048	37937	63904	45766	66134	75470	66520	34693	90449
30	91921	26418	64117	94305	26766	25940	39972	22209	71500	64568	91402	42416	07844	69618
31	00582	04711	87917	77341	42206	35126	74087	99547	81817	42607	43808	76655	62028	76630
32	00725	69884	62797	56170	86324	88072	76222	36086	84637	93161	76038	65855	77919	88006
33	69011	65795	95876	55293	18988	27354	26575	08625	40801	59920	29841	80150	12777	48501
34	25976	57948	29888	88604	67917	48708	18912	82271	65424	69774	33611	54262	85963	03547
35	09763	83473	73577	12908	30883	18317	28290	35797	05998	41688	34952	37888	38917	88050
36	91576	42595	27958	30134	04024	86385	29880	99730	55536	84855	29080	09250	79656	73211
37	17955	56349	90999	49127	20044	59931	06115	20542	18059	02008	73708	83517	36103	42791
38	46503	18584	18845	49618	02304	51038	20655	58727	28168	15475	56942	53389	20562	87338
39	92157	89634	94824	78171	84610	82834	09922	25417	44137	48413	25555	21246	35509	20468
40	14577	62765	35605	81263	39667	47358	56873	56307	61607	49518	89656	20103	77490	18062
41	98427	07523	33362	64270	01638	92477	66969	98420	04880	45585	46565	04102	46880	45709

(continued)

Table I ***(continued)***

Row \ Column	1	2	3	4	5	6	7	8	9	10	11	12	13	14
42	34914	63976	88720	82765	34476	17032	87589	40836	32427	70002	70663	88863	77775	69348
43	70060	28277	39475	46473	23219	53416	94970	25832	69975	94884	19661	72828	00102	66794
44	53976	54914	06990	67245	68350	82948	11398	42878	80287	88267	47363	46634	06541	97809
45	76072	29515	40980	07391	58745	25774	22987	80059	39911	96189	41151	14222	60697	59583
46	90725	52210	83974	29992	65831	38857	50490	83765	55657	14361	31720	57375	56228	41546
47	64364	67412	33339	31926	14883	24413	59744	92351	97473	89286	35931	04110	23726	51900
48	08962	00358	31662	25388	61642	34072	81249	35648	56891	69352	48373	45578	78547	81788
49	95012	68379	93526	70765	10592	04542	76463	54328	02349	17247	28865	14777	62730	92277
50	15664	10493	20492	38391	91132	21999	59516	81652	27195	48223	46751	22923	32261	85653
51	16408	81899	04153	53381	79401	21438	83035	92350	36693	31238	59649	91754	72772	02338
52	18629	81953	05520	91962	04739	13092	97662	24822	94730	06496	35090	04822	86774	98289
53	73115	35101	47498	87637	99016	71060	88824	71013	18735	20286	23153	72924	35165	43040
54	57491	16703	23167	49323	45021	33132	12544	41035	80780	45393	44812	12512	98931	91202
55	30405	83946	23792	14422	15059	45799	22716	19792	09983	74353	68668	30429	70735	25499
56	16631	35006	85900	98275	32388	52390	16815	69290	82732	38480	73817	32523	41961	44437
57	96773	20206	42559	78985	05300	22164	24369	54224	35083	19687	11052	91491	60383	19746
58	38935	64202	14349	82674	66523	44133	00697	35552	35970	19124	63318	29686	03387	59846
59	31624	76384	17403	53363	44167	64486	64758	75366	76554	31601	12614	33072	60332	92325
60	78919	19474	23632	27889	47914	02584	37680	20801	72152	39339	34806	08930	85001	87820
61	03931	33309	57047	74211	63445	17361	62825	39908	05607	91284	68833	25570	38818	46920
62	74426	33278	43972	10110	89917	15665	52872	73823	73144	88662	88970	74492	51805	99378
63	09066	00903	20795	95452	92648	45454	09552	88815	16553	51125	79375	97596	16296	66092
64	42238	12426	87025	14267	20979	04508	64535	31355	86064	29472	47689	05974	52468	16834
65	16153	08002	26504	41744	81959	65642	74240	56302	00033	67107	77510	70625	28725	34191
66	21457	40742	29820	96783	29400	21840	15035	34537	33310	06116	95240	15957	16572	06004
67	21581	57802	02050	89728	17937	37621	47075	42080	97403	48626	68995	43805	33386	21597
68	55612	78095	83197	33732	05810	24813	86902	60397	16489	03264	88525	42786	05269	92532
69	44657	66999	99324	51281	84463	60563	79312	93454	68876	25471	93911	25650	12682	73572
70	91340	84979	46949	81973	37949	61023	43997	15263	80644	43942	89203	71795	99533	50501
71	91227	21199	31935	27022	84067	05462	35216	14486	29891	68607	41867	14951	91696	85065
72	50001	38140	66321	19924	72163	09538	12151	06878	91903	18749	34405	56087	82790	70925
73	65390	05224	72958	28609	81406	39147	25549	48542	42627	45233	57202	94617	23772	07896
74	27504	96131	83944	41575	10573	08619	64482	73923	36152	05184	94142	25299	84387	34925
75	37169	94851	39117	89632	00959	16487	65536	49071	39782	17095	02330	74301	00275	48280
76	11508	70225	51111	38351	19444	66499	71945	05422	13442	78675	84081	66938	93654	59894
77	37449	30362	06694	54690	04052	53115	62757	95348	78662	11163	81651	50245	34971	52924
78	46515	70331	85922	38329	57015	15765	97161	17869	45349	61796	66345	81073	49106	79860
79	30986	81223	42416	58353	21532	30502	32305	86482	05174	07901	54339	58861	74818	46942
80	63798	64995	46583	09785	44160	78128	83991	42865	92520	83531	80377	35909	81250	54238
81	82486	84846	99254	67632	43218	50076	21361	64816	51202	88124	41870	52689	51275	83556
82	21885	32906	92431	09060	64297	51674	64126	62570	26123	05155	59194	52799	28225	85762

(continued)

Table I ***(continued)***

Row \ Column	1	2	3	4	5	6	7	8	9	10	11	12	13	14
83	60336	98782	07408	53458	13564	59089	26445	29789	85205	41001	12535	12133	14645	23541
84	43937	46891	24010	25560	86355	33941	25786	54990	71899	15475	95434	98227	21824	19585
85	97656	63175	89303	16275	07100	92063	21942	18611	47348	20203	18534	03862	78095	50136
86	03299	01221	05418	38982	55758	92237	26759	86367	21216	98442	08303	56613	91511	75928
87	79626	06486	03574	17668	07785	76020	79924	25651	83325	88428	85076	72811	22717	50585
88	85636	68335	47539	03129	65651	11977	02510	26113	99447	68645	34327	15152	55230	93448
89	18039	14367	64337	06177	12143	46609	32989	74014	64708	00533	35398	58408	13261	47908
90	08362	15656	60627	36478	65648	16764	53412	09013	07832	41574	17639	82163	60859	75567
91	79556	29068	04142	16268	15387	12856	66227	38358	22478	73373	88732	09443	82558	05250
92	92608	82674	27072	32534	17075	27698	98204	63863	11951	34648	88022	56148	34925	57031
93	23982	25835	40055	67006	12293	02753	14827	23235	35071	99704	37543	11601	35503	85171
94	09915	96306	05908	97901	28395	14186	00821	80703	70426	75647	76310	88717	37890	40129
95	59037	33300	26695	62247	69927	76123	50842	43834	86654	70959	79725	93872	28117	19233
96	42488	78077	69882	61657	34136	79180	97526	43092	04098	73571	80799	76536	71255	64239
97	46764	86273	63003	93017	31204	36692	40202	35275	57306	55543	53203	18098	47625	88684
98	03237	45430	55417	63282	90816	17349	88298	90183	36600	78406	06216	95787	42579	90730
99	86591	81482	52667	61582	14972	90053	89534	76036	49199	43716	97548	04379	46370	28672
100	38534	01715	94964	87288	65680	43772	39560	12918	86537	62738	19636	51132	25739	56947

Source: Abridged from W. H. Beyer (ed.). *CRC Standard Mathematical Tables* 24th edition (Cleveland: The Chemical Rubber Company), 1976. Reproduced by permission of the publisher.

Table II Binomial Probabilities

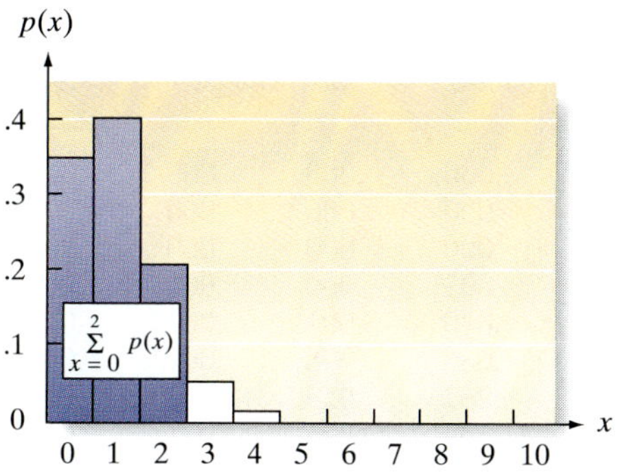

Tabulated values are $\sum_{x=0}^{k} p(x)$. *(Computations are rounded at the third decimal place.)*

a. $n = 5$

k \ p	.01	.05	.10	.20	.30	.40	.50	.60	.70	.80	.90	.95	.99
0	.951	.774	.590	.328	.168	.078	.031	.010	.002	.000	.000	.000	.000
1	.999	.977	.919	.737	.528	.337	.188	.087	.031	.007	.000	.000	.000
2	1.000	.999	.991	.942	.837	.683	.500	.317	.163	.058	.009	.001	.000
3	1.000	1.000	1.000	.993	.969	.913	.812	.663	.472	.263	.081	.023	.001
4	1.000	1.000	1.000	1.000	.998	.990	.969	.922	.832	.672	.410	.226	.049

b. $n = 6$

k \ p	.01	.05	.10	.20	.30	.40	.50	.60	.70	.80	.90	.95	.99
0	.941	.735	.531	.262	.118	.047	.016	.004	.001	.000	.000	.000	.000
1	.999	.967	.886	.655	.420	.233	.109	.041	.011	.002	.000	.000	.000
2	1.000	.998	.984	.901	.744	.544	.344	.179	.070	.017	.001	.000	.000
3	1.000	1.000	.999	.983	.930	.821	.656	.456	.256	.099	.016	.002	.000
4	1.000	1.000	1.000	.998	.989	.959	.891	.767	.580	.345	.114	.033	.001
5	1.000	1.000	1.000	1.000	.999	.996	.984	.953	.882	.738	.469	.265	.059

c. $n = 7$

k \ p	.01	.05	.10	.20	.30	.40	.50	.60	.70	.80	.90	.95	.99
0	.932	.698	.478	.210	.082	.028	.008	.002	.000	.000	.000	.000	.000
1	.998	.956	.850	.577	.329	.159	.063	.019	.004	.000	.000	.000	.000
2	1.000	.996	.974	.852	.647	.420	.227	.096	.029	.005	.000	.000	.000
3	1.000	1.000	.997	.967	.874	.710	.500	.290	.126	.033	.003	.000	.000
4	1.000	1.000	1.000	.995	.971	.904	.773	.580	.353	.148	.026	.004	.000
5	1.000	1.000	1.000	1.000	.996	.981	.937	.841	.671	.423	.150	.044	.002
6	1.000	1.000	1.000	1.000	1.000	.998	.992	.972	.918	.790	.522	.302	.068

d. $n = 8$

k \ p	.01	.05	.10	.20	.30	.40	.50	.60	.70	.80	.90	.95	.99
0	.923	.663	.430	.168	.058	.017	.004	.001	.000	.000	.000	.000	.000
1	.997	.943	.813	.503	.255	.106	.035	.009	.001	.000	.000	.000	.000
2	1.000	.994	.962	.797	.552	.315	.145	.050	.011	.001	.000	.000	.000
3	1.000	1.000	.995	.944	.806	.594	.363	.174	.058	.010	.000	.000	.000
4	1.000	1.000	1.000	.990	.942	.826	.637	.406	.194	.056	.005	.000	.000
5	1.000	1.000	1.000	.999	.989	.950	.855	.685	.448	.203	.038	.006	.000
6	1.000	1.000	1.000	1.000	.999	.991	.965	.894	.745	.497	.187	.057	.003
7	1.000	1.000	1.000	1.000	1.000	.999	.996	.983	.942	.832	.570	.337	.077

(continued)

Table II (continued)

e. $n = 9$

k \ p	.01	.05	.10	.20	.30	.40	.50	.60	.70	.80	.90	.95	.99
0	.914	.630	.387	.134	.040	.010	.002	.000	.000	.000	.000	.000	.000
1	.997	.929	.775	.436	.196	.071	.020	.004	.000	.000	.000	.000	.000
2	1.000	.992	.947	.738	.463	.232	.090	.025	.004	.000	.000	.000	.000
3	1.000	.999	.992	.914	.730	.483	.254	.099	.025	.003	.000	.000	.000
4	1.000	1.000	.999	.980	.901	.733	.500	.267	.099	.020	.001	.000	.000
5	1.000	1.000	1.000	.997	.975	.901	.746	.517	.270	.086	.008	.001	.000
6	1.000	1.000	1.000	1.000	.996	.975	.910	.768	.537	.262	.053	.008	.000
7	1.000	1.000	1.000	1.000	1.000	.996	.980	.929	.804	.564	.225	.071	.003
8	1.000	1.000	1.000	1.000	1.000	1.000	.998	.990	.960	.866	.613	.370	.086

f. $n = 10$

k \ p	.01	.05	.10	.20	.30	.40	.50	.60	.70	.80	.90	.95	.99
0	.904	.599	.349	.107	.028	.006	.001	.000	.000	.000	.000	.000	.000
1	.996	.914	.736	.376	.149	.046	.011	.002	.000	.000	.000	.000	.000
2	1.000	.988	.930	.678	.383	.167	.055	.012	.002	.000	.000	.000	.000
3	1.000	.999	.987	.879	.650	.382	.172	.055	.011	.001	.000	.000	.000
4	1.000	1.000	.998	.967	.850	.633	.377	.166	.047	.006	.000	.000	.000
5	1.000	1.000	1.000	.994	.953	.834	.623	.367	.150	.033	.002	.000	.000
6	1.000	1.000	1.000	.999	.989	.945	.828	.618	.350	.121	.013	.001	.000
7	1.000	1.000	1.000	1.000	.998	.988	.945	.833	.617	.322	.070	.012	.000
8	1.000	1.000	1.000	1.000	1.000	.998	.989	.954	.851	.624	.264	.086	.004
9	1.000	1.000	1.000	1.000	1.000	1.000	.999	.994	.972	.893	.651	.401	.096

g. $n = 15$

k \ p	.01	.05	.10	.20	.30	.40	.50	.60	.70	.80	.90	.95	.99
0	.860	.463	.206	.035	.005	.000	.000	.000	.000	.000	.000	.000	.000
1	.990	.829	.549	.167	.035	.005	.000	.000	.000	.000	.000	.000	.000
2	1.000	.964	.816	.398	.127	.027	.004	.000	.000	.000	.000	.000	.000
3	1.000	.995	.944	.648	.297	.091	.018	.002	.000	.000	.000	.000	.000
4	1.000	.999	.987	.838	.515	.217	.059	.009	.001	.000	.000	.000	.000
5	1.000	1.000	.998	.939	.722	.403	.151	.034	.004	.000	.000	.000	.000
6	1.000	1.000	1.000	.982	.869	.610	.304	.095	.015	.001	.000	.000	.000
7	1.000	1.000	1.000	.996	.950	.787	.500	.213	.050	.004	.000	.000	.000
8	1.000	1.000	1.000	.999	.985	.905	.696	.390	.131	.018	.000	.000	.000
9	1.000	1.000	1.000	1.000	.996	.966	.849	.597	.278	.061	.002	.000	.000
10	1.000	1.000	1.000	1.000	.999	.991	.941	.783	.485	.164	.013	.001	.000
11	1.000	1.000	1.000	1.000	1.000	.998	.982	.909	.703	.352	.056	.005	.000
12	1.000	1.000	1.000	1.000	1.000	1.000	.996	.973	.873	.602	.184	.036	.000
13	1.000	1.000	1.000	1.000	1.000	1.000	1.000	.995	.965	.833	.451	.171	.010
14	1.000	1.000	1.000	1.000	1.000	1.000	1.000	1.000	.995	.965	.794	.537	.140

(continued)

Table II *(continued)*

h. $n = 20$

k \ p	.01	.05	.10	.20	.30	.40	.50	.60	.70	.80	.90	.95	.99
0	.818	.358	.122	.012	.001	.000	.000	.000	.000	.000	.000	.000	.000
1	.983	.736	.392	.069	.008	.001	.000	.000	.000	.000	.000	.000	.000
2	.999	.925	.677	.206	.035	.004	.000	.000	.000	.000	.000	.000	.000
3	1.000	.984	.867	.411	.107	.016	.001	.000	.000	.000	.000	.000	.000
4	1.000	.997	.957	.630	.238	.051	.006	.000	.000	.000	.000	.000	.000
5	1.000	1.000	.989	.804	.416	.126	.021	.002	.000	.000	.000	.000	.000
6	1.000	1.000	.998	.913	.608	.250	.058	.006	.000	.000	.000	.000	.000
7	1.000	1.000	1.000	.968	.772	.416	.132	.021	.001	.000	.000	.000	.000
8	1.000	1.000	1.000	.990	.887	.596	.252	.057	.005	.000	.000	.000	.000
9	1.000	1.000	1.000	.997	.952	.755	.412	.128	.017	.001	.000	.000	.000
10	1.000	1.000	1.000	.999	.983	.872	.588	.245	.048	.003	.000	.000	.000
11	1.000	1.000	1.000	1.000	.995	.943	.748	.404	.113	.010	.000	.000	.000
12	1.000	1.000	1.000	1.000	.999	.979	.868	.584	.228	.032	.000	.000	.000
13	1.000	1.000	1.000	1.000	1.000	.994	.942	.750	.392	.087	.002	.000	.000
14	1.000	1.000	1.000	1.000	1.000	.998	.979	.874	.584	.196	.011	.000	.000
15	1.000	1.000	1.000	1.000	1.000	1.000	.994	.949	.762	.370	.043	.003	.000
16	1.000	1.000	1.000	1.000	1.000	1.000	.999	.984	.893	.589	.133	.016	.000
17	1.000	1.000	1.000	1.000	1.000	1.000	1.000	.996	.965	.794	.323	.075	.001
18	1.000	1.000	1.000	1.000	1.000	1.000	1.000	.999	.992	.931	.608	.264	.017
19	1.000	1.000	1.000	1.000	1.000	1.000	1.000	1.000	.999	.988	.878	.642	.182

i. $n = 25$

k \ p	.01	.05	.10	.20	.30	.40	.50	.60	.70	.80	.90	.95	.99
0	.778	.277	.072	.004	.000	.000	.000	.000	.000	.000	.000	.000	.000
1	.974	.642	.271	.027	.002	.000	.000	.000	.000	.000	.000	.000	.000
2	.998	.873	.537	.098	.009	.000	.000	.000	.000	.000	.000	.000	.000
3	1.000	.966	.764	.234	.033	.002	.000	.000	.000	.000	.000	.000	.000
4	1.000	.993	.902	.421	.090	.009	.000	.000	.000	.000	.000	.000	.000
5	1.000	.999	.967	.617	.193	.029	.002	.000	.000	.000	.000	.000	.000
6	1.000	1.000	.991	.780	.341	.074	.007	.000	.000	.000	.000	.000	.000
7	1.000	1.000	.998	.891	.512	.154	.022	.001	.000	.000	.000	.000	.000
8	1.000	1.000	1.000	.953	.677	.274	.054	.004	.000	.000	.000	.000	.000
9	1.000	1.000	1.000	.983	.811	.425	.115	.013	.000	.000	.000	.000	.000
10	1.000	1.000	1.000	.994	.902	.586	.212	.034	.002	.000	.000	.000	.000
11	1.000	1.000	1.000	.998	.956	.732	.345	.078	.006	.000	.000	.000	.000
12	1.000	1.000	1.000	1.000	.983	.846	.500	.154	.017	.000	.000	.000	.000
13	1.000	1.000	1.000	1.000	.994	.922	.655	.268	.044	.002	.000	.000	.000
14	1.000	1.000	1.000	1.000	.998	.966	.788	.414	.098	.006	.000	.000	.000
15	1.000	1.000	1.000	1.000	1.000	.987	.885	.575	.189	.017	.000	.000	.000
16	1.000	1.000	1.000	1.000	1.000	.996	.946	.726	.323	.047	.000	.000	.000
17	1.000	1.000	1.000	1.000	1.000	.999	.978	.846	.488	.109	.002	.000	.000
18	1.000	1.000	1.000	1.000	1.000	1.000	.993	.926	.659	.220	.009	.000	.000
19	1.000	1.000	1.000	1.000	1.000	1.000	.998	.971	.807	.383	.033	.001	.000
20	1.000	1.000	1.000	1.000	1.000	1.000	1.000	.991	.910	.579	.098	.007	.000
21	1.000	1.000	1.000	1.000	1.000	1.000	1.000	.998	.967	.766	.236	.034	.000
22	1.000	1.000	1.000	1.000	1.000	1.000	1.000	1.000	.991	.902	.463	.127	.002
23	1.000	1.000	1.000	1.000	1.000	1.000	1.000	1.000	.998	.973	.729	.358	.026
24	1.000	1.000	1.000	1.000	1.000	1.000	1.000	1.000	1.000	.996	.928	.723	.222

Table III Poisson Probabilities

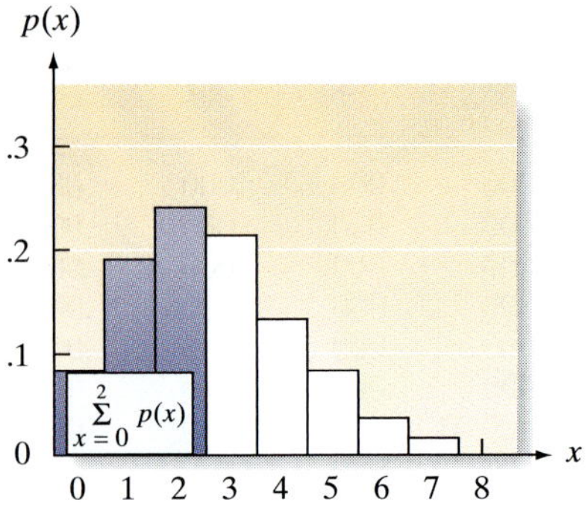

Tabulated values are $\sum_{x=0}^{k} p(x)$. *(Computations are rounded at the third decimal place.)*

λ \ k	0	1	2	3	4	5	6	7	8	9
.02	.980	1.000								
.04	.961	.999	1.000							
.06	.942	.998	1.000							
.08	.923	.997	1.000							
.10	.905	.995	1.000							
.15	.861	.990	.999	1.000						
.20	.819	.982	.999	1.000						
.25	.779	.974	.998	1.000						
.30	.741	.963	.996	1.000						
.35	.705	.951	.994	1.000						
.40	.670	.938	.992	.999	1.000					
.45	.638	.925	.989	.999	1.000					
.50	.607	.910	.986	.998	1.000					
.55	.577	.894	.982	.998	1.000					
.60	.549	.878	.977	.997	1.000					
.65	.522	.861	.972	.996	.999	1.000				
.70	.497	.844	.966	.994	.999	1.000				
.75	.472	.827	.959	.993	.999	1.000				
.80	.449	.809	.953	.991	.999	1.000				
.85	.427	.791	.945	.989	.998	1.000				
.90	.407	.772	.937	.987	.998	1.000				
.95	.387	.754	.929	.981	.997	1.000				
1.00	.368	.736	.920	.981	.996	.999	1.000			
1.1	.333	.699	.900	.974	.995	.999	1.000			
1.2	.301	.663	.879	.966	.992	.998	1.000			
1.3	.273	.627	.857	.957	.989	.998	1.000			
1.4	.247	.592	.833	.946	.986	.997	.999	1.000		
1.5	.223	.558	.809	.934	.981	.996	.999	1.000		
1.6	.202	.525	.783	.921	.976	.994	.999	1.000		
1.7	.183	.493	.757	.907	.970	.992	.998	1.000		
1.8	.165	.463	.731	.891	.964	.990	.997	.999	1.000	
1.9	.150	.434	.704	.875	.956	.987	.997	.999	1.000	
2.0	.135	.406	.677	.857	.947	.983	.995	.999	1.000	
2.2	.111	.355	.623	.819	.928	.975	.993	.998	1.000	
2.4	.091	.308	.570	.779	.904	.964	.988	.997	.999	1.000
2.6	.074	.267	.518	.736	.877	.951	.983	.995	.999	1.000
2.8	.061	.231	.469	.692	.848	.935	.976	.992	.998	.999
3.0	.050	.199	.423	.647	.815	.916	.966	.988	.996	.999
3.2	.041	.171	.380	.603	.781	.895	.955	.983	.994	.998
3.4	.033	.147	.340	.558	.744	.871	.942	.977	.992	.997
3.6	.027	.126	.303	.515	.706	.844	.927	.969	.988	.996
3.8	.022	.107	.269	.473	.668	.816	.909	.960	.984	.994

(continued)

Table III ***(continued)***

λ \ k	0	1	2	3	4	5	6	7	8	9
4.0	.018	.092	.238	.433	.629	.785	.889	.949	.979	.992
4.2	.015	.078	.210	.395	.590	.753	.867	.936	.972	.989
4.4	.012	.066	.185	.359	.551	.720	.844	.921	.964	.985
4.6	.010	.056	.163	.326	.513	.686	.818	.905	.955	.980
4.8	.008	.048	.143	.294	.476	.651	.791	.887	.944	.975
5.0	.007	.040	.125	.265	.440	.616	.762	.867	.932	.968
5.2	.006	.034	.109	.238	.406	.581	.732	.845	.918	.960
5.4	.005	.029	.095	.213	.373	.546	.702	.822	.903	.951
5.6	.004	.024	.082	.191	.342	.512	.670	.797	.886	.941
5.8	.003	.021	.072	.170	.313	.478	.638	.771	.867	.929
6.0	.002	.017	.062	.151	.285	.446	.606	.744	.847	.916

λ \ k	10	11	12	13	14	15	16
2.8	1.000						
3.0	1.000						
3.2	1.000						
3.4	.999	1.000					
3.6	.999	1.000					
3.8	.998	.999	1.000				
4.0	.997	.999	1.000				
4.2	.996	.999	1.000				
4.4	.994	.998	.999	1.000			
4.6	.992	.997	.999	1.000			
4.8	.990	.996	.999	1.000			
5.0	.986	.995	.998	.999	1.000		
5.2	.982	.993	.997	.999	1.000		
5.4	.977	.990	.996	.999	1.000		
5.6	.972	.988	.995	.998	.999	1.000	
5.8	.965	.984	.993	.997	.999	1.000	
6.0	.957	.980	.991	.996	.999	.999	1.000

λ \ k	0	1	2	3	4	5	6	7	8	9
6.2	.002	.015	.054	.134	.259	.414	.574	.716	.826	.902
6.4	.002	.012	.046	.119	.235	.384	.542	.687	.803	.886
6.6	.001	.010	.040	.105	.213	.355	.511	.658	.780	.869
6.8	.001	.009	.034	.093	.192	.327	.480	.628	.755	.850
7.0	.001	.007	.030	.082	.173	.301	.450	.599	.729	.830
7.2	.001	.006	.025	.072	.156	.276	.420	.569	.703	.810
7.4	.001	.005	.022	.063	.140	.253	.392	.539	.676	.788
7.6	.001	.004	.019	.055	.125	.231	.365	.510	.648	.765
7.8	.000	.004	.016	.048	.112	.210	.338	.481	.620	.741
8.0	.000	.003	.014	.042	.100	.191	.313	.453	.593	.717
8.5	.000	.002	.009	.030	.074	.150	.256	.386	.523	.653
9.0	.000	.001	.006	.021	.055	.116	.207	.324	.456	.587
9.5	.000	.001	.004	.015	.040	.089	.165	.269	.392	.522
10.0	.000	.000	.003	.010	.029	.067	.130	.220	.333	.458

λ \ k	10	11	12	13	14	15	16	17	18	19
6.2	.949	.975	.989	.995	.998	.999	1.000			
6.4	.939	.969	.986	.994	.997	.999	1.000			
6.6	.927	.963	.982	.992	.997	.999	.999	1.000		
6.8	.915	.955	.978	.990	.996	.998	.999	1.000		
7.0	.901	.947	.973	.987	.994	.998	.999	1.000		
7.2	.887	.937	.967	.984	.993	.997	.999	.999	1.000	
7.4	.871	.926	.961	.980	.991	.996	.998	.999	1.000	
7.6	.854	.915	.954	.976	.989	.995	.998	.999	1.000	
7.8	.835	.902	.945	.971	.986	.993	.997	.999	1.000	

(continued)

Table III *(continued)*

λ \ k	10	11	12	13	14	15	16	17	18	19
8.0	.816	.888	.936	.966	.983	.992	.996	.998	.999	1.000
8.5	.763	.849	.909	.949	.973	.986	.993	.997	.999	.999
9.0	.706	.803	.876	.926	.959	.978	.989	.995	.998	.999
9.5	.645	.752	.836	.898	.940	.967	.982	.991	.996	.998
10.0	.583	.697	.792	.864	.917	.951	.973	.986	.993	.997

λ \ k	20	21	22
8.5	1.000		
9.0	1.000		
9.5	.999	1.000	
10.0	.998	.999	1.000

λ \ k	0	1	2	3	4	5	6	7	8	9
10.5	.000	.000	.002	.007	.021	.050	.102	.179	.279	.397
11.0	.000	.000	.001	.005	.015	.038	.079	.143	.232	.341
11.5	.000	.000	.001	.003	.011	.028	.060	.114	.191	.289
12.0	.000	.000	.001	.002	.008	.020	.046	.090	.155	.242
12.5	.000	.000	.000	.002	.005	.015	.035	.070	.125	.201
13.0	.000	.000	.000	.001	.004	.011	.026	.054	.100	.166
13.5	.000	.000	.000	.001	.003	.008	.019	.041	.079	.135
14.0	.000	.000	.000	.000	.002	.006	.014	.032	.062	.109
14.5	.000	.000	.000	.000	.001	.004	.010	.024	.048	.088
15.0	.000	.000	.000	.000	.001	.003	.008	.018	.037	.070

λ \ k	10	11	12	13	14	15	16	17	18	19
10.5	.521	.639	.742	.825	.888	.932	.960	.978	.988	.994
11.0	.460	.579	.689	.781	.854	.907	.944	.968	.982	.991
11.5	.402	.520	.633	.733	.815	.878	.924	.954	.974	.986
12.0	.347	.462	.576	.682	.772	.844	.899	.937	.963	.979
12.5	.297	.406	.519	.628	.725	.806	.869	.916	.948	.969
13.0	.252	.353	.463	.573	.675	.764	.835	.890	.930	.957
13.5	.211	.304	.409	.518	.623	.718	.798	.861	.908	.942
14.0	.176	.260	.358	.464	.570	.669	.756	.827	.883	.923
14.5	.145	.220	.311	.413	.518	.619	.711	.790	.853	.901
15.0	.118	.185	.268	.363	.466	.568	.664	.749	.819	.875

λ \ k	20	21	22	23	24	25	26	27	28	29
10.5	.997	.999	.999	1.000						
11.0	.995	.998	.999	1.000						
11.5	.992	.996	.998	.999	1.000					
12.0	.988	.994	.987	.999	.999	1.000				
12.5	.983	.991	.995	.998	.999	.999	1.000			
13.0	.975	.986	.992	.996	.998	.999	1.000			
13.5	.965	.980	.989	.994	.997	.998	.999	1.000		
14.0	.952	.971	.983	.991	.995	.997	.999	.999	1.000	
14.5	.936	.960	.976	.986	.992	.996	.998	.999	.999	1.000
15.0	.917	.947	.967	.981	.989	.994	.997	.998	.999	1.000

(continued)

Table III ***(continued)***

λ \ k	4	5	6	7	8	9	10	11	12	13
16	.000	.001	.004	.010	.022	.043	.077	.127	.193	.275
17	.000	.001	.002	.005	.013	.026	.049	.085	.135	.201
18	.000	.000	.001	.003	.007	.015	.030	.055	.092	.143
19	.000	.000	.001	.002	.004	.009	.018	.035	.061	.098
20	.000	.000	.000	.001	.002	.005	.011	.021	.039	.066
21	.000	.000	.000	.000	.001	.003	.006	.013	.025	.043
22	.000	.000	.000	.000	.001	.002	.004	.008	.015	.028
23	.000	.000	.000	.000	.000	.001	.002	.004	.009	.017
24	.000	.000	.000	.000	.000	.000	.001	.003	.005	.011
25	.000	.000	.000	.000	.000	.000	.001	.001	.003	.006

λ \ k	14	15	16	17	18	19	20	21	22	23
16	.368	.467	.566	.659	.742	.812	.868	.911	.942	.963
17	.281	.371	.468	.564	.655	.736	.805	.861	.905	.937
18	.208	.287	.375	.469	.562	.651	.731	.799	.855	.899
19	.150	.215	.292	.378	.469	.561	.647	.725	.793	.849
20	.105	.157	.221	.297	.381	.470	.559	.644	.721	.787
21	.072	.111	.163	.227	.302	.384	.471	.558	.640	.716
22	.048	.077	.117	.169	.232	.306	.387	.472	.556	.637
23	.031	.052	.082	.123	.175	.238	.310	.389	.472	.555
24	.020	.034	.056	.087	.128	.180	.243	.314	.392	.473
25	.012	.022	.038	.060	.092	.134	.185	.247	.318	.394

λ \ k	24	25	26	27	28	29	30	31	32	33
16	.978	.987	.993	.996	.998	.999	.999	1.000		
17	.959	.975	.985	.991	.995	.997	.999	.999	1.000	
18	.932	.955	.972	.983	.990	.994	.997	.998	.999	1.000
19	.893	.927	.951	.969	.980	.988	.993	.996	.998	.999
20	.843	.888	.922	.948	.966	.978	.987	.992	.995	.997
21	.782	.838	.883	.917	.944	.963	.976	.985	.991	.994
22	.712	.777	.832	.877	.913	.940	.959	.973	.983	.989
23	.635	.708	.772	.827	.873	.908	.936	.956	.971	.981
24	.554	.632	.704	.768	.823	.868	.904	.932	.953	.969
25	.473	.553	.629	.700	.763	.818	.863	.900	.929	.950

λ \ k	34	35	36	37	38	39	40	41	42	43
19	.999	1.000								
20	.999	.999	1.000							
21	.997	.998	.999	.999	1.000					
22	.994	.996	.998	.999	.999	1.000				
23	.988	.993	.996	.997	.999	.999	1.000			
24	.979	.987	.992	.995	.997	.998	.999	.999	1.000	
25	.966	.978	.985	.991	.991	.997	.998	.999	.999	1.000

Table IV Normal Curve Areas

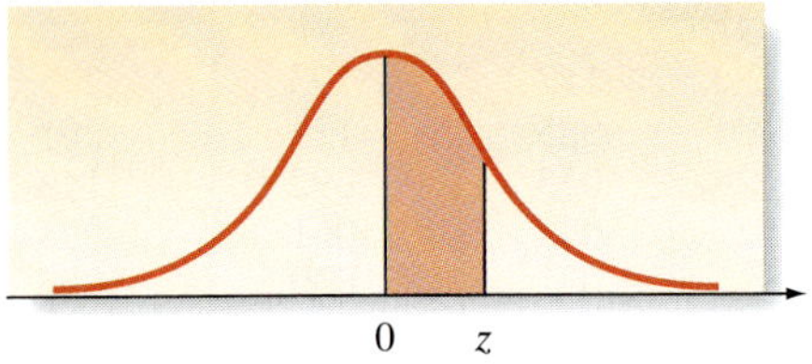

z	.00	.01	.02	.03	.04	.05	.06	.07	.08	.09
.0	.0000	.0040	.0080	.0120	.0160	.0199	.0239	.0279	.0319	.0359
.1	.0398	.0438	.0478	.0517	.0557	.0596	.0636	.0675	.0714	.0753
.2	.0793	.0832	.0871	.0910	.0948	.0987	.1026	.1064	.1103	.1141
.3	.1179	.1217	.1255	.1293	.1331	.1368	.1406	.1443	.1480	.1517
.4	.1554	.1591	.1628	.1664	.1700	.1736	.1772	.1808	.1844	.1879
.5	.1915	.1950	.1985	.2019	.2054	.2088	.2123	.2157	.2190	.2224
.6	.2257	.2291	.2324	.2357	.2389	.2422	.2454	.2486	.2517	.2549
.7	.2580	.2611	.2642	.2673	.2704	.2734	.2764	.2794	.2823	.2852
.8	.2881	.2910	.2939	.2967	.2995	.3023	.3051	.3078	.3106	.3133
.9	.3159	.3186	.3212	.3238	.3264	.3289	.3315	.3340	.3365	.3389
1.0	.3413	.3438	.3461	.3485	.3508	.3531	.3554	.3577	.3599	.3621
1.1	.3643	.3665	.3686	.3708	.3729	.3749	.3770	.3790	.3810	.3830
1.2	.3849	.3869	.3888	.3907	.3925	.3944	.3962	.3980	.3997	.4015
1.3	.4032	.4049	.4066	.4082	.4099	.4115	.4131	.4147	.4162	.4177
1.4	.4192	.4207	.4222	.4236	.4251	.4265	.4279	.4292	.4306	.4319
1.5	.4332	.4345	.4357	.4370	.4382	.4394	.4406	.4418	.4429	.4441
1.6	.4452	.4463	.4474	.4484	.4495	.4505	.4515	.4525	.4535	.4545
1.7	.4554	.4564	.4573	.4582	.4591	.4599	.4608	.4616	.4625	.4633
1.8	.4641	.4649	.4656	.4664	.4671	.4678	.4686	.4693	.4699	.4706
1.9	.4713	.4719	.4726	.4732	.4738	.4744	.4750	.4756	.4761	.4767
2.0	.4772	.4778	.4783	.4788	.4793	.4798	.4803	.4808	.4812	.4817
2.1	.4821	.4826	.4830	.4834	.4838	.4842	.4846	.4850	.4854	.4857
2.2	.4861	.4864	.4868	.4871	.4875	.4878	.4881	.4884	.4887	.4890
2.3	.4893	.4896	.4898	.4901	.4904	.4906	.4909	.4911	.4913	.4916
2.4	.4918	.4920	.4922	.4925	.4927	.4929	.4931	.4932	.4934	.4936
2.5	.4938	.4940	.4941	.4943	.4945	.4946	.4948	.4949	.4951	.4952
2.6	.4953	.4955	.4956	.4957	.4959	.4960	.4961	.4962	.4963	.4964
2.7	.4965	.4966	.4967	.4968	.4969	.4970	.4971	.4972	.4973	.4974
2.8	.4974	.4975	.4976	.4977	.4977	.4978	.4979	.4979	.4980	.4981
2.9	.4981	.4982	.4982	.4983	.4984	.4984	.4985	.4985	.4986	.4986
3.0	.4987	.4987	.4987	.4988	.4988	.4989	.4989	.4989	.4990	.4990
3.1	.49903	.49906	.49910	.49913	.49916	.49918	.49921	.49924	.49926	.48829
3.2	.49931	.49934	.49936	.49938	.49940	.49942	.49944	.49946	.49948	.49950
3.3	.49952	.49953	.49955	.49957	.49958	.49960	.49961	.49962	.49964	.49965
3.4	.49966	.49968	.49969	.49970	.49971	.49972	.49973	.49974	.49975	.49976
3.5	.49977	.49978	.49978	.49979	.49980	.49981	.49981	.49982	.49983	.49983
3.6	.49984	.49985	.49985	.49986	.49986	.49987	.49987	.49988	.49988	.49989
3.7	.49989	.49990	.49990	.49990	.49991	.49991	.49992	.49992	.49992	.49992
3.8	.49993	.49993	.49993	.49994	.49994	.49994	.49994	.49995	.49995	.49995
3.9	.49995	.49995	.49996	.49996	.49996	.49996	.49996	.49996	.49997	.49997

Source: Abridged from Table I of A. Hald. *Statistical Tables and Formulas* (New York: Wiley), 1952. Reproduced by permission of A. Hald.

Table V Critical Values of *t*

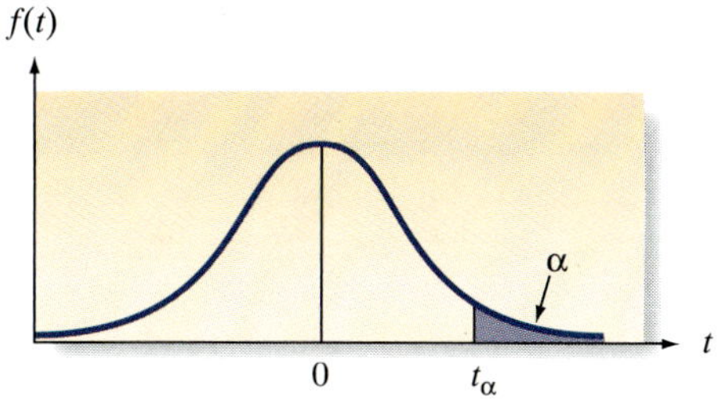

Degrees of Freedom	$t_{.100}$	$t_{.050}$	$t_{.025}$	$t_{.010}$	$t_{.005}$	$t_{.001}$	$t_{.0005}$
1	3.078	6.314	12.706	31.821	63.657	318.31	636.62
2	1.886	2.920	4.303	6.965	9.925	22.326	31.598
3	1.638	2.353	3.182	4.541	5.841	10.213	12.924
4	1.533	2.132	2.776	3.747	4.604	7.173	8.610
5	1.476	2.015	2.571	3.365	4.032	5.893	6.869
6	1.440	1.943	2.447	3.143	3.707	5.208	5.959
7	1.415	1.895	2.365	2.998	3.499	4.785	5.408
8	1.397	1.860	2.306	2.896	3.355	4.501	5.041
9	1.383	1.833	2.262	2.821	3.250	4.297	4.781
10	1.372	1.812	2.228	2.764	3.169	4.144	4.587
11	1.363	1.796	2.201	2.718	3.106	4.025	4.437
12	1.356	1.782	2.179	2.681	3.055	3.930	4.318
13	1.350	1.771	2.160	2.650	3.012	3.852	4.221
14	1.345	1.761	2.145	2.624	2.977	3.787	4.140
15	1.341	1.753	2.131	2.602	2.947	3.733	4.073
16	1.337	1.746	2.120	2.583	2.921	3.686	4.015
17	1.333	1.740	2.110	2.567	2.898	3.646	3.965
18	1.330	1.734	2.101	2.552	2.878	3.610	3.922
19	1.328	1.729	2.093	2.539	2.861	3.579	3.883
20	1.325	1.725	2.086	2.528	2.845	3.552	3.850
21	1.323	1.721	2.080	2.518	2.831	3.527	3.819
22	1.321	1.717	2.074	2.508	2.819	3.505	3.792
23	1.319	1.714	2.069	2.500	2.807	3.485	3.767
24	1.318	1.711	2.064	2.492	2.797	3.467	3.745
25	1.316	1.708	2.060	2.485	2.787	3.450	3.725
26	1.315	1.706	2.056	2.479	2.779	3.435	3.707
27	1.314	1.703	2.052	2.473	2.771	3.421	3.690
28	1.313	1.701	2.048	2.467	2.763	3.408	3.674
29	1.311	1.699	2.045	2.462	2.756	3.396	3.659
30	1.310	1.697	2.042	2.457	2.750	3.385	3.646
40	1.303	1.684	2.021	2.423	2.704	3.307	3.551
60	1.296	1.671	2.000	2.390	2.660	3.232	3.460
120	1.289	1.658	1.980	2.358	2.617	3.160	3.373
∞	1.282	1.645	1.960	2.326	2.576	3.090	3.291

Source: This table is reproduced with the kind permission of the Trustees of Biometrika from E. S. Pearson and H. O. Hartley (eds.). *The Biometrika Tables for Statisticians,* Vol. 1, 3rd ed., Biometrika, 1966.

Table VI Critical Values of χ^2

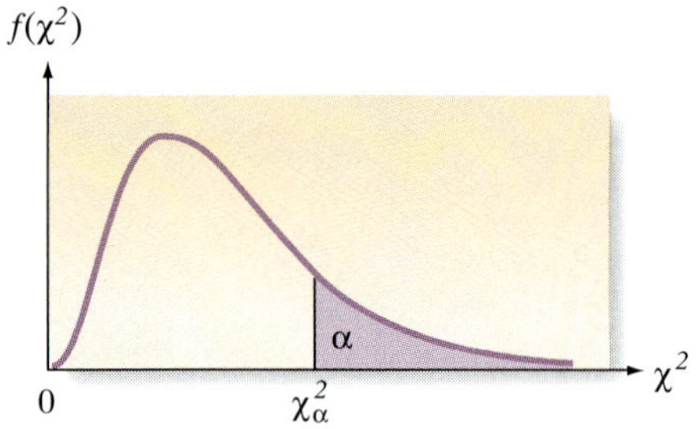

Degrees of Freedom	$\chi^2_{.995}$	$\chi^2_{.990}$	$\chi^2_{.975}$	$\chi^2_{.950}$	$\chi^2_{.900}$
1	.0000393	.0001571	.0009821	.0039321	.0157908
2	.0100251	.0201007	.0506356	.102587	.210720
3	.0717212	.114832	.215795	.351846	.584375
4	.206990	.297110	.484419	.710721	1.063623
5	.411740	.554300	.831211	1.145476	1.61031
6	.675727	.872085	1.237347	1.63539	2.20413
7	.989265	1.239043	1.68987	2.16735	2.83311
8	1.344419	1.646482	2.17973	2.73264	3.48954
9	1.734926	2.087912	2.70039	3.32511	4.16816
10	2.15585	2.55821	3.24697	3.94030	4.86518
11	2.60321	3.05347	3.81575	4.57481	5.57779
12	3.07382	3.57056	4.40379	5.22603	6.30380
13	3.56503	4.10691	5.00874	5.89186	7.04150
14	4.07468	4.66043	5.62872	6.57063	7.78953
15	4.60094	5.22935	6.26214	7.26094	8.54675
16	5.14224	5.81221	6.90766	7.96164	9.31223
17	5.69724	6.40776	7.56418	8.67176	10.0852
18	6.26481	7.01491	8.23075	9.39046	10.8649
19	6.84398	7.63273	8.90655	10.1170	11.6509
20	7.43386	8.26040	9.59083	10.8508	12.4426
21	8.03366	8.89720	10.28293	11.5913	13.2396
22	8.64272	9.54249	10.9823	12.3380	14.0415
23	9.26042	10.19567	11.6885	13.0905	14.8479
24	9.88623	10.8564	12.4011	13.8484	15.6587
25	10.5197	11.5240	13.1197	14.6114	16.4734
26	11.1603	12.1981	13.8439	15.3791	17.2919
27	11.8076	12.8786	14.5733	16.1513	18.1138
28	12.4613	13.5648	15.3079	16.9279	18.9392
29	13.1211	14.2565	16.0471	17.7083	19.7677
30	13.7867	14.9535	16.7908	18.4926	20.5992
40	20.7065	22.1643	24.4331	26.5093	29.0505
50	27.9907	29.7067	32.3574	34.7642	37.6886
60	35.5346	37.4848	40.4817	43.1879	46.4589
70	43.2752	45.4418	48.7576	51.7393	55.3290
80	51.1720	53.5400	57.1532	60.3915	64.2778
90	59.1963	61.7541	65.6466	69.1260	73.2912
100	67.3276	70.0648	74.2219	77.9295	82.3581

(continued)

Table VI ***(continued)***

Degrees of Freedom	$\chi^2_{.100}$	$\chi^2_{.050}$	$\chi^2_{.025}$	$\chi^2_{.010}$	$\chi^2_{.005}$
1	2.70554	3.84146	5.02389	6.63490	7.87944
2	4.60517	5.99147	7.37776	9.21034	10.5966
3	6.25139	7.81473	9.34840	11.3449	12.8381
4	7.77944	9.48773	11.1433	13.2767	14.8602
5	9.23635	11.0705	12.8325	15.0863	16.7496
6	10.6446	12.5916	14.4494	16.8119	18.5476
7	12.0170	14.0671	16.0128	18.4753	20.2777
8	13.3616	15.5073	17.5346	20.0902	21.9550
9	14.6837	16.9190	19.0228	21.6660	23.5893
10	15.9871	18.3070	20.4831	23.2093	25.1882
11	17.2750	19.6751	21.9200	24.7250	26.7569
12	18.5494	21.0261	23.3367	26.2170	28.2995
13	19.8119	22.3621	24.7356	27.6883	29.8194
14	21.0642	23.6848	26.1190	29.1413	31.3193
15	22.3072	24.9958	27.4884	30.5779	32.8013
16	23.5418	26.2962	28.8454	31.9999	34.2672
17	24.7690	27.5871	30.1910	33.4087	35.7185
18	25.9894	28.8693	31.5264	34.8053	37.1564
19	27.2036	30.1435	32.8523	36.1908	38.5822
20	28.4120	31.4104	34.1696	37.5662	39.9968
21	29.6151	32.6705	35.4789	38.9321	41.4010
22	30.8133	33.9244	36.7807	40.2894	42.7956
23	32.0069	35.1725	38.0757	41.6384	44.1813
24	33.1963	36.4151	39.3641	42.9798	45.5585
25	34.3816	37.6525	40.6465	44.3141	46.9278
26	35.5631	38.8852	41.9232	45.6417	48.2899
27	36.7412	40.1133	43.1944	46.9630	49.6449
28	37.9159	41.3372	44.4607	48.2782	50.9933
29	39.0875	42.5569	45.7222	49.5879	52.3356
30	40.2560	43.7729	46.9792	50.8922	53.6720
40	51.8050	55.7585	59.3417	63.6907	66.7659
50	63.1671	67.5048	71.4202	76.1539	79.4900
60	74.3970	79.0819	83.2976	88.3794	91.9517
70	85.5271	90.5312	95.0231	100.425	104.215
80	96.5782	101.879	106.629	112.329	116.321
90	107.565	113.145	118.136	124.116	128.299
100	118.498	124.342	129.561	135.807	140.169

Source: From Thompson, C. M. "Tables of the percentage points of the χ^2-distribution," *Biometrika,* 1941, 32, 188–189. Reproduced by permission of the *Biometrika* Trustees.

Table VII **Percentage Points of the F-Distribution, $\alpha = .10$**

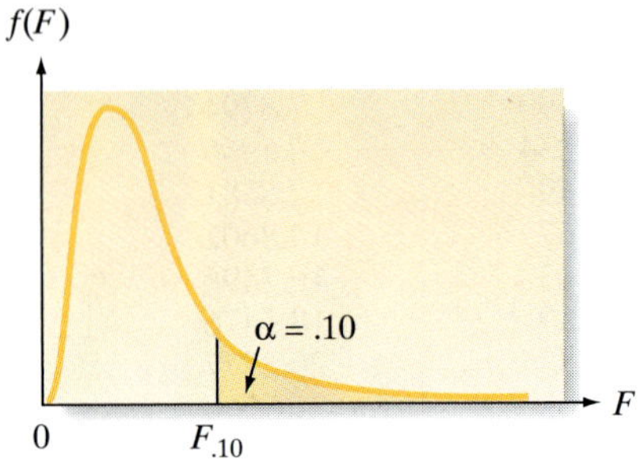

ν_2 \ ν_1 (Numerator Degrees of Freedom)	1	2	3	4	5	6	7	8	9
1	39.86	49.50	53.59	55.83	57.24	58.20	58.91	59.44	59.86
2	8.53	9.00	9.16	9.24	9.29	9.33	9.35	9.37	9.38
3	5.54	5.46	5.39	5.34	5.31	5.28	5.27	5.25	5.24
4	4.54	4.32	4.19	4.11	4.05	4.01	3.98	3.95	3.94
5	4.06	3.78	3.62	3.52	3.45	3.40	3.37	3.34	3.32
6	3.78	3.46	3.29	3.18	3.11	3.05	3.01	2.98	2.96
7	3.59	3.26	3.07	2.96	2.88	2.83	2.78	2.75	2.72
8	3.46	3.11	2.92	2.81	2.73	2.67	2.62	2.59	2.56
9	3.36	3.01	2.81	2.69	2.61	2.55	2.51	2.47	2.44
10	3.29	2.92	2.73	2.61	2.52	2.46	2.41	2.38	2.35
11	3.23	2.86	2.66	2.54	2.45	2.39	2.34	2.30	2.27
12	3.18	2.81	2.61	2.48	2.39	2.33	2.28	2.24	2.21
13	3.14	2.76	2.56	2.43	2.35	2.28	2.23	2.20	2.16
14	3.10	2.73	2.52	2.39	2.31	2.24	2.19	2.15	2.12
15	3.07	2.70	2.49	2.36	2.27	2.21	2.16	2.12	2.09
16	3.05	2.67	2.46	2.33	2.24	2.18	2.13	2.09	2.06
17	3.03	2.64	2.44	2.31	2.22	2.15	2.10	2.06	2.03
18	3.01	2.62	2.42	2.29	2.20	2.13	2.08	2.04	2.00
19	2.99	2.61	2.40	2.27	2.18	2.11	2.06	2.02	1.98
20	2.97	2.59	2.38	2.25	2.16	2.09	2.04	2.00	1.96
21	2.96	2.57	2.36	2.23	2.14	2.08	2.02	1.98	1.95
22	2.95	2.56	2.35	2.22	2.13	2.06	2.01	1.97	1.93
23	2.94	2.55	2.34	2.21	2.11	2.05	1.99	1.95	1.92
24	2.93	2.54	2.33	2.19	2.10	2.04	1.98	1.94	1.91
25	2.92	2.53	2.32	2.18	2.09	2.02	1.97	1.93	1.89
26	2.91	2.52	2.31	2.17	2.08	2.01	1.96	1.92	1.88
27	2.90	2.51	2.30	2.17	2.07	2.00	1.95	1.91	1.87
28	2.89	2.50	2.29	2.16	2.06	2.00	1.94	1.90	1.87
29	2.89	2.50	2.28	2.15	2.06	1.99	1.93	1.89	1.86
30	2.88	2.49	2.28	2.14	2.05	1.98	1.93	1.88	1.85
40	2.84	2.44	2.23	2.09	2.00	1.93	1.87	1.83	1.79
60	2.79	2.39	2.18	2.04	1.95	1.87	1.82	1.77	1.74
120	2.75	2.35	2.13	1.99	1.90	1.82	1.77	1.72	1.68
∞	2.71	2.30	2.08	1.94	1.85	1.77	1.72	1.67	1.63

Denominator Degrees of Freedom (ν_2)

(continued)

Table VII ***(continued)***

ν_2 \ ν_1	Numerator Degrees of Freedom									
Denominator Degrees of Freedom	10	12	15	20	24	30	40	60	120	∞
1	60.19	60.71	61.22	61.74	62.00	62.26	62.53	62.79	63.06	63.33
2	9.39	9.41	9.42	9.44	9.45	9.46	9.47	9.47	9.48	9.49
3	5.23	5.22	5.20	5.18	5.18	5.17	5.16	5.15	5.14	5.13
4	3.92	3.90	3.87	3.84	3.83	3.82	3.80	3.79	3.78	3.76
5	3.30	3.27	3.24	3.21	3.19	3.17	3.16	3.14	3.12	3.10
6	2.94	2.90	2.87	2.84	2.82	2.80	2.78	2.76	2.74	2.72
7	2.70	2.67	2.63	2.59	2.58	2.56	2.54	2.51	2.49	2.47
8	2.54	2.50	2.46	2.42	2.40	2.38	2.36	2.34	2.32	2.29
9	2.42	2.38	2.34	2.30	2.28	2.25	2.23	2.21	2.18	2.16
10	2.32	2.28	2.24	2.20	2.18	2.16	2.13	2.11	2.08	2.06
11	2.25	2.21	2.17	2.12	2.10	2.08	2.05	2.03	2.00	1.97
12	2.19	2.15	2.10	2.06	2.04	2.01	1.99	1.96	1.93	1.90
13	2.14	2.10	2.05	2.01	1.98	1.96	1.93	1.90	1.88	1.85
14	2.10	2.05	2.01	1.96	1.94	1.91	1.89	1.86	1.83	1.80
15	2.06	2.02	1.97	1.92	1.90	1.87	1.85	1.82	1.79	1.76
16	2.03	1.99	1.94	1.89	1.87	1.84	1.81	1.78	1.75	1.72
17	2.00	1.96	1.91	1.86	1.84	1.81	1.78	1.75	1.72	1.69
18	1.98	1.93	1.89	1.84	1.81	1.78	1.75	1.72	1.69	1.66
19	1.96	1.91	1.86	1.81	1.79	1.76	1.73	1.70	1.67	1.63
20	1.94	1.89	1.84	1.79	1.77	1.74	1.71	1.68	1.64	1.61
21	1.92	1.87	1.83	1.78	1.75	1.72	1.69	1.66	1.62	1.59
22	1.90	1.86	1.81	1.76	1.73	1.70	1.67	1.64	1.60	1.57
23	1.89	1.84	1.80	1.74	1.72	1.69	1.66	1.62	1.59	1.55
24	1.88	1.83	1.78	1.73	1.70	1.67	1.64	1.61	1.57	1.53
25	1.87	1.82	1.77	1.72	1.69	1.66	1.63	1.59	1.56	1.52
26	1.86	1.81	1.76	1.71	1.68	1.65	1.61	1.58	1.54	1.50
27	1.85	1.80	1.75	1.70	1.67	1.64	1.60	1.57	1.53	1.49
28	1.84	1.79	1.74	1.69	1.66	1.63	1.59	1.56	1.52	1.48
29	1.83	1.78	1.73	1.68	1.65	1.62	1.58	1.55	1.51	1.47
30	1.82	1.77	1.72	1.67	1.64	1.61	1.57	1.54	1.50	1.46
40	1.76	1.71	1.66	1.61	1.57	1.54	1.51	1.47	1.42	1.38
60	1.71	1.66	1.60	1.54	1.51	1.48	1.44	1.40	1.35	1.29
120	1.65	1.60	1.55	1.48	1.45	1.41	1.37	1.32	1.26	1.19
∞	1.60	1.55	1.49	1.42	1.38	1.34	1.30	1.24	1.17	1.00

Source: From Merrington, M., and Thompson, C. M. "Tables of percentage points of the inverted beta (F)-distribution," *Biometrika,* 1943, 33, 73–88. Reproduced by permission of the *Biometrika* Trustees.

Table VIII Percentage Points of the *F*-Distribution, $\alpha = .05$

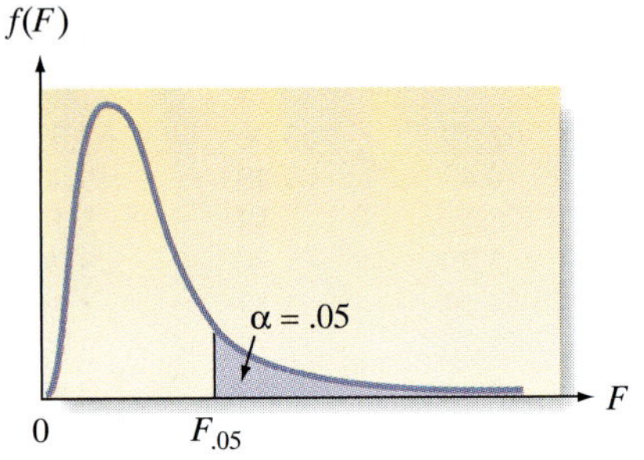

ν_2 \ ν_1	Numerator Degrees of Freedom								
Denominator Degrees of Freedom	1	2	3	4	5	6	7	8	9
1	161.4	199.5	215.7	224.6	230.2	234.0	236.8	238.9	240.5
2	18.51	19.00	19.16	19.25	19.30	19.33	19.35	19.37	19.38
3	10.13	9.55	9.28	9.12	9.01	8.94	8.89	8.85	8.81
4	7.71	6.94	6.59	6.39	6.26	6.16	6.09	6.04	6.00
5	6.61	5.79	5.41	5.19	5.05	4.95	4.88	4.82	4.77
6	5.99	5.14	4.76	4.53	4.39	4.28	4.21	4.15	4.10
7	5.59	4.74	4.35	4.12	3.97	3.87	3.79	3.73	3.68
8	5.32	4.46	4.07	3.84	3.69	3.58	3.50	3.44	3.39
9	5.12	4.26	3.86	3.63	3.48	3.37	3.29	3.23	3.18
10	4.96	4.10	3.71	3.48	3.33	3.22	3.14	3.07	3.02
11	4.84	3.98	3.59	3.36	3.20	3.09	3.01	2.95	2.90
12	4.75	3.89	3.49	3.26	3.11	3.00	2.91	2.85	2.80
13	4.67	3.81	3.41	3.18	3.03	2.92	2.83	2.77	2.71
14	4.60	3.74	3.34	3.11	2.96	2.85	2.76	2.70	2.65
15	4.54	3.68	3.29	3.06	2.90	2.79	2.71	2.64	2.59
16	4.49	3.63	3.24	3.01	2.85	2.74	2.66	2.59	2.54
17	4.45	3.59	3.20	2.96	2.81	2.70	2.61	2.55	2.49
18	4.41	3.55	3.16	2.93	2.77	2.66	2.58	2.51	2.46
19	4.38	3.52	3.13	2.90	2.74	2.63	2.54	2.48	2.42
20	4.35	3.49	3.10	2.87	2.71	2.60	2.51	2.45	2.39
21	4.32	3.47	3.07	2.84	2.68	2.57	2.49	2.42	2.37
22	4.30	3.44	3.05	2.82	2.66	2.55	2.46	2.40	2.34
23	4.28	3.42	3.03	2.80	2.64	2.53	2.44	2.37	2.32
24	4.26	3.40	3.01	2.78	2.62	2.51	2.42	2.36	2.30
25	4.24	3.39	2.99	2.76	2.60	2.49	2.40	2.34	2.28
26	4.23	3.37	2.98	2.74	2.59	2.47	2.39	2.32	2.77
27	4.21	3.35	2.96	2.73	2.57	2.46	2.37	2.31	2.25
28	4.20	3.34	2.95	2.71	2.56	2.45	2.36	2.29	2.24
29	4.18	3.33	2.93	2.70	2.55	2.43	2.35	2.28	2.22
30	4.17	3.32	2.92	2.69	2.53	2.42	2.33	2.27	2.21
40	4.08	3.23	2.84	2.61	2.45	2.34	2.25	2.18	2.12
60	4.00	3.15	2.76	2.53	2.37	2.25	2.17	2.10	2.04
120	3.92	3.07	2.68	2.45	2.29	2.17	2.09	2.02	1.96
∞	3.84	3.00	2.60	2.37	2.21	2.10	2.01	1.94	1.88

(continued)

Table VIII ***(continued)***

ν_2 \ ν_1 (Denominator Degrees of Freedom)	Numerator Degrees of Freedom									
	10	12	15	20	24	30	40	60	120	∞
1	241.9	243.9	245.9	248.0	249.1	250.1	251.1	252.2	253.3	254.3
2	19.40	19.41	19.43	19.45	19.45	19.46	19.47	19.48	19.49	19.50
3	8.79	8.74	8.70	8.66	8.64	8.62	8.59	8.57	8.55	8.53
4	5.96	5.91	5.86	5.80	5.77	5.75	5.72	5.69	5.66	5.63
5	4.74	4.68	4.62	4.56	4.53	4.50	4.46	4.43	4.40	4.36
6	4.06	4.00	3.94	3.87	3.84	3.81	3.77	3.74	3.70	3.67
7	3.64	3.57	3.51	3.44	3.41	3.38	3.34	3.30	3.27	3.23
8	3.35	3.28	3.22	3.15	3.12	3.08	3.04	3.01	2.97	2.93
9	3.14	3.07	3.01	2.94	2.90	2.86	2.83	2.79	2.75	2.71
10	2.98	2.91	2.85	2.77	2.74	2.70	2.66	2.62	2.58	2.54
11	2.85	2.79	2.72	2.65	2.61	2.57	2.53	2.49	2.45	2.40
12	2.75	2.69	2.62	2.54	2.51	2.47	2.43	2.38	2.34	2.30
13	2.67	2.60	2.53	2.46	2.42	2.38	2.34	2.30	2.25	2.21
14	2.60	2.53	2.46	2.39	2.35	2.31	2.27	2.22	2.18	2.13
15	2.54	2.48	2.40	2.33	2.29	2.25	2.20	2.16	2.11	2.07
16	2.49	2.42	2.35	2.28	2.24	2.19	2.15	2.11	2.06	2.01
17	2.45	2.38	2.31	2.23	2.19	2.15	2.10	2.06	2.01	1.96
18	2.41	2.34	2.27	2.19	2.15	2.11	2.06	2.02	1.97	1.92
19	2.38	2.31	2.23	2.16	2.11	2.07	2.03	1.98	1.93	1.88
20	2.35	2.28	2.20	2.12	2.08	2.04	1.99	1.95	1.90	1.84
21	2.32	2.25	2.18	2.10	2.05	2.01	1.96	1.92	1.87	1.81
22	2.30	2.23	2.15	2.07	2.03	1.98	1.94	1.89	1.84	1.78
23	2.27	2.20	2.13	2.05	2.01	1.96	1.91	1.86	1.81	1.76
24	2.25	2.18	2.11	2.03	1.98	1.94	1.89	1.84	1.79	1.73
25	2.24	2.16	2.09	2.01	1.96	1.92	1.87	1.82	1.77	1.71
26	2.22	2.15	2.07	1.99	1.95	1.90	1.85	1.80	1.75	1.69
27	2.20	2.13	2.06	1.97	1.93	1.88	1.84	1.79	1.73	1.67
28	2.19	2.12	2.04	1.96	1.91	1.87	1.82	1.77	1.71	1.65
29	2.18	2.10	2.03	1.94	1.90	1.85	1.81	1.75	1.70	1.64
30	2.16	2.09	2.01	1.93	1.89	1.84	1.79	1.74	1.68	1.62
40	2.08	2.00	1.92	1.84	1.79	1.74	1.69	1.64	1.58	1.51
60	1.99	1.92	1.84	1.75	1.70	1.65	1.59	1.53	1.47	1.39
120	1.91	1.83	1.75	1.66	1.61	1.55	1.50	1.43	1.35	1.25
∞	1.83	1.75	1.67	1.57	1.52	1.46	1.39	1.32	1.22	1.00

Source: From Merrington, M., and Thompson, C. M. "Tables of percentage points of the inverted beta (*F*)-distribution," *Biometrika,* 1943, 33, 73–88. Reproduced by permission of the *Biometrika* Trustees.

Table IX Percentage Points of the F-Distribution, $\alpha = .025$

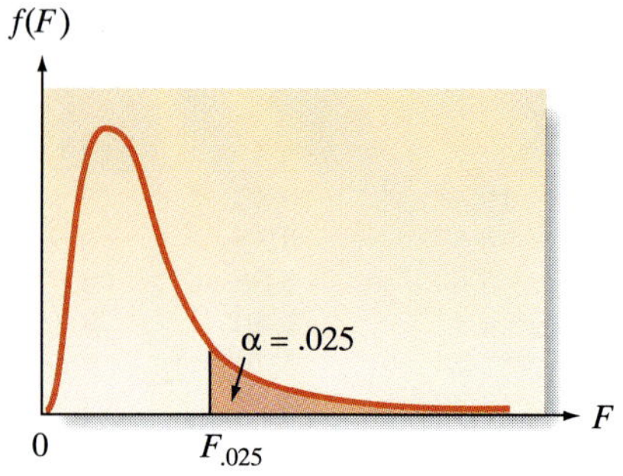

ν_2 \ ν_1 (Numerator Degrees of Freedom)	1	2	3	4	5	6	7	8	9
Denominator Degrees of Freedom									
1	647.8	799.5	864.2	899.6	921.8	937.1	948.2	956.7	963.3
2	38.51	39.00	39.17	39.25	39.30	39.33	39.36	39.37	39.39
3	17.44	16.04	15.44	15.10	14.88	14.73	14.62	14.54	14.47
4	12.22	10.65	9.98	9.60	9.36	9.20	9.07	8.98	8.90
5	10.01	8.43	7.76	7.39	7.15	6.98	6.85	6.76	6.68
6	8.81	7.26	6.60	6.23	5.99	5.82	5.70	5.60	5.52
7	8.07	6.54	5.89	5.52	5.29	5.12	4.99	4.90	4.82
8	7.57	6.06	5.42	5.05	4.82	4.65	4.53	4.43	4.36
9	7.21	5.71	5.08	4.72	4.48	4.32	4.20	4.10	4.03
10	6.94	5.46	4.83	4.47	4.24	4.07	3.95	3.85	3.78
11	6.72	5.26	4.63	4.28	4.04	3.88	3.76	3.66	3.59
12	6.55	5.10	4.47	4.12	3.89	3.73	3.61	3.51	3.44
13	6.41	4.97	4.35	4.00	3.77	3.60	3.48	3.39	3.31
14	6.30	4.86	4.24	3.89	3.66	3.50	3.38	3.29	3.21
15	6.20	4.77	4.15	3.80	3.58	3.41	3.29	3.20	3.12
16	6.12	4.69	4.08	3.73	3.50	3.34	3.22	3.12	3.05
17	6.04	4.62	4.01	3.66	3.44	3.28	3.16	3.06	2.98
18	5.98	4.56	3.95	3.61	3.38	3.22	3.10	3.01	2.93
19	5.92	4.51	3.90	3.56	3.33	3.17	3.05	2.96	2.88
20	5.87	4.46	3.86	3.51	3.29	3.13	3.01	2.91	2.84
21	5.83	4.42	3.82	3.48	3.25	3.09	2.97	2.87	2.80
22	5.79	4.38	3.78	3.44	3.22	3.05	2.93	2.84	2.76
23	5.75	4.35	3.75	3.41	3.18	3.02	2.90	2.81	2.73
24	5.72	4.32	3.72	3.38	3.15	2.99	2.87	2.78	2.70
25	5.69	4.29	3.69	3.35	3.13	2.97	2.85	2.75	2.68
26	5.66	4.27	3.67	3.33	3.10	2.94	2.82	2.73	2.65
27	5.63	4.24	3.65	3.31	3.08	2.92	2.80	2.71	2.63
28	5.61	4.22	3.63	3.29	3.06	2.90	2.78	2.69	2.61
29	5.59	4.20	3.61	3.27	3.04	2.88	2.76	2.67	2.59
30	5.57	4.18	3.59	3.25	3.03	2.87	2.75	2.65	2.57
40	5.42	4.05	3.46	3.13	2.90	2.74	2.62	2.53	2.45
60	5.29	3.93	3.34	3.01	2.79	2.63	2.51	2.41	2.33
120	5.15	3.80	3.23	2.89	2.67	2.52	2.39	2.30	2.22
∞	5.02	3.69	3.12	2.79	2.57	2.41	2.29	2.19	2.11

(continued)

Table IX ***(continued)***

ν_2 \ ν_1 (Denominator Degrees of Freedom \ Numerator Degrees of Freedom)	10	12	15	20	24	30	40	60	120	∞
1	968.6	976.7	984.9	993.1	997.2	1,001	1,006	1,010	1,014	1,018
2	39.40	39.41	39.43	39.45	39.46	39.46	39.47	39.48	39.49	39.50
3	14.42	14.34	14.25	14.17	14.12	14.08	14.04	13.99	13.95	13.90
4	8.84	8.75	8.66	8.56	8.51	8.46	8.41	8.36	8.31	8.26
5	6.62	6.52	6.43	6.33	6.28	6.23	6.18	6.12	6.07	6.02
6	5.46	5.37	5.27	5.17	5.12	5.07	5.01	4.96	4.90	4.85
7	4.76	4.67	4.57	4.47	4.42	4.36	4.31	4.25	4.20	4.14
8	4.30	4.20	4.10	4.00	3.95	3.89	3.84	3.78	3.73	3.67
9	3.96	3.87	3.77	3.67	3.61	3.56	3.51	3.45	3.39	3.33
10	3.72	3.62	3.52	3.42	3.37	3.31	3.26	3.20	3.14	3.08
11	3.53	3.43	3.33	3.23	3.17	3.12	3.06	3.00	2.94	2.88
12	3.37	3.28	3.18	3.07	3.02	2.96	2.91	2.85	2.79	2.72
13	3.25	3.15	3.05	2.95	2.89	2.84	2.78	2.72	2.66	2.60
14	3.15	3.05	2.95	2.84	2.79	2.73	2.67	2.61	2.55	2.49
15	3.06	2.96	2.86	2.76	2.70	2.64	2.59	2.52	2.46	2.40
16	2.99	2.89	2.79	2.68	2.63	2.57	2.51	2.45	2.38	2.32
17	2.92	2.82	2.72	2.62	2.56	2.50	2.44	2.38	2.32	2.25
18	2.87	2.77	2.67	2.56	2.50	2.44	2.38	2.32	2.26	2.19
19	2.82	2.72	2.62	2.51	2.45	2.39	2.33	2.27	2.20	2.13
20	2.77	2.68	2.57	2.46	2.41	2.35	2.29	2.22	2.16	2.09
21	2.73	2.64	2.53	2.42	2.37	2.31	2.25	2.18	2.11	2.04
22	2.70	2.60	2.50	2.39	2.33	2.27	2.21	2.14	2.08	2.00
23	2.67	2.57	2.47	2.36	2.30	2.24	2.18	2.11	2.04	1.97
24	2.64	2.54	2.44	2.33	2.27	2.21	2.15	2.08	2.01	1.94
25	2.61	2.51	2.41	2.30	2.24	2.18	2.12	2.05	1.98	1.91
26	2.59	2.49	2.39	2.28	2.22	2.16	2.09	2.03	1.95	1.88
27	2.57	2.47	2.36	2.25	2.19	2.13	2.07	2.00	1.93	1.85
28	2.55	2.45	2.34	2.23	2.17	2.11	2.05	1.98	1.91	1.83
29	2.53	2.43	2.32	2.21	2.15	2.09	2.03	1.96	1.89	1.81
30	2.51	2.41	2.31	2.20	2.14	2.07	2.01	1.94	1.87	1.79
40	2.39	2.29	2.18	2.07	2.01	1.94	1.88	1.80	1.72	1.64
60	2.27	2.17	2.06	1.94	1.88	1.82	1.74	1.67	1.58	1.48
120	2.16	2.05	1.94	1.82	1.76	1.69	1.61	1.53	1.43	1.31
∞	2.05	1.94	1.83	1.71	1.64	1.57	1.48	1.39	1.27	1.00

Source: From Merrington, M., and Thompson, C. M. "Tables of percentage points of the inverted beta (*F*)-distribution," *Biometrika,* 1943, 33, 73–88. Reproduced by permission of the *Biometrika* Trustees.

Table X **Percentage Points of the *F*-Distribution, $\alpha = .01$**

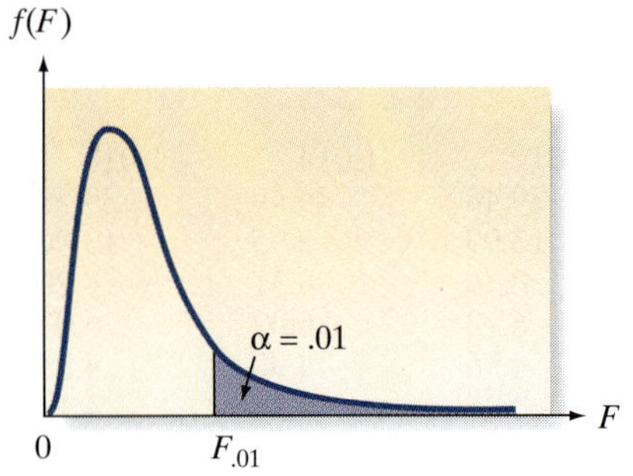

ν_2 \ ν_1	Numerator Degrees of Freedom								
Denominator Degrees of Freedom	1	2	3	4	5	6	7	8	9
1	4,052	4,999.5	5,403	5,625	5,764	5,859	5,928	5,982	6,022
2	98.50	99.00	99.17	99.25	99.30	99.33	99.36	99.37	99.39
3	34.12	30.82	29.46	28.71	28.24	27.91	27.67	27.49	27.35
4	21.20	18.00	16.69	15.98	15.52	15.21	14.98	14.80	14.66
5	16.26	13.27	12.06	11.39	10.97	10.67	10.46	10.29	10.16
6	13.75	10.92	9.78	9.15	8.75	8.47	8.26	8.10	7.98
7	12.25	9.55	8.45	7.85	7.46	7.19	6.99	6.84	6.72
8	11.26	8.65	7.59	7.01	6.63	6.37	6.18	6.03	5.91
9	10.56	8.02	6.99	6.42	6.06	5.80	5.61	5.47	5.35
10	10.04	7.56	6.55	5.99	5.64	5.39	5.20	5.06	4.94
11	9.65	7.21	6.22	5.67	5.32	5.07	4.89	4.74	4.63
12	9.33	6.93	5.95	5.41	5.06	4.82	4.64	4.50	4.39
13	9.07	6.70	5.74	5.21	4.86	4.62	4.44	4.30	4.19
14	8.86	6.51	5.56	5.04	4.69	4.46	4.28	4.14	4.03
15	8.68	6.36	5.42	4.89	4.56	4.32	4.14	4.00	3.89
16	8.53	6.23	5.29	4.77	4.44	4.20	4.03	3.89	3.78
17	8.40	6.11	5.18	4.67	4.34	4.10	3.93	3.79	3.68
18	8.29	6.01	5.09	4.58	4.25	4.01	3.84	3.71	3.60
19	8.18	5.93	5.01	4.50	4.17	3.94	3.77	3.63	3.52
20	8.10	5.85	4.94	4.43	4.10	3.87	3.70	3.56	3.46
21	8.02	5.78	4.87	4.37	4.04	3.81	3.64	3.51	3.40
22	7.95	5.72	4.82	4.31	3.99	3.76	3.59	3.45	3.35
23	7.88	5.66	4.76	4.26	3.94	3.71	3.54	3.41	3.30
24	7.82	5.61	4.72	4.22	3.90	3.67	3.50	3.36	3.26
25	7.77	5.57	4.68	4.18	3.85	3.63	3.46	3.32	3.22
26	7.72	5.53	4.64	4.14	3.82	3.59	3.42	3.29	3.18
27	7.68	5.49	4.60	4.11	3.78	3.56	3.39	3.26	3.15
28	7.64	5.45	4.57	4.07	3.75	3.53	3.36	3.23	3.12
29	7.60	5.42	4.54	4.04	3.73	3.50	3.33	3.20	3.09
30	7.56	5.39	4.51	4.02	3.70	3.47	3.30	3.17	3.07
40	7.31	5.18	4.31	3.83	3.51	3.29	3.12	2.99	2.89
60	7.08	4.98	4.13	3.65	3.34	3.12	2.95	2.82	2.72
120	6.85	4.79	3.95	3.48	3.17	2.96	2.79	2.66	2.56
∞	6.63	4.61	3.78	3.32	3.02	2.80	2.64	2.51	2.41

(continued)

Table X ***(continued)***

ν_2 \ ν_1 (Denominator Degrees of Freedom)	Numerator Degrees of Freedom: 10	12	15	20	24	30	40	60	120	∞
1	6,056	6,106	6,157	6,209	6,235	6,261	6,287	6,313	6,339	6,366
2	99.40	99.42	99.43	99.45	99.46	99.47	99.47	99.48	99.49	99.50
3	27.23	27.05	26.87	26.69	26.60	26.50	26.41	26.32	26.22	26.13
4	14.55	14.37	14.20	14.02	13.93	13.84	13.75	13.65	13.56	13.46
5	10.05	9.89	9.72	9.55	9.47	9.38	9.29	9.20	9.11	9.02
6	7.87	7.72	7.56	7.40	7.31	7.23	7.14	7.06	6.97	6.88
7	6.62	6.47	6.31	6.16	6.07	5.99	5.91	5.82	5.74	5.65
8	5.81	5.67	5.52	5.36	5.28	5.20	5.12	5.03	4.95	4.86
9	5.26	5.11	4.96	4.81	4.73	4.65	4.57	4.48	4.40	4.31
10	4.85	4.71	4.56	4.41	4.33	4.25	4.17	4.08	4.00	3.91
11	4.54	4.40	4.25	4.10	4.02	3.94	3.86	3.78	3.69	3.60
12	4.30	4.16	4.01	3.86	3.78	3.70	3.62	3.54	3.45	3.36
13	4.10	3.96	3.82	3.66	3.59	3.51	3.43	3.34	3.25	3.17
14	3.94	3.80	3.66	3.51	3.43	3.35	3.27	3.18	3.09	3.00
15	3.80	3.67	3.52	3.37	3.29	3.21	3.13	3.05	2.96	2.87
16	3.69	3.55	3.41	3.26	3.18	3.10	3.02	2.93	2.84	2.75
17	3.59	3.46	3.31	3.16	3.08	3.00	2.92	2.83	2.75	2.65
18	3.51	3.37	3.23	3.08	3.00	2.92	2.84	2.75	2.66	2.57
19	3.43	3.30	3.15	3.00	2.92	2.84	2.76	2.67	2.58	2.49
20	3.37	3.23	3.09	2.94	2.86	2.78	2.69	2.61	2.52	2.42
21	3.31	3.17	3.03	2.88	2.80	2.72	2.64	2.55	2.46	2.36
22	3.26	3.12	2.98	2.83	2.75	2.67	2.58	2.50	2.40	2.31
23	3.21	3.07	2.93	2.78	2.70	2.62	2.54	2.45	2.35	2.26
24	3.17	3.03	2.89	2.74	2.66	2.58	2.49	2.40	2.31	2.21
25	3.13	2.99	2.85	2.70	2.62	2.54	2.45	2.36	2.27	2.17
26	3.09	2.96	2.81	2.66	2.58	2.50	2.42	2.33	2.23	2.13
27	3.06	2.93	2.78	2.63	2.55	2.47	2.38	2.29	2.20	2.10
28	3.03	2.90	2.75	2.60	2.52	2.44	2.35	2.26	2.17	2.06
29	3.00	2.87	2.73	2.57	2.49	2.41	2.33	2.23	2.14	2.03
30	2.98	2.84	2.70	2.55	2.47	2.39	2.30	2.21	2.11	2.01
40	2.80	2.66	2.52	2.37	2.29	2.20	2.11	2.02	1.92	1.80
60	2.63	2.50	2.35	2.20	2.12	2.03	1.94	1.84	1.73	1.60
120	2.47	2.34	2.19	2.03	1.95	1.86	1.76	1.66	1.53	1.38
∞	2.32	2.18	2.04	1.88	1.79	1.70	1.59	1.47	1.32	1.00

Source: From Merrington, M., and Thompson, C. M. "Tables of percentage points of the inverted beta (F)-distribution," *Biometrika,* 1943, 33, 73–88. Reproduced by permission of the *Biometrika* Trustees.

Table XI **Control Chart Constants**

Number of Observations in Subgroup, n	A_2	d_2	d_3	D_3	D_4
2	1.880	1.128	.853	.000	3.267
3	1.023	1.693	.888	.000	2.574
4	.729	2.059	.880	.000	2.282
5	.577	2.326	.864	.000	2.114
6	.483	2.534	.848	.000	2.004
7	.419	2.704	.833	.076	1.924
8	.373	2.847	.820	.136	1.864
9	.337	2.970	.808	.184	1.816
10	.308	3.078	.797	.223	1.777
11	.285	3.173	.787	.256	1.744
12	.266	3.258	.778	.283	1.717
13	.249	3.336	.770	.307	1.693
14	.235	3.407	.762	.328	1.672
15	.223	3.472	.755	.347	1.653
16	.212	3.532	.749	.363	1.637
17	.203	3.588	.743	.378	1.622
18	.194	3.640	.738	.391	1.608
19	.187	3.689	.733	.403	1.597
20	.180	3.735	.729	.415	1.585
21	.173	3.778	.724	.425	1.575
22	.167	3.819	.720	.434	1.566
23	.162	3.858	.716	.443	1.557
24	.157	3.895	.712	.451	1.548
25	.153	3.931	.709	.459	1.541

Source: ASTM Manual on the Presentation of Data and Control Chart Analysis, Philadelphia, PA: American Society for Testing Materials, pp. 134–136, 1976.

Table XII Critical Values for the Durbin-Watson *d* Statistic, $\alpha = .05$

	$k = 1$		$k = 2$		$k = 3$		$k = 4$		$k = 5$	
n	d_L	d_U	d_L	d_U	d_L	d_U	d_L	d_U	d_L	d_U
15	1.08	1.36	.95	1.54	.82	1.75	.69	1.97	.56	2.21
16	1.10	1.37	.98	1.54	.86	1.73	.74	1.93	.62	2.15
17	1.13	1.38	1.02	1.54	.90	1.71	.78	1.90	.67	2.10
18	1.16	1.39	1.05	1.53	.93	1.69	.92	1.87	.71	2.06
19	1.18	1.40	1.08	1.53	.97	1.68	.86	1.85	.75	2.02
20	1.20	1.41	1.10	1.54	1.00	1.68	.90	1.83	.79	1.99
21	1.22	1.42	1.13	1.54	1.03	1.67	.93	1.81	.83	1.96
22	1.24	1.43	1.15	1.54	1.05	1.66	.96	1.80	.96	1.94
23	1.26	1.44	1.17	1.54	1.08	1.66	.99	1.79	.90	1.92
24	1.27	1.45	1.19	1.55	1.10	1.66	1.01	1.78	.93	1.90
25	1.29	1.45	1.21	1.55	1.12	1.66	1.04	1.77	.95	1.89
26	1.30	1.46	1.22	1.55	1.14	1.65	1.06	1.76	.98	1.88
27	1.32	1.47	1.24	1.56	1.16	1.65	1.08	1.76	1.01	1.86
28	1.33	1.48	1.26	1.56	1.18	1.65	1.10	1.75	1.03	1.85
29	1.34	1.48	1.27	1.56	1.20	1.65	1.12	1.74	1.05	1.84
30	1.35	1.49	1.28	1.57	1.21	1.65	1.14	1.74	1.07	1.83
31	1.36	1.50	1.30	1.57	1.23	1.65	1.16	1.74	1.09	1.83
32	1.37	1.50	1.31	1.57	1.24	1.65	1.18	1.73	1.11	1.82
33	1.38	1.51	1.32	1.58	1.26	1.65	1.19	1.73	1.13	1.81
34	1.39	1.51	1.33	1.58	1.27	1.65	1.21	1.73	1.15	1.81
35	1.40	1.52	1.34	1.58	1.28	1.65	1.22	1.73	1.16	1.80
36	1.41	1.52	1.35	1.59	1.29	1.65	1.24	1.73	1.18	1.80
37	1.42	1.53	1.36	1.59	1.31	1.66	1.25	1.72	1.19	1.80
38	1.43	1.54	1.37	1.59	1.32	1.66	1.26	1.72	1.21	1.79
39	1.43	1.54	1.38	1.60	1.33	1.66	1.27	1.72	1.22	1.79
40	1.44	1.54	1.39	1.60	1.34	1.66	1.29	1.72	1.23	1.79
45	1.48	1.57	1.43	1.62	1.38	1.67	1.34	1.72	1.29	1.78
50	1.50	1.59	1.46	1.63	1.42	1.67	1.38	1.72	1.34	1.77
55	1.53	1.60	1.49	1.64	1.45	1.68	1.41	1.72	1.38	1.77
60	1.55	1.62	1.51	1.65	1.48	1.69	1.44	1.73	1.41	1.77
65	1.57	1.63	1.54	1.66	1.50	1.70	1.47	1.73	1.44	1.77
70	1.58	1.64	1.55	1.67	1.52	1.70	1.49	1.74	1.46	1.77
75	1.60	1.65	1.57	1.68	1.54	1.71	1.51	1.74	1.49	1.77
80	1.61	1.66	1.59	1.69	1.56	1.72	1.53	1.74	1.51	1.77
85	1.62	1.67	1.60	1.70	1.57	1.72	1.55	1.75	1.52	1.77
90	1.63	1.68	1.61	1.70	1.59	1.73	1.57	1.75	1.54	1.78
95	1.64	1.69	1.62	1.71	1.60	1.73	1.58	1.75	1.56	1.78
100	1.65	1.69	1.63	1.72	1.61	1.74	1.59	1.76	1.57	1.78

Source: From Durbin, J., and Watson, G. S. "Testing for serial correlation in least squares regression, II," *Biometrika,* 1951, 30, 159–178. Reproduced by permission of the *Biometrika* Trustees.

Table XIII Critical Values for the Durbin-Watson *d* Statistic, $\alpha = .01$

n	$k = 1$		$k = 2$		$k = 3$		$k = 4$		$k = 5$	
	d_L	d_U	d_L	d_U	d_L	d_U	d_L	d_U	d_L	d_U
15	.81	1.07	.70	1.25	.59	1.46	.49	1.70	.39	1.96
16	.84	1.09	.74	1.25	.63	1.44	.53	1.66	.44	1.90
17	.87	1.10	.77	1.25	.67	1.43	.57	1.3	.48	1.85
18	.90	1.12	.80	1.26	.71	1.42	.61	1.60	.52	1.80
19	.93	1.13	.83	1.26	.74	1.41	.65	1.58	.56	1.77
20	.95	1.15	.86	1.27	.77	1.41	.68	1.57	.60	1.74
21	.97	1.16	.89	1.27	.80	1.41	.72	1.55	.63	1.71
22	1.00	1.17	.91	1.28	.83	1.40	.75	1.54	.66	1.69
23	1.02	1.19	.94	1.29	.86	1.40	.77	1.53	.70	1.67
24	1.04	1.20	.96	1.30	.88	1.41	.80	1.53	.72	1.66
25	1.05	1.21	.98	1.30	.90	1.41	.83	1.52	.75	1.65
26	1.07	1.22	1.00	1.31	.93	1.41	.85	1.52	.78	1.64
27	1.09	1.23	1.02	1.32	.95	1.41	.88	1.51	.81	1.63
28	1.10	1.24	1.04	1.32	.97	1.41	.90	1.51	.83	1.62
29	1.12	1.25	1.05	1.33	.99	1.42	.92	1.51	.85	1.61
30	1.13	1.26	1.07	1.34	1.01	1.42	.94	1.51	.88	1.61
31	1.15	1.27	1.08	1.34	1.02	1.42	.96	1.51	.90	1.60
32	1.16	1.28	1.10	1.35	1.04	1.43	.98	1.51	.92	1.60
33	1.17	1.29	1.11	1.36	1.05	1.43	1.00	1.51	.94	1.59
34	1.18	1.30	1.13	1.36	1.07	1.43	1.01	1.51	.95	1.59
35	1.19	1.31	1.14	1.27	1.08	1.44	1.03	1.51	.97	1.59
36	1.21	1.32	1.15	1.38	1.10	1.44	1.04	1.51	.99	1.59
37	1.22	1.32	1.16	1.38	1.11	1.45	1.06	1.51	1.00	1.59
38	1.23	1.33	1.18	1.39	1.12	1.45	1.07	1.52	1.02	1.58
39	1.24	1.34	1.19	1.39	1.14	1.45	1.09	1.52	1.03	1.58
40	1.25	1.34	1.20	1.40	1.15	1.46	1.10	1.52	1.05	1.58
45	1.29	1.38	1.24	1.42	1.20	1.48	1.16	1.53	1.11	1.58
50	1.32	1.40	1.28	1.45	1.24	1.49	1.20	1.54	1.16	1.59
55	1.36	1.43	1.32	1.47	1.28	1.51	1.25	1.55	1.21	1.59
60	1.38	1.45	1.35	1.48	1.32	1.52	1.28	1.56	1.25	1.60
65	1.41	1.47	1.38	1.50	1.35	1.53	1.31	1.57	1.28	1.61
70	1.43	1.49	1.40	1.52	1.37	1.55	1.34	1.58	1.31	1.61
75	1.45	1.50	1.42	1.53	1.39	1.56	1.37	1.59	1.34	1.62
80	1.47	1.52	1.44	1.54	1.42	1.57	1.39	1.60	1.36	1.62
85	1.48	1.53	1.46	1.55	1.43	1.58	1.41	1.60	1.39	1.63
90	1.50	1.54	1.47	1.56	1.45	1.59	1.43	1.61	1.41	1.64
95	1.51	1.55	1.49	1.57	1.47	1.60	1.45	1.62	1.42	1.64
100	1.52	1.56	1.50	1.58	1.48	1.60	1.46	1.63	1.44	1.65

Source: From Durbin, J., and Watson, G. S. "Testing for serial correlation in least squares regression, II," *Biometrika,* 1951, 30, 159–178. Reproduced by permission of the *Biometrika* Trustees.

Table XIV Critical Values of T_L and T_U for the Wilcoxon Rank Sum Test: Independent Samples

Test statistic is the rank sum associated with the smaller sample (if equal sample sizes, either rank sum can be used).

a. $\alpha = .025$ one-tailed; $\alpha = .05$ two-tailed

n_2 \ n_1	3		4		5		6		7		8		9		10	
	T_L	T_U	T_L	T_U	T_L	T_U	T_L	T_U	T_L	T_U	T_L	T_U	T_L	T_U	T_L	T_U
3	5	16	6	18	6	21	7	23	7	26	8	28	8	31	9	33
4	6	18	11	25	12	28	12	32	13	35	14	38	15	41	16	44
5	6	21	12	28	18	37	19	41	20	45	21	49	22	53	24	56
6	7	23	12	32	19	41	26	52	28	56	29	61	31	65	32	70
7	7	26	13	35	20	45	28	56	37	68	39	73	41	78	43	83
8	8	28	14	38	21	49	29	61	39	73	49	87	51	93	54	98
9	8	31	15	41	22	53	31	65	41	78	51	93	63	108	66	114
10	9	33	16	44	24	56	32	70	43	83	54	98	66	114	79	131

b. $\alpha = .05$ one-tailed; $\alpha = .10$ two-tailed

n_2 \ n_1	3		4		5		6		7		8		9		10	
	T_L	T_U	T_L	T_U	T_L	T_U	T_L	T_U	T_L	T_U	T_L	T_U	T_L	T_U	T_L	T_U
3	6	15	7	17	7	20	8	22	9	24	9	27	10	29	11	31
4	7	17	12	24	13	27	14	30	15	33	16	36	17	39	18	42
5	7	20	13	27	19	36	20	40	22	43	24	46	25	50	26	54
6	8	22	14	30	20	40	28	50	30	54	32	58	33	63	35	67
7	9	24	15	33	22	43	30	54	39	66	41	71	43	76	46	80
8	9	27	16	36	24	46	32	58	41	71	52	84	54	90	57	95
9	10	29	17	39	25	50	33	63	43	76	54	90	66	105	69	111
10	11	31	18	42	26	54	35	67	46	80	57	95	69	111	83	127

Source: From Wilcoxon, F., and Wilcox, R. A. "Some rapid approximate statistical procedures," 1964, 20–23. Courtesy of Lederle Laboratories Division of American Cyanamid Company, Madison, NJ.

Table XV Critical Values of T_0 in the Wilcoxon Paired Difference Signed Rank Test

One-Tailed	Two-Tailed	$n = 5$	$n = 6$	$n = 7$	$n = 8$	$n = 9$	$n = 10$
$\alpha = .05$	$\alpha = .10$	1	2	4	6	8	11
$\alpha = .025$	$\alpha = .05$		1	2	4	6	8
$\alpha = .01$	$\alpha = .02$			0	2	3	5
$\alpha = .005$	$\alpha = .01$				0	2	3
		$n = 11$	$n = 12$	$n = 13$	$n = 14$	$n = 15$	$n = 16$
$\alpha = .05$	$\alpha = .10$	14	17	21	26	30	36
$\alpha = .025$	$\alpha = .05$	11	14	17	21	25	30
$\alpha = .01$	$\alpha = .02$	7	10	13	16	20	24
$\alpha = .005$	$\alpha = .01$	5	7	10	13	16	19
		$n = 17$	$n = 18$	$n = 19$	$n = 20$	$n = 21$	$n = 22$
$\alpha = .05$	$\alpha = .10$	41	47	54	60	68	75
$\alpha = .025$	$\alpha = .05$	35	40	46	52	59	66
$\alpha = .01$	$\alpha = .02$	28	33	38	43	49	56
$\alpha = .005$	$\alpha = .01$	23	28	32	37	43	49
		$n = 23$	$n = 24$	$n = 25$	$n = 26$	$n = 27$	$n = 28$
$\alpha = .05$	$\alpha = .10$	83	92	101	110	120	130
$\alpha = .025$	$\alpha = .05$	73	81	90	98	107	117
$\alpha = .01$	$\alpha = .02$	62	69	77	85	93	102
$\alpha = .005$	$\alpha = .01$	55	61	68	76	84	92
		$n = 29$	$n = 30$	$n = 31$	$n = 32$	$n = 33$	$n = 34$
$\alpha = .05$	$\alpha = .10$	141	152	163	175	188	201
$\alpha = .025$	$\alpha = .05$	127	137	148	159	171	183
$\alpha = .01$	$\alpha = .02$	111	120	130	141	151	162
$\alpha = .005$	$\alpha = .01$	100	109	118	128	138	149
		$n = 35$	$n = 36$	$n = 37$	$n = 38$	$n = 39$	
$\alpha = .05$	$\alpha = .10$	214	228	242	256	271	
$\alpha = .025$	$\alpha = .05$	195	208	222	235	250	
$\alpha = .01$	$\alpha = .02$	174	186	198	211	224	
$\alpha = .005$	$\alpha = .01$	160	171	183	195	208	
		$n = 40$	$n = 41$	$n = 42$	$n = 43$	$n = 44$	$n = 45$
$\alpha = .05$	$\alpha = .10$	287	303	319	336	353	371
$\alpha = .025$	$\alpha = .05$	264	279	295	311	327	344
$\alpha = .01$	$\alpha = .02$	238	252	267	281	297	313
$\alpha = .005$	$\alpha = .01$	221	234	248	262	277	292
		$n = 46$	$n = 47$	$n = 48$	$n = 49$	$n = 50$	
$\alpha = .05$	$\alpha = .10$	389	408	427	446	466	
$\alpha = .025$	$\alpha = .05$	361	379	397	415	434	
$\alpha = .01$	$\alpha = .02$	329	345	362	380	398	
$\alpha = .005$	$\alpha = .01$	307	323	339	356	373	

Source: From Wilcoxon, F., and Wilcox, R. A. "Some rapid approximate statistical procedures," 1964, p. 28. Courtesy of Lederle Laboratories Division of American Cyanamid Company, Madison, NJ.

Table XVI Critical Values of Spearman's Rank Correlation Coefficient

The values correspond to a one-tailed test of H_0: $p = 0$. The value should be doubled for two-tailed tests.

n	$\alpha = .05$	$\alpha = .025$	$\alpha = .01$	$\alpha = .005$	n	$\alpha = .05$	$\alpha = .025$	$\alpha = .01$	$\alpha = .005$
5	.900	—	—	—	18	.399	.476	.564	.625
6	.829	.886	.943	—	19	.388	.462	.549	.608
7	.714	.786	.893	—	20	.377	.450	.534	.591
8	.643	.738	.833	.881	21	.368	.438	.521	.576
9	.600	.683	.783	.833	22	.359	.428	.508	.562
10	.564	.648	.745	.794	23	.351	.418	.496	.549
11	.523	.623	.736	.818	24	.343	.409	.485	.537
12	.497	.591	.703	.780	25	.336	.400	.475	.526
13	.475	.566	.673	.745	26	.329	.392	.465	.515
14	.457	.545	.646	.716	27	.323	.385	.456	.505
15	.441	.525	.623	.689	28	.317	.377	.448	.496
16	.425	.507	.601	.666	29	.311	.370	.440	.487
17	.412	.490	.582	.645	30	.305	.364	.432	.478

Source: From Olds, E. G. "Distribution of sums of squares of rank differences for small samples," *Annals of Mathematical Statistics,* 1938, 9. Reproduced with the permission of the editor, *Annals of Mathematical Statistics*.

Table XVII Critical Values of the Studentized Range, $\alpha = .05$

ν \ k	2	3	4	5	6	7	8	9	10	11
1	17.97	26.98	32.82	37.08	40.41	43.12	45.40	47.36	49.07	50.59
2	6.08	8.33	9.80	10.88	11.74	12.44	13.03	13.54	13.99	14.39
3	4.50	5.91	6.82	7.50	8.04	8.48	8.85	9.18	9.46	9.72
4	3.93	5.04	5.76	6.29	6.71	7.05	7.35	7.60	7.83	8.03
5	3.64	4.60	5.22	5.67	6.03	6.33	6.58	6.80	6.99	7.17
6	3.46	4.34	4.90	5.30	5.63	5.90	6.12	6.32	6.49	6.65
7	3.34	4.16	4.68	5.06	5.36	5.61	5.82	6.00	6.16	6.30
8	3.26	4.04	4.53	4.89	5.17	5.40	5.60	5.77	5.92	6.05
9	3.20	3.95	4.41	4.76	5.02	5.24	5.43	5.59	5.74	5.87
10	3.15	3.88	4.33	4.65	4.91	5.12	5.30	5.46	5.60	5.72
11	3.11	3.82	4.26	4.57	4.82	5.03	5.20	5.35	5.49	5.61
12	3.08	3.77	4.20	4.51	4.75	4.95	5.12	5.27	5.39	5.51
13	3.06	3.73	4.15	4.45	4.69	4.88	5.05	5.19	5.32	5.43
14	3.03	3.70	4.11	4.41	4.64	4.83	4.99	5.13	5.25	5.36
15	3.01	3.67	4.08	4.37	4.60	4.78	4.94	5.08	5.20	5.31
16	3.00	3.65	4.05	4.33	4.56	4.74	4.90	5.03	5.15	5.26
17	2.98	3.63	4.02	4.30	4.52	4.70	4.86	4.99	5.11	5.21
18	2.97	3.61	4.00	4.28	4.49	4.67	4.82	4.96	5.07	5.17
19	2.96	3.59	3.98	4.25	4.47	4.65	4.79	4.92	5.04	5.14
20	2.95	3.58	3.96	4.23	4.45	4.62	4.77	4.90	5.01	5.11
24	2.92	3.53	3.90	4.17	4.37	4.54	4.68	4.81	4.92	5.01
30	2.89	3.49	3.85	4.10	4.30	4.46	4.60	4.72	4.82	4.92
40	2.86	3.44	3.79	4.04	4.23	4.39	4.52	4.63	4.73	4.82
60	2.83	3.40	3.74	3.98	4.16	4.31	4.44	4.55	4.65	4.73
120	2.80	3.36	3.68	3.92	4.10	4.24	4.36	4.47	4.56	4.64
∞	2.77	3.31	3.63	3.86	4.03	4.17	4.29	4.39	4.47	4.55

ν \ k	12	13	14	15	16	17	18	19	20
1	51.96	53.20	54.33	55.36	56.32	57.22	58.04	58.83	59.56
2	14.75	15.08	15.38	15.65	15.91	16.14	16.37	16.57	16.77
3	9.95	10.15	10.35	10.52	10.69	10.84	10.98	11.11	11.24
4	8.21	8.37	8.52	8.66	8.79	8.91	9.03	9.13	9.23
5	7.32	7.47	7.60	7.72	7.83	7.93	8.03	8.12	8.21
6	6.79	6.92	7.03	7.14	7.24	7.34	7.43	7.51	7.59
7	6.43	6.55	6.66	6.76	6.85	6.94	7.02	7.10	7.17
8	6.18	6.29	6.39	6.48	6.57	6.65	6.73	6.80	6.87
9	5.98	6.09	6.19	6.28	6.36	6.44	6.51	6.58	6.64
10	5.83	5.93	6.03	6.11	6.19	6.27	6.34	6.40	6.47
11	5.71	5.81	5.90	5.98	6.06	6.13	6.20	6.27	6.33
12	5.61	5.71	5.80	5.88	5.95	6.02	6.09	6.15	6.21
13	5.53	5.63	5.71	5.79	5.86	5.93	5.99	6.05	6.11
14	5.46	5.55	5.64	5.71	5.79	5.85	5.91	5.97	6.03
15	5.40	5.49	5.57	5.65	5.72	5.78	5.85	5.90	5.96
16	5.35	5.44	5.52	5.59	5.66	5.73	5.79	5.84	5.90
17	5.31	5.39	5.47	5.54	5.61	5.67	5.73	5.79	5.84
18	5.27	5.35	5.43	5.50	5.57	5.63	5.69	5.74	5.79
19	5.23	5.31	5.39	5.46	5.53	5.59	5.65	5.70	5.75
20	5.20	5.28	5.36	5.43	5.49	5.55	5.61	5.66	5.71
24	5.10	5.18	5.25	5.32	5.38	5.44	5.49	5.55	5.59
30	5.00	5.08	5.15	5.21	5.27	5.33	5.38	5.43	5.47
40	4.90	4.98	5.04	5.11	5.16	5.22	5.27	5.31	5.36
60	4.81	4.88	4.94	5.00	5.06	5.11	5.15	5.20	5.24
120	4.71	4.78	4.84	4.90	4.95	5.00	5.04	5.09	5.13
∞	4.62	4.68	4.74	4.80	4.85	4.89	4.93	4.97	5.01

Source: Biometrika Tables for Statisticians, Vol. I, 3rd ed., edited by E. S. Pearson and H. O. Hartley (Cambridge University Press, 1966). Reproduced by permission of Professor E. S. Pearson and the *Biometrika* Trustees.

Appendix C: Calculation Formulas for Analysis of Variance

C.1 Formulas for the Calculations in the Completely Randomized Design

$$\text{CM} = \text{Correction for mean}$$

$$= \frac{(\text{Total of all observations})^2}{\text{Total number of observations}} = \frac{\left(\sum_{i=1}^{n} y_i\right)^2}{n}$$

$$\text{SS(Total)} = \text{Total sum of squares}$$

$$= (\text{Sum of squares of all observations}) - \text{CM} = \sum_{i=1}^{n} y_i^2 - \text{CM}$$

$$\text{SST} = \text{Sum of squares for treatments}$$

$$= \begin{pmatrix} \text{Sum of squares of treatment totals with} \\ \text{each square divided by the number of} \\ \text{observations for that treatment} \end{pmatrix} - \text{CM}$$

$$= \frac{T_1^2}{n_1} + \frac{T_2^2}{n_2} + \cdots + \frac{T_k^2}{n_k} - \text{CM}$$

$$\text{SSE} = \text{Sum of squares for error} = \text{SS(Total)} - \text{SST}$$

$$\text{MST} = \text{Mean square for treatments} = \frac{\text{SST}}{k - 1}$$

$$\text{MSE} = \text{Mean square for error} = \frac{\text{SSE}}{n - k}$$

$$F = \text{Test statistic} = \frac{\text{MST}}{\text{MSE}}$$

where

n = Total number of observations

k = Number of treatments

T_i = Total for treatment i ($i = 1, 2, \ldots, k$)

C.2 Formulas for the Calculations in the Randomized Block Design

$$\text{CM} = \text{Correction for mean}$$

$$= \frac{(\text{Total of all observations})^2}{\text{Total number of observations}} = \frac{\left(\sum y_i\right)^2}{n}$$

$$\text{SS(Total)} = \text{Total sum of squares}$$
$$= (\text{Sum of squares of all observations}) - \text{CM} = \sum y_i^2 - \text{CM}$$

$$\text{SST} = \text{Sum of squares for treatments}$$
$$= \left(\begin{array}{c}\text{Sum of squares of treatment totals with} \\ \text{each square divided by } b\text{, the number of} \\ \text{observations for that treatment}\end{array}\right) - \text{CM}$$
$$= \frac{T_1^2}{b} + \frac{T_2^2}{b} + \cdots + \frac{T_p^2}{b} - \text{CM}$$

$$\text{SST} = \text{Sum of squares for blocks}$$
$$= \left(\begin{array}{c}\text{Sum of squares of block totals with} \\ \text{each square divided by } k\text{, the number} \\ \text{of observations in that block}\end{array}\right) - \text{CM}$$
$$= \frac{B_1^2}{k} + \frac{B_2^2}{k} + \cdots + \frac{B_b^2}{k} - \text{CM}$$

$$\text{SSE} = \text{Sum of squares for error} = \text{SS(Total)} - \text{SST} - \text{SSB}$$

$$\text{MST} = \text{Mean square for treatments} = \frac{\text{SST}}{k - 1}$$

$$\text{MSB} = \text{Mean square for blocks} = \frac{\text{SSB}}{b - 1}$$

$$\text{MSE} = \text{Mean square for error} = \frac{\text{SSE}}{n - k - b + 1}$$

$$F = \text{Test statistic} = \frac{\text{MST}}{\text{MSE}}$$

where

n = Total number of observations

b = Number of blocks

k = Number of treatments

T_i = Total for treatment i ($i = 1, 2, \ldots, k$)

B_i = Total for block i ($i = 1, 2, \ldots, b$)

C.3 Formulas for the Calculations for a Two-Factor Factorial Experiment

$$\text{CM} = \text{Correction for mean}$$
$$= \frac{(\text{Total of all } n \text{ measurements})^2}{n} = \frac{\left(\sum_{i=1}^{n} y_i\right)^2}{n}$$

$$\text{SS(Total)} = \text{Total sum of squares}$$
$$= (\text{Sum of squares of all } n \text{ measurements}) - \text{CM} = \sum_{i=1}^{n} y_i^2 - \text{CM}$$

$$\text{SS}(A) = \text{Sum of squares for main effects, factor } A$$

$$= \begin{pmatrix} \text{Sum of squares of the totals } A_1, A_2, \ldots, A_a \\ \text{divided by the number of measurements} \\ \text{in a single total, namely } br \end{pmatrix} - \text{CM}$$

$$= \frac{\sum_{i=1}^{a} A_i^2}{br} - \text{CM}$$

$$\text{SS}(B) = \text{Sum of squares for main effects, factor } B$$

$$= \begin{pmatrix} \text{Sum of squares of the totals } B_1, B_2, \ldots, B_b \\ \text{divided by the number of measurements} \\ \text{in a single total, namely } ar \end{pmatrix} - \text{CM}$$

$$= \frac{\sum_{i=1}^{b} B_i^2}{ar} - \text{CM}$$

$$\text{SS}(AB) = \text{Sum of squares for } AB \text{ interaction}$$

$$= \begin{pmatrix} \text{Sum of squares of the cell totals} \\ AB_{11}, AB_{12}, \ldots, AB_{ab} \text{ divided by} \\ \text{the number of measurements} \\ \text{in a single total, namely } r \end{pmatrix} - \text{SS}(A) - \text{SS}(B) - \text{CM}$$

$$= \frac{\sum_{j=1}^{b}\sum_{i=1}^{a} AB_{ij}^2}{r} - \text{SS}(A) - \text{SS}(B) - \text{CM}$$

where

a = Number of levels of factor A

b = Number of levels of factor B

r = Number of replicates (observations per treatment)

A_i = Total for level i of factor A ($i = 1, 2, \ldots, a$)

B_j = Total for level j of factor B ($j = 1, 2, \ldots, b$)

AB_{ij} = Total for treatment (i, j), i.e., for ith level of factor A and jth level of factor B

C.4 Tukey's Multiple Comparisons Procedure (Equal Sample Sizes)

Step 1 Select the desired experimentwise error rate, α

Step 2 Calculate

$$\omega = q_\alpha(k, v)\frac{s}{\sqrt{n_t}}$$

where

k = Number of sample means (i.e., number of treatments)

$s = \sqrt{\text{MSE}}$

v = Number of degress of freedom associated with MSE

n_t = Number of observations in each of the k samples (i.e., number of observations per treatment)

$q_\alpha(k, v)$ = Critical value of the Studentized range (Table XVII in Appendix B)

Step 3 Calculate and rank the k sample means.

Step 4 Place a bar over those pairs of treatment means that differ by less than ω. A pair of treatments not connected by an overbar (i.e., differing by more than ω) implies a difference in the corresponding population means.

Note: The confidence level associated with all inferences drawn from the analysis is $(1 - \alpha)$.

C.5 Bonferroni Multiple Comparisons Procedure (Pairwise Comparisons)

Step 1 Calculate for each treatment pair (i, j)

$$B_{ij} = t_{\alpha/(2c)} s \sqrt{\frac{1}{n_i} + \frac{1}{n_j}}$$

where

k = Number of sample (treatment) means in the experiment

c = Number of pairwise comparisons

[*Note:* If all pairwise comparisons are to be made, then $c = k(k - 1)/2$]

$s = \sqrt{\text{MSE}}$

v = Number of degrees of freedom associated with MSE

n_i = Number of observations in sample for treatment i

n_j = Number of observations in sample for treatment j

$t_{\alpha/(2c)}$ = Critical value of t distribution with v df and tail area $\alpha/(2c)$ (Table V in Appendix B)

Step 2 Rank the sample means and place a bar over any treatment pair (i, j) whose sample means differ by less than B_{ij}. Any pair of means not connected by an overbar implies a difference in the corresponding population means.

Note: The level of confidence associated with all inferences drawn from the analysis is at least $(1 - \alpha)$.

C.6 Scheffé's Multiple Comparisons Procedure (Pairwise Comparisons)

Step 1 Calculate Scheffé's critical difference for each pair of treatments (i, j):

$$S_{ij} = \sqrt{(k - 1)(F_\alpha)(\text{MSE})\left(\frac{1}{n_1} + \frac{1}{n_j}\right)}$$

where

k = Number of sample (treatment) means

MSE = Mean squared error

n_i = Number of observations in sample for treatment i

n_j = Number of observations in sample for treatment j

F_α = Critical value of F distribution with $k - 1$ numerator df and v denominator df (Tables XII, XIII, IX, and X in Appendix B)

v = Number of degrees of freedom associated with MSE

Step 2 Rank the k sample means and place a bar over any treatment pair (i, j) that differs by less than S_{ij}. Any pair of sample means not connected by an overbar implies a difference in the corresponding population means.

Answers to Selected Exercises

Chapter 1

1.3 population; variables; summary tools; conclusions **1.5** published source; designed experiment; survey; observationally **1.13** qualitative; qualitative **1.15 a.** all U.S. citizens **b.** president's job performance; qualitative **c.** 2000 polled individuals **d.** Estimate the proportion of all U.S. citizens who believe the president is doing a good job. **e.** survey **f.** not very likely **1.17** I. qualitative II. quantitative III. qualitative IV. qualitative V. qualitative VI. quantitative **1.19 a.** sample, if interested in CEOs of all U.S companies; population, if interested in only the 500 CEOs in the 2008 scoreboard **b.** (1) qualitative; (2) quantitative; (3) quantitative; (4) quantitative; (5) quantitative; (6) quantitative **1.21 a.** all satellite radio subscribers **b.** satellite receiver in car status **c.** qualitative **d.** 501 satellite subscribers surveyed; yes **e.** estimate of proportion of all satellite subscribers that have a car receiver is $396/501 = .79$ **1.23 a.** quantitative **b.** quantitative **c.** qualitative **d.** quantitative **e.** qualitative **f.** quantitative **g.** qualitative **1.25 b.** speed of the deliveries; accuracy of the invoices; quality of the packaging **c.** total numbers of questionnaires received **1.27 a.** all accounting alumni of the large university **b.** quantitative: age, income, job satisfaction score, Machiavellian rating; qualitative: gender, education level **c.** 198 alumni who returned the questionnaire **d.** survey **e.** Machiavellian behavior is not required to achieve success in business **f.** Nonrespondents may be more Machiavellian in nature, biasing the results **1.31 a.** all persons over 14 in U.S. **b.** employment status; qualitative **c.** inferential

Chapter 2

2.1 16; .18; .45; .15; .14 **2.3 a.** pie chart **b.** task category **c.** material handling (34%) **d.** 58,880 **e.** 52% **2.5 a.** qualitative **2.7 a.** all non-cash transactions made during the year **b.** 21.46 billion **c.** 50% **2.9** companies and employees created 73% of blogs/forums **2.11** Most often: F color (26.6%), clarity VS1 (26.3%); least often: D color (5.2%), IF clarity (14.3%) **2.13 a.** response time **c.** 3,570 **d.** No **2.15. d.** most aquifers are bedrock; most MTBE levels are below the limit; 80% of private wells and 60% of public wells are not contaminated **2.17** 50, 75, 125, 100, 25, 50, 50, 25 **2.19 a.** frequency histogram **b.** 14 **c.** 49 **2.21 a.** approximately 30% **b.** approximately 80% **2.23 b.** years 1996–2000 had the highest number of firms with at least one acquisition **2.25 d.** HRD group **2.27 c.** penalties for CAA tend to be smaller **2.29 a.** 2007 SAT scores are shifted to the right of the scores from 2000 **c.** similar conclusion **d.** Colorado ($\approx$ 50 point improvement) **2.31** "inside job" not likely because histogram appears similar to frequency distribution **2.33 a.** 12 **b.** 40 **c.** 7 **d.** 21 **e.** 144 **2.35 a.** 11.2 **b.** 12 **c.** 30 **2.37 a.** $\bar{x} = 2.717, m = 2.65$ **2.41** mode $= 15; \bar{x} = 14.545; m = 15$ **2.43 a.** mean less than median **b.** mean greater than median **c.** equal **2.45 a.** mean = 653; average amount of sparkling wine exported by the 30 countries is $653,000 **b.** median = 231; half the 30 export values are above $231,000 **c.** mean = 481; average change in amount exported by the 30 countries is 481% **d.** median = 156; half the 30 percentage changes are above 156% **2.47 a.** 11 **b.** 9 **c.** 7 **d.** honey dosage leads to greatest improvement **2.49 a.** $\bar{x} = .63$; average number of carats of 308 diamonds is .63 **b.** $m = .62$; 50% of the diamonds weigh less than .62 carat **c.** mode = 1.0; carat value of 1.0 occurred the most often **d.** mean or median **2.51** data are probably not skewed but close to symmetric **2.53** joint: $= 2.6545, m = 1.5$; no prefiling: $= 4.2364, m = 3.2$; prepack: $= 1.8185, m = 1.4$; three centers **2.55 a.** median **b.** mean **2.57 a.** $R = 4, s^2 = 2.3, s = 1.52$ **b.** $R = 6, s^2 = 3.619, s = 1.90$ **c.** $R = 10, s^2 = 7.111, s = 2.67$ **d.** $R = 4, s^2 = 1.395, s = 1.18$ **2.59 a.** $\bar{x} = 5.6, s^2 = 17.3, s = 4.1593$ **b.** $\bar{x} = 13.75$ feet, $s^2 = 152.25$ square feet, $s = 12.339$ feet **c.** $\bar{x} = -2.5, s^2 = 4.3, s = 2.0736$ **d.** $\bar{x} = .33$ ounce, $s^2 = .0587$ square ounce, $s = .2422$ ounce **2.61** Data set 1: 0,1,2,3,4,5,6,7,8,9; data set 2: 0,0,1,1,2,2,3,3,9,9 **2.63 a.** $R = 3, s^2 = 1.3, s = 1.14$ **b.** $R = 3, s^2 = 1.3, s = 1.14$ **c.** $R = 3, s^2 = 1.3, s = 1.14$ **d.** no effect **2.65 a.** R = $4,882 thousand **b.** s = $1,113 thousand **c.** 1,238,769 **2.67 a.** .92 **b.** .0768 **c.** .2772 **d.** standard deviation **2.69 a.** $R = 10, s^2 = 7.10, s = 2.67$ **b.** $R = 8, s^2 = 4.67, s = 2.16$ **c.** $R = 8, s^2 = 4.50, s = 2.12$ **2.71 a.** dollars; quantitative **b.** at least $3/4$; at least $8/9$; nothing; nothing **2.73** approx. 68%; approx. 95%; essentially all **2.75** between $R/6 = 104.17$ and $R/4 = 156.25$; not feasible **2.77 a.** (105.77, 176.85) **b.** at least $3/4$ **c.** nothing **2.79 a.** at most 25% **b.** $\approx$ 2.5% **2.81.** $\bar{x} = 234.74, s = 9.91$; (205.0, 264.5) **2.83 a.** at least $8/9$ of the velocities fall within (906, 966) **b.** no **2.85** do not buy **2.87** 11:30 and 4:00 **2.89 a.** 25%, 75% **b.** 50%, 50% **c.** 80%, 20% **d.** 16%, 84% **2.91 a.** $z = 2$ **b.** $z = -3$ **c.** $z = -2$ **d.** $z = 1.67$ **2.93** average score is 279; 10% of 8th graders score below 231; 25% score below 255; 75% score below 304; 90% score below 324 **2.95** half the graduates have starting salaries less than $41,100; half the graduates have mid-career salaries less than $71,100; 90% of the graduates have mid-career salaries less than $131,000 **2.97** no **2.99 a.** 0 **b.** 21 **c.** $\bar{x} = 5.24, s = 7.24; z = 5.90$ **d.** yes **2.101** not necessarily because the standard deviation will be large for right-skewed data **2.103 a.** $z = .727$, no **b.** $z = -3.273$, yes **c.** $z = 1.364$, no **d.** $z = 3.727$, yes **2.107 a.** 50% of expenditures are less than $6,232; 25% are less than $5,309; 75% are less than $7,216 **b.** $1,907 **c.** $.75 - .25 = .50$ **2.109. a.** $z = 1.05$, no **b.** greater than 194.6 or less than 88 **2.111 b.** Joint: 1.5; None firms: 3.2; Prepack: 1.4 **d.** no **e.** yes **2.113 a.** 8 outliers: 56, 72, 74, 78, 83, 83, 84, and 84 **b.** 4 outliers: 56, 72, 74, and 78 **c.** no **2.115 b.** customers 268, 269, and 264 **c.** 2.06, 2.13, and 3.14 **2.119** no **2.121** increasing **2.123 a.** yes **b.** USAID and State **2.125** moderate positive trend; yes **2.129 a.** $-1, 1, 2$ **b.** 0, 4, 6 **c.** 1, 3, 4 **d.** .1, .3, .4 **2.131 a.** 6, 27, 5.20 **b.** 6.25, 28.25, 5.32 **c.** 7, 37.67, 6.14 **d.** 3, 0, 0 **2.137 a.** .082, .173, .167, .174, .139, .098, .074, .046, .048; 16.7% of articles published were in the Buyer Behavior area **b.** lower percentage for Marketing Research **2.139** 60% of cars have 4-star rating **2.141 a.** body defect **b.** paint or dents **2.143** $\bar{x} = 52.04; m = 56.7$; mode = 56.7 **2.145 a.** $\bar{x} = 144.5; m = 102.5$; mode = 70 **b.** $\bar{x} = 5.23; m = 5$; mode = 6 **c.** median **d.** mean **e.** increasing **2.147 a.** grounding and fire **b.** $\bar{x} = 66.19, s = 56.05$; $(-101.96, 234.34)$ **2.149 c.** approved brands tend to have lower costs **2.151 a.** no scale on vertical axis **b.** add vertical axis scale **2.153 a.** skewed right **c.** $\approx$ .38 **d.** no, $z = 3.333$ **2.155 a.** both height and width of bars change **2.157** Laid off: $m = 40.5$; not laid off: $m = 40$; company probably not vulnerable **2.159** yes; no observations recorded in interval just below interval centered at 1.000 cm

Chapter 3

3.1 a. .5 **b.** .3 **c.** .6 **3.3** $P(A) = .55, P(B) = .50, P(C) = .70$ **3.5 a.** 10 **b.** 20 **c.** 15,504 **3.7 a.** $(B_1, B_2), (B_1, R_1), (B_1, R_2), (B_1, R_3), (B_2, R_1), (B_2, R_2), (B_2, R_3), (R_1, R_2), (R_1, R_3), (R_2, R_3)$ **b.** $P(E_i) = 1/10$ **c.** $P(A) = 1/10, P(B) = 3/5, P(C) = 3/10$ **3.9 a.** Brown, yellow, red, blue, orange, and green **b.** $P(\text{Br}) = 0.13, P(\text{Y}) = 0.14, P(\text{R}) = 0.13, P(\text{Bl}) = 0.24, P(\text{O}) = 0.2, P(\text{G}) = 0.16$ **c.** .13 **d.** .43 **e.** .76 **3.11** .0057 **3.13 a.** None, 1-2, 3-5, 6-9, 10 or more **b.** $P(\text{none}) = 0.25, P(1\text{–}2) = 0.31, P(3\text{–}5) = 0.25, P(6\text{–}9) = 0.05, P(10 \text{ or more}) = 0.14$ **c.** .44 **3.15 a.** 1/100 = .01 **b.** .0095; yes **3.17 a.** .325 **b.** .711 **3.19** 455 **3.21 a.** 6; $(G_1G_2G_3), (G_1G_2G_4), (G_1G_2G_5), (G_2G_3G_4), (G_2G_3G_5), (G_2G_4G_5)$ **b.** .282, .065, .339, .032, .008, .274 **c.** .686 **3.23 a.** 1 to 2 **b.** 1/2 **c.** 2/5 **3.25 a.** 32 **b.** 1/32 **c.** 12 **d.** 12/32 = .375 **e.** .5 **3.27 b.** $P(A) = 7/8, P(B) = 1/2, P(A \cup B) = 7/8, P(A^c) = 1/8, P(A \cap B) = 1/2$ **d.** no **3.29 a.** 3/4 **b.** 13/20 **c.** 1 **d.** 2/5 **e.** 1/4 **f.** 7/20 **g.** 1 **h.** 1/4 **3.31 a.** .65 **b.** .72 **c.** .25 **d.** .08 **e.** .35 **f.** .72 **g.** 0 **h.** *A* and *C*, *B* and *C*, *C* and *D* **3.33 a.** $A \cap B$ **b.** A^c **c.** $C \cup B$ **d.** $B^c \cap A^c$ **3.35 a.** .43 **b.** .84 **3.37 a.** AC, AW, AF, IC, IW, and IF **b.** .148, .066, .426, .176, .052, .132 **c.** .640 **d.** .118 **e.** .176 **f.** .786 **g.** .676 **3.39 a.** $P \cap S \cap A$ **b.** 1/5 **c.** $A \cup S$; 4/5 **d.** $P \cap S$; 3/10 **3.41 a.** .684 **b.** .124 **c.** no **d.** .316 **e.** .717 **f.** .091 **3.43 a.** $P(A) = .281, P(B) = .276, P(C) = .044, P(D) = .079, P(E) = .044$ **b.** 0 **c.** .557 **d.** 0 **e.** .325 **f.** *A* and *B*, *A* and *C*, *A* and *D*, *A* and *E* **3.45 a.** .09 **b.** .09 **c.** .84 **d.** no **e.** column events not mutually exclusive **3.47 a.** .5 **b.** .25 **c.** no **3.49 a.** .08 **b.** .4 **c.** .52 **3.51 a.** .8, .7, .6 **b.** .25, .375, .375 **d.** no **3.53 a.** .37 **b.** .68 **c.** .15 **d.** .2206 **e.** 0 **f.** 0 **g.** no **3.55 a.** *A and C*, *B and C* **b.** none **c.** .65, .90 **3.57** 3/5 **3.59 a.** .26 **b.** .013 **3.61 a.** .1875 **b.** .60 **3.63 a.** $P(R) = .68, P(A|R) = .07$ **b.** $P(A \cap R) = .048$ **3.65 a.** .222 **b.** .183 **3.67 a.** $P(A \mid I) = .9, P(B \mid I) = .95, P(A \mid I^c) = .2, P(B \mid I^c) = .1$ **b.** .855 **c.** .02 **d.** .995 **3.69 a.** .23 **b.** .729 **3.71.** .60 **3.73 a.** $(.5)^{10} = .000977$ **b.** .00195 **c.** .99805 **3.75** random samples are likely to be representative **3.77 a.** 35,820,200 **b.** 1/35,820,200 **3.81 a.** .000186 **c.** no **3.85 a.** .158 **b.** .074 **c.** .768 **3.87** .0063 **3.89 a.** .52, .39, .09 **b.** .516 **3.91** .6982 **3.93 a.** Supplier 4; $P(S_4 \mid D) = .7147$ **b.** Suppliers 4 or 6 **3.95 b.** .6 **3.97 a.** 0 **b.** no **3.99** .5 **3.101 a.** 0, .2, .9, 1, .7, .3, .4, 0 **3.103 a.** 2,118,760 **3.105** .73 **3.107 a.** false **b.** true **c.** true **d.** false **3.109 a.** $B \cap C$ **b.** A^c **c.** $C \cup B$ **d.** $A \cap C^c$ **3.111 b.** .95 **c.** .25 **d.** .5 **3.113. a.** .5 **b.** .034 **c.** .058 **3.115 a.** .00000625 **b.** .0135 **c.** doubt validity of the manufacturer's claim **d.** no **3.117 a.** .006 **b.** .012 **c.** .018 **3.119 a.** .006 **b.** .0022 **c.** .4133 **d.** .3601 **e.** no **3.121** .79 **3.123 a.** LLLL, LLLU, LLUL, LULL, ULLL, LLUU, LULU, LUUL, ULLU, ULUL, UULL, LUUU, ULUU, UULU, UUUL, UUUU **b.** 1/16 **c.** 5/16 **3.125 a.** .7127 **b.** .2873 **c.** .9639 **d.** .3078 **e.** .0361 **f.** 3 **3.127** .526 **3.129 a.** .0362 **b.** .0352 **3.131** Marilyn

Chapter 4

4.1 a. discrete **b.** continuous **c.** continuous **d.** discrete **e.** discrete **4.3** discrete **4.5** discrete **4.11 a.** $p(x) = 1/6$ for all x values **4.13 a.** .25 **b.** .40 **c.** .75 **4.15 a.** $p(0) = p(3) = 1/8, p(1) = p(2) = 3/8$ **b.** 1/2 **4.17 a.** $\mu = 0, \sigma^2 = 2.94, \sigma = 1.71$ **c.** .96 **4.19 a.** $p(1) = 0, p(2) = .0408, p(3) = .1735, p(4) = .6020, p(5) = .1837$ **b.** .1837 **c.** .0408 **d.** 3.93 **4.21 a.** yes **b.** .06 **c.** .28 **d.** .82 **4.23 b.** $p(0) = .1160, p(1) = .3124, p(2) = .3364, p(3) = .1811, p(4) = .0488, p(5) = .0053$ **d.** .0541 **4.25 a.** .23 **b.** .0809 **c.** .77 **4.27 a.** .508 **b.** .391 **c.** .094 **d.** .007 **4.29 b.** .85 **c.** .6, 0, 0, 0, .7 **d.** a_1: 2.4; a_2: 1.5; $a_3 - a_5$: .90; a_6: 1.65 **e.** a_1: .86, (.68, 4.12); a_2: .67, (.16, 2.84); $a_3 - a_5$: .3, (.3, 1.5); a_6: .57, (.51, 2.79) **4.31 a.** p($300,000) = .3, p($0) = .7 **b.** $90,000 **4.33 a.** $p(x) = .05$ for all x-values **b.** 52.5 **c.** (−5.16, 110.16) **f.** 33.25, 38.3577 **g.** .525 **i.** .20 **j.** .65 **4.35** $.25 **4.37 a.** discrete **b.** binomial **d.** 3.5; 1.02 **4.39 a.** .4096 **b.** .3456 **c.** .027 **d.** .0081 **e.** .3456 **f.** .027 **4.41 a.** 12.5, 6.25, 2.5 **b.** 16, 12.8, 3.578 **c.** 60, 24, 4.899 **d.** 63, 6.3, 2.510 **e.** 48, 9.6, 3.098 **f.** 40, 38.4, 6.197 **4.43 b.** .20 **c.** 20 **4.45 a.** 54 **b.** .0806, .1056 **4.47 b.** $n = 20, p = .8$. **c.** .174 **d.** .804 **e.** 16 **4.49 a.** .055 **b.** in 2020, more than 9% of Denver bridges will have a rating of 4 or below **4.51 a.** .01; .09 **b.** .048; .001 **c.** .3086; .0674 **d.** .8116 **4.53 a.** $\mu = 560, \sigma = 12.96$ **b.** no, $z = -12.35$ **4.55 b.** $\mu = 2.4, \sigma = 1.47$ **c.** $p = .90, q = .10, n = 24, \mu = 21.60, \sigma = 1.47$ **4.57 a.** discrete **b.** Poisson **c.** $\mu = 3, \sigma = 1.73$ **4.59 a.** .3 **b.** .119 **c.** .167 **d.** .167 **4.61 a.** .920 **b.** .677 **c.** .423 **d.** decreases **4.63 a.** hypergeometric **b.** binomial **4.65 a.** .383 **b.** .0002 **4.67 a.** .0111 **b.** .05 **c.** $\mu = 4.5, \sigma = 2.12$ **4.69** .2693 **4.71 a.** .125 **b.** 5 **4.73** .25 **4.75 a.** ≈0 **b.** ≈0 **c.** yes **4.77** No; probability of only 5 females selected is ≈ 0 **4.79 a.** .0721 **b.** .0594 **c.** .2434 **d.** .3457 **e.** .5 **f.** .9233 **4.81 a.** .6826 **b.** .9500 **c.** .90 **d.** .9544 **4.83 a.** 0 **b.** .8413 **c.** .8413 **d.** .1587 **4.85 a.** −.81 **b.** .55 **c.** 1.43 **d.** .21 **e.** −2.05 **f.** .50 **4.87 a.** −2.5 **b.** 0 **c.** −.625 **d.** −3.75 **e.** 1.25 **f.** −1.25 **4.89 a.** .3830 **b.** .3023 **c.** .1525 **d.** .7333 **e.** .1314 **f.** .9545 **4.91 a.** .1558 **b.** .2586 **c.** .0062 **d.** .9525 **4.93 a.** .2743 **b.** .008 **4.95 a.** .8413 **b.** .7528 **4.97 a.** 69.2 **b.** 42.2 **4.99 a.** 25.14% **b.** 90.375 **4.101 a.** .8106 **b.** .1894 **c.** $1,246,100 **4.103** 5.1 **4.105 a.** .68 **b.** .95 **c.** 1.00 **4.107** Plot c **4.109 a.** 7; yes **b.** 6.444 **c.** IQR/s = 1.09 **4.111** no **4.113** IQR/s = 1.3 **4.115** none **4.117** $z = -.75$ for minimum value of 128 **4.119 a.** no **b.** yes **c.** no **d.** yes **e.** yes **f.** yes **4.121 a.** .1788 **b.** .5236 **c.** .6950 **4.123** .0537 **4.125 a.** 280 **b.** ≈0 **4.127** .1762 **4.129** .2676; no **4.131 a.** .4681 **b.** .0436 **c.** .9822 **4.133 a.** $f(x) = .04$ $(20 \leq x \leq 45)$, 0 otherwise **b.** 32.5, 7.22 **4.135 a.** $f(x) = 1/4$ $(3 \leq x \leq 7)$, 0 otherwise **b.** 5, 1.155 **c.** .578 **4.137 a.** .367879 **b.** .950213 **c.** .223130 **d.** .993262 **4.139 a.** 0 **b.** 1 **c.** 1 **4.141 a.** .133; .571 **b.** .267; 0 **4.143 a.** .449329 **b.** .864665 **4.145** yes **4.147 a.** continuous **c.** 7, .2887, (6.422, 7.577) **d.** .5 **e.** 0 **f.** .75 **g.** .0002 **4.149 a.** 17 **b.** .5862 **4.151 a.** .753403 **b.** .667 **c.** .811 **4.153** .4444 **4.159** 2.7 **4.163** no **4.165 a.** 100, 5 **b.** 100, 2 **c.** 100, 1 **d.** 100, 1.414 **e.** 100, .447 **f.** 100, .316 **4.167 a.** .0228 **b.** .0668 **c.** .0062 **d.** .8185 **e.** .0013 **4.169 a.** .8944 **b.** .0228 **c.** .1292 **d.** .9699 **4.171 a.** 141 **b.** 1.8 **c.** ≈ normal **d.** 0.56 **e.** .2877 **4.173** .0838 **4.175 a.** ≈ normal with $\mu_{\bar{x}} = .53$ and $\sigma_{\bar{x}} = .0273$ **b.** .0336 **c.** before: $p = .0139$; after: $p = .3557$; after **4.177 a.** .0034 **b.** ≈0; true mean is larger than 6 ppb **4.179 a.** ≈ normal **b.** .0091 **c.** .9544 **4.181** Handrubbing: $p = .2743$; handwashing: $p = .0047$; handrubbing **4.183 a.** .2734 **b.** .4096 **c.** .3432 **4.185 a.** .192 **b.** .228 **c.** .772 **d.** .987 **e.** .960 **f.** 14, 4.2, 2.049 **g.** .975 **4.187 a.** Poisson **b.** binomial **c.** binomial **4.189 a.** discrete **b.** continuous **c.** continuous **d.** continuous **4.191 a.** .9821 **b.** .0179 **c.** .9505 **d.** .3243 **e.** .9107 **f.** .0764 **4.193 a.** exponential **b.** uniform **c.** normal **4.195 a.** .6915 **b.** .1587 **c.** .1915 **d.** .3085 **e.** 0 **f.** 1 **4.197 a.** 47.68 **b.** 47.68 **c.** 30.13 **d.** 41.5 **e.** 30.13 **4.199 a.** .5 **b.** .0606 **c.** .0985 **d.** .8436 **4.203 b.** .2592 **c.** .0870 **d.** .6826 **e.** 3.00 **4.205 a.** .5, .289 **b.** .2 **c.** no, $p = .005$ **4.207 a.** 89.34, 1.31 **c.** .8461 **d.** .0367 **4.209 a.** .221199 **b.** .002394 **c.** .082085 **4.211 a.** .367869 **b.** .606531 **c.** .135335; .135335 **d.** .814046 **4.213 b.** 20, 4.47 **c.** no, $z = -3.58$ **d.** 0 **4.215** no **4.217 a.** .8264 **b.** 17 times **c.** .6217 **d.** 0, −157 **4.219 a.** 1.25, 1.09; no **b.** .007 **c.** not applicable **4.221 a.** $\mu = 113.24, \sigma = 4.19$ **b.** $z = 6.38$ **4.223** 292 **4.225** .9332 **4.227** $\bar{x} = 5.935$ minutes; $P(\bar{x} < 6 \mid \mu = 15) \approx 0$, therefore, appears drug is effective

Chapter 5

5.1 a. 1.645 **b.** 2.575 **c.** 1.96 **d.** 1.28 **5.3 a.** 28 ± .784 **b.** 102 ± .65 **c.** 15 ± .0588 **d.** 4.05 ± .163 **e.** no **5.5 a.** 26.2 ± .96 **b.** In repeated sampling, 95% of all confidence intervals constructed will include μ. **c.** 26.2 ± 1.26 **d.** increases **e.** yes **5.9** yes **5.11** 6,563 ± 97.65 **5.13 a.** 4.25 ± 4.14 **b.** 99% confident that true mean number of blogs/forums falls between .11 and 8.39 **c.** no; apply the Central Limit Theorem **5.15 a.** μ = mean salary of all 500 CEOs **5.17 a.** 99.6 **b.** 99.6 ± 2.24 **c.** 95% confident that true Mach rating score of all purchasing managers is between 97.36 and 101.84. **d.** yes **5.19** (12.43, 25.57); (2.05, 11.95); SAT–Math **5.21 a.** $z_{.10} = 1.28, t_{.10} = 1.533$ **b.** $z_{.05} = 1.645, t_{.05} = 2.132$ **c.** $z_{.025} = 1.96, t_{.025} = 2.776$ **d.** $z_{.01} = 2.33, t_{.01} = 3.747$ **e.** $z_{.005} = 2.575, t_{.005} = 4.604$ **5.23 a.** 2.228 **b.** 2.228 **c.** −1.812 **d.** 1.725 **e.** 4.032 **5.25 a.** 5 ± 1.88 **b.** 5 ± 2.39 **c.** 5 ± 3.75 **d.** 5 ± .78, 5 ± .94, 5 ± 1.28; decreased width **5.27** (71.67, 79.13) **5.29 a.** (−320, 5,922) **b.** normally distributed **5.31 a.** (303.4, 413.6) **c.** ≈ normal **d.** no **5.33 a.** $\bar{x} = 1.857, s = 1.195$ **c.** 1.857 ± .45 **d.** 90% confident that true mean number of suicide bombings/attacks per transgression against United States is between 1.407 and 2.307 **e.** .90 **5.35 a.** all 441 firms in *Forbe*s Largest Private Companies list **b.** 4.61 ± 3.06 **d.** ≈ normal **e.** yes **5.37 a.** yes **b.** no **c.** no **d.** no **5.39 a.** yes **b.** .46 ± .065 **5.41 a.** all American adults **b.** 1,000 adults surveyed **c.** proportion of all American adults who think Starbucks coffee is overpriced **d.** .73 ± .028 **5.43 a.** .46 **b.** (.433, .487) **c.** 90% confident that true proportion of contractors with Web site falls between .433 and .487 **d.** 90% of all similarly constructed intervals contain the true proportion **5.45 a.** .03 **b.** .03 ± .01 **5.47 a.** .42 **b.** .42 ± .015 **c.** yes **5.49** .269 ± .014; 90% confident that true percentage of all firms with acquisitions is between 25.5% and 28.3% **5.51 a.** .867 **b.** (.781, .953) **d.** not believable **5.53 a.** true proportion of all fillets that are red snapper **c.** .27 ± .17 **d.** 95% confident that true proportion of all fillets that are red snapper is between .10 and .44 **5.55** .85 ± .002 **5.57 a.** 68 **b.** 31 **5.59** 34 **5.61 a.** 0.98, 0.784, 0.56, 0.392, 0.196 **5.63** 21 **5.65** 1,692 **5.67** 14,735 **5.69** 43; 171; 385 **5.71** no **5.73 a.** .7746 **b.** .8944 **c.** .9487 **d.** .995 **5.75 a.** 1.00 **b.** .6124 **c.** 0 **d.** As n increases, standard error decreases. **5.77** .42 ± .011 **5.79 a.** 36.03 ± 3.40 **b.** .7 ± .159 **5.81 a.** .56 ± .012 **c.** 95% confident that true proportion of active NFL players who select a professional coach as most influential is between .548 and .572 **5.83 a.** 156.46 **b.** 18.70 **c.** 156.46 ± 37.405 **d.** not reasonable **5.85** No; .086 ± .041 **5.87 a.** −1.725 **b.** 3.250 **c.** 1.860 **d.** 2.898 **5.89 a.** 32.5 ± 5.15 **b.** 23,871 **5.91** (1) p = proportion with excellent health; (2), (3), and (4) μ = mean number of days health not good **5.93 b.** .29 ± .028 **5.95** 97 **5.97** .694 ± .092 **5.99** .667 ± .065 **5.101 a.** 49.3 ± 8.6 **b.** 99% confident that mean amount removed from all soil specimens is between 40.70% and 57.90% **c.** normal distribution **d.** possible **5.103 a.** 66.83 ± 6.17 **b.** 45.39 ± 4.14 **5.105** 154 **5.107 a.** .833 ± .149 **b.** no **c.** 1,337 **5.111** 818 **5.115 a.** yes **b.** missing measure of reliability **c.** 95% CI for μ: .932 ± .037

Chapter 6

6.1 null; alternative **6.3** α **6.5** Reject H_0 when H_0 is true; accept H_0 when H_0 is true; Reject H_0 when H_0 is false; accept H_0 when H_0 is false **6.7** no **6.9 a.** H_0: $p = .60$ **b.** $|z| > 2.576$ **6.11** H_0: $p = .045$, H_a: $p < .045$ **6.13** H_0: $\mu = 863$, H_a: $\mu < 863$ **6.15 a.** unsafe; safe **c.** α **6.17 c.** α **e.** decrease **f.** increase **6.19 a.** $z > 2.14$ **b.** .0162 **6.21 a.** $z = -1.61$, reject H_0 **b.** $z = -1.61$, do not reject H_0 **6.23 a.** $z < -2.33$ **b.** $z = -.40$ **c.** do not reject H_0 **6.25** yes; $z = 2.50$, do not reject H_0: $\mu = 220$ **6.27 a.** H_0: $\mu = .250$, H_a: $\mu \neq .250$ **b.** yes, $z = 7.02$ **6.29** $z = -1.86$, do not reject H_0 **6.31 a.** no **b.** $z = .61$, do not reject H_0 **d.** no **e.** $z = -.83$, do not reject H_0 **6.33 a.** $z = 2.73$, reject H_0 **c.** no **6.35 a.** do not reject H_0 **b.** reject H_0 **c.** reject H_0 **d.** do not reject H_0 **e.** do not reject H_0 **6.37** .0150 **6.39** .03 **6.41** .06 **6.43 a.** .057 **b.** do not reject H_0 **6.45 a.** .0693 **b.** do not reject H_0 **6.47 a.** $z = 26.15$, p-value ≈ 0, reject H_0 **b.** $z = -14.24$, p-value ≈ 0, reject H_0 **6.49 a.** small n, data normal **b.** mound-shaped and symmetric; t is flatter than z **6.51 a.** $|t| > 2.160$ **b.** $t > 2.500$ **c.** $t > 1.397$ **d.** $t < -2.718$ **e.** $|t| > 1.729$ **f.** $t < -2.353$ **6.53 a.** population is normally distributed **b.** reject H_0 at $\alpha = .05$ **c.** .076 **6.55 a.** H_0: $\mu = 2$, H_a: $\mu \neq 2$ **b.** $t = -1.02$ **c.** $|t| > 2.093$ **d.** do not reject H_0 **e.** .322 **6.57 a.** yes, p-value = .0229 **b.** 90% CI: (1.41, 2.31); evidence that $\mu \neq 2.5$ **c.** yes **d.** population is normal **e.** validity of inference is suspect **6.59** yes, $t = -2.53$ **6.61** no, $t = 2.97$ **6.63** yes, using $\alpha = .10$; Plant 1: p-value = .197; Plant 2: p-value = .094 **6.65 a.** −2.33 **c.** reject H_0 **d.** .0099 **6.67 a.** $z = 1.13$, do not reject H_0 **b.** .1292 **6.69 a.** .42 **b.** H_0: $p = .3$, H_a: $p > .3$ **c.** $z = 16.56$ **d.** $z > 2.33$ **e.** reject H_0 **f.** ≈ 0 **6.71** yes, $z = -3.54$ **6.73 a.** H_0: $p = .5$, H_a: $p < .5$ **b.** .231 **c.** do not reject H_0 **6.75 a.** no, $z = 1.49$ **b.** .0681 **6.77 a.** no, $z = 1.37$ **b.** yes, $z = -4.58$ **6.79** $z = 1.20$, do not reject H_0 **6.81 b.** 532.9 **d.** .1949 **e.** .8051 **6.83 c.** .1469 **d.** .8531 **6.85 c.** .5359 **d.** .0409 **6.87 a.** .1949, Type II error **b.** .05, Type I error **c.** .8051 **6.89** .1075 **6.91 a.** $\chi^2 < 6.26214$ or $\chi^2 > 27.4884$ **b.** $\chi^2 > 40.2894$ **c.** $\chi^2 > 21.0642$ **d.** $\chi^2 < 3.57056$ **e.** $\chi^2 < 1.63539$ or $\chi^2 > 12.5916$ **f.** $\chi^2 < 13.8484$ **6.93 a.** $\chi^2 = 479.16$, reject H_0 **6.95 a.** $\chi^2 < 24.3$ or $\chi^2 > 73.1$ **b.** $\chi^2 = 63.72$ **c.** do not reject H_0 **6.97 a.** H_0: $\sigma^2 = .000004$, H_a: $\sigma^2 \neq .000004$ **b.** $\chi^2 = 48.49$, do not reject H_0 **c.** tee weights ≈ normal **6.99** at $\alpha = .05$, yes; $\chi^2 = 133.90$ **6.101 b.** 1.76 **c.** yes; $\chi^2 = 99$ **6.103** alternative **6.105** big **6.107 a.** $t = -7.51$, reject H_0 **b.** $t = -7.51$, reject H_0 **6.109 a.** $z = -1.78$, reject H_0 **b.** $z = -1.78$, do not reject H_0 **6.111 a.** $\chi^2 = 63.48$, reject H_0 **b.** $\chi^2 = 63.48$, reject H_0 **6.113 a.** .52 **b.** H_0: $p = .62$, H_a: $p \neq .62$ **c.** $z = -5.16$ **d.** $|z| > 1.96$ **e.** reject H_0 **f.** ≈ 0 **6.115 a.** H_0: $\mu = 1$, H_a: $\mu > 1$ **b.** reject H_0 if $t > 1.345$ **d.** t = 2.41 **e.** .01 < p-value < .025, reject H_0 **f.** $\chi^2 = 12.78$, do not reject H_0 **6.117** p-value = .257, do not reject H_0 **6.119 a.** H_0: No disease, H_a: Disease **6.121** $\chi^2 = 187.9$, do not reject H_0 **6.123 a.** $z = -3.22$, do not reject H_0 **b.** $p \approx 1$ **c.** .8159 **6.125 a.** concluding percentage of shoplifters turned over to police is 50% when, in fact, the percentage is higher than 50% **b.** .8461 **c.** decreases to .7389 **6.127 a.** no, $z = 1.41$ **b.** small **c.** .0793 **6.129 a.** no **b.** $\beta = .5910$, power = .4090 **c.** increases **6.131 a.** $z = -1.29$, do not reject H_0 **b.** .0985 **c.** Small **6.133 a.** $z = 2.73$, reject H_0 **6.135** $z = 7.83$, reject H_0

Chapter 7

7.1 a. 150 ± 6 **b.** 150 ± 8 **c.** 0; 5 **d.** 0 ± 10 **e.** variability of the difference is greater **7.3 a.** 35 ± 24.5 **b.** $z = 2.8$, p-value = .0052, reject H_0 **c.** p-value = .0026 **d.** $z = .8$, p-value = .4238, do not reject H_0 **e.** independent random samples **7.5 a.** no **b.** no **c.** no **d.** yes **e.** no **7.7 a.** .5989 **b.** yes, $t = -2.39$ **c.** −1.24 ± .98 **d.** confidence interval **7.9 a.** do not reject H_0 **b.** .0575 **7.11 a.** $t = -1.646$, do not reject H_0 **b.** -2.50 ± 3.12 **7.13 a.** H_0: $\mu_1 - \mu_2 = 0$, H_a: $\mu_1 - \mu_2 \neq 0$ **b.** $t = .62$ **c.** $|t| > 1.684$ **d.** do not reject H_0 **e.** do not reject H_0 **f.** independent random samples, both populations normal, $\sigma_1^2 = \sigma_2^2$ **7.15 a.** $.90 \pm .064$ **b.** Buy-side **c.** large, independent random samples **7.17 a.** $t = 1.56$, do not reject H_0: $\mu_T = \mu_I$ **b.** $t = -.50$, do not reject H_0: $\mu_W = \mu_I$ **c.** no differences **7.19** $z = -2.81$, reject H_0: $\mu_C = \mu_R$ **7.21** yes, $z = 3.20$, sufficient evidence to conclude $\mu_{Honey} > \mu_{DM}$ **7.23 a.** no, $t = -1.22$ **b.** yes, $t = -4.19$ **7.25 a.** $t > 1.796$ **b.** $t > 1.319$ **c.** $t > 3.182$ **d.** $t > 2.374$ **7.27 a.** H_0: $\mu_d = 0$, H_a: $\mu_d < 0$ **b.** $t = -5.29$, p-value $= \approx 0$, reject H_0 **c.** $(-4.98, -2.42)$ **d.** population of differences is normal **7.29 a.** $z = 1.79$, do not reject H_0 **b.** .0734 **c.** no **7.31 a.** H_0: $\mu_d = 0$, H_a: $\mu_d < 0$, where $\mu_d = \mu_{2003} - \mu_{2005}$ **c.** $\bar{d} = -1.98$, $s_d = 3.60$ **d.** $t = -1.98$ **e.** $t < -1.356$ **f.** reject H_0 **g.** population of differences normal; yes **7.33 a.** response rates observed twice for each survey **b.** no, $t = .53$ **c.** reject H_0 **7.35 a.** yes, $t = 2.864$ **7.37** (.036, .224); Huffman-coding **7.39** $t = .46$, p-value $= 0.65$, do not reject H_0 **7.41 a.** binomial distributions **b.** distribution is approx. normal for large n's **7.43 a.** no **b.** no **c.** no **d.** no **e.** no **7.45 a.** $z = -4.02$, reject H_0 **b.** $z = -4.02$, reject H_0 **c.** $z = -4.02$, reject H_0 **d.** $-.10 \pm .04$ **7.47 a.** .153 **b.** .215 **c.** (−.132, .008) **d.** no evidence of a difference **7.49 a.** $p_1 - p_2$ **b.** H_0: $p_1 - p_2 = 0$, H_a: $p_1 - p_2 \neq 0$ **c.** $z = -5.16$ **d.** $|z| > 2.58$ **e.** yes **f.** reject H_0 **7.51 a.** all male and all female cell phone users **b.** .32; .25 **c.** $.07 \pm .04$ **d.** yes **e.** .71; .77 **f.** $z = -2.45$, reject H_0 **7.53 a.** $z = 94.35$, reject H_0 **b.** $(-.155, -.123)$ **7.55** yes; 95% CI for $(p_{AA} - p_{White}) = (.089, .131)$ **7.57 a.** 500, 46 **b.** sample may not be representative **7.59 a.** 29,954 **b.** 2,165 **c.** 1,113 **7.61** 34 **7.63** 54 **7.65** 13,531 for each survey **7.67** 1,729 **7.69 a.** 5,053 **7.71 a.** 4.10 **b.** 3.57 **c.** 8.81 **d.** 3.21 **7.73 a.** $F > 1.74$ **b.** $F > 2.04$ **c.** $F > 2.35$ **d.** $F > 2.78$ **7.75 a.** $F > 2.11$ **b.** $F > 3.01$ **c.** $F > 1.93$ **d.** $F > 2.30$ **e.** $F > 3.41$ **7.77 a.** $F = 4.29$, do not reject H_0 **b.** $.05 < p$-value $< .10$ **7.79 a.** H_0: $\sigma_M^2 = \sigma_F^2$, H_a: $\sigma_M^2 < \sigma_F^2$ **b.** 1.06 **c.** $F > 1.26$ **d.** p-value $> .10$ **e.** do not reject H_0 **7.81 a.** no, $F = 2.27$ **b.** $.05 < p$-value $< .10$ **7.83 a.** yes, $F = 8.29$ **b.** no **7.85** no, $F = 1.30$ **7.87 a.** $t = .78$, do not reject H_0 **b.** 2.50 ± 8.99 **c.** 225 **7.89 a.** $3.90 \pm .31$ **b.** $z = 20.60$, reject H_0 **c.** 346 **7.91 a.** $t = 5.73$, reject H_0 **b.** 3.8 ± 1.84 **7.93 a.** $\mu_1 - \mu_2$ **b.** $p_1 - p_2$ **c.** μ_d **7.95 a.** yes **b.** $z = 4.58$, reject H_0 **7.97 b.** H_0: $\mu_d = 0$, H_a: $\mu_d \neq 0$ **c.** do not reject H_0 **7.99 a.** H_0: $\mu_1 = \mu_2$, H_a: $\mu_1 \neq \mu_2$ **b.** reject H_0 **c.** do not reject H_0 **d.** no practical difference **7.101** yes, $z = -2.25$ **7.103** $.0308 \pm .0341$ **7.105 a.** H_0: $\mu_1 = \mu_2$, H_a: $\mu_1 \neq \mu_2$ **b.** $z = 7.17$, reject H_0 **c.** (.37, .65) **7.107 a.** yes, $t = 1.9557$ **c.** .0504 **d.** -7.4 ± 6.38 **7.109 a.** 6.57 **b.** population mean for all 50 states is known **c.** $z = 2.09$, reject H_0 **7.111** 4,802 **7.113 a.** H_0: $\mu_d = 0$, H_a: $\mu_d \neq 0$ **b.** $t = 5.76$, reject H_0 **c.** yes **7.115** Aad: no, at $\alpha = .05$; Ab: yes, at $\alpha = .05$; Intention: no, at $\alpha = .05$ **7.117** yes; 95% confidence interval for μ_d: (−242.29, −106.96)

Chapter 8

8.1 A, B, C, D **8.5 a.** observational **b.** designed **c.** observational **d.** observational **e.** observational **8.7 a.** age **b.** smokers **c.** screening method **d.** CT and X-ray **8.9 a.** student **b.** yes **c.** class standing and study group **d.** class: low, medium, high; study: review session, practice test **e.** 6 **f.** final exam score **8.11 a.** 4 **b.** (Within-store/home), (Within-store/in store), (Between-store/home), (Between-store/in store) **8.13 a.** dissolution time **b.** binding agent (gum, PVP); binding concentration (.5%, 4%); relative density (low, high) **c.** 8; (gum/.5/low), (gum/.5/high), (gum/4/low), (gum/4/high), (PVP/.5/low), (PVP/.5/high), (PVP/4/low), (PVP/4/high) **8.15 a.** 6.39 **b.** 15.98 **c.** 1.54 **d.** 3.18 **8.17 a.** Plot 2 **b.** $\bar{x}_1 = 9$ and $\bar{x}_2 = 14$ for both plots **c.** SST $= 75$ for both plots **d.** Plot 1: SSE $= 20$; Plot 2: SSE $= 144$ **e.** Plot 1: SS(Total) $= 95$ (78.95%); Plot 2: SS(Total) $= 219$ (34.25%) **f.** Plot 1: $F = 37.5$; Plot 2: $F = 5.21$ **g.** reject H_0 for both plots **h.** both populations normal with equal variances

8.19 Plot 1:

Source	df	SS	MS	F
Treatment	1	75	75	37.5
Error	10	20	2	
Total	11	95		

Plot 2:

Source	df	SS	MS	F
Treatment	1	75	75	5.21
Error	10	144	14.4	
Total	11	219		

8.21 a.

Source	df	SS	MS	F
Treatments	2	12.30	6.15	2.93
Error	9	18.89	2.10	
Total	11	31.19		

b. do not reject H_0 **8.23 a.** H_0: $\mu_1 = \mu_2 = \mu_3 = \mu_4 = \mu_5 = \mu_6 = \mu_7$, H_a: At least two treatment means differ **b.** p-value $= .174$, do not reject H_0 **8.25 a.** completely randomized design **b.** dependent variable = energy expenditure, treatments = colony sizes **c.** H_0: $\mu_1 = \mu_2 = \mu_3 = \mu_4$ **d.** reject H_0 **8.27** yes, $F = 3.90$ **8.29 a.** Factor = Group; Treatments = 5 levels of Group; response variable = ethics score;

experimental units = employees **b.** H_0: $\mu_1 = \mu_2 = \mu_3 = \mu_4 = \mu_5$, H_a: At least two treatment means differ **c.** yes; $F = 9.85$, p-value = 0 **d.** assumption of constant variance is violated **8.31 a.** H_0: $\mu_1 = \mu_2 = \mu_3 = \mu_4 = \mu_5 = \mu_6$ **b.** no **d.** designed **8.33** reject H_0: $\mu_1 = \mu_2 = \mu_3 = \mu_4$ for shell thickness and whipping capacity **8.35 a.** 3 **b.** 10 **c.** 6 **d.** 45 **8.37** P(Type I error) for a single comparison **8.39** $(\mu_1, \mu_4) > (\mu_2, \mu_3)$ **8.41** $\mu_I > (\mu_{II}, \mu_{III})$ **8.43 b.** no **c.** yes **d.** no **e.** $\mu_{Large} > \mu_{Small}$; no other significant differences **f.** 95% confidence **8.45 a.** probability of claiming at least 2 means are different when they are not is .01 **b.** 3 **d.** $(\mu_{Under30} - \mu_{30to60})$; $(\mu_{Under30} - \mu_{Over60})$ **8.47** $\mu_{NC} > (\mu_{AC}, \mu_{EO})$; $\mu_{IE} > \mu_{EO}$ **8.49 a.** $(-.125, -.032)$; $\mu_B > \mu_C$ **b.** $(-.1233, -.0307)$; $\mu_F > \mu_C$ **c.** $(-.105, -.0123)$; $\mu_O > \mu_C$ **d.** $(-.0501, .0535)$; no difference **e.** $(-.0318, .0718)$; no difference **f.** $(-.0335, .0701)$; no difference **g.** $(\mu_B, \mu_F, \mu_O) > \mu_C$; .05

8.51 a.

Source	df	SS	MS	F
Treatments	2	21.5555	10.7778	5.54
Blocks	2	.8889	.4445	.23
Error	4	7.7778	1.9445	
Total	8	30.2222		

b. H_0: $\mu_1 = \mu_2 = \mu_3$ **c.** $F = 5.54$ **e.** do not reject H_0 **8.53 a.** $F = 3.20$; $F = 1.80$ **b.** $F = 13.33$; $F = 2.00$ **c.** $F = 5.33$; $F = 5.00$ **d.** $F = 16.00$; $F = 6.00$ **e.** $F = 2.67$; $F = 1.00$ **8.55 a.** Blocks = employees, treatments = time periods **b.** p-values provided **c.** H_0: $\mu_B = \mu_{2M} = \mu_{2D}$ **d.** mean competence levels differ for three time periods **e.** $\mu_B < (\mu_{2M}, \mu_{2D})$ **8.57 b.** California, Utah, Alaska **c.** Nov 2000, Oct 2001, Nov 2001 **d.** H_0: $\mu_{CAL} = \mu_{UT} = \mu_{AL}$ **e.** $F = 38.07$, p-value = 0.002 **f.** California **8.59** $F = 2.00$, fail to reject H_0: $\mu_M = \mu_T = \mu_W = \mu_R = \mu_F$ **8.61** yes, $F = 34.12$; S4 and S1

8.63 a.

Source	df	SS	MS	F
A	2	.8	.4	3.69
B	3	5.3	1.7667	16.31
AB	6	9.6	1.6	14.77
Error	12	1.3	.1083	
Total	23	17.00		

b. SSA + SSB + SSAB; yes, $F = 13.18$ **c.** yes **e.** $F = 14.77$, reject H_0 **f.** no **8.65 a.** (1,1), (1,2), (1,3), (2,1), (2,2), (2,3) **b.** yes; $F = 21.62$ **c.** yes; $F = 36.62$, reject H_0 **d.** no **8.67 a.** $F(AB) = .75$, $F(A) = 3.00$, $F(B) = 1.50$ **b.** $F(AB) = 7.50$, $F(A) = 3.00$, $F(B) = 3.00$ **c.** $F(AB) = 3.00$, $F(A) = 12.00$, $F(B) = 3.00$ **d.** $F(AB) = 4.50$, $F(A) = 36.00$, $F(B) = 36.00$ **8.69 a.** complete 6 × 6 factorial design **b.** Factors: Coagulant (5, 10, 20, 50, 100, and 200), pH level (4.0, 5.0, 6.0, 7.0, 8.0, and 9.0); 6 × 6 = 36 treatments **8.71 a.** 2 × 2 factorial; factors: color and question; treatments: (red/simple), (red/difficult), (blue/simple), (blue/difficult) **b.** difference between red and blue exam means depends on question difficulty **8.73 a.** factors = housing system, weight class; treatments = Cage, M), (Cage, L), (Free, M), (Free, L), (Barn, M), (Barn, L), (Organic, M), (Organic, L) **c.** no; $F = .76$ **d.** $F = 33.06$, reject H_0 **e.** $F = 2.95$, do not reject H_0 **8.75** yes **8.77 a.** 450; 152.5; 195; 157.5 **b.** 9,120.25 **c.** 1,122.25; 625; 676 **d.** 225; 90.25; 90.25; 100 **e.** 12,132 **f.** 14,555.25

g.

Source	df	SS	MS	F
Load	1	1,122.25	1,122.25	8.88
Name	1	625.00	625.00	4.95
Load x Name	1	676.00	676.00	5.35
Error	96	12,132.00	126.375	
Total	99	14,555.25		

h. yes **i.** sufficient evidence of interaction **j.** normal distributions for each treatment, with equal variances

8.83 a.

Source	df	SS	MS	F
Treatment	3	11.334	3.778	157.42
Block	4	10.688	2.672	111.33
Error	12	0.288	0.024	
Total	19	22.310		

b. yes, $F = 157.42$ **c.** yes, 6 **d.** yes, $F = 111.33$ **8.85 b.** yes **c.** no **8.87 a.** randomized block **b.** experimental units: electronic commerce/Internet-based companies; response: rate of return; treatments: e-companies, Internet software/service, Internet hardware, and Internet communication; blocks: 1 year, 3 year, and 5 year **8.89 a.** yes; $F = 30.4$ **b.** $\mu_A > (\mu_B, \mu_C, \mu_D)$; $\mu_B > \mu_D$ **8.91 a.** observational **b.** day

(7 levels), time (24 levels) **c.** $a = 7, b = 24$ **d.** $F = 25.06$, reject H_0 **e.** H_0: Day and Time do not interact **f.** $F = 1.22$, do not reject H_0 **g.** Day: $F = 68.39$, reject H_0; Time: $F = 156.8$, reject H_0
8.93 a.

Source	df	SS	MS	F
Groups	2	128.70	64.35	0.16
Error	68	27,124.52	398.89	

b. no, $F = .16$ **c.** p-value $> .10$ **d.** no **8.95 a.** quality **b.** temperature and pressure **c.** $3 \times 5 = 15$ combinations of temperature and pressure **d.** steel ingots **8.97 a.** completely randomized; 5 education levels **b.** $F = 3.298$, reject H_0 **c.** $\mu_P > (\mu_{CG}, \mu_{HS}, \mu_{SC}, \mu_{NH})$ **8.99 a.** complete 6×5 factorial design **b.** cylinders (6 levels) and batches (5 levels) **c.** 30
d.

Source	df	SS	MS	F
B	4	62.444	15.611	8.31
C	5	55.789	11.158	5.94
BC	20	48.489	2.424	1.29
Error	60	112.667	1.878	
Total	89	279.389		

f. $F = 1.29$, no evidence of BC interaction **g.** $F = 8.31$, evidence of B main effect; $F = 5.94$, evidence of C main effect **8.101 a.** 2×2 factorial experiment **b.** factors: tent type and location; 4 treatments: (treated/inside), (treated/outside), (untreated/inside), and (untreated/outside) **c.** number of mosquito bites received in a 20-minute interval **8.103** $F = 1.77$, no evidence of interaction; $F = 2.17$, no evidence of Class main effect; $F = 14.40$, evidence of Preparation main effect

Chapter 9

9.1 a. $\chi^2 > 5.99147$ **b.** $\chi^2 > 7.77944$ **c.** $\chi^2 > 11.3449$ **9.3** $E_i \geq 5$ **9.5 a.** $\chi^2 = 3.293$ **c.** (.226, .350) **9.7 a.** Levels: 100%, 75%–99%, 50%–74%, less than 50% **b.** .50, .25. .20, .05 **c.** H_0: $p_1 = .5, p_2 = .25, p_3 = .20, p_4 = .05$ **d.** 4.68 **e.** $\chi^2 > 6.25139$ **f.** do not reject H_0 **g.** (.526, .682) **9.9 a.** 107.2 **b.** H_0: $p_1 = p_2 = \cdots = p_{10} = .10$ **c.** 93.15 **d.** $\chi^2 > 14.6837$ **e.** reject H_0 **9.11 b.** H_0: $p_1 = .5, p_2 = .1, p_3 = .1, p_4 = .1, p_5 = .1, p_6 = .05, p_7 = .05$ **c.** $\chi^2 = 452.48$, reject H_0 **9.13** yes, $\chi^2 = 8.04$ **9.15** $\chi^2 = 12.734$, do not reject H_0 **9.17 a.** $\chi^2 > 26.2962$ **b.** $\chi^2 > 15.9871$ **c.** $\chi^2 > 9.21034$ **9.19 a.** Column 1: 36%, 64%; column 2: 53%, 47%; column 3: 68%, 32%; row 1: 57%; row 2: 43% **c.** yes **9.21 a.** B_1: 30%, 47%, 23%; B_2: 44%, 33%, 23%; B_3: 30%, 49%, 21% **b.** A_1: 35%; A_2: 42%; A_3: 23% **c.** yes **9.23 a.** .282; .511 **b.** .563; .368 **c.** .155; .121 **d.** yes **e.** $\chi^2 = 14.21$, reject H_0 **f.** (−.325, −.133) **9.25 a.** .028, .050, .045, .038, .046, .039 **b.** yes, $\chi^2 = 48.09$ **9.27** no, $\chi^2 = 1.188$ **9.29** $\chi^2 = 12.47$, reject H_0 **9.31** yes, $\chi^2 = 7.38$, p-value $= .025$; no **9.33 a.** $\chi^2 = 4.41$, reject H_0 **b.** no **c.** .04378 **d.** .00571, .00027 **e.** .04976 **9.35 a.** no, $\chi^2 = 2.133$ **b.** $.233 \pm .057$ **9.37 a.** H_0: $p_1 = p_2 = p_3 = 1/3$ **b.** reject H_0, $\chi^2 = 87.74$, p-value $= 0$ **c.** $.539 \pm .047$ **9.39** yes, $\chi^2 = 256.336$ **9.41 a.** no, $\chi^2 = .078$ **b.** no **9.43** $\chi^2 = 16$, p-value $= .003$, reject H_0

9.45 a.

	Committee Accept	Committee Reject
Inspector Accept	101	23
Inspector Reject	10	19

b. yes **c.** $\chi^2 = 26.034$, reject H_0
9.47 a. yes, $\chi^2 = 880.52$ **c.** p-value ≈ 0
9.49 a.

	Low Flight Response	High Flight Response
< 300	85	105
300–600	77	121
≥ 600	17	59

b. $\chi^2 = 11.48$, reject H_0
c.

	Low Flight Response	High Flight Response
< 1,000	37	243
1,000–2,000	68	37
2,000–3,000	44	4
≥ 3,000	30	1

d. $\chi^2 = 207.81$, reject H_0 **e.** maximum height: 300 m; minimum distance: 3,000 m **9.51 a.** 9.65 **b.** 11.0705 **c.** no **d.** $.05 < p\text{-value} < .10$ **9.53** yes, $\chi^2 = 2.28$, do not reject H_0; yes, implies rigged election

Chapter 10

10.3 $\beta_1 = 1/3$, $\beta_0 = 14/3$, $y = 14/3 + 1/3x$ **10.9** no **10.11 a.** $\Sigma(y - \hat{y}) = 0.02$, SSE = 1.2204 **c.** SSE = 108.00 **10.13 b.** negative linear relationship **c.** 8.54; −.994 **10.15 a.** $y = \beta_0 + \beta_1 x + \varepsilon$ **b.** $\hat{y} = -146.856 + 1.144x$ **c.** no practical interpretation **d.** for every 1-point increase in 2000 SAT score, estimate 2007 SAT score to increase 1.144 points **10.17 a.** $\hat{y} = 7.21 + 35.92x$ **b.** y-intercept: no practical interpretation; slope: for each unit change in surface area to volume, estimate drug release rate to increase by 35.92 **c.** 25.17 **d.** not reliable; $x = 50$ is outside range **10.19 a.** $E(y) = \beta_0 + \beta_1 x$ **b.** $\hat{y} = 250.14 - .629x$ **c.** no practical interpretation **d.** for every 1-yard increase in driving distance, driving accuracy to decrease .629% **e.** slope **10.21 a.** $E(y) = \beta_0 + \beta_1 x$ **b.** $\hat{y} = 488.09 + 0.034x$ **d.** y-intercept: no practical interpretation; slope: for each additional \$1 million of net worth, estimate amount pledged to increase by \$.034 million **10.25** Plot b **10.27 a.** .3475 **b.** 1.179 **10.29 a.** SSE = 20,608.39, $s^2 = 420.58$, $s = 20.51$ **b.** about 95% of the observed SAT scores fall within 41 points of their least squares predicted values **10.31 a.** $s = 2.01$ **b.** about 95% of the observed release rates fall within 4% of their least squares predicted values **10.33 a.** SSE = 1.017, $s^2 = .0462$, $s = .215$ **c.** about 95% of the observed sweetness index values fall within .43 units of their least squares predicted values **10.35 a.** $s = 5.37$; about 95% of the observed FCAT-Math scores fall within 10.74 points of their least squares predicted values **b.** $s = 3.42$; about 95% of the observed FCAT-Reading scores fall within 6.84 points of their least squares predicted values **c.** reading score **10.37 a.** 95% CI: 31 ± 1.17; 90% CI: 31 ± .94 **b.** 95% CI: 64 ± 5.08; 90% CI: 64 ± 4.15 **c.** 95% CI: −8.4 ± .75; 90% CI: −8.4 ± .62 **10.39** 80% CI: .82 ± .33; 98% CI: .82 ± .76 **10.41 a.** H_0: $\beta_1 = 0$, H_a: $\beta_1 > 0$ **b.** reject H_0, p-value ≈ 0 **c.** (1.06, 1.23) **10.43** (30.17, 41.67); 90% confident that for each additional unit increase in surface area to volume, the increase in the drug release rate is between 30.17 and 41.67 **10.45 a.** H_0: $\beta_1 = 0$, H_a: $\beta_1 < 0$ **b.** $t = -13.23$, p-value ≈ 0 **c.** reject H_0 **10.47** no, $t = .66$ **10.49 a.** increasing trend **b.** positive **c.** support; $t = 4.60$, reject H_0 **10.51 a.** $\hat{y} = .515 + .00002074x$ **b.** yes, p-value = .008/2 = .004 **c.** possible outlier **d.** do not reject H_0 **10.53 a.** positive **b.** negative **c.** 0 slope **d.** positive or negative **10.55 a.** .9438 **b.** .8020 **10.57 a.** $E(y) = \beta_0 + \beta_1 x$ **b.** moderately strong positive linear relationship between RMP and SET ratings **c.** positive **d.** reject H_0: $p = 0$ **e.** .4624 **10.59 a.** moderately strong positive linear relationship between skill level and goal-setting ability **b.** reject H_0 at $\alpha = .01$ **c.** .49 **10.61 a.** strong positive linear relationship between number of females in managerial positions and number of females with college degree **b.** very weak positive linear relationship between number of females in managerial positions and number of female HS graduates with no college **c.** moderately strong positive linear relationship between number of males in managerial positions and number of males with college degree **d.** moderate positive linear relationship between number of males in managerial positions and number of male HS graduates with no college degree **10.63** $r^2 = .2286$; $r = -.478$ **10.65 a.** weak/moderate positive linear relationship between height and average earnings **b.** $r^2 = .168$; 16.8% of the total sample variation in average earnings is explained by the linear relationship with height **c.** H_0: $p = 0$, H_a: $p > 0$ **d.** $t = 4.82$ **e.** reject H_0 **10.67** $r = .243$, $r^2 = .059$ **10.69 a.** .57 **c.** no **10.71 c.** 4.64 ± 1.12 **d.** 2.28 ± .63; −.41 ± 1.17 **10.73 a.** $\hat{y} = 1.375 \pm .875x$ **c.** 1.5 **d.** .1875 **e.** (3.23, 3.89) **f.** (3.81, 5.94) **10.75 a.** (3,598.1, 3,868.1); 95% confident that mean price of all diamonds with .52 carat falls between \$3,598 and \$3,868 **b.** (1,529.8, 5,936.3); 95% confident that price of a diamond with .52 carat falls between \$1,530 and \$5,936 **10.77** for run = 1: 90% confident that mean sweetness index of all runs with pectin value of 220 falls between 5.65 and 5.84 **10.79 a.** (2.955, 4.066); 99% confident that the mean mass of all spills with elapsed time of 15 minutes is between 2.995 and 4.066 **b.** (1.02, 6.00); 99% confident that the actual mass of a single spill with elapsed time of 15 minutes is between 1.02 and 6.0 **c.** prediction interval for y; yes **10.81 a.** yes, $t = -5.91$; negative **b.** (.656, 2.829) **c.** (1.467, 2.018) **10.83** $t = 12.56$; p-value = 0; reject H_0; $r^2 = .07$ **10.85 a.** $\hat{y} = 37.08 - 1.6x$ **c.** 57.2 **d.** 4.4 **e.** −1.6 ± .5 **f.** 13.08 ± 6.93 **g.** 13.08 ± 7.86 **10.87 a.** yes **b.** 1.39 ± .10 **c.** 90% confident that for each 1 mm increase in Winter runoff, rainfall erosivity index increases between 1.29 and 1.49 **10.89 b.** $\hat{y} = 569.58 - .0019x$ **e.** range of x: \$15,100 to \$70,000 **f.** $t = -.62$; do not reject H_0 **10.91 a.** moderate positive linear relationship between rates of returns for the two countries **10.93 a.** 57.14 ± 34.82 **b.** 110 is outside range of x **c.** $\bar{x} = 44$ **10.95** $\hat{y} = -92.46 + 8.347x$, $t = 3.25$, reject H_0 **10.97 c.** $r_1 = .965$, $r_2 = .996$ **d.** yes **10.99 a.** $\hat{y} = 2.7803 + 0.01696x$ **b.** $t = 4.02$, reject H_0; $r^2 = .402$ **c.** 8 **10.101** machine-hours: $t = 3.30$, $p = .008$, reject H_0, $r^2 = .521$; labor-hours: $t = 1.43$, $p = .183$, do not reject H_0, $r^2 = .170$

Chapter 11

11.1 a. $E(y = \beta_0 + \beta_1 x_1 + \beta_2 x_2$ **b.** $E(y) = \beta_0 + \beta_1 x_1 + \beta_2 x_2 + \beta_3 x_3 + \beta_4 x_4$ **c.** $E(y) = \beta_0 + \beta_1 x_1 + \beta_2 x_2 + \beta_3 x_3 + \beta_4 x_4 + \beta_5 x_5$ **11.3 a.** $t = 1.45$, do not reject H_0 **b.** $t = 3.21$, reject H_0 **11.5** $n - (k + 1)$ **11.7 a.** Yes, $R^2 = .92$, $s = .117$ **b.** Yes, $F = 55.2$ **11.9 a.** $F = 4.74$, p-value $< .01$, reject H_0 **b.** 13% of sample variation in Mach score can be explained by the model **c.** no **11.11 a.** $\hat{y} = 1.81231 + 0.10875x_1 + 0.00017x_2$ **c.** (.025, .192) **d.** (.00009, .00025) **e.** $\hat{y} = 1.20785 + 0.06343x_1 + 0.00056x_2$; (.016, .111); (.00024, .00088) **11.13 a.** β_0: no practical interpretation; β_1: for every 1-unit increase in proportion of block with low-density areas, estimate density to increase 2.006 units; β_2: for every 1-unit increase in proportion of block with high-density areas, estimate density to increase 5.006 units **b.** 68.6% of sample variation in population density can be explained by the model **c.** $H_0: \beta_1 = \beta_2 = 0$ **d.** $F = 133.3$ **e.** reject H_0 **11.15 a.** $E(y) = \beta_0 + \beta_1 x_1 + \beta_2 x_2 + \beta_3 x_3 + \beta_4 x_4$ **b.** reject H_0 **c.** yes **d.** $H_0: \beta_1 = \beta_2 = \beta_3 = \beta_4 = 0$ **e.** $F > 3.37$ **f.** Thrill: reject H_0; Change: do not reject H_0; Surprise: do not reject H_0 **11.17 a.** $E(y) = \beta_0 + \beta_1 x_1 + \beta_2 x_2$ **b.** $\hat{y} = -20.4 + 13.350x_1 + 243.71x_2$ **d.** no, $t = 1.74$ **e.** (105.32, 382.10) **f.** .582 **g.** .513 **h.** $F = 8.36$, $p = .005$, reject H_0 **11.19 a.** $E(y) = \beta_0 + \beta_1 x_1 + \beta_2 x_2 + \beta_3 x_3 + \beta_4 x_4 + \beta_5 x_5$ **b.** $\hat{y} = 13{,}614.5 + 0.0888x_1 - 9.201x_2 + 14.394x_3 + 0.35x_4 - 0.848x_5$ **d.** 458.83 **e.** $F = 147.3$, $p < .001$, reject H_0 **f.** .917 **g.** yes, $t = -6.14$ **11.21 b.** $F = 5.11$, reject H_0 **11.23 a.** Model 1: $t = 2.58$, reject $H_0: \beta_1 = 0$; Model 2: $t = 3.32$, reject $H_0: \beta_1 = 0$, $t = 6.47$, reject $H_0: \beta_2 = 0$, $t = -4.77$, reject $H_0: \beta_3 = 0$, $t = 0.24$, do not reject $H_0: \beta_4 = 0$; Model 3: $t = 3.21$, reject $H_0: \beta_1 = 0$, $t = 5.24$, reject $H_0: \beta_2 = 0$, $t = -4.00$, reject $H_0: \beta_3 = 0$, $t = 2.28$, reject $H_0: \beta_4 = 0$, $t = .014$, do not reject $H_0: \beta_5 = 0$ **c.** Model 2 **11.25 a.** 3.78 **b.** 4.68 **11.27 a.** (1,759.75, 4,275.38) **b.** (2,620.25, 3,414.87) **c.** yes **11.29 a.** (11,599.6, 13,665.5) **b.** (12,157.9, 13,107.1) **c.** yes **11.31 a.** $\hat{y} = .704 + .180x_1 + .073x_2 - .120x_3$ **b.** $F = 7.19$, reject H_0 **c.** (−7.82, 28.02) **11.33 a.** $\hat{y} = -3{,}783 + .00875x_1 + 1.93x_2 + 3{,}444x_3 + 2{,}093x_4$ **b.** $F = 72.11$, reject H_0 **c.** (1,449, 2,424) **11.35 c.** interaction is present **11.37** $\hat{y} = -2.55 + 3.82x_1 + 2.63x_2 - 1.29x_1x_2$ **b.** twisted plane **c.** $x_2 = 1$: $\hat{y} = .08 + 2.53x_1$; $x_2 = 3$: $\hat{y} = 5.34 - .05x_1$; $x_2 = 5$: $\hat{y} = 10.6 - 2.63x_1$ **e.** $H_0: \beta_3 = 0$ vs. $H_a: \beta_3 \neq 0$ **f.** $t = -8.06$, reject H_0 **11.39 a.** $E(y) = \beta_0 + \beta_1 x_1 + \beta_2 x_2 + \beta_3 x_1 x_2$ **b.** $\beta_3 < 0$ **11.41 a.** $\hat{y} = 1{,}042 - 13.24x_1 + 103.3x_2 + 3.621x_1x_2$ **b.** 22.97 **c.** 284.14 **d.** $H_0: \beta_3 = 0$ **e.** .366 **f.** do not reject H_0 **11.43 a.** $E(y) = \beta_0 + \beta_1 x_1 + \beta_2 x_2 + \beta_3 x_3 + \beta_4 x_1 x_3 + \beta_5 x_2 x_3$ **b.** $\hat{y} = 10{,}845 - 1280x_1 + 217.4x_2 - 1549.2x_3 - 11x_1x_3 + 19.98x_2x_3$ **c.** $t = -.93$, do not reject H_0 **d.** $t = 1.78$, do not reject H_0 **11.45 b.** yes, higher R^2 and smaller s **11.47 a.** $E(y) = \beta_0 + \beta_1 x + \beta_2 x^2$ **b.** $E(y) = \beta_0 + \beta_1 x_1 + \beta_2 x_2 + \beta_3 x_1 x_2 + \beta_4 x_1^2 + \beta_5 x_2^2$ **c.** $E(y) = \beta_0 + \beta_1 x_1 + \beta_2 x_2 + \beta_3 x_3 + \beta_4 x_1 x_2 + \beta_5 x_1 x_3 + \beta_6 x_2 x_3 + \beta_7 x_1^2 + \beta_8 x_2^2 + \beta_9 x_3^2$ **11.49 a.** yes, $F = 85.94$ **b.** $H_0: \beta_2 = 0$ vs. $H_a: \beta_2 > 0$ **c.** $H_0: \beta_2 = 0$ vs. $H_a: \beta_2 < 0$ **11.51 a.** yes, $F = 25.93$ **b.** $t = -10.74$, reject H_0 **c.** $t = .60$, do not reject H_0 **11.53 b.** first-order model; first-order model; second-order model **11.55 a.** $E(y) = \beta_0 + \beta_1 x_1 + \beta_2 x_2 + \beta_3 x_1 x_2 + \beta_4 x_1^2 + \beta_5 x_2^2$ **b.** 14% of sample variation in attitude toward improving efficiency can be explained by the model **c.** downward curvature in the relationship between attitude and congruence **d.** evidence of interaction **11.57 a.** $E(y) = \beta_0 + \beta_1 x$ **b.** $E(y) = \beta_0 + \beta_1 x + \beta_2 x^2$ **c.** first-order model **d.** $F = 12.28$, overall model is statistically useful; $t = 1.04$, insufficient evidence of curvature **e.** first-order model **11.59 a.** curvilinear **b.** $\hat{y} = 154{,}243 - 1{,}908.9x + 5.93x^2$ **c.** $t = 5.66$, reject H_0 **11.61** $E(y) = \beta_0 + \beta_1 x + \beta_2 x^2$ **11.63** $E(y) = \beta_0 + \beta_1 x$, $x = \{1$ if level 2, 0 if level 1$\}$ **11.65 a.** 10.2, 6.2, 22.2, 12.2 **b.** $H_0: \beta_1 = \beta_2 = \beta_3 = 0$ **11.67 a.** Method: $x_1 = \{1$ manual, 0 if automated$\}$; Soil: $x_2 = \{1$ if clay, 0 if not$\}$, $x_3 = \{1$ if gravel, 0 if not$\}$; Slope: $x_4 = \{1$ if East, 0 if not$\}$, $x_5 = \{1$ if South, 0 if not$\}$, $x_6 = \{1$ if West, 0 if not$\}$, $x_7 = \{1$ if Southeast, 0 if not$\}$ **b.** $E(y) = \beta_0 + \beta_1 x_1$ **c.** $E(y) = \beta_0 + \beta_1 x_2 + \beta_2 x_3$ **d.** $E(y) = \beta_0 + \beta_1 x_4 + \beta_2 x_5 + \beta_3 x_6 + \beta_4 x_7$ **11.69 a.** $E(y) = \beta_0 + \beta_1 x$, $x = \{1$ if developer, 0 if not$\}$ **b.** $E(y) = \beta_0 + \beta_1 x_1 + \beta_2 x_2$, $x_1 = \{1$ if low, 0 if not$\}$, $x_2 = \{1$ if medium, 0 if not$\}$ **c.** $E(y) = \beta_0 + \beta_1 x$, $x = \{1$ if fixed price, 0 if not$\}$ **d.** $E(y) = \beta_0 + \beta_1 x_1 + \beta_2 x_2$, $x_1 = \{1$ if time of delivery, 0 if not$\}$, $x_2 = \{1$ if cost, 0 if not$\}$ **11.71 b.** (13.12, 24.88) **c.** coaching is effective **11.73 a.** T-Bills: $t = 8.14$, reject $H_0: \beta_1 = 0$; REIT: $t = -3.46$, reject $H_0: \beta_1 = 0$ **c.** .00281, .01863 **11.75 a.** $E(y) = \beta_0 + \beta_1 x$, where $x = \{1$ if Lotion/Cream, 0 if not$\}$ **b.** $\hat{y} = .7775 + .1092x$ **c.** $H_0: \beta_1 = 0$ **d.** $t = .24$, do not reject H_0 **e.** $\hat{y} = 7.56 - 1.65x$; $t = -.46$, do not reject H_0 **11.77 a.** $E(y) = \beta_0 + \beta_1 x_1 + \beta_2 x_2$, $x_1 = \{1$ if group V, 0 if not$\}$, $x_2 = \{1$ if group S, 0 if not$\}$ **b.** $\hat{y} = 3.1667 - 1.0833x_1 - 1.4537x_2$ **c.** $F = 20.45$, reject H_0 **11.79 a.** $E(y) = \beta_0 + \beta_1 x_1 + \beta_2 x_1^2$ **b.** $E(y) = \beta_0 + \beta_1 x_1 + \beta_2 x_1^2 + \beta_3 x_2 + \beta_4 x_3$, where x_2 and x_3 are dummy variables **c.** add terms: $\beta_5 x_1 x_2 + \beta_6 x_1 x_3 + \beta_7 x_1^2 x_2 + \beta_8 x_1^2 x_3$ **d.** $\beta_5 = \beta_6 = \beta_7 = \beta_8 = 0$ **e.** $\beta_2 = \beta_5 = \beta_6 = \beta_7 = \beta_8 = 0$ **f.** $\beta_3 = \beta_4 = \beta_5 = \beta_6 = \beta_7 = \beta_8 = 0$ **11.81 b.** $\hat{y} = 48.8 - 3.4x_1 + .07x_1^2$; $\hat{y} = 46.4 + .3x_1 + .05x_1^2$; $\hat{y} = 41.3 - .7x_1 + .03x_1^2$ **11.83 a.** model is statistically useful; reject $H_0: \beta_1 = \beta_2 = \ldots = \beta_{12} = 0$ **b.** no significant impact; do not reject $H_0: \beta_1 = 0$ **c.** positive impact; reject $H_0: \beta_3 = 0$ **d.** $E(y) = \beta_0 + \beta_1 x_3 + \beta_2 x_5 + \beta_3 x_6 + \ldots + \beta_9 x_{12} + \beta_{10} x_3 x_5 + \beta_{11} x_3 x_6 + \ldots + \beta_{17} x_3 x_{12}$ **11.85 a.** $E(y) = \beta_0 + \beta_1 x_1$; β_1 **b.** $E(y) = (\beta_0 + \beta_2) + (\beta_1 + \beta_3)x_1$; $\beta_1 + \beta_3$ **c.** no evidence of interaction at $\alpha = .01$ **11.87 a.** $x_1 = \{1$ if channel catfish, 0 if not$\}$, $x_2 = \{1$ if largemouth bass, 0 if not$\}$ **b.** $E(y) = \beta_0 + \beta_1 x_1 + \beta_2 x_2 + \beta_3 x_3$, where x_3 = weight **c.** $E(y) = \beta_0 + \beta_1 x_1 + \beta_2 x_2 + \beta_3 x_3 + \beta_4 x_1 x_3 + \beta_5 x_2 x_3$ **d.** $\hat{y} = 3.1 + 26.5x_1 - 4.1x_2 + 0.0037x_3$ **f.** $\hat{y} = 3.5 + 25.6x_1 - 3.5x_2 + 0.0034x_3 + .0008x_1x_3 - .0013x_2x_3$ **g.** .0042 **11.89 a.** $E(y) = \beta_0 + \beta_1 x_1 + \beta_2 x_1^2 + \beta_3 x_2 + \beta_4 x_3 + \beta_5 x_4 + \beta_6 x_1 x_2 + \beta_7 x_1 x_3 + \beta_8 x_1 x_4 + \beta_9 x_1^2 x_2 + \beta_{10} x_1^2 x_3 + \beta_{11} x_1^2 x_4$, where x_1 = sales volume and $x_2 - x_4$ are dummy variables for region **b.** $E(y) = (\beta_0 + \beta_5) + (\beta_1 + \beta_8)x_1 + (\beta_2 + \beta_{11})x_1^2$ **c.** $E(y) = (\beta_0 + \beta_3) + (\beta_1 + \beta_6)x_1 + (\beta_2 + \beta_9)x_1^2$ **d.** β_3 through β_{11} **e.** yes, $F = 8.21$, $p = .000$ **11.91** a and b, a and d, a and e, b and c, b and d, b and e, c and e, d and e **11.93 a.** 5; 3 **b.** $H_0: \beta_3 = \beta_4 = 0$ **c.** $F = .38$, do not reject H_0 **11.95 a.** $H_0: \beta_4 = \beta_5 = 0$ **b.** complete model: $E(y) = \beta_0 + \beta_1 x_1 + \beta_2 x_2 + \beta_3 x_1 x_2 + \beta_4 x_2^2 + \beta_5 x_1 x_2^2$; reduced model: $E(y) = \beta_0 + \beta_1 x_1 + \beta_2 x_2 + \beta_3 x_1 x_2$ **c.** $H_0: \beta_3 = \beta_5 = 0$ **d.** complete model: $E(y) = \beta_0 + \beta_1 x_1 + \beta_2 x_2 + \beta_3 x_1 x_2 + \beta_4 x_2^2 + \beta_5 x_1 x_2^2$; reduced model: $E(y) = \beta_0 + \beta_1 x_1 + \beta_2 x_2 + \beta_4 x_2^2$ **e.** $H_0: \beta_1 = \beta_3 = \beta_5 = 0$ **f.** complete model: $E(y) = \beta_0 + \beta_1 x_1 + \beta_2 x_2 + \beta_3 x_1 x_2 + \beta_4 {x_2}^2 + \beta_5 x_1 {x_2}^2$; reduced model: $E(y) = \beta_0 + \beta_2 x_2 + \beta_4 {x_2}^2$ **11.97 a.** $E(y) = \beta_0 + \beta_1 x_1 + \beta_2 x_2 + \beta_3 x_1^2 + \beta_4 x_2^2 + \beta_5 x_1 x_2$ **b.** $H_0: \beta_3 = \beta_4 = 0$ **c.** complete model: $E(y) = \beta_0 + \beta_1 x_1 + \beta_2 x_2 + \beta_3 x_1^2 + \beta_4 x_2^2 + \beta_5 x_1 x_2$; reduced model: $E(y) = \beta_0 + \beta_1 x_1 + \beta_2 x_2 + \beta_5 x_1 x_2$ **d.** $SSE_R = 25{,}310{,}639$, $SSE_C = 19{,}370{,}350$, $MSE_C = 317{,}547$ **e.** $F = 9.35$ **f.** $F > 2.39$ **g.** reject H_0 **11.99 a.** $E(y) = \beta_0 + \beta_1 x_1 + \beta_2 x_2 + \beta_3 x_3 + \beta_4 x_4 + \beta_5 x_5 + \beta_6 x_6 + \beta_7 x_7 + \beta_8 x_8 + \beta_9 x_9 + \beta_{10} x_{10}$ **b.** $H_0: \beta_3 = \beta_4 = \ldots = \beta_{10} = 0$ **c.** reject H_0 **e.** (8.12, 19.88) **f.** yes **g.** $E(y) = \beta_0 + \beta_1 x_1 + \beta_2 x_2 + \beta_3 x_3 + \beta_4 x_4 + \beta_5 x_5 + \beta_6 x_6 + \beta_7 x_7 + \beta_8 x_8 + \beta_9 x_9 + \beta_{10} x_{10} + \beta_{11} x_1 x_2 + \beta_{12} x_3 x_2 + \beta_{13} x_4 x_2 + \beta_{14} x_5 x_2 + \beta_{15} x_6 x_2 + \beta_{16} x_7 x_2 + \beta_{17} x_8 x_2 + \beta_{18} x_9 x_2 + \beta_{19} x_{10} x_2$ **h.** Test $H_0: \beta_{11} = \beta_{12} = \ldots = \beta_{19} = 0$ using a nested model F-test **11.101 a.** $E(y) = \beta_0 + \beta_1 x_1 + \beta_2 x_2 + \beta_3 x_3$ **b.** add terms: $\beta_4 x_1 x_2 + \beta_5 x_1 x_3$ **c.** AL: β_1; TDS-3A: $\beta_1 + \beta_4$; FE: $\beta_1 + \beta_5$ **d.** Test $H_0: \beta_4 = \beta_5 = 0$ using a nested model F-test **11.103 b.** $F = 38.24$, reject H_0 **c.** no **11.105 a.** 7; $E(y) = \beta_0 + \beta_1 x_i$ **b.** 6; $E(y) = \beta_0 + \beta_1 x_1 + \beta_2 x_i$ **c.** 5; $E(y) = \beta_0 + \beta_1 x_1 + \beta_2 x_2 + \beta_3 x_i$ **11.107 a.** 11 **b.** 10 **c.** 1 **d.** $E(y) = \beta_0 + \beta_1 x_{11} + \beta_2 x_4 + \beta_3 x_2 + \beta_4 x_7 + \beta_5 x_{10} + \beta_6 x_1 + \beta_7 x_9 + \beta_8 x_3$ **e.** 67.7% of sample variation in overall satisfaction can be explained by the model **f.** no interactions or squared terms in model; high probability of making at least one Type I error **11.109** yes **11.111 a.** yes, extreme multicollinearity **b.** no, little multicollinearity **c.** possibly, moderate multicollinearity **d.** possibly, moderate multicollinearity **11.113** multicollinearity **11.115 a.** normal probability plot; yes **b.** plot of residuals versus predicted values; yes **11.117 a.** satisfied **b.** violated **c.** some outliers **d.** violated **e.** no **11.119 a.** $\hat{y} = 30{,}856 - 191.57x$ **b.** −1,212.07, −1,162.07 **c.** yes, curvilinear trend **d.** yes **11.121** confidence interval **11.125 a.** $\hat{y} = 90.1 - 1.836x_1 + .285x_2$ **b.** .916 **c.** yes, $F = 64.91$ **d.** $t = -5.01$, reject H_0 **e.** 10.68 **11.127** $E(y) = \beta_0 + \beta_1x_1 + \beta_2x_2 + \beta_3x_3$, where x_1 = {1 if level 2, 0 otherwise}, x_2 = {1 if level 3, 0 otherwise}, x_3 = { 1 if level 4, 0 otherwise} **11.131** no degrees of freedom for error **11.133 a.** confidence interval for $(\mu_1 - \mu_2)$ **b.** $E(y) = \beta_0 + \beta_1x$, x = {1 if public college, 0 if private college} **11.135 a.** type of extractor is qualitative; size is quantitative **b.** $E(y) = \beta_0 + \beta_1x_1 + \beta_2x_2$, where x_1 = diameter of orange, x_2 = {1 if Brand B, 0 if not} **c.** $E(y) = \beta_0 + \beta_1x_1 + \beta_2x_2 + \beta_3x_1x_2$ **e.** H_0: $\beta_3 = 0$ **11.137** x_4 and x_5 **11.139 a.** 71.2% of sample variation in fee charged can be explained by the model **b.** $F = 111.1$, reject H_0 **c.** negative **11.141** Importance and Support are correlated at .6991; no **11.143 a.** $E(y) = \beta_0 + \beta_1x_1 + \beta_2x_2$, where x_2 = {1 if intervention group, 0 if control group} **b.** slope = β_1 **c.** $E(y) = \beta_0 + \beta_1x_1 + \beta_2x_2 + \beta_3x_1x_2$ **d.** intervention slope = $(\beta_1 + \beta_3)$; control slope = β_1 **11.145 a.** possibly **b.** yes **c.** no **11.147 b.** $E(y) = \beta_0 + \beta_1x_1 + \beta_2x_1^2 + \beta_3x_2 + \beta_4x_1x_2 + \beta_5x_1^2x_2$, where x_2 = {1 if I-35W, 0 if not} **c.** yes, $F = 383.76$ **d.** assumptions satisfied **11.149 a.** $E(y) = \beta_0 + \beta_1x_1 + \beta_2x_2 + \beta_3x_3$, where x_1 = {1 if VH, 0 otherwise}, x_2 = {1 if H, 0 otherwise}, x_3 = {1 if M, 0 otherwise} **b.** no **c.** $\hat{y} = 10.2 + .5x_1 + 2.02x_2 + .683x_3$ **d.** yes, $F = 63.09$ **11.151 b.** yes, $F = 16.10$ **c.** yes, $t = 2.5$ **d.** 945 **11.153 a.** $E(y) = \beta_0 + \beta_1x_1 + \beta_2x_6 + \beta_3x_7$, where x_6 = {1 if good, 0 otherwise}, x_7 = {1 if fair, 0 otherwise} **c.** excellent: $\hat{y} = 188{,}875 + 15{,}617x_1$; good: $\hat{y} = 85{,}829 + 15{,}617x_1$; fair: $\hat{y} = 36{,}388 + 15{,}617x_1$ **e.** yes, $F = 8.43$ **f.** (x_1 and x_3), (x_1 and x_5), (x_3 and x_5) are highly correlated **g.** assumptions are satisfied **11.155 a.** H_0: $\beta_1 = \beta_2 = \beta_3 = \beta_4 = \beta_5 = 0$ **b.** $F = 18.24$, reject H_0 **c.** H_0: $\beta_3 = \beta_4 = \beta_5 = 0$ **d.** $F = 8.46$, reject H_0 **e.** second-order model **11.157** $\hat{y} = -11.5 + 0.189x_1 + 0.159x_2 - 0.00114x_1^2 - 0.000871x_2^2$

Chapter 12

12.7 out of control **12.9 a.** 1.023 **b.** 0.308 **c.** 0.167 **12.11 b.** $\bar{\bar{x}} = 20.11625$, $\bar{R} = 3.31$ **c.** UCL = 22.529, LCL = 17.703 **d.** Upper A–B: 21.725, Lower A–B: 18.507, Upper B–C: 20.920, Lower B–C: 19.312 **e.** yes **12.13 a.** UCL = .9175, LCL = .4291, Upper A–B: .8361, Lower A–B: .5105, Upper B–C: .7547, Lower B–C: .5919 **b.** out of control **c.** no **12.15 a.** $\bar{\bar{x}} = 13.05$, Upper A–B: 13.55, Lower A–B: 12.55, Upper B–C: 13.3, Lower B–C: 12.8 **b.** out of control c. under-reporting likely **12.17 a.** $\bar{\bar{x}} = 100.08$, $\bar{R} = .8065$, UCL = 100.91, LCL = 99.26, Upper A–B: 100.63, Lower A–B: 99.53, Upper B–C: 100.36, Lower B–C: 99.81; in control **b.** $\bar{\bar{x}} = 100.09$, $\bar{R} = .75$, UCL = 100.86, LCL = 99.33, Upper A–B: 100.61, Lower A–B: 99.58, Upper B–C: 100.35, Lower B–C: 99.84; out of control **12.19** $\bar{\bar{x}} = 52.6467$, $\bar{R} = .755$, UCL = 53.419, LCL = 51.874, Upper A–B: 53.162, Lower A–B: 52.132, Upper B–C: 52.904, Lower B–C: 52.389 **b.**out of control **d.** no **12.23 a.** UCL = 16.802 **b.** Upper A–B: 13.853, Lower A–B: 2.043, Upper B–C: 10.900, Lower B–C: 4.996 **c.** in control **12.25** *R*-chart: $\bar{R} = 4.03$, UCL = 7.754, LCL = 0.306, Upper A–B: 6.513, Lower A–B: 1.547, Upper B–C: 5.271, Lower B–C: 2.789, in control; $\bar{x}$-chart: $\bar{\bar{x}} = 21.728$, UCL = 23.417, LCL = 20.039, Upper A–B: 22.854, Lower A–B: 20.602, Upper B–C: 22.291, Lower B–C: 21.165, out of control **12.27 a.** UCL = .7645 **b.** process in control **c.** yes **d.** 8 **12.29 a.** $\bar{R} = .0238$, UCL = .0778, LCL = 0, Upper A–B: .0598, Lower A–B: 0, Upper B–C: .0418, Lower B–C: .0058; in control **b.** $\bar{\bar{x}} = .2214$, UCL = .2661, LCL = .1767, Upper A–B: .2512, Lower A–B: .1916, Upper B–C: .2363, Lower B–C: .2065; in control **c.** in control; .2214 **12.31 a.** $\bar{R} = 2.756$, UCL = 5.826, Upper A–B: 4.803, Lower A–B: .709, Upper B–C: 3.780, Lower B–C: 1.732 **b.** process variation **c.** in control **12.33** proportion **12.35** 104 **12.37** $\bar{p} = .0575$, UCL = .1145, LCL = .0005, Upper A–B: .0955, Lower A–B: .0195, Upper B–C: .0765, Lower B–C: .0385 **d.** no **e.** no **12.39 a.** $\bar{p} = .04$, UCL = .099, LCL = 0, Upper A–B: .079, Lower A–B: .001, Upper B–C: .060, Lower B–C: .020 **b.** no **c.** no **12.41** $\bar{p} = .06$, UCL = .092, LCL = .028, Upper A–B: .081, Lower A–B: .039, Upper B–C: .071, Lower B–C: .049; out of control **12.49 a.** 126 **b.** 31.2 **c.** 660.36 **d.** .0144 **12.51** 6σ **12.53** $C_p = .866$; no **12.55 a.** LSL = 3.12, USL = 3.72 **b.** .152 **c.** $C_p = .505$ **12.71 a.** $\bar{x} = 6.4$ **b.** increasing variance **12.73** out of control (Rule 2) **12.75 a.** $\bar{R} = 7.4$, UCL = 24.1758, Upper A–B: 18.5918, Lower A–B: 0, Upper B–C: 12.9959, Lower B–C: 1.8041; out of control **b.** $\bar{\bar{x}} = 344.15$, UCL = 358.062, LCL = 330.238, Upper A–B: 353.425, Lower A–B: 334.875, Upper B–C: 348.787, Lower B–C: 339.513; out of control **c.** no **d.** .25 **12.77** $\bar{R} = 5.455$, UCL = 11.532, Upper A–B: 9.508, Lower A–B: 1.402, Upper B–C: 7.481, Lower B–C: 3.429 **b.** in control **d.** $\bar{\bar{x}} = 3.867$, UCL = 7.015, LCL = .719, Upper A–B: 5.965, Lower A–B: 1.769, Upper B–C: 4.916, Lower B–C: 2.818 **e.** in control **f.** yes **12.79 a.** $n > 141$ **b.** $\bar{p} = .063$, UCL = .123, LCL = .003, Upper A–B: .103, Lower A–B: .023, Upper B–C: .083, Lower B–C: .043 **c.** out of control **e.** no

Chapter 13

13.5 a. 127.63 **b.** 149.74 **13.7 a.** 100, 103.19, 103.19, 103.72, 102.66, 102.66, 103.72, 103.72, 105.32, 106.38, 108.51, 107.98, 107.45, 107.98, 107.45, 105.85, 106.91, 105.85, 105.32, 105.32, 105.85, 105.85, 106.38, 103.72, 105.32, 104.79, 105.32, 105.85 **b.** quantity **c.** 95.16, 95.10, 95.10, 95.59, 94.61, 94.61, 95.59, 95.59, 97.06, 98.04, 100, 99.51, 99.02, 99.51, 99.02, 97.55, 98.53, 97.55, 97.06, 97.06, 97.55, 97.55, 98.04, 95.59, 97.06, 96.57, 97.06, 97.55 **13.9 a.** 100.00, 157.61, 158.15, 160.05, 167.39, 174.18, 164.67,172.28, 188.59, 185.33, 181.79, 210.87, 261.68, 214.40, 261.68, 292.12, 345.11, 373.64, 353.53 **c.** price **13.11 a.** 51.43, 68.77, 100.00, 158.52, 270.39, 412.59, 581.78, 768.60, 1042.43, 1344.80 **b.** 19.02, 25.43, 36.98, 58.62, 100.00, 152.59, 215.16, 284.26, 385.53, 497.35 **c.** flattens the graph **13.13 a.** Manufacturing: 100.00, 150.52, 197.52, 224.22, 256.11, 297.72; transportation: 100.00, 150.85, 193.88, 220.58, 242.01, 275.85 **b.** Earnings: 100.00, 149.64, 195.01, 224.17, 252.95, 296.82; hours: 100.00, 99.92, 100.51, 100.00, 101.36, 101.61 **13.15** $\omega = .2$ **13.17 a.** 188.0, 189.2, 190.2,…, 198.5, 198.4, 198.1, 198.1, 198.3 **b.** 188.0, 192.8, 193.8,…, 196.0, 197.6, 197.1, 197.8, 198.8 **c.** $\omega = .2$ series **13.19 a.** 384.00, 366.40, 348.48, 357.70, 378.74, 382.95, 386.99, 342.20, 303.64, 283.93, 279.99, 272.80, 302.56, 350.91, 398.18, 435.64, 569.53, 669.91, 831.58 **13.21 a.** $\omega = .1$: 1283.0, 1278.0,…, 1490.7, 1515.4, 1538.4, 1581.5; $\omega = .9$: 1283.0, 1238.0,…, 1918.8, 1756.1, 1746.1, 1946.7 **b.** $\omega = .9$ series **13.25 a.** $\omega = .3$: 198.03; $\omega = .7$: 197.52 **b.** $\omega = .7$ and $\nu = .3$: $F_{2005} = 196.97$, $F_{2006} = 196.65$, $F_{2007} = 196.33$; $\omega = .3$ and $\nu = .7$: $F_{2005} = 197.43$, $F_{2006} = 197.51$, $F_{2007} = 197.59$ **13.27 a.** yes **b.** $F_{2009} = 205.41$ **c.** $F_{2009} = 219.63$ **13.29 a.** forecast for all 4 quarters = 1481.3 **b.** forecast for all 4 quarters = 1443.8

13.31 a. forecast for all 12 months = 783.56 **b.** forecasts: 783.56, 836.58, 879.44, 923.92, 916.81, 902.75, 896.13, 917.96, 878.48, 854.19, 830.40, 795.65 **c.** forecasts: 844.56, 879.16, 913.76, 948.36, 982.96, 1017.56, 1052.16, 1086.76, 1121.36, 1155.96, 1190.56, 1225.16 ; one-step-ahead forecasts: 844.56, 912.94, 965.81, 1015.95, 985.11, 935.08, 899.07, 916.40, 855.32, 813.87, 779.68, 735.04 **13.33 a.** .03, 1.35, 2.67 **b.** −.43, .49, 1.41 **c.** MAD = 1.35, MAPE = .68, RMSE = 1.73 **d.** MAD = .78, MAPE = .39, RMSE = .90 **13.35 a.** MAD = 461.0, MAPE = 43.1, RMSE = 503.1 **b.** MAD = 351.9, MAPE = 33.2, RMSE = 391.2 **c.** Holt series with $w = .7$ and $v = .5$ **13.37 a.** forecasts for all 3 years = 71,893.3 **b.** F_{2006} = 72,387.1, F_{2007} = 72,675.9, F_{2008} = 72,964.7 **c.** Holt: MAD = 1081.1, MAPE = 1.38, RMSE = 1020.9 **13.39 a.** $E(Y_t) = \beta_0 + \beta_1 t + \beta_2 x_1 + \beta_3 x_2 + \beta_4 x_3$, where x_1 = {1 if Qtr. 1, 0 otherwise}, x_2 = {1 if Qtr. 2, 0 otherwise}, x_3 = {1 if Qtr. 3, 0 otherwise} **b.** $\hat{Y}_t = 11.49 + .51t - 3.95x_1 - 2.09x_2 - 4.52x_3$; $F = 1275.44$, reject H_0 **c.** Qtr. 1: (27.22, 29.67); Qtr. 2: (29.59, 32.04); Qtr. 3: (27.67, 30.12); Qtr. 4: (32.70, 35.15) **13.41 a.** $\hat{Y}_t = 11.40 - .278t$ **b.** 4.44; (2.32, 6.57) **13.43 a.** $\hat{Y}_t = 393.88 - 1.447t$ **b.** 2007: 361.8; 2008: 360.7 **c.** 2007: (338.5, 385.2); 2008: (337.2, 384.2) **13.47 a.** inconclusive **b.** inconclusive **c.** reject H_0 **d.** fail to reject H_0 **13.49 a.** Models statistically useful for Banks 1, 2, 3, 4, and 7 **b.** No evidence of positive autocorrelation for all 9 banks **13.51 a.** possibly **b.** $d = 1.29$, do not reject H_0 **c.** valid **13.53 a.** yes **b.** yes **c.** no **13.55 a.** 100, 101.4, 102.6, 103.6, 104.5, 105.6, 107.1, 108.6, 110.0 **b.** quantity **13.57 a.** $\hat{Y}_t = 38.17 + 7.32t$; forecasts: 118.68, 126.00, 133.32 **b.** Year 11: (100.61, 136.75); Year 12: (107.06, 144.94); Year 13: (113.40, 153.24) **13.59** F_{2008} = 5.00, F_{2009} = 4.72, F_{2010} = 4.44; Holt forecast errors are larger **13.61** $\hat{Y}_t = 42.63 - .379t$ **c.** F_{2009} = 50.2, F_{2010} = 50.6 **d.** 2009: (24.7, 75.7); 2010: (24.7, 76.4) **e.** $d = 1.74$, do not reject H_0 **13.63** $F_{2009,1}$ = 14,836.8; $F_{2009,2}$ = 14,874.6; $F_{2009,3}$ = 14,912.4; $F_{2009,4}$ = 14,950.2 **13.67 a.** \$39,745; \$44,124; 2008 **b.** \$34,229

Chapter 14

14.3. a. .035 **b.** .363 **c.** .004 **d.** .151; .151 **e.** .2122; .2119 **14.5.** p-value = .054; reject H_0 **14.7. a.** $H_0\colon \eta = 125{,}000$, $H_a\colon \eta > 125{,}000$ **b.** $S = 9$, p-value = .304, do not reject H_0 **c.** random sample from continuous distribution **14.9. a.** $H_0\colon \eta = 200$, $H_a\colon \eta > 200$ **b.** .188 **c.** do not reject H_0 **d.** nonrandom sample; validity in question **14.11. a.** data not normally distributed **b.** $H_0\colon \eta = 5{,}000$, $H_a\colon \eta < 5{,}000$ **c.** $S = 10$, p-value = .0192 , reject H_0 **14.13.** S = 8, p-value = .110, do not reject H_0 **14.15.** yes, $z = -2.47$ **14.17. a.** $T_1 = 62.5$, reject H_0 **b.** $T_1 = 62.5$, reject H_0 **14.19. b.** $T_1 = 104$ **c.** $T_2 = 106$ **d.** either T_1 or T_2 **e.** do not reject H_0 **14.21. a.** Wilcoxon rank sum test **b.** H_0: CMC and FTF groups have identical probability distributions **c.** $z < -1.28$ **d.** $z = -.21$, do not reject H_0 **14.23. a.** H_0: Two sampled populations have identical probability distributions **b.** $z = -.66$ **c.** $|z| > 1.645$ **d.** do not reject H_0 **14.25. b.** $z = 2.43$, reject H_0 **c.** sample sizes large, Central Limit Theorem applies **14.27. b.** p-value = .3431, do not reject H_0 **14.29. a.** H_0: Two sampled populations have identical probability distributions **b.** $T_- = 3.5$, reject H_0 **14.31. a.** H_0: Two sampled populations have identical probability distributions **b.** $z = 2.50$, reject H_0 **c.** .0062 **14.33. a.** data likely not normal **b.** H_a: Probability distribution of scores at 3rd meeting shifted to the right of probability distribution of scores at 1st meeting **c.** $T_- \leq 92$ **d.** H_a: Probability distribution of scores at 3rd meeting shifted to the right or left of probability distribution of scores at 1st meeting **e.** $T_- \leq 81$ **14.35.** p-value = .011, reject H_0 at $\alpha = .05$; enforcement program is effective **14.37.** $T_- = 3.5$, reject H_0 **14.39. a.** $T_- = 4$, reject H_0 **b.** yes **14.43. a.** completely randomized **b.** H_0: 3 probability distributions are identical **c.** $H > 9.21034$ **d.** $H = 13.85$, reject H_0 **14.45. a.** H_0: 3 probability distributions are identical **b.** $H = 36.04$, p-value = .000 **c.** reject H_0 **14.47. a.** H_0: 4 probability distributions are identical **c.** $R_1 = 52.5$, $R_2 = 125.0$, $R_3 = 88.5$, $R_4 = 34.0$; $H = 16.23$ **d.** $H > 7.81473$ **e.** reject H_0 **14.49. b.** normal data and equal variances **c.** $H = .29$, do not reject H_0 **14.51.** $H = 26.82$, reject H_0 **14.53. a.** H_0: Distributions for 3 treatments are identical **b.** $F_r > 4.60517$ **c.** $F_r = 6.93$, reject H_0 **14.55. a.** $R_1 = 35$, $R_2 = 32$, $R_3 = 23$ **b.** $F_r = 5.20$ **c.** p-value = .074 **d.** reject H_0 at $\alpha = .10$ **14.57. a.** H_0: Distributions for 3 states are identical **b.** $R_1 = 9$, $R_2 = 6$, $R_3 = 3$ **c.** $F_r = 6$ **d.** $F_r > 5.99147$ **e.** reject H_0 **14.59.** No, $F_r = .20$ **14.61.** Yes, $F_r = 6.35$ **14.63. a.** .01 **b.** .01 **c.** .975 **d.** .05 **14.65. a.** .4 **b.** −.9 **c.** −.2 **d.** .2 **14.67. b.** $r_s = .943$ **c.** do not reject H_0 **14.69. b.** Navigability: do not reject H_0; transactions: reject H_0; locatability: reject H_0; information: do not reject H_0; files: do not reject H_0 **14.71.** yes; $r_s = -.829$, reject H_0 **14.73.** moderate positive rank correlation between Methods I and III; all other pairs have weak positive rank correlation **14.75.** $r_s = .341$, do not reject H_0 **14.77. a.** no, $r_s = .40$ **b.** yes, $T_- = 1.5$ **14.79.** Yes, $F_r = 14.9$ **14.81. b.** $H > 15.0863$ **c.** reject H_0 **d.** yes **14.83. a.** 43 **b.** $H_0\colon \eta = 37$, $H_a\colon \eta > 37$ **c.** $S = 11$, p-value = .059, reject H_0 at $\alpha = .10$ **14.85. a.** no, at $\alpha = .05$; S = 20, p-value = .999 **14.87. a.** 27.0, 32.5, 29.0, 31.5 **b.** $F_r = .93$ **c.** p-value = .819 **d.** do not reject H_0 **14.89.** yes, $T_- = 3$ **14.91. a.** $z = 1.77$, do not reject H_0 **14.93.** yes, $T_{\text{before}} = 132.5$ **14.95.** $F_r = 6.21$, do not reject H_0 **14.97 a.** $S = 14$, p-value = .058, do not reject H_0 **b.** $S = 12$, p-value = .252, do not reject H_0 **c.** $T = 50$, do not reject H_0 **d.** $r_s = .774$; yes **14.99.** Evidence of difference in distributions (at $\alpha = .05$) for creative ideas and good use of skills

Index

[Chapters 13 and 14 are included in this Index and appear on the CD.]

Photo Credits

Table of Contents © Talaj/Courtesy of www.istockphoto.com; © IBI/Shutterstock; © Simon Askham/iStockphoto; © Ken Babione/iStockphoto; © Newscom; © Lisa Poole/AP Wide World; © Andres/Shutterstock; © sigh/Shutterstock; © David P. Smith/Shutterstock; © JustASC/Shutterstock; © Silvrshootr/iStockphoto; © olly/Shutterstock; © Varuka/Shutterstock; © Liz Van Steenburgh/Shutterstock

Chapter 1
pp. 1, 2, 10, 17 © Talaj/Courtesy of www.istockphoto.com; **p. 7** © Denis Pepin/Shutterstock; **p. 12** © Kevin Russ/iStockphoto

Chapter 2
pp. 30, 31, 37, 47, 74, 87, 92 © IBI/Shutterstock; **p. 35** © David Sucsy/Courtesy of www.istockphoto.com; **p. 85** © Jeffrey Smith/Courtesy of www.istockphoto.com; **p. 114** © Elena Sherengovskaya/Shutterstock

Chapter 3
p. 116 © Simon Askham/Courtesy of www.istockphoto.com; **p. 117** © Damon Higgins/The Palm Beach Post; **p. 121** © Robert Van Beets/Courtesy of www.istockphoto.com; **pp. 127, 136, 150** © Simon Askham/iStockphoto; **p. 131** © Scott Rothstein/Shutterstock; **p. 142** Tracy Hebden/Courtesy of www.istockphoto.com

Chapter 4
p. 173 © Dieter Spears/iStockphoto; **pp. 174, 214, 221, 250** © Ken Babione/iStockphoto; **p. 175** © Segundo/iStockphoto; **p. 189** © CostinT/iStockphoto; **p. 194** © Stephen Coburn/Shutterstock; **p. 202** © Yuri Arcurs/Shutterstock; **p. 211** © Adrian Baras/Shutterstock; **p. 213** © Sebastian Duda/iStockphoto; **p. 233** © Stephen Finn/Shutterstock; **p. 240** © Marek Szumlas/Shutterstock; **p. 268** © Cary Ulrich Photography/Alamy

Chapter 5
pp. 269, 270, 286, 294, 301 © Newscom; **p. 275** © Fred Goldstein/Shutterstock; **p. 290** © Food Drink and Diet/Mark Sykes/Alamy; **p. 292** © Sean Prior/Shutterstock; **p. 299** © Visions of America, LLC/Alamy

Chapter 6
pp. 319, 320, 329, 340, 353 © Lisa Poole/AP Wide World; **p. 328** © Ellen Creager/Newscom; **p. 344** © Michael-John Wolfe/Shutterstock; **p. 351** © Christine Kublanski/iStockphoto

Chapter 7
pp. 380, 381, 391, 412 © Andres/Shutterstock; **p. 382** © Ron Chapple/Thinkstock RF/Jupiter Unlimited; **p. 402** © Chris Schmidt/iStockphoto; **p. 416** Courtesy of www.istockphoto.com; **p. 421** © Kenneth Sponsler/Shutterstock; **p. 446** © Elena Sherengovskaya/Shutterstock

Chapter 8
p. 447 © Julia Freeman/Dreamstime; **pp. 448, 465, 476, 502** © sjgh/Shutterstock; **p. 451** © isogood/iStockphoto; **p. 455** © Bridget Zawitoski/Shutterstock; **p. 460** © Ben Blankenburg/iStockphoto; **p. 485** © Kim Reinick/Dreamstime

Chapter 9
pp. 521, 522, 527, 537 © David P. Smith/Shutterstock; **p. 523** © zennie/iStockphoto; **p. 526** © Ian Jeffery/iStockphoto; **p. 557** © Hywit Dimyadi/Shutterstock

Chapter 10
pp. 559, 560, 570, 585, 604 © JustASC/Shutterstock; **p. 567** © Scott Leigh/iStockphoto

Chapter 11

pp. 624, 625, 646, 693, 710 © Silvrshootr/iStockphoto; **p. 627** David Stockman/iStockphoto; **p. 632** © Tina Fineberg /AP Wide World; **p. 657** © Trevor Fisher/iStockphoto; **p. 668** © Oytun Karadayi/iStockphoto; **p. 699** © Harry Hu/iStockphoto; **p. 736** © Roberto A. Sanchez/iStockphoto

Chapter 12

pp. 738, 739, 765, 776 © olly/Shutterstock; **p. 762** © GWImages/Shutterstock; **p. 783** © Ed Hidden/iStockphoto; **p. 806** © Collard/Wikipedia

Chapter 13 (on CD only)

pp. 13-1, 13-2, 13-29, 13-32, 13-36 © Varuka/Shutterstock; **p. 13-3** © The Granger Collection, New York

Chapter 14 (on CD only)

pp. 14-1, 14-2, 14-7, 14-14, 14-28, 14-42 © Liz Van Steenburgh/Shutterstock; **p. 14-6** © Alex Kotlov/Shutterstock; **p. 14-9** Courtesy of the Department of Statistics, Florida State University; **p. 14-20** © George Peters/iStockphoto; **p. 14-38** University College London Dept of Psychology

Normal Curve Areas

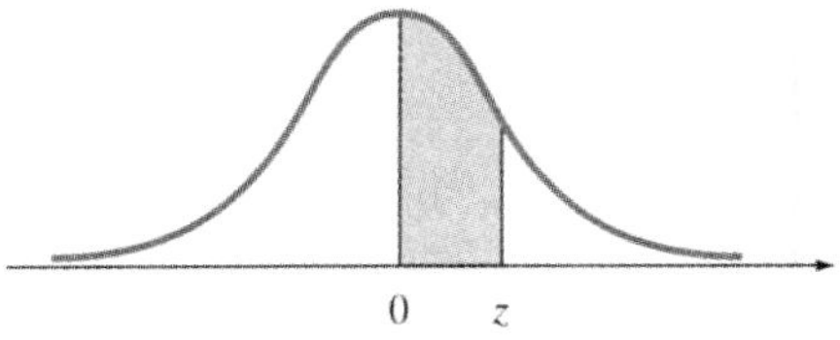

z	.00	.01	.02	.03	.04	.05	.06	.07	.08	.09
.0	.0000	.0040	.0080	.0120	.0160	.0199	.0239	.0279	.0319	.0359
.1	.0398	.0438	.0478	.0517	.0557	.0596	.0636	.0675	.0714	.0753
.2	.0793	.0832	.0871	.0910	.0948	.0987	.1026	.1064	.1103	.1141
.3	.1179	.1217	.1255	.1293	.1331	.1368	.1406	.1443	.1480	.1517
.4	.1554	.1591	.1628	.1664	.1700	.1736	.1772	.1808	.1844	.1879
.5	.1915	.1950	.1985	.2019	.2054	.2088	.2123	.2157	.2190	.2224
.6	.2257	.2291	.2324	.2357	.2389	.2422	.2454	.2486	.2517	.2549
.7	.2580	.2611	.2642	.2673	.2704	.2734	.2764	.2794	.2823	.2852
.8	.2881	.2910	.2939	.2967	.2995	.3023	.3051	.3078	.3106	.3133
.9	.3159	.3186	.3212	.3238	.3264	.3289	.3315	.3340	.3365	.3389
1.0	.3413	.3438	.3461	.3485	.3508	.3531	.3554	.3577	.3599	.3621
1.1	.3643	.3665	.3686	.3708	.3729	.3749	.3770	.3790	.3810	.3830
1.2	.3849	.3869	.3888	.3907	.3925	.3944	.3962	.3980	.3997	.4015
1.3	.4032	.4049	.4066	.4082	.4099	.4115	.4131	.4147	.4162	.4177
1.4	.4192	.4207	.4222	.4236	.4251	.4265	.4279	.4292	.4306	.4319
1.5	.4332	.4345	.4357	.4370	.4382	.4394	.4406	.4418	.4429	.4441
1.6	.4452	.4463	.4474	.4484	.4495	.4505	.4515	.4525	.4535	.4545
1.7	.4554	.4564	.4573	.4582	.4591	.4599	.4608	.4616	.4625	.4633
1.8	.4641	.4649	.4656	.4664	.4671	.4678	.4686	.4693	.4699	.4706
1.9	.4713	.4719	.4726	.4732	.4738	.4744	.4750	.4756	.4761	.4767
2.0	.4772	.4778	.4783	.4788	.4793	.4798	.4803	.4808	.4812	.4817
2.1	.4821	.4826	.4830	.4834	.4838	.4842	.4846	.4850	.4854	.4857
2.2	.4861	.4864	.4868	.4871	.4875	.4878	.4881	.4884	.4887	.4890
2.3	.4893	.4896	.4898	.4901	.4904	.4906	.4909	.4911	.4913	.4916
2.4	.4918	.4920	.4922	.4925	.4927	.4929	.4931	.4932	.4934	.4936
2.5	.4938	.4940	.4941	.4943	.4945	.4946	.4948	.4949	.4951	.4952
2.6	.4953	.4955	.4956	.4957	.4959	.4960	.4961	.4962	.4963	.4964
2.7	.4965	.4966	.4967	.4968	.4969	.4970	.4971	.4972	.4973	.4974
2.8	.4974	.4975	.4976	.4977	.4977	.4978	.4979	.4979	.4980	.4981
2.9	.4981	.4982	.4982	.4983	.4984	.4984	.4985	.4985	.4986	.4986
3.0	.4987	.4987	.4987	.4988	.4988	.4989	.4989	.4989	.4990	.4990
3.1	.49903	.49906	.49910	.49913	.49916	.49918	.49921	.49924	.49926	.48829
3.2	.49931	.49934	.49936	.49938	.49940	.49942	.49944	.49946	.49948	.49950
3.3	.49952	.49953	.49955	.49957	.49958	.49960	.49961	.49962	.49964	.49965
3.4	.49966	.49968	.49969	.49970	.49971	.49972	.49973	.49974	.49975	.49976
3.5	.49977	.49978	.49978	.49979	.49980	.49981	.49981	.49982	.49983	.49983
3.6	.49984	.49985	.49985	.49986	.49986	.49987	.49987	.49988	.49988	.49989
3.7	.49989	.49990	.49990	.49990	.49991	.49991	.49992	.49992	.49992	.49992
3.8	.49993	.49993	.49993	.49994	.49994	.49994	.49994	.49995	.49995	.49995
3.9	.49995	.49995	.49996	.49996	.49996	.49996	.49996	.49996	.49997	.49997

Source: Abridged from Table I of A. Hald. *Statistical Tables and Formulas* (New York: Wiley), 1952. Reproduced by permission of A. Hald.

Critical Values of *t*

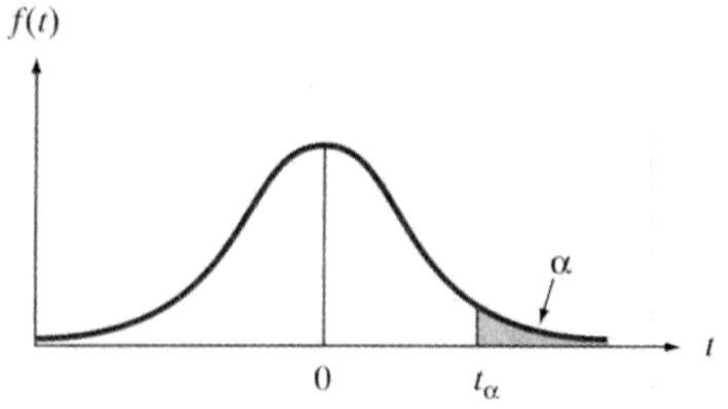

Degrees of Freedom	$t_{.100}$	$t_{.050}$	$t_{.025}$	$t_{.010}$	$t_{.005}$	$t_{.001}$	$t_{.0005}$
1	3.078	6.314	12.706	31.821	63.657	318.31	636.62
2	1.886	2.920	4.303	6.965	9.925	22.326	31.598
3	1.638	2.353	3.182	4.541	5.841	10.213	12.924
4	1.533	2.132	2.776	3.747	4.604	7.173	8.610
5	1.476	2.015	2.571	3.365	4.032	5.893	6.869
6	1.440	1.943	2.447	3.143	3.707	5.208	5.959
7	1.415	1.895	2.365	2.998	3.499	4.785	5.408
8	1.397	1.860	2.306	2.896	3.355	4.501	5.041
9	1.383	1.833	2.262	2.821	3.250	4.297	4.781
10	1.372	1.812	2.228	2.764	3.169	4.144	4.587
11	1.363	1.796	2.201	2.718	3.106	4.025	4.437
12	1.356	1.782	2.179	2.681	3.055	3.930	4.318
13	1.350	1.771	2.160	2.650	3.012	3.852	4.221
14	1.345	1.761	2.145	2.624	2.977	3.787	4.140
15	1.341	1.753	2.131	2.602	2.947	3.733	4.073
16	1.337	1.746	2.120	2.583	2.921	3.686	4.015
17	1.333	1.740	2.110	2.567	2.898	3.646	3.965
18	1.330	1.734	2.101	2.552	2.878	3.610	3.922
19	1.328	1.729	2.093	2.539	2.861	3.579	3.883
20	1.325	1.725	2.086	2.528	2.845	3.552	3.850
21	1.323	1.721	2.080	2.518	2.831	3.527	3.819
22	1.321	1.717	2.074	2.508	2.819	3.505	3.792
23	1.319	1.714	2.069	2.500	2.807	3.485	3.767
24	1.318	1.711	2.064	2.492	2.797	3.467	3.745
25	1.316	1.708	2.060	2.485	2.787	3.450	3.725
26	1.315	1.706	2.056	2.479	2.779	3.435	3.707
27	1.314	1.703	2.052	2.473	2.771	3.421	3.690
28	1.313	1.701	2.048	2.467	2.763	3.408	3.674
29	1.311	1.699	2.045	2.462	2.756	3.396	3.659
30	1.310	1.697	2.042	2.457	2.750	3.385	3.646
40	1.303	1.684	2.021	2.423	2.704	3.307	3.551
60	1.296	1.671	2.000	2.390	2.660	3.232	3.460
120	1.289	1.658	1.980	2.358	2.617	3.160	3.373
∞	1.282	1.645	1.960	2.326	2.576	3.090	3.291

Source: This table is reproduced with the kind permission of the Trustees of Biometrika from E. S. Pearson and H. O. Hartley (eds.). *The Biometrika Tables for Statisticians,* Vol. 1, 3rd ed., Biometrika, 1966.

Selected Formulas

CHAPTER 2

Relative Frequency = (frequency)/n

$$\bar{x} = \frac{\Sigma x}{n}$$

$$s^2 = \frac{\Sigma(x - \bar{x})^2}{n - 1} = \frac{\Sigma x^2 - \frac{(\Sigma x)^2}{n}}{n - 1}$$

$$s = \sqrt{s^2}$$

$$z = \frac{x - \mu}{\sigma} = \frac{x - \bar{x}}{s}$$

Chebyshev = At least $\left(1 - \frac{1}{k^2}\right)100\%$

IQR = $Q_U - Q_L$

CHAPTER 3

$$P(A^c) = 1 - P(A)$$

$$P(A \cup B) = P(A) + P(B) - P(A \cap B)$$

$$= P(A) + P(B) \text{ if } A \text{ and } B \text{ mutually exclusive}$$

$$P(A \cap B) = P(A|B) \cdot P(B) = P(B|A) \cdot P(A)$$

$$= P(A) \cdot P(B) \text{ if } A \text{ and } B \text{ independent}$$

$$P(A|B) = \frac{P(A \cap B)}{P(B)}$$

$$\binom{N}{n} = \frac{N!}{n!(N - n)!}$$

Bayes's: $P(S_i|A) =$

$$\frac{P(S_i)P(A|S_i)}{P(S_1)P(A|S_1) + P(S_2)P(A|S_2) + \cdots + P(S_k)P(A|S_k)}$$

CHAPTER 5

CI for μ: $\bar{x} \pm (z_{\alpha/2})\sigma/\sqrt{n}$ (large n)

$\bar{x} \pm (t_{\alpha/2})s/\sqrt{n}$ (small n, σ unknown)

CI for p: $\hat{p} \pm z_{\alpha/2}\sqrt{\frac{\hat{p}\hat{q}}{n}}$

Estimating μ: $n = (z_{\alpha/2})^2(\sigma^2)/(\text{SE})^2$

Estimating p: $n = (z_{\alpha/2})^2(pq)/(\text{SE})^2$

CHAPTER 4

Key Formulas

Random Variable	Prob. Dist'n	Mean	Variance
General Discrete:	Table, formula, or graph for $p(x)$	$\sum_{\text{all } x} x \cdot p(x)$	$\sum_{\text{all } x} (x - \mu)^2 \cdot p(x)$
Binomial:	$p(x) = \binom{n}{x} p^x q^{n-x}$ $x = 0, 1, 2, \ldots, n$	np	npq
Poisson:	$p(x) = \frac{\lambda^x e^{-\lambda}}{x!}$ $x = 0, 1, 2, \ldots$	λ	λ
Uniform:	$f(x) = 1/(d - c)$ $(c \leq x \leq d)$	$(c + d)/2$	$(d - c)^2/12$
Normal:	$f(x) = \frac{1}{\sigma\sqrt{2\pi}} e^{-\frac{1}{2}[(x-\mu)/\sigma]^2}$	μ	σ^2
Standard Normal:	$f(z) = \frac{1}{\sqrt{2\pi}} e^{-\frac{1}{2}(z)^2}$ $z = (x - \mu)/\sigma$	$\mu = 0$	$\sigma^2 = 1$
Sample Mean: **(large *n*)**	$f(\bar{x}) = \frac{1}{\sigma_{\bar{x}}\sqrt{2\pi}} e^{-\frac{1}{2}[(\bar{x}-\mu)/\sigma_{\bar{x}}]^2}$	$\mu_{\bar{x}} = \mu$	$\sigma_{\bar{x}}^2 = \sigma^2/n$

CHAPTER 6

Test for μ: $z = \dfrac{\bar{x} - \mu}{\sigma/\sqrt{n}}$ (large n)

$t = \dfrac{\bar{x} - \mu}{s/\sqrt{n}}$ (small n, σ unknown)

Test for p: $z = \dfrac{\hat{p} - p_0}{\sqrt{p_0 q_0/n}}$

Test for σ^2: $\chi^2 = (n - 1)s^2/(\sigma_0)^2$

CHAPTER 7

CI for $\mu_1 - \mu_2$:

$$(\bar{x}_1 - \bar{x}_2) \pm z_{\alpha/2}\sqrt{\frac{\sigma_1^2}{n_1} + \frac{\sigma_2^2}{n_2}} \quad \text{(large } n_1 \text{ and } n_2\text{)}$$

Test for $\mu_1 - \mu_2$:

$$z = \frac{(\bar{x}_1 - \bar{x}_2) - (\mu_1 - \mu_2)}{\sqrt{\dfrac{\sigma_1^2}{n_1} + \dfrac{\sigma_2^2}{n_2}}} \quad \text{(large } n_1 \text{ and } n_2\text{)}$$

$$s_p^2 = \frac{(n_1 - 1)s_1^2 + (n_2 - 1)s_2^2}{n_1 + n_2 - 2}$$

CI for $\mu_1 - \mu_2$:

$$(\bar{x}_1 - \bar{x}_2) \pm t_{\alpha/2}\sqrt{s_p^2\left(\frac{1}{n_1} + \frac{1}{n_2}\right)} \quad \text{(small } n_1 \text{ and/or } n_2\text{)}$$

Test for $\mu_1 - \mu_2$:

$$t = \frac{(\bar{x}_1 - \bar{x}_2) - (\mu_1 - \mu_2)}{\sqrt{s_p^2\left(\dfrac{1}{n_1} + \dfrac{1}{n_2}\right)}} \quad \text{(small } n_1 \text{ and/or } n_2\text{)}$$

CI for μ_d: $\bar{x}_d \pm t_{\alpha/2}\dfrac{s_d}{\sqrt{n}}$

Test for μ_d: $t = \dfrac{\bar{x}_d - \mu_d}{s_d/\sqrt{n}}$

CI for $p_1 - p_2$: $(\hat{p}_1 - \hat{p}_2) \pm z_{\alpha/2}\sqrt{\dfrac{\hat{p}_1\hat{q}_1}{n_1} + \dfrac{\hat{p}_2\hat{q}_2}{n_2}}$

Test for $p_1 - p_2$: $z = \dfrac{(\hat{p}_1 - \hat{p}_2) - (p_1 - p_2)}{\sqrt{\hat{p}\hat{q}\left(\dfrac{1}{n_1} + \dfrac{1}{n_2}\right)}}$

$$\hat{p} = \frac{x_1 + x_2}{n_1 + n_2}$$

Test for (σ_1^2/σ_2^2): $F = (s_1^2/s_2^2)$

Estimating $\mu_1 - \mu_2$: $n_1 = n_2 = (z_{\alpha/2})^2(\sigma_1^2 + \sigma_2^2)/(\text{ME})^2$

Estimating $p_1 - p_2$: $n_1 = n_2 = (z_{\alpha/2})^2(p_1q_1 + p_2q_2)/(\text{ME})^2$

CHAPTER 8

ANOVA Test for completely randomized design:
$F = \text{MST/MSE}$

ANOVA Test for randomized block design:
$F = \text{MST/MSE}$

ANOVA Test for factorial design interaction:
$F = \text{MS(A} \times \text{B)/MSE}$

Pairwise comparisons: $c = k(k - 1)/2$

CHAPTER 9

Multinomial test: $\chi^2 = \Sigma \dfrac{(n_i - E_i)^2}{E_i}$

$E_i = n(p_{i0})$

Contingency table test: $\chi^2 = \Sigma \dfrac{(n_{ij} - E_{ij})^2}{E_{ij}}$

$$E_{ij} = \frac{R_i C_j}{n}$$

CHAPTER 10

$$\text{SS}_{xx} = \Sigma x^2 - \frac{(\Sigma x)^2}{n}$$

$$\text{SS}_{yy} = \Sigma y^2 - \frac{(\Sigma y)^2}{n}$$

$$\text{SS}_{xy} = \Sigma xy - \frac{(\Sigma x)(\Sigma y)}{n}$$

$$\hat{y} = \hat{\beta}_0 + \hat{\beta}_1 x$$

$$\hat{\beta}_1 = \frac{\text{SS}_{xy}}{\text{SS}_{xx}}$$

$$\hat{\beta}_0 = \bar{y} - \hat{\beta}_1\bar{x}$$

$$r = \frac{\text{SS}_{xy}}{\sqrt{\text{SS}_{xx}}\sqrt{\text{SS}_{yy}}}$$